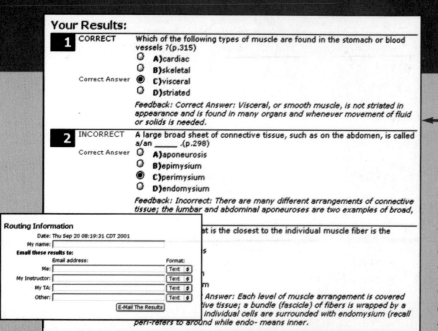

Your Results:

1 CORRECT

Which of the following types of muscle are found in the stomach or blood vessels ?(p.315)

- ○ A)cardiac
- ○ B)skeletal
- **Correct Answer** ● C)visceral
- ○ D)striated

Feedback: Correct Answer: Visceral, or smooth muscle, is not striated in appearance and is found in many organs and whenever movement of fluid or solids is needed.

2 INCORRECT

A large broad sheet of connective tissue, such as on the abdomen, is called a/an _____ .(p.298)

- **Correct Answer** ○ A)aponeurosis
- ○ B)epimysium
- ● C)perimysium
- ○ D)endomysium

Feedback: Incorrect: There are many different arrangements of connective tissue; the lumbar and abdominal aponeuroses are two examples of broad,

Routing Information

Date: Thu Sep 20 08:19:31 CDT 2001

My name:

Email these results to:

Email address: Format:

Me: Text

My Instructor: Text

My TA: Text

Other: Text

[E-Mail The Results]

...t is the closest to the individual muscle fiber is the

Answer: Each level of muscle arrangement is covered ...tive tissue; a bundle (fascicle) of fibers is wrapped by a ...individual cells are surrounded with endomysium (recall peri-refers to around while endo- means inner.

Practice

Take a quiz at the *Human Physiology* Online Learning Center to gauge your mastery of chapter content. Each chapter quiz is specially constructed to test your comprehension of key concepts. Immediate feedback on your responses explains why an answer is correct or incorrect. You can even e-mail your quiz results to your professor!

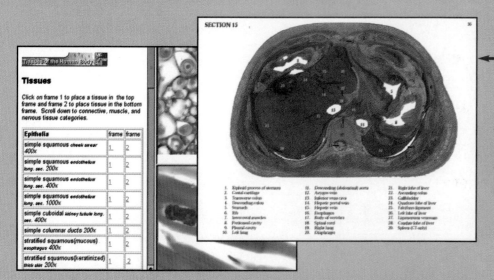

Course Tools

Here you'll find a wealth of online reference material such as histology and anatomy atlases, a scientific encyclopedia, and general study tips. Use these resources to supplement your study of physiology.

Access to Premium Learning Materials

The *Human Physiology* Online Learning Center is your portal to exclusive interactive study tools like McGraw-Hill's Essential Study Partner. Your log in privileges also allow you unlimited use of the 24-hour student tutorial service—help is just an e-mail away!

Visit www.mhhe.com/fox8 today!

McGraw Hill | **Higher Education**

(See next page for site registration instructions.)

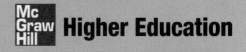

IMPORTANT:

HERE IS YOUR REGISTRATION CODE TO ACCESS
YOUR PREMIUM McGRAW-HILL ONLINE RESOURCES.

For key premium online resources you need THIS CODE to gain access. Once the code is entered, you will be able to use the Web resources for the length of your course.

If your course is using WebCT or Blackboard, you'll be able to use this code to access the McGraw-Hill content within your instructor's online course.

Access is provided if you have purchased a new book. If the registration code is missing from this book, the registration screen on our Website, and within your WebCT or Blackboard course, will tell you how to obtain your new code.

Registering for McGraw-Hill Online Resources

TO gain access to your MCGraw-Hill web resources simply follow the steps below:

1. USE YOUR WEB BROWSER TO GO TO: **http://www.mhhe.com/fox8**
2. CLICK ON **FIRST TIME USER**.
3. ENTER THE REGISTRATION CODE* PRINTED ON THE TEAR-OFF BOOKMARK ON THE RIGHT.
4. AFTER YOU HAVE ENTERED YOUR REGISTRATION CODE, CLICK **REGISTER**.
5. FOLLOW THE INSTRUCTIONS TO SET-UP YOUR PERSONAL UserID AND PASSWORD.
6. WRITE YOUR UserID AND PASSWORD DOWN FOR FUTURE REFERENCE. KEEP IT IN A SAFE PLACE.

TO GAIN ACCESS to the McGraw-Hill content in your instructor's WebCT or Blackboard course simply log in to the course with the UserID and Password provided by your instructor. Enter the registration code exactly as it appears in the box to the right when prompted by the system. You will only need to use the code the first time you click on McGraw-Hill content.

Thank you, and welcome to your MCGraw-Hill online Resources!

*YOUR REGISTRATION CODE CAN BE USED ONLY ONCE TO ESTABLISH ACCESS. IT IS NOT TRANSFERABLE.

0-07-291927-2 FOX: HUMAN PHYSIOLOGY, 8E

MCGRAW-HILL
ONLINE RESOURCES

REGISTRATION CODE

CL48-BJQI-WDC8-9433-S028

HUMAN PHYSIOLOGY

Eighth Edition

Stuart Ira Fox

PIERCE COLLEGE

McGraw Hill **Higher Education**

Boston   Burr Ridge, IL   Dubuque, IA   Madison, WI   New York   San Francisco   St. Louis
Bangkok   Bogotá   Caracas   Kuala Lumpur   Lisbon   London   Madrid   Mexico City
Milan   Montreal   New Delhi   Santiago   Seoul   Singapore   Sydney   Taipei   Toronto

The **McGraw·Hill** Companies

HUMAN PHYSIOLOGY, EIGHTH EDITION

3 4 5 6 7 8 9 0 VNH/VNH 0 9 8 7 6 5 4

ISBN 0-07-291928-0

Publisher: *Martin J. Lange*
Sponsoring editor: *Michelle Watnick*
Developmental editor: *Kristine A. Queck*
Director of development: *Kristine Tibbetts*
Marketing manager: *James F. Connely*
Project manager: *Sheila M. Frank*
Production supervisor: *Sherry L. Kane*
Senior media project manager: *Stacy A. Patch*
Senior media technology producer: *Barbara R. Block*
Coordinator of freelance design: *Michelle D. Whitaker*
Cover designer: *Diane Beasley*
Cover illustration: *William Westwood*
Senior photo research coordinator: *John C. Leland*
Photo research: *Chris Hammond/PhotoFind, LLC*
Supplement producer: *Brenda A. Ernzen*
Compositor: *Shepherd-Imagineering Media Services, Inc.*
Typeface: *10.2/12 Times Roman*
Printer: *Von Hoffmann Corporation*

The credits section for this book begins on page 706 and is considered an extension of the copyright page.

Library of Congress Cataloging-in-Publication Data

Fox, Stuart Ira.
   Human physiology / Stuart Ira Fox.—8th ed.
      p.   cm.
   Includes index.
   ISBN 0-07-291928-0
   1. Human physiology. I. Title.

QP34.5 .F68   2004
612—dc21   2002011254
                                      CIP

www.mhhe.com

To Laura, Eric and Jacob

Brief Contents

Contents

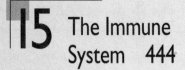

Preface

Human physiology provides the scientific foundation for the field of medicine and all other professions related to human health and physical performance. The scope of topics included in a human physiology course is therefore wide-ranging, yet each topic must be covered in sufficient detail to provide a firm basis for future expansion and application. The rigor of the course, however, need not diminish the student's initial fascination with how the body works. On the contrary, a basic understanding of physiological mechanisms can instill a deeper appreciation for the complexity and beauty of the human body and motivate the student to learn still more.

This text is designed to serve the needs of students in an undergraduate physiology course. The beginning chapters introduce basic chemical and biological concepts to provide these students—many of whom do not have extensive science backgrounds—with the framework they need to comprehend physiological principles. In the chapters that follow, the material is presented in such a way as to promote conceptual understanding rather than rote memorization of facts. Every effort has been made to help students integrate related concepts and to understand the relationships between anatomical structures and their functions.

Abundant summary flowcharts and tables serve as aids for review. Beautifully rendered figures, with a functional use of color, are designed to enhance learning. Health applications are discussed often to heighten interest, deepen understanding of physiological concepts, and help students relate the material they have learned to their individual career goals. In addition, various other pedagogical devices are used extensively (but not intrusively) to add to the value of the text as a comprehensive learning tool.

Changes to the Eighth Edition

Before I began writing this new edition, my editors at McGraw-Hill repeated a successful technique introduced at the last revision cycle: they requested users of the previous edition to send in their suggestions and comments, focusing on their chapter of particular interest. Thus, every chapter was reviewed several times over by people who had experience using the book in their own classrooms. The eighth edition benefited enormously by this input. It also benefited greatly through the reviews provided by faculty who previously used other texts.

Updates and Additions

The eighth edition incorporates a number of new and recently modified physiological concepts. This may surprise people who are unfamiliar with the subject; indeed, I'm sometimes asked if the field really changes much from one edition to the next. It does; that's one of the reasons physiology is so much fun to study. I've tried to impart this sense of excitement and fun in the book by indicating, in a manner appropriate for this level of text, where knowledge is new and where gaps in our knowledge remain. Following is a *partial* list of the topical additions and updates made to the eighth edition. New figures added to support the coordinating text discussion are also indicated.

Chapter 1: The Study of Body Function
- Animal models of human diseases
- Use of measurements and controls in physiology
- Use of statistics in physiology
- Homeostasis of blood glucose as example of negative feedback mechanisms
- Functions of different epithelial membranes

Chapter 3: Cell Structure and Genetic Control
- Human Genome Project
- Capsases and apoptosis
- Telomeres and life expectancy
- Chromatin structure affects gene expression (fig. 3.17)

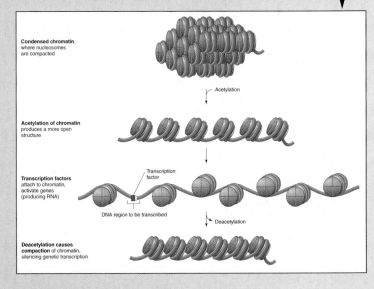

- Chromosomes and spindle fibers (fig. 3.31)

Chapter 4: Enzymes and Energy

- Structural formulas for NAD$^+$, NADH, FAD, and FADH$_2$ (fig. 4.17)

Chapter 6: Interactions Between Cells and the Extracellular Environment

- Discussion of integrins
- Transport across epithelial membranes
- Gated ion channels (fig. 6.4)

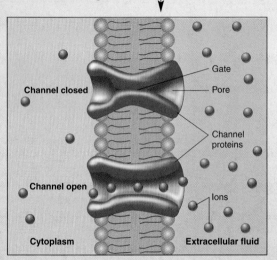

- Red blood cells in isotonic, hypotonic, and hypertonic solutions (fig. 6.11)
- Concentrations of ions in the intracellular and extracellular fluids (fig. 6.23)

Chapter 7: The Nervous System: Neurons and Synapses

- Astrocytes needed for the formation of synapses
- The action of local anesthetics
- The two types of channel inactivation mechanisms
- The function of endocannabinoid neurotransmitters
- Different types of neuroglial cells (fig. 7.5)

Chapter 8: The Central Nervous System

- Technology for visualizing brain function
- Role of neural stem cells in learning and memory
- Synaptic effects of abused drugs
- Glutamate receptors in long-term potentiation (fig. 8.15)

Chapter 9: The Autonomic Nervous System

- Cocaine as a sympathomimetic drug
- *Synapses en passant*

- Sympathetic and parasympathetic axons release different neurotransmitters (fig. 9.9)

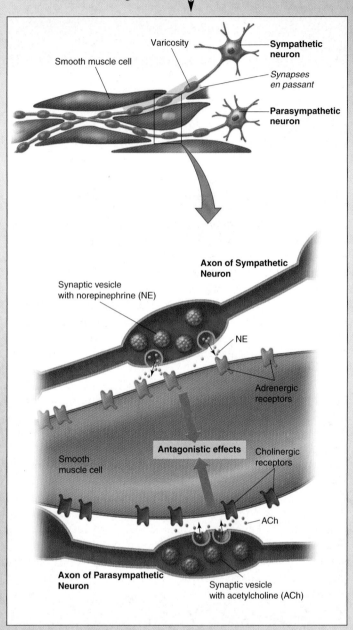

- The receptors involved in autonomic regulation (fig. 9.10)
- Comparison of nicotinic and muscarinic receptors (fig. 9.11)

Chapter 10: Sensory Physiology

- Mechanisms of taste cell activation (fig. 10.8)
- Functions of the retinal pigment epithelium
- Macular degeneration
- Effects of light on retinal cells (fig. 10.39)
- Light causes closing of Na$^+$ channels (fig. 10.38)
- Ganglion cell receptive fields (fig. 10.45)

Chapter 11: Endocrine Glands: Secretion and Action of Hormones
- Steroid hormone receptors
- cGMP as a second messenger and the action of Viagra

Chapter 12: Muscle: Mechanisms of Contraction and Neural Control
- Muscle structure: M lines and titin
- Eccentric muscle contractions
- Type IIX muscle fibers
- Mechanisms of muscle fatigue
- Genetic differences in muscle fiber types
- Role of troponin T, C, and I, and their use in diagnosing myocardial infarction
- A single motor unit (fig. 12.4*b*)
- Power stroke of the cross-bridge (fig. 12.11)

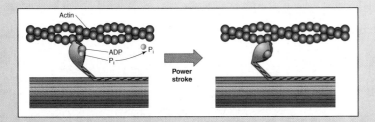

- Incomplete and complete tetanus (fig. 12.18)
- Relative abundance of different muscle fiber types (fig. 12.25)

Chapter 13: Heart and Circulation
- The capillary filtration barrier
- Calcium-stimulated calcium release in cardiac muscle
- Correlation of the ECG with the action potential (fig. 13.21)

Chapter 14: Cardiac Output, Blood Flow, and Blood Pressure
- Length-tension relationship in cardiac compared to skeletal muscle
- Comparison of cardiac and skeletal muscle length-tension relationships (fig. 14.4)

Chapter 15: The Immune System
- AIDS incidence and treatments
- Mechanisms of allergy and asthma
- Stages in the migration of white blood cells out of capillaries (fig. 15.1)
- Antigens on the surface of a bacterium (fig. 15.8)

- Migration of dendritic cells to lymphoid organs to activate T cells (fig. 15.15)

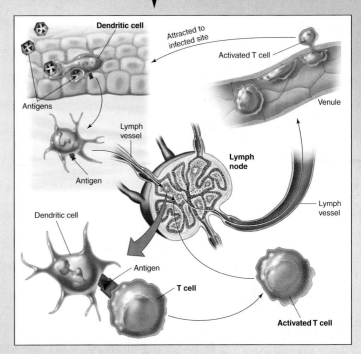

Chapter 16: Respiratory Physiology
- Role of nitric oxide in the hypoxic ventilatory response

Chapter 17: Physiology of the Kidneys
- Tubular secretion of drugs and organic anion transporters
- Use of drugs to inhibit renal tubular secretion of antibiotics
- How homeostasis is maintained by the action of ADH (fig. 17.20)

Chapter 18: The Digestive System
- Regulation of swallowing
- Stomach secretion of ghrelin
- Slow wave conduction by interstitial cells of Cajal
- Amounts of bile salts recirculated and excreted
- Transporters that secrete xenobiotics into bile
- Drugs, SXR nuclear receptors, and cytochrome P450 enzymes
- Slow waves in the intestine (fig. 18.15)
- Pathway for the metabolism of heme and bilirubin (fig. 18.23)
- A pancreatic acinus (fig. 18.28*b*)

Chapter 19: Regulation of Metabolism

- Free radicals, oxidative stress, antioxidants and homeostasis
- Role of ghrelin in the regulation of hunger
- Drugs that bind PPAR (nuclear receptors for treating type 2 diabetes)
- Factors that affect calorie expenditures
- Regulation of adaptive thermogenesis
- Impaired glucose tolerance and oral glucose tolerance test
- Lifestyle changes and impaired glucose tolerance
- Bone resorption and deposition
- Mechanisms of osteoclast activity
- Role of estrogens in bone mineralization
- Reactive oxygen species production and defense (fig. 19.1)
- The action of leptin (fig. 19.3)
- The regulation of insulin secretion (fig. 19.7)
- Resorption of bone by osteoclasts (fig. 19.18b)

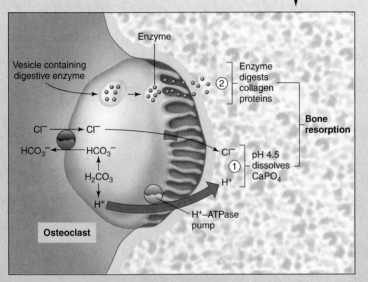

Chapter 20: Reproduction

- Role of estrogens in spermatogenesis
- Role of nitric oxide in penile erection
- Production of weak estrogens in postmenopausal women
- Embryonic stem cells and cloning technology
- Totipotency, pluripotency, and transdifferentiation
- Genetic screening of neonates
- Umbilical cord blood banking
- Passive immunization of fetus and baby by maternal antibodies
- Role of nitric oxide in erection and the action of Viagra fig. 20.23)

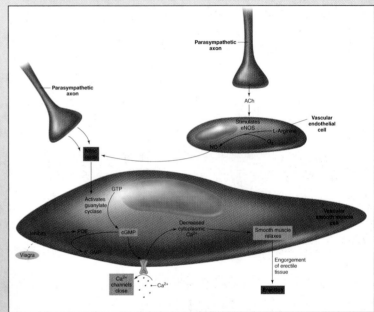

- Implantation of the blastocyst (fig. 20.45b)
- Maternal antibodies that protect the baby (fig. 20.56)

Art Upgrades

The most immediately apparent changes in this edition are in the art program. Although previous editions were praised for the high quality of the figures—their clarity, pedagogical usefulness, and beauty—the art in the current edition represents a marked improvement.

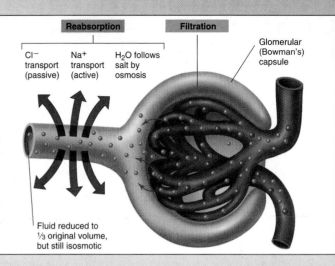

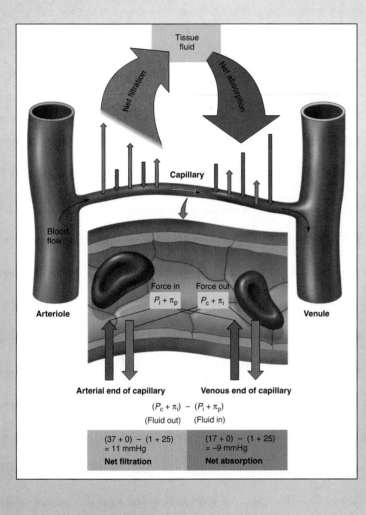

Three-Dimensional Art Brings Concepts to Life

Virtually all of the figures from the previous edition have been revised with a view toward improving the clarity with which they depict physiological concepts. In some cases, this involved changes in labeling; in other cases, changes in the content or balance of the figure components. In most cases, the revisions included making the art more three-dimensional and using more vibrant colors.

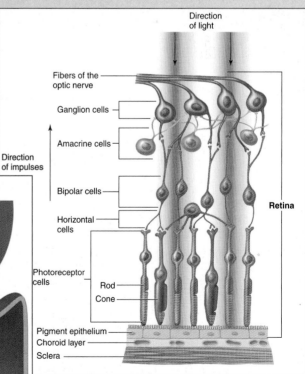

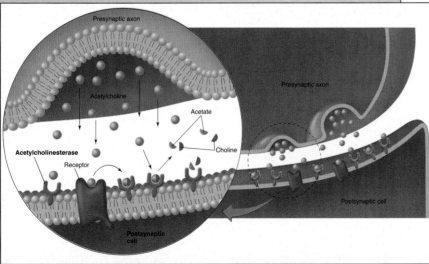

Consistent Colors Promote Understanding

The complete revision of the art program allowed us to standardize the appearance of particular structures so that structures are presented consistently across figures in all chapters of the book. This continuity makes it easier for students to interpret each figure, thereby improving the clarity of the total presentation. This key shows a sampling of some of the structures that have been standardized.

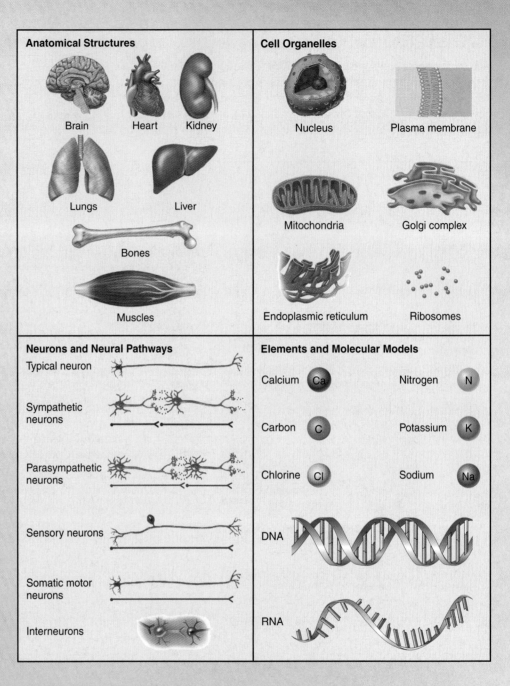

In addition to updating the existing artwork to achieve more dimension and continuity, many entirely new figures have been added to the eighth edition. Despite the many figure changes, the philosophy of the art program remains the same as in previous editions: the art *supports* the text explanation; it does *not substitute* for text explanation. This allows students to learn difficult concepts by following detailed explanations, rather than by trying to decipher overly complex figures. Thus, although the newly enhanced art program attracts the eye, its purpose is not to dazzle but to better illustrate the physiological concepts described in the text.

Unique Text Features Promote Active Learning

Each chapter has a consistent organization to help students learn the concepts explained in the text and illustrated in the figures. The numerous pedagogical features can aid students in their quest to master new terminology, learn new concepts, analyze and understand physiological principles, and finally apply this knowledge in practical ways. The chapter organization, and the learning devices built into each chapter, facilitates this growth by providing mechanisms for active learning of the chapter contents. Because mastery of the content of a human physiology course requires such active learning, students are advised to make use of the learning aids in each chapter.

Chapter Openers Set the Stage

Chapter Objectives
Students can look over the objectives before reading the chapter to get a feeling for the material covered, and check off the objectives as each major section is completed.

Chapter at a Glance
Students can use the chapter outline to get an overview of the chapter, or to find specific topics.

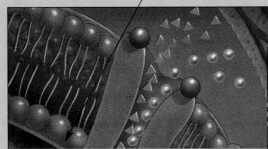

10 Sensory Physiology

Objectives
After studying this chapter, you should be able to . . .

1. explain how sensory receptors are categorized, give examples of functional categories, and explain how tonic and phasic receptors differ.

2. explain the law of specific nerve energies.

3. describe the characteristics of the generator potential.

4. give examples of different types of cutaneous receptors and describe the neural pathways for the cutaneous senses.

5. explain the concepts of receptive [fields and latera]l inhibition.

[explain how t]aste cells are [stimulated by] foods that are salty, [sweet, sour, an]d bitter.

[describe the s]tructure and function [of olfacto]ry receptors and [explain how o]dor discrimination [is accomp]lished.

8. describe the structure of the vestibular apparatus and explain how it provides information about acceleration of the body in different directions.

9. describe the functions of the outer and middle ear.

10. describe the structure of the cochlea and explain how movements of the stapes against the oval window result in vibrations of the basilar membrane.

11. explain how mechanical energy is converted into nerve impulses by the organ of Corti and how pitch perception is accomplished.

12. describe the structure of the eye and explain how images are brought to a focus on the retina.

13. explain how visual accommodation is achieved and describe the defects associated with myopia, hyperopia, and astigmatism.

14. describe the architecture of the retina and trace the pathways of light and nerve activity through the retina.

15. describe the function of rhodopsin in the rods and explain how dark adaptation is achieved.

16. explain how light affects the electrical activity of rods and their synaptic input to bipolar cells.

17. explain the trichromatic theory of color vision.

18. compare rods and cones with respect to their locations, synaptic connections, and functions.

19. describe the neural pathways from the retina, explaining the differences in pathways from different regions of the visual field.

Refresh Your Memory
Before you begin this chapter, you may want to review these concepts from previous chapters:

☑ Cerebral Cortex 193
☑ Ascending Tracts 209
☑ Cranial and Spinal Nerves 212

Take Advantage of the Technology
Visit the Online Learning Center for these additional study resources.

☐ Interactive quizzing
☐ Online study guide
☐ Current news feeds
☐ Crossword puzzles
☐ Vocabulary flashcards
☐ Labeling activities

www.mhhe.com/fox8

Refresh Your Memory
This boxed information at the beginning of each chapter alerts students to material from previous chapters that they may need to review as they begin a new chapter.

Take Advantage of the Technology
Also located in the opening spread of each chapter, these short sections acquaint students with the resources available at the Online Learning Center www.mhhe.com/fox8

In-text Aids Keep Students Focused

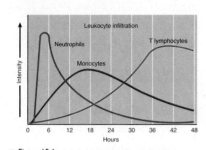

Figure 15.6 Infiltration of an inflamed site by leukocytes. Different types of leukocytes infiltrate the site of a local inflammation. Neutrophils arrive first, followed by monocytes and then lymphocytes.

anticoagulant ability (chapter 13). However, mast cells produce a variety of other molecules that play important roles in inflammation (and in allergy, discussed in a later section).

Mast cells release **histamine** which is stored in intracellular granules and secreted during inflammation and allergy. Histamine binds to its H_1 histamine receptors in the smooth muscle of bronchioles to stimulate bronchiolar constriction (as in asthma), but produces relaxation of the smooth muscles in blood vessels (causing vasodilation). Histamine also promotes increased capillary permeability, bringing more leukocytes to the infected area.

With a time delay, mast cells release inflammatory prostaglandins and leukotrienes (chapter 11), as well as a variety of cytokines that promote inflammation. In addition, mast cells secrete *tumor necrosis factor$_\alpha$* (TNF$_\alpha$), which acts as a chemokine to recruit neutrophils to the infected site.

These effects produce the characteristic symptoms of a local inflammation: *redness* and *warmth* (due to histamine-stimulated vasodilation); *swelling* (edema) and *pus* (the accumulation of dead leukocytes); and *pain*. If the infection continues, the release of endogenous pyrogen from leukocytes and macrophages may also produce a fever, as previously discussed.

Test Yourself Before You Continue

1. List the phagocytic cells found in blood and lymph, and indicate which organs contain fixed phagocytes.
2. Describe the actions of interferons.
3. Distinguish between innate and adaptive immunity, and describe the properties of antigens.
4. Distinguish between B and T lymphocytes in terms of their origins and immune functions.
5. Identify their fu
6. Descri

Functions of B Lymphocytes

B lymphocytes secrete antibodies that can bind to antigens in a specific fashion. This bonding stimulates a cascade of reactions whereby a system of plasma proteins called complement is activated. Some of the activated complement proteins kill the cells containing the antigen; others promote phagocytosis, resulting in a more effective defense against pathogens.

Exposure of a B lymphocyte to the appropriate antigen results in cell growth followed by many cell divisions. Some of the progeny become **memory cells;** these are visually indistinguishable from the original cell and are important in active immunity. Others are transformed into **plasma cells** (fig. 15.7). Plasma cells are protein factories that produce about 2,000 antibody proteins per second.

The antibodies that are produced by plasma cells when B lymphocytes are exposed to a particular antigen react specifically with that antigen. Such antigens may be isolated molecules, as illustrated in figure 15.7, or they may be molecules at the surface of an invading foreign cell (fig. 15.8). The specific bonding of antibodies to antigens serves to identify the enemy and to activate defense mechanisms that lead to the invader's destruction.

Antibodies

Antibody proteins are also known as **immunoglobulins.** They are found in the gamma globulin class of plasma proteins, as identified by a technique called *electrophoresis* in which different types of plasma proteins are separated by their movement in an electric field (fig. 15.9). The five distinct bands of proteins that appear are albumin, alpha-1 globulin, alpha-2 globulin, beta globulin, and gamma globulin.

The gamma globulin band is wide and diffuse because it represents a heterogeneous class of molecules. Since antibodies are specific in their actions, it follows that different types of antibodies should have different structures. An antibody against smallpox, for example, does not confer immunity to poliomyelitis and, therefore, must have a slightly different structure than an antibody against polio. Despite these differences, antibodies are structurally related and form only a few classes.

There are five immunoglobulin (abbreviated Ig) subclasses: *IgG, IgA, IgM, IgD,* and *IgE.* Most of the antibodies in serum are in the IgG subclass, whereas most of the antibodies in external secretions (saliva and milk) are IgA (table 15.6). Antibodies in the IgE subclass are involved in certain allergic reactions.

ed polypep-
re joined to
at these four

Test Yourself Before You Continue

1. List the phagocytic cells found in blood and lymph, and indicate which organs contain fixed phagocytes.
2. Describe the actions of interferons.
3. Distinguish between innate and adaptive immunity, and describe the properties of antigens.
4. Distinguish between B and T lymphocytes in terms of their origins and immune functions.
5. Identify the primary and secondary lymphoid organs and describe their functions.
6. Describe the events that occur during a local inflammation.

Perspectives
Immediately following each major section heading is a concise statement of the section's central concepts, or organizing themes, that will be illustrated in detail in the text that follows. These brief introductions are designed to help students place the sections in perspective, before getting involved with the specifics.

Test Yourself Before You Continue
Each major chapter section ends with a set of learning activities and essay questions that relate only to the material presented in the section. Students are encouraged to answer the essay questions, draw the outlines and flowcharts requested, and otherwise actively participate in their learning of this material. Thus, these sections serve as both a "reality check" for the student and as a mechanism for active learning.

Clinical Content Adds Interest

Clinical information is presented throughout the text to underscore the real-life importance of understanding human physiology and to provide concrete examples that demonstrate the application of complex physiological concepts.

Clinical Investigation
Clinical Investigations are diagnostic puzzles provided at the very beginning of each chapter. These thought-provoking cases are designed to engage students' interest and motivate them to delve into the content of each chapter. Students must read the chapter, understand the concepts, and look for clues in order to arrive at the correct diagnosis.

Clinical Investigation

Jason is a 19-year-old college student who goes to the doctor complaining of chronic fatigue. The doctor palpates Jason's radial pulse and discovers that it is fast and weak. An echocardiogram and later coronary arteriograph reveal that he has a ventricular septal defect and mitral stenosis. His electrocardiogram (ECG) indicates that he has sinus tachycardia. When laboratory test results are returned, they indicate that Jason has a very high plasma cholesterol concentration with a high LDL/HDL ratio.

What can be concluded from these findings, and how are they related to Jason's complaint of chronic fatigue?

Clinical Investigation Clues
Scattered within each chapter, these short boxes remind students of the ongoing clinical investigation puzzle and provide clues to the solution. Clues are carefully placed so they always relate to the information presented in the preceding text. These clues help reinforce comprehension of the text material and spur students to continue reading so they can gather all of the pertinent information needed to solve the puzzle. After attempting to diagnose the case, students can find the solution to each Clinical Investigation in Appendix A.

Clinical Investigation Clues

Remember that Brenda experienced muscle pain and fatigue during her training, and that she had an episode where she experienced severe pain in her left pectoral region following an intense workout.

What produced her muscle pain and fatigue?
What might have caused the severe pain in her left pectoral region?
Which of these effects are normal?

Boxed Clinical and Fitness Applications
Applications—in clinical medicine general health, and physical fitness—of basic physiological principles are found intermittently throughout the body of the text. Placement of these applications is precise—they always relate to concepts that have been presented *immediately preceding* the application. As such, they provide immediate reinforcement for students learning the fundamental principles on which the applications are based. This is preferable to longer but fewer magazine-article-type applications that are separated from the text information. The immediate reinforcement allows students to see the practical importance of learning the material they have just studied.

The saturated fat content (expressed as a percentage of total fat) for some food items is as follows: canola, or rapeseed, oil (6%); olive oil (14%); margarine (17%); chicken fat (31%); palm oil (51%); beef fat (52%); butter fat (66%); and coconut oil (77%). Health authorities recommend that a person's total fat intake not exceed 30% of the total energy int... ...saturated fat contribute less than 10% of the... the diet may co... icant risk factor... mal fats, which... saturated than...

Many people with dangerously high LDL-cholesterol concentrations take drugs known as **statins.** These drugs function as inhibitors of the enzyme *HMG-coenzyme A reductase,* which catalyzes the rate-limiting step in cholesterol synthesis. The statins therefore decrease the ability of the liver to produce its own cholesterol. The lowered intracellular cholesterol then stimulates the production of LDL receptors, allowing the liver cells to engulf more LDL-cholesterol. When a person takes a statin drug, therefore, the liver cells remove more LDL-cholesterol from the blood and thus decrease the amount of blood LDL-cholesterol that can enter the endothelial cells of arteries.

HPer Links Establish Connections

Interactions: HPer Links ————

Interactions page can be found at the end of each chapter or group of chapters relating to a particular body system, and also at the ends of chapters 3, 5, and 6. These resource pages list the many ways a major concept applies to the study of different body systems, and the ways that a given system interacts with other body systems. Each application or interaction includes a page reference to related material in the textbook.

The term *HPer Links* is a hybrid of "hyperlinks" and the initials of "Human Physiology." On the Internet, a hyperlink is a reference that you can click with a mouse to jump from one part of a document or web page to another. Students can use the cross-references offered on the Interactions pages in a similar way to find interrelated topics in the textbook.

► INTERACTIONS

HPer Links of Metabolism Concepts to the Body Systems

Integumentary System
- The skin synthesizes vitamin D from a derivative of cholesterol ....(p. 625)
- The metabolic rate of the skin varies greatly, depending upon ambient temperature ....(p. 428)

Nervous System
- The aerobic respiration of glucose serves most of the energy needs of the brain ....(p. 119)
- Regions of the brain with a faster metabolic rate, resulting from increased brain activity, receive a more abundant blood supply than regions with a slower metabolic rate ....(p. 427)

Endocrine System
- Hormones that bind to receptors in the plasma membrane of their target cells activate enzymes in the target cell cytoplasm ....(p. 294)
- Hormones that bind to nuclear receptors in their target cells alter the target cell metabolism by regulating gene expression ....(p. 292)
- Hormonal secretions from adipose cells regulate hunger and metabolism ..(p. 606)
- Anabolism and catabolism are regulated by a number of hormones ....(p. 609)
- Insulin stimulates the synthesis of glycogen and fat ....(p. 611)
- The adrenal hormones stimulate the breakdown of glycogen, fat, and protein ....(p. 619)
- Thyroxine stimulates the production of a protein that uncouples oxidative phosphorylation. This helps to increase the body's metabolic rate ....(p. 620)
- Growth hormone stimulates protein synthesis ....(p. 621)

Muscular System
- The intensity of exercise that can be performed aerobically depends on a person's maximal oxygen uptake and lactate threshold ....(p. 343)

- The body consumes extra oxygen for a period of time after exercise has ceased. This extra oxygen is used to repay the oxygen debt incurred during exercise ....(p. 344)
- Glycogenolysis and gluconeogenesis by the liver help to supply glucose for exercising muscles ....(p. 343)
- Trained athletes obtain a higher proportion of skeletal muscle energy from the aerobic respiration of fatty acids than do nonathletes ....(p. 346)
- Muscle fatigue is associated with anaerobic respiration and the production of lactic acid ....(p. 346)
- The proportion of energy derived from carbohydrates or lipids by exercising skeletal muscles depends on the intensity of the exercise ....(p. 343)

Circulatory System
- Metabolic acidosis may result from excessive production of either ketone bodies or lactic acid ....(p. 377)
- The metabolic rate of skeletal muscles determines the degree of blood vessel dilation, and thus the rate of blood flow to the organ ....(p. 424)
- Atherosclerosis of coronary arteries can force a region of the heart to metabolize anaerobically and produce lactic acid. This is associated with angina pectoris .(p. 397)

Respiratory System
- Ventilation oxygenates the blood going to the cells for aerobic cell respiration and removes the carbon dioxide produced by the cells ....(p. 480)
- Breathing is regulated primarily by the effects of carbon dioxide produced by aerobic cell respiration ....(p. 500)

Urinary System
- The kidneys eliminate urea and other waste products of metabolism from the blood plasma ....(p. 539)

Digestive System
- The liver contains enzymes needed for many metabolic reactions involved in regulating the blood glucose and lipid concentrations ....(p. 579)
- The pancreas produces many enzymes needed for the digestion of food in the small intestine ....(p. 582)
- The digestion and absorption of carbohydrates, lipids, and proteins provides the body with the substrates used in cell metabolism ....(p. 587)
- Vitamins A and D help to regulate metabolism through the activation of nuclear receptors, which bind to regions of DNA ....(p. 601)

Reproductive System
- The sperm do not contribute mitochondria to the fertilized oocyte ....(p. 58)
- The endometrium contains glycogen that nourishes the developing embryo .(p. 663)

Chapter Review Pages Summarize and Challenge

Chapter Summaries ──────➤
At the end of each chapter, the material is summarized in outline form. This outline summary is organized by major section headings with page references, followed by the key points in the section. Students may read the summary after studying the chapter to be sure that they haven't missed any points, and can use the chapter summaries to help review for examinations.

Summary

Introduction to Physiology 4

I. Physiology is the study of how cells, tissues, and organs function.
   A. In the study of physiology, cause-and-effect sequences are emphasized.
   B. Knowledge of physiological mechanisms is deduced from data obtained experimentally.

II. The science of physiology overlaps with chemistry and physics and shares knowledge with the related sciences of pathophysiology and comparative physiology.
   A. Pathophysiology is concerned with the functions of diseased or injured body systems and is based on knowledge of how normal systems function, which is the focus of physiology.
   B. Comparative physiology is concerned with the physiology of animals other than humans and shares much information with human physiology.

III. All of the information in this book has been gained by applications of the scientific method. This method has three essential characteristics.
   A. It is assumed that the subject under study can ultimately be explained in terms we can understand.
   B. Descriptions and explanations are honestly based on observations of the natural world and can be changed as warranted by new observations.
   C. Humility is an important characteristic of the scientific method; the scientist must be willing to change his or her theories when warranted by the weight of the evidence.

The Primary Tissues 9

I. The body is composed of four primary tissues: muscle, nervous, epithelial, and connective tissues.
   A. There are three types of muscle tissue: skeletal, cardiac, and smooth muscle.
      1. Skeletal and cardiac muscle are striated.
      2. Smooth muscle is found in the walls of the internal organs.
   B. Nervous tissue is composed of neurons and supporting cells.
      1. Neurons are specialized for the generation and conduction of electrical impulses.
      2. Supporting cells provide the neurons with anatomical and functional support.
   C. Epithelial tissue includes membranes and glands.
      1. Epithelial membranes cover and line the body surfaces, and their cells are tightly joined by junctional complexes.
      2. Epithelial membranes may be simple or stratified and their cells may be squamous, cuboidal, or columnar.
      3. Exocrine glands, which secrete into ducts, and endocrine glands, which lack ducts and secrete hormones into the blood, are derived from epithelial membranes.

2. In a negative feedback loop, the effector acts to cause changes in the internal environment that compensate for the initial deviations that were detected by the sensor.
   B. Positive feedback loops serve to amplify changes and may be part of the action of an overall negative feedback mechanism.
   C. The nervous and endocrine systems provide extrinsic regulation of other body systems and act to maintain homeostasis.
   D. The secretion of hormones is stimulated by specific chemicals and is inhibited by negative feedback mechanisms.

II. Effectors act antagonistically to defend the set point against deviations in any direction.

D. Connective tissue is characterized by large intercellular spaces that contain extracellular material.
   1. Connective tissue proper is categorized into subtypes, including loose, dense fibrous, adipose, and others.
   2. Cartilage, bone, and blood are classified as connective tissues because their cells are widely spaced with abundant extracellular material between them.

Organs and Systems 17

I. Organs are units of structure and function that are composed of at least two, and usually all four, primary tissues.
   A. The skin is a good example of an organ.
      1. The epidermis is a stratified squamous keratinized epithelium that protects underlying structures and produces vitamin D.
      2. The dermis is an example of loose connective tissue.
      3. Hair follicles, sweat glands, and sebaceous glands are exocrine glands located within the dermis.
      4. Sensory and motor nerve fibers enter the spaces within the dermis to innervate sensory organs and smooth muscles.
      5. The arrector pili muscles that attach to the hair follicles are composed of smooth muscle.
   B. Organs that are located in different regions of the body and that perform related functions are grouped into systems. These include, among others, the circulatory system, digestive system, and endocrine system.

II. The fluids of the body are divided into two major compartments.
   A. The intracellular compartment refers to the fluid within cells.
   B. The extracellular compartment refers to the fluid outside of cells; extracellular fluid is subdivided into plasma (the fluid portion of the blood) and tissue (interstitial) fluid.

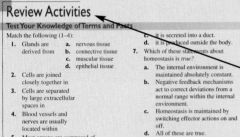

Review Activities

Test Your Knowledge of Terms and Facts

Match the following (1–4):
1. Glands are derived from
   a. nervous tissue
   b. connective tissue
   c. muscular tissue
   d. epithelial tissue
2. Cells are joined closely together in
3. Cells are separated by large extracellular spaces in
4. Blood vessels and nerves are usually located within
5. Most organs are composed of
   a. epithelial tissue.
   b. muscle tissue.
   c. connective tissue.
   d. all of these.
6. Sweat is secreted by exocrine glands. This means that
   a. it is produced by epithelial cells.
   b. it is a hormone.

c. it is secreted into a duct.
d. it is produced outside the body.
7. Which of these statements about homeostasis is *true*?
   a. The internal environment is maintained absolutely constant.
   b. Negative feedback mechanisms act to correct deviations from a normal range within the internal environment.
   c. Homeostasis is maintained by switching effector actions on and off.
   d. All of these are true.
8. In a negative feedback loop, the effector organ produces changes that are
   a. in the same direction as the change produced by the initial stimulus.
   b. opposite in direction to the change produced by the initial stimulus.
   c. unrelated to the initial stimulus.
9. A hormone called parathyroid hormone acts to help raise the blood

calcium concentration. According to the principles of negative feedback, an effective stimulus for parathyroid hormone secretion would be
   a. a fall in blood calcium.
   b. a rise in blood calcium.
10. Which of these consists of dense parallel arrangements of collagen fibers?
   a. skeletal muscle tissue
   b. nervous tissue
   c. tendons
   d. dermis of the skin
11. The act of breathing raises the blood oxygen level, lowers the blood carbon dioxide concentration, and raises the blood pH. According to the principles of negative feedback, sensors that regulate breathing should respond to
   a. a rise in blood oxygen.
   b. a rise in blood pH.
   c. a rise in blood carbon dioxide concentration.
   d. all of these.

Test Your Understanding of Concepts and Principles

1. Describe the structure of the various epithelial membranes and explain how their structures relate to their functions.[1]
2. Compare bone, blood, and the dermis of the skin in terms of their similarities. What are the major structural differences between these tissues?

3. Describe the role of antagonistic negative feedback processes in the maintenance of homeostasis.
4. Using insulin as an example, explain how the secretion of a hormone is controlled by the effects of that hormone's actions.

5. Describe the steps in the development of pharmaceutical drugs and evaluate the role of animal research in this process.
6. Why is Claude Bernard considered the father of modern physiology? Why is the concept he introduced so important in physiology and medicine?

Test Your Ability to Analyze and Apply Your Knowledge

1. What do you think would happen if most of your physiological regulatory mechanisms were to operate by positive feedback rather than by negative feedback? Would life even be possible?

2. Examine figure 1.5 and determine when the compensatory physiological responses began to act, and how many minutes they required to restore the initial set point of blood glucose concentration. Comment on the

importance of quantitative measurements in physiology.
3. Why are interactions between the body-fluid compartments essential for sustaining life?

Related Websites

Check out the Links Library at www.mhhe.com/fox8 for links to sites containing resources related to the study of body function. These links are monitored to ensure current URLs.

[1]*Note:* This question is answered in the chapter 1 Study Guide found on the Online Learning Center at www.mhhe.com/fox8.

Review Activities
A battery of questions collectively titled Review Activities follows each chapter summary. These self-examinations are organized into three increasingly difficult learning levels to help students progress from simple memorization to higher levels of understanding.

- *Test Your Knowledge of Terms and Facts* is a series of multiple-choice questions that prompt students to recall key terms and facts presented in the chapter. Answers to these questions are found in Appendix B.
- *Test Your Understanding of Concepts and Principles* consists of brief essay questions that require students to demonstrate their understanding of chapter material.
- *Test Your Ability to Analyze and Apply Your Knowledge* questions stimulate critical thinking by challenging students to utilize chapter concepts to solve a problem.

Teaching and Learning Supplements

McGraw-Hill offers various tools and technology products to support the eighth edition of *Human Physiology*. Students can order supplemental study materials by contacting their campus bookstore. Instructors can obtain teaching aids by calling the McGraw-Hill Customer Service Department at 1-800-338-3987, visiting our A&P catalog at www.mhhe.com/ap, or contacting your local McGraw-Hill sales representative.

For Instructors

Digital Content Manager

This multimedia collection of visual resources allows instructors to utilize artwork from the text in multiple formats to create customized classroom presentations, visually based tests and quizzes, dynamic course website content, or attractive printed support materials. The digital assets on this cross-platform CD-ROM are grouped by chapter within easy-to-use folders.

Art Library Full-color digital files of all illustrations in the book, plus grayscale versions of the same artwork, are housed in the Art Library. These files can be readily incorporated into lecture presentations, exams, or custom-made classroom materials.

Photo Library The Photo Library folder contains digital files of instructionally significant photographs from the text, which can be reproduced for multiple classroom uses.

Table Library Every table that appears in the text is saved in electronic form.

PowerPoint The Digital Content Manager supplies two types of PowerPoint files for each of the 20 chapters of the text. PowerPoint Lectures are ready-made lecture presentations that combine art with lecture outlines. These lectures can be used as they are, or can be tailored to reflect your preferred lecture topics and sequences. PowerPoint Image Slides present all art, photos, and tables from each chapter pre-inserted into blank PowerPoint slides for ease of use.

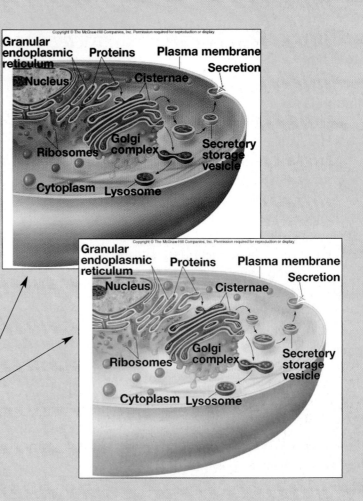

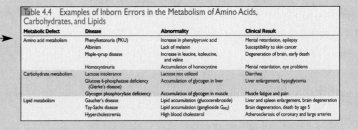

Table 4.4   Examples of Inborn Errors in the Metabolism of Amino Acids, Carbohydrates, and Lipids

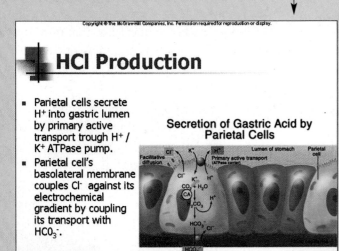

In addition to the assets found within each chapter, the Digital Content Manager for *Human Physiology* contains the following multimedia instructional materials, organized by topic.

Active Art Library Active Art consists of art files that have been converted to a format that allows the artwork to be edited inside of Microsoft PowerPoint. Each piece of art inside an Active Art presentation can be broken down to its core elements, grouped or ungrouped, and edited to create customized illustrations.

Animations Library Numerous full-color animations illustrating physiological processes are provided. Harness the visual impact of processes in motion by importing these files into classroom presentations or online course materials. ⟶

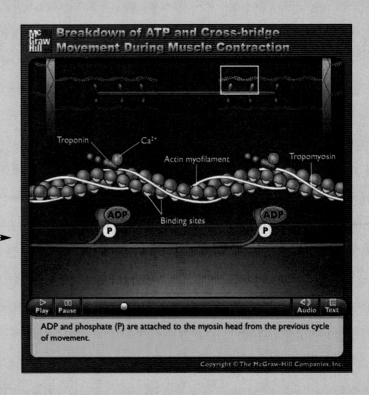

Instructor's Testing and Resource CD

This cross-platform CD provides a wealth of resources for the instructor in one convenient place. Supplements featured on this CD include a computerized test bank utilizing Brownstone Diploma® testing software to quickly create customized exams. This user-friendly program allows instructors to search for questions by topic, format, or difficulty level; edit existing questions or add new ones; and scramble questions and answer keys for multiple versions of the same test. Microsoft Word files of the test bank are included for those instructors who prefer to work outside of the test-generator software.

Other assets on the Instructor's Testing and Resource CD include the Instructor's Manual, Instructor's Manual for Laboratory Guide, Student Study Guide, and answers to the end-of-chapter questions from the text.

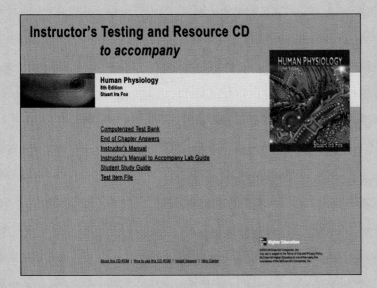

Transparencies

This set of 350 transparency overheads includes key pieces of line art from the textbook. Images are printed with better visibility and contrast than ever before, and labels are large and bold for clear projection.

Laboratory Manual

The **Laboratory Guide to Human Physiology: Concepts and Clinical Applications,** Tenth Edition, by Stuart I. Fox, is self-contained so students can prepare for laboratory exercises and quizzes without having to bring the textbook to the laboratory. The introduction to each exercise contains cross-references to the pages in this textbook where related information can be found. Similarly, those figures in the lab manual that correspond to full-color figures in the textbook are also cross-referenced. Both of these mechanisms help students better integrate the lecture and laboratory portions of their course. The manual provides laboratory exercises, classroom tested for a number of years, that reinforce many of the topics covered in this textbook and in the human physiology course.

Online Learning Center

The *Human Physiology* Online Learning Center at www.mhhe.com/fox8 is a comprehensive website created for instructors and students using the Fox textbook. For details about the student assets included on this site, please refer to the inside front cover of this book. The Online Learning Center allows instructors complete access to all student features, as well as exclusive access to a separate Instructor Center that houses downloadable and printable versions of traditional ancillaries, plus additional instructor content. Contact your McGraw-Hill sales representative for your instructor user name and password.

Course Delivery Systems

With help from our partners WebCT, Blackboard, Top-Class, eCollege, and other course management systems, professors can take complete control over their course content. Course cartridges containing Online Learning Center content, online testing, and powerful student tracking features are readily available for use within these platforms.

For Students

MediaPhys 2.0

This interactive tutorial CD-ROM offers detailed explanations, high-quality illustrations, and animations to provide students with a thorough introduction to the world of physiology—giving them a virtual tour of physiological processes. MediaPhys is filled with interactive activities and quizzes to help reinforce physiology concepts that are often difficult to understand.

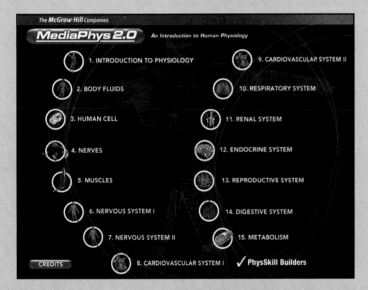

Online Learning Center

The *Human Physiology* Online Learning Center at www.mhhe.com/fox8 offers access to a vast array of premium online content to fortify the learning experience.

Essential Study Partner A collection of interactive study modules that contains hundreds of animations, learning activities, and quizzes designed to help students grasp complex concepts.

Live News Feeds The OLC offers course-specific real-time news articles to help students and instructors stay current with the latest topics in physiology.

Online Tutoring A 24-hour tutorial service moderated by qualified instructors. Help with difficult concepts is only an email away!

In addition to these outstanding online tools, the OLC features quizzes, interactive learning games, and study tools tailored to coincide with each chapter of the text. Turn to the inside front cover to learn more about the exciting features found on this student resource.

Physiology Interactive Lab Simulations (Ph.I.L.S.)

The Ph.I.L.S. CD-ROM is the perfect supplement or replacement for wet labs. Eleven laboratory simulations allow students to perform experiments without using expensive lab equipment or live animals. This easy-to-use software offers students the flexibility to change the parameters of every lab experiment, with no limit to the amount of times a student can repeat experiments or modify variables. This power to manipulate each experiment reinforces key physiology concepts by helping students to view outcomes, make predictions, and draw conclusions.

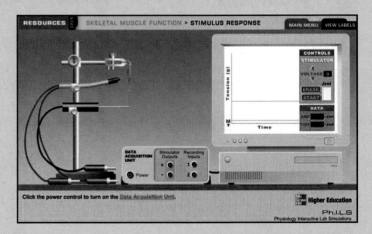

Acknowledgments

As mentioned previously, many users of the seventh edition contributed individual chapter reviews. I am extremely grateful to all of them and have endeavored to incorporate their suggestions wherever possible. In addition to the many professors who contributed chapter reviews of the seventh edition, several colleagues reviewed the entire seventh edition. I am grateful to them for taking on this arduous task, and know that their efforts resulted in a greatly improved eighth edition. I would also like to thank my colleagues at Pierce College, particularly Dr. Lawrence G. Thouin Jr., Dr. James Rikel, and Ms. Karen Gebhardt for their input and support.

Reviewers

William D. Blaker
Furman University

Maureen Burton
University of South Dakota School of Medicine

Robert G. Carroll
Brody School of Medicine, East Carolina University

Paul V. Cupp, Jr.
Eastern Kentucky University

Stephen Dodd
University of Florida

Jeff Engel
Western Illinois University

John Erickson
Ivy Tech State College

Michael T. Griffin
Angelo State University

Eric S. Hall
Rhode Island College

John P. Harley
Eastern Kentucky University

Jean A. S. Helgeson
Collin County Community College

Brent J.F. Hill
Augustana College

William F. Jackson
Western Michigan University

Paul E. Jarrell
Pasadena City College

Kiyomi Koizumi
State University of New York-Health Science Center at Brooklyn

Joan Esterline Lafuze
Indiana University East

Charles K. Levy
Boston University

Harvey N. Mayrovitz
College of Medical Sciences, Nova Southeastern University

Kip L. McGilliard
Eastern Illinois University

William F. Nicholson
University of Arkansas, Monticello

Colleen J. Nolan
St. Mary's University

Robert K. Okazaki
Weber State University

Larry A. Reichard
Maple Woods Community College

Laurel Roberts
University of Pittsburgh

Sarmad Saman
Quinsigamond Community College

David J. Saxon
Morehead State University

Joel D. Schiff
New York University Dental Center

Michael Edward Soulsby
University of Arkansas for Medical Sciences
University of Arkansas at Little Rock

Philip J. Stephens
Villanova University

Leeann S. Sticker
Northwestern State University of Louisiana

Julianna E. Szilagyi
University of Houston, College of Pharmacy

David L. Tauck
Santa Clara University

Deborah L. Taylor
Kansas City Kansas Community College

Suzanne Sawyer Vincent
Oral Roberts University

Margaret A. Weck
St. Louis College of Pharmacy

Cindy Martinez Wedig
University of Texas-Pan American

John Welborn
Mississippi State University

MaryJo A. Witz
Monroe Community College

HUMAN PHYSIOLOGY

The Study of Body Function

Objectives

After studying this chapter, you should be able to . . .

1. describe, in a general way, the topics studied in physiology and explain the importance of physiology in modern medicine.

2. describe the characteristics of the scientific method.

3. define *homeostasis* and explain how this concept is used in physiology and medicine.

4. describe the nature of negative feedback loops and explain how these mechanisms act to maintain homeostasis.

5. explain how antagonistic effectors help to maintain homeostasis.

6. describe the nature of positive feedback loops and explain how these mechanisms function in the body.

7. distinguish between intrinsic and extrinsic regulation and describe, in a general way, the roles of the nervous and endocrine systems in body regulation.

8. explain how negative feedback inhibition helps to regulate the secretion of hormones, using insulin as an example.

9. list the four primary tissues and their subtypes and describe the distinguishing features of each primary tissue.

10. relate the structure of each primary tissue to its functions.

11. describe how the primary tissues are grouped into organs, using the skin as an example.

12. describe the nature of the extracellular and intracellular compartments of the body and explain the significance of this compartmentalization.

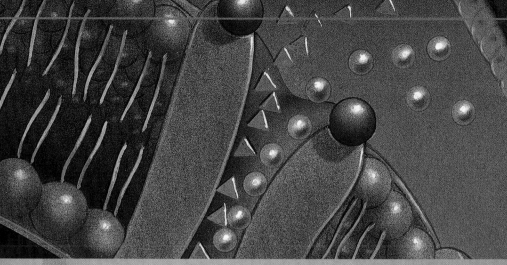

Chapter at a Glance

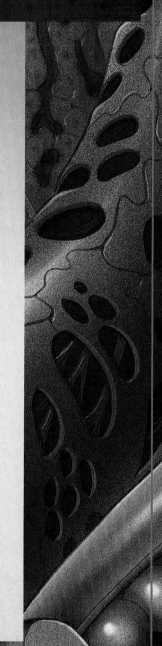

Introduction to Physiology

Human physiology is the study of how the human body functions, with emphasis on specific cause-and-effect mechanisms. Knowledge of these mechanisms has been obtained experimentally through applications of the scientific method.

Physiology (from the Greek *physis* = nature; *logos* = study) is the study of biological function—of how the body works, from cell to tissue, tissue to organ, organ to system, and of how the organism as a whole accomplishes particular tasks essential for life. In the study of physiology, the emphasis is on mechanisms—with questions that begin with the word *how* and answers that involve cause-and-effect sequences. These sequences can be woven into larger and larger stories that include descriptions of the structures involved (anatomy) and that overlap with the sciences of chemistry and physics.

The separate facts and relationships of these cause-and-effect sequences are derived empirically from experimental evidence. Explanations that seem logical are not necessarily true; they are only as valid as the data on which they are based, and they can change as new techniques are developed and further experiments are performed. The ultimate objective of physiological research is to understand the normal functioning of cells, organs, and systems. A related science—*pathophysiology*—is concerned with how physiological processes are altered in disease or injury.

Pathophysiology and the study of normal physiology complement one another. For example, a standard technique for investigating the functioning of an organ is to observe what happens when it is surgically removed from an experimental animal or when its function is altered in a specific way. This study is often aided by "experiments of nature"—diseases—that involve specific damage to the functioning of an organ. The study of disease processes has thus aided our understanding of normal functioning, and the study of normal physiology has provided much of the scientific basis of modern medicine. This relationship is recognized by the Nobel Prize committee, whose members award prizes in the category "Physiology or Medicine."

The physiology of invertebrates and of different vertebrate groups is studied in the science of *comparative physiology*. Much of the knowledge gained from comparative physiology has benefited the study of human physiology. This is because animals, including humans, are more alike than they are different. This is especially true when comparing humans with other mammals. The small differences in physiology between humans and other mammals can be of crucial importance in the development of pharmaceutical drugs (discussed later in this section), but these differences are relatively slight in the overall study of physiology.

Scientific Method

All of the information in this text has been gained by application of the **scientific method.** Although many different techniques are involved in the scientific method, all share three attributes: (1) confidence that the natural world, including ourselves, is ul-

timately explainable in terms we can understand; (2) descriptions and explanations of the natural world that are honestly based on observations and that could be modified or refuted by other observations; and (3) humility, or the willingness to accept the fact that we could be wrong. If further study should yield conclusions that refuted all or part of an idea, the idea would have to be modified accordingly. In short, the scientific method is based on a confidence in our rational ability, honesty, and humility. Practicing scientists may not always display these attributes, but the validity of the large body of scientific knowledge that has been accumulated—as shown by the technological applications and the predictive value of scientific hypotheses—are ample testimony to the fact that the scientific method works.

The scientific method involves specific steps. After making certain observations regarding the natural world, a **hypothesis** is formulated. In order for this hypothesis to be scientific, it must be capable of being refuted by experiments or other observations of the natural world. For example, one might hypothesize that people who exercise regularly have a lower resting pulse rate than other people. Experiments are conducted, or other observations are made, and the results are analyzed. Conclusions are then drawn as to whether the new data either refute or support the hypothesis. If the hypothesis survives such testing, it might be incorporated into a more general **theory.** Scientific theories are statements about the natural world that incorporate a number of proven hypotheses. They serve as a logical framework by which these hypotheses can be interrelated and provide the basis for predictions that may as yet be untested.

The hypothesis in the preceding example is scientific because it is *testable;* the pulse rates of 100 athletes and 100 sedentary people could be measured, for example, to see if there were statistically significant differences. If there were, the statement that athletes, on the average, have lower resting pulse rates than other people would be justified *based on these data.* One must still be open to the fact that this conclusion could be wrong. Before the discovery could become generally accepted as fact, other scientists would have to consistently replicate the results. Scientific theories are based on *reproducible* data.

It is quite possible that when others attempt to replicate the experiment their results will be slightly different. They may then construct scientific hypotheses that the differences in resting pulse rate also depend on other factors—for example, the nature of the exercise performed. When scientists attempt to test these hypotheses, they will likely encounter new problems, requiring new explanatory hypotheses, which then must be tested by additional experiments.

In this way, a large body of highly specialized information is gradually accumulated, and a more generalized explanation (a scientific theory) can be formulated. This explanation will almost always be different from preconceived notions. People who follow the scientific method will then appropriately modify their concepts, realizing that their new ideas will probably have to be changed again in the future as additional experiments are performed.

Use of Measurements, Controls, and Statistics

Suppose you wanted to test the hypothesis that a regular exercise program causes people to have a lower resting heart rate.

First, you would have to decide on the nature of the exercise program. Then, you would have to decide how the heart rate (or pulse rate) would be measured. This is a typical problem in physiology research, because the testing of most physiological hypotheses requires quantitative **measurements.**

The group that is subject to the testing condition—in this case, exercise—is called the **experimental group.** A measurement of the heart rate for this group would only be meaningful if it is compared to that of another group, known as the **control group.** How shall this control group be chosen? Perhaps the subjects could serve as their own controls—that is, a person's resting heart rate could be measured before and after the exercise regimen. If this isn't possible, a control group could be other people who do not follow the exercise program. The choice of control groups is often a controversial aspect of physiology studies. In this example, did the people in the control group really refrain from *any* exercise? Were they comparable to the people in the experimental group with regard to age, sex, ethnicity, body weight, health status, and so on? You can see how difficult it could be in practice to get a control group that could satisfy any potential criticism.

Another potential criticism could be bias in the way that the scientists perform the measurements. This bias could be completely unintentional; scientists are human, after all, and they may have invested months or years in this project! Thus, the person doing the measurements often does not know if a subject is part of the experimental or the control group. This is known as a *blind measurement.*

Now suppose the data are in, and it looks like the experimental group indeed has a lower average resting heart rate than the control group. But there is overlap—some people in the control group have measurements that are lower than some people in the experimental group. Now, is the difference in the average measurements of the groups due to a real, physiological difference, or is it due to chance variations in the measurements? Scientists attempt to test the *null hypothesis* (the hypothesis that the difference is due to chance) by employing the mathematical tools of **statistics.** If the statistical results so warrant, the null hypothesis can be rejected and the experimental hypothesis can be deemed to be supported by this study.

The statistical test chosen will depend upon the design of the experiment, and it can also be a source of contention among scientists in evaluating the validity of the results. Because of the nature of the scientific method, "proof" in science is always provisional. Some other researchers, employing the scientific method in a different way (with different measuring techniques, experimental procedures, choice of control groups, statistical tests, and so on) may later obtain different results. The scientific method is thus an ongoing enterprise.

The results of the scientific enterprise are written up as research articles, and these must be reviewed by other scientists who work in the same field before they can be published in **peer-reviewed journals.** More often than not, the reviewers will suggest that certain changes be made in the articles before they can be accepted for publication.

Examples of such peer-reviewed journals in which articles pertaining to many scientific fields are published include *Science* (www.sciencemag.org/), *Nature* (www.nature.com/nature/), and *Proceedings of the National Academy of Sciences* (www.pnas.org/).

Review articles in physiology can be found in *Annual Review of Physiology* (physiol.annualreviews.org/), *Physiological Reviews* (physrev.physiology.org/), and *News in Physiological Sciences* (www.the-aps.org/publication/journals/ pub_nips_home.htm). Medical research journals, such as the *New England Journal of Medicine* (content.nejm.org/) and *Nature Medicine* (www.nature.com/nm/), also publish articles of physiological interest. There are also a great number of specialty journals in areas of physiology such as neurophysiology, endocrinology, cardiovascular physiology, and so on.

Students who wish to look online for scientific articles published in peer-reviewed journals that relate to a particular subject can do so at the National Library of Medicine website, *PubMed* (www.ncbi.nlm.nih.gov/entrez/query.fcgi).

Development of Pharmaceutical Drugs

The development of new pharmaceutical drugs can serve as an example of how the scientific method is used in physiology and its health applications. The process usually starts with basic physiological research, often at cellular and molecular levels. Perhaps a new family of drugs is developed using cells in tissue culture (in vitro, or outside the body). For example, cell physiologists, studying membrane transport, may discover that a particular family of compounds blocks membrane channels for calcium ions (Ca^{2+}). Because of their knowledge of physiology, other scientists may predict that a drug of this nature might be useful in the treatment of hypertension (high blood pressure). This drug may then be tried in experimental animals.

If a drug is effective at extremely low concentrations in vitro, (in cells cultured outside of the body), there is a chance that it may work in vivo (in the body) at concentrations low enough not to be toxic (poisonous). This possibility must be thoroughly tested utilizing experimental animals, primarily rats and mice. More than 90% of drugs tested in experimental animals are too toxic for further development. Only in those rare cases when the toxicity is low enough may development progress to human/clinical trials.

Biomedical research is often aided by **animal models** of particular diseases. These are strains of laboratory rats and mice that are genetically susceptible to particular diseases that resemble human diseases. Research utilizing laboratory animals typically takes several years and always precedes human (clinical) trials of promising drugs. It should be noted that this length of time does not include all of the years of "basic" physiological research (involving laboratory animals) that provided the scientific foundation for the specific medical application.

In **phase I clinical trials,** the drug is tested on healthy human volunteers. This is done to test its toxicity in humans and to study how the drug is "handled" by the body: how it is metabolized, how rapidly it is removed from the blood by the liver and kidneys, how it can be most effectively administered, and so on. If no toxic effects are observed, the drug can proceed to the next stage. In **phase II clinical trials,** the drug is tested on the target human population (for example, those with hypertension). Only in those exceptional cases where the drug seems to be effective but has minimal toxicity does testing move to the next phase. **Phase III trials** occur in many research centers across the country to maximize the number of test participants. At this point, the test population must include a sufficient number of subjects of both sexes, as well as people of different ethnic groups. In addition, people are tested who have other health problems besides the one that the drug is intended to benefit. For example, those who have diabetes in addition to hypertension would be included in this phase. If the drug passes phase III trials, it goes to the Food and Drug Administration (FDA) for approval. **Phase IV trials** test other potential uses of the drug.

The percentage of drugs that make it all the way through these trials to eventually become approved and marketed is very low. Notice the crucial role of basic research, using experimental animals, in this process. Virtually every prescription drug on the market owes its existence to such research.

Test Yourself Before You Continue

1. How has the study of physiology aided, and been aided by, the study of diseases?
2. Describe the steps involved in the scientific method. What would qualify a statement as unscientific?
3. Describe the different types of trials a new drug must undergo before it is "ready for market."

Homeostasis and Feedback Control

The regulatory mechanisms of the body can be understood in terms of a single shared function: that of maintaining constancy of the internal environment. A state of relative constancy of the internal environment is known as homeostasis, and it is maintained by effectors that are regulated by sensory information from the internal environment.

History of Physiology

The Greek philosopher Aristotle (384–322 B.C.) speculated on the function of the human body, but another ancient Greek, Erasistratus (304–250? B.C.), is considered the father of physiology because he attempted to apply physical laws to the study of human function. Galen (A.D. 130–201) wrote widely on the subject and was considered the supreme authority until the advent of the Renaissance. Physiology became a fully experimental science with the revolutionary work of the English physician William Harvey (1578–1657), who demonstrated that the heart pumps blood through a closed system of vessels.

However, the father of modern physiology is the French physiologist Claude Bernard (1813–1878), who observed that the *milieu interieur* ("internal environment") remains remarkably constant despite changing conditions in the external environment. In a book entitled *The Wisdom of the Body,* published in 1932, the

American physiologist Walter Cannon (1871–1945) coined the term **homeostasis** to describe this internal constancy. Cannon further suggested that the many mechanisms of physiological regulation have but one purpose—the maintenance of internal constancy.

Most of our present knowledge of human physiology has been gained in the twentieth century. Further, new knowledge is being added at an ever more rapid pace, fueled in more recent decades by the revolutionary growth of molecular genetics and its associated biotechnology, and by the availability of ever more powerful computers and other equipment. A very brief history of twentieth-century physiology, limited by space to only two citations per decade, is provided in table 1.1.

Negative Feedback Loops

The concept of homeostasis has been of immense value in the study of physiology because it allows diverse regulatory mechanisms to be understood in terms of their "why" as well as their "how." The concept of homeostasis also provides a major foundation for medical diagnostic procedures. When a particular measurement of the internal environment, such as a blood measurement (table 1.2), deviates significantly from the normal range of values, it can be concluded that homeostasis is not being maintained and that the person is sick. A number of such measurements, combined with clinical observations, may allow the particular defective mechanism to be identified.

In order for internal constancy to be maintained, the body must have sensors that are able to detect deviations from a **set point.** The set point is analogous to the temperature set on a house thermostat. In a similar manner, there is a set point for body temperature, blood glucose concentration, the tension on a tendon, and so on. When a sensor detects a deviation from a particular set point, it must relay this information to an **integrating center,** which usually receives information from many different sensors. The integrating center is often a particular region of the brain or spinal cord, but in some cases it can also be a group of cells in an endocrine gland. The relative strengths of different sensory inputs are weighed in the integrating center, which responds by either increasing or decreasing the activity of particular **effectors**—generally, muscles or glands.

The thermostat of a house can serve as a simple example. Suppose you set the thermostat at a set point of 70° F. If the temperature of the house rises sufficiently above the set point, a sensor within the thermostat will detect the deviation. This will then act, via the thermostat's equivalent of an integrating center, to activate the effector. The effector in this case may be an air conditioner, which acts to reverse the deviation from the set point.

If the body temperature exceeds the set point of 37° C, sensors in a part of the brain detect this deviation and, acting via an integrating center (also in the brain), stimulate activities of effectors (including sweat glands) that lower the temperature. If, as another example, the blood glucose concentration falls below normal, the effectors act to increase the blood glucose. One can think of the effectors as "defending" the set points against deviations. Since the activity of the effectors is influenced by the effects they produce, and since this regulation is in a negative, or

Table 1.1 History of Twentieth-Century Physiology (limited to two citations per decade)

1900	Karl Landsteiner discovers the A, B, and O blood groups.
1904	Ivan Pavlov wins the Nobel Prize for his work on the physiology of digestion.
1910	Sir Henry Dale describes properties of histamine.
1918	Earnest Starling describes how the force of the heart's contraction relates to the amount of blood in it.
1921	John Langley describes the functions of the autonomic nervous system.
1923	Sir Frederick Banting, Charles Best, and John Macleod win the Nobel Prize for the discovery of insulin.
1932	Sir Charles Sherrington and Lord Edgar Adrian win the Nobel Prize for discoveries related to the functions of neurons.
1936	Sir Henry Dale and Otto Loewi win the Nobel Prize for discovery of acetylcholine in synaptic transmission.
1939–47	Albert von Szent-Georgi explains the role of ATP and contributes to the understanding of actin and myosin in muscle contraction.
1949	Hans Selye discovers the common physiological responses to stress.
1953	Sir Hans Krebs wins the Nobel Prize for his discovery of the citric acid cycle.
1954	Hugh Huxley, Jean Hanson, R. Niedergerde, and Andrew Huxley propose the sliding filament theory of muscle contraction.
1962	Francis Crick, James Watson, and Maurice Wilkins win the Nobel Prize for determining the structure of DNA.
1963	Sir John Eccles, Sir Alan Hodgkin, and Sir Andrew Huxley win the Nobel Prize for their discoveries relating to the nerve impulse.
1971	Earl Sutherland wins the Nobel Prize for his discovery of the mechanism of hormone action.
1977	Roger Guillemin and Andrew Schally win the Nobel Prize for discoveries of the peptide hormone production by the brain.
1981	Roger Sperry wins the Nobel Prize for his discoveries regarding the specializations of the right and left cerebral hemispheres.
1986	Stanley Cohen and Rita Levi-Montalcini win the Nobel Prize for their discoveries of growth factors regulating the nervous system.
1994	Alfred Gilman and Martin Rodbell win the Nobel Prize for their discovery of the functions of G-proteins in signal transduction in cells.
1998	Robert Furchgott, Louis Ignarro, and Ferid Murad win the Nobel Prize for discovering the role of nitric oxide as a signaling molecule in the cardiovascular system.

Table 1.2 Approximate Normal Ranges for Measurements of Some Fasting Blood Values

Measurement	Normal Range
Arterial pH	7.35–7.45
Bicarbonate	24–28 mEq/L
Sodium	135–145 mEq/L
Calcium	4.5–5.5 mEq/L
Oxygen content	17.2–22.0 ml/100 ml
Urea	12–35 mg/100 ml
Amino acids	3.3–5.1 mg/100 ml
Protein	6.5–8.0 g/100 ml
Total lipids	400–800 mg/100 ml
Glucose	75–110 mg/100 ml

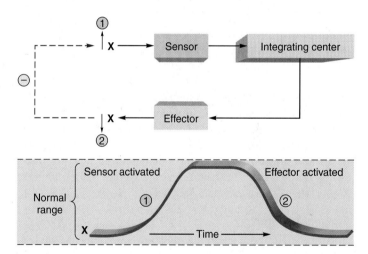

■ **Figure 1.1** A rise in some factor of the internal environment (⇑X) **is detected by a sensor.** This information is relayed to an integrating center, which causes an effector to produce a change in the opposite direction (⇓X). The initial deviation is thus reversed, completing a negative feedback loop (shown by the dashed arrow and negative sign). The numbers indicate the sequence of changes.

reverse, direction, this type of control system is known as a **negative feedback loop** (fig. 1.1). (Notice that in fig. 1.1 and in all subsequent figures, negative feedback is indicated by a dashed line and a negative sign.)

The nature of the negative feedback loop can be understood by again referring to the analogy of the thermostat and air conditioner. After the air conditioner has been on for some time, the room temperature may fall significantly below the set point of the thermostat. When this occurs, the air conditioner will be turned off. The effector (air conditioner) is turned on by a high temperature and, when activated, produces a negative change (lowering of the temperature) that ultimately causes the effector to be turned off. In this way, constancy is maintained.

It is important to realize that these negative feedback loops are continuous, ongoing processes. Thus, a particular nerve fiber that is part of an effector mechanism may always display some

activity, and a particular hormone, which is part of another effector mechanism, may always be present in the blood. The nerve activity and hormone concentration may decrease in response to deviations of the internal environment in one direction (fig. 1.1), or they may increase in response to deviations in the opposite direction (fig. 1.2). Changes from the normal range in either direction are thus compensated for by reverse changes in effector activity.

Since negative feedback loops respond after deviations from the set point have stimulated sensors, the internal environment is

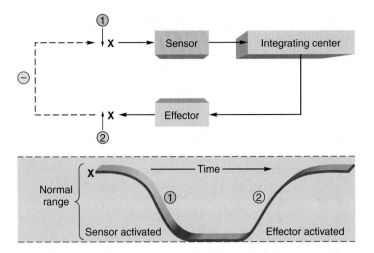

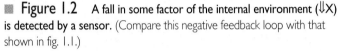

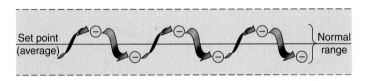

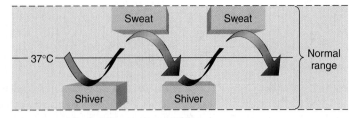

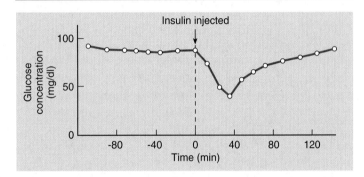

Figure 1.2 **A fall in some factor of the internal environment (⇓X) is detected by a sensor.** (Compare this negative feedback loop with that shown in fig. 1.1.)

Figure 1.3 Negative feedback loops maintain a state of dynamic constancy within the internal environment. The completion of the negative feedback loop is indicated by negative signs.

Figure 1.4 **How body temperature is maintained within the normal range.** The body temperature normally has a set point of 37° C. This is maintained, in part, by two antagonistic mechanisms—shivering and sweating. Shivering is induced when the body temperature falls too low, and it gradually subsides as the temperature rises. Sweating occurs when the body temperature is too high, and it diminishes as the temperature falls. Most aspects of the internal environment are regulated by the antagonistic actions of different effector mechanisms.

Figure 1.5 Homeostasis of the blood glucose concentration. Average blood glucose concentrations of five healthy individuals are graphed before and after a rapid intravenous injection of insulin. The "0" indicates the time of the injection. Notice that, following injection of insulin, the blood glucose is brought back up to the normal range. This occurs as a result of the action of hormones antagonistic to insulin, which cause the liver to secrete glucose into the blood. In this way, homeostasis is maintained.

never absolutely constant. Homeostasis is best conceived as a state of **dynamic constancy,** in which conditions are stabilized above and below the set point. These conditions can be measured quantitatively, in degrees Celsius for body temperature, for example, or in milligrams per deciliter (one-tenth of a liter) for blood glucose. The set point can be taken as the average value within the normal range of measurements (fig. 1.3).

Antagonistic Effectors

Most factors in the internal environment are controlled by several effectors, which often have antagonistic actions. Control by antagonistic effectors is sometimes described as "push-pull," where the increasing activity of one effector is accompanied by decreasing activity of an antagonistic effector. This affords a finer degree of control than could be achieved by simply switching one effector on and off.

Room temperature can be maintained for example, by simply turning an air conditioner on and off, or by just turning a heater on and off. A much more stable temperature, however, can be achieved if the air conditioner and heater are both controlled by a thermostat. Then the heater is turned on when the air conditioner is turned off, and vice versa. Normal body temperature is maintained about a set point of 37° C by the antagonistic effects of sweating, shivering, and other mechanisms (fig. 1.4).

The blood concentrations of glucose, calcium, and other substances are regulated by negative feedback loops involving hormones that promote opposite effects. While insulin, for example, lowers blood glucose, other hormones raise the blood

glucose concentration. The heart rate, similarly, is controlled by nerve fibers that produce opposite effects: stimulation of one group of nerve fibers increases heart rate; stimulation of another group slows the heart rate.

Quantitative Measurements

Normal ranges and deviations from the set point must be known quantitatively in order to study physiological mechanisms. For these and other reasons, quantitative measurements are basic to the science of physiology. One example of this, and of the actions of antagonistic mechanisms in maintaining homeostasis, is shown in figure 1.5. Blood glucose concentrations were measured in five healthy people before and after an injection of insulin, a hormone that acts to lower the blood glucose concentration. A graph of the data reveals that the blood glucose concentration decreased rapidly but was brought back up to normal levels within 80 minutes after the injection. This demonstrates that negative feedback mechanisms acted to restore homeostasis in this experiment. These mechanisms involve the action of hormones whose effects are antagonistic to that of insulin—that is, they promote the secretion of glucose from the liver (see chapter 19).

Positive Feedback

Constancy of the internal environment is maintained by effectors that act to compensate for the change that served as the stimulus for their activation; in short, by negative feedback loops. A thermostat, for example, maintains a constant temperature by increasing heat production when it is cold and decreasing heat production when it is warm. The opposite occurs during **positive feedback**—in this case, the action of effectors *amplifies* those changes that stimulated the effectors. A thermostat that works by positive feedback, for example, would increase heat production in response to a rise in temperature.

It is clear that homeostasis must ultimately be maintained by negative rather than by positive feedback mechanisms. The effectiveness of some negative feedback loops, however, is increased by positive feedback mechanisms that amplify the actions of a negative feedback response. Blood clotting, for example, occurs as a result of a sequential activation of clotting factors; the activation of one clotting factor results in activation of many in a positive feedback cascade. In this way, a single change is amplified to produce a blood clot. Formation of the clot, however, can prevent further loss of blood, and thus represents the completion of a negative feedback loop that restores homeostasis.

Neural and Endocrine Regulation

Homeostasis is maintained by two general categories of regulatory mechanisms: (1) those that are **intrinsic,** or "built-in," to the organs being regulated and (2) those that are **extrinsic,** as in regulation of an organ by the nervous and endocrine systems. The endocrine system functions closely with the nervous system in regulating and integrating body processes and maintaining homeostasis. The nervous system controls the secretion of many endocrine glands, and some hormones in turn affect the function of the nervous system. Together, the nervous and endocrine systems regulate the activities of most of the other systems of the body.

Regulation by the endocrine system is achieved by the secretion of chemical regulators called **hormones** into the blood. Since hormones are secreted into the blood, they are carried by the blood to all organs in the body. Only specific organs can respond to a particular hormone, however; these are known as the **target organs** of that hormone.

Nerve fibers are said to *innervate* the organs that they regulate. When stimulated, these fibers produce electrochemical nerve impulses that are conducted from the origin of the fiber to its end point in the target organ innervated by the fiber. These target organs can be muscles or glands that may function as effectors in the maintenance of homeostasis.

Feedback Control of Hormone Secretion

The nature of the endocrine glands, the interaction of the nervous and endocrine systems, and the actions of hormones will be discussed in detail in later chapters. For now, it is sufficient to describe the regulation of hormone secretion very broadly, since it so superbly illustrates the principles of homeostasis and negative feedback regulation.

Hormones are secreted in response to specific chemical stimuli. A rise in the plasma glucose concentration, for example, stimulates insulin secretion from structures in the pancreas known as the pancreatic islets, or islets of Langerhans. Hormones are also secreted in response to nerve stimulation and to stimulation by other hormones.

The secretion of a hormone can be inhibited by its own effects, in a negative feedback manner. Insulin, as previously described, produces a lowering of blood glucose. Since a rise in blood glucose stimulates insulin secretion, a lowering of blood glucose caused by insulin's action inhibits further insulin secretion. This closed-loop control system is called **negative feedback inhibition** (fig. 1.6*a*).

Homeostasis of blood glucose is too important—the brain uses blood glucose as its primary source of energy—to entrust to the regulation of only one hormone, insulin. So, during fasting, when blood glucose falls, it is prevented from falling too far by several mechanisms (fig. 1.6*b*). First, insulin secretion decreases, preventing muscle, liver, and adipose cells from taking too much glucose from the blood. Second, the secretion of a hormone antagonistic to insulin, called *glucagon,* increases. Glucagon stimulates processes in the liver (breakdown of a stored, starchlike molecule called glycogen—see chapter 2) that cause it to secrete glucose into the blood. Through these and other antagonistic negative feedback mechanisms, the blood glucose is maintained within a homeostatic range.

Test Yourself Before You Continue

1. Define *homeostasis* and describe how this concept can be used to explain physiological control mechanisms.

2. Define the term *negative feedback* and explain how it contributes to homeostasis. Illustrate this concept by drawing a negative feedback loop.

3. Describe *positive feedback* and explain how this process functions in the body.

4. Explain how the secretion of a hormone is controlled by negative feedback inhibition. Use the control of insulin secretion as an example.

The Primary Tissues

The organs of the body are composed of four different primary tissues, each of which has its own characteristic structure and function. The activities and interactions of these tissues determine the physiology of the organs.

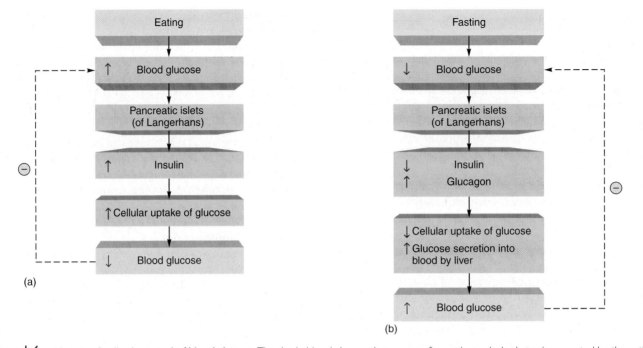

■ Figure 1.6 Negative feedback control of blood glucose. The rise in blood glucose that occurs after eating carbohydrates is corrected by the action of insulin, which is secreted in increasing amounts (*a*) at that time. During fasting, when blood glucose falls, insulin secretion is inhibited and the secretion of an antagonistic hormone, glucagon, is increased (*b*). This stimulates the liver to secrete glucose into the blood, helping to prevent blood glucose from continuing to fall. In this way, blood glucose concentrations are maintained within a homeostatic range following eating and during fasting.

Although physiology is the study of function, it is difficult to properly understand the function of the body without some knowledge of its anatomy, particularly at a microscopic level. Microscopic anatomy constitutes a field of study known as *histology.* The anatomy and histology of specific organs will be discussed together with their functions in later chapters. In this section, the common "fabric" of all organs is described.

Cells are the basic units of structure and function in the body. Cells that have similar functions are grouped into categories called *tissues.* The entire body is composed of only four major types of tissues. These **primary tissues** include (1) muscle, (2) nervous, (3) epithelial, and (4) connective tissues. Groupings of these four primary tissues into anatomical and functional units are called **organs.** Organs, in turn, may be grouped together by common functions into **systems.** The systems of the body act in a coordinated fashion to maintain the entire organism.

Muscle Tissue

Muscle tissue is specialized for contraction. There are three types of muscle tissue: **skeletal, cardiac,** and **smooth.** Skeletal muscle is often called *voluntary muscle* because its contraction is consciously controlled. Both skeletal and cardiac muscles are **striated;** they have striations, or stripes, that extend across the width of the muscle cell (figs. 1.7 and 1.8). These striations are produced by a characteristic arrangement of contractile proteins, and for this reason skeletal and cardiac muscle have similar mechanisms of contraction. Smooth muscle (fig. 1.9) lacks these striations and has a different mechanism of contraction.

Skeletal Muscle

Skeletal muscles are generally attached to bones at both ends by means of tendons; hence, contraction produces movements of the skeleton. There are exceptions to this pattern, however. The tongue, superior portion of the esophagus, anal sphincter, and diaphragm are also composed of skeletal muscle, but they do not cause movements of the skeleton.

Beginning at about the fourth week of embryonic development, separate cells called *myoblasts* fuse together to form **skeletal muscle fibers,** or **myofibers** (from the Greek *myos,* meaning "muscle"). Although myofibers are often referred to as skeletal muscle cells, each is actually a *syncytium,* or multinucleate mass formed from the union of separate cells. Despite their unique origin and structure, each myofiber contains mitochondria and other organelles (described in chapter 3) common to all cells.

The muscle fibers within a skeletal muscle are arranged in bundles, and within these bundles the fibers extend in parallel from one end to the other of the bundle. The parallel arrangement of muscle fibers (shown in fig. 1.7) allows each fiber to be controlled individually: one can thus contract fewer or more muscle fibers and, in this way, vary the strength of contraction of the whole muscle. The ability to vary, or "grade," the strength of skeletal muscle contraction is obviously needed for precise control of skeletal movements.

Cardiac Muscle

Although cardiac muscle is striated, it differs markedly from skeletal muscle in appearance. Cardiac muscle is found only in the heart, where the **myocardial cells** are short, branched, and intimately interconnected to form a continuous fabric. Special areas of contact between adjacent cells stain darkly to show *intercalated discs* (fig. 1.8), which are characteristic of heart muscle.

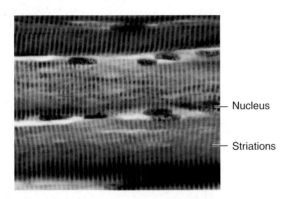

Figure 1.7 Three skeletal muscle fibers showing the characteristic light and dark cross striations. Because of this feature, skeletal muscle is also called striated muscle.

Nucleus

Striations

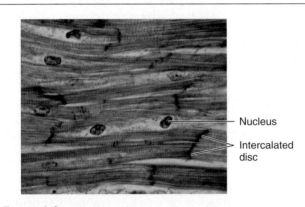

Figure 1.8 Human cardiac muscle. Notice the striated appearance and dark-staining intercalated discs.

Nucleus

Intercalated disc

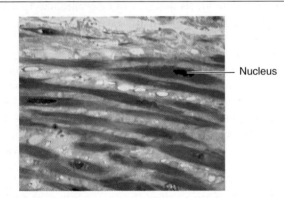

Figure 1.9 A photomicrograph of smooth muscle cells. Notice that these cells contain single, centrally located nuclei and lack striations.

Nucleus

The intercalated discs couple myocardial cells together mechanically and electrically. Unlike skeletal muscles, therefore, the heart cannot produce a graded contraction by varying the number of cells stimulated to contract. Because of the way it is constructed, the stimulation of one myocardial cell results in the stimulation of all other cells in the mass and a "wholehearted" contraction.

Smooth Muscle

As implied by the name, smooth muscle cells (fig. 1.9) do not have the striations characteristic of skeletal and cardiac muscle. Smooth muscle is found in the digestive tract, blood vessels,

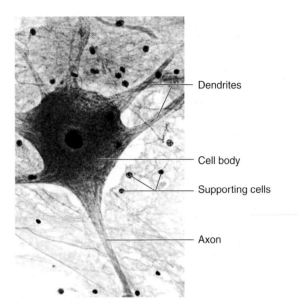

Dendrites

Cell body

Supporting cells

Axon

Figure 1.10 A photomicrograph of nerve tissue. A single neuron and numerous smaller supporting cells can be seen.

bronchioles (small air passages in the lungs), and in the ducts of the urinary and reproductive systems. Circular arrangements of smooth muscle in these organs produce constriction of the *lumen* (cavity) when the muscle cells contract. The digestive tract also contains longitudinally arranged layers of smooth muscle. The series of wavelike contractions of circular and longitudinal layers of muscle known as *peristalsis* pushes food from one end of the digestive tract to the other.

The three types of muscle tissue are discussed further in chapter 12.

Nervous Tissue

Nervous tissue consists of nerve cells, or **neurons,** which are specialized for the generation and conduction of electrical events, and of **supporting cells,** which provide the neurons with anatomical and functional support. Supporting cells in the brain and spinal cord are referred to as *neuroglial cells,* or often simply as *glial cells.*

Each neuron consists of three parts: (1) a *cell body,* (2) *dendrites,* and (3) an *axon* (fig. 1.10). The cell body contains the nucleus and serves as the metabolic center of the cell. The dendrites (literally, "branches") are highly branched cytoplasmic extensions of the cell body that receive input from other neurons or from receptor cells. The axon is a single cytoplasmic extension of the cell body that can be quite long (up to a few feet in length). It is specialized for conducting nerve impulses from the cell body to another neuron or to an effector (muscle or gland) cell.

The supporting cells do not conduct impulses but instead serve to bind neurons together, modify the extracellular environment of the nervous system, and influence the nourishment and electrical activity of neurons. Supporting cells are about five times more abundant than neurons in the nervous system and, unlike neurons, maintain a limited ability to divide by mitosis throughout life.

Neurons and supporting cells are discussed in detail in chapter 7.

Table 1.3   Summary of Epithelial Membranes

Type	Structure and Function	Location
Simple Epithelia	Single layer of cells; function varies with type	Covering visceral organs; linings of body cavities, tubes, and ducts
Simple squamous epithelium	Single layer of flattened, tightly bound cells; diffusion and filtration	Capillary walls; pulmonary alveoli of lungs; covering visceral organs; linings of body cavities
Simple cuboidal epithelium	Single layer of cube-shaped cells; excretion, secretion, or absorption	Surface of ovaries; linings of kidney tubules, salivary ducts, and pancreatic ducts
Simple columnar epithelium	Single layer of nonciliated, tall, column-shaped cells; protection, secretion, and absorption	Lining of most of digestive tract
Simple ciliated columnar epithelium	Single layer of ciliated, column-shaped cells; transportive role through ciliary motion	Lining of uterine tubes
Pseudostratified ciliated columnar epithelium	Single layer of ciliated, irregularly shaped cells; many goblet cells; protection, secretion, ciliary movement	Lining of respiratory passageways
Stratified Epithelia	Two or more layers of cells; function varies with type	Epidermal layer of skin; linings of body openings, ducts, and urinary bladder
Stratified squamous epithelium (keratinized)	Numerous layers containing keratin, with outer layers flattened and dead; protection	Epidermis of skin
Stratified squamous epithelium (nonkeratinized)	Numerous layers lacking keratin, with outer layers moistened and alive; protection and pliability	Linings of oral and nasal cavities, vagina, and anal canal
Stratified cuboidal epithelium	Usually two layers of cube-shaped cells; strengthening of luminal walls	Large ducts of sweat glands, salivary glands, and pancreas
Transitional epithelium	Numerous layers of rounded, nonkeratinized cells; distension	Walls of ureters, part of urethra, and urinary bladder

Epithelial Tissue

Epithelial tissue consists of cells that form **membranes,** which cover and line the body surfaces, and of **glands,** which are derived from these membranes. There are two categories of glands. *Exocrine glands* (*exo* = outside) secrete chemicals through a duct that leads to the outside of a membrane, and thus to the outside of a body surface. *Endocrine glands* (from the Greek *endon* = within) secrete chemicals called *hormones* into the blood. Endocrine glands are discussed in chapter 11.

Epithelial Membranes

Epithelial membranes are classified according to the number of their layers and the shape of the cells in the upper layer (table 1.3). Epithelial cells that are flattened in shape are **squamous;** those that are taller than they are wide are **columnar;** and those that are as wide as they are tall are **cuboidal** (fig. 1.11*a–c*). Those epithelial membranes that are only one cell layer thick are known as **simple membranes;** those that are composed of a number of layers are **stratified membranes.**

Epithelial membranes cover all body surfaces and line the cavity (lumen) of every hollow organ. Thus, epithelial membranes provide a barrier between the external environment and the internal environment of the body. Stratified epithelial membranes are specialized to provide protection. Simple epithelial membranes, in contrast, provide little protection; instead, they are specialized for transport of substances between the internal and external environments. In order for a substance to get into the body, it must pass through an epithelial membrane, and simple epithelia are specialized for this function For example, a simple squamous epithelium in the lungs allows the rapid passage of oxygen and carbon dioxide between the air (external environment) and blood (internal environment). A simple columnar epithelium in the small intestine, as another example, allows digestion products to pass from the intestinal lumen (external environment) to the blood (internal environment).

Dispersed among the columnar epithelial cells are specialized unicellular glands called *goblet cells* that secrete mucus. The columnar epithelial cells in the uterine (fallopian) tubes of females and in the respiratory passages contain numerous *cilia* (hairlike structures, described in chapter 3) that can move in a coordinated fashion and aid the functions of these organs.

The epithelial lining of the esophagus and vagina that provides protection for these organs is a stratified squamous epithelium (fig. 1.12). This is a *nonkeratinized* membrane, and all layers consist of living cells. The *epidermis* of the skin, by contrast, is *keratinized,* or *cornified* (fig. 1.13). Since the epidermis is dry and exposed to the potentially desiccating effects of the air, the surface is covered with dead cells that are filled with a water-resistant protein known as *keratin.* This protective layer is constantly flaked off from the surface of the skin and therefore must be constantly replaced by the division of cells in the deeper layers of the epidermis.

The constant loss and renewal of cells is characteristic of epithelial membranes. The entire epidermis is completely replaced every 2 weeks; the stomach lining is renewed every 2 to 3 days. Examination of the cells that are lost, or "exfoliated," from the outer layer of epithelium lining the female reproductive tract is a common procedure in gynecology (as in the Pap smear).

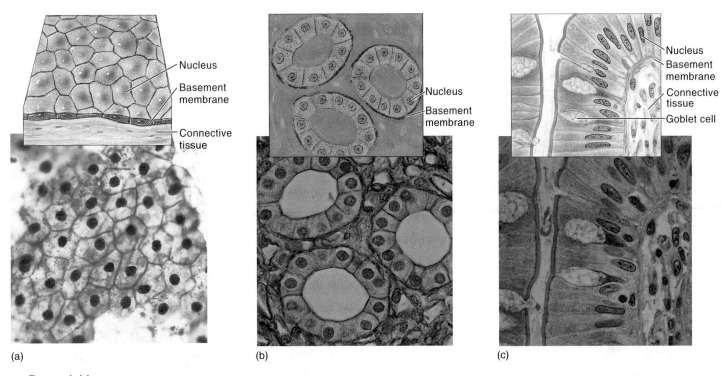

Figure 1.11 **Different types of simple epithelial membranes.** (*a*) Simple squamous, (*b*) simple cuboidal, and (*c*) simple columnar epithelial membranes. The tissue beneath each membrane is connective tissue.

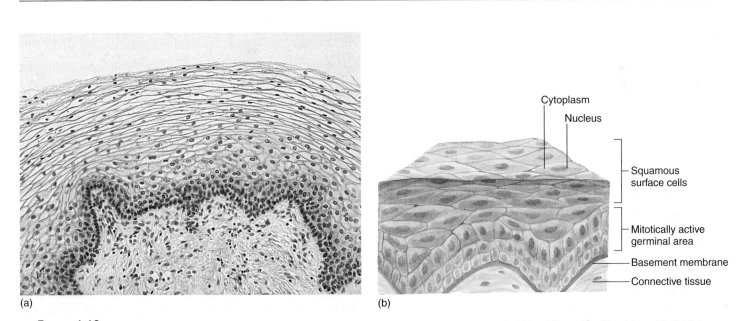

Figure 1.12 **A stratified squamous nonkeratinized epithelial membrane.** This is a photomicrograph (*a*) and illustration (*b*) of the epithelial lining of the vagina.

In order to form a strong membrane that is effective as a barrier at the body surfaces, epithelial cells are very closely packed and are joined together by structures collectively called **junctional complexes.** There is no room for blood vessels between adjacent epithelial cells. The epithelium must therefore receive nourishment from the tissue beneath, which has large intercellular spaces that can accommodate blood vessels and nerves. This underlying tissue is called *connective tissue.* Epithelial membranes are attached to the underlying connective tissue by a layer of proteins and polysaccharides known as the **basement membrane.** This layer can be observed only under the microscope using specialized staining techniques.

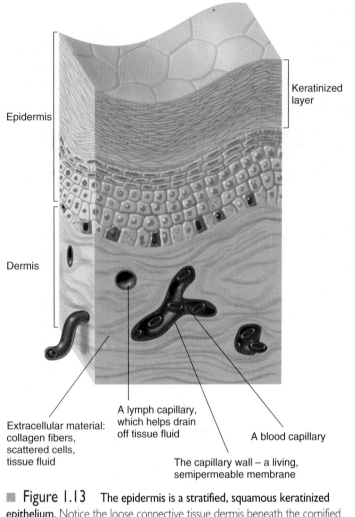

Epidermis

Keratinized layer

Dermis

Extracellular material: collagen fibers, scattered cells, tissue fluid

A lymph capillary, which helps drain off tissue fluid

A blood capillary

The capillary wall – a living, semipermeable membrane

Figure 1.13   The epidermis is a stratified, squamous keratinized epithelium. Notice the loose connective tissue dermis beneath the cornified epidermis. Loose connective tissue contains scattered collagen fibers in a matrix of protein-rich fluid. The intercellular spaces also contain cells and blood vessels.

Exocrine Glands

Exocrine glands are derived from cells of epithelial membranes. The secretions of these cells are passed to the outside of the epithelial membranes (and hence to the surface of the body) through *ducts.* This is in contrast to *endocrine glands,* which lack ducts and which therefore secrete into capillaries within the body (fig. 1.14). The structure of endocrine glands will be described in chapter 11.

The secretory units of exocrine glands may be simple tubes, or they may be modified to form clusters of units around branched ducts (fig. 1.15). These clusters, or **acini,** are often surrounded by tentacle-like extensions of *myoepithelial cells* that contract and squeeze the secretions through the ducts. The rate of secretion and the action of myoepithelial cells are subject to neural and endocrine regulation.

Examples of exocrine glands in the skin include the lacrimal (tear) glands, sebaceous glands (which secrete oily sebum into hair follicles), and sweat glands. There are two types

of sweat glands. The more numerous, the *eccrine* (or *merocrine*) *sweat glands,* secrete a dilute salt solution that serves in thermoregulation (evaporation cools the skin). The *apocrine sweat glands,* located in the axillae (underarms) and pubic region, secrete a protein-rich fluid. This provides nourishment for bacteria that produce the characteristic odor of this type of sweat.

All of the glands that secrete into the digestive tract are also exocrine. This is because the lumen of the digestive tract is a part of the external environment, and secretions of these glands go to the outside of the membrane that lines this tract. Mucous glands are located throughout the length of the digestive tract. Other relatively simple glands of the tract include salivary glands, gastric glands, and simple tubular glands in the intestine.

The *liver* and *pancreas* are exocrine (as well as endocrine) glands, derived embryologically from the digestive tract. The exocrine secretion of the pancreas—pancreatic juice—contains digestive enzymes and bicarbonate and is secreted into the small intestine via the pancreatic duct. The liver produces and secretes bile (an emulsifier of fat) into the small intestine via the gallbladder and bile duct.

Exocrine glands are also prominent in the reproductive system. The female reproductive tract contains numerous mucus-secreting exocrine glands. The male accessory sex organs—the *prostate* and *seminal vesicles*—are exocrine glands that contribute to semen. The testes and ovaries (the gonads) are both endocrine and exocrine glands. They are endocrine because they secrete sex steroid hormones into the blood; they are exocrine because they release gametes (ova and sperm) into the reproductive tracts.

Connective Tissue

Connective tissue is characterized by large amounts of extracellular material in the spaces between the connective tissue cells. This extracellular material may be of various types and arrangements and, on this basis, several types of connective tissues are recognized: (1) connective tissue proper, (2) cartilage, (3) bone, and (4) blood. **Blood** is usually classified as connective tissue because about half its volume is composed of an extracellular fluid known as *plasma.*

Connective tissue proper includes a variety of subtypes. An example of *loose connective tissue* (or *areolar tissue*) is the dermis of the skin (see fig. 1.13). This connective tissue consists of scattered fibrous proteins, called *collagen,* and tissue fluid, which provides abundant space for the entry of blood and lymphatic vessels and nerve fibers. Another type of connective tissue proper, *dense fibrous connective tissue,* contains densely packed fibers of collagen that may be irregularly or regularly arranged. Dense irregular connective tissue (fig. 1.16) contains a meshwork of randomly oriented collagen fibers that resist forces applied from many directions. This tissue forms the tough capsules and sheaths surrounding organs. Tendons, which connect muscle to bone, and ligaments, which connect bones together at joints, are examples of dense regular connective tissue. The collagen fibers of this tissue are oriented in the same direction (fig. 1.17).

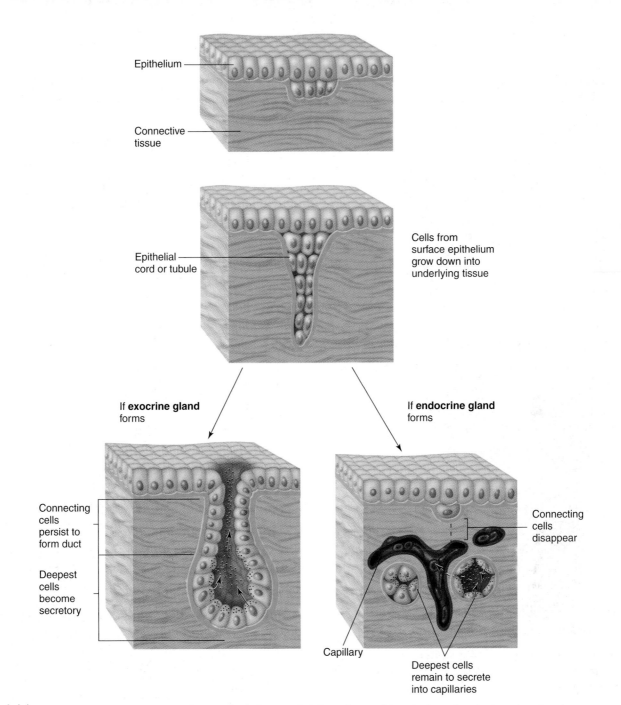

Epithelium

Connective tissue

Epithelial cord or tubule

Cells from surface epithelium grow down into underlying tissue

If **exocrine gland** forms

If **endocrine gland** forms

Connecting cells persist to form duct

Deepest cells become secretory

Connecting cells disappear

Capillary

Deepest cells remain to secrete into capillaries

■ Figure 1.14 **The formation of exocrine and endocrine glands from epithelial membranes.** Note that exocrine glands retain a duct that can carry their secretion to the surface of the epithelial membrane, whereas endocrine glands are ductless.

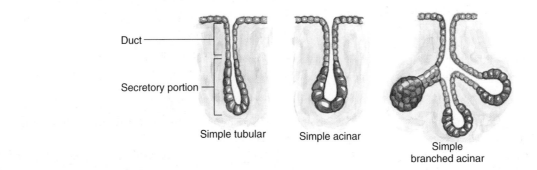

Duct

Secretory portion

Simple tubular

Simple acinar

Simple branched acinar

■ Figure 1.15 **The structure of exocrine glands.** Exocrine glands may be simple invaginations of epithelial membranes, or they may be more complex derivatives.

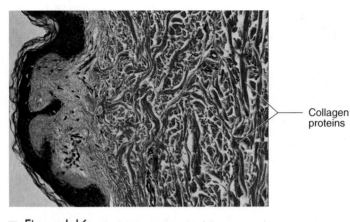

■ **Figure 1.16**   A photomicrograph of dense irregular connective **tissue.** Notice the tightly packed, irregularly arranged collagen proteins.

Collagen proteins

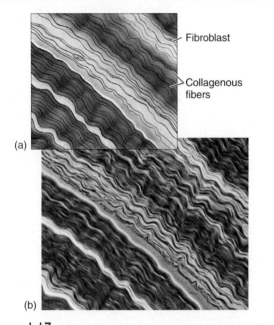

Fibroblast

Collagenous fibers

(a)

(b)

■ **Figure 1.17**   Dense regular connective tissue. (*a*) Labeled diagram and (*b*) photomicrograph of a tendon. Notice the dense regular arrangement of collagenous fibers.

Adipose tissue is a specialized type of loose connective tissue. Each adipose cell, or *adipocyte,* has its cytoplasm stretched around a central globule of fat (fig. 1.18). The synthesis and breakdown of fat are accomplished by enzymes within the cytoplasm of the adipocytes.

Cartilage consists of cells, called *chondrocytes,* surrounded by a semisolid ground substance that imparts elastic properties to the tissue. Cartilage is a type of supportive and protective tissue commonly called "gristle." It forms the precursor to many bones that develop in the fetus and persists at the articular (joint) surfaces on the bones at all movable joints in adults.

Bone is produced as concentric layers, or *lamellae,* of calcified material laid around blood vessels. The bone-forming cells, or *osteoblasts,* surrounded by their calcified products, become trapped within cavities called *lacunae.* The trapped cells, which are now called *osteocytes,* remain alive because they are

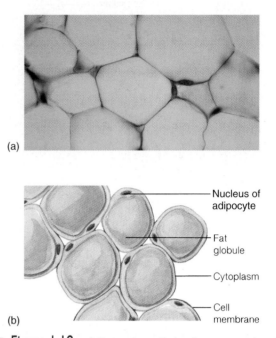

(a)

Nucleus of adipocyte

Fat globule

Cytoplasm

Cell membrane

(b)

■ **Figure 1.18**   Adipose tissue. Each adipocyte contains a large, central globule of fat surrounded by the cytoplasm of the adipocyte. (*a*) Photomicrograph and (*b*) illustration of adipose tissue.

nourished by "lifelines" of cytoplasm that extend from the cells to the blood vessels in *canaliculi* (little canals). The blood vessels lie within central canals, surrounded by concentric rings of bone lamellae with their trapped osteocytes. These units of bone structure are called *haversian systems* (fig. 1.19).

The *dentin* of a tooth (fig. 1.20) is similar in composition to bone, but the cells that form this calcified tissue are located in the pulp (composed of loose connective tissue). These cells send cytoplasmic extensions, called *dentinal tubules,* into the dentin. Dentin, like bone, is thus a living tissue that can be remodeled in response to stresses. The cells that form the outer *enamel* of a tooth, by contrast, are lost as the tooth erupts. Enamel is a highly calcified material, harder than bone or dentin, that cannot be regenerated; artificial "fillings" are therefore required to patch holes in the enamel.

Test Yourself Before You Continue

1. List the four primary tissues and describe the distinguishing features of each type.

2. Compare and contrast the three types of muscle tissue.

3. Describe the different types of epithelial membranes and state their locations in the body.

4. Explain why exocrine and endocrine glands are considered epithelial tissues and distinguish between these two types of glands.

5. Describe the different types of connective tissues and explain how they differ from one another in their content of extracellular material.

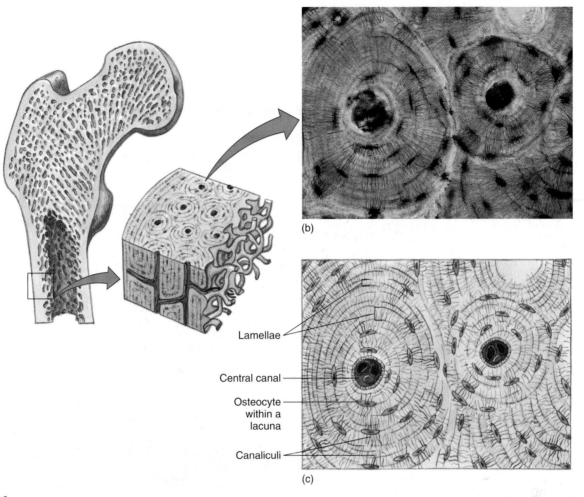

(a)

(b)

Lamellae

Central canal

Osteocyte
within a
lacuna

Canaliculi

(c)

■ **Figure 1.19 The structure of bone.** (*a*) A diagram of a long bone, (*b*) a photomicrograph showing haversian systems, and (*c*) a diagram of haversian systems. Within each central canal, an artery (red), vein (blue), and nerve (yellow) is illustrated.

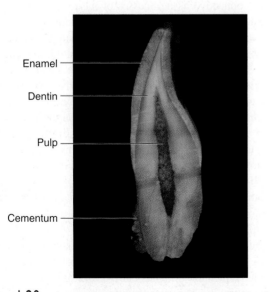

Enamel

Dentin

Pulp

Cementum

■ **Figure 1.20** **A cross section of a tooth showing pulp, dentin, and enamel.** The root of the tooth is covered by cementum, a calcified connective tissue that helps to anchor the tooth in its bony socket.

Organs and Systems

Organs are composed of two or more primary tissues that serve the different functions of the organ. The skin is an organ that has numerous functions provided by its constituent tissues.

An **organ** is a structure composed of at least two, and usually all four, primary tissues. The largest organ in the body, in terms of surface area, is the skin (fig. 1.21). In this section, the numerous functions of the skin serve to illustrate how primary tissues cooperate in the service of organ physiology.

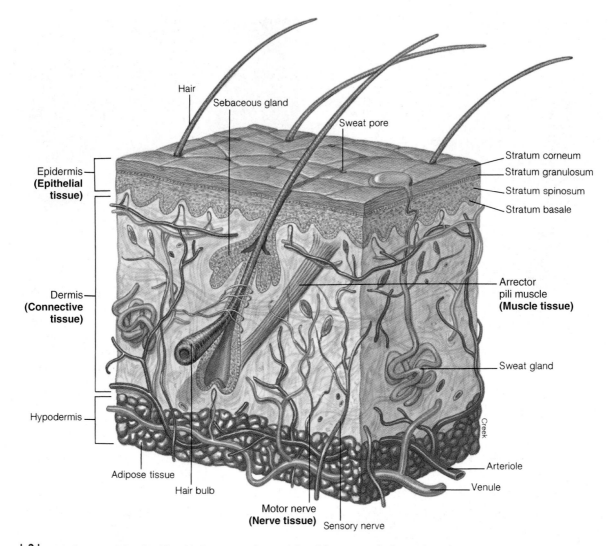

Hair
Sebaceous gland
Sweat pore
Stratum corneum
Stratum granulosum
Stratum spinosum
Stratum basale
Epidermis **(Epithelial tissue)**
Dermis **(Connective tissue)**
Arrector pili muscle **(Muscle tissue)**
Sweat gland
Hypodermis
Adipose tissue
Hair bulb
Arteriole
Venule
Motor nerve **(Nerve tissue)**
Sensory nerve
Creek

■ Figure 1.21   **A diagram of the skin.** The skin is an organ that contains all four types of primary tissues.

An Example of an Organ: The Skin

The cornified *epidermis* protects the skin against water loss and against invasion by disease-causing organisms. Invaginations of the epithelium into the underlying connective tissue *dermis* create the exocrine glands of the skin. These include hair follicles (which produce the hair), sweat glands, and sebaceous glands. The secretion of sweat glands cools the body by evaporation and produces odors that, at least in lower animals, serve as sexual attractants. Sebaceous glands secrete oily sebum into hair follicles, which transport the sebum to the surface of the skin. Sebum lubricates the cornified surface of the skin, helping to prevent it from drying and cracking.

The skin is nourished by blood vessels within the dermis. In addition to blood vessels, the dermis contains wandering white blood cells and other types of cells that protect against invading disease-causing organisms. It also contains nerve fibers and fat cells; however, most of the fat cells are grouped together to form the *hypodermis* (a layer beneath the dermis). Although fat cells are a type of connective tissue, masses of fat deposits throughout the body—such as subcutaneous fat—are referred to as *adipose tissue.*

Sensory nerve endings within the dermis mediate the cutaneous sensations of touch, pressure, heat, cold, and pain. Some of these sensory stimuli directly affect the sensory nerve endings. Others act via sensory structures derived from nonneural primary tissues. The pacinian (lamellated) corpuscles in the dermis of the skin (fig. 1.22), for example, monitor sensations of pressure. Motor nerve fibers in the skin stimulate effector organs, resulting in, for example, the secretions of exocrine glands and contractions of the arrector pili muscles, which attach to hair follicles and surrounding connective tissue (producing goose bumps). The degree of constriction or dilation of cutaneous blood vessels—and therefore the rate of blood flow—is also regulated by motor nerve fibers.

The epidermis itself is a dynamic structure that can respond to environmental stimuli. The rate of its cell division—and consequently the thickness of the cornified layer—increases under the stimulus of constant abrasion. This produces calluses.

Sensory neuron

Figure 1.22 **A diagram of a pacinian corpuscle.** This receptor for deep pressure consists of epithelial cells and connective tissue proteins that form concentric layers around the ending of a sensory neuron.

The skin also protects itself against the dangers of ultraviolet light by increasing its production of *melanin* pigment, which absorbs ultraviolet light while producing a tan. In addition, the skin is an endocrine gland; it synthesizes and secretes vitamin D (derived from cholesterol under the influence of ultraviolet light), which functions as a hormone.

The architecture of most organs is similar to that of the skin. Most are covered by an epithelium that lies immediately over a connective tissue layer. The connective tissue contains blood vessels, nerve endings, scattered cells for fighting infection, and possibly glandular tissue as well. If the organ is hollow—as with the digestive tract or blood vessels—the lumen is also lined with an epithelium overlying a connective tissue layer. The presence, type, and distribution of muscle tissue and nervous tissue vary in different organs.

Systems

Organs that are located in different regions of the body and that perform related functions are grouped into **systems.** These include the integumentary system, nervous system, endocrine system, skeletal system, muscular system, circulatory system, immune system, respiratory system, urinary system, digestive system, and reproductive system (table 1.4). By means of numerous regulatory mechanisms, these systems work together to maintain the life and health of the entire organism.

Body-Fluid Compartments

Tissues, organs, and systems can all be divided into two major parts, or compartments. The **intracellular compartment** is that part inside the cells; the **extracellular compartment** is that part outside the cells. Both compartments consist primarily of water—

Table 1.4 Organ Systems of the Body

System	Major Organs	Primary Functions
Integumentary	Skin, hair, nails	Protection, thermoregulation
Nervous	Brain, spinal cord, nerves	Regulation of other body systems
Endocrine	Hormone-secreting glands, such as the pituitary, thyroid, and adrenals	Secretion of regulatory molecules called hormones
Skeletal	Bones, cartilages	Movement and support
Muscular	Skeletal muscles	Movements of the skeleton
Circulatory	Heart, blood vessels, lymphatic vessels	Movement of blood and lymph
Immune	Bone marrow, lymphoid organs	Defense of the body against invading pathogens
Respiratory	Lungs, airways	Gas exchange
Urinary	Kidneys, ureters, urethra	Regulation of blood volume and composition
Digestive	Mouth, stomach, intestine, liver, gallbladder, pancreas	Breakdown of food into molecules that enter the body
Reproductive	Gonads, external genitalia, associated glands and ducts	Continuation of the human species

they are said to be *aqueous.* The two compartments are separated by the cell membrane surrounding each cell (see chapter 3).

The extracellular compartment is subdivided into two parts. One part is the *blood plasma,* the fluid portion of the blood. The other is the fluid that bathes the cells within the organs of the body. This is called *tissue fluid,* or *interstitial fluid.* In most parts of the body, blood plasma and tissue fluid communicate freely through blood capillaries. The kidneys regulate the volume and composition of the blood plasma, and thus, indirectly, the fluid volume and composition of the entire extracellular compartment.

There is also selective communication between the intracellular and extracellular compartments through the movement of molecules and ions through the cell membrane, as described in chapter 6. This is how cells obtain the molecules they need for life and how they eliminate waste products.

Test Yourself Before You Continue

1. State the location of each type of primary tissue in the skin.
2. Describe the functions of nervous, muscle, and connective tissue in the skin.
3. Describe the functions of the epidermis and explain why this tissue is called "dynamic."
4. Distinguish between the intracellular and extracellular compartments and explain their significance.

Summary

Introduction to Physiology 4

I. Physiology is the study of how cells, tissues, and organs function.

  A. In the study of physiology, cause-and-effect sequences are emphasized.

  B. Knowledge of physiological mechanisms is deduced from data obtained experimentally.

II. The science of physiology overlaps with chemistry and physics and shares knowledge with the related sciences of pathophysiology and comparative physiology.

  A. Pathophysiology is concerned with the functions of diseased or injured body systems and is based on knowledge of how normal systems function, which is the focus of physiology.

  B. Comparative physiology is concerned with the physiology of animals other than humans and shares much information with human physiology.

III. All of the information in this book has been gained by applications of the scientific method. This method has three essential characteristics.

  A. It is assumed that the subject under study can ultimately be explained in terms we can understand.

  B. Descriptions and explanations are honestly based on observations of the natural world and can be changed as warranted by new observations.

  C. Humility is an important characteristic of the scientific method; the scientist must be willing to change his or her theories when warranted by the weight of the evidence.

Homeostasis and Feedback Control 6

I. Homeostasis refers to the dynamic constancy of the internal environment.

  A. Homeostasis is maintained by mechanisms that act through negative feedback loops.

    1. A negative feedback loop requires (1) a sensor that can detect a change in the internal environment and (2) an effector that can be activated by the sensor.

    2. In a negative feedback loop, the effector acts to cause changes in the internal environment that compensate for the initial deviations that were detected by the sensor.

  B. Positive feedback loops serve to amplify changes and may be part of the action of an overall negative feedback mechanism.

  C. The nervous and endocrine systems provide extrinsic regulation of other body systems and act to maintain homeostasis.

  D. The secretion of hormones is stimulated by specific chemicals and is inhibited by negative feedback mechanisms.

II. Effectors act antagonistically to defend the set point against deviations in any direction.

The Primary Tissues 9

I. The body is composed of four primary tissues: muscle, nervous, epithelial, and connective tissues.

  A. There are three types of muscle tissue: skeletal, cardiac, and smooth muscle.

    1. Skeletal and cardiac muscle are striated.

    2. Smooth muscle is found in the walls of the internal organs.

  B. Nervous tissue is composed of neurons and supporting cells.

    1. Neurons are specialized for the generation and conduction of electrical impulses.

    2. Supporting cells provide the neurons with anatomical and functional support.

  C. Epithelial tissue includes membranes and glands.

    1. Epithelial membranes cover and line the body surfaces, and their cells are tightly joined by junctional complexes.

    2. Epithelial membranes may be simple or stratified and their cells may be squamous, cuboidal, or columnar.

    3. Exocrine glands, which secrete into ducts, and endocrine glands, which lack ducts and secrete hormones into the blood, are derived from epithelial membranes.

  D. Connective tissue is characterized by large intercellular spaces that contain extracellular material.

    1. Connective tissue proper is categorized into subtypes, including loose, dense fibrous, adipose, and others.

    2. Cartilage, bone, and blood are classified as connective tissues because their cells are widely spaced with abundant extracellular material between them.

Organs and Systems 17

I. Organs are units of structure and function that are composed of at least two, and usually all four, primary tissues.

  A. The skin is a good example of an organ.

    1. The epidermis is a stratified squamous keratinized epithelium that protects underlying structures and produces vitamin D.

    2. The dermis is an example of loose connective tissue.

    3. Hair follicles, sweat glands, and sebaceous glands are exocrine glands located within the dermis.

    4. Sensory and motor nerve fibers enter the spaces within the dermis to innervate sensory organs and smooth muscles.

    5. The arrector pili muscles that attach to the hair follicles are composed of smooth muscle.

  B. Organs that are located in different regions of the body and that perform related functions are grouped into systems. These include, among others, the circulatory system, digestive system, and endocrine system.

II. The fluids of the body are divided into two major compartments.

  A. The intracellular compartment refers to the fluid within cells.

  B. The extracellular compartment refers to the fluid outside of cells; extracellular fluid is subdivided into plasma (the fluid portion of the blood) and tissue (interstitial) fluid.

Review Activities

Test Your Knowledge of Terms and Facts

Match the following (1–4):

1. Glands are derived from
   a. nervous tissue
   b. connective tissue
   c. muscular tissue
   d. epithelial tissue

2. Cells are joined closely together in

3. Cells are separated by large extracellular spaces in

4. Blood vessels and nerves are usually located within

5. Most organs are composed of
   a. epithelial tissue.
   b. muscle tissue.
   c. connective tissue.
   d. all of these.

6. Sweat is secreted by exocrine glands. This means that
   a. it is produced by epithelial cells.
   b. it is a hormone.
   c. it is secreted into a duct.
   d. it is produced outside the body.

7. Which of these statements about homeostasis is *true?*
   a. The internal environment is maintained absolutely constant.
   b. Negative feedback mechanisms act to correct deviations from a normal range within the internal environment.
   c. Homeostasis is maintained by switching effector actions on and off.
   d. All of these are true.

8. In a negative feedback loop, the effector organ produces changes that are
   a. in the same direction as the change produced by the initial stimulus.
   b. opposite in direction to the change produced by the initial stimulus.
   c. unrelated to the initial stimulus.

9. A hormone called parathyroid hormone acts to help raise the blood calcium concentration. According to the principles of negative feedback, an effective stimulus for parathyroid hormone secretion would be
   a. a fall in blood calcium.
   b. a rise in blood calcium.

10. Which of these consists of dense parallel arrangements of collagen fibers?
    a. skeletal muscle tissue
    b. nervous tissue
    c. tendons
    d. dermis of the skin

11. The act of breathing raises the blood oxygen level, lowers the blood carbon dioxide concentration, and raises the blood pH. According to the principles of negative feedback, sensors that regulate breathing should respond to
    a. a rise in blood oxygen.
    b. a rise in blood pH.
    c. a rise in blood carbon dioxide concentration.
    d. all of these.

Test Your Understanding of Concepts and Principles

1. Describe the structure of the various epithelial membranes and explain how their structures relate to their functions.[1]

2. Compare bone, blood, and the dermis of the skin in terms of their similarities. What are the major structural differences between these tissues?

3. Describe the role of antagonistic negative feedback processes in the maintenance of homeostasis.

4. Using insulin as an example, explain how the secretion of a hormone is controlled by the effects of that hormone's actions.

5. Describe the steps in the development of pharmaceutical drugs and evaluate the role of animal research in this process.

6. Why is Claude Bernard considered the father of modern physiology? Why is the concept he introduced so important in physiology and medicine?

Test Your Ability to Analyze and Apply Your Knowledge

1. What do you think would happen if most of your physiological regulatory mechanisms were to operate by positive feedback rather than by negative feedback? Would life even be possible?

2. Examine figure 1.5 and determine when the compensatory physiological responses began to act, and how many minutes they required to restore the initial set point of blood glucose concentration. Comment on the importance of quantitative measurements in physiology.

3. Why are interactions between the body-fluid compartments essential for sustaining life?

Related Websites

Check out the Links Library at www.mhhe.com/fox8 for links to sites containing resources related to the study of body function. These links are monitored to ensure current URLs.

---

[1]*Note:* This question is answered in the chapter 1 Study Guide found on the Online Learning Center at www.mhhe.com/fox8.

2 Chemical Composition of the Body

Objectives

After studying this chapter, you should be able to . . .

1. describe the structure of an atom and define the terms *atomic mass* and *atomic number*.

2. explain how covalent bonds are formed and distinguish between nonpolar and polar covalent bonds.

3. describe the structure of an ion and explain how ionic bonds are formed.

4. describe the nature of hydrogen bonds and explain their significance.

5. describe the structure of a water molecule and explain why some compounds are hydrophilic and others are hydrophobic.

6. define the terms *acid* and *base* and explain what is meant by the pH scale.

7. explain how the pH of the blood is stabilized by bicarbonate buffer and define the terms *acidosis* and *alkalosis*.

8. describe the various types of carbohydrates and give examples of each type.

9. describe the mechanisms of dehydration synthesis and hydrolysis reactions and explain their significance.

10. state the common characteristic of lipids and describe the different categories of lipids.

11. describe how peptide bonds are formed and discuss the different orders of protein structure.

12. list some of the functions of proteins and explain why proteins can provide the specificity required to perform these functions.

13. describe the structure of DNA and RNA, and explain the law of complementary base pairing.

Chapter at a Glance

Clinical Investigation

George decides that it is immoral to eat plants or animals, and so resolves to eat only artificial food. After raiding the storeroom of his freshman chemistry lab, he places himself on a diet that consists only of the D-amino acids and L-sugars that he obtained in his raid. He feels very weak after several days and seeks medical attention.

Laboratory analysis of his urine reveals very high concentrations of ketone bodies (a condition call *ketonuria*). What might be the cause of his weakness and ketonuria?

Atoms, Ions, and Chemical Bonds

The study of physiology requires some familiarity with the basic concepts and terminology of chemistry. A knowledge of atomic and molecular structure, the nature of chemical bonds, and the nature of pH and associated concepts provides the foundation for much of human physiology.

The structures and physiological processes of the body are based, to a large degree, on the properties and interactions of atoms, ions, and molecules. Water is the major constituent of the body and accounts for 65% to 75% of the total weight of an average adult. Of this amount, two-thirds is contained within the body cells, or in the *intracellular compartment;* the remainder is contained in the *extracellular compartment,* a term that refers to the blood and tissue fluids. Dissolved in this water are many organic molecules (carbon-containing molecules such as carbohydrates, lipids, proteins, and nucleic acids), as well as inorganic molecules and ions (atoms with a net charge). Before describing the structure and function of organic molecules within the body, it would be useful to consider some basic chemical concepts, terminology, and symbols.

Atoms

Atoms are the smallest units of matter that can undergo chemical change. They are much too small to be seen individually, even with the most powerful electron microscope. Through the efforts of generations of scientists, however, atomic structure is now well understood. At the center of an atom is its **nucleus.** The nucleus contains two types of particles—**protons,** which bear a positive charge, and **neutrons,** which carry no charge (are

neutral). The mass of a proton is equal to the mass of a neutron, and the sum of the protons and neutrons in an atom is equal to the **atomic mass** of the atom. For example, an atom of carbon, which contains six protons and six neutrons, has an atomic mass of 12 (table 2.1). Note that the mass of electrons is not considered when calculating the atomic mass, because it is insignificantly small compared to the mass of protons and neutrons.

The number of protons in an atom is given as its **atomic number.** Carbon has six protons and thus has an atomic number of 6. Outside the positively charged nucleus are negatively charged subatomic particles called **electrons.** Since the number of electrons in an atom is equal to the number of protons, atoms have a net charge of zero.

Although it is often convenient to think of electrons as orbiting the nucleus like planets orbiting the sun, this simplified model of atomic structure is no longer believed to be correct. A given electron can occupy any position in a certain volume of space called the *orbital* of the electron. The orbitals form a "shell," or energy level, beyond which the electron usually does not pass.

There are potentially several such shells surrounding a nucleus, with each successive shell being farther from the nucleus. The first shell, closest to the nucleus, can contain only two electrons. If an atom has more than two electrons (as do all atoms except hydrogen and helium), the additional electrons must occupy shells that are more distant from the nucleus. The second shell can contain a maximum of eight electrons and higher shells can contain still more electrons that possess more energy the farther they are from the nucleus. Most elements of biological significance (other than hydrogen), however, require eight electrons to complete the outermost shell. The shells are filled from the innermost outward. Carbon, with six electrons, has two electrons in its first shell and four electrons in its second shell (fig. 2.1).

It is always the electrons in the outermost shell, if this shell is incomplete, that participate in chemical reactions and form chemical bonds. These outermost electrons are known as the **valence electrons** of the atom.

Isotopes

A particular atom with a given number of protons in its nucleus may exist in several forms that differ from one another in their number of neutrons. The atomic number of these forms is thus the same, but their atomic mass is different. These different forms are called **isotopes.** All of the isotopic forms of a given

Table 2.1 Atoms Commonly Present in Organic Molecules

Atom	Symbol	Atomic Number	Atomic Mass	Shell 1	Shell 2	Shell 3	Number of Chemical Bonds
Hydrogen	H	1	1	1	0	0	1
Carbon	C	6	12	2	4	0	4
Nitrogen	N	7	14	2	5	0	3
Oxygen	O	8	16	2	6	0	2
Sulfur	S	16	32	2	8	6	2

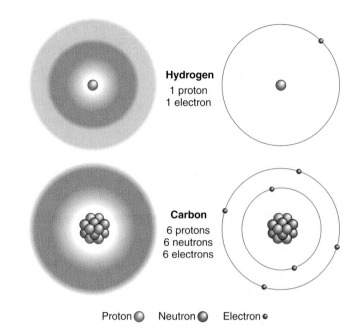

Proton ⬤   Neutron ⬤   Electron ●

Hydrogen
1 proton
1 electron

Carbon
6 protons
6 neutrons
6 electrons

■ **Figure 2.1**   Diagrams of the hydrogen and carbon atoms. The electron shells on the left are represented by shaded spheres indicating probable positions of the electrons. The shells on the right are represented by concentric circles.

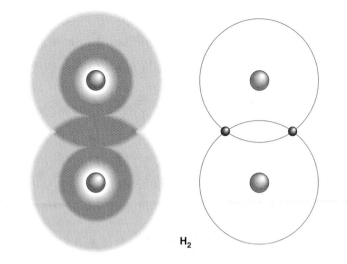

H_2

■ **Figure 2.2**   A hydrogen molecule showing the covalent bonds between hydrogen atoms. These bonds are formed by the equal sharing of electrons.

atom are included in the term **chemical element.** The element hydrogen, for example, has three isotopes. The most common of these has a nucleus consisting of only one proton. Another isotope of hydrogen (called *deuterium*) has one proton and one neutron in the nucleus, whereas the third isotope (*tritium*) has one proton and two neutrons. Tritium is a radioactive isotope that is commonly used in physiological research and in many clinical laboratory procedures.

Chemical Bonds, Molecules, and Ionic Compounds

Molecules are formed through interaction of the valence electrons between two or more atoms. These interactions, such as the sharing of electrons, produce **chemical bonds** (fig. 2.2). The number of bonds that each atom can have is determined by the number of electrons needed to complete the outermost shell. Hydrogen, for example, must obtain only one more electron—and can thus form only one chemical bond—to complete the first shell of two electrons. Carbon, by contrast, must obtain four more electrons—and can thus form four chemical bonds—to complete the second shell of eight electrons (fig. 2.3, *left*).

Covalent Bonds

Covalent bonds result when atoms share their valence electrons. Covalent bonds that are formed between identical atoms, as in oxygen gas (O_2) and hydrogen gas (H_2), are the strongest because their electrons are equally shared. Since the electrons are equally distributed between the two atoms, these molecules are said to be **nonpolar** and the bonds between them are non-

polar covalent bonds. Such bonds are also important in living organisms. The unique nature of carbon atoms and the organic molecules formed through covalent bonds between carbon atoms provides the chemical foundation of life.

When covalent bonds are formed between two different atoms, the electrons may be pulled more toward one atom than the other. The end of the molecule toward which the electrons are pulled is electrically negative compared to the other end. Such a molecule is said to be **polar** (has a positive and negative "pole"). Atoms of oxygen, nitrogen, and phosphorus have a particularly strong tendency to pull electrons toward themselves when they bond with other atoms; thus, they tend to form polar molecules.

Water is the most abundant molecule in the body and serves as the solvent for body fluids. Water is a good solvent because it is polar; the oxygen atom pulls electrons from the two hydrogens toward its side of the water molecule, so that the oxygen side is more negatively charged than the hydrogen side of the molecule (fig. 2.4). The significance of the polar nature of water in its function as a solvent is discussed in the next section.

Ionic Bonds

Ionic bonds result when one or more valence electrons from one atom are completely transferred to a second atom. Thus, the electrons are not shared at all. The first atom loses electrons, so that its number of electrons becomes smaller than its number of protons; it becomes positively charged. Atoms or molecules that have positive or negative charges are called **ions.** Positively charged ions are called *cations* because they move toward the negative pole, or cathode, in an electric field. The second atom now has more electrons than it has protons and becomes a negatively charged ion, or *anion* (so called because it moves toward the positive pole, or anode, in an electric field). The cation and anion then attract each other to form an **ionic compound.**

Common table salt, sodium chloride (NaCl), is an example of an ionic compound. Sodium, with a total of eleven electrons,

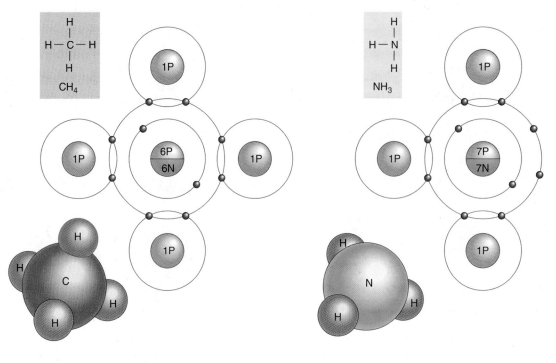

Methane (CH₄)   **Ammonia (NH₃)**

■ **Figure 2.3**   The molecules methane and ammonia represented in three different ways. Notice that a bond between two atoms consists of a pair of shared electrons (the electrons from the outer shell of each atom).

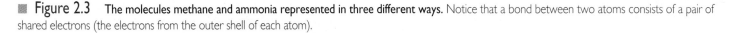

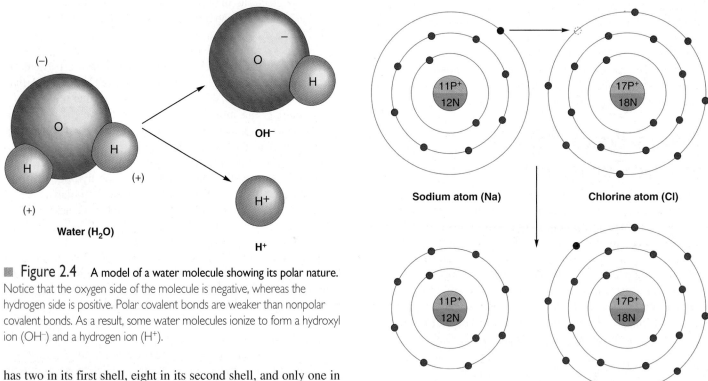

Water (H₂O)   **H⁺**

■ **Figure 2.4**   A model of a water molecule showing its polar nature. Notice that the oxygen side of the molecule is negative, whereas the hydrogen side is positive. Polar covalent bonds are weaker than nonpolar covalent bonds. As a result, some water molecules ionize to form a hydroxyl ion (OH⁻) and a hydrogen ion (H⁺).

has two in its first shell, eight in its second shell, and only one in its third shell. Chlorine, conversely, is one electron short of completing its outer shell of eight electrons. The lone electron in sodium's outer shell is attracted to chlorine's outer shell. This creates a chloride ion (represented as Cl⁻) and a sodium ion (Na⁺). Although table salt is shown as NaCl, it is actually composed of Na⁺Cl⁻ (fig. 2.5).

Sodium atom (Na)   **Chlorine atom (Cl)**

Sodium ion (Na⁺)   **Chloride ion (Cl⁻)**

■ **Figure 2.5**   The reaction of sodium with chlorine to produce sodium and chloride ions. The positive sodium and negative chloride ions attract each other, producing the ionic compound sodium chloride (NaCl).

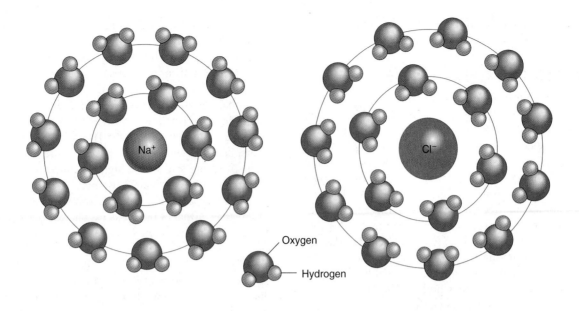

Water molecule

■ **Figure 2.6** **How NaCl dissolves in water.** The negatively charged oxygen-ends of water molecules are attracted to the positively charged Na⁺, whereas the positively charged hydrogen-ends of water molecules are attracted to the negatively charged Cl⁻. Other water molecules are attracted to this first concentric layer of water, forming hydration spheres around the sodium and chloride ions.

Ionic bonds are weaker than polar covalent bonds, and therefore ionic compounds easily separate (dissociate) when dissolved in water. Dissociation of NaCl, for example, yields Na⁺ and Cl⁻. Each of these ions attracts polar water molecules; the negative ends of water molecules are attracted to the Na⁺, and the positive ends of water molecules are attracted to the Cl⁻ (fig. 2.6). The water molecules that surround these ions in turn attract other molecules of water to form *hydration spheres* around each ion.

The formation of hydration spheres makes an ion or a molecule soluble in water. Glucose, amino acids, and many other organic molecules are water-soluble because hydration spheres can form around atoms of oxygen, nitrogen, and phosphorus, which are joined by polar covalent bonds to other atoms in the molecule. Such molecules are said to be **hydrophilic.** By contrast, molecules composed primarily of nonpolar covalent bonds, such as the hydrocarbon chains of fat molecules, have few charges and thus cannot form hydration spheres. They are insoluble in water, and in fact are repelled by water molecules. For this reason, nonpolar molecules are said to be **hydrophobic.**

Hydrogen Bonds

When a hydrogen atom forms a polar covalent bond with an atom of oxygen or nitrogen, the hydrogen gains a slight positive charge as the electron is pulled toward the other atom. This other atom is thus described as being *electronegative.* Since the hydrogen has a slight positive charge, it will have a weak attraction for a second electronegative atom (oxygen or nitrogen) that may be located near it. This weak attraction is called a **hydrogen bond.** Hydrogen bonds are usually shown with dashed or dotted lines (fig. 2.7) to distinguish them from strong covalent bonds, which are shown with solid lines.

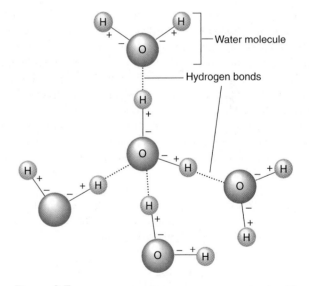

■ **Figure 2.7** **Hydrogen bonds between water molecules.** The oxygen atoms of water molecules are weakly joined together by the attraction of the electronegative oxygen for the positively charged hydrogen. These weak bonds are called hydrogen bonds.

Although each hydrogen bond is relatively weak, the sum of their attractive forces is largely responsible for the folding and bending of long organic molecules such as proteins and for the holding together of the two strands of a DNA molecule (described later in this chapter). Hydrogen bonds can also be formed between adjacent water molecules (fig. 2.7). The hydrogen bonding between water molecules is responsible for many of the biologically important properties of water, including its *surface tension* (see chapter 16) and its ability to be pulled as a column through narrow channels in a process called *capillary action.*

Acids, Bases, and the pH Scale

The bonds in water molecules joining hydrogen and oxygen atoms together are, as previously discussed, polar covalent bonds. Although these bonds are strong, a small proportion of them break as the electron from the hydrogen atom is completely transferred to oxygen. When this occurs, the water molecule ionizes to form a *hydroxyl ion* (OH^-) and a hydrogen ion (H^+), which is simply a free proton (see fig. 2.4). A proton released in this way does not remain free for long, however, because it is attracted to the electrons of oxygen atoms in water molecules. This forms a *hydronium ion,* shown by the formula H_3O^+. For the sake of clarity in the following discussion, however, H^+ will be used to represent the ion resulting from the ionization of water.

Ionization of water molecules produces equal amounts of OH^- and H^+. Since only a small proportion of water molecules ionize, the concentrations of H^+ and OH^- are each equal to only 10^{-7} molar (the term *molar* is a unit of concentration, described in chapter 6; for hydrogen, one molar equals one gram per liter). A solution with 10^{-7} molar hydrogen ion, which is produced by the ionization of water molecules in which the H^+ and OH^- concentrations are equal, is said to be **neutral.**

A solution that has a higher H^+ concentration than that of water is called *acidic;* one with a lower H^+ concentration is called *basic,* or *alkaline.* An **acid** is defined as a molecule that can release protons (H^+) into a solution; it is a "proton donor." A **base** can be a molecule such as ammonia (NH_3) that can combine with H^+ (to form NH_4^+, ammonium ion). More commonly, it is a molecule such as NaOH that can ionize to produce a negatively charged ion (hydroxyl, OH^-), which in turn can combine with H^+ (to form H_2O, water). A base thus removes H^+ from solution; it is a "proton acceptor," thus lowering the H^+ concentration of the solution. Examples of common acids and bases are shown in table 2.2.

pH

The H^+ concentration of a solution is usually indicated in pH units on a pH scale that runs from 0 to 14. The pH value is equal to the logarithm of 1 over the H^+ concentration:

$$pH = \log \frac{1}{[H^+]}$$

where $[H^+]$ = molar H^+ concentration. This can also be expressed as $pH = -\log [H^+]$.

Table 2.2 Common Acids and Bases

Acid	Symbol	Base	Symbol
Hydrochloric acid	HCl	Sodium hydroxide	NaOH
Phosphoric acid	H_3PO_4	Potassium hydroxide	KOH
Nitric acid	HNO_3	Calcium hydroxide	$Ca(OH)_2$
Sulfuric acid	H_2SO_4	Ammonium hydroxide	NH_4OH
Carbonic acid	H_2CO_3		

Pure water has a H^+ concentration of 10^{-7} molar at 25° C, and thus has a pH of 7 (neutral). Because of the logarithmic relationship, a solution with 10 times the hydrogen ion concentration (10^{-6} M) has a pH of 6, whereas a solution with one-tenth the H^+ concentration (10^{-8} M) has a pH of 8. The pH value is easier to write than the molar H^+ concentration, but it is admittedly confusing because it is *inversely related* to the H^+ concentration— that is, a solution with a higher H^+ concentration has a lower pH value, and one with a lower H^+ concentration has a higher pH value. A strong acid with a high H^+ concentration of 10^{-2} molar, for example, has a pH of 2, whereas a solution with only 10^{-10} molar H^+ has a pH of 10. **Acidic solutions,** therefore, have a pH of less than 7 (that of pure water), whereas **basic (alkaline) solutions** have a pH between 7 and 14 (table 2.3).

Buffers

A **buffer** is a system of molecules and ions that acts to prevent changes in H^+ concentration and thus serves to stabilize the pH of a solution. In blood plasma, for example, the pH is stabilized by the following reversible reaction involving the bicarbonate ion (HCO_3^-) and carbonic acid (H_2CO_3):

$$HCO_3^- + H^+ \rightleftharpoons H_2CO_3$$

The double arrows indicate that the reaction could go either to the right or to the left; the net direction depends on the concentration of molecules and ions on each side. If an acid (such as lactic acid) should release H^+ into the solution, for example, the increased concentration of H^+ would drive the equilibrium to the right and the following reaction would be promoted:

$$HCO_3^- + H^+ \rightarrow H_2CO_3$$

Table 2.3 The pH Scale

	H^+ Concentration (Molar)*	pH	OH^- Concentration (Molar)*
Acids	1.0	0	10^{-14}
	0.1	1	10^{-13}
	0.01	2	10^{-12}
	0.001	3	10^{-11}
	0.0001	4	10^{-10}
	10^{-5}	5	10^{-9}
	10^{-6}	6	10^{-8}
Neutral	10^{-7}	7	10^{-7}
Bases	10^{-8}	8	10^{-6}
	10^{-9}	9	10^{-5}
	10^{-10}	10	0.0001
	10^{-11}	11	0.001
	10^{-12}	12	0.01
	10^{-13}	13	0.1
	10^{-14}	14	1.0

*Molar concentration is the number of moles of a solute dissolved in one liter. One mole is the atomic or molecular weight of the solute in grams. Since hydrogen has an atomic weight of one, one molar hydrogen is one gram of hydrogen per liter of solution.

Notice that, in this reaction, H⁺ is taken out of solution. Thus, the H⁺ concentration is prevented from rising (and the pH prevented from falling) by the action of bicarbonate buffer.

Blood pH

Lactic acid and other organic acids are produced by the cells of the body and secreted into the blood. Despite the release of H⁺ by these acids, the arterial blood pH normally does not decrease but remains remarkably constant at pH 7.40 ± 0.05. This constancy is achieved, in part, by the buffering action of bicarbonate shown in the preceding equation. Bicarbonate serves as the major buffer of the blood.

Certain conditions could cause an opposite change in pH. For example, excessive vomiting that results in loss of gastric acid could cause the concentration of free H⁺ in the blood to fall and the blood pH to rise. In this case, the reaction previously described could be reversed:

$$H_2CO_3 \rightarrow H^+ + HCO_3^-$$

The dissociation of carbonic acid yields free H⁺, which helps to prevent an increase in pH. Bicarbonate ions and carbonic acid thus act as a *buffer pair* to prevent either decreases or increases in pH, respectively. This buffering action normally maintains the blood pH within the narrow range of 7.35 to 7.45.

If the arterial blood pH falls below 7.35, the condition is called *acidosis*. A blood pH of 7.20, for example, represents significant acidosis. Notice that acidotic blood need not be acidic. An increase in blood pH above 7.45, conversely, is known as *alkalosis*. Acidosis and alkalosis are normally prevented by the action of the bicarbonate/carbonic acid buffer pair and by the functions of the lungs and kidneys. Regulation of blood pH is discussed in more detail in chapters 13, 16, and 17.

Organic Molecules

Organic molecules are those molecules that contain the atoms carbon and hydrogen. Since the carbon atom has four electrons in its outer shell, it must share four additional electrons by covalently bonding with other atoms to fill its outer shell with eight electrons. The unique bonding requirements of carbon enable it to join with other carbon atoms to form chains and rings while still allowing the carbon atoms to bond with hydrogen and other atoms.

Most organic molecules in the body contain hydrocarbon chains and rings, as well as other atoms bonded to carbon. Two adjacent carbon atoms in a chain or ring may share one or two pairs of electrons. If the two carbon atoms share one pair of electrons, they are said to have a *single covalent bond;* this leaves each carbon atom free to bond with as many as three other atoms. If the two carbon atoms share two pairs of electrons, they have a *double covalent bond,* and each carbon atom can bond with a maximum of only two additional atoms (fig. 2.8).

The ends of some hydrocarbons are joined together to form rings. In the shorthand structural formulas for these molecules, the carbon atoms are not shown but are understood to be located at the corners of the ring. Some of these cyclic molecules have a double bond between two adjacent carbon atoms. Benzene and related molecules are shown as a six-sided ring with alternating double bonds. Such compounds are called **aromatic.** Since all of the carbons in an aromatic ring are equivalent, double bonds can be shown between any two adjacent carbons in the ring (fig. 2.9), or even as a circle within the hexagonal structure of carbons.

The hydrocarbon chain or ring of many organic molecules provides a relatively inactive molecular "backbone" to which more reactive groups of atoms are attached. Known as *functional groups*

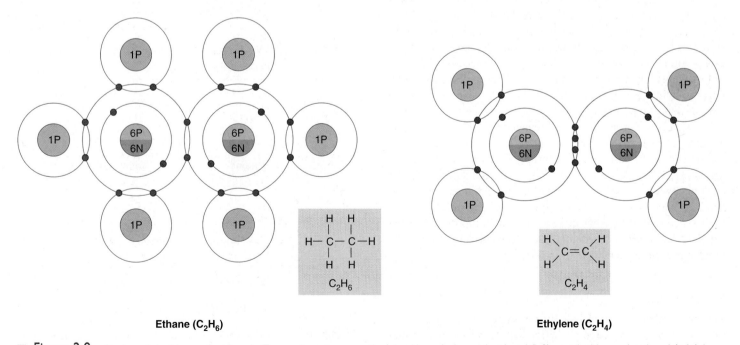

Ethane (C₂H₆)

Ethylene (C₂H₄)

■ **Figure 2.8** **Single and double covalent bonds.** Two carbon atoms may be joined by a single covalent bond *(left)* or a double covalent bond *(right).* In both cases, each carbon atom shares four pairs of electrons (has four bonds) to complete the eight electrons required to fill its outer shell.

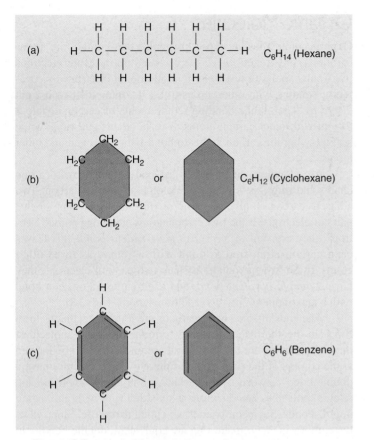

■ **Figure 2.9** Different shapes of hydrocarbon molecules.
Hydrocarbon molecules can be (a) linear, (b) cyclic, or (c) have aromatic rings.

of the molecule, these reactive groups usually contain atoms of oxygen, nitrogen, phosphorus, or sulfur. They are largely responsible for the unique chemical properties of the molecule (fig. 2.10).

Classes of organic molecules can be named according to their functional groups. **Ketones,** for example, have a *carbonyl group* within the carbon chain. An organic molecule is an **alcohol** if it has a *hydroxyl group* bound to a hydrocarbon chain. All **organic acids** (acetic acid, citric acids, lactic acid, and others) have a *carboxyl group* (fig. 2.11).

A carboxyl group can be abbreviated COOH. This group is an acid because it can donate its proton (H^+) to the solution. Ionization of the OH part of COOH forms COO^- and H^+ (fig. 2.12). The ionized organic acid is designated with the suffix *-ate.* For example, when the carboxyl group of lactic acid ionizes, the molecule is called *lactate.* Since both ionized and unionized forms of the molecule exist together in a solution (the proportion of each depends on the pH of the solution), one can correctly refer to the molecule as either lactic acid or lactate.

Stereoisomers

Two molecules may have exactly the same atoms arranged in exactly the same sequence yet differ with respect to the spatial orientation of a key functional group. Such molecules are called **stereoisomers** of each other. Depending upon the direction in

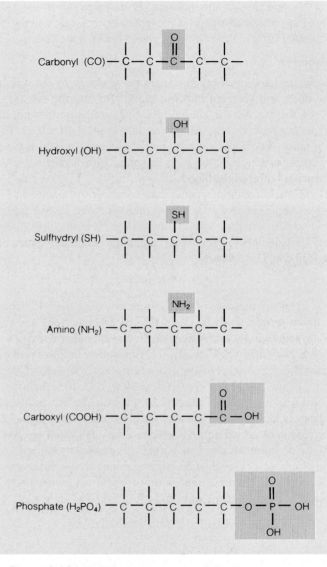
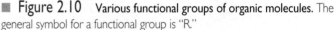

■ **Figure 2.10** Various functional groups of organic molecules. The general symbol for a functional group is "R."

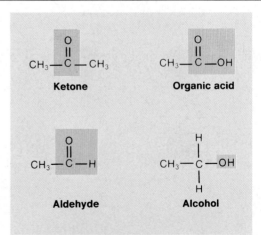
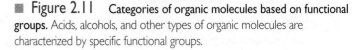

■ **Figure 2.11** Categories of organic molecules based on functional groups. Acids, alcohols, and other types of organic molecules are characterized by specific functional groups.

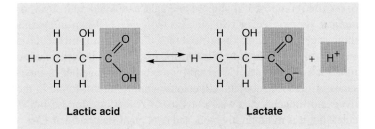

Figure 2.12 **The carboxyl group of an organic acid.** This group can ionize to yield a free proton, which is a hydrogen ion (H^+). This process is shown for lactic acid, with the double arrows indicating that the reaction is reversible.

Severe birth defects often resulted when pregnant women used the sedative **thalidomide** in the early 1960s to alleviate morning sickness. The drug available at the time contained a mixture of both right-handed (D) and left-handed (L) forms. This tragic circumstance emphasizes the clinical importance of stereoisomers. It has since been learned that the L-stereoisomer is a potent tranquilizer, but the right-handed version causes disruption of fetal development and the re-sulting birth defects. Interestingly, thalidomide is now being used in the treatment of people with AIDS, leprosy, and *cachexia* (pro-longed ill health and malnutrition).

which the key functional group is oriented with respect to the molecules, stereoisomers are called either *D-isomers* (for *dextro*, or right-handed) or *L-isomers* (for *levo*, or left-handed). Their relationship is similar to that of a right and left glove—if the palms are both pointing in the same direction, the two cannot be superimposed.

These subtle differences in structure are extremely impor-tant biologically. They ensure that enzymes—which interact with such molecules in a stereo-specific way in chemical reactions—cannot combine with the "wrong" stereoisomer. The enzymes of all cells (human and others) can combine only with L-amino acids and D-sugars, for example. The opposite stereoisomers (D-amino acids and L-sugars) cannot be used by any enzyme in metabolism.

Clinical Investigation Clues

Remember that George ate only the D-amino acids and L-sugars he obtained in the chemistry storeroom.

Could his body absorb and use these molecules?

What would be his nutritional status as a result of this diet?

Test Yourself Before You Continue

1. List the components of an atom and explain how they are organized. Explain why different atoms are able to form characteristic numbers of chemical bonds.

2. Describe the nature of nonpolar and polar covalent bonds, ionic bonds, and hydrogen bonds. Why are ions and polar molecules soluble in water?

3. Define the terms *acidic, basic, acid,* and *base.* Also define *pH* and describe the relationship between pH and the H^+ concentration of a solution.

4. Using chemical equations, explain how bicarbonate ion and carbonic acid function as a buffer pair.

5. Explain how carbon atoms can bond with each other and with atoms of hydrogen, oxygen, and nitrogen.

Carbohydrates and Lipids

Carbohydrates are a class of organic molecules that includes monosaccharides, disaccharides, and polysaccharides. All of these molecules are based on a characteristic ratio of carbon, hydrogen, and oxygen atoms. Lipids constitute a category of diverse organic molecules that share the physical property of being nonpolar, and thus insoluble in water.

Carbohydrates and lipids are similar in many ways. Both groups of molecules consist primarily of the atoms carbon, hydro-gen, and oxygen, and both serve as major sources of energy in the body (accounting for most of the calories consumed in food). Car-bohydrates and lipids differ, however, in some important aspects of their chemical structures and physical properties. Such differences significantly affect the functions of these molecules in the body.

Carbohydrates

Carbohydrates are organic molecules that contain carbon, hy-drogen, and oxygen in the ratio described by their name—*carbo* (carbon) and *hydrate* (water, H_2O). The general formula for a carbohydrate molecule is thus $C_nH_{2n}O_n$; the molecule contains twice as many hydrogen atoms as carbon or oxygen atoms (the number of each is indicated by the subscript n).

Monosaccharides, Disaccharides, and Polysaccharides

Carbohydrates include simple sugars, or **monosaccharides,** and longer molecules that contain a number of monosaccharides joined together. The suffix *-ose* denotes a sugar molecule; the term *hexose,* for example, refers to a six-carbon monosaccharide with the formula $C_6H_{12}O_6$. This formula is adequate for some purposes, but it does not distinguish between related hexose sug-ars, which are *structural isomers* of each other. The structural isomers glucose, fructose, and galactose, for example, are mono-saccharides that have the same ratio of atoms arranged in slightly different ways (fig. 2.13).

Two monosaccharides can be joined covalently to form a **di-saccharide,** or double sugar. Common disaccharides include table sugar, or *sucrose* (composed of glucose and fructose); milk sugar,

or *lactose* (composed of glucose and galactose); and malt sugar, or *maltose* (composed of two glucose molecules). When numerous monosaccharides are joined together, the resulting molecule is called a **polysaccharide.** *Starch,* for example, a polysaccharide found in many plants, is formed by the bonding together of thousands of glucose subunits. **Glycogen** (animal starch), found in the liver and muscles, likewise consists of repeating glucose molecules, but it is more highly branched than plant starch (fig. 2.14).

Many cells store carbohydrates for use as an energy source, as described in chapter 5. If a cell were to store many thousands of separate monosaccharide molecules, however, their high concentration would draw an excessive amount of water into the cell, damaging or even killing it. The net movement of water through membranes is called osmosis, and is discussed in chapter 6. Cells that store carbohydrates for energy minimize this osmotic damage by instead joining the glucose molecules together to form the polysaccharides starch or glycogen. Since there are fewer of these larger molecules, less water is drawn into the cell by osmosis (see chapter 6).

Dehydration Synthesis and Hydrolysis

In the formation of disaccharides and polysaccharides, the separate subunits (monosaccharides) are bonded together covalently by a type of reaction called **dehydration synthesis,** or **condensation.** In this reaction, which requires the participation of specific enzymes (chapter 4), a hydrogen atom is removed from one monosaccharide and a hydroxyl group (OH) is removed from another. As a covalent bond is formed between the two monosaccharides, water (H_2O) is produced. Dehydration synthesis reactions are illustrated in figure 2.15.

When a person eats disaccharides or polysaccharides, or when the stored glycogen in the liver and muscles is to be used by tissue cells, the covalent bonds that join monosaccharides to form disaccharides and polysaccharides must be broken. These *digestion reactions* occur by means of **hydrolysis.** Hydrolysis (from the Greek *hydro* = water; *lysis* = break) is the reverse of dehydration synthesis. When a covalent bond joining two monosaccharides is broken, a water molecule provides the atoms needed to complete their structure. The water molecule is split, and the resulting hydrogen atom is added to one of the free glucose molecules as the hydroxyl group is added to the other (fig. 2.16).

When a potato is eaten, the starch within it is hydrolyzed into separate glucose molecules within the small intestine. This glucose is absorbed into the blood and carried to the tissues. Some tissue cells may use this glucose for energy. Liver and muscles, however, can store excess glucose in the form of glycogen by dehydration synthesis reactions in these cells. During fasting or prolonged exercise, the liver can add glucose to the blood through hydrolysis of its stored glycogen.

Dehydration synthesis and hydrolysis reactions do not occur spontaneously; they require the action of specific enzymes. Similar reactions, in the presence of other enzymes, build and break down lipids, proteins, and nucleic acids. In general, therefore, hydrolysis reactions digest molecules into their subunits, and dehydration synthesis reactions build larger molecules by the bonding together of their subunits.

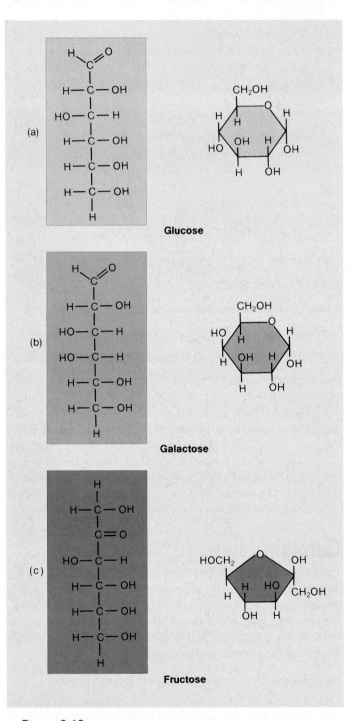

Figure 2.13 Structural formulas for three hexose sugars. These are (a) glucose, (b) galactose, and (c) fructose. All three have the same ratio of atoms—$C_6H_{12}O_6$. The representations on the left more clearly show the atoms in each molecule, while the ring structures on the right more accurately reflect the way these atoms are arranged.

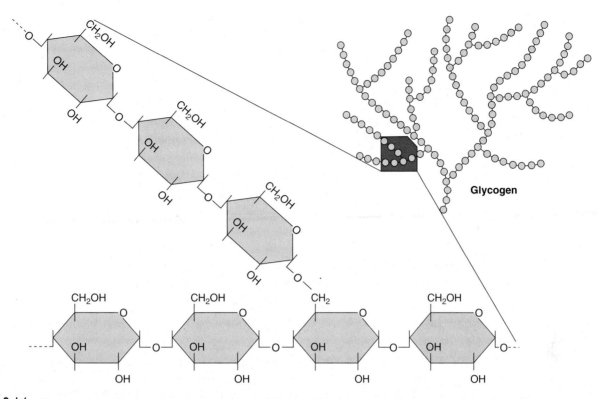

■ **Figure 2.14** **The structure of glycogen.** Glycogen is a polysaccharide composed of glucose subunits joined together to form a large, highly branched molecule.

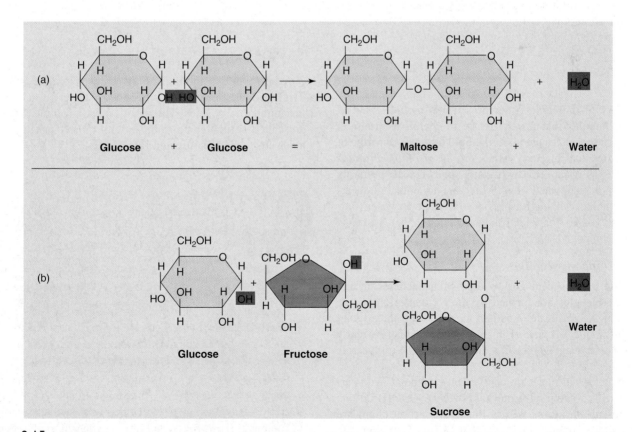

■ **Figure 2.15** **Dehydration synthesis of disaccharides.** The two disaccharides formed here are (a) maltose and (b) sucrose (table sugar). Notice that a molecule of water is produced as the disaccharides are formed.

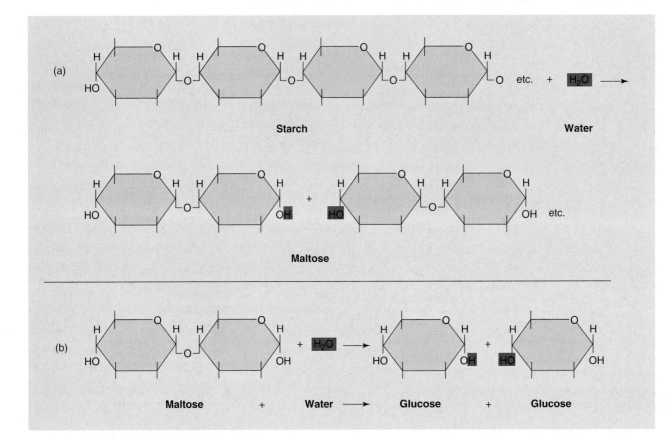

Figure 2.16 **The hydrolysis of starch.** The polysaccharide is first hydrolyzed into (*a*) disaccharides (maltose) and then into (*b*) monosaccharides (glucose). Notice that as the covalent bond between the subunits breaks, a molecule of water is split. In this way, the hydrogen atom and hydroxyl group from the water are added to the ends of the released subunits.

Lipids

The category of molecules known as **lipids** includes several types of molecules that differ greatly in chemical structure. These diverse molecules are all in the lipid category by virtue of a common physical property—they are all *insoluble in polar solvents* such as water. This is because lipids consist primarily of hydrocarbon chains and rings, which are nonpolar and therefore hydrophobic. Although lipids are insoluble in water, they can be dissolved in nonpolar solvents such as ether, benzene, and related compounds.

Triglyceride (Triacylglycerol)

Triglyceride is the subcategory of lipids that includes fat and oil. These molecules are formed by the condensation of one molecule of *glycerol* (a three-carbon alcohol) with three molecules of *fatty acids*. Because of this structure, chemists currently prefer the name **triacylglycerol,** although the name triglyceride is still in wide use.

Each fatty acid molecule consists of a nonpolar hydrocarbon chain with a carboxyl group (abbreviated COOH) on one end. If the carbon atoms within the hydrocarbon chain are joined by single covalent bonds so that each carbon atom can also bond with two hydrogen atoms, the fatty acid is said to be *saturated.* If there are a number of double covalent bonds within the hydrocarbon chain so that each carbon atom can bond with

only one hydrogen atom, the fatty acid is said to be *unsaturated.* Triglycerides contain combinations of different saturated and unsaturated fatty acids. Those with mostly saturated fatty acids are called **saturated fats;** those with mostly unsaturated fatty acids are called **unsaturated fats** (fig. 2.17).

The saturated fat content (expressed as a percentage of total fat) for some food items is as follows: canola, or rapeseed, oil (6%); olive oil (14%); margarine (17%); chicken fat (31%); palm oil (51%); beef fat (52%); butter fat (66%); and coconut oil (77%). Health authorities recommend that a person's total fat intake not exceed 30% of the total energy intake per day, and that saturated fat contribute less than 10% of the daily energy intake. This is because saturated fat in the diet may contribute to high blood cholesterol, which is a significant risk factor in heart disease and stroke (see chapter 13). Animal fats, which are solid at room temperature, are generally more saturated than vegetable oils because the hardness of the triglyceride is determined partly by the degree of saturation. Palm and coconut oil, however, are notable exceptions. Though very saturated, they nonetheless remain liquid at room temperature because they have short fatty acid chains.

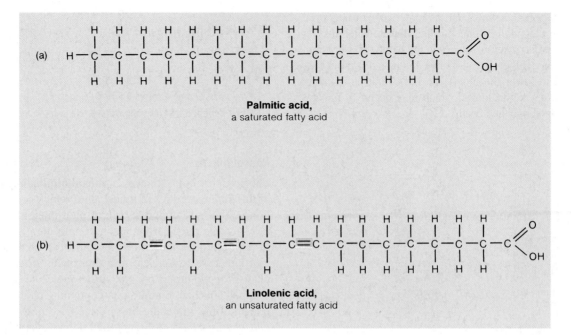

Figure 2.17 Structural formulas for fatty acids. (*a*) The formula for saturated fatty acids and (*b*) the formula for unsaturated fatty acids. Double bonds, which are points of unsaturation, are highlighted in yellow.

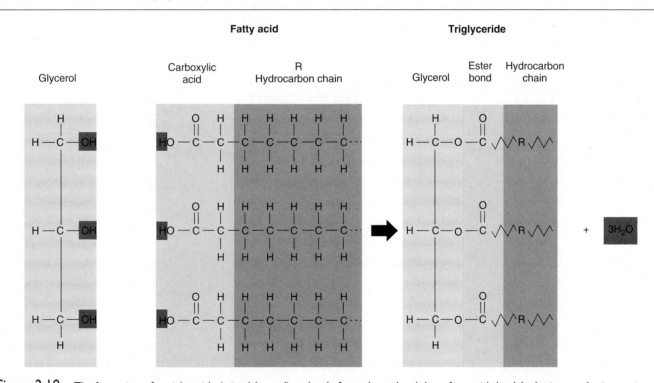

Figure 2.18 The formation of a triglyceride (triacylglycerol) molecule from glycerol and three fatty acids by dehydration synthesis reactions. A molecule of water is produced as an ester bond forms between each fatty acid and the glycerol. Sawtooth lines represent hydrocarbon chains, which are symbolized by an *R*.

Within the adipose cells of the body, triglycerides are formed as the carboxyl ends of fatty acid molecules condense with the hydroxyl groups of a glycerol molecule (fig. 2.18). Since the hydrogen atoms from the carboxyl ends of fatty acids form water molecules during dehydration synthesis, fatty acids that are combined with glycerol can no longer release H⁺ and function as acids. For this reason, triglycerides are described as *neutral fats*.

Ketone Bodies

Hydrolysis of triglycerides within adipose tissue releases *free fatty acids* into the blood. Free fatty acids can be used as an immediate source of energy by many organs; they can also be converted by the liver into derivatives called **ketone bodies** (fig. 2.19). These include four-carbon-long acidic molecules (acetoacetic acid and ß-hydroxybutyric acid) and acetone (the solvent in nail-polish remover). A rapid breakdown of fat, as may occur during strict low-carbohydrate diets and in uncontrolled diabetes mellitus, results in elevated levels of ketone bodies in the blood. This is a condition called **ketosis.** If there are sufficient amounts of ketone bodies in the blood to lower the blood pH, the condition is called **ketoacidosis.** Severe ketoacidosis, which may occur in diabetes mellitus, can lead to coma and death.

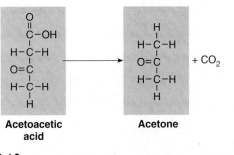

■ **Figure 2.19** Ketone bodies. Acetoacetic acid, an acidic ketone body, can spontaneously decarboxylate (lose carbon dioxide) to form acetone. Acetone is a volatile ketone body that escapes in the exhaled breath, thereby lending a "fruity" smell to the breath of people with ketosis (elevated blood ketone bodies).

Phospholipids

The group of lipids known as **phospholipids** includes a number of different categories of lipids, all of which contain a phosphate group. The most common type of phospholipid molecule is one in which the three-carbon alcohol molecule glycerol is attached to two fatty acid molecules; the third carbon atom of the glycerol molecule is attached to a phosphate group, and the phosphate group in turn is bound to other molecules. If the phosphate group is attached to a nitrogen-containing choline molecule, the phospholipid molecule thus formed is known as **lecithin** (or *phosphatidylcholine*). Figure 2.20 shows a simple way of illustrating the structure of a phospholipid—the parts of the molecule capable of ionizing (and thus becoming charged) are shown as a circle, whereas the nonpolar parts of the molecule are represented by sawtooth lines.

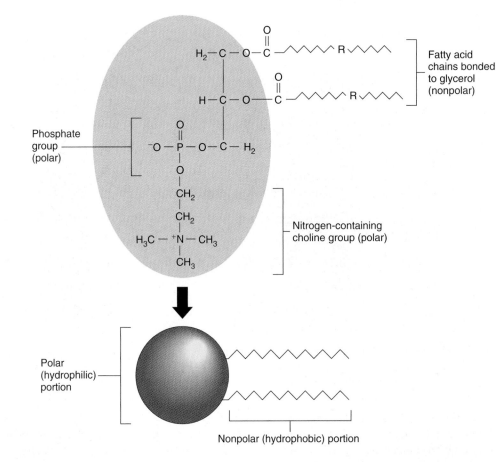

■ **Figure 2.20** The structure of lecithin. Lecithin is also called phosphatidylcholine, where choline is the nitrogen-containing portion of the molecule (interestingly, choline is also part of an important neurotransmitter known as acetylcholine, discussed in chapter 7). The detailed structure of the phospholipid (*top*) is usually shown in simplified form (*bottom*), where the circle represents the polar portion and the saw-toothed lines the nonpolar portion of the molecule.

Since the nonpolar ends of phospholipids are hydrophobic, they tend to group together when mixed in water. This allows the hydrophilic parts (which are polar) to face the surrounding water molecules (fig. 2.21). Such aggregates of molecules are called **micelles.** The dual nature of phospholipid molecules (part polar, part nonpolar) allows them to alter the interaction of water molecules and thus decrease the surface tension of water. This function of phospholipids makes them **surfactants** (surface-active agents). The surfactant effect of phospholipids prevents the lungs from collapsing due to surface tension forces (see chapter 16). Phospholipids are also the major component of cell membranes, as will be described in chapter 3.

Steroids

In terms of structure, **steroids** differ considerably from triglycerides or phospholipids, yet steroids are still included in the lipid category of molecules because they are nonpolar and insoluble in water. All steroid molecules have the same basic structure: three six-carbon rings joined to one five-carbon ring (fig. 2.22). However, different kinds of steroids have different functional groups attached to this basic structure, and they vary in the number and position of the double covalent bonds between the carbon atoms in the rings.

Cholesterol is an important molecule in the body because it serves as the precursor (parent molecule) for the steroid hormones produced by the gonads and adrenal cortex. The testes and ovaries (collectively called the *gonads*) secrete **sex steroids,** which include estradiol and progesterone from the ovaries and testosterone from the testes. The adrenal cortex secretes the **corticosteroids,** including hydrocortisone and aldosterone, as well as weak androgens (including dehydroepiandrosterone, or DHEA). Cholesterol is also an important component of cell membranes, and serves as the precursor molecule for bile salts and vitamin D_3.

Prostaglandins

Prostaglandins are a type of fatty acid with a cyclic hydrocarbon group. Although their name is derived from the fact that they were originally noted in the semen as a secretion of the prostate, it has since been shown that they are produced by and are active in almost

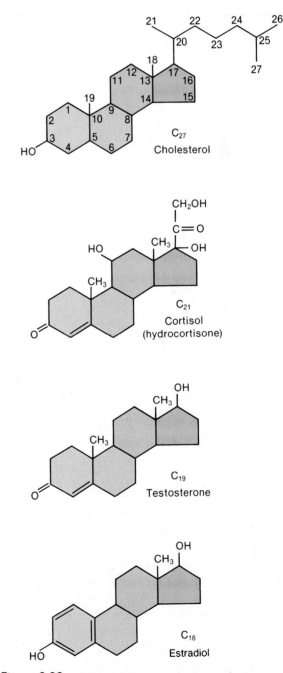

Figure 2.22 Cholesterol and some of the steroid hormones derived from cholesterol. The steroid hormones are secreted by the gonads and the adrenal cortex.

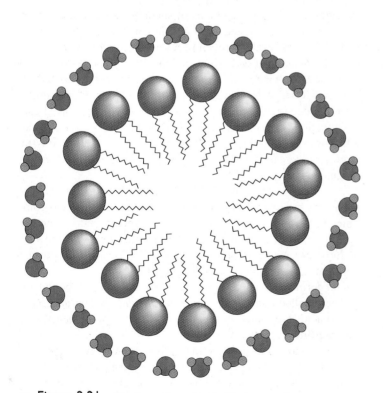

Figure 2.21 The formation of a micelle structure by phospholipids such as lecithin. The hydrophilic outer layer of the micelle faces the aqueous environment.

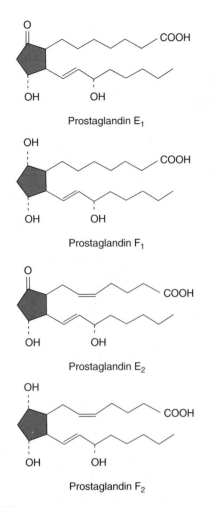

Figure 2.23 Structural formulas for various prostaglandins. Prostaglandins are a family of regulatory compounds derived from a membrane lipid known as arachidonic acid.

all organs, where they serve a variety of regulatory functions. Prostaglandins are implicated in the regulation of blood vessel diameter, ovulation, uterine contraction during labor, inflammation reactions, blood clotting, and many other functions. Structural formulas for different types of prostaglandins are shown in figure 2.23.

Test Yourself Before You Continue

1. Describe the structure characteristic of all carbohydrates and distinguish between monosaccharides, disaccharides, and polysaccharides.
2. Using dehydration synthesis and hydrolysis reactions, explain how disaccharides and monosaccharides can be interconverted and how triglycerides can be formed and broken down.
3. Describe the characteristics of a lipid and discuss the different subcategories of lipids.
4. Relate the functions of phospholipids to their structure and explain the significance of the prostaglandins.

Proteins

Proteins are large molecules composed of amino acid subunits. Since there are twenty different types of amino acids that can be used in constructing a given protein, the variety of protein structures is immense. This variety allows each type of protein to perform very specific functions.

The enormous diversity of protein structure results from the fact that there are twenty different building blocks—the *amino acids*—that can be used to form a protein. These amino acids, as will be described in the next section, are joined together to form a chain. Because of chemical interactions between the amino acids, the chain can twist and fold in a specific manner. The sequence of amino acids in a protein, and thus the specific structure of the protein, is determined by genetic information. This genetic information for protein synthesis is contained in another category of organic molecules, the *nucleic acids,* which includes the macromolecules DNA and RNA. The structure of nucleic acids is described in the next section, and the mechanisms by which the genetic information they encode directs protein synthesis are described in chapter 3.

Structure of Proteins

Proteins consist of long chains of subunits called **amino acids.** As the name implies, each amino acid contains an *amino group* (NH$_2$) on one end of the molecule and a *carboxyl group* (COOH) on another end. There are about twenty different amino acids, each with a distinct structure and chemical properties, that are used to build proteins. The differences between the amino acids are due to differences in their *functional groups.* "R" is the abbreviation for *functional group* in the general formula for an amino acid (fig. 2.24). The R symbol actually stands for the word *residue,* but it can be thought of as indicating the "*rest* of the molecule."

When amino acids are joined together by dehydration synthesis, the hydrogen from the amino end of one amino acid combines with the hydroxyl group of the carboxyl end of another amino acid. As a covalent bond is formed between the two amino acids, water is produced (fig. 2.25). The bond between adjacent amino acids is called a **peptide bond,** and the compound formed is called a *peptide.* Two amino acids bound together is called a *dipeptide;* three, a *tripeptide.* When numerous amino acids are joined in this way, a chain of amino acids, or a **polypeptide,** is produced.

The lengths of polypeptide chains vary widely. A hormone called *thyrotropin-releasing hormone,* for example, is only three amino acids long, whereas myosin, a muscle protein, contains about 4,500 amino acids. When the length of a polypeptide

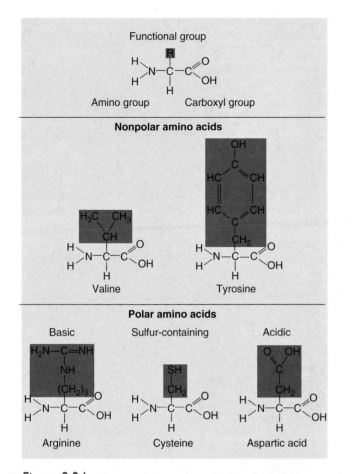

Figure 2.24 **Representative amino acids.** The figure depicts different types of functional (R) groups. Each amino acid differs from other amino acids in the number and arrangement of its functional groups.

chain becomes very long (containing more than about 100 amino acids), the molecule is called a **protein.**

The structure of a protein can be described at four different levels. At the first level, the sequence of amino acids in the protein is described; this is called the **primary structure** of the protein. Each type of protein has a different primary structure. All of the billions of *copies* of a given type of protein in a person have the same structure, however, because the structure of a given protein is coded by the person's genes. The primary structure of a protein is illustrated in figure 2.26a.

Weak hydrogen bonds may form between the hydrogen atom of an amino group and an oxygen atom from a different amino acid nearby. These weak bonds cause the polypeptide chain to assume a particular shape, known as the **secondary structure** of the protein (fig. 2.26b,c). This can be the shape of an *alpha* (α) *helix,* or alternatively, the shape of what is called a *beta* (β) *pleated sheet.*

Most polypeptide chains bend and fold upon themselves to produce complex three-dimensional shapes called the **tertiary structure** of the protein (fig. 2.26d). Each type of protein has its own characteristic tertiary structure. This is because the folding and bending of the polypeptide chain is produced by chemical interactions between particular amino acids located in different regions of the chain.

Most of the tertiary structure of proteins is formed and stabilized by weak chemical bonds (such as hydrogen bonds) between the functional groups of widely spaced amino acids. Since most of the tertiary structure is stabilized by weak bonds, this structure can easily be disrupted by high temperature or by changes in pH. Irreversible changes in the tertiary structure of proteins that occur by these means are referred to as *denaturation* of the proteins. The tertiary structure of some

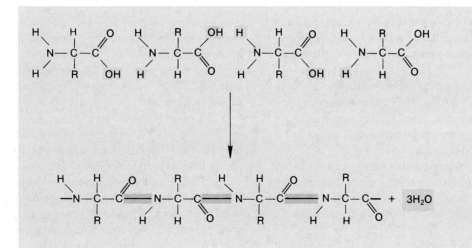

Figure 2.25 **The formation of peptide bonds by dehydration synthesis reactions.** Water molecules are split off as the peptide bonds (highlighted in green) are produced between the amino acids.

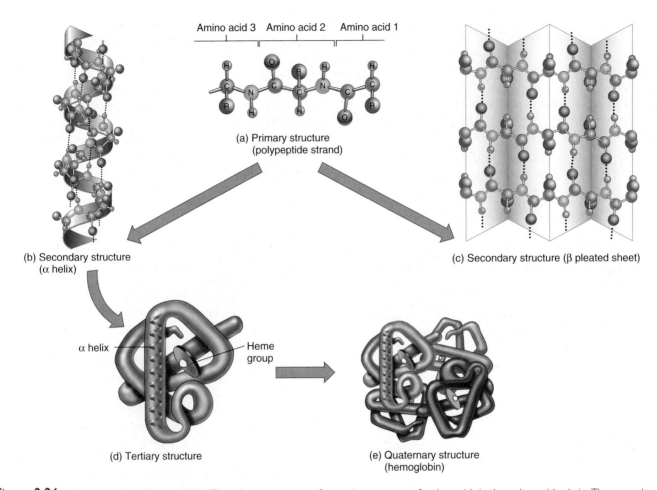

Amino acid 3 Amino acid 2 Amino acid 1

(a) Primary structure
(polypeptide strand)

(b) Secondary structure
(α helix)

(c) Secondary structure (β pleated sheet)

α helix — Heme group

(d) Tertiary structure

(e) Quaternary structure
(hemoglobin)

■ **Figure 2.26** **The structure of proteins.** (*a*) The primary structure refers to the sequence of amino acids in the polypeptide chain. The secondary structure refers to the conformation of the chain created by hydrogen bonding between amino acids; this can be either an alpha helix (*b*) or a beta pleated sheet (*c*). The tertiary structure (*d*) is the three-dimensional structure of the protein. The formation of a protein by the bonding together of two or more polypeptide chains is the quaternary structure (*e*) of the protein.

proteins, however, is made more stable by strong covalent bonds between sulfur atoms (called *disulfide bonds* and abbreviated S—S) in the functional group of an amino acid known as cysteine (fig. 2.27).

Denatured proteins retain their primary structure (the peptide bonds are not broken) but have altered chemical properties. Cooking a pot roast, for example, alters the texture of the meat proteins—it doesn't result in an amino acid soup. Denaturation is most dramatically demonstrated by frying an egg. Egg albumin proteins are soluble in their native state, in which they form the clear, viscous fluid of a raw egg. When denatured by cooking, these proteins change shape, cross-bond with each other, and by this means form an insoluble white precipitate— the egg white.

Hemoglobin and insulin are composed of a number of polypeptide chains covalently bonded together. This is the **quaternary structure** of these molecules. Insulin, for exam- ple, is composed of two polypeptide chains—one that is twenty-one amino acids long, the other that is thirty amino acids long. Hemoglobin (the protein in red blood cells that carries oxygen) is composed of four separate polypeptide chains (see fig. 2.26*e*). The composition of various body proteins is shown in table 2.4.

Many proteins in the body are normally found combined, or *conjugated,* with other types of molecules. **Glycoproteins** are proteins conjugated with carbohydrates. Examples of such molecules include certain hormones and some proteins found in the cell membrane. **Lipoproteins** are proteins conjugated with lipids. These are found in cell membranes and in the plasma (the fluid portion of the blood). Proteins may also be conjugated with pigment molecules. These include hemoglobin, which transports oxygen in red blood cells, and the cytochromes, which are needed for oxygen utilization and energy production within cells.

Table 2.4 Composition of Selected Proteins Found in the Body

Protein	Number of Polypeptide Chains	Nonprotein Component	Function
Hemoglobin	4	Heme pigment	Carries oxygen in the blood
Myoglobin	1	Heme pigment	Stores oxygen in muscle
Insulin	2	None	Hormonal regulation of metabolism
Blood group proteins	1	Carbohydrate	Produces blood types
Lipoproteins	1	Lipids	Transports lipids in blood

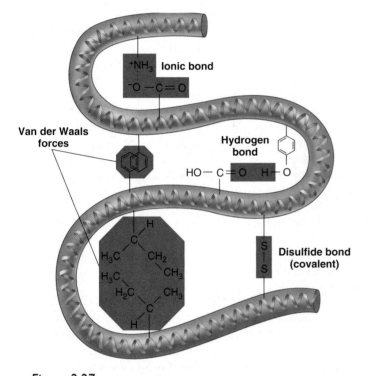

Figure 2.27 The bonds responsible for the tertiary structure of a protein. The tertiary structure of a protein is held in place by a variety of bonds. These include relatively weak bonds, such as hydrogen bonds, ionic bonds, and Van der Waals (hydrophobic) forces, as well as the strong covalent disulfide bonds.

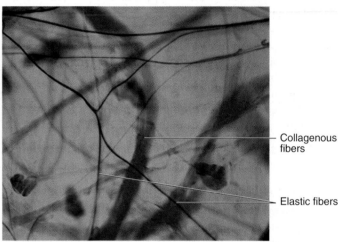

Figure 2.28 A photomicrograph of collagenous fibers within connective tissue. Collagen proteins strengthen the connective tissues.

Many proteins play a more active role in the body, where specificity of structure and function is required. *Enzymes* and *antibodies,* for example, are proteins—no other type of molecule could provide the vast array of different structures needed for their tremendously varied functions. As another example, proteins in cell membranes may serve as *receptors* for specific regulator molecules (such as hormones) and as *carriers* for transport of specific molecules across the membrane. Proteins provide the diversity of shape and chemical properties required by these functions.

Functions of Proteins

Because of their tremendous structural diversity, proteins can serve a wider variety of functions than any other type of molecule in the body. Many proteins, for example, contribute significantly to the structure of different tissues and in this way play a passive role in the functions of these tissues. Examples of such *structural proteins* include collagen (fig. 2.28) and keratin. Collagen is a fibrous protein that provides tensile strength to connective tissues, such as tendons and ligaments. Keratin is found in the outer layer of dead cells in the epidermis, where it prevents water loss through the skin.

Test Yourself Before You Continue

1. Write the general formula for an amino acid and describe how amino acids differ from one another.

2. Describe and account for the different levels of protein structure.

3. Describe the different categories of protein function in the body and explain why proteins can serve functions that are so diverse.

Nucleic Acids

Nucleic acids include the macromolecules DNA and RNA which are critically important in genetic regulation, and the subunits from which these molecules are formed. These subunits are known as nucleotides.

Nucleotides are the subunits of nucleic acids, bonded together in dehydration synthesis reactions to form long polynucleotide chains. Each nucleotide, however, is itself composed of three smaller subunits: a five-carbon (*pentose*) sugar, a phosphate group attached to one end of the sugar, and a *nitrogenous base* attached to the other end of the sugar (fig. 2.29). The nitrogenous bases are nitrogen-containing molecules of two kinds: pyrimidines and purines. The *pyrimidines* contain a single ring of carbon and nitrogen, whereas the *purines* have two such rings.

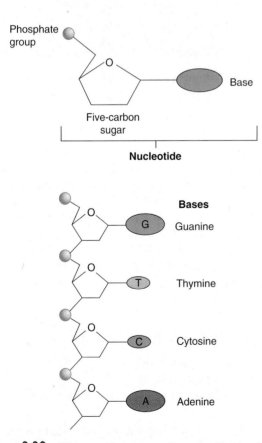

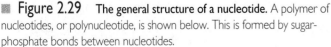

■ Figure 2.29   **The general structure of a nucleotide.** A polymer of nucleotides, or polynucleotide, is shown below. This is formed by sugar-phosphate bonds between nucleotides.

Deoxyribonucleic Acid

The structure of **DNA (deoxyribonucleic acid)** serves as the basis for the genetic code. For this reason, it might seem logical that DNA should have an extremely complex structure. DNA is indeed larger than any other molecule in the cell, but its structure is actually simpler than that of most proteins. This simplicity of structure deceived some early investigators into believing that the protein content of chromosomes, rather than their DNA content, provided the basis for the genetic code.

Sugar molecules in the nucleotides of DNA are a type of pentose (five-carbon) sugar called **deoxyribose.** Each deoxyribose can be covalently bonded to one of four possible bases. These bases include the two purines **(guanine and adenine)** and the two pyrimidines **(cytosine and thymine)** (fig. 2.30). There are thus four different types of nucleotides that can be used to produce the long DNA chains. If you remember that there are twenty different amino acids used to produce proteins, you can now understand why many scientists were deceived into thinking that genes were composed of proteins rather than nucleic acids.

When nucleotides combined to form a chain, the phosphate group of one condenses with the deoxyribose sugar of another nucleotide. This forms a sugar-phosphate chain as water is removed in dehydration synthesis. Since the nitrogenous bases are attached to the sugar molecules, the sugar-phosphate chain looks like a "backbone" from which the bases project. Each of these bases can form hydrogen bonds with other bases, which are in turn joined to a different chain of nucleotides. Such hydrogen bonding between bases thus produces a *double-stranded* DNA molecule; the two strands are like a staircase, with the paired bases as steps (fig. 2.30).

Actually, the two chains of DNA twist about each other to form a **double helix,** so that the molecule resembles a spiral staircase (fig. 2.31). It has been shown that the number of purine bases in DNA is equal to the number of pyrimidine bases. The reason for this is explained by the **law of complementary base pairing:** *adenine can pair only with thymine* (through two hydrogen bonds), whereas *guanine can pair only with cytosine* (through three hydrogen bonds). With knowledge of this rule, we could predict the base sequence of one DNA strand if we knew the sequence of bases in the complementary strand.

Although we can be certain of which base is opposite a given base in DNA, we cannot predict which bases will be above or below that particular pair within a single polynucleotide chain. Although there are only four bases, the number of possible base sequences along a stretch of several thousand nucleotides (the length of most genes) is almost infinite. To gain perspective, it is useful to realize that the total human **genome** (all of the genes in a cell) consists of over 3 billion base pairs that would extend over a meter if the DNA molecules were unraveled and stretched out.

Yet, even with this amazing variety of possible base sequences, almost all of the billions of copies of a particular gene in a person are identical. The mechanisms by which identical DNA copies are made and distributed to the daughter cells when a cell divides will be described in chapter 3.

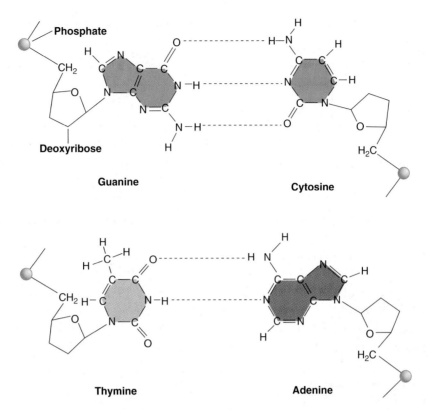

■ Figure 2.30 The four nitrogenous bases in deoxyribonucleic acid (DNA). Notice that hydrogen bonds can form between guanine and cytosine and between thymine and adenine.

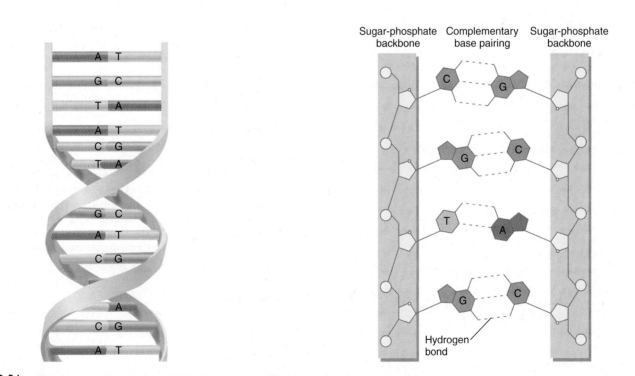

■ Figure 2.31 The double-helix structure of DNA. The two strands are held together by hydrogen bonds between complementary bases in each strand.

Ribonucleic Acid

DNA can direct the activities of the cell only by means of another type of nucleic acid—**RNA (ribonucleic acid).** Like DNA, RNA consists of long chains of nucleotides joined together by sugar-phosphate bonds. Nucleotides in RNA, however, differ from those in DNA (fig. 2.32) in three ways: (1) a **ribonucleotide** contains the sugar **ribose** (instead of deoxyribose), (2) the base **uracil** is found in place of thymine, and (3) RNA is composed of a single polynucleotide strand (it is not double-stranded like DNA).

There are three types of RNA molecules that function in the cytoplasm of cells: *messenger RNA (mRNA), transfer RNA (tRNA),* and *ribosomal RNA (rRNA).* All three types are made within the cell nucleus by using information contained in DNA as a guide. The functions of RNA are described in chapter 3.

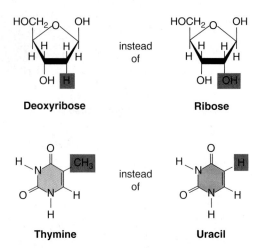

Figure 2.32 Differences between the nucleotides and sugars in DNA and RNA. DNA has deoxyribose and thymine; RNA has ribose and uracil. The other three bases are the same in DNA and RNA.

Test Yourself Before You Continue

1. What are nucleotides, and of what are they composed?
2. Describe the structure of DNA and explain the law of complementary base pairing.
3. List the types of RNA, and explain how the structure of RNA differs from the structure of DNA.

Summary

Atoms, Ions, and Chemical Bonds 24

I. Covalent bonds are formed by atoms that share electrons. They are the strongest type of chemical bond.
  A. Electrons are equally shared in nonpolar covalent bonds and unequally shared in polar covalent bonds.
  B. Atoms of oxygen, nitrogen, and phosphorus strongly attract electrons and become electrically negative compared to the other atoms sharing electrons with them.

II. Ionic bonds are formed by atoms that transfer electrons. These weak bonds join atoms together in an ionic compound.
  A. If one atom in this compound takes an electron from another atom, it gains a net negative charge and the other atom becomes positively charged.

  B. Ionic bonds easily break when the ionic compound is dissolved in water. Dissociation of the ionic compound yields charged atoms called ions.

III. When hydrogen bonds with an electronegative atom, it gains a slight positive charge and is weakly attracted to another electronegative atom. This weak attraction is a hydrogen bond.

IV. Acids donate hydrogen ions to solution, whereas bases lower the hydrogen ion concentration of a solution.
  A. The pH scale is a negative function of the logarithm of the hydrogen ion concentration.
  B. In a neutral solution, the concentration of H^+ is equal to the concentration of OH^-, and the pH is 7.

  C. Acids raise the H^+ concentration and thus lower the pH below 7; bases lower the H^+ concentration and thus raise the pH above 7.

V. Organic molecules contain atoms of carbon and hydrogen joined together by covalent bonds. Atoms of nitrogen, oxygen, phosphorus, or sulfur may be present as specific functional groups in the organic molecule.

Carbohydrates and Lipids 31

I. Carbohydrates contain carbon, hydrogen, and oxygen, usually in a ratio of 1:2:1.
  A. Carbohydrates consist of simple sugars (monosaccharides), disaccharides, and polysaccharides (such as glycogen).
  B. Covalent bonds between monosaccharides are formed by

dehydration synthesis, or condensation. Bonds are broken by hydrolysis reactions.

II. Lipids are organic molecules that are insoluble in polar solvents such as water.

   A. Triglycerides (fat and oil) consist of three fatty acid molecules joined to a molecule of glycerol.

   B. Ketone bodies are smaller derivatives of fatty acids.

   C. Phospholipids (such as lecithin) are phosphate-containing lipids that have a hydrophilic polar group. The rest of the molecule is hydrophobic.

   D. Steroids (including the hormones of the adrenal cortex and gonads) are lipids with a characteristic four-ring structure.

   E. Prostaglandins are a family of cyclic fatty acids that serve a variety of regulatory functions.

Proteins 38

I. Proteins are composed of long chains of amino acids bound together by covalent peptide bonds.

   A. Each amino acid contains an amino group, a carboxyl group, and a functional group. Differences in the functional groups give each of the more than twenty different amino acids an individual identity.

   B. The polypeptide chain may be twisted into a helix (secondary structure) and bent and folded to form the tertiary structure of the protein.

   C. Proteins that are composed of two or more polypeptide chains are said to have a quaternary structure.

   D. Proteins may be combined with carbohydrates, lipids, or other molecules.

   E. Because they are so diverse structurally, proteins serve a wider variety of specific functions than any other type of molecule.

Nucleic Acids 42

I. DNA is composed of four nucleotides, each of which contains the sugar deoxyribose.

   A. Two of the bases contain the purines adenine and guanine; two contain the pyrimidines cytosine and thymine.

   B. DNA consists of two polynucleotide chains joined together by hydrogen bonds between their bases.

   C. Hydrogen bonds can only form between the bases adenine and thymine, and between the bases guanine and cytosine.

   D. This complementary base pairing is critical for DNA synthesis and for genetic expression.

II. RNA consists of four nucleotides, each of which contains the sugar ribose.

   A. The nucleotide bases are adenine, guanine, cytosine, and uracil (in place of the DNA base thymine).

   B. RNA consists of only a single polynucleotide chain.

   C. There are different types of RNA, which have different functions in genetic expression.

Review Activities

Test Your Knowledge of Terms and Facts

1. Which of these statements about atoms is *true?*
   a. They have more protons than electrons.
   b. They have more electrons than protons.
   c. They are electrically neutral.
   d. They have as many neutrons as they have electrons.

2. The bond between oxygen and hydrogen in a water molecule is
   a. a hydrogen bond.
   b. a polar covalent bond.
   c. a nonpolar covalent bond.
   d. an ionic bond.

3. Which of these is a nonpolar covalent bond?
   a. bond between two carbons
   b. bond between sodium and chloride
   c. bond between two water molecules
   d. bond between nitrogen and hydrogen

4. Solution A has a pH of 2, and solution B has a pH of 10. Which of these statements about these solutions is *true?*
   a. Solution A has a higher H^+ concentration than solution B.
   b. Solution B is basic.
   c. Solution A is acidic.
   d. All of these are true.

5. Glucose is
   a. a disaccharide.
   b. a polysaccharide.
   c. a monosaccharide.
   d. a phospholipid.

6. Digestion reactions occur by means of
   a. dehydration synthesis.
   b. hydrolysis.

7. Carbohydrates are stored in the liver and muscles in the form of
   a. glucose.
   b. triglycerides.
   c. glycogen.
   d. cholesterol.

8. Lecithin is
   a. a carbohydrate.
   b. a protein.
   c. a steroid.
   d. a phospholipid.

9. Which of these lipids have regulatory roles in the body?
   a. steroids
   b. prostaglandins
   c. triglycerides
   d. both *a* and *b*
   e. both *b* and *c*

10. The tertiary structure of a protein is *directly* determined by
    a. genes.
    b. the primary structure of the protein.
    c. enzymes that "mold" the shape of the protein.
    d. the position of peptide bonds.

11. The type of bond formed between two molecules of water is
    a. a hydrolytic bond.
    b. a polar covalent bond.
    c. a nonpolar covalent bond.
    d. a hydrogen bond.
12. The carbon-to-nitrogen bond that joins amino acids together is called
    a. a glycosidic bond.
    b. a peptide bond.
    c. a hydrogen bond.
    d. a double bond.
13. The RNA nucleotide base that pairs with adenine in DNA is
    a. thymine.
    b. uracil.
    c. guanine.
    d. cytosine.
14. If four bases in one DNA strand are A (adenine), G (guanine), C (cytosine), and T (thymine), the complementary bases in the RNA strand made from this region are
    a. T,C,G,A.
    b. C,G,A,U.
    c. A,G,C,U.
    d. U,C,G,A.

Test Your Understanding of Concepts and Principles

1. Compare and contrast nonpolar covalent bonds, polar covalent bonds, and ionic bonds.[1]
2. Define *acid* and *base* and explain how acids and bases influence the pH of a solution.
3. Using dehydration synthesis and hydrolysis reactions, explain the relationships between starch in an ingested potato, liver glycogen, and blood glucose.
4. "All fats are lipids, but not all lipids are fats." Explain why this is an accurate statement.
5. What are the similarities and differences between a fat and an oil? Comment on the physiological and clinical significance of the degree of saturation of fatty acid chains.
6. Explain how one DNA molecule serves as a template for the formation of another DNA molecule and why DNA synthesis is said to be semiconservative.

Test Your Ability to Analyze and Apply Your Knowledge

1. Explain the relationship between the primary structure of a protein and its secondary and tertiary structures. What do you think would happen to the tertiary structure if some amino acids were substituted for others in the primary structure? What physiological significance might this have?
2. Suppose you try to discover a hormone by homogenizing an organ in a fluid, filtering the fluid to eliminate the solid material, and then injecting the extract into an animal to see the effect. If an aqueous (water) extract does not work but one using benzene as the solvent does have an effect, what might you conclude about the chemical nature of the hormone? Explain.
3. From the ingredients listed on a food wrapper, it would appear that the food contains high amounts of fat. Yet on the front of the package is the large slogan, "Cholesterol Free!" In what sense is this slogan chemically correct? In what way is it misleading?

Related Websites

Check out the Links Library at www.mhhe.com/fox8 for links to sites containing resources related to the chemical composition of the body. These links are monitored to ensure current URLs.

---

[1]*Note:* This question is answered in the chapter 2 Study Guide found on the Online Learning Center at www.mhhe.com/fox8.

3 Cell Structure and Genetic Control

Objectives

After studying this chapter, you should be able to . . .

1. describe the structure of the plasma membrane and explain its functional significance.

2. state which cells in the human body transport themselves by amoeboid movement and explain how they perform this movement.

3. describe the structure of cilia and flagella, and state some of their functions.

4. describe the processes of phagocytosis, pinocytosis, receptor-mediated endocytosis, and exocytosis.

5. state the functions of the cytoskeleton, lysosomes, mitochondria, and the endoplasmic reticulum.

6. describe the structure of the cell nucleus and explain its significance.

7. explain how RNA is produced according to the genetic information in DNA and distinguish between the different types of RNA.

8. describe how proteins are produced according to the information contained in messenger RNA.

9. describe the structure of the rough endoplasmic reticulum and Golgi complex and explain how they function together in the secretion of proteins.

10. explain what is meant by the semiconservative mechanism of DNA replication.

11. describe the different stages of the cell cycle and list the events that occur in the different phases of mitosis.

12. define the terms *hypertrophy* and *hyperplasia* and explain their physiological importance.

13. describe the events that occur in meiosis, compare them to those that occur in mitosis, and discuss the significance of meiotic cell division in human physiology.

Take Advantage of the Technology

Visit the Online Learning Center for these additional study resources.

■ Interactive quizzing

■ Online study guide

■ Current news feeds

■ Crossword puzzles and vocabulary flashcards

■ Labeling activities

www.mhhe.com/fox8

Clinical Investigation

Timothy is only eighteen years old, but appears to have liver disease. A liver biopsy is performed, and different microscopic techniques are employed for viewing the samples. The biopsy reveals an unusually extensive smooth endoplasmic reticulum. In addition, an abnormally large amount of glycogen granules are found, and many intact glycogen granules are seen within secondary lysosomes.

Upon questioning, Timothy admits that he has a history of drug abuse, but claims that he is now in recovery. Laboratory analysis reveals that he has an abnormally low amount of the enzyme that hydrolyzes glycogen. What is the relationship between these observations?

Plasma Membrane and Associated Structures

The cell is the basic unit of structure and function in the body. Many of the functions of cells are performed by particular subcellular structures known as organelles. The plasma (cell) membrane allows selective communication between the intracellular and extracellular compartments and aids cellular movement.

Cells look so small and simple when viewed with the ordinary (light) microscope that it is difficult to think of each one as a living entity unto itself. Equally amazing is the fact that the physiology of our organs and systems derives from the complex functions of the cells of which they are composed. Complexity of function demands complexity of structure, even at the subcellular level.

As the basic functional unit of the body, each cell is a highly organized molecular factory. Cells come in a wide variety of shapes and sizes. This great diversity, which is also apparent in the subcellular structures within different cells, reflects the diversity of function of different cells in the body. All cells, however, share certain characteristics; for example, they are all surrounded by a plasma membrane, and most of them possess the structures listed in table 3.1. Thus, although no single cell can be considered "typical," the general structure of cells can be indicated by a single illustration (fig. 3.1).

For descriptive purposes, a cell can be divided into three principal parts:

1. **Plasma (cell) membrane.** The selectively permeable plasma membrane surrounds the cell, gives it form, and separates the cell's internal structures from the extracellular environment. The plasma membrane also participates in intercellular communication.

Table 3.1 Cellular Components: Structure and Function

Component	Structure	Function
Plasma (cell) membrane	Membrane composed of double layer of phospholipids in which proteins are embedded	Gives form to cell and controls passage of materials into and out of cell
Cytoplasm	Fluid, jellylike substance between the cell membrane and the nucleus in which organelles are suspended	Serves as matrix substance in which chemical reactions occur
Endoplasmic reticulum	System of interconnected membrane-forming canals and tubules	Agranular (smooth) endoplasmic reticulum metabolizes nonpolar compounds and stores Ca^{2+} in striated muscle cells, granular (rough) endoplasmic reticulum assists in protein synthesis
Ribosomes	Granular particles composed of protein and RNA	Synthesize proteins
Golgi complex	Cluster of flattened membranous sacs	Synthesizes carbohydrates and packages molecules for secretion, secretes lipids and glycoproteins
Mitochondria	Membranous sacs with folded inner partitions	Release energy from food molecules and transform energy into usable ATP
Lysosomes	Membranous sacs	Digest foreign molecules and worn and damaged organelles
Peroxisomes	Spherical membranous vesicles	Contain enzymes that detoxify harmful molecules and break down hydrogen peroxide
Centrosome	Nonmembranous mass of two rodlike centrioles	Helps to organize spindle fibers and distribute chromosomes during mitosis
Vacuoles	Membranous sacs	Store and release various substances within the cytoplasm
Microfilaments and microtubules	Thin, hollow tubes	Support cytoplasm and transport materials within the cytoplasm
Cilia and flagella	Minute cytoplasmic projections that extend from the cell surface	Move particles along cell surface or move the cell
Nuclear envelope	Double-layered membrane that surrounds the nucleus, composed of protein and lipid molecules	Supports nucleus and controls passage of materials between nucleus and cytoplasm
Nucleolus	Dense nonmembranous mass composed of protein and RNA molecules	Produces ribosomal RNA for ribosomes
Chromatin	Fibrous strands composed of protein and DNA	Contains genetic code that determines which proteins (including enzymes) will be manufactured by the cell

2. **Cytoplasm and organelles.** The cytoplasm is the aqueous content of a cell inside the cell membrane but outside the nucleus. Organelles (excluding the nucleus) are subcellular structures within the cytoplasm that perform specific functions. The term **cytosol** is frequently used to describe the fluid portion of the cytoplasm; that is, the part that cannot be removed by centrifugation.

3. **Nucleus.** The nucleus is a large, generally spheroid body within a cell. The largest of the organelles, it contains the DNA, or genetic material, of the cell and thus directs the cell's activities. The nucleus also contains one or more *nucleoli*. Nucleoli are centers for the production of ribosomes, which are the sites of protein synthesis.

Structure of the Plasma Membrane

Because both the intracellular and extracellular environments (or "compartments") are aqueous, a barrier must be present to prevent the loss of enzymes, nucleotides, and other cellular molecules that are water-soluble. Since this barrier surrounding the cell cannot itself be composed of water-soluble molecules, it is instead composed of lipids.

The **plasma membrane** (also called the **cell membrane**), and indeed all of the membranes surrounding organelles within the cell, are composed primarily of phospholipids and proteins. Phospholipids, described in chapter 2, are polar (and hydrophilic) in the region that contains the phosphate group and nonpolar (and hydrophobic) throughout the rest of the molecule. Since the environment on each side of the membrane is aqueous, the hydrophobic parts of the molecules "huddle together" in the center of the membrane, leaving the polar parts exposed to water on both surfaces. This results in the formation of a double layer of phospholipids in the cell membrane.

The hydrophobic middle of the membrane restricts the passage of water and water-soluble molecules and ions. Certain of these polar compounds, however, do pass through the membrane. The specialized functions and selective transport properties of the membrane are believed to be due to its protein content. Membrane proteins are described as peripheral or integral. *Peripheral proteins* are only partially embedded in one face of the membrane, whereas *integral proteins* span the membrane from one side to the other. Since the membrane is not solid—phospholipids and proteins are free to

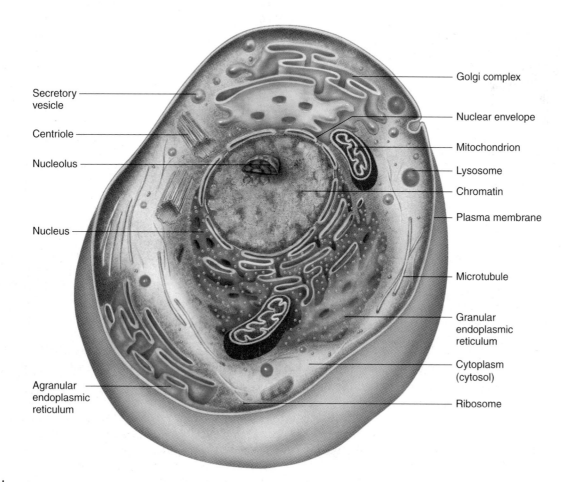

Labels (left): Secretory vesicle; Centriole; Nucleolus; Nucleus; Agranular endoplasmic reticulum

Labels (right): Golgi complex; Nuclear envelope; Mitochondrion; Lysosome; Chromatin; Plasma membrane; Microtubule; Granular endoplasmic reticulum; Cytoplasm (cytosol); Ribosome

■ **Figure 3.1** A generalized human cell showing the principal organelles. Since most cells of the body are highly specialized, they have structures that differ from those shown here.

Extracellular side

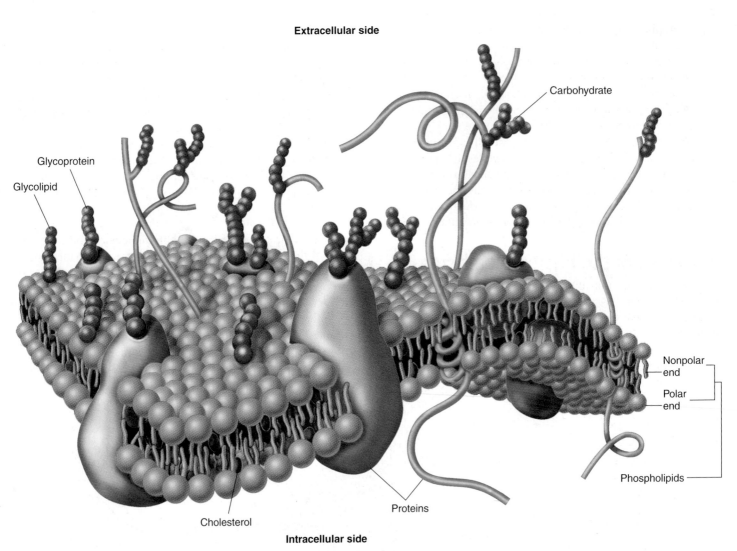

Intracellular side

■ **Figure 3.2**   The fluid-mosaic model of the plasma membrane.   The membrane consists of a double layer of phospholipids, with the polar regions (shown by spheres) oriented outward and the nonpolar hydrocarbons (wavy lines) oriented toward the center. Proteins may completely or partially span the membrane. Carbohydrates are attached to the outer surface.

move laterally—the proteins within the phospholipid "sea" are not uniformly distributed. Rather, they present a constantly changing mosaic pattern, an arrangement known as the **fluid-mosaic model** of membrane structure (fig. 3.2).

The proteins found in the plasma membrane serve a variety of functions, including structural support, transport of molecules across the membrane, and enzymatic control of chemical reactions at the cell surface. Some proteins function as receptors for hormones and other regulatory molecules that arrive at the outer surface of the membrane. Receptor proteins are usually specific for one particular messenger much like an enzyme that is specific for a single substrate. Other cellular proteins serve as "markers" (antigens) that identify the blood and tissue type of an individual.

The plasma membranes of all higher organisms contain cholesterol. The cells in the body with the highest content of cholesterol are the Schwann cells, which form insulating layers by wrapping around certain nerve fibers (see chapter 7). Their high cholesterol content is believed to be important in this insulating function. The ratio of cholesterol to phospholipids also helps to determine the flexibility of a plasma membrane. When there is an inherited defect in this ratio, the flexibility of the cell may be reduced. This could result, for example, in the inability of red blood cells to flex at the middle when passing through narrow blood channels, thereby causing occlusion of these small vessels.

Pseudopod

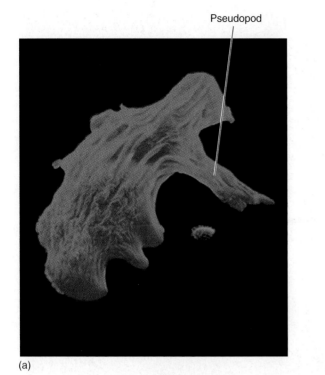

(a)

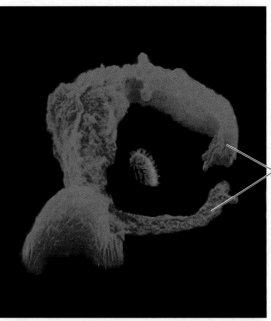

Pseudopods
forming food
vacuole

(b)

Figure 3.3   Scanning electron micrographs of phagocytosis. (*a*) The formation of pseudopods and (*b*) the entrapment of the prey within a food vacuole.

In addition to lipids and proteins, the plasma membrane also contains carbohydrates, which are primarily attached to the outer surface of the membrane as glycoproteins and glycolipids. These surface carbohydrates have numerous negative charges and, as a result, affect the interaction of regulatory molecules with the membrane. The negative charges at the surface also affect interactions between cells—they help keep red blood cells apart, for example. Stripping the carbohydrates from the outer red blood cell surface results in their more rapid destruction by the liver, spleen, and bone marrow.

Phagocytosis

Most of the movement of molecules and ions between the intracellular and extracellular compartments involves passage through the plasma membrane (see chapter 6). However, the plasma membrane also participates in the **bulk transport** of larger portions of the extracellular environment. Bulk transport includes the processes of *phagocytosis* and *endocytosis.*

Some body cells—including certain white blood cells and macrophages in connective tissues—are able to move in the manner of an amoeba (a single-celled organism). They perform this **amoeboid movement** by extending parts of their cytoplasm to form *pseudopods,* which attach to a substrate and pull the cell along. This process depends on the bonding of membrane-spanning proteins called *integrins* with proteins outside the membrane in the *extracellular matrix* (generally, an extracellular gel of proteins and carbohydrates).

Cells that exhibit amoeboid motion—as well as certain liver cells, which are not mobile—use pseudopods to surround and engulf particles of organic matter (such as bacteria). This process is a type of cellular "eating" called **phagocytosis.** It serves to protect the body from invading microorganisms and to remove extracellular debris.

Phagocytic cells surround their victim with pseudopods, which join together and fuse (fig. 3.3). After the inner membrane of the pseudopods has become a continuous membrane surrounding the ingested particle, it pinches off from the plasma membrane. The ingested particle is now contained in an organelle called a *food vacuole* within the cell. The food vacuole will subsequently fuse with an organelle called a lysosome (described later), and the particle will be digested by lysosomal enzymes.

Endocytosis

Endocytosis is a process in which the plasma membrane furrows inward, instead of extending outward with pseudopods. One form of endocytosis, **pinocytosis,** is a nonspecific process performed by many cells. The plasma membrane invaginates to produce a deep, narrow furrow. The membrane near the surface of this furrow then fuses, and a small vesicle containing the extracellular fluid is pinched off and enters the cell. Pinocytosis allows a cell to engulf large molecules such as proteins, as well as any other molecules that may be present in the extracellular fluid.

Another type of endocytosis involves a smaller area of plasma membrane, and it occurs only in response to specific molecules in

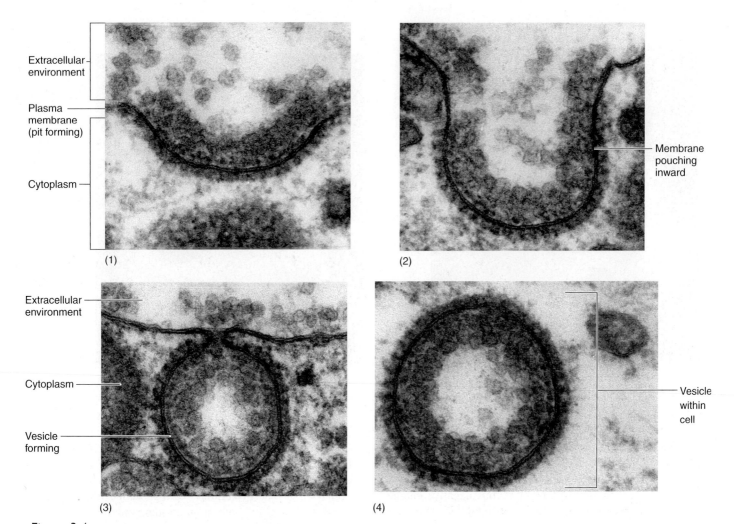

Extracellular environment

Plasma membrane (pit forming)

Cytoplasm

(1)

Membrane pouching inward

(2)

Extracellular environment

Cytoplasm

Vesicle forming

(3)

Vesicle within cell

(4)

Figure 3.4 **Receptor-mediated endocytosis.** In stages 1 through 4 shown here, specific bonding of extracellular particles with membrane receptor proteins results in the formation of endocytotic vesicles.

the extracellular environment. Since the extracellular molecules must bind to very specific *receptor proteins* in the plasma membrane, this process is known as **receptor-mediated endocytosis.**

In receptor-mediated endocytosis, the interaction of specific molecules in the extracellular fluid with specific membrane receptor proteins causes the membrane to invaginate, fuse, and pinch off to form a vesicle (fig. 3.4). Vesicles formed in this way contain extracellular fluid and molecules that could not have passed by other means into the cell. Cholesterol attached to specific proteins, for example, is taken up into artery cells by receptor-mediated endocytosis. This is in part responsible for atherosclerosis, as described in chapter 13. Hepatitis, polio, and AIDS viruses also exploit the process of receptor-mediated endocytosis to invade cells.

Exocytosis

Exocytosis is a process by which cellular products are secreted into the extracellular environment. Proteins and other molecules produced within the cell that are destined for export (secretion) are packaged within vesicles by an organelle known as the Golgi complex. In the process of exocytosis, these secretory vesicles fuse with the plasma membrane and release their contents into the extracellular environment (see fig. 3.13). Nerve endings, for example, release their chemical neurotransmitters in this manner (see chapter 7).

When the vesicle containing the secretory products of the cell fuses with the plasma membrane during exocytosis, the total surface area of the cell membrane is increased. This process replaces material that was lost from the plasma membrane during endocytosis.

Cilia and Flagella

Cilia are tiny hairlike structures that project from the surface of a cell and, like the coordinated action of rowers in a boat, stroke in unison. Cilia in the human body are found on the apical surface (the surface facing the lumen, or cavity) of stationary epithelial cells in the respiratory and female reproductive tracts. In the respiratory system, the cilia transport strands of mucus to the pharynx (throat), where the mucus can either be swallowed or expectorated. In the female reproductive tract, ciliary movements in the epithelial lining of the uterine tube draw the ovum (egg) into the tube and move it toward the uterus.

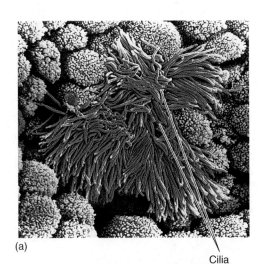

(a)

Cilia

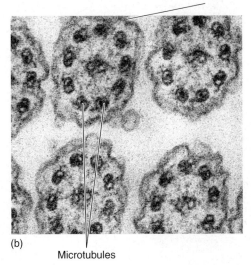

(b)

Microtubules

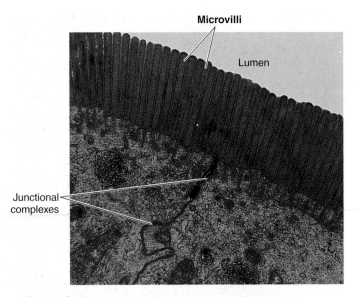

Microvilli

Lumen

Junctional
complexes

■ **Figure 3.6**   Microvilli in the small intestine. Microvilli are seen in this colorized electron micrograph, which shows two adjacent cells joined together by junctional complexes.

■ **Figure 3.5**   **Electron micrographs of cilia.** The cilia can be seen in (*a*) a scanning electron micrograph and (*b*) cross sections in a transmission electron micrograph. Notice the characteristic "9 + 2" arrangement of microtubules in the cross sections.

Sperm cells are the only cells in the human body that have **flagella.** The flagellum is a single whiplike structure that propels the sperm cell through its environment. Both cilia and flagella are composed of *microtubules* (thin cylinders formed from proteins) arranged in a characteristic way. One pair of microtubules in the center of a cilium or flagellum is surrounded by nine other pairs of microtubules, to produce what is often described as a "9 + 2" arrangement (fig. 3.5).

Microvilli

In areas of the body that are specialized for rapid diffusion, the surface area of the cell membranes may be increased by numerous folds called **microvilli.** The rapid passage of the products of digestion across the epithelial membranes in the intestine, for example, is aided by these structural adaptations. The surface area of the apical membranes (the part facing the lumen) in the intestine is increased by the numerous tiny fingerlike projections (fig. 3.6). Similar microvilli are found in the epithelium of the kidney tubule, which must reabsorb various molecules that are filtered out of the blood.

Test Yourself Before You Continue

1. Describe the structure of the plasma membrane.
2. Describe the different ways that cells can engulf materials in the extracellular fluid.
3. Explain the process of exocytosis.
4. Describe the structure and function of cilia, flagella, and microvilli.

Cytoplasm and Its Organelles

Many of the functions of a cell that are performed in the cytoplasmic compartment result from the activity of specific structures called organelles. Among these are the lysosomes, which contain digestive enzymes, and the mitochondria, where most of the cellular energy is produced. Other organelles participate in the synthesis and secretion of cellular products.

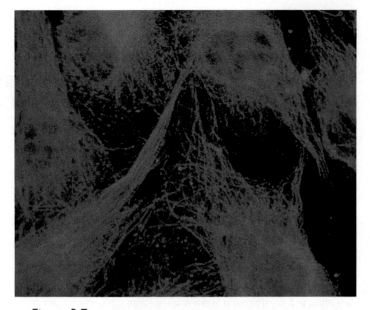

■ Figure 3.7   An immunofluorescence photograph of microtubules. The microtubules in this photograph are visualized with the aid of fluorescent antibodies against tubulin, the major protein component of the microtubules.

Cytoplasm and Cytoskeleton

The jellylike matrix within a cell (exclusive of that within the nucleus) is known as **cytoplasm.** Cytoplasm includes structures called **organelles** that are visible under the microscope, and the fluidlike **cytosol** that surrounds the organelles. When viewed in a microscope without special techniques, the cytoplasm appears to be uniform and unstructured. According to modern evidence, however, the cytosol is not a homogenous solution; it is, rather, a highly organized structure in which protein fibers—in the form of *microtubules* and *microfilaments*—are arranged in a complex latticework surrounding the membrane-bound organelles. Using fluorescence microscopy, these structures can be visualized with the aid of antibodies against their protein components (fig. 3.7). The interconnected microfilaments and microtubules are believed to provide structural organization for cytoplasmic enzymes and support for various organelles.

The latticework of microfilaments and microtubules is said to function as a **cytoskeleton** (fig. 3.8). The structure of this "skeleton" is not rigid; it is capable of quite rapid movement and reorganization. Contractile proteins—including actin and myosin, which are responsible for muscle contraction—are microtubules found in most cells. Such microtubules aid in amoeboid movement, for example, so that the cytoskeleton is also the cell's "musculature." Microtubules, as another example, form the *spindle apparatus* that pulls chromosomes away from each other in cell division. Microtubules also form the central parts of cilia and flagella and contribute to the structure and movements of these projections from the cells.

The cytoplasm of some cells contains stored chemicals in aggregates called **inclusions.** Examples are *glycogen granules* in the liver, striated muscles, and some other tissues; *melanin granules* in the melanocytes of the skin; and *triglycerides* within adipose cells.

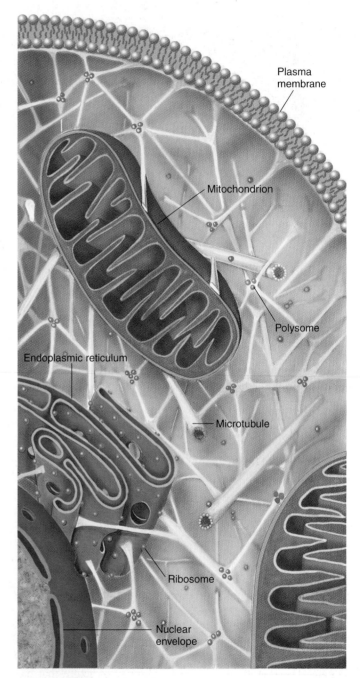

■ Figure 3.8   The formation of the cytoskeleton by microtubules. Microtubules are also important in the motility (movement) of the cell, and movement of materials within the cell.

Lysosomes

After a phagocytic cell has engulfed the proteins, polysaccharides, and lipids present in a particle of "food" (such as a bacterium), these molecules are still kept isolated from the cytoplasm by the membranes surrounding the food vacuole. The large molecules of proteins, polysaccharides, and lipids must first be digested into their smaller subunits (including amino acids, monosaccharides, and fatty acids) before they can cross the vacuole membrane and enter the cytoplasm.

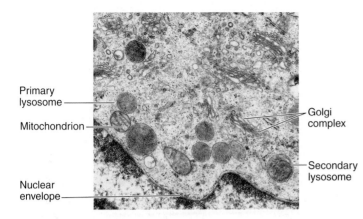

Primary
lysosome

Mitochondrion

Golgi
complex

Secondary
lysosome

Nuclear
envelope

Figure 3.9 An electron micrograph of lysosomes. This photograph shows primary and secondary lysosomes, mitochondria, and the Golgi complex.

The digestive enzymes of a cell are isolated from the cytoplasm and concentrated within membrane-bound organelles called **lysosomes** (fig. 3.9). A *primary lysosome* is one that contains only digestive enzymes (about forty different types) within an environment that is considerably more acidic than the surrounding cytoplasm. A primary lysosome may fuse with a food vacuole (or with another cellular organelle) to form a *secondary lysosome* in which worn-out organelles and the products of phagocytosis can be digested. Thus, a secondary lysosome contains partially digested remnants of other organelles and ingested organic material. A lysosome that contains undigested wastes is called a *residual body.* Residual bodies may eliminate their waste by exocytosis, or the wastes may accumulate within the cell as the cell ages.

Partly digested membranes of various organelles and other cellular debris are often observed within secondary lysosomes. This is a result of **autophagy,** a process that destroys worn-out organelles so that they can be continuously replaced. Lysosomes are thus aptly characterized as the "digestive system" of the cell.

Lysosomes have also been called "suicide bags" because a break in their membranes would release their digestive enzymes and thus destroy the cell. This happens normally in *programmed cell death* (or *apoptosis*), described later in the discussion of the cell cycle. An example is the loss of tissues that must accompany embryonic development, when earlier structures (such as gill pouches) are remodeled or replaced as the embryo matures.

Most, if not all, molecules in the cell have a limited life span. They are continuously destroyed and must be continuously replaced. Glycogen and some complex lipids in the brain, for example, are normally digested at a particular rate by lysosomes. If a person, because of some genetic defect, does not have the proper amount of these lysosomal enzymes, the resulting abnormal accumulation of glycogen and lipids could destroy the tissues. Examples of such defects include **Tay Sach's disease** and **Gaucher's disease.**

Clinical Investigation Clues

Remember that Timothy has large amounts of glycogen granules, with many intact granules seen within his secondary lysosomes.

Could his apparent liver disease be caused by another disorder?

What condition may Timothy have that would explain the presence of intact glycogen granules in his lysosomes?

Peroxisomes

Peroxisomes are membrane-enclosed organelles containing several specific enzymes that promote oxidative reactions. Although peroxisomes are present in most cells, they are particularly large and active in the liver.

All peroxisomes contain one or more enzymes that promote reactions in which hydrogen is removed from particular organic molecules and transferred to molecular oxygen (O_2), thereby oxidizing the molecule and forming hydrogen peroxide (H_2O_2) in the process. The oxidation of toxic molecules by peroxisomes in this way is an important function of liver and kidney cells. For example, much of the alcohol ingested in alcoholic drinks is oxidized into acetaldehyde by liver peroxisomes.

The enzyme *catalase* within the peroxisomes prevents the excessive accumulation of hydrogen peroxide by catalyzing the reaction $2H_2O_2 \rightarrow 2\ H_2O + O_2$. Catalase is one of the fastest acting enzymes known (see chapter 4), and it is this reaction that produces the characteristic fizzing when hydrogen peroxide is poured on a wound.

Mitochondria

All cells in the body, with the exception of mature red blood cells, have from a hundred to a few thousand organelles called **mitochondria** (singular, **mitochondrion**). Mitochondria serve as sites for the production of most of the energy of cells (see chapter 5).

Mitochondria vary in size and shape, but all have the same basic structure (fig. 3.10). Each mitochondrion is surrounded by an inner and outer membrane, separated by a narrow intermembranous space. The outer mitochondrial membrane is smooth, but the inner membrane is characterized by many folds, called *cristae,* which project like shelves into the central area (or *matrix*) of the mitochondrion. The cristae and the matrix compartmentalize the space within the mitochondrion and have different roles in the generation of cellular energy. The structure and functions of mitochondria will be described in more detail in the context of cellular metabolism in chapter 5.

Mitochondria can migrate through the cytoplasm of a cell and are able to reproduce themselves. Indeed, mitochondria contain their own DNA. This is a more primitive form of DNA (consisting of a circular, relatively small, double-stranded molecule) than that found within the cell nucleus. For this and other reasons, many scientists believe that mitochondria evolved from separate organisms, related to bacteria, that invaded the ancestors of animal cells and remained in a state of symbiosis.

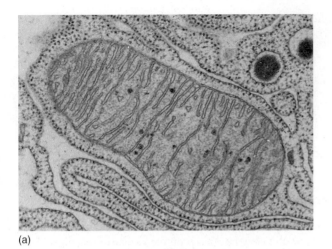

(a)

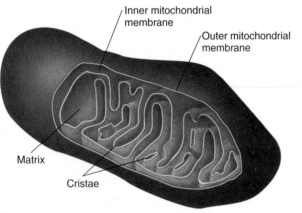

(b)

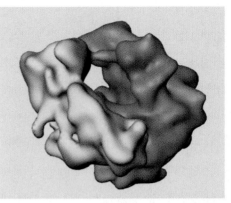

Figure 3.11 **A ribosome is composed of two subunits.** This is a model of the structure of a ribosome, showing the smaller (lighter) and larger (darker) subunits. The space between the two subunits accommodates a molecule of transfer RNA, needed to bring amino acids to the growing polypeptide chain.

Figure 3.10 **The structure of a mitochondrion.** (*a*) An electron micrograph of a mitochondrion. The outer mitochondrial membrane and the infoldings of the inner membrane—the cristae—are clearly seen. The fluid in the center is the matrix. (*b*) A diagram of the structure of a mitochondrion.

Ribosomes

Ribosomes are often called the "protein factories" of the cell, because it is here that proteins are produced according to the genetic information contained in messenger RNA (discussed in a later section). The ribosomes are quite tiny, about 25 nanometers in size, and can be found both free in the cytoplasm and located on the surface of an organelle called the endoplasmic reticulum (discussed in the next section).

Each ribosome consists of two subunits (fig. 3.11) that are designated 30S and 50S, after their sedimentation rate in a centrifuge (this is measured in Svedberg units, from which the "S" is derived). Each of the subunits is composed of both ribosomal RNA and proteins. Contrary to earlier expectations of most scientists, it now appears that the ribosomal RNA molecules serve as enzymes (called *ribozymes*) for many of the reactions in the ribosomes that are required for protein synthesis. Protein synthesis is covered later in this chapter, and the general subject of enzymes and catalysis is discussed in chapter 4.

Endoplasmic Reticulum

Most cells contain a system of membranes known as the **endoplasmic reticulum,** or **ER.** The ER may be either of two types: (1) a **granular,** or **rough, endoplasmic reticulum** and (2) an **agranular,** or **smooth, endoplasmic reticulum** (fig. 3.12). A granular endoplasmic reticulum bears ribosomes on its surface, whereas an agranular endoplasmic reticulum does not. The agranular endoplasmic reticulum serves a variety of purposes in different cells; it provides a site for enzyme reactions in steroid hormone

An unfertilized ovum (egg cell) contains numerous mitochondria, and upon fertilization, gains few if any mitochondria from the sperm. The mitochondrial DNA replicates itself and the mitochondria subsequently divide by pinching off, so that mitochondria can enter the proliferating cells of the embryo and fetus. Thus, all (or nearly all) of the mitochondria in a person are ultimately inherited from that person's mother. This provides a unique form of inheritance that is passed only from mother to child. A rare cause of blindness known as **Leber's hereditary optic neuropathy,** as well as several other disorders, are inherited only along the maternal lineage and are known to be caused by defective mitochondrial DNA.

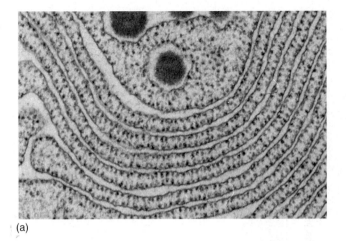

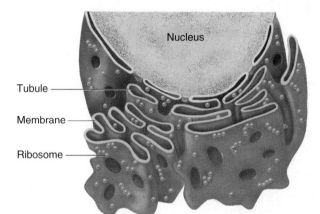

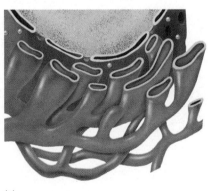

Figure 3.12 **The endoplasmic reticulum.** (*a*) An electron micrograph of a granular endoplasmic reticulum (about 100,000×). The granular endoplasmic reticulum (*b*) has ribosomes attached to its surface, whereas the agranular endoplasmic reticulum (*c*) lacks ribosomes.

production and inactivation, for example, and a site for the storage of Ca^{2+} in striated muscle cells. The granular endoplasmic reticulum is abundant in cells that are active in protein synthesis and secretion, such as those of many exocrine and endocrine glands.

The agranular endoplasmic reticulum in liver cells contains enzymes used for the inactivation of steroid hormones and many drugs. This inactivation is generally achieved by reactions that convert these compounds to more water-soluble and less active forms, which can be more easily excreted by the kidneys. When people take certain drugs (such as alcohol and phenobarbital) for a long period of time, increasingly large doses of these compounds are required to achieve the effect produced initially. This phenomenon, called **tolerance,** is accompanied by growth of the agranular endoplasmic reticulum, and thus an increase in the amount of enzymes charged with inactivation of these drugs.

Clinical Investigation Clues

Remember that Timothy's liver cells have an unusually extensive smooth endoplasmic reticulum.

■ *Why is his endoplasmic reticulum so well developed, and what beneficial function might this serve?*

■ *What could he do to determine if this is the cause of his liver problems?*

Golgi Complex

The **Golgi complex,** also called the **Golgi apparatus,** consists of a stack of several flattened sacs (fig. 3.13). This is something like a stack of pancakes, but the Golgi sac "pancakes" are hollow, with cavities called *cisternae* within each sac. One side of the stack faces the endoplasmic reticulum and serves as a site of entry for vesicles from the endoplasmic reticulum that contain cellular products. These products are passed from one sac to the next, probably by means of vesicles that are budded from one sac and fuse with the next, though other mechanisms may also be involved.

The opposite side of the Golgi stack of sacs faces toward the plasma membrane. As the cellular product passes toward that side it is chemically modified, and then released within vesicles that are budded off the sac. Depending on the nature of the specific product, the vesicles that leave the Golgi complex may become lysosomes, storage granules, secretory vesicles, or additions to the plasma membrane.

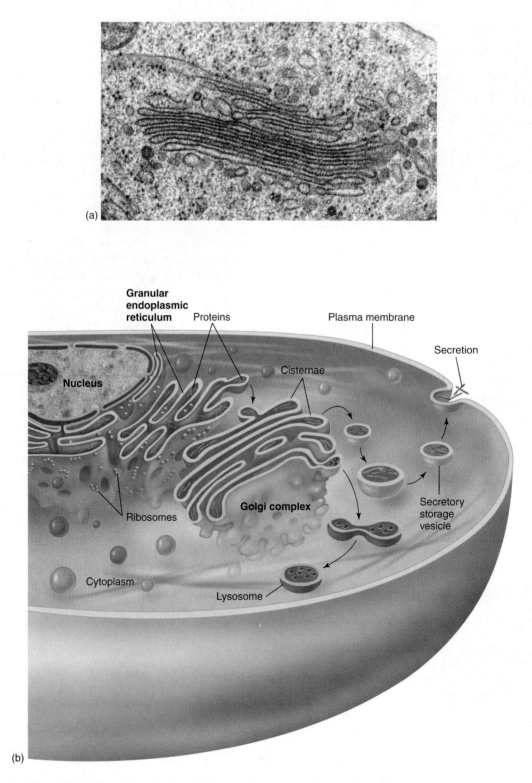

(a)

(b)

■ **Figure 3.13** **The Golgi complex.** (*a*) An electron micrograph of a Golgi complex. Notice the formation of vesicles at the ends of some of the flattened sacs. (*b*) An illustration of the processing of proteins by the granular endoplasmic reticulum and Golgi complex.

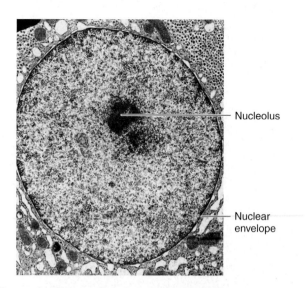

Figure 3.14 **The structure of a nucleus.** The nucleus of a liver cell, with its nuclear envelope and nucleolus, is shown in this electron micrograph.

Cell Nucleus and Gene Expression

The nucleus is the organelle that contains the DNA of a cell. A gene is a length of DNA that codes for the production of a specific polypeptide chain. In order for genes to be expressed, they must first direct the production of complementary RNA molecules. That process is called genetic transcription.

Most cells in the body have a single **nucleus.** Exceptions include skeletal muscle cells, which have two or more nuclei, and mature red blood cells, which have none. The nucleus is enclosed by two membranes—an inner membrane and an outer membrane—that together are called the **nuclear envelope** (fig. 3.14). The outer membrane is continuous with the endoplasmic reticulum in the cytoplasm. At various points, the inner and outer membranes are fused together by structures called *nuclear pore complexes.* These structures function as rivets, holding the two membranes together. Each nuclear pore complex has a central opening, the *nuclear pore* (fig. 3.15), surrounded by interconnected rings and columns of proteins. Small molecules may pass through the complexes by diffusion, but movement of protein and RNA through the nuclear pores is a selective, energy-requiring process.

Transport of specific proteins from the cytoplasm into the nucleus through the nuclear pores may serve a variety of functions, including regulation of gene expression by hormones (see chapter 11). Transport of RNA out of the nucleus, where it is formed, is required for gene expression. As described in this section, *genes* are regions of the DNA within the nucleus. Each gene contains the code for the production of a particular type of RNA called messenger RNA (mRNA). As an mRNA molecule is transported through the nuclear pore, it becomes associated with ribosomes that are either free in the cytoplasm or associated with the granular endoplasmic reticulum. The mRNA then provides the code for the production of a specific type of protein.

The primary structure of the protein (its amino acid sequence) is determined by the sequence of bases in mRNA. The base sequence of mRNA has been previously determined by the sequence of bases in the region of the DNA (the gene) that codes for the mRNA. **Genetic expression** therefore occurs in two stages: first **genetic transcription** (synthesis of RNA) and then **genetic translation** (synthesis of protein).

Each nucleus contains one or more dark areas (see fig. 3.14). These regions, which are not surrounded by membranes, are called **nucleoli.** The DNA within the nucleoli contains the genes that code for the production of ribosomal RNA (rRNA).

The **Human Genome Project** began in 1990 as an international effort to sequence the human genome. In February of 2001, two versions were published: one sponsored by public agencies that was published in the journal *Science,* and one produced by a private company that was published in the journal *Nature.* It soon became apparent that human DNA is 99.9% similar among people; a mere 0.1% is responsible for human genetic variation. It also seems that humans have only about 30,000 to 40,000 genes (segments that code for polypeptide chains), rather than the 100,000 genes that scientists had previously believed.

Nuclear pores

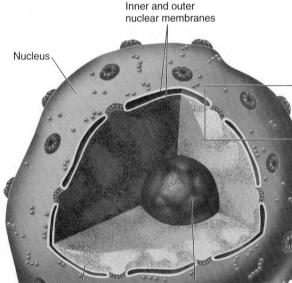

(a)

Chromatin

DNA is composed of four different nucleotide subunits that contain the nitrogenous bases adenine, guanine, cytosine, and thymine. These nucleotides form two polynucleotide chains, joined by complementary base pairing and twisted to form a double helix. This structure is discussed in chapter 2 and illustrated in figures 2.30 and 2.31.

The DNA within the cell nucleus is combined with protein to form **chromatin,** the threadlike material that makes up the chromosomes. Much of the protein content of chromatin is of a type known as *histones.* Histone proteins are positively charged and organized to form spools, about which the negatively charged strands of DNA are wound. Each spool consists of two turns of DNA, comprising 146 base pairs, wound around a core of histone proteins. This spooling creates particles known as **nucleosomes** (fig. 3.16).

Chromatin that is active in genetic transcription (RNA synthesis) is in a relatively extended form known as **euchromatin.** Chromatin regions called **heterochromatin,** in contrast, are highly condensed and form blotchy-looking areas in the nucleus. The condensed heterochromatin contains genes that are said to be "silenced," which means that they are permanently inactivated.

In the euchromatin, genes may be activated or repressed at different times. This is believed to be accomplished by chemical changes in the histones. Such changes include acetylation (the addition of two-carbon-long chemical groups), which turns on genetic transcription, and deacetylation (the removal of those groups), which stops the gene from being transcribed. The acetylation of histone proteins produces a less condensed, more open configuration of the chromatin in specific locations (fig. 3.17), allowing the DNA to be "read" by transcription factors (those that promote RNA synthesis, described in the next section).

Inner and outer nuclear membranes

Nucleus

Chromatin

Pore

Nucleolus

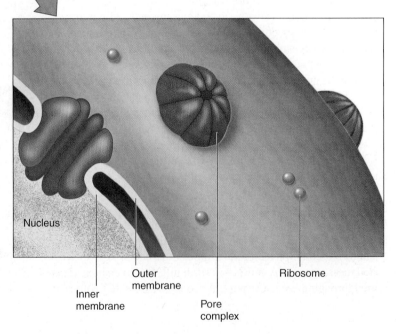

Nucleus

Inner membrane

Outer membrane

Pore complex

Ribosome

(b)

Figure 3.15 **The nuclear pores.** (*a*) An electron micrograph of a freeze-fractured nuclear membrane showing the nuclear pores. (*b*) A diagram showing the nuclear pore complexes.

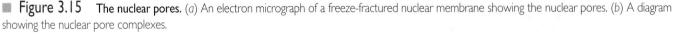

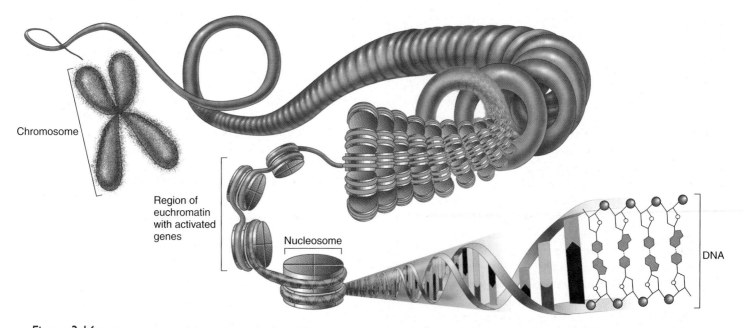

Chromosome

Region of
euchromatin
with activated
genes

Nucleosome

DNA

Figure 3.16 **The structure of chromatin.** Part of the DNA is wound around complexes of histone proteins, forming particles known as nucleosomes.

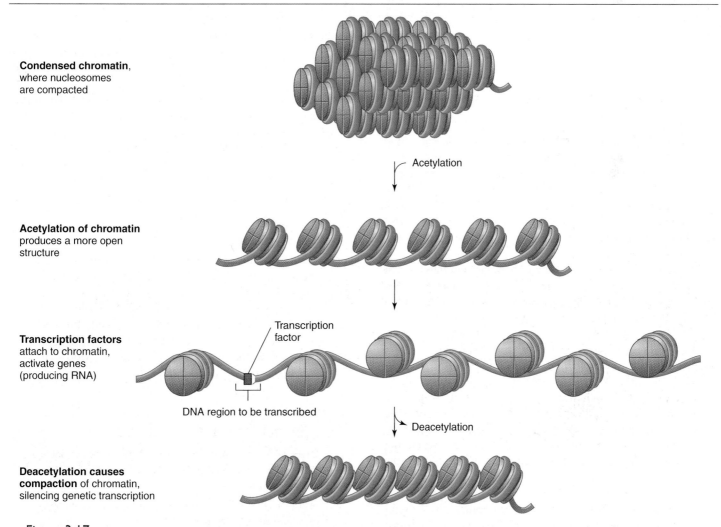

Condensed chromatin, where nucleosomes are compacted

Acetylation

Acetylation of chromatin produces a more open structure

Transcription factors attach to chromatin, activate genes (producing RNA)

Transcription factor

DNA region to be transcribed

Deacetylation

Deacetylation causes compaction of chromatin, silencing genetic transcription

Figure 3.17 **Chromatin structure affects gene expression.** The ability of DNA to be transcribed into messenger RNA is affected by the structure of the chromatin. The genes are silenced when the chromatin is condensed. Acetylation (addition of two-carbon groups) produces a more open chromatin structure that can be activated by transcription factors, producing mRNA. Deacetylation (removal of the acetyl groups) silences genetic transcription.

It is estimated that only about 300 genes out of a total of 30,000 are active in any given cell. This is because each cell becomes specialized for particular functions, in a process called *differentiation*. The differentiated cells of an adult are derived, or "stem from," those of the embryo. Early **embryonic stem cells** can become any cell in the body—they are said to be *totipotent*. As development proceeds, most genes are silenced as cells become more differentiated. *Adult stem cells* can differentiate into a range of specific cell types, but are not normally totipotent. For example, the bone marrow of an adult contains such stem cells (also described in chapter 13, p. 371). These include **hematopoietic stem cells,** which can form the blood cells, and **mesenchymal stem cells,** which can differentiate into osteocytes (bone cells), chondrocytes (cartilage cells), adipocytes (fat cells), and others. **Neural stem cells** (also described in chapter 8, p. 203) have been identified in the adult nervous system. These can migrate to particular locations and differentiate into specific neuron and glial cell types in these locations. Many scientists hope that stem cells grown in tissue culture might someday be used to grow transplantable tissues and organs.

RNA Synthesis

One gene codes for one polypeptide chain. Each gene is a stretch of DNA that is several thousand nucleotide pairs long. The DNA in a human cell contains over 3 billion base pairs—enough to code for at least 3 million proteins. Since the average human cell contains less than this amount (30,000 to 150,000 different proteins), it follows that only a fraction of the DNA in each cell is used to code for proteins. The remainder of the DNA may be inactive or redundant. Also, some segments of DNA serve to regulate those regions that do code for proteins.

In order for the genetic code to be translated into the synthesis of specific proteins, the DNA code first must be copied onto a strand of RNA. This is accomplished by DNA-directed RNA synthesis—the process of **genetic transcription.**

In RNA synthesis, the enzyme **RNA polymerase** breaks the weak hydrogen bonds between paired DNA bases. This does not occur throughout the length of DNA, but only in the regions that are to be transcribed. There are base sequences that code for "start" and "stop," and there are regions of DNA that function as *promoters.* Specific regulatory molecules, such as hormones, act as **transcription factors** by binding to the promoter region of a particular gene and thereby activating the gene. The double-stranded DNA separates in the region to be transcribed, so that the freed bases can pair with the complementary RNA nucleotide bases in the nucleoplasm.

This pairing of bases, like that which occurs in DNA replication (described in a later section), follows the law of complementary base pairing: *guanine bonds with cytosine* (and vice versa), and *adenine bonds with uracil* (because uracil in RNA is

equivalent to thymine in DNA). Unlike DNA replication, however, only *one* of the two freed strands of DNA serves as a guide for RNA synthesis (fig. 3.18). Once an RNA molecule has been produced, it detaches from the DNA strand on which it was formed. This process can continue indefinitely, producing many thousands of RNA copies of the DNA strand that is being transcribed. When the gene is no longer to be transcribed, the separated DNA strands can then go back together again.

Types of RNA

There are four types of RNA produced within the nucleus by transcription: (1) **precursor messenger RNA (pre-mRNA),** which is altered within the nucleus to form mRNA; (2) **messenger RNA (mRNA),** which contains the code for the synthesis of specific proteins; (3) **transfer RNA (tRNA),** which is needed for decoding the genetic message contained in mRNA; and (4) **ribosomal RNA (rRNA),** which forms part of the structure of ribosomes. The DNA that codes for rRNA synthesis is located in the part of the nucleus called the nucleolus. The DNA that codes for pre-mRNA and tRNA synthesis is located elsewhere in the nucleus.

In bacteria, where the molecular biology of the gene is best understood, a gene that codes for one type of protein produces an mRNA molecule that begins to direct protein synthesis as soon as it is transcribed. This is not the case in higher organisms, including humans. In higher cells, a pre-mRNA is produced that must be modified within the nucleus before it can enter the cytoplasm as mRNA and direct protein synthesis.

Precursor mRNA is much larger than the mRNA it forms. Surprisingly, this large size of pre-mRNA is not due to excess bases at the ends of the molecule that must be trimmed; rather, the excess bases are located *within* the pre-mRNA. The genetic code for a particular protein, in other words, is split up by stretches of base pairs that do not contribute to the code. These regions of noncoding DNA within a gene are called *introns;* the coding regions are known as *exons.* Consequently, pre-mRNA must be cut and spliced to make mRNA (fig. 3.19). This cutting and splicing can be quite extensive—a single gene may contain up to 50 introns, which must be removed from the pre-mRNA in order to convert it to mRNA.

Introns are cut out of the pre-mRNA, and the ends of the exons spliced, by macromolecules called *snRNPs* (pronounced "snurps"), producing the functional mRNA that leaves the nucleus and enters the cytoplasm. SnRNPs stands for *small nuclear ribonucleoproteins.* These are small, ribosome-like aggregates of RNA and protein that form a body called a *spliceosome* that splices the exons together.

Test Yourself Before You Continue

1. Describe the appearance and composition of chromatin and the structure of nucleosomes. Comment on the significance of histone proteins.
2. Explain how RNA is produced within the nucleus according to the information contained in DNA.
3. Explain how precursor mRNA is modified to produce mRNA.

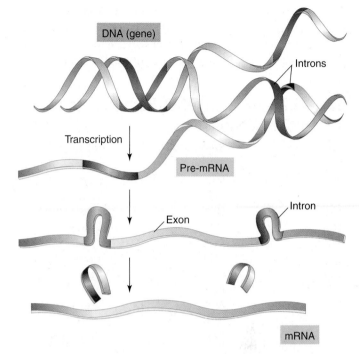

Figure 3.19 The processing of pre-mRNA into mRNA. Noncoding regions of the genes, called introns, produce excess bases within the pre-mRNA. These excess bases are removed, and the coding regions of mRNA are spliced together.

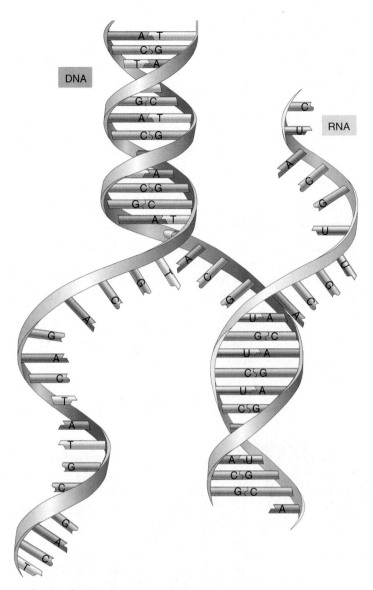

Figure 3.18 RNA synthesis (transcription). Notice that only one of the two DNA strands is used to form a single-stranded molecule of RNA.

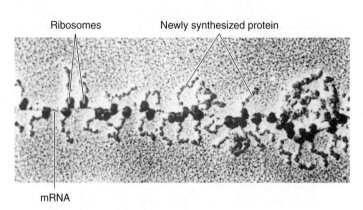

Figure 3.20 An electron micrograph of a polyribosome. An RNA strand joins the ribosomes together.

Protein Synthesis and Secretion

In order for a gene to be expressed, it first must be used as a guide, or template, in the production of a complementary strand of messenger RNA. This mRNA is then itself used as a guide to produce a particular type of protein whose sequence of amino acids is determined by the sequence of base triplets (codons) in the mRNA.

When mRNA enters the cytoplasm, it attaches to **ribosomes,** which appear in the electron microscope as numerous small particles. A ribosome is composed of four molecules of ribosomal RNA and eighty-two proteins, arranged to form two

subunits of unequal size. The mRNA passes through a number of ribosomes to form a "string-of-pearls" structure called a *poly-ribosome* (or *polysome,* for short), as shown in figure 3.20. The association of mRNA with ribosomes is needed for the process of **genetic translation**—the production of specific proteins according to the code contained in the mRNA base sequence.

Each mRNA molecule contains several hundred or more nucleotides, arranged in the sequence determined by complementary base pairing with DNA during transcription (RNA synthesis). Every three bases, or *base triplet,* is a code word—called a **codon**—for a specific amino acid. Sample codons and their amino acid "translations" are listed in table 3.2 and illustrated in figure 3.21. As mRNA moves through the ribosome, the sequence of codons is translated into a sequence of specific amino acids within a growing polypeptide chain.

Table 3.2   Selected DNA Base Triplets and mRNA Codons

DNA Triplet	RNA Codon	Amino Acid
TAC	AUG	"Start" (Methionine)
ATC	UAG	"Stop"
AAA	UUU	Phenylalanine
AGG	UCC	Serine
ACA	UGU	Cysteine
GGG	CCC	Proline
GAA	CUU	Leucine
GCT	CGA	Arginine
TTT	AAA	Lysine
TGC	ACG	Threonine
CCG	GGC	Glycine
CTC	GAG	Glutamic acid

Transfer RNA

Translation of the codons is accomplished by tRNA and particular enzymes. Each tRNA molecule, like mRNA and rRNA, is single-stranded. Although tRNA is single-stranded, it bends in on itself to form a cloverleaf structure (fig. 3.22*a*), which is believed to be further twisted into an upside down "L" shape (fig. 3.22*b*). One end of the "L" contains the **anticodon**—three nucleotides that are complementary to a specific codon in mRNA.

Enzymes in the cell cytoplasm called *aminoacyl-tRNA synthetase enzymes* join specific amino acids to the ends of tRNA, so that a tRNA with a given anticodon can bind to only one specific amino acid. There are twenty different varieties of synthetase enzymes, one for each type of amino acid. Not only must each synthetase recognize its specific amino acid, it also must be able to attach this amino acid to the particular tRNA that has the correct anticodon for that amino acid. The cytoplasm of a cell thus contains tRNA molecules that are each bonded to a specific amino acid, and each of these tRNA molecules is capable of bonding with a specific codon in mRNA via its anticodon base triplet.

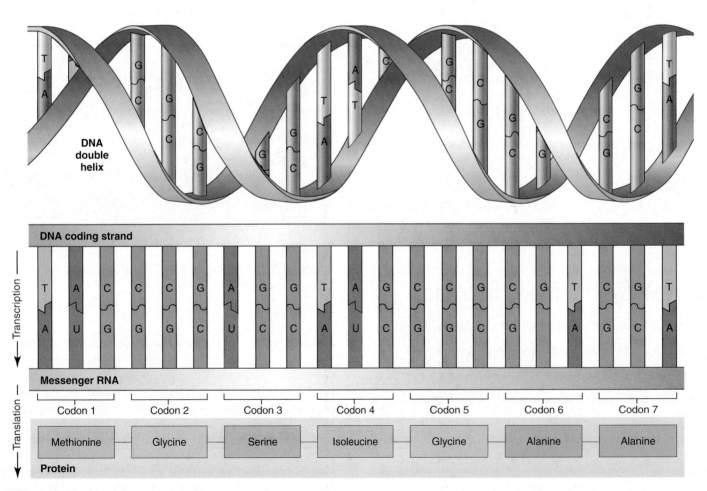

■ **Figure 3.21**   **Transcription and translation.** The genetic code is first transcribed into base triplets (codons) in mRNA and then translated into a specific sequence of amino acids in a polypeptide.

Formation of a Polypeptide

The anticodons of tRNA bind to the codons of mRNA as the mRNA moves through the ribosome. Since each tRNA molecule carries a specific amino acid, the joining together of these amino acids by peptide bonds creates a polypeptide whose amino acid sequence has been determined by the sequence of codons in mRNA.

The first and second tRNA bring the first and second amino acids close together. The first amino acid then detaches from its tRNA and is enzymatically transferred to the amino acid on the second tRNA, forming a dipeptide. When the third tRNA binds to the third codon, the amino acid it brings forms a peptide bond with the second amino acid (which detaches from its tRNA). A tripeptide is now attached by the third amino acid to the third tRNA. The polypeptide chain thus grows as new amino acids are added to its growing tip (fig. 3.23). This growing polypeptide chain is always attached by means of only one tRNA to the strand of mRNA, and this tRNA molecule is always the one that has added the latest amino acid to the growing polypeptide.

As the polypeptide chain grows in length, interactions between its amino acids cause the chain to twist into a helix (secondary structure) and to fold and bend upon itself (tertiary structure). At the end of this process, the new protein detaches from the tRNA as the last amino acid is added. Many proteins are further modified after they are formed; these modifications occur in the rough endoplasmic reticulum and Golgi complex.

Functions of the Endoplasmic Reticulum and Golgi Complex

Proteins that are to be used within the cell are likely to be produced by polyribosomes that float freely in the cytoplasm, unattached to other organelles. If the protein is to be secreted by the cell, however, it is made by mRNA-ribosome complexes that are located on the granular endoplasmic reticulum. The membranes of this system enclose fluid-filled spaces called *cisternae,* into which the newly formed proteins may enter. Once in the cisternae, the structure of these proteins is modified in specific ways.

When proteins destined for secretion are produced, the first thirty or so amino acids are primarily hydrophobic. This *leader sequence* is attracted to the lipid component of the membranes of the endoplasmic reticulum. As the polypeptide chain elongates, it is "injected" into the cisterna within the endoplasmic reticulum. The leader sequence is, in a sense, an "address" that directs secretory proteins into the endoplasmic reticulum. Once the proteins are in the cisterna, the leader sequence is enzymatically removed so that the protein cannot reenter the cytoplasm (fig. 3.24).

The processing of the hormone insulin can serve as an example of the changes that occur within the endoplasmic reticulum. The original molecule enters the cisterna as a single polypeptide composed of 109 amino acids. This molecule is called *preproinsulin.* The first twenty-three amino acids serve as a leader sequence that allows the molecule to be injected into the cisterna within the endoplasmic reticulum. The leader sequence is then quickly removed, producing a molecule called *proinsulin.* The remaining chain folds within the cisterna so that the first and last amino acids in the polypeptide are brought close together. Enzymatic removal of the central region produces two chains—one of them, twenty-one amino acids long; the other, thirty amino acids long—that are subsequently joined together by disulfide bonds (fig. 3.25). This is the form of insulin that is normally secreted from the cell.

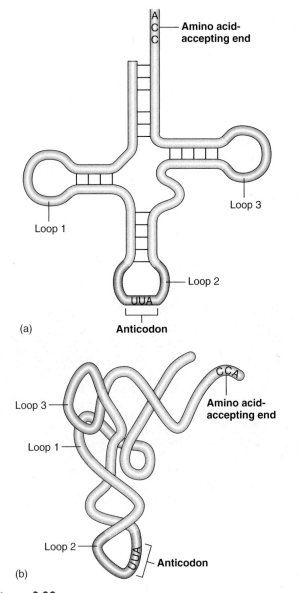

■ Figure 3.22 The structure of transfer RNA (tRNA). (*a*) A simplified cloverleaf representation and (*b*) the three-dimensional structure of tRNA.

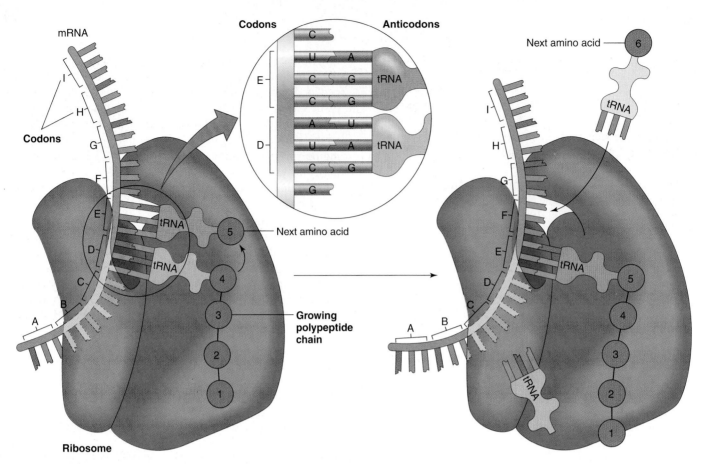

Figure 3.23 **The translation of messenger RNA (mRNA).** As the anticodon of each new aminoacyl-tRNA bonds with a codon on the mRNA, new amino acids are joined to the growing tip of the polypeptide chain.

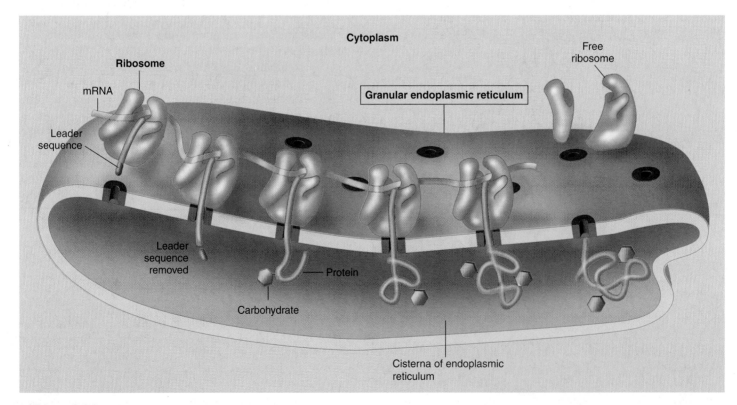

Figure 3.24 **How secretory proteins enter the endoplasmic reticulum.** A protein destined for secretion begins with a leader sequence that enables it to be inserted into the cisterna (cavity) of the endoplasmic reticulum. Once it has been inserted, the leader sequence is removed and carbohydrate is added to the protein.

Advances in the identification of human genes, methods of cloning (replicating) isolated genes, and other technologies have made **gene therapy** a realistic possibility. Although attempts at gene therapy were made as early as 1990, it was not until 2000 that children with the less severe form of *Severe Combined Immunodeficiency,* or *SCID,* were successfully treated by gene therapy. Then, in 2002, two children with the more severe form of SCID were cured of their condition. In this case, the children lack the gene for a specific enzyme, *adenine deaminase,* and this lack prevents the development of a functioning immune system. By inserting genes that code for ADA into the children's blood-forming stem cells, and getting these cells to proliferate in the bone marrow, scientists have apparently restored the immune system of these children. Prior to this new gene therapy, children with SCID had to be kept isolated in sterile environments (the "boy in the bubble"), because even common infections could be fatal.

The Cell Cycle

Unlike the life of an organism, which can be viewed as a linear progression from birth to death, the life of a cell follows a cyclical pattern. Each cell is produced as a part of its "parent" cell; when the daughter cell divides, it in turn becomes two new cells. In a sense, then, each cell is potentially immortal as long as its progeny can continue to divide. Some cells in the body divide frequently; the epidermis of the skin, for example, is renewed approximately every 2 weeks, and the stomach lining is renewed every 2 or 3 days. Other cells, such as striated muscle cells in the adult, do not divide at all. All cells in the body, of course, live only as long as the person lives (some cells live longer than others, but eventually all cells die when vital functions cease).

The nondividing cell is in a part of its life cycle known as interphase (fig. 3.27), which is subdivided into G_1, S, and G_2 phases, as will be described shortly. The chromosomes are in their extended form, and their genes actively direct the synthesis of RNA. Through their direction of RNA synthesis, genes control the metabolism of the cell. The cell may be growing during this time, and this part of interphase is known as the G_1 phase (G stands for *gap*). Although sometimes described as "resting," cells in the G_1 phase perform the physiological functions characteristic of the tissue in which they are found. The DNA of resting cells in the G_1 phase thus produces mRNA and proteins as previously described.

If a cell is going to divide, it replicates its DNA in a part of interphase known as the S phase (S stands for *synthesis*). Once DNA has replicated in the S phase, the chromatin condenses in the G_2 phase to form short, thick, structures by the end of G_2. Though condensed, the chromosomes are not yet in their more familiar, visible form in the ordinary (light) microscope; these will first make their appearance at prophase of mitosis (fig. 3.28).

Cyclins and p53

A group of proteins known as the cyclins promote different phases of the cell cycle. During the G_1 phase of the cycle, for example, an increase in the concentration of *cyclin D* proteins

within the cell acts to move the cell quickly through this phase. Cyclin D proteins do this by activating a group of otherwise inactive enzymes known as *cyclin-dependant kinases.* Therefore, overactivity of a gene that codes for a cyclin D might be predicted to cause uncontrolled cell division, as occurs in a cancer. Indeed, overexpression of the gene for cyclin D1 has been shown to occur in some cancers, including those of the breast and esophagus. Genes that contribute to cancer are called **oncogenes.** Oncogenes are mutated forms of normal genes, called *proto-oncogenes,* that are functional in normal, healthy cells.

While oncogenes promote cancer, other genes—called **tumor suppressor genes**—inhibit its development. One very important tumor suppressor gene is known as **p53.** This name refers to the protein coded by the gene, which has a molecular weight of 53,000. The normal gene protects against cancer by indirectly blocking the ability of cyclins to stimulate cell division. In part, p53 accomplishes this by inducing the expression of another gene, called *p21,* which produces a protein that binds to and inactivates the cyclin-dependant kinases. The p21 protein thus inhibits cell division as it promotes cell differentiation (specialization).

For these reasons, cancer is likely to develop if the p53 gene becomes mutated and therefore ineffective as a tumor suppressor gene. Indeed, mutated p53 genes are found in over 50% of all cancers. Mice whose p53 genes were "knocked out" all developed tumors. (**Knockout mice** are strains of mice in which a specific

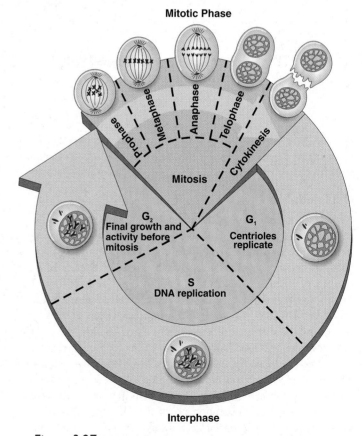

Figure 3.27 **The life cycle of a cell.** The different stages of mitotic division are shown; it should be noted, however, that not all cells undergo mitosis.

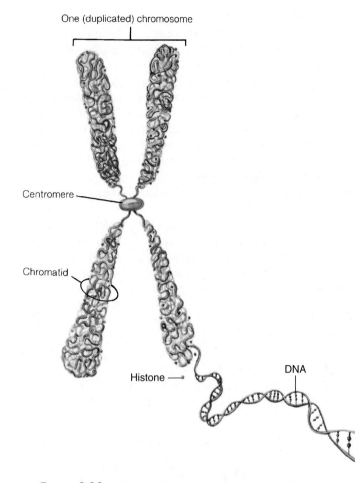

One (duplicated) chromosome

Centromere

Chromatid

Histone ⟶

DNA

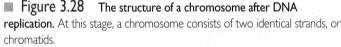

■ **Figure 3.28**   The structure of a chromosome after DNA replication. At this stage, a chromosome consists of two identical strands, or chromatids.

Apoptosis has been implicated in many disease processes, but it also occurs normally as part of programmed cell death—a process described previously in the section on lysosomes. Programmed cell death refers to the physiological process responsible for the remodeling of tissues during embryonic development and for tissue turnover in the adult body. As mentioned earlier, the epithelial cells lining the digestive tract are programmed to die 2 to 3 days after they are produced, and epidermal cells of the skin live only for about 2 weeks until they die and become completely cornified. Apoptosis is also important in the functioning of the immune system. A neutrophil (a type of white blood cell), for example, is programmed to die by apoptosis 24 hours after its creation in the bone marrow. A killer T lymphocyte (another type of white blood cell) destroys targeted cells by triggering their apoptosis.

Using mice with their gene for p53 knocked out, scientists have learned that p53 is needed for the apoptosis that occurs when a cell's DNA is damaged. The damaged DNA, if not repaired, activates p53, which in turn causes the cell to be destroyed. If the p53 gene has mutated to an ineffective form, however, the cell will not be destroyed by apoptosis as it should; rather, it will divide to produce daughter cells with damaged DNA. This may be one mechanism responsible for the development of a cancer.

There are three forms of **skin cancer**—squamous cell carcinoma, basal cell carcinoma, and melanoma, depending on the type of epidermal cell involved—all of which are promoted by the damaging effects of the ultraviolet portion of sunlight. Ultraviolet light promotes a characteristic type of DNA mutation in which either of two pyrimidines (cytosine or thymine) is affected. In squamous cell and basal cell carcinoma (but not melanoma), the cancer is believed to involve mutations that affect the p53 gene, among others. Whereas cells with normal p53 genes may die by apoptosis when their DNA is damaged, and are thus prevented from replicating themselves and perpetuating the damaged DNA, those damaged cells with a mutated p53 gene survive and divide to produce the cancer.

Mitosis

At the end of the G_2 phase of the cell cycle, which is generally shorter than G_1, each chromosome consists of two strands called **chromatids** that are joined together by a *centromere* (see fig. 3.28). The two chromatids within a chromosome contain identical DNA base sequences because each is produced by the semiconservative replication of DNA. Each chromatid, therefore, contains a complete double-helix DNA molecule that is a copy of the single DNA molecule existing prior to replication. Each chromatid will become a separate chromosome once mitotic cell division has been completed.

The G_2 phase completes interphase. The cell next proceeds through the various stages of cell division, or **mitosis.** This is the *M phase* of the cell cycle. Mitosis is subdivided into four stages: *prophase, metaphase, anaphase,* and *telophase* (fig. 3.29).

targeted gene has been inactivated by developing the mice from embryos injected with specifically mutated cells.) These important discoveries have obvious relevance to cancer diagnosis and treatment.

Cell Death

Cell death occurs both pathologically and naturally. Pathologically, cells deprived of a blood supply may swell, rupture their membranes, and burst. Such cellular death, leading to tissue death, is known as **necrosis.** In certain cases, however, a different pattern is observed. Instead of swelling, the cells shrink. The membranes remain intact but become bubbled, and the nuclei condense. This process was named **apoptosis** (from a Greek term describing the shedding of leaves from a tree), and its discoverers were awarded the 2002 Nobel prize in Physiology or Medicine.

The machinery of cell death is set in motion by a family of enzymes called *caspases,* which are normally inactive within the cell but become activated during apoptosis. These enzymes have thus been called the "executioners" of the cell. Mitochondria may play an essential role in the activation of caspases and resulting apoptosis. This occurs when certain stimuli cause the outer and inner mitochondrial membranes to become permeable to proteins and other products that do not normally leak into the cell cytoplasm.

(a) Interphase
- The chromosomes are in an extended form and seen as chromatin in the electron microscope.
- The nucleus is visible

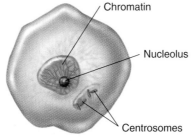

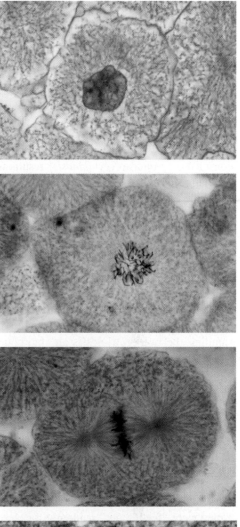

(b) Prophase
- The chromosomes are seen to consist of two chromatids joined by a centromere.
- The centrioles move apart toward opposite poles of the cell.
- Spindle fibers are produced and extend from each centrosome.
- The nuclear membrane starts to disappear.
- The nucleolus is no longer visible.

(c) Metaphase
- The chromosomes are lined up at the equator of the cell.
- The spindle fibers from each centriole are attached to the centromeres of the chromosomes.
- The nuclear membrane has disappeared.

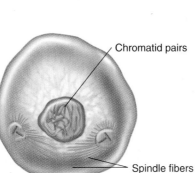

(d) Anaphase
- The centromere split, and the sister chromatids separate as each is pulled to an opposite pole.

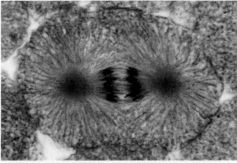

(e) Telophase
- The chromosomes become longer, thinner, and less distinct.
- New nuclear membranes form.
- The nucleolus reappears.
- Cell division is nearly complete.

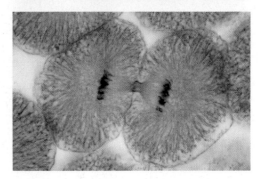

Figure 3.29 **The stages of mitosis.** The events that occur in each stage are indicated in the figure.

In prophase, chromosomes become visible as distinctive structures. In metaphase of mitosis, the chromosomes line up single file along the equator of the cell. This aligning of chromosomes at the equator is believed to result from the action of **spindle fibers,** which are attached to a protein structure called the *kinetochore* at the centromere of each chromosome (fig. 3.29).

Anaphase begins when the centromeres split apart and the spindle fibers shorten, pulling the two chromatids in each chromosome to opposite poles. Each pole therefore gets one copy of each of the forty-six chromosomes. During early telophase, division of the cytoplasm (*cytokinesis*) results in the production of two daughter cells that are genetically identical to each other and to the original parent cell.

Role of the Centrosome

All animal cells have a **centrosome,** located near the nucleus in a nondividing cell. At the center of the centrosome are two **centrioles,** which are positioned at right angles to each other. Each centriole is composed of nine evenly spaced bundles of microtubules, with three microtubules per bundle (fig. 3.30). Surrounding the two centrioles is an amorphous mass of material called the *pericentriolar material*. Microtubules grow out of the pericentriolar material, which is believed to function as the center for the organization of microtubules in the cytoskeleton.

Through a mechanism that is still incompletely understood, the centrosome replicates itself during interphase if a cell is going to divide. The two identical centrosomes then move away from each other during prophase of mitosis and take up positions at opposite poles of the cell by metaphase. At this time, the centrosomes produce new microtubules. These new microtubules are very dynamic, rapidly growing and shrinking as if they were "feeling out" randomly for chromosomes. A microtubule becomes stabilized when it finally binds to the proper region of a chromosome. In this way, the microtubules from both centrosomes form the spindle fibers that are attached to each of the replicated chromosomes at metaphase (fig 3.31).

The spindle fibers pull the chromosomes to opposite poles of the cell during anaphase, so that at telophase, when the cell pinches inward, two identical daughter cells will be produced. This also requires the centrosomes, which somehow organize a ring of contractile filaments halfway between the two poles. These filaments are attached to the cell membrane, and when they contract, the cell is pinched in two. The filaments consist of actin and myosin proteins, the same contractile proteins present in muscle.

Telomeres and Cell Division

Certain types of cells can be removed from the body and grown in nutrient solutions (outside the body, or *in vitro*). Under these artificial conditions, the potential longevity of different cell lines can be studied. For unknown reasons, normal connective tissue cells (called fibroblasts) stop dividing in vitro after a certain number of population doublings. Cells from a newborn will divide 80 to 90 times, while those from a 70-year-old will stop after 20 to 30 divisions. The decreased ability to divide is thus an indicator of senescence (aging). Cells that become transformed into cancer, however, apparently do not age and continue dividing indefinitely in culture.

This senescent decrease in the ability of cells to replicate may be related to a loss of DNA sequences at the ends of chromosomes, in regions called **telomeres** (from the Greek *telos* = end). The telomeres serve as caps on the ends of DNA, preventing enzymes from mistaking the normal ends for broken DNA and doing damage by trying to "repair" them.

The DNA polymerase enzyme does not fully copy the DNA at the end-regions. Each time a chromosome replicates, it loses 50 to 100 base pairs in its telomeres. Cell division may ultimately stop when there is too much loss of DNA in the telomeres, and the cell dies because of damage it sustains in the

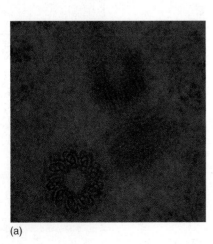

(a)

(b)

■ **Figure 3.30** **The centrioles.** (*a*) A micrograph of the two centrioles in a centrosome. (*b*) A diagram showing that the centrioles are positioned at right angles to each other.

course of aging. Interestingly, Dolly (the famous cloned sheep) had short telomeres, because her DNA was older than she was. For reasons not presently clear, however, cloned cattle seem to have long telomeres, despite the short telomeres of the donors. Will Dolly's life be shorter and the cloned cattle's longer because of this? It is too soon to tell.

Germinal cells that give rise to gametes (sperm cells and ova) can continue to divide indefinitely, perhaps because they produce an enzyme called **telomerase,** which duplicates the telomere DNA. Telomerase is also found in hematopoietic stem cells (those in bone marrow that produce blood cells) and other stem cells that must divide continuously. Similarly, telomerase is produced by most cancer cells, and there is evidence to suggest that telomerase may be responsible for their ability to divide indefinitely.

Hypertrophy and Hyperplasia

The growth of an individual from a fertilized egg into an adult involves an increase in the number of cells and an increase in the size of cells. Growth that is due to an increase in cell number results from an increased rate of mitotic cell division and is termed **hyperplasia.** Growth of a tissue or organ due to an increase in cell size is termed **hypertrophy.**

Most growth is due to hyperplasia. A callus on the palm of the hand, for example, involves thickening of the skin by hyperplasia due to frequent abrasion. An increase in skeletal muscle size as a result of exercise, by contrast, is produced by hypertrophy.

Skeletal muscle and cardiac (heart) muscle can grow only by hypertrophy. When growth occurs in skeletal muscles in response to an increased workload—during weight training, for example—it is called **compensatory hypertrophy.** The heart muscle may also demonstrate compensatory hypertrophy when its workload increases because of hypertension (high blood pressure). The opposite of hypertrophy is **atrophy,** the wasting or decrease in size of a cell, tissue, or organ. This may result from the disuse of skeletal muscles, as occurs in prolonged bed rest, various diseases, or advanced age.

Meiosis

When a cell is going to divide, either by mitosis or meiosis, the DNA is replicated (forming chromatids) and the chromosomes become shorter and thicker, as previously described. At this point the cell has forty-six chromosomes, each of which consists of two duplicate chromatids.

The short, thick chromosomes seen at the end of the G₂ phase can be matched as pairs, the members of each pair appearing to be structurally identical. These matched chromosomes are called **homologous chromosomes.** One member of each homologous pair is derived from a chromosome

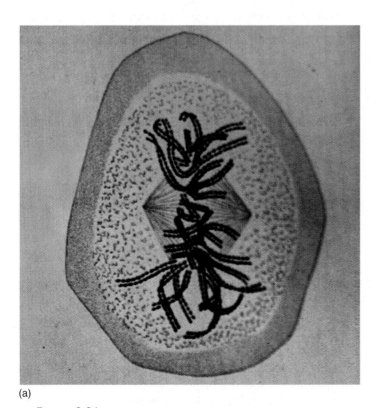

(a)

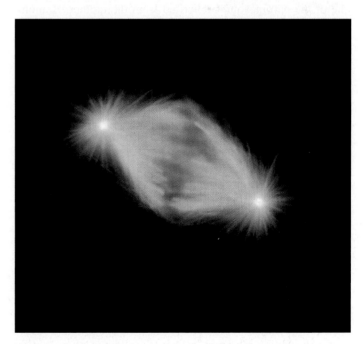

(b)

 Figure 3.31   **Chromosomes and spindle fibers.** The duplicate chromatids are clearly seen in (*a*), though the spindle fibers are just barely visible. In a technique callled immunofluorescence, the spindle fibers shine in (*b*) due to a reaction with microtubules, the major constituent of the spindles.

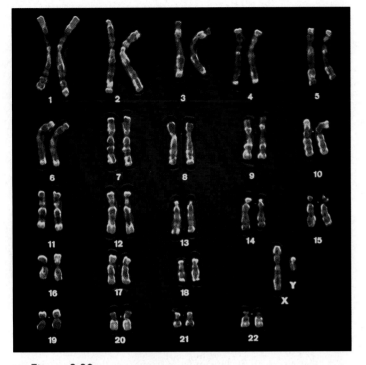

Figure 3.32 A karyotype, in which chromosomes are arranged in homologous pairs. A false-color light micrograph of chromosomes from a male arranged in numbered homologous pairs, from the largest to the smallest.

Table 3.3 Stages of Meiosis

Stage	Events
First Meiotic Division	
Prophase I	Chromosomes appear double-stranded.
	Each strand, called a chromatid, contains duplicate DNA joined together by a structure known as a centromere.
	Homologous chromosomes pair up side by side.
Metaphase I	Homologous chromosome pairs line up at equator.
	Spindle apparatus is complete.
Anaphase I	Homologous chromosomes separate; the two members of a homologous pair move to opposite poles.
Telophase I	Cytoplasm divides to produce two haploid cells.
Second Meiotic Division	
Prophase II	Chromosomes appear, each containing two chromatids.
Metaphase II	Chromosomes line up single file along equator as spindle formation is completed.
Anaphase II	Centromeres split and chromatids move to opposite poles.
Telophase II	Cytoplasm divides to produce two haploid cells from each of the haploid cells formed at telophase I.

inherited from the father, and the other member is a copy of one of the chromosomes inherited from the mother. Homologous chromosomes do not have identical DNA base sequences; one member of the pair may code for blue eyes, for example, and the other for brown eyes. There are twenty-two homologous pairs of *autosomal chromosomes* and one pair of *sex chromosomes,* described as X and Y. Females have two X chromosomes, whereas males have one X and one Y chromosome (fig. 3.32).

Meiosis, which has two divisional sequences, is a special type of cell division that occurs only in the gonads (testes and ovaries), where it is used only in the production of gametes—sperm cells and ova. (Gamete production is described in detail in chapter 20.) In the first division of meiosis, the homologous chromosomes line up side by side, rather than single file, along the equator of the cell. The spindle fibers then pull one member of a homologous pair to one pole of the cell, and the other member of the pair to the other pole. Each of the two daughter cells thus acquires only one chromosome from each of the twenty-three homologous pairs contained in the parent. The daughter cells, in other words, contain twenty-three rather than forty-six chromosomes. For this reason, meiosis (from the Greek *meion* = less) is also known as **reduction division.**

At the end of this cell division, each daughter cell contains twenty-three chromosomes—but *each of these consists of two chromatids.* (Since the two chromatids per chromosome are identical, this does not make forty-six chromosomes; there are still only twenty-three *different* chromosomes per cell at this point.) The chromatids are separated by a second meiotic division. Each of the daughter cells from the first cell division itself divides, with the duplicate chromatids going to each of two new daughter cells. A grand total of four daughter cells can thus be produced from the meiotic cell division of one parent cell. This occurs in the testes, where one parent cell produces four sperm cells. In the ovaries, one parent cell also produces four daughter cells, but three of these die and only one progresses to become a mature egg cell (as will be described in chapter 20).

The stages of meiosis are subdivided according to whether they occur in the first or the second meiotic cell division. These stages are designated as prophase I, metaphase I, anaphase I, telophase I; and then prophase II, metaphase II, anaphase II, and telophase II (table 3.3 and fig. 3.33).

The reduction of the chromosome number from forty-six to twenty-three is obviously necessary for sexual reproduction, where the sex cells join and add their content of chromosomes together to produce a new individual. The significance of meiosis, however, goes beyond the reduction of chromosome number. At metaphase I, the pairs of homologous chromosomes can line up with either member facing a given pole of the cell. (Recall that each member of a homologous pair came from a different parent.) Maternal and paternal members of homologous pairs are thus randomly shuffled. Hence, when the first meiotic

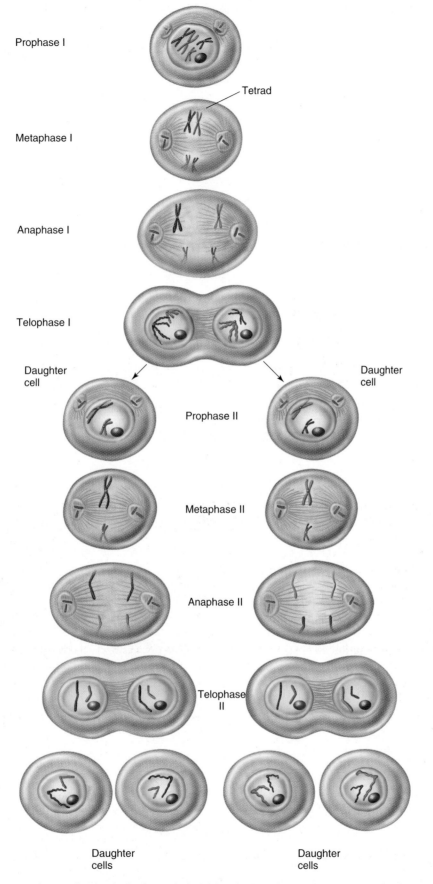

Prophase I

Tetrad

Metaphase I

Anaphase I

Telophase I

Daughter cell

Prophase II

Daughter cell

Metaphase II

Anaphase II

Telophase II

Daughter cells

Daughter cells

■ **Figure 3.33** **Meiosis, or reduction division.** In the first meiotic division, the homologous chromosomes of a diploid parent cell are separated into two haploid daughter cells. Each of these chromosomes contains duplicate strands, or chromatids. In the second meiotic division, these chromosomes are distributed to two new haploid daughter cells.

(a) First meiotic prophase   Chromosomes pairing   Chromosomes crossing-over

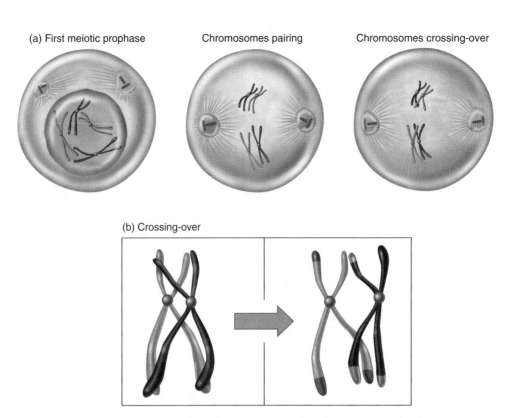

(b) Crossing-over

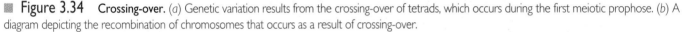

■ **Figure 3.34** **Crossing-over.** (*a*) Genetic variation results from the crossing-over of tetrads, which occurs during the first meiotic prophose. (*b*) A diagram depicting the recombination of chromosomes that occurs as a result of crossing-over.

division occurs, each daughter cell will obtain a complement of twenty-three chromosomes that are randomly derived from the maternal or paternal contribution to the homologous pairs of chromosomes of the parent cell.

In addition to this "shuffling of the deck" of chromosomes, exchanges of parts of homologous chromosomes can occur at prophase I. That is, pieces of one chromosome of a homologous pair can be exchanged with the other homologous chromosome in a process called *crossing-over* (fig. 3.34). These events together result in **genetic recombination** and ensure that the gametes produced by meiosis are genetically unique. This provides additional genetic diversity for organisms that reproduce sexually, and genetic diversity is needed to promote survival of species over evolutionary time.

Test Yourself Before You Continue

1. Draw a simple diagram of the semiconservative replication of DNA using stick figures and two colors.

2. Describe the cell cycle using the proper symbols to indicate the different stages of the cycle.

3. List the phases of mitosis and briefly describe the events that occur in each phase.

4. Distinguish between mitosis and meiosis in terms of their final result and their functional significance.

5. Summarize the events that occur during the two meiotic cell divisions and explain the mechanisms by which genetic recombination occurs during meiosis.

HPer Links of Basic Cell Concepts to the Body Systems

Nervous System

- Regeneration of neurons is regulated by several different chemicals . . . .(p. 157)
- Different forms (alleles) of a gene produce different forms of receptors for particular neurotransmitter chemicals . . . . .(p. 178)
- Microglia, located in the brain and spinal cord, are cells that transport themselves by amoeboid movement . . . .(p. 155)
- The insulating material around nerve fibers, called a myelin sheath, is derived from the cell membrane of certain cells in the nervous system . . . .(p. 156)
- Cytoplasmic transport processes are important for the movement of neurotransmitters and other substances within neurons . . . .(p. 153)

Endocrine System

- Many hormones act on their target cells by regulating gene expression . . . . . .(p. 292)
- Other hormones bind to receptor proteins located on the outer surface of the cell membrane of the target cells . . . . .(p. 294)
- The endoplasmic reticulum of some cells stores Ca^{2+}, which is released in response to hormone action . . . .(p. 296)
- Chemical regulators called prostaglandins are derived from a type of lipid associated with the cell membrane . . . .(p. 317)
- Liver and adipose cells store glycogen and triglycerides, respectively, which can be mobilized for energy needs by the action of particular hormones . . . .(p. 609)
- The sex of an individual is determined by the presence of a particular region of DNA in the Y chromosome . . . .(p. 635)

Muscular System

- Muscle cells have cytoplasmic proteins called actin and myosin that are needed for contraction . . . .(p. 330)
- The endoplasmic reticulum of skeletal muscle fibers stores Ca^{2+}, which is needed for muscle contraction . . . .(p. 336)

Circulatory System

- Blood cells are formed in the bone marrow . . . .(p. 370)
- Mature red blood cells lack nuclei and mitochondria . . . .(p. 368)
- The different white blood cells are distinguished by the shape of their nuclei and the presence of cytoplasmic granules . . . .(p. 369)

Immune System

- The carbohydrates outside the cell membrane of many bacteria help to target these cells for immune attack . . . .(p. 446)
- Some white blood cells and tissue macrophages destroy bacteria by phagocytosis . . . .(p. 446)
- When a B lymphocyte is stimulated by a foreign molecule (antigen), its endoplasmic reticulum becomes more developed and produces more antibody proteins (p. 453)
- Apoptosis is responsible for the destruction of T lymphocytes after an infection has been cleared . . . .(p. 462)

Respiratory System

- The air sacs (alveoli) of the lungs are composed of cells that are very thin, minimizing the separation between air and blood . . . .(p. 480)
- The epithelial cells lining the airways of the conducting zone have cilia that move mucus . . . .(p. 483)

Urinary System

- Parts of the renal tubules have microvilli that increase the rate of reabsorption . .(p. 526)
- Some regions of the renal tubules have water channels; these are produced by the Golgi complex and inserted by means of vesicles into the cell membrane . .(p. 536)

Digestive System

- The mucosa of the digestive tract has unicellular glands called goblet cells that secrete mucus . . . .(p. 566)
- The cells of the small intestine have microvilli that increase the rate of absorption . . . .(p. 570)
- The liver contains phagocytic cells . .(p. 575)

Reproductive System

- Males have an X and a Y chromosome, whereas females have two X chromosomes per diploid cell . .(p. 634)
- Gametes are produced by meiotic cell division . . . .(p. 634)
- Follicles degenerate (undergo atresia) in the ovaries by means of apoptosis (p. 656)
- Sperm cells are motile through the action of flagella . . . .(p. 650)
- The uterine tubes are lined with cilia that help to move the ovulated egg toward the uterus . . . .(p. 654)

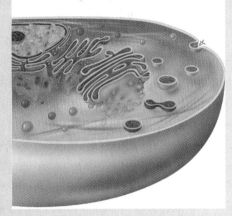

Summary

Plasma Membrane and Associated Structures 50

I. The structure of the cell (plasma) membrane is described by a fluid-mosaic model.

  A. The membrane is composed predominately of a double layer of phospholipids.

  B. The membrane also contains proteins, most of which span its entire width.

II. Some cells move by extending pseudopods; cilia and flagella protrude from the cell membrane of some specialized cells.

III. In the process of endocytosis, invaginations of the plasma membrane allow the cells to take up molecules from the external environment.

  A. In phagocytosis, the cell extends pseudopods that eventually fuse together to create a food vacuole; pinocytosis involves the formation of a narrow furrow in the membrane, which eventually fuses.

  B. Receptor-mediated endocytosis requires the interaction of a specific molecule in the extracellular environment with a specific receptor protein in the cell membrane.

  C. Exocytosis, the reverse of endocytosis, is a process that allows the cell to secrete its products.

Cytoplasm and Its Organelles 55

I. Microfilaments and microtubules produce a cytoskeleton that aids movements of organelles within a cell.

II. Lysosomes contain digestive enzymes and are responsible for the elimination of structures and molecules within the cell and for digestion of the contents of phagocytic food vacuoles.

III. Mitochondria serve as the major sites for energy production within the cell. They have an outer membrane with a smooth contour and an inner membrane with infoldings called cristae.

IV. Ribosomes are small protein factories composed of ribosomal RNA and protein arranged into two subunits.

V. The endoplasmic reticulum is a system of membranous tubules in the cell.

  A. The granular endoplasmic reticulum is covered with ribosomes and is involved in protein synthesis.

  B. The agranular endoplasmic reticulum provides a site for many enzymatic reactions and, in skeletal muscles, serves to store Ca^{2+}.

VI. The Golgi complex is a series of membranous sacs that receive products from the endoplasmic reticulum, modify those products, and release the products within vesicles.

Cell Nucleus and Gene Expression 61

I. The cell nucleus is surrounded by a double-layered nuclear envelope. At some points, the two layers are fused by nuclear pore complexes that allow for the passage of molecules.

II. Genetic expression occurs in two stages: transcription (RNA synthesis) and translation (protein synthesis).

  A. The DNA in the nucleus is combined with proteins to form the threadlike material known as chromatin.

  B. In chromatin, DNA is wound around regulatory proteins known as histones to form particles called nucleosomes.

  C. Chromatin that is active in directing RNA synthesis is euchromatin; the highly condensed, inactive chromatin is heterochromatin.

III. RNA is single-stranded. Four types are produced within the nucleus: ribosomal RNA, transfer RNA, precursor messenger RNA, and messenger RNA.

IV. Active euchromatin directs the synthesis of RNA in a process called transcription.

  A. The enzyme RNA polymerase causes separation of the two strands of DNA along the region of the DNA that constitutes a gene.

  B. One of the two separated strands of DNA serves as a template for the production of RNA. This

occurs by complementary base pairing between the DNA bases and ribonucleotide bases.

Protein Synthesis and Secretion 65

I. Messenger RNA leaves the nucleus and attaches to the ribosomes.

II. Each transfer RNA, with a specific base triplet in its anticodon, binds to a specific amino acid.

  A. As the mRNA moves through the ribosomes, complementary base pairing between tRNA anticodons and mRNA codons occurs.

  B. As each successive tRNA molecule binds to its complementary codon, the amino acid it carries is added to the end of a growing polypeptide chain.

III. Proteins destined for secretion are produced in ribosomes located on the granular endoplasmic reticulum and enter the cisternae of this organelle.

IV. Secretory proteins move from the granular endoplasmic reticulum to the Golgi complex.

  A. The Golgi complex modifies the proteins it contains, separates different proteins, and packages them in vesicles.

  B. Secretory vesicles from the Golgi complex fuse with the plasma membrane and release their products by exocytosis.

DNA Synthesis and Cell Division 69

I. Replication of DNA is semiconservative; each DNA strand serves as a template for the production of a new strand.

  A. The strands of the original DNA molecule gradually separate along their entire length and, through complementary base pairing, form a new complementary strand.

  B. In this way, each DNA molecule consists of one old and one new strand.

II. During the G_1 phase of the cell cycle, the DNA directs the synthesis of RNA, and hence that of proteins.

III. During the S phase of the cycle, DNA directs the synthesis of new DNA and replicates itself.

IV. After a brief time gap (G_2), the cell begins mitosis (the M stage of the cycle).

    A. Mitosis consists of the following phases: interphase, prophase, metaphase, anaphase, and telophase.

    B. In mitosis, the homologous chromosomes line up single file and are pulled by spindle fibers to opposite poles.

    C. This results in the production of two daughter cells, each containing forty-six chromosomes, just like the parent cell.

V. Meiosis is a special type of cell division that results in the production of gametes in the gonads.

    A. The homologous chromosomes line up side by side, so that only one of each pair is pulled to each pole.

    B. This results in the production of two daughter cells, each containing only twenty-three chromosomes, which are duplicated.

    C. The duplicate chromatids are separated into two new daughter cells during the second meiotic cell division.

Review Activities

Test Your Knowledge of Terms and Facts

1. According to the fluid-mosaic model of the plasma membrane

    a. protein and phospholipids form a regular, repeating structure.

    b. the membrane is a rigid structure.

    c. phospholipids form a double layer, with the polar parts facing each other.

    d. proteins are free to move within a double layer of phospholipids.

2. After the DNA molecule has replicated itself, the duplicate strands are called

    a. homologous chromosomes.

    b. chromatids.

    c. centromeres.

    d. spindle fibers.

3. Nerve and skeletal muscle cells in the adult, which do not divide, remain in the

    a. G_1 phase.

    b. S phase.

    c. G_2 phase.

    d. M phase.

4. The phase of mitosis in which the chromosomes line up at the equator of the cell is called

    a. interphase.

    b. prophase.

    c. metaphase.

    d. anaphase.

    e. telophase.

5. The phase of mitosis in which the chromatids separate is called

    a. interphase.

    b. prophase.

    c. metaphase.

    d. anaphase.

    e. telophase.

6. Chemical modifications of histone proteins are believed to directly influence

    a. genetic transcription.

    b. genetic translation.

    c. both transcription and translation.

    d. posttranslational changes in the newly synthesized proteins.

7. Which of these statements about RNA is *true?*

    a. It is made in the nucleus.

    b. It is double-stranded.

    c. It contains the sugar deoxyribose.

    d. It is a complementary copy of the entire DNA molecule.

8. Which of these statements about mRNA is *false?*

    a. It is produced as a larger pre-mRNA.

    b. It forms associations with ribosomes.

    c. Its base triplets are called anticodons.

    d. It codes for the synthesis of specific proteins.

9. The organelle that combines proteins with carbohydrates and packages them within vesicles for secretion is

    a. the Golgi complex.

    b. the granular endoplasmic reticulum.

    c. the agranular endoplasmic reticulum.

    d. the ribosome.

10. The organelle that contains digestive enzymes is

    a. the mitochondrion.

    b. the lysosome.

    c. the endoplasmic reticulum.

    d. the Golgi complex.

11. Which of these descriptions of rRNA is *true?*

    a. It is single-stranded.

    b. It catalyzes steps in protein synthesis.

    c. It forms part of the structure of both subunits of a ribosome.

    d. It is produced in the nucleolus.

    e. All of these are true.

12. Which of these statements about tRNA is *true?*

    a. It is made in the nucleus.

    b. It is looped back on itself.

    c. It contains the anticodon.

    d. There are over twenty different types.

    e. All of these are true.

13. The step in protein synthesis during which tRNA, rRNA, and mRNA are all active is known as

    a. transcription.

    b. translation.

    c. replication.

    d. RNA polymerization.

14. The anticodons are located in

    a. tRNA.

    b. rRNA.

    c. mRNA.

    d. ribosomes.

    e. endoplasmic reticulum.

Test Your Understanding of Concepts and Principles

1. Give some specific examples that illustrate the dynamic nature of the plasma membrane.[1]

2. Describe the structure of nucleosomes, and explain the role of histone proteins in chromatin structure and function.

3. What is the genetic code, and how does it affect the structure and function of the body?

4. Why may tRNA be considered the "interpreter" of the genetic code?

5. Compare the processing of cellular proteins with that of proteins secreted by a cell.

6. Explain the interrelationship between the endoplasmic reticulum and the Golgi complex. What becomes of vesicles released from the Golgi complex?

7. Explain the functions of centrioles in nondividing and dividing cells.

8. Describe the phases of the cell cycle and explain how this cycle may be regulated.

9. Distinguish between oncogenes and tumor suppressor genes and give examples of how such genes may function.

10. Define *apoptosis* and explain the physiological significance of this process.

Test Your Ability to Analyze and Apply Your Knowledge

1. Discuss the role of chromatin proteins in regulating gene expression. How does the three-dimensional structure of the chromatin affect genetic regulation? How do hormones influence genetic regulation?

2. Explain how p53 functions as a tumor suppressor gene. How can mutations in p53 lead to cancer, and how might gene therapy or other drug interventions inhibit the growth of a tumor?

3. Release of lysosomal enzymes from white blood cells during a local immune attack can contribute to the symptoms of inflammation. Suppose, to alleviate inflammation, you develop a drug that destroys all lysosomes. Would this drug have negative side effects? Explain.

4. Antibiotics can have different mechanisms of action. An antibiotic called puromycin blocks genetic translation. One called actinomycin D blocks genetic transcription. These drugs can be used to determine how regulatory molecules, such as hormones, work. For example, if a hormone's effects on a tissue were blocked immediately by puromycin but not by actinomycin D, what would that tell you about the mechanism of action of the hormone?

Related Websites

Check out the Links Library at www.mhhe.com/fox8 for links to sites containing resources related to cell structure and genetic control. These links are monitored to ensure current URLs.

---

[1]*Note:* This question is answered in the chapter 3 Study Guide found on the Online Learning Center at www.mhhe.com/fox8.

4 Enzymes and Energy

Objectives

After studying this chapter, you should be able to . . .

1. state the principles of catalysis and explain how enzymes function as catalysts.

2. explain how the names of enzymes are derived and comment on the significance of isoenzymes.

3. describe the effects of pH and temperature on the rate of enzyme-catalyzed reactions and explain how these effects are produced.

4. describe the roles of cofactors and coenzymes in enzymatic reactions.

5. explain how the law of mass action helps to account for the direction of reversible reactions.

6. explain how enzymes work together to produce a metabolic pathway and how this pathway may be affected by end-product inhibition and inborn errors of metabolism.

7. explain how the first and second laws of thermodynamics can be used to predict whether metabolic reactions will be endergonic or exergonic.

8. describe the production of ATP and explain the significance of ATP as the universal energy carrier.

9. define the terms *oxidation, reduction, oxidizing agent,* and *reducing agent.*

10. describe the use of NAD and FAD in oxidation-reduction reactions and explain the functional significance of these two molecules.

Chapter at a Glance

Chapter Four

Clinical Investigation

Tom is a 77-year-old man who was brought to the hospital because of severe chest pain. He also complained that he had difficulty urinating and that he "got the runs" when he ate ice cream.

Laboratory tests were performed and demonstrated an abnormally high plasma concentration of the MB isoform of creatine phosphokinase. The tests also demonstrated a high blood level of acid phosphatase. What might be responsible for Tom's symptoms?

Enzymes as Catalysts

Enzymes are biological catalysts that increase the rate of chemical reactions. Most enzymes are proteins, and their catalytic action results from their complex structure. The great diversity of protein structure allows different enzymes to be specialized in their action.

The ability of yeast cells to make alcohol from glucose (a process called *fermentation*) had been known since antiquity, yet even as late as the mid-nineteenth century no scientist had been able to duplicate this process in the absence of living yeast. Also, a vast array of chemical reactions occurred in yeast and other living cells at body temperature that could not be duplicated in the chemistry laboratory without adding substantial amounts of heat energy. These observations led many mid-nineteenth-century scientists to believe that chemical reactions in living cells were aided by a "vital force" that operated beyond the laws of the physical world. This *vitalist concept* was squashed along with the yeast cells when a pioneering biochemist, Eduard Buchner, demonstrated that juice obtained from yeast could ferment glucose to alcohol. The yeast juice was not alive—evidently some chemicals in the cells were responsible for fermentation. Buchner didn't know what these chemicals were, so he simply named them **enzymes** (Greek for "in yeast").

Chemically, enzymes are a subclass of proteins. The only known exceptions are the few special cases in which RNA demonstrates enzymatic activity; in these cases they are called *ribozymes.* Ribozymes function as enzymes in reactions involving remodeling of the RNA molecules themselves, and in the formation of a growing polypeptide in ribosomes.

Functionally, enzymes (and ribozymes) are biological **catalysts.** A catalyst is a chemical that (1) increases the rate of a reaction, (2) is not itself changed at the end of the reaction, and (3) does not change the nature of the reaction or its final result. The same reaction would have occurred to the same degree in the absence of the catalyst, but it would have progressed at a much slower rate.

In order for a given reaction to occur, the reactants must have sufficient energy. The amount of energy required for a reaction to proceed is called the **activation energy.** By analogy, a match will not burn and release heat energy unless it is first "activated" by striking the match or by placing it in a flame.

In a large population of molecules, only a small fraction will possess sufficient energy for a reaction. Adding heat will raise the energy level of all the reactant molecules, thus increasing the percentage of the population that has the activation energy. Heat makes reactions go faster, but it also produces undesirable side effects in cells. Catalysts make reactions go faster at lower temperatures by lowering the activation energy required, thus ensuring that a larger percentage of the population of reactant molecules will have sufficient energy to participate in the reaction (fig. 4.1).

Since a small fraction of the reactants will have the activation energy required for a reaction even in the absence of a catalyst, the reaction could theoretically occur spontaneously at a slow rate. This rate, however, would be much too slow for the needs of a cell. So, from a biological standpoint, the presence or absence of a specific enzyme catalyst acts as a switch—the reaction will occur if the enzyme is present and will not occur if the enzyme is absent.

Mechanism of Enzyme Action

The ability of enzymes to lower the activation energy of a reaction is a result of their structure. Enzymes are large proteins with complex, highly ordered, three-dimensional shapes produced by physical and chemical interactions between their amino acid subunits. Each type of enzyme has a characteristic three-dimensional shape, or *conformation,* with ridges, grooves, and pockets lined with specific amino acids. The particular pockets that are active in catalyzing a reaction are called the *active sites* of the enzyme.

The reactant molecules, which are called the **substrates** of the enzyme, have specific shapes that allow them to fit into the active sites. The enzyme can thus be thought of as a lock into which only a specifically shaped key—the substrate—can fit. This **lock-and-key model** of enzyme activity is illustrated in figure 4.2.

In some cases, the fit between an enzyme and its substrate may not be perfect at first. A perfect fit may be induced, however, as the substrate gradually slips into the active site. This induced fit, together with temporary bonds that form between the substrate and the amino acids lining the active sites of the enzyme, weaken the existing bonds within the substrate molecules and allows them to be more easily broken. New bonds are more easily formed as substrates are brought close together in the proper orientation. This model of enzyme activity, in which the enzyme undergoes a slight structural change to better fit the substrate, is called the **induced-fit model.** The *enzyme-substrate complex,* formed temporarily in the course of the reaction, then dissociates to yield *products* and the free unaltered enzyme.

Since enzymes are very specific as to their substrates and activity, the concentration of a specific enzyme in a sample of fluid can be measured relatively easily. This is usually done by measuring the rate of conversion of the enzyme's substrates into products under specified conditions. The presence of an enzyme in a sample can thus be detected by the job it does, and its concentration can be measured by how rapidly it performs its job.

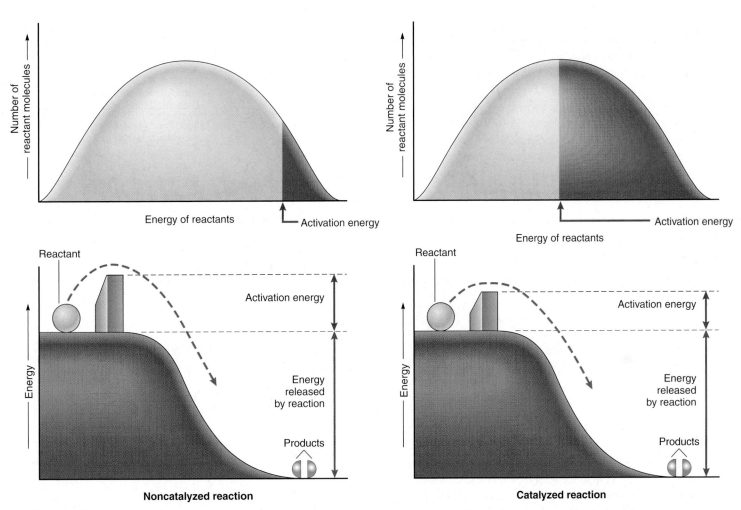

Figure 4.1 **A comparison of noncatalyzed and catalyzed reactions.** The upper figures compare the proportion of reactant molecules that have sufficient activation energy to participate in the reaction (blue = insufficient energy; green = sufficient energy). This proportion is increased in the enzyme-catalyzed reaction because enzymes lower the activation energy required for the reaction (shown as a barrier on top of an energy "hill" in the lower figures). Reactants that can overcome this barrier are able to participate in the reaction, as shown by arrows pointing to the bottom of the energy hill.

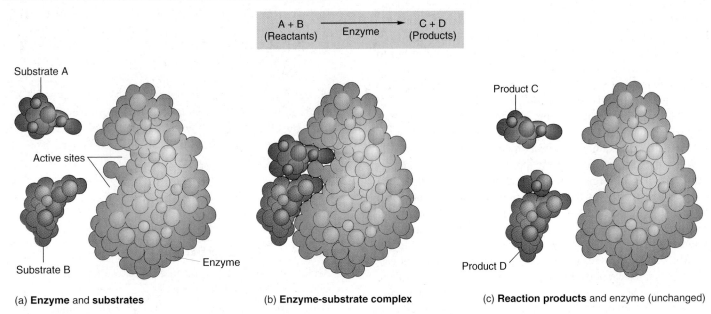

Figure 4.2 **The lock-and-key model of enzyme action.** (a) Substrates A and B fit into active sites in the enzyme, forming an enzyme-substrate complex. (b) This complex then dissociates (c), releasing the products of the reaction and the free enzyme.

When tissues become damaged as a result of diseases, some of the dead cells disintegrate and release their enzymes into the blood. Most of these enzymes are not normally active in the blood for lack of their specific substrates, but their enzymatic activity can be measured in a test tube by the addition of the appropriate substrates to samples of plasma. Such measurements are clinically useful because abnormally high plasma concentrations of particular enzymes are characteristic of certain diseases (table 4.1).

Clinical Investigation Clues

Remember that Tom had elevated blood levels of acid phosphatase and creatine phosphokinase.

How might these laboratory results help to explain his difficulty in urination?

What two different conditions might cause the elevated creatine phosphokinase?

Naming of Enzymes

In the past, enzymes were given names that were somewhat arbitrary. The modern system for naming enzymes, established by an international committee, is more orderly and informative. With the exception of some older enzyme names (such as pepsin, trypsin, and renin), all enzyme names end with the suffix *-ase* (table 4.2), and classes of enzymes are named according to their activity, or "job category." *Hydrolases,* for example, promote hydrolysis reactions. Other enzyme categories include *phosphatases,* which catalyze the removal of phosphate groups; *synthases* and *synthetases,* which catalyze dehydration synthesis reactions; *dehydrogenases,* which remove hydrogen atoms from their substrates; and *kinases,* which add a phosphate group to (phosphorylate) particular molecules. Enzymes called *isomerases* rearrange atoms within their substrate molecules to form structural isomers, such as glucose and fructose.

The names of many enzymes specify both the substrate of the enzyme and the job category of the enzyme. Lactic acid dehydrogenase, for example, removes hydrogens from lactic acid. Enzymes that do exactly the same job (that catalyze the same reaction) in different organs have the same name, since the name describes the activity of the enzyme. Different organs, however, may make slightly different "models" of the enzyme that differ in one or a few amino acids. These different models of the same enzyme are called **isoenzymes.** The differences in structure do not affect the active sites (otherwise the enzymes would not catalyze the same reaction), but they do alter the structure of the enzymes at other locations, so that the different isoenzymatic forms can be separated by standard biochemical procedures. These techniques are useful in the diagnosis of diseases.

Table 4.1   Examples of the Diagnostic Value of Some Enzymes Found in Plasma

Enzyme	Diseases Associated with Abnormal Plasma Enzyme Concentrations
Alkaline phosphatase	Obstructive jaundice, Paget's disease (osteitis deformans), carcinoma of bone
Acid phosphatase	Benign hypertrophy of prostate, cancer of prostate
Amylase	Pancreatitis, perforated peptic ulcer
Aldolase	Muscular dystrophy
Creatine kinase (or creatine phosphokinase-CPK)	Muscular dystrophy, myocardial infarction
Lactate dehydrogenase (LDH)	Myocardial infarction, liver disease, renal disease, pernicious anemia
Transaminases (AST and ALT)	Myocardial infarction, hepatitis, muscular dystrophy

Table 4.2   Selected Enzymes and the Reactions They Catalyze

Enzyme	Reaction Catalyzed
Catalase	$2 H_2O_2 \rightarrow 2 H_2O + O_2$
Carbonic anhydrase	$H_2CO_3 \rightarrow H_2O + CO_2$
Amylase	starch $+ H_2O \rightarrow$ maltose
Lactate dehydrogenase	lactic acid $\rightarrow$ pyruvic acid $+ H_2$
Ribonuclease	RNA $+ H_2O \rightarrow$ ribonucleotides

Different organs, when they are diseased, may liberate different isoenzymatic forms of an enzyme that can be measured in a clinical laboratory. For example, the enzyme **creatine phosphokinase,** abbreviated either **CPK** or **CK,** exists in three isoenzymatic forms. These forms are identified by two letters that indicate two components of this enzyme. One form is identified as MM and is liberated from diseased skeletal muscle; the second is BB, released by a damaged brain; and the third is MB, released from a diseased heart. Clinical tests utilizing antibodies that can bind to the M and B components are now available to specifically measure the level of the MB form in the blood when heart disease is suspected.

Clinical Investigation Clues

Remember that Tom had elevated blood levels of the MB isoform of creatine phosphokinase.

What condition might produce this, and explain his chest pain?

Control of Enzyme Activity

The rate of an enzyme-catalyzed reaction depends on numerous factors, including the concentration of the enzyme and the pH and temperature of the solution. Genetic control of enzyme concentration, for example, affects the rate of progress along particular metabolic pathways and thus regulates cellular metabolism.

The activity of an enzyme, as measured by the rate at which its substrates are converted to products, is influenced by such factors as (1) the temperature and pH of the solution; (2) the concentration of cofactors and coenzymes, which are needed by many enzymes as "helpers" for their catalytic activity; (3) the concentration of enzyme and substrate molecules in the solution; and (4) the stimulatory and inhibitory effects of some products of enzyme action on the activity of the enzymes that helped to form these products.

Effects of Temperature and pH

An increase in temperature will increase the rate of non-enzyme-catalyzed reactions. A similar relationship between temperature and reaction rate occurs in enzyme-catalyzed reactions. At a temperature of 0° C the reaction rate is immeasurably slow. As the temperature is raised above 0° C the reaction rate increases, but only up to a point. At a few degrees above body temperature (which is 37° C) the reaction rate reaches a plateau; further increases in temperature actually *decrease* the rate of the reaction (fig. 4.3). This decrease is due to the fact that the tertiary structure of enzymes becomes altered at higher temperatures.

A similar relationship is observed when the rate of an enzymatic reaction is measured at different pH values. Each enzyme characteristically exhibits peak activity in a very narrow pH range, which is the **pH optimum** for the enzyme. If the pH is changed so that it is no longer within the enzyme's optimum range, the reaction rate will decrease (fig. 4.4). This decreased enzyme activity is due to changes in the conformation of the enzyme and in the charges of the R groups of the amino acids lining the active sites.

The pH optimum of an enzyme usually reflects the pH of the body fluid in which the enzyme is found. The acidic pH optimum of the protein-digesting enzyme *pepsin,* for example, allows it to be active in the strong hydrochloric acid of gastric juice. Similarly, the neutral pH optimum of *salivary amylase* and the alkaline pH optimum of *trypsin* in pancreatic juice allow these enzymes to digest starch and protein, respectively, in other parts of the digestive tract.

Although the pH of other body fluids shows less variation than that of the fluids of the digestive tract, the pH optima of different enzymes found throughout the body do show significant differences (table 4.3). Some of these differences can be exploited for diagnostic purposes. Disease of the prostate, for example, may be associated with elevated blood levels of a prostatic phosphatase with an acidic pH optimum (descriptively called *acid phosphatase*). Bone disease, on the other hand, may be associated with elevated blood levels of *alkaline phosphatase,* which has a higher pH optimum than the similar enzyme released from the diseased prostate.

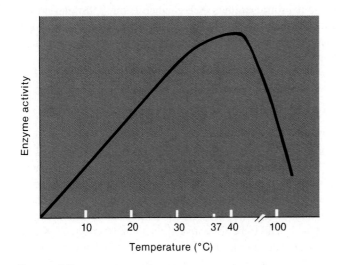

Figure 4.3   The effect of temperature on enzyme activity. This effect is measured by the rate of the enzyme-catalyzed reaction under standardized conditions as the temperature of the reaction is varied.

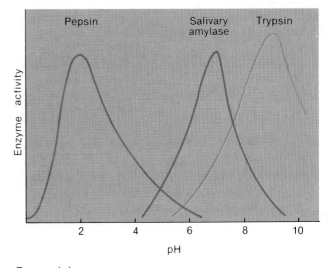

Figure 4.4   The effect of pH on the activity of three digestive enzymes. Salivary amylase is found in saliva, which has a pH close to neutral; pepsin is found in acidic gastric juice, and trypsin is found in alkaline pancreatic juice.

Table 4.3 pH Optima of Selected Enzymes

Enzyme	Reaction Catalyzed	pH Optimum
Pepsin (stomach)	Digestion of protein	2.0
Acid phosphatase (prostate)	Removal of phosphate group	5.5
Salivary amylase (saliva)	Digestion of starch	6.8
Lipase (pancreatic juice)	Digestion of fat	7.0
Alkaline phosphatase (bone)	Removal of phosphate group	9.0
Trypsin (pancreatic juice)	Digestion of protein	9.5
Monoamine oxidase (nerve endings)	Removal of amine group from norepinephrine	9.8

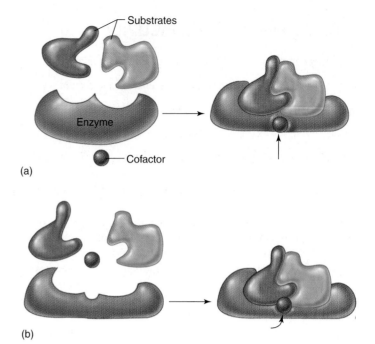

(a)

(b)

■ **Figure 4.5** The roles of cofactors in enzyme function. In (*a*) the cofactor changes the conformation of the active site, allowing for a better fit between the enzyme and its substrates. In (*b*) the cofactor participates in the temporary bonding between the active site and the substrates.

Cofactors and Coenzymes

Many enzymes are completely inactive when isolated in a pure state. Evidently some of the ions and smaller organic molecules that are removed in the purification procedure play an essential role in enzyme activity. These ions and smaller organic molecules needed for the activity of specific enzymes are called *cofactors* and *coenzymes*.

Cofactors include metal ions such as Ca^{2+}, Mg^{2+}, Mn^{2+}, Cu^{2+}, Zn^{2+}, and selenium. Some enzymes with a cofactor requirement do not have a properly shaped active site in the absence of the cofactor. In these enzymes, the attachment of cofactors causes a conformational change in the protein that allows it to combine with its substrate. The cofactors of other enzymes participate in the temporary bonds between the enzyme and its substrate when the enzyme-substrate complex is formed (fig. 4.5).

Coenzymes are organic molecules, derived from water-soluble vitamins such as niacin and riboflavin, that are needed for the function of particular enzymes. Coenzymes participate in enzyme-catalyzed reactions by transporting hydrogen atoms and small molecules from one enzyme to another. Examples of the actions of cofactors and coenzymes in specific reactions will be given in the context of their roles in cellular metabolism later in this chapter.

Enzyme Activation

There are a number of important cases in which enzymes are produced as inactive forms. In the cells of the pancreas, for example, many digestive enzymes are produced as inactive *zymogens,* which are activated after they are secreted into the intestine. Activation of zymogens in the intestinal lumen (cavity) protects the pancreatic cells from self-digestion.

In liver cells, as another example, the enzyme that catalyzes the hydrolysis of stored glycogen is inactive when it is produced, and must later be activated by the addition of a phosphate group. A different enzyme, called a *protein kinase,* catalyzes the addition of the phosphate group to that enzyme. At a later time, enzyme inactivation is achieved by another enzyme that catalyzes the removal of the phosphate group. The activation/inactivation of this enzyme (and many others) is thus achieved by the processes of *phosphorylation/dephosphorylation.*

Going back a step, the protein kinase itself may be produced as an inactive enzyme. In this case, activation of the protein kinase requires that it bind to a particular **ligand** (smaller molecule). Such ligands serve as intracellular regulators that are called **second messengers.** In many cases, this ligand is a molecule called *cyclic AMP (cAMP).* Cyclic AMP activates the protein kinase by promoting the dissociation of an inhibitory subunit from the active enzyme. Since the production of cyclic AMP within cells is stimulated by regulatory molecules that include neurotransmitters (see chapter 7, fig. 7.28) and hormones (see chapter 11, fig. 11.8), this topic will be discussed more completely in the context of neural and endocrine regulation.

Substrate Concentration and Reversible Reactions

At a given level of enzyme concentration, the rate of product formation will increase as the substrate concentration increases. Eventually, however, a point will be reached where additional increases in substrate concentration do not result in comparable increases in reaction rate. When the relationship between substrate concentration and reaction rate reaches a plateau of maximum velocity, the enzyme is said to be *saturated.* If we think of enzymes as workers in a plant that converts a raw material (say, metal ore) into a product (say, iron), then enzyme saturation is like the plant working at full capacity, with no idle time for the workers. Increasing the amount of raw material (substrate) at this point cannot increase the rate of product formation. This concept is illustrated in figure 4.6.

Some enzymatic reactions within a cell are reversible, with both the forward and backward reactions catalyzed by the

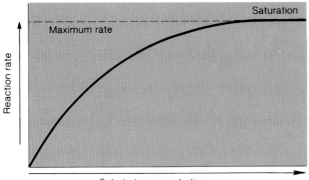

Figure 4.6 The effect of substrate concentration on the rate of an enzyme-catalyzed reaction. When the reaction rate is at a maximum, the enzyme is said to be saturated.

same enzyme. The enzyme *carbonic anhydrase,* for example, is named because it can catalyze the following reaction:

$$H_2CO_3 \rightarrow H_2O + CO_2$$

The same enzyme, however, can also catalyze the reverse reaction:

$$H_2O + CO_2 \rightarrow H_2CO_3$$

The two reactions can be more conveniently illustrated by a single equation with double arrows:

$$H_2O + CO_2 \rightleftarrows H_2CO_3$$

The direction of the reversible reaction depends, in part, on the relative concentrations of the molecules to the left and right of the arrows. If the concentration of CO_2 is very high (as it is in the tissues), the reaction will be driven to the right. If the concentration of CO_2 is low and that of H_2CO_3 is high (as it is in the lungs), the reaction will be driven to the left. The principle that reversible reactions will be driven from the side of the equation where the concentration is higher to the side where the concentration is lower is known as the **law of mass action.**

Although some enzymatic reactions are not directly reversible, the net effects of the reactions can be reversed by the action of different enzymes. Some of the enzymes that convert glucose to pyruvic acid, for example, are different from those that reverse the pathway and produce glucose from pyruvic acid. Likewise, the formation and breakdown of glycogen (a polymer of glucose) are catalyzed by different enzymes.

Metabolic Pathways

The many thousands of different types of enzymatic reactions within a cell do not occur independently of each other. They are, rather, all linked together by intricate webs of interrelationships, the total pattern of which constitutes cellular metabolism. A sequence of enzymatic reactions that begins with an *initial substrate,* progresses through a number of *intermediates,* and ends with a *final product* is known as a **metabolic pathway.**

The enzymes in a metabolic pathway cooperate in a manner analogous to workers on an assembly line, where each contributes a small part to the final product. In this process, the

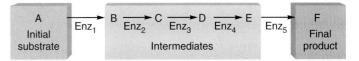

Figure 4.7 The general pattern of a metabolic pathway. In metabolic pathways, the product of one enzyme becomes the substrate of the next.

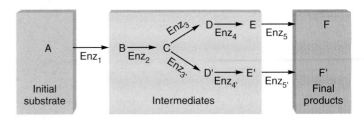

Figure 4.8 A branched metabolic pathway. Two or more different enzymes can work on the same substrate at the branch point of the pathway, catalyzing two or more different reactions.

product of one enzyme in the line becomes the substrate of the next enzyme, and so on (fig. 4.7).

Few metabolic pathways are completely linear. Most are branched so that one intermediate at the branch point can serve as a substrate for two different enzymes. Two different products can thus be formed that serve as intermediates of two pathways (fig. 4.8).

End-Product Inhibition

The activities of enzymes at the branch points of metabolic pathways are often regulated by a process called **end-product inhibition,** which is a form of negative feedback inhibition. In this process, one of the final products of a divergent pathway inhibits the activity of the branch-point enzyme that began the path toward the production of this inhibitor. This inhibition prevents that final product from accumulating excessively and results in a shift toward the final product of the alternate pathway (fig. 4.9).

The mechanism by which a final product inhibits an earlier enzymatic step in its pathway is known as **allosteric inhibition.** The allosteric inhibitor combines with a part of the enzyme at a location other than the active site. This causes the active site to change shape so that it can no longer combine properly with its substrate.

Inborn Errors of Metabolism

Since each different polypeptide in the body is coded by a different gene (chapter 3), each enzyme protein that participates in a metabolic pathway is coded by a different gene. An inherited defect in one of these genes may result in a disease known as an **inborn error of metabolism.** In this type of disease, the quantity of intermediates formed *prior* to the defective enzymatic step *increases,* and the quantity of intermediates and final products formed *after* the defective step *decreases.* Diseases may result from deficiencies of the normal end product or from excessive accumulation of intermediates formed prior to the defective step. If the defective enzyme is active at a step that follows a

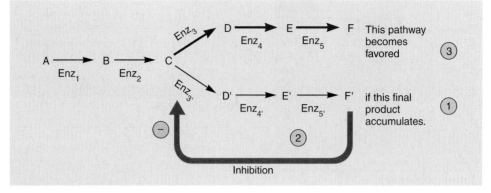

■ **Figure 4.9**   End-product inhibition in a branched metabolic pathway. Inhibition is shown by the arrow in step 2.

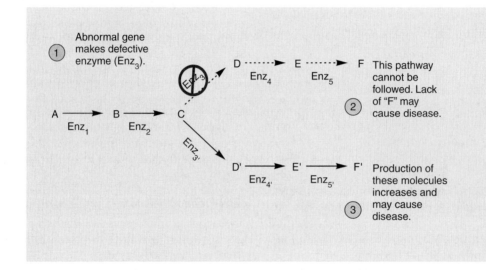

■ **Figure 4.10**   The effects of an inborn error of metabolism on a branched metabolic pathway. The defective gene produces a defective enzyme, indicated here by a line through its symbol.

branch point in a pathway, the intermediates and final products of the alternate pathway will increase (fig. 4.10). An abnormal increase in the production of these products can be the cause of some metabolic diseases.

One of the conversion products of phenylalanine is a molecule called DOPA, an acronym for dihydroxyphenylalanine. DOPA is a precursor of the pigment molecule *melanin,* which gives skin, eyes, and hair their normal coloration. The condition of *albinism* results from an inherited defect in the enzyme that catalyzes the formation of melanin from DOPA (fig. 4.11). Besides PKU and albinism, there are many other inborn errors of amino acid metabolism, as well as errors in carbohydrate and lipid metabolism. Some of these are described in table 4.4.

The branched metabolic pathway that begins with phenylalanine as the initial substrate is subject to a number of inborn errors of metabolism (fig. 4.11). When the enzyme that converts this amino acid to the amino acid tyrosine is defective, the final product of a divergent pathway accumulates and can be detected in the blood and urine. This disease—**phenylketonuria (PKU)**—can result in severe mental retardation and a shortened life span. PKU occurs often enough (although no inborn error of metabolism is common) to warrant the testing of all newborn babies for the defect. If the disease is detected early, brain damage can be prevented by placing the child on an artificial diet low in the amino acid phenylalanine.

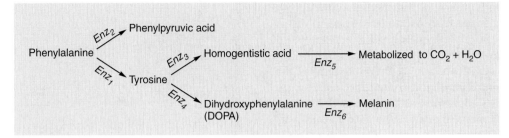

■ **Figure 4.11** Metabolic pathways for the degradation of the amino acid phenylalanine. Defective enzyme₁ produces phenylketonuria (PKU), defective enzyme₅ produces alcaptonuria (not a clinically significant condition), and defective enzyme₆ produces albinism.

Table 4.4 Examples of Inborn Errors in the Metabolism of Amino Acids, Carbohydrates, and Lipids

Metabolic Defect	Disease	Abnormality	Clinical Result
Amino acid metabolism	Phenylketonuria (PKU)	Increase in phenylpyruvic acid	Mental retardation, epilepsy
	Albinism	Lack of melanin	Susceptibility to skin cancer
	Maple-syrup disease	Increase in leucine, isoleucine, and valine	Degeneration of brain, early death
	Homocystinuria	Accumulation of homocystine	Mental retardation, eye problems
Carbohydrate metabolism	Lactose intolerance	Lactose not utilized	Diarrhea
	Glucose 6-phosphatase deficiency (Gierke's disease)	Accumulation of glycogen in liver	Liver enlargement, hypoglycemia
	Glycogen phosphorylase deficiency	Accumulation of glycogen in muscle	Muscle fatigue and pain
Lipid metabolism	Gaucher's disease	Lipid accumulation (glucocerebroside)	Liver and spleen enlargement, brain degeneration
	Tay-Sachs disease	Lipid accumulation (ganglioside G_{M2})	Brain degeneration, death by age 5
	Hypercholestremia	High blood cholesterol	Atherosclerosis of coronary and large arteries

Test Yourself Before You Continue

1. Draw graphs to represent the effects of changes in temperature, pH, and enzyme and substrate concentration on the rate of enzymatic reactions. Explain the mechanisms responsible for the effects you have graphed.

2. Using arrows and letters of the alphabet, draw a flowchart of a metabolic pathway with one branch point.

3. Describe a reversible reaction and explain how the law of mass action affects this reaction.

4. Define *end-product inhibition* and use your diagram of a branched metabolic pathway to explain how this process will affect the concentrations of different intermediates.

5. Because of an inborn error of metabolism, suppose that the enzyme that catalyzed the third reaction in your pathway (question no. 2) was defective. Describe the effects this would have on the concentrations of the intermediates in your pathway.

Bioenergetics

Living organisms require the constant expenditure of energy to maintain their complex structures and processes. Central to life processes are chemical reactions that are coupled, so that the energy released by one reaction is incorporated into the products of another reaction.

Bioenergetics refers to the flow of energy in living systems. Organisms maintain their highly ordered structure and life-sustaining activities through the constant expenditure of energy obtained ultimately from the environment. The energy flow in living systems obeys the first and second laws of a branch of physics known as *thermodynamics*.

According to the **first law of thermodynamics,** energy can be transformed (changed from one form to another), but it can neither be created nor destroyed. This is sometimes called the *law of conservation of energy.* As a result of energy transformations, according to the **second law of thermodynamics,** the universe and its parts (including living systems) become increasingly disorganized. The term *entropy* is used to describe the degree of disorganization of a system. Energy transformations thus increase the amount of entropy of a system. Only energy that is in an organized state—called *free energy*—can be used to do work. Since entropy increases in every energy transformation, the amount of free energy available to do work decreases. As a result of the increased entropy described by the second law, systems tend to go from states of higher free energy to states of lower free energy.

The chemical bonding of atoms into molecules obeys the laws of thermodynamics. A complex organic molecule such as glucose has more free energy (less entropy) than six separate molecules each of carbon dioxide and water. Therefore, in order to convert carbon dioxide and water to glucose, energy must be added. Plants perform this feat using energy from the sun in the process of *photosynthesis* (fig. 4.12).

Endergonic and Exergonic Reactions

Chemical reactions that require an input of energy are known as **endergonic reactions.** Since energy is added to make these reactions "go," the products of endergonic reactions must contain more free energy than the reactants. A portion of the energy added, in other words, is contained within the product molecules. This follows from the fact that energy cannot be created or destroyed (first law of thermodynamics) and from the fact that a more-organized state of matter contains more free energy, or less entropy, than a less-organized state (second law of thermodynamics).

The fact that glucose contains more free energy than carbon dioxide and water can easily be proven by combusting glucose to CO_2 and H_2O. This reaction releases energy in the form of heat. Reactions that convert molecules with more free energy to molecules with less—and, therefore, that release energy as they proceed—are called **exergonic reactions.**

As illustrated in figure 4.13, the total amount of energy released by a molecule in a combustion reaction can be released in smaller portions, by enzymatically controlled exergonic reactions within cells. This allows the cells to use the energy to "drive" other processes, as described in the next section. Since the energy obtained by the body from the cellular oxidation of a molecule is the same as the amount released when the molecule is combusted, the energy in food molecules can conveniently be measured by the heat released when the molecules are combusted.

Heat is measured in units called *calories.* One calorie is defined as the amount of heat required to raise the temperature of one cubic centimeter of water one degree on the Celsius scale. The caloric value of food is usually indicated in *kilocalories* (one kilocalorie = 1,000 calories), which are often called large calories and spelled with a capital C.

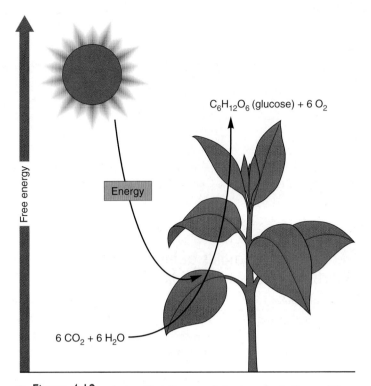

■ **Figure 4.12**   **A simplified diagram of photosynthesis.** Some of the sun's radiant energy is captured by plants and used to produce glucose from carbon dioxide and water. As the product of this endergonic reaction, glucose has more free energy than the initial reactants.

Coupled Reactions: ATP

In order to remain alive, a cell must maintain its highly organized, low-entropy state at the expense of free energy in its environment. Accordingly, the cell contains many enzymes that catalyze exergonic reactions using substrates that come ultimately from the environment. The energy released by these exergonic reactions is used to drive the energy-requiring processes (endergonic reactions) in the cell. Since cells cannot use heat energy to drive energy-requiring processes, the chemical-bond energy that is released in exergonic reactions must be directly transferred to chemical-bond energy in the products of endergonic reactions. Energy-liberating reactions are thus *coupled* to energy-requiring reactions. This relationship is like that of two meshed gears; the turning of one (the energy-releasing exergonic gear) causes turning of the other (the energy-requiring endergonic gear). This relationship is illustrated in figure 4.14.

The energy released by most exergonic reactions in the cell is used, either directly or indirectly, to drive one particular endergonic reaction (fig. 4.15): the formation of **adenosine triphosphate (ATP)** from adenosine diphosphate (ADP) and inorganic phosphate (abbreviated P_i).

The formation of ATP requires the input of a fairly large amount of energy. Since this energy must be conserved (first law of thermodynamics), the bond produced by joining P_i to ADP must contain a part of this energy. Thus, when enzymes reverse this reaction and convert ATP to ADP and P_i, a large amount of energy is released. Energy released from the breakdown of ATP

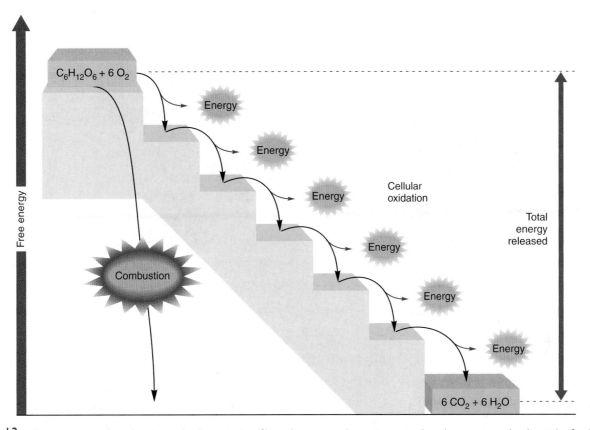

■ **Figure 4.13** **A comparison of combustion and cell respiration.** Since glucose contains more energy than six separate molecules each of carbon dioxide and water, the combustion of glucose is an exergonic reaction. The same amount of energy is released when glucose is broken down stepwise within the cell.

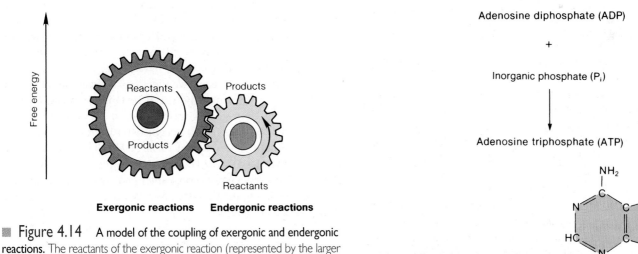

■ **Figure 4.14** **A model of the coupling of exergonic and endergonic reactions.** The reactants of the exergonic reaction (represented by the larger gear) have more free energy than the products of the endergonic reaction because the coupling is not 100% efficient—some energy is lost as heat.

is used to power the energy-requiring processes in all cells. As the **universal energy carrier,** ATP serves to more efficiently couple the energy released by the breakdown of food molecules to the energy required by the diverse endergonic processes in the cell (fig. 4.16).

■ **Figure 4.15** **The formation and structure of adenosine triphosphate (ATP).** ATP is the universal energy carrier of the cell. High-energy bonds are indicated by a squiggle (**~**).

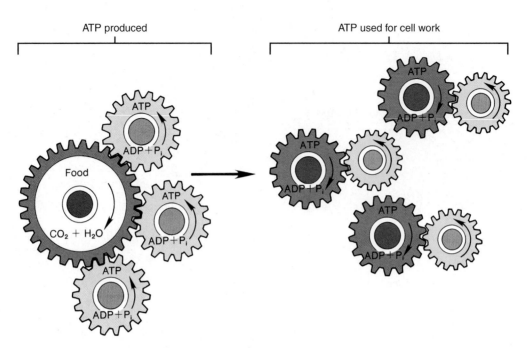

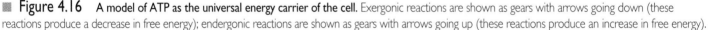

■ **Figure 4.16**   A model of ATP as the universal energy carrier of the cell. Exergonic reactions are shown as gears with arrows going down (these reactions produce a decrease in free energy); endergonic reactions are shown as gears with arrows going up (these reactions produce an increase in free energy).

Coupled Reactions: Oxidation-Reduction

When an atom or a molecule gains electrons, it is said to become **reduced;** when it loses electrons, it is said to become **oxidized.** Reduction and oxidation are always coupled reactions: an atom or a molecule cannot become oxidized unless it donates electrons to another, which therefore becomes reduced. The atom or molecule that donates electrons *to* another is a **reducing agent,** and the one that accepts electrons *from* another is an **oxidizing agent.** It is important to understand that a particular atom (or molecule) can play both roles; it may function as an oxidizing agent in one reaction and as a reducing agent in another reaction. When atoms or molecules play both roles, they gain electrons in one reaction and pass them on in another reaction to produce a series of coupled oxidation-reduction reactions—like a bucket brigade, with electrons in the buckets.

Notice that the term *oxidation* does not imply that oxygen participates in the reaction. This term is derived from the fact that oxygen has a great tendency to accept electrons; that is, to act as a strong oxidizing agent. This property of oxygen is exploited by cells; oxygen acts as the final electron acceptor in a chain of oxidation-reduction reactions that provides energy for ATP production.

Oxidation-reduction reactions in cells often involve the transfer of hydrogen atoms rather than free electrons. Since a hydrogen atom contains one electron (and one proton in the nucleus), a molecule that loses hydrogen becomes oxidized, and one that gains hydrogen becomes reduced. In many oxidation-reduction reactions, pairs of electrons—either as free electrons or as a pair of hydrogen atoms—are transferred from the reducing agent to the oxidizing agent.

Two molecules that serve important roles in the transfer of hydrogens are **nicotinamide adenine dinucleotide (NAD),** which is derived from the vitamin niacin (vitamin B_3), and **flavin adenine dinucleotide (FAD),** which is derived from the vitamin riboflavin (vitamin B_2). These molecules (fig. 4.17) are coenzymes that function as *hydrogen carriers* because they accept hydrogens (becoming reduced) in one enzyme reaction and donate hydrogens (becoming oxidized) in a different enzyme reaction (fig. 4.18). The oxidized forms of these molecules are written simply as NAD (or NAD^+) and FAD.

Each FAD can accept two electrons and can bind two protons. Therefore, the reduced form of FAD is combined with the equivalent of two hydrogen atoms and may be written as $FADH_2$. Each NAD can also accept two electrons but can bind only one proton. The reduced form of NAD is therefore indicated by $NADH + H^+$ (the H^+ represents a free proton). When the reduced forms of these two coenzymes participate in an oxidation-reduction reaction, they transfer two hydrogen atoms to the oxidizing agent (fig. 4.18).

 Production of the coenzymes NAD and FAD is the major reason that we need the vitamins niacin and riboflavin in our diet. As described in chapter 5, NAD and FAD are required to transfer hydrogen atoms in the chemical reactions that provide energy for the body. Niacin and riboflavin do not themselves provide the energy, although this is often claimed in misleading advertisements for health foods. Nor can eating extra amounts of niacin and riboflavin provide extra energy. Once the cells have obtained sufficient NAD and FAD, the excess amounts of these vitamins are simply eliminated in the urine.

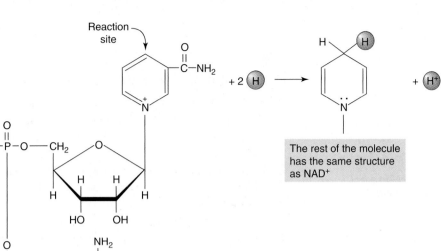

Reaction site

+ 2 (H) →

H

(H)

+ (H⁺)

The rest of the molecule has the same structure as NAD⁺

NAD⁺

(a) Oxidized state

NADH

Reduced state

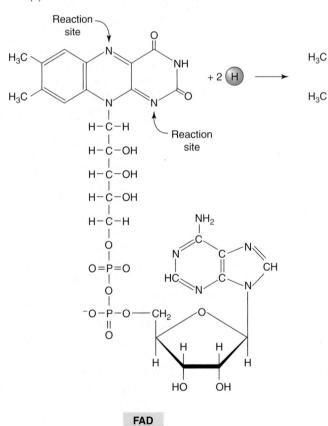

■ **Figure 4.17** Structural formulas for NAD⁺, NADH, FAD, and FADH₂.
(*a*) When NAD⁺ reacts with two hydrogen atoms, it binds to one of them and accepts the electron from the other. This is shown by two dots above the nitrogen ($\ddot{N}$) in the formula for NADH.
(*b*) When FAD reacts with two hydrogen atoms to form FADH₂, it binds each of them to a nitrogen atom at the reaction sites.

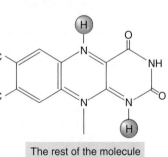

Reaction site

+ 2 (H) →

(H)

(H)

The rest of the molecule has the same structure as FAD

FAD

(b) Oxidized state

FADH₂

Reduced state

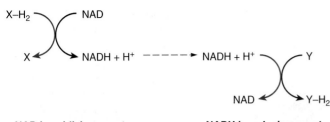

NAD is oxidizing agent
(it becomes reduced)

NADH is reducing agent
(it becomes oxidized)

Figure 4.18   The action of NAD. NAD is a coenzyme that transfers pairs of hydrogen atoms from one molecule to another. In the first reaction, NAD is reduced (acts as an oxidizing agent); in the second reaction, NADH is oxidized (acts as a reducing agent). Oxidation reactions are shown by red arrows, reduction reactions by blue arrows.

Test Yourself Before You Continue

1. Describe the first and second laws of thermodynamics. Use these laws to explain why the chemical bonds in glucose represent a source of potential energy and describe the process by which cells can obtain this energy.

2. Define the terms *exergonic reaction* and *endergonic reaction*. Use these terms to describe the function of ATP in cells.

3. Using the symbols X-H_2 and Y, draw a coupled oxidation-reduction reaction. Designate the molecule that is reduced and the one that is oxidized and state which one is the reducing agent and which is the oxidizing agent.

4. Describe the functions of NAD, FAD, and oxygen (in terms of oxidation-reduction reactions) and explain the meaning of the symbols NAD, $NADH + H^+$, FAD, and $FADH_2$.

Summary

Enzymes as Catalysts   86

I. Enzymes are biological catalysts.
   A. Catalysts increase the rate of chemical reactions.
      1. A catalyst is not altered by the reaction.
      2. A catalyst does not change the final result of a reaction.
   B. Catalysts lower the activation energy of chemical reactions.
      1. The activation energy is the amount of energy needed by the reactant molecules to participate in a reaction.
      2. In the absence of a catalyst, only a small proportion of the reactants possess the activation energy to participate.
      3. By lowering the activation energy, enzymes allow a larger proportion of the reactants to participate in the reaction, thus increasing the reaction rate.

II. Most enzymes are proteins.
   A. Protein enzymes have specific three-dimensional shapes that are determined by the amino acid sequence and, ultimately, by the genes.
   B. The reactants in an enzyme-catalyzed reaction—called the substrates of the enzyme—fit into a specific pocket in the enzyme called the active site.
   C. By forming an enzyme-substrate complex, substrate molecules are brought into proper orientation and existing bonds are weakened. This allows new bonds to be formed more easily.

Control of Enzyme Activity   89

I. The activity of an enzyme is affected by a variety of factors.
   A. The rate of enzyme-catalyzed reactions increases with increasing temperature, up to a maximum.
      1. This is because increasing the temperature increases the energy in the total population of reactant molecules, thus increasing the proportion of reactants that have the activation energy.
      2. At a few degrees above body temperature, however, most enzymes start to denature, which decreases the rate of the reactions that they catalyze.
   B. Each enzyme has optimal activity at a characteristic pH—called the pH optimum for that enzyme.
      1. Deviations from the pH optimum will decrease the reaction rate because the pH affects the shape of the enzyme and charges within the active site.
      2. The pH optima of different enzymes can vary widely— pepsin has a pH optimum of 2, for example, while trypsin is most active at a pH of 9.
   C. Many enzymes require metal ions in order to be active. These ions are therefore said to be cofactors for the enzymes.
   D. Many enzymes require smaller organic molecules for activity. These smaller organic molecules are called coenzymes.
      1. Coenzymes are derived from water-soluble vitamins.
      2. Coenzymes transport hydrogen atoms and small substrate molecules from one enzyme to another.

E. Some enzymes are produced as inactive forms that are later activated within the cell.

    1. Activation may be achieved by phosphorylation of the enzyme, in which case the enzyme can later be inactivated by dephosphorylation.

    2. Phosphorylation of enzymes is catalyzed by an enzyme called protein kinase.

    3. Protein kinase itself may be inactive and require the binding of a second messenger called cyclic AMP in order to become activated.

F. The rate of enzymatic reactions increases when either the substrate concentration or the enzyme concentration is increased.

    1. If the enzyme concentration remains constant, the rate of the reaction increases as the substrate concentration is raised, up to a maximum rate.

    2. When the rate of the reaction does not increase upon further addition of substrate, the enzyme is said to be saturated.

II. Metabolic pathways involve a number of enzyme-catalyzed reactions.

A. A number of enzymes usually cooperate to convert an initial substrate to a final product by way of several intermediates.

B. Metabolic pathways are produced by multienzyme systems in which the product of one enzyme becomes the substrate of the next.

C. If an enzyme is defective due to an abnormal gene, the intermediates that are formed following the step catalyzed by the defective enzyme will decrease, and the intermediates that are formed prior to the defective step will accumulate.

    1. Diseases that result from defective enzymes are called inborn errors of metabolism.

    2. Accumulation of intermediates often results in damage to the organ in which the defective enzyme is found.

D. Many metabolic pathways are branched, so that one intermediate can serve as the substrate for two different enzymes.

E. The activity of a particular pathway can be regulated by end-product inhibition.

    1. In end-product inhibition, one of the products of the pathway inhibits the activity of a key enzyme.

    2. This is an example of allosteric inhibition, in which the product combines with its specific site on the enzyme, changing the conformation of the active site.

Bioenergetics 93

I. The flow of energy in the cell is called bioenergetics.

A. According to the first law of thermodynamics, energy can neither be created nor destroyed but only transformed from one form to another.

B. According to the second law of thermodynamics, all energy transformation reactions result in an increase in entropy (disorder).

    1. As a result of the increase in entropy, there is a decrease in free (usable) energy.

    2. Atoms that are organized into large organic molecules thus contain more free energy than more-disorganized, smaller molecules.

C. In order to produce glucose from carbon dioxide and water, energy must be added.

    1. Plants use energy from the sun for this conversion, in a process called photosynthesis.

    2. Reactions that require the input of energy to produce molecules with higher free energy than the reactants are called endergonic reactions.

D. The combustion of glucose to carbon dioxide and water releases energy in the form of heat.

    1. A reaction that releases energy, thus forming products that contain less free energy than the reactants, is called an exergonic reaction.

    2. The same total amount of energy is released when glucose is converted into carbon dioxide and water within cells, even though this process occurs in many small steps.

E. The exergonic reactions that convert food molecules into carbon dioxide and water in cells are coupled to endergonic reactions that form adenosine triphosphate (ATP).

    1. Some of the chemical-bond energy in glucose is therefore transferred to the "high energy" bonds of ATP.

    2. The breakdown of ATP into adenosine diphosphate (ADP) and inorganic phosphate results in the liberation of energy.

    3. The energy liberated by the breakdown of ATP is used to power all of the energy-requiring processes of the cell. ATP is thus the "universal energy carrier" of the cell.

II. Oxidation-reduction reactions are coupled and usually involve the transfer of hydrogen atoms.

A. A molecule is said to be oxidized when it loses electrons; it is said to be reduced when it gains electrons.

B. A reducing agent is thus an electron donor; an oxidizing agent is an electron acceptor.

C. Although oxygen is the final electron acceptor in the cell, other molecules can act as oxidizing agents.

D. A single molecule can be an electron acceptor in one reaction and an electron donor in another.

    1. NAD and FAD can become reduced by accepting electrons from hydrogen atoms removed from other molecules.

    2. $NADH + H^+$, and $FADH_2$, in turn, donate these electrons to other molecules in other locations within the cells.

    3. Oxygen is the final electron acceptor (oxidizing agent) in a chain of oxidation-reduction reactions that provide energy for ATP production.

Review Activities

Test Your Knowledge of Terms and Facts

1. Which of these statements about enzymes is *true*?
   a. Most proteins are enzymes.
   b. Most enzymes are proteins.
   c. Enzymes are changed by the reactions they catalyze.
   d. The active sites of enzymes have little specificity for substrates.

2. Which of these statements about enzyme-catalyzed reactions is *true*?
   a. The rate of reaction is independent of temperature.
   b. The rate of all enzyme-catalyzed reactions is decreased when the pH is lowered from 7 to 2.
   c. The rate of reaction is independent of substrate concentration.
   d. Under given conditions of substrate concentration, pH, and temperature, the rate of product formation varies directly with enzyme concentration up to a maximum, at which point the rate cannot be increased further.

3. Which of these statements about lactate dehydrogenase is *true*?
   a. It is a protein.
   b. It oxidizes lactic acid.
   c. It reduces another molecule (pyruvic acid).
   d. All of these are true.

4. In a metabolic pathway,
   a. the product of one enzyme becomes the substrate of the next.
   b. the substrate of one enzyme becomes the product of the next.

5. In an inborn error of metabolism,
   a. a genetic change results in the production of a defective enzyme.
   b. intermediates produced prior to the defective step accumulate.
   c. alternate pathways are taken by intermediates at branch points that precede the defective step.
   d. All of these are true.

6. Which of these represents an endergonic reaction?
   a. $ADP + P_i \rightarrow ATP$
   b. $ATP \rightarrow ADP + P_i$
   c. $glucose + O_2 \rightarrow CO_2 + H_2O$
   d. $CO_2 + H_2O \rightarrow glucose$
   e. both *a* and *d*
   f. both *b* and *c*

7. Which of these statements about ATP is *true*?
   a. The bond joining ADP and the third phosphate is a high-energy bond.
   b. The formation of ATP is coupled to energy-liberating reactions.
   c. The conversion of ATP to ADP and P_i provides energy for biosynthesis, cell movement, and other cellular processes that require energy.
   d. ATP is the "universal energy carrier" of cells.
   e. All of these are true.

8. When oxygen is combined with two hydrogens to make water,
   a. oxygen is reduced.
   b. the molecule that donated the hydrogens becomes oxidized.
   c. oxygen acts as a reducing agent.
   d. both *a* and *b* apply.
   e. both *a* and *c* apply.

9. Enzymes increase the rate of chemical reactions by
   a. increasing the body temperature.
   b. decreasing the blood pH.
   c. increasing the affinity of reactant molecules for each other.
   d. decreasing the activation energy of the reactants.

10. According to the law of mass action, which of these conditions will drive the reaction $A + B \rightleftharpoons C$ to the right?
    a. an increase in the concentration of A and B
    b. a decrease in the concentration of C
    c. an increase in the concentration of enzyme
    d. both *a* and *b*
    e. both *b* and *c*

Test Your Understanding of Concepts and Principles

1. Explain the relationship between an enzyme's chemical structure and the function of the enzyme, and describe how both structure and function may be altered in various ways.[1]

2. Explain how the rate of enzymatic reactions may be regulated by the relative concentrations of substrates and products.

3. Explain how end-product inhibition represents a form of negative feedback regulation.

4. Using the first and second laws of thermodynamics, explain how ATP is formed and how it serves as the universal energy carrier.

5. The coenzymes NAD and FAD can "shuttle" hydrogens from one reaction to another. How does this process serve to couple oxidation and reduction reactions?

6. Using albinism and phenylketonuria as examples, explain what is meant by inborn errors of metabolism.

7. Why do we need to eat food containing niacin and riboflavin? How do these vitamins function in the body?

[1]*Note:* This question is answered in the chapter 4 Study Guide found on the Online Learning Center at www.mhhe.com/fox8.

Test Your Ability to Analyze and Apply Your Knowledge

1. Metabolic pathways can be likened to intersecting railroad tracks, with enzymes as the switches. Discuss this analogy.

2. A student, learning that someone has an elevated blood level of lactate dehydrogenase (LDH), wonders how the enzyme got into this person's blood and worries about whether it will digest the blood. What explanation can you give to allay the student's fears?

3. Suppose you come across a bottle of enzyme tablets at your local health food store. The clerk tells you this enzyme will help your digestion, but you notice that it is derived from a plant. What concerns might you have regarding the effectiveness of these tablets?

Related Websites

Check out the Links Library at www.mhhe.com/fox8 for links to sites containing resources related to enzymes and energy. These links are monitored to ensure current URLs.

5 Cell Respiration and Metabolism

Objectives

After studying this chapter, you should be able to . . .

1. describe the steps of glycolysis and discuss the significance of this metabolic pathway.

2. describe how lactic acid is formed and explain the physiological significance of this pathway.

3. define the term *gluconeogenesis* and describe the Cori cycle.

4. describe the pathway for the aerobic respiration of glucose through the steps of the Krebs cycle.

5. explain the functional significance of the Krebs cycle in relation to the electron-transport system.

6. describe the electron-transport system and oxidative phosphorylation.

7. describe the role of oxygen in aerobic respiration.

8. compare the lactic acid pathway and aerobic respiration in terms of initial substrates, final products, cellular locations, and the total number of ATP molecules produced per glucose respired.

9. explain how glucose and glycogen can be interconverted and how the liver can secrete free glucose derived from its stored glycogen.

10. define the terms *lipolysis* and *β-oxidation* and explain how these processes function in cellular energy production.

11. explain how ketone bodies are formed.

12. describe the processes of oxidative deamination and transamination of amino acids and explain how these processes can contribute to energy production.

13. explain how carbohydrates or protein can be converted to fat in terms of the metabolic pathways involved.

14. state the preferred energy sources of different organs.

Take Advantage of the Technology

Visit the Online Learning Center for these additional study resources.

■ Interactive quizzing
■ Online study guide
■ Current news feeds
■ Crossword puzzles
■ Vocabulary flashcards
■ Labeling activities

www.mhhe.com/fox8

Clinical Investigation

Brenda is a second-year college student training to make the swim team. In the early stages of her training, she experienced great fatigue following a workout, and she found herself gasping and panting for air more than her teammates did. Her coach suggested that she eat less protein and more carbohydrates than was her habit, and that she train more gradually. She also complained of pain in her arms and shoulders that began with her training. Following a particularly intense workout, she experienced severe pain in her left pectoral region and sought medical aid.

What might be responsible for Brenda's symptoms?

Glycolysis and the Lactic Acid Pathway

In cellular respiration, energy is released by the stepwise breakdown of glucose and other molecules, and some of this energy is used to produce ATP. The complete combustion of glucose requires the presence of oxygen and yields thirty ATP for each molecule of glucose. However, some energy can be obtained in the absence of oxygen by the pathway that leads to the production of lactic acid. This process results in a net gain of two ATP per glucose.

All of the reactions in the body that involve energy transformation are collectively termed **metabolism.** Metabolism may be divided into two categories: *anabolism* and *catabolism.* Catabolic reactions release energy, usually by the breakdown of larger organic molecules into smaller molecules. Anabolic reactions require the input of energy and include the synthesis of large energy-storage molecules, including glycogen, fat, and protein.

The catabolic reactions that break down glucose, fatty acids, and amino acids serve as the primary sources of energy for the synthesis of ATP. For example, this means that some of the chemical-bond energy in glucose is transferred to the chemical-bond energy in ATP. Since energy transfers can never be 100% efficient (according to the second law of thermodynamics), some of the chemical-bond energy from glucose is lost as heat.

This energy transfer involves oxidation-reduction reactions. As explained in chapter 4, oxidation of a molecule occurs when the molecule loses electrons. This must be coupled to the reduction of another atom or molecule, which accepts the electrons. In the breakdown of glucose and other molecules for energy, some of the electrons initially present in these molecules are transferred to intermediate carriers and then to a *final electron acceptor.* When a molecule is completely broken down to carbon dioxide and water within an animal cell, the final electron acceptor is always an atom of oxygen. Because of the involvement of oxygen, the metabolic pathway that converts molecules such as glucose or fatty acid to carbon dioxide and water (transferring some of the energy to ATP) is called **aerobic cell respiration.** The oxygen for this process is obtained from the blood. The blood, in turn, obtains oxygen from air in the lungs through the process of breathing, or ventilation, as described in chapter 16. Ventilation also serves the important function of eliminating the carbon dioxide produced by aerobic cell respiration.

Unlike the process of burning, or combustion, which quickly releases the energy content of molecules as heat (and which can be measured as kilocalories—see chapter 4), the conversion of glucose to carbon dioxide and water within the cells occurs in small, enzymatically catalyzed steps. Oxygen is used only at the last step. Since a small amount of the chemical-bond energy of glucose is released at early steps in the metabolic pathway, some tissue cells can obtain energy for ATP production in the temporary absence of oxygen. This process is described in the next two sections.

Glycolysis

The breakdown of glucose for energy involves a metabolic pathway in the cytoplasm known as **glycolysis.** This term is derived from the Greek *glykys* = sweet and *lysis* = a loosening, and it refers to the cleavage of sugar. Glycolysis is the metabolic pathway by which glucose—a six-carbon (hexose) sugar (see fig. 2.13)—is converted into two molecules of pyruvic acid, or pyruvate. Even though each pyruvic acid molecule is roughly half the size of a glucose, glycolysis is *not* simply the breaking in half of glucose. Glycolysis is a metabolic pathway involving many enzymatically controlled steps.

Each pyruvic acid molecule contains three carbons, three oxygens, and four hydrogens (see fig. 5.3). The number of carbon and oxygen atoms in one molecule of glucose—$C_6H_{12}O_6$—can thus be accounted for in the two pyruvic acid molecules. Since the two pyruvic acids together account for only eight hydrogens, however, it is clear that four hydrogen atoms are removed from the intermediates in glycolysis. Each pair of these hydrogen atoms is used to reduce a molecule of NAD. In this process, each pair of hydrogen atoms donates two electrons to NAD, thus reducing it. The reduced NAD binds one proton from the hydrogen atoms, leaving one proton unbound as H^+ (chapter 4, fig. 4.17). Starting from one glucose molecule, therefore, glycolysis results in the production of two molecules of NADH and two H^+. The H^+ will follow the NADH in subsequent reactions, so for simplicity we can refer to reduced NAD simply as NADH.

Glycolysis is exergonic, and a portion of the energy that is released is used to drive the endergonic reaction ADP + P_i → ATP. At the end of the glycolytic pathway, there is a net gain of two ATP molecules per glucose molecule, as indicated in the overall equation for glycolysis:

$$\text{Glucose} + 2\ \text{NAD} + 2\ \text{ADP} + 2\ P_i \rightarrow$$

$$2\ \text{pyruvic acid} + 2\ \text{NADH} + 2\ \text{ATP}$$

Although the overall equation for glycolysis is exergonic, glucose must be "activated" at the beginning of the pathway before energy can be obtained. This activation requires the addition

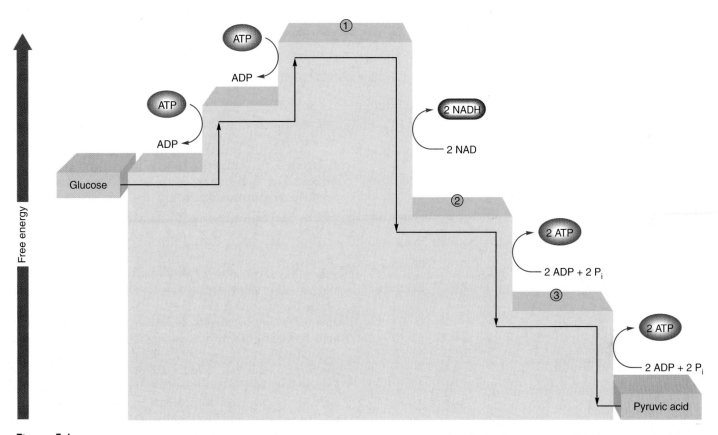

Figure 5.1 **The energy expenditure and gain in glycolysis.** Notice that there is a "net profit" of two ATP and two NADH for every molecule of glucose that enters the glycolytic pathway. Molecules listed by number are (1) fructose 1,6-biphosphate, (2) 1,3-biphosphoglyceric acid, and (3) 3-phosphoglyceric acid (see fig. 5.2).

of two phosphate groups derived from two molecules of ATP. Energy from the reaction ATP → ADP + P_i is therefore consumed at the beginning of glycolysis. This is shown as an "up-staircase" in figure 5.1. Notice that the P_i is not shown in these reactions in figure 5.1; this is because the phosphate is not released, but instead is added to the intermediate molecules of glycolysis. The addition of a phosphate group is known as *phosphorylation*. Besides being essential for glycolysis, the phosphorylation of glucose (to glucose 6-phosphate) has an important side benefit: it traps the glucose within the cell. This is because phosphorylated organic molecules cannot cross cell membranes.

At later steps in glycolysis, four molecules of ATP are produced (and two molecules of NAD are reduced) as energy is liberated (the "down-staircase" in fig. 5.1). The two molecules of ATP used in the beginning, therefore, represent an energy investment; the net gain of two ATP and two NADH molecules by the end of the pathway represents an energy profit. The overall equation for glycolysis obscures the fact that this is a metabolic pathway consisting of nine separate steps. The individual steps in this pathway are shown in figure 5.2.

In figure 5.2, glucose is phosphorylated to glucose 6-phosphate using ATP at step 1, and then is converted into its isomer, fructose 6-phosphate, in step 2. Another ATP is used to form fructose 1, 6-biphosphate at step 3. Notice that the six-carbon-long molecule is split into two separate three-carbon-long molecules at

step 4. At step 5, two pairs of hydrogens are removed and used to reduce two NAD to two NADH + H^+. These reduced coenzymes are important products of glycolysis. Then, at step 6, a phosphate group is removed from each 1,3-biphosphoglyceric acid, forming two ATP and two molecules of 3-phosphoglyceric acid. Steps 7 and 8 are isomerizations. Then, at step 9, the last phosphate group is removed from each intermediate; this forms another two ATP (for a net gain of two ATP), and two molecules of pyruvic acid.

Lactic Acid Pathway

In order for glycolysis to continue, there must be adequate amounts of NAD available to accept hydrogen atoms. Therefore, the NADH produced in glycolysis must become oxidized by donating its electrons to another molecule. (In aerobic respiration this other molecule is located in the mitochondria and ultimately passes its electrons to oxygen.)

When oxygen is not available in sufficient amounts, the NADH (+ H^+) produced in glycolysis is oxidized in the cytoplasm by donating its electrons to pyruvic acid. This results in the re-formation of NAD and the addition of two hydrogen atoms to pyruvic acid, which is thus reduced. This addition of two hydrogen atoms to pyruvic acid produces lactic acid (fig. 5.3).

The metabolic pathway by which glucose is converted to lactic acid is frequently referred to by physiologists as

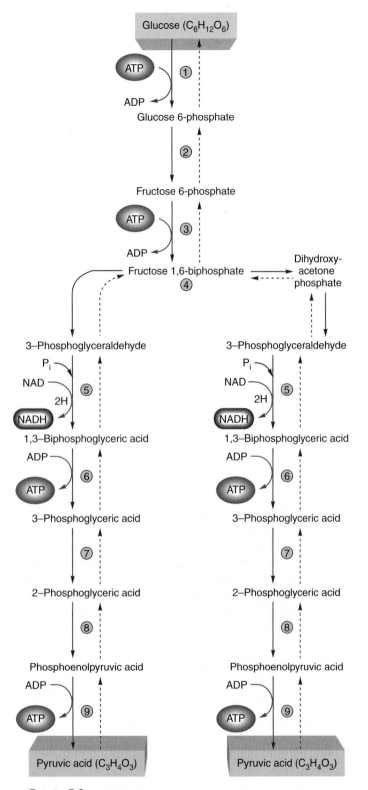

Figure 5.2 Glycolysis. In glycolysis, one glucose is converted into two pyruvic acids in nine separate steps. In addition to two pyruvic acids, the products of glycolysis include two NADH and four ATP. Since two ATP were used at the beginning, however, the net gain is two ATP per glucose. Dashed arrows indicate reverse reactions that may occur under other conditions.

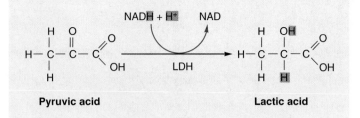

Figure 5.3 The formation of lactic acid. The addition of two hydrogen atoms (colored boxes) from reduced NAD to pyruvic acid produces lactic acid and oxidized NAD. This reaction is catalyzed by lactic acid dehydrogenase (LDH) and is reversible under the proper conditions.

anaerobic respiration. "Anaerobic" describes the fact that oxygen is not used in the process. This is the term that will be used in this text for the pathway leading to lactic acid production. Many biologists, however, prefer the name **lactic acid fermentation** for this pathway. This is because the lactic acid pathway is basically similar to the way yeast cells convert glucose into ethyl alcohol, a process universally known as *fermentation.* In both lactic acid and alcohol production, the last electron acceptor is an organic molecule (as opposed to an atom of oxygen, as will be described for aerobic respiration).

The lactic acid pathway yields a net gain of two ATP molecules (produced by glycolysis) per glucose molecule. A cell can survive without oxygen as long as it can produce sufficient energy for its needs in this way and as long as lactic acid concentrations do not become excessive. Some tissues are better adapted to anaerobic conditions than others—skeletal muscles survive longer than cardiac muscle, which in turn survives under anaerobic conditions longer than the brain.

Red blood cells, which lack mitochondria, can use only the lactic acid pathway; therefore (for reasons described in the next section), they cannot use oxygen. This spares the oxygen they carry for delivery to other cells. Except for red blood cells, anaerobic respiration occurs for only a limited period of time in tissues that have energy requirements in excess of their aerobic ability. Anaerobic respiration occurs in the skeletal muscles and heart when the *ratio of oxygen supply to oxygen need* (related to the concentration of NADH) falls below a critical level. Anaerobic respiration is, in a sense, an emergency procedure that provides some ATP until the emergency (oxygen deficiency) has passed.

It should be noted, though, that there is no real "emergency" in the case of skeletal muscles, where anaerobic respiration is a normal, daily occurrence that does not harm muscle tissue or the individual. Excessive lactic acid production by muscles, however, is associated with pain and muscle fatigue. (The metabolism of skeletal muscles is discussed in chapter 12.) In contrast to skeletal muscles, the heart normally respires only aerobically. If anaerobic conditions do occur in the heart, a potentially dangerous situation may be present.

Ischemia refers to inadequate blood flow to an organ, such that the rate of oxygen delivery is insufficient to maintain aerobic respiration. Inadequate blood flow to the heart, or *myocardial ischemia*, may occur if the coronary blood flow is occluded by atherosclerosis, a blood clot, or by an artery spasm. People with myocardial ischemia often experience *angina pectoris*—severe pain in the chest and left (or sometimes, right) arm area. This pain is associated with increased blood levels of lactic acid which are produced by the ischemic heart muscle. If the ischemia is prolonged, the cells may die and produce an area called an *infarct*. The degree of ischemia and angina can be decreased by vasodilator drugs such as nitroglycerin, which improve blood flow to the heart and also decrease the work of the heart by dilating peripheral blood vessels.

Clinical Investigation Clues

Remember that Brenda experienced muscle pain and fatigue during her training, and that she had an episode where she experienced severe pain in her left pectoral region following an intense workout.

What produced her muscle pain and fatigue?

What might have caused the severe pain in her left pectoral region?

Which of these effects are normal?

Glycogenesis and Glycogenolysis

Cells cannot accumulate very many separate glucose molecules, because an abundance of these would exert an osmotic pressure (see chapter 6) that would draw a dangerous amount of water into the cells. Instead, many organs, particularly the liver, skeletal muscles, and heart, store carbohydrates in the form of glycogen.

The formation of glycogen from glucose is called **glycogenesis.** In this process, glucose is converted to glucose 6-phosphate by utilizing the terminal phosphate group of ATP. Glucose 6-phosphate is then converted into its isomer, glucose 1-phosphate. Finally, the enzyme *glycogen synthase* removes these phosphate groups as it polymerizes glucose to form glycogen.

The reverse reactions are similar. The enzyme *glycogen phosphorylase* catalyzes the breakdown of glycogen to glucose 1-phosphate. (The phosphates are derived from inorganic phosphate, not from ATP, so glycogen breakdown does not require metabolic energy.) Glucose 1-phosphate is then converted to glucose 6-phosphate. The conversion of glycogen to glucose 6-phosphate is called **glycogenolysis.** In most tissues, glucose 6-phosphate can then be respired for energy (through glycolysis) or used to resynthesize glycogen. Only in the liver, for reasons that will now be explained, can the glucose 6-phosphate also be used to produce free glucose for secretion into the blood.

As mentioned earlier, organic molecules with phosphate groups cannot cross cell membranes. Since the glucose derived from glycogen is in the form of glucose 1-phosphate and then

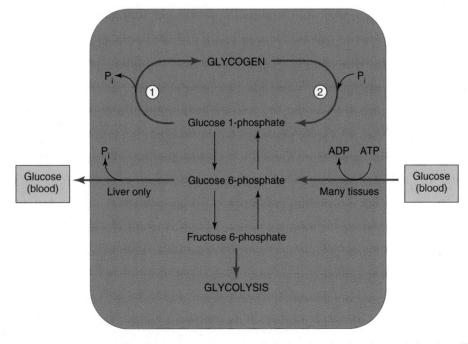

Figure 5.4 **Glycogenesis and glycogenolysis.** Blood glucose entering tissue cells is phosphorylated to glucose 6-phosphate. This intermediate can be metabolized for energy in glycolysis, or it can be converted to glycogen (*1*) in a process called *glycogenesis.* Glycogen represents a storage form of carbohydrates that can be used as a source for new glucose 6-phosphate (*2*) in a process called *glycogenolysis.* The liver contains an enzyme that can remove the phosphate from glucose 6-phosphate; liver glycogen thus serves as a source for new blood glucose.

glucose 6-phosphate, it cannot leak out of the cell. Similarly, glucose that enters the cell from the blood is "trapped" within the cell by conversion to glucose 6-phosphate. Skeletal muscles, which have large amounts of glycogen, can generate glucose 6-phosphate for their own glycolytic needs, but they cannot secrete glucose into the blood because they lack the ability to remove the phosphate group.

Unlike skeletal muscles, the liver contains an enzyme—known as *glucose 6-phosphatase*—that can remove the phosphate groups and produce free glucose (fig. 5.4). This free glucose can then be transported through the cell membrane. The liver, then, can secrete glucose into the blood, whereas skeletal muscles cannot. Liver glycogen can thus supply blood glucose for use by other organs, including exercising skeletal muscles that may have depleted much of their own stored glycogen during exercise.

Clinical Investigation Clues

Remember that Brenda's coach advised her to eat more carbohydrates during her training.

What will happen to the extra carbohydrates she eats?
What benefits might be derived from such "carbohydrate loading"?

Cori Cycle

In humans and other mammals, much of the lactic acid produced in anaerobic respiration is later eliminated by aerobic respiration of the lactic acid to carbon dioxide and water. However, some of the lactic acid produced by exercising skeletal muscles is delivered by the blood to the liver. Within the liver cells under these conditions, the enzyme *lactic acid dehydrogenase (LDH)* converts lactic acid to pyruvic acid. This is the reverse of the step of anaerobic respiration shown in figure 5.3, and in the process NAD is reduced to NADH + H$^+$. Unlike most other organs, the liver contains the enzymes needed to take pyruvic acid molecules and convert them to glucose 6-phosphate, a process that is essentially the reverse of glycolysis.

Glucose 6-phosphate in liver cells can then be used as an intermediate for glycogen synthesis, or it can be converted to free glucose that is secreted into the blood. The conversion of noncarbohydrate molecules (not just lactic acid, but also amino acids and glycerol) through pyruvic acid to glucose is an extremely important process called **gluconeogenesis.** The significance of this process in conditions of fasting will be discussed in a later section on amino acid metabolism.

During exercise, some of the lactic acid produced by skeletal muscles may be transformed through gluconeogenesis in the liver to blood glucose. This new glucose can serve as an energy source during exercise and can be used after exercise to help replenish the depleted muscle glycogen. This two-way traffic between skeletal muscles and the liver is called the **Cori cycle** (fig. 5.5). Through

the Cori cycle, gluconeogenesis in the liver allows depleted skeletal muscle glycogen to be restored within 48 hours.

Test Yourself Before You Continue

1. Define the term *glycolysis* in terms of its initial substrates and products. Explain why there is a net gain of two molecules of ATP in this process.
2. Discuss the two meanings of the term *anaerobic respiration*. As the term is used in this text, what are its initial substrates and final products?
3. Describe the physiological functions of anaerobic respiration. In which tissue(s) is anaerobic respiration normal? In which tissue is it abnormal?
4. Describe the pathways by which glucose and glycogen can be interconverted. Explain why only the liver can secrete glucose derived from its stored glycogen.
5. Define the term *gluconeogenesis* and explain how this process replenishes the glycogen stores of skeletal muscles following exercise.

Aerobic Respiration

In the aerobic respiration of glucose, pyruvic acid is formed by glycolysis and then converted into acetyl coenzyme A. This begins a cyclic metabolic pathway called the Krebs cycle. As a result of these pathways, a large amount of reduced NAD and FAD (NADH and FADH$_2$) is generated. These reduced coenzymes provide electrons for an energy-generating process that drives the formation of ATP.

Aerobic respiration is equivalent to combustion in terms of its final products (CO_2 and H_2O) and in terms of the total amount of energy liberated. In aerobic respiration, however, the energy is released in small, enzymatically controlled oxidation reactions, and a portion (38% to 40%) of the energy released is captured in the high-energy bonds of ATP.

The aerobic respiration of glucose begins with glycolysis. Glycolysis in both anaerobic and aerobic respiration results in the production of two molecules of pyruvic acid, two ATP, and two NADH + H$^+$ per glucose molecule. In aerobic respiration, however, the electrons in NADH are *not* donated to pyruvic acid and lactic acid is not formed, as happens in anaerobic respiration. Instead, the pyruvic acids will move to a different cellular location and undergo a different reaction; the NADH produced by glycolysis will eventually be oxidized, but that occurs later in the story.

In aerobic respiration, pyruvic acid leaves the cell cytoplasm and enters the interior (the matrix) of mitochondria. Once pyruvic acid is inside a mitochondrion, carbon dioxide is enzymatically removed from each three-carbon-long pyruvic acid to

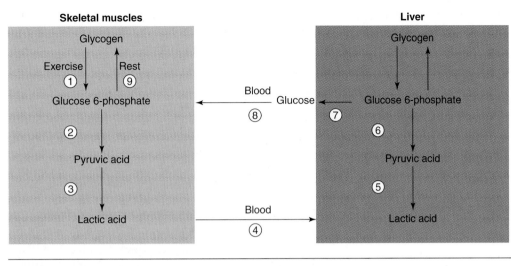

■ **Figure 5.5** The Cori cycle. The sequence of steps is indicated by numbers 1 through 9.

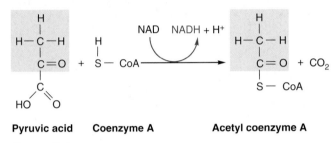

■ **Figure 5.6** The formation of acetyl coenzyme A in aerobic respiration. Notice that NAD is reduced to NADH in this process.

form a two-carbon-long organic acid—acetic acid. The enzyme that catalyzes this reaction combines the acetic acid with a coenzyme (derived from the vitamin pantothenic acid) called *coenzyme A*. The combination thus produced is called **acetyl coenzyme A,** abbreviated **acetyl CoA** (fig. 5.6).

Glycolysis converts one glucose molecule into two molecules of pyruvic acid. Since each pyruvic acid molecule is converted into one molecule of acetyl CoA and one CO_2, two molecules of acetyl CoA and two molecules of CO_2 are derived from each glucose. These acetyl CoA molecules serve as substrates for mitochondrial enzymes in the aerobic pathway, while the carbon dioxide is a waste product that is carried by the blood to the lungs for elimination. It is important to note that the oxygen in CO_2 is derived from pyruvic acid, not from oxygen gas.

Krebs Cycle

Once acetyl CoA has been formed, the acetic acid subunit (two carbons long) combines with oxaloacetic acid (four carbons long) to form a molecule of citric acid (six carbons long). Coenzyme A acts only as a transporter of acetic acid from one enzyme to another (similar to the transport of hydrogen by NAD). The formation of citric acid begins a cyclic metabolic pathway known as the **citric acid cycle,** or **TCA cycle** (for tricarboxylic acid; citric acid has three carboxylic acid groups). Most commonly, however, this cyclic pathway is called the **Krebs cycle,** after its principal discoverer, Sir Hans Krebs. A simplified illustration of this pathway is shown in figure 5.7.

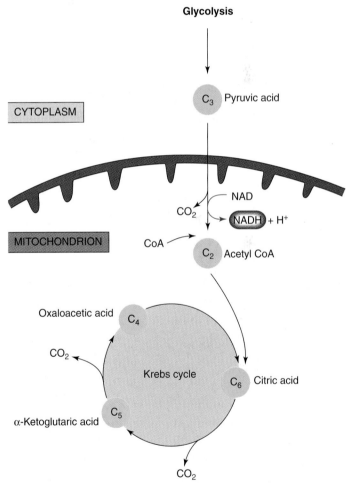

■ **Figure 5.7** A simplified diagram of the Krebs cycle. This diagram shows how the original four-carbon-long oxaloacetic acid is regenerated at the end of the cyclic pathway. Only the numbers of carbon atoms in the Krebs cycle intermediates are shown; the numbers of hydrogens and oxygens are not accounted for in this simplified scheme.

Through a series of reactions involving the elimination of two carbons and four oxygens (as two CO_2 molecules) and the removal of hydrogens, citric acid is eventually converted to ox-aloacetic acid, which completes the cyclic metabolic pathway (fig. 5.8). In this process, these events occur:

1. One guanosine triphosphate (GTP) is produced (step 5 of fig. 5.8), which donates a phosphate group to ADP to produce one ATP.
2. Three molecules of NAD are reduced to NADH (steps 4, 5, and 8 of fig. 5.8).
3. One molecule of FAD is reduced to $FADH_2$ (step 6).

The production of NADH and $FADH_2$ by each "turn" of the Krebs cycle is far more significant, in terms of energy production, than the single GTP (converted to ATP) produced directly by the cycle. This is because NADH and $FADH_2$ eventually donate their electrons to an energy-transferring process that results in the formation of a large number of ATP.

Electron Transport and Oxidative Phosphorylation

Built into the foldings, or cristae, of the inner mitochondrial membrane are a series of molecules that serve as an **electron-transport system** during aerobic respiration. This electron-transport chain of molecules consists of a protein containing *flavin mononucleotide* (abbreviated *FMN* and derived from the vitamin riboflavin), *coenzyme Q,* and a group of iron-containing pigments called *cytochromes.* The last of these cytochromes is cytochrome a_3, which donates electrons to oxygen in the final oxidation-reduction reaction (as will be described shortly). These molecules of the electron-transport system are fixed in position within the inner mitochondrial membrane in such a way that they can pick up electrons from NADH and $FADH_2$ and transport them in a definite sequence and direction.

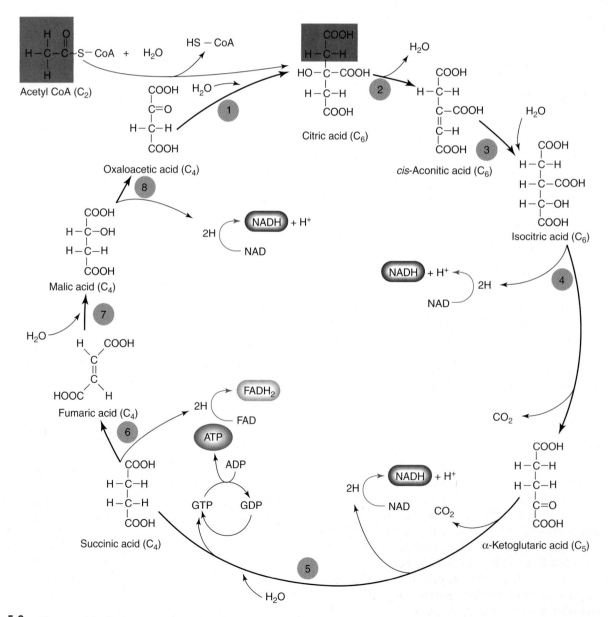

■ **Figure 5.8** **The complete Krebs cycle.** Notice that, for each "turn" of the cycle, one ATP, three NADH, and one $FADH_2$ are produced.

Free radicals are molecules with unpaired electrons, in contrast to molecules that are not free radicals because they have two electrons per orbital. A *superoxide radical* is an oxygen molecule with an extra, unpaired electron. These can be generated in mitochondria through the accidental leakage of electrons from the electron-transport system. Superoxide radicals have some known, physiological functions; for example, they are produced in phagocytic white blood cells where they are needed for the destruction of bacteria. However, the production of free radicals and other molecules classified as *reactive oxygen species* (including the superoxide, hydroxyl, and nitric oxide free radicals, and hydrogen peroxide) have been implicated in many disease processes, including atherosclerosis (hardening of the arteries—see chapter 13). Accordingly, reactive oxygen species have been described as exerting an *oxidative stress* on the body. **Antioxidants** are molecules that scavenge free radicals and protect the body from reactive oxygen species. Antioxidants produced in the body cells include the enzyme superoxide dismutase, which converts superoxide radicals to hydrogen peroxide, and a tripeptide called glutathione, which functions as the major cellular scavenger of free radicals. Those ingested in the diet include ascorbic acid (vitamin C), α-tocopherol (vitamin E), and many other molecules found in different fruits and vegetables.

In aerobic respiration, NADH and $FADH_2$ become oxidized by transferring their pairs of electrons to the electron-transport system of the cristae. It should be noted that the protons (H^+) are not transported together with the electrons; their fate will be described a little later. The oxidized forms of NAD and FAD are thus regenerated and can continue to "shuttle" electrons from the Krebs cycle to the electron-transport chain. The first molecule of the electron-transport chain in turn becomes reduced when it accepts the electron pair from NADH. When the cytochromes receive a pair of electrons, two ferric ions (Fe^{3+}) become reduced to two ferrous ions (Fe^{2+}).

The electron-transport chain thus acts as an oxidizing agent for NAD and FAD. Each element in the chain, however, also functions as a reducing agent; one reduced cytochrome transfers its electron pair to the next cytochrome in the chain (fig. 5.9). In this way, the iron ions in each cytochrome alternately become reduced (from Fe^{3+} to Fe^{2+}) and oxidized (from Fe^{2+} to Fe^{3+}). This is an exergonic process, and the energy derived is used to phosphorylate ADP to ATP. The production of ATP in this manner is thus appropriately termed **oxidative phosphorylation.**

The coupling is not 100% efficient between the energy released by electron transport (the "oxidative" part of oxidative phosphorylation) and the energy incorporated into the chemical bonds of ATP (the "phosphorylation" part of the term). This difference in energy escapes the body as heat. Metabolic heat production is needed to maintain our internal body temperature.

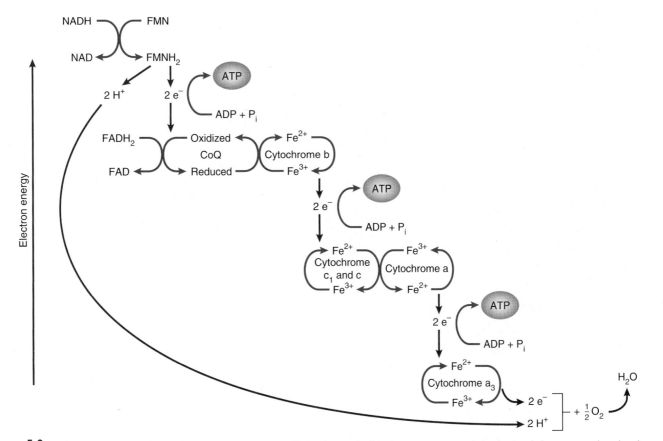

■ **Figure 5.9** Electron transport and oxidative phosphorylation. Each element in the electron-transport chain alternately becomes reduced and oxidized as it transports electrons to the next member of the chain. This process provides energy for the formation of ATP. At the end of the electron-transport chain, the electrons are donated to oxygen, which becomes reduced (by the addition of two hydrogen atoms) to water.

Coupling of Electron Transport to ATP Production

According to the **chemiosmotic theory,** the electron-transport system, powered by the transport of electrons, pumps protons (H^+) from the mitochondrial matrix into the space between the inner and outer mitochondrial membranes. The electron-transport system is grouped into three complexes that serve as **proton pumps** (fig. 5.10). The first pump (the NADH-coenzyme Q reductase complex) transports four H^+ from the matrix to the intermembrane space for every pair of electrons moved along the electron-transport system. The second pump (the cytochrome c reductase complex) also transports four protons into the intermembrane space, and the third pump (the cytochrome c oxidase complex) transports two protons into the intermembrane space. As a result, there is a higher concentration of H^+ in the intermembrane space than in the matrix, favoring the diffusion of H^+ back out into the matrix. The inner mitochondrial membrane, however, does not permit diffusion of H^+, except through structures called *respiratory assemblies.*

The respiratory assemblies consist of a group of proteins that form a "stem" and a globular subunit. The stem contains a channel through the inner mitochondrial membrane that permits the passage of protons (H^+). The globular subunit, which protrudes into the matrix, contains an **ATP synthase** enzyme that is capable of catalyzing the reaction $ADP + P_i \rightarrow ATP$ when it is activated by the diffusion of protons through the respiratory assemblies and into the matrix (fig. 5.10). In this way, phosphorylation (the addition of phosphate to ADP) is coupled to oxidation (the transport of electrons) in oxidative phosphorylation.

Function of Oxygen

If the last cytochrome remained in a reduced state, it would be unable to accept more electrons. Electron transport would then progress only to the next-to-last cytochrome. This process would continue until all of the elements of the electron-transport chain remained in the reduced state. At this point, the electron-transport system would stop functioning and no ATP could be produced in the mitochondria. With the electron-transport system incapacitated, NADH and $FADH_2$ could not become oxidized by donating their electrons to the chain and, through inhibition of Krebs cycle enzymes, no more NADH and $FADH_2$ could be produced in the mitochondria. The Krebs cycle would stop and respiration would become anaerobic.

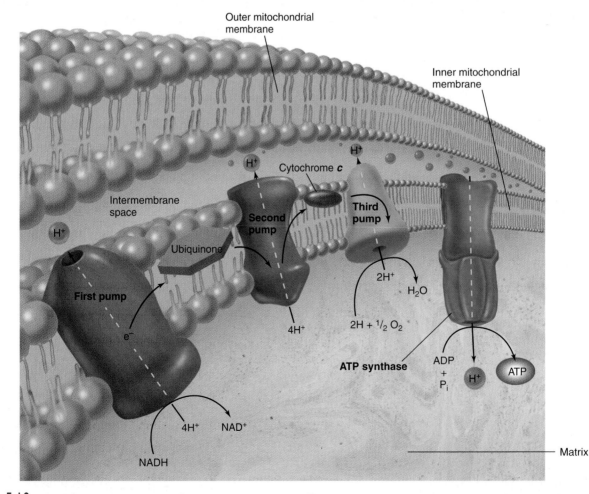

■ **Figure 5.10** **A schematic representation of the chemiosmotic theory.** The matrix and the compartment between the inner and outer mitochondrial membranes showing how the electron-transport system functions to pump H^+ from the matrix to the intermembrane space. This results in a steep H^+ gradient between the intermembrane space and the cytoplasm of the cell. The diffusion of H^+ through ATP synthase results in the production of ATP.

Oxygen, from the air we breathe, allows electron transport to continue by functioning as the **final electron acceptor** of the electron-transport chain. This oxidizes cytochrome a_3, allowing electron transport and oxidative phosphorylation to continue. At the very last step of aerobic respiration, therefore, oxygen becomes reduced by the two electrons that were passed to the chain from NADH and $FADH_2$. This reduced oxygen binds two protons, and a molecule of water is formed. Since the oxygen atom is part of a molecule of oxygen gas (O_2), this last reaction can be shown as follows:

$$O_2 + 4\,e^- + 4\,H^+ \rightarrow 2\,H_2O$$

 Cyanide is a fast-acting lethal poison that produces such symptoms as rapid heart rate, tiredness, seizures, and headache. Cyanide poisoning can result in coma, and ultimately death, in the absence of quick treatment. The reason that cyanide is so deadly is that it has one very specific action: it blocks the transfer of electrons from cytochrome a_3 to oxygen. The effects are thus the same as would occur if oxygen were completely removed—aerobic cell respiration and the production of ATP by oxidative phosphorylation comes to a halt.

ATP Balance Sheet

Overview

There are two different methods of ATP formation in cell respiration. One method is the **direct** (also called **substrate-level**) **phosphorylation** that occurs in glycolysis (producing a net gain of 2 ATP) and the Krebs cycle (producing 1 ATP per cycle). These numbers are certain and constant. In the second method of ATP formation, **oxidative phosphorylation,** the numbers of ATP molecules produced vary under different conditions and for different kinds of cells. For many years, it was believed that 1 NADH yielded 3 ATP and that 1 $FADH_2$ yielded 2 ATP by oxidative phosphorylation. This gave a grand total of 36 to 38 molecules of ATP per glucose through cell respiration (see the footnote in table 5.1). Newer biochemical information, however,

suggests that these numbers may be overestimates, because, of the 36 to 38 ATP produced per glucose in the mitochondrion, only 30 to 32 ATP actually enter the cytoplasm of the cell.

Roughly three protons must pass through the respiratory assemblies and activate ATP synthase to produce 1 ATP. However, the newly formed ATP is in the mitochondrial matrix and must be moved into the cytoplasm; this transport also uses the proton gradient and costs one more proton. The ATP and H^+ are transported into the cytoplasm in exchange for ADP and P_i, which are transported into the mitochondrion. Thus, it effectively takes four protons to produce 1 ATP that enters the cytoplasm.

To summarize: The **theoretical ATP yield** is 36 to 38 ATP per glucose. The **actual ATP yield,** allowing for the costs of transport, is about 30 to 32 ATP per glucose. The details of how these numbers are obtained are described in the following section.

Detailed Accounting

Each NADH formed in the mitochondrion donates two electrons to the electron transport system at the first proton pump (fig. 5.10). The electrons are then passed to the second and third proton pumps, activating each of them in turn until the two electrons are ultimately passed to oxygen. The first and second pumps transport four protons each, and the third pump transports two protons, for a total of ten. Dividing ten protons by the four it takes to produce an ATP gives 2.5 ATP that are produced for every pair of electrons donated by an NADH. (There is no such thing as half an ATP; the decimal fraction simply indicates an average.)

Three molecules of NADH are formed with each Krebs cycle, and 1 NADH is also produced when pyruvate is converted into acetyl CoA (see fig. 5.6). Starting from one glucose, two Krebs cycles (producing 6 NADH) and two pyruvates converted to acetyl CoA (producing 2 NADH) yield 8 NADH. Multiplying by 2.5 ATP per NADH gives 20 ATP.

Electrons from $FADH_2$ are donated later in the electron-transport system than those donated by NADH; consequently, these electrons activate only the second and third proton pumps. Since the first proton pump is bypassed, the electrons passed from $FADH_2$ result in the pumping of only six protons (four by the second pump and two by the third pump). Since 1 ATP is

Table 5.1 ATP Yield per Glucose in Aerobic Respiration

Phases of Respiration	ATP Made Directly	Reduced Coenzymes	ATP Made by Oxidative Phosphorylation*
Glucose to pyruvate (in cytoplasm)	**2 ATP** (net gain)	2 NADH, but usually goes into mitochondria as 2 $FADH_2$	1.5 ATP per $FADH_2$ × 2 = **3 ATP**
Pyruvate to acetyl CoA (× 2 because one glucose yields 2 pyruvates)	None	1 NADH (× 2) = 2 NADH	2.5 ATP per NADH × 2 = **5 ATP**
Krebs cycle (× 2 because one glucose yields 2 Krebs cycles)	1 ATP (× 2) = **2 ATP**	3 NADH (× 2) 1 $FADH_2$ (× 2)	2.5 ATP per NADH × 3 = 7.5 ATP × 2 = **15 ATP** 1.5 ATP per $FADH_2$ × 2 = **3 ATP**
Subtotals	4 ATP		26 ATP
Grand Total		**30 ATP**	

*Theoretical estimates of ATP production from oxidative phosphorylation are 2 ATP per $FADH_2$ and 3 ATP per NADH. If these numbers are used, a total of 32 ATP will be calculated as arising from oxidative phosphorylation. This is increased to 34 ATP if the cytoplasmic NADH remains as NADH when it is shuttled into the mitochondrion. Adding these numbers to the ATP made directly gives a total of 38 ATP produced from a molecule of glucose. Estimates of the actual number of ATP obtained by the cell are lower because of the costs of transporting ATP out of the mitochondria.

produced for every four protons pumped, electrons derived from FADH$_2$ result in the formation of $6 \div 4 = 1.5$ ATP. Each Krebs cycle produces 1 FADH$_2$ and we get two Krebs cycles from one glucose, so there are 2 FADH$_2$ that give 2×1.5 ATP = 3 ATP.

The 23 ATP subtotal from oxidative phosphorylation we have at this point includes only the NADH and FADH$_2$ produced in the mitochondrion. Remember that glycolysis, which occurs in the cytoplasm, also produces 2 NADH. These cytoplasmic NADH cannot directly enter the mitochondrion, but there is a process by which their electrons can be "shuttled" in. The net effect of the most common shuttle is that a molecule of NADH in the cytoplasm is translated into a molecule of FADH$_2$ in the mitochondrion. The 2 NADH produced in glycolysis, therefore, usually become 2 FADH$_2$ and yield 2×1.5 ATP = 3 ATP by oxidative phosphorylation. (An alternative pathway, where the cytoplasmic NADH is transformed into mitochondrial NADH and produces 2×2.5 ATP = 5 ATP, is less common; however, this is the dominant pathway in the liver and heart, which are metabolically highly active.)

We now have a total 26 ATP (or, less commonly, 28 ATP) produced by oxidative phosphorylation from glucose. We can add the 2 ATP made by direct (substrate-level) phosphorylation in glycolysis and the 2 ATP made directly by the two Krebs cycles to give a grand total of 30 ATP (or, less commonly, 32 ATP) produced by the aerobic respiration of glucose (table 5.1).

Test Yourself Before You Continue

1. Compare the fate of pyruvic acid in aerobic respiration with its fate in anaerobic respiration.

2. Draw a simplified Krebs cycle using C$_2$ for acetic acid, C$_4$ for oxaloacetic acid, C$_5$ for alpha-ketoglutaric acid, and C$_6$ for citric acid. List the high-energy products that are produced at each turn of the Krebs cycle.

3. Using a diagram, show how electrons from NADH and FADH$_2$ are transferred by the cytochromes. Represent the oxidized and reduced forms of the cytochromes with Fe^{3+} and Fe^{2+}, respectively.

4. Explain how ATP molecules are produced in the process of oxidative phosphorylation.

5. Explain why a cell gets an average of 2.5 ATP from NADH in the mitochondrion and 1.5 ATP from FADH$_2$.

Metabolism of Lipids and Proteins

Triglycerides can be hydrolyzed into glycerol and fatty acids. The latter are of particular importance because they can be converted into numerous molecules of acetyl CoA that can enter Krebs cycles and generate a large amount of ATP. Amino acids derived from proteins also may be used for energy. This involves deamination (removal of the amine group) and the conversion of the remaining molecule into either pyruvic acid or one of the Krebs cycle molecules.

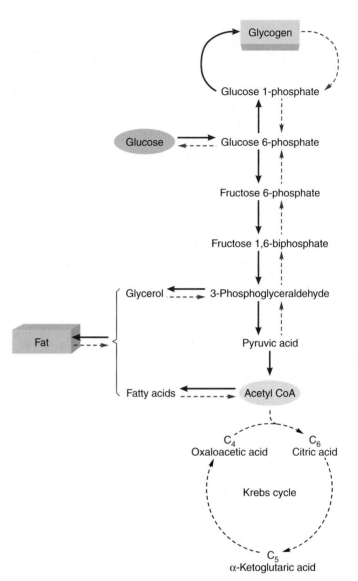

■ **Figure 5.11** **The conversion of glucose into glycogen and fat.** This occurs as a result of inhibition of respiratory enzymes when the cell has adequate amounts of ATP. Favored pathways are indicated by blue arrows.

Energy can be derived by the cellular respiration of lipids and proteins using the same aerobic pathway previously described for the metabolism of pyruvic acid. Indeed, some organs preferentially use molecules other than glucose as an energy source. Pyruvic acid and the Krebs cycle acids also serve as common intermediates in the interconversion of glucose, lipids, and amino acids.

When food energy is taken into the body faster than it is consumed, the concentration of ATP within body cells rises. Cells, however, do not store extra energy in the form of extra ATP. When cellular ATP concentrations rise because more energy (from food) is available than can be immediately used, ATP production is inhibited and glucose is instead converted into glycogen and fat (fig. 5.11).

Lipid Metabolism

When glucose is going to be converted into fat, glycolysis occurs and pyruvic acid is converted into acetyl CoA. Some of the glycolytic intermediates—phosphoglyceraldehyde and dihydroxyacetone phosphate—do not complete their conversion to pyruvic acid, however, and acetyl CoA does not enter a Krebs cycle. The acetic acid subunits of these acetyl CoA molecules can instead be used to produce a variety of lipids, including cholesterol (used in the synthesis of bile salts and steroid hormones), ketone bodies, and fatty acids (fig. 5.12). Acetyl CoA may thus be considered a branch point from which a number of different possible metabolic pathways may progress.

In the formation of fatty acids, a number of acetic acid (two-carbon) subunits are joined together to form the fatty acid chain. Six acetyl CoA molecules, for example, will produce a fatty acid that is twelve carbons long. When three of these fatty acids condense with one glycerol (derived from phosphoglyceraldehyde), a *triglyceride* (also called *triacylglycerol*) molecule is produced. The formation of fat, or **lipogenesis,** occurs primarily in adipose tissue and in the liver when the concentration of blood glucose is elevated following a meal.

Fat represents the major form of energy storage in the body. One gram of fat contains 9 kilocalories of energy, compared to 4 kilocalories for a gram of carbohydrates or protein. In a nonobese 70-kilogram (155-pound) man, 80% to 85% of the body's energy is stored as fat, which amounts to about 140,000 kilocalories. Stored glycogen, by contrast, accounts for less than 2,000 kilocalories, most of which (about 350 g) is stored in skeletal muscles and is available for use only by the muscles. The liver contains between 80 and 90 grams of glycogen, which can be converted to glucose and used by other organs. Protein accounts for 15% to 20% of the stored calories in the body, but protein is usually not used extensively as an energy source because that would involve the loss of muscle mass.

The ingestion of excessive calories increases fat production. The rise in blood glucose that follows carbohydrate-rich meals stimulates insulin secretion, and this hormone, in turn, promotes the entry of blood glucose into adipose cells. Increased availability of glucose within adipose cells, under conditions of high insulin secretion, promotes the conversion of glucose to fat (see figs. 5.11 and 5.12). The lowering of insulin secretion, conversely, promotes the breakdown of fat. This is exploited for weight reduction by low-carbohydrate diets.

Breakdown of Fat (Lipolysis)

When fat stored in adipose tissue is going to be used as an energy source, *lipase* enzymes hydrolyze triglycerides into glycerol and free fatty acids in a process called **lipolysis.** These molecules (primarily the free fatty acids) serve as *blood-borne energy carriers* that can be used by the liver, skeletal muscles, and other organs for aerobic respiration.

A few organs can utilize glycerol for energy by virtue of an enzyme that converts glycerol to phosphoglyceraldehyde. Free fatty acids, however, serve as the major energy source derived from triglycerides. Most fatty acids consist of a long hydrocarbon chain with a carboxyl, or carboxylic acid group (COOH) at one end. In a process known as **β-oxidation** (β is the Greek letter *beta*), enzymes

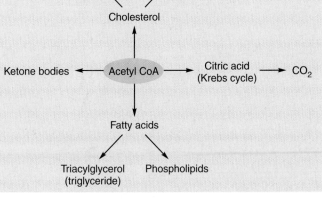

Figure 5.12 Divergent metabolic pathways for acetyl coenzyme A. Acetyl CoA is a common substrate that can be used to produce a number of chemically related products.

remove two-carbon acetic acid molecules from the acid end of a fatty acid chain. This results in the formation of acetyl CoA, as the third carbon from the end becomes oxidized to produce a new carboxyl group. The fatty acid chain is thus decreased in length by two carbons. The process of oxidation continues until the entire fatty acid molecule is converted to acetyl CoA (fig. 5.13).

A sixteen-carbon-long fatty acid, for example, yields eight acetyl CoA molecules. Each of these can enter a Krebs cycle and produce ten ATP per turn of the cycle, so that eight times ten, or eighty, ATP are produced. In addition, each time an acetyl CoA molecule is formed and the end carbon of the fatty acid chain is oxidized, one NADH and one $FADH_2$ are produced. Oxidative phosphorylation produces 2.5 ATP per NADH and 1.5 ATP per $FADH_2$. For a sixteen-carbon-long fatty acid, these four ATP molecules would be formed seven times (producing four times seven, or twenty-eight, ATP). Not counting the single ATP used to start β-oxidation (fig. 5.13), this fatty acid could yield a grand total of 28 + 80, or 108 ATP molecules!

Clinical Investigation Clues

Remember that Brenda's coach advised her to exercise more gradually. Under these conditions, skeletal muscles utilize a higher proportion of fatty acids for energy.

If her skeletal muscles used fatty acids more for energy, how would this help to alleviate her pain and fatigue?

Function of Brown Fat

The amount of **brown fat** in the body is greatest at the time of birth. Brown fat is the major site for thermogenesis (heat production) in the newborn, and is especially prominent around the kidneys and adrenal glands. Smaller amounts are also found around the blood vessels of the chest and neck. In response to regulation by thyroid hormone (see chapter 11) and norepinephrine from sympathetic nerves (see chapter 9), brown fat produces a unique

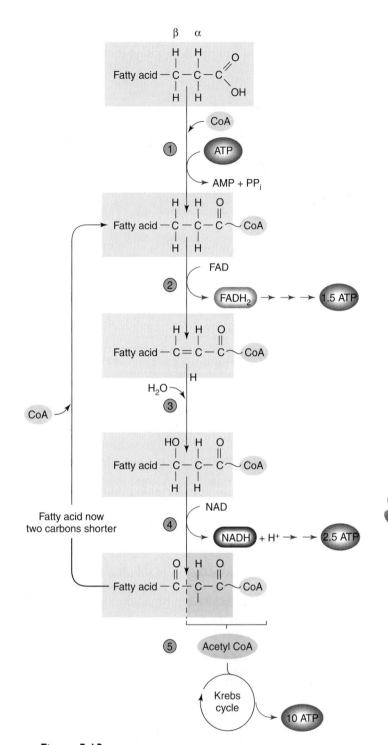

β   α

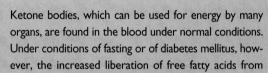

Figure 5.13 *Beta-oxidation of a fatty acid.* After the attachment of coenzyme A to the carboxyl group (*step 1*), a pair of hydrogens is removed from the fatty acid and used to reduce one molecule of FAD (*step 2*). When this electron pair is donated to the cytochrome chain, 1.5 ATP are produced. The addition of a hydroxyl group from water (*step 3*), followed by the oxidation of the β-carbon (*step 4*), results in the production of 2.5 ATP from the electron pair donated by NADH. The bond between the α and β carbons in the fatty acid is broken (*step 5*), releasing acetyl coenzyme A and a fatty acid chain that is two carbons shorter than the original. With the addition of a new coenzyme A to the shorter fatty acid, the process begins again (*step 2*), as acetyl CoA enters the Krebs cycle and generates ten ATP.

uncoupling protein. This protein causes H⁺ to leak out of the inner mitochondrial membrane, so that less H⁺ is available to pass through the respiratory assemblies and drive ATP synthase activity. Therefore, less ATP is made by the electron-transport system than would otherwise be the case. Lower ATP concentrations cause the electron-transport system to be more active and generate more heat from the respiration of fatty acids. This extra heat may be needed to prevent hypothermia (low body temperature) in newborns.

Ketone Bodies

Even when a person is not losing weight, the triglycerides in adipose tissue are continuously being broken down and resynthesized. New triglycerides are produced, while others are hydrolyzed into glycerol and fatty acids. This turnover ensures that the blood will normally contain a sufficient level of fatty acids for aerobic respiration by skeletal muscles, the liver, and other organs. When the rate of lipolysis exceeds the rate of fatty acid utilization—as it may in starvation, dieting, and in diabetes mellitus—the blood concentration of fatty acids increases.

If the liver cells contain sufficient amounts of ATP so that further production of ATP is not needed, some of the acetyl CoA derived from fatty acids is channeled into an alternate pathway. This pathway involves the conversion of two molecules of acetyl CoA into four-carbon-long acidic derivatives, *acetoacetic acid* and β-*hydroxybutyric acid.* Together with *acetone,* which is a three-carbon-long derivative of acetoacetic acid, these products are known as **ketone bodies** (see chapter 2, fig. 2.19).

> **CLINICAL**
>
> Ketone bodies, which can be used for energy by many organs, are found in the blood under normal conditions. Under conditions of fasting or of diabetes mellitus, however, the increased liberation of free fatty acids from adipose tissue results in the increased production of ketone bodies by the liver. The secretion of abnormally high amounts of ketone bodies into the blood produces **ketosis,** which is one of the signs of fasting or an uncontrolled diabetic state. A person in this condition may also have a sweet-smelling breath due to the presence of acetone, which is volatile and leaves the blood in the exhaled air.

Amino Acid Metabolism

Nitrogen is ingested primarily as proteins, enters the body as amino acids, and is excreted mainly as urea in the urine. In childhood, the amount of nitrogen excreted may be less than the amount ingested because amino acids are incorporated into proteins during growth. Growing children are thus said to be in a state of *positive nitrogen balance.* People who are starving or suffering from prolonged wasting diseases, by contrast, are in a state of *negative nitrogen balance;* they excrete more nitrogen than they ingest because they are breaking down their tissue proteins.

Healthy adults maintain a state of nitrogen balance, in which the amount of nitrogen excreted is equal to the amount ingested. This does not imply that the amino acids ingested are unnecessary; on the contrary, they are needed to replace the protein that is "turned over" each day. When more amino acids are ingested than are needed to replace proteins, the excess amino acids are not stored as additional protein (one cannot build muscles simply by eating large amounts of protein). Rather, the amine groups can be removed, and the "carbon skeletons" of the organic acids that are left can be used for energy or converted to carbohydrate and fat.

Transamination

An adequate amount of all twenty amino acids is required to build proteins for growth and to replace the proteins that are turned over. However, only eight of these (nine in children) cannot be produced by the body and must be obtained in the diet. These are the **essential amino acids** (table 5.2). The remaining amino acids are "nonessential" only in the sense that the body can produce them if provided with a sufficient amount of carbohydrates and the essential amino acids.

Pyruvic acid and the Krebs cycle acids are collectively termed *keto acids* because they have a ketone group; these should not be confused with the ketone bodies (derived from acetyl CoA) discussed in the previous section. Keto acids can be converted to amino acids by the addition of an amine (NH_2) group. This amine group is usually obtained by "cannibalizing" another amino acid; in this process, a new amino acid is formed as the one that was cannibalized is converted to a new keto acid. This type of reaction, in which the amine group is transferred from one amino acid to form another, is called **transamination** (fig. 5.14).

Each transamination reaction is catalyzed by a specific enzyme (a transaminase) that requires vitamin B_6 (pyridoxine) as a coenzyme. The amine group from glutamic acid, for example, may be transferred to either pyruvic acid or oxaloacetic acid. The former reaction is catalyzed by the enzyme alanine transaminase (ALT); the latter reaction is catalyzed by aspartate transaminase

(AST). These enzyme names reflect the fact that the addition of an amine group to pyruvic acid produces the amino acid alanine; the addition of an amine group to oxaloacetic acid produces the amino acid known as aspartic acid (fig. 5.14).

Oxidative Deamination

As shown in figure 5.15, glutamic acid can be formed through transamination by the combination of an amine group with α-ketoglutaric acid. Glutamic acid is also produced in the liver from the ammonia that is generated by intestinal bacteria and carried to the liver in the hepatic portal vein. Since free ammonia is very toxic, its removal from the blood and incorporation into glutamic acid is an important function of a healthy liver.

Table 5.2 The Essential and Nonessential Amino Acids

Essential Amino Acids	Nonessential Amino Acids
Lysine	Aspartic acid
Tryptophan	Glutamic acid
Phenylalanine	Proline
Threonine	Glycine
Valine	Serine
Methionine	Alanine
Leucine	Cysteine
Isoleucine	Arginine
Histidine (in children)	Asparagine
	Glutamine
	Tyrosine

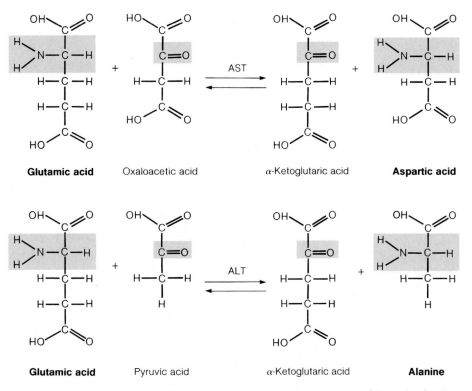

Glutamic acid   Oxaloacetic acid   α-Ketoglutaric acid   **Aspartic acid**

Glutamic acid   Pyruvic acid   α-Ketoglutaric acid   **Alanine**

Figure 5.14 **Two important transamination reactions.** The areas shaded in blue indicate the parts of the molecules that are changed. (AST = aspartate transaminase; ALT = alanine transaminase. The amino acids are identified in boldface.)

If there are more amino acids than are needed for protein synthesis, the amine group from glutamic acid may be removed and excreted as *urea* in the urine (fig. 5.15). The metabolic pathway that removes amine groups from amino acids—leaving a keto acid and ammonia (which is converted to urea)—is known as **oxidative deamination.**

A number of amino acids can be converted into glutamic acid by transamination. Since glutamic acid can donate amine groups to urea (through deamination), it serves as a channel through which other amino acids can be used to produce keto

acids (pyruvic acid and Krebs cycle acids). These keto acids may then be used in the Krebs cycle as a source of energy (fig. 5.16).

Depending upon which amino acid is deaminated, the keto acid left over may be either pyruvic acid or one of the Krebs cycle acids. These can be respired for energy, converted to fat, or converted to glucose. In the last case, the amino acids are eventually changed to pyruvic acid, which is used to form glucose. This process—the formation of glucose from amino acids or other noncarbohydrate molecules—is called *gluconeogenesis,* as mentioned previously in connection with the Cori cycle.

The main substrates for gluconeogenesis are the three-carbon-long molecules of alanine (an amino acid), lactic acid, and glycerol. This illustrates the interrelationship between amino acids, carbohydrates, and fat, as shown in figure 5.17. Recent experiments in humans have suggested that, even in the early stages of fasting, most of the glucose secreted by the liver is derived through gluconeogenesis. Findings indicate that hydrolysis of liver glycogen (glycogenolysis) contributes only 36% of the glucose secreted during the early stages of a fast. At 42 hours of fasting, all of the glucose secreted by the liver is being produced by gluconeogenesis.

Uses of Different Energy Sources

The blood serves as a common trough from which all the cells in the body are fed. If all cells used the same energy source, such as glucose, this source would quickly be depleted and cellular starvation would occur. Normally however, the blood contains

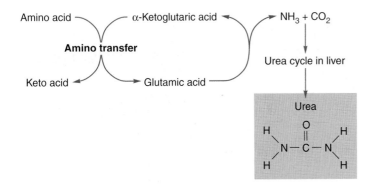

Figure 5.15 Oxidative deamination. Glutamic acid is converted to α-ketoglutaric acid as it donates its amine group to the metabolic pathway that results in the formation of urea.

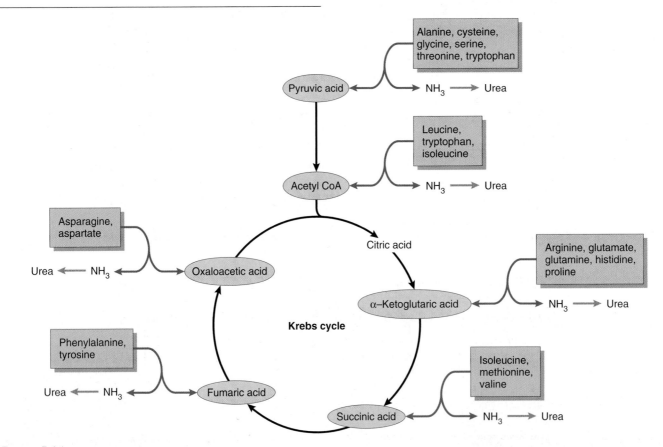

Figure 5.16 Pathways by which amino acids can be catabolized for energy. These pathways are indirect for some amino acids, which first must be transaminated into other amino acids before being converted into keto acids by deamination.

a variety of energy sources from which to draw: glucose and ketone bodies that come from the liver, fatty acids from adipose tissue, and lactic acid and amino acids from muscles. Some organs preferentially use one energy source more than the others, so that each energy source is "spared" for organs with strict energy needs.

The brain uses blood glucose as its major energy source. Under fasting conditions, blood glucose is supplied primarily by the liver through glycogenolysis and gluconeogenesis. In addition, the blood glucose concentration is maintained because many organs spare glucose by using fatty acids, ketone bodies, and lactic acid as energy sources (table 5.3). During severe starvation, the brain also gains some ability to metabolize ketone bodies for energy.

As mentioned earlier, lactic acid produced anaerobically during exercise can be used for energy following the cessation of exercise. The lactic acid, under aerobic conditions, is reconverted to pyruvic acid, which then enters the aerobic respiratory pathway. The extra oxygen required to metabolize lactic acid contributes to the *oxygen debt* following exercise (see chapter 12).

Clinical Investigation Clues

Remember that Brenda found herself gasping and panting for air more than her teammates.

What is the term for the extra oxygen she needs following exercise? What function does it serve?

What would cause her to need less, and thus to gasp and pant less following exercise?

Test Yourself Before You Continue

1. Construct a flowchart to show the metabolic pathway by which glucose can be converted to fat. Indicate only the major intermediates involved (not all of the steps of glycolysis).

2. Define the terms *lipolysis* β*-oxidation* and explain, in general terms, how fat can be used for energy.

3. Describe transamination and deamination and explain their functional significance.

4. List five blood-borne energy carriers and explain, in general terms, how these are used as sources of energy.

Table 5.3 Relative Importance of Different Molecules in the Blood with Respect to the Energy Requirements of Different Organs

Organ	Glucose	Fatty Acids	Ketone Bodies	Lactic Acid
Brain	+++	−	+	−
Skeletal muscles (resting)	+	+++	+	−
Liver	+	+++	++	+
Heart	+	++	+	+

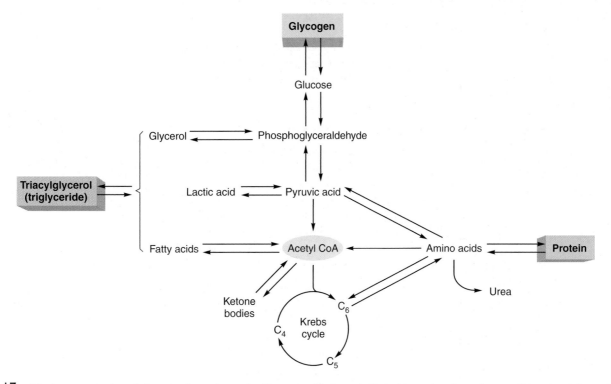

Figure 5.17 **The interconversion of glycogen, fat, and protein.** These simplified metabolic pathways show how glycogen, fat, and protein can be interconverted. Note that while most reactions are reversible, the reaction from pyruvic acid to acetyl CoA is not. This is because a CO_2 is removed in the process. (Only plants, in a phase of photosynthesis called the dark reaction, can use CO_2 to produce glucose.)

HPer Links of Metabolism Concepts to the Body Systems

Integumentary System

- The skin synthesizes vitamin D from a derivative of cholesterol ....(p. 625)
- The metabolic rate of the skin varies greatly, depending upon ambient temperature ....(p. 428)

Nervous System

- The aerobic respiration of glucose serves most of the energy needs of the brain ....(p. 119)
- Regions of the brain with a faster metabolic rate, resulting from increased brain activity, receive a more abundant blood supply than regions with a slower metabolic rate ....(p. 427)

Endocrine System

- Hormones that bind to receptors in the plasma membrane of their target cells activate enzymes in the target cell cytoplasm ....(p. 294)
- Hormones that bind to nuclear receptors in their target cells alter the target cell metabolism by regulating gene expression ....(p. 292)
- Hormonal secretions from adipose cells regulate hunger and metabolism ..(p. 606)
- Anabolism and catabolism are regulated by a number of hormones ....(p. 609)
- Insulin stimulates the synthesis of glycogen and fat ....(p. 611)
- The adrenal hormones stimulate the breakdown of glycogen, fat, and protein ....(p. 619)
- Thyroxine stimulates the production of a protein that uncouples oxidative phosphorylation. This helps to increase the body's metabolic rate ....(p. 620)
- Growth hormone stimulates protein synthesis ....(p. 621)

Muscular System

- The intensity of exercise that can be performed aerobically depends on a person's maximal oxygen uptake and lactate threshold ....(p. 343)

- The body consumes extra oxygen for a period of time after exercise has ceased. This extra oxygen is used to repay the oxygen debt incurred during exercise ....(p. 344)
- Glycogenolysis and gluconeogenesis by the liver help to supply glucose for exercising muscles ....(p. 343)
- Trained athletes obtain a higher proportion of skeletal muscle energy from the aerobic respiration of fatty acids than do nonathletes ....(p. 346)
- Muscle fatigue is associated with anaerobic respiration and the production of lactic acid ....(p. 346)
- The proportion of energy derived from carbohydrates or lipids by exercising skeletal muscles depends on the intensity of the exercise ....(p. 343)

Circulatory System

- Metabolic acidosis may result from excessive production of either ketone bodies or lactic acid ....(p. 377)
- The metabolic rate of skeletal muscles determines the degree of blood vessel dilation, and thus the rate of blood flow to the organ ....(p. 424)
- Atherosclerosis of coronary arteries can force a region of the heart to metabolize anaerobically and produce lactic acid. This is associated with angina pectoris .(p. 397)

Respiratory System

- Ventilation oxygenates the blood going to the cells for aerobic cell respiration and removes the carbon dioxide produced by the cells ....(p. 480)
- Breathing is regulated primarily by the effects of carbon dioxide produced by aerobic cell respiration ....(p. 500)

Urinary System

- The kidneys eliminate urea and other waste products of metabolism from the blood plasma ....(p. 539)

Digestive System

- The liver contains enzymes needed for many metabolic reactions involved in regulating the blood glucose and lipid concentrations ....(p. 579)
- The pancreas produces many enzymes needed for the digestion of food in the small intestine ....(p. 582)
- The digestion and absorption of carbohydrates, lipids, and proteins provides the body with the substrates used in cell metabolism ....(p. 587)
- Vitamins A and D help to regulate metabolism through the activation of nuclear receptors, which bind to regions of DNA ....(p. 601)

Reproductive System

- The sperm do not contribute mitochondria to the fertilized oocyte ....(p. 58)
- The endometrium contains glycogen that nourishes the developing embryo .(p. 663)

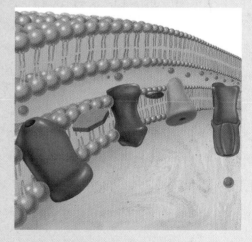

Summary

Glycolysis and the Lactic Acid Pathway   104

I. Glycolysis refers to the conversion of glucose to two molecules of pyruvic acid.

   A. In the process, two molecules of ATP are consumed and four molecules of ATP are formed. Thus, there is a net gain of two ATP.

   B. In the steps of glycolysis, two pairs of hydrogens are released. Electrons from these hydrogens reduce two molecules of NAD.

II. When respiration is anaerobic, reduced NAD is oxidized by pyruvic acid, which accepts two hydrogen atoms and is thereby reduced to lactic acid.

   A. Skeletal muscles use anaerobic respiration and thus produce lactic acid during exercise. Heart muscle respires anaerobically for just a short time, under conditions of ischemia.

   B. Lactic acid can be converted to glucose in the liver by a process called gluconeogenesis.

Aerobic Respiration   108

I. The Krebs cycle begins when coenzyme A donates acetic acid to an enzyme that adds it to oxaloacetic acid to form citric acid.

   A. Acetyl CoA is formed from pyruvic acid by the removal of carbon dioxide and two hydrogens.

   B. The formation of citric acid begins a cyclic pathway that ultimately forms a new molecule of oxaloacetic acid.

   C. As the Krebs cycle progresses, one molecule of ATP is formed, and three molecules of NAD and one of FAD are reduced by hydrogens from the Krebs cycle.

II. Reduced NAD and FAD donate their electrons to an electron-transport chain of molecules located in the cristae.

   A. The electrons from NAD and FAD are passed from one cytochrome of the electron-transport chain to the next in a series of coupled oxidation-reduction reactions.

   B. As each cytochrome ion gains an electron, it becomes reduced; as it passes the electron to the next cytochrome, it becomes oxidized.

   C. The last cytochrome becomes oxidized by donating its electron to oxygen, which functions as the final electron acceptor.

   D. When one oxygen atom accepts two electrons and two protons, it becomes reduced to form water.

   E. The energy provided by electron transport is used to form ATP from ADP and P_i in the process known as oxidative phosphorylation.

III. Thirty to thirty-two molecules of ATP are produced by the aerobic respiration of one glucose molecule. Of these, two are produced in the cytoplasm by glycolysis and the remainder are produced in the mitochondria.

IV. The formation of glycogen from glucose is called glycogenesis; the breakdown of glycogen is called glycogenolysis.

   A. Glycogenolysis yields glucose 6-phosphate, which can enter the pathway of glycolysis.

   B. The liver contains an enzyme (which skeletal muscles do not) that can produce free glucose from glucose 6-phosphate. Thus, the liver can secrete glucose derived from glycogen.

V. Carbohydrate metabolism is influenced by the availability of oxygen and by a negative feedback effect of ATP on glycolysis and the Krebs cycle.

Metabolism of Lipids and Proteins   114

I. In lipolysis, triglycerides yield glycerol and fatty acids.

   A. Glycerol can be converted to phosphoglyceraldehyde and used for energy.

   B. In the process of β-oxidation of fatty acids, a number of acetyl CoA molecules are produced.

   C. Processes that operate in the reverse direction can convert glucose to triglycerides.

II. Amino acids derived from the hydrolysis of proteins can serve as sources of energy.

   A. Through transamination, a particular amino acid and a particular keto acid (pyruvic acid or one of the Krebs cycle acids) can serve as substrates to form a new amino acid and a new keto acid.

   B. In oxidative deamination, amino acids are converted into keto acids as their amino group is incorporated into urea.

III. Each organ uses certain blood-borne energy carriers as its preferred energy source.

   A. The brain has an almost absolute requirement for blood glucose as its energy source.

   B. During exercise, the needs of skeletal muscles for blood glucose can be met by glycogenolysis and by gluconeogenesis in the liver.

Review Activities

Test Your Knowledge of Terms and Facts

1. The net gain of ATP per glucose molecule in anaerobic respiration (lactic acid fermentation) is _____; the net gain in aerobic respiration is generally _____.
   a. 2;4
   b. 2;30
   c. 30;2
   d. 24;38

2. In anaerobic respiration in humans, the oxidizing agent for NADH (that is, the molecule that removes electrons from NADH) is
   a. pyruvic acid.
   b. lactic acid.
   c. citric acid.
   d. oxygen.

3. When skeletal muscles lack sufficient oxygen, there is an increased blood concentration of
   a. pyruvic acid.
   b. glucose.
   c. lactic acid.
   d. ATP.

4. The conversion of lactic acid to pyruvic acid occurs
   a. in anaerobic respiration.
   b. in the heart, where lactic acid is aerobically respired.
   c. in the liver, where lactic acid can be converted to glucose.
   d. in both a and b.
   e. in both b and c.

5. Which of these statements about the oxygen in the air we breathe is true?
   a. It functions as the final electron acceptor of the electron-transport chain.
   b. It combines with hydrogen to form water.
   c. It combines with carbon to form CO_2.
   d. Both a and b are true.
   e. Both a and c are true.

6. In terms of the number of ATP molecules directly produced, the major energy-yielding process in the cell is
   a. glycolysis.
   b. the Krebs cycle.
   c. oxidative phosphorylation.
   d. gluconeogenesis.

7. Ketone bodies are derived from
   a. fatty acids.
   b. glycerol.
   c. glucose.
   d. amino acids.

8. The conversion of glycogen to glucose 6-phosphate occurs in
   a. the liver.
   b. skeletal muscles.
   c. both a and b.

9. The conversion of glucose 6-phosphate to free glucose, which can be secreted into the blood, occurs in

   a. the liver.
   b. skeletal muscles.
   c. both a and b.

10. The formation of glucose from pyruvic acid derived from lactic acid, amino acids, or glycerol is called
    a. glycogenesis.
    b. glycogenolysis.
    c. glycolysis.
    d. gluconeogenesis.

11. Which of these organs has an almost absolute requirement for blood glucose as its energy source?
    a. liver
    b. brain
    c. skeletal muscles
    d. heart

12. When amino acids are used as an energy source,
    a. oxidative deamination occurs.
    b. pyruvic acid or one of the Krebs cycle acids (keto acids) is formed.
    c. urea is produced.
    d. all of these occur.

13. Intermediates formed during fatty acid metabolism can enter the Krebs cycle as
    a. keto acids.
    b. acetyl CoA.
    c. Krebs cycle molecules.
    d. pyruvic acid.

Test Your Understanding of Concepts and Principles

1. State the advantages and disadvantages of anaerobic respiration.[1]

2. What purpose is served by the formation of lactic acid during anaerobic respiration? How is this accomplished during aerobic respiration?

3. Describe the effect of cyanide on oxidative phosphorylation and on the Krebs cycle. Why is cyanide deadly?

4. Describe the metabolic pathway by which glucose can be converted into fat. How can end-product inhibition by ATP favor this pathway?

5. Describe the metabolic pathway by which fat can be used as a source of energy and explain why the metabolism of fatty acids can yield more ATP than the metabolism of glucose.

6. Explain how energy is obtained from the metabolism of amino acids. Why does a starving person have a high concentration of urea in the blood?

7. Explain why the liver is the only organ able to secrete glucose into the blood. What are the possible sources of hepatic glucose?

8. Explain the two possible meanings of the term anaerobic respiration. Why is

the production of lactic acid sometimes termed a "fermentation" pathway?

9. Explain the function of brown fat. What does its mechanism imply about the effect of ATP concentrations on the rate of cell respiration?

10. What three molecules serve as the major substrates for gluconeogenesis? Describe the situations in which each one would be involved in this process. Why can't fatty acids be used as a substrate for gluconeogenesis? (Hint: Count the carbons in acetyl CoA and pyruvic acid.)

---

[1]Note: This question is answered in the chapter 5 Study Guide found on the Online Learning Center at www.mhhe.com/fox8.

Test Your Ability to Analyze and Apply Your Knowledge

1. A friend, wanting to lose weight, eliminates all fat from her diet. How would this help her to lose weight? Could she possibly gain weight on this diet? How? Discuss the health consequences of such a diet.

2. Suppose a drug is developed that promotes the channeling of H^+ out of the intermembrane space into the matrix of the mitochondria of adipose cells. How could this drug affect the production of ATP, body temperature, and body weight?

3. For many years, the total number of molecules of ATP produced for each molecule of glucose in aerobic respiration was given as 38. Later, it was estimated to be closer to 36, and now it is believed to be closer to 30. What factors must be considered in estimating the yield of ATP molecules? Why are the recent numbers considered to be approximate values?

Related Websites

Check out the Links Library at www.mhhe.com/fox8 for links to sites containing resources related to cell respiration and metabolism. These links are monitored to ensure current URLs.

6

Interactions Between Cells and the Extracellular Environment

Objectives

After studying this chapter, you should be able to . . .

1. describe the composition of the extracellular environment.

2. describe diffusion and explain its physical basis.

3. explain how nonpolar molecules, inorganic ions, and water can diffuse through a cell membrane.

4. state the factors that influence the rate of diffusion through cell membranes.

5. define the term *osmosis* and describe the conditions required for osmosis to occur.

6. define the terms *osmolality* and *osmotic pressure* and explain how these factors relate to osmosis.

7. define the term *tonicity* and distinguish between isotonic, hypertonic, and hypotonic solutions.

8. describe the characteristics of carrier-mediated transport.

9. describe the facilitated diffusion of glucose through cell membranes and give examples of its occurrence in the body.

10. explain what is meant by active transport and describe how the Na^+/K^+ pumps work.

11. explain how an equilibrium potential is produced when only one ion is able to diffuse through a cell membrane.

12. explain why the resting membrane potential is slightly different than the potassium equilibrium potential and describe the effect of the extracellular potassium concentration on the resting membrane potential.

13. discuss the role of the Na^+/K^+ pumps in the maintenance of the resting membrane potential.

14. distinguish between the different types of cell signaling.

Chapter at a Glance

Take Advantage of the Technology

Visit the Online Learning Center for these additional study resources.

- Interactive quizzing
- Online study guide
- Current news feeds
- Crossword puzzles
- Vocabulary flashcards
- Labeling activities

www.mhhe.com/fox8

Clinical Investigation

Jessica, a student of physiology, is constantly drinking from a water bottle yet claims to be constantly thirsty. During her physiology laboratory exercise involving urinalysis, she discovers that she has a significant amount of glucose in her urine. Alarmed, because urine normally should contain little or no glucose, she seeks medical attention. As a result of a later medical examination, she learns that she has hyperglycemia, hyperkalemia, and a high plasma osmolality. When she shows the doctor her EKG that she recorded in the physiology lab, he remarks that it has some abnormalities.

How might Jessica's symptoms and medical findings be related?

Extracelluar Environment

The extracellular environment surrounding cells consists of a fluid compartment, in which molecules are dissolved, and a matrix of polysaccharides and proteins that give form to the tissues. Interactions between the intracellular and extracellular environment occur across the plasma membrane.

The extracellular environment includes all constituents of the body located outside of the cells. The cells of our body must receive nourishment from, and release their waste products into, the extracellular environment. Further, the different cells of a tissue, the cells of different tissues within an organ, and the cells of different organs interact with each other through chemical regulators secreted into the extracellular environment.

Body Fluids

The water content of the body is divided into two compartments. Approximately 67% of the total body water is contained within cells, in the **intracellular compartment.** The remaining 33% of the total body water comprises the **extracellular compartment.**

About 20% of this extracellular fluid is contained within the vessels of the cardiovascular system, where it comprises the fluid portion of the blood, or **blood plasma.**

The blood transports oxygen from the lungs to the body cells, and carbon dioxide from the body cells to the lungs. It also transports nutrients derived from food in the intestine to the body cells; other nutrients between organs (such as glucose from the liver to the brain, or lactic acid from muscles to the liver); metabolic wastes from the body cells to the liver and kidneys for elimination in the bile and urine, respectively; and regulatory molecules called hormones from endocrine glands to the cells of their target organs.

The remaining 80% of the extracellular fluid is located outside of the vascular system, and comprises **tissue fluid,** also called **interstitial fluid.** This fluid is contained in the gel-like extracellular matrix, as described in the next section. Body fluid distribution is illustrated in figure 14.8, p. 413, in conjunction with a discussion of the cardiovascular system. This is because the interstitial fluid is formed continuously from blood plasma, and it continuously returns to the blood plasma through mechanisms described in chapter 14 (see fig. 14.9). Oxygen, nutrients, and regulatory molecules traveling in the blood must first pass into the interstitial fluid before reaching the body cells; waste products and hormone secretions from the cells must first pass into the interstitial fluid before reaching the blood plasma (fig. 6.1).

Extracellular Matrix

The cells that comprise the organs of our body are embedded within the extracellular material of connective tissues. This material is called the **extracellular matrix,** and it consists of the protein fibers *collagen* and *elastin* (see chapter 2, fig. 2.28), as well as gel-like *ground substance.* The interstitial fluid referred to previously exists primarily in the hydrated gel of the ground substance.

Although the ground substance seemingly lacks form (is amorphous) when viewed under a microscope, it is actually a

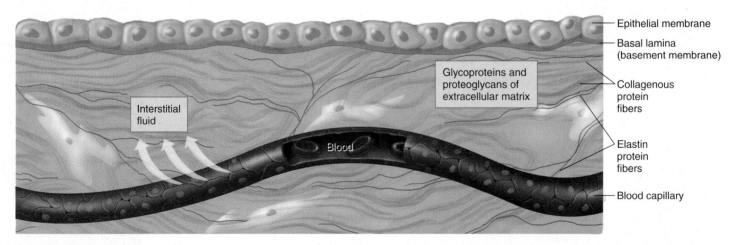

Epithelial membrane

Basal lamina (basement membrane)

Glycoproteins and proteoglycans of extracellular matrix

Collagenous protein fibers

Interstitial fluid

Elastin protein fibers

Blood

Blood capillary

■ **Figure 6.1   The extracellular environment.** The extracellular environment contains fluid, as interstitial, or tissue, fluid, within a matrix of glycoproteins and proteoglycans. This fluid, derived from blood plasma, provides nutrients and regulatory molecules to the cells. The extracellular environment is supported by collagen and elastin protein fibers, which also form the basal lamina below epithelial membranes.

highly functional, complex organization of molecules chemically linked to the extracellular protein fibers of collagen and elastin, as well as to the carbohydrates that cover the outside surface of the cell's plasma membrane (see chapter 3, fig. 3.2). The gel is composed of *glycoproteins* (proteins with numerous side chains of sugars) and molecules called *proteoglycans.* These molecules (formerly called mucopolysaccharides) are composed primarily of polysaccharides and have a high content of bound water molecules.

The collagen and elastin fibers have been likened to the reinforcing iron bars in concrete—they provide structural strength to the connective tissues. One type of collagen (there are about fifteen different types known) constitutes the *basal lamina* (or *basement membrane*) underlying epithelial membranes (see chapter 1, fig. 1.11). By forming chemical bonds between the carbohydrates on the outside surface of the plasma membrane of the epithelial cells, and the glycoproteins and proteoglycans of the matrix in the connective tissues, the basal lamina helps to wed the epithelium to its underlying connective tissues (fig. 6.1)

There is an important family of enzymes that can break down extracellular matrix proteins. These enzymes are called **matrix metalloproteinases (MMPs)** because of their need for a zinc ion cofactor. MMPs are required for tissue remodeling (for example, during embryonic development and wound healing), and for migration of phagocytic cells and other white blood cells during the fight against infection. MMPs are secreted as inactive enzymes and then activated extracellularly. They can contribute to disease processes, however, if they are produced or activated inappropriately. For example, cancer cells that become invasive (that metastasize, or spread to different locations) produce active MMPs, which break down the collagen of the basal lamina and allow the cancerous cells to migrate. The destruction of cartilage protein in arthritis may also involve the action of these enzymes, and MMPs have been implicated in the pathogenesis of such neural diseases as multiple sclerosis, Alzheimer's disease, and others. Therefore, scientists are attempting to develop drugs that may be able to treat these and other diseases by selectively blocking different matrix metalloproteinases.

Integrins are a class of glycoproteins that extend from the cytoskeleton within a cell, through its plasma membrane, and into the extracellular matrix. By binding to components within the matrix, they serve as a sort of "glue" (or adhesion molecule) between cells and the extracellular matrix. Moreover, by physically joining the intracellular to the extracellular compartments, they serve to relay signals between these two compartments (or integrate these two compartments—hence the origin of the term integrin.). Interestingly, certain snake venoms slow blood clotting by blocking integrin-binding sites on blood platelets, preventing them from sticking together (see chapter 13 for a discussion of blood clotting).

Categories of Transport Across the Plasma Membrane

The plasma (cell) membrane separates the intracellular environment from the extracellular environment. Molecules that move from the blood to the interstitial fluid, or molecules that move within the interstitial fluid between different cells, must eventually come into contact with the plasma membrane surrounding the cells. Some of these molecules may be able to penetrate the membrane, while others may not. Similarly, some intracellular molecules can penetrate, or "permeate," the plasma membrane and some cannot. The plasma membrane is thus said to be **selectively permeable.**

The plasma membrane is generally not permeable to proteins, nucleic acids, and other molecules needed for the structure and function of the cell. It is, however, permeable to many other molecules, permitting the two-way traffic of nutrients and wastes needed to sustain metabolism. The plasma membrane is also selectively permeable to certain ions; this permits electrochemical currents across the membrane used for production of impulses in nerve and muscle cells.

The mechanisms involved in the transport of molecules and ions through the cell membrane may be divided into two categories: (1) transport that requires the action of specific carrier proteins in the membrane, called **carrier-mediated transport;** and (2) transport through the membrane that is not carrier mediated. Carrier-mediated transport may be further subdivided into *facilitated diffusion* and *active transport,* both of which will be described later. Membrane transport that does not use carrier proteins involves the *simple diffusion* of ions, lipid-soluble molecules, and water through the membrane. *Osmosis* is the net diffusion of solvent (water) through a membrane.

Membrane transport processes may also be categorized by their energy requirements. **Passive transport** is the net movement of molecules and ions across a membrane from higher to lower concentration (down a concentration gradient); it does not require metabolic energy. Passive transport includes simple diffusion, osmosis, and facilitated diffusion. **Active transport** is net movement across a membrane that occurs against a concentration gradient (to the region of higher concentration). Active transport requires the expenditure of metabolic energy (ATP) and involves specific carrier proteins.

Test Yourself before You Continue

1. Describe the distribution of fluid in the body.
2. Describe the composition of the extracellular matrix and explain the importance of the matrix metalloproteinases.
3. List the subcategories of passive transport and distinguish between passive transport and active transport.

Diffusion and Osmosis

Net diffusion of a molecule or ion through a cell membrane always occurs in the direction of its lower concentration. Nonpolar molecules can penetrate the phospholipid barrier, and small inorganic ions can pass through channels in the membrane. The net diffusion of water through a membrane is known as osmosis.

A *solution* consists of the *solvent,* water, and *solute* molecules that are dissolved in the water. The molecules of a solution (solvent and solute) are in a constant state of random motion as a result of their thermal (heat) energy. If there is a *concentration difference,* or *concentration gradient,* between two regions of a solution, this random motion tends to eliminate the concentration difference as the molecules become more diffusely spread out (fig. 6.2). Hence, this random molecular motion is known as **diffusion.** In terms of the second law of thermodynamics, the concentration difference represents an unstable state of high organization (low entropy) that changes to produce a uniformly distributed solution with maximum disorganization (high entropy).

As a result of random molecular motion, molecules in the part of the solution with a higher concentration will enter the area of lower concentration. Molecules will also move in the opposite direction, but not as frequently. As a result, there will be a *net movement* from the region of higher to the region of lower concentration until the concentration difference no longer exists. This net movement is called **net diffusion.** Net diffusion is a physical process that occurs whenever there is a concentration difference across a membrane and the membrane is permeable to the diffusing substance.

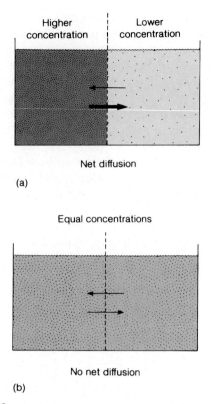

Net diffusion

(a)

Equal concentrations

No net diffusion

(b)

■ **Figure 6.2** **Diffusion of a solute.** *(a)* Net diffusion occurs when there is a concentration difference (or concentration gradient) between two regions of a solution, provided that the membrane separating these regions is permeable to the diffusing substance. *(b)* Diffusion tends to equalize the concentrations of these regions, and thus to eliminate the concentration differences.

In the kidneys, blood is filtered through pores in capillary walls to produce a filtrate that will become urine. Wastes and other dissolved molecules can pass through the pores, but blood cells and proteins are held back. Then, the molecules needed by the body are reabsorbed from the filtrate back into the blood by transport processes. Wastes generally remain in the filtrate and are thus excreted in the urine. When the kidneys fail to perform this function, the wastes must be removed from the blood artificially by means of **dialysis.** In this process, waste molecules are removed from the blood by having them diffuse through an artificial porous membrane. The wastes pass into a solution (called a dialysate) surrounding the dialysis membrane. Molecules needed by the body, however, are kept in the blood by including them in the dialysate. This prevents their net diffusion by abolishing their concentration gradients.

Diffusion Through the Plasma Membrane

Since the plasma (cell) membrane consists primarily of a double layer of phospholipids, molecules that are nonpolar, and thus lipid-soluble, can easily pass from one side of the membrane to the other. The plasma membrane, in other words, does not present a barrier to the diffusion of nonpolar molecules such as oxygen gas (O_2) or steroid hormones. Small molecules that have polar covalent bonds, but which are uncharged, such as CO_2 (as well as ethanol and urea), are also able to penetrate the phospholipid bilayer. Net diffusion of these molecules can thus easily occur between the intracellular and extracellular compartments when concentration gradients exist.

The oxygen concentration is relatively high, for example, in the extracellular fluid because oxygen is carried from the lungs to the body tissues by the blood. Since oxygen is combined with hydrogen to form water in aerobic cell respiration, the oxygen concentration within the cells is lower than in the extracellular fluid. The concentration gradient for carbon dioxide is in the opposite direction because cells produce CO_2. *Gas exchange* thus

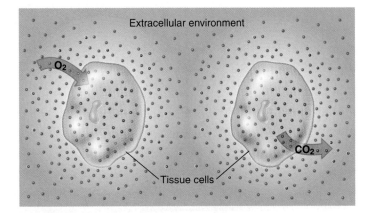

■ **Figure 6.3** Gas exchange occurs by diffusion. The colored dots, which represent oxygen and carbon dioxide molecules, indicate relative concentrations inside the cell and in the extracellular environment. Gas exchange between the intracellular and extracellular compartments thus occurs by diffusion.

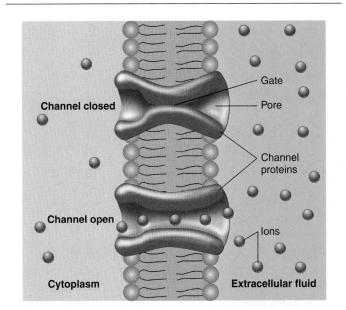

■ **Figure 6.4** Ions pass through membrane channels. These channels are composed of integral proteins that span the thickness of the membrane. Although some channels are always open, many others have structures known as "gates" than can open or close the channel. This figure depicts a generalized ion channel; most, however, are relatively selective—they allow only particular ions to pass.

occurs by diffusion between the cells and their extracellular environments (fig. 6.3).

Although water is not lipid-soluble, water molecules can diffuse through the plasma membrane to a limited degree because of their small size and lack of net charge. In certain membranes, however, the passage of water is aided by specific channels that are inserted into the membrane in response to physiological regulation. The net diffusion of water molecules (the solvent) across the membrane is known as *osmosis.* Since

osmosis is the simple diffusion of solvent instead of solute, a unique terminology (discussed shortly) is used to describe it.

Larger polar molecules, such as glucose, cannot pass through the double layer of phospholipid molecules and thus require special *carrier proteins* in the membrane for transport. The phospholipid portion of the membrane is similarly impermeable to charged inorganic ions, such as Na^+ and K^+. However, tiny **ion channels** through the membrane, which are too small to be seen even with an electron microscope, permit passage of these ions. The ion channels are provided by some of the proteins that span the thickness of the membrane (fig. 6.4).

Some ion channels are always open, so that diffusion of the ion through the plasma membrane is an ongoing process. Many ion channels, however, are **gated**—they have structures ("gates") that can open or close the channel (fig. 6.4). In this way, particular physiological stimuli (such as binding of the channel to a specific chemical regulator) can open an otherwise closed channel. In the production of nerve and muscle impulses, specific channels for Na^+ and others for K^+ open and close in response to membrane voltage (discussed in chapter 7).

Cystic fibrosis occurs about once in every 2,500 births in the Caucasian population. As a result of a genetic defect, abnormal NaCl and water movement occurs across wet epithelial membranes. Where such membranes line the pancreatic ductules and small respiratory airways, they produce a dense, viscous mucus that cannot be properly cleared, which may lead to pancreatic and pulmonary disorders. The genetic defect involves a particular glycoprotein that forms chloride (Cl^-) channels in the apical membrane of the epithelial cells. This protein, known as *CFTR* (for cystic fibrosis transmembrane conductance regulator), is formed in the usual manner in the endoplasmic reticulum. It does not move into the Golgi complex for processing, however, and therefore, it doesn't get correctly processed and inserted into vesicles that would introduce it into the cell membrane (chapter 3). The gene for CFTR has been identified and cloned. More research is required, however, before gene therapy for cystic fibrosis becomes an effective therapy.

Rate of Diffusion

The speed at which diffusion occurs, measured by the number of diffusing molecules passing through a membrane per unit time, depends on (1) the magnitude of the concentration difference across the membrane (the "steepness" of the concentration gradient), (2) the permeability of the membrane to the diffusing substances, (3) the temperature of the solution, and (4) the surface area of the membrane through which the substances are diffusing.

The magnitude of the concentration difference across a membrane serves as the driving force for diffusion. Regardless of this concentration difference, however, the diffusion of a substance across a membrane will not occur if the membrane is not permeable to that substance. With a given concentration difference, the speed at which a substance diffuses through a membrane will depend on how permeable the membrane is to it. In a resting neuron, for example, the plasma (cell) membrane is about twenty times more permeable to potassium (K^+) than to sodium (Na^+); consequently, K^+ diffuses much more rapidly than does Na^+. Changes in the protein structure of the membrane channels, however, can change the permeability of the membrane. This occurs during the production of a nerve impulse (see chapter 7), when specific stimulation opens Na^+ channels temporarily and allows a faster diffusion rate for Na^+ than for K^+.

In areas of the body that are specialized for rapid diffusion, the surface area of the cell membranes may be increased by numerous folds. The rapid passage of the products of digestion across the epithelial membranes in the small intestine, for example, is aided by tiny fingerlike projections called *microvilli* (discussed in chapter 3). Similar microvilli are found in the kidney tubule epithelium, which must reabsorb various molecules that are filtered out of the blood.

Osmosis

Osmosis is the net diffusion of water (the solvent) across the membrane. For osmosis to occur, the membrane must be *selectively permeable;* that is, it must be more permeable to water molecules than to at least one species of solute. There are thus two requirements for osmosis: (1) there must be a difference in the concentration of a solute on the two sides of a selectively permeable membrane; and (2) the membrane must be relatively impermeable to the solute. Solutes that cannot freely pass through the membrane are said to be **osmotically active.**

Like the diffusion of solute molecules, the diffusion of water occurs when the water is more concentrated on one side of the membrane than on the other side; that is, when one solution is more dilute than the other (fig. 6.5). The more dilute solution has a higher concentration of water molecules and a lower concentration of solute. Although the terminology associated with osmosis can be awkward (because we are describing water instead of solute), the principles of osmosis are the same as those governing the diffusion of solute molecules through a membrane. Remember that, during osmosis, there is a net movement of water molecules from the side of higher water concentration to the side of lower water concentration.

Imagine a cylinder divided into two equal compartments by an artificial membrane partition that can freely move. One compartment initially contains 180 g/L (grams per liter) of glucose and the other compartment contains 360 g/L of glucose. If the membrane is permeable to glucose, glucose will diffuse from the 360-g/L compartment to the 180-g/L compartment until both compartments contain 270 g/L of glucose. If the membrane is not permeable to glucose but is permeable to water, the same result (270-g/L solutions on both sides of the membrane) will be achieved by the diffusion of water. As water diffuses from the 180-g/L compartment to the 360-g/L compartment (from the higher to the lower water concentration), the for-

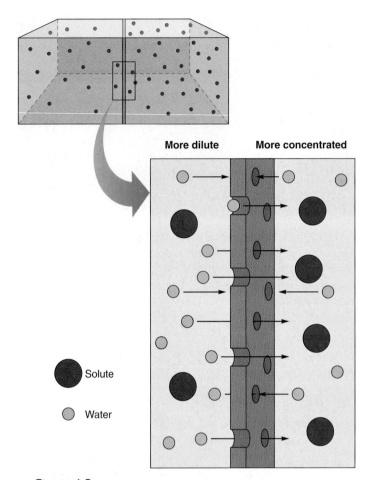

More dilute   **More concentrated**

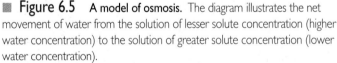

● Solute

○ Water

■ **Figure 6.5** **A model of osmosis.** The diagram illustrates the net movement of water from the solution of lesser solute concentration (higher water concentration) to the solution of greater solute concentration (lower water concentration).

mer solution becomes more concentrated while the latter becomes more dilute. This is accompanied by volume changes, as illustrated in figure 6.6. Osmosis ceases when the concentrations become equal on both sides of the membrane.

Cell membranes behave in a similar manner because water is able to move to some degree through the lipid component of most cell membranes. The membranes of some cells, however, have special water channels that allow water to move through more rapidly. These channels are known as **aquaporins.** In some cells, the plasma membrane always has aquaporin channels; in others, the aquaporin channels are inserted into the plasma membrane in response to regulatory molecules. Such regulation is particularly important in the functioning of the kidneys, as will be described in chapter 17.

Osmotic Pressure

Osmosis and the movement of the membrane partition could be prevented by an opposing force. If one compartment contained 180 g/L of glucose and the other compartment contained pure water, the osmosis of water into the glucose solution could be prevented by pushing against the membrane with a certain force. This concept is illustrated in figure 6.7.

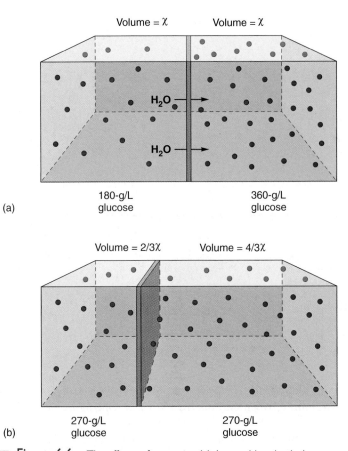

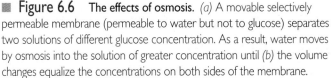

Figure 6.6 **The effects of osmosis.** (a) A movable selectively permeable membrane (permeable to water but not to glucose) separates two solutions of different glucose concentration. As a result, water moves by osmosis into the solution of greater concentration until (b) the volume changes equalize the concentrations on both sides of the membrane.

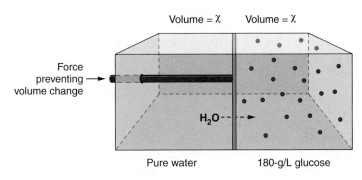

Figure 6.7 A model illustrating osmotic pressure. If a selectively permeable membrane separates pure water from a 180-g/L glucose solution, water will tend to move by osmosis into the glucose solution, thus creating a hydrostatic pressure that will push the membrane to the left and expand the volume of the glucose solution. The amount of pressure that must be applied to just counteract this volume change is equal to the osmotic pressure of the glucose solution.

The force that would have to be exerted to prevent osmosis in the situation just described is the **osmotic pressure** of the solution. This backward measurement indicates how strongly the solution "draws" water into it by osmosis. The greater the

solute concentration of a solution, the greater its osmotic pressure. Pure water thus has an osmotic pressure of zero, and a 360-g/L glucose solution has twice the osmotic pressure of a 180-g/L glucose solution.

Clinical Investigation Clues

Remember that Jessica has glucose in her urine, a solute that is normally absent from urine.

What would the presence of the extra solute, glucose, do to the osmotic pressure of the urine?

How might this be the cause of her frequent urination?

Molarity and Molality

Glucose is a monosaccharide with a molecular weight of 180 (the sum of its atomic weights). Sucrose is a disaccharide of glucose and fructose, which have molecular weights of 180 each. When glucose and fructose join together by dehydration synthesis to form sucrose, a molecule of water (molecular weight = 18) is split off. Therefore, sucrose has a molecular weight of 342 (180 + 180 − 18). Since the molecular weights of sucrose and glucose are in a ratio of 342/180, it follows that 342 grams of sucrose must contain the same number of molecules as 180 grams of glucose.

Notice that an amount of any compound equal to its molecular weight in grams must contain the same number of molecules as an amount of any other compound equal to its molecular weight in grams. This unit of weight, *a mole,* always contains 6.02×10^{23} molecules (**Avogadro's number**). One mole of solute dissolved in water to make one liter of solution is described as a **one-molar solution** (abbreviated 1.0 *M*). Although this unit of measurement is commonly used in chemistry, it is not completely desirable in discussions of osmosis because the exact ratio of solute to water is not specified. For example, more water is needed to make a 1.0 *M* NaCl solution (where a mole of NaCl weighs 58.5 grams) than is needed to make a 1.0 *M* glucose solution, since 180 grams of glucose takes up more volume than 58.5 grams of salt.

Since the ratio of solute to water molecules is of critical importance in osmosis, a more desirable measurement of concentration is **molality.** In a one-molal solution (abbreviated

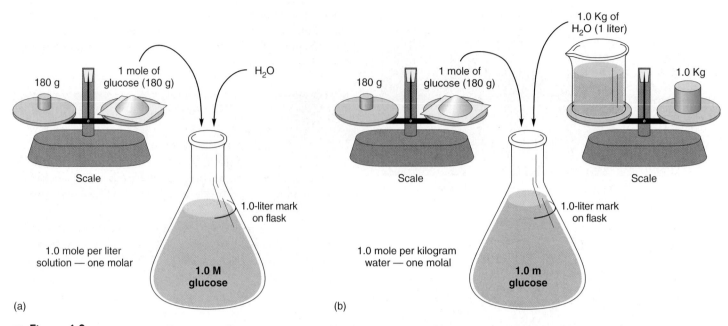

■ Figure 6.8 Molar and molal solutions. The diagrams illustrate the difference between (a) a one-molar (1.0 M) and (b) a one-molal (1.0 m) glucose solution.

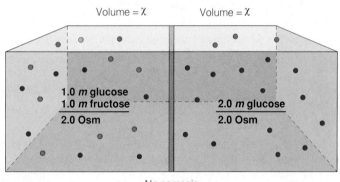

■ Figure 6.9 The osmolality of a solution. The osmolality (Osm) is equal to the sum of the molalities of each solute in the solution. If a selectively permeable membrane separates two solutions with equal osmolalities, no osmosis will occur.

1.0 m), 1 mole of solute (180 grams of glucose, for example) is dissolved in 1 kilogram of water (equal to 1 liter at 4° C). A 1.0 m NaCl solution and a 1.0 m glucose solution therefore both contain a mole of solute dissolved in exactly the same amount of water (fig. 6.8).

Osmolality

If 180 grams of glucose and 180 grams of fructose were dissolved in the same kilogram of water, the osmotic pressure of the solution would be the same as that of a 360-g/L glucose solution. Osmotic pressure depends on the ratio of solute to solvent, *not* on the chemical nature of the solute molecules. The expression for the total molality of a solution is **osmolality (Osm).** Thus, the solution of 1.0 m glucose plus 1.0 m fructose has a total molality, or *osmolality,* of 2.0 osmol/L (abbreviated 2.0 Osm). This osmolality is the same as that of the 360-g/L glucose solution, which has a concentration of 2.0 m and 2.0 Osm (fig. 6.9).

Unlike glucose, fructose, and sucrose, electrolytes such as NaCl ionize when they dissolve in water. One molecule of NaCl dissolved in water yields two ions (Na^+ and Cl^-); 1 mole of NaCl ionizes to form 1 mole of Na^+ and 1 mole of Cl^-. Thus, a 1.0 m NaCl solution has a total concentration of 2.0 Osm. The effect of this ionization on osmosis is illustrated in figure 6.10.

Measurement of Osmolality

Plasma and other biological fluids contain many organic molecules and electrolytes. The osmolality of such complex solutions only can be estimated by calculations. Fortunately, however, there is a relatively simple method for measuring osmolality. This method is based on the fact that the freezing point of a solution, like its osmotic pressure, is affected by the total concentration of the solution and not by the chemical nature of the solute.

One mole of solute per liter depresses the freezing point of water by $-1.86°$ C. Accordingly, a 1.0 m glucose solution freezes at a temperature of $-1.86°$ C, and a 1.0 m NaCl solution freezes at a temperature of $2 \times -1.86 = -3.72°$ C because of ionization. Thus, the *freezing-point depression* is a measure of the osmolality. Since plasma freezes at about $-0.56°$ C, its osmolality is equal to $0.56 \div 1.86 = 0.3$ Osm, which is more commonly indicated as 300 milliosmolal (or 300 mOsm).

Clinical Investigation Clues

Remember that Jessica's plasma has a higher than normal osmolality.

■ *What is the normal osmolality of plasma?*

■ *What is the relationship between the glucose in Jessica's urine, her frequent urination, and her high plasma osmolality?*

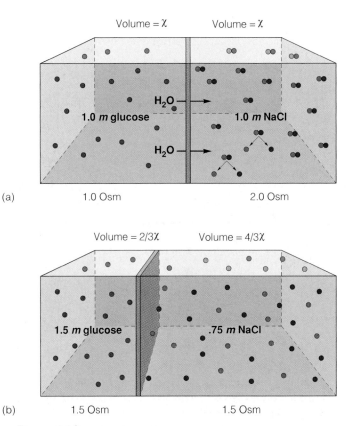

Volume = X   Volume = X

H₂O →

1.0 *m* glucose   1.0 *m* NaCl

H₂O →

(a)   1.0 Osm   2.0 Osm

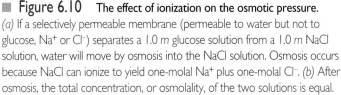

Volume = 2/3X   Volume = 4/3X

1.5 *m* glucose   .75 *m* NaCl

(b)   1.5 Osm   1.5 Osm

Figure 6.10   The effect of ionization on the osmotic pressure. *(a)* If a selectively permeable membrane (permeable to water but not to glucose, Na⁺ or Cl⁻) separates a 1.0 *m* glucose solution from a 1.0 *m* NaCl solution, water will move by osmosis into the NaCl solution. Osmosis occurs because NaCl can ionize to yield one-molal Na⁺ plus one-molal Cl⁻. *(b)* After osmosis, the total concentration, or osmolality, of the two solutions is equal.

Tonicity

A 0.3 *m* glucose solution, which is 0.3 Osm, or 300 milliosmolal (300 mOsm), has the same osmolality and osmotic pressure as plasma. The same is true of a 0.15 *m* NaCl solution, which ionizes to produce a total concentration of 300 mOsm. Both of these solutions are used clinically as intravenous infusions, labeled 5% *dextrose* (5 g of glucose per 100 ml, which is 0.3 *m*) and *normal saline* (0.9 g of NaCl per 100 ml, which is 0.15 *m*). Since 5% dextrose and normal saline have the same osmolality as plasma, they are said to be **isosmotic** to plasma.

The term **tonicity** is used to describe the effect of a solution on the osmotic movement of water. For example, if an isosmotic glucose or saline solution is separated from plasma by a membrane that is permeable to water, but not to glucose or NaCl, osmosis will not occur. In this case, the solution is said to be **isotonic** (from the Greek *isos* = equal; *tonos* = tension) to plasma.

Red blood cells placed in an isotonic solution will neither gain nor lose water. It should be noted that a solution may be isosmotic but not isotonic; such is the case whenever the solute in the isosmotic solution can freely penetrate the membrane. A 0.3 *m* urea solution, for example, is isosmotic but not isotonic because the cell membrane is permeable to urea. When red blood cells are placed in a 0.3 *m* urea solution, the urea diffuses into the cells until its concentration on both sides of the cell membranes becomes equal.

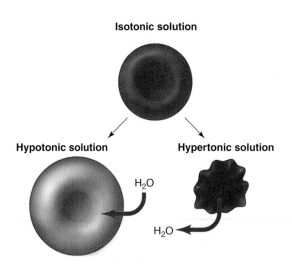

Isotonic solution

Hypotonic solution   Hypertonic solution

H₂O

H₂O

Figure 6.11   Red blood cells in isotonic, hypotonic, and hypertonic solutions.   In each case, the external solution has an equal, lower, or higher osmotic pressure, respectively, than the intracellular fluid. As a result, water moves by osmosis into the red blood cells placed in hypotonic solutions, causing them to swell and even to burst. Similarly, water moves out of red blood cells placed in a hypertonic solution, causing them to shrink and become crenated.

Meanwhile, the solutes within the cells that cannot exit—and which are therefore osmotically active—cause osmosis of water into the cells. Red blood cells placed in 0.3 *m* urea will thus eventually burst.

Solutions that have a lower total concentration of solutes than that of plasma, and therefore a lower osmotic pressure, are **hypo-osmotic** to plasma. If the solute is osmotically active, such solutions are also **hypotonic** to plasma. Red blood cells placed in hypotonic solutions gain water and may burst—a process called *hemolysis*. When red blood cells are placed in a **hypertonic** solution (such as sea water), which contains osmotically active solutes at a higher osmolality and osmotic pressure than plasma, they shrink because of the osmosis of water out of the cells. This process is called *crenation* (*crena* = notch) because the cell surface takes on a scalloped appearance (fig. 6.11).

Intravenous fluids must be isotonic to blood in order to maintain the correct osmotic pressure and prevent cells from either expanding or shrinking from the gain or loss of water. Common fluids used for this purpose are *normal saline* and *5% dextrose*, which, as previously described, have about the same osmolality as normal plasma (approximately 300 mOsm). Another isotonic solution frequently used in hospitals is *Ringer's lactate*. This solution contains glucose and lactic acid in addition to a number of different salts.

Regulation of Blood Osmolality

The osmolality of the blood plasma is normally maintained within very narrow limits by a variety of regulatory mechanisms. When a person becomes dehydrated, for example, the blood becomes more concentrated as the total blood volume is reduced. The increased blood osmolality and osmotic pressure stimulate *osmoreceptors,* which are neurons located in a part of the brain called the hypothalamus.

As a result of increased osmoreceptor stimulation, the person becomes thirsty and, if water is available, drinks. Along with increased water intake, a person who is dehydrated excretes a lower volume of urine. This occurs as a result of the following sequence of events:

1. Increased plasma osmolality stimulates osmoreceptors in the hypothalamus of the brain.
2. The osmoreceptors in the hypothalamus then stimulate a tract of axons that terminate in the posterior pituitary; this causes the posterior pituitary to release **antidiuretic hormone (ADH)** into the blood.
3. ADH acts on the kidneys to promote water retention, so that a lower volume of more concentrated urine is excreted.

Clinical Investigation Clues

Remember that Jessica is constantly thirsty, despite drinking large amounts of water.

What is stimulating Jessica's sense of thirst?

How is this related to the glucose in her urine and her frequent urination?

A person who is dehydrated, therefore, drinks more and urinates less. This represents a negative feedback loop (fig. 6.12), which acts to maintain homeostasis of the plasma concentration (osmolality) and, in the process, helps to maintain a proper blood volume.

A person with a normal blood volume who eats salty food will also get thirsty, and more ADH will be released from the posterior pituitary. By drinking more and excreting less water in the urine, the salt from the food will become diluted to restore the normal blood concentration, but at a higher blood volume. The opposite occurs in salt deprivation. With a lower plasma osmolality, the osmoreceptors are not stimulated as much, and the posterior pituitary releases less ADH. Consequently, more water is excreted in the urine to again restore the proper range of plasma concentration, but at a lower blood volume. Low blood volume and pressure as a result of prolonged salt deprivation can be fatal (refer to the discussion of blood volume and pressure in chapter 14).

Test Yourself Before You Continue

1. Explain what is meant by simple diffusion and list the factors that influence the diffusion rate.
2. Define the terms *osmosis, osmolality,* and *osmotic pressure,* and state the conditions that are needed for osmosis to occur.
3. Define the terms *isotonic, hypotonic,* and *hypertonic,* and explain why hospitals use 5% dextrose and normal saline as intravenous infusions.
4. Explain how the body detects changes in the osmolality of plasma and describe the regulatory mechanisms by which a proper range of plasma osmolality is maintained.

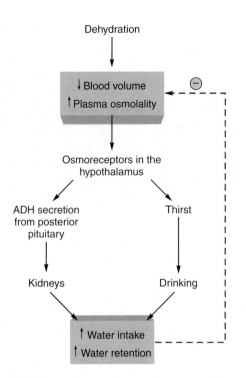

■ **Figure 6.12** **Homeostasis of plasma concentration.** An increase in plasma osmolality (increased concentration and osmotic pressure) due to dehydration stimulates thirst and increased ADH secretion. These effects cause the person to drink more and urinate less. The blood volume, as a result, is increased while the plasma osmolality is decreased. These effects help to bring the blood volume back to the normal range and complete the negative feedback loop (indicated by a negative sign).

Carrier-Mediated Transport

Molecules such as glucose are transported across plasma membranes by special protein carriers. Carrier-mediated transport in which the net movement is down a concentration gradient, and which is therefore passive, is called facilitated diffusion. Carrier-mediated transport that occurs against a concentration gradient, and which therefore requires metabolic energy, is called active transport.

In order to sustain metabolism, cells must take up glucose, amino acids, and other organic molecules from the extracellular environment. Molecules such as these, however, are too large and polar to pass through the lipid barrier of the plasma membrane by a process of simple diffusion. The transport of such molecules is mediated by **protein carriers** within the membrane. Although such carriers cannot be directly observed, their presence has been inferred by the observation that this transport has characteristics in common with enzyme activity. These characteristics include (1) *specificity,* (2) *competition,* and (3) *saturation.*

Like enzyme proteins, carrier proteins interact only with specific molecules. Glucose carriers, for example, can interact only with glucose and not with closely related monosaccharides.

As a further example of specificity, particular carriers for amino acids transport some types of amino acids but not others. Two amino acids that are transported by the same carrier compete with each other, so that the rate of transport for each is lower when they are present together than it would be if each were present alone (fig. 6.13).

As the concentration of a transported molecule is increased, its rate of transport will also be increased—but only up to a maximum. Beyond this rate, called the *transport maximum* (T_m), further increases in concentration do not further increase the transport rate. This indicates that the carriers have become saturated (fig. 6.13).

As an example of saturation, imagine a bus stop that is serviced once an hour by a bus that can hold a maximum of forty people (its "transport maximum"). If there are ten people waiting at the bus stop, ten will be transported each hour. If twenty people are waiting, twenty will be transported each hour. This linear relationship will hold up to a maximum of forty people; if there are eighty people at the bus stop, the transport rate will still be forty per hour.

The kidneys transport a number of molecules from the blood filtrate (which will become urine) back into the blood. Glucose, for example, is normally completely reabsorbed so that urine is normally free of glucose. If the glucose concentration of the blood and filtrate is too high (a condition called **hyperglycemia**), however, the transport maximum will be exceeded. In this case, glucose will be found in the urine (a condition called *glycosuria*). This may result from the consumption of too much sugar or from inadequate action of the hormone insulin in the disease *diabetes mellitus*.

Facilitated Diffusion

The transport of glucose from the blood across plasma membranes occurs by **facilitated diffusion.** Facilitated diffusion, like simple diffusion, is powered by the thermal energy of the diffusing molecules and involves net transport from the side of higher to the side of lower concentration. ATP is not required for either facilitated or simple diffusion.

Unlike simple diffusion of nonpolar molecules, water, and inorganic ions through a membrane, the diffusion of glucose through the plasma membrane displays the properties of carrier-mediated transport: specificity, competition, and saturation. The diffusion of glucose through a plasma membrane must therefore be mediated by carrier proteins. In the conceptual model shown in figure 6.14, each transport carrier is composed of two protein subunits that interact with glucose in such a way as to create a channel through the membrane, thus enabling the movement of glucose down its concentration gradient.

Like the isoenzymes described in chapter 4, carrier proteins that do the same job may exist in various tissues in slightly different forms. The transport carriers for the facilitative diffusion of glucose are designated with the letters **GLUT,** followed by a number for the isoform. The carrier for glucose in skeletal muscles, for example, is designated *GLUT4.*

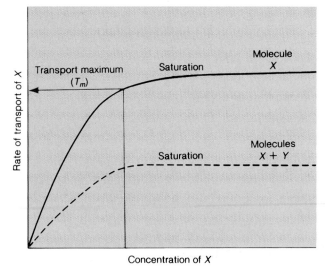

Figure 6.13 **Characteristics of carrier-mediated transport.** Carrier-mediated transport displays the characteristics of saturation (illustrated by the transport maximum) and competition. Since molecules X and Y compete for the same carrier, the rate of transport of each is lower when they are both present than when either is present alone.

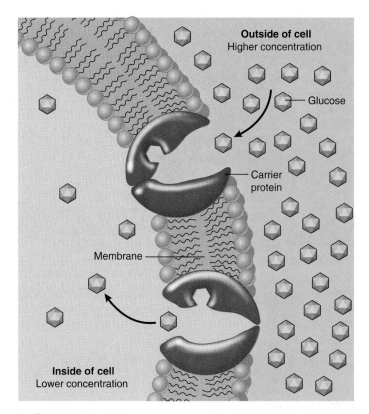

Figure 6.14 **A model of the facilitated diffusion of glucose.** A carrier—with characteristics of specificity and saturation—is required for this transport, which occurs from the blood into cells such as muscle, liver, and fat cells. This is passive transport because the net movement is to the region of lower concentrations, and ATP is not required.

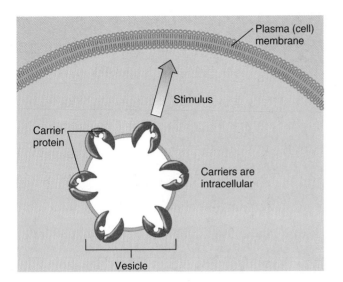

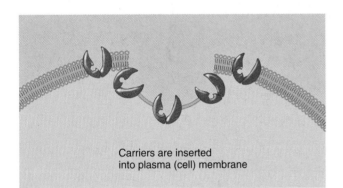

■ **Figure 6.15** The insertion of carrier proteins into the plasma (cell) membrane. In the unstimulated state, carrier proteins (such as those for glucose) may be located in the membrane of intracellular vesicles. In response to stimulation, the vesicle fuses with the plasma membrane and the carriers are thereby inserted into the membrane.

In unstimulated muscles, the GLUT4 proteins are within the membrane enclosing cytoplasmic vesicles. Exercise—and stimulation by insulin—causes these vesicles to fuse with the plasma membrane. This process is similar to exocytosis (chapter 3; also see fig. 6.20), except that no cellular product is secreted. Instead, the transport carriers are inserted into the plasma membrane (fig. 6.15). During exercise and insulin stimulation, therefore, more glucose is able to enter the skeletal muscle cells from the blood plasma.

Transport of glucose by GLUT carriers is a form of passive transport, where glucose is always transported down its concentration gradient. However, in certain cases (such as the epithelial cells of the kidney tubules and small intestine), glucose is transported against its concentration gradient by a different kind of carrier, one that is dependent on simultaneous transport of Na^+. Since this is a type of active transport, it will be described shortly in a different section.

The rate of the facilitated diffusion of glucose into tissue cells depends directly on the plasma glucose concentration. When the plasma glucose concentration is abnormally low—a condition called **hypoglycemia**—the rate of transport of glucose into brain cells may be too slow for the metabolic needs of the brain. Severe hypoglycemia, as may be produced in a diabetic person by an overdose of insulin, can thus result in loss of consciousness or even death.

Active Transport

Some aspects of cell transport cannot be explained by simple or facilitated diffusion. The epithelial linings of the small intestine and kidney tubules, for example, move glucose from the side of lower to the side of higher concentration—from the space within the tube (*lumen*) to the blood. Similarly, all cells extrude Ca^{2+} into the extracellular environment and, by this means, maintain an intracellular Ca^{2+} concentration that is 1,000 to 10,000 times lower than the extracellular Ca^{2+} concentration. This steep concentration gradient sets the stage for Ca^{2+} to be used as a regulatory signal. The opening of plasma membrane Ca^{2+} channels, and the rapid diffusion of Ca^{2+} that results, provides a signal for neurotransmitter release, muscle contraction, and many other cellular activities.

Active transport is the movement of molecules and ions against their concentration gradients, from lower to higher concentrations. This transport requires the expenditure of cellular energy obtained from ATP; if a cell is poisoned with cyanide (which inhibits oxidative phosphorylation), active transport will stop. Passive transport, by contrast, can continue even if metabolic poisons kill the cell by preventing the formation of ATP.

Primary Active Transport

Primary active transport occurs when the hydrolysis of ATP is directly required for the function of the carriers. These carriers are composed of proteins that span the thickness of the membrane. The following sequence of events is believed to occur: (1) the molecule or ion to be transported binds to a specific "recognition site" on one side of the carrier protein; (2) this bonding stimulates the breakdown of ATP, which in turn results in phosphorylation of the carrier protein; (3) as a result of phosphorylation, the carrier protein undergoes a conformational (shape) change; and (4) a hinge-like motion of the carrier protein releases the transported molecule or ion on the opposite side of the membrane. This model of active transport is illustrated in figure 6.16.

The Sodium-Potassium Pump

Primary active transport carriers are often referred to as *pumps*. Although some of these carriers transport only one molecule or ion at a time, others exchange one molecule or ion for another. The most important of the latter type of carrier is the **Na^+/K^+ pump**. This carrier protein, which is also an ATPase enzyme that converts ATP to ADP and P_i, actively extrudes three sodium ions (Na^+) from the cell as it transports two potassium ions (K^+) into the cell. This transport is energy dependent because Na^+ is

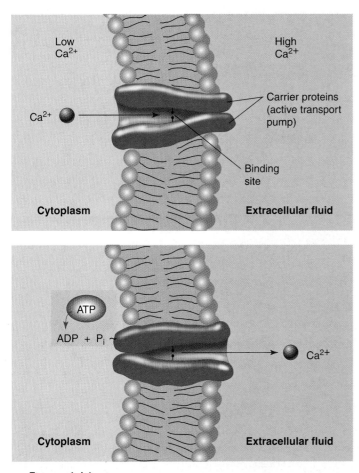

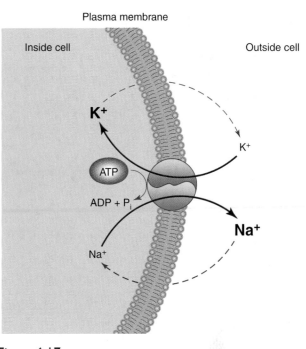

Figure 6.16 A model of active transport. This model (a mental construct, consistent with the scientific evidence) features a hingelike motion of the integral protein subunits.

Figure 6.17 The exchange of intracellular Na^+ for K^+ by the Na^+/K^+ pump. The active transport carrier itself is an ATPase that breaks down ATP for energy. Dashed arrows indicate the direction of passive transport (diffusion); solid arrows indicate the direction of active transport.

more highly concentrated outside the cell and K^+ is more concentrated within the cell. Both ions, in other words, are moved against their concentration gradients (fig. 6.17).

Most cells have numerous Na^+/K^+ pumps that are constantly active. For example, there are about 200 Na^+/K^+ pumps per red blood cell, about 35,000 per white blood cell, and several million per cell in a part of the tubules within the kidney. This represents an enormous expenditure of energy used to maintain a steep gradient of Na^+ and K^+ across the cell membrane. This steep gradient serves four functions:

1. The steep Na^+ gradient is used to provide energy for the "coupled transport" of other molecules.
2. The activity of the Na^+/K^+ pumps can be adjusted (primarily by thyroid hormones) to regulate the resting calorie expenditure and basal metabolic rate of the body.
3. The gradients for Na^+ and K^+ concentrations across the plasma membranes of nerve and muscle cells are used to produce electrochemical impulses needed for functions of the nerve and muscles, including the heart muscle.
4. The active extrusion of Na^+ is important for osmotic reasons; if the pumps stop, the increased Na^+ concentrations within cells promote the osmotic inflow of water, damaging the cells.

Secondary Active Transport (Coupled Transport)

In **secondary active transport,** or **coupled transport,** the energy needed for the "uphill" movement of a molecule or ion is obtained from the "downhill" transport of Na^+ into the cell. Hydrolysis of ATP by the action of the Na^+/K^+ pumps is required indirectly, in order to maintain low intracellular Na^+ concentrations. The diffusion of Na^+ down its concentration gradient into the cell can then power the movement of a different ion or molecule against its concentration gradient. If the other molecule or ion is moved in the same direction as Na^+ (that is, into the cell), the coupled transport is called either *cotransport* or *symport*. If the other molecule or ion is moved in the opposite direction (out of the cell), the process is called either *countertransport* or *antiport*.

An example of symport is the cotransport of glucose and Na^+ from the extracellular fluid into the epithelial cells of the small intestine and kidney tubules. In these cases, a carrier protein simultaneously binds to glucose and Na^+ in the extracellular fluid. The downhill transport of Na^+ (from higher to lower concentration) into the cell furnishes the energy for the uphill transport of glucose (fig. 6.18). Notice that, in order for this secondary active transport to work, a steep gradient for Na^+ must have already been established by the activity of the Na^+/K^+ pumps.

An example of countertransport is the uphill extrusion of Ca^{2+} from a cell by a type of pump that is coupled to the passive diffusion of Na^+ into the cell. Cellular energy, obtained from ATP, is not used to move Ca^{2+} directly out of the cell in this case, but energy is constantly required to maintain the steep Na^+ gradient. Another example of countertransport is the exchange of chloride (Cl^-) for bicarbonate (HCO_3^-) across the red blood

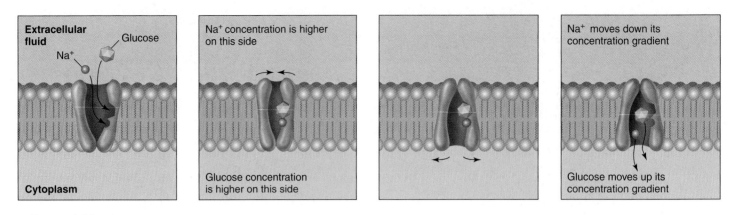

Figure 6.18 **A model for the cotransport of Na⁺ and glucose into a cell.** The sequence of events is illustrated left-to-right. This is secondary active transport because it is dependent upon the diffusion gradient for Na⁺ created by the Na⁺/K⁺ pumps.

cell membrane. Diffusion of bicarbonate out of the cell powers the entry of chloride (this is discussed in connection with red blood cell function in chapter 16).

Transport Across Epithelial Membranes

As discussed in chapter 1, epithelial membranes cover all body surfaces and line the cavities of all hollow organs. Therefore, in order for a molecule or ion to move from the external environment into the blood (and from there to the body organs), it must first pass through an epithelial membrane. The transport of digestion products (such as glucose) across the intestinal epithelium into the blood is called **absorption.** The transport of molecules out of the urinary filtrate (originally derived from blood) back into the blood is called **reabsorption.**

The cotransport of Na⁺ and glucose described in the last section can serve as an example. The cotransport carriers for Na⁺ and glucose are located in the apical (top) plasma membrane of the epithelial cells, which faces the lumen of the intestine or kidney tubule. The Na⁺/K⁺ pumps, and the carriers for the facilitated diffusion of glucose, are on the opposite side of the epithelial cell (facing the location of blood capillaries). As a result of these active and passive transport processes, glucose is moved from the lumen, through the cell, and then to the blood (fig. 6.19).

The membrane transport mechanisms described in this section move materials through the cytoplasm of the epithelial cells, a process termed **transcellular transport.** However, diffusion and osmosis may also occur to a limited extent in the very tiny spaces between epithelial cells, a process termed **paracellular transport.** Such passive transport processes that do occur are limited by the *tight junctions* (regions where the plasma membranes of adjacent cells are fused all the way around) and *desmosomes* (buttonlike points of fusion of plasma membranes) between epithelial cells of the intestine and kidney tubules.

It should be noted that there are additional processes that cause movements of materials across various epithelial membranes. For example, the epithelial cells that comprise the walls of many blood capillaries (the thinnest of blood vessels) have pores between them that can be relatively large, permitting filtration of water and dissolved molecules out of the capillaries through the

paracellular route. In the capillaries of the brain, however, such filtration is prevented by tight junctions, so molecules must be transported transcellularly. This involves the cell transport mechanisms previously described, as well as the processes of endocytosis and exocytosis, as described in the next section.

Severe diarrhea is responsible for almost half of all deaths worldwide of children under the age of 4 (amounting to about 4 million deaths per year). Because rehydration through intravenous therapy is often not practical, the World Health Organization (WHO) developed a simpler, more economical treatment called **oral rehydration therapy.** The therapy is effective because (1) the absorption of water by osmosis across the intestine is proportional to the absorption of Na⁺ and (2) the intestinal epithelium cotransports Na⁺ and glucose. The WHO provides those in need with a mixture (which can be diluted with tap water in the home) containing both glucose and Na⁺ as well as other ions. The glucose in the mixture promotes the cotransport of Na⁺ and the Na⁺ transport promotes the osmotic movement of water from the intestine into the blood. It has been estimated that oral rehydration therapy saves the lives of more than a million small children each year.

Bulk Transport

Polypeptides and proteins, as well as many other molecules, are too large to be transported through a membrane by the carriers described in previous sections. Yet many cells do secrete these molecules—for example, as hormones or neurotransmitters—by the process of **exocytosis.** As described in chapter 3, this involves the fusion of a membrane-bound vesicle that contains these cellular products with the plasma membrane, so that the membranes become continuous (fig. 6.20).

The process of **endocytosis** resembles exocytosis in reverse. In receptor-mediated endocytosis (see fig. 3.4), specific molecules,

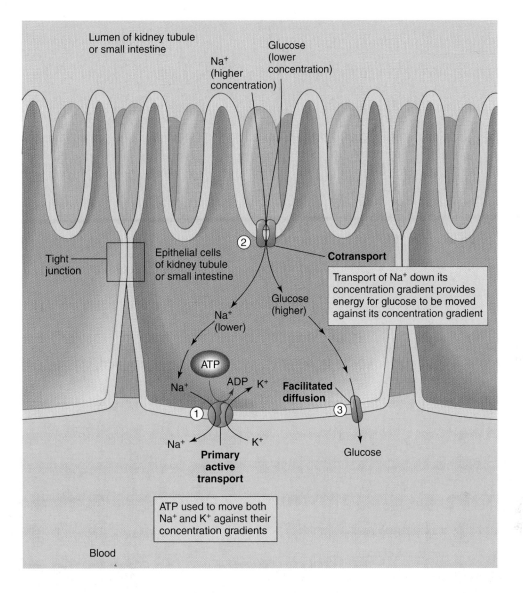

Figure 6.19 Transport processes involved in the epithelial absorption of glucose. When glucose is to be absorbed across the epithelial membranes of the kidney tubules or the small intestine, several processes are involved. (1) Primary active transport (the Na⁺/K⁺ pumps) in the basal membrane use ATP to maintain a low intracellular concentration of Na⁺. (2) Secondary active transport uses carriers in the apical membrane to transport glucose up its concentration gradient, using the energy from the "downhill" flow of Na⁺ into the cell. Finally, (3) facilitated diffusion of glucose using carriers in the basal membrane allows the glucose to leave the cells and enter the blood.

such as protein-bound cholesterol, can be taken into the cell because of the interaction between the cholesterol transport protein and a protein receptor on the plasma membrane. Cholesterol is removed from the blood by the liver and by the walls of blood vessels through this mechanism.

Exocytosis and endocytosis together provide **bulk transport** out of and into the cell, respectively. (The term "bulk" is used because many molecules are moved at the same time). It should be noted that molecules taken into a cell by endocytosis are still separated from the cytoplasm by the membrane of the endocytotic vesicle. Some of these molecules, such as membrane receptors, will be moved back to the plasma membrane, while the rest will end up in lysosomes.

Test Yourself Before You Continue

1. List the three characteristics of facilitated diffusion that distinguish it from simple diffusion.

2. Draw a figure that illustrates two of the characteristics of carrier-mediated transport and explain how this type of movement differs from simple diffusion.

3. Describe active transport, including primary and secondary active transport in your description. Explain how active transport differs from facilitated diffusion.

4. Discuss the physiological significance of the Na⁺/K⁺ pumps.

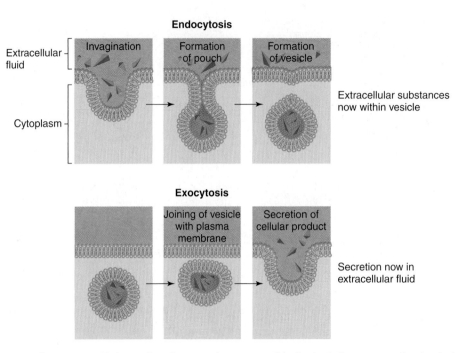

Figure 6.20 **Endocytosis and exocytosis.** Endocytosis and exocytosis are responsible for the bulk transport of molecules into and out of a cell.

The Membrane Potential

As a result of the permeability properties of the plasma membrane, the presence of nondiffusible negatively charged molecules inside the cell, and the action of the Na⁺/K⁺ pumps, there is an unequal distribution of charges across the membrane. As a result, the inside of the cell is negatively charged compared to the outside. This difference in charge, or potential difference, is known as the membrane potential.

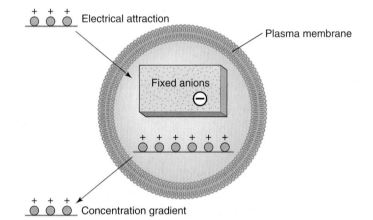

Figure 6.21 **The effect of fixed anions on the distribution of cations.** Proteins, organic phosphates, and other organic anions that cannot leave the cell create a fixed negative charge on the inside of the membrane. This negative charge attracts positively charged inorganic ions (cations), which therefore accumulate within the cell at a higher concentration than is found in the extracellular fluid. The amount of cations that accumulates within the cell is limited by the fact that a concentration gradient builds up, which favors the diffusion of the cations out of the cell.

In the preceding section, the action of the Na⁺/K⁺ pumps was discussed in conjunction with the topic of active transport, and it was noted that these pumps move Na⁺ and K⁺ against their concentration gradients. This action alone would create and amplify a difference in the concentration of these ions across the plasma membrane. There is, however, another reason why the concentration of Na⁺ and K⁺ would be unequal across the membrane.

Cellular proteins and the phosphate groups of ATP and other organic molecules are negatively charged at the pH of the cell cytoplasm. These negative ions (anions) are "fixed" within the cell because they cannot penetrate the plasma membrane. As a result, these anions attract positively charged inorganic ions (cations) from the extracellular fluid that are small enough to diffuse through the membrane pores. The distribution of small inorganic cations (mainly K⁺, Na⁺, and Ca²⁺) between the intracellular and extracellular compartments is thus influenced by the negatively charged fixed ions within the cell.

Since the plasma membrane is more permeable to K⁺ than to any other cation, K⁺ accumulates within the cell more than the others as a result of its electrical attraction for the fixed an-

ions (fig. 6.21). So, instead of being evenly distributed between the intracellular and extracellular compartments, K⁺ becomes more highly concentrated within the cell. The intracellular K⁺ concentration is 150 mEq/L in the human body compared to an extracellular concentration of 5 mEq/L (mEq = milliequivalents, which is the millimolar concentration multiplied by the valence of the ion—in this case, by one).

As a result of the unequal distribution of charges between the inside and outside of cells, each cell acts as a tiny battery with the positive pole outside the plasma membrane and the negative pole inside. The magnitude of this charge difference is measured in *voltage*. Although the voltage of this battery is very small (less than a tenth of a volt), it is of critical importance in such physiological processes as muscle contraction, the regulation of the heartbeat, and the generation of nerve impulses. In order to understand these processes, then, we must first examine the electrical properties of cells.

Equilibrium Potentials

An equilibrium potential is a theoretical voltage that would be produced across a plasma membrane if only one ion were able to diffuse through the membrane. Since the membrane is most permeable to K^+, we can construct a theoretical approximation by determining what would happen if K^+ were the *only* ion able to cross the membrane. If this were the case, K^+ would diffuse until its concentration inside and outside of a cell became stable, thus establishing an *equilibrium*. In this condition, if a certain amount of K^+ were to move inside the cell (by electrical attraction for the fixed anions), an identical amount of K^+ would diffuse out of the cell (down its concentration gradient). At equilibrium, the forces of electrical attraction and of the diffusion gradient are equal and opposite.

At this equilibrium, the concentration of K^+ would be higher inside the cell than outside the cell; a concentration difference would exist across the plasma membrane that was stabilized by the attraction of K^+ to the fixed anions. At this point we could ask, Are the fixed anions neutralized . . . are the charges balanced? The answer depends on how much K^+ gets into the cell, which in turn depends on the K^+ concentration in the extracellular fluid. At the K^+ concentrations that are, in fact, found in the body, the answer to our question is no. Not enough K^+ is present in the cell to neutralize the fixed anions (fig. 6.22).

At equilibrium, therefore, the inside of the cell membrane would have a higher concentration of negative charges than the outside of the membrane. There is a difference in charge, as well as a difference in concentration, across the membrane. The magnitude of the difference in charge, or **potential difference,** on the two sides of the membrane under these conditions is 90 millivolts (mV). A sign (+ or –) placed in front of this number indicates the polarity within the cell. This is shown with a negative sign (as –90 mV) to indicate that the inside of the cell is the negative pole. The potential difference of –90 mV, which would be developed if K^+ were the only diffusible ion, is called the **K^+ equilibrium potential** (abbreviated E_K).

Nernst Equation

There is another way to look at the equilibrium potential: it is the membrane potential that would *exactly balance* the diffusion gradient and prevent the net movement of a particular ion. Since the diffusion gradient depends on the difference in concentration of the ion, the value of the equilibrium potential must depend on the ratio of the concentrations of the ion on the two sides of the membrane. The **Nernst equation** allows this theoretical equilibrium potential to be calculated for a particular ion when its con-

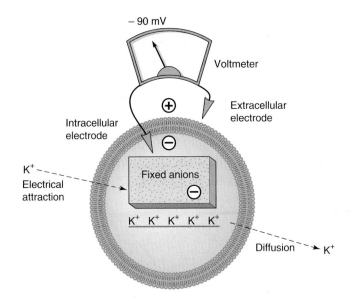

Figure 6.22 Potassium equilibrium potential. If K^+ were the only ion able to diffuse through the plasma membrane, it would distribute itself between the intracellular and extracellular compartments until an equilibrium was established. At equilibrium, the K^+ concentration within the cell would be higher than outside the cell because of the attraction of K^+ for the fixed anions. Not enough K^+ would accumulate within the cell to neutralize these anions, however, so the inside of the cell would be –90 millivolts compared to the outside of the cell. This membrane voltage is the equilibrium potential (E_K) for potassium.

centrations are known. The following simplified form of the equation is valid at a temperature of 37° C:

$$E_x = \frac{61}{z} \log \frac{[X_o]}{[X_i]}$$

where

E_x = equilibrium potential in millivolts (mV) for ion x
X_o = concentration of the ion outside the cell
X_i = concentration of the ion inside the cell
z = valence of the ion (+1 for Na^+ or K^+)

Note that, using the Nernst equation, the equilibrium potential for a cation has a negative value when X_i is greater than X_o. If we substitute K^+ for X, this is indeed the case. As a hypothetical example, if the concentration of K^+ were ten times higher inside compared to outside the cell, the equilibrium potential would be 61 mV (log 1/10) = 61 × (–1) = –61 mV. In reality, the concentration of K^+ inside the cell is actually thirty times greater than outside (150 mEq/L inside compared to 5 mEq/L outside). Thus,

$$E_K = 61 \text{ mV} \log \frac{5 \text{ mEq / L}}{150 \text{ mEq / L}} = -90 \text{ mV}$$

This means that a membrane potential of 90 mV, with the inside of the cell negative, would be required to prevent the diffusion of K^+ out of the cell.

If we wish to calculate the equilibrium potential for Na^+, different values must be used. The concentration of Na^+ in the extracellular fluid is 145 mEq/L, whereas its concentration inside cells is only 12 mEq/L. The diffusion gradient thus promotes the movement of Na^+ into the cell, and, in order to oppose this diffusion, the membrane potential would have to have a positive polarity on the inside of the cell. This is indeed what the Nernst equation would provide. Thus,

$$E_{Na} = 61 \text{ mV log} \frac{145 \text{ mEq / L}}{12 \text{ mEq / L}} = +60 \text{ mV}$$

This means that a membrane potential of 60 mV, with the inside of the cell positive, would be required to prevent the diffusion of Na^+ into the cell.

Resting Membrane Potential

A membrane potential of +60 mV would prevent the diffusion of Na^+ into the cell, while a membrane potential of –90 mV would prevent the diffusion of K^+ out of the cell. It is clear that the membrane potential cannot be both values at the same time; indeed, it is seldom either value but instead is somewhere between these two extremes. We will call this the **resting membrane potential** to distinguish it from the theoretical equilibrium potentials.

The actual value of the resting membrane potential depends on two factors:

1. The *ratio of the concentrations* (X_o/X_i) of each ion on the two sides of the plasma membrane.
2. The *specific permeability* of the membrane to each different ion.

Many ions—including K^+, Na^+, Ca^{2+}, and Cl^-—contribute to the resting membrane potential. Their individual contributions are determined by (a) the differences in their concentrations across the membrane (fig. 6.23), and (b) by their membrane permeabilities.

This has two important implications:

1. For any given ion, a change in its concentration in the extracellular fluid will change the resting membrane potential—but only to the extent that the membrane is permeable to that ion. Because *the resting membrane is most permeable to K^+*, a change in the extracellular concentration of K^+ has the greatest effect on the resting membrane potential. This is the mechanism behind the fact that "lethal injections" are of KCl (raising the extracellular K^+ concentrations and depolarizing cardiac cells.).
2. A change in the membrane permeability to any given ion will change the membrane potential. This fact is central to the production of nerve and muscle impulses, as will be described in chapter 7. Most often, it is the opening and closing of Na^+ and K^+ channels that are involved, but gated channels for Ca^{2+} and Cl^- are also very important in physiology.

The resting membrane potential of most cells in the body ranges from –65 mV to –85 mV (in neurons it averages –70 mV). This value is close to the E_K, because the resting plasma membrane is more permeable to K^+ than to other ions. During nerve and muscle impulses, however, the permeability properties change, as will be described in chapter 7. An in-

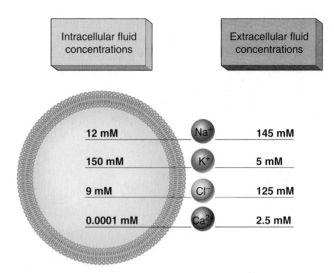

Figure 6.23 Concentrations of ions in the intracellular and extracellular fluids. This distribution of ions, and the different permeabilities of the plasma membrane to these ions, affects the membrane potential and other physiological processes.

creased membrane permeability to Na^+ drives the membrane potential toward E_{Na} (+60 mV) for a short time. This is the reason that the term *resting* is used to describe the membrane potential when it is not producing impulses.

The resting membrane potential is particularly sensitive to changes in plasma potassium concentration. Since the maintenance of a particular membrane potential is critical for the generation of electrical events in the heart, mechanisms that act primarily through the kidneys maintain plasma K^+ concentrations within very narrow limits. An abnormal increase in the blood concentration of K^+ is called **hyperkalemia**. When hyperkalemia occurs, more K^+ can enter the cell. In terms of the Nernst equation, the ratio $[K^+_o]/[K^+_i]$ is decreased. This reduces the membrane potential (brings it closer to zero) and thus interferes with the proper function of the heart. For these reasons, the blood electrolyte concentrations are monitored very carefully in patients with heart or kidney disease.

Clinical Investigation Clues

Remember that Jessica's medical tests revealed hyperkalemia.
What is hyperkalemia and why might Jessica have this condition? What is the relationship between the hyperkalemia and her abnormal EKG?

Role of the Na^+/K^+ Pumps

Since the resting membrane potential is less negative than E_K, some K^+ leaks out of the cell (fig. 6.24). The cell is *not* at equilibrium with respect to K^+ and Na^+ concentrations. Nonetheless,

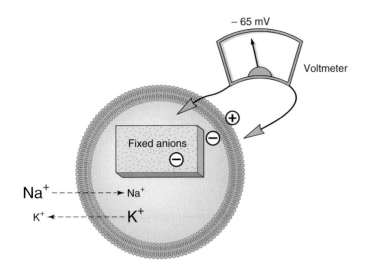

Figure 6.24 **The resting membrane potential.** Because some Na⁺ leaks into the cell by diffusion, the actual resting membrane potential is lower than the K⁺ equilibrium potential. As a result, some K⁺ diffuses out of the cell, as indicated by the dashed lines.

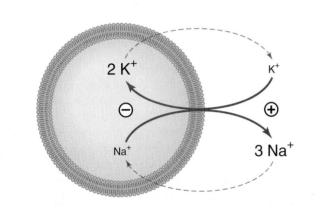

Figure 6.25 **The contribution of the Na⁺/K⁺ pumps to the membrane potential.** The concentrations of Na⁺ and K⁺ both inside and outside the cell do not change as a result of diffusion (dashed arrows) because of active transport (solid arrows) by the Na⁺/K⁺ pump. Since the pump transports three Na⁺ for every two K⁺, the pump itself helps to create a charge separation (a potential difference, or voltage) across the membrane.

Test Yourself Before You Continue

1. Define the term *membrane potential* and explain how it is measured.
2. Explain how an equilibrium potential is produced when potassium is the only diffusible cation. State how the value of the equilibrium potential is affected by the potassium concentrations outside and inside the cell.
3. Explain why the resting membrane potential is close to, but different from, the potassium equilibrium potential.
4. Suppose a person has hyperkalemia such that the extracellular K⁺ concentration increases from 5 mM to 10 mM (a potentially fatal condition). Use the Nernst equation to calculate the new E_K, and then verbally describe how the resting membrane potential would be changed.
5. Describe the role of the Na⁺/K⁺ pumps in the generation and maintenance of the resting membrane potential.

Cell Signaling

Cells communicate by signaling each other chemically. These chemical signals are regulatory molecules released by neurons and endocrine glands, and by different cells within an organ.

The membrane potential and the permeability properties of the plasma membrane to ions discussed in the previous section set the stage for the discussion of nerve impulses in chapter 7. Nerve impulses are a type of signal that is conducted along the axon of a neuron. When the impulses reach the end of the axon, however, the signal must somehow be transmitted to the next cell.

Cell signaling refers to how cells communicate with each other. In certain specialized cases, the signal can travel directly from one cell to the next because their plasma membranes are fused together and their cytoplasm is continuous through tiny **gap junctions** in the fused membranes (see chapter 7, fig. 7.19). In these cases, ions and regulatory molecules can travel by diffusion through the cytoplasm of adjoining cells. In most cases, however, cells signal each other by releasing chemicals into the extracellular environment. In these cases, cell signaling can be divided into three general categories: (1) paracrine signaling; (2) synaptic signaling; and (3) endocrine signaling.

In **paracrine** signaling (fig. 6.26), cells within an organ secrete regulatory molecules that diffuse through the extracellular matrix to nearby *target cells* (those that respond to the regulatory molecule). Paracrine regulation is considered to be *local*, because it involves the cells of a particular organ. Numerous paracrine regulators have been discovered that regulate organ growth and coordinate the activities of the different cells and tissues within an organ.

Synaptic signaling refers to the means by which neurons regulate their target cells. The axon of a neuron (see chapter 1, fig. 1.10) is said to *innervate* its target organ through a functional

the concentrations of K⁺ and Na⁺ are maintained constant because of the constant expenditure of energy in active transport by the Na⁺/K⁺ pumps. The Na⁺/K⁺ pumps act to counter the leaks and thus maintain the membrane potential.

Actually, the Na⁺/K⁺ pump does more than simply work against the ion leaks; since it transports *three* Na⁺ out of the cell for every *two* K⁺ that it moves in, it has the net effect of contributing to the negative intracellular charge (fig. 6.25). This *electrogenic effect* of the pumps adds approximately 3 mV to the membrane potential. As a result of all of these activities, a real cell has (1) a relatively constant intracellular concentration of Na⁺ and K⁺ and (2) a constant membrane potential (in the absence of stimulation) in nerves and muscles of –65 mV to –85 mV.

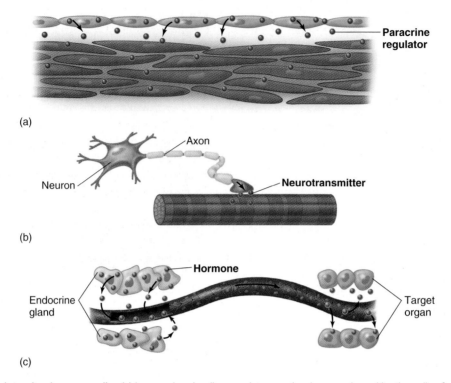

(a)

(b)

(c)

■ **Figure 6.26** **Chemical signaling between cells.** *(a)* In paracrine signaling, regulatory molecules are released by the cells of an organ and target other cells in the same organ. *(b)* In synaptic signaling, the axon of a neuron releases a chemical neurotransmitter, which regulates a target cell. *(c)* In endocrine signaling, an endocrine gland secretes hormones into the blood, which carries the hormones to the target organs.

connection, or *synapse,* between the axon ending and the target cell. There is a small synaptic gap, or cleft, between the two cells, and chemical regulators called *neurotransmitters* are released by the axon endings (fig. 6.26).

In **endocrine** signaling, the cells of endocrine glands secrete chemical regulators called *hormones* into the extracellular fluid. The hormones enter the blood and are carried by the blood to all the cells in the body. Only the target cells for a particular hormone, however, can respond to the hormone.

In order for a target cell to respond to a hormone, neurotransmitter, or paracrine regulator, it must have specific **receptor proteins** for these molecules. These receptor proteins may be located on the outer surface of the plasma membrane of the target cells, or they may be located intracellularly, in either the cytoplasm or nucleus. The location of the receptor proteins depends on whether the regulatory molecule can penetrate the plasma membrane of the target cell.

If the regulatory molecule is nonpolar, it can diffuse through the cell membrane and enter the target cell. Such non-polar regulatory molecules include steroid hormones, thyroid hormones, and nitric oxide gas (a paracrine regulator). In these cases, the receptor proteins are intracellular in location. Regulatory molecules that are large or polar—such as epinephrine (an amine hormone), acetylcholine (an amine neurotransmitter), and insulin (a polypeptide hormone)—cannot enter their target cells. In these cases, the receptor proteins are located on the outer surface of the plasma membrane. The details of how these signals influence their target cells are described in conjunction with neural and endocrine regulation in the next several chapters.

Test Yourself Before You Continue

1. Distinguish between synaptic, endocrine, and paracrine regulation.
2. Identify the location of the receptor proteins for different regulatory molecules.

HPer Links of Membrane Transport Concepts to the Body Systems

Skeletal System

- Osteoblasts secrete Ca^{2+} and PO_4^{3-} into the extracellular matrix, forming calcium phosphate crystals that account for the hardness of bone . . . .**(p. 623)**

Nervous System

- Glucose enters neurons by facilitative diffusion . . . .**(p. 135)**
- Voltage-gated ion channels produce action potentials, or nerve impulses . . . . .**(p. 161)**
- Ion channels in particular regions of a neuron open in response to binding to a chemical ligand known as a neurotransmitter . . . .**(p. 169)**
- Neurotransmitters are released by axons through the process of exocytosis .**(p. 169)**
- Sensory stimuli generally cause the opening of ion channels and depolarization of receptor cells . . . .**(p. 242)**

Endocrine System

- Lipophilic hormones pass through the cell membrane of their target cells, where they then bind to receptors in the cytoplasm or nucleus . . . .**(p. 292)**
- Active transport Ca^+ pumps and the passive diffusion of Ca^+ are important in mediating the actions of some hormones . . . .**(p. 296)**
- Insulin stimulates the facilitative diffusion of glucose into skeletal muscle cells .**(p. 611)**

Muscular System

- Exercise increases the number of carriers for the facilitative diffusion of glucose in the muscle cell membrane . . . .**(p. 343)**
- Ca^{2+} transport processes in the endoplasmic reticulum of skeletal muscle fibers are important in the regulation of muscle contraction . . . .**(p. 336)**
- Voltage-gated Ca^{2+} channels in the cell membrane of smooth muscle open in response to depolarization, producing contraction of the muscle . . . .**(p. 355)**

Circulatory System

- Transport processes through the capillary endothelial cells of the brain are needed in order for molecules to cross the blood-brain barrier and enter the brain . .**(p. 159)**

- Ion diffusion across the plasma membrane of myocardial cells is responsible for the electrical activity of the heart . . . .**(p. 385)**
- The LDL carriers for blood cholesterol are taken into arterial smooth muscle cells by receptor-mediated endocytosis . . .**(p. 396)**

Immune System

- B lymphocytes secrete antibody proteins that function in humoral (antibody-mediated) immunity . . . .**(p. 453)**
- T lymphocytes secrete polypeptides called cytokines that promote the cell-mediated immune response . . . .**(p. 458)**
- Antigen-presenting cells engulf foreign proteins by pinocytosis, modify these proteins, and present them to T lymphocytes . . . .**(p. 460)**

Respiratory System

- Oxygen and carbon dioxide pass through the cells of the pulmonary alveoli (air sacs) by simple diffusion . . . .**(p. 480)**
- Surfactant is secreted into pulmonary alveoli by exocytosis . . . .**(p. 486)**

Urinary System

- Urine is produced as a filtrate of blood plasma, but most of the filtered water is reabsorbed back into the blood by osmosis . . . .**(p. 534)**

- Osmosis across the wall of the renal tubules is promoted by membrane pores known as aquaporins . . . .**(p. 536)**
- Transport of urea occurs passively across particular regions of the renal tubules . . . .**(p. 536)**
- Antidiuretic hormone stimulates the permeability of the renal tubule to water . . . .**(p. 536)**
- Aldosterone stimulates Na^+ transport in a region of the renal tubule . . . .**(p. 544)**
- Glucose and amino acids are reabsorbed by secondary active transport . . . . . .**(p. 543)**

Digestive System

- Cells in the stomach have a membrane H^+/K^+ ATPase active transport pump that creates an extremely acidic gastric juice . . . .**(p. 566)**
- Water is absorbed in the intestine by osmosis following the absorption of sodium chloride . . . .**(p. 574)**
- An intestinal membrane carrier protein transports dipeptides and tripeptides from the intestinal lumen into the epithelial cells . . . .**(p. 589)**

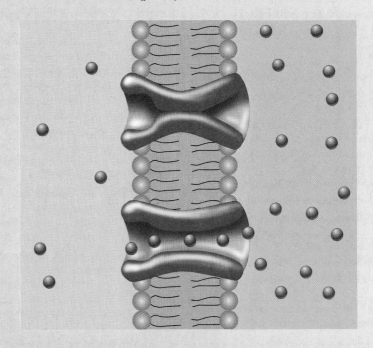

Summary

Extracellular Environment   126

I. Body fluids are divided into an intracellular compartment and an extracellular compartment.

    A. The extracellular compartment consists of blood plasma and interstitial, or tissue, fluid.

    B. Interstitial fluid is derived from plasma and returns to plasma.

II. The extracellular matrix consists of protein fibers of collagen and elastin and an amorphorus ground substance.

    A. The collagen and elastin fibers provide structural support.

    B. The ground substance contains glycoproteins and proteoglycans forming a hydrated gel, which contains most of the interstitial fluid.

Diffusion and Osmosis   128

I. Diffusion is the net movement of molecules or ions from regions of higher to regions of lower concentration.

    A. This is a type of passive transport—energy is provided by the thermal energy of the molecules, not by cellular metabolism.

    B. Net diffusion stops when the concentration is equal on both sides of the membrane.

II. The rate of diffusion is dependent on a variety of factors.

    A. The rate of diffusion depends on the concentration difference across the two sides of the membrane.

    B. The rate depends on the permeability of the plasma membrane to the diffusing substance.

    C. The rate depends on the temperature of the solution.

    D. The rate of diffusion through a membrane is also directly proportional to the surface area of the membrane, which can be increased by such adaptations as microvilli.

III. Simple diffusion is the type of passive transport in which small molecules and inorganic ions move through the plasma membrane.

    A. Inorganic ions such as Na^+ and K^+ pass through specific channels in the membrane.

    B. Steroid hormones and other lipids can pass directly through the phospholipid layers of the membrane by simple diffusion.

IV. Osmosis is the simple diffusion of solvent (water) through a membrane that is more permeable to the solvent than it is to the solute.

    A. Water moves from the solution that is more dilute to the solution that has a higher solute concentration.

    B. Osmosis depends on a difference in total solute concentration, not on the chemical nature of the solute.

        1. The concentration of total solute, in moles per kilogram (liter) of water, is measured in osmolality units.

        2. The solution with the higher osmolality has the higher osmotic pressure.

        3. Water moves by osmosis from the solution of lower osmolality and osmotic pressure to the solution of higher osmolality and osmotic pressure.

    C. Solutions containing osmotically active solutes that have the same osmotic pressure as plasma (such as 0.9% NaCl and 5% glucose) are said to be isotonic to plasma.

        1. Solutions with a lower osmotic pressure are hypotonic; those with a higher osmotic pressure are hypertonic.

        2. Cells in a hypotonic solution gain water and swell; those in a hypertonic solution lose water and shrink (crenate).

    D. The osmolality and osmotic pressure of the plasma is detected by osmoreceptors in the hypothalamus of the brain and maintained within a normal range by the action of antidiuretic hormone (ADH) released from the posterior pituitary.

        1. Increased osmolality of the blood stimulates the osmoreceptors.

        2. Stimulation of the osmoreceptors causes thirst and triggers the release of antidiuretic hormone (ADH) from the posterior pituitary.

        3. ADH promotes water retention by the kidneys, which serves to maintain a normal blood volume and osmolality.

Carrier-Mediated Transport   134

I. The passage of glucose, amino acids, and other polar molecules through the plasma membrane is mediated by carrier proteins in the cell membrane.

    A. Carrier-mediated transport exhibits the properties of specificity, competition, and saturation.

    B. The transport rate of molecules such as glucose reaches a maximum when the carriers are saturated. This maximum rate is called the transport maximum (T_m).

II. The transport of molecules such as glucose from the side of higher to the side of lower concentration by means of membrane carriers is called facilitated diffusion.

    A. Like simple diffusion, facilitated diffusion is passive transport—cellular energy is not required.

    B. Unlike simple diffusion, facilitated diffusion displays the properties of specificity, competition, and saturation.

III. The active transport of molecules and ions across a membrane requires the expenditure of cellular energy (ATP).

    A. In active transport, carriers move molecules or ions from the side of lower to the side of higher concentration.

    B. One example of active transport is the action of the Na^+/K^+ pump.

        1. Sodium is more concentrated on the outside of the cell, whereas potassium is more concentrated on the inside of the cell.

2. The Na$^+$/K$^+$ pump helps to maintain these concentration differences by transporting Na$^+$ out of the cell and K$^+$ into the cell.

The Membrane Potential   140

I. The cytoplasm of the cell contains negatively charged organic ions (anions) that cannot leave the cell—they are "fixed" anions.

A. These fixed anions attract K$^+$, which is the inorganic ion that can pass through the plasma membrane most easily.

B. As a result of this electrical attraction, the concentration of K$^+$ within the cell is greater than the concentration of K$^+$ in the extracellular fluid.

C. If K$^+$ were the only diffusible ion, the concentrations of K$^+$ on the inside and outside of the cell would reach an equilibrium.

1. At this point, the rate of K$^+$ entry (due to electrical attraction) would equal the rate of K$^+$ exit (due to diffusion).

2. At this equilibrium, there would still be a higher concentration of negative charges within the cell (because of the fixed anions) than outside the cell.

3. At this equilibrium, the inside of the cell would be 90 millivolts negative (–90 mV) compared to the outside of the cell. This potential difference is called the K$^+$ equilibrium potential (E_K).

D. The resting membrane potential is less than E_K (usually –65 mV to –85 mV) because some Na$^+$ can also enter the cell.

1. Na$^+$ is more highly concentrated outside than inside the cell, and the inside of the cell is negative. These forces attract Na$^+$ into the cell.

2. The rate of Na$^+$ entry is generally slow because the membrane is usually not very permeable to Na$^+$.

II. The slow rate of Na$^+$ entry is accompanied by a slow rate of K$^+$ leakage out of the cell.

A. The Na$^+$/K$^+$ pump counters this leakage, thus maintaining constant concentrations and a constant resting membrane potential.

B. Most cells in the body contain numerous Na$^+$/K$^+$ pumps that require a constant expenditure of energy.

C. The Na$^+$/K$^+$ pump itself contributes to the membrane potential because it pumps more Na$^+$ out than it pumps K$^+$ in (by a ratio of three to two).

Cell Signaling   143

I. Cells signal each other generally by secreting regulatory molecules into the extracellular fluid.

II. There are three categories of chemical regulation between cells.

A. Paracrine signaling refers to the release of regulatory molecules that act within the organ in which they are made.

B. Synaptic signaling refers to the release of chemical neurotransmitters by axon endings.

C. Endocrine signaling refers to the release of regulatory molecules called hormones, which travel in the blood to their target cells.

Review Activities

Test Your Knowledge of Terms and Facts

1. The movement of water across a plasma membrane occurs by
   a   active transport.
   b. facilitated diffusion.
   c. simple diffusion (osmosis).
   d. all of these.

2. Which of these statements about the facilitated diffusion of glucose is *true?*
   a   There is a net movement from the region of lower to the region of higher concentration.
   b. Carrier proteins in the cell membrane are required for this transport.
   c. This transport requires energy obtained from ATP.
   d. It is an example of cotransport.

3. If a poison such as cyanide stopped the production of ATP, which of the following transport processes would cease?
   a. the movement of Na$^+$ out of a cell
   b. osmosis
   c. the movement of K$^+$ out of a cell
   d. all of these

4. Red blood cells crenate in
   a. a hypotonic solution.
   b. an isotonic solution.
   c. a hypertonic solution.

5. Plasma has an osmolality of about 300 mOsm. The osmolality of isotonic saline is equal to
   a. 150 mOsm.
   b. 300 mOsm.

   c. 600 mOsm.
   d. none of these.

6. Which of these statements comparing a 0.5 *m* NaCl solution and a 1.0 *m* glucose solution is *true?*
   a. They have the same osmolality.
   b. They have the same osmotic pressure.
   c. They are isotonic to each other.
   d. All of these are true.

7. The most important diffusible ion in the establishment of the membrane potential is
   a   K$^-$.
   b. Na$^+$.
   c. Ca^{2+}.
   d. Cl$^-$.

8. Which of these statements regarding an increase in blood osmolality is *true?*
    a   It can occur as a result of dehydration.
    b. It causes a decrease in blood osmotic pressure.
    c. It is accompanied by a decrease in ADH secretion.
    d. All of these are true.

9. In hyperkalemia, the resting membrane potential
    a. moves farther from 0 millivolts.
    b. moves closer to 0 millivolts.
    c. remains unaffected.

10. Which of these statements about the Na^+/K^+ pump is *true?*
    a. Na^+ is actively transported into the cell.
    b. K^- is actively transported out of the cell.
    c. An equal number of Na^+ and K^+ ions are transported with each cycle of the pump.
    d. The pumps are constantly active in all cells.

11. Which of these statements about carrier-mediated facilitated diffusion is *true?*
    a. It uses cellular ATP.
    b. It is used for cellular uptake of blood glucose.

    c. It is a form of active transport.
    d. None of these are true.

12. Which of these is *not* an example of cotransport?
    a. movement of glucose and Na^+ through the apical epithelial membrane in the intestinal epithelium
    b. movement of Na^+ and K^+ through the action of the Na^+/K^+ pumps
    c. movement of Na^+ and glucose across the kidney tubules
    d. movement of Na^+ into a cell while Ca^{2+} moves out

Test Your Understanding of Concepts and Principles

1. Describe the conditions required to produce osmosis and explain why osmosis occurs under these conditions.[1]

2. Explain how simple diffusion can be distinguished from facilitated diffusion and how active transport can be distinguished from passive transport.

3. Compare the theoretical membrane potential that occurs at K^+ equilibrium with the true resting membrane

potential. Explain why these values differ.

4. Explain how the Na^+/K^+ pump contributes to the resting membrane potential.

5. Describe the cause-and-effect sequence whereby a genetic defect results in improper cellular transport and the symptoms of cystic fibrosis.

6. Using the principles of osmosis, explain why movement of Na^+ through a plasma membrane is followed by movement of water. Use this concept to explain the rationale on which oral rehydration therapy is based.

7. Distinguish between primary active transport and secondary active transport, and between cotransport and countertransport. Give examples of each.

Test Your Ability to Analyze and Apply Your Knowledge

1. Mannitol is a sugar that does not pass through the walls of blood capillaries in the brain (does not cross the "blood-brain barrier," as described in chapter 7). It also does not cross the walls of kidney tubules, the structures that transport blood filtrate to become urine (see chapter 17). Explain why mannitol can be described as osmotically active. How might its clinical administration

help to prevent swelling of the brain in head trauma? Also, explain the effect it might have on the water content of urine.

2. Discuss carrier-mediated transport. How could you experimentally distinguish between the different types of carrier-mediated transport?

3. Remembering the effect of cyanide (described in chapter 5), explain how

you might determine the extent to which the Na^+/K^+ pumps contribute to the resting membrane potential. Using a measurement of the resting membrane potential as your guide, how could you experimentally determine the relative permeability of the plasma membrane to Na^+ and K^+?

Related Websites

Check out the Links Library at www.mhhe.com/fox8 for links to sites containing resources related to cells and the extracellular environment. These links are monitored to ensure current URLs.

---

[1]*Note:* This question is answered in the chapter 6 Study Guide found on the Online Learning Center at www.mhhe.com/fox8.

7

The Nervous System: Neurons and Synapses

Objectives *After studying this chapter, you should be able to . . .*

1. describe the structure of a neuron and explain the functional significance of its principal regions.

2. classify neurons on the basis of their structure and function.

3. describe the locations and functions of the different types of supporting cells.

4. explain what is meant by the blood-brain barrier and discuss its significance.

5. describe the neurilemma and explain how it functions in the regeneration of cut peripheral nerve fibers.

6. explain how a myelin sheath is formed.

7. define *depolarization, repolarization, and hyperpolarization.*

8. explain the actions of voltage-regulated Na^+ and K^+ channels and describe the events that occur during the production of an action potential.

9. describe the properties of action potentials and explain the significance of the all-or-none law and the refractory periods.

10. explain how action potentials are regenerated along myelinated and nonmyelinated axons.

11. describe the events that occur in the interval between the electrical excitation of an axon and the release of neurotransmitter.

12. describe the two general categories of chemically regulated ion channels and explain how these channels operate using nicotinic and muscarinic ACh receptors as examples.

13. explain how ACh produces EPSPs and IPSPs, and discuss the significance of these processes.

14. compare the characteristics of EPSPs and action potentials.

15. compare the mechanisms that inactivate ACh with those that inactivate monoamine neurotransmitters.

16. explain the role of cyclic AMP in the action of monoamine neurotransmitters and describe some of the actions of monoamines in the nervous system.

17. explain the significance of the inhibitory effects of glycine and GABA in the central nervous system.

18. list some of the polypeptide neurotransmitters and explain the significance of the endogenous opioids in the nervous system.

19. discuss the significance of nitric oxide as a neurotransmitter.

20. explain how EPSPs and IPSPs can interact and discuss the significance of spatial and temporal summation and of presynaptic and postsynaptic inhibition.

21. describe the nature of long-term potentiation and discuss its significance.

Take Advantage of the Technology

Visit the Online Learning Center for
these additional study resources.

■ Interactive quizzing

■ Online study guide

■ Current news feeds

■ Crossword puzzles

■ Vocabulary flashcards

■ Labeling activities

www.mhhe.com/fox8

Clinical Investigation

Sandra, whose clinical depression was causing her grades to fall, decides to treat herself to dinner at a seafood restaurant. After eating a meal of mussels and clams, which were gathered from the local shore, she falls to the floor. Paramedics quickly arrive at the scene and notice that she has flaccid paralysis of her muscles and is having difficulty breathing. Fortunately, their emergency care saves her life. While the emergency care is being administered, a prescription bottle containing a monoamine oxidase (MAO) inhibitor is found in her purse.

Laboratory tests later reveal that her blood contained amounts of the MAO inhibitor that were consistent with its therapeutic use. What might have caused Sandra's medical emergency?

Neurons and Supporting Cells

The nervous system is composed of neurons, which produce and conduct electrochemical impulses, and supporting cells, which assist the functions of neurons. Neurons are classified functionally and structurally; the various types of supporting cells perform specialized functions.

The nervous system is divided into the **central nervous system (CNS),** which includes the brain and spinal cord, and the **peripheral nervous system (PNS),** which includes the *cranial nerves* arising from the brain and the *spinal nerves* arising from the spinal cord.

The nervous system is composed of only two principal types of cells—neurons and supporting cells. **Neurons** are the basic structural and functional units of the nervous system. They are specialized to respond to physical and chemical stimuli, conduct electrochemical impulses, and release chemical regulators. Through these activities, neurons enable the perception of sensory stimuli, learning, memory, and the control of muscles and glands. Most neurons cannot divide by mitosis, although many can regenerate a severed portion or sprout small new branches under certain conditions.

Supporting cells aid the functions of neurons and are about five times more abundant than neurons. In the CNS, supporting cells are collectively called **neuroglia,** or simply **glial cells** (*glia* = glue). Unlike neurons, which do not divide mitotically (except for particular ones, discussed in a clinical box on neural stem cells in chapter 8), glial cells are able to divide by mitosis. This helps to explain why brain tumors in adults are usually composed of glial cells rather than of neurons.

Neurons

Although neurons vary considerably in size and shape, they generally have three principal regions: (1) a cell body, (2) dendrites, and (3) an axon (figs. 7.1 and 7.2). Dendrites and axons can be referred to generically as *processes,* or extensions from the cell body.

The **cell body** is the enlarged portion of the neuron that contains the nucleus. It is the "nutritional center" of the neuron where macromolecules are produced. The cell body also contains densely staining areas of rough endoplasmic reticulum known as *Nissl bodies* that are not found in the dendrites or axon. The cell bodies within the CNS are frequently clustered into groups called

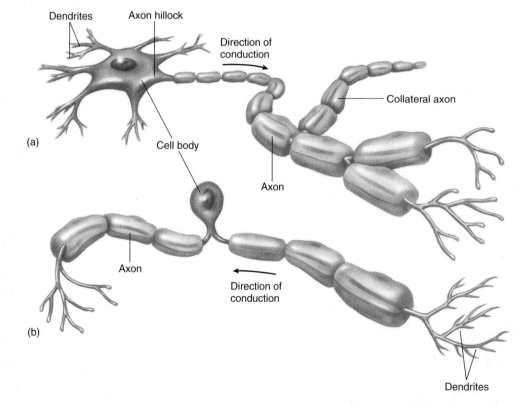

■ Figure 7.1   The structure of two kinds of neurons. (*a*) A motor neuron and (*b*) a sensory neuron are depicted here.

nuclei (not to be confused with the nucleus of a cell). Cell bodies in the PNS usually occur in clusters called *ganglia* (table 7.1).

Dendrites (*dendron* = tree branch) are thin, branched processes that extend from the cytoplasm of the cell body. Dendrites provide a receptive area that transmits electrical impulses to the cell body. The **axon** is a longer process that conducts impulses away from the cell body. Axons vary in length from only a millimeter long to up to a meter or more (for those that extend from the CNS to the foot). The origin of the axon near the cell

body is an expanded region called the *axon hillock;* it is here that nerve impulses originate. Side branches called *axon collaterals* may extend from the axon.

Proteins and other molecules are transported through the axon at faster rates than could be achieved by simple diffusion. This rapid movement is produced by two different mechanisms: axoplasmic flow and axonal transport (table 7.2). **Axoplasmic flow,** the slower of the two, results from rhythmic waves of contraction that push the cytoplasm from the axon hillock to the nerve endings.

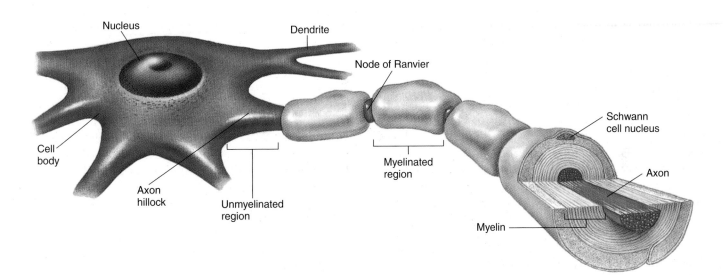

■ **Figure 7.2** Parts of a neuron. The axon of this neuron is wrapped by Schwann cells, which form a myelin sheath.

Table 7.1 Terminology Pertaining to the Nervous System

Term	Definition
Central nervous system (CNS)	Brain and spinal cord
Peripheral nervous system (PNS)	Nerves, ganglia, and nerve plexuses (outside of the CNS)
Association neuron (interneuron)	Multipolar neuron located entirely within the CNS
Sensory neuron (afferent neuron)	Neuron that transmits impulses from a sensory receptor into the CNS
Motor neuron (efferent neuron)	Neuron that transmits impulses from the CNS to an effector organ, for example, a muscle
Nerve	Cablelike collection of many axons, may be "mixed" (contain both sensory and motor fibers)
Somatic motor nerve	Nerve that stimulates contraction of skeletal muscles
Autonomic motor nerve	Nerve that stimulates contraction (or inhibits contraction) of smooth muscle and cardiac muscle and that stimulates glandular secretion
Ganglion	Grouping of neuron cell bodies located outside the CNS
Nucleus	Grouping of neuron cell bodies within the CNS
Tract	Grouping of nerve fibers that interconnect regions of the CNS

Table 7.2 Comparison of Axoplasmic Flow and Axonal Transport

Axoplasmic Flow	Axonal Transport
Transport rate comparatively slow (1–2 mm/day)	Transport rate comparatively fast (200–400 mm/day)
Molecules transported only from cell body	Molecules transported from cell body to axon endings and in reverse direction
Bulk movement of proteins in axoplasm, including microfilaments and tubules	Transport of specific proteins, mainly of membrane proteins and acetylcholinesterase
Transport accompanied by peristaltic waves of axon membrane	Transport dependent on cagelike microtubule structure within axon and on actin and Ca^{2+}

Axonal transport, which employs microtubules and is more rapid and more selective, may occur in a reverse (retrograde) direction as well as in a forward (orthograde) direction. Indeed, retrograde transport may be responsible for the movement of herpes virus, rabies virus, and tetanus toxin from the nerve terminals into cell bodies.

Classification of Neurons and Nerves

Neurons may be classified according to their function or structure. The functional classification is based on the direction in which they conduct impulses, as indicated in figure 7.3. **Sensory,** or **afferent, neurons** conduct impulses from sensory receptors *into* the CNS. **Motor,** or **efferent, neurons** conduct impulses *out* of the CNS to effector organs (muscles and glands). **Association neurons,** or **interneurons,** are located entirely within the CNS and serve the associative, or integrative, functions of the nervous system.

There are two types of motor neurons: somatic and autonomic. **Somatic motor neurons** are responsible for both reflex and voluntary control of skeletal muscles. **Autonomic motor neurons** innervate (send axons to) the involuntary effectors—smooth muscle, cardiac muscle, and glands. The cell bodies of the autonomic neurons that innervate these organs are located outside the CNS in autonomic ganglia (fig. 7.3). There are two subdivisions of autonomic neurons: *sympathetic* and *parasympathetic*. Autonomic motor neurons, together with their central control centers, constitute the *autonomic nervous system,* the focus of chapter 9.

The structural classification of neurons is based on the number of processes that extend from the cell body of the neuron (fig. 7.4). **Pseudounipolar neurons** have a single short process that branches like a T to form a pair of longer processes. They are called pseudounipolar (*pseudo* = false) because, though they originate with two processes, during early embryonic development their two processes converge and partially fuse. Sensory neurons are pseudounipolar—one of the branched processes receives sensory stimuli and produces nerve impulses; the other delivers these impulses to synapses within the brain or spinal cord. Anatomically, the part of the process that conducts impulses toward the cell body can be considered a dendrite, and the part that conducts impulses away from the cell body can be considered an axon. Functionally, however, the two branched processes behave as a single long axon; only the small projections at the receptive end of the process function as typical dendrites. **Bipolar neurons** have two processes, one at either end; this type is found in the retina of the eye. **Multipolar neurons,** the most common type, have several dendrites and one axon extending from the cell body; motor neurons are good examples of this type.

A **nerve** is a bundle of axons located outside the CNS. Most nerves are composed of both motor and sensory fibers and are thus called *mixed nerves.* Some of the cranial nerves, however, contain sensory fibers only. These are the nerves that serve the special senses of sight, hearing, taste, and smell.

Supporting Cells

Unlike other organs that are "packaged" in connective tissue derived from mesoderm (the middle layer of embryonic tissue), the supporting cells of the nervous system are derived from the same embryonic tissue layer (ectoderm) that produces neurons.

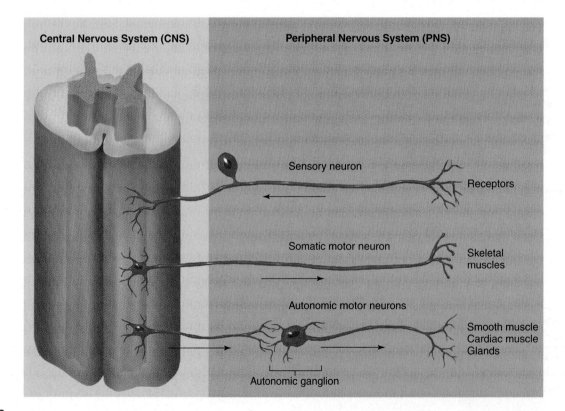

■ Figure 7.3 **The relationship between the CNS and PNS.** Sensory and motor neurons of the peripheral nervous system carry information into and out of, respectively, the central nervous system (brain and spinal cord).

There are two types of supporting cells in the peripheral nervous system:

1. **Schwann cells,** which form myelin sheaths around peripheral axons; and
2. **satellite cells,** or **ganglionic gliocytes,** which support neuron cells bodies within the ganglia of the PNS.

There are four types of supporting cells, called neuroglial (or glial) cells, in the central nervous system (fig. 7.5):

1. **oligodendrocytes,** which form myelin sheaths around axons of the CNS;
2. **microglia,** which migrate through the CNS and phagocytose foreign and degenerated material;

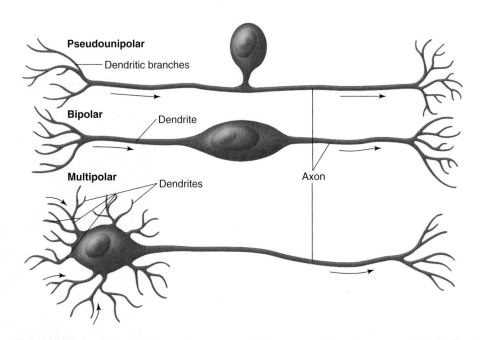

Figure 7.4 **Three different types of neurons.** Pseudounipolar neurons, which are sensory, have one process that splits. Bipolar neurons, found in the retina and cochlea, have two processes. Multipolar neurons, which are motor and association neurons, have many dendrites and one axon.

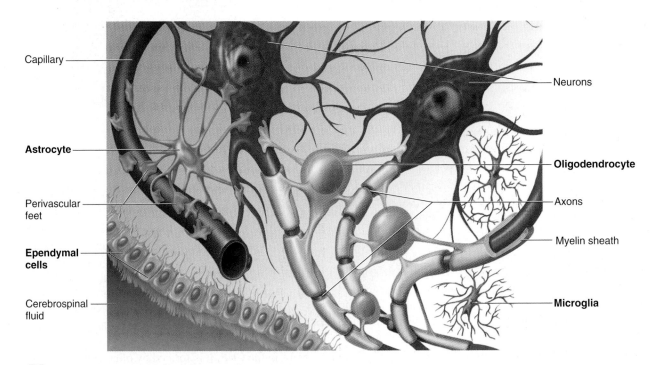

Figure 7.5 **The different types of neuroglial cells.** Myelin sheaths around axons are formed in the CNS by oligodendrocytes. Astrocytes have extensions that surround both blood capillaries and neurons. Microglia are phagocytic, and ependymal cells line the brain ventricles and central canal of the spinal cord.

Table 7.3 Supporting Cells and Their Functions*

Cell Type	Location	Functions
Schwann cells	PNS	Surround axons of all peripheral nerve fibers, forming a neurilemmal sheath, or sheath of Schwann; wrap around many peripheral fibers to form myelin sheaths; also called neurolemmocytes
Satellite cells	PNS	Support functions of neurons within sensory and autonomic ganglia; also called ganglionic gliocytes
Oligodendrocytes	CNS	Form myelin sheaths around central axons, producing "white matter" of the CNS
Microglia	CNS	Phagocytose pathogens and cellular debris in the CNS
Astrocytes	CNS	Cover capillaries of the CNS and induce the blood-brain barrier; interact metabolically with neurons and modify the extracellular environment of neurons
Ependymal cells	CNS	Form the epithelial lining of brain cavities (ventricles) and the central canal of the spinal cord; cover tufts of capillaries to form choroid plexuses—structures that produce cerebrospinal fluid

*Supporting cells in the CNS are known as neuroglia.

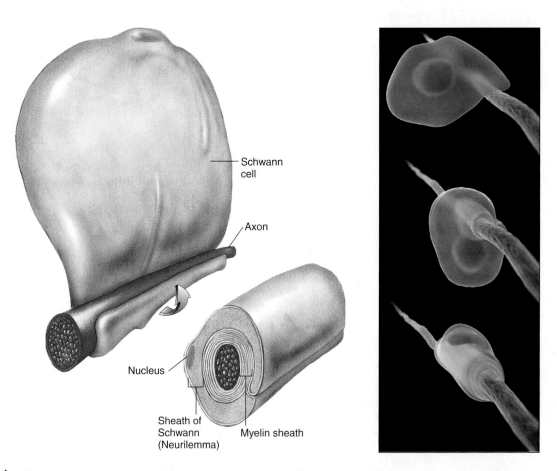

Schwann cell

Axon

Nucleus

Sheath of Schwann (Neurilemma)

Myelin sheath

■ **Figure 7.6** **The formation of a myelin sheath around a peripheral axon.** The myelin sheath is formed by successive wrappings of the Schwann cell membranes, leaving most of the Schwann cell cytoplasm outside the myelin. The sheath of Schwann is thus external to the myelin sheath.

3. **astrocytes,** which help to regulate the external environment of neurons in the CNS; and
4. **ependymal cells,** which line the ventricles (cavities) of the brain and the central canal of the spinal cord.

A summary of the supporting cells is presented in table 7.3.

Recent evidence suggests a more exciting role for the ependymal cells that line the ventricles of the brain, and also for the astrocytes immediately adjacent to this region—they can function as **neural stem cells.** That is, they can divide and their progeny can differentiate (specialize) along different lines, to become new neurons and neuroglial cells. Reptile and bird brains have been known to generate new neurons throughout life, but only recently has this ability been demonstrated in mammalian (including human) brains.

Neurilemma and Myelin Sheath

All axons in the PNS (myelinated and unmyelinated) are surrounded by a continuous, living sheath of Schwann cells, known as the **neurilemma,** or **sheath of Schwann.** The axons of the CNS, by contrast, lack a neurilemma (Schwann cells are only found in the PNS). This is significant in terms of regeneration of damaged axons, as will be described shortly.

Some axons in the PNS and CNS are surrounded by a **myelin sheath.** In the PNS, this insulating covering is formed by successive wrappings of the cell membrane of Schwann cells; in the CNS, it is formed by oligodendrocytes. Those axons smaller than 2 micrometers (2 μm) in diameter are usually *un-myelinated* (have no myelin sheath), whereas those that are

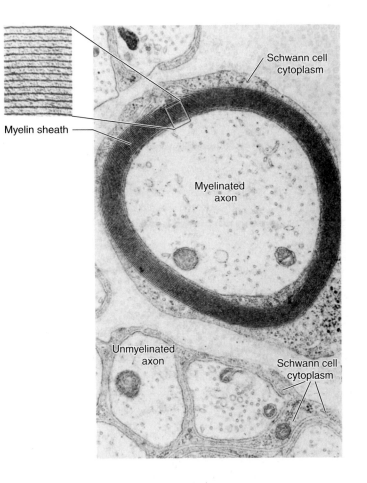

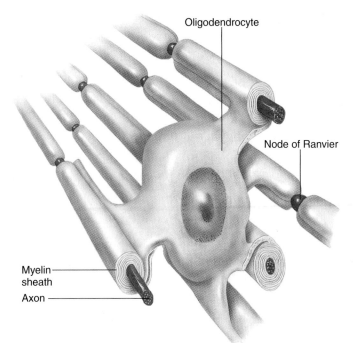

Figure 7.8 The formation of myelin sheaths in the CNS by an **oligodendrocyte.** One oligodendrocyte forms myelin sheaths around several axons.

Figure 7.7 An electron micrograph of unmyelinated and myelinated axons. Notice that myelinated axons have Schwann cell cytoplasm to the outside of their myelin sheath, and that Schwann cell cytoplasm also surrounds unmyelinated axons.

larger are likely to be *myelinated.* Myelinated axons conduct impulses more rapidly than those that are unmyelinated.

Myelin Sheath in PNS

In the process of myelin formation in the PNS, Schwann cells roll around the axon, much like a roll of electrician's tape is wrapped around a wire. Unlike electrician's tape, however, the Schwann cell wrappings are made in the same spot, so that each wrapping overlaps the previous layers. The cytoplasm, meanwhile, is forced into the outer region of the Schwann cell, much as toothpaste is squeezed to the top of the tube as the bottom is rolled up (fig. 7.6). Each Schwann cell wraps only about a millimeter of axon, leaving gaps of exposed axon between the adjacent Schwann cells. These gaps in the myelin sheath are known as the **nodes of Ranvier.** The successive wrappings of Schwann cell membrane provide insulation around the axon, leaving only the nodes of Ranvier exposed to produce nerve impulses.

The Schwann cells remain alive as their cytoplasm is forced to the outside of the myelin sheath. As a result, myelinated axons of the PNS are surrounded by a living sheath of Schwann cells, or neurilemma (fig. 7.7). Unmyelinated axons are

also surrounded by a neurilemma, but they differ from myelinated axons in that they lack the multiple wrappings of Schwann cell plasma membrane that comprise the myelin sheath.

Myelin Sheath in CNS

As mentioned earlier, the myelin sheaths of the CNS are formed by oligodendrocytes. This process occurs mostly postnatally (after birth). Unlike a Schwann cell, which forms a myelin sheath around only one axon, each oligodendrocyte has extensions, like the tentacles of an octopus, that form myelin sheaths around several axons (fig. 7.8). The myelin sheaths around axons of the CNS give this tissue a white color; areas of the CNS that contain a high concentration of axons thus form the **white matter.** The **gray matter** of the CNS is composed of high concentrations of cell bodies and dendrites, which lack myelin sheaths.

Regeneration of a Cut Axon

When an axon in a peripheral nerve is cut, the distal portion of the axon that was severed from the cell body degenerates and is phagocytosed by Schwann cells. The Schwann cells, surrounded by the basement membrane, then form a *regeneration tube* (fig. 7.9) as the part of the axon that is connected to the cell body begins to grow and exhibit amoeboid movement. The Schwann cells of the regeneration tube are believed to secrete chemicals that attract the growing axon tip, and the regeneration tube helps to guide the regenerating axon to its proper destination. Even a severed major nerve may be surgically reconnected—and the function of the nerve largely reestablished—if the surgery is performed before tissue death occurs.

■ **Figure 7.9** The process of peripheral neuron regeneration. *(a)* If a neuron is severed through a myelinated axon, the proximal portion may survive, but *(b)* the distal portion will degenerate through phagocytosis. The myelin sheath provides a pathway *(c)* and *(d)* for the regeneration of an axon, and *(e)* innervation is restored.

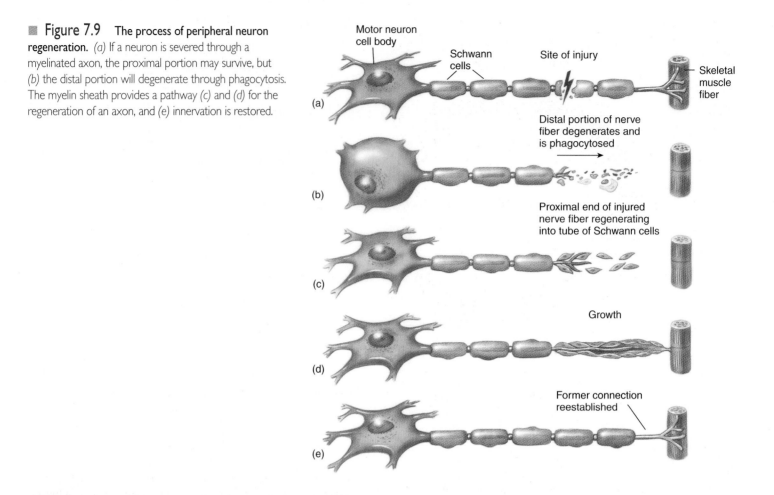

Multiple sclerosis (MS) is a neurological disease usually diagnosed in people between the ages of 20 and 40. It is a chronic, degenerating, remitting, and relapsing disease that progressively destroys the myelin sheaths of neurons in multiple areas of the CNS. Initially, lesions form on the myelin sheaths and soon develop into hardened *scleroses,* or scars (from the Greek word *sklerosis,* meaning "hardened"). Destruction of the myelin sheaths prohibits the normal conduction of impulses, resulting in a progressive loss of functions. Because myelin degeneration is widespread and affects different areas of the nervous system in different people, MS has a wider variety of symptoms than any other neurological disease. Although the causes of MS are not fully known, there is evidence that the disease involves a genetic susceptibility combined with an immune attack on the oligodendrocytes and myelin, perhaps triggered by viruses. Inflammation and demyelination then occur, leading to the symptoms of MS.

Injury in the CNS stimulates growth of axon collaterals, but central axons have a much more limited ability to regenerate than peripheral axons. This may be due in part to the absence of a continuous neurilemma (as is present in the PNS), which precludes the formation of a regeneration tube, and to inhibitory molecules produced by oligodendrocytes and astrocytes in the injured CNS. In addition to the limited ability of CNS neurons to regenerate, injury to the spinal cord has recently been shown to actually evoke apoptosis (cell suicide—chapter 3) in neurons that were not directly damaged by the injury.

Neurotrophins

In a developing fetal brain, chemicals called **neurotrophins** promote neuron growth. *Nerve growth factor (NGF)* was the first neurotrophin to be identified; others include *brain-derived neurotrophic factor (BDNF); glial-derived neurotrophic factor (GDNF); neurotrophin-3;* and *neurotrophin-4/5* (the number depends on the species). NGF and neurotrophin-3 are known to be particularly important in the embryonic development of sensory neurons and sympathetic ganglia.

Neurotrophins also have important functions in the adult nervous system. NGF is required for the maintenance of sympathetic ganglia, and there is evidence that neurotrophins are required for mature sensory neurons to regenerate after injury. In addition, GDNF may be needed in the adult to maintain spinal motor neurons and to sustain neurons in the brain that use the chemical dopamine as a neurotransmitter.

Experiments suggest that neurons of the CNS can regenerate if they are provided with the appropriate environment. While neurotrophins promote neuron growth, some chemicals, including *myelin-associated inhibitory proteins,* have been shown to inhibit axon regeneration. Research in this area, with its important implications for the repair of spinal cord and brain injury, is ongoing.

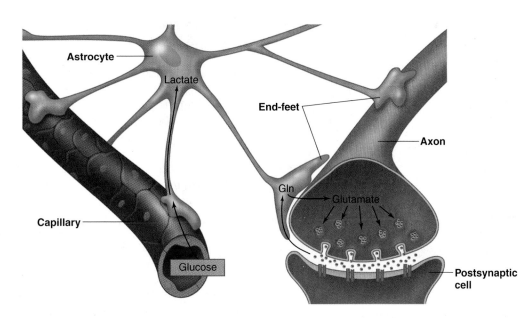

■ Figure 7.10 Astrocytes have processes that end on capillaries and neurons. Astrocyte end-feet take up glucose from blood capillaries and use this to help supply energy substrates for neurons. Astrocytes also take up the neurotransmitter glutamate from synapses and convert it to glutamine (Gln), which is then recycled to the neurons.

Functions of Astrocytes

Astrocytes (*aster* = star) are large stellate cells with numerous cytoplasmic processes that radiate outward. They are the most abundant of the glial cells in the CNS, constituting up to 90% of the nervous tissue in some areas of the brain.

Astrocytes (fig. 7.10) have processes that terminate in *end-feet* surrounding the capillaries of the CNS; indeed, the entire surface of these capillaries is covered by the astrocyte end-feet. In addition, astrocytes have other extensions adjacent to the synapses (connections) between the axon terminal of one neuron and the dendrite or cell body of another neuron. The astrocytes are thus ideally situated to influence the interactions between neurons and between neurons and the blood.

Here are some of the proposed functions of astrocytes:

1. **Astrocytes take up K$^+$ from the extracellular fluid.** Since K$^+$ diffuses out of neurons during the production of nerve impulses (described later), this function may be important in maintaining the proper ionic environment for neurons.
2. **Astrocytes take up some neurotransmitters released from the axon terminals of neurons.** For example, the neurotransmitter glutamate is taken into astrocytes and transformed into glutamine (fig. 7.10). The glutamine is then released back to the neurons, which can use it to reform the neurotransmitter glutamate.
3. **The astrocyte end-feet surrounding blood capillaries take up glucose from the blood.** The glucose is metabolized into lactic acid, or lactate (fig. 7.10). The lactate is then released and use as an energy source by neurons, which metabolize it aerobically into CO$_2$ and H$_2$O for the production of ATP.
4. **Astrocytes appear to be needed for the formation of synapses in the CNS.** Few synapses form in the absence of astrocytes, and those that do are defective. Normal synapses in the CNS are ensheathed by astrocytes (fig. 7.10).
5. **Astrocytes induce the formation of the blood-brain barrier.** The nature of the blood-brain barrier is described in the next section.

Blood-Brain Barrier

Capillaries in the brain, unlike those of most other organs, do not have pores between adjacent endothelial cells (the cells that compose the walls of capillaries). Instead, the endothelial cells of brain capillaries are joined together by tight junctions. Unlike other organs, therefore, the brain cannot obtain molecules from the blood plasma by a nonspecific filtering process. Instead, molecules within brain capillaries must be moved through the endothelial cells by diffusion and active transport, as well as by endocytosis and exocytosis. This feature of brain capillaries imposes a very selective **blood-brain barrier.** There is evidence to suggest that the development of tight junctions between adjacent endothelial cells in brain capillaries, and thus the development of the blood-brain barrier, results from the effects of astrocytes on the brain capillaries.

The blood-brain barrier presents difficulties in the chemotherapy of brain diseases because drugs that could enter other organs may not be able to enter the brain. In the treatment of **Parkinson's disease,** for example, patients who need a chemical called dopamine in the brain are often given a precursor molecule called levodopa (L-dopa) because L-dopa can cross the blood-brain barrier but dopamine cannot. Some antibiotics also cannot cross the blood-brain barrier; therefore, in treating infections such as meningitis, only those antibiotics that can cross the blood-brain barrier are used.

Test Yourself Before You Continue

1. Draw a neuron, label its parts, and describe the functions of these parts.
2. Distinguish between sensory neurons, motor neurons, and association neurons in terms of structure, location, and function.
3. Describe the structure of the sheath of Schwann, or neurilemma, and explain how it promotes nerve regeneration. Explain how a myelin sheath is formed in the PNS.
4. Explain how myelin sheaths are formed in the CNS. How does the presence or absence of myelin sheaths in the CNS determine the color of this tissue?
5. Explain what is meant by the blood-brain barrier. Describe its structure and discuss its clinical significance.

Electrical Activity in Axons

The permeability of the axon membrane to Na$^+$ and K$^+$ is regulated by gates, which open in response to stimulation. Net diffusion of these ions occurs in two stages: first Na$^+$ moves into the axon, then K$^+$ moves out. This flow of ions, and the changes in the membrane potential that result, constitute an event called an action potential.

All cells in the body maintain a potential difference (voltage) across the membrane, or **resting membrane potential,** in which the inside of the cell is negatively charged in comparison to the outside of the cell (for example, in neurons it is –70 mV). As explained in chapter 6, this potential difference is largely the result of the permeability properties of the plasma membrane. The membrane traps large, negatively charged organic molecules within the cell and permits only limited diffusion of positively charged inorganic ions. These properties result in an unequal distribution of these ions across the membrane. The action of the Na$^+$/K$^+$ pumps also helps to maintain a potential difference because they pump out three sodium ions (Na$^+$) for every two potassium ions (K$^+$) that they transport into the cell. Partly as a result of these pumps, Na$^+$ is more highly concentrated in the extracellular fluid than inside the cell, whereas K$^+$ is more highly concentrated within the cell.

Although all cells have a membrane potential, only a few types of cells have been shown to alter their membrane potential in response to stimulation. Such alterations in membrane potential are achieved by varying the membrane permeability to specific ions in response to stimulation. A central aspect of the physiology of neurons and muscle cells is their ability to produce and conduct these changes in membrane potential. Such an ability is termed *excitability* or *irritability*.

An increase in membrane permeability to a specific ion results in the diffusion of that ion down its concentration gradient, either into or out of the cell. These *ion currents* occur only across limited patches of membrane (located fractions of a millimeter apart), where specific ion channels are located. Changes

in the potential difference across the membrane at these points can be measured by the voltage developed between two electrodes—one placed inside the cell and the other placed outside the plasma membrane at the region being recorded. The voltage between these two recording electrodes can be visualized by connecting them to an oscilloscope (fig. 7.11).

In an oscilloscope, electrons from a cathode-ray "gun" are sprayed across a fluorescent screen, producing a line of light. This line deflects upward or downward in response to a potential difference between the two electrodes. The oscilloscope can be calibrated in such a way that an upward deflection of the line indicates that the inside of the membrane has become less negative (or more positive) compared to the outside of the membrane. A downward deflection of the line, conversely, indicates that the inside of the cell has become more negative. The oscilloscope can thus function as a voltmeter with an ability to display voltage changes as a function of time.

If both recording electrodes are placed outside of the cell, the potential difference between the two will be zero (because there is no charge separation). When one of the two electrodes penetrates the cell membrane, the oscilloscope will indicate that the intracellular electrode is electrically negative with respect to the extracellular electrode; a membrane potential is recorded. We will call this the *resting membrane potential (rmp)* to distinguish it from events described in later sections. All cells have a resting membrane potential, but its magnitude can be different in different

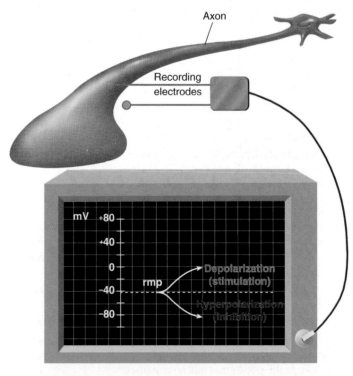

■ **Figure 7.11**   Observing depolarization and hyperpolarization. The difference in potential (in millivolts [mV]) between an intracellular and extracellular recording electrode is displayed on an oscilloscope screen. The resting membrane potential (rmp) of the axon may be reduced (depolarization) or increased (hyperpolarization). Depolarization is seen as a line deflecting upward from the rmp, and hyperpolarization by a line deflecting downward from the rmp.

types of cells. Neurons maintain an average rmp of –70 mV, for example, whereas heart muscle cells may have an rmp of –85 mV.

If appropriate stimulation causes positive charges to flow into the cell, the line will deflect upward. This change is called **depolarization,** since the potential difference between the two recording electrodes is reduced. A return to the resting membrane potential is known as **repolarization.** If stimulation causes the inside of the cell to become more negative than the resting membrane potential, the line on the oscilloscope will deflect downward. This change is called **hyperpolarization** (fig. 7.11). Hyperpolarization can be caused either by positive charges leaving the cell or by negative charges entering the cell.

Depolarization of a dendrite or cell body is *excitatory,* whereas hyperpolarization is *inhibitory,* in terms of their effects on the production of nerve impulses. The reasons for this relate to the nature of nerve impulses (action potentials), as will be explained shortly.

Ion Gating in Axons

The changes in membrane potential just described—depolarization, repolarization, and hyperpolarization—are caused by changes in the net flow of ions through ion channels in the membrane. Ions such as Na^+, K^+, and others pass through ion channels in the plasma membrane that are said to be *gated channels.* The "gates" are part of the proteins that comprise the channels, and can open or close the ion channels in response to particular changes. When ion channels are closed, the plasma membrane is less permeable, and when the channels are open, the membrane is more permeable to an ion (fig. 7.12).

The ion channels for Na^+ and K^+ are fairly specific for each of these ions. It is believed that there are two types of channels for K^+; one type is always open, whereas the other type is closed in the resting cell. Channels for Na^+, by contrast, are always closed in the resting cell. The resting cell is thus more permeable to K^+ than to Na^+. (As described in chapter 6, some Na^+ does leak into the cell; this leakage may occur in a nonspecific manner through open K^+ channels.) Because of the slight inward leakage of Na^+, the resting membrane potential is a little less negative than the equilibrium potential for K^+.

Depolarization of a small region of an axon can be experimentally induced by a pair of stimulating electrodes that act as if they were injecting positive charges into the axon. If two recording electrodes are placed in the same region (one electrode within the axon and one outside), an upward deflection of the oscilloscope line will be observed as a result of this depolarization. If a certain level of depolarization is achieved (from –70 mV to –55 mV, for example) by this artificial stimulation, a sudden and very rapid change in the membrane potential will be observed. This is because *depolarization to a threshold level causes the Na^+ channels to open.* Now the permeability properties of the membrane are changed, and Na^+ diffuses down its concentration gradient into the cell.

A fraction of a second after the Na^+ channels open, they close again. Just before they do, *the depolarization stimulus causes the K^+ gates to open.* This makes the membrane more permeable to K^+ than it is at rest, and K^+ diffuses down its con-

centration gradient out of the cell. The K^+ gates will then close and the permeability properties of the membrane will return to what they were at rest.

Since opening of the gated Na^+ and K^+ channels is stimulated by depolarization, these ion channels in the axon membrane are said to be **voltage regulated.** The channel gates are closed at the resting membrane potential of –70 mV and open in response to depolarization of the membrane to a threshold value.

Action Potentials

We will now consider the events that occur at one point in an axon, when a small region of axon membrane is stimulated artificially and responds with changes in ion permeabilities. The resulting changes in membrane potential at this point are detected by recording electrodes placed in this region of the axon. The

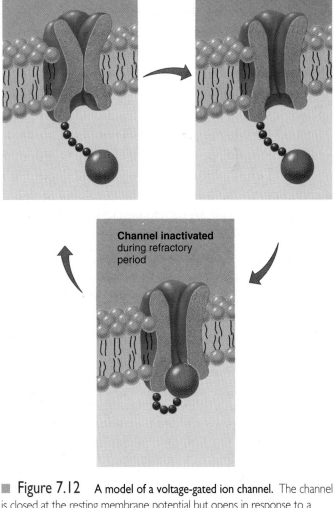

Channel closed at resting membrane potential

Channel open by depolarization (action potential)

Channel inactivated during refractory period

■ **Figure 7.12** **A model of a voltage-gated ion channel.** The channel is closed at the resting membrane potential but opens in response to a threshold level of depolarization. This permits the diffusion of ions required for action potentials. After a brief period of time, the channel is inactivated by the "ball and chain" portion of a polypeptide chain (discussed in the section on refractory periods in the text).

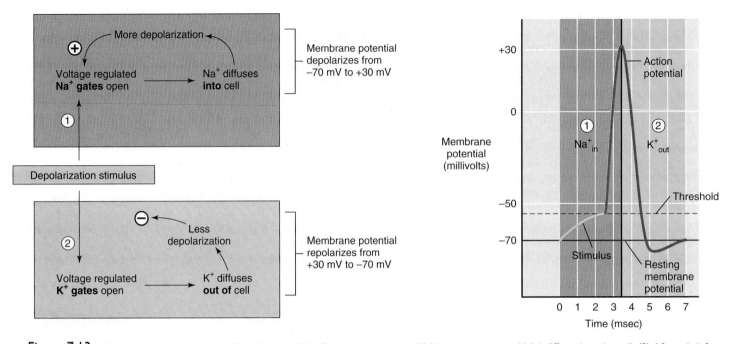

Figure 7.13   Depolarization of an axon affects Na$^+$ and K$^+$ diffusion in sequence. *(1)* Na$^+$ gates open and Na$^+$ diffuses into the cell. *(2)* After a brief period, K$^+$ gates open and K$^+$ diffuses out of the cell. An inward diffusion of Na$^+$ causes further depolarization, which in turn causes further opening of Na$^+$ gates in a positive feedback (+) fashion. The opening of K$^+$ gates and outward diffusion of K$^+$ makes the inside of the cell more negative, and thus has a negative feedback effect (−) on the initial depolarization.

nature of the stimulus in vivo (in the body), and the manner by which electrical events are conducted to different points along the axon, will be described in later sections.

When the axon membrane has been depolarized to a threshold level—in the previous example, by stimulating electrodes—the Na$^+$ gates open and the membrane becomes permeable to Na$^+$. This permits Na$^+$ to enter the axon by diffusion, which further depolarizes the membrane (makes the inside less negative, or more positive). Since the gates for the Na$^+$ channels of the axon membrane are voltage regulated, this additional depolarization opens more Na$^+$ channels and makes the membrane even more permeable to Na$^+$. As a result, more Na$^+$ can enter the cell and induce a depolarization that opens even more voltage-regulated Na$^+$ gates. A *positive feedback loop* (fig. 7.13) is thus created, causing the rate of Na$^+$ entry and depolarization to accelerate in an explosive fashion.

The explosive increase in Na$^+$ permeability results in a rapid reversal of the membrane potential in that region from −70 mV to +30 mV (fig. 7.13). At that point in time, the channels for Na$^+$ close (they actually become inactivated, as illustrated in figure 7.12), causing a rapid decrease in Na$^+$ permeability. Also at this time, as a result of a time-delayed effect of the depolarization, voltage-gated K$^+$ channels open and K$^+$ diffuses rapidly out of the cell.

Since K$^+$ is positively charged, the diffusion of K$^+$ out of the cell makes the inside of the cell less positive, or more negative, and acts to restore the original resting membrane potential of −70 mV. This process is called **repolarization** and represents the completion of a *negative feedback loop* (fig. 7.13). These changes in Na$^+$ and K$^+$ diffusion and the resulting changes in membrane potential they produce constitute an event called the **action potential,** or **nerve impulse.**

Local anesthetics block the conduction of action potentials in sensory axons. They do this by reversibly binding to specific sites within the voltage-gated Na$^+$ channels, reducing the ability of membrane depolarization to produce action potentials. *Cocaine* was the first local anesthetic to be used, but because of its toxicity and potential for abuse, alternatives have been developed. The first synthetic analog of cocaine used for local anesthesia, *procaine,* was produced in 1905. Other local anesthetics of this type include *lidocaine* and *tetracaine.*

The correlation between ion movements and changes in membrane potential is shown in figure 7.14. The bottom portion of this figure illustrates the movement of Na$^+$ and K$^+$ through the axon membrane in response to a depolarization stimulus. Notice that the explosive increase in Na$^+$ diffusion causes rapid depolarization to 0 mV and then *overshoot* of the membrane potential so that the inside of the membrane actually becomes positively charged (almost +30 mV) compared to the outside (top portion of fig. 7.14). The greatly increased permeability to Na$^+$ thus drives the membrane potential toward the equilibrium potential for Na$^+$ (chapter 6). The Na$^+$ permeability then rapidly decreases and the diffusion of K$^+$ increases, resulting in repolarization to the resting membrane potential.

Once an action potential has been completed, the Na$^+$/K$^+$ pumps will extrude the extra Na$^+$ that has entered the axon and recover the K$^+$ that has diffused out of the axon. This active transport of ions occurs very quickly because the events described occur across only a very small area of membrane. Only

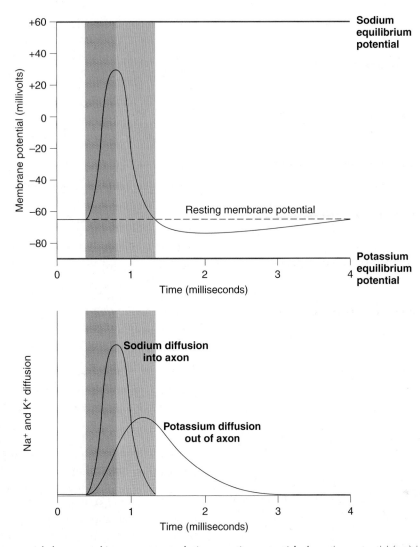

Figure 7.14 **Membrane potential changes and ion movements during an action potential.** An action potential (*top*) is produced by an increase in sodium diffusion that is followed, after a short delay, by an increase in potassium diffusion (*bottom*). This drives the membrane potential first toward the sodium equilibrium potential and then toward the potassium equilibrium potential.

a relatively small amount of Na+ and K+ actually diffuse through the membrane during the production of an action potential, and so the total concentrations of Na+ and K+ in the axon and in the extracellular fluid are not significantly changed.

Notice that active transport processes are not directly involved in the production of an action potential; both depolarization and repolarization are produced by the diffusion of ions down their concentration gradients. A neuron poisoned with cyanide, so that it cannot produce ATP, can still produce action potentials for a period of time. After awhile, however, the lack of ATP for active transport by the Na+/K+ pumps will result in a decline in the concentration gradients, and therefore in the ability of the axon to produce action potentials. This shows that the Na+/K+ pumps are not directly involved; rather, they are required to maintain the concentration gradients needed for the diffusion of Na+ and K+ during action potentials.

All-or-None Law

Once a region of axon membrane has been depolarized to a threshold value, the positive feedback effect of depolarization on

Na+ permeability and of Na+ permeability on depolarization causes the membrane potential to shoot toward about +30 mV. It does not normally become more positive than +30 mV because the Na+ channels quickly close and the K+ channels open. The length of time that the Na+ and K+ channels stay open is independent of the strength of the depolarization stimulus.

The amplitude (size) of action potentials is therefore **all or none.** When depolarization is below a threshold value, the voltage-regulated gates are closed; when depolarization reaches threshold, a maximum potential change (the action potential) is produced. Since the change from −70 mV to +30 mV and back to −70 mV lasts only about 3 msec, the image of an action potential on an oscilloscope screen looks like a spike. Action potentials are therefore sometimes called *spike potentials.*

The channels are only open for a fixed period of time because they are soon *inactivated,* a process different from simply closing the gates. Inactivation occurs automatically and lasts until the membrane potential has repolarized. Because of this automatic inactivation, all action potentials have about the same duration. Likewise, since the concentration gradient for Na+ is

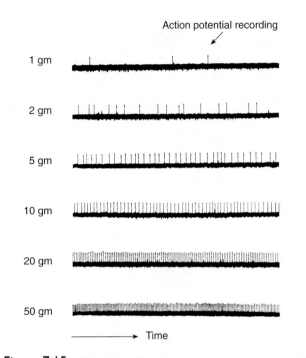

Action potential recording

1 gm

2 gm

5 gm

10 gm

20 gm

50 gm

→ Time

■ **Figure 7.15**   The effect of stimulus strength on action potential **frequency.** These are recordings from a single sensory fiber of the sciatic nerve of a frog stimulated by varying degrees of stretch of the gastrocnemius muscle. Notice that increasing degrees of stretch (indicated by increasing weights attached to the muscle) result in a higher frequency of action potentials.

relatively constant, the amplitudes of the action potentials are about equal in all axons at all times (from −70 mV to +30 mV, or about 100 mV in total amplitude).

Coding for Stimulus Intensity

Because action potentials are all-or-none events, a stronger stimulus cannot produce an action potential of greater amplitude. The code for stimulus strength in the nervous system is not amplitude modulated (AM). When a greater stimulus strength is applied to a neuron, identical action potentials are produced more frequently (more are produced per second). Therefore, the code for stimulus strength in the nervous system is frequency modulated (FM). This concept is illustrated in figure 7.15.

When an entire collection of axons (in a nerve) is stimulated, different axons will be stimulated at different stimulus intensities. A weak stimulus will activate only those few axons with low thresholds, whereas stronger stimuli can activate axons with higher thresholds. As the intensity of stimulation increases, more and more axons will become activated. This process, called **recruitment,** represents another mechanism by which the nervous system can code for stimulus strength.

Refractory Periods

If a stimulus of a given intensity is maintained at one point of an axon and depolarizes it to threshold, action potentials will be produced at that point at a given frequency (number per second). As the stimulus strength is increased, the frequency of action potentials produced at that point will increase accordingly. As action

potentials are produced with increasing frequency, the time between successive action potentials will decrease—but only up to a minimum time interval. The interval between successive action potentials will never become so short as to allow a new action potential to be produced before the preceding one has finished.

During the time that a patch of axon membrane is producing an action potential, it is incapable of responding—or *refractory*—to further stimulation. If a second stimulus is applied during most of the time that an action potential is being produced, the second stimulus will have no effect on the axon membrane. The membrane is thus said to be in an **absolute refractory period;** it cannot respond to any subsequent stimulus.

The cause of the absolute refractory period is now understood at a molecular level. In addition to the voltage-regulated gates that open and close the channel, an ion channel may have a polypeptide that functions as a "ball and chain" apparatus dangling from its cytoplasmic side (see fig. 7.12). After a voltage-regulated channel is opened by depolarization for a set time, it enters an *inactive state*. The inactivated channel cannot be opened by depolarization. The reason for its inactivation depends on the type of voltage-gated channel. In the type of channel shown in fig. 7.12, the channel becomes blocked by a molecular ball attached to a chain. In a different type of voltage-gated channel, the channel shape becomes altered through molecular rearrangements. The inactivation ends after a fixed period of time in both cases, either because the ball leaves the mouth of the channel, or because molecular rearrangements restore the resting form of the channel. In the resting state, unlike the inactivated state, the channel is closed but it can be opened in response to a depolarization stimulus of sufficient strength.

If a second stimulus is applied while the K^+ gates are open (and the membrane is in the process of repolarizing), the membrane is said to be in a **relative refractory period.** During this time, only a very strong depolarization can overcome the repolarization effects of the open K^+ channels and produce a second action potential (fig. 7.16).

Because the cell membrane is refractory during the time it is producing an action potential, each action potential remains a separate, all-or-none event. In this way, as a continuously applied stimulus increases in intensity, its strength can be coded strictly by the frequency of the action potentials it produces at each point of the axon membrane.

After a large number of action potentials have been produced, one might think that the relative concentrations of Na^+ and K^+ would be changed in the extracellular and intracellular compartments. This is not the case. In a typical mammalian axon that is 1 mm in diameter, for example, only one intracellular K^+ in 3,000 would be exchanged for a Na^+ to produce an action potential. Since a typical neuron has about 1 million Na^+/K^+ pumps that can transport nearly 200 million ions per second, these small changes can be quickly corrected.

Cable Properties of Neurons

If a pair of stimulating electrodes produces a depolarization that is too weak to cause the opening of voltage-regulated Na^+ gates—that is, if the depolarization is below threshold (about −55 mV)—the change in membrane potential will be *localized*

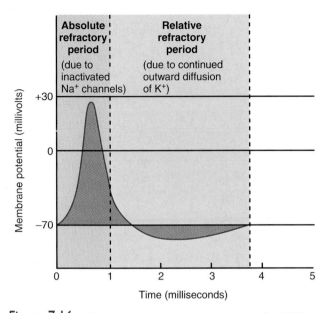

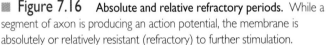

Figure 7.16 Absolute and relative refractory periods. While a segment of axon is producing an action potential, the membrane is absolutely or relatively resistant (refractory) to further stimulation.

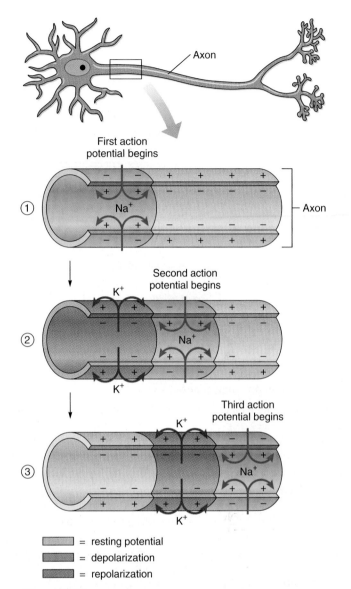

Figure 7.17 The conduction of action potentials in an unmyelinated axon. Each action potential "injects" positive charges that spread to adjacent regions. The region that has just produced an action potential is refractory. The next region, not having been stimulated previously, is partially depolarized. As a result, its voltage-regulated Na⁺ gates open and the process is repeated. Successive segments of the axon thereby regenerate, or "conduct," the action potential.

to within 1 to 2 mm of the point of stimulation. For example, if the stimulus causes depolarization from -70 mV to -60 mV at one point, and the recording electrodes are placed only 3 mm away from the stimulus, the membrane potential recorded will remain at -70 mV (the resting potential). The axon is thus a very poor conductor compared to a metal wire.

The term **cable properties** refers to the ability of a neuron to transmit charges through its cytoplasm. These cable properties are quite poor because there is a high internal resistance to the spread of charges and because many charges leak out of the axon through its membrane. If an axon had to conduct only through its cable properties, therefore, no axon could be more than a millimeter in length. The fact that some axons are a meter or more in length suggests that the conduction of nerve impulses does not rely on the cable properties of the axon.

Conduction of Nerve Impulses

When stimulating electrodes artificially depolarize one point of an axon membrane to a threshold level, voltage-regulated channels open and an action potential is produced at that small region of axon membrane containing those gates. For about the first millisecond of the action potential, when the membrane voltage changes from -70 mV to $+30$ mV, a current of Na⁺ enters the cell by diffusion because of the opening of the Na⁺ gates. Each action potential thus "injects" positive charges (sodium ions) into the axon (fig. 7.17).

These positively charged sodium ions are conducted, by the cable properties of the axon, to an adjacent region that still has a membrane potential of -70 mV. Within the limits of the cable properties of the axon (1 to 2 mm), this helps to depolarize the adjacent region of axon membrane. When this adjacent region of

membrane reaches a threshold level of depolarization, it too produces an action potential as its voltage-regulated gates open.

Each action potential thus acts as a stimulus for the production of another action potential at the next region of membrane that contains voltage-regulated gates. In the description of action potentials earlier in this chapter, the stimulus for their production was artificial—depolarization produced by a pair of stimulating electrodes. Now it can be seen that each action potential is produced by depolarization that results from the preceding action potential. This explains how all action potentials along an axon are produced after the first action potentials are generated at the initial segment of the axon.

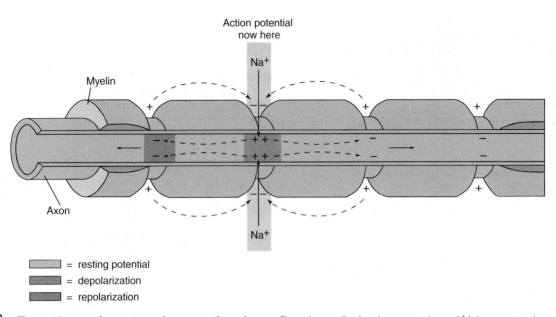

Figure 7.18 **The conduction of a nerve impulse in a myelinated axon.** Since the myelin sheath prevents inward Na⁺ current, action potentials can be produced only at gaps in the myelin sheath called the nodes of Ranvier. This "leaping" of the action potential from node to node is known as saltatory conduction.

Conduction in an Unmyelinated Axon

In an unmyelinated axon, every patch of membrane that contains Na⁺ and K⁺ gates can produce an action potential. Action potentials are thus produced along the entire length of the axon. The cablelike spread of depolarization induced by the influx of Na⁺ during one action potential helps to depolarize the adjacent regions of membrane—a process that is also aided by movements of ions on the outer surface of the axon membrane (fig. 7.17). This process would depolarize the adjacent membranes on each side of the region to produce an action potential, but the area that had previously produced one cannot produce another at this time because it is still in its refractory period.

It is important to recognize that action potentials are not really "conducted," although it is convenient to use that word. Each action potential is a separate, complete event that is repeated, or *regenerated*, along the axon's length. This is analogous to the "wave" performed by spectators in a stadium. One person after another gets up (depolarization) and then sits down (repolarization); it is thus the "wave" (spread of action potentials) that travels, not the people (individual action potentials).

The action potential produced at the end of the axon is thus a completely new event that was produced in response to depolarization from the previous action potential. The last action potential has the same amplitude as the first. Action potentials are thus said to be **conducted without decrement** (without decreasing in amplitude).

The spread of depolarization by the cable properties of an axon is fast compared to the time it takes to produce an action potential. Thus, the more action potentials along a given stretch of axon that have to be produced, the slower the conduction. Since action potentials must be produced at every fraction of a micrometer in an unmyelinated axon, the conduction rate is relatively slow. This conduction rate is somewhat faster if the unmyelinated axon is thicker, because thicker axons have less resistance to the flow of charges (so conduction of charges by cable properties is faster). The conduction rate is substantially faster if the axon is myelinated because fewer action potentials are produced along a given length of myelinated axon.

Conduction in a Myelinated Axon

The myelin sheath provides insulation for the axon, preventing movements of Na⁺ and K⁺ through the membrane. If the myelin sheath were continuous, therefore, action potentials could not be produced. The myelin thus has interruptions—the *nodes of Ranvier*, as previously described.

Because the cable properties of axons can conduct depolarizations only over a very short distance (1 to 2 mm), the nodes of Ranvier cannot be separated by more than this distance. Studies have shown that Na⁺ channels are highly concentrated at the nodes (estimated at 10,000 per square micrometer) and almost absent in the regions of axon membrane between the nodes. Action potentials, therefore, occur only at the nodes of Ranvier (fig. 7.18) and seem to "leap" from node to node—a process called **saltatory conduction** (*saltario* = leap). The leaping is, of course, just a metaphor; the action potential at one node depolarizes the membrane at the next node to threshold, so that a new action potential is produced at the next node of Ranvier.

Since the cablelike spread of depolarization between the nodes is very fast and fewer action potentials need to be produced per given length of axon, saltatory conduction allows a faster rate of conduction than is possible in an unmyelinated fiber. Conduction rates in the human nervous system vary from 1.0 m/sec—in thin, unmyelinated fibers that mediate slow, visceral responses—to faster than 100 m/sec (225 miles per hour)—in thick, myelinated fibers involved in quick stretch reflexes in skeletal muscles (table 7.4).

In summary, the speed of action potential conduction is increased by (1) increased diameter of the axon, because this

Table 7.4 Conduction Velocities and Functions of Mammalian Nerves of Different Diameters

Diameter (μm)	Conduction Velocity (m/sec)	Examples of Functions Served
12–22	70–120	Sensory: muscle position
5–13	30–90	Somatic motor fibers
3–8	15–40	Sensory: touch, pressure
1–5	12–30	Sensory: pain, temperature
1–3	3–15	Autonomic fibers to ganglia
0.3–1.3	0.7–2.2	Autonomic fibers to smooth and cardiac muscles

reduces the resistance to the spread of charges by cable properties; and (2) myelination, because the myelin sheath results in saltatory conduction of action potentials. These methods of affecting conduction speed are generally combined in the nervous system: the thinnest axons tend to be unmyelinated and the thickest tend to be myelinated.

Test Yourself Before You Continue

1. Define the terms *depolarization* and *repolarization,* and illustrate these processes graphically.

2. Describe how the permeability of the axon membrane to Na^+ and K^+ is regulated and how changes in permeability to these ions affect the membrane potential.

3. Describe how gating of Na^+ and K^+ in the axon membrane results in the production of an action potential.

4. Explain the all-or-none law of action potentials and describe the effect of increased stimulus strength on action potential production. How do the refractory periods affect the frequency of action potential production?

5. Describe how action potentials are conducted by unmyelinated nerve fibers. Why is saltatory conduction in myelinated fibers more rapid?

The Synapse

Axons end close to, or in some cases at the point of contact with, another cell. Once action potentials reach the end of an axon, they directly or indirectly stimulate (or inhibit) the other cell. In specialized cases, action potentials can directly pass from one cell to another. In most cases, however, the action potentials stop at the axon ending, where they stimulate the release of a chemical neurotransmitter that affects the next cell.

A **synapse** is the functional connection between a neuron and a second cell. In the CNS, this other cell is also a neuron. In the PNS, the other cell may be either a neuron or an *effector cell* within a muscle or gland. Although the physiology of neuron-neuron synapses and neuron-muscle synapses is similar, the latter synapses are often called **myoneural,** or **neuromuscular, junctions.**

Neuron-neuron synapses usually involve a connection between the axon of one neuron and the dendrites, cell body, or axon of a second neuron. These are called, respectively, *axodendritic, axosomatic,* and *axoaxonic synapses.* In almost all synapses, transmission is in one direction only—from the axon of the first (or **presynaptic**) neuron to the second (or **postsynaptic**) neuron. Most commonly, the synapse occurs between the axon of the presynaptic neuron and the dendrites or cell body of the postsynaptic neuron.

In the early part of the twentieth century, most physiologists believed that synaptic transmission was *electrical*—that is, that action potentials were conducted directly from one cell to the next. This was a logical assumption given that nerve endings appeared to touch the postsynaptic cells and that the delay in synaptic conduction was extremely short (about 0.5 msec). Improved histological techniques, however, revealed tiny gaps in the synapses, and experiments demonstrated that the actions of autonomic nerves could be duplicated by certain chemicals. This led to the hypothesis that synaptic transmission might be *chemical*—that the presynaptic nerve endings might release chemicals called **neurotransmitters** that stimulated action potentials in the postsynaptic cells.

In 1921, a physiologist named Otto Loewi published the results of experiments suggesting that synaptic transmission was indeed chemical, at least at the junction between a branch of the vagus nerve (see chapter 9) and the heart. He had isolated the heart of a frog and, while stimulating the branch of the vagus that innervates the heart, perfused the heart with an isotonic salt solution. Stimulation of this nerve slowed the heart rate, as expected. More importantly, application of this salt solution to the heart of a second frog caused the second heart also to slow its rate of beat.

Loewi concluded that the nerve endings of the vagus must have released a chemical—which he called *Vagusstoff*—that inhibited the heart rate. This chemical was subsequently identified as **acetylcholine,** or **ACh.** In the decades following Loewi's discovery, many other examples of chemical synapses were discovered, and the theory of electrical synaptic transmission fell into disrepute. More recent evidence, ironically, has shown that electrical synapses do exist in the nervous system (though they are the exception), within smooth muscles, and between cardiac cells in the heart.

Electrical Synapses: Gap Junctions

In order for two cells to be electrically coupled, they must be approximately equal in size and they must be joined by areas of contact with low electrical resistance. In this way, impulses can be regenerated from one cell to the next without interruption. Adjacent cells that are electrically coupled are joined together by **gap junctions.** In gap junctions, the membranes of the two

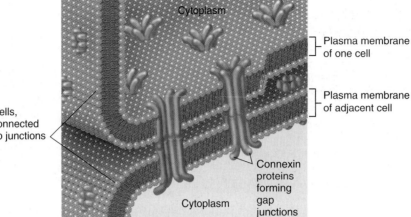

■ **Figure 7.19** **The structure of gap junctions.** Gap junctions are water-filled channels through which ions can pass from one cell to another. This permits impulses to be conducted directly from one cell to another. Each gap junction is composed of connexin proteins. Six connexin proteins in one plasma membrane line up with six connexin proteins in the other plasma membrane to form each gap junction.

cells are separated by only 2 nanometers (1 nanometer = 10^{-9} meter). A surface view of gap junctions in the electron microscope reveals hexagonal arrays of particles that function as channels through which ions and molecules may pass from one cell to the next (fig. 7.19). Each gap junction is now known to be composed of twelve proteins known as *connexins,* which are arranged like staves of a barrel to form a water-filled pore.

Gap junctions are present in cardiac muscle and some smooth muscles, where they allow excitation and rhythmic contraction of large masses of muscle cells. Gap junctions have also been observed in various regions of the brain. Although their functional significance in the brain is unknown, it has been speculated that they may allow a two-way transmission of impulses (in contrast to chemical synapses, which are always one-way). Gap junctions also have been observed between glial cells; these may act as channels for the passage of informational molecules between cells. It is interesting in this regard that gap junctions are present in many embryonic tissues, and that these gap junctions disappear as the tissue becomes more specialized.

Chemical Synapses

Transmission across the majority of synapses in the nervous system is one-way and occurs through the release of chemical neurotransmitters from presynaptic axon endings. These presynaptic endings, called **terminal boutons** (*bouton* = button) because of their swollen appearance, are separated from the postsynaptic cell by a **synaptic cleft** so narrow (about 10 nm) that it can be seen clearly only with an electron microscope (fig. 7.20).

Neurotransmitter molecules within the presynaptic neuron endings are contained within many small, membrane-enclosed **synaptic vesicles.** In order for the neurotransmitter within these vesicles to be released into the synaptic cleft, the vesicle membrane must fuse with the axon membrane in the process of *exocytosis* (chapter 3). The neurotransmitter is released in multiples of the amount contained in one vesicle, and the number of vesicles that undergo exocytosis depends on the frequency of action potentials produced at the presynaptic axon ending. Therefore,

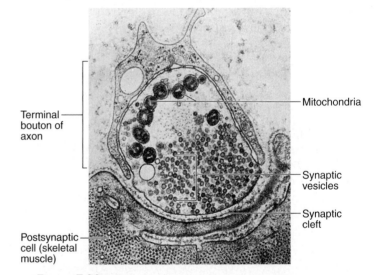

■ **Figure 7.20** **An electron micrograph of a chemical synapse.** This synapse between the axon of a somatic motor neuron and a skeletal muscle cell shows the synaptic vesicles at the end of the axon and the synaptic cleft. The synaptic vesicles contain the neurotransmitter chemical.

when stimulation of the presynaptic axon is increased, more of its vesicles will release their neurotransmitters to more greatly affect the postsynaptic cell.

Action potentials that arrive at the end of the axon trigger the release of neurotransmitter quite rapidly. The release is rapid because many synaptic vesicles are already "docked" at the correct areas of the presynaptic membrane before the arrival of the action potentials. At these docking sites, the vesicles are attached by proteins to form a *fusion complex* associated with the presynaptic membrane. The fusion complex attaches the vesicle to the docking site, but actual fusion of the vesicle membrane and the axon membrane is prevented until the arrival of action potentials.

Voltage-regulated calcium (Ca^{2+}) channels are located in the axon terminal adjacent to the docking sites. The arrival of action potentials at the axon terminal opens these voltage-regulated calcium channels, and it is the inward diffusion of Ca^{2+} that triggers

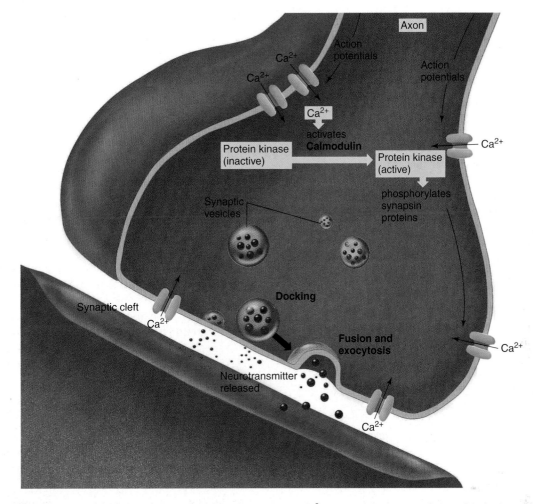

■ **Figure 7.21** **The release of neurotransmitter.** Action potentials, by opening Ca^{2+} channels, stimulate the fusion of docked synaptic vesicles with the cell membrane of the axon terminals. This leads to exocytosis and the release of neurotransmitter. The activation of protein kinase by Ca^{2+} may also contribute to this process.

the rapid fusion of the synaptic vesicle with the axon membrane and the release of neurotransmitter through exocytosis (fig. 7.21).

In addition, Ca^{2+} diffusing into the axon terminal activates a regulatory protein within the cytoplasm known as **calmodulin,** which in turn activates an enzyme called **protein kinase.** This enzyme phosphorylates (adds a phosphate group to) specific proteins known as *synapsins* in the membrane of the synaptic vesicle. This action may aid the fusion of synaptic vesicles with the plasma membrane. The Ca^{2+}-calmodulin-protein kinase regulatory mechanism is also important in the action of some hormones, and is therefore discussed in more detail in chapter 11.

Tetanus toxin and **botulinum toxin** are bacterial products that cause paralysis by preventing neurotransmission. These neurotoxins function as *proteases* (protein-digesting enzymes), digesting particular components of the fusion complex and thereby inhibiting the exocytosis of synaptic vesicles and preventing the release of neurotransmitter. Botulinum toxin prevents the release of ACh, causing flaccid paralysis; tetanus toxin blocks inhibitory synapses (discussed later), causing spastic paralysis.

Once the neurotransmitter molecules have been released from the presynaptic axon terminals, they diffuse rapidly across the synaptic cleft and reach the membrane of the postsynaptic cell. The neurotransmitters then bind to specific **receptor proteins** that are part of the postsynaptic membrane. Receptor proteins have high specificity for their neurotransmitter, which is the **ligand** of the receptor protein. The term *ligand* in this case refers to a smaller molecule (the neurotransmitter) that binds to and forms a complex with a larger protein molecule (the receptor). Binding of the neurotransmitter ligand to its receptor protein causes ion channels to open in the postsynaptic membrane. The gates that regulate these channels, therefore, can be called **chemically regulated** (or **ligand-regulated**) **gates** because they open in response to the binding of a chemical ligand to its receptor in the postsynaptic plasma membrane.

Note that two broad categories of gated ion channels have been described: *voltage-regulated* and *chemically regulated.* Voltage-regulated channels are found primarily in the axons; chemically regulated channels are found in the postsynaptic membrane. Voltage-regulated channels open in response to depolarization; chemically regulated channels open in response to the binding of postsynaptic receptor proteins to their neurotransmitter ligands.

The chemically regulated channels are opened by a number of different mechanisms, and the effects of opening these channels vary. Opening of ion channels often produces a depolarization—the inside of the postsynaptic membrane becomes less negative. This depolarization is called an **excitatory postsynaptic potential (EPSP)** because the membrane potential moves toward threshold. In other cases, a hyperpolarization occurs—the inside of the postsynaptic membrane becomes more negative. This hyperpolarization is called an **inhibitory postsynaptic potential (IPSP)** because the membrane potential moves farther from threshold. The mechanisms by which EPSPs and IPSPs are produced will be described in the sections that deal with different types of neurotransmitters.

Excitatory postsynaptic potentials, as their name implies, stimulate the postsynaptic cell to produce action potentials, and inhibitory postsynaptic potentials antagonize this effect. In synapses between the axon of one neuron and the dendrites of another, the EPSPs and IPSPs are produced at the dendrites and must propagate to the initial segment of the axon to influence action potential production (fig. 7.22). The total depolarization produced by the summation of EPSPs at the initial segment of the axon will determine whether the axon will fire action potentials, and the frequency with which it fires action potentials. Once the first action potentials are produced, they will regenerate themselves along the axon as previously described.

In summary, the following sequence of events occurs:

1. **An excitatory neurotransmitter produces a depolarization.** This occurs when the neurotransmitter binds to its receptor and causes the opening of chemically regulated ion channels in the postsynaptic membrane. (An inhibitory neurotransmitter has the opposite effect—it causes a hyperpolarization.)
2. **The depolarization causes the opening of voltage-regulated ion channels.** This occurs if the depolarization reaches threshold.
3. **Opening of voltage-regulated channels produces action potentials.** This occurs in the first region of the postsynaptic membrane that contains voltage-regulated channels. In neurons, this is the initial segment of the axon.
4. **The action potential is regenerated along the axon or muscle cell.** An action potential in one region serves as the depolarization stimulus for the next region.

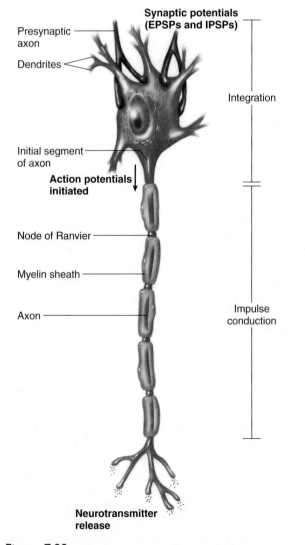

Synaptic potentials (EPSPs and IPSPs)

Presynaptic axon

Dendrites

Integration

Initial segment of axon

Action potentials initiated

Node of Ranvier

Myelin sheath

Axon

Impulse conduction

Neurotransmitter release

■ **Figure 7.22** The functional specialization of different regions in a multipolar neuron. Integration of input (EPSPs and IPSPs) generally occurs in the dendrites and cell body, with the axon serving to conduct action potentials.

Test Yourself Before You Continue

1. Describe the structure, locations, and functions of gap junctions.
2. Describe the location of neurotransmitters within an axon and explain the relationship between presynaptic axon activity and the amount of neurotransmitters released.
3. Describe the sequence of events by which action potentials stimulate the release of neurotransmitters from presynaptic axons.
4. Distinguish between voltage-regulated and chemically regulated ion channels.

Acetylcholine as a Neurotransmitter

When acetylcholine (ACh) binds to its receptor, it directly or indirectly causes the opening of chemically regulated gates. In many cases, this produces a depolarization called an excitatory postsynaptic potential, or EPSP. In some cases, however, ACh causes a hyperpolarization known as an inhibitory postsynaptic potential, or IPSP.

Acetylcholine (ACh) is used as an excitatory neurotransmitter by some neurons in the CNS and by somatic motor neurons at the neuromuscular junction. At autonomic nerve endings, ACh may be either excitatory or inhibitory, depending on the organ involved.

The varying responses of postsynaptic cells to the same chemical can be explained, in part, by the fact that different postsynaptic cells have different subtypes of ACh receptors. These receptor subtypes can be specifically stimulated by particular toxins, and they are named for these toxins. The stimulatory effect of ACh on skeletal muscle cells is produced by the binding of ACh to **nicotinic ACh receptors,** so named because they can also be activated by nicotine. Effects of ACh on other cells occur when ACh binds to **muscarinic ACh receptors;** these effects can also be produced by muscarine (a drug derived from certain poisonous mushrooms).

An overview of the distribution of the two types of ACh receptors demonstrates that this terminology and its associated concepts will be important in understanding the physiology of different body systems.

1. *Nicotinic ACh receptors.* These are found in specific regions of the brain (chapter 8), in autonomic ganglia (chapter 9), and in skeletal muscle fibers (chapter 12). The release of ACh from somatic motor neurons and its subsequent binding to nicotinic receptors, for example, stimulates muscle contraction.
2. *Muscarinic ACh receptors.* These are found in the plasma membrane of smooth muscle cells, cardiac muscle cells, and the cells of particular glands (chapter 9). Thus, the activation of muscarinic ACh receptors by ACh released from autonomic axons is required for the regulation of the cardiovascular system (chapter 14), digestive system (chapter 18), and others.

Chemically Regulated Channels

The binding of a neurotransmitter to its receptor protein can cause the opening of ion channels through two different mechanisms. These two mechanisms can be illustrated by the actions of ACh on the nicotinic and muscarinic subtypes of the ACh receptors.

Ligand-Operated Channels

This is the most direct mechanism by which chemically regulated gates can be opened. In this case, the ion channel runs through the receptor itself. The ion channel is opened by the binding of the receptor to the neurotransmitter ligand.

Such is the case when ACh binds to its nicotinic ACh receptor. This receptor consists of five polypeptide subunits that enclose the ion channel. Two of these subunits contain ACh-binding sites, and the channel opens when both sites bind to ACh (fig. 7.23). The opening of this channel permits the simultaneous diffusion of Na^+ into and K^+ out of the postsynaptic cell. The effects of the inward flow of Na^+ predominate, however, because of its steeper electrochemical gradient. This produces the depolarization of an excitatory postsynaptic potential (EPSP).

Although the inward diffusion of Na^+ predominates in an EPSP, the simultaneous outward diffusion of K^+ prevents the

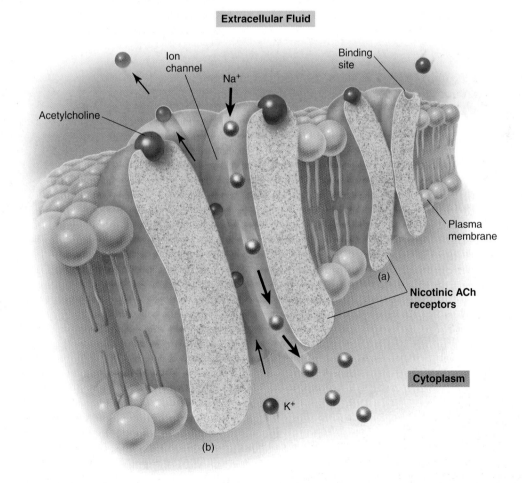

Extracellular Fluid

Ion channel

Na^+

Binding site

Acetylcholine

Plasma membrane

(a)

Nicotinic ACh receptors

Cytoplasm

K^+

(b)

Figure 7.23 Nicotinic acetylcholine (ACh) receptors also function as ion channels. The nicotinic acetylcholine receptor contains a channel that is closed (*a*) until the receptor binds to ACh. (*b*) Na^+ and K^+ diffuse simultaneously, and in opposite directions, through the open ion channel. The electrochemical gradient for Na^+ is greater than for K^+, so that the effect of the inward diffusion of Na^+ predominates, resulting in a depolarization known as an excitatory postsynaptic potential (EPSP).

Table 7.5   Comparison of Action Potentials and Excitatory Postsynaptic Potentials (EPSPs)

Characteristic	Action Potential	Excitatory Postsynaptic Potential
Stimulus for opening of ionic gates	Depolarization	Acetylcholine (ACh)
Initial effect of stimulus	Na$^+$ channels open	Common channels for Na$^+$ and K$^+$ open
Cause of repolarization	Opening of K$^+$ gates	Loss of intracellular positive charges with time and distance
Conduction distance	Regenerated over length of the axon	1–2 mm; a localized potential
Positive feedback between depolarization and opening of Na$^+$ gates	Yes	No
Maximum depolarization	+40 mV	Close to zero
Summation	No summation—all-or-none event	Summation of EPSPs, producing graded depolarizations
Refractory period	Yes	No
Effect of drugs	Inhibited by tetrodotoxin, not by curare	Inhibited by curare, not by tetrodotoxin

Table 7.6   Drugs That Affect the Neural Control of Skeletal Muscles

Drug	Origin	Effects
Botulinum toxin	Produced by *Clostridium botulinum* (bacteria)	Inhibits release of acetylcholine (Ach)
Curare	Resin from a South American tree	Prevents interaction of ACh with the postsynaptic receptor protein
α-Bungarotoxin	Venom of *Bungarus* snakes	Binds to ACh receptor proteins and prevents ACh from binding
Saxitoxin	Red tide (*Gonyaulax*) algae	Blocks voltage-gated Na$^+$ channels
Tetrodotoxin	Pufferfish	Blocks voltage-gated Na$^+$ channels
Nerve gas	Artificial	Inhibits acetylcholinesterase in postsynaptic membrane
Neostigmine	Nigerian bean	Inhibits acetylcholinesterase in postsynaptic membrane
Strychnine	Seeds of an Asian tree	Prevents IPSPs in spinal cord that inhibit contraction of antagonistic muscles

depolarization from overshooting 0 mV. Therefore, the membrane polarity does not reverse in an EPSP as it does in an action potential. (Remember that action potentials are produced by separate voltage-gated channels for Na$^+$ and K$^+$, where the channel for K$^+$ opens only after the Na$^+$ channel has closed.)

A comparison of EPSPs and action potentials is provided in table 7.5. Action potentials occur in axons, where the voltage-gated channels are located, whereas EPSPs occur in the dendrites and cell body. Unlike action potentials, EPSPs have *no threshold;* the ACh released from a single synaptic vesicle produces a tiny depolarization of the postsynaptic membrane. When more vesicles are stimulated to release their ACh, the depolarization is correspondingly greater. EPSPs are therefore *graded* in magnitude, unlike all-or-none action potentials. Since EPSPs can be graded, and have *no refractory period,* they are capable of *summation.* That is, the depolarizations of several different EPSPs can be added together. Action potentials are prevented from summating by their all-or-none nature and by the refractory periods they exhibit.

Muscle weakness in the disease **myasthenia gravis** is due to the fact that ACh receptors are blocked and destroyed by antibodies secreted by the immune system of the affected person. Paralysis in people who eat shellfish poisoned with saxitoxin, or pufferfish containing tetrodotoxin, results from the blockage of Na$^+$ channels. The effects of these and other poisons on neuromuscular transmission are summarized in table 7.6.

Clinical Investigation Clue

Remember that Sandra had flaccid paralysis and difficulty breathing after eating mussels and clams gathered from the local shore.

Mussels and clams are filter feeders that can concentrate the poison in the organisms responsible for the red tide. How might eating these mussels and clams cause her flaccid paralysis?

G-Protein-Operated Channels

The muscarinic ACh receptors are formed from only a single subunit, which can bind to one ACh molecule. Unlike the nicotinic receptors, these receptors do not contain ion channels. The ion channels are separate proteins located at some distance from the muscarinic receptors. Binding of ACh (the ligand) to the muscarinic receptor causes it to activate a complex of proteins in the cell membrane known as **G-proteins**—so named because their activity is influenced by guanosine nucleotides (GDP and GTP).

There are three G-protein subunits, designated alpha, beta, and gamma. In response to the binding of ACh to its receptor, the alpha subunit dissociates from the other two subunits, which stick together to form a beta-gamma complex. Depending on the specific case, either the alpha subunit or the beta-gamma complex then diffuses through the membrane until it binds to an ion channel, causing the channel to open (fig. 7.24). A short time later, the G-protein alpha subunit (or beta-gamma complex) dissociates

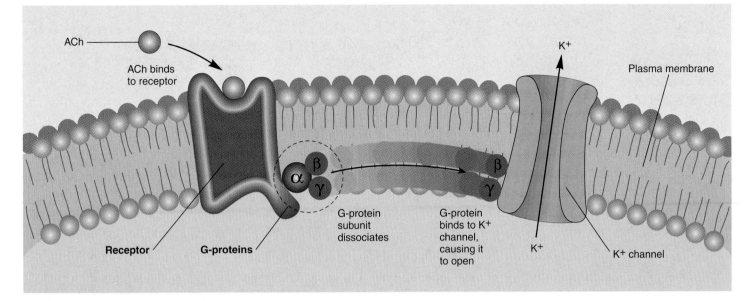

■ **Figure 7.24** Muscarinic ACh receptors require the mediation of G-proteins. The figure depicts the effects of ACh on the pacemaker cells of the heart. Binding of ACh to its muscarinic receptor causes the beta-gamma subunits to dissociate from the alpha subunit. The beta-gamma complex of G-proteins then binds to a K⁺ channel, causing it to open. Outward diffusion of K⁺ results, slowing the heart rate.

Table 7.7 Steps in the Activation and Inactivation of G-Proteins

Step 1	The alpha, beta, and gamma G-proteins are joined together and bind to GDP before the arrival of the neurotransmitter.
Step 2	The ligand (neurotransmitter chemical) binds to its receptor in the membrane.
Step 3	GDP is released, and the alpha subunit of the G-proteins binds GTP.
Step 4	This causes the dissociation of the alpha subunit from the beta-gamma subunits.
Step 5	In different cases, either the alpha subunit, or the beta-gamma complex, can interact with membrane ion channels or membrane-bound enzymes.
Step 6	Deactivation is initiated by the hydrolysis of GTP to GDP by the alpha subunit.
Step 7	Bound to GDP again, the alpha subunit comes back together with the beta-gamma complex to reassemble the alpha-beta-gamma G-proteins.

from the channel and moves back to its previous position. This causes the ion channel to close. The steps of this process are summarized in table 7.7.

The binding of ACh to its muscarinic receptors indirectly affects the permeability of K⁺ channels. This can produce hyperpolarization in some organs (if the K⁺ channels are opened) and depolarization in other organs (if the K⁺ channels are closed). Specific examples should help to clarify this point.

Scientists have learned that it is the beta-gamma complex that binds to the K⁺ channels in the heart muscle cells and causes these channels to open (fig. 7.24). This leads to the diffusion of K⁺ out of the postsynaptic cell (because that is the direction of its concentration gradient). As a result, the cell becomes hyperpolar-

ized, producing an inhibitory postsynaptic potential (IPSP). Such an effect is produced in the heart, for example, when autonomic nerve fibers (part of the vagus nerve) synapse with pacemaker cells and slow the rate of beat. It should be noted that inhibition also occurs in the CNS in response to other neurotransmitters, but those IPSPs are produced by a different mechanism.

There are cases in which the alpha subunit is the effector, and examples where its effects are substantially different from the one shown in figure 7.24. In the smooth muscle cells of the stomach, the binding of ACh to its muscarinic receptors causes a different type of G-protein alpha subunit to dissociate and bind to the K⁺ channels. In this case, however, the binding of the G-protein subunit to the K⁺ channels causes the channels to close rather than to open. As a result, the outward diffusion of K⁺, which occurs at an ongoing rate in the resting cell, is reduced to below resting levels. Since the resting membrane potential is maintained by a balance between cations flowing into the cell and cations flowing out, a reduction in the outward flow of K⁺ produces a depolarization. This depolarization produced in these smooth muscle cells results in contractions of the stomach (see chapter 12).

Acetylcholinesterase (AChE)

The bond between ACh and its receptor protein exists for only a brief instant. The ACh-receptor complex quickly dissociates but can be quickly re-formed as long as free ACh is in the vicinity. In order for activity in the postsynaptic cell to be stopped, free ACh must be inactivated very soon after it is released. The inactivation of ACh is achieved by means of an enzyme called **acetylcholinesterase,** or **AChE,** which is present on the postsynaptic membrane or immediately outside the membrane, with its active site facing the synaptic cleft (fig. 7.25).

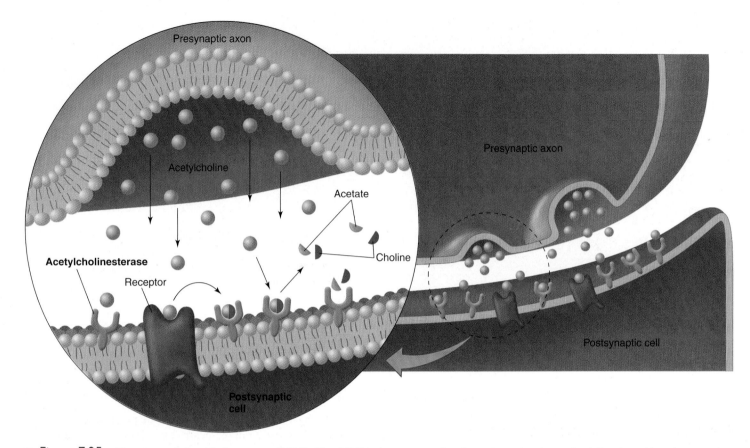

Figure 7.25 **The action of acetylcholinesterase (AChE).** The AChE in the postsynaptic cell membrane inactivates the ACh released into the synaptic cleft. This prevents continued stimulation of the postsynaptic cell unless more ACh is released by the axon.

 Nerve gas exerts its odious effects by inhibiting AChE in skeletal muscles. Since ACh is not degraded, it can continue to combine with receptor proteins and can continue to stimulate the postsynaptic cell, leading to spastic paralysis. Clinically, cholinesterase inhibitors (such as neostigmine) are used to enhance the effects of ACh on muscle contraction when neuromuscular transmission is weak, as in the disease *myasthenia gravis*.

Acetylcholine in the PNS

Somatic motor neurons form synapses with skeletal muscle cells (muscle fibers). At these synapses, or **neuromuscular junctions,** the postsynaptic membrane of the muscle fiber is known as a *motor end plate*. Therefore, the EPSPs produced by ACh in skeletal muscle fibers are often called **end-plate potentials.** This depolarization opens voltage-regulated channels that are adjacent to the end plate. Voltage-regulated channels produce action potentials in the muscle fiber, and these are reproduced by other voltage-regulated channels along the muscle

plasma membrane. This conduction is analogous to conduction of action potentials by axons; it is significant because action potentials produced by muscle fibers stimulate muscle contraction (as described in chapter 12).

Clinical Investigation Clue

Remember that Sandra had flaccid paralysis and difficulty breathing after eating mussels and clams gathered from the local shore.

What caused her difficulty in breathing?

If any stage in the process of neuromuscular transmission is blocked, muscle weakness—sometimes leading to paralysis and death—may result. The drug *curare,* for example, competes with ACh for attachment to the nicotinic ACh receptors and thus reduces the size of the end-plate potentials (see table 7.6). This drug was first used on blow-gun darts by South American Indians because it produced flaccid paralysis in their victims. Clinically, curare is used in surgery as a muscle relaxant and in electroconvulsive shock therapy to prevent muscle damage.

Autonomic motor neurons innervate cardiac muscle, smooth muscles in blood vessels and visceral organs, and glands. As previously mentioned, there are two classifications of autonomic nerves: sympathetic and parasympathetic. Most of the parasympathetic axons that innervate the effector organs use ACh as their neurotransmitter. In some cases, these axons have an inhibitory effect on the organs they innervate through the binding of ACh to muscarinic ACh receptors. The action of the vagus nerve in slowing the heart rate is an example of this inhibitory effect. In other cases, ACh released by autonomic neurons produces stimulatory effects as previously described. The structures and functions of the autonomic system are described in chapter 9.

Acetylcholine in the CNS

There are many **cholinergic neurons** (those that use ACh as a neurotransmitter) in the CNS, where the axon terminals of one neuron typically synapse with the dendrites or cell body of another. The dendrites and cell body thus serve as the receptive area of the neuron, and it is in these regions that receptor proteins for neurotransmitters and chemically regulated gated channels are located. The first voltage-regulated gated channels are located at the *axon hillock,* a cone-shaped elevation on the cell body from which the axon arises. The *initial segment* of the axon, which is the unmyelinated region of the axon around the axon hillock, has a high concentration of voltage-regulated gated channels. It is here that action potentials are first produced (see fig. 7.22).

Depolarizations—EPSPs—in the dendrites and cell body spread by cable properties to the initial segment of the axon in order to stimulate action potentials. If the depolarization is at or above threshold by the time it reaches the initial segment of the axon, the EPSP will stimulate the production of action potentials, which can then regenerate themselves along the axon. If, however, the EPSP is below threshold at the initial segment, no action potentials will be produced in the postsynaptic cell (fig. 7.26). Gradations in the strength of the EPSP above threshold determine the frequency with which action potentials will be produced at the axon hillock, and at each point in the axon where the impulse is conducted. The action potentials that begin at the initial segment of the axon are conducted without loss of amplitude toward the axon terminals.

Earlier in this chapter, the action potential was introduced by describing the events that occurred when a depolarization stimulus was artificially produced by stimulating electrodes. Now it is apparent that EPSPs, conducted from the dendrites and cell body, serve as the normal stimuli for the production of action potentials at the axon hillock, and that the action potentials at this point serve as the depolarization stimuli for the next region, and so on. This chain of events ends at the terminal boutons of the axon, where neurotransmitter is released.

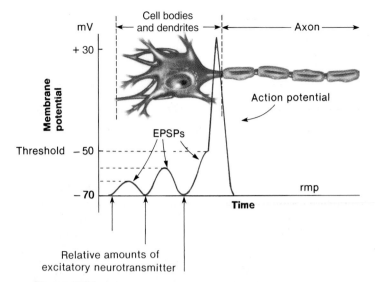

Figure 7.26 The graded nature of excitatory postsynaptic potentials (EPSPs). Stimuli of increasing strength produce increasing amounts of depolarization. When a threshold level of depolarization is produced, action potentials are generated in the axon.

Alzheimer's disease, the most common cause of senile dementia, often begins in middle age and produces progressive mental deterioration. Brain lesions develop that consist of dense extracellular deposits of an insoluble protein called *amyloid beta protein,* and degenerating nerve fibers. Twisted fibrils, called *neurofibrillar tangles,* form within the dead or dying neurons. Alzheimer's is associated with a loss of cholinergic neurons that terminate in the hippocampus and cerebral cortex of the brain (areas concerned with memory storage). Treatments for Alzheimer's disease currently include the use of cholinesterase (AChE) inhibitors to augment cholinergic transmission in the brain, and the use of vitamin E and other antioxidants to limit the oxidative stress produced by free radicals (see chapter 5), which may contribute to neural damage.

Test Yourself Before You Continue

1. Distinguish between the two types of chemically regulated channels and explain how ACh opens each type.
2. State a location at which ACh has stimulatory effects. Where does it exert inhibitory effects? How are stimulation and inhibition accomplished?
3. Describe the function of acetylcholinesterase and discuss its physiological significance.
4. Compare the properties of EPSPs and action potentials and state where these events occur in a postsynaptic neuron.
5. Explain how EPSPs produce action potentials in the postsynaptic neuron.

Monoamines as Neurotransmitters

A variety of chemicals in the CNS function as neurotransmitters. Among these are the monoamines, a chemical family that includes dopamine, norepinephrine, and serotonin. Although these molecules have similar mechanisms of action, they are used by different neurons for different functions.

The regulatory molecules epinephrine, norepinephrine, dopamine, and serotonin are in the chemical family known as **monoamines.** Serotonin is derived from the amino acid tryptophan. Epinephrine, norepinephrine, and dopamine are derived from the amino acid tyrosine and form a subfamily of monoamines called the **catecholamines** (see fig. 9.8, p. 229). Epinephrine (also called adrenaline) is a hormone secreted by the adrenal gland, not a neurotransmitter, while the closely related norepinephrine functions both as a hormone and a neurotransmitter.

Like ACh, monoamine neurotransmitters are released by exocytosis from presynaptic vesicles, diffuse across the synaptic cleft, and interact with specific receptor proteins in the membrane of the postsynaptic cell. The stimulatory effects of these monoamines, like those of ACh, must be quickly inhibited so as to maintain proper neural control. The inhibition of monoamine action is due to (1) reuptake of monoamines into the presynaptic neuron endings, (2) enzymatic degradation of monoamines in the presynaptic neuron endings by *monoamine oxidase (MAO),* and (3) the enzymatic degradation of catecholamines in the postsynaptic neuron by *catechol-O-methyltransferase (COMT).* This process is illustrated in figure 7.27.

Monoamine oxidase (MAO) inhibitors are drugs that block monoamine oxidase, the enzyme in presynaptic endings that breaks down catecholamines and serotonin after they have been taken up from the synaptic cleft. These drugs thus promote transmission at synapses that use monoamines as neurotransmitters. Such drugs have proven useful in the treatment of clinical depression, suggesting that a deficiency in monoamine transmission contributes to that disorder. An MAO inhibitor is also used to treat Parkinson's disease, because it increases the ability of dopamine to function as a neurotransmitter.

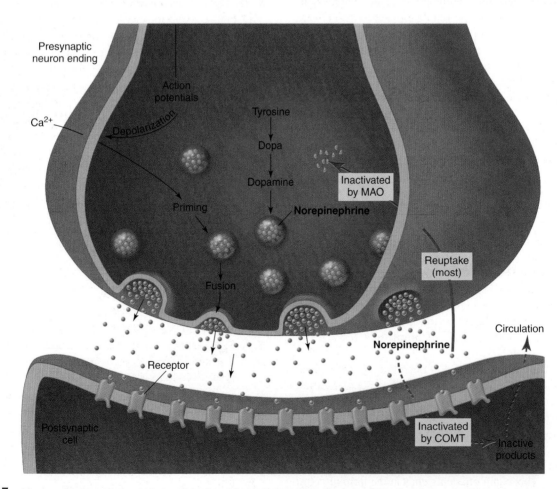

■ **Figure 7.27** **The production, release, and reuptake of catecholamine neurotransmitters.** The transmitters combine with receptor proteins in the postsynaptic membrane. (COMT = catechol-O-methyltransferase; MAO = monoamine oxidase.)

Clinical Investigation Clues

Remember that Sandra was taking an MAO inhibitor, and that her blood levels of this drug were not unduly high.

- *Why was Sandra taking an MAO inhibitor drug?*
- *Why might the paramedics suspect that she might have a neuromuscular disorder?*

The monoamine neurotransmitters do not directly cause opening of ion channels in the postsynaptic membrane. Instead, these neurotransmitters act by means of an intermediate regulator, known as a **second messenger.** In the case of some synapses that use catecholamines for synaptic transmission, this second messenger is a compound known as **cyclic adenosine monophosphate (cAMP).** Although other synapses can use other second messengers, only the function of cAMP as a second messenger will be considered here. Other second-messenger systems are discussed in conjunction with hormone action in chapter 11.

Binding of norepinephrine, for example, with its receptor in the postsynaptic membrane stimulates the dissociation of the G-protein alpha subunit from the others in its complex (fig. 7.28). This subunit diffuses in the membrane until it binds to an enzyme known as *adenylate cyclase* (also called *adenylyl cyclase*). This enzyme converts ATP to cyclic AMP (cAMP) and pyrophosphate (two inorganic phosphates) within the postsynaptic cell cyto-

plasm. Cyclic AMP in turn activates another enzyme, *protein kinase,* which phosphorylates (adds a phosphate group to) other proteins (fig. 7.28). Through this action, ion channels are opened in the postsynaptic membrane.

Serotonin as a Neurotransmitter

Serotonin, or *5-hydroxytryptamine (5-HT),* is used as a neurotransmitter by neurons with cell bodies in what are called the *raphe nuclei* that are located along the midline of the brain stem (see chapter 8). Serotonin is derived from the amino acid L-tryptophan, and variations in the amount of this amino acid in the diet (tryptophan-rich foods include milk and turkey) can affect the amount of serotonin produced by the neurons. Physiological functions attributed to serotonin include a role in the regulation of mood and behavior, appetite, and cerebral circulation.

Since LSD (a powerful hallucinogen) mimics the structure, and thus likely the function, of serotonin, scientists have long suspected that serotonin should influence mood and emotion. This suspicion is confirmed by the actions of the antidepressant drugs *Prozac, Paxil, Zoloft,* and *Luvox,* which act as **serotonin-specific reuptake inhibitors (SSRIs).** By blocking the reuptake of serotonin into presynaptic endings, and thereby increasing the effectiveness of serotonin transmission at synapses, these drugs have proven effective in the treatment of depression.

Serotonin's diverse functions are related to the fact that there are a large number of different subtypes of serotonin receptors— over a dozen are currently known. Thus, while Prozac may be

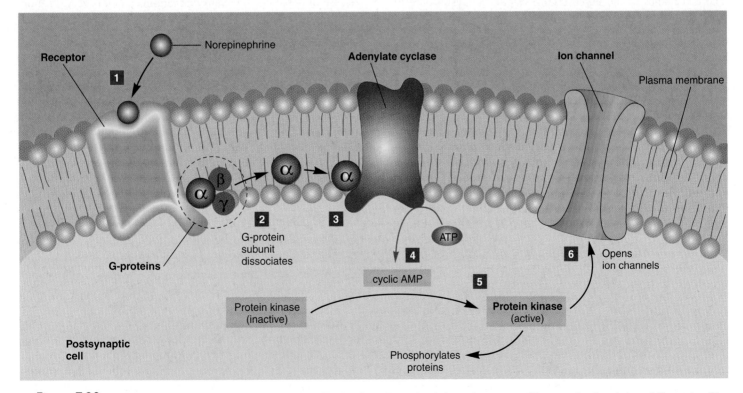

■ **Figure 7.28   Norepinephrine action requires G-proteins.** The binding of norepinephrine to its receptor (*1*) causes the dissociation of G-proteins (*2*). Binding of the alpha G-protein subunit to the enzyme adenylate cyclase (*3*) activates this enzyme, leading to the production of cyclic AMP (*4*). Cyclic AMP, in turn, activates protein kinase (*5*), which can open ion channels (*6*) and produce other effects.

given to relieve depression, another drug that promotes serotonin action is sometimes given to reduce the appetite of obese patients. A different drug that may activate a different serotonin receptor is used to treat anxiety, and yet another drug that promotes serotonin action is given to relieve migraine headaches. It should be noted that the other monoamine neurotransmitters, dopamine and norepinephrine, also influence mood and behavior in a way that complements the actions of serotonin.

Dopamine as a Neurotransmitter

Neurons that use **dopamine** as a neurotransmitter are called **dopaminergic neurons.** Neurons that have dopamine receptor proteins on the postsynaptic membrane, and that therefore respond to dopamine, have been identified in postmortem brain tissue. More recently, the location of these receptors has been observed in the living brain using the technique of *positron emission tomography (PET)* (see chapter 8). These investigations have been spurred by the great clinical interest in the effects of dopaminergic neurons.

The cell bodies of dopaminergic neurons are highly concentrated in the midbrain. Their axons project to different parts of the brain and can be divided into two systems: the *nigrostriatal dopamine system,* involved in motor control, and the *mesolimbic dopamine system,* involved in emotional reward (see chapter 8, fig. 8.18).

Nigrostriatal Dopamine System

The cell bodies of the **nigrostriatal dopamine system** are located in a part of the midbrain called the *substantia nigra* ("dark substance") because it contains melanin pigment. Neurons in the substantia nigra send fibers to a group of nuclei known collectively as the *corpus striatum* because of its striped appearance—hence the term *nigrostriatal system.* These regions are part of the *basal nuclei*—large masses of neuron cell bodies deep in the cerebrum involved in the initiation of skeletal movements (chapter 8). There is much evidence that **Parkinson's disease** is caused by degeneration of the dopaminergic neurons in the substantia nigra. Parkinson's disease is the second most common neuro-degenerative disease (after Alzheimer's disease), and is associated with such symptoms as muscle tremors and rigidity, difficulty in initiating movements and speech, and other severe motor problems. Patients are often treated with L-dopa and MAO inhibitors in an attempt to increase dopaminergic transmission.

The cause of the degeneration of dopaminergic neurons in Parkinson's disease is not well understood. Some scientists believe that neural destruction might be caused by free radicals (superoxide and nitric oxide), perhaps released by overactive microglia, that produce oxidative damage.

Mesolimbic Dopamine System

The **mesolimbic dopamine system** involves neurons that originate in the midbrain and send axons to structures in the forebrain that are part of the limbic system (see fig. 8.18). The dopamine released by these neurons may be involved in behavior and reward. For example, several studies involving human twins separated at birth and reared in different environments, and other studies involving the use of rats, have implicated the gene that codes for one subtype of dopamine receptor (designated D_2) in alcoholism. Other addictive drugs, including cocaine, morphine, and amphetamines, are also known to activate dopaminergic pathways.

Cocaine—a stimulant related to the amphetamines in its action—is currently widely abused in the United States. Although early use of this drug produces feelings of euphoria and social adroitness, continued use leads to social withdrawal, depression, dependence upon ever-higher dosages, and serious cardiovascular and renal disease that can result in heart and kidney failure. The numerous effects of cocaine on the central nervous system appear to be mediated by one primary mechanism: cocaine binds to the reuptake transporters for dopamine, norepinephrine, and serotonin, and blocks their reuptake into the presynaptic axon endings. This results in overstimulation of those neural pathways that use dopamine as a neurotransmitter.

Recent studies demonstrate that alcohol, amphetamines, cocaine, marijuana, and morphine promote the activity of dopaminergic neurons that arise in the midbrain and terminate in a particular location, the *nucleus accumbens,* of the forebrain. Interestingly, nicotine also has recently been shown to promote the release of dopamine by axons that terminate in this very location. This suggests that the physiological mechanism for nicotine addiction in smokers is similar to that for other abused drugs.

All drugs used to treat schizophrenia (drugs called *neuroleptics*) act as antagonists of the D_2 subtype of dopamine receptor. This suggests that overactivity of the mesolimbic dopamine pathways contributes to schizophrenia, a concept that helps to explain why people with Parkinson's disease may develop symptoms of schizophrenia if treated with too much L-dopa. It should be noted that abnormalities in other neurotransmitters (including norepinephrine and glutamate) may also contribute to schizophrenia.

Norepinephrine as a Neurotransmitter

Norepinephrine, like ACh, is used as a neurotransmitter in both the PNS and the CNS. Sympathetic neurons of the PNS use norepinephrine as a neurotransmitter at their synapse with smooth muscles, cardiac muscle, and glands. Some neurons in the CNS also use norepinephrine as a neurotransmitter; these neurons seem to be involved in general behavioral arousal. This would help to explain the mental arousal elicited by *amphetamines,* which stimulate pathways in which norepinephrine is used as a neurotransmitter. Such drugs also stimulate the PNS pathways that use norepinephrine, however, and this duplicates the effects of sympathetic nerve activation. A rise in blood pressure, constriction of arteries, and other effects similar to the deleterious consequences of cocaine use can thereby be produced.

Other Neurotransmitters

A surprisingly large number of diverse molecules appear to function as neurotransmitters. These include some amino acids and their derivatives, many polypeptides, and even the gas nitric oxide.

Amino Acids as Neurotransmitters

Excitatory Neurotransmitters

The amino acids **glutamic acid** and **aspartic acid** function as excitatory neurotransmitters in the CNS. Glutamic acid (or *glutamate*), indeed, is the major excitatory neurotransmitter in the brain, producing excitatory postsynaptic potentials (EPSPs). Research has revealed that each of the glutamate receptors encloses an ion channel, similar to the arrangement seen in the nicotinic ACh receptors (see fig. 7.23).

Among these EPSP-producing glutamate receptors, three subtypes can be distinguished. These are named according to the molecules (other than glutamate) that they bind, and include: (1) **NMDA receptors** (named for N-methyl-D-aspartate); (2) *AMPA receptors;* and (3) *kainate* receptors. NMDA and AMPA receptors are illustrated in chapter 8, figure 8.15.

The NMDA receptors for glutamate are involved in memory storage, as will be discussed more fully in the section on long-term potentiation. These receptors are quite complex, because the ion channel will not open simply by the binding of glutamate to its receptor. Instead, two other conditions must be met at the same time: (1) the NMDA receptor must also bind to glycine (or D-serine, which has recently been shown to be produced by astrocytes); and (2) the membrane must be partially depolarized at this time by a different neurotransmitter molecule that binds to a different receptor (for example, by glutamate binding to the AMPA receptors). Once open, the NMDA receptor channels permit the entry of Ca^{2+} and Na^+ (and exit of K^+) into the dendrites of the postsynaptic neuron.

Inhibitory Neurotransmitters

The amino acid **glycine** is inhibitory; instead of depolarizing the postsynaptic membrane and producing an EPSP, it hyperpolarizes the postsynaptic membrane and produces an inhibitory postsynaptic potential (IPSP). The binding of glycine to its receptor proteins causes the opening of chloride (Cl^-) channels in the postsynaptic membrane. As a result, Cl^- diffuses into the postsynaptic neuron and produces the hyperpolarization. This inhibits the neuron by making the membrane potential even more negative than it is at rest, and therefore farther from the threshold depolarization required to stimulate action potentials.

The inhibitory effects of glycine are very important in the spinal cord, where they help in the control of skeletal movements. Flexion of an arm, for example, involves stimulation of the flexor muscles by motor neurons in the spinal cord. The motor neurons that innervate the antagonistic extensor muscles are inhibited by IPSPs produced by glycine released from other neurons. The importance of the inhibitory actions of glycine is revealed by the deadly effects of *strychnine,* a poison that causes spastic paralysis by specifically blocking the glycine receptor proteins. Animals poisoned with strychnine die from asphyxiation because they are unable to relax the diaphragm.

The neurotransmitter **gamma-aminobutyric acid (GABA)** is a derivative of another amino acid, glutamic acid. GABA is the most prevalent neurotransmitter in the brain; in fact, as many as one-third of all the neurons in the brain use GABA as a neurotransmitter. Like glycine, GABA is inhibitory—it hyperpolarizes the postsynaptic membrane by opening Cl^- channels. Also, the effects of GABA, like those of glycine, are involved in motor control. For example, the large Purkinje cells mediate the motor functions of the cerebellum by producing IPSPs in their postsynaptic neurons. A deficiency of GABA-releasing neurons is responsible for the uncontrolled movements seen in people with *Huntington's chorea.*

Polypeptides as Neurotransmitters

Many polypeptides of various sizes are found in the synapses of the brain. These are often called **neuropeptides** and are believed to function as neurotransmitters. Interestingly, some of the polypeptides that function as hormones secreted by the small intestine and other endocrine glands are also produced in the brain and may function there as neurotransmitters (table 7.8). For example, *cholecystokinin (CCK),* which is secreted as a hormone from the small intestine, is also released from neurons and used as a neurotransmitter in the brain. Recent evidence suggests that CCK, acting as a neurotransmitter, may promote feelings of satiety in the brain following meals. Another polypeptide found in

Table 7.8 Examples of Chemicals That Are Either Proven or Suspected Neurotransmitters

Category	Chemicals
Amines	Acetylcholine
	Histamine
	Serotonin
Catecholamines	Dopamine
	(Epinephrine—a hormone)
	Norepinephrine
Amino acids	Aspartic acid
	GABA (gamma-aminobutyric acid)
	Glutamic acid
	Glycine
Polypeptides	Glucagon
	Insulin
	Somatostatin
	Substance P
	ACTH (adrenocorticotrophic hormone)
	Angiotensin II
	Endogenous opioids (enkephalins and endorphins)
	LHRH (luteinizing hormone-releasing hormone)
	TRH (thyrotrophin-releasing hormone)
	Vasopressin (antidiuretic hormone)
	CCK (cholecystokinin)
Lipids	Endocannabinoids
Gases	Nitric oxide
	Carbon monoxide

many organs, *substance P,* functions as a neurotransmitter in pathways in the brain that mediate sensations of pain.

Synaptic Plasticity

Although some of the polypeptides released from neurons may function as neurotransmitters in the traditional sense (that is, by stimulating the opening of ionic gates and causing changes in the membrane potential), others may have more subtle and poorly understood effects. **Neuromodulators** has been proposed as a name for compounds with such alternative effects. An exciting recent discovery is that some neurons in both the PNS and CNS produce both a classical neurotransmitter (ACh or a catecholamine) and a polypeptide neurotransmitter. These are contained in different synaptic vesicles that can be distinguished using the electron microscope. The neuron can thus release either the classical neurotransmitter or the polypeptide neurotransmitter under different conditions.

Discoveries such as the one just described indicate that synapses have a greater capacity for alteration at the molecular level than was previously believed. This attribute has been termed **synaptic plasticity.** Synapses are also more plastic at the cellular level. There is evidence that sprouting of new axon branches can occur over short distances to produce a turnover of synapses, even in the mature CNS. This breakdown and re-forming

of synapses may occur within a time span of only a few hours. These events may play a role in learning and conditioning.

Endogenous Opioids

The ability of opium and its analogues—that is, the **opioids**—to relieve pain (promote analgesia) has been known for centuries. Morphine, for example, has long been used for this purpose. The discovery in 1973 of opioid receptor proteins in the brain suggested that the effects of these drugs might be due to the stimulation of specific neuron pathways. This implied that opioids—along with LSD, mescaline, and other mind-altering drugs—might mimic the actions of neurotransmitters produced by the brain.

The analgesic effects of morphine are blocked in a specific manner by a drug called *naloxone.* In the same year that opioid receptor proteins were discovered, it was found that naloxone also blocked the analgesic effect of electrical brain stimulation. Subsequent evidence suggested that the analgesic effects of hypnosis and acupuncture could also be blocked by naloxone. These experiments indicated that the brain might be producing its own endogenous morphinelike analgesic compounds that served as the natural ligands of the opioid receptors in the brain.

These compounds have been identified as a family of polypeptides produced by the brain and pituitary gland. One member is called β-*endorphin* (for "endogenously produced morphinelike compound"). Another consists of a group of five-amino-acid peptides called *enkephalins,* and a third is a polypeptide neurotransmitter called *dynorphin.*

The endogenous opioid system is inactive under normal conditions, but when activated by stressors it can block the transmission of pain. For example, a burst in β-endorphin secretion was shown to occur in pregnant women during parturition (childbirth).

Exogenous opioids such as opium and morphine can produce euphoria, and so endogenous opioids may mediate reward or positive reinforcement pathways. This is consistent with the observation that overeating in genetically obese mice can be blocked by naloxone. It has also been suggested that the feeling of well-being and reduced anxiety following exercise (the "joggers high") may be an effect of endogenous opioids. Blood levels of β-endorphin increase when exercise is performed at greater than 60% of the maximal oxygen uptake (see chapter 12) and peak 15 minutes after the exercise has ended. Although obviously harder to measure, an increased level of opioids in the brain and cerebrospinal fluid has also been found to result from exercise. The opioid antagonist drug naloxone, however, does not block the exercise-induced euphoria, suggesting that the joggers high is not primarily an opioid effect. Use of naloxone, however, does demonstrate that the endogenous opioids are involved in the effects of exercise on blood pressure, and that they are responsible for the ability of exercise to raise the pain threshold.

Neuropeptide Y

Neuropeptide Y is the most abundant neuropeptide in the brain. It has been shown to have a variety of physiological effects, including a role in the response to stress, in the regulation of circadian rhythms, and in the control of the cardiovascular system.

Neuropeptide Y has been shown to inhibit the release of the excitatory neurotransmitter glutamate in a part of the brain called the hippocampus. This is significant because excessive glutamate released in this area can cause convulsions. Indeed, frequent seizures were a symptom of a recently developed strain of mice with the gene for neuropeptide Y "knocked out." (Knockout strains of mice have specific genes inactivated, as described in chapter 3.)

Neuropeptide Y is a powerful stimulator of appetite. When injected into a rat's brain, it can cause the rat to eat until it becomes obese. Conversely, inhibitors of neuropeptide Y that are injected into the brain inhibit eating. This research has become particularly important in light of the recent discovery of *leptin,* a satiety factor secreted by adipose tissue. Leptin suppresses appetite by acting, at least in part, to inhibit neuropeptide Y release. This topic is discussed in more detail in chapter 19.

Endocannabinoids as Neurotransmitters

In addition to producing endogenous opioids, the brain also produces compounds with effects similar to the active ingredient in marijuana—Δ^9-tetrahydrocannabinol (THC). These endogenous cannabinoids, or **endocannabinoids,** are neurotransmitters that bind to the same receptor proteins in the brain as does THC from marijuana.

The endocannabinoids, like the endogenous opioids, are believed to act as analgesics. Unlike the polypeptide opioids, however, the endocannabinoids are lipids. As such, they are the only lipid neurotransmitters currently identified. The endocannabinoids are also distinguished by evidence that they may function as backward, or retrograde, neurotransmitters. That is, they are produced in the postsynaptic neuron when it is depolarized, and then they diffuse backward to the presynaptic neuron to inhibit the release of the neurotransmitter (for example, GABA) from the presynaptic axon terminal. The physiological significance of these actions is presently unclear.

Nitric Oxide and Carbon Monoxide as Neurotransmitters

Nitric oxide (NO) was the first gas to be identified as a neurotransmitter. Produced by nitric oxide synthetase in the cells of many organs from the amino acid L-arginine, nitric oxide's actions are very different from those of the more familiar nitrous oxide (N_2O), or laughing gas, sometimes used as a mild anesthetic in dentistry.

Nitric oxide has a number of different roles in the body. Within blood vessels, it acts as a local tissue regulator that causes the smooth muscles of those vessels to relax, so that the blood vessels dilate. This role will be described in conjunction with the circulatory system in chapter 14. Within macrophages and other cells, nitric oxide helps to kill bacteria. This activity is described

in conjunction with the immune system in chapter 15. In addition, nitric oxide is a neurotransmitter of certain neurons in both the PNS and CNS. It diffuses out of the presynaptic axon and into neighboring cells by simply passing through the lipid portion of the cell membranes. Once in the target cells, NO exerts its effects by stimulating the production of cyclic guanosine monophosphate (cGMP), which acts as a second messenger.

In the PNS, nitric oxide is released by some neurons that innervate the gastrointestinal tract, penis, respiratory passages, and cerebral blood vessels. These are autonomic neurons that cause smooth muscle relaxation in their target organs. This can produce, for example, the engorgement of the spongy tissue of the penis with blood. In fact, scientists now believe that erection of the penis results from the action of nitric oxide, and indeed the drug *Viagra* works by increasing this action of nitric oxide (as described in chapter 20; see fig. 20.23). Nitric oxide is also released as a neurotransmitter in the brain, and has been implicated in the processes of learning and memory. This will be discussed in more detail later in this chapter.

In addition to nitric oxide, another gas—**carbon monoxide (CO)**—may function as a neurotransmitter. Certain neurons, including those of the cerebellum and olfactory epithelium, have been shown to produce carbon monoxide (derived from the conversion of one pigment molecule, heme, to another, biliverdin; see fig. 18.23). Also, carbon monoxide, like nitric oxide, has been shown to stimulate the production of cGMP within the neurons. Experiments suggest that carbon monoxide may promote odor adaptation in olfactory neurons, contributing to the regulation of olfactory sensitivity. Other physiological functions of neuronal carbon monoxide have also been suggested, including neuroendocrine regulation in the hypothalamus.

Although its importance in the body was recognized only recently, nitric oxide has already been exploited for medical use. The hypotension (low blood pressure) of *septic shock,* for example, is apparently due to vasodilation caused by nitric oxide and has been successfully treated with drugs that inhibit nitric oxide synthetase. Conversely inhalation of nitric oxide has been used to treat *pulmonary hypertension,* as well as *respiratory distress syndrome* (discussed in chapter 16).

Test Yourself Before You Continue

1. Explain the significance of glutamate in the brain and of NMDA receptors.
2. Describe the mechanism of action of glycine and GABA as neurotransmitters and discuss their significance.
3. Give examples of endogenous opioid polypeptides and discuss their significance.
4. Explain how nitric acid is produced in the body and describe its functions.

Synaptic Integration

The summation of numerous EPSPs may be needed to produce a depolarization of sufficient magnitude to stimulate the postsynaptic cell. The net effect of EPSPs on the postsynaptic neuron is reduced by hyperpolarization (IPSPs), which is produced by inhibitory neurotransmitters. The activity of neurons within the central nervous system is thus the net result of both excitatory and inhibitory effects.

Unlike action potentials, synaptic potentials are graded and can add together, or summate. **Spatial summation** occurs because numerous presynaptic nerve fibers (up to a thousand, in some cases) converge on a single postsynaptic neuron. In spatial summation, synaptic depolarizations (EPSPs) produced at different synapses summate in the postsynaptic dendrites and cell body (fig 7.29). In **temporal summation,** the successive activity of a presynaptic axon terminal causes successive waves of transmitter release, resulting in the summation of EPSPs in the postsynaptic neuron. The summation of EPSPs helps to determine if the depolarization that reaches the axon hillock will be of sufficient magnitude to generate new action potentials in the postsynaptic neuron.

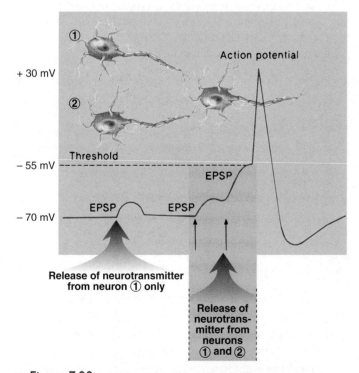

■ **Figure 7.29**   Spatial summation.   When only one presynaptic neuron releases excitatory neurotransmitter, the EPSP produced may not be sufficiently strong to stimulate action potentials in the postsynaptic neuron. When more than one presynaptic neuron produces EPSPs at the same time, however, the EPSPs can summate at the axon hillock to produce action potentials.

Long-Term Potentiation

When a presynaptic neuron is experimentally stimulated at a high frequency, even for just a few seconds, the excitability of the synapse is enhanced—or potentiated—when this neuron pathway is subsequently stimulated. The improved efficacy of synaptic transmission may last for hours or even weeks and is called **long-term potentiation (LTP).** Long-term potentiation may favor transmission along frequently used neural pathways and thus may represent a mechanism of neural "learning." It is interesting in this regard that LTP has been observed in the hippocampus of the brain, which is an area implicated in memory storage (see chapter 8).

Most of the neural pathways in the hippocampus use glutamate as a neurotransmitter that activates NMDA receptors. This implicates glutamate and its NMDA receptors in learning and memory, and indeed, in a recent experiment, it was demonstrated that genetically altered mice with enhanced NMDA expression were smarter when tested in a maze. The association of NMDA receptors with synaptic changes during learning and memory is discussed more fully in chapter 8.

Although glutamate-mediated neurotransmission is necessary for normal brain function, excessive release of glutamate can cause epilepsy and neuronal cell death, a process termed **excitotoxicity.** This process has been implicated in the neuronal damage that occurs in stroke and traumatic damage to the CNS, and in the loss of neurons in various neurodegenerative diseases. Interestingly, the street drug known as *PCP* or *angel dust* blocks NMDA receptors, suggesting that the aberrant schizophrenia-like effects of this drug are produced by a reduction in glutamate stimulation of NMDA receptors.

Synaptic Inhibition

Although many neurotransmitters depolarize the postsynaptic membrane (produce EPSPs), some transmitters do just the opposite. The neurotransmitters glycine and GABA hyperpolarize the postsynaptic membrane; that is, they make the inside of the membrane more negative than it is at rest (fig. 7.30). Since hyperpolarization (from –70 mV to, for example, –85 mV) drives the membrane potential farther from the threshold depolarization required to stimulate action potentials, this inhibits the activity of the postsynaptic neuron. Hyperpolarizations produced by neurotransmitters are therefore called *inhibitory postsynaptic potentials (IPSPs),* as previously described. The inhibition produced in this way is called **postsynaptic inhibition.** Postsynaptic inhibition in the brain is produced by GABA, while in the spinal cord it is mainly produced by glycine (although GABA is also involved).

Excitatory and inhibitory inputs (EPSPs and IPSPs) to a postsynaptic neuron can summate in an algebraic fashion. The effects of IPSPs in this way reduce, or may even eliminate, the ability of

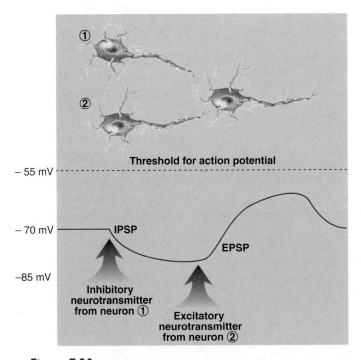

Figure 7.30 An IPSP hyperpolarizes the postsynaptic membrane. An inhibitory postsynaptic potential (IPSP) makes the inside of the postsynaptic membrane more negative than the resting potential—it hyperpolarizes the membrane. Subsequent or simultaneous excitatory postsynaptic potentials (EPSPs), which are depolarizations, must thus be stronger to reach the threshold required to generate action potentials at the axon hillock.

EPSPs to generate action potentials in the postsynaptic cell. Considering that a given neuron may receive as many as 1,000 presynaptic inputs, the interactions of EPSPs and IPSPs can vary greatly.

In **presynaptic inhibition** (fig. 7.31), the amount of an excitatory neurotransmitter released at the end of an axon is decreased by the effects of a second neuron, whose axon makes a synapse with the axon of the first neuron (an axoaxonic synapse). The neurotransmitter exerting this presynaptic inhibition may be GABA or excitatory neurotransmitters, such as ACh and glutamate.

Excitatory neurotransmitters can cause presynaptic inhibition by producing depolarization of the axon terminals, leading to inactivation of Ca^{2+} channels. This decreases the inflow of Ca^{2+} into the axon terminals and thus inhibits the release of neurotransmitter. The ability of the opiates to promote analgesia (reduce pain) is an example of such presynaptic inhibition. By reducing Ca^{2+} flow into axon terminals containing substance P, the opioids inhibit the release of the neurotransmitter involved in pain transmission.

Test Yourself Before You Continue

1. Define spatial summation and temporal summation and explain their functional importance.

2. Describe long-term potentiation, explain how it is produced, and discuss its significance.

3. Explain how postsynaptic inhibition is produced and how IPSPs and EPSPs can interact.

4. Describe the mechanism of presynaptic inhibition.

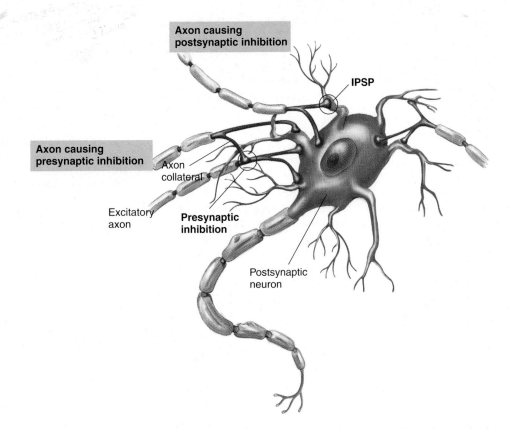

Figure 7.31 A diagram illustrating postsynaptic and presynaptic inhibition. These and other processes permit extensive integration within the CNS.

Summary

Neurons and Supporting Cells   152

I.   The nervous system is divided into the central nervous system (CNS) and the peripheral nervous system (PNS).

   A.   The central nervous system includes the brain and spinal cord, which contain nuclei and tracts.

   B.   The peripheral nervous system consists of nerves, ganglia, and nerve plexuses.

II.   A neuron consists of dendrites, a cell body, and an axon.

   A.   The cell body contains the nucleus, Nissl bodies, neurofibrils, and other organelles.

   B.   Dendrites receive stimuli, and the axon conducts nerve impulses away from the cell body.

III.   A nerve is a collection of axons in the PNS.

   A.   A sensory, or afferent, neuron is pseudounipolar and conducts impulses from sensory receptors into the CNS.

   B.   A motor, or efferent, neuron is multipolar and conducts impulses from the CNS to effector organs.

   C.   Interneurons, or association neurons, are located entirely within the CNS.

   D.   Somatic motor nerves innervate skeletal muscle; autonomic nerves innervate smooth muscle, cardiac muscle, and glands.

IV.   Supporting cells include Schwann cells and satellite cells in the PNS; in the CNS they include the various types of glial cells: oligodendrocytes, microglia, astrocytes, and ependymal cells.

   A.   Schwann cells form a sheath of Schwann, or neurilemma, around axons of the PNS.

   B.   Some neurons are surrounded by successive wrappings of supporting cell membrane called a myelin sheath. This sheath is formed by Schwann cells in the PNS and by oligodendrocytes in the CNS.

   C.   Astrocytes in the CNS may contribute to the blood-brain barrier.

Electrical Activity in Axons   160

I.   The permeability of the axon membrane to Na^+ and K^+ is regulated by gated ion channels.

   A.   At the resting membrane potential of –70 mV, the membrane is relatively impermeable to Na^+ and only slightly permeable to K^+.

   B.   The voltage-regulated Na^+ and K^+ channels open in response to the stimulus of depolarization.

   C.   When the membrane is depolarized to a threshold level, the Na^+ channels open first, followed quickly by opening of the K^+ channels.

II.   The opening of voltage-regulated channels produces an action potential.

   A.   The opening of Na^+ channels in response to depolarization allows Na^+ to diffuse into the axon, thus further depolarizing the membrane in a positive feedback fashion.

   B.   The inward diffusion of Na^+ causes a reversal of the membrane potential from –70 mV to +30 mV.

   C.   The opening of K^+ channels and outward diffusion of K^+ causes the reestablishment of the resting membrane potential. This is called repolarization.

   D.   Action potentials are all-or-none events.

   E.   The refractory periods of an axon membrane prevent action potentials from running together.

   F.   Stronger stimuli produce action potentials with greater frequency.

III.   One action potential serves as the depolarization stimulus for production of the next action potential in the axon.

   A.   In unmyelinated axons, action potentials are produced fractions of a micrometer apart.

   B.   In myelinated axons, action potentials are produced only at the nodes of Ranvier. This saltatory conduction is faster than conduction in an unmyelinated nerve fiber.

The Synapse   167

I.   Gap junctions are electrical synapses found in cardiac muscle, smooth muscle, and some regions of the brain.

II.   In chemical synapses, neurotransmitters are packaged in synaptic vesicles and released by exocytosis into the synaptic cleft.

   A.   The neurotransmitter can be called the ligand of the receptor.

   B.   Binding of the neurotransmitter to the receptor causes the opening of chemically regulated gates of ion channels.

Acetylcholine as a Neurotransmitter   170

I.   There are two subtypes of ACh receptors: nicotinic and muscarinic.

   A.   Nicotinic receptors enclose membrane channels and open when ACh binds to the receptor. This causes a depolarization called an excitatory postsynaptic potential (EPSP).

   B.   The binding of ACh to muscarinic receptors opens ion channels indirectly, through the action of G-proteins. This can cause a hyperpolarization called an inhibitory postsynaptic potential (IPSP).

   C.   After ACh acts at the synapse, it is inactivated by the enzyme acetylcholinesterase (AChE).

II.   EPSPs are graded and capable of summation. They decrease in amplitude as they are conducted.

III.   ACh is used in the PNS as the neurotransmitter of somatic motor neurons, which stimulate skeletal muscles to contract, and by some autonomic neurons.

IV.   ACh in the CNS produces EPSPs at synapses in the dendrites or cell body. These EPSPs travel to the axon hillock, stimulate opening of voltage-regulated channels, and generate action potentials in the axon.

Monoamines as Neurotransmitters   176

I.   Monoamines include serotonin, dopamine, norepinephrine, and epinephrine. The last three are included in the subcategory known as catecholamines.

  A. These neurotransmitters are inactivated after being released, primarily by reuptake into the presynaptic nerve endings.

  B. Catecholamines may activate adenylate cyclase in the postsynaptic cell, which catalyzes the formation of cyclic AMP.

II. Dopaminergic neurons (those that use dopamine as a neurotransmitter) are implicated in the development of Parkinson's disease and schizophrenia. Norepinephrine is used as a neurotransmitter by sympathetic neurons in the PNS and by some neurons in the CNS.

Other Neurotransmitters 179

I. The amino acids glutamate and aspartate are excitatory in the CNS.

  A. The subclass of glutamate receptor designated as NMDA receptors are implicated in learning and memory.

  B. The amino acids glycine and GABA are inhibitory. They produce hyperpolarizations, causing IPSPs by opening Cl$^-$ channels.

II. Numerous polypeptides function as neurotransmitters, including the endogenous opioids.

III. Nitric oxide functions as both a local tissue regulator and a neurotransmitter in the PNS and CNS. It promotes smooth muscle relaxation and is implicated in memory.

Synaptic Integration 182

I. Spatial and temporal summation of EPSPs allows a depolarization of sufficient magnitude to cause the stimulation of action potentials in the postsynaptic neuron.

  A. IPSPs and EPSPs from different synaptic inputs can summate.

  B. The production of IPSPs is called postsynaptic inhibition.

II. Long-term potentiation is a process that improves synaptic transmission as a result of the use of the synaptic pathway. This process thus may be a mechanism for learning.

Review Activities

Test Your Knowledge of Terms and Facts

1. The supporting cells that form myelin sheaths in the peripheral nervous system are
   a. oligodendrocytes.
   b. satellite cells.
   c. Schwann cells.
   d. astrocytes.
   e. microglia.

2. A collection of neuron cell bodies located outside the CNS is called
   a. a tract.
   b. a nerve.
   c. a nucleus.
   d. a ganglion.

3. Which of these neurons are pseudounipolar?
   a. sensory neurons
   b. somatic motor neurons
   c. neurons in the retina
   d. autonomic motor neurons

4. Depolarization of an axon is produced by
   a. inward diffusion of Na$^+$.
   b. active extrusion of K$^+$.
   c. outward diffusion of K$^+$.
   d. inward active transport of Na$^+$.

5. Repolarization of an axon during an action potential is produced by
   a. inward diffusion of Na$^+$.
   b. active extrusion of K$^+$.
   c. outward diffusion of K$^+$.
   d. inward active transport of Na$^+$.

6. As the strength of a depolarizing stimulus to an axon is increased,
   a. the amplitude of action potentials increases.
   b. the duration of action potentials increases.
   c. the speed with which action potentials are conducted increases.
   d. the frequency with which action potentials are produced increases.

7. The conduction of action potentials in a myelinated nerve fiber is
   a. saltatory.
   b. without decrement.
   c. faster than in an unmyelinated fiber.
   d. all of the these.

8. Which of these is *not* a characteristic of synaptic potentials?
   a. They are all or none in amplitude.
   b. They decrease in amplitude with distance.
   c. They are produced in dendrites and cell bodies.
   d. They are graded in amplitude.
   e. They are produced by chemically regulated gates.

9. Which of these is *not* a characteristic of action potentials?
   a. They are produced by voltage-regulated gates.
   b. They are conducted without decrement.
   c. Na$^+$ and K$^+$ gates open at the same time.
   d. The membrane potential reverses polarity during depolarization.

10. A drug that inactivates acetylcholinesterase
   a. inhibits the release of ACh from presynaptic endings.
   b. inhibits the attachment of ACh to its receptor protein.
   c. increases the ability of ACh to stimulate muscle contraction.
   d. does all of the these.

11. Postsynaptic inhibition is produced by
   a. depolarization of the postsynaptic membrane.
   b. hyperpolarization of the postsynaptic membrane.
   c. axoaxonic synapses.
   d. long-term potentiation.

12. Hyperpolarization of the postsynaptic membrane in response to glycine or GABA is produced by the opening of
    a. Na$^+$ channels.
    b. K$^+$ channels.
    c. Ca^{2+} channels.
    d. Cl$^-$ channels.

13. The absolute refractory period of a neuron
    a. is due to the high negative polarity of the inside of the neuron.
    b. occurs only during the repolarization phase.
    c. occurs only during the depolarization phase.
    d. occurs during depolarization and the first part of the repolarization phase.

14. Which of these statements about catecholamines is *false?*
    a. They include norepinephrine, epinephrine, and dopamine.

b. Their effects are increased by action of the enzyme catechol-O-methyltransferase.
    c. They are inactivated by monoamine oxidase.
    d. They are inactivated by reuptake into the presynaptic axon.
    e. They may stimulate the production of cyclic AMP in the postsynaptic axon.

15. The summation of EPSPs from numerous presynaptic nerve fibers converging onto one postsynaptic neuron is called
    a. spatial summation.
    b. long-term potentiation.
    c. temporal summation.
    d. synaptic plasticity.

16. Which of these statements about ACh receptors is *false?*
    a. Skeletal muscles contain nicotinic ACh receptors.
    b. The heart contains muscarinic ACh receptors.

c. G-proteins are needed to open ion channels for nicotinic receptors.
    d. Stimulation of nicotinic receptors results in the production of EPSPs.

17. Hyperpolarization is caused by all of these neurotransmitters *except*
    a. glutamic acid in the CNS.
    b. ACh in the heart.
    c. glycine in the spinal cord.
    d. GABA in the brain.

18. Which of these may be produced by the action of nitric oxide?
    a. dilation of blood vessels
    b. erection of the penis
    c. relaxation of smooth muscles in the digestive tract
    d. long-term potentiation (LTP) among neighboring synapses in the brain
    e. all of the these

Test Your Understanding of Concepts and Principles

1. Compare the characteristics of action potentials with those of synaptic potentials.[1]

2. Explain how voltage-regulated channels produce an all-or-none action potential.

3. Explain how action potentials are regenerated along an axon.

4. Explain why conduction in a myelinated axon is faster than in an unmyelinated axon.

5. Describe the structure of nicotinic ACh receptors. Explain how ACh causes the production of an EPSP and relate this

process to the neural stimulation of skeletal muscle contraction.

6. Describe the nature of muscarinic ACh receptors and the function of G-proteins in the action of these receptors. How does stimulation of these receptors cause the production of a hyperpolarization or a depolarization?

7. Trace the course of events in the interval between the production of an EPSP and the generation of action potentials at the axon hillock. Describe

the effect of spatial and temporal summation on this process.

8. Explain how an IPSP is produced and how IPSPs can inhibit activity of the postsynaptic neuron.

9. List the endogenous opioids in the brain and describe some of their proposed functions.

10. Explain what is meant by long-term potentiation and discuss the significance of this process. What may account for LTP and what role might nitric oxide play?

Test Your Ability to Analyze and Apply Your Knowledge

1. Grafting peripheral nerves onto the two parts of a cut spinal cord in rats was found to restore some function in the hind limbs. Apparently, when the white matter of the peripheral nerve was joined to the gray matter of the spinal cord, some regeneration of central neurons occurred across the two spinal cord sections. What

component of the peripheral nerve probably contributed to the regeneration? Discuss the factors that promote and inhibit central neuron regeneration.

2. Discuss the different states of a voltage-gated ion channel and distinguish between these states. How has molecular biology/biochemistry

aided our understanding of the physiology of the voltage-gated channels?

3. Suppose you are provided with an isolated nerve-muscle preparation in order to study synaptic transmission. In one of your experiments, you give this preparation a drug that blocks voltage-regulated Ca$^+$ channels; in another, you

---

[1]*Note:* This question is answered in the chapter 7 Study Guide found on the Online Learning Center at www.mhhe.com/fox8.

give tetanus toxin to the preparation. How will synaptic transmission be affected in each experiment?

4. What functions do G-proteins serve in synaptic transmission? Speculate on the advantages of having G-proteins mediate the effects of a neurotransmitter.

5. Studies indicate that alcoholism may be associated with a particular allele (form of a gene) for the D_2 dopamine receptor. Suggest some scientific investigations that might further explore these possible genetic and physiological relationships.

Related Websites

Check out the Links Library at www.mhhe.com/fox8 for links to sites containing resources related to neurons and synapses. These links are monitored to ensure current URLs.

8

The Central Nervous System

Objectives

After studying this chapter, you should be able to . . .

1. locate the major brain regions and describe the structures within each of these regions.

2. describe the organization of the cerebrum and the primary roles of its lobes.

3. describe the location and functions of the sensory cortex and motor cortex.

4. explain the lateralization of functions in the right and left cerebral hemispheres.

5. describe the structures involved in the control of speech and explain their interrelationships.

6. describe the different types of aphasias that result from damage to specific regions of the brain.

7. describe the structures included in the limbic system and discuss the possible role of this system in emotion.

8. distinguish between different types of memory and describe the roles of different brain regions in memory.

9. describe the location of the thalamus and explain the significance of this region.

10. describe the location of the hypothalamus and explain the significance of this region.

11. describe the structures located in the midbrain and hindbrain, and explain the role of the medulla oblongata in the control of visceral functions.

12. explain how the spinal cord is organized and how ascending and descending tracts are named.

13. describe the origin and pathways of the pyramidal motor tracts and explain the significance of these descending tracts.

14. explain the role of the basal nuclei and cerebellum in motor control via the extrapyramidal system and describe the pathways of this system.

15. describe the structures and pathways involved in a reflex arc.

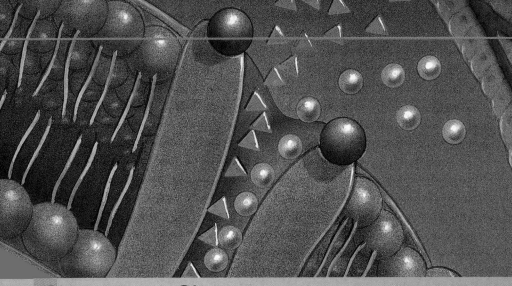

Refresh Your Memory

Before you begin this chapter, you may want to review these concepts from previous chapters:

- Neurons and Supporting Cells   152
- Dopamine as a Neurotransmitter   178
- Synaptic Integration   182

Chapter at a Glance

Take Advantage of the Technology

Visit the Online Learning Center for these additional study resources.

- Interactive quizzing
- Online study guide
- Current news feeds
- Crossword puzzles
- Vocabulary flashcards
- Labeling activities

www.mhhe.com/fox8

Clinical Investigation

Frank, a 72-year-old man, is brought to the hospital by his wife. As he leans on her for support, she explains to the doctor that her husband has suddenly become partially paralyzed and has difficulty speaking. During a neurological exam, the doctor determines that Frank is paralyzed on the right side of his body, but despite this, the doctor is able to elicit a knee-jerk reflex. Frank doesn't voluntarily speak to the doctor, and when questioned, he answers slowly and with great difficulty. His answers, however, are coherent.

Magnetic resonance imaging (MRI) of his brain reveals a blockage of blood flow in the middle cerebral artery. What might explain Frank's symptoms?

ciate appropriate motor responses with sensory stimuli, and thus to maintain homeostasis in the internal environment and the continued existence of the organism in a changing external environment. Further, the central nervous systems of all vertebrates (and most invertebrates) are capable of at least rudimentary forms of learning and memory. This capability—most highly developed in the human brain—permits behavior to be modified by experience and is thus of obvious benefit to survival. Perceptions, learning, memory, emotions, and perhaps even the self-awareness that forms the basis of consciousness, are creations of the brain. Whimsical though it seems, the study of brain physiology is the process of the brain studying itself.

The study of the structure and function of the central nervous system requires a knowledge of its basic "plan," which is established during the course of embryonic development. The early embryo contains an embryonic tissue layer known as *ectoderm* on its surface; this will eventually form the epidermis of the skin, among other structures. As development progresses, a groove appears in this ectoderm along the dorsal midline of the embryo's body. This groove deepens, and by the twentieth day after conception, has fused to form a **neural tube.** The part of the ectoderm where the fusion occurs becomes a separate structure called the **neural crest,** which is located between the neural tube and the surface ectoderm (fig. 8.2). Eventually, the neural tube will become the central nervous system, and the neural crest will become the ganglia of the peripheral nervous system, among other structures.

By the middle of the fourth week after conception, three distinct swellings are evident on the anterior end of the neural tube, which is going to form the brain: the **forebrain** (*prosencephalon*),

Structural Organization of the Brain

The brain is composed of an enormous number of association neurons and accompanying neuroglia, arranged in regions and subdivisions. These neurons receive sensory information, direct the activity of motor neurons, and perform such higher brain functions as learning and memory.

The **central nervous system (CNS),** consisting of the brain and spinal cord (fig. 8.1), receives input from *sensory neurons* and directs the activity of *motor neurons* that innervate muscles and glands. The *association neurons* within the brain and spinal cord are in a position, as their name implies, to asso-

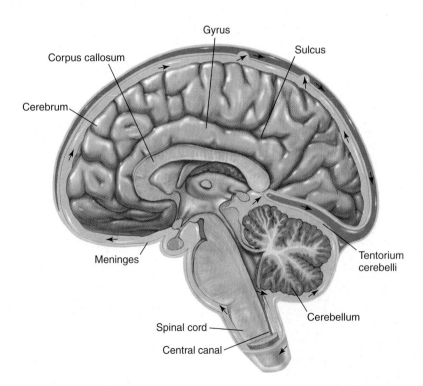

Figure 8.1 **The CNS consists of the brain and the spinal cord.** Both of these structures are covered with meninges and bathed in cerebrospinal fluid.

midbrain (*mesencephalon*), and hindbrain (*rhomben-cephalon*). During the fifth week, these areas become modified to form five regions. The forebrain divides into the *telen-cephalon* and *diencephalon;* the mesencephalon remains unchanged; and the hindbrain divides into the *metencephalon* and *myelencephalon* (fig. 8.3). These regions subsequently become greatly modified, but the terms described here are still used to indicate general regions of the adult brain.

The basic structural plan of the CNS can now be understood. The telencephalon (refer to fig. 8.3) grows disproportion-

ately in humans, forming the two enormous hemispheres of the *cerebrum* that cover the diencephalon, the midbrain, and a portion of the hindbrain. Also, notice that the CNS begins as a hollow tube, and indeed remains hollow as the brain regions are formed. The cavities of the brain are known as *ventricles* and become filled with cerebrospinal fluid (CSF). The cavity of the spinal cord is called the *central canal,* and is also filled with CSF (fig. 8.4).

The CNS is composed of gray and white matter, as described in chapter 7. The gray matter, consisting of neuron cell bodies and dendrites, is found in the *cortex* (surface layer) of the

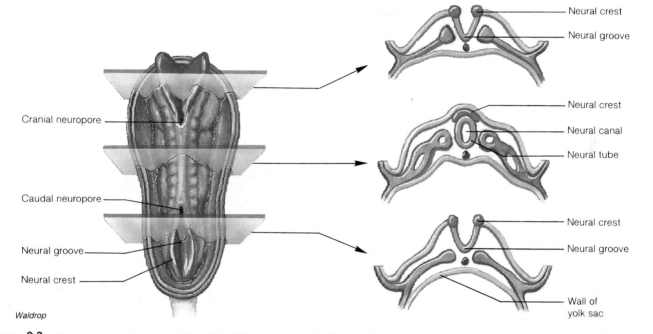

Figure 8.2 **Embryonic development of the CNS.** This dorsal view of a 22-day-old embryo shows transverse sections at three levels of the developing central nervous system.

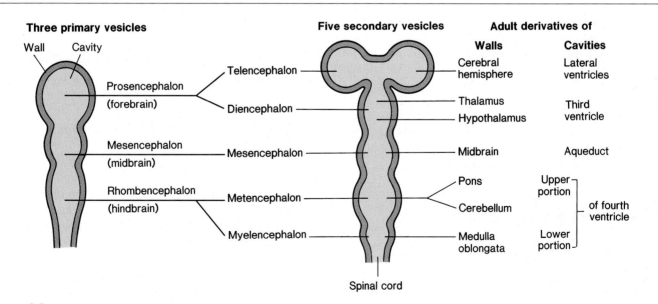

Figure 8.3 **The developmental sequence of the brain.** (*a*) During the fourth week, three principal regions of the brain are formed. (*b*) During the fifth week, a five-regioned brain develops and specific structures begin to form.

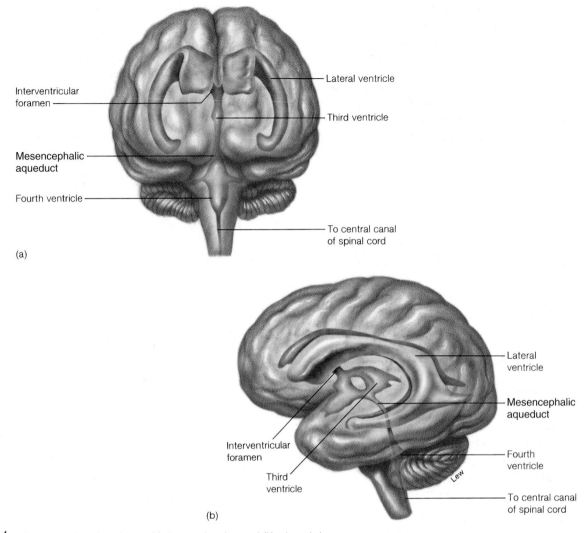

Interventricular foramen

Mesencephalic aqueduct

Fourth ventricle

Lateral ventricle

Third ventricle

To central canal of spinal cord

(a)

Lateral ventricle

Mesencephalic aqueduct

Fourth ventricle

To central canal of spinal cord

Interventricular foramen

Third ventricle

(b)

Figure 8.4 **The ventricles of the brain.** (*a*) An anterior view and (*b*) a lateral view.

brain and deeper within the brain in aggregations known as *nuclei*. White matter consists of axon tracts (the myelin sheaths produce the white color) that underlie the cortex and surround the nuclei. The adult brain contains an estimated 100 billion (10^{11}) neurons, weighs approximately 1.5 kg (3 to 3.5 lb), and receives about 20% of the total blood flow to the body per minute. This high rate of blood flow is a consequence of the high metabolic requirements of the brain; it is not, as Aristotle believed, because the brain's function is to cool the blood. (This fanciful notion—completely incorrect—is a striking example of prescientific thought, having no basis in experimental evidence.)

Test Yourself Before You Continue

1. Identify the three brain regions formed by the middle of the fourth week of gestation and the five brain regions formed during the fifth week.
2. Describe the embryonic origin of the brain ventricles. Where are they located and what do they contain?

Cerebrum

The cerebrum, consisting of five paired lobes within two convoluted hemispheres, contains gray matter in its cortex and in deeper cerebral nuclei. Most of what are considered to be the higher functions of the brain are performed by the cerebrum.

The **cerebrum** (fig. 8.5), which is the only structure of the telencephalon, is the largest portion of the brain (accounting for about 80% of its mass) and is the brain region primarily responsible for higher mental functions. The cerebrum consists of *right* and *left hemispheres,* which are connected internally by a large fiber tract called the *corpus callosum* (see fig. 8.1). The corpus callosum is the major tract of axons that functionally interconnects the right and left cerebral hemispheres.

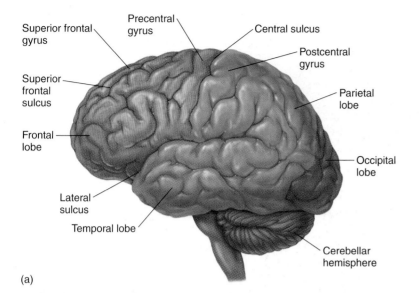

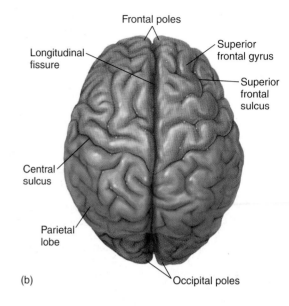

Figure 8.5 **The cerebrum.** (*a*) A lateral view and (*b*) a superior view.

Cerebral Cortex

The cerebrum consists of an outer **cerebral cortex,** composed of 2 to 4 mm of gray matter and underlying white matter. The cerebral cortex is characterized by numerous folds and grooves called *convolutions.* The elevated folds of the convolutions are called *gyri,* and the depressed grooves are the *sulci.* Each cerebral hemisphere is subdivided by deep sulci, or *fissures,* into five lobes, four of which are visible from the surface (fig. 8.6). These lobes are the *frontal, parietal, temporal,* and *occipital,* which are visible from the surface, and the deep *insula,* which is covered by portions of the frontal, parietal, and temporal lobes (table 8.1).

The **frontal lobe** is the anterior portion of each cerebral hemisphere. A deep fissure, called the *central sulcus,* separates the frontal lobe from the **parietal lobe.** The *precentral gyrus* (figs. 8.5 and 8.6), involved in motor control, is located in the frontal lobe, just in front of the central sulcus. The neuron cell bodies located here are called *upper motor neurons,* because of

their role in muscle regulation (chapter 12). The *postcentral gyrus,* which is located just behind the central sulcus in the parietal lobe, is the primary area of the cortex responsible for the perception of *somatesthetic sensation*—sensation arising from cutaneous, muscle, tendon, and joint receptors. This neural pathway is described in chapter 10.

The precentral (motor) and postcentral (sensory) gyri have been mapped in conscious patients undergoing brain surgery. Electrical stimulation of specific areas of the precentral gyrus causes specific movements, and stimulation of different areas of the postcentral gyrus evokes sensations in specific parts of the body. Typical maps of these regions (fig. 8.7) show an upside-down picture of the body, with the superior regions of cortex devoted to the toes and the inferior regions devoted to the head.

A striking feature of these maps is that the areas of cortex responsible for different parts of the body do not correspond to the size of the body parts being served. Instead, the body regions with the highest densities of receptors are represented by the largest areas of the sensory cortex, and the body regions with the greatest number of motor innervations are represented by the largest areas of motor cortex. The hands and face, therefore, which have a high density of sensory receptors and motor innervation, are served by larger areas of the precentral and postcentral gyri than is the rest of the body.

The **temporal lobe** contains auditory centers that receive sensory fibers from the cochlea of each ear. This lobe is also involved in the interpretation and association of auditory and visual information. The **occipital lobe** is the primary area responsible for vision and for the coordination of eye movements. The functions of the temporal and occipital lobes will be considered in more detail in chapter 10, in conjunction with the physiology of hearing and vision.

The **insula** is implicated in memory encoding and in the integration of sensory information (principally pain) with visceral responses. In particular, the insula seems to be involved in coordinating the cardiovascular responses to stress.

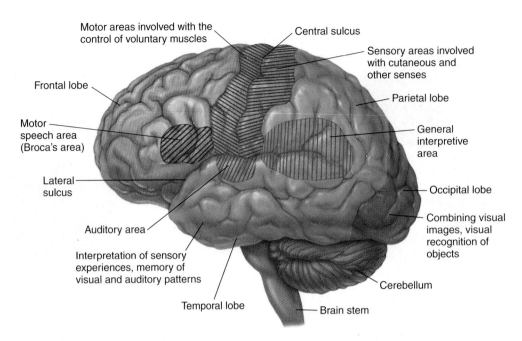

Figure 8.6   **The lobes of the left cerebral hemisphere.** This diagram shows the principal motor and sensory areas of the cerebral cortex.

Table 8.1   Functions of the Cerebral Lobes

Lobe	Functions
Frontal	Voluntary motor control of skeletal muscles; personality; higher intellectual processes (e.g., concentration, planning, and decision making); verbal communication
Parietal	Somatesthetic interpretation (e.g., cutaneous and muscular sensations); understanding speech and formulating words to express thoughts and emotions; interpretation of textures and shapes
Temporal	Interpretation of auditory sensations; storage (memory) of auditory and visual experiences
Occipital	Integration of movements in focusing the eye; correlation of visual images with previous visual experiences and other sensory stimuli; conscious perception of vision
Insula	Memory; sensory (principally pain) and visceral integration

People with **Alzheimer's disease** have (1) a loss of neurons; (2) an accumulation of intracellular proteins forming *neurofibrillar tangles;* and (3) an accumulation of extracellular protein deposits called *amyloid plaques.* The major constituent of the plaques is a polypeptide called *amyloid β-peptide (Aβ).* Aβ is formed by cleavage of a precursor protein by an enzyme called *secretase.* One isoform of the enzyme, γ-secretase, is activated by *presenilin* proteins, which are defective in some people with an inherited type of Alzheimer's. The structure of another isoform of the enzyme, β-secretase, has recently been characterized. Scientists hope that this will help them to develop a drug that will block secretase action and perhaps thereby slow the progression of Alzheimer's disease.

Visualizing the Brain

Several relatively new imaging techniques permit the brains of living people to be observed in detail for medical and research purposes. The first of these to be developed was **x-ray computed tomography (CT).** CT involves complex computer manipulation of data obtained from x-ray absorption by tissues of different densities. Using this technique, soft tissues such as the brain can be observed at different depths.

The next technique to be developed was **positron-emission tomography (PET).** In this technique, radioisotopes that emit positrons are injected into the bloodstream. Positrons are like electrons but carry a positive charge. The collision of a positron and an electron results in their mutual annihilation and the emission of gamma rays, which can be detected and used to pinpoint brain cells that are most active. Scientists have used PET to study brain metabolism, drug distribution in the brain, and changes in blood flow as a result of brain activity.

A newer technique for visualizing the living brain is **magnetic resonance imaging (MRI).** This technique is based on the concept that protons (H^+) respond to a magnetic field. The magnetic field is used to align the protons, which emit a detectable radio-wave signal when appropriately stimulated. With this technique, excellent images can be obtained (figs. 8.8 and 8.9) without subjecting the person to any known danger. Scientists are now using MRI together with other techniques to study the function of the brain (see fig. 8.8) in a technique called *functional magnetic resonance imaging (fMRI).* Various techniques for visualizing the functioning brain are summarized in table 8.2.

Electroencephalogram

The synaptic potentials (discussed in chapter 7) produced at the cell bodies and dendrites of the cerebral cortex create electrical currents that can be measured by electrodes placed on the scalp. A

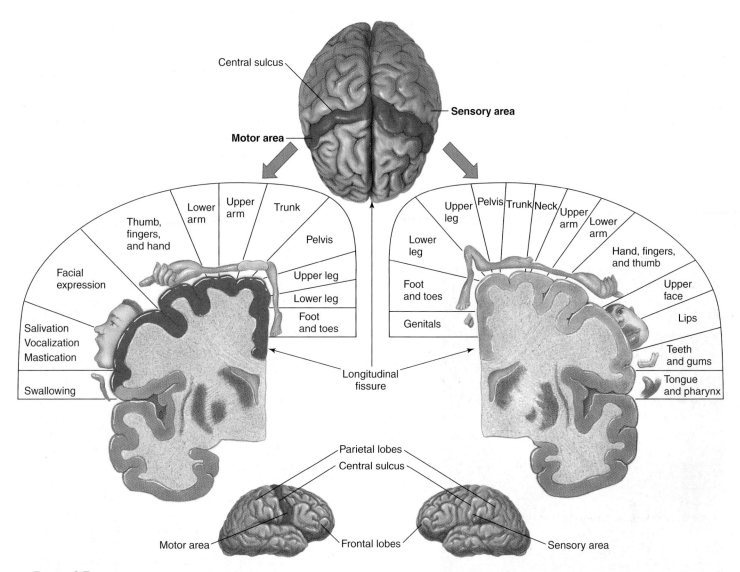

Figure 8.7 **Motor and sensory areas of the cerebral cortex.** (*a*) Motor areas that control skeletal muscles and (*b*) sensory areas that receive somatesthetic sensations.

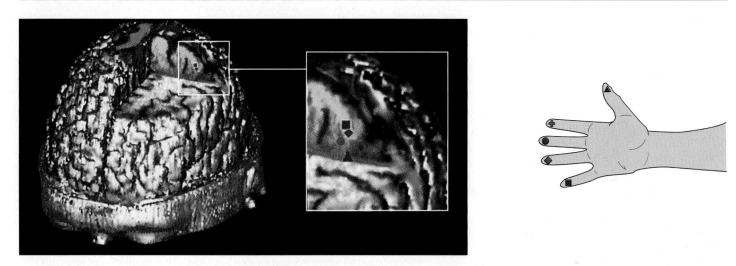

Figure 8.8 **An MRI image of the brain reveals the sensory cortex.** The integration of MRI and EEG information shows the location on the sensory cortex that corresponds to each of the digits of the hand.

record of these electrical currents is called an **electroencephalogram,** or **EEG.** Deviations from normal EEG patterns can be used clinically to diagnose epilepsy and other abnormal states, and the absence of an EEG can be used to signify brain death.

There are normally four types of EEG patterns (fig. 8.10). **Alpha waves** are best recorded from the parietal and occipital regions while a person is awake and relaxed but with the eyes closed. These waves are rhythmic oscillations of 10 to 12 cycles/second. The alpha rhythm of a child under the age of 8 occurs at a slightly lower frequency of 4 to 7 cycles/second.

Beta waves are strongest from the frontal lobes, especially the area near the precentral gyrus. These waves are produced by visual stimuli and mental activity. Because they respond to stimuli from receptors and are superimposed on the continuous activity patterns, they constitute *evoked activity.* Beta waves occur at a frequency of 13 to 25 cycles per second.

Theta waves are emitted from the temporal and occipital lobes. They have a frequency of 5 to 8 cycles/second and are common in newborn infants. The recording of theta waves in adults generally indicates severe emotional stress and can be a forewarning of a nervous breakdown.

Delta waves are seemingly emitted in a general pattern from the cerebral cortex. These waves have a frequency of 1 to 5 cycles/second and are common during sleep and in an awake

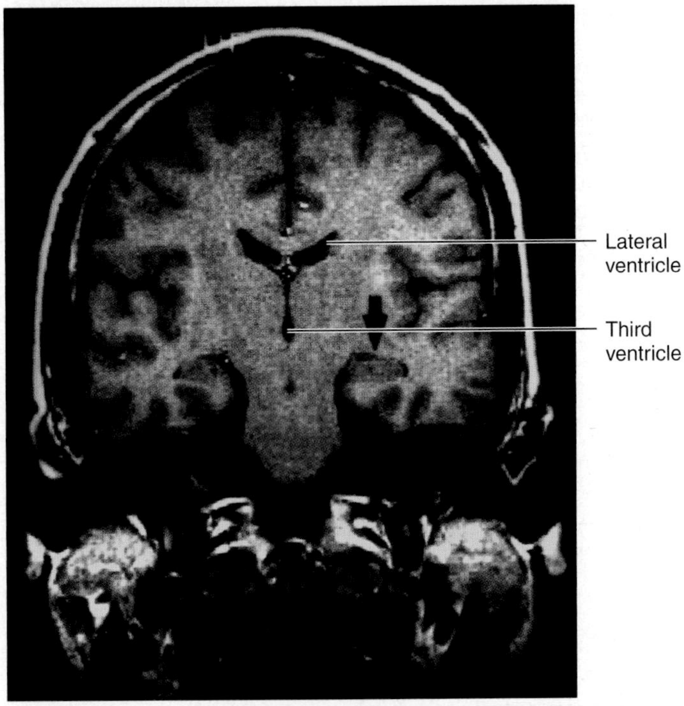

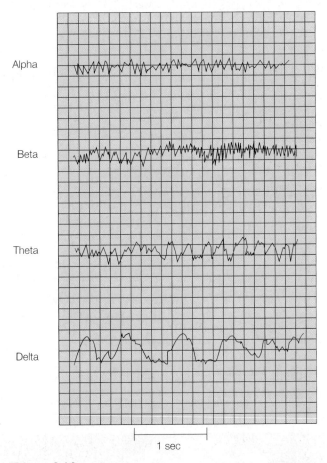

■ Figure 8.9 **An MRI scan of a normal brain.** In this coronal view of the brain, the lateral and third ventricles can be clearly seen. The arrow points to a part of the hippocampus.

From W. T. Carpenter and R. W. Buchanan, "Medical Progress: Schizophrenia" in *New England Journal of Medicine,* 330:685, 1994, fig 1A.

■ Figure 8.10 **Different types of waves in an electroencephalogram (EEG).** Notice that the delta waves (bottom) have the highest amplitude and lowest frequency.

Table 8.2 Techniques for Visualizing Brain Function

Abbreviation	Technique Name	Principle Behind Technique
EEG	Electroencephalogram	Neuronal activity is measured as maps with scalp electrodes.
fMRI	Functional magnetic resonance imaging	Increased neuronal activity increases cerebral blood flow and oxygen consumption in local areas. This is detected by effects of changes in blood oxyhemoglobin/deoxyhemoglobin ratios.
MEG	Magnetoencephalogram	Neuronal magnetic activity is measured using magnetic coils and mathematical plots.
PET	Positron emission tomography	Increased neuronal activity increases cerebral blood flow and metabolite consumption in local areas. This is measured using radioactively labeled deoxyglucose.
SPECT	Single photon emission computed tomography	Increased neuronal activity increases cerebral blood flow. This is measured using emitters of single photons, such as technetium.

Source: Burkhart Bromm "Brain images of pain." *News in Physiological Sciences 16* (Feb. 2001): 244–249.

infant. The presence of delta waves in an awake adult indicates brain damage.

Two different types of EEG patterns are seen during sleep, corresponding to the two phases of sleep: **rapid eye movement (REM)** sleep, when dreams occur, and **non-REM,** or **resting,** sleep. During non-REM sleep the EEG displays large, slow delta waves (high amplitude, low-frequency waves). Superimposed on these are *sleep spindles,* which are waxing and waning bursts of 7 to 14 cycles per second that last for 1 to 3-second periods. During REM sleep, when the eyes move about rapidly, the EEG waves are similar to that of wakefulness. That is, they are lower in amplitude and display high-frequency oscillations.

Basal Nuclei

The **basal nuclei** (or **basal ganglia**) are masses of gray matter composed of neuron cell bodies located deep within the white matter of the cerebrum (fig. 8.11). The most prominent of the basal nuclei is the **corpus striatum,** which consists of several masses of nuclei (a *nucleus* is a collection of cell bodies in the CNS). The upper mass, called the *caudate nucleus,* is separated from two lower masses, collectively called the *lentiform nucleus.* The lentiform nucleus consists of a lateral portion, the *putamen,* and a medial portion, the *globus pallidus.* The basal nuclei function in the control of voluntary movements.

Degeneration of the caudate nucleus (as in *Huntington's disease*) produces **chorea**—a hyperkinetic disorder characterized by rapid, uncontrolled, jerky movements. Degeneration of dopaminergic neurons to the caudate nucleus from the substantia nigra, a small nucleus in the midbrain, produces most of the symptoms of **Parkinson's disease.** As discussed in chapter 7, this disease is associated with rigidity, resting tremor, and difficulty in initiating voluntary movements.

Cerebral Lateralization

By way of motor fibers originating in the precentral gyrus, each cerebral cortex controls movements of the contralateral (opposite) side of the body. At the same time, somatesthetic sensation from each side of the body projects to the contralateral postcentral gyrus as a result of *decussation* (crossing over) of fibers. In a similar manner, images falling in the left half of each retina project to the right occipital lobe, and images in the right half of each retina project to the left occipital lobe. Each cerebral hemisphere, however, receives information from both sides of the body because the two hemispheres communicate with each other via the **corpus callosum,** a large tract composed of about 200 million fibers.

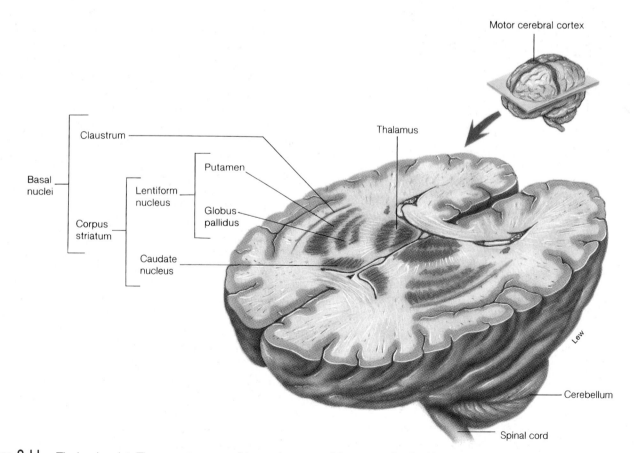

Figure 8.11 **The basal nuclei.** These are structures of the cerebrum containing neurons involved in the control of skeletal muscles (higher motor neurons). The thalamus is a relay center between the motor cerebral cortex and other brain areas.

The corpus callosum has been surgically cut in some people with severe epilepsy as a way of alleviating their symptoms. These *split-brain procedures* isolate each hemisphere from the other, but, surprisingly, to a casual observer split-brain patients do not show evidence of disability as a result of the surgery. However, in specially designed experiments in which each hemisphere is separately presented with sensory images and the patient is asked to perform tasks (speech or writing or drawing with the contralateral hand), it has been learned that each hemisphere is good at certain categories of tasks and poor at others (fig. 8.12).

In a typical experiment, the image of an object may be presented to either the right or left hemisphere (by presenting it to either the left or right visual field only) and the person may be asked to name the object. Findings indicate that, in most people, the task can be performed successfully by the left hemisphere but not by the right. Similar experiments have shown that the left hemisphere is generally the one in which most of the language and analytical abilities reside.

Clinical Investigation Clues

Remember that Frank had paralysis of the right side of his body and suffered speech impairment.

What is the most likely explanation for the paralysis on the right side of his body?

How does this relate to his speech impairment?

These findings have led to the concept of **cerebral dominance,** which is analogous to the concept of handedness—people generally have greater motor competence with one hand than with the other. Since most people are right-handed, and the right hand is also controlled by the left hemisphere, the left hemisphere was naturally considered to be the dominant hemisphere in most people. Further experiments have shown, however, that the right hemisphere is specialized along different, less obvious lines—rather than one hemisphere being dominant and the other subordinate, the two hemispheres appear to have complementary functions. The term **cerebral lateralization,** or specialization of function in one hemisphere or the other, is thus now preferred to the term *cerebral dominance,* although both terms are currently used.

Experiments have shown that the right hemisphere does have limited verbal ability; more noteworthy is the observation that the right hemisphere is most adept at *visuospatial tasks.* The right hemisphere, for example, can recognize faces better than the left, but it cannot describe facial appearances as well as the left. Acting through its control of the left hand, the right hemisphere is better than the left (controlling the right hand) at arranging blocks or drawing cubes. Patients with damage to the right hemisphere, as might be predicted from the results of split-brain research, have difficulty finding their way around a house and reading maps.

Perhaps as a result of the role of the right hemisphere in the comprehension of patterns and part-whole relationships, the ability to compose music, but not to critically understand it, appears to depend on the right hemisphere. Interestingly, damage to the left hemisphere may cause severe speech problems while leaving the ability to sing unaffected.

The lateralization of functions just described—with the left hemisphere specialized for language and analytical ability, and the right hemisphere specialized for visuospatial ability—is true for 97% of all people. It is true for all right-handers (who account for 90% of all people) and for 70% of all left-handers. The remaining left-handers are split about equally into those who have language-analytical ability in the right hemisphere and those in whom this ability is present in both hemispheres.

It is interesting to speculate that the creative ability of a person may be related to the interaction of information between the right and left hemispheres. The finding of one study—that the number of left-handers among college art students is disproportionately higher than the number of left-handers in the general population—suggests that this interaction may be greater in left-handed people. The observation that Leonardo da Vinci and Michelangelo were both left-handed is interesting in this regard, but obviously does not constitute scientific proof of any hypothesis.

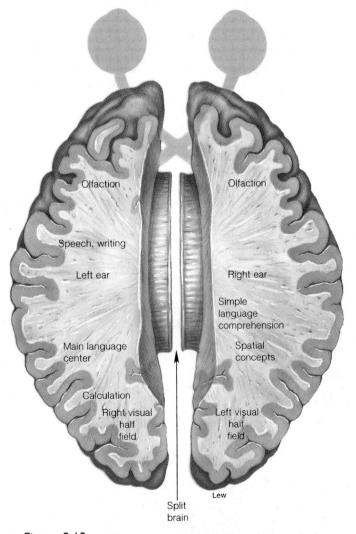

■ Figure 8.12   Different functions of the right and left cerebral hemispheres. These differences were revealed by experiments with people whose corpus callosum—the tract connecting the two hemispheres—was surgically split.

Further research on the lateralization of function of the cerebral hemispheres may reveal much more about brain function and the creative process.

Language

Knowledge of the brain regions involved in language has been gained primarily by the study of *aphasias*—speech and language disorders caused by damage to the brain through head injury or stroke. In most people, the language areas of the brain are primarily located in the left hemisphere of the cerebral cortex, as previously described. Even in the nineteenth century, two areas of the cortex—Broca's area and Wernicke's area (fig. 8.13)—were recognized as areas of particular importance in the production of aphasias.

Broca's aphasia is the result of damage to **Broca's area,** located in the left inferior frontal gyrus and surrounding areas. Common symptoms include weakness in the right arm and the right side of the face. People with Broca's aphasia are reluctant to speak, and when they try, their speech is slow and poorly articulated. Their comprehension of speech is unimpaired, however. People with this aphasia can understand a sentence but have difficulty repeating it. It should be noted that this is not simply due to a problem in motor control, since the neural control over the musculature of the tongue, lips, larynx, and so on is unaffected.

Wernicke's aphasia is caused by damage to **Wernicke's area,** located in the superior temporal gyrus of the left hemisphere (in most people). This results in speech that is rapid and fluid but without meaning. People with Wernicke's aphasia produce speech that has been described as a "word salad." The words used may be real words that are chaotically mixed together, or they may be made-up words. Language comprehension is destroyed; people with Wernicke's aphasia cannot understand either spoken or written language.

It appears that the concept of words originates in Wernicke's area. Thus, in order to understand words that are read, information from the visual cortex (in the occipital lobe) must project to Wernicke's area. Similarly, in order to understand spoken words, the auditory cortex (in the temporal lobe) must send information to Wernicke's area.

To speak intelligibly, the concept of words originating in Wernike's area must be communicated to Broca's area; this is accomplished by a fiber tract called the **arcuate fasciculus.** Broca's area, in turn, sends fibers to the motor cortex (precentral gyrus), which directly controls the musculature of speech. Damage to the arcuate fasciculus produces *conduction aphasia,* which is fluent but nonsensical speech as in Wernicke's aphasia, even though both Broca's and Wernicke's areas are intact.

The **angular gyrus,** located at the junction of the parietal, temporal, and occipital lobes, is believed to be a center for the integration of auditory, visual, and somatesthetic information. Damage to the angular gyrus produces aphasias, which suggests that this area projects to Wernicke's area. Some patients with damage to the left angular gyrus can speak and understand spoken language but cannot read or write. Other patients can write a sentence but cannot read it, presumably because of damage to the projections from the occipital lobe (involved in vision) to the angular gyrus.

Clinical Investigation Clues

Remember that Frank had difficulty speaking, but his speech was coherent.

Which type of aphasia did Frank most likely have?

Which part of the brain sustained the damage?

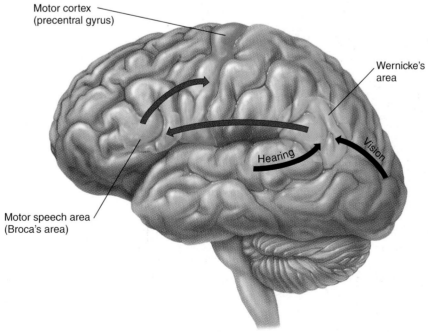

Motor cortex
(precentral gyrus)

Wernicke's area

Hearing

Vision

Motor speech area
(Broca's area)

■ **Figure 8.13** **Brain areas involved in the control of speech.** Damage to these areas produces speech deficits, known as aphasias. Wernicke's area, required for language comprehension, receives information from many areas of the brain, including the auditory cortex (for heard words), the visual cortex (for read words), and other brain areas. In order for a person to be able to speak intelligibly, Wernicke's area must send messages to Broca's area, which controls the motor aspects of speech by way of its input to the motor cortex.

Emotion and Motivation

The parts of the brain that appear to be of paramount importance in the neural basis of emotional states are the hypothalamus (in the diencephalon) and the **limbic system.** The limbic system consists of a group of forebrain nuclei and fiber tracts that form a ring around the brain stem (*limbus* = ring). Among the components of the limbic system are the *cingulate gyrus* (part of the cerebral cortex), the *amygdaloid nucleus* (or *amygdala*), the *hippocampus,* and the *septal nuclei* (fig. 8.14).

The limbic system was once called the *rhinencephalon,* or "smell brain," because it is involved in the central processing of olfactory information. This may be its primary function in lower vertebrates, whose limbic system may constitute the entire forebrain. It is now known however, that the limbic system in humans is a center for basic emotional drives. The limbic system was derived early in the course of vertebrate evolution, and its tissue is phylogenetically older than the cerebral cortex. There are thus few synaptic connections between the cerebral cortex and the structures of the limbic system, which perhaps helps to explain why we have so little conscious control over our emotions.

There is a closed circuit of information flow between the limbic system and the thalamus and hypothalamus (fig. 8.14) called the *Papez circuit.* (The thalamus and hypothalamus are part of the diencephalon, described in a later section.) In the Papez circuit, a fiber tract, the *fornix,* connects the hippocampus to the mammillary bodies of the hypothalamus, which in turn project to the anterior nuclei of the thalamus. The thalamic nuclei, in turn, send fibers to the cingulate gyrus, which then completes the circuit by sending fibers to the hippocampus. Through these interconnections, the limbic system and the hypothalamus appear to cooperate in the neural basis of emotional states.

Studies of the functions of these regions include electrical stimulation of specific locations, destruction of tissue (producing *lesions*) in particular sites, and surgical removal, or *ablation,* of specific structures. These studies suggest that the hypothalamus and limbic system are involved in the following feelings and behaviors:

1. **Aggression.** Stimulation of certain areas of the amygdala produces rage and aggression, and lesions of the amygdala can produce docility in experimental animals. Stimulation of particular areas of the hypothalamus can produce similar effects.

2. **Fear.** Fear can be produced by electrical stimulation of the amygdala and hypothalamus, and surgical removal of the limbic system can result in an absence of fear. Monkeys are normally terrified of snakes, for example, but they will handle snakes without fear if their limbic system is removed. Humans with damage to their amygdala have demonstrated an impaired ability to recognize facial expressions of fear and anger.

3. **Feeding.** The hypothalamus contains both a *feeding center* and a *satiety center.* Electrical stimulation of the former

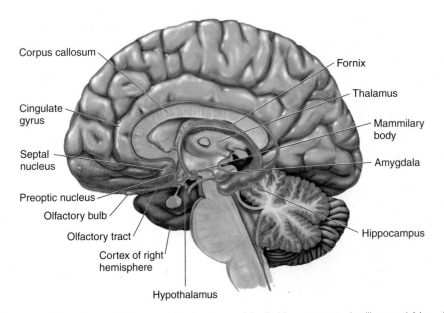

Corpus callosum

Cingulate gyrus

Septal nucleus

Preoptic nucleus

Olfactory bulb

Olfactory tract

Cortex of right hemisphere

Hypothalamus

Fornix

Thalamus

Mammilary body

Amygdala

Hippocampus

■ **Figure 8.14**   **The limbic system.** The pathways that connect the structures of the limbic system are also illustrated. Note that the left temporal lobe of the cerebral cortex has been removed to make these structures visible.

causes overeating, and stimulation of the latter will stop feeding behavior in experimental animals.

4. **Sex.** The hypothalamus and limbic system are involved in the regulation of the sexual drive and sexual behavior, as shown by stimulation and ablation studies in experimental animals. The cerebral cortex, however, is also critically important for the sex drive in lower animals, and the role of the cerebrum is even more important for the sex drive in humans.

5. **Goal-directed behavior (reward and punishment system).** Electrodes placed in particular sites between the frontal cortex and the hypothalamus can deliver shocks that function as a reward. In rats, this reward is more powerful than food or sex in motivating behavior. Similar studies have been done in humans, who report feelings of relaxation and relief from tension, but not of ecstasy. Electrodes placed in slightly different positions apparently stimulate a punishment system in experimental animals, who stop their behavior when stimulated in these regions.

One of the most dramatic examples of the role of higher brain areas in personality and emotion is the famous crowbar accident of 1848. A 25-year-old railroad foreman, Phineas P. Gage, was tamping blasting powder into a hole in a rock with a metal rod when the blasting powder suddenly exploded. The rod—three feet, seven inches long and one and one-fourth inches thick—was driven above his left eye and through his brain, finally emerging through the top of his skull.

After a few minutes of convulsions, Gage got up, rode a horse three-quarters of a mile into town, and walked up a long flight of stairs to see a doctor. He recovered well, with no noticeable sensory or motor deficits. His associates, however, noted striking personality changes. Before the accident Gage was a responsible, capable, and financially prudent man. Afterward, he appeared to have lost his social inhibitions, engaging, for example, in gross profanity (which he had never done before the accident). He also seemed to be tossed about by chance whims. He was eventually fired from his job, and his old friends remarked that he was "no longer Gage."

Memory

Brain Regions in Memory

Clinical studies of *amnesia* (loss of memory) suggest that several different brain regions are involved in memory storage and retrieval. Amnesia has been found to result from damage to the temporal lobe of the cerebral cortex, the hippocampus, the head of the caudate nucleus (in Huntington's disease), or the dorsomedial thalamus (in alcoholics suffering from Korsakoff's syndrome with thiamine deficiency). A number of researchers now believe that there are several different systems of information storage in the brain. One system relates to the simple learning of stimulus-response that even invertebrates can do to some degree. This, together with skill learning and different kinds of conditioning and habits, are retained in people with amnesia.

People with amnesia have an impaired ability to remember facts and events, which some scientists have called "declarative memory." This system of memory can be divided into two major categories: **short-term memory** and **long-term memory.** People with head trauma, for example, and patients who undergo *electroconvulsive shock (ECS) therapy* may lose their memory of recent events but retain their older memories. Recent evidence suggests that the consolidation of long-term memory requires the activation of genes, leading to altered protein synthesis and synaptic connections. The consolidation of short-term memory into long-term memory is the function of the **medial temporal lobe,** an area that includes the hippocampus, amygdaloid nucleus, and adjacent areas of the cerebral cortex (fig. 8.14). Once the memory is put into long-term storage, however, it is independent of the medial temporal lobe.

Using functional magnetic resonance imaging (fMRI) of subjects asked to remember words, scientists detected more brain activity in the left medial temporal lobe and left frontal lobe for words that were remembered compared to words that were subsequently forgotten. When pictures of scenes rather than words were used, the scenes that were remembered evoked more fMRI activity in left and right medial temporal lobes and right frontal lobe compared to that evoked by scenes that were subsequently forgotten. The increased fMRI activity in these brain regions seems to indicate the encoding of the memories. Indeed, lesions of the left medial temporal lobe impairs verbal memory, while lesions of the right medial temporal lobe impairs nonverbal memories, such as the ability to remember faces.

Surgical removal of the right and left medial temporal lobes was performed in one patient, designated "H.M.," in an effort to treat his epilepsy. After the surgery he was unable to consolidate any short-term memory. He could repeat a phone number and carry out a normal conversation; he could not remember the phone number if momentarily distracted, however, and if the person to whom he was talking left the room and came back a few minutes later, H.M. would have no recollection of seeing that person or of having had a conversation with that person before. Although his memory of events that occurred before the operation was intact, all subsequent events in his life seemed as if they were happening for the first time.

The effects of bilateral removal of H.M.'s medial temporal lobes were due to the fact that the hippocampus and amygdaloid nucleus (fig. 8.14) were also removed in the process. Surgical removal of the left medial temporal lobe impairs the consolidation of short-term verbal memories into long-term memory, and removal of the right medial temporal lobe impairs the consolidation of nonverbal memories.

On the basis of additional clinical experience, it appears that the **hippocampus** is a critical component of the memory system. Magnetic resonance imaging (MRI) reveals that the hippocampus is often shrunken in living amnesic patients. However, the degree of memory impairment is increased when other structures, as well as the hippocampus, are damaged. The hippocampus and associated structures of the medial temporal lobe are thus needed for the acquisition of new information about facts and events, and for the consolidation of short-term into long-term memory, which is stored in the cerebral cortex. Emotional arousal, acting via the structures of the limbic system, can enhance or inhibit long-term memory storage. The **amygdala** appears to be particularly important in the memory of fear responses. Studies demonstrate increased neural activity of the human amygdala during visual processing of fearful faces, and patients with bilateral damage to the amygdala were unable to read danger when shown threatening pictures.

The cerebral cortex is thought to store factual information, with verbal memories lateralized to the left hemisphere and visuospatial information to the right hemisphere. The neurosurgeon Wilder Penfield was the first to electrically stimulate various brain regions of awake patients, often evoking visual or auditory memories that were extremely vivid. Electrical stimulation of specific points in the temporal lobe evoked specific memories so detailed that the patients felt as if they were reliving the experience. The medial regions of the temporal lobes, however, cannot be the site where long-term memory is stored, since destruction of these areas in patients being treated for epilepsy did not destroy the memory of events prior to the surgery. The **inferior temporal lobes,** on the other hand, do appear to be sites for the storage of long-term visual memories.

The **left inferior frontal lobe** has recently been shown to participate in performing exact mathematical calculations. Scientists have speculated that this brain region may be involved because it stores verbally coded facts about numbers. Using fMRI, researchers have recently demonstrated that complex, problem-solving and planning activities involve the most anterior portion of the frontal lobes, an area called the **prefrontal cortex.** There is evidence that signals are sent from the prefrontal cortex to the inferior temporal lobes, where visual long-term memories are stored. Lesions of the prefrontal cortex interfere with memory in a less dramatic way than lesions of the medial temporal lobe.

The amount of memory destroyed by ablation (removal) of brain tissue seems to depend more on the amount of brain tissue removed than on the location of the surgery. On the basis of these observations, it was formerly believed that the memory was diffusely located in the brain; stimulation of the correct location of the cortex then retrieved the memory. According to current thinking, however, particular aspects of the memory—visual, auditory, olfactory, spatial, and so on—are stored in particular areas, and the cooperation of all of these areas is required to elicit the complete memory.

Synaptic Changes in Memory

Since long-term memory is not destroyed by electroconvulsive shock, it seems reasonable to conclude that the consolidation of memory depends on relatively permanent changes in the chemical structure of neurons and their synapses. Experiments suggest that protein synthesis is required for the consolidation of the "memory trace." The nature of the synaptic changes involved in memory storage has been studied using the phenomenon of long-term potentiation (LTP) in the hippocampus, as described in chapter 7.

Long-term potentiation is a type of synaptic learning, in that synapses that are first stimulated at high frequency will subsequently exhibit increased excitability. Long-term potentiation has been studied extensively in the hippocampus, where most of the axons use glutamate as a neurotransmitter. Here, the induction of LTP requires activation of the NMDA receptors for glutamate (described in chapter 7). Activation of NMDA receptors—where the receptor channels for Ca^{2+} and Na^+ open—requires not only binding by glutamate, but also binding by another ligand (glycine or D-serine) and a simultaneous partial depolarization of the postsynaptic membrane by different membrane channels. This can involve the binding of glutamate to different receptors, known as AMPA receptors. It is interesting in this regard that AMPA receptors move into the postsynaptic membrane during LTP. Once glutamate is able to activate its NMDA receptors, their channels for Ca^{2+} are opened in the dendritic plasma membrane. Long-term potentiation is thus characterized by the diffusion of Ca^{2+} into the dendrites of the postsynaptic neuron (fig 8.15).

Morphological (structural) changes also occur in the postsynaptic neuron as a result of LTP. *Dendritic spines,* which are tiny spikelike extensions from the dendrites, grow as a consequence of LTP. Recent evidence suggests that, as a result of LTP, the growth of new dendritic spines results in increased area of contact between the presynaptic axon terminal and the postsynaptic membrane.

The induction of LTP may also involve presynaptic changes, so that there is increased release of neurotransmitter. This may involve a "retrograde messenger," sent from the postsynaptic neuron to the presynaptic axon. Some scientists have proposed that nitric oxide plays this role. In this proposed sequence of events:

1. The binding of glutamate to its NMDA receptors and simultaneous depolarization of the postsynaptic membrane causes the NMDA receptor channels to open.
2. This opening of the NMDA receptor channels allows Ca^{2+} to enter.
3. The entry of Ca^{2+} into the postsynaptic neuron causes long-term potentiation in that neuron.
4. The entry of Ca^{2+} into the postsynaptic neuron also activates nitric oxide synthase, causing nitric oxide production.
5. The nitric oxide then acts as a retrograde messenger, diffusing into the presynaptic neuron and somehow causing it to release more neurotransmitter.

In these ways, synaptic transmission is strengthened through frequent use. Although the mechanisms by which LTP

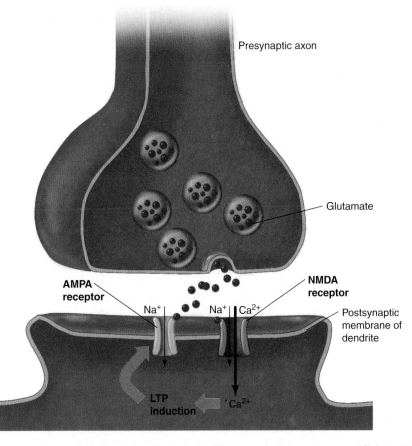

Figure 8.15 Role of glutamate receptors in long-term potentiation (LTP). The neurotransmitter glutamate (Glu) can bind to two different receptors, designated AMPA and NMDA. The activation of the NMDA receptors promotes an increased concentration of Ca^{2+} in the cytoplasm, which is needed in order for LTP to be induced. LTP is believed to be a mechanism of learning at the level of the single synapse.

is produced are still incompletely understood, and the causal association between LTP and learning still unproven, the evidence suggests that LTP is involved in the changes that occur when memories are made.

Neural Stem Cells in Learning and Memory

As mentioned previously, mammalian brains have recently been demonstrated to contain *neural stem cells*—cells that both renew themselves through mitosis and produce differentiated (specialized) neurons and neuroglia. It is particularly exciting that one of the brain regions shown to contain stem cells, the hippocampus, is required for the consolidation of long-term memory and for spatial learning.

Given this observation, it is natural to wonder if the production of new neurons, called **neurogenesis,** is involved in learning and memory. There is now evidence, at least in rats, that this is the case for the learning and retention of a particular type of task. There is also indirect evidence linking neurogenesis in the hippocampus with learning and memory. For example, conditions of stress inhibit neurogenesis in the hippocampus (and retard hippocampus-dependent forms of learning), while increased environmental complexity has the opposite effects on both neurogenesis and learning. Mitotically active neural stem

cells have recently been surgically isolated from the human hippocampus of adult patients, in the hope that cells obtained in this way may someday be useful in treating people with a damaged or degenerated hippocampus.

Test Yourself Before You Continue

1. Describe the locations of the sensory and motor areas of the cerebral cortex and explain how these areas are organized.

2. Describe the locations and functions of the basal nuclei. Of what structures are the basal nuclei composed?

3. Identify the structures of the limbic system and explain the functional significance of this system.

4. Explain the difference in function of the right and left cerebral hemispheres.

5. List the areas of the brain believed to be involved in the production of speech and describe the different types of aphasias produced by damage to these areas.

6. Describe the different forms of memory, list the brain structures shown to be involved in memory, and discuss some of the experimental evidence on which this information is based.

Diencephalon

The diencephalon is the part of the forebrain that contains such important structures as the thalamus, hypothalamus, and part of the pituitary gland. The hypothalamus performs numerous vital functions, most of which relate directly or indirectly to the regulation of visceral activities by way of other brain regions and the autonomic nervous system.

The **diencephalon,** together with the telencephalon (cerebrum) previously discussed, constitutes the forebrain and is almost completely surrounded by the cerebral hemispheres. The *third ventricle* is a narrow midline cavity within the diencephalon.

Thalamus and Epithalamus

The **thalamus** composes about four-fifths of the diencephalon and forms most of the walls of the third ventricle (fig. 8.16). It consists of paired masses of gray matter, each positioned immediately below the lateral ventricle of its respective cerebral hemisphere. The thalamus acts primarily as a relay center through which all sensory information (except smell) passes on the way to the cerebrum. For example, the *lateral geniculate nuclei* relay visual information, and the *medial geniculate nuclei* relay auditory information, from the thalamus to the occipital and temporal lobes, respectively, of the cerebral cortex. The *intralaminar nuclei* of the thalamus are activated by many different sensory modalities and in turn project to many areas of the cerebral cortex. This is part of the system that promotes a state of alertness and causes arousal from sleep in response to any sufficiently strong sensory stimulus.

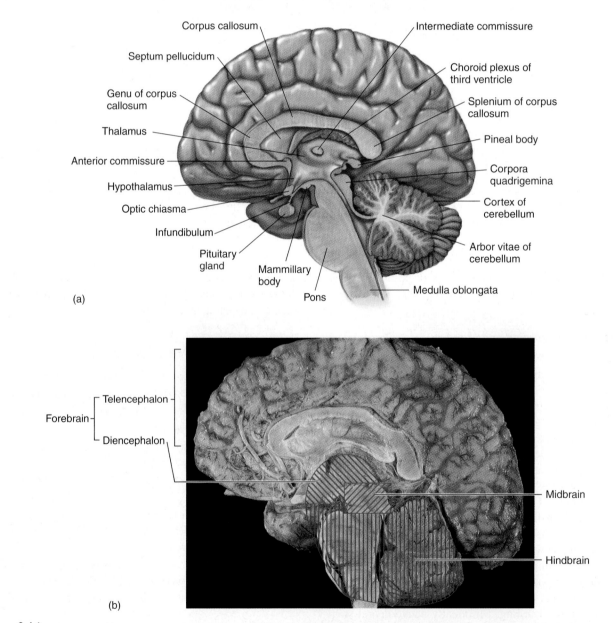

Figure 8.16   **A midsagittal section through the brain.** (*a*) A diagram and (*b*) a photograph. Areas of the diencephalon, midbrain (mesencephalon), and hindbrain (rhombencephalon) are shaded. All of the brain outside of the these shaded areas is included in the telencephalon.

The **epithalamus** is the dorsal segment of the diencephalon containing a *choroid plexus* over the third ventricle, where cerebrospinal fluid is formed, and the *pineal gland (epiphysis).* The pineal gland secretes the hormone *melatonin,* which may play a role in the endocrine control of reproduction (discussed in chapter 20).

Hypothalamus and Pituitary Gland

The **hypothalamus** is the most inferior portion of the diencephalon. Located below the thalamus, it forms the floor and part of the lateral walls of the third ventricle. This small but extremely important brain region contains neural centers for hunger and thirst and for the regulation of body temperature and hormone secretion from the pituitary gland (fig. 8.17). In addition, centers in the hypothalamus contribute to the regulation of sleep, wakefulness, sexual arousal and performance, and such emotions as anger, fear, pain, and pleasure. Acting through its connections with the medulla oblongata of the brain stem, the hypothalamus helps to evoke the visceral responses to various emotional states. In its regulation of emotion, the hypothalamus works together with the limbic system, as was discussed in the previous section.

Experimental stimulation of different areas of the hypothalamus can evoke the autonomic responses characteristic of aggression, sexual behavior, hunger, or satiety. Chronic stimulation of the lateral hypothalamus, for example, can make an animal eat and become obese, whereas stimulation of the medial hypothalamus inhibits eating. Other areas contain osmoreceptors that stimulate thirst and the release of antidiuretic hormone (ADH) from the posterior pituitary.

The hypothalamus is also where the body's "thermostat" is located. Experimental cooling of the preoptic-anterior hypothalamus causes shivering (a somatic motor response) and nonshivering thermogenesis (a sympathetic motor response). Experimental heating of this hypothalamic area results in hyperventilation (stimulated by somatic motor nerves), vasodilation, salivation, and sweat-gland secretion (regulated by sympathetic nerves). These responses serve to correct the temperature deviations in a negative feedback fashion.

The coordination of sympathetic and parasympathetic reflexes is thus integrated with the control of somatic and endocrine responses by the hypothalamus. The activities of the hypothalamus are in turn influenced by higher brain centers.

The **pituitary gland** is located immediately inferior to the hypothalamus. Indeed, the posterior pituitary derives embryonically from a downgrowth of the diencephalon, and the entire pituitary remains connected to the diencephalon by means of a stalk (a relationship described in more detail in chapter 11). Neurons within the *supraoptic* and *paraventricular nuclei* of the hypothalamus (fig. 8.17) produce two hormones—**antidiuretic hormone (ADH),** which is also known as *vasopressin,* and **oxytocin.** These two hormones are transported in axons of the *hypothalamo-hypophyseal tract* to the **neurohypophysis** (posterior pituitary), where they are stored and released in response to hypothalamic stimulation. Oxytocin stimulates contractions of the uterus during labor, and ADH stimulates the kidneys to reabsorb water and thus to excrete a smaller volume of urine. Neurons in the hypothalamus also produce hormones known as **releasing hormones** and **inhibiting hormones** that are transported by the blood to the **adenohypophysis**

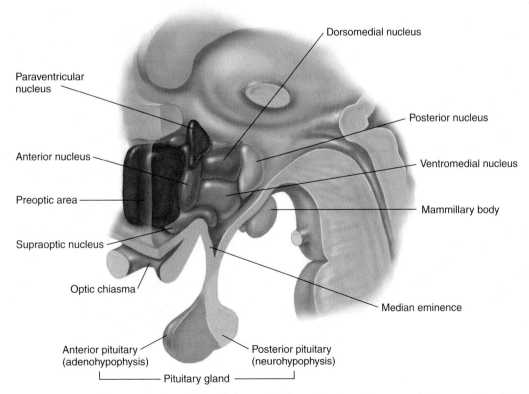

Figure 8.17 A diagram of some of the nuclei within the hypothalamus. The hypothalamic nuclei, composed of neuron cell bodies have different functions.

(anterior pituitary). These hypothalamic releasing and inhibiting hormones regulate the secretions of the anterior pituitary and, by this means, regulate the secretions of other endocrine glands (as described in chapter 11).

Midbrain and Hindbrain

The midbrain and hindbrain contain many important relay centers for sensory and motor pathways, and are particularly important in the control of skeletal movements by the brain. The medulla oblongata, a vital region of the hindbrain, contains centers for the control of breathing and cardiovascular function.

Midbrain

The *mesencephalon,* or **midbrain,** is located between the diencephalon and the pons. The **corpora quadrigemina** are four rounded elevations on the dorsal surface of the midbrain (see fig. 8.16). The two upper mounds, the *superior colliculi,* are involved in visual reflexes; the *inferior colliculi,* immediately below, are relay centers for auditory information.

The midbrain also contains the cerebral peduncles, red nucleus, substantia nigra, and other nuclei. The **cerebral peduncles** are a pair of structures composed of ascending and descending fiber tracts. The **red nucleus,** an area of gray matter deep in the midbrain, maintains connections with the cerebrum and cerebellum and is involved in motor coordination.

As discussed in chapter 7, the midbrain has two systems of dopaminergic (dopamine-releasing) neurons that project to other areas of the brain. The *nigrostriatal system* projects from the **substantia nigra** to the corpus striatum of the basal nuclei; this system is required for motor coordination, and it is the degeneration of these fibers that produces Parkinson's disease. Other dopaminergic neurons that are part of the *mesolimbic system* project from nuclei adjacent to the substantia nigra to the limbic system of the forebrain (fig. 8.18). This system is in-

volved in behavior and reward, and the release of dopamine from these neurons is promoted by abused drugs.

Hindbrain

The *rhombencephalon,* or **hindbrain,** is composed of two regions: the metencephalon and the myelencephalon. Each of these regions will be discussed separately.

Metencephalon

The *metencephalon* is composed of the pons and the cerebellum. The **pons** can be seen as a rounded bulge on the underside of the brain, between the midbrain and the medulla oblongata (fig. 8.19). Surface fibers in the pons connect to the cerebellum, and deeper fibers are part of motor and sensory tracts that pass from the medulla oblongata, through the pons, and on to the midbrain. Within the pons are several nuclei associated with specific cranial nerves—the trigeminal (V), abducens (VI), facial (VII), and vestibulocochlear (VIII). Other nuclei of the pons cooperate with nuclei in the medulla oblongata to regulate breathing. The two respiratory control centers in the pons are known as the *apneustic* and the *pneumotaxic centers.*

The **cerebellum,** containing over a hundred billion neurons, is the second largest structure of the brain. Like the cerebrum, it contains outer gray and inner white matter. Fibers from the cerebellum pass through the red nucleus to the thalamus, and then to the motor areas of the cerebral cortex. Other fiber tracts connect the cerebellum with the pons, medulla oblongata, and spinal cord. The cerebellum receives input from *proprioceptors* (joint, tendon, and muscle receptors) and, working together with the basal nuclei and motor areas of the cerebral cortex, participates in the coordination of movement.

The cerebellum is needed for motor learning and for coordinating the movement of different joints during a movement. It is also required for the proper timing and force required for limb movements. The cerebellum, for example, is needed in order to touch your nose with your finger, bring a fork of food to your mouth, or find keys by touch in your pocket or purse.

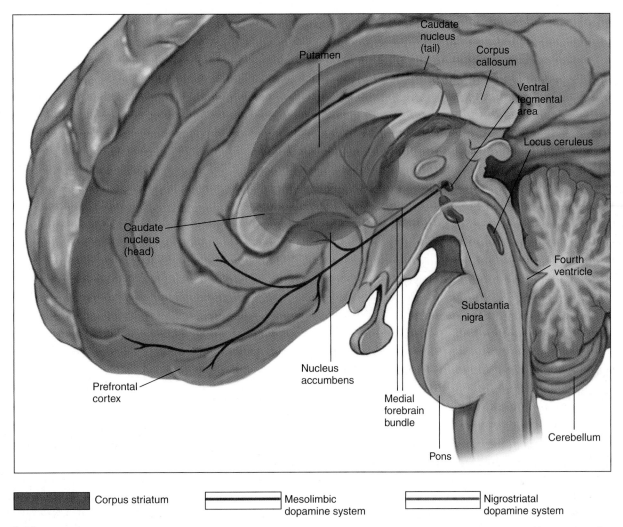

Corpus striatum   Mesolimbic dopamine system   Nigrostriatal dopamine system

▦ **Figure 8.18   Dopaminergic pathways in the brain.** Axons that use dopamine as a neurotransmitter (that are dopaminergic) leave the substantia nigra of the midbrain and synapse in the corpus striatum. This is the nigrostriatal system, used for motor control. Dopaminergic axons from the midbrain to the nucleus accumbens and prefrontal cortex constitute the mesolimbic system, which functions in emotional reward.

Table 8.3   Synaptic Effects of Some Abused Drugs

Drug	Action	How Synaptic Transmission Is Affected
Opiates	Stimulates opioid receptors	Exogenous opioids bind to and stimulate the G-protein-coupled receptors for the endogenous opioids.
Cocaine	Inhibits transporter needed for reuptake of dopamine (and serotonin and norepinephrine) into presynaptic axon terminals	Receptors for monoamines are stimulated indirectly because more neurotransmitters remain in the synaptic cleft.
Amphetamines	Stimulates the release of dopamine from dopaminergic neurons	Receptors for dopamine are stimulated indirectly because more dopamine is released into the synaptic cleft.
Ethanol (alcohol)	Facilitates GABA receptor function (promoting inhibition) and inhibits NMDA glutamate receptor function (decreasing excitation)	Receptors for GABA and the NMDA receptors for glutamate are ligand-gated channels, opened directly by binding to these neurotransmitters.
Nicotine	Stimulates nicotinic acetylcholine receptors	Nicotinic ACh receptors are ligand-gated channels, opened directly by binding to ACh or nicotine.

Source: Reprinted by permission from *Nature Reviews Neuroscience: Vol. 2, No. 2,* p. 120 (2001). Copyright 2001 Macmillan Magazines Ltd.

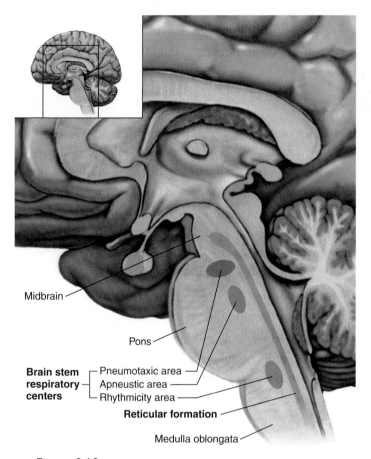

Midbrain

Pons

Brain stem respiratory centers ┌ Pneumotaxic area
                                    ├ Apneustic area
                                    └ Rhythmicity area

Reticular formation

Medulla oblongata

■ **Figure 8.19** Respiratory control centers in the brain stem. These are nuclei within the pons and medulla oblongata that control the motor nerves required for breathing. The location of the reticular formation is also shown.

Damage to the cerebellum produces **ataxia**—lack of coordination resulting from errors in the speed, force, and direction of movement. The movements and speech of people afflicted with ataxia may resemble those of someone who is intoxicated. This condition is also characterized by intention tremor, which differs from the resting tremor of Parkinson's disease in that it occurs only when intentional movements are made. People with cerebellar damage may reach for an object and miss it by placing their hand too far to the left or right; then, they will attempt to compensate by moving their hand in the opposite direction. This back-and-forth movement can result in oscillations of the limb.

Myelencephalon

The *myelencephalon* is composed of only one structure, the **medulla oblongata,** often simply called the *medulla.* About 3 cm (1 in.) long, the medulla is continuous with the pons superiorly and the spinal cord inferiorly. All of the descending and ascending fiber tracts that provide communication between the spinal cord and the brain must pass through the medulla. Many of these fiber tracts cross to the contralateral side in elevated triangular structures in the medulla called the **pyramids.** Thus, the left side of the brain receives sensory information from the right side of the body and vice versa. Similarly, because of the decussation of fibers, the right side of the brain controls motor activity in the left side of the body and vice versa.

Many important nuclei are contained within the medulla. Several nuclei are involved in motor control, giving rise to axons within cranial nerves VIII, IX, X, XI, and XII. The *vagus nuclei* (there is one on each lateral side of the medulla), for example, give rise to the highly important vagus (X) nerves. Other nuclei relay sensory information to the thalamus and then to the cerebral cortex.

The medulla contains groupings of neurons required for the regulation of breathing and of cardiovascular responses; hence, they are known as the *vital centers.* The **vasomotor center** controls the autonomic innervation of blood vessels; the **cardiac control center,** closely associated with the vasomotor center, regulates the autonomic nerve control of the heart; and the **respiratory center** of the medulla acts together with centers in the pons to control breathing.

Reticular Formation

The **reticular formation** (fig. 8.19) is a complex network of nuclei and nerve fibers within the medulla, pons, midbrain, thalamus, and hypothalamus that functions as the **reticular activating system,** or **RAS.** Because of its many interconnections, the RAS is activated in a nonspecific fashion by any modality of sensory information. Nerve fibers from the RAS, in turn, project diffusely to the cerebral cortex; this results in *nonspecific arousal* of the cerebral cortex to incoming sensory information.

The RAS, through its nonspecific arousal of the cortex, helps to maintain a state of alert consciousness. Not surprisingly, there is evidence that general anesthetics may produce unconsciousness by depressing the RAS. Similarly, the ability to fall asleep may be due to the action of specific neurotransmitters that inhibit activity of the RAS.

Test Yourself Before You Continue

1. List the structures of the midbrain and describe their functions.
2. Describe the functions of the medulla oblongata and pons.
3. Locate the reticular formation in the brain. What is the primary function of the reticular activating system and how is this function accomplished?

Spinal Cord Tracts

Sensory information from receptors throughout most of the body is relayed to the brain by means of ascending tracts of fibers that conduct impulses up the spinal cord. When the brain directs motor activities, these directions are in the form of nerve impulses that travel down the spinal cord in descending tracts of fibers.

The spinal cord extends from the level of the foramen magnum of the skull to the first lumbar vertebra. Unlike the brain, in which the gray matter forms a cortex over white matter, the gray matter of the spinal cord is located centrally, surrounded by white matter. The central gray matter of the spinal cord is arranged in the form of an H, with two *dorsal horns* and two *ventral horns* (also called posterior and anterior horns, respectively). The white matter of the spinal cord is composed of ascending and descending fiber tracts. These are arranged into six columns of white matter called *funiculi*.

The fiber tracts within the white matter of the spinal cord are named to indicate whether they are ascending (sensory) or descending (motor) tracts. The names of the ascending tracts usually start with the prefix *spino-* and end with the name of the brain region where the spinal cord fibers first synapse. The anterior spinothalamic tract, for example, carries impulses conveying the sense of touch and pressure, and synapses in the thalamus. From there it is relayed to the cerebral cortex. The names of descending motor tracts, conversely, begin with a prefix denoting the brain region that gives rise to the fibers and end with the suffix *-spinal*. The lateral corticospinal tracts, for example, begin in the cerebral cortex and descend the spinal cord.

Ascending Tracts

The ascending fiber tracts convey sensory information from cutaneous receptors, proprioceptors (muscle and joint receptors), and visceral receptors (table 8.4). Most of the sensory information that originates in the right side of the body crosses over to eventually reach the region on the left side of the brain that analyzes this information. Similarly, the information arising in the left side of the body is ultimately analyzed by the right side of the brain. For some sensory modalities, this decussation occurs in the medulla oblongata (fig. 8.20); for others, it occurs in the spinal cord. These neural pathways are discussed in more detail in chapter 10.

Descending Tracts

The descending fiber tracts that originate in the brain consist of two major groups: the **corticospinal, or pyramidal tracts,** and the **extrapyramidal tracts** (table 8.5). The pyramidal tracts descend directly, without synaptic interruption, from the cerebral cortex to the spinal cord. The cell bodies that contribute fibers to these pyramidal tracts are located primarily in the *precentral gyrus* (also called the *motor cortex*). Other areas of the cerebral cortex, however, also contribute to these tracts.

From 80% to 90% of the corticospinal fibers decussate in the pyramids of the medulla oblongata (hence the name "pyramidal tracts") and descend as the *lateral corticospinal tracts.* The remaining uncrossed fibers form the *anterior corticospinal tracts,* which decussate in the spinal cord. Because of the crossing over of fibers, the right cerebral hemisphere controls the musculature on the left side of the body (fig. 8.21), whereas the left hemisphere controls the right musculature. The corticospinal tracts are primarily concerned with the control of fine movements that require dexterity.

Table 8.4   Principal Ascending Tracts of Spinal Cord

Tract	Origin	Termination	Function
Anterior spinothalamic	Posterior horn on one side of cord but crosses to opposite side	Thalamus, then cerebral cortex	Conducts sensory impulses for crude touch and pressure
Lateral spinothalamic	Posterior horn on one side of cord but crosses to opposite side	Thalamus, then cerebral cortex	Conducts pain and temperature impulses that are interpreted within cerebral cortex
Fasciculus gracilis and fasciculus cuneatus	Peripheral afferent neurons; ascends on ipsilateral side of spinal cord but crosses over in medulla	Nucleus gracilis and nucleus cuneatus of medulla; eventually thalamus, then cerebral cortex	Conducts sensory impulses from skin, muscles, tendons, and joints, which are interpreted as sensations of fine touch, precise pressures, and body movements
Posterior spinocerebellar	Posterior horn; does not cross over	Cerebellum	Conducts sensory impulses from one side of body to same side of cerebellum; necessary for coordinated muscular contractions
Anterior spinocerebellar	Posterior horn; some fibers cross, others do not	Cerebellum	Conducts sensory impulses from both sides of body to cerebellum; necessary for coordinated muscular contractions

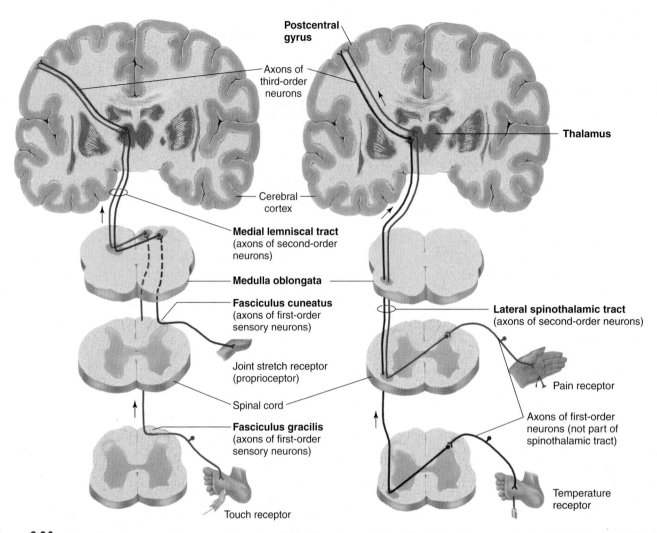

Figure 8.20 **Ascending tracts carrying sensory information.** This information is delivered by third-order neurons to the cerebral cortex. (*a*) Medial lemniscal tract; (*b*) lateral spinothalamic tract.

Table 8.5 Descending Motor Tracts to Spinal Interneurons and Motor Neurons

Tract	Category	Origin	Crossed/Uncrossed
Lateral corticospinal	Pyramidal	Cerebral cortex	Crossed
Anterior corticospinal	Pyramidal	Cerebral cortex	Uncrossed
Rubrospinal	Extrapyramidal	Red nucleus (midbrain)	Crossed
Tectospinal	Extrapyramidal	Superior colliculus (midbrain)	Crossed
Vestibulospinal	Extrapyramidal	Vestibular nuclei (medulla oblongata)	Uncrossed
Reticulospinal	Extrapyramidal	Reticular formation (medulla and pons)	Crossed

Clinical Investigation Clue

Remember that Frank was paralyzed on the right side of his body.

Damage to which descending motor tract would account for Frank's paralysis?

The remaining descending tracts are extrapyramidal motor tracts, which originate in the midbrain and brain stem regions (table 8.5). If the pyramidal tracts of an experimental animal are cut, electrical stimulation of the cerebral cortex, cerebellum, and basal nuclei can still produce movements. The descending fibers that produce these movements must, by definition, be extrapyramidal motor tracts. The regions of the cerebral cortex, basal nuclei, and cerebellum that participate in this motor control have

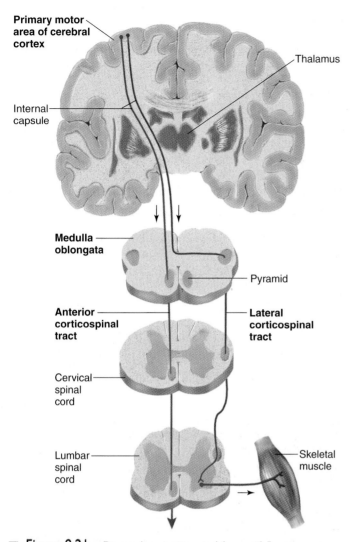

Figure 8.21 Descending corticospinal (pyramidal) motor tracts. These tracts contain axons that pass from the precentral gyrus of the cerebral cortex down the spinal cord to make synapses with spinal interneurons and lower motor neurons.

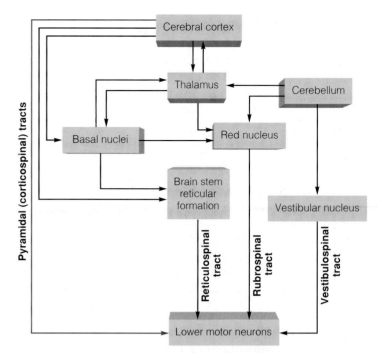

Figure 8.22 The higher motor neuron control of skeletal muscles. The pyramidal (corticospinal) tracts are shown in pink and the extrapyramidal tracts are shown in black.

numerous synaptic interconnections, and they can influence movement only indirectly by means of stimulation or inhibition of the nuclei that give rise to the extrapyramidal tracts. Notice that this motor control differs from that exerted by the neurons of the precentral gyrus, which send fibers directly down to the spinal cord in the pyramidal tracts.

The *reticulospinal tracts* are the major descending pathways of the extrapyramidal system. These tracts originate in the reticular formation of the brain stem, which receives either stimulatory or inhibitory input from the cerebrum and the cerebellum. There are no descending tracts from the cerebellum; the cerebellum can influence motor activity only indirectly by its effect on the vestibular nuclei, red nucleus, and basal nuclei (which send axons to the reticular formation). These nuclei, in turn, send axons down the spinal cord via the *vestibulospinal tracts, rubrospinal tracts,* and reticulospinal tracts, respectively (fig. 8.22). Neural control of skeletal muscle is explained in more detail in chapter 12.

The corticospinal tracts appear to be particularly important in voluntary movements that require complex interactions between sensory input and the motor cortex. Speech, for example, is impaired when the corticospinal tracts are damaged in the thoracic region of the spinal cord, whereas involuntary breathing continues. Damage to the pyramidal motor system can be detected clinically by the presence of **Babinski's reflex,** in which stimulation of the sole of the foot causes extension of the great toe upward and fanning of the other toes. (Normally, in adults, such stimulation causes the plantar reflex, a downward flexion, or curling, of the toes.) Babinski's reflex is normally present in infants because neural control is not yet fully developed.

Test Yourself Before You Continue

1. Explain why each cerebral hemisphere receives sensory input from and directs motor output to the contralateral side of the body.
2. List the tracts of the pyramidal motor system and describe the function of the pyramidal system.
3. List the tracts of the extrapyramidal system and explain how this system differs from the pyramidal motor system.

Cranial and Spinal Nerves

The central nervous system communicates with the body by means of nerves that exit the CNS from the brain (cranial nerves) and spinal cord (spinal nerves). These nerves, together with aggregations of cell bodies located outside the CNS, constitute the peripheral nervous system.

As mentioned in chapter 7, the *peripheral nervous system (PNS)* consists of nerves (collections of axons) and their associated ganglia (collections of cell bodies). Although this chapter is devoted to the CNS, the CNS cannot function without the PNS. This section thus serves to complete our discussion of the CNS and introduces concepts pertaining to the PNS that will be explored more thoroughly in later chapters (particularly chapters 9, 10, and 12).

Cranial Nerves

Of the twelve pairs of **cranial nerves,** two pairs arise from neuron cell bodies located in the forebrain and ten pairs arise from the midbrain and hindbrain. The cranial nerves are designated by Roman numerals and by names. The Roman numerals refer to the order in which the nerves are positioned from the front of the brain to the back. The names indicate the structures innervated by these nerves (e.g., facial) or the principal function of the nerves (e.g., oculomotor). A summary of the cranial nerves is presented in table 8.6.

Most cranial nerves are classified as *mixed nerves.* This term indicates that the nerve contains both sensory and motor fibers. Those cranial nerves associated with the special senses (e.g., olfactory, optic), however, consist of sensory fibers only. The cell bodies of these sensory neurons are not located in the brain, but instead are found in ganglia near the sensory organ.

Table 8.6 Summary of Cranial Nerves

Number and Name	Composition	Function
I Olfactory	Sensory	Olfaction
II Optic	Sensory	Vision
III Oculomotor	Motor	Motor impulses to levator palpebrae superioris and extrinsic eye muscles, except superior oblique and lateral rectus; innervation to muscles that regulate amount of light entering eye and that focus the lens
	Sensory: proprioception	Proprioception from muscles innervated with motor fibers
IV Trochlear	Motor	Motor impulses to superior oblique muscle of eyeball
	Sensory: proprioception	Proprioception from superior oblique muscle of eyeball
V Trigeminal		
Ophthalmic division	Sensory	Sensory impulses from cornea, skin of nose, forehead, and scalp
Maxillary division	Sensory	Sensory impulses from nasal mucosa, upper teeth and gums, palate, upper lip, and skin of cheek
Mandibular division	Sensory	Sensory impulses from temporal region, tongue, lower teeth and gums, and skin of chin and lower jaw
	Sensory: proprioception	Proprioception from muscles of mastication
	Motor	Motor impulses to muscles of mastication and muscle that tenses the tympanum
VI Abducens	Motor	Motor impulses to lateral rectus muscle of eyeball
	Sensory: proprioception	Proprioception from lateral rectus muscle of eyeball
VII Facial	Motor	Motor impulses to muscles of facial expression and muscle that tenses the stapes
	Motor: parasympathetic	Secretion of tears from lacrimal gland and salivation from sublingual and submandibular salivary glands
	Sensory	Sensory impulses from taste buds on anterior two-thirds of tongue; nasal and palatal sensation.
	Sensory: proprioception	Proprioception from muscles of facial expression
VIII Vestibulocochlear	Sensory	Sensory impulses associated with equilibrium
		Sensory impulses associated with hearing
IX Glossopharyngeal	Motor	Motor impulses to muscles of pharynx used in swallowing
	Sensory: proprioception	Proprioception from muscles of pharynx
	Sensory	Sensory impulses from pharynx, middle-ear cavity, carotid sinus, and taste buds on posterior one-third of tongue
	Parasympathetic	Salivation from parotid salivary gland
X Vagus	Motor	Contraction of muscles of pharynx (swallowing) and larynx (phonation)
	Sensory: proprioception	Proprioception from visceral muscles
	Sensory	Sensory impulses from taste buds on rear of tongue; sensations from auricle of ear; general visceral sensations
	Motor: parasympathetic	Regulation of many visceral functions
XI Accessory	Motor	Laryngeal movement; soft palate
		Motor impulses to trapezius and sternocleidomastoid muscles for movement of head, neck, and shoulders
	Sensory: proprioception	Proprioception from muscles that move head, neck, and shoulders
XII Hypoglossal	Motor	Motor impulses to intrinsic and extrinsic muscles of tongue and infrahyoid muscles
	Sensory: proprioception	Proprioception from muscles of tongue

Spinal Nerves

There are thirty-one pairs of spinal nerves. These nerves are grouped into eight cervical, twelve thoracic, five lumbar, five sacral, and one coccygeal according to the region of the vertebral column from which they arise (fig. 8.23).

Each spinal nerve is a mixed nerve composed of sensory and motor fibers. These fibers are packaged together in the nerve, but they separate near the attachment of the nerve to the spinal cord. This produces two "roots" to each nerve. The **dorsal root** is composed of sensory fibers, and the **ventral root** is composed of motor fibers (fig. 8.24). An enlargement of the dorsal root, the **dorsal root ganglion,** contains the cell bodies of the sensory neurons. The motor neuron shown in figure 8.24 is a somatic motor neuron that innervates skeletal muscles; its cell body is not located in a ganglion, but instead is contained within the gray matter of the spinal cord. The cell bodies of some autonomic motor neurons (which innervate involuntary effectors), however, are located in ganglia outside the spinal cord (the autonomic system is discussed separately in chapter 9).

Reflex Arc

The functions of the sensory and motor components of a spinal nerve can be understood most easily by examining a simple reflex; that is, an unconscious motor response to a sensory stimulus. Figure 8.24 demonstrates the neural pathway involved in a **reflex arc.** Stimulation of sensory receptors evokes action potentials that are conducted into the spinal cord by sensory

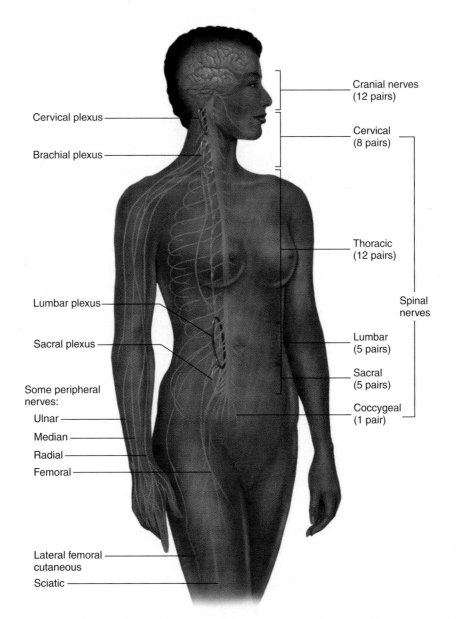

Cranial nerves
(12 pairs)

Cervical
(8 pairs)

Thoracic
(12 pairs)

Spinal
nerves

Lumbar
(5 pairs)

Sacral
(5 pairs)

Coccygeal
(1 pair)

Cervical plexus

Brachial plexus

Lumbar plexus

Sacral plexus

Some peripheral
nerves:

Ulnar

Median

Radial

Femoral

Lateral femoral
cutaneous

Sciatic

■ **Figure 8.23**　Distribution of the spinal nerves.　These interconnect at plexuses (shown on the left) and form specific peripheral nerves.

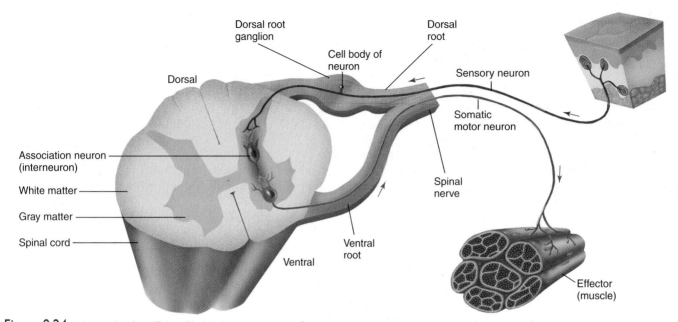

Figure 8.24 **A spinal reflex.** This reflex involves three types of neurons: a sensory neuron, an association neuron (interneuron), and a somatic motor neuron at the spinal cord level.

neurons. In the example shown, a sensory neuron synapses with an association neuron (or interneuron), which in turn synapses with a somatic motor neuron. The somatic motor neuron then conducts impulses out of the spinal cord to the muscle and stimulates a reflex contraction. Notice that the brain is not directly involved in this reflex response to sensory stimulation. Some reflex arcs are even simpler than this; in a muscle stretch reflex (the knee-jerk reflex, for example) the sensory neuron synapses directly with a motor neuron. Other reflexes are more complex, involving a number of association neurons and resulting in motor responses on both sides of the spinal cord at different levels. These skeletal muscle reflexes are described together with muscle control in chapter 12, and autonomic reflexes, involving smooth and cardiac muscle, are described in chapter 9.

Clinical Investigation Clues

Remember that Frank displayed a knee-jerk reflex despite his paralysis.

Why was the knee-jerk reflex present?

What is the most likely cause of Frank's symptoms?

Test Yourself Before You Continue

1. Define the terms *dorsal root, dorsal root ganglion, ventral root,* and *mixed nerve.*

2. Describe the neural pathways and structures involved in a reflex arc.

Summary

Structural Organization of the Brain   190

I. During embryonic development, five regions of the brain are formed: the telencephalon, diencephalon, mesencephalon, metencephalon, and myelencephalon.

    A. The telencephalon and diencephalon constitute the forebrain; the mesencephalon is the midbrain, and the hindbrain is composed of the metencephalon and the myelencephalon.

    B. The CNS begins as a hollow tube, and thus the brain and spinal cord are hollow. The cavities of the brain are known as ventricles.

Cerebrum   192

I. The cerebrum consists of two hemispheres connected by a large fiber tract called the corpus callosum.

    A. The outer part of the cerebrum, the cerebral cortex, consists of gray matter.

    B. Under the gray matter is white matter, but nuclei of gray matter, known as the basal nuclei, lie deep within the white matter of the cerebrum.

    C. Synaptic potentials within the cerebral cortex produce the electrical activity seen in an electroencephalogram (EEG).

II. The two cerebral hemispheres exhibit some specialization of function, a phenomenon called cerebral lateralization.

    A. In most people, the left hemisphere is dominant in language and analytical ability, whereas the right hemisphere is more important in pattern recognition, musical composition, singing, and the recognition of faces.

    B. The two hemispheres cooperate in their functions; this cooperation is aided by communication between the two via the corpus callosum.

III. Particular regions of the left cerebral cortex appear to be important in language ability; when these areas are damaged, characteristic types of aphasias result.

    A. Wernicke's area is involved in speech comprehension, whereas Broca's area is required for the mechanical performance of speech.

    B. Wernicke's area is believed to control Broca's area by means of the arcuate fasciculus.

    C. The angular gyrus is believed to integrate different sources of sensory information and project to Wernicke's area.

IV. The limbic system and hypothalamus are regions of the brain that have been implicated as centers for various emotions.

V. Memory can be divided into short-term and long-term categories.

    A. The medial temporal lobes—in particular the hippocampus and perhaps the amygdaloid nucleus— appear to be required for the consolidation of short-term memory into long-term memory.

    B. Particular aspects of a memory may be stored in numerous brain regions.

    C. Long-term potentiation is a phenomenon that may be involved in some aspects of memory.

Diencephalon   204

I. The diencephalon is the region of the forebrain that includes the thalamus, epithalamus, hypothalamus, and pituitary gland.

    A. The thalamus serves as an important relay center for sensory information, among its other functions.

    B. The epithalamus contains a choroid plexus, where cerebrospinal fluid is formed. The pineal gland, which secretes the hormone melatonin, is also part of the epithalamus.

    C. The hypothalamus forms the floor of the third ventricle, and the pituitary gland is located immediately inferior to the hypothalamus.

II. The hypothalamus is the main control center for visceral activities.

    A. The hypothalamus contains centers for the control of thirst, hunger, body temperature, and (together with the limbic system) various emotions.

    B. The hypothalamus regulates the secretions of the pituitary gland. It controls the posterior pituitary by means of a fiber tract, and it controls the anterior pituitary by means of hormones.

Midbrain and Hindbrain   206

I. The midbrain contains the superior and inferior colliculi, which are involved in visual and auditory reflexes, respectively, and nuclei that contain dopaminergic neurons that project to the corpus striatum and limbic system of the forebrain.

II. The hindbrain consists of two regions: the metencephalon and the myelencephalon.

    A. The metencephalon contains the pons and cerebellum. The pons contains nuclei for four pairs of cranial nerves, and the cerebellum plays an important role in the control of skeletal movements.

    B. The myelencephalon consists of only one region, the medulla oblongata. The medulla contains centers for the regulation of such vital functions as breathing and the control of the cardiovascular system.

Spinal Cord Tracts   209

I. Ascending tracts carry sensory information from sensory organs up the spinal cord to the brain.

II. Descending tracts are motor tracts and are divided into two groups: the pyramidal and the extrapyramidal systems.

    A. Pyramidal tracts are the corticospinal tracts. They begin in the precentral gyrus and descend, without synapsing, into the spinal cord.

B. Most of the corticospinal fibers decussate in the pyramids of the medulla oblongata.

C. Regions of the cerebral cortex, the basal nuclei, and the cerebellum control movements indirectly by synapsing with other regions that give rise to descending extrapyramidal fiber tracts.

D. The major extrapyramidal motor tract is the reticulospinal tract, which has its origin in the reticular formation of the midbrain.

Cranial and Spinal Nerves 212

I. There are twelve pairs of cranial nerves. Most of these are mixed, but some are exclusively sensory in function.

II. There are thirty-one pairs of spinal nerves. Each pair contains both sensory and motor fibers.

A. The dorsal root of a spinal nerve contains sensory fibers, and the cell bodies of these neurons are contained in the dorsal root ganglion.

B. The ventral root of a spinal nerve contains motor fibers.

III. A reflex arc is a neural pathway involving a sensory neuron and a motor neuron. One or more association neurons also may be involved in some reflexes.

Review Activities

Test Your Knowledge of Terms and Facts

1. Which of these statements about the precentral gyrus is *true?*
   a. It is involved in motor control.
   b. It is involved in sensory perception.
   c. It is located in the frontal lobe.
   d. Both *a* and *c* are true.
   e. Both *b* and *c* are true.

2. In most people, the right hemisphere controls movement of
   a. the right side of the body primarily.
   b. the left side of the body primarily.
   c. both the right and left sides of the body equally.
   d. the head and neck only.

3. Which of these statements about the basal nuclei is *true?*
   a. They are located in the cerebrum.
   b. They contain the caudate nucleus.
   c. They are involved in motor control.
   d. They are part of the extrapyramidal system.
   e. All of these are true.

4. Which of these acts as a relay center for somatesthetic sensation?
   a. the thalamus
   b. the hypothalamus
   c. the red nucleus
   d. the cerebellum

5. Which of these statements about the medulla oblongata is *false?*
   a. It contains nuclei for some cranial nerves.
   b. It contains the apneustic center.
   c. It contains the vasomotor center.
   d. It contains ascending and descending fiber tracts.

6. The reticular activating system
   a. is composed of neurons that are part of the reticular formation.
   b. is a loose arrangement of neurons with many interconnecting synapses.
   c. is located in the brain stem and midbrain.
   d. functions to arouse the cerebral cortex to incoming sensory information.
   e. is described correctly by all of these.

7. In the control of emotion and motivation, the limbic system works together with
   a. the pons.
   b. the thalamus.
   c. the hypothalamus.
   d. the cerebellum.
   e. the basal nuclei.

8. Verbal ability predominates in
   a. the left hemisphere of right-handed people.
   b. the left hemisphere of most left-handed people.
   c. the right hemisphere of 97% of all people.
   d. both *a* and *b*.
   e. both *b* and *c*.

9. The consolidation of short-term memory into long-term memory appears to be a function of

   a. the substantia nigra.
   b. the hippocampus.
   c. the cerebral peduncles.
   d. the arcuate fasciculus.
   e. the precentral gyrus.

For questions 10–12, match the nature of the aphasia with its cause (choices are listed under question 12).

10. Comprehension good; can speak and write, but cannot read (although can see).

11. Comprehension good; speech is slow and difficult (but motor ability is not damaged).

12. Comprehension poor; speech is fluent but meaningless.
    a. damage to Broca's area
    b. damage to Wernicke's area
    c. damage to angular gyrus
    d. damage to precentral gyrus

13. Antidiuretic hormone (ADH) and oxytocin are synthesized by supraoptic and paraventricular nuclei, which are located in
    a. the thalamus.
    b. the pineal gland.
    c. the pituitary gland.
    d. the hypothalamus.
    e. the pons.

14. The superior colliculi are twin bodies within the corpora quadrigemina of the midbrain that are involved in
    a. visual reflexes.
    b. auditory reflexes.
    c. relaying of cutaneous information.
    d. release of pituitary hormones.

Test Your Understanding of Concepts and Principles

1. Define the term *decussation* and explain its significance in terms of the pyramidal motor system.[1]

2. Electrical stimulation of the basal nuclei or cerebellum can produce skeletal movements. Describe the pathways by which these brain regions control motor activity.

3. Define the term *ablation*. Give two examples of how this experimental technique has been used to learn about the function of particular brain regions.

4. Explain how "split-brain" patients have contributed to research on the function of the cerebral hemispheres. Propose some experiments that would reveal the lateralization of function in the two hemispheres.

5. What evidence do we have that Wernicke's area may control Broca's area? What evidence do we have that the angular gyrus has input to Wernicke's area?

6. State two reasons why researchers distinguish between short-term and long-term memory.

7. Describe evidence showing that the hippocampus is involved in the consolidation of short-term memory. After long-term memory is established, why may there be no need for hippocampal involvement?

8. Can we be aware of a reflex action involving our skeletal muscles? Is this awareness necessary for the response? Explain, identifying the neural pathways involved in the reflex response and the conscious awareness of a stimulus.

Test Your Ability to Analyze and Apply Your Knowledge

1. Fetal alcohol syndrome, produced by excessive alcohol consumption during pregnancy, affects different aspects of embryonic development. Two brain regions known to be particularly damaged in this syndrome are the corpus callosum and the basal nuclei. Speculate on what effects damage to these areas may produce.

2. Recent studies suggest that medial temporal lobe activity is needed for memory retrieval. What is the difference between memory storage and retrieval, and what scientific evidence might allow them to be distinguished?

3. Much has been made (particularly by left-handers) of the fact that Leonardo da Vinci was left-handed. Do you think his accomplishments are in any way related to his left-handedness? Why or why not?

Related Websites

Check out the Links Library at www.mhhe.com/fox8 for links to sites containing resources related to the central nervous system. These links are monitored to ensure current URLs.

---

[1]*Note:* This question is answered in the chapter 8 Study Guide found on the Online Learning Center at www.mhhe.com/fox8.

9

The Autonomic Nervous System

Objectives

After studying this chapter, you should be able to . . .

1. compare the structures and pathways of the autonomic system with those involved in the control of skeletal muscle.

2. explain how autonomic innervation of involuntary effectors differs from the innervation of skeletal muscle.

3. describe the structure and general functions of the sympathetic division of the autonomic system.

4. describe the structure and general functions of the parasympathetic division of the autonomic system.

5. list the neurotransmitters of the preganglionic and postganglionic neurons of the sympathetic and parasympathetic systems.

6. describe the structural and functional relationships between the sympathetic system and the adrenal medulla.

7. distinguish between the different types of adrenergic receptors and explain the physiological and clinical significance of these receptors.

8. explain how the cholinergic receptors are categorized and describe the effects produced by stimulation of these receptors.

9. explain the antagonistic, complementary, and cooperative effects of sympathetic and parasympathetic innervation on different organs.

10. explain how the autonomic system is controlled by the brain.

Take Advantage of the Technology

Visit the Online Learning Center for these additional study resources.

■ Interactive quizzing
■ Online study guide
■ Current news feeds
■ Crossword puzzles
■ Vocabulary flashcards
■ Labeling activities

www.mhhe.com/fox8

Chapter Nine

Clinical Investigation

Cathy stayed up through the night studying for a big examination. She felt very on edge and found that she frequently had to use the inhaler to treat her asthma. In the physiology lab that afternoon, she found that her pulse rate and blood pressure were higher than usual. In the physiology lab exercise the following week, Cathy handled a number of drugs (epinephrine, atropine, and others) that she administered to a frog heart. Later that day, she developed a severe headache and had a very dry mouth. When she looked at her face in the mirror, she noticed that her pupils were dilated.

What may have been responsible for Cathy's fast pulse and high blood pressure the day of her exam, and for her headache and other symptoms the day of the frog lab?

Neural Control of Involuntary Effectors

The autonomic nervous system helps to regulate the activities of cardiac muscle, smooth muscles, and glands. In this regulation, impulses are conducted from the CNS by an axon that synapses with a second autonomic neuron. It is the axon of this second neuron in the pathway that innervates the involuntary effectors.

Autonomic motor nerves innervate organs whose functions are not usually under voluntary control. The effectors that respond to autonomic regulation include **cardiac muscle** (the heart), **smooth muscles,** and **glands.** These effectors are part of the *visceral organs* (organs within the body cavities) and of blood vessels. The involuntary effects of autonomic innervation contrast with the voluntary control of skeletal muscles by way of somatic motor neurons.

Autonomic Neurons

As discussed in chapter 7, neurons of the peripheral nervous system (PNS) that conduct impulses away from the central nervous system (CNS) are known as *motor,* or *efferent, neurons.* There are two major categories of motor neurons: somatic and autonomic. Somatic motor neurons have their cell bodies within the CNS and send axons to **skeletal muscles,** which are usually under voluntary control. This was briefly described in chapter 8 (see fig. 8.23), in the section on the reflex arc, and is reviewed in figure 9.1a. The control of skeletal muscles by somatic motor neurons is discussed in depth in chapter 12.

Unlike somatic motor neurons, which conduct impulses along a single axon from the spinal cord to the neuromuscular junction, autonomic motor control involves two neurons in the efferent pathway (table 9.1). The first of these neurons has its cell body in the gray matter of the brain or spinal cord. The

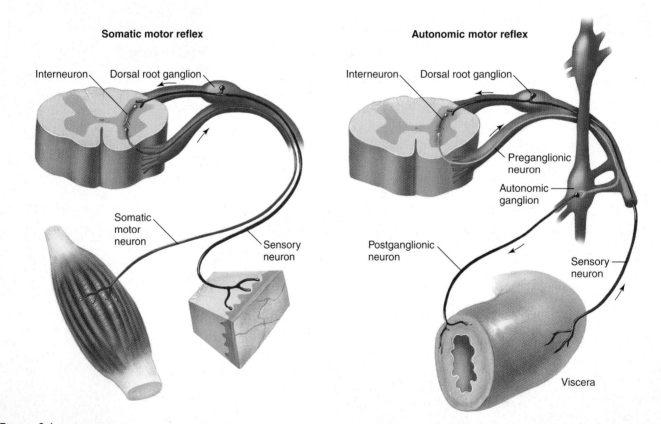

Somatic motor reflex

Interneuron · Dorsal root ganglion

Somatic motor neuron

Sensory neuron

Autonomic motor reflex

Interneuron · Dorsal root ganglion

Preganglionic neuron

Autonomic ganglion

Postganglionic neuron

Sensory neuron

Viscera

Figure 9.1 Comparison of a somatic motor reflex and an autonomic motor reflex. In a skeletal muscle reflex, a single somatic motor neuron passes from the CNS to the skeletal muscle. In an autonomic reflex, a preganglionic neuron passes from the CNS to an autonomic ganglion, where it synapses with a second autonomic neuron. It is that second, or postganglionic, neuron that innervates the smooth muscle, cardiac muscle, or gland.

axon of this neuron does not directly innervate the effector organ but instead synapses with a second neuron within an *autonomic ganglion* (a ganglion is a collection of cell bodies outside the CNS). The first neuron is thus called a **preganglionic neuron.** The second neuron in this pathway, called a **postganglionic neuron,** has an axon that extends from the autonomic ganglion to an effector organ, where it synapses with its target tissue (fig. 9.1*b*).

Preganglionic autonomic fibers originate in the midbrain and hindbrain and in the upper thoracic to the fourth sacral levels of the spinal cord. Autonomic ganglia are located in the head, neck, and abdomen; chains of autonomic ganglia also parallel the right and left sides of the spinal cord. The origin of the preganglionic fibers and the location of the autonomic ganglia help to distinguish the *sympathetic* and *parasympathetic divisions* of the autonomic system, discussed in later sections of this chapter.

Visceral Effector Organs

Since the autonomic nervous system helps to regulate the activities of glands, smooth muscles, and cardiac muscle, autonomic control is an integral aspect of the physiology of most of the body systems. Autonomic regulation, then, partly explains endocrine regulation (chapter 11), smooth muscle function (chapter 12), functions of the heart and circulation (chapters 13 and 14), and, in fact, all the remaining systems to be discussed. Although the functions of the target organs of autonomic innervation are described in subsequent chapters, at this point we will consider some of the common features of autonomic regulation.

Unlike skeletal muscles, which enter a state of flaccid paralysis and atrophy when their motor nerves are severed, the involuntary effectors are somewhat independent of their innervation. Smooth muscles maintain a resting tone (tension) in the absence of nerve stimulation, for example. In fact, damage to an autonomic nerve makes its target tissue more sensitive than normal to stimulating agents. This phenomenon is called **denervation hypersensitivity.** Such compensatory changes can explain why, for example, the ability of the stomach mucosa to secrete acid may be restored after its neural supply from the vagus nerve has been severed. (This procedure is called vagotomy, and is sometimes performed as a treatment for ulcers.)

In addition to their intrinsic ("built-in") muscle tone, cardiac muscle and many smooth muscles take their autonomy a step further. These muscles can contract rhythmically, even in the absence of nerve stimulation, in response to electrical waves of depolarization initiated by the muscles themselves. Autonomic innervation simply increases or decreases this intrinsic activity. Autonomic nerves also maintain a resting tone in the sense that they maintain a baseline firing rate that can be either increased or decreased. A decrease in the excitatory input to the heart, for example, will slow its rate of beat.

The release of acetylcholine (ACh) from somatic motor neurons always stimulates the effector organ (skeletal muscles). By contrast, some autonomic nerves release transmitters that inhibit the activity of their effectors. An increase in the activity of the vagus, a nerve that supplies inhibitory fibers to the heart, for example, will slow the heart rate, whereas a decrease in this inhibitory input will increase the heart rate.

Test Yourself Before You Continue

1. Describe the preganglionic and postganglionic neurons in the autonomic system. Use a diagram to illustrate the difference in efferent outflow between somatic and autonomic nerves.

2. Compare the control of cardiac muscle and smooth muscles with that of skeletal muscles. How is each type of muscle tissue affected by cutting its innervation?

Table 9.1   Comparison of the Somatic Motor System and the Autonomic Motor System

Feature	Somatic Motor	Autonomic Motor
Effector organs	Skeletal muscles	Cardiac muscle, smooth muscle, and glands
Presence of ganglia	No ganglia	Cell bodies of postganglionic autonomic fibers located in paravertebral, prevertebral (collateral), and terminal ganglia
Number of neurons from CNS to effector	One	Two
Type of neuromuscular junction	Specialized motor end plate	No specialization of postsynaptic membrane; all areas of smooth muscle cells contain receptor proteins for neurotransmitters
Effect of nerve impulse on muscle	Excitatory only	Either excitatory or inhibitory
Type of nerve fibers	Fast-conducting, thick (9–13 μm), and myelinated	Slow-conducting; preganglionic fibers lightly myelinated but thin (3 μm); postganglionic fibers unmyelinated and very thin (about 1.0 μm)
Effect of denervation	Flaccid paralysis and atrophy	Muscle tone and function persist; target cells show denervation hypersensitivity

Divisions of the Autonomic Nervous System

Preganglionic neurons of the sympathetic division of the autonomic system originate in the thoracic and lumbar levels of the spinal cord and send axons to sympathetic ganglia, which parallel the spinal cord. Preganglionic neurons of the parasympathetic division, by contrast, originate in the brain and in the sacral level of the spinal cord, and send axons to ganglia located in or near the effector organs.

The sympathetic and parasympathetic divisions of the autonomic system have some structural features in common. Both consist of preganglionic neurons that originate in the CNS and postganglionic neurons that originate outside of the CNS in ganglia. However, the specific origin of the preganglionic fibers and the location of the ganglia differ in the two divisions of the autonomic system.

Sympathetic Division

The **sympathetic division** is also called the *thoracolumbar division* of the autonomic system because its preganglionic fibers exit the spinal cord from the first thoracic (T1) to the second lumbar (L2) levels. Most sympathetic nerve fibers, however, separate from the somatic motor fibers and synapse with postganglionic neurons within a double row of sympathetic ganglia, called **paravertebral ganglia,** located on either side of the spinal cord (fig. 9.2). Ganglia within each row are interconnected, forming a **sympathetic chain of ganglia** that parallels the spinal cord on each lateral side.

The myelinated preganglionic sympathetic axons exit the spinal cord in the ventral roots of spinal nerves, but they soon diverge from the spinal nerves within short pathways called *white rami communicantes.* The axons within each ramus enter the sympathetic chain of ganglia, where they can travel to ganglia at different levels and synapse with postganglionic sympathetic neurons. The axons of the postganglionic sympathetic neurons are unmyelinated and form the *gray rami communicantes* as they return to the spinal nerves and travel as part of the spinal nerves to their effector organs (fig. 9.3). Since sympathetic axons form a component of spinal nerves, they are widely distributed to the skeletal muscles and skin of the body, where they innervate blood vessels and other involuntary effectors.

Divergence occurs within the sympathetic chain of ganglia as preganglionic fibers branch to synapse with numerous postganglionic neurons located in ganglia at different levels in the chain. *Convergence* also occurs here when a postganglionic neuron receives synaptic input from a large number of preganglionic fibers. The divergence of impulses from the spinal cord to the ganglia and the convergence of impulses within the ganglia results in the **mass activation** of almost all of the postganglionic sympathetic neurons. This explains why the sympathetic system is usually activated as a unit, affecting all of its effector organs at the same time.

Collateral Ganglia

Many preganglionic fibers that exit the spinal cord below the level of the diaphragm pass through the sympathetic chain of ganglia without synapsing. Beyond the sympathetic chain, these preganglionic fibers form *splanchnic nerves.* Preganglionic fibers in the splanchnic nerves synapse in **collateral,** or **prevertebral ganglia.** These include the *celiac, superior mesenteric,* and *inferior mesenteric ganglia* (fig. 9.4). Postganglionic

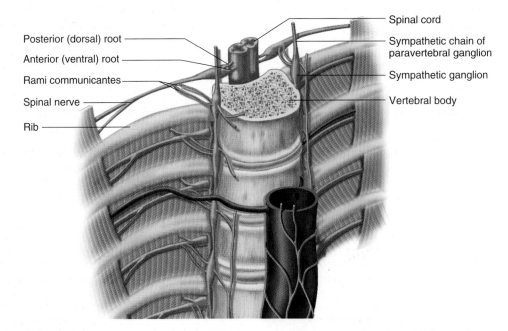

Posterior (dorsal) root
Anterior (ventral) root
Rami communicantes
Spinal nerve
Rib

Spinal cord
Sympathetic chain of paravertebral ganglion
Sympathetic ganglion
Vertebral body

■ **Figure 9.2**   **The sympathetic chain of paravertebral ganglia.** This diagram shows the anatomical relationship between the sympathetic ganglia and the vertebral column and spinal cord.

fibers that arise from the collateral ganglia innervate organs of the digestive, urinary, and reproductive systems.

Adrenal Glands

The paired **adrenal glands** are located above each kidney. Each adrenal is composed of two parts: an outer **cortex** and an inner **medulla.** These two parts are really two functionally different glands with different embryonic origins, different hormones, and different regulatory mechanisms. The adrenal cortex secretes steroid hormones; the adrenal medulla secretes the hormone **epinephrine** (adrenaline) and, to a lesser degree, **norepinephrine,** when it is stimulated by the sympathetic system.

The adrenal medulla can be likened to a modified sympathetic ganglion; its cells are derived from the same embryonic tissue (the neural crest, chapter 8) that forms postganglionic sympathetic neurons. Like a sympathetic ganglion, the cells of the adrenal medulla are innervated by preganglionic sympathetic fibers. The adrenal medulla secretes epinephrine into the blood in response to this neural stimulation. The effects of epinephrine are complementary to those of the neurotransmitter norepinephrine, which is released from postganglionic sympathetic nerve endings. For this reason, and because the adrenal medulla is stimulated as part of the mass activation of the sympathetic system, the two are often grouped together as a single **sympathoadrenal system.**

Parasympathetic Division

The **parasympathetic division** is also known as the *craniosacral division* of the autonomic system. This is because its preganglionic fibers originate in the brain (specifically, in the midbrain, medulla oblongata, and pons) and in the second through fourth sacral levels of the spinal column. These preganglionic parasympathetic fibers synapse in ganglia that are located next to—or actually within—the organs innervated. These parasympathetic ganglia, called **terminal ganglia,** supply the postganglionic fibers that synapse with the effector cells.

The comparative structures of the sympathetic and parasympathetic divisions are listed in tables 9.2 and 9.3. It should be noted that most parasympathetic fibers do not travel within spinal nerves, as do sympathetic fibers. As a result, cutaneous effectors (blood vessels, sweat glands, and arrector pili muscles) and blood vessels in skeletal muscles receive sympathetic but not parasympathetic innervation.

Four of the twelve pairs of cranial nerves (described in chapter 8) contain preganglionic parasympathetic fibers. These are the oculomotor (III), facial (VII), glossopharyngeal (IX), and vagus (X) nerves. Parasympathetic fibers within the first three of these cranial nerves synapse in ganglia located in the head; fibers in the vagus nerve synapse in terminal ganglia located in widespread regions of the body.

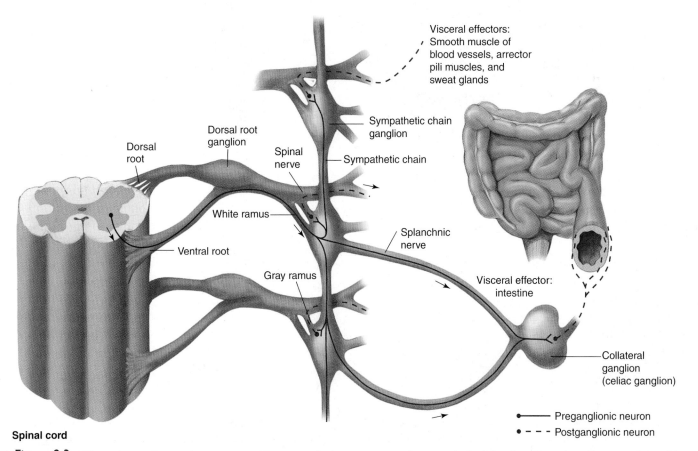

■ Figure 9.3   **The pathway of sympathetic neurons.** The preganglionic neurons enter the sympathetic chain of ganglia on the white ramus (one of the two rami communicantes). Some synapse there, and the postganglionic axon leaves on the grey ramus to rejoin a spinal nerve. Others pass through the ganglia without synapsing. These ultimately synapse in a collateral ganglion, such as the celiac ganglion.

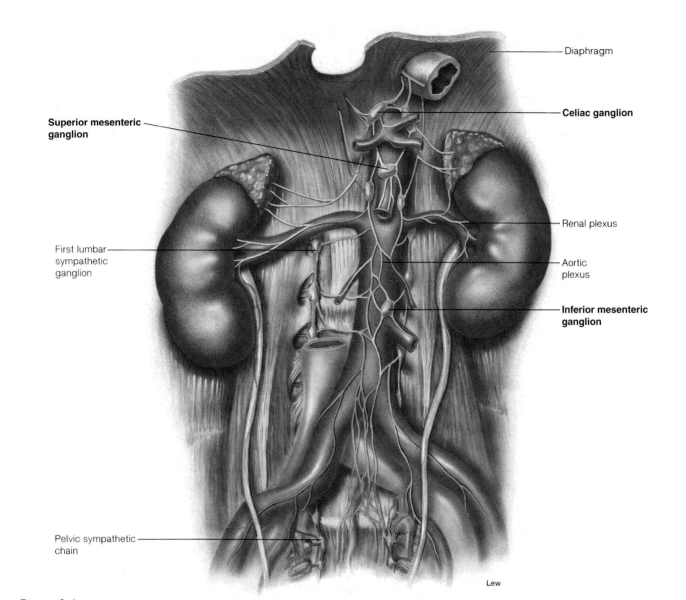

Lew

■ **Figure 9.4**   **The collateral sympathetic ganglia.** These include the celiac ganglion and the superior and inferior mesenteric ganglia.

Table 9.2   The Sympathetic Division

Parts of Body Innervated	Spinal Origin of Preganglionic Fibers	Origin of Postganglionic Fibers
Eye	C8 and T1	Cervical ganglia
Head and neck	T1 to T4	Cervical ganglia
Heart and lungs	T1 to T5	Upper thoracic (paravertebral) ganglia
Upper extremities	T2 to T9	Lower cervical and upper thoracic (paravertebral) ganglia
Upper abdominal viscera	T4 to T9	Celiac and superior mesenteric (collateral) ganglia
Adrenal	T10 and T11	Not applicable
Urinary and reproductive systems	T12 to L2	Celiac and interior mesenteric (collateral) ganglia
Lower extremities	T9 to L2	Lumbar and upper sacral (paravertebral) ganglia

The oculomotor nerve contains somatic motor and parasympathetic fibers that originate in the oculomotor nuclei of the midbrain. These parasympathetic fibers synapse in the *ciliary ganglion,* whose postganglionic fibers innervate the ciliary muscle and constrictor fibers in the iris of the eye. Preganglionic fibers that originate in the pons travel in the facial nerve to the *pterygopalatine ganglion,* which sends postganglionic fibers to the nasal mucosa, pharynx, palate, and lacrimal glands. Another group of fibers in the facial nerve terminates in the *submandibular ganglion,* which sends postganglionic fibers to the submandibular and sublingual salivary glands. Preganglionic fibers of the glossopharyngeal nerve synapse in the *otic ganglion,*

Table 9.3   The Parasympathetic Division

Nerve	Origin of Preganglionic Fibers	Location of Terminal Ganglia	Effector Organs
Oculomotor (third cranial) nerve	Midbrain (cranial)	Ciliary ganglion	Eye (smooth muscle in iris and ciliary body)
Facial (seventh cranial)	Pons (cranial	Pterygopalatine and submandibular ganglia	Lacrimal, mucous, and salivary glands
Glossopharyngeal (ninth cranial) nerve	Medulla oblongata (cranial)	Otic ganglion	Parotid gland
Vagus (tenth cranial) nerve	Medulla oblongata (cranial)	Terminal ganglia in or near organ Terminal ganglia near organs	Heart, lungs, gastrointestinal tract, liver, pancreas
Pelvic spinal nerves	S2 to S4 (sacral)		Lower half of large intestine, rectum, urinary bladder, and reproductive organs

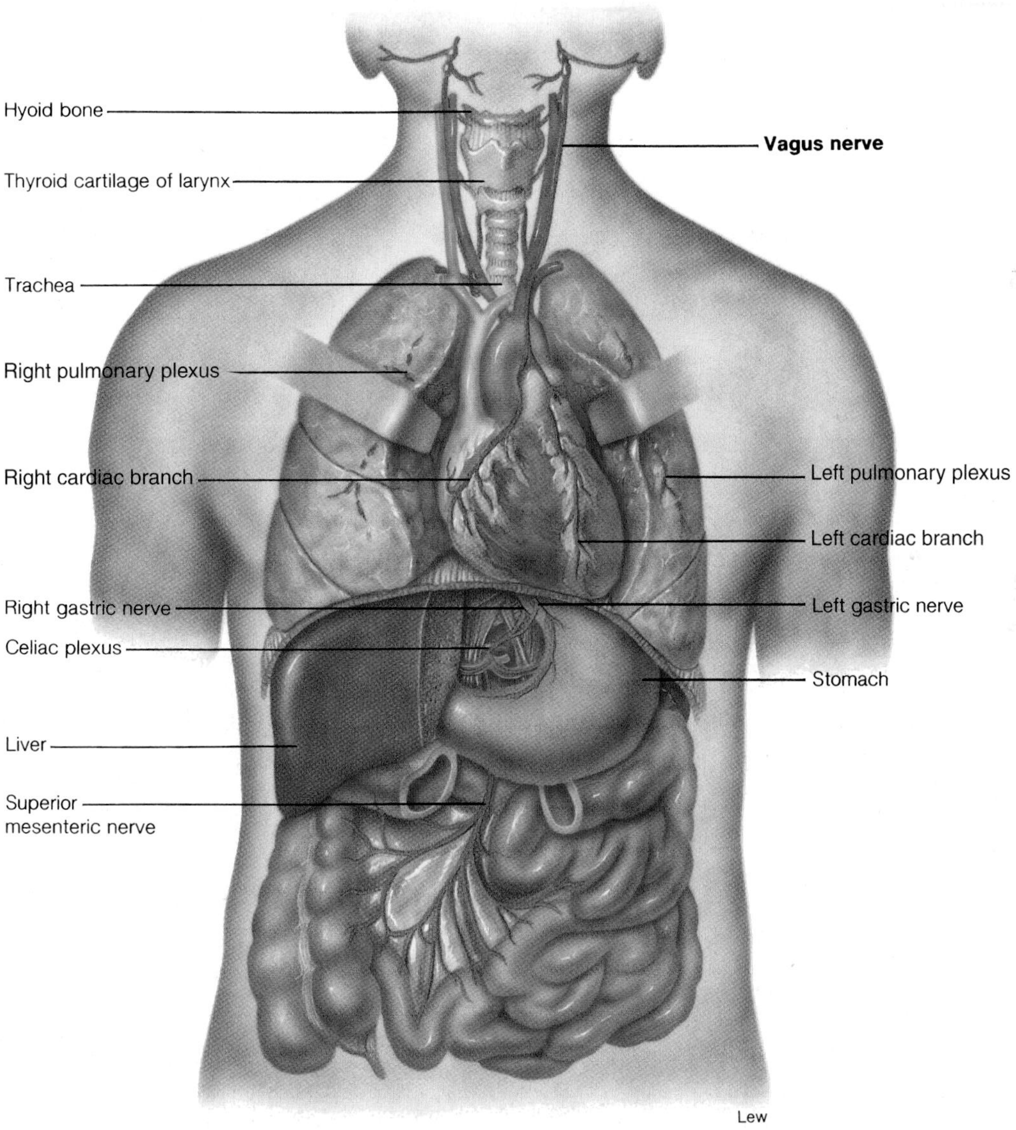

■ **Figure 9.5**   **The path of the vagus nerves.** The vagus nerves and their branches provide parasympathetic innervation to most organs within the thoracic and abdominal cavities.

which sends postganglionic fibers to innervate the parotid salivary gland.

Nuclei in the medulla oblongata contribute preganglionic fibers to the very long *tenth cranial,* or *vagus nerves* (the "vagrant" or "wandering" nerves), which provide the major parasympathetic innervation in the body. These preganglionic fibers travel through the neck to the thoracic cavity and through the esophageal opening in the diaphragm to the abdominal cavity (fig. 9.5). In each region, some of these preganglionic fibers branch from the main trunks of the vagus nerves and synapse with postganglionic neurons located *within* the innervated organs. The preganglionic vagus fibers are thus quite long; they provide parasympathetic innervation to the heart, lungs, esophagus, stomach, pancreas, liver, small intestine, and the upper half of the large intestine. Postganglionic parasympathetic fibers arise from terminal ganglia within these organs and synapse with effector cells (smooth muscles and glands).

Preganglionic fibers from the sacral levels of the spinal cord provide parasympathetic innervation to the lower half of the large intestine, the rectum, and to the urinary and reproductive systems. These fibers, like those of the vagus, synapse with terminal ganglia located within the effector organs.

Parasympathetic nerves to the visceral organs thus consist of preganglionic fibers, whereas sympathetic nerves to these organs contain postganglionic fibers. A composite view of the sympathetic and parasympathetic systems is provided in figure 9.6.

Test Yourself Before You Continue

1. Using a simple line diagram, illustrate the sympathetic pathway (a) from the spinal cord to the heart and (b) from the spinal cord to the adrenal gland. Label the preganglionic and postganglionic fibers and the ganglion.

2. Explain what is meant by the mass activation of the sympathetic system and discuss the significance of the term *sympathoadrenal system*.

3. Using a simple line diagram, illustrate the parasympathetic pathway from the brain to the heart. Compare the parasympathetic and sympathetic divisions in terms of the locations of the pre- and postganglionic fibers and their ganglia.

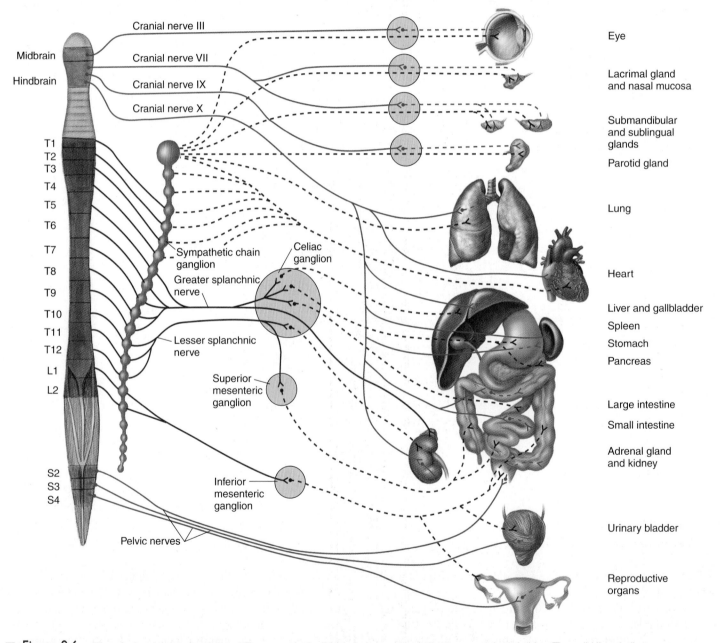

■ **Figure 9.6**   **The autonomic nervous system.** The sympathetic division is shown in red; the parasympathetic in blue. The solid lines indicate preganglionic fibers, and the dashed lines indicate postganglionic fibers.

Functions of the Autonomic Nervous System

The sympathetic division of the autonomic system activates the body to "fight or flight," largely through the release of norepinephrine from postganglionic fibers and the secretion of epinephrine from the adrenal medulla. The parasympathetic division often produces antagonistic effects through the release of acetylcholine from its postganglionic fibers. The actions of the two divisions must be balanced in order to maintain homeostasis.

The sympathetic and parasympathetic divisions of the autonomic system affect the visceral organs in different ways. Mass activation of the sympathetic system prepares the body for intense physical activity in emergencies; the heart rate increases, blood glucose rises, and blood is diverted to the skeletal muscles (away from the visceral organs and skin). These and other effects are listed in table 9.4. The theme of the sympathetic system has been aptly summarized in a phrase: **"fight or flight."** An examination of the first two columns in table 9.4 will reveal how each organ responds to sympathetic nerve stimulation during the fight-or-flight response.

Cocaine blocks the reuptake of dopamine and norepinephrine into the presynaptic axon terminals. This causes an excessive amount of these neurotransmitters to remain in the synaptic cleft and stimulate their target cells. Since sympathetic nerve effects are produced mainly by the action of norepinephrine, cocaine is a *sympathomimetic drug* (a drug that promotes sympathetic nerve effects). This can result in vasoconstriction of coronary arteries, leading to heart damage (myocardial ischemia, myocardial infarction, and left ventricular hypertrophy). The combination of cocaine with alcohol is more deadly than either drug taken separately, and is a common cause of death from substance abuse.

Table 9.4   Effects of Autonomic Nerve Stimulation on Various Effector Organs

Effector Organ	Sympathetic Effect	Parasympathetic Effect
Eye		
Iris (radial muscle)	Dilation of pupil	—
Iris (sphincter muscle)	—	Constriction of pupil
Ciliary muscle	Relaxation (for far vision)	Contraction (for near vision)
Glands		
Lacrimal (tear)	—	Stimulation of secretion
Sweat	Stimulation of secretion	—
Salivary	Decreased secretion; saliva becomes thick	Increased secretion; saliva becomes thin
Stomach	—	Stimulation of secretion
Intestine	—	Stimulation of secretion
Adrenal medulla	Stimulation of hormone secretion	—
Heart		
Rate	Increased	Decreased
Conduction	Increased rate	Decreased rate
Strength	Increased	—
Blood Vessels	Mostly constriction; affects all organs	Dilation in a few organs (e.g., penis)
Lungs		
Bronchioles (tubes)	Dilation	Constriction
Mucous glands	Inhibition of secretion	Stimulation of secretion
Gastrointestinal Tract		
Motility	Inhibition of movement	Stimulation of movement
Sphincters	Closing stimulated	Closing inhibited
Liver	Stimulation of glycogen hydrolysis	—
Adipose (Fat) Cells	Stimulation of fat hydrolysis	—
Pancreas	Inhibition of exocrine secretions	Stimulation of exocrine secretions
Spleen	Contraction	—
Urinary Bladder	Muscle tone aided	Contraction
Arrector Pili Muscles	Erection of hair and goose bumps	—
Uterus	If pregnant: contraction; if not pregnant: relaxation	—
Penis	Ejaculation	Erection (due to vasodilation)

The effects of parasympathetic nerve stimulation are in many ways opposite to those produced by sympathetic stimulation. The parasympathetic system, however, is not normally activated as a whole. Stimulation of separate parasympathetic nerves can result in slowing of the heart, dilation of visceral blood vessels, and increased activity of the digestive tract (table 9.4). Visceral organs respond differently to sympathetic and parasympathetic nerve activity because the postganglionic fibers of these two divisions release different neurotransmitters.

Adrenergic and Cholinergic Synaptic Transmission

Acetylcholine (ACh) is the neurotransmitter of all preganglionic fibers (both sympathetic and parasympathetic). Acetylcholine is also the transmitter released by most parasympathetic post-

ganglionic fibers at their synapses with effector cells (fig. 9.7). Transmission at these synapses is thus said to be **cholinergic.**

The neurotransmitter released by most postganglionic sympathetic nerve fibers is **norepinephrine** (*noradrenaline*). Transmission at these synapses is thus said to be **adrenergic.** There are a few exceptions, however. Some sympathetic fibers that innervate blood vessels in skeletal muscles, as well as sympathetic fibers to sweat glands, release ACh (are cholinergic).

In view of the fact that the cells of the adrenal medulla are embryologically related to postganglionic sympathetic neurons, it is not surprising that the hormones they secrete should consist of epinephrine (about 85%) and norepinephrine (about 15%). Epinephrine differs from norepinephrine only in that the former has an additional methyl (CH_3) group, as shown in figure 9.8. Epinephrine, norepinephrine, and dopamine (a transmitter within the CNS) are all derived from

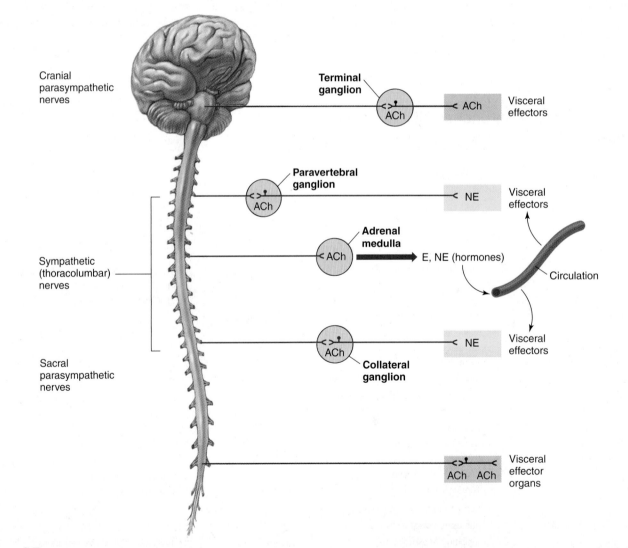

■ **Figure 9.7**   Neurotransmitters of the autonomic motor system. ACh = acetylcholine; NE = norepinephrine; E = epinephrine. Those nerves that release ACh are called cholinergic; those nerves that release NE are called adrenergic. The adrenal medulla secretes both epinephrine (85%) and norepinephrine (15%) as hormones into the blood.

the amino acid tyrosine and are collectively termed **catecholamines** (fig. 9.8).

Where the axons of postganglionic autonomic neurons enter into their target organs, they have numerous swellings, called *varicosities,* that contain the neurotransmitter molecules. Neurotransmitters can thereby be released along a length of axon, rather than just at the axon terminal. Thus, autonomic neurons are said to form *synapses en passant* ("synapses in passing") with their target cells (fig. 9.9). Sympathetic and parasympathetic axons often innervate the same target cells, where they release different neurotransmitters that promote different (and usually antagonistic effects).

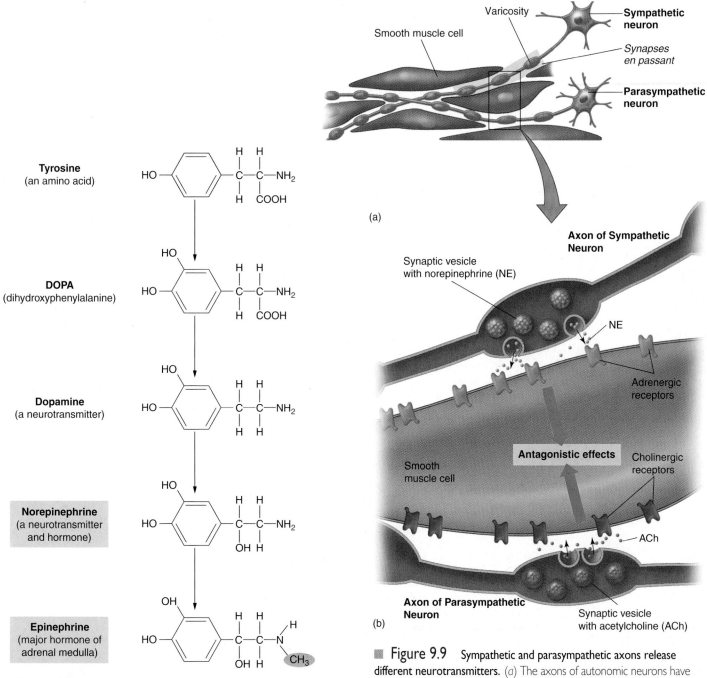

■ **Figure 9.8**   The catecholamine family of molecules. Catecholamines are derived from the amino acid tyrosine, and include both neurotransmitters (dopamine and norepinephrine) and a hormone (epinephrine). Notice that epinephrine has an additional methyl (CH_3) group compared to norepinephrine.

■ **Figure 9.9**   Sympathetic and parasympathetic axons release different neurotransmitters. (*a*) The axons of autonomic neurons have varicosities that form *synapses en passant* with the target cells. (*b*) In general, sympathetic axons release norepinephrine, which binds to its adrenergic receptors, while parasympathetic neurons release acetylcholine, which binds to its cholinergic receptors (discussed in chapter 7). In most cases, these two neurotransmitters elicit antagonistic responses from smooth muscles.

Responses to Adrenergic Stimulation

Adrenergic stimulation—by epinephrine in the blood and by norepinephrine released from sympathetic nerve endings—has both excitatory and inhibitory effects. The heart, dilatory muscles of the iris, and the smooth muscles of many blood vessels are stimulated to contract. The smooth muscles of the bronchioles and of some blood vessels, however, are inhibited from contracting; adrenergic chemicals, therefore, cause these structures to dilate.

Since excitatory and inhibitory effects can be produced in different tissues by the same neurotransmitter, the responses must depend on the characteristics of the cells. To some degree, this is due to the presence of different membrane *receptor proteins* for the catecholamine neurotransmitters. (The interaction of neurotransmitters and receptor proteins in the postsynaptic membrane was described in chapter 7.) The two major classes of these receptor proteins are designated **alpha-** (α) and **beta-** (β) **adrenergic receptors.**

Experiments have revealed that each class of adrenergic receptor has two major subtypes. These are designated by subscripts: α_1 and α_2; β_1 and β_2. Compounds have been developed that selectively bind to one or the other type of adrenergic receptor and, by this means, either promote or inhibit the normal action produced when epinephrine or norepinephrine binds to the receptor. As a result of its binding to an adrenergic receptor, a drug may either promote or inhibit the adrenergic effect. Also, by using these selective compounds, it has been possible to determine which subtype of adrenergic receptor is present in each organ (table 9.5). An additional subtype of adrenergic receptor, designated β_3, has been demonstrated in adipose tissue, but its physiological significance has not yet been established.

All adrenergic receptors act via G-proteins. The action of G-proteins was described in chapter 7, and can be reviewed by reference to fig. 7.28 and table 7.7. In short, the binding of epinephrine and norepinephrine to their receptors causes the group of three G-proteins (designated α, β, and γ) to dissociate into an α subunit and a $\beta\gamma$ complex. In different cases, either the α subunit or the $\beta\gamma$ complex causes the opening or closing of an ion channel in the plasma membrane, or the activation of an enzyme in the membrane. This begins the sequence of events that culminates in the effects of epinephrine and norepinephrine on the target cells.

All subtypes of beta receptors produce their effects by stimulating the production of cyclic AMP (discussed in chapter 7) within the target cells. The response of a target cell when norepinephrine binds to the α_1 receptors is mediated by a different second-messenger system—a rise in the cytoplasmic concentration of Ca^{2+}. This Ca^{2+} second-messenger system is similar, in many ways, to the cAMP system and is discussed together with endocrine regulation in chapter 11. It should be remembered that each of the intracellular changes following the binding of norepinephrine to its receptor ultimately results in the characteristic response of the tissue to the neurotransmitter.

The physiology of α_2-adrenergic receptors is complex. These receptors are located on *presynaptic* axon terminals, and when stimulated, cause a decreased release of norepinephrine. This may represent a form of negative feedback control. On the other hand, vascular smooth muscle cells also have α_2-adrenergic receptors on the *postsynaptic* membrane, where they can be activated to produce vasoconstriction. This action would cause a rise in blood pressure. However, drugs that activate α_2-adrenergic receptors are used to *lower* blood pressure. This is because they stimulate α_2-adrenergic receptors in the

Table 9.5   Selected Adrenergic Effects in Different Organs

Organ	Adrenergic Effects of Sympathoadrenal System	Adrenergic Receptor
Eye	Contraction of radial fibers of the iris dilates the pupils	α_1
Heart	Increase in heart rate and contraction strength	β_1 primarily
Skin and visceral vessels	Arterioles constrict due to smooth muscle contraction	α_1
Skeletal muscle vessels	Arterioles constrict due to sympathetic nerve activity	α_1
	Arterioles dilate due to hormone epinephrine	β_2
Lungs	Bronchioles (airways) dilate due to smooth muscle relaxation	β_2
Stomach and intestine	Contraction of sphincters slows passage of food	α_1
Liver	Glycogenolysis and secretion of glucose	α_1, β_2

Source: Simplified from table 6-1, pp. 110–111 of Goodman and Gilman's *The Pharmacological Basis of Therapeutics.* Ninth edition. J.E. Hardman et al., eds. 1996. McGraw-Hill.

brain, and this somehow reduces the activity of the entire sympathetic nervous system!

A review of table 9.5 reveals certain generalities about the actions of adrenergic receptors. The stimulation of α_1-adrenergic receptors consistently causes contraction of smooth muscles. We can thus state that the vasoconstrictor effect of sympathetic nerves always results from the activation of alpha-adrenergic receptors. The effects of beta-adrenergic activation are more complex; stimulation of beta-adrenergic receptors promotes the relaxation of smooth muscles (in the digestive tract, bronchioles, and uterus, for example) but increases the force of contraction of cardiac muscle and promotes an increase in cardiac rate.

The diverse effects of epinephrine and norepinephrine can be understood in terms of the "fight-or-flight" theme.

Adrenergic stimulation wrought by activation of the sympathetic division produces an increase in cardiac pumping (a β_1 effect), vasoconstriction and thus reduced blood flow to the visceral organs (an α_1 effect), dilation of pulmonary bronchioles (a β_2 effect), and so on, preparing the body for physical exertion (fig. 9.10).

A drug that binds to the receptors for a neurotransmitter and that promotes the processes that are stimulated by that neurotransmitter is said to be an *agonist* of that neurotransmitter. A drug that blocks the action of a neurotransmitter, by contrast, is said to be an *antagonist*. The use of specific drugs that selectively stimulate or block α_1, α_2, β_1, and β_2 receptors has proven extremely useful in many medical applications (see the boxed information).

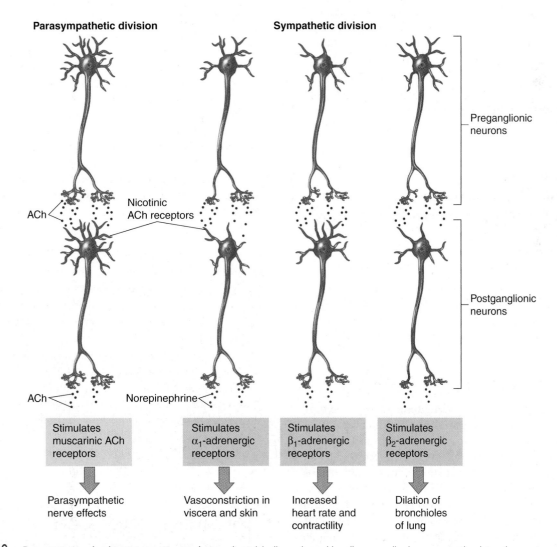

Figure 9.10 Receptors involved in autonomic regulation. Acetylcholine released by all preganglionic neurons stimulates the postganglionic neurons by means of nicotinic ACh receptors. Postganglionic parasympathetic axons regulate their target organs using muscarinic ACh receptors. Postganglionic sympathetic axons provide adrenergic regulation of their target organs by binding of norepinephrine to α_1, β_1, and β_2-adrenergic receptors.

Many people with hypertension were once treated with a beta-blocking drug known as *propranolol.* This drug blocks β_1 receptors, which are located in the heart, and thus produces the desired effect of lowering the cardiac rate and blood pressure. Propranolol, however, also blocks β_2 receptors, which are located in the bronchioles of the lungs. This reduces the bronchodilation effect of epinephrine, producing bronchoconstriction and asthma in susceptible people. A more selective β_1 antagonist, *atenolol,* is now used instead to slow the cardiac rate and lower blood pressure. At one time, asthmatics inhaled an epinephrine spray, which stimulates β_1 receptors in the heart as well as β_2 receptors in the airways. Now, drugs such as *terbutaline* that selectively function as β_2 agonists are more commonly used.

Drugs such as *phenylephrine,* which function as α_1 agonists, are often included in nasal sprays because they promote vasoconstriction in the nasal mucosa. *Clonidine* is a drug that selectively stimulates α_2 receptors located on neurons in the brain. As a consequence of its action, clonidine suppresses the activation of the sympathoadrenal system and thereby helps to lower the blood pressure. For reasons that are poorly understood, this drug is also helpful in treating patients with an addiction to opiates who are experiencing withdrawal symptoms.

Clinical Investigation Clues

Remember that Cathy had a rapid pulse and higher than usual blood pressure after staying up studying for an exam and taking her asthma inhaler.

Why did Cathy have a rapid pulse and higher blood pressure than usual?

Was there more than one factor that contributed to these symptoms?

Responses to Cholinergic Stimulation

All somatic motor neurons, all preganglionic neurons (sympathetic and parasympathetic), and most postganglionic parasympathetic neurons are cholinergic—they release acetylcholine (ACh) as a neurotransmitter. The effects of ACh released by somatic motor neurons, and by preganglionic autonomic neurons, are always excitatory. The effects of ACh released by postganglionic parasympathetic axons are usually excitatory, but in some cases they are inhibitory. For example, the cholinergic effect of the postganglionic parasympathetic axons innervating the heart (a part of the vagus nerve) slows the heart rate. It is useful to remember that, in general, the effects of parasympathetic innervation are opposite to the effects of sympathetic innervation.

The effects of ACh in an organ depend on the nature of the cholinergic receptor (fig. 9.11). As may be recalled from chapter 7, there are two types of cholinergic receptors—nicotinic and muscarinic. Nicotine (derived from the tobacco plant), as well as ACh, stimulates the nicotinic ACh receptors. These are located in the neuromuscular junction of skeletal muscle fibers and in the autonomic ganglia. Nicotinic receptors are thus stimulated by ACh released by somatic motor neurons and by preganglionic autonomic neurons. Muscarine (derived from some poisonous mushrooms), as well as ACh, stimulates the ACh receptors in the visceral organs. Muscarinic receptors are thus stimulated by ACh released by postganglionic parasympathetic axons to produce the parasympathetic effects. Nicotinic and muscarinic receptors are further distinguished by the action of the drugs *curare (tubocurarine),* which specifically blocks the nicotinic ACh receptors, and *atropine* (or *belladonna*), which specifically blocks the muscarinic ACh receptors.

As described in chapter 7, the nicotinic ACh receptors are ligand-gated ion channels. That is, binding to ACh causes the ion channel to open within the receptor protein. This allows Na^+ to diffuse inward, causing depolarization. As a result, nicotinic ACh receptors are always excitatory. In contrast, muscarinic ACh receptors are coupled to G-proteins, which can then close or open different membrane channels and activate different membrane enzymes. As a result, their effects can be either excitatory or inhibitory (fig. 9.11).

Scientists have identified five different subtypes of muscarinic receptors (M_1 through M_5; table 9.6). Some of these cause contraction of smooth muscles and secretion of glands, while others cause the inhibition that results in a slowing of the heart rate. These actions are mediated by second-messenger systems that will be discussed in more detail in conjunction with hormone action in chapter 11.

The muscarinic effects of ACh are specifically inhibited by the drug **atropine,** derived from the deadly nightshade plant (*Atropa belladonna*). Indeed, extracts of this plant were used by women during the Middle Ages to dilate their pupils (atropine inhibits parasympathetic stimulation of the iris). This was thought to enhance their beauty (in Italian, *bella* = beautiful, *donna* = woman). Atropine is used clinically today to dilate pupils during eye examinations, to reduce secretions of the respiratory tract prior to general anesthesia, to inhibit spasmodic contractions of the lower digestive tract, and to inhibit stomach acid secretion in a person with gastritis.

Clinical Investigation Clues

Remember that Cathy developed a headache, dry mouth, and dilated pupils following the use of various drugs in the frog heart lab.

Which drug likely produced these effects in Cathy?

How did the drug have these effects?

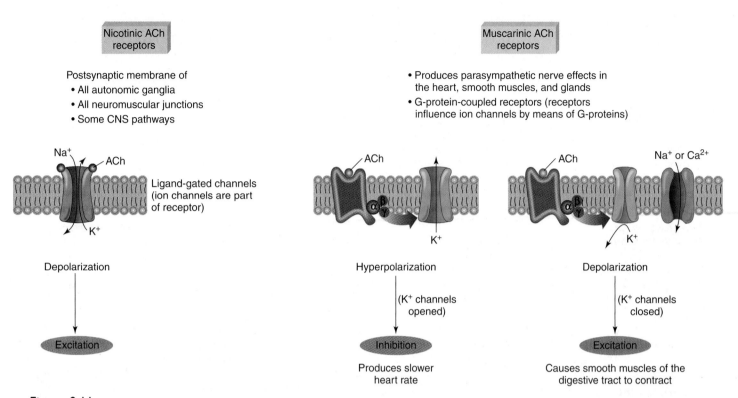

Figure 9.11 Comparison of nicotinic and muscarinic acetylcholine receptors. Nicotinic receptors are ligand-gated, meaning that the ion channel (which runs through the receptor) is opened by binding to the neurotransmitter molecule (the ligand). The muscarinic ACh receptors are G-protein coupled receptors, meaning that the binding of ACh to its receptor indirectly opens or closes ion channels through the action of G-proteins.

Table 9.6 Cholinergic Receptors and Responses to Acetylcholine

Receptor	Tissue	Response	Mechanisms
Nicotinic	Skeletal muscle	Depolarization, producing action potentials and muscle contraction	ACh opens cation channel in receptor
Nicotinic	Autonomic ganglia	Depolarization, causing activation of postganglionic neurons	ACh opens cation channel in receptor
Muscarinic (M_3, M_5)	Smooth muscle, glands	Depolarization and contraction of smooth muscle, secretion of glands	ACh activates G-protein coupled receptor, opening Ca^{2+} channels and increasing cytosolic Ca^{2+}
Muscarinic (M_2)	Heart	Hyperpolarization, slowing rate of spontaneous depolarization	ACh activates G-protein coupled receptor, opening channels for K^+

Source: Simplified from table 6-2, p. 119 of Goodman and Gilman's *The Pharmacological Basis of Therapeutics.* Ninth edition. J.E. Hardman et al., eds. 1996. McGraw-Hill.

Other Autonomic Neurotransmitters

Certain postganglionic autonomic axons produce their effects through mechanisms that do not involve either norepinephrine or acetylcholine. This can be demonstrated experimentally by the inability of drugs that block adrenergic and cholinergic effects from inhibiting the actions of those autonomic axons. These axons, consequently, have been termed "nonadrenergic, noncholinergic fibers." Proposed neurotransmitters for these axons include ATP, a polypeptide called vasoactive intestinal peptide (VIP), and nitric oxide (NO).

The nonadrenergic, noncholinergic parasympathetic axons that innervate the blood vessels of the penis cause the smooth muscles of these vessels to relax, thereby producing vasodilation and a consequent erection of the penis (chapter 20, fig. 20.23).

These parasympathetic axons have been shown to use the gas nitric oxide (chapter 7) as their neurotransmitter. In a similar manner, nitric oxide appears to function as the autonomic neurotransmitter that causes vasodilation of cerebral arteries. Studies suggest that nitric oxide is not stored in synaptic vesicles, as are other neurotransmitters, but instead is produced immediately when Ca^{2+} enters the axon terminal in response to action potentials. This Ca^{2+} indirectly activates nitric oxide synthetase, the enzyme that forms nitric oxide from the amino acid L-arginine. Nitric oxide then diffuses across the synaptic cleft and promotes relaxation of the postsynaptic smooth muscle cells.

Nitric oxide can produce relaxation of smooth muscles in many organs, including the stomach, small intestine, large intestine, and urinary bladder. There is some controversy, however, about whether the nitric oxide functions as a neurotransmitter in

Table 9.7   Adrenergic and Cholinergic Effects of Sympathetic and Parasympathetic Nerves

| | Effect of | | | |
| Organ | Sympathetic | | Parasympathetic | |
	Action	Receptor*	Action	Receptor*
Eye				
Iris				
Radial muscle	Contracts	α_1	—	—
Circular muscle	—	—	Contracts	M
Heart				
Sinoatrial node	Accelerates	β_1	Decelerates	M
Contractility	Increases	β_1	Decreases (atria)	M
Vascular Smooth Muscle				
Skin, splanchnic vessels	Contracts	α, β	—	—
Skeletal muscle vessels	Relaxes	β_2	—	—
	Relaxes	M**	—	—
Bronchiolar Smooth Muscle	Relaxes	β_2	Contracts	M
Gastrointestinal Tract				
Smooth Muscle				
Walls	Relaxes	β_2	Contracts	M
Sphincters	Constricts	α_1	Relaxes	M
Secretion	Decreases	α_1	Increases	M
Myenteric plexus	Inhibits	α_1	—	—
Genitourinary Smooth Muscle				
Bladder wall	Relaxes	β_2	Contracts	M
Urethral sphincter	Constricts	α_1	Relaxes	M
Uterus, pregnant	Relaxes	β_2	—	—
	Contracts	α_1	—	—
Penis	Ejaculation	α_1	Erection	M
Skin				
Pilomotor smooth muscle	Contracts	α_1	—	—
Sweat glands				
Thermoregulatory	Increases	M	—	—
Apocrine (stress)	Increases	α_1	—	—

Source: Reproduced and modified, with permission, from Katzung, B.G.: *Basic and Clinical Pharmacology,* 6th edition, copyright Appleton & Lange, Norwalk, CT, 1995.

*Adrenergic receptors are indicated as alpha (α) or beta (β); cholinergic receptors are indicated as muscarinic (M).

**Vascular smooth muscle in skeletal muscle has sympathetic cholinergic dilator fibers.

each case. It has been argued that, in some cases, nitric oxide could be produced in the organ itself in response to autonomic stimulation. The fact that different tissues, such as the endothelium of blood vessels, can produce nitric oxide lends support to this argument. Indeed, nitric oxide is a member of a class of local tissue regulatory molecules called *paracrine regulators* (see chapter 11). Regulation can therefore be a complex process involving the interacting effects of different neurotransmitters, hormones, and paracrine regulators.

Organs with Dual Innervation

Most visceral organs receive **dual innervation**—they are innervated by both sympathetic and parasympathetic fibers. In this condition, the effects of the two divisions of the autonomic system may be antagonistic, complementary, or cooperative (table 9.7).

Antagonistic Effects

The effects of sympathetic and parasympathetic innervation of the pacemaker region of the heart is the best example of the antagonism of these two systems. In this case, sympathetic and parasympathetic fibers innervate the same cells. Adrenergic stimulation from sympathetic fibers increases the heart rate, whereas the release of acetylcholine from parasympathetic fibers decreases the heart rate. A reverse of this antagonism is seen in the digestive tract, where sympathetic nerves inhibit and parasympathetic nerves stimulate intestinal movements and secretions.

The effects of sympathetic and parasympathetic stimulation on the diameter of the pupil of the eye are analogous to the reciprocal innervation of flexor and extensor skeletal muscles by somatic motor neurons (see chapter 12). This is because the iris contains antagonistic muscle layers. Contraction of the radial muscles, which are innervated by sympathetic nerves, causes dilation;

contraction of the circular muscles, which are innervated by parasympathetic nerve endings, causes constriction of the pupils (chapter 10, fig. 10.27).

Complementary and Cooperative Effects

The effects of sympathetic and parasympathetic nerves are generally antagonistic; in a few cases, however, they can be complementary or cooperative. The effects are complementary when sympathetic and parasympathetic stimulation produce similar effects. The effects are cooperative, or synergistic, when sympathetic and parasympathetic stimulation produce different effects that work together to promote a single action.

The effects of sympathetic and parasympathetic stimulation on salivary gland secretion are complementary. The secretion of watery saliva is stimulated by parasympathetic nerves, which also stimulate the secretion of other exocrine glands in the digestive tract. Sympathetic nerves stimulate the constriction of blood vessels throughout the digestive tract. The resultant decrease in blood flow to the salivary glands causes the production of a thicker, more viscous saliva.

The effects of sympathetic and parasympathetic stimulation on the reproductive and urinary systems are cooperative. Erection of the penis, for example, is due to vasodilation resulting from parasympathetic nerve stimulation; ejaculation is due to stimulation through sympathetic nerves. The two divisions of the autonomic system thus cooperate to enable sexual function in the male. They also cooperate in the female; clitoral erection and vaginal secretions are stimulated by parasympathetic nerves, whereas orgasm is a sympathetic nerve response, as it is in the male.

There is also cooperation between the two divisions in the micturition (urination) reflex. Although the contraction of the urinary bladder is largely independent of nerve stimulation, it is promoted in part by the action of parasympathetic nerves. This reflex is also enhanced by sympathetic nerve activity, which increases the tone of the bladder muscles. Emotional states that are accompanied by high sympathetic nerve activity (such as extreme fear) may thus result in reflex urination at bladder volumes that are normally too low to trigger this reflex.

Organs without Dual Innervation

Although most organs are innervated by both sympathetic and parasympathetic nerves, some—including the adrenal medulla, arrector pili muscles, sweat glands, and most blood vessels—receive only sympathetic innervation. In these cases, regulation is achieved by increases or decreases in the tone (firing rate) of the sympathetic fibers. Constriction of cutaneous blood vessels, for example, is produced by increased sympathetic activity that stimulates alpha-adrenergic receptors, and vasodilation results from decreased sympathetic nerve stimulation.

The sympathoadrenal system is required for *nonshivering thermogenesis:* animals deprived of their sympathetic system and adrenals cannot tolerate cold stress. The sympathetic system itself is required for proper thermoregulatory responses to heat. In a hot room, for example, decreased sympathetic stimulation produces dilation of the blood vessels in the skin, which increases cutaneous blood flow and provides better heat radiation. During exercise, by contrast, sympathetic activity increases, causing constriction of the blood vessels in the skin of the limbs and stimulation of sweat glands in the trunk.

Autonomic dysreflexia, a serious condition producing rapid elevations in blood pressure that can lead to stroke (cerebrovascular accident), occurs in 85% of people with quadriplegia and others with spinal cord lesions above the sixth thoracic level. Lesions to the spinal cord first produce the symptoms of *spinal shock,* characterized by the loss of both skeletal muscle and autonomic reflexes. After a period of time, both types of reflexes return in an exaggerated state. The skeletal muscles may become spastic in the absence of higher inhibitory influences, and the visceral organs experience denervation hypersensitivity. Patients in this condition have difficulty emptying their urinary bladders and often must be catheterized.

Noxious stimuli, such as overdistension of the urinary bladder, can result in reflex activation of the sympathetic nerves below the spinal cord lesion. This produces goose bumps, cold skin, and vasoconstriction in the regions served by the spinal cord below the level of the lesion. The rise in blood pressure resulting from this vasoconstriction activates pressure receptors that transmit impulses along sensory nerve fibers to the medulla oblongata. In response to this sensory input, the medulla directs a reflex slowing of the heart and vasodilation. Since descending impulses are blocked by the spinal lesion, however, the skin above the lesion is warm and moist (due to vasodilation and sweat gland secretion), but it is cold below the level of spinal cord damage.

The sweat glands in the trunk secrete a watery fluid in response to cholinergic sympathetic stimulation. Evaporation of this dilute sweat helps to cool the body. The sweat glands also secrete a chemical called *bradykinin* in response to sympathetic stimulation. Bradykinin stimulates dilation of the surface blood vessels near the sweat glands, helping to radiate some heat despite the fact that other cutaneous blood vessels are constricted. At the conclusion of exercise, sympathetic stimulation is reduced, causing cutaneous blood vessels to dilate. This increases blood flow to the skin, which helps to eliminate metabolic heat. Notice that all of these thermoregulatory responses are achieved without the direct involvement of the parasympathetic system.

Table 9.8 Effects Resulting from Sensory Input from Afferent Fibers in the Vagus, Which Transmit This Input to Centers in the Medulla Oblongata

Organs	Type of Receptors	Reflex Effects
Lungs	Stretch receptors	Further inhalation inhibited; increase in cardiac rate and vasodilation stimulated
	Type J receptors	Stimulated by pulmonary congestion—produces feelings of breathlessness and causes a reflex fall in cardiac rate and blood pressure
Aorta	Chemoreceptors	Stimulated by rise in CO_2 and fall in O_2—produces increased rate of breathing, rise in heart rate, and vasoconstriction
	Baroreceptors	Stimulated by increased blood pressure—produces a reflex decrease in heart rate
Heart	Atrial stretch receptors	Antidiuretic hormone secretion inhibited, thus increasing the volume of urine excreted
	Stretch receptors in ventricles	Produces a reflex decrease in heart rate and vasodilation
Gastrointestinal tract	Stretch receptors	Feelings of satiety, discomfort, and pain

Control of the Autonomic Nervous System by Higher Brain Centers

Visceral functions are largely regulated by autonomic reflexes. In most autonomic reflexes, sensory input is transmitted to brain centers that integrate this information and respond by modifying the activity of preganglionic autonomic neurons. The neural centers that directly control the activity of autonomic nerves are influenced by higher brain areas, as well as by sensory input.

The **medulla oblongata** of the brain stem is the area that most directly controls the activity of the autonomic system. Almost all autonomic responses can be elicited by experimental stimulation of the medulla, where centers for the control of the cardiovascular, pulmonary, urinary, reproductive, and digestive systems are located. Much of the sensory input to these centers travels in the afferent fibers of the vagus nerve—a mixed nerve containing both sensory and motor fibers. The reflexes that result are listed in table 9.8.

Although it directly regulates the activity of autonomic motor fibers, the medulla itself is responsive to regulation by higher brain areas. One of these areas is the hypothalamus, the brain region that contains centers for the control of body temperature, hunger, and thirst; for regulation of the pituitary gland; and (together with the limbic system and cerebral cortex) for various emotional states.

As described in chapter 8, the limbic system is a group of fiber tracts and nuclei that form a ring around the brain stem. It includes the cingulate gyrus of the cerebral cortex, the hypothalamus, the fornix (a fiber tract), the hippocampus, and the amygdaloid nucleus (see fig. 8.14). The limbic system is involved in basic emotional drives, such as anger, fear, sex, and hunger. The involvement of the limbic system with the control of autonomic function is responsible for the visceral responses that are characteristic of these emotional states. Blushing, pallor, fainting, breaking out in a cold sweat, a racing heartbeat, and "butterflies in the stomach" are only some of the many visceral reactions that accompany emotions as a result of autonomic activation.

The autonomic correlates of motion sickness—nausea, sweating, and cardiovascular changes—are eliminated by cutting the motor tracts of the cerebellum. This demonstrates that

impulses from the cerebellum to the medulla oblongata influence activity of the autonomic nervous system. Experimental and clinical observations have also demonstrated that the frontal and temporal lobes of the cerebral cortex influence lower brain areas as part of their involvement in emotion and personality.

Traditionally, the distinction between the somatic system and the autonomic nervous system was drawn on the basis that the former is under conscious control whereas the latter is not. Recently, however, we have learned that conscious processes in the cerebrum can influence autonomic activity. In **biofeedback** techniques, data obtained from devices that detect and amplify changes in blood pressure and heart rate, for example, are "fed back" to patients in the form of light signals or audible tones. The patients can often be trained to consciously reduce the frequency of the signals and, eventually, to control visceral activities without the aid of a machine. Biofeedback has been used successfully to treat hypertension, stress, and migraine headaches.

Test Yourself Before You Continue

1. Define *adrenergic* and *cholinergic* and use these terms to describe the neurotransmitters of different autonomic nerve fibers.

2. List the effects of sympathoadrenal stimulation on different effector organs. In each case, indicate whether the effect is due to alpha- or beta-receptor stimulation.

3. Describe the effects of the drug atropine and explain these effects in terms of the actions of the parasympathetic system.

4. Explain how the effects of the sympathetic and parasympathetic systems can be antagonistic, cooperative, or complementary. Include specific examples of these different types of effects in your explanation.

5. Explain the mechanisms involved when a person blushes. What structures are involved in this response?

HPer Links of the Nervous System with Other Body Systems

Integumentary System

- The skin houses receptors for heat, cold, pain, pressure, and vibration ..... **(p. 244)**
- Afferent neurons conduct impulses from cutaneous receptors .... **(p. 245)**
- Sympathetic neurons to the skin help to regulate cutaneous blood flow ... **(p. 428)**

Skeletal System

- The skeleton supports and protects the brain and spinal cord .... **(p. 190)**
- Bones store calcium needed for neural function .... **(p. 623)**
- Afferent neurons from sensory receptors monitor movements of joints .... **(p. 242)**

Muscular System

- Muscle contractions generate body heat to maintain constant temperature for neural function .... **(p. 608)**
- Afferent neurons from muscle spindles transmit impulses to the CNS .... **(p. 348)**
- Somatic motor neurons innervate skeletal muscles .... **(p. 347)**
- Autonomic motor neurons innervate cardiac and smooth muscles ..... **(p. 220)**

Endocrine System

- Many hormones, including sex steroids, act on the brain .... **(p. 304)**
- Hormones and neurotransmitters, such as epinephrine and norepinephrine, can have synergistic actions on a target tissue .... **(p. 290)**
- Autonomic neurons innervate endocrine glands such as the pancreatic islets .... **(p. 613)**
- The brain controls anterior pituitary function .... **(p. 301)**
- The brain controls posterior pituitary function .... **(p. 301)**

Circulatory System

- The circulatory system transports O_2 and CO_2, nutrients, and fluids to and from all organs, including the brain and spinal cord .... **(p. 366)**

- Autonomic nerves help to regulate cardiac output .... **(p. 411)**
- Autonomic nerves promote constriction and dilation of blood vessels, helping to regulate blood flow and blood pressure .... **(p. 420)**

Immune System

- Chemical factors called cytokines, released by cells of the immune system, act on the brain to promote a fever .... **(p. 448)**
- Cytokines from the immune system act on the brain to modify its regulation of pituitary gland secretion .... **(p. 462)**
- The nervous system plays a role in regulating the immune response ..**(p. 462)**

Respiratory System

- The lungs provide oxygen for all body systems and eliminate carbon dioxide .... **(p. 480)**
- Neural centers within the brain control breathing .... **(p. 499)**

Urinary System

- The kidneys eliminate metabolic wastes and help to maintain homeostasis of the blood plasma .... **(p. 524)**
- The kidneys regulate plasma concentrations of Na^+, K^+, and other ions needed for the functioning of neurons .... **(p. 544)**
- The nervous system innervates organs of the urinary system to control urination .... **(p. 525)**
- Autonomic nerves help to regulate renal blood flow .... **(p. 531)**

Digestive System

- The GI tract provides nutrients for all body organs, including those of nervous system .... **(p. 561)**
- Autonomic nerves innervate digestive organs .... **(p. 563)**

- The GI tract contains a complex neural system, called an enteric brain, that regulates its motility and secretions .... **(p. 585)**
- Secretions of gastric juice can be stimulated through activation of brain regions .... **(p. 583)**
- Hunger is controlled by centers in the hypothalamus of the brain .... **(p. 606)**

Reproductive System

- Gonads produce sex hormones that influence brain development ..... **(p. 640)**
- The brain helps to regulate secretions of gonadotropic hormones from the anterior pituitary .... **(p. 640)**
- Autonomic nerves regulate blood flow into the external genitalia, contributing to the male and female sexual response .... **(p. 643)**
- The nervous and endocrine systems cooperate in the control of lactation .... **(p. 677)**

Summary

Neural Control of Involuntary Effectors 220

I. Preganglionic autonomic neurons originate in the brain or spinal cord; postganglionic neurons originate in ganglia located outside the CNS.

II. Smooth muscle, cardiac muscle, and glands receive autonomic innervation.

    A. The involuntary effectors are somewhat independent of their innervation and become hypersensitive when their innervation is removed.

    B. Autonomic nerves can have either excitatory or inhibitory effects on their target organs.

Divisions of the Autonomic Nervous System 222

I. Preganglionic neurons of the sympathetic division originate in the spinal cord, between the thoracic and lumbar levels.

    A. Many of these fibers synapse with postganglionic neurons whose cell bodies are located in a double chain of sympathetic (paravertebral) ganglia outside the spinal cord.

    B. Some preganglionic fibers synapse in collateral (prevertebral) ganglia. These are the celiac, superior mesenteric, and inferior mesenteric ganglia.

    C. Some preganglionic fibers innervate the adrenal medulla, which secretes epinephrine (and some norepinephrine) into the blood in response to stimulation.

II. Preganglionic parasympathetic fibers originate in the brain and in the sacral levels of the spinal cord.

    A. Preganglionic parasympathetic fibers contribute to cranial nerves III, VII, IX, and X.

    B. The long preganglionic fibers of the vagus (X) nerve synapse in terminal ganglia located next to or within the innervated organ. Short postganglionic fibers then innervate the effector cells.

    C. The vagus provides parasympathetic innervation to the heart, lungs, esophagus, stomach, liver, small intestine, and upper half of the large intestine.

    D. Parasympathetic outflow from the sacral levels of the spinal cord innervates terminal ganglia in the lower half of the large intestine, in the rectum, and in the urinary and reproductive systems.

Functions of the Autonomic Nervous System 227

I. The sympathetic division of the autonomic system activates the body to "fight or flight" through adrenergic effects. The parasympathetic division often exerts antagonistic actions through cholinergic effects.

II. All preganglionic autonomic nerve fibers are cholinergic (use ACh as a neurotransmitter).

    A. All postganglionic parasympathetic fibers are cholinergic.

    B. Most postganglionic sympathetic fibers are adrenergic (use norepinephrine as a neurotransmitter).

    C. Sympathetic fibers that innervate sweat glands and those that innervate blood vessels in skeletal muscles are cholinergic.

III. Adrenergic effects include stimulation of the heart, vasoconstriction in the viscera and skin, bronchodilation, and glycogenolysis in the liver.

A. The two main classes of adrenergic receptor proteins are alpha and beta.

B. Some organs have only alpha or only beta receptors; other organs (such as the heart) have both types of receptors.

C. There are two subtypes of alpha receptors (α_1 and α_2) and two subtypes of beta receptors (β_1 and β_2). These subtypes can be selectively stimulated or blocked by therapeutic drugs.

IV. Cholinergic effects of parasympathetic nerves are promoted by the drug muscarine and inhibited by atropine.

V. In organs with dual innervation, the effects of the sympathetic and parasympathetic divisions can be antagonistic, complementary, or cooperative.

    A. The effects are antagonistic in the heart and pupils of the eyes.

    B. The effects are complementary in the regulation of salivary gland secretion and are cooperative in the regulation of the reproductive and urinary systems.

VI. In organs without dual innervation (such as most blood vessels), regulation is achieved by variations in sympathetic nerve activity.

VII. The medulla oblongata of the brain stem is the area that most directly controls the activity of the autonomic system.

    A. The medulla oblongata is in turn influenced by sensory input and by input from the hypothalamus.

    B. The hypothalamus is influenced by input from the limbic system, cerebellum, and cerebrum. These interconnections provide an autonomic component to some of the visceral responses that accompany emotions.

Review Activities

Test Your Knowledge of Terms and Facts

1. When a visceral organ is denervated,
    a. it ceases to function.
    b. it becomes less sensitive to subsequent stimulation by neurotransmitters.
    c. it becomes hypersensitive to subsequent stimulation.

2. Parasympathetic ganglia are located
    a. in a chain parallel to the spinal cord.
    b. in the dorsal roots of spinal nerves.
    c. next to or within the organs innervated.
    d. in the brain.

3. The neurotransmitter of preganglionic sympathetic fibers is
    a. norepinephrine.
    b. epinephrine.
    c. acetylcholine.
    d. dopamine.

4. Which of these results from stimulation of alpha-adrenergic receptors?
   a. constriction of blood vessels
   b. dilation of bronchioles
   c. decreased heart rate
   d. sweat gland secretion

5. Which of these fibers release norepinephrine?
   a. preganglionic parasympathetic fibers
   b. postganglionic parasympathetic fibers
   c. postganglionic sympathetic fibers in the heart
   d. postganglionic sympathetic fibers in sweat glands
   e. all of these

6. The effects of sympathetic and parasympathetic fibers are cooperative in
   a. the heart.
   b. the reproductive system.

   c. the digestive system.
   d. the eyes.

7. Propranolol is a beta blocker. It would therefore cause
   a. vasodilation.
   b. slowing of the heart rate.
   c. increased blood pressure.
   d. secretion of saliva.

8. Atropine blocks parasympathetic nerve effects. It would therefore cause
   a. dilation of the pupils.
   b. decreased mucus secretion.
   c. decreased movements of the digestive tract.
   d. increased heart rate.
   e. all of these.

9. Which area of the brain is most directly involved in the reflex control of the autonomic system?
   a. hypothalamus
   b. cerebral cortex

   c. medulla oblongata
   d. cerebellum

10. The two subtypes of cholinergic receptors are
    a. adrenergic and nicotinic.
    b. dopaminergic and muscarinic.
    c. nicotinic and muscarinic.
    d. nicotinic and dopaminergic.

11. A fall in cyclic AMP within the target cell occurs when norepinephrine binds to which of adrenergic receptors?
    a. α_1
    b. α_2
    c. β_1
    d. β_2

12. A drug that serves as an agonist for α_2 receptors can be used to
    a. increase the heart rate.
    b. decrease the heart rate.
    c. dilate the bronchioles.
    d. constrict the bronchioles.
    e. constrict the blood vessels.

Test Your Understanding of Concepts and Principles

1. Compare the sympathetic and parasympathetic systems in terms of the location of their ganglia and the distribution of their nerves.[1]

2. Explain the anatomical and physiological relationship between the sympathetic nervous system and the adrenal glands.

3. Compare the effects of adrenergic and cholinergic stimulation on the cardiovascular and digestive systems.

4. Explain how effectors that receive only sympathetic innervation are regulated by the autonomic system.

5. Distinguish between the different types of adrenergic receptors and state where these receptors are located in the body.

6. Give examples of drugs that selectively stimulate or block different adrenergic receptors and explain how these drugs are used clinically.

7. Explain what is meant by nicotinic and muscarinic ACh receptors and describe the distribution of these receptors in the body.

8. Give examples of drugs that selectively stimulate and block the nicotinic and muscarinic receptors and explain how these drugs are used clinically.

Test Your Ability to Analyze and Apply Your Knowledge

1. Shock is the medical condition that results when body tissues do not receive enough oxygen-carrying blood. It is characterized by low blood flow to the brain, leading to decreased levels of consciousness. Why would a patient with a cervical spinal cord injury be at risk of going into shock?

2. A person in shock may have pale, cold, and clammy skin and a rapid and weak pulse. What is the role of the autonomic nervous system in producing these symptoms? Discuss how drugs that influence autonomic

activity might be used to treat someone in shock.

3. Imagine yourself at the starting block of the 100-meter dash of the Olympics. The gun is about to go off in the biggest race of your life. What is the autonomic nervous system doing at this point? How are your organs responding?

4. Some patients with hypertension (high blood pressure) are given beta-blocking drugs to lower their blood pressure. How does this effect occur?

Explain why these drugs are not administered to patients with a history of asthma. Why might drinking coffee help asthma?

5. Why do many cold medications contain an alpha-adrenergic agonist and atropine (belladonna)? Why is there a label warning for people with hypertension? Why would a person with gastritis be given a prescription for atropine? Explain how this drug might affect the ability to digest and absorb food.

Related Websites

Check out the Links Library at www.mhhe.com/fox8 for links to sites containing resources related to the autonomic nervous system. These links are monitored to ensure current URLs.

[1]*Note:* This question is answered in the chapter 9 Study Guide found on the Online Learning Center at www.mhhe.com/fox8.

10 Sensory Physiology

Objectives

After studying this chapter, you should be able to . . .

1. explain how sensory receptors are categorized, give examples of functional categories, and explain how tonic and phasic receptors differ.

2. explain the law of specific nerve energies.

3. describe the characteristics of the generator potential.

4. give examples of different types of cutaneous receptors and describe the neural pathways for the cutaneous senses.

5. explain the concepts of receptive fields and lateral inhibition.

6. Explain how taste cells are stimulated by foods that are salty, sour, sweet, and bitter.

7. describe the structure and function of the olfactory receptors and explain how odor discrimination might be accomplished.

8. describe the structure of the vestibular apparatus and explain how it provides information about acceleration of the body in different directions.

9. describe the functions of the outer and middle ear.

10. describe the structure of the cochlea and explain how movements of the stapes against the oval window result in vibrations of the basilar membrane.

11. explain how mechanical energy is converted into nerve impulses by the organ of Corti and how pitch perception is accomplished.

12. describe the structure of the eye and explain how images are brought to a focus on the retina.

13. explain how visual accommodation is achieved and describe the defects associated with myopia, hyperopia, and astigmatism.

14. describe the architecture of the retina and trace the pathways of light and nerve activity through the retina.

15. describe the function of rhodopsin in the rods and explain how dark adaptation is achieved.

16. explain how light affects the electrical activity of rods and their synaptic input to bipolar cells.

17. explain the trichromatic theory of color vision.

18. compare rods and cones with respect to their locations, synaptic connections, and functions.

19. describe the neural pathways from the retina, explaining the differences in pathways from different regions of the visual field.

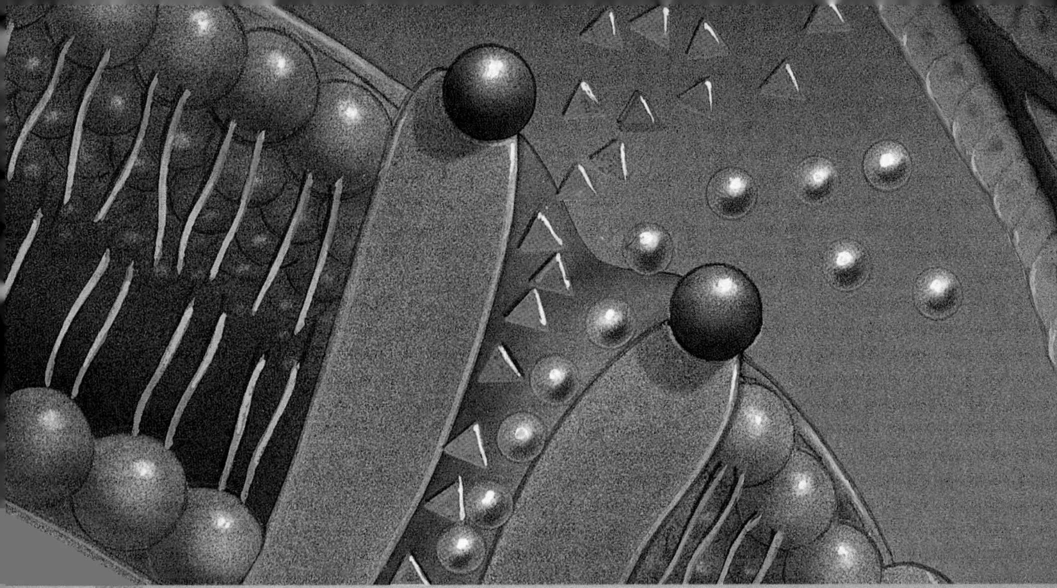

Chapter at a Glance

Take Advantage of the Technology

*Visit the Online Learning Center for
these additional study resources.*

▪ Interactive quizzing

▪ Online study guide

▪ Current news feeds

▪ Crossword puzzles

▪ Vocabulary flashcards

▪ Labeling activities

www.mhhe.com/fox8

Clinical Investigation

Ed is a 45-year-old man who goes to the doctor complaining of severe ear pain and reduced hearing immediately after disembarking from an international flight. It is apparent that Ed has a bad head cold, and the doctor recommends that he take a decongestant. He further recommends that Ed come back after the cold is better for an audiology test, if his hearing has not improved by then. While talking to the doctor, Ed complains that he can't see print very clearly anymore, even though he's never worn glasses. However, he tells the doctor that his distant vision, and ability to drive, are still fine.

What may have caused Ed's ear pain and reduced hearing? What may be responsible for his impaired ability to see print?

Characteristics of Sensory Receptors

Each type of sensory receptor responds to a particular modality of environmental stimulus by causing the production of action potentials in a sensory neuron. These impulses are conducted to parts of the brain that provide the proper interpretation of the sensory information when that particular neural pathway is activated.

Our perceptions of the world—its textures, colors, and sounds; its warmth, smells, and tastes—are created by the brain from electrochemical nerve impulses delivered to it from sensory receptors. These receptors **transduce** (change) different forms of energy in the "real world" into the energy of nerve impulses that are conducted into the central nervous system by sensory neurons. Different *modalities* (forms) of sensation—sound, light, pressure, and so forth—result from differences in neural pathways and synaptic connections. The brain thus interprets impulses arriving from the auditory nerve as sound and from the optic nerve as sight, even though the impulses themselves are identical in the two nerves.

We know, through the use of scientific instruments, that our senses act as energy filters that allow us to perceive only a narrow range of energy. Vision, for example, is limited to light in the visible spectrum; ultraviolet and infrared light, X rays and radio waves, which are the same type of energy as visible light, cannot normally excite the photoreceptors in the eyes. The perception of cold is entirely a product of the nervous system—there is no such thing as cold in the physical world, only varying degrees of heat. The perception of cold, however, has obvious survival value. Although filtered and distorted by the limitations of sensory function, our perceptions of the world allow us to interact effectively with the environment.

Categories of Sensory Receptors

Sensory receptors can be categorized on the basis of structure or various functional criteria. Structurally, the sensory receptors may be the dendritic endings of sensory neurons. These dendritic endings may be free—such as those that respond to pain and temperature—or encapsulated within nonneural structures—such as those that respond to pressure (see fig. 10.4). The photoreceptors in the retina of the eyes (rods and cones) are highly specialized neurons that synapse with other neurons in the retina. In the case of taste buds and of hair cells in the inner ears, modified epithelial cells respond to an environmental stimulus and activate sensory neurons.

Functional Categories

Sensory receptors can be grouped according to the type of stimulus energy they transduce. These categories include (1) **chemoreceptors,** which sense chemical stimuli in the environment or the blood (e.g., the taste buds, olfactory epithelium, and the aortic and carotid bodies); (2) **photoreceptors**—the rods and cones in the retina of the eye; (3) **thermoreceptors,** which respond to heat and cold; and (4) **mechanoreceptors,** which are stimulated by mechanical deformation of the receptor cell membrane (e.g., touch and pressure receptors in the skin and hair cells within the inner ear).

Nociceptors—or pain receptors—have a higher threshold for activation than do the other cutaneous receptors; thus, a more intense stimulus is required for their activation. Their firing rate then increases with stimulus intensity. Receptors that subserve other sensations may also become involved in pain transmission when the stimulus is prolonged, particularly when tissue damage occurs.

Receptors also can be grouped according to the type of sensory information they deliver to the brain. **Proprioceptors** include the muscle spindles, Golgi tendon organs, and joint receptors. These provide a sense of body position and allow fine control of skeletal movements (as discussed in chapter 12). **Cutaneous** (skin) **receptors** include (1) touch and pressure receptors, (2) heat and cold receptors, and (3) pain receptors. The receptors that mediate sight, hearing, and equilibrium are grouped together as the **special senses.**

Tonic and Phasic Receptors: Sensory Adaptation

Some receptors respond with a burst of activity when a stimulus is first applied, but then quickly decrease their firing rate—adapt to the stimulus—if the stimulus is maintained. Receptors with this response pattern are called *phasic receptors.* Receptors that produce a relatively constant rate of firing as long as the stimulus is maintained are known as *tonic receptors* (fig. 10.1).

Phasic receptors alert us to changes in sensory stimuli and are in part responsible for the fact that we can cease paying attention to constant stimuli. This ability is called **sensory adaptation.** Odor, touch, and temperature, for example, adapt rapidly; bathwater feels hotter when we first enter it. Sensations of pain, by contrast, adapt little if at all.

Law of Specific Nerve Energies

Stimulation of a sensory nerve fiber produces only one sensation—touch, cold, pain, and so on. According to the **law of specific nerve energies,** the sensation characteristic of each sensory neuron is that produced by its normal stimulus, or *adequate stimulus* (table 10.1).

Also, although a variety of different stimuli may activate a receptor, the adequate stimulus requires the least amount of energy to do so. The adequate stimulus for the photoreceptors of the eye, for example, is light, where a single photon can have a measurable effect. If these receptors are stimulated by some other means—such as by the high pressure produced by a punch to the eye—a flash of light (the adequate stimulus) may be perceived.

The effect of *paradoxical cold* provides another example of the law of specific nerve energies. When the tip of a cold metal rod is touched to the skin, the perception of cold gradually disappears as the rod warms to body temperature. Then, when the tip of a rod heated to 45° C is applied to the same spot, the sensation of cold is perceived once again. This paradoxical cold is produced because the heat slightly damages receptor endings, and by this means produces an "injury current" that stimulates the receptor.

Regardless of how a sensory neuron is stimulated, therefore, only one sensory modality will be perceived. This specificity is due to the synaptic pathways within the brain that are activated by the sensory neuron. The ability of receptors to function as sensory filters so that they are stimulated by only one type of stimulus (the adequate stimulus) allows the brain to perceive the stimulus accurately under normal conditions.

Generator (Receptor) Potential

The electrical behavior of sensory nerve endings is similar to that of the dendrites of other neurons. In response to an environmental stimulus, the sensory endings produce local graded changes in the membrane potential. In most cases, these potential changes are depolarizations that are analogous to the excitatory postsynaptic potentials (EPSPs) described in chapter 7. In the sensory endings, however, these potential changes in response to environmental stimulation are called **receptor,** or **generator, potentials** because they serve to generate action potentials in response to the sensory stimulation. Since sensory neurons are pseudounipolar (chapter 7), the action potentials produced in response to the generator potential are conducted continuously from the periphery into the CNS.

The *pacinian,* or *lamellated, corpuscle,* a cutaneous receptor for pressure (see fig. 10.4), can serve as an example of sensory transduction. When a light touch is applied to the receptor, a small depolarization (the generator potential) is produced. Increasing the pressure on the pacinian corpuscle increases the magnitude of the generator potential until it reaches the threshold depolarization required to produce an action potential (fig. 10.2). The pacinian corpuscle, however, is a phasic receptor; if the pressure is maintained, the size of the generator potential produced quickly diminishes. It is interesting to note that this phasic response is a result of the onionlike covering around the dendritic nerve ending; if the layers are peeled off and the nerve ending is stimulated directly, it will respond in a tonic fashion.

When a tonic receptor is stimulated, the generator potential it produces is proportional to the intensity of the stimulus. After a threshold depolarization is produced, increases in the amplitude of the generator potential result in increases in the *frequency* with which action potentials are produced (fig. 10.3). In this way, the frequency of action potentials that are conducted into the central nervous system serves as the code for the strength of the stimulus. As described in chapter 7, this frequency code is needed because the amplitude of action potentials is constant (all or none). Acting through changes in action potential frequency, tonic receptors thus provide information about the relative intensity of a stimulus.

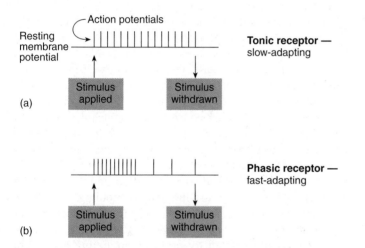

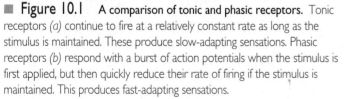

Figure 10.1 A comparison of tonic and phasic receptors. Tonic receptors (a) continue to fire at a relatively constant rate as long as the stimulus is maintained. These produce slow-adapting sensations. Phasic receptors (b) respond with a burst of action potentials when the stimulus is first applied, but then quickly reduce their rate of firing if the stimulus is maintained. This produces fast-adapting sensations.

Table 10.1 Classification of Receptors Based on Their Normal (or "Adequate") Stimulus

Receptor	Normal Stimulus	Mechanisms	Examples
Mechanoreceptors	Mechanical force	Deforms cell membranes of sensory dendrites or deforms hair cells that activate sensory nerve endings	Cutaneous touch and pressure receptors; vestibular apparatus and cochlea
Pain receptors	Tissue damage	Damaged tissues release chemicals that excite sensory endings	Cutaneous pain receptors
Chemoreceptors	Dissolved chemicals	Chemical interaction affects ionic permeability of sensory cells	Smell and taste (exteroceptors) osmoreceptors and carotid body chemoreceptors (interoceptors)
Photoreceptors	Light	Photochemical reaction affects ionic permeability of receptor cell	Rods and cones in retina of eye

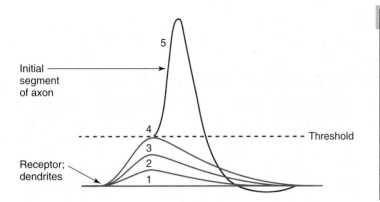

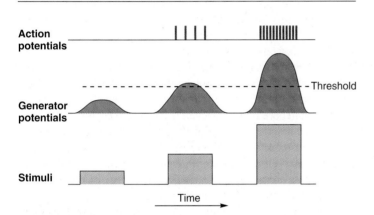

Figure 10.2   The receptor (generator) potential. Sensory stimuli result in the production of local graded potential changes known as receptor, or generator, potentials (numbers 1–4). If the receptor potential reaches a threshold value of depolarization, it generates action potentials (number 5) in the sensory neuron.

Figure 10.3   The response of tonic receptors to stimuli. Three successive stimuli of increasing strengths are delivered to a receptor. The increasing amplitude of the generator potential results in increases in the frequency of action potentials, which persist as long as the stimulus is maintained.

Test Yourself before You Continue

1. Our perceptions are products of our brains; they relate to physical reality only indirectly and incompletely. Explain this statement, using examples of vision and the perception of cold.

2. Explain what is meant by the law of specific nerve energies and the adequate stimulus, and relate these concepts to your answer for question no. 1.

3. Describe sensory adaptation in olfactory and pain receptors. Using a line drawing, relate sensory adaptation to the responses of phasic and tonic receptors.

4. Explain how the magnitude of a sensory stimulus is transduced into a receptor potential and how the magnitude of the receptor potential is coded in the sensory nerve fiber.

Cutaneous Sensations

There are several different types of sensory receptors in the skin, each of which is specialized to be maximally sensitive to one modality of sensation. A receptor will be activated when a given area of the skin is stimulated; this area is the receptive field of that receptor. A process known as lateral inhibition helps to sharpen the perceived location of the stimulus on the skin.

The **cutaneous sensations** of touch, pressure, heat and cold, and pain are mediated by the dendritic nerve endings of different sensory neurons. The receptors for heat, cold, and pain are simply the naked endings of sensory neurons. Sensations of touch are mediated by naked dendritic endings surrounding hair follicles and by expanded dendritic endings, called Ruffini endings and Merkel's discs. The sensations of touch and pressure are also mediated by dendrites that are encapsulated within various structures (table 10.2); these include Meissner's corpuscles and pacinian (lamellated) corpuscles. In pacinian corpuscles, for example, the dendritic endings are encased within thirty to fifty onionlike layers of connective tissue (fig. 10.4). These layers absorb some of the pressure when a stimulus is maintained, which helps to accentuate the phasic response of this receptor. The encapsulated touch receptors thus adapt rapidly, in contrast to the more slowly adapting Ruffini endings and Merkel's discs.

There are far more free dendritic endings that respond to cold than to warm. The receptors for cold are located in the upper region of the dermis, just below the epidermis. These receptors are stimulated by cooling and inhibited by warming. The warm receptors are located somewhat deeper in the dermis and are excited by warming and inhibited by cooling. Nociceptors are also free sensory nerve endings of either myelinated or unmyelinated fibers. The initial sharp sensation of pain, as from a pin-prick, is transmitted by rapidly conducting myelinated axons, whereas a dull, persistent ache is transmitted by slower conducting unmyelinated axons. These afferent neurons synapse in the spinal cord, using substance P (an eleven-amino-acid polypeptide) and glutamate as neurotransmitters.

Hot temperatures produce sensations of pain through the action of a particular membrane protein in sensory dendrites. This protein, called a *capsaicin receptor*, serves as both an ion channel and a receptor for capsaicin—the molecule in chili peppers that causes sensations of heat and pain. In response to a noxiously high temperature, or to capsaicin in chili peppers, these ion channels open. This allows Ca^{2+} and Na^+ to diffuse into the neuron, producing depolarization and resulting action potentials that are transmitted to the CNS and perceived as heat and pain.

While the capsaicin receptor for pain is activated by intense heat, other nociceptors may be activated by mechanical stimuli that cause cellular damage. There is evidence that ATP released from damaged cells can cause pain, as can a local fall in pH produced during infection and inflammation.

Table 10.2   Cutaneous Receptors

Receptor	Structure	Sensation	Location
Free nerve endings	Unmyelinated dendrites of sensory neurons	Light touch; hot; cold; nociception (pain)	Around hair follicles; throughout skin
Merkel's discs	Expanded dendritic endings	Sustained touch and pressure	Base of epidermis (stratum basale)
Ruffini corpuscles (endings)	Enlarged dendritic endings with open, elongated capsule	Sustained pressure	Deep in dermis and hypodermis
Meissner's corpuscles	Dendrites encapsulated in connective tissue	Changes in texture; slow vibrations	Upper dermis (papillary layer)
Pacinian corpuscles	Dendrites encapsulated by concentric lamellae of connective tissue structures	Deep pressure; fast vibrations	Deep in dermis

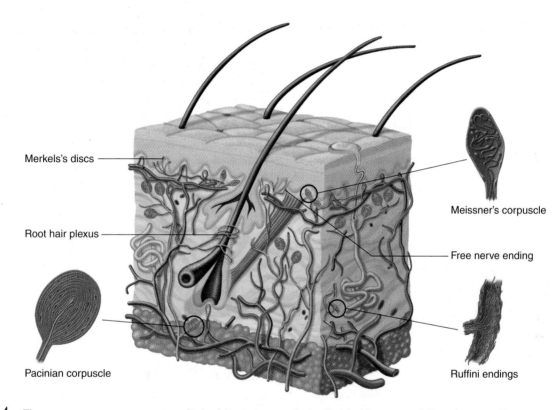

Merkels's discs

Root hair plexus

Pacinian corpuscle

Meissner's corpuscle

Free nerve ending

Ruffini endings

Figure 10.4   **The cutaneous sensory receptors.** Each of these structures is associated with a sensory (afferent) neuron. Free nerve endings are naked, dendritic branches that serve a variety of cutaneous sensations, including that of heat. Some cutaneous receptors are dendritic branches encapsulated within associated structures. Examples of this type include the pacinian (lamellated) corpuscles, which provide a sense of deep pressure, and the Meissner's corpuscles, which provide cutaneous information related to changes in texture.

Neural Pathways for Somatesthetic Sensations

The conduction pathways for the **somatesthetic senses**—a term that includes sensations from cutaneous receptors and proprioceptors—are shown in chapter 8 (fig. 8.20). These pathways involve three orders of neurons in series. Sensory information from proprioceptors and pressure receptors is first carried by large, myelinated nerve fibers that ascend in the dorsal columns of the spinal cord on the same (ipsilateral) side. These fibers do not synapse until they reach the medulla oblongata of the brain stem; hence, fibers that carry these sensations from the feet are remarkably long. After the fibers synapse in the medulla with other second-order sensory neurons, information in the latter neurons crosses over to the contralateral side as it ascends via a fiber tract, called the **medial lemniscus,** to the thalamus (chapter 8, fig. 8.20). Third-order sensory neurons in the thalamus that receive this input in turn project to the **postcentral gyrus** (the sensory cortex, fig. 8.7).

Sensations of heat, cold, and pain are carried into the spinal cord mostly by thin, unmyelinated sensory neurons. Within the spinal cord, these neurons synapse with second-order association neurons that cross over to the contralateral side and ascend to the brain in the **lateral spinothalamic tract.** Fibers that mediate touch and pressure ascend in the **anterior spinothalamic tract.** Fibers of both spinothalamic tracts synapse with third-order neurons in the thalamus, which in turn project to the postcentral gyrus. Notice that somatesthetic information is always carried to the postcentral gyrus in third-order neurons. Also, because of crossing-over, somatesthetic information from each side of the body is projected to the postcentral gyrus of the contralateral cerebral hemisphere.

Since all somatesthetic information from the same area of the body projects to the same area of the postcentral gyrus, a "map" of the body can be drawn on the postcentral gyrus to represent sensory projection points (see fig. 8.7). This map is distorted, however, because it shows larger areas of cortex devoted to sensation in the face and hands than in other areas in the body. This disproportionately large area of the cortex devoted to the face and hands reflects the fact that the density of sensory receptors is higher in these regions.

Receptive Fields and Sensory Acuity

The **receptive field** of a neuron serving cutaneous sensation is the area of skin whose stimulation results in changes in the firing rate of the neuron. Changes in the firing rate of primary sensory neurons affect the firing of second- and third-order neurons, which in turn affects the firing of those neurons in the postcentral gyrus that receive input from the third-order neurons. Indirectly, therefore, neurons in the postcentral gyrus can be said to have receptive fields in the skin.

The area of each receptive field in the skin varies inversely with the density of receptors in the region. In the back and legs, where a large area of skin is served by relatively few sensory endings, the receptive field of each neuron is correspondingly large. In the fingertips, where a large number of cutaneous receptors serve a small area of skin, the receptive field of each sensory neuron is correspondingly small.

Two-Point Touch Threshold

The approximate size of the receptive fields serving light touch can be measured by the *two-point touch threshold test.* In this procedure, the two points of a pair of calipers are lightly touched to the skin at the same time. If the distance between the points is sufficiently great, each point will stimulate a different receptive field and a different sensory neuron—two separate points of touch will thus be felt. If the distance is sufficiently small, both points will touch the receptive field of only one sensory neuron, and only one point of touch will be felt (fig. 10.5).

The **two-point touch threshold,** which is the minimum distance at which two points of touch can be perceived as separate, is a measure of the distance between receptive fields. If the distance between the two points of the calipers is less than this minimum distance, only one "blurred" point of touch can be felt. The two-point touch threshold is thus an indication of *tactile acuity* (*acus* = needle), or the sharpness of touch perception.

The tactile acuity of the fingertips is exploited in the reading of braille. Braille symbols are formed by raised dots on the page that are separated from each other by 2.5 mm, which is slightly greater than the two-point touch threshold in the finger-

The phenomenon of the **phantom limb** was first described by a neurologist during the Civil War. In this account, a veteran with amputated legs asked for someone to massage his cramped leg muscle. It is now known that this phenomenon is common in amputees, who may experience complete sensations from the missing limbs. These sensations are sometimes useful; for example, in fitting prostheses into which the phantom has seemingly entered. However, pain in the phantom is experienced by 70% of amputees, and the pain can be severe and persistent.

One explanation for phantom limbs is that the nerves remaining in the stump can grow into nodules called neuromas, and these may generate nerve impulses that are transmitted to the brain and interpreted as arising from the missing limb. However, a phantom limb may occur in cases where the limb has not been amputated, but the nerves that normally enter from the limb have been severed. Or it may occur in individuals with spinal cord injuries above the level of the limb, so that sensations from the limb do not enter the brain. Current theories propose that the phantom may be produced by brain reorganization caused by the absence of the sensations that would normally arise from the missing limb. Such brain reorganization has been demonstrated in the thalamus and in the representational map of the body in the postcentral gyrus of the cerebral cortex.

tips (table 10.3). Experienced braille readers can scan words at about the same speed that a sighted person can read aloud—a rate of about 100 words per minute.

Lateral Inhibition

When a blunt object touches the skin, a number of receptive fields are stimulated—some more than others. The receptive fields in the center areas where the touch is strongest will be stimulated more than those in the neighboring fields where the touch is lighter. Stimulation will gradually diminish from the point of greatest contact, without a clear, sharp boundary. What we perceive, however, is not the fuzzy sensation that might be predicted. Instead, only a single touch with well-defined borders is felt. This sharpening of sensation is due to a process called **lateral inhibition** (fig. 10.6).

Lateral inhibition and the resultant sharpening of sensation occur within the central nervous system. Those sensory neurons whose receptive fields are stimulated most strongly inhibit—via interneurons that pass "laterally" within the CNS—sensory neurons that serve neighboring receptive fields.

Lateral inhibition is a common theme in sensory physiology, though the mechanisms involved are different for each sense. In hearing, lateral inhibition helps to more sharply tune the ability of the brain to distinguish sounds of different pitches. In vision, it helps the brain to more sharply distinguish borders of light and darkness; and in olfaction, it helps the brain to more clearly distinguish closely related odors.

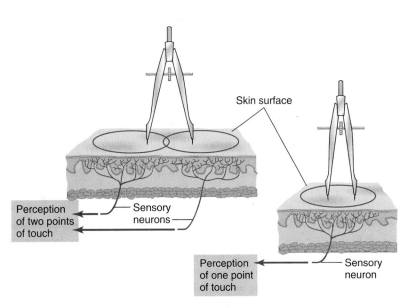

■ **Figure 10.5** **The two-point touch threshold test.** If each point touches the receptive fields of different sensory neurons, two separate points of touch will be felt. If both caliper points touch the receptive field of one sensory neuron, only one point of touch will be felt.

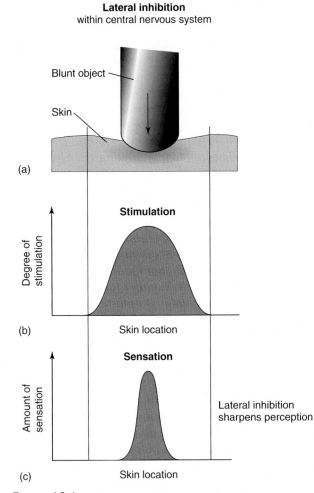

■ **Figure 10.6** **Lateral inhibition.** When an object touches the skin (a), receptors in the central area of the touched skin are stimulated more than neighboring receptors (b). Lateral inhibition within the central nervous system reduces the input from these neighboring sensory neurons. Sensation, as a result, is sharpened within the area of skin that was stimulated the most (c).

Table 10.3 The Two-Point Touch Threshold for Different Regions of the Body

Body Region	Two-Point Touch Threshold (mm)
Big toe	10
Sole of foot	22
Calf	48
Thigh	46
Back	42
Abdomen	36
Upper arm	47
Forehead	18
Palm of hand	13
Thumb	3
First finger	2

Source: From S. Weinstein and D. R. Kenshalo, editors, *The Skin Senses,* © 1968. Courtesy of Charles C. Thomas, Publisher, Ltd., Springfield, Illinois.

Test Yourself Before You Continue

1. Using a flow diagram, describe the neural pathways leading from cutaneous pain and pressure receptors to the postcentral gyrus. Indicate where crossing-over occurs.

2. Define the term *sensory acuity* and explain how acuity is related to the density of receptive fields in different parts of the body.

3. Explain the mechanism of lateral inhibition in cutaneous sensory perception and discuss its significance.

Taste and Smell

The receptors for taste and smell respond to molecules that are dissolved in fluid; hence, they are classified as chemoreceptors. Although there are only four basic modalities of taste, they combine in various ways and are influenced by the sense of smell, thus permitting a wide variety of different sensory experiences.

Chemoreceptors that respond to chemical changes in the internal environment are called **interoceptors;** those that respond to chemical changes in the external environment are **exteroceptors.** Included in the latter category are *taste (gustatory) receptors,* which respond to chemicals dissolved in food or drink, and *smell (olfactory) receptors,* which respond to gaseous molecules in the air. This distinction is somewhat arbitrary, however, because odorant molecules in air must first dissolve in fluid within the olfactory mucosa before the sense of smell can be stimulated. Also, the sense of olfaction strongly influences the sense of taste, as can easily be verified by eating an onion (or almost anything else) with the nostrils pinched together.

Taste

Gustation, the sense of taste, is evoked by receptors that consist of barrel-shaped **taste buds** (fig. 10.7). Located primarily on the dorsal surface of the tongue, each taste bud consists of 50 to 100 specialized epithelial cells with long microvilli that extend through a pore in the taste bud to the external environment, where they are bathed in saliva. Although these sensory epithelial cells are not neurons, they behave like neurons; they become depolarized when stimulated appropriately, produce action potentials, and release neurotransmitters that stimulate sensory neurons associated with the taste buds.

Taste buds in the anterior two-thirds of the tongue are innervated by the *facial nerve (VII),* and those in the posterior third of the tongue by the *glossopharyngeal nerve (IX).* Dendritic endings of the *facial nerve (VII)* are located around the taste buds and relay sensations of touch and temperature. Taste sensations are passed to the medulla oblongata, where the neurons synapse with second-order neurons that project to the thalamus. From here, third-order neurons project to the area of the postcentral gyrus of the cerebral cortex that is devoted to sensations from the tongue.

The specialized epithelial cells of the taste bud are known as **taste cells.** The different categories of taste are produced by different chemicals that come into contact with the microvilli of these cells (figure 10.8). Four different categories of taste are traditionally recognized: *salty, sour, sweet,* and *bitter.* There may also be a fifth category of taste, termed *umami* (a Japanese term related to a meaty flavor), for the amino acid glutamate (and stimulated by the flavor-enhancer monosodium glutamate). Although scientists long believed that different regions of the tongue were specialized for different tastes, this is no longer believed to be true. Indeed, is seems that each taste bud contains taste cells responsive to each of the different taste categories! It also appears

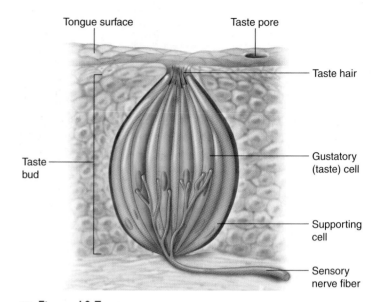

Figure 10.7 **A taste bud.** Chemicals dissolved in the fluid at the pore bind to receptor proteins in the microvilli of the sensory cells. This ultimately leads to the release of neurotransmitter, which activates the associated sensory neuron.

that a given sensory neuron may be stimulated by more than one taste cell in a number of different taste buds, and so one sensory fiber may not transmit information specific for only one category of taste. The brain interprets the pattern of stimulation of these sensory neurons, together with the nuances provided by the sense of smell, as the complex tastes that we are capable of perceiving.

The salty taste of food is due to the presence of sodium ions (Na^+), or some other cations, which activate specific receptor cells for the salty taste. Different substances taste salty to the degree that they activate these particular receptor cells. The Na^+ passes into the sensitive receptor cells through channels in the apical membranes. This depolarizes the cells, causing them to release their transmitter. The anion associated with the Na^+, however, modifies the perceived saltiness to a surprising degree: NaCl tastes much saltier than other sodium salts (such as sodium acetate). There is evidence to suggest that the anions can pass through the tight junctions between the receptor cells, and that the Cl^- anion passes through this barrier more readily than the other anions. This is presumably related to the ability of Cl^- to impart a saltier taste to the Na^+ than do the other anions.

Sour taste, like salty taste, is produced by ion movement through membrane channels. Sour taste, however, is due to the presence of hydrogen ions (H^+); all acids therefore taste sour. In contrast to the salty and sour tastes, the sweet and bitter tastes are produced by interaction of taste molecules with specific membrane receptor proteins.

Most organic molecules, particularly sugars, taste sweet to varying degrees. Bitter taste is evoked by quinine and seemingly unrelated molecules. It is the most acute taste sensation and is generally associated with toxic molecules (although not all toxins taste bitter). Both sweet and bitter sensations are mediated by receptors that are coupled to G-proteins (chapter 7). The particular type of G-protein involved in taste has recently been identified

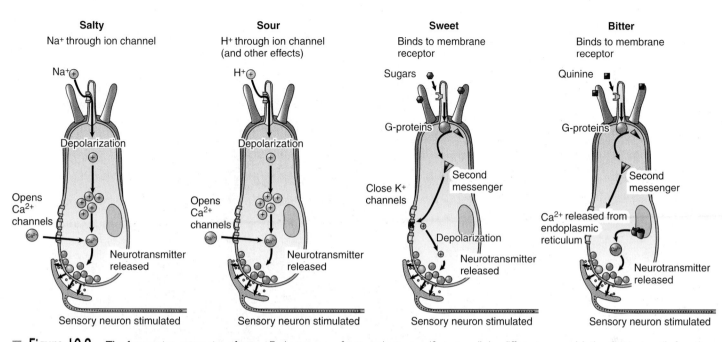

Figure 10.8 The four major categories of taste. Each category of taste activates specific taste cells by different means. Notice that taste cells for salty and sour are depolarized by ions (Na^+ and H^+, respectively) in the food, whereas taste cells for sweet and bitter are depolarized by sugars and quinine, respectively, by means of G-protein-coupled receptors and the actions of second messengers.

and termed **gustducin.** This term is used to emphasize the similarity to a related group of G-proteins, of a type called *transducin,* associated with the photoreceptors in the eye. Dissociation of the gustducin G-protein subunit activates second-messenger systems, leading to depolarization of the receptor cell (fig. 10.8). The stimulated receptor cell, in turn, activates an associated sensory neuron that transmits impulses to the brain, where they are interpreted as the corresponding taste perception.

Although all sweet and bitter taste receptors act via G-proteins, the second-messenger systems activated by the G-proteins depend on the molecule tasted. In the case of the sweet taste of sugars, for example, the G-proteins activate adenylate cyclase, producing cyclic AMP (cAMP; see chapter 7). The cAMP, in turn, produces depolarization by closing K^+ channels that were previously open. On the other hand, the sweet taste of the amino acids phenylalanine and tryptophan, as well as of the artificial sweeteners saccharin and cyclamate, may enlist different second-messenger systems. These involve the activation of a membrane enzyme that produces the second messengers inositol triphosphate (IP_3) and diacylglycerol (DAG). These second-messenger systems are described in chapter 11.

Smell

The receptors responsible for **olfaction,** the sense of smell, are located in the olfactory epithelium. The olfactory apparatus consists of receptor cells (which are bipolar neurons), supporting (sustentacular) cells, and basal (stem) cells. The basal cells generate new receptor cells every 1 to 2 months to replace the neurons damaged by exposure to the environment. The supporting cells are epithe-

lial cells rich in enzymes that oxidize hydrophobic, volatile odorants, thereby making these molecules less lipid-soluble and thus less able to penetrate membranes and enter the brain.

Each bipolar sensory neuron has one dendrite that projects into the nasal cavity, where it terminates in a knob containing cilia (figs. 10.9 and 10.10). The bipolar sensory neuron also has a single unmyelinated axon that projects through holes in the cribriform plate of the ethmoid bone into the olfactory bulb of the cerebrum, where it synapses with second-order neurons. Therefore, unlike other sensory modalities that are relayed to the cerebrum from the thalamus, the sense of smell is transmitted directly to the cerebral cortex. The processing of olfactory information begins in the olfactory bulb, where the bipolar sensory neurons synapse with neurons located in spherically shaped arrangements called *glomeruli* (fig. 10.9). Evidence suggests that each glomerulus receives input from one type of olfactory receptor. The smell of a flower, which releases many different molecular odorants, may be identified by the pattern of excitation it produces in the glomeruli of the olfactory bulb. Identification of an odor is improved by lateral inhibition in the olfactory bulb, which appears to involve dendrodendritic synapses between neurons of adjacent glomeruli.

Neurons in the olfactory bulb project to the olfactory cortex in the medial temporal lobes, and to the associated hippocampus and amygdaloid nuclei. These structures are part of the limbic system, which was described in chapter 8 as having important roles in both emotion and memory. The human amygdala, in particular, has been implicated in the emotional responses to olfactory stimulation. Perhaps this explains why the smell of a particular odor can so powerfully evoke emotionally charged memories.

The molecular basis of olfaction is complex. At least in some cases, odorant molecules bind to receptors and act through

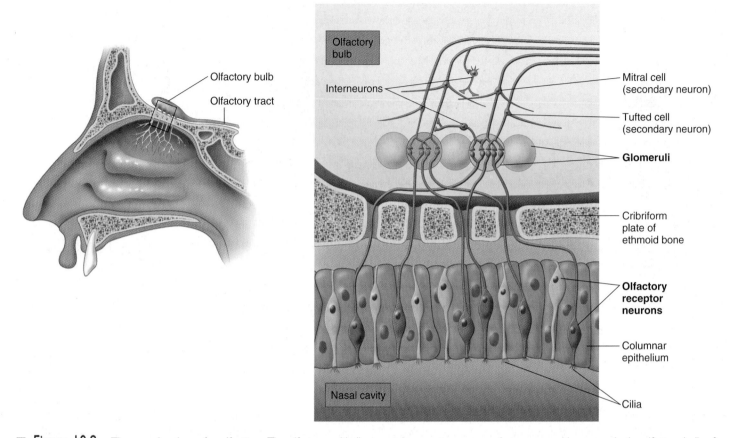

■ Figure 10.9 **The neural pathway for olfaction.** The olfactory epithelium contains receptor neurons that synapse with neurons in the olfactory bulb of the cerebral cortex. The synapses occur in rounded structures called glomeruli. Secondary neurons, known as tufted cells and mitral cells, transmit impulses from the olfactory bulb to the olfactory cortex in the medial temporal lobes. Notice that each glomerulus receives input from only one type of olfactory receptor, regardless of where those receptors are located in the olfactory epithelium.

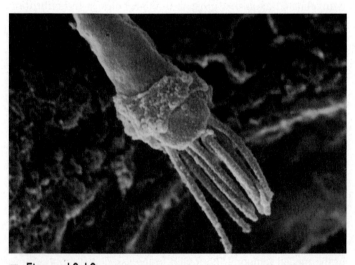

■ Figure 10.10 A scanning electron micrograph of an olfactory neuron. The tassel of cilia is clearly visible.

G-proteins to increase the cyclic AMP within the cell. This, in turn, opens membrane channels and causes the depolarization of the generator potential, which then stimulates the production of action potentials. Up to fifty G-proteins may be associated with a single receptor protein. Dissociation of these G-proteins re-

leases many G-protein subunits, thereby amplifying the effect many times. This amplification could account for the extreme sensitivity of the sense of smell: the human nose can detect a billionth of an ounce of perfume in air. Even at that, our sense of smell is not nearly as keen as that of many other mammals.

A family of genes that codes for the olfactory receptor proteins has been discovered. This is a large family that may include as many as a thousand genes. The large number may reflect the importance of the sense of smell to mammals in general. Even a thousand different genes coding for a thousand different receptor proteins, however, cannot account for the fact that humans can distinguish up to 10,000 different odors. Clearly, the brain must integrate the signals from several sensory neurons that have different olfactory receptor proteins and then interpret the pattern as a characteristic "fingerprint" for a particular odor.

Test Yourself Before You Continue

1. Explain how the mechanisms for sour and salty tastes are similar to each other, and how these differ from the mechanisms responsible for sweet and bitter tastes.

2. Explain how odorant molecules stimulate the olfactory receptors. Why is it that our sense of smell is so keen?

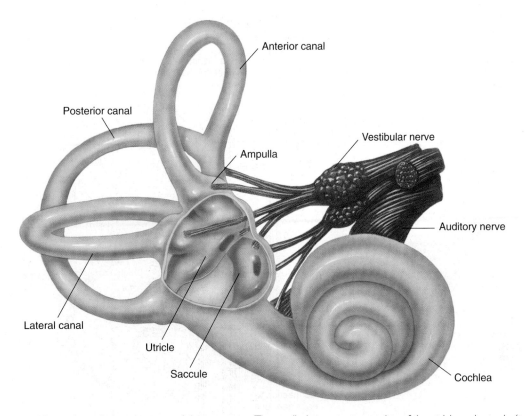

Anterior canal

Posterior canal

Vestibular nerve

Ampulla

Auditory nerve

Lateral canal

Utricle

Saccule

Cochlea

■ **Figure 10.11** **The cochlea and vestibular apparatus of the inner ear.** The vestibular apparatus consists of the utricle and saccule (together called the otolith organs) and the three semicircular canals. The base of each semicircular canal is expanded into an ampulla that contains sensory hair cells.

Vestibular Apparatus and Equilibrium

The sense of equilibrium is provided by structures in the inner ear, collectively known as the vestibular apparatus. Movements of the head cause fluid within these structures to bend extensions of sensory hair cells, and this bending results in the production of action potentials.

The sense of equilibrium, which provides orientation with respect to gravity, is due to the function of an organ called the **vestibular apparatus.** The vestibular apparatus and a snail-like structure called the *cochlea,* which is involved in hearing, form the *inner ear* within the temporal bones of the skull. The vestibular apparatus consists of two parts: (1) the *otolith organs,* which include the *utricle* and *saccule,* and (2) the *semicircular canals* (fig. 10.11).

The sensory structures of the vestibular apparatus and cochlea are located within the **membranous labyrinth** (fig. 10.12), a tubular structure that is filled with a fluid similar in composition to intracellular fluid. This fluid is called *endolymph.* The membranous labyrinth is located within a bony cavity in the skull, the **bony labyrinth.** Within this cavity, between the membranous labyrinth and the bone, is a fluid called *perilymph.* Perilymph is similar in composition to cerebrospinal fluid.

Sensory Hair Cells of the Vestibular Apparatus

The utricle and saccule provide information about *linear acceleration*—changes in velocity when traveling horizontally or vertically. We therefore have a sense of acceleration and deceleration when riding in a car or when skipping rope. A sense of *rotational,* or *angular, acceleration* is provided by the semicircular canals, which are oriented in three planes like the faces of a cube. This helps us maintain balance when turning the head, spinning, or tumbling.

The receptors for equilibrium are modified epithelial cells. They are known as **hair cells** because each cell contains twenty to fifty hairlike extensions. All but one of these hairlike extensions are **stereocilia**—processes containing filaments of protein surrounded by part of the cell membrane. One larger extension has the structure of a true cilium (chapter 3), and it is known as a **kinocilium** (fig. 10.13). When the stereocilia are bent in the direction of the kinocilium, the cell membrane is depressed and becomes depolarized. This causes the hair cell to release a synaptic transmitter that stimulates the dendrites of sensory neurons that are part of the vestibulocochlear nerve (VIII). When the stereocilia are bent in the opposite direction, the membrane of the hair cell becomes hyperpolarized (fig. 10.13) and, as a result, releases less synaptic transmitter. In this way, the frequency of action potentials in the sensory neurons that innervate the hair cells carries information about movements that cause the hair cell processes to bend.

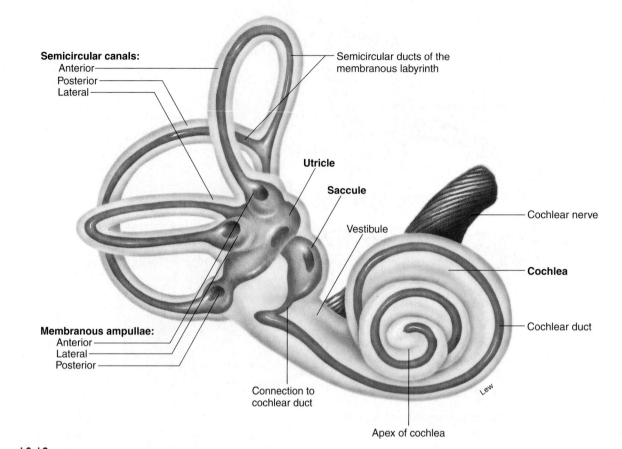

Semicircular canals:
Anterior
Posterior
Lateral

Semicircular ducts of the
membranous labyrinth

Utricle

Saccule

Vestibule

Cochlear nerve

Cochlea

Cochlear duct

Membranous ampullae:
Anterior
Lateral
Posterior

Connection to
cochlear duct

Lew

Apex of cochlea

■ **Figure 10.12**   **The labyrinths of the inner ear.** The membranous labyrinth (darker blue) is contained within the bony labyrinth.

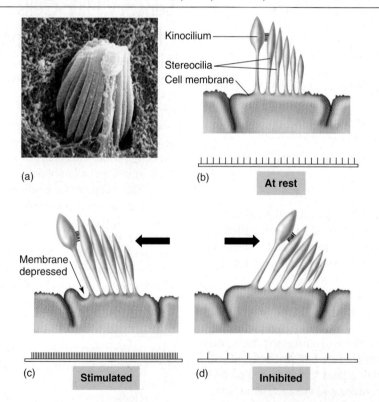

Kinocilium

Stereocilia

Cell membrane

(a)

(b)   **At rest**

Membrane
depressed

(c)   **Stimulated**

(d)   **Inhibited**

■ **Figure 10.13**   **Sensory hair cells within the vestibular apparatus.** (a) A scanning electron photograph of a kinocilium and stereocilia. (b) Each sensory hair cell contains a single kinocilium and several stereocilia. (c) When stereocilia are displaced toward the kinocilium (arrow), the cell membrane is depressed and the sensory neuron innervating the hair cell is stimulated. (d) When the stereocilia are bent in the opposite direction, away from the kinocilium, the sensory neuron is inhibited.

Utricle and Saccule

The otolith organs, the **utricle** and **saccule,** each have a patch of specialized epithelium called a *macula* that consists of hair cells and supporting cells. The hair cells project into the endolymph-filled membranous labyrinth, with their hairs embedded in a **gelatinous otolithic membrane** (fig. 10.14). The otolithic membrane contains microscopic crystals of calcium carbonate (otoliths) from which it derives its name (*oto* = ear; *lith* = stone). These stones increase the mass of the membrane, which results in a higher inertia (resistance to change in movement).

Because of the orientation of their hair cell processes into the otolithic membrane, the utricle is more sensitive to horizontal acceleration and the saccule is more sensitive to vertical acceleration. During forward acceleration, the otolithic membrane lags behind the hair cells, so the hairs of the utricle are pushed backward. This is similar to the backward thrust of the body when a car quickly accelerates forward. The inertia of the otolithic membrane similarly causes the hairs of the saccule to be pushed upward when a person descends rapidly in an elevator. These effects, and the opposite ones that occur when a person accelerates backward or upward, produce a changed pattern of action potentials in sensory nerve fibers that allows us to maintain our equilibrium with respect to gravity during linear acceleration.

Semicircular Canals

The three **semicircular canals** project in three different planes at nearly right angles to each other. Each canal contains an inner extension of the membranous labyrinth called a *semicircular duct,* and at the base of each duct is an enlarged swelling called the *ampulla.* The *crista ampullaris,* an elevated area of the ampulla, is where the sensory hair cells are located. The processes of these cells are embedded in a gelatinous membrane, the **cupula** (fig. 10.15), which has a higher density than that of the surrounding endolymph. Like a sail in the wind, the cupula can be pushed in one direction or the other by movements of the endolymph.

The endolymph of the semicircular canals serves a function analogous to that of the otolithic membrane—it provides inertia so that the sensory processes will be bent in a direction opposite to that of the angular acceleration. As the head rotates to the right, for example, the endolymph causes the cupula to be bent toward the left, thereby stimulating the hair cells. Hair cells in the anterior semicircular canal are stimulated when doing a somersault, those in the posterior semicircular canal are stimulated when performing a cartwheel, and those in the lateral semicircular canal are stimulated when spinning around the long axis of the body.

Neural Pathways

Stimulation of hair cells in the vestibular apparatus activates sensory neurons of the *vestibulocochlear nerve (VIII).* These

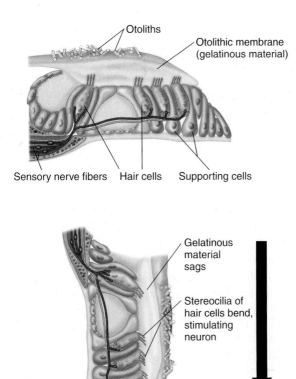

Figure 10.14 **The otolith organ.** *(a)* When the head is in an upright position, the weight of the otoliths applies direct pressure to the sensitive cytoplasmic extensions of the hair cells. *(b)* As the head is tilted forward, the extensions of the hair cells bend in response to gravitational force and cause the sensory nerve fibers to be stimulated.

Otoliths

Otolithic membrane (gelatinous material)

Sensory nerve fibers   Hair cells   Supporting cells

(a) **Head upright**

Gelatinous material sags

Stereocilia of hair cells bend, stimulating neuron

Gravitational force

(b) **Head bent forward**

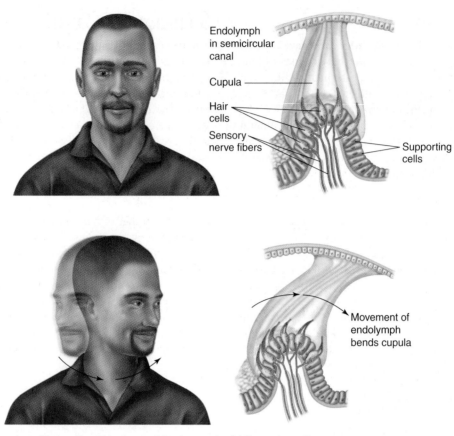

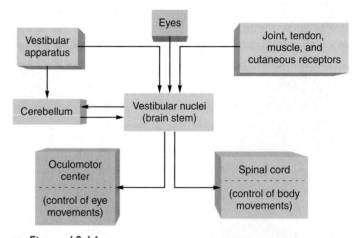

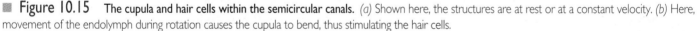

■ **Figure 10.15**   **The cupula and hair cells within the semicircular canals.** *(a)* Shown here, the structures are at rest or at a constant velocity. *(b)* Here, movement of the endolymph during rotation causes the cupula to bend, thus stimulating the hair cells.

fibers transmit impulses to the cerebellum and to the vestibular nuclei of the medulla oblongata. The vestibular nuclei, in turn, send fibers to the oculomotor center of the brain stem and to the spinal cord (fig. 10.16). Neurons in the oculomotor center control eye movements, and neurons in the spinal cord stimulate movements of the head, neck, and limbs. Movements of the eyes and body produced by these pathways serve to maintain balance and "track" the visual field during rotation.

Nystagmus and Vertigo

When a person first begins to spin, the inertia of endolymph within the semicircular ducts causes the cupula to bend in the opposite direction. As the spin continues, however, the inertia of the endolymph is overcome and the cupula straightens. At this time, the endolymph and the cupula are moving in the same direction and at the same speed. If movement is suddenly stopped, the greater inertia of the endolymph causes it to continue moving in the previous direction of spin and to bend the cupula in that direction.

Bending of the cupula affects muscular control of the eyes and body through the neural pathways previously discussed. During a spin, this produces smooth movements of the eyes in a direction opposite to that of the head movement so that a stable visual fixation point can be maintained. When the spin is abruptly stopped, the eyes continue to move smoothly in a direction opposite to that of the spin (because of the continued bending of the cupula) and then are jerked rapidly back to the midline position. This produces involuntary oscillations of the eyes called **vestibular nystagmus.** People

■ **Figure 10.16**   **Neural pathways involved in the maintenance of equilibrium and balance.** Sensory input enters the vestibular nuclei and the cerebellum, which coordinate motor responses.

experiencing this effect may feel that they, or the room, are spinning. The loss of equilibrium that results is called **vertigo.**

Vertigo as a result of spinning is a natural response of the vestibular apparatus. Pathologically, vertigo may be caused by anything that alters the firing rate of one of the vestibulocochlear nerves (right or left) compared to the other. This is usually due to a viral infection causing vestibular neuritis. Severe vertigo is often accompanied by dizziness, pallor, sweating, nausea, and vomiting due to involvement of the autonomic nervous system, which is activated by vestibular input to the brain stem.

Vestibular nystagmus is one of the symptoms of an inner-ear disease called **Ménière's disease.** The early symptom of this disease is often "ringing in the ears," or *tinnitus*. Since the endolymph of the cochlea and the endolymph of the vestibular apparatus are continuous through a tiny canal, the duct of Hensen, vestibular symptoms of vertigo and nystagmus often accompany hearing problems in this disease.

Test Yourself Before You Continue

1. Describe the structure of the utricle and saccule and explain how linear acceleration results in stimulation of the hair cells within these organs.
2. Describe the structure of the semicircular canals and explain how they provide a sense of angular acceleration.

The Ears and Hearing

Sound causes vibrations of the tympanic membrane. These vibrations, in turn, produce movements of the middle-ear ossicles, which press against a membrane called the oval window in the cochlea. Movements of the oval window produce pressure waves within the fluid of the cochlea, which in turn cause movements of a membrane called the basilar membrane. Sensory hair cells are located on the basilar membrane, and the movements of this membrane in response to sound result in the bending of the hair cell processes. This stimulates action potentials that are transmitted to the brain in sensory fibers and interpreted as sound.

Sound waves are alternating zones of high and low pressure traveling in a medium, usually air or water. (Thus, sound waves cannot travel in space.) Sound waves travel in all directions from their source, like ripples in a pond where a stone has been dropped. These waves are characterized by their frequency and intensity. The **frequency** is measured in *hertz (Hz),* which is the modern designation for *cycles per second (cps).* The *pitch* of a sound is directly related to its frequency—the greater the frequency of a sound, the higher its pitch.

The **intensity,** or loudness, of a sound is directly related to the amplitude of the sound waves and is measured in units called *decibels (dB).* A sound that is barely audible—at the threshold of hearing—has an intensity of zero decibels. Every 10 decibels indicates a tenfold increase in sound intensity; a sound is ten times louder than threshold at 10 dB, 100 times louder at 20 dB, a million times louder at 60 dB, and 10 billion times louder at 100 dB.

The ear of a trained, young individual can hear sound over a frequency range of 20 to 20,000 Hz, yet still can distinguish between two pitches that have only a 0.3% difference in frequency. The human ear can detect differences in sound intensities of only 0.1 to 0.5 dB, while the range of audible intensities covers twelve orders of magnitude (10^{12}), from the barely audible to the limits of painful loudness.

Outer Ear

Sound waves are funneled by the *pinna,* or *auricle,* into the *external auditory meatus* (fig. 10.17). These two structures form the **outer ear.** The external auditory meatus channels the sound waves (while increasing their intensity) to the eardrum, or **tympanic membrane.** Sound waves in the external auditory meatus produce extremely small vibrations of the tympanic membrane; movements of the eardrum during speech (with an average sound intensity of 60 dB) are estimated to be about the diameter of a molecule of hydrogen!

Middle Ear

The **middle ear** is the cavity between the tympanic membrane on the outer side and the cochlea on the inner side (fig. 10.18). Within this cavity are three **middle-ear ossicles**—the *malleus* (hammer), *incus* (anvil), and *stapes* (stirrup). The malleus is attached to the tympanic membrane, so that vibrations of this membrane are transmitted via the malleus and incus to the stapes. The stapes, in turn, is attached to a membrane in the cochlea called the *oval window,* which thus vibrates in response to vibrations of the tympanic membrane.

The **auditory (eustachian) tube** is a passageway leading from the middle ear to the nasopharynx (a cavity positioned behind the nasal cavity and extending down to the soft palate). The auditory tube is usually collapsed, so that debris and infectious agents are prevented from traveling from the oral cavity to the middle ear. In order to open the auditory tube, the *tensor tympani muscle,* attaching to the auditory tube and the malleus (fig. 10.18), must contract. This occurs during swallowing, yawning, and sneezing. People sense a "popping" sensation in their ears as they swallow when driving up a mountain because the opening of the auditory canal permits air to move from the region of higher pressure in the middle ear to the region of lower pressure in the nasopharynx.

Clinical Investigation Clues

Remember that Ed experienced severe ear pain and reduced hearing immediately after disembarking from an international flight. Remember also that he had a bad head cold.

What may have caused his pain and hearing impairment?
How could this be helped by taking a decongestant?

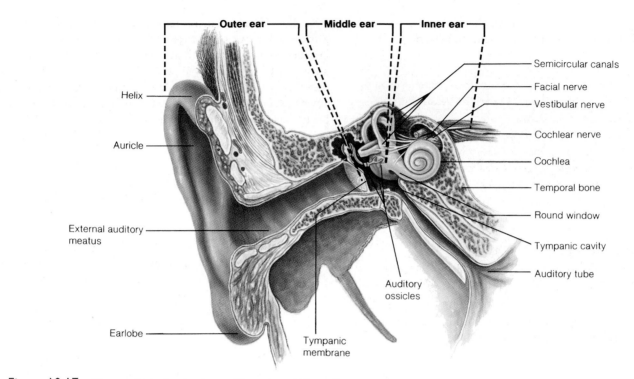

Outer ear   **Middle ear**   **Inner ear**

Helix

Auricle

External auditory
meatus

Earlobe

Tympanic
membrane

Auditory
ossicles

Semicircular canals

Facial nerve

Vestibular nerve

Cochlear nerve

Cochlea

Temporal bone

Round window

Tympanic cavity

Auditory tube

■ Figure 10.17   **The ear.** Note the structures of the outer, middle, and inner ear.

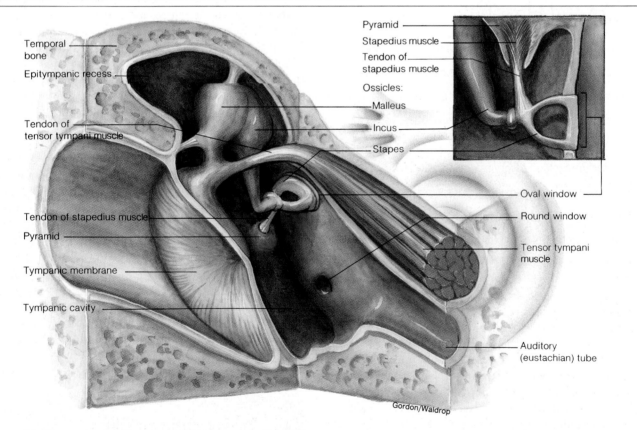

Temporal
bone

Epitympanic recess

Tendon of
tensor tympani muscle

Tendon of stapedius muscle

Pyramid

Tympanic membrane

Tympanic cavity

Pyramid

Stapedius muscle

Tendon of
stapedius muscle

Ossicles:

Malleus

Incus

Stapes

Oval window

Round window

Tensor tympani
muscle

Auditory
(eustachian) tube

Gordon/Waldrop

■ Figure 10.18   A medial view of the middle ear. The locations of auditory muscles, attached to the middle-ear ossicles, are indicated.

The fact that vibrations of the tympanic membrane are transferred through three bones instead of just one affords protection. If the sound is too intense, the ossicles may buckle. This protection is increased by the action of the *stapedius muscle,* which attaches to the neck of the stapes (fig. 10.18). When sound becomes too loud, the stapedius muscle contracts and dampens the movements of the stapes against the oval window. This action helps to prevent nerve damage within the cochlea. If sounds reach high amplitudes very quickly, however—as in gunshots—the stapedius muscle may not respond soon enough to prevent nerve damage.

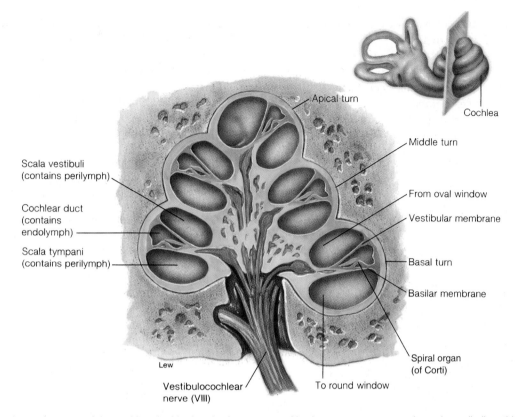

Apical turn

Cochlea

Middle turn

Scala vestibuli
(contains perilymph)

From oval window

Cochlear duct
(contains
endolymph)

Vestibular membrane

Scala tympani
(contains perilymph)

Basal turn

Basilar membrane

Spiral organ
(of Corti)

Lew

Vestibulocochlear
nerve (VIII)

To round window

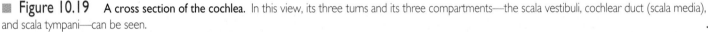

■ **Figure 10.19** **A cross section of the cochlea.** In this view, its three turns and its three compartments—the scala vestibuli, cochlear duct (scala media), and scala tympani—can be seen.

Cochlea

Encased within the dense temporal bone of the skull is an organ called the **cochlea,** about the size of a pea and shaped like the shell of a snail. Together with the vestibular apparatus (previously described), it composes the **inner ear.**

Damage to the tympanic membrane or middle-ear ossicles produces **conduction deafness.** This impairment can result from a variety of causes, including otitis media and otosclerosis. In *otitis media,* which sometimes follows allergic reactions or respiratory disease, inflammation produces an excessive accumulation of fluid within the middle ear. This, in turn, can result in the excessive growth of epithelial tissue and damage to the eardrum. In otosclerosis, bone is resorbed and replaced by "sclerotic bone" that grows over the oval window and immobilizes the footplate of the stapes. In conduction deafness, these pathological changes hinder the transmission of sound waves from the air to the cochlea of the inner ear.

Vibrations of the stapes and oval window displace perilymph fluid within a part of the bony labyrinth known as the **scala vestibuli,** which is the upper of three chambers within the cochlea. The lower of the three chambers is also a part of the bony labyrinth and is known as the **scala tympani.** The middle chamber of the cochlea is a part of the membranous labyrinth called the **cochlear duct,** or **scala media.** Like the cochlea as a whole, the cochlear duct coils to form three turns (fig. 10.19), similar to the basal, middle, and apical portions of a snail shell. Since the cochlear duct is a part of the membranous labyrinth, it contains endolymph rather than perilymph.

The perilymph of the scala vestibuli and scala tympani is continuous at the apex of the cochlea because the cochlear duct ends blindly, leaving a small space called the *helicotrema* between the end of the cochlear duct and the wall of the cochlea. Vibrations of the oval window produced by movements of the stapes cause pressure waves within the scala vestibuli, which pass to the scala tympani. Movements of perilymph within the scala tympani, in turn, travel to the base of the cochlea where they cause displacement of a membrane called the *round window* into the middle-ear cavity (see fig. 10.18). This occurs because fluid, such as perilymph, cannot be compressed; an inward movement of the oval window is thus compensated for by an outward movement of the round window.

When the sound frequency (pitch) is sufficiently low, there is adequate time for the pressure waves of perilymph within the upper scala vestibuli to travel through the helicotrema to the scala tympani. As the sound frequency increases, however, pressure waves of perilymph within the scala vestibuli do not have time to travel all the way to the apex of the cochlea. Instead, they are transmitted through the *vestibular membrane,* which separates the scala vestibuli from the cochlear duct, and through the **basilar membrane,** which separates the cochlear duct from the scala tympani, to the perilymph of the scala tympani (fig. 10.19).

The distance that these pressure waves travel, therefore, decreases as the sound frequency increases.

Sound waves transmitted through perilymph from the scala vestibuli to the scala tympani thus produce displacement of the vestibular membrane and the basilar membrane. Although the movement of the vestibular membrane does not directly contribute to hearing, displacement of the basilar membrane is central to pitch discrimination. Each sound frequency produces maximum vibrations at a different region of the basilar membrane. Sounds of higher frequency (pitch) cause maximum vibrations of the basilar membrane closer to the stapes, as illustrated in figure 10.20.

Spiral Organ (Organ of Corti)

The sensory *hair cells* are located on the basilar membrane, with their "hairs" (actually stereocilia) projecting into the endolymph of the cochlear duct. These hair cells are arranged to form one row of inner cells, which extends the length of the basilar membrane, and multiple rows of outer hair cells: three rows in the basal turn, four in the middle turn, and five in the apical turn of the cochlea (fig. 10.21).

The stereocilia of the outer hair cells are embedded in a gelatinous **tectorial membrane** (*tectum* = roof, covering), which overhangs the hair cells within the cochlear duct (fig. 10.22). The

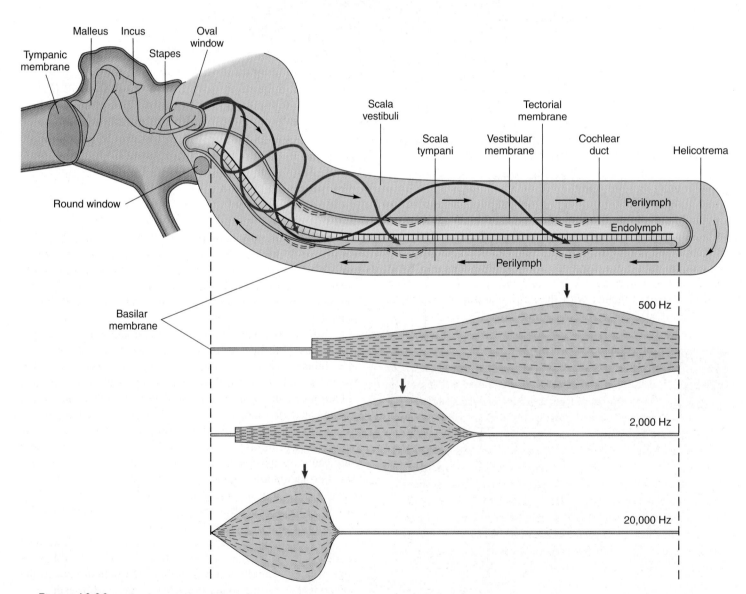

Figure 10.20 **The effect of sounds of different frequency on the basilar membrane.** The cochlea is shown "unwound" in this diagram. Sounds of low frequency cause pressure waves of perilymph to pass through the helicotrema. Sounds of higher frequency cause pressure waves to "shortcut" through the cochlear duct. This causes displacement of the basilar membrane, which is central to the transduction of sound waves into nerve impulses. Maximum displacement of the basilar membrane occurs closer to its base as the sound frequency is increased. (The frequency of sound waves is measured in hertz [Hz], or cycles per second.)

Outer hair cells

■ **Figure 10.21** A scanning electron micrograph of the hair cells of the spiral organ (organ of Corti).

association of the basilar membrane, hair cells with sensory fibers, and tectorial membrane forms a functional unit called the **spiral organ,** or **organ of Corti** (fig. 10.22). When the cochlear duct is displaced by pressure waves of perilymph, a shearing force is created between the basilar membrane and the tectorial membrane. This causes the stereocilia to move and bend. Such movement causes ion channels in the membrane to open, which in turn depolarizes the hair cells. Each depolarized hair cell then releases a transmitter chemical, believed to be glutamate, that stimulates an associated sensory neuron.

The greater the displacement of the basilar membrane and the bending of the stereocilia, the greater the amount of transmitter released by the hair cell, and therefore the greater the generator potential produced in the sensory neuron. In other

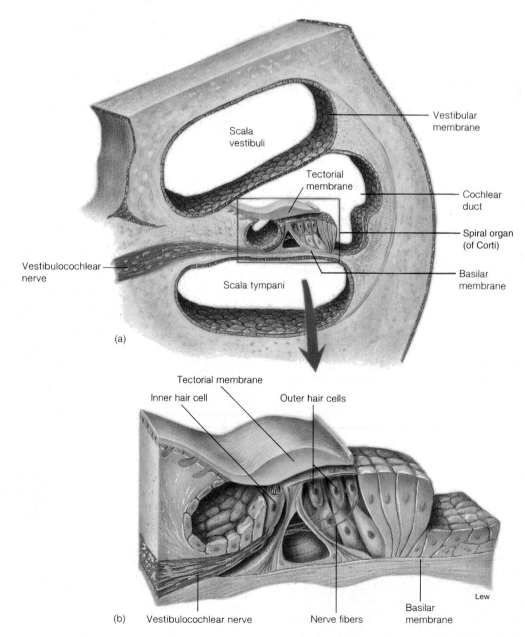

■ **Figure 10.22** **The spiral organ (organ of Corti).** This functional unit of hearing is depicted *(a)* within the cochlear duct and *(b)* isolated to show greater detail.

words, a greater bending of the stereocilia will increase the frequency of action potentials produced by the fibers of the cochlear nerve that are stimulated by the hair cells. Experiments suggest that the stereocilia need bend only 0.3 nanometers to be detected at the threshold of hearing! A greater bending will result in a higher frequency of action potentials, which will be perceived as a louder sound.

As mentioned earlier, traveling waves in the basilar membrane reach a peak in different regions, depending on the pitch of the sound. High-pitched sounds produce a peak displacement closer to the base, while sounds of lower pitch cause peak displacement further toward the apex (see fig. 10.20). Those neurons that originate in hair cells located where the displacement is greatest will be stimulated more than neurons that originate in other regions. This mechanism provides a neural code for **pitch discrimination.**

There is evidence that the outer hair cells shorten and stiffen in the peak stimulated region of the basilar membrane, and thereby contribute to pitch discrimination. This active response of the hair cells may serve to more sharply tune the frequency response of the basilar membrane. Lateral inhibition by neurons in the CNS accentuates the response of the spiral organ to different frequencies of sound, and thereby serves to sharpen pitch discrimination.

Neural Pathways for Hearing

Sensory neurons in the vestibulocochlear nerve (VIII) synapse with neurons in the medulla oblongata that project to the inferior colliculus of the midbrain (fig. 10.23). Neurons in this area, in turn, project to the thalamus, which sends axons to the auditory cortex of the temporal lobe. By means of this pathway, neurons in different regions of the basilar membrane stimulate neurons in corresponding areas of the auditory cortex. Each area of this cortex thus represents a different part of the basilar membrane and a different pitch (fig. 10.24).

Hearing Impairments

There are two major categories of deafness: (1) **conduction deafness,** in which the transmission of sound waves through the middle ear to the oval window is impaired, and (2) **sensorineural,** or **perceptive, deafness,** in which the transmission of nerve impulses anywhere from the cochlea to the auditory cortex is impaired. Conduction deafness can be caused by middle-ear damage from otitis media or otosclerosis (discussed in the previous clinical applications box, p. 257). Sensorineural deafness may result from a wide variety of pathological processes and from exposure to extremely loud sounds. Unfortunately, the hair cells in the inner ears of mammals cannot regenerate once they are destroyed. Experiments have shown, however, that the hair cells of reptiles and birds can regenerate by cell division when they are damaged. Scientists are currently trying to determine if mammalian sensory hair cells might be made to respond in a similar fashion.

Conduction deafness impairs hearing at all sound frequencies. Sensorineural deafness, by contrast, often impairs the ability to hear some pitches more than others. This may be due to

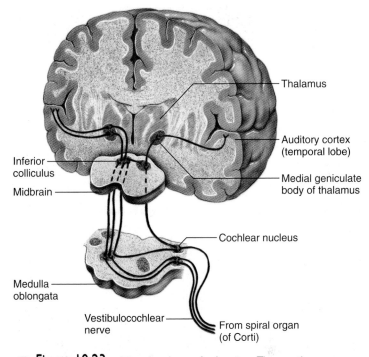

Figure 10.23   Neural pathways for hearing. These pathways extend from the spiral organ in the cochlea to the auditory cortex.

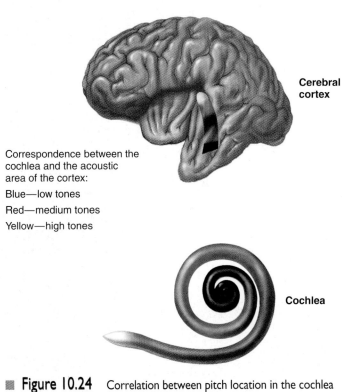

Correspondence between the cochlea and the acoustic area of the cortex:

Blue—low tones
Red—medium tones
Yellow—high tones

Figure 10.24   Correlation between pitch location in the cochlea and auditory cortex. Sounds of different frequencies (pitches) cause vibration of different parts of the basilar membrane, exciting different sensory neurons in the cochlea. These in turn send their input to different regions of the auditory cortex.

pathological processes or to changes that occur during aging. Age-related hearing impairment—called *presbycusis*—begins after age 20 when the ability to hear high frequencies (18,000 to 20,000 Hz) diminishes. Men are affected to a greater degree than women, and although the progression is variable, the deficits may gradually extend into the 4,000-to-8,000-Hz range. These impairments can be detected by *audiometry*, a technique in which the threshold intensity of different pitches is determined. The ability to hear speech is particularly affected by hearing loss in the higher frequencies.

People with conduction deafness can be helped by **hearing aids**—devices that amplify sounds and conduct the sound waves through bone to the inner ear. People with sensorineural deafness sometimes choose to have **cochlear implants,** which electrically stimulate the fibers of the vestibulocochlear nerve in response to sounds. Experiments with animals suggest that such devices produce a reorganization of the auditory cortex, demonstrating a plasticity similar to that described previously in the somatosensory cortex (the postcentral gyrus) in people with amputated limbs.

Clinical Investigation Clues

Remember that the doctor suggested that Ed have an audiology exam if he still experienced a hearing impairment after his cold got better.

What type of hearing loss could be detected by an audiology exam?

What might cause such a hearing loss?

Test Yourself Before You Continue

1. Use a flowchart to describe how sound waves in air within the external auditory meatus are transduced into movements of the basilar membrane.

2. Explain how movements of the basilar membrane affect hair cells, and how hair cells can stimulate associated sensory neurons.

3. Explain how sounds of different intensities affect the function of the cochlea. How are different pitches of sounds distinguished by the cochlea?

The Eyes and Vision

Light from an observed object is focused by the cornea and lens onto the photoreceptive retina at the back of the eye. The focus is maintained on the retina at different distances between the object and the eyes by muscular contractions that change the thickness and degree of curvature of the lens.

The eyes transduce energy in the electromagnetic spectrum (fig. 10.25) into nerve impulses. Only a limited part of this spectrum can excite the photoreceptors—electromagnetic energy with wavelengths between 400 and 700 nanometers (1 nm = 10^{-9} m, or one-billionth of a meter) constitutes *visible light*. Light of longer

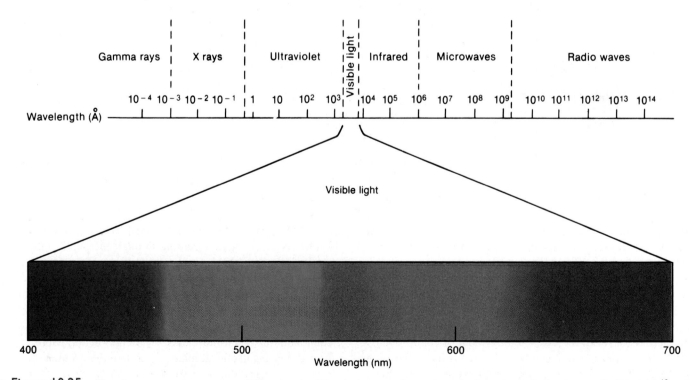

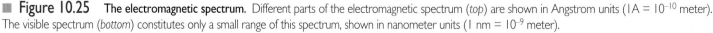

Figure 10.25 **The electromagnetic spectrum.** Different parts of the electromagnetic spectrum (*top*) are shown in Angstrom units (1 A = 10^{-10} meter). The visible spectrum (*bottom*) constitutes only a small range of this spectrum, shown in nanometer units (1 nm = 10^{-9} meter).

Table 10.4   Structures of the Eyeball

Tunic and Structure	Location	Composition	Function
Fibrous tunic	Outer layer of eyeball	Avascular connective tissue	Gives shape to the eyeball
Sclera	Posterior outer layer; white of the eye	Tightly bound elastic and collagen fibers	Supports and protects the eyeball
Cornea	Anterior surface of eyeball	Tightly packed dense connective tissue—transparent and convex	Transmits and refracts light
Vascular tunic (uvea)	Middle layer of eyeball	Highly vascular pigmented tissue	Supplies blood; prevents reflection
Choroid	Middle layer in posterior portion of eyeball	Vascular layer	Supplies blood to eyeball
Ciliary body	Anterior portion of vascular tunic	Smooth muscle fibers and glandular epithelium	Supports the lens through suspensory ligament and determines its thickness; secretes aqueous humor
Iris	Anterior portion of vascular tunic; continuous with ciliary body	Pigment cells and smooth muscle fibers	Regulates the diameter of the pupil, and hence the amount of light entering the vitreous chamber
Internal tunic	Inner layer of eyeball	Tightly packed photoreceptors, neurons, blood vessels, and connective tissue	Provides location and support for rods and cones
Retina	Principal portion of internal tunic	Photoreceptor neurons (rods and cones), bipolar neurons, and ganglion neurons	Photoreception; transmits impulses
Lens (not part of any tunic)	Between posterior and vitreous chambers; supported by suspensory ligament of ciliary body	Tightly arranged protein fibers; transparent	Refracts light and focuses onto fovea centralis

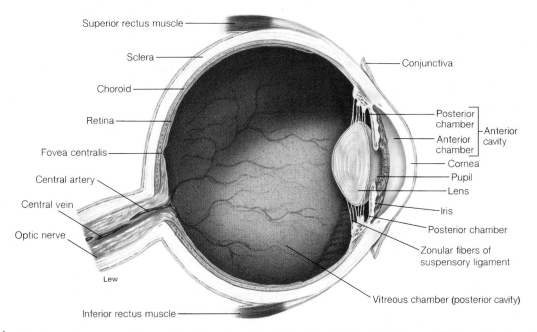

■ Figure 10.26   **The internal anatomy of the eyeball.** Light enters the eye from the right side of this figure and is focused on the retina.

wavelengths in the infrared regions of the spectrum is felt as heat but does not have sufficient energy to excite the receptors. Ultraviolet light, which has shorter wavelengths and more energy than visible light, is filtered out by the yellow color of the eye's lens. Honeybees—and people who have had their lenses removed—can see light in the ultraviolet range.

The structures of the eyeball are summarized in table 10.4. The outermost layer of the eye is a tough coat of connective tissue called the *sclera,* which can be seen externally as the white of the eyes. The tissue of the sclera is continuous with the transparent *cornea.* Light passes through the cornea to enter the *anterior chamber* of the eye. Light then passes through an opening

called the *pupil,* which is surrounded by a pigmented muscle known as the *iris.* After passing through the pupil, light enters the *lens* (fig. 10.26).

The iris is like the diaphragm of a camera; it can increase or decrease the diameter of its aperture (the pupil) to admit more or less light. Constriction of the pupils is produced by contraction of circular muscles within the iris; dilation is produced by contraction of radial muscles. Constriction of the pupils results from parasympathetic stimulation through the occulomotor (III) nerve, whereas dilation results from sympathetic stimulation (fig. 10.27). Variations in the diameter of the pupil are similar in effect to variations in the f-stop of a camera.

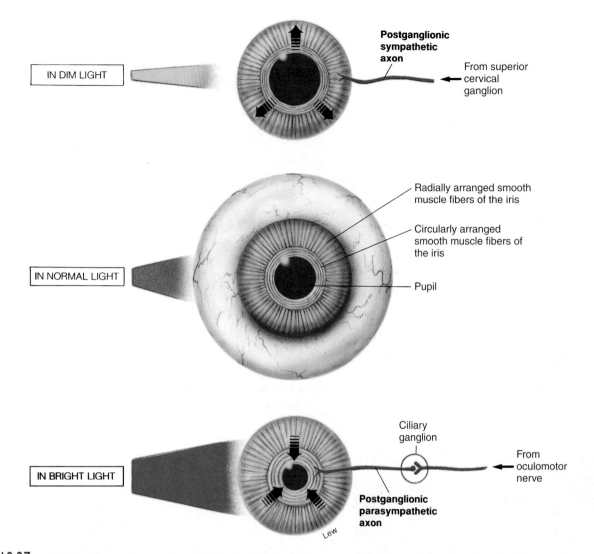

IN DIM LIGHT

**Postganglionic
sympathetic
axon**

From superior
cervical
ganglion

Radially arranged smooth
muscle fibers of the iris

Circularly arranged
smooth muscle fibers of
the iris

IN NORMAL LIGHT

Pupil

IN BRIGHT LIGHT

Ciliary
ganglion

From
oculomotor
nerve

**Postganglionic
parasympathetic
axon**

Lew

■ Figure 10.27   **Dilation and constriction of the pupil.** In dim light, the radially arranged smooth muscle fibers are stimulated to contract by sympathetic neurons, dilating the pupil. In bright light, the circularly arranged smooth muscle fibers are stimulated to contract by parasympathetic neurons, constricting the pupil.

The posterior part of the iris contains a pigmented epithelium that gives the eye its color. The color of the eye is determined by the amount of pigment—blue eyes have the least pigment, brown eyes have more, and black eyes have the greatest amount of pigment. In the condition of *albinism*—a congenital absence of normal pigmentation caused by an inability to produce melanin pigment—the eyes appear pink because the absence of pigment allows blood vessels to be seen.

The lens is suspended from a muscular process called the **ciliary body,** which connects the sclera and encircles the lens. *Zonular fibers* (*zon* = girdle) suspend the lens from the ciliary body, forming a **suspensory ligament** that supports the lens. The space between the cornea and iris is the *anterior chamber,* and the space between the iris and the ciliary body and lens is the *posterior chamber* (fig. 10.28).

The anterior and posterior chambers are filled with a fluid called **aqueous humor.** This fluid is secreted by the ciliary body into the posterior chamber and passes through the pupil into the anterior chamber, where it provides nourishment to the avascular lens

and cornea. The aqueous humor is drained from the anterior chamber into the *scleral venous sinus (canal of Schlemm),* which returns it to the venous blood (fig. 10.28). Inadequate drainage of aqueous humor can lead to excessive accumulation of fluid, which in turn results in increased intraocular pressure. This condition, called *glaucoma,* may produce serious damage to the retina and loss of vision.

The portion of the eye located behind the lens is filled with a thick, viscous substance known as the **vitreous body,** or **vitreous humor.** Light from the lens that passes through the vitreous body enters the neural layer, which contains photoreceptors, at the back of the eye. This neural layer is called the **retina.** Light that passes through the retina is absorbed by a darkly pigmented *choroid layer* underneath. While passing through the retina, some of this light stimulates photoreceptors, which in turn activate other neurons. Neurons in the retina contribute fibers that are gathered together at a region called the *optic disc* (fig. 10.29), where they exit the retina as the optic nerve. This region lacks photoreceptors, and is therefore known as the *blind spot.* The optic disc is also the site of entry and exit of blood vessels.

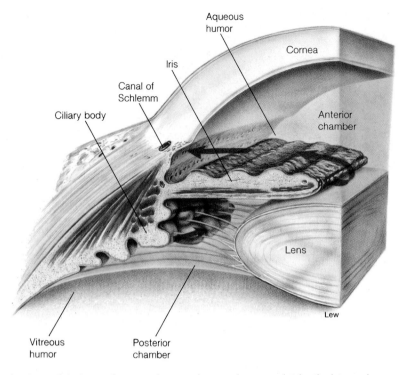

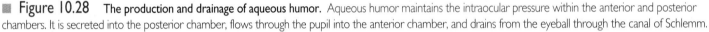

Figure 10.28   The production and drainage of aqueous humor. Aqueous humor maintains the intraocular pressure within the anterior and posterior chambers. It is secreted into the posterior chamber, flows through the pupil into the anterior chamber, and drains from the eyeball through the canal of Schlemm.

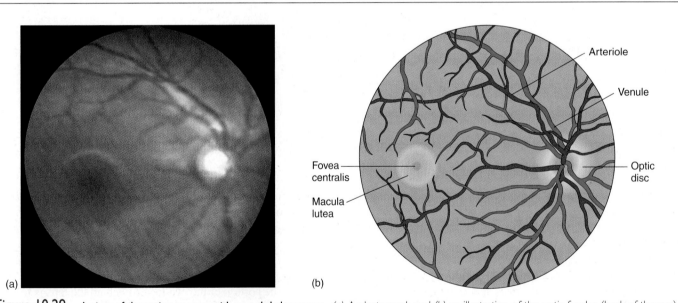

Figure 10.29   A view of the retina as seen with an ophthalmoscope. (a) A photograph and (b) an illustration of the optic fundus (back of the eye). Optic nerve fibers leave the eyeball at the optic disc to form the optic nerve. (Note the blood vessels that can be seen entering the eyeball at the optic disc.)

Refraction

Light that passes from a medium of one density into a medium of a different density is *refracted,* or bent. The degree of refraction depends on the comparative densities of the two media, as indicated by their *refractive index.* The refractive index of air is set at 1.00; the refractive index of the cornea, by comparison, is 1.38; and the refractive indices of the aqueous humor and lens are 1.33 and 1.40, respectively. Since the greatest difference in refractive index occurs at the air-cornea interface, the light is refracted most at the cornea.

The degree of refraction also depends on the curvature of the interface between two media. The curvature of the cornea is constant, but the curvature of the lens can be varied. The

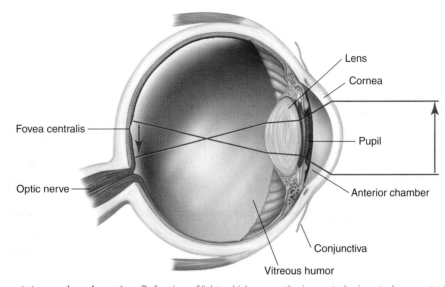

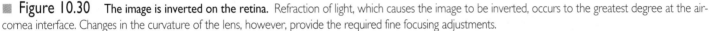

Figure 10.30 **The image is inverted on the retina.** Refraction of light, which causes the image to be inverted, occurs to the greatest degree at the air-cornea interface. Changes in the curvature of the lens, however, provide the required fine focusing adjustments.

refractive properties of the lens can thus provide fine control for focusing light on the retina. As a result of light refraction, the image formed on the retina is upside down and right to left (fig. 10.30).

The *visual field*—which is the part of the external world projected onto the retina—is thus reversed in each eye. The cornea and lens focus the right part of the visual field on the left half of the retina of each eye, while the left half of the visual field is focused on the right half of each retina (fig. 10.31). The medial (or nasal) half-retina of the left eye therefore receives the same image as the lateral (or temporal) half-retina of the right eye. The nasal half-retina of the right eye receives the same image as the temporal half-retina of the left eye.

Accommodation

When a normal eye views an object, parallel rays of light are refracted to a point, or *focus,* on the retina (see fig. 10.34). If the degree of refraction remained constant, movement of the object closer to or farther from the eye would cause corresponding movement of the focal point, so that the focus would either be behind or in front of the retina.

The ability of the eyes to keep the image focused on the retina as the distance between the eyes and object varies is called **accommodation.** Accommodation results from contraction of the ciliary muscle, which is like a sphincter muscle that can vary its aperture (fig. 10.32). When the ciliary muscle is relaxed, its aperture is wide. Relaxation of the ciliary muscle thus places tension on the zonular fibers of the suspensory ligament and pulls the lens taut. These are the conditions that prevail when viewing an object that is 20 feet or more from a normal eye; the image is focused on the retina and the lens is in its most flat, least convex form. As the object moves closer to the eyes,

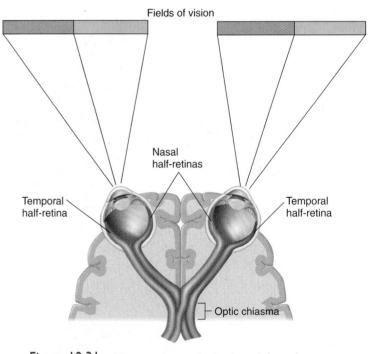

Figure 10.31 **The image is switched right-to-left on the retina.** The left side of the visual field is projected to the right half of each retina, while the right side of each visual field is projected to the left half of each retina.

the muscles of the ciliary body contract. This muscular contraction narrows the aperture of the ciliary body and thus reduces the tension on the zonular fibers that suspend the lens. When the tension is reduced, the lens becomes more rounded and convex as a result of its inherent elasticity (fig. 10.33).

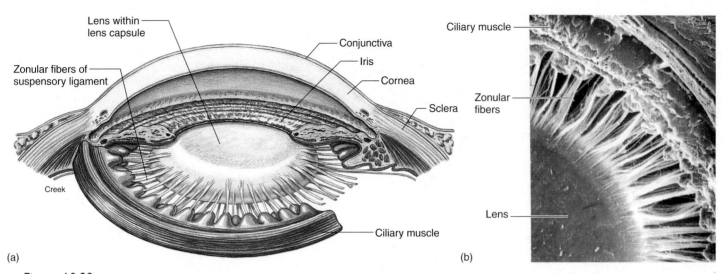

(a)   (b)

Figure 10.32   **The relationship between the ciliary muscle and the lens.** *(a)* A diagram, and *(b)* a scanning electron micrograph (from the eye of a 17-year-old boy) showing the relationship between the lens, zonular fibers, and ciliary muscle of the eye.

Part (b) from "How the Eye Focuses" by James F. Koretz and George H. Handleman. Copyright © 1988 by *Scientific American,* Inc. All rights reserved.

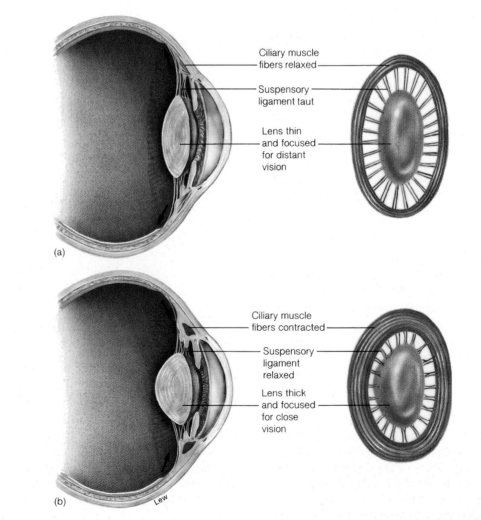

Figure 10.33   **Changes in the shape of the lens permit accommodation.** *(a)* The lens is flattened for distant vision when the ciliary muscle fibers are relaxed and the suspensory ligament is taut. *(b)* The lens is more spherical for close-up vision when the ciliary muscle fibers are contracted and the suspensory ligament is relaxed.

The ability of a person's eyes to accommodate can be measured by the near-point-of-vision test. The *near point of vision* is the minimum distance from the eyes at which an object can be brought into focus. This distance increases with age; indeed, accommodation in almost everyone over the age of 45 is significantly impaired. Loss of accommodating ability with age is known as **presbyopia** (*presby* = old). This loss appears to have a number of causes, including reduced flexibility of the lens and a forward movement of the attachments of the zonular fibers to the lens. As a result of these changes, the zonular fibers and lens are pulled taut even when the ciliary muscle contracts. The lens is thus not able to thicken and increase its refraction when, for example, a printed page is brought close to the eyes.

Clinical Investigation Clue

Remember that Ed had difficulty seeing print, although he never needed glasses before and his distance vision was still good.

What condition was most likely responsible for Ed's vision impairment?

Visual Acuity

Visual acuity refers to the sharpness of vision. The sharpness of an image depends on the *resolving power* of the visual system—that is, on the ability of the visual system to distinguish (resolve) two closely spaced dots. The better the resolving power of the system, the closer together these dots can be and still be seen as

separate. When the resolving power of the system is exceeded, the dots blur and are perceived as a single image.

Myopia and Hyperopia

When a person with normal visual acuity stands 20 feet from a *Snellen eye chart* (so that accommodation is not a factor influencing acuity), the line of letters marked "20/20" can be read. If a person has **myopia** (nearsightedness), this line will appear blurred because the image will be brought to a focus in front of the retina. This is usually due to the fact that the eyeball is too long. Myopia is corrected by glasses with concave lenses that cause the light rays to diverge, so that the point of focus is farther from the lens and is thus pushed back to the retina (fig. 10.34).

If the eyeball is too short, the line marked "20/20" will appear blurred because the focal length of the lens is longer than the distance to the retina. Thus, the focus of the image would have been behind the retina, and the object will have to be placed farther from the eyes to be seen clearly. This condition is called **hyperopia** (farsightedness). Hyperopia is corrected by glasses with convex lenses that increase the convergence of light, so that the point of focus is brought closer to the lens and falls on the retina.

Astigmatism

Because the curvature of the cornea and lens is not perfectly symmetrical, light passing through some parts of these structures may be refracted to a different degree than light passing through other parts. When the asymmetry of the cornea and/or lens is significant, the person is said to have **astigmatism.** If a person with astigmatism views a circle of lines radiating from the center, like the spokes of a wheel, the image of these lines will not appear clear in all 360 degrees. The parts of the circle that appear blurred can thus be used to map the astigmatism. This condition is corrected by cylindrical lenses that compensate for the asymmetry in the cornea or lens of the eye.

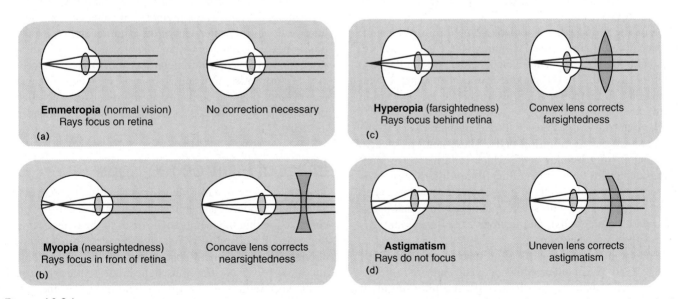

Figure 10.34   Problems of refraction and how they are corrected. In a normal eye (*a*), parallel rays of light are brought to a focus on the retina by refraction in the cornea and lens. If the eye is too long, as in myopia (*b*), the focus is in front of the retina. This can be corrected by a concave lens. If the eye is too short, as in hyperopia (*c*), the focus is behind the retina. This is corrected by a convex lens. In astigmatism (*d*), light refraction is uneven because of irregularities in the shape of the cornea or lens.

Retina

There are two types of photoreceptor neurons: rods and cones. Both receptor cell types contain pigment molecules that undergo dissociation in response to light, and it is this photochemical reaction that eventually results in the production of action potentials in the optic nerve. Rods provide black-and-white vision under conditions of low light intensities, whereas cones provide sharp color vision when light intensities are greater.

The **retina** consists of a single-cell-thick pigmented epithelium, photoreceptor neurons called **rods** and **cones,** and layers of other neurons. The neural layers of the retina are actually a forward extension of the brain. In this sense, the optic nerve can be considered a tract, and indeed the myelin sheaths of its fibers are derived from oligodendrocytes (like other CNS axons) rather than from Schwann cells.

Since the retina is an extension of the brain, the neural layers face outward, toward the incoming light. Light, therefore, must pass through several neural layers before striking the photoreceptors (fig. 10.35). The photoreceptors then synapse with other neurons, so that nerve impulses are conducted outward in the retina.

The outer layers of neurons that contribute axons to the optic nerve are called **ganglion cells.** These neurons receive synaptic input from **bipolar cells,** which in turn receive input from rods and cones. In addition to the flow of information from photoreceptors to bipolar cells to ganglion cells, neurons called *horizontal cells* synapse with several photoreceptors (and possibly also with bipolar cells), and neurons called *amacrine cells* synapse with several ganglion cells.

Each rod and cone consists of an inner and an outer segment (fig. 10.36). Each outer segment contains hundreds of flattened membranous sacs, or discs, with the photopigment molecules required for vision. The photoreceptor cells continuously add new discs at the base of the outer segment as the tip regions are removed by the cells of the **retinal pigment epithelium** (see fig. 10.35) through a process of phagocytosis.

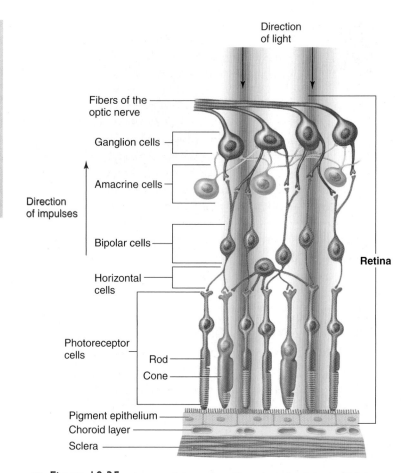

Figure 10.35 **Layers of the retina.** Since the retina is inverted, light must pass through various layers of nerve cells before reaching the photoreceptors (rods and cones).

The photoreceptors are in close proximity to the retinal pigment epithelium. These pigment epithelial cells are needed to remove the old tip segments of the photoreceptors; if that function is not performed properly, vision can be impaired (see the discussion of macular degeneration, p. 274). The pigment cells also perform other functions important in vision. For example, the melanin pigment contained in these cells absorbs light that might otherwise reflect back to the photoreceptors and reduce the clarity of vision.

Effect of Light on the Rods

The photoreceptors—rods and cones (fig. 10.36)—are activated when light produces a chemical change in molecules of pigment contained within the membranous **discs** of the outer segments of the receptor cells. Rods contain a purple pigment known as **rhodopsin.** The pigment appears purple (a combination of red and blue) because it transmits light in the red and blue regions of the spectrum, while absorbing light energy in the green region. The wavelength of light that is absorbed best—the *absorption maximum*—is about 500 nm (blue–green light).

Green cars (and other green objects) are seen more easily at night—when rods are used for vision—than are red objects. This is

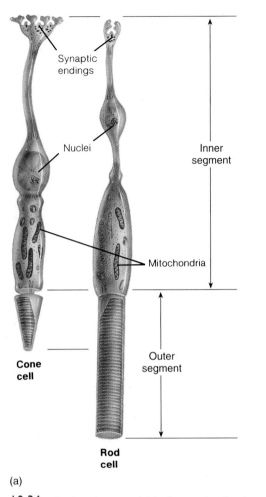

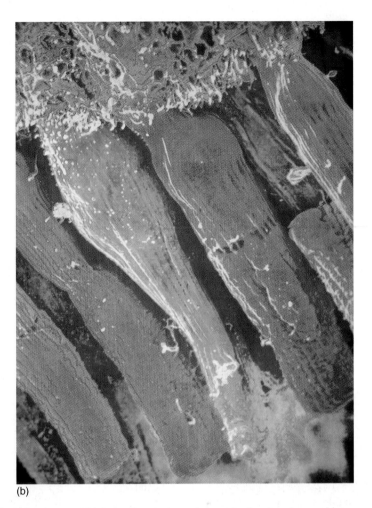

(a)

(b)

Figure 10.36 Rods and cones. *(a)* A diagram showing the structure of a rod and a cone. *(b)* A scanning electron micrograph of rods and cones. Note that each photoreceptor contains an outer and inner segment.

because red light is not absorbed well by rhodopsin, and only absorbed light can produce the photochemical reaction that results in vision. In response to absorbed light, rhodopsin dissociates into its two components: the pigment **retinaldehyde** (also called **retinene** or **retinal**), which is derived from vitamin A, and a protein called **opsin.** This reaction is known as the **bleaching reaction.**

Scientists have recently discovered the genetic basis for blindness in the disease **dominant retinitis pigmentosa.** People with this disease inherit a gene for the opsin protein in which a single base change in the gene (substitution of adenine for cytosine) causes the amino acid histidine to be substituted for proline at a specific point in the polypeptide chain. This abnormal opsin leads to degeneration of the photoreceptors.

Retinene can exist in two possible configurations (shapes)—one known as the all-*trans* form and one called the 11-*cis* form (fig. 10.37). The all-*trans* form is more stable, but only the 11-*cis* form is found attached to opsin. In response to

absorbed **light energy,** the 11-*cis*-retinene is converted to the all-*trans* isomer, causing it to dissociate from the opsin. This dissociation reaction in response to light initiates changes in the ionic permeability of the rod plasma membrane and ultimately results in the production of nerve impulses in the ganglion cells. As a result of these effects, rods provide black-and-white vision under conditions of low light intensity.

Dark Adaptation

The bleaching reaction that occurs in the light results in a lowered amount of rhodopsin in the rods and lowered amounts of visual pigments in the cones. When a light-adapted person first enters a darkened room, therefore, sensitivity to light is low and vision is poor. A gradual increase in photoreceptor sensitivity, known as **dark adaptation,** then occurs, reaching maximal sensitivity in about 20 minutes. The increased sensitivity to low light intensity is partly due to increased amounts of visual pigments produced in the dark. Increased pigments in the cones produce a slight dark adaptation in the first 5 minutes. Increased rhodopsin in the rods produces a much greater increase in sensitivity to low light levels and is partly responsible for the adaptation that occurs after about 5 minutes in the dark. In addition to the increased concentration of rhodopsin, other more subtle (and less well understood) changes occur in the

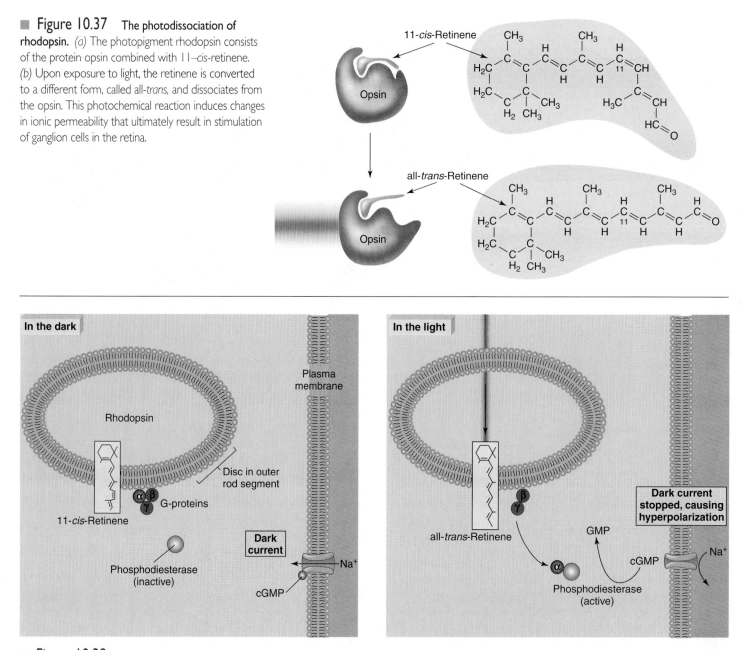

■ **Figure 10.37**   The photodissociation of
rhodopsin. *(a)* The photopigment rhodopsin consists
of the protein opsin combined with 11–*cis*-retinene.
(b) Upon exposure to light, the retinene is converted
to a different form, called all-*trans*, and dissociates from
the opsin. This photochemical reaction induces changes
in ionic permeability that ultimately result in stimulation
of ganglion cells in the retina.

■ **Figure 10.38**   **Light striking the photoreceptors causes Na⁺ channels to close.** Light causes the photoisomerization of 11-*cis* to all-*trans*-retinene
and its dissociation from opsin. This releases α subunits of G-proteins, which activate phosphodiesterase. This activated enzyme converts cyclic GMP
(cGMP) to GMP. Since the cGMP is needed to keep the Na⁺ channels of the plasma membrane open, conversion of cGMP to GMP causes these channels
to close. This stops the dark current produced by Na⁺ entry in the dark, leading to hyperpolarization of the photoreceptor.

rods that ultimately result in a 100,000-fold increase in light sensitivity in dark-adapted as compared to light-adapted eyes.

Electrical Activity of Retinal Cells

The only neurons in the retina that produce all-or-none action potentials are ganglion cells and amacrine cells. The photoreceptors,
bipolar cells, and horizontal cells instead produce only graded depolarizations or hyperpolarizations, analogous to EPSPs and IPSPs.

The transduction of light energy into nerve impulses follows
a cause-and-effect sequence that is the inverse of the usual way in

which sensory stimuli are detected. This is because, in the dark, the
photoreceptors release an inhibitory neurotransmitter that hyperpolarizes the bipolar neurons. Thus inhibited, the bipolar neurons do
not release excitatory neurotransmitter to the ganglion cells. Light
inhibits the photoreceptors from releasing their inhibitory neurotransmitter and by this means *stimulates* the bipolar cells, and thus
the ganglion cells that transmit action potentials to the brain.

A rod or cone contains many Na⁺ channels in the plasma
membrane of its outer segment (see fig. 10.38), and in the dark,
many of these channels are open. As a consequence, Na⁺ continuously diffuses into the outer segment and across the narrow

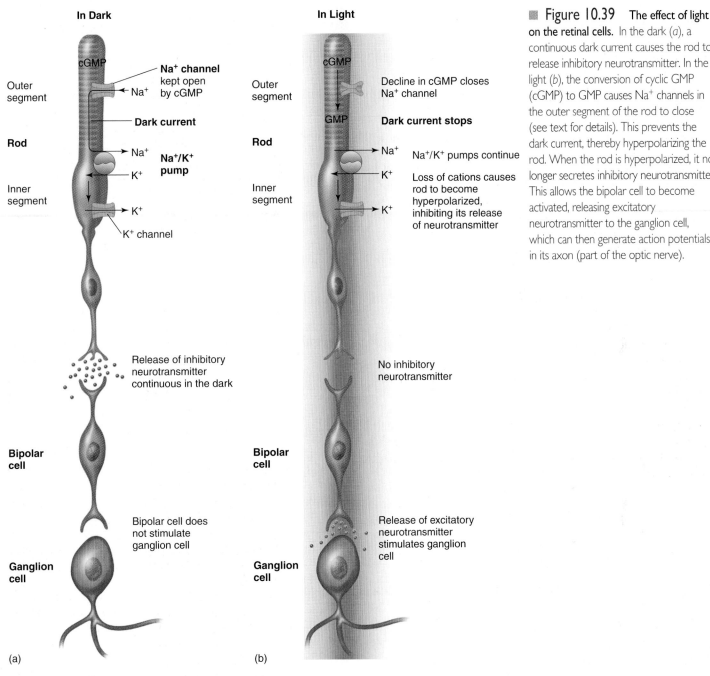

In Dark

Outer segment

cGMP

Na⁺ channel kept open by cGMP

← Na⁺

Dark current

Rod

→ Na⁺   **Na⁺/K⁺ pump**

← K⁺

Inner segment

→ K⁺

K⁺ channel

Release of inhibitory neurotransmitter continuous in the dark

Bipolar cell

Bipolar cell does not stimulate ganglion cell

Ganglion cell

(a)

In Light

cGMP

Outer segment

Decline in cGMP closes Na⁺ channel

GMP

Dark current stops

Rod

→ Na⁺   Na⁺/K⁺ pumps continue

← K⁺

Loss of cations causes rod to become hyperpolarized, inhibiting its release of neurotransmitter

Inner segment

→ K⁺

No inhibitory neurotransmitter

Bipolar cell

Release of excitatory neurotransmitter stimulates ganglion cell

Ganglion cell

(b)

■ **Figure 10.39** The effect of light on the retinal cells. In the dark (*a*), a continuous dark current causes the rod to release inhibitory neurotransmitter. In the light (*b*), the conversion of cyclic GMP (cGMP) to GMP causes Na⁺ channels in the outer segment of the rod to close (see text for details). This prevents the dark current, thereby hyperpolarizing the rod. When the rod is hyperpolarized, it no longer secretes inhibitory neurotransmitter. This allows the bipolar cell to become activated, releasing excitatory neurotransmitter to the ganglion cell, which can then generate action potentials in its axon (part of the optic nerve).

stalk to the inner segment. This small flow of Na⁺ that occurs in the absence of light stimulation is called the **dark current,** and it causes the membrane of a photoreceptor to be somewhat depolarized in the dark. The Na⁺ channels in the outer segment rapidly close in response to light, reducing the dark current and causing the photoreceptor to hyperpolarize.

It has been discovered that cyclic GMP (cGMP) is required to keep the Na⁺ channels open, and that the channels will close if the cGMP is converted into GMP. Light causes this conversion and consequent closing of the Na⁺ channels. When a photopigment absorbs light, 11-*cis*-retinene is converted into its isomer, all-*trans*-retinene (fig. 10.38) and dissociates from the opsin, causing the opsin protein to change shape. Each opsin is

associated with over a hundred regulatory *G-proteins* (see chapter 7) known as **transducins,** and the change in the opsin induced by light causes the alpha subunits of the G-proteins to dissociate. These G-protein subunits then bind to and activate hundreds of molecules of the enzyme *phosphodiesterase.* This enzyme converts cGMP to GMP, thus closing the Na⁺ channels at a rate of about 1,000 per second and inhibiting the dark current (fig. 10.38). The absorption of a single photon of light can block the entry of more than a million Na⁺, thereby causing the photoreceptor to hyperpolarize and release less inhibitory neurotransmitter. Freed from inhibition, the bipolar cells activate ganglion cells, and the ganglion cells transmit action potentials to the brain so that light can be perceived (fig. 10.39).

Cones and Color Vision

Cones are less sensitive than rods to light, but the cones provide color vision and greater visual acuity, as described in the next section. During the day, therefore, the high light intensity bleaches out the rods, and color vision with high acuity is provided by the cones. Humans and other primates have **trichromatic color vision** (are *trichromats*). This means that our perception of a multitude of colors is produced by stimulation of only three types of cones. This fact is exploited by television screens and computer monitors, which display only red, green, and blue pixels. Interestingly, other mammals that are able to see colors get by with only two types of cones (they are *dichromats*).

The three different cones responsible for human color vision are designated *blue, green,* and *red,* according to the region of the visible spectrum in which each cone pigment absorbs light best (fig. 10.40). This is the cone's *absorption maximum,* and corresponds to wavelengths of 420 nanometers (nm) for the blue cones (also called short wavelengths, or *S cones*), 530 nm for the green cones (also called medium wavelength, or *M cones*), and 562 nm for red cones (also called long wavelengths, or *L cones*). The gene for the S cones is located on chromosome number 7, whereas the genes for the M and L cones are located on the long arm of the X chromosomes.

Each type of cone contains retinene, as in rhodopsin, but the retinene in the cones is associated with proteins called **photopsins,** which are different from the opsin in rods. It is the three different photopsin proteins (coded by three different genes) that gives each type of cone its unique absorption maximum.

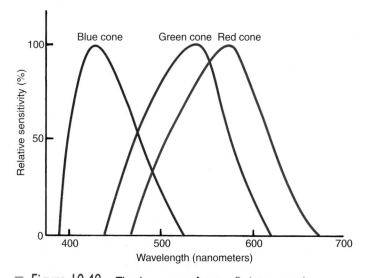

Figure 10.40 **The three types of cones.** Each type contains retinene, but the protein with which the retinene is combined is different in each case. Thus, each different pigment absorbs light maximally at a different wavelength. Color vision is produced by the activity of these blue cones, green cones, and red cones.

Color blindness is caused by a congenital lack of one or more types of cones, usually the absence of either the L (red) or M (green) cones. Since such people have only two functioning types of cones, they are dichromats. The absence of functioning M cones, a condition called *deuteranopia,* is the most common form of color blindness. The absence of L cones (*protanopia*) is less common, and the absence of S cones (*tritanopia*) is the least common. People who have only one cone in the middle to long wavelength region (M or L) have difficulty distinguishing reds from greens. Since the M and L cone pigments (photopsins) are coded on the X chromosome, and since men have only one X chromosome (and therefore cannot carry the trait in a recessive state—see chapter 20), such red-green color blindness is far more common in men (with an incidence of 8%) than in women (0.5%).

Suppose a person has become dark adapted in a photographic darkroom over a period of 20 minutes or longer, but needs light to examine some prints. Since rods do not absorb red light but red cones do, a red light in a photographic darkroom allows vision (because of the red cones) but does not cause bleaching of the rods. When the light is turned off, therefore, the rods will still be dark adapted and the person will still be able to see.

Visual Acuity and Sensitivity

While reading or similarly viewing objects in daylight, each eye is oriented so that the image falls within a tiny area of the retina called the **fovea centralis.** The fovea is a pinhead-sized pit (*fovea* = pit) within a yellow area of the retina called the *macula lutea.* The pit is formed as a result of the displacement of neural layers around the periphery; therefore, light falls directly on photoreceptors in the center (fig. 10.41). Light falling on other areas, by contrast, must pass through several layers of neurons, as previously described.

There are approximately 120 million rods and 6 million cones in each retina, but only about 1.2 million nerve fibers enter the optic nerve of each eye. This gives an overall convergence ratio of photoreceptors on ganglion cells of about 105 to 1. This is misleading, however, because the degree of convergence is much lower for cones than for rods. In the fovea, the ratio is 1 to 1.

The photoreceptors are distributed in such a way that the fovea contains only cones, whereas more peripheral regions of the retina contain a mixture of rods and cones. Approximately 4,000 cones in the fovea provide input to approximately 4,000 ganglion cells; each ganglion cell in this region, therefore, has a private line to the visual field. Each ganglion cell in the fovea thus receives input from an area of retina corresponding to the diameter of one cone (about 2 μm). Peripheral to the fovea, however, many rods synapse with a single bipolar cell, and many bipolar cells synapse with a single ganglion cell. A single ganglion cell outside the fovea thus may receive input from large numbers of rods, corresponding to an area of about 1 mm² on the retina (fig. 10.42).

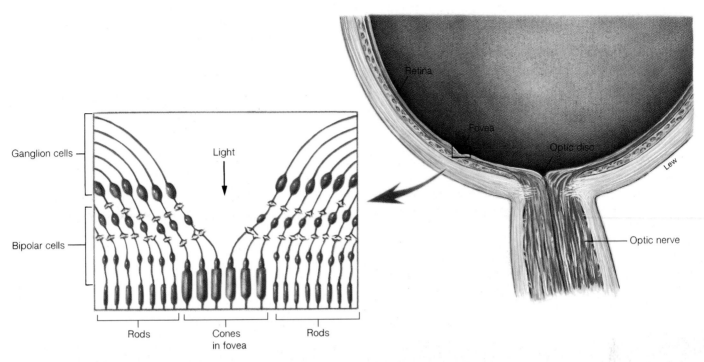

Figure 10.41 **The fovea centralis.** When the eyes "track" an object, the image is cast upon the fovea centralis of the retina. The fovea is literally a "pit" formed by parting of the neural layers. In this region, light thus falls directly on the photoreceptors (cones).

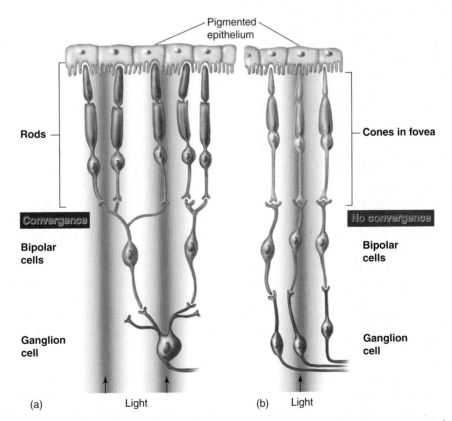

Figure 10.42 **Convergence in the retina and light sensitivity.** Since bipolar cells receive input from the convergence of many rods (a), and since a number of such bipolar cells converge on a single ganglion cell, rods maximize sensitivity to low levels of light at the expense of visual acuity. By contrast, the 1:1:1 ratio of cones to bipolar cells to ganglion cells in the fovea (b) provides high visual acuity, but sensitivity to light is reduced.

Since each cone in the fovea has a private line to a ganglion cell, and since each ganglion cell receives input from only a tiny region of the retina, visual acuity is greatest and sensitivity to low light is poorest when light falls on the fovea. In dim light only the rods are activated, and vision is best out of the corners of the eye when the image falls away from the fovea. Under these conditions, the convergence of large numbers of rods on a single bipolar cell and the convergence of large numbers of bipolar cells on a single ganglion cell increase sensitivity to dim light at the expense of visual acuity. Night vision is therefore less distinct than day vision.

The difference in visual sensitivity between cones in the fovea centralis and rods in the periphery of the retina can be demonstrated easily using a technique called *averted vision.* If you go out on a clear night and stare hard at a very dim star, it will disappear. This is because the light falls on the fovea and is not sufficiently bright to activate the cones. If you then look slightly off to the side, the star will reappear because the light falls away from the fovea, on the rods.

The only part of the visual field that is actually seen clearly is that tiny part (about 1%) that falls on the fovea centralis. We are unaware of this because rapid eye movements shift different parts of the visual field onto the fovea. A common visual impairment, particularly in older people, is **macular degeneration**—degeneration of the macula lutea and its central fovea. People with macular degeneration lose the clarity of vision provided by the fovea in the central region of the visual field. In most cases, the damage is believed to be related to the loss of retinal pigment epithelium in this region. In some cases, the damage is made worse by growth of new blood vessels (*neovascularization*) from the choroid into the retina.

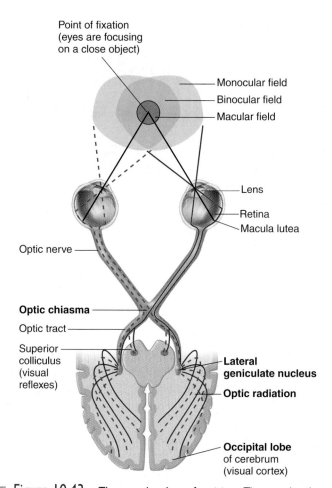

■ **Figure 10.43**   The neural pathway for vision.   The neural pathway leading from the retina to the lateral geniculate body, and then to the visual cortex, is needed for visual perception. As a result of the crossing of optic fibers, the visual cortex of each cerebral hemisphere receives input from the opposite (contralateral) visual field.

Neural Pathways from the Retina

As a result of light refraction by the cornea and lens, the right half of the visual field is projected to the left half of the retina of both eyes (the temporal half of the left retina and the nasal half of the right retina). The left half of the visual field is projected to the right half of the retina of both eyes. The temporal half of the left retina and the nasal half of the right retina therefore see the same image. Axons from ganglion cells in the left (temporal) half of the left retina pass to the left **lateral geniculate nucleus** of the thalamus. Axons from ganglion cells in the nasal half of the right retina cross over (decussate) in the x-shaped *optic chiasma,* also to synapse in the left lateral geniculate body. The left lateral geniculate, therefore, receives input from both eyes that relates to the right half of the visual field (fig. 10.43).

The right lateral geniculate body, similarly, receives input from both eyes relating to the left half of the visual field. Neurons in both lateral geniculate bodies of the thalamus in turn project to the **striate cortex** of the occipital lobe in the cerebral cortex (fig. 10.44). This area is also called area 17, in reference to a numbering system developed by K. Brodmann in 1906. Neurons in area 17 synapse with neurons in areas 18 and 19 of the occipital lobe (fig. 10.44).

Approximately 70% to 80% of the axons from the retina pass to the lateral geniculate bodies and to the striate cortex. This **geniculostriate system** is involved in perception of the visual field. Put another way, the geniculostriate system is needed to answer the question, What is it? Approximately 20% to 30% of the fibers from the retina, however, follow a different path to the *superior colliculus* of the midbrain (also called the *optic tectum*). Axons from the superior colliculus activate motor pathways leading to eye and body movements. The **tectal system,** in other words, is needed to answer the question, Where is it?

Superior Colliculus and Eye Movements

Neural pathways from the superior colliculus to motor neurons in the spinal cord help mediate the startle response to the sight of an unexpected intruder. Other nerve fibers from the superior colliculus stimulate the *extrinsic eye muscles,* which are the striated muscles that move the eyes.

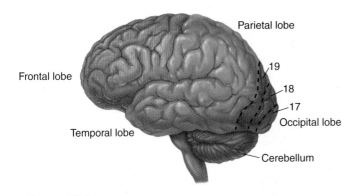

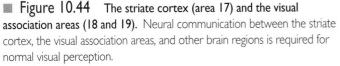

■ **Figure 10.44** The striate cortex (area 17) and the visual association areas (18 and 19). Neural communication between the striate cortex, the visual association areas, and other brain regions is required for normal visual perception.

Two types of eye movements are coordinated by the superior colliculus. **Smooth pursuit movements** track moving objects and keep the image focused on the fovea centralis. **Saccadic eye movements** are quick (lasting 20 to 50 msec), jerky movements of both eyes that occur while the eyes appear to be still. These saccadic movements continuously move the image to different photoreceptors; if they were to stop, the image would disappear as the photoreceptors became bleached. Saccadic eye movements are also responsible for the ability of the eyes to jump from word to word as you read a line, so that the image of each word in succession is focused on the fovea.

The tectal system is also involved in the control of the intrinsic eye muscles—the iris and the muscles of the ciliary body. Shining a light into one eye stimulates the *pupillary reflex,* in which both pupils constrict. This is caused by activation of parasympathetic neurons in the superior colliculus. Postganglionic axons from the ciliary ganglia behind the eyes, in turn, stimulate constrictor fibers in the iris (see fig. 10.27). Contraction of the ciliary body during accommodation also involves parasympathetic stimulation by the superior colliculus.

Test Yourself Before You Continue

1. Describe the different layers of the retina and trace the path of light and of nerve activity through these layers.

2. Describe the photochemical reaction in the rods and explain how dark adaptation occurs.

3. Describe the electrical state of photoreceptors in the dark. Explain how light affects the electrical activity of retinal cells.

4. Explain what is meant by the trichromatic theory of color vision.

5. Compare the architecture of the fovea centralis with more peripheral regions of the retina. How does this architecture relate to visual acuity and sensitivity?

6. Describe how different parts of the visual field are projected onto the retinas of both eyes. Trace the neural pathways for this information in the geniculostriate system.

7. Describe the neural pathways involved in the tectal system. What are the functions of these pathways?

Neural Processing of Visual Information

Electrical activity in ganglion cells of the retina and in neurons of the lateral geniculate nucleus and cerebral cortex is evoked in response to light on the retina. The way in which each type of neuron responds to light at a particular point on the retina provides information about how the brain interprets visual information.

Light cast on the retina directly affects the activity of photoreceptors and indirectly affects the neural activity in bipolar and ganglion cells. The part of the visual field that affects the activity of a particular ganglion cell can be considered its **receptive field.** As previously mentioned, each cone in the fovea has a private line to a ganglion cell, and thus the receptive fields of these ganglion cells are equal to the width of one cone (about 2 µm). By contrast, ganglion cells in more peripheral parts of the retina receive input from hundreds of photoreceptors, and are therefore influenced by a larger area of the retina (about 1 mm in diameter).

Ganglion Cell Receptive Fields

Studies of the electrical activity of ganglion cells have yielded some interesting results. In the dark, each ganglion cell discharges spontaneously at a slow rate. When the room lights are turned on, the firing rate of many (but not all) ganglion cells increases slightly. With some ganglion cells, however, if a small spot of light is directed at the center of their receptive fields, a large increase in firing rate results. Surprisingly, then, a small spot of light can be a more effective stimulus than larger areas of light!

When the spot of light is moved only a short distance away from the center of the receptive field, the ganglion cell responds in the opposite manner. The ganglion cell that was stimulated with light at the center of its receptive field is inhibited by light in the periphery of its field. The responses produced by light in the center and by light in the "surround" of the visual field are *antagonistic.* Those ganglion cells that are stimulated by light at the center of their visual fields are said to have **on-center fields;** those that are inhibited by light in the center and stimulated by light in the surround have **off-center fields** (fig. 10.45).

The reason wide illumination of the retina has a weaker effect than pinpoint illumination is now clear; diffuse illumination gives the ganglion cell conflicting orders—on and off. Because of the antagonism between the center and surround of ganglion cell receptive fields, the activity of each ganglion cell is a result of the *difference in light intensity* between the center and surround of its visual field. This is a form of lateral inhibition that helps to accentuate the contours of images and improve visual acuity.

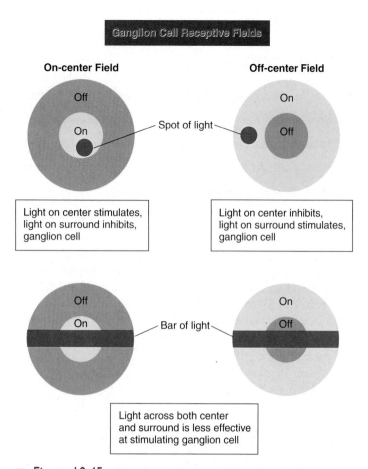

Figure 10.45 *Ganglion cell receptive fields.* Each ganglion cell receives input from photoreceptors in the retina that are part of the ganglion cell's "receptive field." Because of the antagonism between the field's center and its surround, an image that falls across the entire field has less effect than one that only excites just the center or surround. Because of this, edges of an image are enhanced, improving the clarity of vision.

Lateral Geniculate Nuclei

Each lateral geniculate nucleus of the thalamus receives input from ganglion cells in both eyes. The right lateral geniculate receives input from the right half of each retina (corresponding to the left half of the visual field); the left lateral geniculate receives input from the left half of each retina (corresponding to the right half of the visual field). Each neuron in the lateral geniculate, however, is activated by input from only one eye.

The receptive field of each ganglion cell, as previously described, is the part of the retina it "sees" through its photoreceptor input. The receptive field of lateral geniculate neurons, similarly, is the part of the retina it "sees" through its ganglion cell input. Experiments in which the lateral geniculate receptive fields are mapped with a spot of light reveal that they are circular, with an antagonistic center and surround, much like the ganglion cell receptive fields.

Cerebral Cortex

Projections of nerve fibers from the lateral geniculate bodies to area 17 of the occipital lobe form the **optic radiation** (see fig. 10.43). Because these fiber projections give area 17 a striped or striated ap-

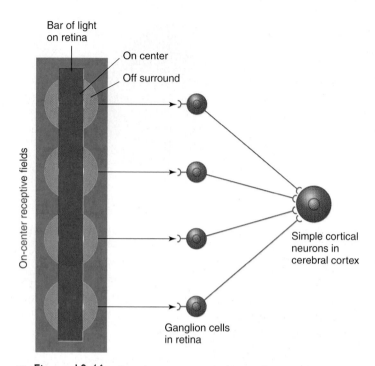

Figure 10.46 *Stimulus requirements for simple cortical neurons.* Cortical neurons called simple cells have rectangular receptive fields that are best stimulated by slits of light of particular orientations. This may be due to the fact that these simple cells receive input from ganglion cells that have circular receptive fields along a particular line.

pearance, this area is also known as the *striate cortex.* As mentioned earlier, neurons in area 17 project to areas 18 and 19 of the occipital lobe. Cortical neurons in areas 17, 18, and 19 are thus stimulated indirectly by light on the retina. On the basis of their stimulus requirements, these cortical neurons are classified as *simple, complex,* and *hypercomplex.*

The receptive fields of **simple neurons** are rectangular rather than circular. This is because they receive input from lateral geniculate neurons whose receptive fields are aligned in a particular way (as illustrated in fig. 10.46). Simple cortical neurons are best stimulated by a slit or bar of light located in a precise part of the visual field (of either eye) at a precise orientation.

The striate cortex (area 17) contains simple, complex, and hypercomplex neurons. The other visual association areas, designated areas 18 and 19, contain only complex and hypercomplex cells. Complex neurons receive input from simple cells, and hypercomplex neurons receive input from complex cells.

Test Yourself Before You Continue

1. Describe the way in which ganglion cells typically respond to light on the retina. Why may a small spot of light be a more effective stimulus than general illumination of the retina?

2. How can the arrangement of the receptive fields of ganglion cells enhance visual acuity?

3. Describe the stimulus requirements of simple cortical neurons.

HPer Links of the Sensory System with Other Body Systems

Integumentary System

- The skin helps to protect the body from pathogens ....**(p. 446)**
- The skin helps to regulate body temperature ....**(p. 428)**
- Cutaneous receptors provide sensations of touch, pressure, pain, heat, and cold ....**(p. 244)**

Skeletal System

- The skull provides protection and support for the eye and ear ....**(p. 255)**
- Proprioceptors provide sensory information about joint movement and the tension of tendons ....**(p. 242)**

Muscular System

- Sensory information from the heart helps to regulate the heartbeat ....**(p. 432)**
- Sensory information from certain arteries helps to regulate the blood pressure ....**(p. 430)**
- Muscle spindles within skeletal muscles monitor the length of the muscle .**(p. 348)**

Nervous System

- Afferent neurons transduce graded receptor potentials into action potentials ....**(p. 243)**
- Afferent neurons conduct action potentials from sensory receptors into the CNS for processing ....**(p. 245)**

Endocrine System

- Stimulation of stretch receptors in the heart causes the secretion of atrial natriuretic hormone ....**(p. 432)**
- Stimulation of receptors in the GI tract causes the secretion of particular hormones ....**(p. 583)**
- Stimulation of sensory endings in the breast by the sucking action of an infant evokes the secretion of hormones involved in lactation ....**(p. 678)**

Circulatory System

- The blood delivers oxygen and nutrients to sensory organs and removes metabolic wastes ....**(p. 366)**
- Sensory stimuli from the heart provide information for neural regulation of the heartbeat ....**(p. 432)**
- Sensory stimuli from certain blood vessels provide information for the neural regulation of blood flow and blood pressure ....**(p. 430)**

Immune System

- The immune system protects against infections of sensory organs .....**(p. 446)**
- Pain sensations may arise from swollen lymph nodes, alerting us to infection ....**(p. 450)**
- The detection of particular chemicals in the brain evokes a fever, which may help to defeat infections ....**(p. 448)**

Respiratory System

- The lungs provide oxygen for the blood and provide for the elimination of carbon dioxide ....**(p. 480)**
- Chemoreceptors in the aorta, carotid arteries, and medulla oblongata provide sensory information for the regulation of breathing ....**(p. 500)**

Urinary System

- The kidneys regulate the volume, pH, and electrolyte balance of the blood and eliminate wastes ....**(p. 524)**
- Stretch receptors in the atria of the heart cause secretion of natriuretic factor, which helps to regulate the kidneys .....**(p. 432)**

- Receptors in renal blood vessels contribute to the regulation of renal blood flow ....**(p. 531)**

Digestive System

- The GI tract provides nutrients for all the body organs, including those of the sensory system ....**(p. 561)**
- Stretch receptors in the GI tract participate in the reflex control of the digestive system ....**(p. 586)**
- Chemoreceptors in the GI tract contribute to the regulation of digestive activities ....**(p. 584)**

Reproductive System

- Gonads produce sex hormones that influence sensations involved in the male and female sexual response .....**(p. 643)**
- Sensory receptors provide information for erection and orgasm, as well as for other aspects of the sexual response ...**(p. 643)**

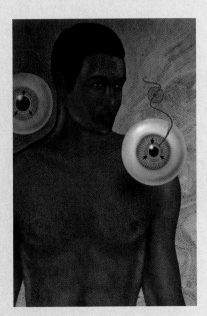

Summary

Characteristics of Sensory Receptors 242

I. Sensory receptors may be categorized on the basis of their structure, the stimulus energy they transduce, or the nature of their response.

   A. Receptors may be dendritic nerve endings, specialized neurons, or specialized epithelial cells associated with sensory nerve endings.

   B. Receptors may be chemoreceptors, photoreceptors, thermoreceptors, mechanoreceptors, or nociceptors.

      1. Proprioceptors include receptors in the muscles, tendons, and joints.

      2. The senses of sight, hearing, taste, olfaction, and equilibrium are grouped as special senses.

   C. Receptors vary in the duration of their firing in response to a constant stimulus.

      1. Tonic receptors continue to fire as long as the stimulus is maintained; they monitor the presence and intensity of a stimulus.

      2. Phasic receptors respond to stimulus changes; they do not respond to a sustained stimulus, and this partly accounts for sensory adaptation.

II. According to the law of specific nerve energies, each sensory receptor responds with lowest threshold to only one modality of sensation.

   A. That stimulus modality is called the adequate stimulus.

   B. Stimulation of the sensory nerve from a receptor by any means is interpreted in the brain as the adequate stimulus modality of that receptor.

III. Generator potentials are graded changes (usually depolarizations) in the membrane potential of the dendritic endings of sensory neurons.

   A. The magnitude of the potential change of the generator potential is directly proportional to the strength of the stimulus applied to the receptor.

   B. After the generator potential reaches a threshold value, increases in the magnitude of the depolarization result in increased frequency of action potential production in the sensory neuron.

Cutaneous Sensations 244

I. Somatesthetic information—from cutaneous receptors and proprioceptors—is carried by third-order neurons to the postcentral gyrus of the cerebrum.

   A. Proprioception and pressure sensations ascend on the ipsilateral side of the spinal cord, synapse in the medulla and cross to the contralateral side, and then ascend in the medial lemniscus to the thalamus; neurons in the thalamus in turn project to the postcentral gyrus.

   B. Sensory neurons from other cutaneous receptors synapse and cross to the contralateral side in the spinal cord and ascend in the lateral and ventral spinothalamic tracts to the thalamus; neurons in the thalamus then project to the postcentral gyrus.

II. The receptive field of a cutaneous sensory neuron is the area of skin that, when stimulated, produces responses in the neuron.

   A. The receptive fields are smaller where the skin has a greater density of cutaneous receptors.

   B. The two-point touch threshold test reveals that the fingertips and tip of the tongue have a greater density of touch receptors, and thus a greater sensory acuity, than other areas of the body.

III. Lateral inhibition acts to sharpen a sensation by inhibiting the activity of sensory neurons coming from areas of the skin around the area that is most greatly stimulated.

Taste and Smell 248

I. The sense of taste is mediated by taste buds.

   A. There are four well-established modalities of taste (salty, sour, sweet, and bitter); a fifth, called umami, which is stimulated by glutamate, is now also recognized.

   B. Salty and sour taste are produced by the movement of sodium and hydrogen ions, respectively, through membrane channels; sweet and bitter tastes are produced by binding of molecules to protein receptors that are coupled to G-proteins.

II. The olfactory receptors are neurons that synapse within the olfactory bulb of the brain.

   A. Odorant molecules bind to membrane protein receptors. There may be as many as 1,000 different receptor proteins responsible for the ability to detect as many as 10,000 different odors.

   B. Binding of an odorant molecule to its receptor causes the dissociation of large numbers of G-protein subunits. The effect is thereby amplified, which may contribute to the extreme sensitivity of the sense of smell.

Vestibular Apparatus and Equilibrium 251

I. The structures for equilibrium and hearing are located in the inner ear, within the membranous labyrinth.

   A. The structure involved in equilibrium, known as the vestibular apparatus, consists of the otolith organs (utricle and saccule) and the semicircular canals.

   B. The utricle and saccule provide information about linear acceleration, whereas the semicircular canals provide information about angular acceleration.

   C. The sensory receptors for equilibrium are hair cells that support numerous stereocilia and one kinocilium.

      1. When the stereocilia are bent in the direction of the kinocilium, the cell membrane becomes depolarized.

      2. When the stereocilia are bent in the opposite direction, the membrane becomes hyperpolarized.

II. The stereocilia of the hair cells in the utricle and saccule project into the endolymph of the membranous labyrinth and are embedded in a gelatinous otolithic membrane.

    A. When a person is upright, the stereocilia of the utricle are oriented vertically; those of the saccule are oriented horizontally.

    B. Linear acceleration produces a shearing force between the hairs of the otolithic membrane, thus bending the stereocilia and electrically stimulating the sensory endings.

III. The three semicircular canals are oriented at nearly right angles to each other, like the faces of a cube.

    A. The hair cells are embedded within a gelatinous membrane called the cupula, which projects into the endolymph.

    B. Movement along one of the planes of a semicircular canal causes the endolymph to bend the cupula and stimulate the hair cells.

    C. Stimulation of the hair cells in the vestibular apparatus activates the sensory neurons of the vestibulocochlear nerve (VIII), which projects to the cerebellum and to the vestibular nuclei of the medulla oblongata.

       1. The vestibular nuclei in turn send fibers to the oculomotor center, which controls eye movements.

       2. Spinning and then stopping abruptly can thus cause oscillatory movements of the eyes (nystagmus).

The Ears and Hearing 255

I. The outer ear funnels sound waves of a given frequency (measured in hertz) and intensity (measured in decibels) to the tympanic membrane, causing it to vibrate.

II. Vibrations of the tympanic membrane cause movement of the middle-ear ossicles—malleus, incus, and stapes— which in turn produces vibrations of the oval window of the cochlea.

III. Vibrations of the oval window set up a traveling wave of perilymph in the scala vestibuli.

    A. This wave can pass around the helicotrema to the scala tympani, or it can reach the scala tympani

by passing through the scala media (cochlear duct).

    B. The scala media is filled with endolymph.

       1. The membrane of the cochlear duct that faces the scala vestibuli is called the vestibular membrane.

       2. The membrane that faces the scala tympani is called the basilar membrane.

IV. The sensory structure of the cochlea is called the spiral organ or organ of Corti.

    A. The organ of Corti rests on the basilar membrane and contains sensory hair cells.

       1. The stereocilia of the hair cells project upward into an overhanging tectorial membrane.

       2. The hair cells are innervated by the vestibulocochlear nerve (VIII).

    B. Sounds of high frequency cause maximum displacement of the basilar membrane closer to its base, near the stapes; sounds of lower frequency produce maximum displacement of the basilar membrane closer to its apex, near the helicotrema.

       1. Displacement of the basilar membrane causes the hairs to bend against the tectorial membrane and stimulate the production of nerve impulses.

       2. Pitch discrimination is thus dependent on the region of the basilar membrane that vibrates maximally to sounds of different frequencies.

       3. Pitch discrimination is enhanced by lateral inhibition.

The Eyes and Vision 261

I. Light enters the cornea of the eye, passes through the pupil (the opening of the iris) and then through the lens, from which point it is projected to the retina in the back of the eye.

    A. Light rays are bent, or refracted, by the cornea and lens.

    B. Because of refraction, the image on the retina is upside down and right to left.

    C. The right half of the visual field is projected to the left half of the retina in each eye, and vice versa.

II. Accommodation is the ability to maintain a focus on the retina as the distance between the object and the eyes is changed.

    A. Accommodation is achieved by changes in the shape and refractive power of the lens.

    B. When the muscles of the ciliary body are relaxed, the suspensory ligament is tight, and the lens is pulled to its least convex form.

       1. This gives the lens a low refractive power for distance vision.

       2. As an object is brought closer than 20 feet from the eyes, the ciliary body contracts, the suspensory ligament becomes less tight, and the lens becomes more convex and more powerful.

III. Visual acuity refers to the sharpness of the image. It depends in part on the ability of the lens to bring the image to a focus on the retina.

    A. People with myopia have an eyeball that is too long, so that the image is brought to a focus in front of the retina; this is corrected by a concave lens.

    B. People with hyperopia have an eyeball that is too short, so that the image is brought to a focus behind the retina; this is corrected by a convex lens.

    C. Astigmatism is the condition in which asymmetry of the cornea and/or lens causes uneven refraction of light around 360 degrees of a circle, resulting in an image that is not sharply focused on the retina.

Retina 268

I. The retina contains rods and cones— photoreceptor neurons that synapse with bipolar cells.

    A. When light strikes the rods, it causes the photodissociation of rhodopsin into retinene and opsin.

       1. This bleaching reaction occurs maximally with a light wavelength of 500 nm.

       2. Photodissociation is caused by the conversion of the 11-*cis* form of retinene to the all-*trans* form that cannot bind to opsin.

    B. In the dark, more rhodopsin can be produced, and increased rhodopsin in the rods makes the eyes more sensitive to light. The increased concentration of rhodopsin in the rods is partly responsible for dark adaptation.

C. The rods provide black-and-white vision under conditions of low light intensity. At higher light intensity, the rods are bleached out and the cones provide color vision.

II. In the dark, a constant movement of Na^+ into the rods produces what is known as a "dark current."

A. When light causes the dissociation of rhodopsin, the Na^+ channels become blocked and the rods become hyperpolarized in comparison to their membrane potential in the dark.

B. When the rods are hyperpolarized, they release less neurotransmitter at their synapses with bipolar cells.

C. Neurotransmitters from rods cause depolarization of bipolar cells in some cases and hyperpolarization of bipolar cells in other cases; thus, when the rods are in light and release less neurotransmitter, these effects are inverted.

III. According to the trichromatic theory of color vision, there are three systems of cones, each of which responds to one of three colors: red, blue, or green.

A. Each type of cone contains retinene attached to a different type of protein.

B. The names for the cones signify the region of the spectrum in which the cones absorb light maximally.

IV. The fovea centralis contains only cones; more peripheral parts of the retina contain both cones and rods.

A. Each cone in the fovea synapses with one bipolar cell, which in turn synapses with one ganglion cell.

1. The ganglion cell that receives input from the fovea thus has a visual field limited to that part of the retina that activated its cone.

2. As a result of this 1:1 ratio of cones to bipolar cells, visual

acuity is high in the fovea but sensitivity to low light levels is lower than in other regions of the retina.

B. In regions of the retina where rods predominate, large numbers of rods provide input to each ganglion cell (there is a great convergence). As a result, visual acuity is impaired, but sensitivity to low light levels is improved.

V. The right half of the visual field is projected to the left half of the retina of each eye.

A. The left half of the left retina sends fibers to the left lateral geniculate body of the thalamus.

B. The left half of the right retina also sends fibers to the left lateral geniculate body. This is because these fibers decussate in the optic chiasma.

C. The left lateral geniculate body thus receives input from the left half of the retina of both eyes, corresponding to the right half of the visual field; the right lateral geniculate receives information about the left half of the visual field.

1. Neurons in the lateral geniculate bodies send fibers to the striate cortex of the occipital lobes.

2. The geniculostriate system is involved in providing meaning to the images that form on the retina.

D. Instead of synapsing in the geniculate bodies, some fibers from the ganglion cells of the retina synapse in the superior colliculus of the midbrain, which controls eye movement.

1. Since this brain region is also called the optic tectum, this pathway is called the tectal system.

2. The tectal system enables the eyes to move and track an object; it is also responsible for the pupillary reflex and the changes in lens shape that are needed for accommodation.

Neural Processing of Visual Information   275

I. The area of the retina that provides input to a ganglion cell is called the receptive field of the ganglion cell.

A. The receptive field of a ganglion cell is roughly circular, with an "on" or "off" center and an antagonistic surround.

1. A spot of light in the center of an "on" receptive field stimulates the ganglion cell, whereas a spot of light in its surround inhibits the ganglion cell.

2. The opposite is true for ganglion cells with "off" receptive cells.

3. Wide illumination that stimulates both the center and the surround of a receptive field affects a ganglion cell to a lesser degree than a pinpoint of light that illuminates only the center or the surround.

B. The antagonistic center and surround of the receptive field of ganglion cells provide lateral inhibition, which enhances contours and provides better visual acuity.

II. Each lateral geniculate body receives input from both eyes relating to the same part of the visual field.

A. The neurons receiving input from each eye are arranged in layers within the lateral geniculate.

B. The receptive fields of neurons in the lateral geniculate are circular, with an antagonistic center and surround—much like the receptive field of ganglion cells.

III. Cortical neurons involved in vision may be simple, complex, or hypercomplex.

A. Simple neurons receive input from neurons in the lateral geniculate; complex neurons receive input from simple cells; and hypercomplex neurons receive input from complex cells.

B. Simple neurons are best stimulated by a slit or bar of light that is located in a precise part of the visual field and that has a precise orientation.

Review Activities

Test Your Knowledge of Terms and Facts

Match the vestibular organ on the left with its correct component on the right.

1. utricle and saccule
2. semicircular canals
3. cochlea

   a. cupula
   b. ciliary body
   c. basilar membrane
   d. otolithic membrane

4. The dissociation of rhodopsin in the rods in response to light causes
   a. the Na^+ channels to become blocked.
   b. the rods to secrete less neurotransmitter.
   c. the bipolar cells to become either stimulated or inhibited.
   d. all of these.

5. Tonic receptors
   a. are fast-adapting.
   b. do not fire continuously to a sustained stimulus.
   c. produce action potentials at a greater frequency as the generator potential is increased.
   d. are described by all of these.

6. Cutaneous receptive fields are smallest in
   a. the fingertips.
   b. the back.
   c. the thighs.
   d. the arms.

7. The process of lateral inhibition
   a. increases the sensitivity of receptors.
   b. promotes sensory adaptation.
   c. increases sensory acuity.
   d. prevents adjacent receptors from being stimulated.

8. The receptors for taste are
   a. naked sensory nerve endings.
   b. encapsulated sensory nerve endings.
   c. specialized epithelial cells.

9. Which of these statements about the utricle and saccule are *true?*
   a. They are otolith organs.
   b. They are located in the middle ear.
   c. They provide a sense of linear acceleration.
   d. Both *a* and *c* are true.
   e. Both *b* and *c* are true.

10. Since fibers of the optic nerve that originate in the nasal halves of each retina cross over at the optic chiasma, each lateral geniculate receives input from
    a. both the right and left sides of the visual field of both eyes.
    b. the ipsilateral visual field of both eyes.
    c. the contralateral visual field of both eyes.
    d. the ipsilateral field of one eye and the contralateral field of the other eye.

11. When a person with normal vision views an object from a distance of at least 20 feet,
    a. the ciliary muscles are relaxed.
    b. the suspensory ligament is tight.
    c. the lens is in its most flat, least convex shape.
    d. all of these apply.

12. Glasses with concave lenses help to correct
    a. presbyopia.
    b. myopia.
    c. hyperopia.
    d. astigmatism.

13. Parasympathetic nerves that stimulate constriction of the iris (in the pupillary reflex) are activated by neurons in
    a. the lateral geniculate.
    b. the superior colliculus.

c. the inferior colliculus.
d. the striate cortex.

14. A bar of light in a specific part of the retina, with a particular length and orientation, is the most effective stimulus for
    a. ganglion cells.
    b. lateral geniculate cells.
    c. simple cortical cells.
    d. complex cortical cells.

15. The ability of the lens to increase its curvature and maintain a focus at close distances is called
    a. convergence.
    b. accommodation.
    c. astigmatism.
    d. ambylopia.

16. Which of these sensory modalities is transmitted directly to the cerebral cortex without being relayed through the thalamus?
    a. taste
    b. sight
    c. smell
    d. hearing
    e. touch

17. Stimulation of membrane protein receptors by binding to specific molecules is *not* responsible for
    a. the sense of smell.
    b. sweet taste sensations.
    c. sour taste sensations.
    d. bitter taste sensations.

18. Epithelial cells release transmitter chemicals that excite sensory neurons in all of these senses *except*
    a. taste.
    b. smell.
    c. equilibrium.
    d. hearing.

Test Your Understanding of Concepts and Principles

1. Explain what is meant by lateral inhibition and give examples of its effects in three sensory systems.[1]

2. Describe the nature of the generator potential and explain its relationship to stimulus intensity and to frequency of action potential production.

3. Describe the phantom limb phenomenon and give a possible explanation for its occurrence.

4. Explain the relationship between smell and taste. How are these senses similar? How do they differ?

5. Explain how the vestibular apparatus provides information about changes in the position of our body in space.

6. Describe the sequence of changes that occur during accommodation. Why is it more of a strain on the eyes to look

---

[1]*Note:* This question is answered in the chapter 10 Study Guide found on the Online Learning Center at www.mhhe.com/fox8.

at a small nearby object than at large objects far away?

7. Describe the effects of light on the photoreceptors and explain how these effects influence the bipolar cells.

8. Explain why images that fall on the fovea centralis are seen more clearly than images that fall on the periphery of the retina. Why are the "corners of the eyes" more sensitive to light than the fovea?

9. Explain why rods provide only black-and-white vision. Include a discussion of different types of color blindness in your answer.

10. Explain why green objects can be seen better at night than objects of other colors. What effect does red light in a darkroom have on a dark-adapted eye?

11. Describe the receptive fields of ganglion cells and explain how the nature of these fields helps to improve visual acuity.

12. How many genes code for the sense of color vision? How many for taste? How many for smell? What does this information say about the level of integration required by the brain for the perception of these senses?

Test Your Ability to Analyze and Apply Your Knowlede

1. You are firing your laser canon from your position on the bridge of your starship. You see the hostile enemy starship explode, but you hear no accompanying sound. Can you explain this? How do receptors for sight and hearing differ?

2. People with conduction deafness often speak quietly. By contrast, people with sensorineural deafness tend to speak louder than normal. Explain these differences.

3. Opioid drugs reduce the sensation of dull, persistent pain but have little effect on the initial sharp pain of a noxious stimulus (e.g., a pin prick). What do these different effects imply? What conclusion can be drawn from the fact that aspirin (a drug that inhibits the formation of prostaglandins) functions as a pain reliever?

4. Compare the role of G-proteins in the senses of taste and sight. What is the advantage of having G-proteins mediate the effect of a stimulus on a receptor cell?

5. Discuss the role that inertia plays in the physiology of the vestibular apparatus. Why is there no sensation of movement in an airplane once it has achieved cruising speed?

Related Websites

Check out the Links Library at www.mhhe.com/fox8 for links to sites containing resources related to sensory physiology. These links are monitored to ensure current URLs.

11

Endocrine Glands

Secretion and Action of Hormones

Objectives

After studying this chapter, you should be able to . . .

1. define the terms *hormone* and *endocrine gland* and describe how chemical transformations in the target cells can activate certain hormones.

2. list the general chemical categories of hormones and give examples of hormones within each category.

3. explain how different hormones can exert synergistic, permissive, or antagonistic effects.

4. explain how hormone concentrations in the blood are regulated and how the effects of a hormone are influenced by its concentration.

5. describe the mechanisms of hormone action for steroid and thyroid hormones.

6. describe the mechanism of hormone action when cAMP is used as a second messenger.

7. describe the mechanism of hormone action when Ca^{2+} is used as a second messenger.

8. describe the structure of the pituitary gland and explain the functional relationship between the pituitary and the hypothalamus.

9. list the hormones released by the posterior pituitary, state the origin of these hormones, and explain how the hypothalamus regulates their release.

10. list the hormones of the anterior pituitary and explain how their secretion is regulated by the hypothalamus.

11. describe the production and actions of the thyroid hormones and explain how thyroid secretion is regulated.

12. describe the location of the parathyroid glands and explain the actions of PTH and the regulation of its secretion.

13. describe the types and actions of corticosteroids and explain how the secretions of the adrenal cortex are regulated.

14. describe the actions of epinephrine and norepinephrine, and explain how the secretions of the adrenal medulla are regulated.

15. explain why the pancreas is both an exocrine and an endocrine gland and describe the structure and functions of the pancreatic islets.

16. describe the actions of insulin and glucagon, and explain the regulation of their secretion.

17. list the hormones secreted by the pineal gland and thymus, and explain the significance of these hormones in general terms.

18. list the hormones secreted by the gonads and placenta.

19. describe autocrine and paracrine regulation with reference to the blood vessels and immune system.

20. describe the chemical nature and physiological roles of the prostaglandins and explain how the nonsteroidal anti-inflammatory drugs work.

Chapter at a Glance

Clinical Investigation

Rosemary, a 32-year-old office worker, discovers after taking a physical that she has hypertension and hyperglycemia. She returns to take an oral glucose tolerance test, which is found to be normal. Blood tests reveal normal blood levels of T_4 and T_3, but more extensive tests show that her blood cortisol levels are abnormally high. Rosemary has a generalized "puffiness," but not myxedema. Upon questioning, she states that she does not have a history of chronic inflammation and has not been taking immunosuppressive drugs.

A later assay of her blood ACTH levels reveals that they are only about one-fiftieth of normal. What might account for Rosemary's symptoms?

Endocrine Glands and Hormones

Hormones are regulatory molecules secreted into the blood by endocrine glands. Chemical categories of hormones include steroids, amines, polypeptides, and glycoproteins. Interactions between the various hormones produce effects that may be synergistic, permissive, or antagonistic.

Endocrine glands lack the ducts that are present in exocrine glands (chapter 1). The endocrine glands secrete their products, which are biologically active molecules called **hormones,** into the blood. The blood carries the hormones to *target cells* that contain specific *receptor proteins* for the hormones, and which therefore can respond in a specific fashion to them. Many endocrine glands are discrete organs (fig. 11.1a) whose primary functions are the production and secretion of hormones. The pancreas functions as both an exocrine and an endocrine gland; the endocrine portion of the pancreas is composed of clusters of cells called the pancreatic islets (islets of Langerhans) (fig. 11.1b). The concept of the **endocrine system,** however, must be extended beyond these organs. In recent years, it has been discovered that many other organs in the body secrete hormones. When these hormones can be demonstrated to have significant physiological functions, the organs that produce them may be categorized as endocrine glands, although they serve other functions as well. It is appropriate, then, that a partial list of the endocrine glands (table 11.1) should include the heart, liver, adipose tissue, and kidneys.

Some specialized neurons, particularly in the hypothalamus, secrete chemical messengers into the blood rather than into a narrow synaptic cleft. In these cases, the chemical that the neurons secrete is sometimes called a *neurohormone.* In addition, a number of chemicals—norepinephrine, for example—are secreted

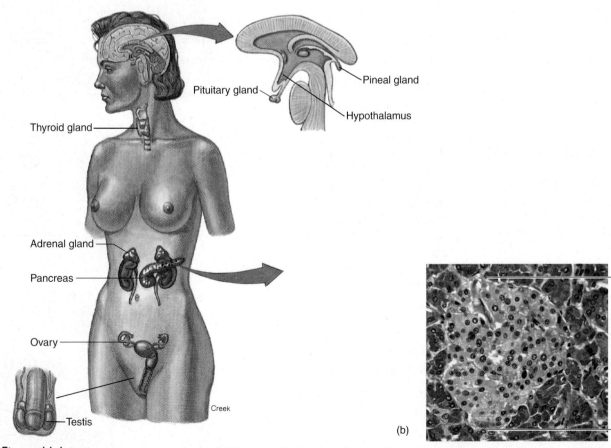

Pituitary gland
Pineal gland
Hypothalamus
Thyroid gland
Adrenal gland
Pancreas
Ovary
Creek
Testis
(a)

Pancreatic islet (of Langerhans)
(b)

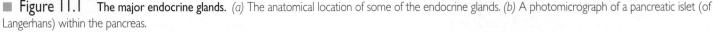

■ **Figure 11.1** **The major endocrine glands.** *(a)* The anatomical location of some of the endocrine glands. *(b)* A photomicrograph of a pancreatic islet (of Langerhans) within the pancreas.

both as a neurotransmitter and a hormone. Thus, a sharp distinction between the nervous and endocrine systems cannot always be drawn on the basis of the chemicals they release.

Hormones affect the metabolism of their target organs and, by this means, help to regulate total body metabolism, growth, and reproduction. The effects of hormones on body metabolism and growth are discussed in chapter 19; the regulation of reproductive functions by hormones is considered in chapter 20.

Chemical Classification of Hormones

Hormones secreted by different endocrine glands vary widely in chemical structure. All hormones, however, can be divided into a few chemical classes.

1. **Amines.** These are hormones derived from the amino acids tyrosine and tryptophan. They include the hormones secreted by the adrenal medulla, thyroid, and pineal glands.

2. **Polypeptides and proteins.** Polypeptide hormones generally contain less than 100 amino acids; an example is antidiuretic hormone (table 11.2). Protein hormones are polypeptides with more than 100 amino acids; growth hormone is an example. The distinction between polypeptide and protein hormones is blurred in the case of insulin, which is composed of two polypeptide chains that are both derived from the same protein precursor.

3. **Glycoproteins.** These molecules consist of a long polypeptide (containing more than 100 amino acids) bound to one or more carbohydrate groups. Examples are follicle-stimulating hormone (FSH) and luteinizing hormone (LH).

4. **Steroids.** These are lipids derived from cholesterol. They include the hormones testosterone, estradiol, progesterone, and cortisol (fig. 11.2).

In terms of their actions in target cells, hormone molecules can be divided into those that are polar, and therefore water-soluble,

Table 11.1 A Partial Listing of the Endocrine Glands

Endocrine Gland	Major Hormones	Primary Target Organs	Primary Effects
Adipose tissue	Leptin	Hypothalamus	Suppresses appetite
Adrenal cortex	Glucocorticoids Aldosterone	Liver and muscles Kidneys	Glucocorticoids influence glucose metabolism; aldosterone promotes Na^+ retention, K^+ excretion
Adrenal medulla	Epinephrine	Heart, bronchioles, and blood vessels	Causes adrenergic stimulation
Heart	Atrial natriuretic hormone	Kidneys	Promotes excretion of Na^+ in the urine
Hypothalamus	Releasing and inhibiting hormones	Anterior pituitary	Regulates secretion of anterior pituitary hormones
Small intestine	Secretin and cholecystokinin	Stomach, liver, and pancreas	Inhibits gastric motility and stimulates bile and pancreatic juice secretion
Islets of Langerhans (pancreas)	Insulin Glucagon	Many organs Liver and adipose tissue	Insulin promotes cellular uptake of glucose and formation of glycogen and fat; glucagon stimulates hydrolysis of glycogen and fat
Kidneys	Erythropoietin	Bone marrow	Stimulates red blood cell production
Liver	Somatomedins	Cartilage	Stimulates cell division and growth
Ovaries	Estradiol-17β and progesterone	Female reproductive tract and mammary glands	Maintains structure of reproductive tract and promotes secondary sex characteristics
Parathyroid glands	Parathyroid hormone	Bone, small intestine, and kidneys	Increases Ca^{2+} concentration in blood
Pineal gland	Melatonin	Hypothalamus and anterior pituitary	Affects secretion of gonadotrophic hormones
Pituitary, anterior	Trophic hormones	Endocrine glands and other organs	Stimulates growth and development of target organs; stimulates secretion of other hormones
Pituitary, posterior	Antidiuretic hormone Oxytocin	Kidneys and blood vessels Uterus and mammary glands	Antidiuretic hormone promotes water retention and vasoconstriction; oxytocin stimulates contraction of uterus and mammary secretory units
Skin	1,25-Dihydroxyvitamin D_3	Small intestine	Stimulates absorption of Ca^{2+}
Stomach	Gastrin	Stomach	Stimulates acid secretion
Testes	Testosterone	Prostate, seminal vesicles, and other organs	Stimulates secondary sexual development
Thymus	Thymopoietin	Lymph nodes	Stimulates white blood cell production
Thyroid gland	Thyroxine (T_4) and triiodothyronine (T_3); calcitonin	Most organs	Thyroxine and triiodothyronine promote growth and development and stimulate basal rate of cell respiration (basal metabolic rate or BMR); calcitonin may participate in the regulation of blood Ca^{2+} levels

Table 11.2 Examples of Polypeptide and Glycoprotein Hormones

Hormone	Structure	Gland	Primary Effects
Antidiuretic hormone	8 amino acids	Posterior pituitary	Water retention and vasoconstriction
Oxytocin	8 amino acids	Posterior pituitary	Uterine and mammary contraction
Insulin	21 and 30 amino acids (double chain)	Beta cells in islets of Langerhans	Cellular glucose uptake, lipogenesis, and glycogenesis
Glucagon	29 amino acids	Alpha cells in islets of Langerhans	Hydrolysis of stored glycogen and fat
ACTH	39 amino acids	Anterior pituitary	Stimulation of adrenal cortex
Parathyroid hormone	84 amino acids	Parathyroid	Increase in blood Ca^{2+} concentration
FSH, LH, TSH	Glycoproteins	Anterior pituitary	Stimulation of growth, development, and secretory activity of target glands

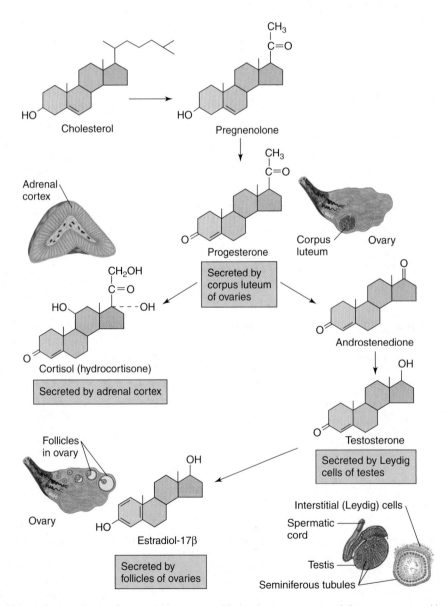

■ **Figure 11.2** **Simplified biosynthetic pathways for steroid hormones.** Notice that progesterone (a hormone secreted by the corpus luteum of the ovaries) is a common precursor of all other steroid hormones and that testosterone (the major androgen secreted by the Leydig cells of the testes) is a precursor of estradiol-17β, the major estrogen secreted by the follicles of the ovaries.

and those that are nonpolar, and thus insoluble in water. Since the nonpolar hormones are soluble in lipids, they are often referred to as **lipophilic hormones.** Unlike the polar hormones, which cannot pass through plasma membranes, lipophilic hormones can gain entry into their target cells. These lipophilic hormones include the steroid hormones and thyroid hormones.

Steroid hormones are secreted by only two endocrine glands: the adrenal cortex and the gonads (fig. 11.2). The gonads secrete *sex steroids;* the adrenal cortex secretes *corticosteroids* (including cortisol and aldosterone) and small amounts of sex steroids.

The major thyroid hormones are composed of two derivatives of the amino acid tyrosine bonded together (fig. 11.3). When the hormone contains four iodine atoms, it is called *tetraiodothyronine* (T_4), or *thyroxine.* When it contains three atoms of iodine, it is called *triiodothyronine* (T_3). Although these hormones are not steroids, they are like steroids in that they are relatively small, nonpolar molecules. Steroid and thyroid hormones are active when taken orally (as a pill). Sex steroids are the active agents in contraceptive pills, and thyroid hormone pills are taken by people whose thyroid is deficient (who are hypothyroid). By contrast, polypeptide and glycoprotein hormones cannot be taken orally because they would be di-

gested into inactive fragments before being absorbed into the blood. Thus, insulin-dependent diabetics must inject themselves with this hormone.

The pineal gland secretes melatonin, a hormone derived from the amino acid tryptophan. Melatonin has properties that are similar in some ways to both the lipophilic hormones and the water-soluble hormones. The adrenal medulla secretes the *catecholamines* epinephrine and norepinephrine (see fig. 9.8), which are derived from the amino acid tyrosine. Like polypeptide and glycoprotein hormones, the catecholamine hormones are too large and polar to pass through plasma membranes.

Prohormones and Prehormones

Hormone molecules that affect the metabolism of target cells are often derived from less active "parent," or *precursor,* molecules. In the case of polypeptide hormones, the precursor may be a longer chained **prohormone** that is cut and spliced together to make the hormone. Insulin, for example, is produced from *proinsulin* within the beta cells of the islets of Langerhans of the pancreas (see fig. 3.25). In some cases, the prohormone itself is derived from an even larger precursor molecule; in the case of insulin, this molecule is called *preproinsulin.* The term *prehormone* is sometimes used to indicate such precursors of prohormones.

In some cases, the molecule secreted by the endocrine gland (and considered to be the hormone of that gland) is actually inactive in the target cells. In order to become active, the target cells must modify the chemical structure of the secreted hormone. Thyroxine (T_4), for example, must be changed into T_3 within the target cells in order to affect the metabolism of these cells. Similarly, testosterone (secreted by the testes) and vitamin D_3 (secreted by the skin) are converted into more active molecules within their target cells (table 11.3). In this text, the term prehormone will be used to designate those molecules secreted by endocrine glands that are inactive until changed by their target cells.

Common Aspects of Neural and Endocrine Regulation

The fact that endocrine regulation is chemical in nature might lead one to believe that it differs fundamentally from neural control systems that depend on the electrical properties of cells. This assumption is incorrect. As explained in chapter 7, electrical nerve impulses are, in fact, chemical events produced by the

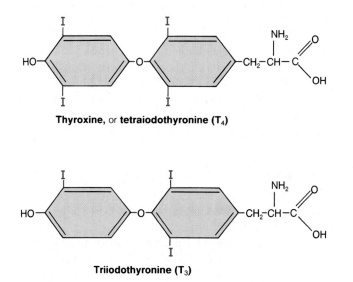

Thyroxine, or tetraiodothyronine (T_4)

Triiodothyronine (T_3)

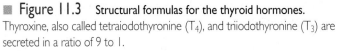

■ **Figure 11.3** Structural formulas for the thyroid hormones. Thyroxine, also called tetraiodothyronine (T_4), and triiodothyronine (T_3) are secreted in a ratio of 9 to 1.

Table 11.3 Conversion of Prehormones into Biologically Active Derivatives

Endocrine Gland	Prehormone	Active Products	Comments
Skin	Vitamin D_3	1,25-Dihydroxyvitamin D_3	Conversion (through hydroxylation reactions) occurs in the liver and the kidneys.
Testes	Testosterone	Dihydrotestosterone (DHT)	DHT and other 5α-reduced androgens are formed in most androgen-dependent tissue.
		Estradiol-17β (E_2)	E_2 is formed in the brain from testosterone, where it is believed to affect both endocrine function and behavior; small amounts of E_2 are also produced in the testes.
Thyroid	Thyroxine (T_4)	Triiodothyronine (T_3)	Conversion of T_4 to T_3 occurs in almost all tissues.

diffusion of ions through the neuron plasma membrane. Interestingly, the action of some hormones (such as insulin) is accompanied by ion diffusion and electrical changes in the target cells, so changes in membrane potential are not unique to the nervous system. Also, most nerve fibers stimulate the cells they innervate through the release of a chemical neurotransmitter. Neurotransmitters do not travel in the blood as do hormones; instead, they diffuse across a narrow synaptic cleft to the membrane of the postsynaptic cell. In other respects, however, the actions of neurotransmitters are very similar to the actions of hormones.

Indeed, many polypeptide hormones, including those secreted by the pituitary gland and by the digestive tract, have been discovered in the brain. In certain locations in the brain, some of these compounds are produced and secreted as hormones. In other brain locations, some of these compounds apparently serve as neurotransmitters. The discovery of polypeptide hormones in unicellular organisms, which of course lack a nervous and endocrine system, suggests that these regulatory molecules appeared early in evolution and were incorporated into the function of nervous and endocrine tissue as these systems evolved. This fascinating theory would help to explain, for example, why insulin, a polypeptide hormone produced in the pancreas of vertebrates, is found in neurons of invertebrates (which lack a distinct endocrine system).

Regardless of whether a particular chemical is acting as a neurotransmitter or as a hormone, in order for it to function in physiological regulation: (1) target cells must have specific **receptor proteins** that combine with the regulatory molecule; (2) the combination of the regulatory molecule with its receptor proteins must cause a specific sequence of changes in the target cells; and (3) there must be a mechanism to quickly turn off the action of the regulator. This mechanism, which involves rapid removal and/or chemical inactivation of the regulator molecules, is essential because without an "off-switch" physiological control would be impossible.

Hormone Interactions

A given target tissue is usually responsive to a number of different hormones. These hormones may antagonize each other or work together to produce effects that are additive or complementary. The responsiveness of a target tissue to a particular hormone is thus affected not only by the concentration of that hormone, but also by the effects of other hormones on that tissue. Terms used to describe hormone interactions include *synergistic, permissive,* and *antagonistic.*

Synergistic and Permissive Effects

When two or more hormones work together to produce a particular result, their effects are said to be **synergistic.** These effects may be additive or complementary. The action of epinephrine and norepinephrine on the heart is a good example of an additive effect. Each of these hormones separately produces an increase in cardiac rate; acting together in the same concentrations, they stimulate an even greater increase in car-

diac rate. The synergistic action of FSH and testosterone is an example of a complementary effect; each hormone separately stimulates a different stage of spermatogenesis during puberty, so that both hormones together are needed at that time to complete sperm development. Likewise, the ability of mammary glands to produce and secrete milk requires the synergistic action of many hormones—estrogen, cortisol, prolactin, oxytocin, and others.

A hormone is said to have a **permissive effect** on the action of a second hormone when it enhances the responsiveness of a target organ to the second hormone or when it increases the activity of the second hormone. Prior exposure of the uterus to estrogen, for example, induces the formation of receptor proteins for progesterone, which improves the response of the uterus when it is subsequently exposed to progesterone. Estrogen thus has a permissive effect on the responsiveness of the uterus to progesterone. Glucocorticoids (a class of corticosteroids including cortisol) exert permissive effects on the actions of catecholamines (epinephrine and norepinephrine). When these permissive effects are not produced because of abnormally low glucocorticoids, the catecholamines will not be as effective as they are normally. One symptom of this condition may be an abnormally low blood pressure.

Vitamin D_3 is a prehormone that must be modified by enzymes in the kidneys and liver, where two hydroxyl (OH^-) groups are added to form the active hormone 1,25-dihydroxyvitamin D_3. This hormone helps to raise blood calcium levels. Parathyroid hormone (PTH) has a permissive effect on the actions of vitamin D_3 because it stimulates the production of the hydroxylating enzymes in the kidneys and liver. By this means, an increased secretion of PTH has a permissive effect on the ability of vitamin D_3 to stimulate the intestinal absorption of calcium.

Antagonistic Effects

In some situations, the actions of one hormone antagonize the effects of another. Lactation during pregnancy, for example, is inhibited because the high concentration of estrogen in the blood inhibits the secretion and action of prolactin. Another example of antagonism is the action of insulin and glucagon (two hormones from the pancreatic islets) on adipose tissue; the formation of fat is promoted by insulin, whereas glucagon promotes fat breakdown.

Effects of Hormone Concentrations on Tissue Response

The concentration of hormones in the blood primarily reflects the rate of secretion by the endocrine glands. Hormones do not generally accumulate in the blood because they are rapidly removed by target organs and by the liver. The **half-life** of a hormone—the time required for the plasma concentration of a given amount of the hormone to be reduced to half its reference level—ranges from minutes to hours for most hormones (thyroid hormone,

however, has a half-life of several days). Hormones removed from the blood by the liver are converted by enzymatic reactions into less active products. Steroids, for example, are converted into more water-soluble polar derivatives that are released into the blood and excreted in the urine and bile.

The effects of hormones are very dependent on concentration. Normal tissue responses are produced only when the hormones are present within their normal, or *physiological,* range of concentrations. When some hormones are taken in abnormally high, or *pharmacological,* concentrations (as when they are taken as drugs), their effects may be different from those produced by lower, more physiological, concentrations. The fact that abnormally high concentrations of a hormone may cause the hormone to bind to tissue receptor proteins of different but related hormones may account, in part, for these different effects. Also, since some steroid hormones can be converted by their target cells into products that have different biological effects (as in the conversion of androgens into estrogens), the administration of large quantities of one steroid can result in the production of a significant quantity of other steroids with different effects.

Anabolic steroids are synthetic androgens (male hormones) that promote protein synthesis in muscles and other organs. Use of these drugs by bodybuilders, weightlifters, and others is prohibited by most athletic organizations. Although administration of exogenous androgens does promote muscle growth, it can also cause a number of undesirable side effects. Since the liver and adipose tissue can change androgens into estrogens, male athletes who take exogenous androgens often develop *gynecomastia*—an abnormal growth of femalelike mammary tissue. High levels of exogenous androgens also inhibit the secretion of FSH and LH from the pituitary, causing atrophy of the testes and erectile dysfunction. The exogenous androgens also promote acne, aggressive behavior, male pattern baldness, and premature closure of the epiphyseal discs (growth plates in bones), stunting the growth of adolescents. Female users of exogenous androgens display masculinization and antisocial behavior. In both sexes, the anabolic steroids raise blood levels of LDL cholesterol (the "bad cholesterol") and triglycerides, while lowering the levels of HDL cholesterol (the "good cholesterol"), thus predisposing users to increased risk of heart disease and stroke.

Pharmacological doses of hormones, particularly of steroids, can thus have widespread and often damaging side effects. People with inflammatory diseases who are treated with high doses of cortisone over long periods of time, for example, may develop osteoporosis and characteristic changes in soft tissue structure. Contraceptive pills, which contain sex steroids, have a number of potential side effects that could not have been predicted in 1960, when "the pill" was first introduced. At that time, the concentrations of sex steroids were much higher than they are in the pills presently being marketed.

Priming Effects

Variations in hormone concentration within the normal, physiological range can affect the responsiveness of target cells. This is due in part to the effects of polypeptide and glycoprotein hormones on the number of their receptor proteins in target cells. More receptors may be formed in the target cells in response to particular hormones. Small amounts of gonadotropin-releasing hormone (GnRH) secreted by the hypothalamus, for example, increase the sensitivity of anterior pituitary cells to further GnRH stimulation. This is a *priming effect,* sometimes also called **upregulation.** Subsequent stimulation by GnRH thus causes a greater response from the anterior pituitary.

Desensitization and Downregulation

Prolonged exposure to high concentrations of polypeptide hormones has been found to *desensitize* the target cells. Subsequent exposure to the same concentration of the same hormone thus produces less of a target tissue response. This desensitization may be partly due to the fact that high concentrations of these hormones cause a decrease in the number of receptor proteins in their target cells—a phenomenon called **downregulation.** Such desensitization and downregulation of receptors has been shown to occur, for example, in adipose cells exposed to high concentrations of insulin and in testicular cells exposed to high concentrations of luteinizing hormone (LH).

In order to prevent desensitization from occurring under normal conditions, many polypeptide and glycoprotein hormones are secreted in spurts rather than continuously. This *pulsatile secretion* is an important aspect, for example, in the hormonal control of the reproductive system. The pulsatile secretion of GnRH and LH is needed to prevent desensitization; when these hormones are artificially presented in a continuous fashion, they produce a decrease (rather than the normal increase) in gonadal function. This effect has important clinical implications, as will be described in chapter 20.

Test Yourself Before You Continue

1. Compare the four chemical classes of hormones with reference to hormones within each class.
2. Define *prohormone* and *prehormone,* and give examples of each of these molecules.
3. Describe the common characteristics of hormones and neurotransmitters.
4. List the terms used to describe hormone interactions and give examples of these effects.
5. Explain how the response of the body to a given hormone can be affected by the concentration of that hormone in the blood.

Mechanisms of Hormone Action

Each hormone exerts its characteristic effects on target organs by acting on the cells of these organs. Hormones of the same chemical class have similar mechanisms of action. Lipid-soluble hormones pass through the target cell's plasma membrane, bind to intracellular receptor proteins, and act directly within the target cell. Polar hormones do not enter the target cells, but instead bind to receptors on the plasma membrane. This results in the activation of intracellular second-messenger systems that mediate the actions of the hormone.

Although each hormone exerts its own characteristic effects on specific target cells, hormones that are in the same chemical category have similar mechanisms of action. These similarities involve the location of cellular receptor proteins and the events that occur in the target cells after the hormone has combined with its receptor protein.

Hormones are delivered by the blood to every cell in the body, but only the **target cells** are able to respond to these hormones. In order to respond to any given hormone, a target cell must have specific receptor proteins for that hormone. Receptor protein–hormone interaction is highly specific. In addition to this property of *specificity*, hormones bind to receptors with a *high affinity* (high bond strength) and a *low capacity*. The latter characteristic refers to the possibility of saturating receptors with hormone molecules because of the limited number of receptors per target cell (usually a few thousand). Notice that the characteristics of specificity and saturation that apply to receptor proteins are similar to the characteristics of enzyme and carrier proteins discussed in previous chapters.

The location of a hormone's receptor proteins in its target cells depends on the chemical nature of the hormone. Since the lipophilic hormones (steroids and thyroxine) can pass through the plasma membrane and enter their target cells, the receptor proteins for lipophilic hormones are located within the cytoplasm and nucleus. Since the water-soluble hormones (catecholamines, polypeptides, and glycoproteins) cannot pass through the plasma membrane, their receptors are located on the outer surface of the membrane. In these cases, hormone action requires the activation of second messengers within the cell.

Hormones That Bind to Nuclear Receptor Proteins

Unlike the water-soluble hormones, the lipophilic steroid and thyroid hormones do not travel dissolved in the aqueous portion of the plasma; rather, they are transported to their target cells attached to plasma *carrier proteins*. These hormones must then dissociate from their carrier proteins in the blood in order to pass through the lipid component of the plasma membrane and enter the target cell, within which their receptor proteins are located (fig 11.4).

The receptors for the lipophilic hormones are known as **nuclear hormone receptors** because they function within the

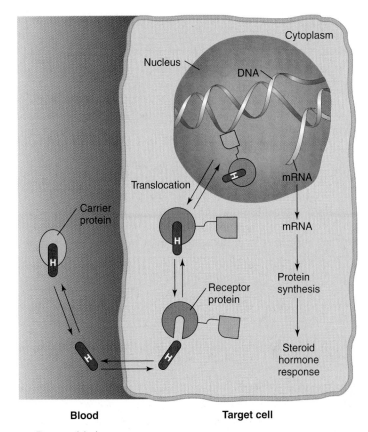

■ Figure 11.4 The mechanism of action of a steroid hormone on the target cells. Some steroids bind to a cytoplasmic receptor, which then translocates to the nucleus. Other steroid hormones enter the nucleus and then bind to their receptor. In both cases, the steroid-receptor complex can then bind to a specific area of DNA and activate specific genes.

cell nucleus to activate genetic transcription (production of mRNA). The nuclear hormone receptors thus function as **transcription factors** that first must be activated by binding to their hormone ligands. The newly formed mRNA produced by the activated genes directs the synthesis of specific enzyme proteins that change the metabolism of the target cell in ways that are characteristic of the effects of that hormone on that target cell.

Each nuclear hormone receptor has two regions, or *domains:* a *ligand (hormone)-binding domain* and a *DNA-binding domain* (fig. 11.5). The receptor must be activated by binding to its hormone ligand before it can bind to a specific region of the DNA, which is called a **hormone-response element.** This is a short DNA span, composed of characteristic nucleotide bases, located adjacent to the gene that will be transcribed when the nuclear receptor binds to the hormone-response element.

The nuclear hormone receptors are said to constitute a superfamily composed of two major families: the *steroid family* and the *thyroid hormone* (or *nonsteroid*) *family*. In addition to the receptor for thyroid hormone, the latter family also includes the receptors for the active form of vitamin D and for retinoic acid (derived from vitamin A, or retinol). Vitamin D and retinoic acid, like the steroid and thyroid hormones, are lipophilic molecules that play important roles in the regulation of cell function and organ physiology.

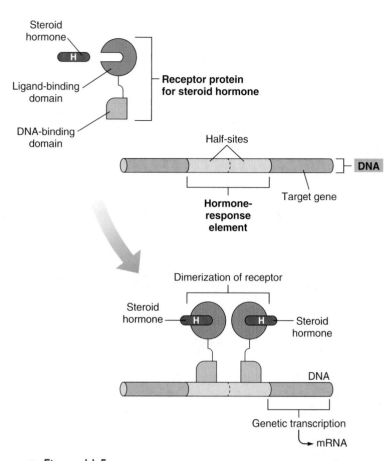

Figure 11.5 **Receptors for steroid hormones.** (a) Each nuclear hormone receptor protein has a ligand-binding domain, which binds to a hormone molecule, and a DNA-binding domain, which binds to the hormone-response element of DNA. (b) Binding to the hormone causes the receptor to dimerize on the half-sites of the hormone-response element. This stimulates genetic transcription (synthesis of RNA).

Modern molecular biology has ushered in a new era in endocrine research, where nuclear receptors can be identified and their genes cloned before their hormone ligands are known. In fact, scientists have currently identified the hormone ligand for only about half of the approximately seventy different nuclear receptors that are now known. The receptors for unknown hormone ligands are called **orphan receptors.** For example, the receptor known as the retinoid X receptor (abbreviated RXR) was an orphan until its ligand, 9-*cis*-retinoic acid (a vitamin A derivative) was discovered. The significance of this receptor will be described shortly.

Mechanism of Steroid Hormone Action

There is some controversy regarding the location of the steroid hormone receptors prior to their binding to steroid hormones. They have been found in both the cytoplasm and the nucleus, and it appears that a particular steroid hormone receptor is distributed in a characteristic way between both compartments. When the cytoplasmic receptor binds to its specific steroid hormone ligand, however, it *translocates* (moves) into the nucleus, where its DNA-binding domain binds to the specific hormone-response element of the DNA (see fig. 11.4).

As illustrated in figure 11.5, the hormone-response element of DNA consists of two *half-sites,* each six nucleotide bases long, separated by a three-nucleotide spacer segment. One steroid receptor, bound to one molecule of the steroid hormone, attaches as a single unit to one of the half-sites. Another steroid receptor, bound to another steroid hormone, attaches to the other half-site of the hormone-response element. The process of two receptor units coming together at the two half-sites is called **dimerization** (fig 11.5). Since both receptor units of the pair are the same, the steroid receptor is said to form a *homodimer.* (The situation is different for the nonsteroid family of receptors, as will be described.) Once dimerization has occurred, the activated nuclear hormone receptor stimulates transcription of particular genes, and thus hormonal regulation of the target cell (see fig. 11.4).

Mechanism of Thyroid Hormone Action

As previously discussed, the major hormone secreted by the thyroid gland is thyroxine, or tetraiodothyronine (T_4). Like steroid hormones, thyroxine travels in the blood attached to carrier proteins (primarily to *thyroxine-binding globulin,* or *TBG*). The thyroid also secretes a small amount of triiodothyronine, or T_3. The carrier proteins have a higher affinity for T_4 than for T_3, however, and, as a result, the amount of unbound (or "free") T_3 in the plasma is about ten times greater than the amount of free T_4.

Approximately 99.96% of the thyroxine in the blood is attached to carrier proteins in the plasma; the rest is free. Only the free thyroxine and T_3 can enter target cells; the protein-bound thyroxine serves as a reservoir of this hormone in the blood (this is why it takes a couple of weeks after surgical removal of the thyroid for the symptoms of hypothyroidism to develop). Once the free thyroxine passes into the target cell cytoplasm, it is enzymatically converted into T_3. As previously discussed, it is the T_3 rather than T_4 that is active within the target cells.

Unlike many of the steroid receptors, the inactive receptor proteins for T_3 are located in the nucleus. Until they bind to T_3, however, the receptors are incapable of binding to DNA and stimulating transcription. The T_3 may enter the cell from the plasma, or it may be produced in the cell by conversion from T_4. In either case, it uses some nonspecific binding proteins as "stepping stones" to enter the nucleus, where it binds to the ligand-binding domain of the receptor (fig. 11.6). Once the receptor binds to T_3, its DNA-binding domain can attach to the half-site of the DNA hormone-response element.

The other half-site, however, does *not* bind to another T_3 receptor protein. Unlike the steroid hormone receptors, the nuclear receptors in the nonsteroid family bind to DNA as *heterodimers.* The thyroid hormone receptor (abbreviated *TR*) is one partner in the heterodimer; the other partner is a receptor (abbreviated *RXR*) for the vitamin A derivative 9-*cis*-retinoic acid. Once bound to their different ligands, the two partners in the heterodimer can bind to the DNA to activate the hormone-response element for thyroid hormone (fig. 11.7). In this way, thyroid hormones stimulate transcription of genes, production of specific mRNA, and therefore the production of specific enzymes (see fig. 11.6).

Interestingly, the receptor for 1,25-dihydroxyvitamin D_3, the active form of vitamin D, also forms heterodimers with the receptor for 9-*cis*-retinoic acid (the RXR receptor) when it binds

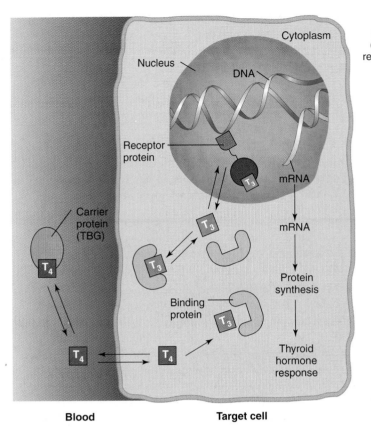

Figure 11.6 The mechanism of action of thyroid hormones on the target cells. T_4 is first converted into T_3 within the cytoplasm of the target cell. T_3 then enters the nucleus and binds to its nuclear receptor. The hormone-receptor complex can then bind to a specific area of DNA and activate specific genes.

to DNA and activates genes. The RXR receptor and its vitamin A derivative ligand thus form a link between the mechanisms of action of thyroid hormone, vitamin A, and vitamin D, along with those of some other molecules that are important regulators of genetic expression.

Hormones That Use Second Messengers

Hormones that are catecholamines (epinephrine and norepinephrine), polypeptides, and glycoproteins cannot pass through the lipid barrier of the target cell's plasma membrane. Although some of these hormones may enter the cell by pinocytosis, most of their effects result from their binding to receptor proteins on the outer surface of the target cell membrane. Since they exert their effects without entering the target cells, the actions of these hormones must be mediated by other molecules within the target cells. If you think of hormones as "messengers" from the endocrine glands, the intracellular mediators of the hormone's action can be called **second messengers.** (The concept of second messengers was introduced in connection with synaptic transmission in chapter 7.) Second messengers are thus a component of *signal-transduction mechanisms,* since extracellular signals (hormones) are transduced into intracellular signals (second messengers).

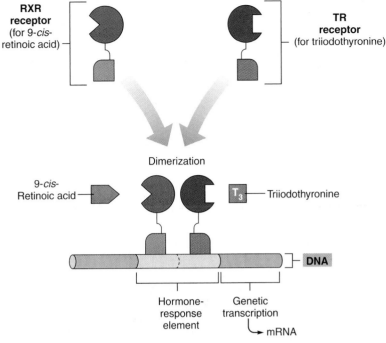

Figure 11.7 The receptor for triiodothyronine (T_3). The nuclear receptor protein for T_3 forms a dimer with the receptor protein for 9-*cis*-retinoic acid, a derivative of vitamin A. This occurs when each binds to its ligand and to the hormone-response element of DNA. Thus, 9-*cis*-retinoic acid is required for the action of T_3. The heterodimer formed on the DNA stimulates genetic transcription.

When these hormones bind to membrane receptor proteins, they must activate specific proteins in the plasma membrane in order to produce the second messengers required to exert their effects. On the basis of the membrane enzyme activated, we can distinguish second-messenger systems that involve the activation of (1) adenylate cyclase, (2) phospholipase C, and (3) tyrosine kinase.

Adenylate Cyclase–Cyclic AMP Second-Messenger System

Cyclic adenosine monophosphate (abbreviated **cAMP**) was the first "second messenger" to be discovered and is the best understood. When epinephrine and norepinephrine bind to their β-adrenergic receptors (chapter 9), the effects of these hormones are due to cAMP production within the target cells. It was later discovered that the effects of many (but not all) polypeptide and glycoprotein hormones are also mediated by cAMP.

When one of these hormones binds to its receptor protein, it causes the dissociation of a subunit from the complex of G-proteins (discussed in chapter 7; see table 7.7). This G-protein subunit moves through the membrane until it reaches the enzyme **adenylate** (or *adenylyl*) **cyclase** (fig 11.8). The G-protein subunit then binds to and activates this enzyme, which catalyzes the following reaction within the cytoplasm of the cell:

$$ATP \rightarrow cAMP + PP_i$$

Adenosine triphosphate (ATP) is thus converted into cyclic AMP (cAMP) and two inorganic phosphates (*pyrophosphate,* abbreviated PP_i). As a result of the interaction of the hormone with its receptor and the activation of adenylate cyclase,

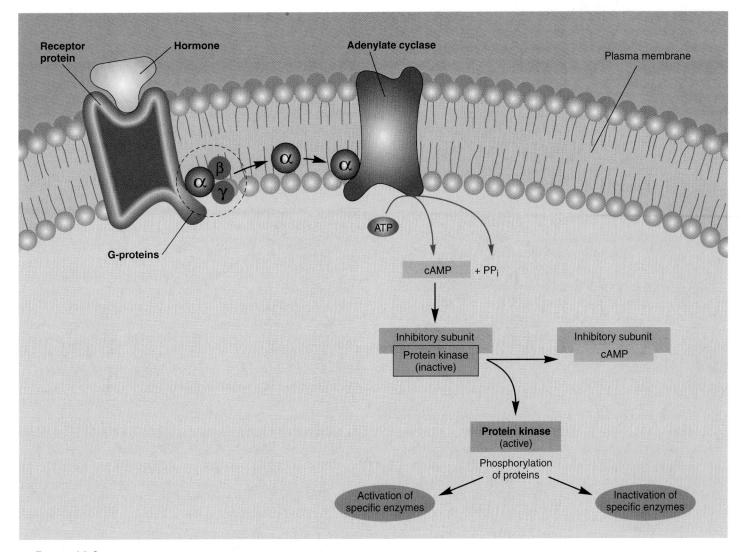

Figure 11.8 **The adenylate cyclase-cyclic AMP second-messenger system.** The hormone causes the production of cAMP within the target cell cytoplasm, and cAMP activates protein kinase. The activated protein kinase then causes the activation or inactivation of a number of specific enzymes. These changes lead to the characteristic effects of the hormone on the target cell.

therefore, the intracellular concentration of cAMP is increased. Cyclic AMP activates a previously inactive enzyme in the cytoplasm called **protein kinase.** The inactive form of this enzyme consists of two subunits: a catalytic subunit and an inhibitory subunit. The enzyme is produced in an inactive form and becomes active only when cAMP attaches to the inhibitory subunit. Binding of cAMP to the inhibitory subunit causes it to dissociate from the catalytic subunit, which then becomes active (fig. 11.8). In summary, the hormone—acting through an increase in cAMP production—causes an increase in protein kinase enzyme activity within its target cells.

Active protein kinase catalyzes the phosphorylation of (attachment of phosphate groups to) different proteins in the target cells. This causes some enzymes to become activated and others to become inactivated. Cyclic AMP, acting through protein kinase, thus modulates the activity of enzymes that are already present in the target cell. This alters the metabolism of the target tissue in a manner characteristic of the actions of that specific hormone (table 11.4).

Table 11.4 Sequence of Events Involving Cyclic AMP as a Second Messenger

1. The hormone binds to its receptor on the outer surface of the target cell's plasma membrane.
2. Hormone-receptor interaction acts by means of G-proteins to stimulate the activity of adenylate cyclase on the cytoplasmic side of the membrane.
3. Activated adenylate cyclase catalyzes the conversion of ATP to cyclic AMP (cAMP) within the cytoplasm.
4. Cyclic AMP activates protein kinase enzymes that were already present in the cytoplasm in an inactive state.
5. Activated cAMP-dependent protein kinase transfers phosphate groups to (phosphorylates) other enzymes in the cytoplasm.
6. The activity of specific enzymes is either increased or inhibited by phosphorylation.
7. Altered enzyme activity mediates the target cell's response to the hormone.

Like all biologically active molecules, cAMP must be rapidly inactivated for it to function effectively as a second messenger in hormone action. This inactivation is accomplished by **phosphodiesterase,** an enzyme within the target cells that hydrolyzes cAMP into inactive fragments. Through the action of phosphodiesterase, the stimulatory effect of a hormone that uses cAMP as a second messenger depends upon the continuous generation of new cAMP molecules, and thus depends on the level of secretion of the hormone.

Drugs that inhibit the activity of phosphodiesterase prevent the breakdown of cAMP and thus result in increased concentrations of cAMP within the target cells. The drug **theophylline** and its derivatives, for example, are used clinically to raise cAMP levels within bronchiolar smooth muscle. This duplicates and enhances the effect of epinephrine on the bronchioles (producing dilation) in people who suffer from asthma. **Caffeine,** a compound related to theophylline, is also a phosphodiesterase inhibitor, and thus exerts its effects by raising the cAMP concentrations within cells.

In addition to cyclic AMP, **cyclic guanosine monophosphate** (cGMP) functions as a second messenger in certain cases. For example, the regulatory molecule nitric oxide (discussed in chapter 7 and later in this chapter) exerts its effects on smooth muscle by stimulating the production of cGMP in its target cells. One example of this is the vascular smooth muscle relaxation that produces erection of the penis (see chapter 20, fig. 20.23). Indeed, as illustrated in this figure, the drug *Viagra* helps treat erectile dysfunction by inhibiting the phosphodiesterase enzyme that breaks down cGMP.

Phospholipase C–Ca²⁺ Second-Messenger System

The concentration of Ca^{2+} in the cytoplasm is kept very low by the action of active transport carriers—calcium pumps—in the plasma membrane. Through the action of these pumps, the concentration of calcium is about 10,000 times lower in the cytoplasm than in the extracellular fluid. In addition, the endoplasmic reticulum (chapter 3) of many cells contains calcium pumps that actively transport Ca^{2+} from the cytoplasm into the cisternae of the endoplasmic reticulum. The steep concentration gradient for Ca^{2+} that results allows various stimuli to evoke a rapid, though brief, diffusion of Ca^{2+} into the cytoplasm, which can serve as a signal in different control systems.

At the terminal boutons of axons, for example, the entry of Ca^{2+} through voltage-regulated Ca^{2+} channels in the plasma membrane serves as a signal for the release of neurotransmitters (chapter 7; see fig. 7.21). Similarly, when muscles are stimulated to contract, Ca^{2+} couples electrical excitation of the muscle cell to the mechanical processes of contraction (see chapter 12). Additionally, it is now known that Ca^{2+} serves as a part of a second-messenger system in the action of a number of hormones.

When epinephrine stimulates its target organs, it must first bind to adrenergic receptor proteins in the membrane of its target cells. As discussed in chapter 9, there are two types of adrenergic receptors—alpha and beta (see fig. 9.10). Stimulation of the beta-adrenergic receptors by epinephrine results in activation of adenylate cyclase and the production of cAMP. Stimulation of alpha-adrenergic receptors by epinephrine, in contrast, activates the target cell via the Ca^{2+} second-messenger system (see fig. 11.10).

The binding of epinephrine to its alpha-adrenergic receptor activates, via a G-protein intermediate, an enzyme in the plasma membrane known as **phospholipase C.** The substrate of this enzyme, a particular membrane phospholipid, is split by the active enzyme into **inositol triphosphate (IP₃)** and another derivative, **diacylglycerol (DAG).** Both derivatives serve as second messengers, but the action of IP₃ is somewhat better understood and will be discussed in this section.

The IP₃ leaves the plasma membrane and diffuses through the cytoplasm to the endoplasmic reticulum. The membrane of the endoplasmic reticulum contains receptor proteins for IP₃, so that the IP₃ is a second messenger in its own right, carrying the hormone's message from the plasma membrane to the endoplasmic reticulum. Binding of IP₃ to its receptors causes specific Ca^{2+} channels to open, so that Ca^{2+} diffuses out of the endoplasmic reticulum and into the cytoplasm (fig. 11.9).

As a result of these events, there is a rapid and transient rise in the cytoplasmic Ca^{2+} concentration. This signal is augmented, through mechanisms that are incompletely understood, by the opening of Ca^{2+} channels in the plasma membrane. This may be due to the action of yet a different (and currently unknown) messenger sent from the endoplasmic reticulum to the plasma membrane. The Ca^{2+} that enters the cytoplasm binds to a protein called **calmodulin.** Once Ca^{2+} binds to calmodulin, the now-active calmodulin in turn activates specific protein kinase enzymes (those that add phosphate groups to proteins) that modify the actions of other enzymes in the cell (fig. 11.10). Activation of specific calmodulin-dependent enzymes is analogous to the activation of enzymes by cAMP-dependent protein kinase. The steps of the Ca^{2+} second-messenger system are summarized in table 11.5.

Tyrosine Kinase Second-Messenger System

Insulin promotes glucose and amino acid transport and stimulates glycogen, fat, and protein synthesis in its target organs—primarily the liver, skeletal muscles, and adipose tissue. These effects are achieved by means of a mechanism of action that is quite complex, and in some ways still incompletely understood. Nevertheless, it is known that insulin's mechanism of action bears similarities to the mechanism of action of other regulatory molecules known as **growth factors.** These growth factors, examples of which are *epidermal growth factor (EGF), platelet-derived growth factor (PDGF), and insulin-like growth factors (IGFs)* are autocrine regulators (described at the end of this chapter).

In the case of insulin and the growth factors, the receptor protein is located in the plasma membrane and is itself a kind of enzyme known as a **tyrosine kinase.** A *kinase* is an enzyme that adds phosphate groups to proteins, and a *tyrosine* kinase

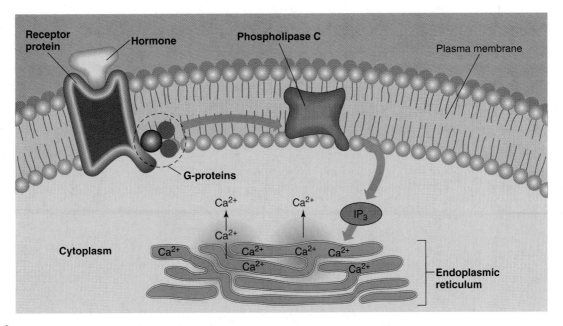

Figure 11.9 The phospholipase C-Ca^{2+} second-messenger system. Some hormones, when they bind to their membrane receptors, activate phospholipase C (PLC). This enzyme catalyzes the formation of inositol triphosphate (IP$_3$), which causes Ca^{2+} channels to open in the endoplasmic reticulum. Ca^{2+} is thus released and acts as a second messenger in the action of the hormone.

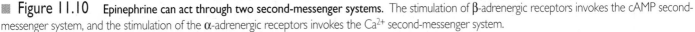

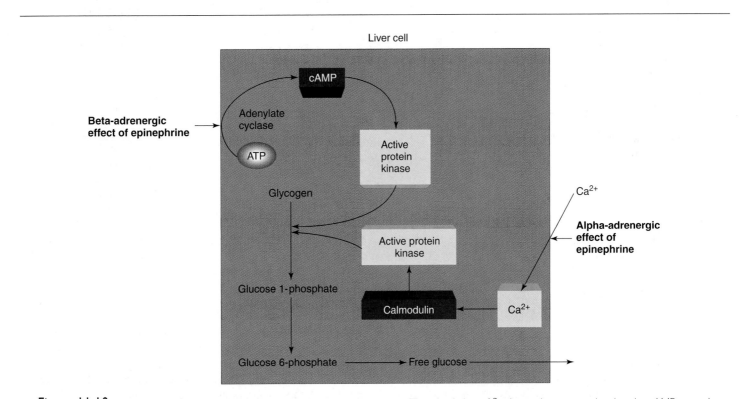

Figure 11.10 Epinephrine can act through two second-messenger systems. The stimulation of β-adrenergic receptors invokes the cAMP second-messenger system, and the stimulation of the α-adrenergic receptors invokes the Ca^{2+} second-messenger system.

specifically adds these phosphate groups to the amino acid tyrosine within the proteins. The insulin receptor consists of two units that come together (dimerize) when they bind with insulin to form an active tyrosine kinase enzyme (fig. 11.11). Each unit of the receptor contains a site on the outside of the cell that binds to insulin (termed the *ligand-binding site*) and a part that spans the plasma membrane, with an *enzymatic site* in the cytoplasm. The enzymatic site is inactive until insulin binds to the ligand-binding site and causes dimerization of the receptor. When insulin binding and dimerization occur, the enzymatic

Table 11.5 Sequence of Events Involving the Ca^{2+} Second-Messenger System

1. The hormone binds to its receptor on the outer surface of the target cell's plasma membrane.
2. Hormone-receptor interaction stimulates the activity of a membrane enzyme, phospholipase C.
3. Activated phospholipase C catalyzes the conversion of particular phospholipids in the membrane to inositol triphosphate (IP$_3$) and another derivative, diacylglycerol.
4. Inositol triphosphate enters the cytoplasm and diffuses to the endoplasmic reticulum, where it binds to its receptor proteins and causes the opening of Ca^{2+} channels.
5. Since the endoplasmic reticulum accumulates Ca^{2+} by active transport, there exists a steep Ca^{2+} concentration gradient favoring the diffusion of Ca^{2+} into the cytoplasm.
6. Ca^{2+} that enters the cytoplasm binds to and activates a protein called calmodulin.
7. Activated calmodulin, in turn, activates protein kinase, which phosphorylates other enzyme proteins.
8. Altered enzyme activity mediates the target cell's response to the hormone.

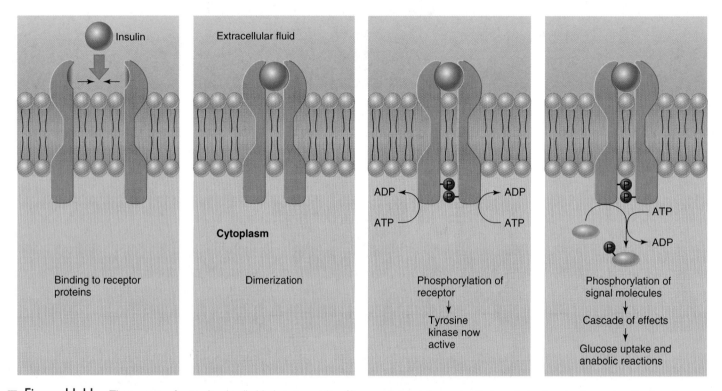

■ **Figure 11.11**   **The receptor for insulin.** Insulin binds to two units of its receptor protein, causing these units to dimerize (come together) on the plasma membrane. This activates the tyrosine kinase enzyme portion of the receptor. As a result, the receptor phosphorylates itself, thereby making the enzyme even more active. The receptor then phosphorylates a number of cytoplasmic "signal molecules" that exert a cascade of effects in the target cell.

site is activated in each unit of the receptor, and one unit phosphorylates the other. This process, termed *autophosphorylation,* increases the tyrosine kinase activity of the dimerized receptor.

The activated tyrosine kinase receptor then phosphorylates other proteins that serve as **signaling molecules.** Some of these signaling molecules are themselves kinase enzymes that phosphorylate and activate other second-messenger systems. As a result of a complex series of activations, insulin and the different growth factors regulate the metabolism of their target cells.

For example, insulin indirectly stimulates the insertion of GLUT-4 carrier proteins (for the facilitated diffusion of glucose; see chapter 6, fig. 6.15) into the plasma membrane of skeletal muscle, adipose, and liver cells. In this way, insulin stimulates

the uptake of plasma glucose into these organs. Also, the binding of insulin to its receptor indirectly causes the activation of glycogen synthetase, the enzyme in liver and skeletal muscles that catalyzes the production of glycogen in these organs.

The complexity of different second-messenger systems is needed so that different signaling molecules can have varying effects. For example, insulin uses the tyrosine kinase second-messenger system to stimulate glucose uptake into the liver and its synthesis into glycogen, whereas glucagon (another hormone secreted by the pancreatic islets) promotes opposite effects—the hydrolysis of hepatic glycogen and subsequent secretion of glucose—by activating a different second-messenger system that involves the production of cAMP.

Pituitary Gland

The pituitary gland includes the anterior pituitary and the posterior pituitary. The posterior pituitary stores and releases hormones that are actually produced by the hypothalamus, whereas the anterior pituitary produces and secretes its own hormones. The anterior pituitary, however, is regulated by hormones secreted by the hypothalamus, as well as by feedback from the target gland hormones.

The **pituitary gland,** or **hypophysis,** is located on the inferior aspect of the brain in the region of the diencephalon (chapter 8). Roughly the size of a pea—about 1.3 cm (0.5 in.) in diameter—it is attached to the hypothalamus by a stalklike structure called the *infundibulum* (fig. 11.12).

The pituitary gland is structurally and functionally divided into an anterior lobe, or **adenohypophysis,** and a posterior lobe called the **neurohypophysis.** These two parts have different embryonic origins. The adenohypophysis is derived from a pouch of epithelial tissue (*Rathke's pouch*) that migrates upward from the embryonic mouth, whereas the neurohypophysis is formed as a downgrowth of the brain. The adenohypophysis consists of two parts in adults: (1) the *pars distalis,* also known as the **anterior pituitary,** is the rounded portion and the major endocrine part of the gland, and (2) the *pars tuberalis* is the thin extension in contact with the infundibulum. These parts are illustrated in figure 11.12. A *pars intermedia,* a strip of tissue between the anterior and posterior lobes, exists in the fetus. During fetal development, its cells mingle with those of the anterior lobe, and in adults they no longer constitute a separate structure.

The neurohypophysis is the neural part of the pituitary gland. It consists of the *pars nervosa,* also called the **posterior pituitary,** which is in contact with the adenohypophysis and the infundibulum. Nerve fibers extend through the infundibulum along with small neuroglia-like cells called *pituicytes.*

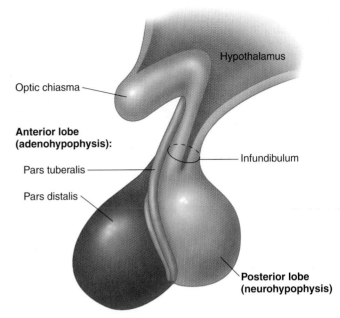

Figure 11.12 **The structure of the pituitary gland.** The anterior lobe is composed of glandular tissue, whereas the posterior lobe is composed largely of neuroglia and nerve fibers.

Pituitary Hormones

The hormones secreted by the anterior pituitary (the pars distalis of the adenohypophysis) are called **trophic hormones.** The term *trophic* means "feed." Although the anterior pituitary hormones are not food for their target organs, this term is used because high concentrations of the anterior pituitary hormones cause their target organs to hypertrophy, while low levels cause their target organs to atrophy. When names are applied to the hormones of the anterior pituitary, "trophic" (conventionally shortened to *tropic,* meaning "attracted to") is incorporated into them. This is why the shortened forms of the names for the anterior pituitary hormones end in the suffix *-tropin.* The hormones of the anterior pituitary, listed here, are summarized in table 11.6.

1. **Growth hormone (GH,** or **somatotropin).** GH promotes the movement of amino acids into cells and the incorporation of these amino acids into proteins, thus promoting overall tissue and organ growth.
2. **Thyroid-stimulating hormone (TSH,** or **thyrotropin).** TSH stimulates the thyroid gland to produce and secrete thyroxine (tetraiodothyronine, or T_4) and triiodothyronine (T_3).
3. **Adrenocorticotropic hormone (ACTH,** or **corticotropin).** ACTH stimulates the adrenal cortex to secrete the glucocorticoids, such as hydrocortisone (cortisol).
4. **Follicle-stimulating hormone (FSH,** or **folliculotropin).** FSH stimulates the growth of ovarian follicles in females and the production of sperm cells in the testes of males.

Table 11.6   Anterior Pituitary Hormones

Hormone	Target Tissue	Principal Actions	Regulation of Secretion
ACTH (adrenocorticotropic hormone)	Adrenal cortex	Stimulates secretion of glucocorticoids	Stimulated by CRH (corticotropin-releasing hormone); inhibited by glucocorticoids
TSH (thyroid-stimulating hormone)	Thyroid gland	Stimulates secretion of thyroid hormones	Stimulated by TRH (thyrotropin-releasing hormone); inhibited by thyroid hormones
GH (growth hormone)	Most tissue	Promotes protein synthesis and growth; lipolysis and increased blood glucose	Inhibited by somatostatin; stimulated by growth hormone–releasing hormone
FSH (follicle-stimulating hormone)	Gonads	Promotes gamete production and stimulates estrogen production in females	Stimulated by GnRH (gonadotropin-releasing hormone); inhibited by sex steroids and inhibin
PRL (prolactin)	Mammary glands and other sex accessory organs	Promotes milk production in lactating females; additional actions in other organs	Inhibited by PIH (prolactin-inhibiting hormone)
LH (luteinizing hormone)	Gonads	Stimulates sex hormone secretion; ovulation and corpus luteum formation in females; stimulates testosterone secretion in males	Stimulated by GnRH; inhibited by sex steroids

5. **Luteinizing hormone (LH,** or **luteotropin).** This hormone and FSH are collectively called **gonadotropic hormones.** In females, LH stimulates ovulation and the conversion of the ovulated ovarian follicle into an endocrine structure called a corpus luteum. In males, LH is sometimes called *interstitial cell stimulating hormone,* or *ICSH;* it stimulates the secretion of male sex hormones (mainly testosterone) from the interstitial cells (Leydig cells) in the testes.

6. **Prolactin (PRL).** This hormone is secreted in both males and females. Its best known function is the stimulation of milk production by the mammary glands of women after the birth of a baby. Prolactin plays a supporting role in the regulation of the male reproductive system by the gonadotropins (FSH and LH) and acts on the kidneys to help regulate water and electrolyte balance.

As mentioned earlier, the pars intermedia of the adenohypophysis ceases to exist as a separate lobe in the adult human pituitary, but it is present in the human fetus and in other animals. Until recently, it was thought to secrete **melanocyte-stimulating hormone (MSH),** as it does in fish, amphibians, and reptiles, where it causes darkening of the skin. In humans, however, plasma concentrations of MSH are insignificant. Some cells of the adenohypophysis, derived from the fetal pars intermedia, produce a large polypeptide called *pro-opiomelanocortin (POMC).* POMC is a prohormone whose major products are beta-endorphin (chapter 7), MSH, and ACTH. Since part of the ACTH molecule contains the amino acid sequence of MSH, elevated secretions of ACTH (as in Addison's disease) cause a marked darkening of the skin.

Inadequate growth hormone secretion during childhood causes **pituitary dwarfism.** Hyposecretion of growth hormone in an adult produces a rare condition called *pituitary cachexia (Simmonds' disease).* One of the symptoms of this disease is premature aging caused by tissue atrophy. Oversecretion of growth hormone during childhood, by contrast, causes **gigantism.** Excessive growth hormone secretion in an adult does not cause further growth in length because the cartilaginous epiphyseal discs have already ossified. Hypersecretion of growth hormone in an adult instead causes **acromegaly** (see fig. 19.15), in which the person's appearance gradually changes as a result of thickening of bones and the growth of soft tissues, particularly in the face, hands, and feet.

The posterior pituitary, or pars nervosa, stores and releases two hormones, both of which are produced in the hypothalamus.

1. **Antidiuretic hormone (ADH),** also known as **arginine vasopressin (AVP).** ADH promotes the retention of water by the kidneys so that less water is excreted in the urine and more water is retained in the blood. At high doses, this hormone also has a "pressor" effect; that is, it causes vasoconstriction in experimental animals. The physiological significance of this pressor effect in humans is controversial, however.

2. **Oxytocin.** In females, oxytocin stimulates contractions of the uterus during labor and for this reason is needed for parturition (childbirth). Oxytocin also stimulates

contractions of the mammary gland alveoli and ducts, which result in the milk-ejection reflex in a lactating woman. In men, a rise in oxytocin secretion at the time of ejaculation has been measured, but the physiological significance of this hormone in males remains to be demonstrated.

Injections of oxytocin may be given to a pregnant woman to induce labor if the pregnancy is prolonged or if the fetal membranes have ruptured and there is danger of infection. Labor also may be induced by injections of oxytocin in the case of severe pregnancy-induced hypertension, or **preeclampsia.** Oxytocin administration after delivery causes the uterus to regress in size and squeezes the blood vessels, thus minimizing the danger of hemorrhage.

Hypothalamic Control of the Posterior Pituitary

Both of the posterior pituitary hormones—antidiuretic hormone and oxytocin—are actually produced in neuron cell bodies of the *supraoptic nuclei* and *paraventricular nuclei* of the hypothalamus. These nuclei within the hypothalamus are thus endocrine glands. The hormones they produce are transported along axons of the **hypothalamo-hypophyseal tract** (fig. 11.13) to the posterior pituitary, where they are stored and later released. The posterior pituitary is thus more a storage organ than a true gland.

The release of ADH and oxytocin from the posterior pituitary is controlled by **neuroendocrine reflexes.** In nursing mothers, for example, the mechanical stimulus of suckling acts, via sensory nerve impulses to the hypothalamus, to stimulate the reflex secretion of oxytocin (chapter 20). The secretion of ADH is stimulated by osmoreceptor neurons in the hypothalamus in response to a rise in blood osmotic pressure (chapter 6); its secretion is inhibited by sensory impulses from stretch receptors in the left atrium of the heart in response to a rise in blood volume (chapter 14).

Hypothalamic Control of the Anterior Pituitary

At one time the anterior pituitary was called the "master gland" because it secretes hormones that regulate some other endocrine glands (fig. 11.14 and table 11.6). Adrenocorticotropic hormone (ACTH), thyroid-stimulating hormone (TSH), and the gonadotropic hormones (FSH and LH) stimulate the adrenal cortex, thyroid, and gonads, respectively, to secrete their hormones. The anterior pituitary hormones also have a "trophic" effect on their target glands in that the health of these glands depends on adequate stimulation by anterior pituitary hormones. The ante-

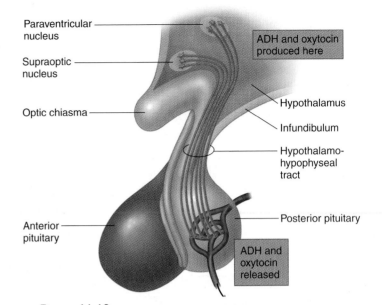

Figure 11.13 *Hypothalamic control of the posterior pituitary.* The posterior pituitary, or neurohypophysis, stores and releases hormones—vasopressin and oxytocin—that are actually produced in neurons within the supraoptic and paraventricular nuclei of the hypothalamus. These hormones are transported to the posterior pituitary by axons in the hypothalamo-hypophyseal tract.

rior pituitary, however, is not really the master gland, since secretion of its hormones is in turn controlled by hormones secreted by the hypothalamus.

Releasing and Inhibiting Hormones

Since axons do not enter the anterior pituitary, hypothalamic control of the anterior pituitary is achieved through hormonal rather than neural regulation. Releasing and inhibiting hormones, produced by neurons in the hypothalamus, are transported to axon endings in the basal portion of the hypothalamus. This region, known as the *median eminence* (fig 11.15), contains blood capillaries that are drained by venules in the stalk of the pituitary.

The venules that drain the median eminence deliver blood to a second capillary bed in the anterior pituitary. Since this second capillary bed is downstream from the capillary bed in the median eminence and receives venous blood from it, the vascular link between the median eminence and the anterior pituitary forms a *portal system.* (This is analogous to the hepatic portal system that delivers venous blood from the intestine to the liver, as described in chapter 18.) The vascular link between the hypothalamus and the anterior pituitary is thus called the **hypothalamo-hypophyseal portal system.**

Regulatory hormones are secreted into the hypothalamo-hypophyseal portal system by neurons of the hypothalamus. These hormones regulate the secretions of the anterior pituitary (fig. 11.15 and table 11.7). **Thyrotropin-releasing hormone (TRH)**

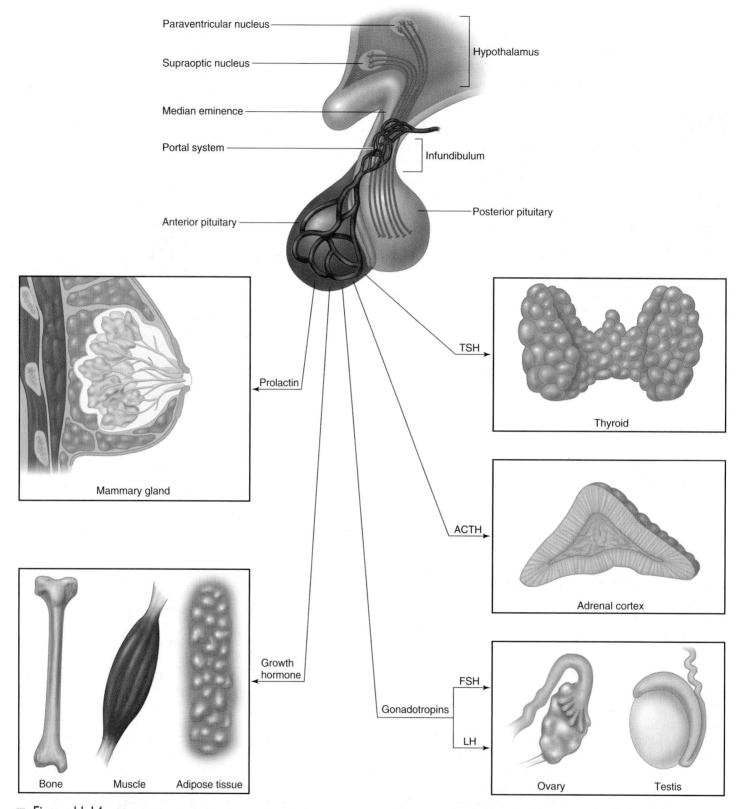

■ Figure 11.14   Hormones secreted by the anterior pituitary and their target organs. Notice that the anterior pituitary controls some (but by no means all) of the other endocrine glands.

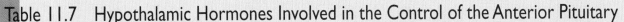

Table 11.7 Hypothalamic Hormones Involved in the Control of the Anterior Pituitary

Hypothalamic Hormone	Structure	Effect on Anterior Pituitary
Corticotropin-releasing hormone (CRH)	41 amino acids	Stimulates secretion of adrenocorticotropic hormone (ACTH)
Gonadotropin-releasing hormone (GnRH)	10 amino acids	Stimulates secretion of follicle-stimulating hormone (FHS) and luteinizing hormone (LH)
Prolactin-inhibiting hormone (PIH)	Dopamine	Inhibits prolactin secretion
Somatostatin	14 amino acids	Inhibits secretion of growth hormone
Thyrotropin-releasing hormone (TRH)	3 amino acids	Stimulates secretion of thyroid-stimulating hormone (TSH)
Growth hormone–releasing hormone (GHRH)	44 amino acids	Stimulates growth hormone secretion

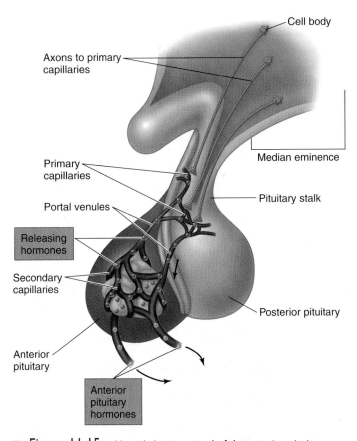

■ **Figure 11.15** Hypothalamic control of the anterior pituitary. Neurons in the hypothalamus secrete releasing hormones (shown as dots) into the blood vessels of the hypothalamo-hypophyseal portal system. These releasing hormones stimulate the anterior pituitary to secrete its hormones into the general circulation.

stimulates the secretion of TSH, and **corticotropin-releasing hormone (CRH)** stimulates the secretion of ACTH from the anterior pituitary. A single releasing hormone, **gonadotropin-releasing hormone,** or **GnRH,** stimulates the secretion of both gonadotropic hormones (FSH and LH) from the anterior pituitary. The secretion of prolactin and of growth hormone from the anterior pituitary is regulated by hypothalamic inhibitory hormones, known as **prolactin-inhibiting hormone (PIH)** and **somatostatin,** respectively.

A specific **growth hormone-releasing hormone (GHRH)** that stimulates growth hormone secretion has been identified as a polypeptide consisting of forty-four amino acids. Experiments suggest that a releasing hormone for prolactin may also exist, but no such specific releasing hormone has yet been discovered.

Feedback Control of the Anterior Pituitary

In view of its secretion of releasing and inhibiting hormones, the hypothalamus might be considered the "master gland." The chain of command, however, is not linear; the hypothalamus and anterior pituitary are controlled by the effects of their own actions. In the endocrine system, to use an analogy, the general takes orders from the private. The hypothalamus and anterior pituitary are not master glands because their secretions are controlled by the target glands they regulate.

Anterior pituitary secretion of ACTH, TSH, and the gonadotropins (FSH and LH) is controlled by **negative feedback inhibition** from the target gland hormones. Secretion of ACTH is inhibited by a rise in corticosteroid secretion, for example, and TSH is inhibited by a rise in the secretion of thyroxine from the thyroid. These negative feedback relationships are easily demonstrated by removal of the target glands. Castration (surgical removal of the gonads), for example, produces a rise in the secretion of FSH and LH. In a similar manner, removal of the adrenals or the thyroid results in an abnormal increase in ACTH or TSH secretion from the anterior pituitary.

Clinical Investigation Clue

Remember that Rosemary has a blood ACTH level that was only about one-fiftieth of normal.

What might account for the low ACTH secretion from the anterior pituitary?

The effects of removal of the target glands demonstrate that, under normal conditions, these glands exert an inhibitory effect on the anterior pituitary. This inhibitory effect can occur at two levels: (1) the target gland hormones can act on the hypothalamus and inhibit the secretion of releasing hormones, and (2) the target gland hormones can act on the anterior pituitary and inhibit its response to the releasing hormones. Thyroxine,

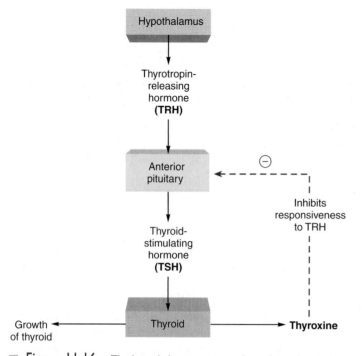

Figure 11.16 The hypothalamus-pituitary-thyroid axis (control system). The secretion of thyroxine from the thyroid is stimulated by thyroid-stimulating hormone (TSH) from the anterior pituitary. The secretion of TSH is stimulated by thyrotropin-releasing hormone (TRH) secreted from the hypothalamus. This stimulation is balanced by negative feedback inhibition (blue arrow) from thyroxine, which decreases the responsiveness of the anterior pituitary to stimulation by TRH.

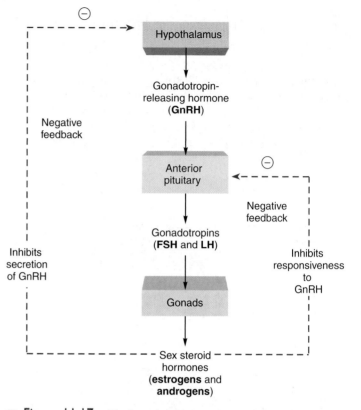

Figure 11.17 The hypothalamus-pituitary-gonad axis (control system). The hypothalamus secretes GnRH, which stimulates the anterior pituitary to secrete the gonadotropins (FSH and LH). These, in turn, stimulate the gonads to secrete the sex steroids. The secretions of the hypothalamus and anterior pituitary are themselves regulated by negative feedback inhibition (blue arrows) from the sex steroids.

for example, appears to inhibit the response of the anterior pituitary to TRH and thus acts to reduce TSH secretion (fig. 11.16). Sex steroids, by contrast, reduce the secretion of gonadotropins by inhibiting both GnRH secretion and the ability of the anterior pituitary to respond to stimulation by GnRH (fig. 11.17).

Evidence suggests that there may be retrograde transport of blood from the anterior pituitary to the hypothalamus. This may permit a *short feedback loop* in which a particular trophic hormone inhibits the secretion of its releasing hormone from the hypothalamus. A high secretion of TSH, for example, may inhibit further secretion of TRH by this means.

In addition to negative feedback control of the anterior pituitary, there is one instance of a hormone from a target organ that actually stimulates the secretion of an anterior pituitary hormone. Toward the middle of the menstrual cycle, the rising secretion of estradiol from the ovaries stimulates the anterior pituitary to secrete a "surge" of LH, which results in ovulation. This is commonly described as a *positive feedback effect* to distinguish it from the more usual negative feedback inhibition of target gland hormones on anterior pituitary secretion. Interestingly, higher levels of estradiol at a later stage of the menstrual cycle exert the opposite effect—negative feedback inhibition—on LH secretion. The control of gonadotropin secretion is discussed in more detail in chapter 20.

Higher Brain Function and Pituitary Secretion

The relationship between the anterior pituitary and a particular target gland is described as an *axis;* the pituitary-gonad axis, for example, refers to the action of gonadotropic hormones on the testes and ovaries. This axis is stimulated by GnRH from the hypothalamus, as previously described. Since the hypothalamus receives neural input from "higher brain centers," however, it is not surprising that the pituitary-gonad axis can be affected by emotions. Indeed, the ability of intense emotions to alter the timing of ovulation or menstruation is well known. Psychological stress, as another example, also stimulates another axis—the pituitary-adrenal axis (described in the next section).

Stressors, as described later in this chapter, produce an increase in CRH secretion from the hypothalamus, which in turn results in elevated ACTH and corticosteroid secretion. In addition, the influence of higher brain centers produces *circadian* ("about a day") *rhythms* in the secretion of many anterior pituitary hormones. The secretion of growth hormone, for example, is highest during sleep and decreases during wakefulness, although its secretion is also stimulated by the absorption of particular amino acids following a meal.

The influence of higher brain centers on the pituitary-gonad axis helps to explain the "dormitory effect"—that is, the tendency for the menstrual cycles of female roommates to synchronize. This synchronization will not occur in a new roommate if her nasal cavity is plugged with cotton, suggesting that the dormitory effect is due to the action of chemicals called **pheromones**. These chemicals are excreted to the outside of the body and act through the olfactory sense to modify the physiology or behavior of another member of the same species. Pheromones are important regulatory molecules in the urine, vaginal fluid, and other secretions of most mammals, and help to regulate their reproductive cycles and behavior. The role of pheromones in humans is difficult to assess. Recently, however, scientists discovered that pheromones produced in the axillae (underarms) of women may contribute to the dormitory effect.

Test Yourself Before You Continue

1. Describe the embryonic origins of the adenohypophysis and neurohypophysis, and list the parts of each. Which of these parts is also called the anterior pituitary? Which is called the posterior pituitary?

2. List the hormones released by the posterior pituitary. Where do these hormones originate and how are their secretions regulated?

3. List the hormones secreted by the anterior pituitary and explain how the hypothalamus controls the secretion of each.

4. Draw a negative feedback loop showing the control of ACTH secretion. Explain how this system would be affected by (a) an injection of ACTH, (b) surgical removal of the pituitary, (c) an injection of corticosteroids, and (d) surgical removal of the adrenal glands.

Adrenal Glands

The adrenal cortex and adrenal medulla are structurally and functionally different. The adrenal medulla secretes catecholamine hormones, which complement the sympathetic nervous system in the "fight-or-flight" reaction. The adrenal cortex secretes steroid hormones that participate in the regulation of mineral and energy balance.

The **adrenal glands** are paired organs that cap the superior borders of the kidneys (fig. 11.18). Each adrenal consists of an outer cortex and inner medulla that function as separate glands. The differences in function of the adrenal cortex and medulla are related to the differences in their embryonic derivation. The adrenal medulla is derived from embryonic neural crest ectoderm (the same tissue that produces the sympathetic

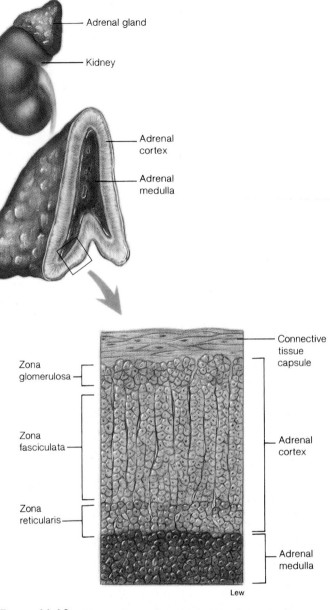

■ **Figure 11.18** **The structure of the adrenal gland, showing the three zones of the adrenal cortex.** The zona glomerulosa secretes the mineralocorticoids (including aldosterone), whereas the other two zones secrete the glucocorticoids (including cortisol).

ganglia), whereas the adrenal cortex is derived from a different embryonic tissue (mesoderm).

As a consequence of its embryonic derivation, the adrenal medulla secretes catecholamine hormones (mainly epinephrine, with lesser amounts of norepinephrine) into the blood in response to stimulation by preganglionic sympathetic nerve fibers (chapter 9). The adrenal cortex does not receive neural innervation, and so must be stimulated hormonally (by ACTH secreted from the anterior pituitary). The cortex consists of three zones: an outer *zona glomerulosa,* a middle *zona fasciculata,* and an inner *zona reticularis* (fig. 11.18). These zones are believed to have different functions.

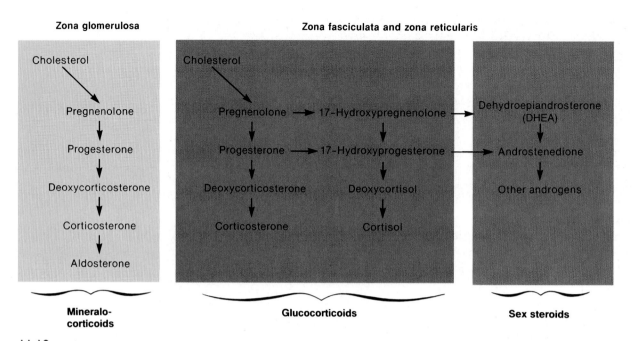

Zona glomerulosa

Zona fasciculata and zona reticularis

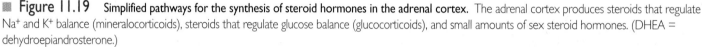

■ **Figure 11.19** Simplified pathways for the synthesis of steroid hormones in the adrenal cortex. The adrenal cortex produces steroids that regulate Na⁺ and K⁺ balance (mineralocorticoids), steroids that regulate glucose balance (glucocorticoids), and small amounts of sex steroid hormones. (DHEA = dehydroepiandrosterone.)

Functions of the Adrenal Cortex

The adrenal cortex secretes steroid hormones called **corticosteroids,** or **corticoids,** for short. There are three functional categories of corticosteroids: (1) **mineralocorticoids,** which regulate Na⁺ and K⁺ balance; (2) **glucocorticoids,** which regulate the metabolism of glucose and other organic molecules; and (3) **sex steroids,** which are weak androgens (including *dehydroepiandrosterone,* or *DHEA*) that supplement the sex steroids secreted by the gonads. These three categories of steroid hormones are derived from the same precursor (parent molecule), cholesterol. The biosynthetic pathways from cholesterol diverges in the different zones of the adrenal cortex, so that a particular category of corticosteroid is produced in a particular zone of the adrenal cortex (fig. 11.19).

Aldosterone is the most potent mineralocorticoid. The mineralocorticoids are produced in the zona glomerulosa and stimulate the kidneys to retain Na⁺ and water while excreting K⁺ in the urine. These actions help to increase the blood volume and pressure (as described in chapter 14), and to regulate blood electrolyte balance (as described in chapter 17).

The predominant glucocorticoid in humans is *cortisol (hydrocortisone),* which is secreted by the zona fasciculata and perhaps also by the zona reticularis. The secretion of cortisol is stimulated by ACTH from the anterior pituitary (fig. 11.20). Cortisol and other glucocorticoids have many effects on metabolism; they stimulate gluconeogenesis (production of glucose from amino acids and lactic acid) and inhibit glucose utilization, which help to raise the blood glucose level; and they promote lipolysis (breakdown of fat) and the consequent release of free

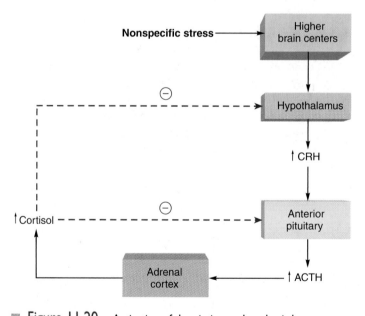

■ **Figure 11.20** Activation of the pituitary-adrenal axis by nonspecific stress. Negative feedback control of the adrenal cortex (blue arrows) is also shown.

fatty acids into the blood. The roles of glucocorticoids and other hormones in metabolic regulation are explained in chapter 19.

Exogenous glucocorticoids (taken as pills, injections, sprays, and topical creams) are used medically to suppress the immune response and inhibit inflammation. Thus, these drugs are very useful in treating inflammatory diseases such as asthma and rheumatoid arthritis. As might be predicted based on their metabolic actions, the side effects of glucocorticoids include hyperglycemia and decreased glucose tolerance. Other negative side effects include decreased synthesis of collagen and other extracellular matrix proteins (chapter 6) and increased bone resorption, leading to osteoporosis.

Hypersecretion of corticosteroids results in **Cushing's syndrome.** This disorder is generally caused by oversecretion of ACTH from the anterior pituitary, but it can also result from a tumor of the adrenal cortex. Cushing's syndrome is characterized by changes in carbohydrate and protein metabolism, hyperglycemia, hypertension, and muscular weakness. Metabolic problems give the body a puffy appearance and can cause structural changes characterized as "buffalo hump" and "moon face."

Addison's disease is caused by inadequate secretion of both glucocorticoids and mineralocorticoids, which results in hypoglycemia, sodium and potassium imbalance, dehydration, hypotension, rapid weight loss, and generalized weakness. A person with this condition who is not treated with corticosteroids will die within a few days because of severe electrolyte imbalance and dehydration. President John F. Kennedy had Addison's disease, but few knew of it because it was well controlled by corticosteroids.

Clinical Investigation Clue

Remember that Rosemary had high blood levels of cortisol together with low ACTH. She also had a puffy appearance.

What disease, due to what cause, is most likely responsible for Rosemary's condition?

Functions of the Adrenal Medulla

The cells of the adrenal medulla secrete **epinephrine** and **norepinephrine** in an approximate ratio of 4 to 1, respectively. The effects of these catecholamine hormones are similar to those caused by stimulation of the sympathetic nervous system, except that the hormonal effect lasts about ten times longer. The hormones from the adrenal medulla increase the cardiac output and heart rate, dilate coronary blood vessels, increase mental alertness, increase the respiratory rate, and elevate the metabolic rate.

The adrenal medulla is innervated by preganglionic sympathetic axons, and secretes its hormones whenever the sympathetic nervous system is activated during "fight or flight" (chapter 9, fig. 9.7). These sympathoadrenal effects are supported by the metabolic actions of epinephrine and norepinephrine: a rise in blood glucose due to stimulation of hepatic

glycogenolysis (breakdown of glycogen) and a rise in blood fatty acids due to stimulation of lipolysis (breakdown of fat). The endocrine regulation of metabolism is described more fully in chapter 19.

A tumor of the adrenal medulla is referred to as a **pheochromocytoma.** This tumor causes hypersecretion of epinephrine and norepinephrine, which produces an effect similar to continuous sympathetic nerve stimulation. The symptoms of this condition are hypertension, elevated metabolism, hyperglycemia and sugar in the urine, nervousness, digestive problems, and sweating. It does not take long for the body to become totally fatigued under these conditions, making the patient susceptible to other diseases.

Stress and the Adrenal Gland

In 1936, a Canadian physiologist, Hans Selye, discovered that injections of a cattle ovary extract into rats (1) stimulated growth of the adrenal cortex; (2) caused atrophy of the lymphoid tissue of the spleen, lymph nodes, and thymus; and (3) produced bleeding peptic ulcers. At first he attributed these effects to the action of a specific hormone in the extract. However subsequent experiments revealed that injections of a variety of substances—including foreign chemicals such as formaldehyde—could produce the same effects. Indeed, the same pattern occurred when Selye subjected rats to cold environments or when he dropped them into water and made them swim until they were exhausted.

The specific pattern of effects produced by these procedures suggested that the effects were due to something the procedures shared in common. Selye reasoned that all of the procedures were stressful. Stress, according to Selye, is the reaction of an organism to stimuli called *stressors,* which may produce damaging effects. The pattern of changes he observed represented a specific response to any stressful agent. He later discovered that stressors produce these effects because they stimulate the pituitary-adrenal axis. Under stressful conditions, there is increased secretion of ACTH from the anterior pituitary, and thus there is increased secretion of glucocorticoids from the adrenal cortex.

On this basis, Selye stated that there is "a nonspecific response of the body to readjust itself following any demand made upon it." A rise in the plasma glucocorticoid levels results from the demands of the stressors. Selye termed this nonspecific response the **general adaptation syndrome (GAS).** Stress, in other words, produces GAS. There are three stages in the response to stress: (1) the *alarm reaction,* when the adrenal glands are activated; (2) the *stage of resistance,* in which readjustment occurs; and (3) if the readjustment is not complete, the *stage of exhaustion,* which may lead to sickness and possibly death.

Glucocorticoids, such as hydrocortisone, can inhibit the immune system. For this reason, these steroids are often administered to treat various inflammatory diseases and to suppress the immune rejection of a transplanted organ. It seems reasonable, therefore, that the elevated glucocorticoid secretion that can accompany stress may inhibit the ability of the immune system to protect against disease. Indeed, studies suggest that prolonged stress results in an increased incidence of cancer and other diseases.

Selye's concept of stress has been refined by subsequent research. These investigations demonstrate that the sympathoadrenal system becomes activated, with increased secretion of epinephrine and norepinephrine, in response to stressors that challenge the organism to respond physically. This is the "fight-or-flight" reaction described in chapter 9. Different emotions, however, are accompanied by different endocrine responses. The pituitary-adrenal axis, with rising levels of glucocorticoids, becomes more active when the stress is of a chronic nature and when the person is more passive and feels less in control.

Test Yourself Before You Continue

1. List the categories of corticosteroids and identify the zone of the adrenal cortex that secretes the hormones within each category.
2. Identify the hormones of the adrenal medulla and describe their effects.
3. Explain how the secretions of the adrenal cortex and adrenal medulla are regulated.
4. Explain how stress affects the secretions of the adrenal cortex and medulla. Why does hypersecretion of the adrenal medullary hormones make a person more susceptible to disease?

Thyroid and Parathyroid Glands

The thyroid secretes thyroxine (T_4) and triiodothyronine (T_3), which are needed for proper growth and development and which are primarily responsible for determining the basal metabolic rate (BMR). The parathyroid glands secrete parathyroid hormone, which helps to raise the blood Ca^{2+} concentration.

The **thyroid gland** is located just below the larynx (fig. 11.21). Its two lobes are positioned on either side of the trachea and are connected anteriorly by a medial mass of thyroid tissue called the *isthmus*. The thyroid is the largest of the pure endocrine glands, weighing between 20 and 25 grams.

On a microscopic level, the thyroid gland consists of numerous spherical hollow sacs called **thyroid follicles** (fig. 11.22).

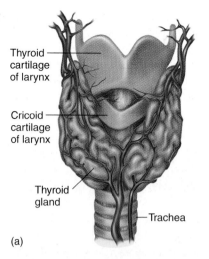

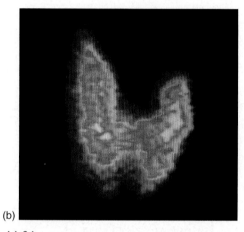

Figure 11.21   The thyroid gland. (*a*) Its relationship to the larynx and trachea. (*b*) A scan of the thyroid gland 24 hours after the intake of radioactive iodine

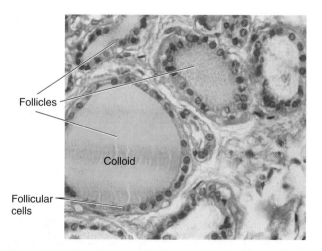

Figure 11.22   A photomicrograph (250×) of a thyroid gland. Numerous thyroid follicles are visible. Each follicle consists of follicular cells surrounding the fluid known as colloid, which contains thyroglobulin.

These follicles are lined with a simple cuboidal epithelium composed of *follicular cells* that synthesize the principal thyroid hormone, *thyroxine*. The interior of the follicles contains *colloid*, a protein-rich fluid. In addition to the follicular cells that secrete thyroxine, the thyroid also contains *parafollicular cells* that secrete a hormone known as *calcitonin* (or *thyrocalcitonin*).

Production and Action of Thyroid Hormones

The thyroid follicles actively accumulate iodide (I^-) from the blood and secrete it into the colloid. Once the iodide has entered the colloid, it is oxidized to iodine and attached to a specific amino acid (tyrosine) within the polypeptide chain of a protein called **thyroglobulin.** The attachment of one iodine to tyrosine produces *monoiodotyrosine (MIT);* the attachment of two iodines produces *diiodotyrosine (DIT).*

Within the colloid, enzymes modify the structure of MIT and DIT and couple them together. When two DIT molecules that are appropriately modified are coupled together, a molecule of **tetraiodothyronine (T_4),** or **thyroxine,** is produced (fig. 11.23). The combination of one MIT with one DIT forms **triiodothyronine (T_3).** Note that at this point T_4 and T_3 are still attached to thyroglobulin. Upon stimulation by TSH, the cells of the follicle take up a small volume of colloid by pinocytosis, hydrolyze the T_3 and T_4 from the thyroglobulin, and secrete the free hormones into the blood.

The transport of thyroid hormones through the blood and their mechanism of action at the cellular level was described earlier in this chapter. Through the activation of genes, thyroid hormones stimulate protein synthesis, promote maturation of the nervous system, and increase the rate of cell respiration in most tissues of the body. Through this action, thyroxine (after it is converted into T_3) elevates the *basal metabolic rate* (*BMR,* discussed in chapter 19), which is the resting rate of calorie expenditure by the body.

Calcitonin, secreted by the parafollicular cells of the thyroid, works in concert with parathyroid hormone (discussed shortly) to regulate the calcium levels of the blood. Calcitonin inhibits the dissolution of the calcium phosphate crystals of bone and stimulates the excretion of calcium in the urine by the kidneys. Both of these actions result in the lowering of blood calcium concentrations.

Diseases of the Thyroid

Thyroid-stimulating hormone (TSH) from the anterior pituitary stimulates the thyroid to secrete thyroxine; however, it also exerts a trophic (growth-stimulating) effect on the thyroid. This trophic effect is evident in people who develop an **iodine-deficiency (endemic) goiter,** or abnormal growth of

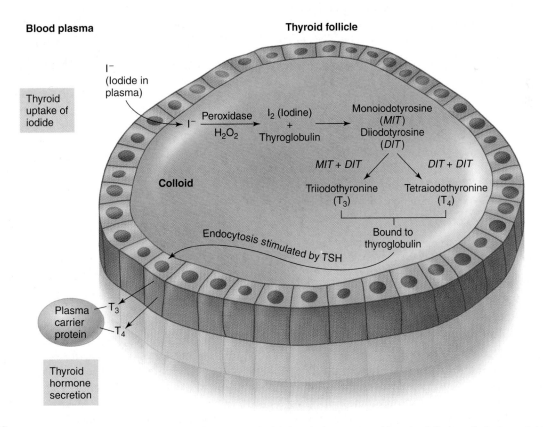

■ **Figure 11.23** **The production and storage of thyroid hormones.** Iodide is actively transported into the follicular cells. In the colloid, it is converted into iodine and attached to tyrosine amino acids within the thyroglobulin protein. MIT (monoiodotyrosine) and DIT (diiodotyrosine) are used to produce T_3 and T_4 within the colloid. Upon stimulation by TSH, the thyroid hormones, bound to thyroglobulin, are taken into the follicular cells by pinocytosis. Hydrolysis reactions within the follicular cells release the free T_4 and T_3, which are secreted.

■ **Figure 11.24** Endemic goiter is caused by insufficient iodine in the diet. A lack of iodine causes hypothyroidism, and the resulting elevation in TSH secretion stimulates the excessive growth of the thyroid.

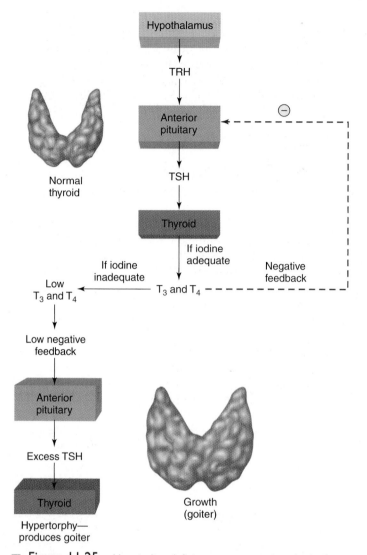

■ **Figure 11.25** How iodine deficiency causes a goiter. Lack of adequate iodine in the diet interferes with the negative feedback control of TSH secretion, resulting in the formation of an endemic goiter.

the thyroid gland (fig. 11.24). In the absence of sufficient dietary iodine, the thyroid cannot produce adequate amounts of T_4 and T_3. The resulting lack of negative feedback inhibition causes abnormally high levels of TSH secretion, which in turn stimulates the abnormal growth of the thyroid. These events are summarized in figure 11.25.

People who have inadequate secretion of thyroid hormones are said to be **hypothyroid.** As might be predicted from the effects of thyroxine, people who are hypothyroid have an abnormally low basal metabolic rate and experience weight gain and lethargy. A thyroxine deficiency also decreases the ability to adapt to cold stress. Hypothyroidism in adults causes **myxedema**—accumulation of mucoproteins and fluid in subcutaneous connective tissues. Symptoms of this disease include swelling of the hands, face, feet, and tissues around the eyes.

Clinical Investigation Clue

Remember that Rosemary's puffiness was determined not to be myxedema, and that her blood T_4 and T_3 levels were normal. *What disorder is ruled out by these observations?*

Hypothyroidism can result from a thyroid gland defect or secondarily from insufficient thyrotropin-releasing hormone (TRH) secretion from the hypothalamus, or insufficient TSH secretion from the anterior pituitary, or insufficient iodine in the diet. In the latter case, excessive TSH secretion stimulates abnormal thyroid growth and the development of an endemic goiter, as described previously. The hypothyroidism and goiter caused by iodine deficiency can be reversed by iodine supplements.

A goiter can also be produced by another mechanism. In **Graves' disease,** autoantibodies (chapter 15) exert TSH-like effects on the thyroid. Since the production of these antibodies is not inhibited by negative feedback, the high secretion of thyroxine that results cannot turn off the excessive stimulation of the thyroid. As a result, the person is **hyperthyroid** (has excessive thyroxine secretion) and develops a goiter. This condition is called *toxic goiter,* or *thyrotoxicosis.* This condition is often accompanied by *exophthalmos,* or bulging eyes, due to edema in the orbits (fig. 11.26). The hyperthyroid state produces a high BMR accompanied by weight loss, nervousness, irritability, and an intolerance to heat. There is also a significant increase in cardiac output and blood pressure (chapter 14). The symptoms of hypothyroidism and hyperthyroidism are compared in table 11.8.

Because of its stimulation of protein synthesis, children need thyroxine for body growth and, most importantly, for the proper development of the central nervous system. The need for thyroxine is particularly great when the brain is undergoing its

Table 11.8   Comparison of Hypothyroidism and Hyperthyroidism

Feature	Hypothyroid	Hyperthyroid
Growth and development	Impaired growth	Accelerated growth
Activity and sleep	Lethargy; increased sleep	Increased activity; decreased sleep
Temperature tolerance	Intolerance to cold	Intolerance to heat
Skin characteristics	Coarse, dry skin	Normal skin
Perspiration	Absent	Excessive
Pulse	Slow	Rapid
Gastrointestinal symptoms	Constipation; decreased appetite; increased weight	Frequent bowel movements; increased appetite; decreased weight
Reflexes	Slow	Rapid
Psychological aspects	Depression and apathy	Nervous, "emotional" state
Plasma T_4 levels	Decreased	Increased

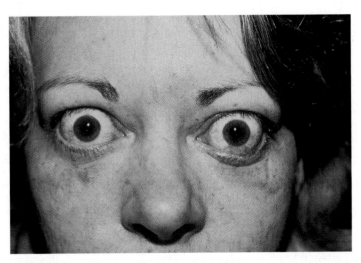

■ **Figure 11.26**   **A symptom of hyperthyroidism.** Hyperthyroidism is characterized by an increased metabolic rate, weight loss, muscular weakness, and nervousness. The eyes may also protrude (exophthalmos) due to edema in the orbits.

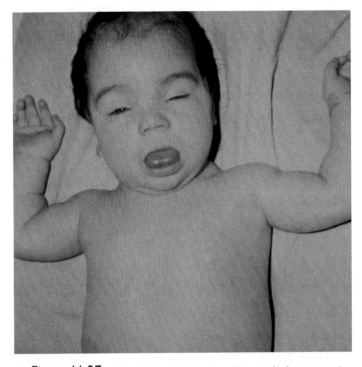

■ **Figure 11.27**   **Cretinism.** Cretinism is a disease of infancy caused by an underactive thyroid gland.

greatest rate of development—from the end of the first trimester of prenatal life to 6 months after birth. Hypothyroidism during this time may result in **cretinism** (fig. 11.27). Unlike people with *dwarfism*, who have inadequate secretion of growth hormone from the anterior pituitary, people with cretinism suffer severe mental retardation. Treatment with thyroxine soon after birth, particularly before 1 month of age, has been found to completely or almost completely restore development of intelligence as measured by IQ tests administered 5 years later.

Parathyroid Glands

The small, flattened **parathyroid glands** are embedded in the posterior surfaces of the lateral lobes of the thyroid gland, as shown in figure 11.28. There are usually four parathyroid glands: a *superior* and an *inferior pair,* although the precise number can vary. Each parathyroid gland is a small yellowish-brown body 3 to 8 mm (0.1 to 0.3 in.) long, 2 to 5 mm (0.07 to 0.2 in.) wide, and about 1.5 mm (0.05 in.) deep.

Parathyroid hormone (PTH) is the only hormone secreted by the parathyroid glands. PTH, however, is the single most important hormone in the control of the calcium levels of the blood. It promotes a rise in blood calcium levels by acting on the bones, kidneys, and intestine (fig. 11.29). Regulation of calcium balance is described in more detail in chapter 19.

Test Yourself Before You Continue

1. Describe the structure of the thyroid gland and list the effects of thyroid hormones.
2. Describe how thyroid hormones are produced and how their secretion is regulated.
3. Explain the consequences of an inadequate dietary intake of iodine.

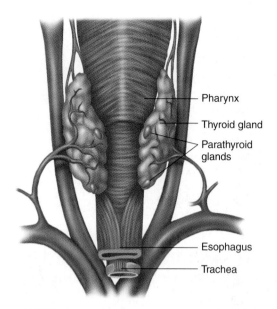

■ **Figure 11.28** A posterior view of the parathyroid glands. The parathyroids are embedded in the tissue of the thyroid gland.

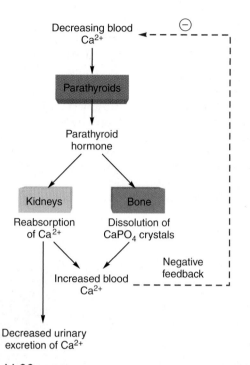

■ **Figure 11.29** The actions of parathyroid hormone and the control of its secretion. An increased level of parathyroid hormone causes the bones to release calcium and the kidneys to conserve calcium that would otherwise be lost through the urine. A rise in blood Ca^+ can then exert negative feedback inhibition on parathyroid hormone secretion.

Pancreas and Other Endocrine Glands

The pancreatic islets secrete two hormones, insulin and glucagon. Insulin promotes the lowering of blood glucose and the storage of energy in the form of glycogen and fat. Glucagon has antagonistic effects that act to raise the blood glucose concentration. Additionally, many other organs secrete hormones that help to regulate digestion, metabolism, growth, immune function, and reproduction.

The **pancreas** is both an endocrine and an exocrine gland. The gross structure of this gland and its exocrine functions in digestion are described in chapter 18. The endocrine portion of the pancreas consists of scattered clusters of cells called the **pancreatic islets** or **islets of Langerhans.** These endocrine structures are most common in the body and tail of the pancreas (fig. 11.30).

Pancreatic Islets (Islets of Langerhans)

On a microscopic level, the most conspicuous cells in the islets are the *alpha* and *beta cells* (fig. 11.30). The alpha cells secrete the hormone **glucagon,** and the beta cells secrete **insulin.**

Alpha cells secrete glucagon in response to a fall in blood glucose concentrations. Glucagon stimulates the liver to hydrolyze glycogen to glucose (*glycogenolysis*), which causes the blood glucose level to rise. This effect represents the completion of a negative feedback loop. Glucagon also stimulates the hydrolysis of stored fat (*lipolysis*) and the consequent release of free fatty acids into the blood. This effect helps to provide energy substrates for the body during fasting, when blood glucose levels decrease. Glucagon, together with other hormones, also stimulates the conversion of fatty acids to ketone bodies, which can be secreted by the liver into the blood and used by other organs as an energy source. Glucagon is thus a hormone that helps to maintain homeostasis during times of fasting, when the body's energy reserves must be utilized.

Beta cells secrete insulin in response to a rise in blood glucose concentrations (fig 11.31). Insulin promotes the entry of glucose into tissue cells, and the conversion of this glucose into energy storage molecules of glycogen and fat. Insulin also aids the entry of amino acids into cells and the production of cellular protein. Thus, insulin promotes the deposition of energy storage molecules (primarily glycogen and fat) following meals, when the blood glucose concentration rises. This action is antagonistic to that of glucagon, and the secretion of glucagon is normally decreased when insulin secretion increases. During times of fasting, conversely, the secretion of insulin is decreased while the secretion of glucagon is increased. The secretion and actions of insulin and glucagon, and the association of these hormones with diabetes mellitus, is more fully explained in chapter 19.

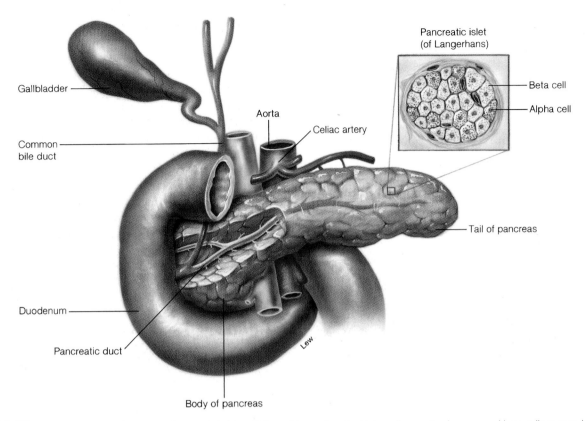

Pancreatic islet
(of Langerhans)

Beta cell

Alpha cell

Gallbladder

Aorta

Celiac artery

Common
bile duct

Tail of pancreas

Duodenum

Pancreatic duct

Lew

Body of pancreas

■ **Figure 11.30** **The pancreas and associated pancreatic islets (islets of Langerhans).** Alpha cells secrete glucagon and beta cells secrete insulin. The pancreas is also exocrine, producing pancreatic juice for transport via the pancreatic duct to the duodenum of the small intestine.

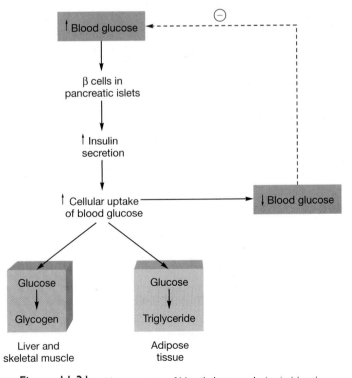

⊖

↑ Blood glucose

β cells in
pancreatic islets

↑ Insulin
secretion

↑ Cellular uptake
of blood glucose

↓ Blood glucose

Glucose
↓
Glycogen

Glucose
↓
Triglyceride

Liver and
skeletal muscle

Adipose
tissue

■ **Figure 11.31** **Homeostasis of blood glucose.** A rise in blood glucose concentration stimulates insulin secretion. Insulin promotes a fall in blood glucose by stimulating the cellular uptake of glucose and the conversion of glucose to glycogen and fat.

Diabetes mellitus is characterized by fasting hyperglycemia and the presence of glucose in the urine. There are two forms of this disease. *Type 1,* or insulin-dependent diabetes mellitus, is caused by destruction of the beta cells and the resulting lack of insulin secretion. *Type 2,* or non-insulin-dependent diabetes mellitus (the more common form), is caused by decreased tissue sensitivity to the effects of insulin, so that larger than normal amounts of insulin are required to produce a normal effect. Both types of diabetes mellitus are also associated with abnormally high levels of glucagon secretion. The causes and symptoms of diabetes mellitus are described in more detail in chapter 19.

Clinical Investigation Clues

Remember that Rosemary has hyperglycemia and hypertension, and that her oral glucose tolerance test (which tests a person's insulin secretion and action) was normal.

Was Rosemary's hyperglycemia and hypertension the result of diabetes mellitus?

If not, what did produce these symptoms?

Pineal Gland

The small, cone-shaped **pineal gland** is located in the roof of the third ventricle of the diencephalon (chapter 8), where it is encapsulated by the meninges covering the brain. The pineal gland of a child weighs about 0.2 g and is 5 to 8 mm (0.2 to 0.3 in.) long and 9 mm wide. The gland begins to regress in size at about age 7 and in the adult appears as a thickened strand of fibrous tissue. Although the pineal gland lacks direct nervous connections to the rest of the brain, it is highly innervated by the sympathetic nervous system from the superior cervical ganglion.

The principal hormone of the pineal gland is **melatonin.** Production and secretion of this hormone is stimulated by activity of the **suprachiasmatic nucleus (SCN)** in the hypothalamus of the brain via activation of sympathetic neurons to the pineal gland (fig. 11.32). The SCN is the primary center for **circadian rhythms** in the body: rhythms of physiological activity that follow a 24-hour pattern. The circadian activity of the SCN is automatic, but environmental light/dark changes are required to entrain (synchronize) this activity to a day/night cycle. Activity of the SCN, and thus secretion of melatonin, begins to increase with darkness and peaks by the middle of the night. During the day, neural pathways

from the retina of the eyes to the hypothalamus (fig. 11.32) act to depress the activity of the SCN, reducing sympathetic stimulation of the pineal and thus decreasing melatonin secretion.

The pineal gland has been implicated in a variety of physiological processes. One of the most widely studied is the ability of melatonin to inhibit the pituitary-gonad axis (inhibiting GnRH secretion or the response of the anterior pituitary to GnRH, depending on the species of animal). Indeed, a decrease in melatonin secretion in many species is required for the maturation of the gonads during the reproductive season of seasonal breeders. Although there is evidence to support an antigonadotropic effect in humans, this possibility has not yet been proven. For example, excessive melatonin secretion in humans is associated with a delay in the onset of puberty. Research findings indicate that melatonin secretion is highest in children between the ages of 1 and 5 and decreases thereafter, reaching its lowest levels at the end of puberty, when concentrations are 75% lower than during early childhood. This suggests a role for melatonin in the onset of human puberty. However, because of much conflicting data, the importance of melatonin in human reproduction is still highly controversial.

The pattern of melatonin secretion is altered when a person works night shifts or flies across different time zones. There

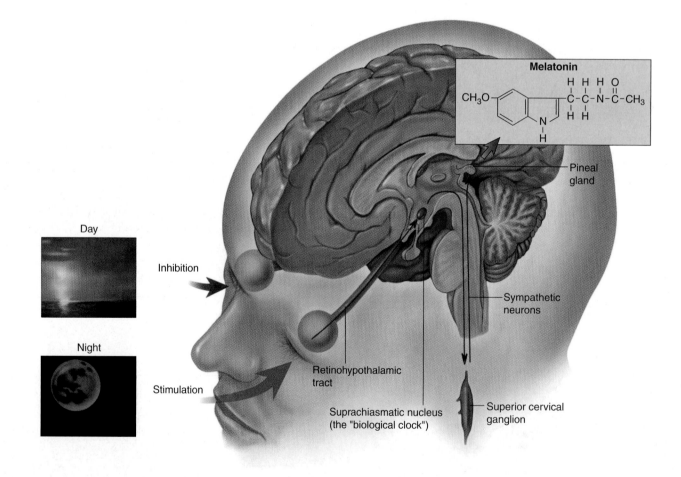

■ **Figure 11.32** **The secretion of melatonin.** The secretion of melatonin by the pineal gland is stimulated by sympathetic axons originating in the superior cervical ganglion. Activity of these neurons is regulated by the cyclic activity of the suprachiasmatic nucleus of the hypothalamus, which sets a circadian rhythm. This rhythm is entrained to light/dark cycles by neurons in the retina.

is evidence that exogenous melatonin (taken as a pill) may be beneficial in the treatment of jet lag, but the optimum dosage is not currently known. Phototherapy using bright fluorescent lamps, which act like sunlight to inhibit melatonin secretion, has been used effectively in the treatment of *seasonal affective disorder (SAD),* or "winter depression."

Melatonin pills decrease the time required to fall asleep and increase the duration of rapid eye movement (REM) sleep; for these reasons, they may be useful in the treatment of insomnia. This is particularly significant for elderly people with insomnia, who have the lowest nighttime levels of endogenous melatonin secretion. Melatonin can also act, much like vitamin E, as a scavenger of hydroxyl and other free radicals that can cause oxidative damage to cells. This antioxidant effect of melatonin, however, only occurs at pharmacological, rather than at normal physiological, doses. The purported beneficial effects of exogenous melatonin (other than for insomnia and jet lag) are not yet proven, and the consensus of current medical opinion is against the uncontrolled use of melatonin pills.

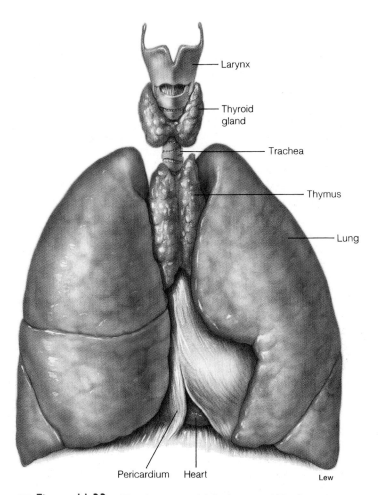

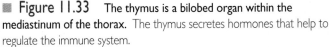

■ **Figure 11.33** **The thymus is a bilobed organ within the mediastinum of the thorax.** The thymus secretes hormones that help to regulate the immune system.

Thymus

The **thymus** is a bilobed organ positioned in front of the aorta and behind the manubrium of the sternum (fig. 11.33). Although the size of the thymus varies considerably from person to person, it is relatively large in newborns and children, and sharply regresses in size after puberty. Besides decreasing in size, the thymus of adults becomes infiltrated with strands of fibrous and fatty connective tissue.

The thymus is the site of production of **T cells** (*thymus-dependent cells),* which are the lymphocytes involved in cell-mediated immunity (see chapter 15). In addition to providing T cells, the thymus secretes a number of hormones that are believed to stimulate T cells after they leave the thymus.

Gastrointestinal Tract

The stomach and small intestine secrete a number of hormones that act on the gastrointestinal tract itself and on the pancreas and gallbladder (chapter 18; the hormone actions are summarized in table 18.6). These hormones, acting in concert with regulation by the autonomic nervous system, coordinate the activities of different regions of the digestive tract and the secretions of pancreatic juice and bile.

Gonads and Placenta

The gonads (**testes** and **ovaries**) secrete sex steroids. These include male sex hormones, or *androgens,* and female sex hormones—*estrogens* and *progesterone.* The androgens and estrogens are families of hormones. The principal androgen secreted by the testes is *testosterone,* and the principal estrogen secreted by the ovaries is *estradiol-17β*. The principal estrogen during pregnancy, however, is a weaker estrogen called *estriol,* secreted by the placenta. After menopause, the principal estrogen is *estrone,* produced primarily by fat cells.

The testes consist of two compartments: *seminiferous tubules,* which produce sperm cells, and *interstitial tissue* between the convolutions of the tubules. Within the interstitial tissue are *Leydig cells,* which secrete testosterone. Testosterone is needed for the development and maintenance of the male genitalia (penis and scrotum) and the male accessory sex organs (prostate, seminal vesicles, epididymides, and vas deferens), as well as for the development of male secondary sex characteristics.

During the first half of the menstrual cycle, *estradiol-17β* is secreted by small structures within the ovary called *ovarian follicles.* These follicles contain the egg cell, or *ovum,* and *granulosa cells* that secrete estrogen. By about midcycle, one of these follicles grows very large and, in the process of ovulation, extrudes its ovum from the ovary. The empty follicle, under the influence of luteinizing hormone (LH) from the anterior pituitary, then becomes a new endocrine structure called a *corpus luteum.* The corpus luteum secretes progesterone as well as estradiol-17β.

The **placenta**—the organ responsible for nutrient and waste exchange between the fetus and mother—is also an endocrine gland that secretes large amounts of estrogens and progesterone. In addition, it secretes a number of polypeptide and protein hormones that are similar to some hormones secreted by the anterior pituitary. These hormones include *human chorionic gonadotropin (hCG),* which is similar to LH, and *somatomammotropin,* which is similar in action to both growth hormone and prolactin. The physiology of the placenta and other aspects of reproductive endocrinology are considered in chapter 20.

Test Yourself Before You Continue

1. Describe the structure of the endocrine pancreas. Which cells secrete insulin and which secrete glucagon?
2. Describe how insulin and glucagon secretion are affected by eating and by fasting and explain the actions of these two hormones.
3. Describe the location of the pineal gland and discuss the possible functions of melatonin.
4. Describe the location and function of the thymus.
5. Explain how the gonadal and placental hormones are categorized and list the hormones secreted by each gland.

Autocrine and Paracrine Regulation

Many regulatory molecules produced throughout the body act within the organs that produce them. These molecules may regulate different cells within one tissue, or they may be produced within one tissue and regulate a different tissue within the same organ.

Thus far in this text, two types of regulatory molecules have been considered—neurotransmitters in chapter 7 and hormones in the present chapter. These two classes of regulatory molecules cannot be defined simply by differences in chemical structure, since on this basis the same molecule (such as norepinephrine) could be included in both categories; rather, they must be defined by function. Neurotransmitters are released by axons, travel across a narrow synaptic cleft, and affect a postsynaptic cell. Hormones are secreted into the blood by an endocrine gland and, through transport in the blood, influence the activities of one or more target organs.

There are yet other classes of regulatory molecules. These molecules are distinguished by the fact that they are produced in many different organs and are active within the organ in which they are produced. Molecules of this type are called **autocrine regulators** if they are produced and act within the same tissue of an organ. They are called **paracrine regulators** if they are produced within one tissue and regulate a different tissue of the same organ (table 11.9). In the following discussion, for the sake of simplicity and because the same chemical can function as an autocrine or a paracrine regulator, the term *autocrine* will be used in a generic sense to refer to both types of local regulation.

Examples of Autocrine Regulation

Many autocrine regulatory molecules are also known as **cytokines,** particularly if they regulate different cells of the immune system, and as **growth factors** if they promote growth and cell division in any organ. This distinction is somewhat blurred, however, because some cytokines may also function as growth factors. Cytokines produced by lymphocytes (the type of white blood cell involved in specific immunity—see chapter 15) are also known as *lymphokines,* and the specific molecules involved are called *interleukins.* The terminology can be confusing because new regulatory molecules, and new functions for previously named regulatory molecules, are being discovered at a rapid pace. As described in chapter 15, cytokines secreted by macrophages (phagocytic cells found in connective tissues) and lymphocytes stimulate proliferation of specific cells involved in the immune response.

Neurotrophins, including *nerve growth factor,* guide regenerating peripheral neurons that have been injured (chapter 7). Nitric oxide, which can function as a neurotransmitter in memory processes (chapters 7 and 8) and in other functions, is also produced by the endothelium of blood vessels. In this context, it is a paracrine regulator because it diffuses to the smooth muscle layer of the blood vessel and promotes relaxation, leading to dilation of the blood vessel. In this action, nitric oxide functions

Table 11.9 Examples of Autocrine and Paracrine Regulators

Autocrine or Paracrine Regulator	Major Sites of Production	Major Actions
Insulin-like growth factors (somatomedins)	Many organs, particularly the liver and cartilages	Growth and cell division
Nitric oxide	Endothelium of blood vessels; neurons; macrophages	Dilation of blood vessels; neural messenger; antibacterial agent
Endothelins	Endothelium of blood vessels; other organs	Constriction of blood vessels; other effects
Platelet-derived growth factor	Platelets; macrophages; vascular smooth muscle cells	Cell division within blood vessels
Epidermal growth factors	Epidermal tissues	Cell division in wound healing
Neurotrophins	Schwann cells; neurons	Regeneration of peripheral nerves
Bradykinin	Endothelium of blood vessels	Dilation of blood vessels
Interleukins (cytokines)	Macrophages; lymphocytes	Regulation of immune system
Prostaglandins	Many tissues	Wide variety (see text)
TNFα (tumor necrosis factor alpha)	Macrophages; adipocytes	Wide variety

as the paracrine regulator previously known as *endothelium-derived relaxation factor.* Neural and paracrine regulation interact in this case, since autonomic axons that release acetylcholine in blood vessels cause dilation by stimulating the synthesis of nitric oxide in those vessels (see chapter 20, fig. 20.23).

The endothelium of blood vessels also produces other paracrine regulators. These include the *endothelins* (specifically *endothelin-1* in humans), which directly promote vasoconstriction, and *bradykinin,* which promotes vasodilation. These regulatory molecules are thus very important in the control of blood flow and blood pressure. They are also involved in the development of atherosclerosis, the leading cause of heart disease and stroke (see chapter 13). In addition, endothelin-1 is produced by the epithelium of the airways and may be important in the embryological development and function of the respiratory system.

All autocrine regulators control gene expression in their target cells to some degree. This is very clearly the case with the various growth factors. These include *platelet-derived growth factor, epidermal growth factor,* and the *insulin-like growth factors* that stimulate cell division and proliferation of their target cells. Regulators in the last group interact with the endocrine system in a number of ways, as will be described in chapter 19.

Prostaglandins

The most diverse group of autocrine regulators are the **prostaglandins.** These twenty-carbon-long fatty acids contain a five-membered carbon ring. Prostaglandins are members of a family called the **eicosanoids** (from the Greek *eicosa* = twenty),

which are molecules derived from the precursor *arachidonic acid.* Upon stimulation by hormones or other agents, arachidonic acid is released from phospholipids in the plasma membrane and may then enter one of two possible metabolic pathways. In one case, the arachidonic acid is converted by the enzyme *cyclooxygenase* into a prostaglandin, which can then be changed by other enzymes into other prostaglandins. In the other case, arachidonic acid is converted by the enzyme *lipoxygenase* into **leukotrienes,** which are eicosanoids that are closely related to the prostaglandins (fig. 11.34).

Prostaglandins are produced in almost every organ and have been implicated in a wide variety of regulatory functions. The study of prostaglandins can be confusing because of the diversity of their actions, and because different prostaglandins may exert antagonistic effects in some tissues. For example, the smooth muscle of blood vessels relaxes (producing vasodilation) in response to prostaglandin E_2 (abbreviated PGE_2) and $PGF_{2\alpha}$; these effects promote reddening and heat during an inflammation reaction. In the smooth muscles of the bronchioles (airways of the lungs), however, $PGF_{2\alpha}$ stimulates contraction, contributing to the symptoms of asthma.

The antagonistic effects of prostaglandins on blood clotting make good physiological sense. Blood platelets, which are required for blood clotting, produce *thromboxane A_2.* This prostaglandin promotes clotting by stimulating platelet aggregation and vasoconstriction. The endothelial cells of blood vessels, by contrast, produce a different prostaglandin, known as PGI_2 or *prostacyclin,* whose effects are the opposite—it inhibits platelet aggregation and causes vasodilation. These antagonistic effects

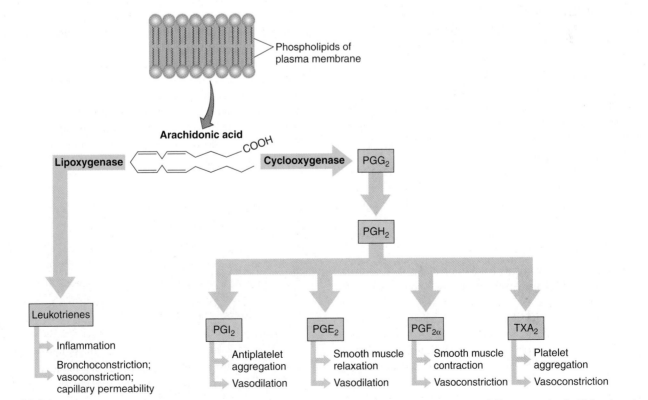

Figure 11.34   **The formation of leukotrienes and prostaglandins.** The actions of these autocrine regulators (PG = prostaglandin; TX = thromboxane) are summarized.

ensure that, while clotting is promoted, the clots will not normally form on the walls of intact blood vessels.

Examples of Prostaglandin Actions

Some of the regulatory functions proposed for prostaglandins in different systems of the body are as listed here:

1. **Immune system.** Prostaglandins promote many aspects of the inflammatory process, including the development of pain and fever. Drugs that inhibit prostaglandin synthesis help to alleviate these symptoms.

2. **Reproductive system.** Prostaglandins may play a role in ovulation and corpus luteum function in the ovaries and in contraction of the uterus. Excessive production of PGE_2 and PGI_2 may be involved in premature labor, endometriosis, dysmenorrhea (painful menstrual cramps), and other gynecological disorders.

3. **Digestive system.** The stomach and intestines produce prostaglandins, which are believed to inhibit gastric secretions and influence intestinal motility and fluid absorption. Since prostaglandins inhibit gastric secretion, drugs that suppress prostaglandin production may make a patient more susceptible to peptic ulcers.

4. **Respiratory system.** Some prostaglandins cause constriction whereas others cause dilation of blood vessels in the lungs and of bronchiolar smooth muscle. The leukotrienes are potent bronchoconstrictors, and these compounds, together with $PGF_{2\alpha}$, may cause respiratory distress and contribute to bronchoconstriction in asthma.

5. **Circulatory system.** Some prostaglandins are vasoconstrictors and others are vasodilators. Thromboxane A_2, a vasoconstrictor, and prostacyclin, a vasodilator, play a role in blood clotting, as previously described. In a fetus, PGE_2 is believed to promote dilation of the *ductus arteriosus*—a short vessel that connects the pulmonary artery with the aorta. After birth, the ductus arteriosus normally closes as a result of a rise in blood oxygen when the baby breathes. If the ductus remains patent (open), however, it can be closed by the administration of drugs that inhibit prostaglandin synthesis.

6. **Urinary system.** Prostaglandins are produced in the renal medulla and cause vasodilation, resulting in increased renal blood flow and increased excretion of water and electrolytes in the urine.

Inhibitors of Prostaglandin Synthesis

Aspirin is the most widely used member of a class of drugs known as **nonsteroidal anti-inflammatory drugs (NSAIDs).** Other members of this class are indomethacin and ibuprofen. These drugs produce their effects because they specifically inhibit the cyclooxygenase enzyme that is needed for prostaglandin synthesis. Through this action, the drugs inhibit inflammation but produce some unwanted side effects, including gastric bleeding, possible kidney problems, and prolonged clotting time.

It is now known that there are two isoenzyme forms (chapter 4) of cyclooxygenase. The type I isoform (*COX1*) is produced constitutively (that is, in a constant fashion) by cells of the stomach and kidneys and by blood platelets, which are cell fragments involved in blood clotting (see chapter 13). The type II isoform of the enzyme (*COX2*) is induced in a number of cells in response to cytokines involved in inflammation, and the prostaglandins produced by this isoenzyme promote the inflammatory condition.

The two isoforms of cyclooxygenase are thus quite distinct. The COX1 isoform is produced continuously by a gene on chromosome 9 and is required for the normal, physiological functioning of different organs, for platelet aggregation in blood clotting, and for the health of the gastric mucosa. The production of the COX2 isoform (by a gene on chromosome 1), is kept at a low level until it is stimulated during an inflammation. Interestingly, the ability of the glucocorticoids (such as hydrocortisone) to inhibit inflammation has been shown to be due to their ability to inhibit the COX2 isoenzyme.

When aspirin and indomethacin inhibit the COX1 isoenzyme, they reduce the synthesis of prostacyclin (PGI_2) and PGE_2 in the gastric mucosa. This is believed to result in the stomach irritation caused by these NSAIDs. Indeed, inhibition of the COX1 isoenzyme may cause serious gastrointestinal and renal toxicity in long-term use. This has spurred research into the development of next-generation NSAIDs that more selectively inhibit the COX2 isoenzyme. These newer COX2-selective drugs, including *celecoxib* and *rofecoxib,* (for example, *Celebrex* and *Vioxx*) thus inhibit inflammation while producing fewer negative side effects in the gastric mucosa.

There is, however, one important benefit derived from the inhibition of the type I isoenzyme by aspirin. The type I isoenzyme is the form of cyclooxygenase present in blood platelets, where it is needed for the production of thromboxane A_2. Since this prostaglandin is needed for platelet aggregation, inhibition of its synthesis by aspirin reduces the ability of the blood to clot. While this can have negative consequences in some circumstances, low doses of aspirin have been shown to significantly reduce the risk of heart attacks and strokes by reducing platelet function. It should be noted that this beneficial effect is produced by lower doses of aspirin than are commonly taken to reduce inflammation.

Test Yourself Before You Continue

1. Explain the nature of autocrine regulation. How does it differ from regulation by hormones and neurotransmitters?

2. List some of the paracrine regulators produced by blood vessels and describe their actions. Also, identify specific growth factors and describe their actions.

3. Describe the chemical nature of prostaglandins. List some of the different forms of prostaglandins and describe their actions.

4. Explain the significance of the isoenzymatic forms of cyclooxygenase in the action of nonsteroidal anti-inflammatory drugs.

HPer Links of the Endocrine System with Other Body Systems

Integumentary System

- The skin helps to protect the body from pathogens . . . .(p. 446)
- The skin produces vitamin D₃, which acts as a prehormone . . . .(p. 625)

Skeletal System

- Bones support and protect the pituitary gland . . . .(p. 299)
- Bones store calcium, which is needed for the action of many hormones . . . .(p. 623)
- Anabolic hormones, including growth hormone, stimulate bone development . . . .(p. 622)
- Parathyroid hormone and calcitonin regulate calcium deposition and resorption in bones . . . .(p. 624)
- Sex hormones help to maintain bone mass in adults . . . .(p. 625)

Muscular System

- Anabolic hormones promote muscle growth . . . .(p. 609)
- Insulin stimulates the uptake of blood glucose into muscles . . . .(p. 613)
- The catabolism of muscle glycogen and proteins is promoted by several hormones . . . .(p. 609)

Nervous System

- The hypothalamus secretes hormones that control the anterior pituitary . . . .(p. 301)
- The hypothalamus produces the hormones released by the posterior pituitary . . . .(p. 301)
- Sympathetic nerves stimulate the secretions of the adrenal medulla . . . .(p. 307)
- Parasympathetic nerves stimulate the secretions of the pancreatic islets . . . .(p. 613)
- Neurons stimulate the secretion of melatonin from the pineal gland, which in turn regulates parts of the brain . .(p. 314)
- Sex hormones from the gonads regulate the hypothalamus . . . .(p. 304)

Circulatory System

- The blood transports oxygen, nutrients, and regulatory molecules to endocrine glands and removes wastes . . . . . .(p. 366)
- The blood transports hormones from endocrine glands to target cells . . .(p. 286)
- Epinephrine and norepinephrine from the adrenal medulla stimulate the heart . . . .(p. 408)

Immune System

- The immune system protects against infections that could damage endocrine glands . . . .(p. 446)
- Autoimmune destruction of the pancreatic islets causes type I diabetes mellitus . . . .(p. 616)
- Hormones from the thymus help to regulate lymphocytes . . . .(p. 315)
- Adrenal corticosteroids have a suppressive effect on the immune system . . . . .(p. 462)

Respiratory System

- The lungs provide oxygen for transport by the blood and eliminate carbon dioxide . . . .(p. 480)
- Thyroxine and epinephrine stimulate the rate of cell respiration in the body . . . .(p. 609)
- Epinephrine promotes bronchodilation, reducing airway resistance . . . .(p. 492)

Urinary System

- The kidneys eliminate metabolic wastes produced by body organs, including endocrine glands . . . .(p. 524)
- The kidneys release renin, which participates in the renin-angiotensin-aldosterone system . . . .(p. 545)
- The kidneys secrete erythropoietin, which serves as a hormone that regulates red blood cell production . . . .(p. 371)
- Antidiuretic hormone, aldosterone, and atrial natriuretic hormone regulate kidney functions . . . .(p. 416)

Digestive System

- The GI tract provides nutrients to the body organs, including those of the endocrine system . . . .(p. 561)
- Hormones of the stomach and small intestine help to coordinate the activities of different regions of the GI tract . .(p. 563)
- Hormones from adipose tissue contribute to the sensation of hunger . . . .(p. 607)

Reproductive System

- Gonadal hormones help to regulate the secretions of the anterior pituitary . . . .(p. 640)
- Pituitary hormones regulate the ovarian cycle . . . .(p. 656)
- Testicular androgens regulate the male accessory sex organs . . . .(p. 644)
- Ovarian hormones regulate the uterus during the menstrual cycle . . . . . .(p. 660)
- Oxytocin plays an essential role in labor and delivery . . . .(p. 675)
- The placenta secretes several hormones that influence the course of pregnancy . . . .(p. 673)
- Several hormones are needed for lactation in a nursing mother . . . .(p. 678)

Summary

Endocrine Glands and Hormones   286

I. Hormones are chemicals that are secreted into the blood by endocrine glands.

    A. The chemical classes of hormones include amines, polypeptides, glycoproteins, and steroids.

    B. Nonpolar hormones, which can pass through the plasma membrane of their target cells, are called lipophilic hormones.

II. Precursors of active hormones may be classified as either prohormones or prehormones.

    A. Prohormones are relatively inactive precursor molecules made in the endocrine cells.

    B. Prehormones are the normal secretions of an endocrine gland that must be converted to other derivatives by target cells in order to be active.

III. Hormones can interact in permissive, synergistic, or antagonistic ways.

IV. The effects of a hormone in the body depend on its concentration.

    A. Abnormally high amounts of a hormone can result in atypical effects.

    B. Target tissues can become desensitized by high hormone concentrations.

Mechanisms of Hormone Action   292

I. The lipophilic hormones (steroids and thyroid hormones) bind to nuclear receptor proteins, which function as ligand-dependent transcription factors.

    A. Some steroid hormones bind to cytoplasmic receptors, which then move into the nucleus. Other steroids and thyroxine bind to receptors already in the nucleus.

    B. Each receptor binds to both the hormone and to a region of DNA called a hormone-response element.

    C. Two units of the nuclear receptor are needed to bind to the hormone-response element to

activate a gene; as a result, the gene is transcribed (makes mRNA).

II. The polar hormones bind to receptors located on the outer surface of the plasma membrane. This activates enzymes that enlist second-messenger molecules.

    A. Many hormones activate adenylate cyclase when they bind to their receptors. This enzyme produces cyclic AMP (cAMP), which activates protein kinase enzymes within the cell cytoplasm.

    B. Other hormones may activate phospholipase C when they bind to their receptors. This leads to the release of inositol triphosphate (IP_3), which stimulates the endoplasmic reticulum to release Ca^{2+} into the cytoplasm, activating calmodulin.

    C. The membrane receptors for insulin and various growth factors are tyrosine kinase enzymes that are activated by binding to the hormone. Once activated, the receptor kinase phosphorylates signaling molecules in the cytoplasm that can have many effects.

Pituitary Gland   299

I. The pituitary gland secretes eight hormones.

    A. The anterior pituitary secretes growth hormone, thyroid-stimulating hormone, adrenocorticotropic hormone, follicle-stimulating hormone, luteinizing hormone, and prolactin.

    B. The posterior pituitary releases antidiuretic hormone (also called vasopressin) and oxytocin, both of which are produced in the hypothalamus and transported to the posterior pituitary by the hypothalamo-hypophyseal tract.

II. The release of posterior pituitary hormones is controlled by neuroendocrine reflexes.

III. Secretions of the anterior pituitary are controlled by hypothalamic hormones that stimulate or inhibit these secretions.

    A. Hypothalamic hormones include TRH, CRH, GnRH, PIH, somatostatin, and a growth hormone-releasing hormone (GHRH).

    B. These hormones are carried to the anterior pituitary by the hypothalamo-hypophyseal portal system.

IV. Secretions of the anterior pituitary are also regulated by the feedback (usually negative feedback) exerted by target gland hormones.

V. Higher brain centers, acting through the hypothalamus, can influence pituitary secretion.

Adrenal Glands   305

I. The adrenal cortex secretes mineralocorticoids (mainly aldosterone), glucocorticoids (mainly cortisol), and sex steroids (primarily weak androgens).

    A. The glucocorticoids help to regulate energy balance. They also can inhibit inflammation and suppress immune function.

    B. The pituitary-adrenal axis is stimulated by stress as part of the general adaptation syndrome.

II. The adrenal medulla secretes epinephrine and lesser amounts of norepinephrine. These hormones complement the action of the sympathetic nervous system.

Thyroid and Parathyroid Glands   308

I. The thyroid follicles secrete tetraiodothyronine (T_4, or thyroxine) and lesser amounts of triiodothyronine (T_3).

    A. These hormones are formed within the colloid of the thyroid follicles.

B. The parafollicular cells of the thyroid secrete the hormone calcitonin, which may act to lower blood calcium levels.

II. The parathyroids are small structures embedded in the thyroid gland. They secrete parathyroid hormone (PTH), which promotes a rise in blood calcium levels.

Pancreas and Other Endocrine Glands 312

I. Beta cells in the islets secrete insulin, and alpha cells secrete glucagon.

A. Insulin lowers blood glucose and stimulates the production of glycogen, fat, and protein.

B. Glucagon raises blood glucose by stimulating the breakdown of liver glycogen. It also promotes lipolysis and the formation of ketone bodies.

C. The secretion of insulin is stimulated by a rise in blood glucose following meals. The secretion of glucagon is stimulated by a fall in blood glucose during periods of fasting.

II. The pineal gland, located on the roof of the third ventricle of the brain, secretes melatonin.

A. Melatonin secretion is regulated by the suprachiasmatic nucleus of the hypothalamus, which is the major center for the control of circadian rhythms.

B. Melatonin secretion is highest at night, and this hormone has a sleep-promoting effect. In many species, it also has an antigonadotropic effect and may play a role in timing the onset of puberty in humans, although this is as yet unproven.

III. The thymus is the site of T cell lymphocyte production and secretes a number of hormones that may help to regulate the immune system.

IV. The gastrointestinal tract secretes a number of hormones that help to regulate digestive functions.

V. The gonads secrete sex steroid hormones.

A. Leydig cells in the interstitial tissue of the testes secrete testosterone and other androgens.

B. Granulosa cells of the ovarian follicles secrete estrogen.

C. The corpus luteum of the ovaries secretes progesterone, as well as estrogen.

VI. The placenta secretes estrogen, progesterone, and a variety of polypeptide and protein hormones that have actions similar to some anterior pituitary hormones.

Autocrine and Paracrine Regulation 316

I. Autocrine regulators are produced and act within the same tissue of an organ, whereas paracrine regulators are produced within one tissue and regulate a different tissue of the same organ. Both types are local regulators—they do not travel in the blood.

II. Prostaglandins are special twenty-carbon-long fatty acids produced by many different organs. They usually have regulatory functions within the organ in which they are produced.

Review Activities

Test Your Knowledge of Terms and Facts

1. Which of these statements about hypothalamic-releasing hormones is *true?*

a. They are secreted into capillaries in the median eminence.

b. They are transported by portal veins to the anterior pituitary.

c. They stimulate the secretion of specific hormones from the anterior pituitary.

d. All of these are true.

2. The hormone primarily responsible for setting the basal metabolic rate and for promoting the maturation of the brain is

a. cortisol.

b. ACTH.

c. TSH.

d. thyroxine.

3. Which of these statements about the adrenal cortex is *true?*

a. It is not innervated by nerve fibers.

b. It secretes some androgens.

c. The zona glomerulosa secretes aldosterone.

d. The zona fasciculata is stimulated by ACTH.

e. All of these are true.

4. Which of these statements about the hormone insulin is *true?*

a. It is secreted by alpha cells in the islets of Langerhans.

b. It is secreted in response to a rise in blood glucose.

c. It stimulates the production of glycogen and fat.

d. Both *a* and *b* are true.

e. Both *b* and *c* are true.

Match the hormone with the primary agent that stimulates its secretion.

5. epinephrine   **a.** TSH

6. thyroxine   **b.** ACTH

7. corticosteroids   **c.** growth hormone

8. ACTH   **d.** sympathetic nerves

     e. CRH

9. Steroid hormones are secreted by

a. the adrenal cortex.

b. the gonads.

c. the thyroid

d. both *a* and *b*.

e. both *b* and *c*.

10. The secretion of which of these hormones would be *increased* in a person with endemic goiter?
    a. TSH
    b. thyroxine
    c. triiodothyronine
    d. all of these

11. Which of these hormones uses cAMP as a second messenger?
    a. testosterone
    b. cortisol
    c. insulin
    d. epinephrine

12. Which of these terms best describes the interactions of insulin and glucagon?
    a. synergistic
    b. permissive
    c. antagonistic
    d. cooperative

13. Which of these correctly describes the role of inositol triphosphate in hormone action?
    a. It activates adenylate cyclase.
    b. It stimulates the release of Ca^{2+} from the endoplasmic reticulum.
    c. It activates protein kinase.
    d. all of these

14. Which of these hormones may have a primary role in many circadian rhythms?
    a. estradiol
    b. insulin
    c. adrenocorticotropic hormone
    d. melatonin

15. Human chorionic gonadotropin (hCG) is secreted by
    a. the anterior pituitary
    b. the posterior pituitary
    c. the placenta
    d. the thymus
    e. the pineal gland

16. What do insulin-like growth factors, neurotrophins, nitric oxide, and lymphokines have in common?
    a. They are hormones.
    b. They are autocrine or paracrine regulators.
    c. They are neurotransmitters.
    d. They all use cAMP as a second messenger.
    e. They all use Ca^{2+} as a second messenger.

Test Your Understanding of Concepts and Principles

1. Explain how regulation of the neurohypophysis and of the adrenal medulla are related to the embryonic origins of these organs.[1]

2. Explain the mechanism of action of steroid hormones and thyroxine.

3. Explain why polar hormones cannot regulate their target cells without using second messengers. Also explain how cyclic AMP is used as a second messenger in hormone action.

4. Describe the sequence of events by which a hormone can cause an increase in the Ca^{2+} concentration within a target cell. How can this increased Ca^{2+} affect the metabolism of the target cell?

5. Explain the significance of the term *trophic* with respect to the actions of anterior pituitary hormones.

6. Suppose a drug blocks the conversion of T_4 to T_3. Explain what the effects of this drug would be on (a) TSH secretion, (b) thyroxine secretion, and (c) the size of the thyroid gland.

7. Explain why the anterior pituitary is sometimes referred to as the "master gland" and why this reference is misleading.

8. Suppose a person's immune system made antibodies against insulin receptor proteins. What effect might this condition have on carbohydrate and fat metabolism?

9. Explain how light affects the function of the pineal gland. What is the relationship between pineal gland function and circadian rhythms?

10. Distinguish between endocrine and autocrine/paracrine regulation. List some of these autocrine/paracrine regulators and describe their functions.

Test Your Ability to Analyze and Apply Your Knowledge

1. Brenda, your roommate, has been having an awful time lately. She can't even muster enough energy to go out on a date. She's been putting on weight, she's always cold, and every time she pops in the work-out video she complains of weakness. When she finally goes to the doctor, he finds her to have a slow pulse and a low blood pressure. Laboratory tests reveal that her T_4 is low and her TSH is high. What is the matter with Brenda? Why are her symptoms typical of this disorder, and what type of treatment will the doctor most likely prescribe?

2. Your friend Bud has the talent to be a star basketball center—if only he weren't five foot eight. Since you're a good friend, you start injecting him with growth hormone as he sleeps each night. You think this is a clever strategy, but after a time you notice that he hasn't grown an inch. Instead, his jaw and forehead seem to have gotten disproportionately large and his hands and feet are swollen. Explain why the growth hormone didn't make Bud grow taller and why it had the effect it did. What disease state do these changes mimic?

3. You see your friend Joe for the first time in over a year. When you last saw him, he had been trying to bulk up by working out daily at the gym, but he was getting discouraged because his progress seemed so slow. Now,

[1]*Note:* This question is answered in the chapter 11 Study Guide found on the Online Learning Center at www.mhhe.com/fox8.

however, he's very muscular. In a frank discussion, he admits that he's been getting into trouble because he's become very aggressive. He also tells you in strict confidence, that his testes have gotten smaller and that he's been developing breasts! What might Joe be doing to cause these changes? Explain how these changes came about.

4. Distinguish between the steroid and nonsteroid group of nuclear hormone receptors. Explain the central role of vitamin A in the actions of the nonsteroid group of receptors.

5. Suppose, in an experiment, that you incubate isolated rat testes with hCG. What would be the effect, if any, of the hCG on the testes? Explain your answer. If there was an effect, discuss its potential significance in research and clinical settings.

Related Websites

Check out the Links Library at www.mhhe.com/fox8 for links to sites containing resources related to the endocrine glands. These links are monitored to ensure current URLs.

12 Muscle

Mechanisms of Contraction and Neural Control

Objectives
After studying this chapter, you should be able to . . .

1. describe the gross and microscopic structure of skeletal muscles.

2. describe the nature of a muscle twitch and explain how summation and tetanus are produced.

3. distinguish among isometric, isotonic, and eccentric contractions.

4. explain how the series-elastic component affects muscle contraction.

5. define the term *motor unit* and explain how motor units are used to control muscle contraction.

6. describe the structure of myofibrils and explain how it accounts for the striated appearance of skeletal muscle fibers.

7. explain what is meant by the sliding filament theory of contraction.

8. list the events that occur during cross-bridge cycles and describe the role of ATP in muscle contraction.

9. explain how tropomyosin and troponin control muscle contraction and relaxation, and describe the role of Ca^{2+} and the sarcoplasmic reticulum in excitation-contraction coupling.

10. describe the structure and functions of muscle spindles and explain the mechanisms involved in a stretch reflex.

11. describe the function of Golgi tendon organs and explain why a slow, gradual muscle stretch could avoid the spasm that may result from a rapid stretch.

12. explain what is meant by reciprocal innervation and describe the neural pathways involved in a crossed-extensor reflex.

13. explain the significance of gamma motoneurons in the neural control of muscle contraction and in the maintenance of muscle tone.

14. describe the neural pathways involved in the pyramidal and extrapyramidal systems.

15. explain the significance of the maximal oxygen uptake and describe the function of phosphocreatine in muscles.

16. explain how slow-twitch, fast-twitch, and intermediate fibers differ in structure and function.

17. describe skeletal muscle metabolism during exercise, and explain how muscles fatigue and how muscle fibers change as a result of physical training.

18. compare cardiac muscle and skeletal muscle in terms of structure and physiology.

19. describe the structure of smooth muscle and explain how its contraction is regulated.

Refresh Your Memory

Before you begin this chapter, you may want to review these concepts from previous chapters:

- Cytoplasm and Its Organelles 55
- Glycolysis and the Lactic Acid Pathway 104
- Aerobic Respiration 108
- Electrical Activity in Axons 160
- Acetylcholine as a Neurotransmitter 170

Take Advantage of the Technology

Visit the Online Learning Center for these additional study resources.

- Interactive quizzing
- Online study guide
- Current news feeds
- Crossword puzzles
- Vocabulary flashcards
- Labeling activities

www.mhhe.com/fox8

Clinical Investigation

Maria, an energetic 40-year-old who plays softball and has been active in athletics most of her life, complains that she is experiencing fatigue and muscle pain and that her body just doesn't seem as limber as it should. Upon exercise testing, she is found to have a high maximal oxygen uptake. Her muscles, though not large, are well toned—perhaps excessively so. Laboratory tests reveal a normal blood level of creatine phosphokinase but an elevated blood Ca^{2+} concentration. She has hypertension, which is well controlled with a calcium channel-blocking drug.

What might be responsible for Maria's fatigue and muscle pain?

Table 12.1   Skeletal Muscle Actions

Category	Action
Extensor	Increases the angle at a joint
Flexor	Decreases the angle at a joint
Abductor	Moves limb away from the midline of the body
Adductor	Moves limb toward the midline of the body
Levator	Moves insertion upward
Depressor	Moves insertion downward
Rotator	Rotates a bone along its axis
Sphincter	Constricts an opening

Skeletal Muscles

Skeletal muscles are composed of individual muscle fibers that contract when stimulated by a motor neuron. Each motor neuron branches to innervate a number of muscle fibers, and all of these fibers contract when their motor neuron is activated. Activation of varying numbers of motor neurons, and thus varying numbers of muscle fibers, results in gradations in the strength of contraction of the whole muscle.

Skeletal muscles are usually attached to bone on each end by tough connective tissue tendons. When a muscle contracts, it shortens, and this places tension on its tendons and attached bones. The muscle tension causes movement of the bones at a joint, where one of the attached bones generally moves more than the other. The more movable bony attachment of the muscle, known as its *insertion,* is pulled toward its less movable attachment known as its *origin.* A variety of skeletal movements are possible, depending on the type of joint involved and the attachments of the muscles (table 12.1). When *flexor muscles* contract, for example, they decrease the angle of a joint. Contraction of *extensor muscles* increases the angle of their attached bones at the joint. The prime mover of any skeletal movement is called the **agonist muscle;** in flexion, for example, the flexor is the agonist muscle. Flexors and extensors that act on the same joint to produce opposite actions are **antagonistic muscles.**

Structure of Skeletal Muscles

The fibrous connective tissue proteins within the tendons extend around the muscle in an irregular arrangement, forming a sheath known as the *epimysium* (*epi* = above; *my* = muscle). Connective tissue from this outer sheath extends into the body of the muscle, subdividing it into columns, or *fascicles* (these are the "strings" in stringy meat). Each of these fascicles is thus surrounded by its own connective tissue sheath, which is known as the *perimysium* (*peri* = around).

Dissection of a muscle fascicle under a microscope reveals that it, in turn, is composed of many **muscle fibers,** or *myofibers.*

Each is surrounded by a plasma membrane, or **sarcolemma,** enveloped by a thin connective tissue layer called an *endomysium* (fig. 12.1). Since the connective tissue of the tendons, epimysium, perimysium, and endomysium is continuous, muscle fibers do not normally pull out of the tendons when they contract.

Duchenne's muscular dystrophy is the most severe of the muscular dystrophies, afflicting 1 out of 3,500 boys each year. This disease, inherited as an X-linked recessive trait, involves progressive muscular wasting and usually results in death by the age of 20. The product of the defective gene is a protein named *dystrophin,* which is associated with the plasma membrane of skeletal muscle fibers (the sarcolemma). Using this information, scientists have recently developed laboratory tests that can detect this disease in fetal cells obtained by amniocentesis. This research has been aided by the development of a strain of mice that exhibit an equivalent form of the disease. When the "good genes" for dystrophin are inserted into mouse embryos of this strain, the mice do not develop the disease. Insertion of the gene into large numbers of mature muscle cells, however, is more difficult, and so far has met with only limited success.

Despite their unusual elongated shape, muscle fibers have the same organelles that are present in other cells: mitochondria, endoplasmic reticulum, glycogen granules, and others. Unlike most other cells in the body, skeletal muscle fibers are multinucleate—that is, they contain multiple nuclei. This is because, as described in chapter 1, each muscle fiber is a syncytial structure. That is, each muscle fiber is formed from the union of several embryonic myoblast cells. The most distinctive feature of skeletal muscle fibers, however, is their **striated** appearance when viewed microscopically (fig. 12.2). The striations (stripes) are produced by alternating dark and light bands that appear to span the width of the fiber.

The dark bands are called **A bands,** and the light bands are called **I bands.** At high magnification in an electron microscope,

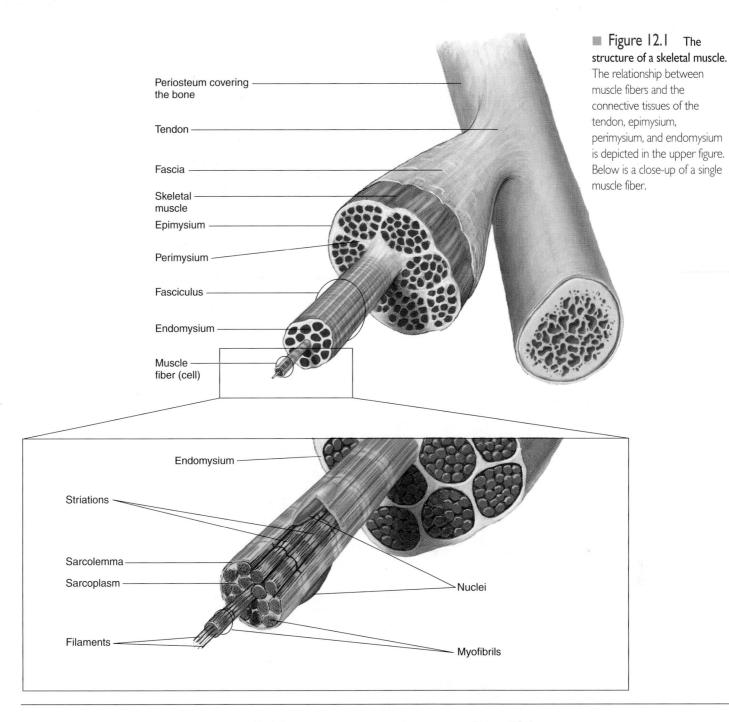

Periosteum covering the bone

Tendon

Fascia

Skeletal muscle

Epimysium

Perimysium

Fasciculus

Endomysium

Muscle fiber (cell)

Figure 12.1 The **structure of a skeletal muscle.** The relationship between muscle fibers and the connective tissues of the tendon, epimysium, perimysium, and endomysium is depicted in the upper figure. Below is a close-up of a single muscle fiber.

Endomysium

Striations

Sarcolemma

Sarcoplasm

Filaments

Nuclei

Myofibrils

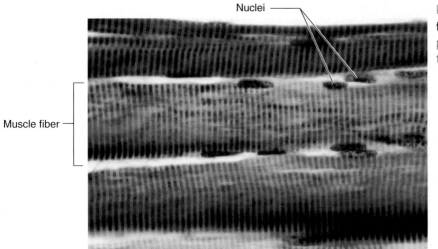

Nuclei

Muscle fiber

Figure 12.2 The appearance of skeletal muscle fibers through the light microscope. The striations are produced by alternating dark A bands and light I bands. (Note the peripheral location of the nuclei.)

thin dark lines can be seen in the middle of the I bands. These are called **Z lines** (see fig. 12.6). The labels A, I, and Z—derived in the course of early muscle research—are useful for describing the functional architecture of muscle fibers. The letters *A* and *I* stand for *anisotropic* and *isotropic,* respectively, which indicate the behavior of polarized light as it passes through these regions; the letter Z comes from the German word *Zwischenscheibe,* which translates to "between disc." These derivations are of historical interest only.

Motor Units

In vivo, each muscle fiber receives a single axon terminal from a somatic motor neuron. The motor neuron stimulates the muscle fiber to contract by liberating acetylcholine at the neuromuscular junction (described in chapter 7). The specialized region of the sarcolemma of the muscle fiber at the neuromuscular junction is known as a **motor end plate** (fig. 12.3).

The acetylcholine (ACh) released by the axon terminals diffuses across the synaptic cleft and binds to ACh receptors in the plasma membrane of the end plate, thereby stimulating the muscle fiber. Prior to its release, the ACh is contained in synaptic vesicles that dock and fuse with the plasma membrane of the axon terminal and undergo exocytosis (see chapter 7, fig. 7.21). The potentially deadly **botulinum toxin,** produced by the bacteria *Clostridium botulinum,* is selectively taken into cholinergic nerve endings and cleaves the proteins needed for the exocytosis of the synaptic vesicles. This blocks nerve stimulation of the muscles, producing a flaccid paralysis. Interestingly, botulinum toxin is now used medically in certain cases to relieve muscle spasms due to excessive nerve stimulation. For example, it is injected into an affected extraocular muscle in order to help correct **strabismus** (deviation of the eye). Intramuscular injections of **Botox** (a brand name for botulinum toxin) are also given for the temporary cosmetic treatment of skin wrinkles.

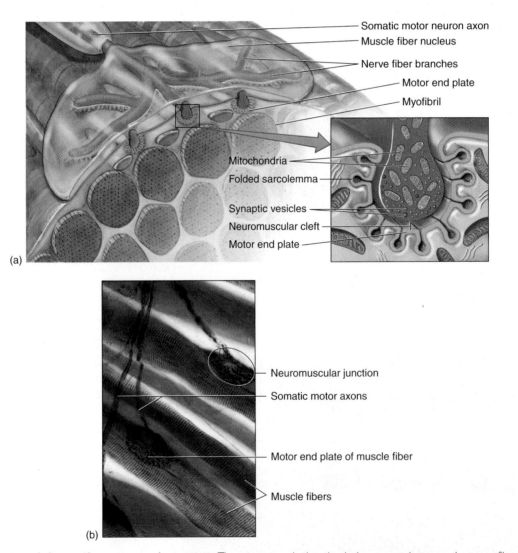

Figure 12.3 **Motor end plates at the neuromuscular junction.** The neuromuscular junction is the synapse between the nerve fiber and muscle fiber. The motor end plate is the specialized portion of the sarcolemma of a muscle fiber surrounding the terminal end of the axon. (*a*) An illustration of the neuromuscular junction. Notice the slight gap between the membrane of the axon and that of the muscle fiber. (*b*) A photomicrograph of muscle fibers and neuromuscular junctions.

The cell body of a somatic motor neuron is located in the ventral horn of the gray matter of the spinal cord and gives rise to a single axon that emerges in the ventral root of a spinal nerve (chapter 8). Each axon, however, can produce a number of collateral branches to innervate an equal number of muscle fibers. Each somatic motor neuron, together with all of the muscle fibers that it innervates, is known as a **motor unit** (fig. 12.4).

Whenever a somatic motor neuron is activated, all of the muscle fibers that it innervates are stimulated to contract. In vivo, graded contractions of whole muscles are produced by variations in the number of motor units that are activated. In order for these graded contractions to be smooth and sustained, different motor units must be activated by rapid, asynchronous stimulation.

Fine neural control over the strength of muscle contraction is optimal when there are many small motor units involved. In the extraocular muscles that position the eyes, for example, the *innervation ratio* (motor neuron:muscle fibers) of an average motor unit is one neuron per twenty-three muscle fibers. This affords a fine degree of control. The innervation ratio of the gastrocnemius, by contrast, averages one neuron per thousand muscle fibers. Stimulation of these motor units results in more powerful contractions at the expense of finer gradations in contraction strength.

All of the motor units controlling the gastrocnemius, however, are not the same size. Innervation ratios vary from 1:100 to 1:2,000. A neuron that innervates fewer muscle fibers has a smaller cell body and is stimulated by lower levels of excitatory input than a larger neuron that innervates a greater number of muscle fibers. The smaller motor units, as a result, are the ones that are used most often. When contractions of greater strength are required, larger and larger motor units are activated in a process known as **recruitment** of motor units.

Test Yourself Before You Continue

1. Describe the actions of muscles when they contract, and define the terms *agonist* and *antagonist* in muscle action.
2. Describe the different levels of muscle structure, explaining how the muscle and its substructures are packaged in connective tissues.
3. Define the terms *motor unit* and *innervation ratio* as they relate to muscle function, and draw a simple diagram of a motor unit with a 1:5 innervation ratio.
4. Using the concept of recruitment, explain how muscle contraction can be graded in its strength.

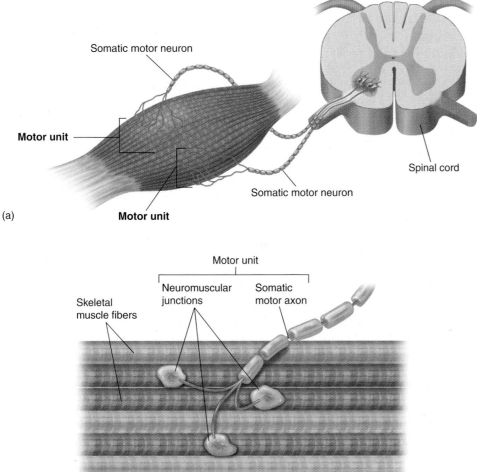

(a)

(b)

Figure 12.4 Motor units. A motor unit consists of a somatic motor neuron and the muscle fibers it innervates. (*a*) Illustration of a muscle containing two motor units. In reality, a muscle would contain many hundreds of motor units, and each motor unit would contain many more muscle fibers than are shown here. (*b*) A single motor unit consisting of a branched motor axon and the three muscle fibers it innervates (the fibers that are highlighted) is depicted. The other muscle fibers would be part of different motor units and would be innervated by different neurons (not shown).

Mechanisms of Contraction

The A bands within each muscle fiber are composed of thick filaments and the I bands contain thin filaments. Movement of cross bridges that extend from the thick to the thin filaments causes sliding of the filaments, and thus muscle tension and shortening. The activity of the cross bridges is regulated by the availability of Ca^{2+}, which is increased by electrical stimulation of the muscle fiber. Electrical stimulation produces contraction of the muscle through the binding of Ca^{2+} to regulatory proteins within the thin filaments.

When muscle cells are viewed in the electron microscope, which can produce images at several thousand times the magnification possible in an ordinary light microscope, each cell is seen to be composed of many subunits known as **myofibrils** (*fibrils* = little fibers) (fig. 12.5). These myofibrils are approximately 1 micrometer (1 μm) in diameter and extend in parallel rows from one end of the muscle fiber to the other. The myofibrils are so densely packed that other organelles, such as mitochondria and intracellular membranes, are restricted to the narrow cytoplasmic spaces that remain between adjacent myofibrils.

With the electron microscope, it can be seen that the muscle fiber does not have striations that extend from one side of the fiber to the other. It is the myofibrils that are striated with dark *A bands* and light *I bands* (fig. 12.6). The striated appearance of the entire muscle fiber when seen with a light microscope is an illusion created by the alignment of the dark and light bands of the myofibrils from one side of the fiber to the other. Since the separate myofibrils are not clearly seen at low magnification, the dark and light bands appear to be continuous across the width of the fiber.

Each myofibril contains even smaller structures called **myofilaments,** or simply **filaments.** When a myofibril is observed at high magnification in longitudinal section (side view), the A bands are seen to contain **thick filaments.** These are about 110 angstroms thick (110 Å, where 1 Å = 10^{-10} m) and are stacked in register. It is these thick filaments that give the A band its dark appearance. The lighter I band, by contrast, contains **thin filaments** (from 50 to 60 Å thick). The thick filaments are primarily composed of the protein **myosin,** and the thin filaments are primarily composed of the protein **actin.**

The I bands within a myofibril are the lighter areas that extend from the edge of one stack of thick filaments to the edge of the next stack of thick filaments. They are light in appearance because they contain only thin filaments. The thin filaments, however, do not end at the edges of the I bands. Instead, each thin filament extends partway into the A bands on each side (between the stack of thick filaments on each side of an I band). Since thick and thin filaments overlap at the edges of each A band, the edges of the A band are darker in appearance than the central region. These central lighter regions of the A bands are called the *H bands* (for *helle,* a German word meaning "bright"). The central H bands thus contain only thick filaments that are not overlapped by thin filaments.

In the center of each I band is a thin dark Z line. The arrangement of thick and thin filaments between a pair of Z lines forms a repeating pattern that serves as the basic subunit of striated muscle contraction. These subunits, from Z to Z, are known as **sarcomeres** (fig. 12.7a). A longitudinal section of a myofibril thus presents a side view of successive sarcomeres.

This side view is, in a sense, misleading; there are numerous sarcomeres within each myofibril that are out of the plane of the section (and out of the picture). A better appreciation of the three-dimensional structure of a myofibril can be obtained by viewing the myofibril in cross section. In this view, it can be seen that the Z lines are actually **Z discs,** and that the thin filaments that penetrate these Z discs surround the thick filaments

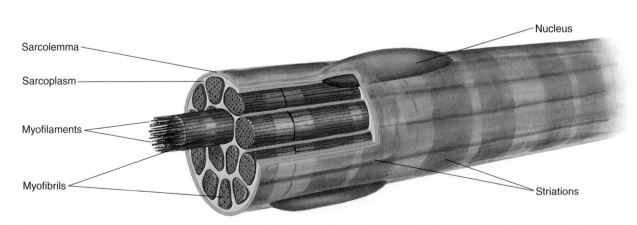

Sarcolemma

Sarcoplasm

Myofilaments

Myofibrils

Nucleus

Striations

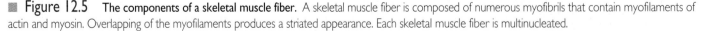

■ Figure 12.5   **The components of a skeletal muscle fiber.** A skeletal muscle fiber is composed of numerous myofibrils that contain myofilaments of actin and myosin. Overlapping of the myofilaments produces a striated appearance. Each skeletal muscle fiber is multinucleated.

in a hexagonal arrangement (fig. 12.7b, right). If we concentrate on a single row of dark thick filaments in this cross section, the alternating pattern of thick and thin filaments seen in longitudinal section becomes apparent.

Figure 12.8 indicates two structures not shown in the previous sarcomere figures. The **M lines** are produced by protein filaments located at the center of the thick filaments (and thus the A band) in a sarcomere. These serve to anchor the thick filaments, helping them to stay together during a contraction. Also shown are filaments of **titin**, a type of elastic protein that runs through the thick filaments from the M lines to the Z discs. Because of its elastic properties, titin is believed to contribute to the elastic recoil of muscles that helps them to return to their resting length.

Sliding Filament Theory of Contraction

When a muscle contracts it decreases in length as a result of the shortening of its individual fibers. Shortening of the muscle fibers, in turn, is produced by shortening of their myofibrils, which occurs as a result of the shortening of the distance from Z line to Z line. As the sarcomeres shorten in length, however, the A bands do *not* shorten but instead move closer together. The I bands—which represent the distance between A bands of successive sarcomeres—decrease in length (table 12.2).

The thin filaments composing the I band, however, do not shorten. Close examination reveals that the thick and thin filaments remain the same length during muscle contraction.

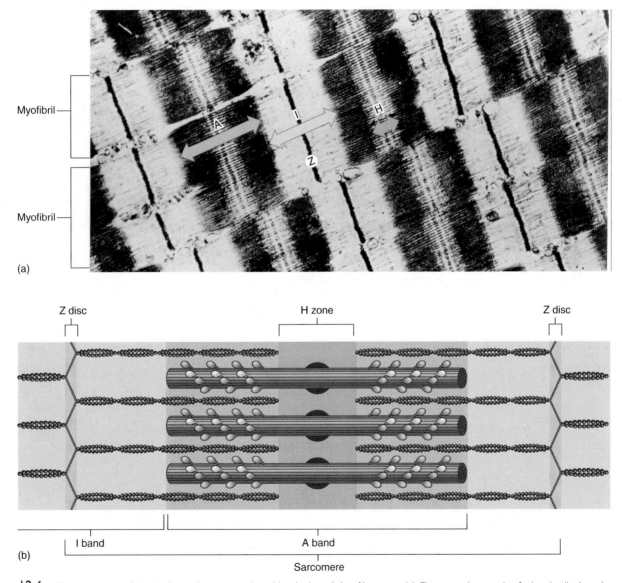

Figure 12.6   The striations of skeletal muscles are produced by thick and thin filaments. (*a*) Electron micrograph of a longitudinal section of myofibrils, showing the banding pattern characteristic of striated muscle. (*b*) Illustration of the arrangement of thick and thin filaments that produces the banding pattern. The colors used in (*a*) to depict different bands and structures correspond to the colors of (*b*).

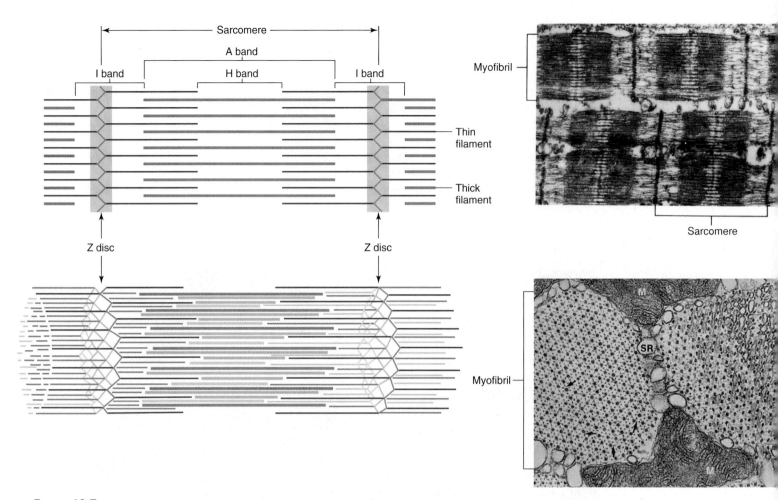

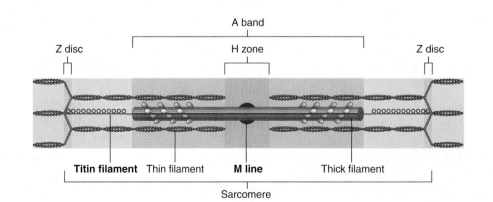

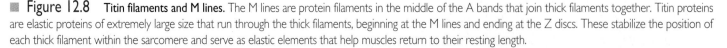

■ **Figure 12.8** **Titin filaments and M lines.** The M lines are protein filaments in the middle of the A bands that join thick filaments together. Titin proteins are elastic proteins of extremely large size that run through the thick filaments, beginning at the M lines and ending at the Z discs. These stabilize the position of each thick filament within the sarcomere and serve as elastic elements that help muscles return to their resting length.

Table 12.2 Summary of the Sliding Filament Theory of Contraction

1. A myofiber, together with all its myofibrils, shortens by movement of the insertion toward the origin of the muscle.
2. Shortening of the myofibrils is caused by shortening of the sarcomeres—the distance between Z lines (or discs) is reduced.
3. Shortening of the sarcomeres is accomplished by sliding of the myofilaments—the length of each filament remains the same during contraction.
4. Sliding of the filaments is produced by asynchronous power strokes of myosin cross bridges, which pull the thin filaments (actin) over the thick filaments (myosin).
5. The A bands remain the same length during contraction, but are pulled toward the origin of the muscle.
6. Adjacent A bands are pulled closer together as the I bands between them shorten.
7. The H bands shorten during contraction as the thin filaments on the sides of the sarcomeres are pulled toward the middle.

Shortening of the sarcomeres is produced not by shortening of the filaments, but rather by the *sliding* of thin filaments over and between the thick filaments. In the process of contraction, the thin filaments on either side of each A band slide deeper and deeper toward the center, producing increasing amounts of overlap with the thick filaments. The I bands (containing only thin filaments) and H bands (containing only thick filaments) thus get shorter during contraction (fig. 12.9).

Cross Bridges

Sliding of the filaments is produced by the action of numerous **cross bridges** that extend out from the myosin toward the actin. These cross bridges are part of the myosin proteins that extend from the axis of the thick filaments to form "arms" that terminate in globular "heads" (fig. 12.10). A myosin protein has two globular heads that serve as cross bridges. The orientation of the myosin heads on one side of a sarcomere is opposite to that of

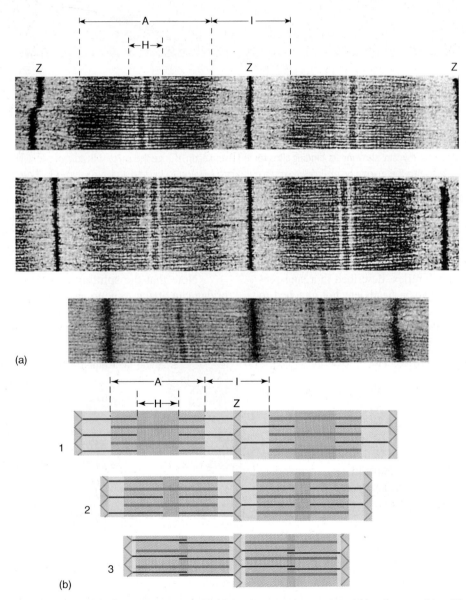

(a)

(b)

■ Figure 12.9 **The sliding filament model of muscle contraction.** (*a*) An electron micrograph and (*b*) a diagram of the sliding filament model of contraction. As the filaments slide, the Z lines are brought closer together and the sarcomeres get shorter. (*1*) Relaxed muscle; (*2*) partially contracted muscle; (*3*) fully contracted muscle.

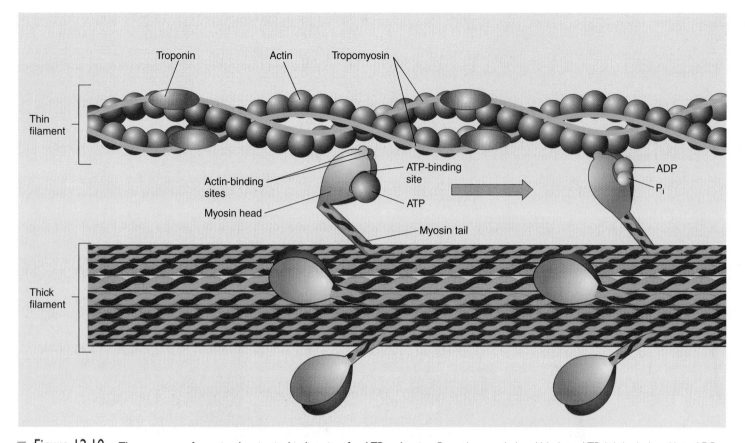

■ **Figure 12.10** **The structure of myosin, showing its binding sites for ATP and actin.** Once the myosin head binds to ATP, it is hydrolyzed into ADP and inorganic phosphate (P_i). This activates the myosin head, "cocking it" to put it into position to bind to attachment sites in the actin molecules.

the other side, so that, when the myosin heads form cross bridges by attaching to actin on each side of the sarcomere, they can pull the actin from each side toward the center.

Isolated muscles are easily stretched (although this is opposed in the body by the stretch reflex, described in a later section), demonstrating that the myosin heads are not attached to actin when the muscle is at rest. Each globular myosin head of a cross bridge contains an ATP-binding site closely associated with an actin-binding site (fig. 12.10, *left*). The globular heads function as **myosin ATPase** enzymes, splitting ATP into ADP and P_i.

This reaction must occur *before* the myosin heads can bind to actin. When ATP is hydrolyzed to ADP and P_i, the myosin head "cocks" (by analogy to the hammer of a gun), putting it into position to bind to actin (fig. 12.10, *right*).

Once the myosin head binds to actin, forming a cross bridge, the bound P_i is released. This results in a conformational change in the myosin, causing the cross bridge to produce a **power stroke** (fig. 12.11). This is the force that pulls the thin filaments toward the center of the A band.

After the power stroke, with the myosin head now in its flexed position, the bound ADP is released as a new ATP molecule binds to the myosin head. This release of ADP and binding to a new ATP is required in order for the myosin head to break its bond with actin after the power stroke is completed. If this process were prevented, the myosin heads would remain bound

to the actin (see the Clinical box discussion of rigor mortis). The myosin head will then split ATP to ADP and P_i, and—if nothing prevents the binding of the myosin head to the actin—a new cross-bridge cycle will occur (fig. 12.12).

Note that the splitting of ATP is required *before* a cross bridge can attach to actin and undergo a power stroke, and that the attachment of a *new ATP* is needed for the cross bridge to release from actin at the end of a power stroke (fig. 12.12).

The detachment of a cross bridge from actin at the end of a power stroke requires that a new ATP molecule bind to the myosin ATPase. The importance of this process is illustrated by the muscular contracture called **rigor mortis** that occurs due to lack of ATP when the muscle dies. Without ATP, the ADP remains bound to the cross bridges, and the cross bridges remain tightly bound to actin. This results in the formation of "rigor complexes" between myosin and actin that cannot detach. In rigor mortis, the muscles remain stiff until the myosin and actin begin to decompose.

Because the cross bridges are quite short, a single contraction cycle and power stroke of all the cross bridges in a muscle would shorten the muscle by only about 1% of its resting length. Since muscles can shorten up to 60% of their resting lengths, it is

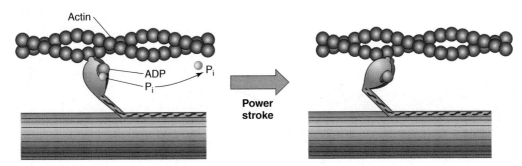

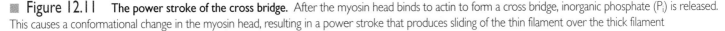

■ **Figure 12.11** **The power stroke of the cross bridge.** After the myosin head binds to actin to form a cross bridge, inorganic phosphate (P_i) is released. This causes a conformational change in the myosin head, resulting in a power stroke that produces sliding of the thin filament over the thick filament.

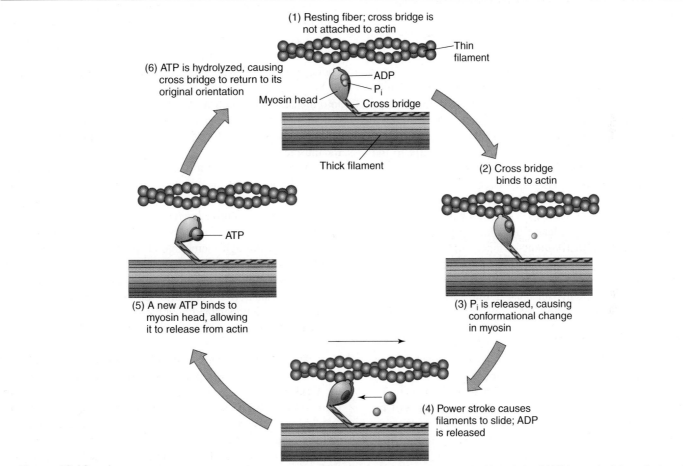

■ **Figure 12.12** **The cross-bridge cycle that causes sliding of the filaments and muscle contraction.** Hydrolysis of ATP is required for activation of the cross bridge, and the binding of a new ATP is required for the cross bridge to release from the actin at the end of a cycle.

obvious that the contraction cycles must be repeated many times. In order for this to occur, the cross bridges must detach from the actin at the end of a power stroke, reassume their resting orientation, and then reattach to the actin and repeat the cycle.

During normal contraction, however, only a portion of the cross bridges are attached at any given time. The power strokes are thus not in synchrony, as the strokes of a competitive rowing team would be. Rather, they are like the actions of a team engaged in tug-of-war, where the pulling action of the members is asynchronous. Some cross bridges are engaged in power strokes at all times during the contraction.

Regulation of Contraction

When the cross bridges attach to actin, they undergo power strokes and cause muscle contraction. In order for a muscle to relax, therefore, the attachment of myosin cross bridges to actin must be prevented. The regulation of cross-bridge attachment to actin is a function of two proteins that are associated with actin in the thin filaments.

The actin filament—or *F-actin*—is a polymer formed of 300 to 400 globular subunits (*G-actin*), arranged in a double row and twisted to form a helix (fig. 12.13). A different type of protein,

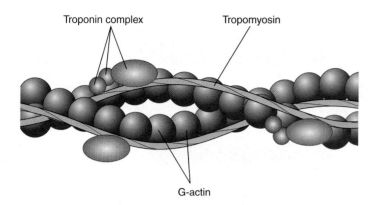

Troponin complex   Tropomyosin

G-actin

■ **Figure 12.13** The structural relationship between troponin, tropomyosin, and actin. The tropomyosin is attached to actin, whereas the troponin complex of three subunits is attached to tropomyosin (not directly to actin).

known as **tropomyosin,** lies within the groove between the double row of G-actin monomers. There are forty to sixty tropomyosin molecules per thin filament, with each tropomyosin spanning a distance of approximately seven actin subunits.

Attached to the tropomyosin, rather than directly to the actin, is a third type of protein called **troponin** (actually a complex of three proteins—see fig. 12.13). Troponin and tropomyosin work together to regulate the attachment of cross bridges to actin, and thus serve as a switch for muscle contraction and relaxation. In a relaxed muscle, the position of the tropomyosin in the thin filaments is such that it physically blocks the cross bridges from bonding to specific attachment sites in the actin. Thus, in order for the myosin cross bridges to attach to actin, the tropomyosin must be moved. This requires the interaction of troponin with Ca^{2+}.

Role of Ca^{2+} in Muscle Contraction

In a relaxed muscle, when tropomyosin blocks the attachment of cross bridges to actin, the concentration of Ca^{2+} in the sarcoplasm (cytoplasm of muscle cells) is very low. When the muscle cell is stimulated to contract, mechanisms that will be discussed shortly cause the concentration of Ca^{2+} in the sarcoplasm to quickly rise. Some of this Ca^{2+} attaches to troponin, causing a conformational change that moves the troponin complex *and* its attached tropomyosin out of the way so that the cross bridges can attach to actin (fig. 12.14). Once the attachment sites on the actin are exposed, the cross bridges can bind to actin, undergo power strokes, and produce muscle contraction.

The position of the troponin-tropomyosin complexes in the thin filaments is thus adjustable. When Ca^{2+} is not attached to troponin, the tropomyosin is in a position that inhibits attachment of cross bridges to actin, preventing muscle contraction. When Ca^{2+} attaches to troponin, the troponin-tropomyosin complexes shift position. The cross bridges can then attach to actin, produce a power stroke, and detach from actin. These contraction cycles can continue as long as Ca^{2+} is attached to troponin.

Clinical Investigation Clues

Remember that Maria has muscle pain and fatigue, and that her body seems stiff. Further, she has an elevated blood concentration of Ca^{2+}.

How might the high blood Ca^{2+} be related to Maria's symptoms? What might cause an elevated blood Ca^{2+} (hint—see chapter 11 or 19).

Excitation-Contraction Coupling

Muscle contraction is turned on when sufficient amounts of Ca^{2+} bind to troponin. This occurs when the Ca^{2+} concentration of the sarcoplasm rises above 10^{-6} molar. In order for muscle relaxation to occur, therefore, the Ca^{2+} concentration of the sarcoplasm must be lowered to below this level. Muscle relaxation is produced by the active transport of Ca^{2+} out of the sarcoplasm into the **sarcoplasmic reticulum** (fig. 12.15). The sarcoplasmic reticulum is a modified endoplasmic reticulum, consisting of interconnected sacs and tubes that surround each myofibril within the muscle cell.

Most of the Ca^{2+} in a relaxed muscle fiber is stored within expanded portions of the sarcoplasmic reticulum known as *terminal cisternae*. When a muscle fiber is stimulated to contract by either a motor neuron *in vivo* or electric shocks *in vitro*, the stored Ca^{2+} is released from the sarcoplasmic reticulum by passive diffusion through membrane channels termed **calcium release channels** (fig. 12.16), so that the Ca^{2+} can attach to troponin. When a muscle fiber is no longer stimulated, the Ca^{2+} is actively transported back into the sarcoplasmic reticulum. Now, in order to understand how the release and uptake of Ca^{2+} is regulated, one more organelle within the muscle fiber must be described.

The terminal cisternae of the sarcoplasmic reticulum are separated only by a very narrow gap from **transverse tubules** (or **T tubules**). These are narrow membranous "tunnels" formed from and continuous with the sarcolemma (muscle plasma membrane). The transverse tubules thus open to the extracellular environment through pores in the cell surface and are able to conduct action potentials. The stage is now set to explain exactly how a motor neuron stimulates a muscle fiber to contract.

The release of acetylcholine from axon terminals at the neuromuscular junctions (motor end plates), as previously described, causes electrical activation of skeletal muscle fibers. End-plate potentials (analogous to EPSPs—chapter 7) are produced that generate action potentials. Action potentials in muscle cells, like those in nerve cells, are all-or-none events that are regenerated along the plasma membrane. It must be remembered that action potentials involve the flow of ions between the extracellular and intracellular environments across a plasma membrane that separates these two compartments. In muscle cells, therefore, action potentials can be conducted into the interior of the fiber across the membrane of the transverse tubules.

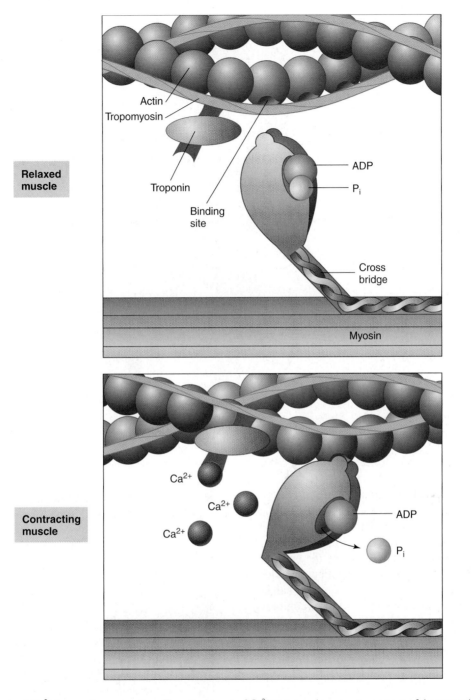

Figure 12.14 **The role of Ca²⁺ in muscle contraction.** The attachment of Ca²⁺ to troponin causes movement of the troponin-tropomyosin complex, which exposes binding sites on the actin. The myosin cross bridges can then attach to actin and undergo a power stroke.

Action potentials in the transverse tubules cause the release of Ca²⁺ from the sarcoplasmic reticulum. This process is known as **excitation-contraction coupling** (table 12.3). Since the transverse tubules are not physically continuous with the sarcoplasmic reticulum, however, there must be some mechanism to permit communication between these two organelles. It is currently believed that there may be a direct coupling, on a molecular level, between voltage-regulated Ca²⁺ channels in the transverse tubules and the Ca²⁺ release channels in the sarcoplasmic reticulum. The Ca²⁺ release channel proteins of the sarcoplasmic reticulum have a part that extends into the cytoplasm. This part, which has a footlike appearance in the electron microscope, may be able to interact directly with the Ca²⁺ channel proteins of the transverse tubules (fig 12.16).

This arrangement has been described as an *electromechanical release* mechanism, because changes in membrane voltage

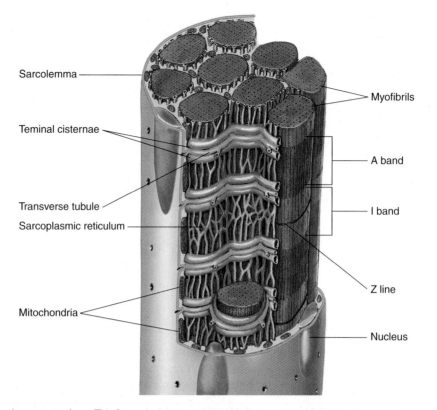

Sarcolemma

Teminal cisternae

Transverse tubule

Sarcoplasmic reticulum

Mitochondria

Myofibrils

A band

I band

Z line

Nucleus

■ **Figure 12.15** **The sarcoplasmic reticulum.** This figure depicts the relationship between myofibrils, the transverse tubules, and the sarcoplasmic reticulum. The sarcoplasmic reticulum *(green)* stores Ca^{2+} and is stimulated to release it by action potentials arriving in the transverse tubules.

(action potentials) in the transverse tubules cause a change in protein conformation of calcium channels, which are mechanically linked to other calcium channels in the sarcoplasmic reticulum. There is also evidence that the Ca^{2+} flow through the channels in the transverse tubules may stimulate the opening of other calcium channels in the sarcoplasmic reticulum. This is termed a *Ca^{2+}-induced Ca^{2+} release mechanism,* and has been shown to be the major mechanism for excitation-contraction coupling in heart muscle. By these mechanisms, Ca^{2+} can be released from the sarcoplasmic reticulum, bind to troponin, and stimulate muscle contraction.

Muscle Relaxation

As long as action potentials continue to be produced—which is as long as neural simulation of the muscle is maintained—the Ca^{2+} release channels in the sarcoplasmic reticulum will remain open, Ca^{2+} will passively diffuse out of the sarcoplasmic reticulum, and the Ca^{2+} concentration of the sarcoplasm will remain high. Thus, Ca^{2+} will remain attached to troponin, and the cross-bridge cycle will continue to maintain contraction.

To stop this action, the production of action potentials must cease, causing the Ca^{2+} release channels to close. When this occurs, the effects of other transport proteins in the sarcoplasmic reticulum becomes unmasked. These are active transport pumps for Ca^{2+}—termed **Ca^{2+}-ATPase pumps,** which move Ca^{2+} from the sarcoplasm into the sarcoplasmic reticulum.

Since these active transport pumps are powered by the hydrolysis of ATP, ATP is needed for muscle relaxation as well as for muscle contraction.

Test Yourself Before You Continue

1. With reference to the sliding filament theory, explain how the lengths of the A, I, and H bands change during contraction.

2. Describe a cycle of cross-bridge activity during contraction and discuss the role of ATP in this cycle.

3. Draw a sarcomere in a relaxed muscle and a sarcomere in a contracted muscle and label the bands in each. What is the significance of the differences in your drawings?

4. Describe the molecular structure of myosin and actin. How are tropomyosin and troponin positioned in the thin filaments and how do they function in the contraction cycle?

5. Use a flowchart to show the sequence of events from the time ACh is released from a nerve ending to the time Ca^{2+} is released from the sarcoplasmic reticulum.

6. Explain the requirements for Ca^{2+} and ATP in muscle contraction and relaxation.

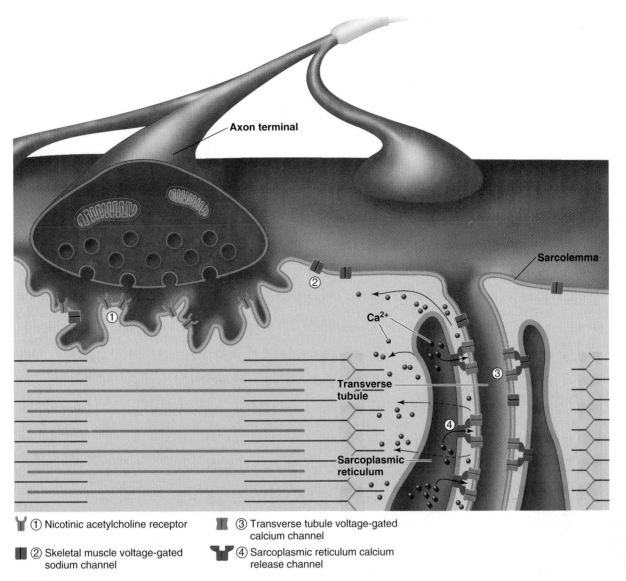

① Nicotinic acetylcholine receptor

③ Transverse tubule voltage-gated calcium channel

② Skeletal muscle voltage-gated sodium channel

④ Sarcoplasmic reticulum calcium release channel

Figure 12.16 **The structures involved in excitation-contraction coupling.** The acetylcholine released from the axon binds to its nicotinic receptors in the motor end plate. This stimulates the production of a depolarization, which causes the opening of voltage-gated Na⁺ channels and the resulting production of action potentials along the sarcolemma. The spread of action potentials into the transverse tubules stimulates the opening of their voltage-gated Ca²⁺ channels, which (directly or indirectly) causes the opening of voltage-gated Ca²⁺ channels in the sarcoplasmic reticulum. Calcium diffuses out of the sarcoplasmic reticulum, binds to troponin, and stimulates contraction.

Table 12.3 Summary of Events in Excitation-Contraction Coupling

1. Action potentials in a somatic motor neuron cause the release of acetylcholine neurotransmitter at the myoneural junction (one myoneural junction per myofiber).
2. Acetylcholine, through its interaction with receptors in the muscle cell membrane (sarcolemma), produces action potentials that are regenerated across the sarcolemma.
3. The membranes of the transverse tubules (T tubules) are continuous with the sarcolemma and conduct action potentials deep into the muscle fiber.
4. Action potentials in the T tubules, acting through a mechanism that is incompletely understood, stimulate the release of Ca²⁺ from the terminal cisternae of the sarcoplasmic reticulum.
5. Ca²⁺ released into the sarcoplasm attaches to troponin, causing a change in its structure.
6. The shape change in troponin causes its attached tropomyosin to shift position in the actin filament, thus exposing binding sites for the myosin cross bridges.
7. Myosin cross bridges, previously activated by the hydrolysis of ATP, attach to actin.
8. Once the previously activated cross bridges attach to actin, they undergo a power stroke and pull the thin filaments over the thick filaments.
9. Attachment of fresh ATP allows the cross bridges to detach from actin and repeat the contraction cycle as long as Ca²⁺ remains attached to troponin.
10. When action potentials stop being produced, the sarcoplasmic reticulum actively accumulates Ca²⁺ and tropomyosin returns to its inhibitory position.

Contractions of Skeletal Muscles

Contraction of muscles generates tension, which allows muscles to shorten and thereby perform work. The contraction strength of skeletal muscles must be sufficiently great to overcome the load on a muscle in order for that muscle to shorten.

The contractions of skeletal muscles generally produce movements of bones at joints, which act as levers to move the loads against which the muscle's force is exerted. The contractile behavior of the muscle, however, is more easily studied *in vitro* (outside the body) than *in vivo* (within the body). When a muscle—for example, the gastrocnemius (calf muscle) of a frog—is studied *in vitro,* it is usually mounted so that one end is fixed and the other is movable. The mechanical force of the muscle contraction is transduced (changed) into an electric current, which can be amplified and displayed as pen deflections in a multichannel recorder (fig. 12.17). In this way, the contractile behavior of the whole muscle in response to experimentally administered electric shocks can be studied.

Twitch, Summation, and Tetanus

When the muscle is stimulated with a single electric shock of sufficient voltage, it quickly contracts and relaxes. This response is called a **twitch.** Increasing the stimulus voltage increases the strength of the twitch, up to a maximum. The strength of a muscle contraction can thus be *graded,* or varied—an obvious requirement for the proper control of skeletal movements. If a second electric shock is delivered immediately after the first, it will produce a second twitch that may partially "ride piggyback" on the first. This response is called **summation.**

Stimulation of fibers within a muscle *in vitro* with an electric stimulator, or *in vivo* by motor axons, usually results in the full contraction of the individual fibers. Stronger muscle contractions are produced by the stimulation of greater numbers of muscle fibers. Skeletal muscles can thus produce **graded contractions,** the strength of which depends on the number of fibers stimulated rather than on the strength of the contractions of individual muscle fibers.

If the stimulator is set to deliver an increasing frequency of electric shocks automatically, the relaxation time between successive twitches will get shorter and shorter as the strength of contraction increases in amplitude. This effect is known as **incomplete tetanus** (fig. 12.18). Finally, at a particular "fusion frequency" of stimulation, there is no visible relaxation between successive twitches. Contraction is smooth and sustained, as it is during normal muscle contraction *in vivo.* This smooth, sustained contraction is called **complete tetanus.** (The term *tetanus* should not be confused with the disease of the same name, which is accompanied by a painful state of muscle contracture, or *tetany.*) The tetanus produced *in vitro* by the asynchronous twitches of muscle fibers simulates the normal, smooth contraction produced *in vivo* by the asynchronous activation of motors units.

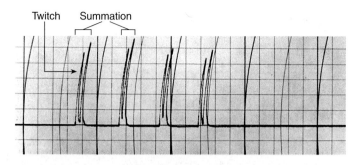

■ **Figure 12.17**   **Recording muscle contractions.** Recorder tracings demonstrating twitch and summation of an isolated frog gastrocnemius muscle.

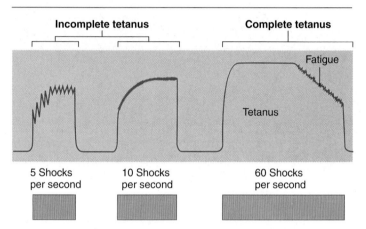

■ **Figure 12.18**   **Incomplete and complete tetanus.** When an isolated muscle is shocked repeatedly, the separate twitches summate to produce a sustained contracton. At a relatively slow rate of simulation (5 or 10 per second), the separate muscle twitches can still be observed. This is incomplete tetanus. When the frequency of stimulation increases to 60 shocks per second, however, complete tetanus—a smooth, sustained contraction—is observed. If the stimulation is continued, the muscle will demonstrate fatigue.

Treppe

If the voltage of the electrical shocks delivered to an isolated muscle *in vitro* is gradually increased from zero, the strength of the muscle twitches will increase accordingly, up to a maximal value at which all of the muscle fibers are stimulated. This demonstrates the graded nature of the muscle contraction. If a series of electrical shocks at this maximal voltage is given to a fresh muscle so that each shock produces a separate twitch, each of the twitches evoked will be successively stronger, up to a higher maximum. This demonstrates **treppe,** or the *staircase effect.* Treppe may represent a warmup effect, and is believed to be due to an increase in intracellular Ca^{2+}, which is needed for muscle contraction.

Isotonic, Isometric, and Eccentric Contractions

In order for muscle fibers to shorten when they contract, they must generate a force that is greater than the opposing forces that act to prevent movement of the muscle's insertion. When a

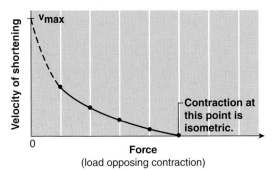

Figure 12.19 **Force-velocity curve.** This graph illustrates the inverse relationship between the force opposing muscle contraction (the load against which the muscle must work) and the velocity of muscle shortening. A force that is sufficiently great prevents muscle shortening, so that the contraction is isometric. If there is no force acting against the muscle contraction, the velocity of shortening is maximal (V_{max}). Since this cannot be measured (because there will always be some load), the estimated position of the curve is shown with a dashed line.

weight is lifted by flexing the elbow joint, for example, the force produced by contraction of the biceps brachii muscle is greater than the force of gravity on the object being lifted. The tension produced by the contraction of each muscle fiber separately is insufficient to overcome the opposing force, but the combined contractions of numerous muscle fibers may be sufficient to overcome the opposing force and flex the forearm. In this case, the muscle and all of its fibers shorten in length.

This process can be seen by examining the **force-velocity curve.** This graph shows the inverse relationship between the force opposing muscle contraction (the load against which the muscle must work) and the velocity of muscle shortening (fig. 12.19). The tension produced by the shortening muscle is just greater than the force (load) at each value, causing the muscle to shorten. Since the contraction strength is constant at each load, a muscle contraction during shortening is called an **isotonic contraction** (*iso* = same; *tonic* = strength).

If the load is zero, a muscle contracts and shortens with its maximum velocity. As the load increases, the velocity of muscle shortening decreases. When the force opposing contraction (the load) becomes sufficiently great, the muscle is unable to shorten when it exerts a given tension. That is, its velocity of shortening is zero. At this point, where muscle tension does not cause muscle shortening, the contraction is called an **isometric** (literally, "same length") **contraction.**

Isometric contraction can be voluntarily produced, for example, by lifting a weight and maintaining the forearm in a partially flexed position. We can then increase the amount of muscle tension produced by recruiting more muscle fibers until the muscle begins to shorten; at this point, isometric contraction is converted to isotonic contraction.

When a force exerted on a muscle to stretch it is greater than the force of muscle contraction, the muscle will lengthen as it contracts. In other words, the muscle lengthens *despite* its contraction. This is known as an **eccentric contraction.** In this case, the muscle can absorb some of the mechanical energy of the external force, and thereby serve as a shock absorber.

For example, when you jump from a height and land in a flexed-leg position, the extensor muscles of your legs (the quadriceps femoris group) contract eccentrically to absorb some of the shock. In this case, most of the energy absorbed by the muscles is dissipated as heat. Less dramatically (and somewhat less painfully), these muscles also contract eccentrically when you jog downhill or hike down a steep mountain trail.

Series-Elastic Component

In order for a muscle to shorten when it contracts, and thus to move its insertion toward its origin, the noncontractile parts of the muscle and the connective tissue of its tendons must first be pulled tight. These structures, particularly the tendons, have elasticity—they resist distension, and when the distending force is released, they tend to spring back to their resting lengths. Tendons provide what is called a **series-elastic component** because they are somewhat elastic and in line (in series) with the force of muscle contraction. The series-elastic component absorbs some of the tension as a muscle contracts, and it must be pulled tight before muscle contraction can result in muscle shortening.

When the gastrocnemius muscle was stimulated with a single electric shock as described earlier, the amplitude of the twitch was reduced because some of the force of contraction was used to stretch the series-elastic component. Quick delivery of a second shock thus produced a greater degree of muscle shortening than the first shock, culminating at the fusion frequency of stimulation with complete tetanus, in which the strength of contraction was much greater than that of individual twitches.

Some of the energy used to stretch the series-elastic component during muscle contraction is released by elastic recoil when the muscle relaxes. This elastic recoil, which helps the muscles return to their resting length, is of particular importance for the muscles involved in breathing. As we will see in chapter 16, inspiration is produced by muscle contraction and expiration is produced by the elastic recoil of the thoracic structures that were stretched during inspiration.

Length-Tension Relationship

The strength of a muscle's contraction is influenced by a variety of factors. These include the number of fibers within the muscle that are stimulated to contract, the frequency of stimulation, the thickness of each muscle fiber (thicker fibers have more myofibrils and thus can exert more power), and the initial length of the muscle fibers when they are at rest.

There is an "ideal" resting length for striated muscle fibers. This is the length at which they can generate maximum force. When the resting length exceeds this ideal, the overlap between actin and myosin is so small that few cross bridges can attach. When the muscle is stretched to the point that there is no overlap of actin with myosin, no cross bridges can attach to the thin filaments and the muscle cannot contract.

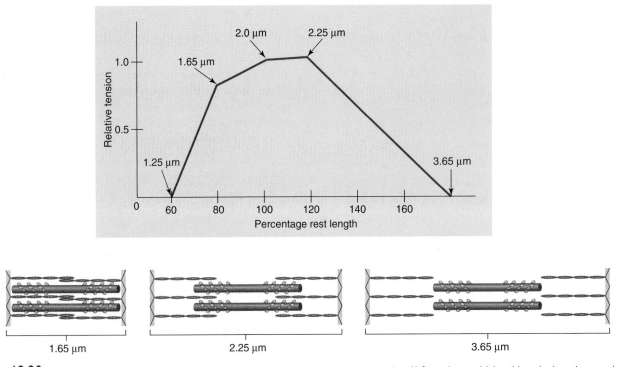

Figure 12.20 **The length-tension relationship in skeletal muscles.** Maximum relative tension (1.0 on the y axis) is achieved when the muscle is 100% to 120% of its resting length (sarcomere lengths from 2.0 to 2.25 μm). Increases or decreases in muscle (and sarcomere) lengths result in rapid decreases in tension.

When the muscle is shortened to about 60% of its resting length, the Z lines abut the thick filaments so that further contraction cannot occur.

The strength of a muscle's contraction can be measured by the force required to prevent it from shortening. Under these isometric conditions, the strength of contraction, or *tension,* can be measured when the muscle length at rest is varied. Maximum tension of skeletal muscle is produced when the muscle is at its normal resting length *in vivo* (fig. 12.20). If the muscle were any shorter or longer than its normal length, in other words, its strength of contraction would be reduced. This resting length is maintained by reflex contraction in response to passive stretching, as described in a later section of this chapter.

Test Yourself Before You Continue

1. Explain how graded contractions and smooth, sustained contractions can be produced *in vitro* and *in vivo*.

2. Distinguish among isotonic, isometric, and eccentric contractions, and describe what factors determine if a contraction will be isometric or isotonic.

3. Identify the nature and physiological significance of the series-elastic component of muscle contraction.

4. Describe the relationship between the resting muscle length and the strength of its contraction.

Energy Requirements of Skeletal Muscles

Skeletal muscles generate ATP through aerobic and anaerobic respiration and through the use of phosphate groups donated by creatine phosphate. The aerobic and anaerobic abilities of skeletal muscle fibers differ according to muscle fiber type, which are described according to their speed of contraction, color, and major mode of energy metabolism.

Skeletal muscles at rest obtain most of their energy from the aerobic respiration of fatty acids. During exercise, muscle glycogen and blood glucose are also used as energy sources (fig. 12.21). Energy obtained by cell respiration is used to make ATP, which serves as the immediate source of energy for (1) the movement of the cross bridges for muscle contraction and (2) the pumping of Ca^{2+} into the sarcoplasmic reticulum for muscle relaxation.

Metabolism of Skeletal Muscles

Skeletal muscles respire anaerobically for the first 45 to 90 seconds of moderate-to-heavy exercises because the cardiopulmonary system requires this amount of time to sufficiently increase the oxygen supply to the exercising muscles. If exercise is moderate, aerobic

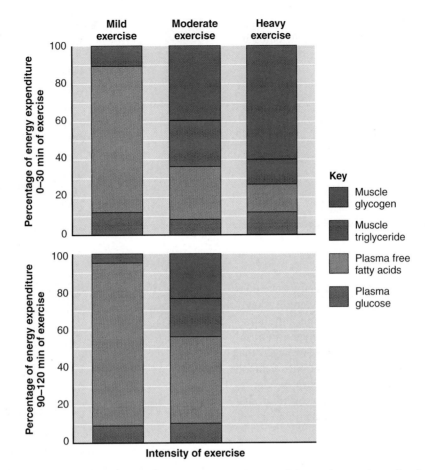

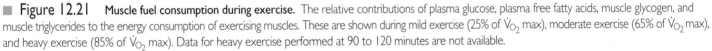

■ Figure 12.21 **Muscle fuel consumption during exercise.** The relative contributions of plasma glucose, plasma free fatty acids, muscle glycogen, and muscle triglycerides to the energy consumption of exercising muscles. These are shown during mild exercise (25% of $\dot{V}_{O_2}$ max), moderate exercise (65% of $\dot{V}_{O_2}$ max), and heavy exercise (85% of $\dot{V}_{O_2}$ max). Data for heavy exercise performed at 90 to 120 minutes are not available.

respiration contributes the major portion of the skeletal muscle energy requirements following the first 2 minutes of exercise.

Maximal Oxygen Uptake

Whether exercise is light, moderate, or heavy for a given person depends on that person's maximal capacity for aerobic exercise. The maximum rate of oxygen consumption (by aerobic respiration) in the body is called the **maximal oxygen uptake,** or the **aerobic capacity,** and is often expressed in abbreviated form as the $\dot{V}_{O_2}$ **max.** The maximal oxygen uptake is determined primarily by a person's age, size, and sex. It is from 15% to 20% higher for males than for females and highest at age 20 for both sexes. The $\dot{V}_{O_2}$ max ranges from about 12 ml of O_2 per minute per kilogram body weight for older, sedentary people to about 84 ml per minute per kilogram for young, elite male athletes. Some world-class athletes have maximal oxygen uptakes that are twice the average for their age and sex—this appears to be due largely to genetic factors, but training can increase the maximum oxygen uptake by about 20%.

The intensity of exercise can also be defined by the **lactate** (or **anaerobic**) **threshold.** This is the percentage of the maximal oxygen uptake at which a significant rise in blood lactate levels occurs. For average healthy people, for example, a

significant amount of blood lactate appears when exercise is performed at about 50% to 70% of the $\dot{V}_{O_2}$ max.

During light exercise (at about 25% of the $\dot{V}_{O_2}$ max), most of the exercising muscle's energy is derived from the aerobic respiration of fatty acids. These are derived mainly from stored fat in adipose tissue, and to a lesser extent from triglycerides stored in the muscle (fig. 12.21). When a person exercises just below the lactate threshold, where the exercise can be described as moderately intense (at 50% to 70% of the $\dot{V}_{O_2}$ max), the energy is derived almost equally from fatty acids and glucose (obtained from stored muscle glycogen and the blood plasma). By contrast, glucose from these sources supplies two-thirds of the energy for muscles during heavy exercise above the lactate threshold.

During exercise, the carrier protein for the facilitated diffusion of glucose (GLUT4—chapter 6) is moved into the muscle fiber's plasma membrane, so that the cell can take up an increasing amount of blood glucose. The uptake of plasma glucose contributes 15% to 30% of the muscle's energy needs during moderate exercise and up to 40% of the energy needs during very heavy exercise. This would produce hypoglycemia if the liver failed to increase its output of glucose. The liver increases its output of glucose primarily through hydrolysis of its stored glycogen, but gluconeogenesis (the production of glucose from amino acids, lactate, and glycerol) contributes increasingly to the liver's glucose production as exercise is prolonged.

The enzyme that transfers phosphate between creatine and ATP is called **creatine kinase,** or **creatine phosphokinase.** Skeletal muscle and heart muscle have two different forms of this enzyme (they have different isoenzymes, as described in chapter 4). The skeletal muscle isoenzyme is found to be elevated in the blood of people with muscular dystrophy (degenerative disease of skeletal muscles). The plasma concentration of the isoenzyme characteristic of heart muscle is elevated as a result of myocardial infarction (damage to heart muscle), and measurements of this enzyme are thus used as a means of diagnosing heart disease.

Clinical Investigation Clue

Remember that Maria has a high maximal oxygen uptake, consistent with her athletic lifestyle.

Is it possible, likely, or unlikely that Maria's muscle pain and fatigue are caused by her playing softball?

Oxygen Debt

When a person stops exercising, the rate of oxygen uptake does not immediately return to pre-exercise levels; it returns slowly (the person continues to breathe heavily for some time afterward). This extra oxygen is used to repay the **oxygen debt** incurred during exercise. The oxygen debt includes oxygen that was withdrawn from savings deposits—hemoglobin in blood and myoglobin in muscle (see chapter 16); the extra oxygen required for metabolism by tissues warmed during exercise; and the oxygen needed for the metabolism of the lactic acid produced during anaerobic respiration.

Phosphocreatine

During sustained muscle activity, ATP may be used faster than it can be produced through cell respiration. At these times, the rapid renewal of ATP is extremely important. This is accomplished by combining ADP with phosphate derived from another high-energy phosphate compound called **phosphocreatine,** or **creatine phosphate.**

Within muscle cells, the phosphocreatine concentration is more than three times the concentration of ATP and represents a ready reserve of high-energy phosphate that can be donated directly to ADP (fig. 12.22). Production of ATP from ADP and phosphocreatine is so efficient that, even though the rate of ATP breakdown rapidly increases from rest to heavy exercise, muscle ATP concentrations hardly change! During times of rest, the depleted reserve of phosphocreatine can be restored by the reverse reaction—phosphorylation of creatine with phosphate derived from ATP.

Clinical Investigation Clue

Remember that Maria had a normal blood level of creatine phosphokinase.

What does this suggest about the health of her muscles and heart?

Slow- and Fast-Twitch Fibers

Skeletal muscle fibers can be divided on the basis of their contraction speed (time required to reach maximum tension) into **slow-twitch,** or **type I, fibers,** and **fast-twitch,** or **type II, fibers.** These differences are associated with different myosin ATPase isoenzymes, which can also be designated as "slow" and "fast." The two fiber types can be distinguished by their ATPase isoenzyme when they are appropriately stained (fig. 12.23). The extraocular muscles that position the eyes, for example, have a high proportion of fast-twitch fibers and reach maximum tension in about 7.3 msec (milliseconds—thousandths of a second). The soleus muscle in the leg, by contrast, has a high proportion of slow-twitch fibers and requires about 100 msec to reach maximum tension (fig. 12.24).

Muscles like the soleus are *postural muscles;* they are able to sustain a contraction for a long period of time without fatigue. The resistance to fatigue demonstrated by these muscles is aided by other characteristics of slow-twitch (type I) fibers that endow them with a high oxidative capacity for aerobic respiration. Hence, the type I fibers are often referred to as **slow oxidative fibers.** These fibers have a rich capillary supply, numerous mitochondria

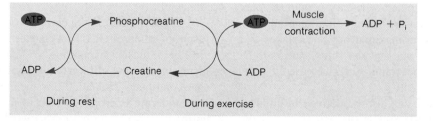

■ **Figure 12.22**   **The production and utilization of phosphocreatine in muscles.** Phosphocreatine serves as a muscle reserve of high-energy phosphate, used for the rapid formation of ATP.

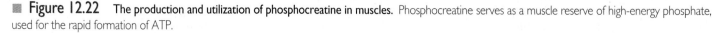

and aerobic respiratory enzymes, and a high concentration of *myoglobin.* Myoglobin is a red pigment, similar to the hemoglobin in red blood cells, that improves the delivery of oxygen to the slow-twitch fibers. Because of their high myoglobin content, slow-twitch fibers are also called **red fibers.**

The thicker, fast-twitch (type II) fibers have fewer capillaries and mitochondria than slow-twitch fibers and not as much myoglobin; hence, these fibers are also called **white fibers.** Fast-twitch fibers are adapted to respire anaerobically by a large store of glycogen and a high concentration of glycolytic enzymes.

In addition to the type I (slow-twitch) and type II (fast-twitch) fibers, human muscles have an intermediate fiber type. These intermediate fibers are fast-twitch but also have a high oxidative capacity; therefore, they are relatively resistant to fatigue. They are called type **IIA fibers,** or **fast oxidative fibers,** because of their aerobic ability. The other fast-twitch fibers are anaerobically adapted; these are called **fast glycolytic fibers** because of their high rate of glycolysis. Not all fast glycolytic fibers are alike, however. There are different fibers in this class, which vary in their contraction speeds and glycolytic abilities. In some animals, the extreme fast glycolytic fibers are of the type designated **type IIB fibers.** In humans, these fast glycolytic fibers are currently designated **type IIX fibers.** The three major fiber types in humans are compared in table 12.4.

People vary tremendously in the proportion of fast- and slow-twitch fibers in their muscles (fig. 12.25). The percent of slow-twitch, type I fibers in the quadriceps femoris muscles of the legs, for example, can vary from under 20% (in people who are excellent sprinters) to as high as 95% (in people who are good marathon runners). These differences are believed to be primarily the result of differences in genetics.

A muscle such as the gastrocnemius contains both fast- and slow-twitch fibers, although fast-twitch fibers predominate.

■ **Figure 12.23**   Skeletal muscle stained to indicate activity of myosin ATPase. ATPase activity is greater in the type II (fast-twitch) fibers than in the type I (slow-twitch) fibers. Among the fast-twitch fibers, ATPase activity is greatest in the fast-glycolytic (IIX) fibers. The fast-oxidative (IIA) fibers show an intermediate level of activity.

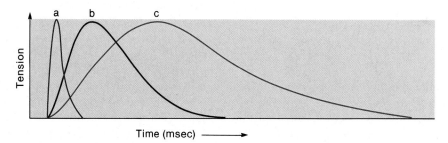

■ **Figure 12.24**   A comparison of the rates at which maximum tension is developed in three muscles.   These are (a) the relatively fast-twitch extraocular and (b) gastrocnemius muscles, and (c) the slow-twitch soleus muscle.

▌Table 12.4   Characteristics of Muscle Fiber Types

Feature	Slow Oxidative/Red (Type I)	Fast Oxidative/White (Type IIA)	Fast Glycolytic/White (Type IIX)
Diameter	Small	Intermediate	Large
Z-line thickness	Wide	Intermediate	Narrow
Glycogen content	Low	Intermediate	High
Resistance to fatigue	High	Intermediate	Low
Capillaries	Many	Many	Few
Myoglobin content	High	High	Low
Respiration	Aerobic	Aerobic	Anaerobic
Oxidative capacity	High	High	Low
Glycolytic ability	Low	High	High
Twitch rate	Slow	Fast	Fast
Myosin ATPase content	Low	High	High

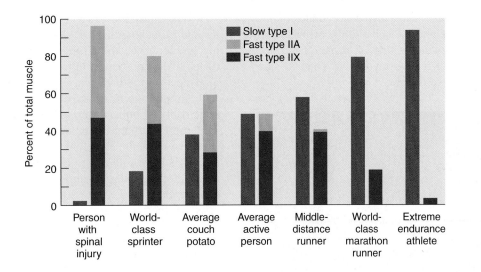

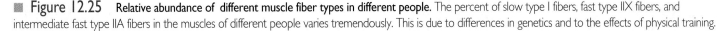

Figure 12.25 **Relative abundance of different muscle fiber types in different people.** The percent of slow type I fibers, fast type IIX fibers, and intermediate fast type IIA fibers in the muscles of different people varies tremendously. This is due to differences in genetics and to the effects of physical training.

A given somatic motor axon, however, innervates muscle fibers of one type only. The sizes of these motor units differ; the motors units composed of slow-twitch fibers tend to be smaller (have fewer fibers) than the motor units of fast-twitch fibers. As mentioned earlier, motor units are recruited from smaller to larger when increasing effort is required; thus, the smaller motor units with slow-twitch fibers would be used most often in routine activities. Larger motor units with fast-twitch fibers, which can exert a great deal of force but which respire anaerobically and thus fatigue quickly, would be used relatively infrequently and for only short periods of time.

Muscle Fatigue

Muscle fatigue may be defined as any exercise-induced reduction in the ability of a muscle to generate force or power. Fatigue during a sustained maximal contraction, when all the motor units are used and the rate of neural firing is maximal—as when lifting an extremely heavy weight—appears to be due to an accumulation of extracellular K^+. (Remember that K^+ leaves axons and muscle fibers during the repolarization phase of action potentials.) This reduces the membrane potential of muscle fibers and interferes with their ability to produce action potentials. Fatigue under these circumstances lasts only a short time, and maximal tension can again be produced after less than a minute's rest.

Muscle fatigue during moderate exercise occurs as the slow-twitch fibers deplete their reserve glycogen and fast-twitch fibers are increasingly recruited. Fast-twitch fibers obtain their energy through anaerobic respiration, converting glucose to lactic acid, and this results in a rise in intracellular H^+ and a fall in pH. The decrease in muscle pH, in turn, promotes muscle fatigue, but the exact physiological mechanisms by which this occurs are not well understood. The increased concentration of H^+ may interfere with cross-bridge formation, but other factors may also be involved. Depletion of glycogen, production of lactic

acid, and other metabolic changes have been shown to somehow interfere with the ability of the sarcoplasmic reticulum to release Ca^{2+} when a muscle fiber is stimulated by a nerve. This interference with excitation-contraction coupling, rather than a depletion of muscle ATP, appears to underlie muscle fatigue.

The foregoing is a description of the reasons that muscle tissue can fatigue during exercise. When humans exercise, however, we often experience fatigue *before* our muscles themselves have fatigued sufficiently to limit exercise. Put another way, our maximum voluntary muscle force is often less than the maximum force the our muscle is itself capable of producing. This demonstrates **central fatigue**—muscle fatigue caused by changes in the CNS rather than by fatigue of the muscles themselves. During exercise, a progressive reduction in the voluntary activation of muscles demonstrates central fatigue.

Evidence suggests that central fatigue is complex. In part, it involves a reduced ability of the "upper motoneurons" (in the brain) to drive the "lower motoneurons" (in the spinal cord). Muscle fatigue thus has two major components: a peripheral component (fatigue in the muscles themselves) and a central component (fatigue in the activation of muscles by motoneurons).

Adaptations of Muscles to Exercise Training

The maximal oxygen uptake, obtained during very strenuous exercise, averages 50 ml of O_2 per minute per kilogram body weight in males between the ages of 20 and 25 (females average 25% lower). For trained endurance athletes (such as swimmers and long-distance runners), maximal oxygen uptakes can be as high as 86 ml of O_2 per minute per kilogram. These considerable differences affect the lactate threshold, and thus the amount of exercise that can be performed before lactic acid production contributes to muscle fatigue. In addition to having a higher aerobic capacity, well-trained athletes also have a lactate threshold that is

Table 12.5 Effects of Endurance Training on Skeletal Muscles

1. Improved ability to obtain ATP from oxidative phosphorylation
2. Increased size and number of mitochondria
3. Less lactic acid produced per given amount of exercise
4. Increased myoglobin content
5. Increased intramuscular triglyceride content
6. Increased lipoprotein lipase (enzyme needed to utilize lipids from blood)
7. Increased proportion of energy derived from fat; less from carbohydrates
8. Lower rate of glycogen depletion during exercise
9. Improved efficiency in extracting oxygen from blood
10. Decreased number of type IIX (fast glycolytic) fibers; increased number of type IIA (fast oxidative) fibers

a higher percentage of their $\dot{V}_{O_2}$ max. The lactate threshold of an untrained person, for example, might be 60% of the $\dot{V}_{O_2}$ max, whereas the lactate threshold of a trained athlete can be up to 80% of the $\dot{V}_{O_2}$ max. These athletes thus produce less lactic acid at a given level of exercise than the average person, and therefore they are less subject to fatigue than the average person.

Since the depletion of muscle glycogen places a limit on exercise, any adaptation that spares muscle glycogen will improve physical endurance. This is achieved in trained athletes by an increased proportion of energy that is derived from the aerobic respiration of fatty acids, resulting in a slower depletion of their muscle glycogen. The greater the level of physical training, the higher the proportion of energy derived from the oxidation of fatty acids during exercise below the $\dot{V}_{O_2}$ max.

All fiber types adapt to endurance training by an increase in mitochondria, and thus in aerobic respiratory enzymes. In fact, the maximal oxygen uptake can be increased by as much as 20% through endurance training. There is a decrease in type IIX (fast glycolytic) fibers, which have a low oxidative capacity, accompanied by an increase in type IIA (fast oxidative) fibers, which have a high oxidative capacity. Although the type IIA fibers are still classified as fast-twitch, they show an increase in the slow myosin ATPase isoenzyme form, indicating that they are in a transitional state between the type II and type I fibers. A summary of the changes that occur as a result of endurance training is presented in table 12.5.

Endurance training does not increase the size of muscles. Muscle enlargement is produced only by frequent periods of high-intensity exercise in which muscles work against a high resistance, as in weightlifting. As a result of resistance training, type II muscle fibers become thicker, and the muscle therefore grows by hypertrophy (an increase in cell size rather than number of cells). This happens first because the myofibrils within a muscle fiber thicken because of the synthesis of actin and myosin proteins and the addition of new sarcomeres. Then, after a myofibril has attained a certain thickness, it may split into two myofibrils, each of which may become thicker as a result of the addition of sarcomeres. Muscle hypertrophy, in short, is associated with an increase in the size of the myofibrils, and then in the number of myofibrils within the muscle fibers.

The decline in physical strength of older people is associated with a reduced muscle mass, which is due to a loss of muscle fibers and to a decrease in the size of fast-twitch muscle fibers. Aging is also associated with a reduced density of blood capillaries surrounding the muscle fibers, leading to a decrease in oxidative capacity. Resistance training can cause the surviving muscle fibers to hypertrophy and become stronger, partially compensating for the decline in the number of muscle fibers in elderly people. Endurance training can increase the density of blood capillaries in the muscles, improving the ability of the blood to deliver oxygen to the muscles. The muscle glycogen of older people can also be increased by endurance training, but it cannot be raised to the levels present in youth.

Test Yourself Before You Continue

1. Draw a figure illustrating the relationship between ATP and creatine phosphate, and explain the physiological significance of this relationship.
2. Describe the characteristics of slow- and fast-twitch fibers (including intermediate fibers). Explain how the fiber types are determined and list the functions of different fiber types.
3. Explain the different causes of muscle fatigue with reference to the various fiber types.
4. Describe the effects of endurance training and resistance training on the fiber characteristics of muscles.

Neural Control of Skeletal Muscles

Skeletal muscles contain stretch receptors called muscle spindles that stimulate the production of impulses in sensory neurons when a muscle is stretched. These sensory neurons can synapse with alpha motoneurons, which stimulate the muscle to contract in response to the stretch. Other motor neurons, called gamma motoneurons, stimulate the tightening of the spindles and thus increase their sensitivity.

Motor neurons in the spinal cord, or **lower motor neurons** (often shortened to *motoneurons*), are those previously described that have cell bodies in the spinal cord and axons within nerves that stimulate muscle contraction (table 12.6). The activity of these neurons is influenced by (1) sensory feedback from the muscles and tendons and (2) facilitory and inhibitory effects from **upper motor neurons** in the brain that contribute axons to descending motor tracts. Lower motor neurons are thus said to be the *final common pathway* by which sensory stimuli and higher brain centers exert control over skeletal movements.

Table 12.6 A Partial Listing of Terms Used to Describe the Neural Control of Skeletal Muscles

Term	Description
1. Lower motoneurons	Neurons whose axons innervate skeletal muscles—also called the "final common pathway" in the control of skeletal muscles
2. Higher motoneurons	Neurons in the brain that are involved in the control of skeletal movements and that act by facilitating or inhibiting (usually by way of interneurons) the activity of the lower motoneurons
3. Alpha motoneurons	Lower motoneurons whose fibers innervate ordinary (extrafusal) muscle fibers
4. Gamma motoneurons	Lower motoneurons whose fibers innervate the muscle spindle fibers (intrafusal fibers)
5. Agonist/antagonist	A pair of muscles or muscle groups that insert on the same bone, the agonist being the muscle of reference
6. Synergist	A muscle whose action facilitates the action of the agonist
7. Ipsilateral/contralateral	Ipsilateral—located on the same side, or the side of reference; contralateral—located on the opposite side
8. Afferent/efferent	Afferent neurons—sensory; efferent neurons—motor

The disease known as **amyotrophic lateral sclerosis (ALS)** involves degeneration of the lower motor neurons, leading to skeletal muscle atrophy and paralysis. This disease is sometimes called Lou Gehrig's disease, after the baseball player who suffered from it, and also includes the famous physicist Steven Hawking among its victims. Scientists have recently learned that the inherited form of this disease is caused by a defect in the gene for a specific enzyme—*superoxide dismutase.* This enzyme is responsible for eliminating superoxide free radicals, which are highly toxic products that can damage the motor neurons. The mutant gene produces an enzyme that has a different, and in fact destructive, action.

The cell bodies of lower motor neurons are located in the ventral horn of the gray matter of the spinal cord (chapter 8). Axons from these cell bodies leave the ventral side of the spinal cord to form the *ventral roots* of spinal nerves (see fig. 8.24). The *dorsal roots* of spinal nerves contain sensory fibers whose cell bodies are located in the *dorsal root ganglia.* Both sensory (*afferent*) and motor (*efferent*) fibers join in a common connective tissue sheath to form the spinal nerves at each segment of the spinal cord. In the lumbar region there are about 12,000 sensory and 6,000 motor fibers per spinal nerve.

About 375,000 cell bodies have been counted in a lumbar segment—a number far larger than can be accounted for by the number of motor neurons. Most of these neurons do not contribute fibers to the spinal nerve. Rather, they serve as *interneurons,* whose fibers conduct impulses up, down, and across the central nervous system. Those fibers that conduct impulses to higher spinal cord segments and the brain form *ascending tracts,* and those that conduct to lower spinal segments contribute to *descending tracts.* Those fibers that cross the midline of the CNS to synapse on the opposite side are part of *commissural tracts.* Interneurons can thus conduct impulses up and down on the same, or *ipsilateral,* side, and can affect neurons on the opposite, or *contralateral,* side of the central nervous system.

Muscle Spindle Apparatus

In order for the nervous system to control skeletal movements properly, it must receive continuous sensory feedback information concerning the effects of its actions. This sensory information includes (1) the tension that the muscle exerts on its tendons, provided by the **Golgi tendon organs,** and (2) muscle length, provided by the **muscle spindle apparatus.** The spindle apparatus, so called because it is wider in the center and tapers toward the ends, functions as a length detector. Muscles that require the finest degree of control, such as the muscles of the hand, have the highest density of spindles.

Each spindle apparatus contains several thin muscle cells, called *intrafusal fibers* (*fusus* = spindle), packaged within a connective tissue sheath. Like the stronger and more numerous "ordinary" muscle fibers outside the spindles—the *extrafusal fibers*—the spindles insert into tendons on each end of the muscle. Spindles are therefore said to be in parallel with the extrafusal fibers.

Unlike the extrafusal fibers, which contain myofibrils along their entire length, the contractile apparatus is absent from the central regions of the intrafusal fibers. The central, noncontracting part of an intrafusal fiber contains nuclei. There are two types of intrafusal fibers. One type, the *nuclear bag fibers,* have their nuclei arranged in a loose aggregate in the central regions of the fibers. The other type of intrafusal fibers, called *nuclear chain fibers,* have their nuclei arranged in rows. Two types of sensory neurons serve these intrafusal fibers. **Primary,** or **annulospiral, sensory endings** wrap around the central regions of the nuclear bag and chain fibers (fig. 12.26), and **secondary,** or **flower-spray, endings** are located over the contracting poles of the nuclear chain fibers.

Since the spindles are arranged in parallel with the extrafusal muscle fibers, stretching a muscle causes its spindles to stretch. This stimulates both the primary and secondary sensory endings. The spindle apparatus thus serves as a length detector because the frequency of impulses produced in the primary and secondary endings is proportional to the length of the muscle. The primary endings, however, are most stimulated at the onset of stretch, whereas the secondary endings respond in a more tonic (sustained) fashion as stretch is maintained. Sudden, rapid stretching of a muscle activates both types of sensory endings, and is thus a more powerful

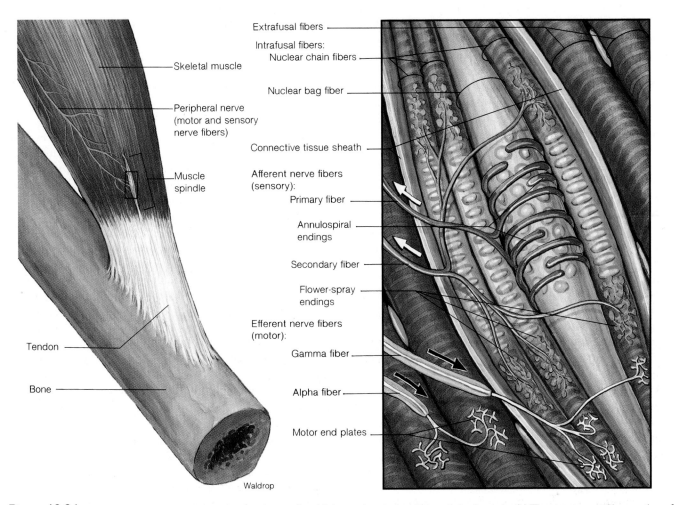

Extrafusal fibers
Intrafusal fibers:
Nuclear chain fibers
Nuclear bag fiber
Connective tissue sheath
Afferent nerve fibers
(sensory):
Primary fiber
Annulospiral
endings
Secondary fiber
Flower-spray
endings
Efferent nerve fibers
(motor):
Gamma fiber
Alpha fiber
Motor end plates

Skeletal muscle

Peripheral nerve
(motor and sensory
nerve fibers)

Muscle
spindle

Tendon

Bone

Waldrop

■ **Figure 12.26** **The location and structure of a muscle spindle.** *(a)* A muscle spindle within a skeletal muscle. *(b)* The structure and innervation of a muscle spindle.

stimulus for the muscle spindles than a slower, more gradual stretching that has less of an effect on the primary sensory endings. Since the activation of the sensory endings in muscle spindles produces a reflex contraction, the force of this reflex contraction is greater in response to rapid stretch than to gradual stretch.

Rapid stretching of skeletal muscles produces very forceful muscle contractions as a result of the activation of primary and secondary endings in the muscle spindles and the monosynaptic stretch reflex. This can result in painful muscle spasms, as may occur, for example, when muscles are forcefully pulled in the process of setting broken bones. Painful muscle spasms may be avoided in physical exercise by stretching slowly and thereby stimulating mainly the secondary endings in the muscle spindles. A slower rate of stretch also allows time for the inhibitory Golgi tendon organ reflex to occur and promote muscle relaxation.

Alpha and Gamma Motoneurons

In the spinal cord, two types of lower motor neurons innervate skeletal muscles. The motor neurons that innervate the extrafusal muscle fibers are called **alpha motoneurons;** those that innervate the intrafusal fibers are called **gamma motoneurons** (fig. 12.26). The alpha motoneurons are faster conducting (60 to 90 meters per second) than the thinner gamma motoneurons (10 to 40 meters per second). Since only the extrafusal muscle fibers are sufficiently strong and numerous to cause a muscle to shorten, only stimulation by the alpha motoneurons can cause muscle contraction that results in skeletal movements.

The intrafusal fibers of the muscle spindle are stimulated to contract by gamma motoneurons, which represent one-third of all efferent fibers in spinal nerves. However, because the intrafusal fibers are too few in number and their contraction too weak to cause a muscle to shorten, stimulation by gamma motoneurons results only in isometric contraction of the spindles. Since myofibrils are present in the poles but absent in the central regions of intrafusal fibers, the more distensible central region of the intrafusal fiber is pulled toward the ends in response to stimulation by gamma motoneurons. As a result, the spindle is tightened. This effect of gamma motoneurons, which is sometimes termed *active stretch* of the spindles, serves to increase the sensitivity of the spindles when the entire muscle is passively stretched by external forces. The activation of gamma motoneurons thus enhances the stretch reflex and is an important factor in the voluntary control of skeletal movements.

Coactivation of Alpha and Gamma Motoneurons

Most of the fibers in the descending motor tracts synapse with interneurons in the spinal cord; only about 10% of the descending fibers synapse directly with the lower motor neurons. It is likely that very rapid movements are produced by direct synapses with the lower motor neurons, whereas most other movements are produced indirectly via synapses with spinal interneurons, which in turn stimulate the motor neurons.

Upper motor neurons—neurons in the brain that contribute fibers to descending motor tracts—usually stimulate alpha and gamma motoneurons simultaneously. Such stimulation is known as **coactivation.** Stimulation of alpha motoneurons results in muscle contraction and shortening; stimulation of gamma motoneurons stimulates contraction of the intrafusal fibers, and thus "takes out the slack" that would otherwise be present in the spindles as the muscles shorten. In this way, the spindles remain under tension and provide information about the length of the muscle even while the muscle is shortening.

Under normal conditions, the activity of gamma motoneurons is maintained at the level needed to keep the muscle spindles under proper tension while the muscles are relaxed. Undue relaxation of the muscles is prevented by stretch and activation of the spindles, which in turn elicits a reflex contraction (described in the next section). This mechanism produces a normal resting muscle length and state of tension, or **muscle tone.**

Skeletal Muscle Reflexes

Although skeletal muscles are often called voluntary muscles because they are controlled by descending motor pathways that are under conscious control, they often contract in an uncon-

scious, reflex fashion in response to particular stimuli. In the simplest type of reflex, a skeletal muscle contracts in response to the stimulus of muscle stretch. More complex reflexes involve inhibition of antagonistic muscles and regulation of a number of muscles on both sides of the body.

The Monosynaptic Stretch Reflex

Reflex contraction of skeletal muscles occurs in response to sensory input and does not depend on the activation of upper motor neurons. The **reflex arc,** which describes the nerve impulse pathway from sensory to motor endings in such reflexes, involves only a few synapses within the CNS. The simplest of all reflexes—the *muscle stretch reflex*—consists of only one synapse within the CNS. The sensory neuron directly synapses with the motor neuron, without involving spinal cord interneurons. The stretch reflex is thus a **monosynaptic reflex** in terms of the individual reflex arcs (although, of course, many sensory neurons are activated at the same time, leading to the activation of many motor neurons). Resting skeletal muscles are maintained at an optimal length, as previously described under the heading "Length-Tension Relationship," by stretch reflexes.

The stretch reflex is present in all muscles, but it is most dramatic in the extensor muscles of the limbs. The **knee-jerk reflex**—the most commonly evoked stretch reflex—is initiated by striking the patellar ligament with a rubber mallet. This stretches the entire body of the muscle, and thus passively stretches the spindles within the muscle so that sensory nerves with primary (annulospiral) endings in the spindles are activated. Axons of these sensory neurons synapse within the ventral gray matter of the spinal cord with *alpha motoneurons.* These large, fast-conducting motor nerve fibers stimulate the extrafusal fibers of the extensor muscle, resulting in isotonic contraction and the knee jerk. This is an example of negative feedback—stretching of the muscles (and spindles) stimulates shortening of the muscles (and spindles). These events are summarized in table 12.7 and illustrated in figure 12.27.

Golgi Tendon Organs

The **Golgi tendon organs** continuously monitor tension in the tendons produced by muscle contraction or passive stretching of a muscle. Sensory neurons from these receptors synapse with interneurons in the spinal cord; these interneurons, in turn, have

Table 12.7   Summary of Events in a Monosynaptic Stretch Reflex

1. Passive stretch of a muscle (produced by tapping its tendon) stretches the spindle (intrafusal) fibers.
2. Stretching of a spindle distorts its central (bag or chain) region, which stimulates dendritic endings of sensory nerves.
3. Action potentials are conducted by afferent (sensory) nerve fibers into the spinal cord on the dorsal roots of spinal nerves.
4. Axons of sensory neurons synapse with dendrites and cell bodies of somatic motor neurons located in the ventral horn gray matter of the spinal cord.
5. Efferent nerve impulses in the axons of somatic motor neurons (which form the ventral roots of spinal nerves) are conducted to the ordinary (extrafusal) muscle fibers. These neurons are alpha motoneurons.
6. Release of acetylcholine from the endings of alpha motoneurons stimulates the contraction of the extrafusal fibers, and thus of the whole muscle.
7. Contraction of the muscle relieves the stretch of its spindles, thus decreasing electrical activity in the spindle afferent nerve fibers.

inhibitory synapses (via IPSPs and postsynaptic inhibition—chapter 7) with motor neurons that innervate the muscle (fig. 12.28). This inhibitory Golgi tendon organ reflex is called a **disynaptic reflex** (because two synapses are crossed in the CNS), and it helps to prevent excessive muscle contractions or excessive passive muscle stretching. Indeed, if a muscle is stretched extensively, it will actually relax as a result of the inhibitory effects produced by the Golgi tendon organs.

Damage to spinal nerves, or to the cell bodies of lower motor neurons (by poliovirus, for example), produces a **flaccid paralysis,** characterized by reduced muscle tone, depressed stretch reflexes, and atrophy. Damage to upper motor neurons or descending motor tracts at first produces spinal shock in which there is a flaccid paralysis. This is followed in a few weeks by **spastic paralysis,** characterized by increased muscle tone, exaggerated stretch reflexes, and other signs of hyperactive lower motor neurons.

The appearance of spastic paralysis suggests that upper motor neurons normally exert an inhibitory effect on lower alpha and gamma motor neurons. When this inhibition is removed, the gamma motoneurons become hyperactive and the spindles thus become overly sensitive to stretch. This can be demonstrated dramatically by forcefully dorsiflecting the patient's foot (pushing it up) and then releasing it. Forced extension stretches the antagonistic flexor muscles, which contract and produce the opposite movement (plantar flexion). Alternative activation of antagonistic stretch reflexes produces a flapping motion known as *clonus.*

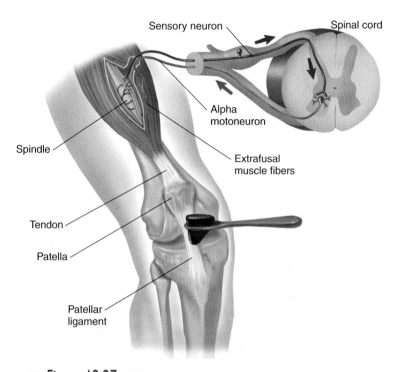

Figure 12.27   **The knee-jerk reflex.** This is an example of a monosynaptic stretch reflex.

Reciprocal Innervation and the Crossed-Extensor Reflex

In the knee-jerk and other stretch reflexes, the sensory neuron that stimulates the motor neuron of a muscle also stimulates interneurons within the spinal cord via collateral branches. These interneurons inhibit the motor neurons of antagonist muscles via inhibitory postsynaptic potentials (IPSPs). This dual stimulatory and inhibitory activity is called **reciprocal innervation** (fig. 12.29).

When a limb is flexed, for example, the antagonistic extensor muscles are passively stretched. Extension of a limb similarly stretches the antagonistic flexor muscles. If the monosynaptic stretch reflexes were not inhibited, reflex contraction of the

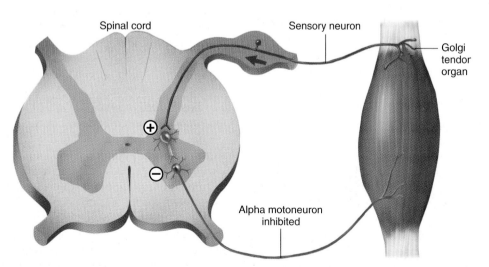

Figure 12.28   **The action of the Golgi tendon organ.** An increase in muscle tension stimulates the activity of sensory nerve endings in the Golgi tendon organ. This sensory input stimulates an interneuron, which in turn inhibits the activity of a motor neuron innervating that muscle. This is therefore a disynaptic reflex.

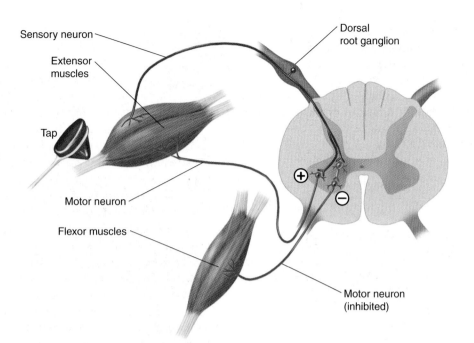

Sensory neuron

Extensor muscles

Tap

Motor neuron

Flexor muscles

Dorsal root ganglion

⊕

⊖

Motor neuron (inhibited)

■ **Figure 12.29**   **A diagram of reciprocal innervation.** Afferent impulses from muscle spindles stimulates alpha motoneurons to the agonists muscle (the extensor) directly, but (via an inhibitory interneuron) they inhibit activity in the alpha motoneuron to the antagonist muscle.

antagonistic muscles would always interfere with the intended movement. Fortunately, whenever the "intended," or agonist muscles, are stimulated to contract, the alpha and gamma motoneurons that stimulate the antagonist muscles are inhibited.

The stretch reflex, with its reciprocal innervations, involves the muscles of one limb only and is controlled by only one segment of the spinal cord. More complex reflexes involve muscles controlled by numerous spinal cord segments and affect muscles on the contralateral side of the cord. Such reflexes involve **double reciprocal innervation** of muscles.

Double reciprocal innervation is illustrated by the **crossed-extensor reflex.** If you step on a tack with your right foot, for example, this foot is withdrawn by contraction of the flexors and relaxation of the extensors of your right leg. The contralateral left leg, by contrast, extends to help support your body during this withdrawal reflex. The extensors of your left leg contract while its flexors relax. These events are illustrated in figure 12.30.

Upper Motor Neuron Control of Skeletal Muscles

As previously described, upper motor neurons are neurons in the brain that influence the control of skeletal muscle by lower motor neurons (alpha and gamma motoneurons). Neurons in the precentral gyrus of the cerebral cortex contribute axons that cross to the contralateral sides in the pyramids of the medulla oblongata; these tracts are thus called **pyramidal tracts** (chapter 8). The pyramidal tracts include the *lateral* and *ventral corticospinal tracts.* Neurons in other areas of the brain produce the **extrapyramidal tracts.** The major extrapyramidal tract is the *reticulospinal tract,* which originates in the reticular formation of the medulla oblongata and pons. Brain areas that influence the activity of extrapyramidal tracts are believed to produce the inhibition of lower motor neurons described in the preceding section.

Cerebellum

The **cerebellum,** like the cerebrum, receives sensory input from muscle spindles and Golgi tendon organs. It also receives fibers from areas of the cerebral cortex devoted to vision, hearing, and equilibrium.

There are no descending tracts from the cerebellum. The cerebellum can influence motor activity only indirectly, through its output to the vestibular nuclei, red nucleus, and basal nuclei. These structures, in turn, affect lower motor neurons via the vestibulospinal tract, rubrospinal tract, and reticulospinal tract. It is interesting that all output from the cerebellum is inhibitory; these inhibitory effects aid motor coordination by eliminating inappropriate neural activity. Damage to the cerebellum interferes with the ability to coordinate movements with spatial judgment. Under- or overreaching for an object may occur, followed by *intention tremor,* in which the limb moves back and forth in a pendulum-like motion.

Basal Nuclei

The **basal nuclei,** sometimes called the **basal ganglia,** include the *caudate nucleus, putamen,* and *globus pallidus* (chapter 8; see fig. 8.11). Often included in this group are other nuclei of the *thalamus, subthalamus, substantia nigra,* and *red nucleus.* Acting directly via the rubrospinal tract and indirectly via synapses in the reticular formation and thalamus, the basal nuclei have profound effects on the activity of lower motor neurons.

In particular, through their synapses in the reticular formation, the basal nuclei exert an inhibitory influence on the activity of lower motor neurons. Damage to the basal nuclei thus results in increased muscle tone, as previously described. People with such damage display *akinesia,* lack of desire to use the affected limb, and *chorea,* sudden and uncontrolled random movements (table 12.8).

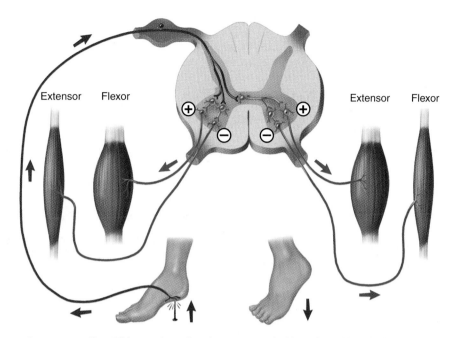

■ Figure 12.30 **The crossed-extensor reflex.** This complex reflex demonstrates double reciprocal innervation.

Table 12.8 Symptoms of Upper Motor Neuron Damage

Babinski's reflex—Extension of the great toe when the sole of the foot is stroked along the lateral border

Spastic paralysis—High muscle tone and hyperactive stretch reflexes; flexion of arms and extension of legs

Hemiplegia—Paralysis of upper and lower limbs on one side—commonly produced by damage to motor tracts as they pass through internal capsule (such as by cerebrovascular accident—stroke)

Paraplegia—Paralysis of the lower limbs on both sides as a result of lower spinal cord damage

Quadriplegia—Paralysis of upper and lower limbs on both sides as a result of damage to the upper region of the spinal cord or brain

Chorea—Random uncontrolled contractions of different muscle groups (as in Saint Vitus' dance) as a result of damage to basal nuclei

Resting tremor—Shaking of limbs at rest; disappears during voluntary movements; produced by damage to basal nuclei

Intention tremor—Oscillations of the arm following voluntary reaching movements; produced by damage to cerebellum

Test Yourself Before You Continue

1. Draw a muscle spindle surrounded by a few extrafusal fibers. Indicate the location of primary and secondary sensory endings and explain how these endings respond to muscle stretch.

2. Describe all of the events that occur from the time the patellar tendon is struck with a mallet to the time the leg kicks.

3. Explain how a Golgi tendon organ is stimulated and describe the disynaptic reflex that occurs.

4. Explain the significance of reciprocal innervation and double reciprocal innervation in muscle reflexes.

5. Describe the functions of gamma motoneurons and explain why they are stimulated at the same time as alpha motoneurons during voluntary muscle contractions.

6. Explain how a person with spinal cord damage might develop clonus.

CLINICAL

Parkinson's disease (or *paralysis agitans*) is a disorder of the basal nuclei involving degeneration of fibers from the substantia nigra. These fibers, which use dopamine as a neurotransmitter, are required to antagonize the effects of other fibers that use acetylcholine (ACh) as a transmitter. The relative deficiency of dopamine compared to ACh is believed to produce the symptoms of Parkinson's disease, including resting tremor. This "shaking" of the limbs tends to disappear during voluntary movements and then reappear when the limb is again at rest.

Cardiac and Smooth Muscles

Cardiac muscle, like skeletal muscle, is striated and contains sarcomeres that shorten by sliding of thin and thick filaments. But while skeletal muscle requires nervous stimulation to contract, cardiac muscle can produce impulses and contract spontaneously. Smooth muscles lack sarcomeres, but they do contain actin and myosin that produce contractions in response to a unique regulatory mechanism.

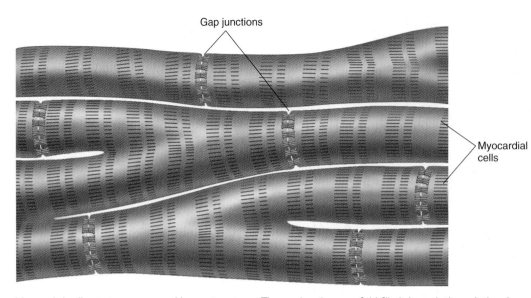

Gap junctions

Myocardial cells

■ Figure 12.31 **Myocardial cells are interconnected by gap junctions.** The gap junctions are fluid-filled channels through the plasma membrane of adjacent cells that permit the conduction of impulses from one cell to the next. The gap junctions are concentrated at the ends of each myocardial cell, and each gap junction is composed of connexin proteins (also see chapter 7, fig. 7.19).

Unlike skeletal muscles, which are voluntary effectors regulated by somatic motor neurons, cardiac and smooth muscles are involuntary effectors regulated by autonomic motor neurons. Although there are important differences between skeletal muscle and cardiac and smooth muscle, there are also significant similarities. All types of muscle are believed to contract by means of sliding of thin filaments over thick filaments. The sliding of the filaments is produced by the action of myosin cross bridges in all types of muscles, and excitation-contraction coupling in all types of muscles involves Ca^{2+}.

Cardiac Muscle

Like skeletal muscle cells, cardiac (heart) muscle cells, or **myocardial cells,** are striated; they contain actin and myosin filaments arranged in the form of sarcomeres, and they contract by means of the sliding filament mechanism. The long, fibrous skeletal muscle cells, however, are structurally and functionally separated from each other, whereas the myocardial cells are short, branched, and interconnected. Each myocardial cell is tubular in structure and joined to adjacent myocardial cells by electrical synapses, or **gap junctions** (see chapter 7, fig. 7.19).

The gap junctions are concentrated at the ends of each myocardial cell (fig. 12.31), which permits electrical impulses to be conducted primarily along the long axis from cell to cell. Gap junctions in cardiac muscle have an affinity for stain that makes them appear as dark lines between adjacent cells when viewed in the light microscope. These dark-staining lines are known as *intercalated discs* (fig. 12.32).

Electrical impulses that originate at any point in a mass of myocardial cells, called a **myocardium,** can spread to all cells in the mass that are joined by gap junctions. Because all cells in a myocardium are electrically joined, a myocardium behaves as a single functional unit. Thus, unlike skeletal muscles that produce contractions that are graded depending on the number of

Nucleus

Intercalated discs

■ Figure 12.32 **Cardiac muscle.** Notice that the cells are short, branched, and striated and that they are interconnected by intercalated discs.

cells stimulated, a myocardium contracts to its full extent each time because all of its cells contribute to the contraction. The ability of the myocardial cells to contract, however, can be increased by the hormone epinephrine and by stretching of the heart chambers. The heart contains two distinct myocardia (atria and ventricles), as will be described in chapter 13.

Cardiac muscle, like skeletal muscle, contains the troponin complex of three proteins (see fig. 12.13). **Troponin I** helps inhibit the binding of the myosin crossbridges to actin; **troponin T** binds to tropomyosin in the thin filaments, and **troponin C** binds to Ca^{2+} for muscle contraction. Damage to myocardial cells, which occurs in **myocardial infarction,** causes troponins to be released into the blood. Fortunately for clinical diagnosis, troponins T and I are slightly different in cardiac muscle than in skeletal muscle. Thus, troponins T and I released by damaged myocardial cells can be distinguished and measured by laboratory tests using specific antibodies. Such tests are now an important tool in the diagnosis of myocardial infarction.

Unlike skeletal muscles, which require external stimulation by somatic motor nerves before they can produce action potentials and contract, cardiac muscle is able to produce action potentials automatically. Cardiac action potentials normally originate in a specialized group of cells called the *pacemaker.* However, the rate of this spontaneous depolarization, and thus the rate of the heartbeat, are regulated by autonomic innervation. Regulation of the cardiac rate is described more fully in chapter 14.

Smooth Muscle

Smooth (visceral) muscles are arranged in circular layers in the walls of blood vessels and bronchioles (small air passages in the lungs). Both circular and longitudinal smooth muscle layers occur in the tubular digestive tract, the ureters (which transport urine), the ductus deferentia (which transport sperm cells), and the uterine tubes (which transport ova). The alternate contraction of circular and longitudinal smooth muscle layers in the intestine produces **peristaltic waves,** which propel the contents of these tubes in one direction.

Although smooth muscle cells do not contain sarcomeres (which produce striations in skeletal and cardiac muscle), they do contain a great deal of actin and some myosin, which produces a ratio of thin to thick filaments of about 16 to 1 (in striated muscles the ratio is 2 to 1). Unlike striated muscles, in which the thin filaments are relatively short (extending from a Z disc into a sarcomere), the thin filaments of smooth muscle cells are quite long. They attach either to regions of the plasma membrane of the smooth muscle cell or to cytoplasmic protein structures called **dense bodies,** which are analogous to the Z discs of striated muscle (fig. 12.33*b*).

In smooth muscle, the myosin proteins of the thick filaments are stacked vertically so that their long axis is perpendicular to the long axis of the thick filament (fig. 12.33*c*). In this way, the myosin heads can form cross bridges with actin all along the length of the thick filaments. This is different from the horizontal arrangement of myosin proteins in the thick filaments of striated muscles (see fig. 12.10), which is required to cause the shortening of sarcomeres.

The arrangement of the contractile apparatus in smooth muscle cells, and the fact that it is not organized into sarcomeres, is required for proper smooth muscle function. Smooth muscles must be able to contract even when greatly stretched—in the urinary bladder, for example, the smooth muscle cells may be stretched up to two and a half times their resting length. The smooth muscle cells of the uterus may be stretched up to eight times their original length by the end of pregnancy. Striated muscles, because of their structure, lose their ability to contract when the sarcomeres are stretched to the point where actin and myosin no longer overlap.

Excitation-Contraction Coupling in Smooth Muscles

As in striated muscles, the contraction of smooth muscles is triggered by a sharp rise in the Ca^{2+} concentration within the cytoplasm of the muscle cells. However, the sarcoplasmic reticulum of smooth muscles is less developed than that of skeletal muscles, and Ca^{2+} released from this organelle may account for

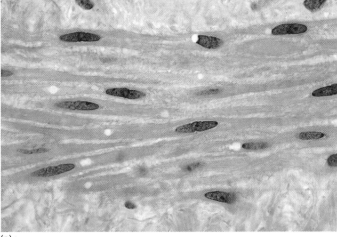

(a)

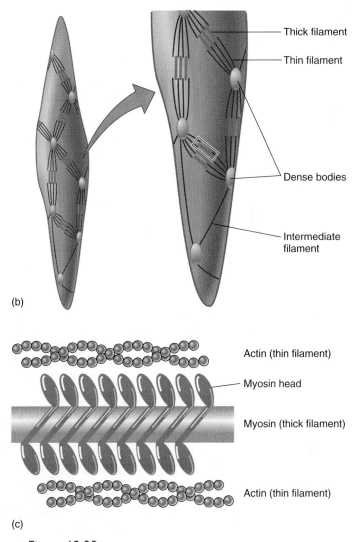

(b)

(c)

■ **Figure 12.33** Smooth muscle and its contractile apparatus.
(a) A photomicrograph of smooth muscle cells in the wall of a blood vessel. *(b)* Arrangement of thick and thin filaments in smooth muscles. Note that dense bodies are also interconnected by intermediate fibers. *(c)* The myosin proteins are stacked in a different arrangement in smooth muscles than in striated muscles.

only the initial phase of smooth muscle contraction. Extracellular Ca^{2+} diffusing into the smooth muscle cell through its plasma membrane is responsible for sustained contractions. This Ca^{2+} enters primarily through voltage-regulated Ca^{2+} channels in the plasma membrane. The opening of these channels is graded by the amount of depolarization; the greater the depolarization, the more Ca^{2+} will enter the cell and the stronger will be the smooth muscle contraction.

The events that follow the entry of Ca^{2+} into the cytoplasm are somewhat different in smooth muscles than in striated muscles. In striated muscles, Ca^{2+} combines with troponin. Troponin, however, is not present in smooth muscle cells. In smooth muscles, Ca^{2+} combines with a protein in the cytoplasm called **calmodulin,** which is structurally similar to troponin. Calmodulin was previously discussed in relation to the function of Ca^{2+} as a second messenger in hormone action (chapter 11). The calmodulin-Ca^{2+} complex thus formed combines with and activates **myosin light-chain kinase (MLCK),** an enzyme that cat-

alyzes the phosphorylation (addition of phosphate groups) of *myosin light chains,* a component of the myosin cross bridges. In smooth muscle (unlike striated muscle), the phosphorylation of myosin cross bridges is the regulatory event that permits them to bind to actin and thereby produce a contraction (fig. 12.34).

Unlike the situation in striated muscle cells, which produce all-or-none action potentials, smooth muscle cells can produce graded depolarizations and contractions without producing action potentials. Indeed, only these graded depolarizations are conducted from cell to cell in many smooth muscles. The greater the depolarization of a smooth muscle cell, the more Ca^{2+} will enter, and the more MLCK enzymes will be activated. With more MLCK enzymes activated, more cross bridges will become phosphorylated and able to bind to actin. In this way, a stronger depolarization of the smooth muscle cell leads to a stronger contraction.

Relaxation of the smooth muscle follows the closing of the Ca^{2+} channels and lowering of the cytoplasmic Ca^{2+} concentrations by the action of Ca^{2+}-ATPase active transport pumps.

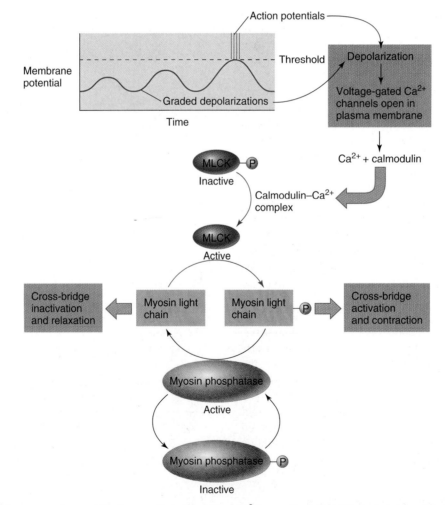

■ **Figure 12.34** Excitation-contraction coupling in smooth muscle. When Ca^{2+} passes through voltage-gated channels in the plasma membrane it enters the cytoplasm and binds to calmodulin. The calmodulin-Ca^{2+} complex then activates myosin light-chain kinase (MLCK) by removing a phosphate group. The activated MLCK, in turn, phosphorylates the myosin light chains, thereby activating the cross bridges to cause contraction. Contraction is ended when myosin phosphatase becomes activated. Upon its activation, myosin phosphatase removes the phosphates from the myosin light chains and thereby inactivates the cross bridges.

Under these conditions, calmodulin dissociates from the myosin light-chain kinase, thereby inactivating this enzyme. The phosphate groups that were added to the myosin are then removed by a different enzyme, a *myosin phosphatase* (fig. 12.34). Dephosphorylation inhibits the cross bridge from binding to actin and producing another power stroke.

In addition to being graded, the contractions of smooth muscle cells are slow and sustained. The slowness of contraction is related to the fact that myosin ATPase in smooth muscle is slower in its action (splitting ATP for the cross-bridge cycle) than it is in striated muscle. The sustained nature of smooth muscle contraction is explained by the theory that cross bridges in smooth muscles can enter a *latch state.*

The latch state allows smooth muscle to maintain its contraction in a very energy-efficient manner, hydrolyzing less ATP than would otherwise be required. This ability is obviously important for smooth muscles, given that they encircle the walls of hollow organs and must sustain contractions for long periods of time. The mechanisms by which the latch state is produced, however, are complex and poorly understood.

The three muscle types—skeletal, cardiac, and smooth—are compared in table 12.9.

Drugs such as *nifedipine (Procardia)* and related newer compounds are **calcium channel blockers.** These drugs block Ca^{2+} channels in the membrane of smooth muscle cells within the walls of blood vessels, causing the muscles to relax and the vessels to dilate. This effect, called vasodilation, may be helpful in treating some cases of hypertension (high blood pressure). Calcium-channel-blocking drugs are also used when spasm of the coronary arteries (vasospasm) produces angina pectoris, which is pain caused by insufficient blood flow to the heart.

Clinical Investigation Clues

Remember that Maria was taking a calcium-channel-blocking drug to treat her hypertension.

How do such drugs help to lower blood pressure?

Is it likely that this drug contributed to Maria's skeletal muscle pain and fatigue?

Could it raise her blood Ca^{2+} levels?

If not, what could raise her blood Ca^{2+}?

Single-Unit and Multiunit Smooth Muscles

Smooth muscles are often grouped into two functional categories: **single-unit** and **multiunit** (fig. 12.35). Single-unit smooth muscles have numerous gap junctions (electrical synapses) between adjacent cells that weld them together electrically; they thus behave as a single unit, much like cardiac muscle. Most smooth muscles—including those in the digestive tract and uterus—are single-unit.

Only some cells of single-unit smooth muscles receive autonomic innervation, but the ACh released by the axon can diffuse to other smooth muscle cells. Binding of ACh to its muscarinic receptors causes depolarization by closing K^+ channels, as described in chapter 9 (see fig. 9.11). Such stimulation, however, only modifies the automatic behavior of single-unit smooth muscles. Single-unit smooth muscles display *pacemaker* activity, in which certain cells stimulate others in the mass. This is similar to the situation in cardiac muscle. Single-unit smooth muscles also display intrinsic, or *myogenic,* electrical activity and contraction in response to stretch. For example, the stretch induced by an increase in the volume of a ureter or a section of the digestive tract can stimulate myogenic contraction. Such contraction does not require stimulation by autonomic nerves.

Contraction of multiunit smooth muscles, by contrast, requires nerve stimulation. Multiunit smooth muscles have few, if

Table 12.9 Comparison of Skeletal, Cardiac, and Smooth Muscle

Skeletal Muscle	Cardiac Muscle	Smooth Muscle
Striated; actin and myosin arranged in sarcomeres	Striated; actin and myosin arranged in sarcomeres	Not striated; more actin than myosin; actin inserts into dense bodies and cell membrane
Well-developed sarcoplasmic reticulum and transverse tubules	Moderately developed sarcoplasmic reticulum and transverse tubules	Poorly developed sarcoplasmic reticulum; no transverse tubules
Contains troponin in the thin filaments	Contains troponin in the thin filaments	Contains calmodulin, a protein that, when bound to Ca^{2+}, activates the enzyme myosin light-chain kinase
Ca^{2+} released into cytoplasm from sarcoplasmic reticulum	Ca^{2+} enters cytoplasm from sarcoplasmic reticulum and extracellular fluid	Ca^{2+} enters cytoplasm from extracellular fluid, sarcoplasmic reticulum, and perhaps mitochondria
Cannot contract without nerve stimulation; denervation results in muscle atrophy	Can contract without nerve stimulation; action potentials originate in pacemaker cells of heart	Maintains tone in absence of nerve stimulation; visceral smooth muscle produces pacemaker potentials; denervation results in hypersensitivity to stimulation
Muscle fibers stimulated independently; no gap junctions	Gap junctions present as intercalated discs	Gap junctions generally present

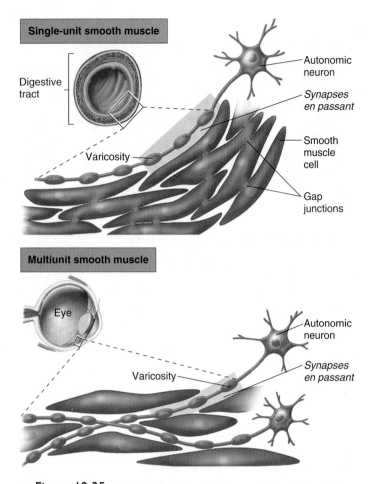

Single-unit smooth muscle

Digestive tract

Varicosity

Autonomic neuron

Synapses en passant

Smooth muscle cell

Gap junctions

Multiunit smooth muscle

Eye

Varicosity

Autonomic neuron

Synapses en passant

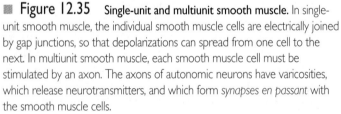

■ Figure 12.35   **Single-unit and multiunit smooth muscle.** In single-unit smooth muscle, the individual smooth muscle cells are electrically joined by gap junctions, so that depolarizations can spread from one cell to the next. In multiunit smooth muscle, each smooth muscle cell must be stimulated by an axon. The axons of autonomic neurons have varicosities, which release neurotransmitters, and which form *synapses en passant* with the smooth muscle cells.

any, gap junctions. The cells must thus be stimulated individually by nerve fibers. Examples of multiunit smooth muscles are the arrector pili muscles in the skin and the ciliary muscles attached to the lens of the eye.

Autonomic Innervation of Smooth Muscles

The neural control of skeletal muscles differs significantly from that of smooth muscles. A skeletal muscle fiber has only one junction with a somatic nerve fiber, and the receptors for the neurotransmitter are located only at the neuromuscular junction. By contrast, the entire surface of smooth muscle cells contains neurotransmitter receptor proteins. Neurotransmitter molecules are released along a stretch of an autonomic nerve fiber that is located some distance from the smooth muscle cells. The regions of the autonomic fiber that release transmitters appear as bulges, or *varicosities,* and the neurotransmitters released from these varicosities stimulate a number of smooth muscle cells. Since there are numerous varicosities along a stretch of an autonomic nerve ending, they form synapses "in passing"—or *synapses en passant*—with the smooth muscle cells. This was described in chapter 9 (see figure 9.9) and is shown in figure 12.35.

Test Yourself Before You Continue

1. Explain how cardiac muscle differs from skeletal muscle in its structure and regulation of contraction.
2. Contrast the structure of a smooth muscle cell with that of a skeletal muscle fiber and discuss the advantages of each type of structure.
3. Describe the events by which depolarization of a smooth muscle cell results in contraction and explain why smooth muscle contractions are slow and sustained.
4. Distinguish between single-unit and multiunit smooth muscles.

HPer Links of the Muscular System with Other Body Systems

Integumentary System

- The skin helps to protect all organs of the body from invasion by pathogens .(p. 446)
- The smooth muscles of cutaneous blood vessels are needed for the regulation of cutaneous blood flow . . . .(p. 427)
- The arrector pili muscles in the skin produce goose bumps . . . .(p. 358)

Skeletal System

- Bones store calcium, which is needed for the control of muscle contraction .(p. 623)
- The skeleton provides attachment sites for muscles . . . .(p. 326)
- Joints of the skeleton provide levers for movement . . . .(p. 326)
- Muscle contractions maintain the health and strength of bone . . . .(p. 623)

Nervous System

- Somatic motor neurons stimulate contraction of skeletal muscles . . .(p. 154)
- Autonomic neurons stimulate smooth muscle contraction or relaxation .(p. 220)
- Autonomic nerves increase cardiac output during exercise . . . .(p. 424)
- Sensory neurons from muscles monitor muscle length and tension . . . . . .(p. 348)

Endocrine System

- Sex hormones promote muscle development and maintenance . . .(p. 609)
- Parathyroid hormone and other hormones regulate blood calcium and phosphate concentrations . . . .(p. 624)

- Epinephrine and norepinephrine influence contractions of cardiac muscle and smooth muscles . . . .(p. 227)
- Insulin promotes glucose entry into skeletal muscles . . . .(p. 611)
- Adipose tissue secretes hormones that regulate the sensitivity of muscles to insulin . . . .(p. 606)

Circulatory System

- Blood transports O_2 and nutrients to muscles and removes CO_2 and lactic acid . . . .(p. 366)
- Contractions of skeletal muscles serve as a pump to assist blood movement within veins . . . .(p. 394)
- Cardiac muscle enables the heart to function as a pump . . . .(p. 378)
- Smooth muscle enables blood vessels to constrict and dilate . . . .(p. 390)

Respiratory System

- The lungs provide oxygen for muscle metabolism and eliminate carbon dioxide . . . .(p. 480)
- Respiratory muscles enable ventilation of the lungs . . . .(p. 488)

Urinary System

- The kidneys eliminate creatinine and other metabolic wastes from muscle . . .(p. 524)
- The kidneys help to regulate the blood calcium and phosphate concentrations . . . .(p. 627)
- Muscles of the urinary tract are needed for the control of urination . . . .(p. 525)

Digestive System

- The GI tract provides nutrients for all body organs, including muscles . . . .(p. 561)
- Smooth muscle contractions push digestion products along the GI tract . . . . . .(p. 564)
- Muscular sphincters of the GI tract help to regulate the passage of food . . . . .(p. 565)

Reproductive System

- Testicular androgen promotes growth of skeletal muscle . . . .(p. 645)
- Muscle contractions contribute to orgasm in both sexes . . . .(p. 643)
- Uterine muscle contractions are required for vaginal delivery of a fetus . . . . .(p. 675)

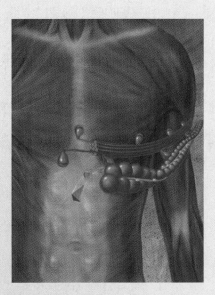

Summary

Skeletal Muscles　326

I. Skeletal muscles are attached to bones by tendons.
  A. Skeletal muscles are composed of separate cells, or fibers, that are attached in parallel to the tendons.
  B. Individual muscle fibers are covered by the endomysium; bundles of fibers, called fascicles, are covered by the perimysium; and the entire muscle is covered by the epimysium.
  C. Skeletal muscle fibers are striated.
    1. The dark striations are called A bands, and the light regions are called I bands.
    2. Z lines are located in the middle of each I band.

II. The contraction of muscle fibers *in vivo* is stimulated by somatic motor neurons.
  A. Each somatic motor axon branches to innervate numerous muscle fibers.
  B. The motor neuron and the muscle fibers it innervates are called a motor unit.
    1. When a muscle is composed of a relatively large number of motor units (such as in the hand), there is fine control of muscle contraction.
    2. The large muscles of the leg have relatively few motor units, which are correspondingly large in size.
    3. Sustained contractions are produced by the asynchronous stimulation of different motor units.

Mechanisms of Contraction　330

I. Skeletal muscle cells, or fibers, contain structures called myofibrils.
  A. Each myofibril is striated with dark (A) and light (I) bands. In the middle of each I band are Z lines.
  B. The A bands contain thick filaments, composed primarily of myosin.
    1. The edges of each A band also contain thin filaments, which overlap the thick filaments.
    2. The central regions of the A bands contain only thick filaments—these regions are the H bands.
  C. The I bands contain only thin filaments, composed primarily of actin.
  D. Thin filaments are composed of globular actin subunits known as G-actin. A protein known as tropomyosin is also located at intervals in the thin filaments. Another protein—troponin—is attached to the tropomyosin.

II. Myosin cross bridges extend out from the thick filaments to the thin filaments.
  A. At rest, the cross bridges are not attached to actin.
    1. The cross-bridge heads function as ATPase enzymes.
    2. ATP is split into ADP and P_i, activating the cross bridge.
  B. When the activated cross bridges attach to actin, they release P_i and undergo a power stroke.
  C. At the end of a power stroke, the cross bridge releases the ADP and binds to a new ATP.
    1. This allows the cross bridge to detach from actin and repeat the cycle.
    2. Rigor mortis is caused by the inability of cross bridges to detach from actin because of a lack of ATP.

III. The activity of the cross bridges causes the thin filaments to slide toward the centers of the sarcomeres.
  A. The filaments slide—they do not shorten—during muscle contraction.
  B. The lengths of the H and I bands decrease, whereas the A bands stay the same length during contraction.

IV. When a muscle is at rest, the Ca^{2+} concentration of the sarcoplasm is very low and cross bridges are prevented from attaching to actin.
  A. The Ca^{2+} is actively transported into the sarcoplasmic reticulum.
  B. The sarcoplasmic reticulum is a modified endoplasmic reticulum that surrounds the myofibrils.

V. Action potentials are conducted by transverse tubules into the muscle fiber.
  A. Transverse tubules are invaginations of the cell membrane that almost touch the sarcoplasmic reticulum.
  B. Action potentials in the transverse tubules stimulate the opening of Ca^{2+}-release channels in the sarcoplasmic reticulum, causing Ca^{2+} to diffuse into the sarcoplasm and stimulate contractions.

VI. When action potentials cease, the Ca^{2+}-release channels in the sarcoplasmic reticulum close.
  A. This allows the active transport Ca^{2+}-ATPase pumps in the sarcoplasmic reticulum to accumulate Ca, removing it from the sarcoplasm and sarcomeres.
  B. As a result of the removal of Ca^{2+} from troponin, the muscle relaxes.

Contractions of Skeletal Muscles　340

I. Muscles *in vitro* can exhibit twitch, summation, and tetanus.
  A. The rapid contraction and relaxation of muscle fibers is called a twitch.
  B. A whole muscle also produces a twitch in response to a single electrical pulse *in vitro*.
    1. The stronger the electric shock, the stronger the muscle twitch—whole muscles can produce graded contractions.
    2. The graded contraction of whole muscles is due to different numbers of fibers participating in the contraction.
  C. The summation of fiber twitches can occur so rapidly that the muscle produces a smooth, sustained contraction known as tetanus.
  D. When a muscle exerts tension without shortening, the contraction is termed isometric; when shortening does occur, the contraction is isotonic.
  E. When a muscle contracts but, despite its contraction, is made to lengthen due to the application of an external force, the contraction is said to be eccentric.

II. The series-elastic component refers to the elastic composition of the muscle and its associated structures, which must be stretched tight before the tension exerted by the muscle can cause movement.

III. The strength of a muscle contraction is dependant upon its resting length.

    A. If the muscle is too short or too long prior to stimulation, the filaments in the sarcomeres will not have an optimum amount of overlap.

    B. At its normal resting length *in vivo,* a muscle is at its optimum length for contraction.

Energy Requirements of Skeletal Muscles   342

I. Aerobic cell respiration is ultimately required for the production of ATP needed for cross-bridge activity.

    A. Resting muscles and muscles performing light exercise obtain most of their energy from fatty acids.

    B. During moderate exercise, just below the lactate threshold, energy is obtained about equally from fatty acids and glucose.

    C. Glucose, from the muscle's stored glycogen and from blood plasma, becomes an increasingly important energy source during heavy exercise.

    D. New ATP can be quickly produced from the combination of ADP with phosphate derived from phosphocreatine.

    E. Muscle fibers are of three types.

        1. Slow-twitch red fibers are adapted for aerobic respiration and are resistant to fatigue.

        2. Fast-twitch white fibers are adapted for anaerobic respiration.

        3. Intermediate fibers are fast-twitch but adapted for aerobic respiration.

II. Muscle fatigue may be caused by a number of mechanisms.

    A. Fatigue during sustained maximal contraction may be produced by the accumulation of extracellular K^+ as a result of high levels of nerve activity.

    B. Fatigue during moderate exercise is primarily a result of anaerobic respiration by fast-twitch fibers.

        1. The productions of lactic acid and consequent fall in pH, the depletion of muscle glycogen, and other metabolic changes interfere with the release of Ca^{2+} from the sarcoplasmic reticulum.

        2. Interference with excitation contraction coupling, rather than depletion of ATP, appears to be responsible for muscle fatigue.

    C. In human exercise, however, fatigue is often caused by changes in the CNS before the muscles themselves fatigue; this central fatigue reduces the force of voluntary contractions.

III. Physical training affects the characteristics of the muscle fibers.

    A. Endurance training increases the aerobic capacity of muscle fibers and their use of fatty acids for energy, so that their reliance on glycogen and anaerobic respiration—and thus their susceptibility to fatigue—is reduced.

    B. Resistance training causes hypertrophy of muscle fibers because of an increase in the size and number of myofibrils.

Neural Control of Skeletal Muscles   347

I. The somatic motor neurons that innervate the muscles are called lower motor neurons.

    A. Alpha motoneurons innervate the ordinary, or extrafusal, muscle fibers. These are the fibers that produce muscle shortening during contraction.

    B. Gamma motoneurons innervate the intrafusal fibers of the muscle spindles.

II. Muscle spindles function as length detectors in muscles.

    A. Spindles consist of several intrafusal fibers wrapped together. The spindles are in parallel with the extrafusal fibers.

    B. Stretching of the muscle stretches the spindles, which excites sensory endings in the spindle apparatus.

        1. Impulses in the sensory neurons travel into the spinal cord in the dorsal roots of spinal nerves.

        2. The sensory neuron synapses directly with an alpha motoneuron within the spinal cord, which produces a monosynaptic reflex.

        3. The alpha motoneuron stimulates the extrafusal muscle fibers to contract, thus relieving the stretch. This is called the stretch reflex.

    C. The activity of gamma motoneurons tightens the spindles, thus making them more sensitive to stretch and better able to monitor the length of the muscle, even during muscle shortening.

III. The Golgi tendon organs monitor the tension that the muscle exerts on its tendons.

    A. As the tension increases, sensory neurons from Golgi tendon organs inhibit the activity of alpha motoneurons.

    B. This is a disynaptic reflex because the sensory neurons synapse with interneurons, which in turn make inhibitory synapses with motoneurons.

IV. A crossed-extensor reflex occurs when a foot steps on a tack.

    A. Sensory input from the injured foot causes stimulation of flexor muscles and inhibition of the antagonistic extensor muscles.

    B. The sensory input also crosses the spinal cord to cause stimulation of extensor and inhibition of flexor muscles in the contralateral leg.

V. Most of the fibers of descending tracts synapse with spinal interneurons, which in turn synapse with the lower motor neurons.

    A. Alpha and gamma motoneurons are usually stimulated at the same time, or coactivated.

    B. The stimulation of gamma motoneurons keeps the muscle spindles under tension and sensitive to stretch.

    C. Upper motor neurons, primarily in the basal nuclei, also exert inhibitory effects on gamma motoneurons.

VI. Neurons in the brain that affect the lower motor neurons are called upper motor neurons.

    A. The fibers of neurons in the precentral gyrus, or motor cortex, descend to the lower motor neurons as the lateral and ventral corticospinal tracts.

        1. Most of these fibers cross to the contralateral side in the brain stem, forming structures called the pyramids; therefore, this system is called the pyramidal system.

2. The left side of the brain thus controls the musculature on the right side, and vice versa.

B. Other descending motor tracts are part of the extrapyramidal system.

    1. The neurons of the extrapyramidal system make numerous synapses in different areas of the brain, including the midbrain, brain stem, basal nuclei, and cerebellum.

    2. Damage to the cerebellum produces intention tremor, and degeneration of dopaminergic neurons in the basal nuclei produces Parkinson's disease.

Cardiac and Smooth Muscles   353

I. Cardiac muscle is striated and contains sarcomeres.

A. In contrast to skeletal muscles, which require neural stimulation to contract, action potentials in the heart originate in myocardial cells; stimulation by neurons is not required.

B. Also unlike the situation in skeletal muscles, action potentials can cross from one myocardial cell to another.

II. The thin and thick filaments in smooth muscles are not organized into sarcomeres.

A. The thin filaments extend from the plasma membrane and from dense bodies in the cytoplasm.

B. The myosin proteins are stacked perpendicular to the long axis of the thick filaments, so they can bind to actin all along the length of the thick filament.

C. Depolarizations are graded and conducted from one smooth muscle cell to the next.

    1. The depolarizations stimulate the entry of Ca^{2+}, which binds to calmodulin; this complex then activates myosin light-chain kinase, which phosphorylates the myosin heads.

    2. Phosphorylation of the myosin heads is needed for them to be able to bind to actin and produce contractions.

D. Smooth muscles are classified as single-unit, if they are interconnected by gap junctions, and as multiunit if they are not so connected.

E. Autonomic neurons have varicosities that release neurotransmitter all along their length of contact with the smooth muscle cells, making *synapses en passant.*

Review Activities

Test Your Knowledge of Terms and Facts

1. A graded whole muscle contraction is produced *in vivo* primarily by variations in
   a. the strength of the fiber's contraction.
   b. the number of fibers that are contracting.
   c. both of these.
   d. neither of these.

2. The series-elastic component of muscle contraction is responsible for
   a. increased muscle shortening to successive twitches.
   b. a time delay between contraction and shortening.
   c. the lengthening of muscle after contraction has ceased.
   d. all of these.

3. Which of these muscles have motor units with the highest innervation ratio?
   a. leg muscles
   b. arm muscles
   c. muscles that move the fingers
   d. muscles of the trunk

4. The stimulation of gamma motoneurons produces
   a. isotonic contraction of intrafusal fibers.
   b. isometric contraction of intrafusal fibers.
   c. either isotonic or isometric contraction of intrafusal fibers.
   d. contraction of extrafusal fibers.

5. In a single reflex arc involved in the knee-jerk reflex, how many synapses are activated within the spinal cord?
   a. thousands
   b. hundreds
   c. dozens
   d. two
   e. one

6. Spastic paralysis may occur when there is damage to
   a. the lower motor neurons.
   b. the upper motor neurons.
   c. either the lower or the upper motor neurons.

7. When a skeletal muscle shortens during contraction, which of these statements is *false?*
   a. The A bands shorten.
   b. The H bands shorten.
   c. The I bands shorten.
   d. The sarcomeres shorten.

8. Electrical excitation of a muscle fiber *most directly* causes
   a. movement of tropomyosin.
   b. attachment of the cross bridges to action.
   c. release of Ca^{2+} from the sarcoplasmic reticulum.
   d. splitting of ATP.

9. The energy for muscle contraction is *most directly* obtained from
   a. phosphocreatine.
   b. ATP.
   c. anaerobic respiration.
   d. aerobic respiration.

10. Which of these statements about cross bridges is *false?*
   a. They are composed of myosin.
   b. They bind to ATP after they detach from actin.
   c. They contain an ATPase.
   d. They split ATP before they attach to actin.

11. When a muscle is stimulated to contract, Ca^{2+} binds to
   a. myosin.
   b. tropomyosin.
   c. actin.
   d. troponin.

12. Which of these statements about muscle fatigue is *false?*
   a. It may result when ATP is no longer available for the cross-bridge cycle.

    b. It may be caused by a loss of muscle cell Ca^{2+}.

    c. It may be caused by the accumulation of extracellular K^+.

    d. It may be a result of lactic acid production.

13. Which of these types of muscle cells *are not* capable of spontaneous depolarization?

    a. single-unit smooth muscle

    b. multiunit smooth muscle

    c. cardiac muscle

    d. skeletal muscle

    e. both *b* and *d*

    f. both *a* and *c*

14. Which of these muscle types is striated and contains gap junctions?

    a. single-unit smooth muscle

    b. multiunit smooth muscle

    c. cardiac muscle

    d. skeletal muscle

15. In an isotonic muscle contraction,

    a. the length of the muscle remains constant.

    b. the muscle tension remains constant.

    c. both muscle length and tension are changed.

    d. movement of bones does not occur.

Test Your Understanding of Concepts and Principles

1. Using the concept of motor units, explain how skeletal muscles *in vivo* produce graded and sustained contractions.[1]

2. Describe how an isometric contraction can be converted into an isotonic contraction using the concepts of motor unit recruitment and the series-elastic component of muscles.

3. Trace the sequence of events in which the cross bridges attach to the thin filaments when a muscle is stimulated by a nerve. Why don't the cross bridges attach to the thin filaments when a muscle is relaxed?

4. Using the sliding filament theory of contraction, explain why the contraction strength of a muscle is maximal at a particular muscle length.

5. Explain why muscle tone is first decreased and then increased when descending motor tracts are damaged. How is muscle tone maintained?

6. Explain the role of ATP in muscle contraction and muscle relaxation.

7. Why are all the muscle fibers of a given motor unit of the same type? Why are smaller motor units and slow-twitch muscle fibers used more frequently than larger motor units and fast-twitch fibers?

8. What changes occur in muscle metabolism as the intensity of exercise is increased? Describe the changes that occur as a result of endurance training and explain how these changes raise the level of exercise that can be performed before the onset of muscle fatigue.

9. Compare the mechanism of excitation-coupling in striated muscle with that in smooth muscle.

10. Compare cardiac muscle, single-unit smooth muscle, and multiunit smooth muscle with respect to the regulation of their contraction.

Test Your Ability to Analyze and Apply Your Knowledge

1. Your friend eats huge helpings of pasta for two days prior to a marathon, claiming such "carbo loading" is of benefit in the race. Is he right? What are some other things he can do to improve his performance?

2. Compare muscular dystrophy and amyotrophic lateral sclerosis (ALS) in terms of their causes and their effects on muscles.

3. Why is it important to have a large amount of stored high-energy phosphates in the form of creatine phosphate for the function of muscles during exercise? What might happen to a muscle in your body if it ever ran out of ATP?

4. How is electrical excitation of a skeletal muscle fiber coupled to muscle contraction? Speculate on why the exact mechanism of this coupling has been difficult to determine.

5. How would a rise in the extracellular Ca^{2+} concentration affect the beating of a heart? Explain the mechanisms involved. Lowering the blood Ca^{2+} concentration can cause muscle spasms. What might be responsible for this effect?

Related Websites

Check out the Links Library at www.mhhe.com/fox8 for links to sites containing resources related to the muscles. These links are monitored to ensure current URLs.

---

[1]*Note:* This question is answered in the chapter 12 Study Guide found on the Online Learning Center at www.mhhe.com/fox8.

13 Heart and Circulation

Objectives

After studying this chapter, you should be able to . . .

1. describe the general functions and major components of the circulatory system.

2. describe the composition of blood plasma and the physical characteristics and functions of the formed elements of the blood.

3. identify the chemical regulators of blood cell production and describe the process of erythropoiesis.

4. describe the ABO system of red blood cell antigens and explain the significance of the blood types.

5. explain how a blood clot is formed and how it is ultimately destroyed.

6. explain how the acid-base balance of the blood is affected by carbon dioxide and bicarbonate, and describe the roles of the lungs and kidneys in maintaining acid-base balance.

7. describe the path of the blood through the heart and the function of the atrioventricular and semilunar valves.

8. describe the structures and pathways of the pulmonary and systemic circulations.

9. describe the structures and pathways of electrical impulse conduction in the heart.

10. describe the electrical activity in the sinoatrial node and explain why this tissue functions as the heart's normal pacemaker.

11. relate the time involved in the production of an action potential to the time involved in the contraction of myocardial cells and explain the significance of this relationship.

12. describe the pressure changes that occur in the ventricles during the cardiac cycle and relate these changes to the action of the valves and the flow of blood.

13. explain the origin of the heart sounds and state when in the cardiac cycle these sounds are produced.

14. explain how electrocardiogram waves are produced and relate these waves to other events in the cardiac cycle.

15. compare the structure of an artery and vein, and explain how the structure of each type of vessel relates to its function.

16. describe the structure of capillaries and explain the physiological significance of this structure.

17. explain how atherosclerosis may develop and comment on the significance of this condition.

18. define *ischemia* and discuss the possible causes of myocardial ischemia.

19. describe some common arrhythmias that can be detected with an ECG.

20. describe the components and functions of the lymphatic system.

Refresh Your Memory

Before you begin this chapter, you may want to review these concepts from previous chapters:

- Action Potentials 161
- Functions of the Autonomic System 227
- Mechanisms of Contraction 330
- Cardiac and Smooth Muscle 353

Take Advantage of the Technology

Visit the Online Learning Center for these additional study resources.

- Interactive quizzing
- Online study guide
- Current news feeds
- Crossword puzzles
- Vocabulary flashcards
- Labeling activities

www.mhhe.com/fox8

Chapter at a Glance

Clinical Investigation

Jason is a 19-year-old college student who goes to the doctor complaining of chronic fatigue. The doctor palpates Jason's radial pulse and discovers that it is fast and weak. An echocardiogram and later coronary arteriograph reveal that he has a ventricular septal defect and mitral stenosis. His electrocardiogram (ECG) indicates that he has sinus tachycardia. When laboratory test results are returned, they indicate that Jason has a very high plasma cholesterol concentration with a high LDL/HDL ratio.

What can be concluded from these findings, and how are they related to Jason's complaint of chronic fatigue?

Functions and Components of the Circulatory System

Blood serves numerous functions, including the transport of respiratory gases, nutritive molecules, metabolic wastes, and hormones. Blood is transported through the body in a system of vessels leading from and returning to the heart.

A unicellular organism can provide for its own maintenance and continuity by performing the wide variety of functions needed for life. By contrast, the complex human body is composed of specialized cells that demonstrate a division of labor. The specialized cells of a multicellular organism depend on one another for the very basics of their existence; since most are firmly implanted in tissues, they must have their oxygen and nutrients brought to them, and their waste products removed. Therefore, a highly effective means of transporting materials within the body is needed.

The blood serves this transportation function. An estimated 60,000 miles of vessels throughout the body of an adult ensure that continued sustenance reaches each of the trillions of living cells. But then, too, the blood can serve to transport disease-causing viruses, bacteria, and their toxins. To guard against this, the circulatory system has protective mechanisms—the white blood cells and the lymphatic system. In order to perform its various functions, the circulatory system works together with the respiratory, urinary, digestive, endocrine, and integumentary systems in maintaining homeostasis.

Functions of the Circulatory System

The functions of the circulatory system can be divided into three broad areas: transportation, regulation, and protection.

1. **Transportation.** All of the substances essential for cellular metabolism are transported by the circulatory system. These substances can be categorized as follows:
   a. *Respiratory.* Red blood cells, or *erythrocytes,* transport oxygen to the cells. In the lungs, oxygen from the inhaled air attaches to hemoglobin molecules within the erythrocytes and is transported to the cells for aerobic respiration. Carbon dioxide produced by cell respiration is carried by the blood to the lungs for elimination in the exhaled air.
   b. *Nutritive.* The digestive system is responsible for the mechanical and chemical breakdown of food so that it can be absorbed through the intestinal wall into the blood and lymphatic vessels. The blood then carries these absorbed products of digestion through the liver and to the cells of the body.
   c. *Excretory.* Metabolic wastes (such as urea), excess water and ions, and other molecules not needed by the body are carried by the blood to the kidneys and excreted in the urine.
2. **Regulation.** The circulatory system contributes to both hormonal and temperature regulation.
   a. *Hormonal.* The blood carries hormones from their site of origin to distant target tissues, where they perform a variety of regulatory functions.
   b. *Temperature.* Temperature regulation is aided by the diversion of blood from deeper to more superficial cutaneous vessels or vice versa. When the ambient temperature is high, diversion of blood from deep to superficial vessels helps to cool the body, and when the ambient temperature is low, the diversion of blood from superficial to deeper vessels helps to keep the body warm.
3. **Protection.** The circulatory system protects against blood loss from injury and against foreign microbes or toxins introduced into the body.
   a. *Clotting.* The clotting mechanism protects against blood loss when vessels are damaged.
   b. *Immune.* The immune function of the blood is performed by the *leukocytes* (white blood cells) that protect against many disease-causing agents (pathogens).

Major Components of the Circulatory System

The **circulatory system** consists of two subdivisions: the cardiovascular system and the lymphatic system. The **cardiovascular system** consists of the heart and blood vessels, and the **lymphatic system** consists of lymphatic vessels and lymphoid tissues within the spleen, thymus, tonsils, and lymph nodes.

The **heart** is a four-chambered double pump. Its pumping action creates the pressure head needed to push blood through the vessels to the lungs and body cells. At rest, the heart of an adult pumps about 5 liters of blood per minute. At this rate, it takes about 1 minute for blood to be circulated to the most distal extremity and back to the heart.

Blood vessels form a tubular network that permits blood to flow from the heart to all the living cells of the body and then back to the heart. *Arteries* carry blood away from the heart, whereas *veins* return blood to the heart. Arteries and veins are continuous with each other through smaller blood vessels.

Arteries branch extensively to form a "tree" of progressively smaller vessels. The smallest of the arteries are called *arterioles*. Blood passes from the arterial to the venous system in microscopic *capillaries*, which are the thinnest and most numerous of the blood vessels. All exchanges of fluid, nutrients, and wastes between the blood and tissues occur across the walls of capillaries. Blood flows through capillaries into microscopic veins called *venules*, which deliver blood into progressively larger veins that eventually return the blood to the heart.

As blood *plasma* (the fluid portion of the blood) passes through capillaries, the hydrostatic pressure of the blood forces some of this fluid out of the capillary walls. Fluid derived from plasma that passes out of capillary walls into the surrounding tissues is called *tissue fluid,* or *interstitial fluid.* Some of this fluid returns directly to capillaries, and some enters into **lymphatic vessels** located in the connective tissues around the blood vessels. Fluid in lymphatic vessels is called *lymph.* This fluid is returned to the venous blood at specific sites. **Lymph nodes,** positioned along the way, cleanse the lymph prior to its return to the venous blood. The lymphatic system is thus considered a part of the circulatory system and is discussed at the end of this chapter.

Test Yourself Before You Continue

1. State the components of the circulatory system that function in oxygen transport, in the transport of nutrients from the digestive system, and in protection.
2. Describe the functions of arteries, veins, and capillaries.
3. Define the terms *interstitial fluid* and *lymph.* How do these fluids relate to blood plasma?

Composition of the Blood

Blood consists of formed elements that are suspended and carried in a fluid called plasma. The formed elements—erythrocytes, leukocytes, and platelets—function, respectively, in oxygen transport, immune defense, and blood clotting. Plasma contains different types of proteins and many water-soluble molecules.

The total blood volume in the average-sized adult is about 5 liters, constituting about 8% of the total body weight. Blood leaving the heart is referred to as *arterial blood.* Arterial blood, with the exception of that going to the lungs, is bright red because of a high concentration of oxyhemoglobin (the combination of oxygen and hemoglobin) in the red blood cells. *Venous blood* is blood returning to the heart. Except for the venous blood from the lungs, it contains less oxygen, and is therefore a darker red than the oxygen-rich arterial blood.

Blood is composed of a cellular portion, called *formed elements,* and a fluid portion, called *plasma.* When a blood sample is centrifuged, the heavier formed elements are packed into the bottom of the tube, leaving plasma at the top (fig. 13.1). The formed elements constitute approximately 45% of the total blood volume (a measurement called the *hematocrit*), and the plasma accounts for the remaining 55%.

Plasma

Plasma is a straw-colored liquid consisting of water and dissolved solutes. The major solute of the plasma in terms of its concentration is Na^+. In addition to Na^+, plasma contains many other ions, as well as organic molecules such as metabolites, hormones,

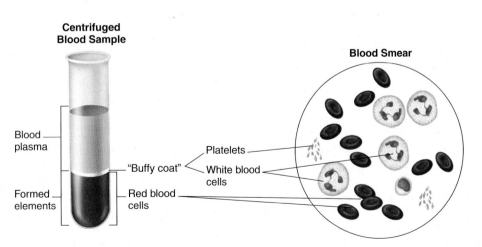

Centrifuged Blood Sample

Blood Smear

Blood plasma

"Buffy coat"

Platelets

White blood cells

Formed elements

Red blood cells

Red blood cells

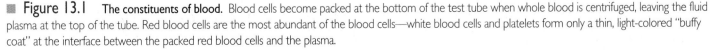

■ **Figure 13.1** **The constituents of blood.** Blood cells become packed at the bottom of the test tube when whole blood is centrifuged, leaving the fluid plasma at the top of the tube. Red blood cells are the most abundant of the blood cells—white blood cells and platelets form only a thin, light-colored "buffy coat" at the interface between the packed red blood cells and the plasma.

Table 13.1 Representative Normal Plasma Values

Measurement	Normal Range
Blood volume	80–85 ml/kg body weight
Blood osmolality	280–296 mOsm
Blood pH	7.35–7.45
Enzymes	
Creatine phosphokinase (CPK)	Female: 10–79 U/L
	Male: 17–148 U/L
Lactic dehydrogenase (LDH)	45–90 U/L
Phosphatase (acid)	Female: 0.01–0.56 Sigma U/ml
	Male: 0.13–0.63 Sigma U/ml
Hematology Values	
Hematocrit	Female: 37%–48%
	Male: 45%–52%
Hemoglobin	Female: 12–16 g/100 ml
	Male: 13–18 g/100 ml
Red blood cell count	4.2–5.9 million/mm³
White blood cell count	4,300–10,800/mm³
Hormones	
Testosterone	Male: 300–1,100 ng/100 ml
	Female: 25–90 ng/100 ml
Adrenocorticotrophic hormone (ACTH)	15–70 pg/ml
Growth hormone	Children: over 10 ng/ml
	Adult male: below 5 ng/ml
Insulin	6–26 μU/ml (fasting)
Ions	
Bicarbonate	24–30 mmol/l
Calcium	2.1–2.6 mmol/l
Chloride	100–106 mmol/l
Potassium	3.5–5.0 mmol/l
Sodium	135–145 mmol/l
Organic Molecules (Other)	
Cholesterol	120–220 mg/100 ml
Glucose	70–110 mg/100 ml (fasting)
Lactic acid	0.6–1.8 mmol/l
Protein (total)	6.0–8.4 g/100 ml
Triglyceride	40–150 mg/100 ml
Urea nitrogen	8–25 mg/100 ml
Uric acid	3–7 mg/100 ml

Source: Excerpted from material appearing in The New England Journal of Medicine, "Case Records of the Massachusetts General Hospital," 302:37–38 and 314:39–49. 1980, 1986.

enzymes, antibodies, and other proteins. The concentrations of some of these plasma constituents are shown in table 13.1.

Plasma Proteins

Plasma proteins constitute 7% to 9% of the plasma. The three types of proteins are albumins, globulins, and fibrinogen. **Albumins** account for most (60% to 80%) of the plasma proteins and are the smallest in size. They are produced by the liver and provide the osmotic pressure needed to draw water from the surrounding tissue fluid into the capillaries. This action is needed

to maintain blood volume and pressure. **Globulins** are grouped into three subtypes: **alpha globulins, beta globulins,** and **gamma globulins.** The alpha and beta globulins are produced by the liver and function in transporting lipids and fat-soluble vitamins. Gamma globulins are antibodies produced by lymphocytes (one of the formed elements found in blood and lymphoid tissues) and function in immunity. **Fibrinogen,** which accounts for only about 4% of the total plasma proteins, is an important clotting factor produced by the liver. During the process of clot formation (described later in this chapter), fibrinogen is converted into insoluble threads of *fibrin.* Thus, the fluid from clotted blood, called **serum,** does not contain fibrinogen, but it is otherwise identical to plasma.

Plasma Volume

A number of regulatory mechanisms in the body maintain homeostasis of the plasma volume. If the body should lose water, the remaining plasma becomes excessively concentrated—its osmolality (chapter 6) increases. This is detected by osmoreceptors in the hypothalamus, resulting in a sensation of thirst and the release of antidiuretic hormone (ADH) from the posterior pituitary (chapter 11). This hormone promotes water retention by the kidneys, which—together with increased intake of fluids—helps to compensate for the dehydration and lowered blood volume. This regulatory mechanism, together with others that influence plasma volume, are very important in maintaining blood pressure as described in chapter 14.

The Formed Elements of Blood

The **formed elements** of blood include two types of blood cells: *erythrocytes,* or *red blood cells,* and *leukocytes,* or *white blood cells.* Erythrocytes are by far the more numerous of the two. A cubic millimeter of blood contains 5.1 million to 5.8 million erythrocytes in males and 4.3 million to 5.2 million erythrocytes in females. The same volume of blood, by contrast, contains only 5,000 to 9,000 leukocytes.

Erythrocytes

Erythrocytes are flattened, biconcave discs, about 7 μm in diameter and 2.2 μm thick. Their unique shape relates to their function of transporting oxygen; it provides an increased surface area through which gas can diffuse (fig. 13.2). Erythrocytes lack nuclei and mitochondria (they obtain energy through anaerobic respiration). Partly because of these deficiencies, erythrocytes have a relatively short circulating life span of only about 120 days. Older erythrocytes are removed from the circulation by phagocytic cells in the liver, spleen, and bone marrow.

Each erythrocyte contains approximately 280 million **hemoglobin** molecules, which give blood its red color. Each hemoglobin molecule consists of four protein chains called *globins,* each of which is bound to one *heme,* a red-pigmented molecule that contains iron. The iron group of heme is able to combine with oxygen in the lungs and release oxygen in the tissues.

■ **Figure 13.2**   A scanning electron micrograph of red blood cells. As seen here, they are clinging to a hypodermic needle. Notice the shape of the red blood cells, sometimes described as a "biconcave disc."

Anemia refers to any condition in which there is an abnormally low hemoglobin concentration and/or red blood cell count. The most common type is **iron-deficiency anemia,** caused by a deficiency of iron, which is an essential component of the hemoglobin molecule. In **pernicious anemia** there is an inadequate amount of vitamin B_{12}, which is needed for red blood cell production. This is usually due to atrophy of the glandular mucosa of the stomach, which normally secretes a protein called *intrinsic factor.* In the absence of intrinsic factor, the vitamin B_{12} obtained in the diet cannot be absorbed by intestinal cells. **Aplastic anemia** is anemia due to destruction of the bone marrow, which may be caused by chemicals (including benzene and arsenic) or by radiation.

with blue-staining granules are called **basophils.** Those with granules that have little affinity for either stain are **neutrophils** (fig. 13.3). Neutrophils are the most abundant type of leukocyte, accounting for 50% to 70% of the leukocytes in the blood. Immature neutrophils have sausage-shaped nuclei and are called *band cells.* As the band cells mature, their nuclei become lobulated, with two to five lobes connected by thin strands. At this stage, the neutrophils are also known as *polymorphonuclear leukocytes (PMNs).*

There are two types of agranular leukocytes: lymphocytes and monocytes. **Lymphocytes** are usually the second most numerous type of leukocyte; they are small cells with round nuclei and little cytoplasm. **Monocytes,** by contrast, are the largest of the leukocytes and generally have kidney- or horseshoe-shaped nuclei. In addition to these two cell types, there are smaller numbers of *plasma cells,* which are derived from lymphocytes. Plasma cells produce and secrete large amounts of antibodies. The immune functions of the different white blood cells are described in more detail in chapter 15.

Blood cell counts are an important source of information in assessing a person's health. An abnormal increase in erythrocytes, for example, is termed **polycythemia** and is indicative of several dysfunctions. As previously mentioned, an abnormally low red blood cell count is termed *anemia.* (Polycythemia and anemia are described in detail in chapter 16.) An elevated leukocyte count, called **leukocytosis,** is often associated with infection (see chapter 15). A large number of immature leukocytes in a blood sample is diagnostic of the disease *leukemia.* A low white blood cell count, called **leukopenia,** may be due to a variety of factors; low numbers of lymphocytes, for example, may result from poor nutrition or from whole-body irradiation treatment for cancer.

Leukocytes

Leukocytes differ from erythrocytes in several respects. Leukocytes contain nuclei and mitochondria and can move in an amoeboid fashion. Because of their amoeboid ability, leukocytes can squeeze through pores in capillary walls and move to a site of infection, whereas erythrocytes usually remain confined within blood vessels. The movement of leukocytes through capillary walls is referred to as *diapedesis* or *extravasation.*

White blood cells are almost invisible under the microscope unless they are stained; therefore, they are classified according to their staining properties. Those leukocytes that have granules in their cytoplasm are called **granular leukocytes;** those without clearly visible granules are called **agranular** (or **nongranular) leukocytes.**

The stain used to identify white blood cells is usually a mixture of a pink-to-red stain called *eosin* and a blue-to-purple stain called a "basic stain." Granular leukocytes with pink-staining granules are therefore called **eosinophils,** and those

Platelets

Platelets, or **thrombocytes,** are the smallest of the formed elements and are actually fragments of large cells called *megakaryocytes,* which are found in bone marrow. (This is why the term *formed elements* is used instead of *blood cells* to describe erythrocytes, leukocytes, and platelets.) The fragments that enter the circulation as platelets lack nuclei but, like leukocytes, are capable of amoeboid movement. The platelet count per cubic millimeter of blood ranges from 130,000 to 400,000, but this count can vary greatly under different physiological conditions. Platelets survive for about 5 to 9 days before being destroyed by the spleen and liver.

Platelets play an important role in blood clotting. They constitute most of the mass of the clot, and phospholipids in their cell membranes activate the clotting factors in plasma that result in threads of fibrin, which reinforce the platelet plug. Platelets that attach together in a blood clot release *serotonin,* a chemical that stimulates constriction of the blood vessels, thus

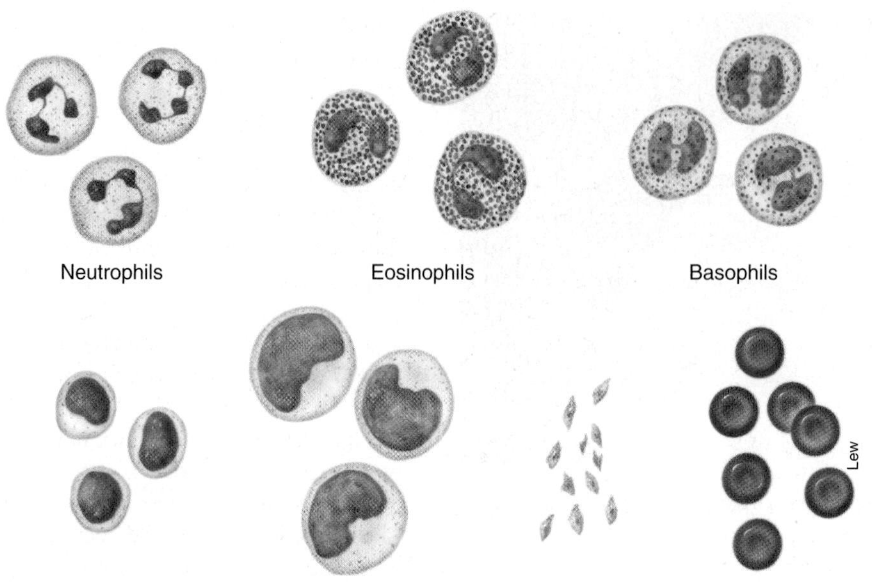

Figure 13.3 **The blood cells and platelets.** The white blood cells depicted above are granular leukocytes; the lymphocytes and monocytes are nongranular leukocytes.

Table 13.2   Formed Elements of the Blood

Component	Description	Number Present	Function
Erythrocyte (red blood cell)	Biconcave disc without nucleus; contains hemoglobin; survives 100 to 120 days	4,000,000 to 6,000,000 / mm³	Transports oxygen and carbon dioxide
Leukocytes (white blood cells)		5,000 to 10,000 / mm³	Aid in defense against infections by microorganisms
Granulocytes	About twice the size of red blood cells; cytoplasmic granules present; survive 12 hours to 3 days		
1. Neutrophil	Nucleus with 2 to 5 lobes; cytoplasmic granules stain slightly pink	54% to 62% of white cells present	Phagocytic
2. Eosinophil	Nucleus bilobed; cytoplasmic granules stain red in eosin stain	1% to 3% of white cells present	Helps to detoxify foreign substances; secretes enzymes that dissolve clots; fights parasitic infections
3. Basophil	Nucleus lobed; cytoplasmic granules stain blue in hematoxylin stain	Less than 1% of white cells present	Releases anticoagulant heparin
Agranulocytes	Cytoplasmic granules not visible; survive 100 to 300 days (some much longer)		
1. Monocyte	2 to 3 times larger than red blood cell; nuclear shape varies from round to lobed	3% to 9% of white cells present	Phagocytic
2. Lymphocyte	Only slightly larger than red blood cell; nucleus nearly fits cell	25% to 33% of white cells present	Provides specific immune response (including antibodies)
Platelet (thrombocyte)	Cytoplasmic fragment; survives 5 to 9 days	130,000 to 400,000 / mm³	Enables clotting; releases serotonin, which causes vasoconstriction

reducing the flow of blood to the injured area. Platelets also secrete growth factors (autocrine regulators—see chapter 11), which are important in maintaining the integrity of blood vessels. These regulators also may be involved in the development of atherosclerosis, as described in a later section.

The formed elements of the blood are illustrated in figure 13.3, and their characteristics are summarized in table 13.2.

Hematopoiesis

Blood cells are constantly formed through a process called **hematopoiesis** (also called **hemopoiesis**). The hematopoietic **stem cells**—those that give rise to blood cells—originate in the yolk sac of the human embryo and then migrate to the liver. Hematopoiesis thus occurs in the liver of the fetus. The stem cells then migrate to the bone marrow, and shortly after birth the liver ceases to be a source of blood cell production.

The term **erythropoiesis** refers to the formation of erythrocytes, and **leukopoiesis** to the formation of leukocytes. These processes occur in two classes of tissues after birth, myeloid and lymphoid. **Myeloid tissue** is the red bone marrow of the long bones, ribs, sternum, pelvis, bodies of the vertebrae, and portions of the skull. **Lymphoid tissue** includes the lymph nodes, tonsils, spleen, and thymus. The bone marrow produces all of the different types of blood cells; the lymphoid tissue produces lymphocytes derived from cells that originated in the bone marrow.

Hematopoiesis begins the same way in both myeloid and lymphoid tissue. A population of undifferentiated (unspecialized) cells gradually differentiate (specialize) to become stem cells, which give rise to the blood cells. At each step along the way the stem cells can duplicate themselves by mitosis, thus ensuring that the parent population will never become depleted. As the cells become differentiated, they develop membrane receptors for chemical signals that cause further development along particular lines. The earliest cells that can be distinguished under a microscope are the *erythroblasts* (which become erythrocytes), *myeloblasts* (which become granular leukocytes), *lymphoblasts* (which form lymphocytes), and *monoblasts* (which form monocytes).

Erythropoiesis is an extremely active process. It is estimated that about 2.5 million erythrocytes are produced every second in order to replace those that are continuously destroyed by the spleen and liver. The life span of an erythrocyte is approximately 120 days. Agranular leukocytes remain functional for 100 to 300 days under normal conditions. Granular leukocytes, by contrast, have an extremely short life span of 12 hours to 3 days.

The production of different subtypes of leukocytes is stimulated by chemicals called **cytokines.** These are autocrine regulators secreted by various cells of the immune system. The particular cytokines involved in leukopoiesis are discussed below. The production of red blood cells is stimulated by the hormone **erythropoietin,** which is secreted by the kidneys. The gene for erythropoietin has been commercially cloned, so that this hormone is now available for the treatment of the anemia that results from kidney disease in patients undergoing dialysis.

Scientists have identified a specific cytokine that stimulates proliferation of megakaryocytes and their maturation into platelets. By analogy with erythropoietin, they named this regulatory molecule **thrombopoietin.** The gene that codes for thrombopoietin also has been cloned, so that recombinant thrombopoietin is now available for medical research and applications. In clinical trials, thrombopoietin has been used to treat the thrombocytopenia (low platelet count) that occurs as a result of bone marrow depletion in patients undergoing chemotherapy for cancer.

Regulation of Leukopoiesis

A variety of cytokines stimulate different stages of leukocyte development. The cytokines known as *multipotent growth factor-1,*

interleukin-1, and *interleukin-3* have general effects, stimulating the development of different types of white blood cells. *Granulocyte colony-stimulating factor (G-CSF)* acts in a highly specific manner to stimulate the development of neutrophils, whereas *granulocyte-monocyte colony-stimulating factor (GM-CSF)* stimulates the development of monocytes and eosinophils. The genes for the cytokines G-CSF and GM-CSF have been cloned, making these cytokines available for medical applications.

Approximately 10,000 **bone marrow transplants** are performed worldwide each year. This procedure generally involves the aspiration of marrow from the iliac crest and separation of the hematopoietic stem cells, which constitute only about 1% of the nucleated cells in the marrow. Stem cells have also been isolated from peripheral blood when the donor is first injected with G-CSF and GM-CSF, which stimulate the marrow to release more stem cells. Another recent technology involves the storage, or "banking," of hematopoietic stem cells obtained from the placenta or umbilical cord blood of a neonate. These cells may then be used later in life if the person needs them for transplantation.

Regulation of Erythropoiesis

The primary regulator of erythropoiesis is erythropoietin, secreted by the kidneys whenever blood oxygen levels are decreased. One of the possible causes of decreased blood oxygen levels is a decreased red blood cell count. Because of erythropoietin stimulation, the daily production of new red blood cells compensates for the daily destruction of old red blood cells, preventing a decrease in the blood oxygen content. An increased secretion of erythropoietin and production of new red blood cells occurs when a person is at a high altitude or has lung disease, which are conditions that reduce the oxygen content of the blood.

Erythropoietin acts by binding to membrane receptors on cells that will become erythroblasts (fig. 13.4). The erythropoietin-stimulated cells undergo cell division and differentiation, leading to the production of erythroblasts. These are transformed into *normoblasts,* which lose their nuclei to become *reticulocytes.* The reticulocytes then change into fully mature erythrocytes. This process takes 3 days; the reticulocyte normally stays in the bone marrow for the first 2 days and then circulates in the blood on the third day. At the end of the erythrocyte life span of 120 days, the old red blood cells are removed by phagocytic cells of the spleen, liver, and bone marrow. Most of the iron contained in the hemoglobin molecules of the destroyed red blood cells is recycled back to the myeloid tissue to be used in the production of hemoglobin for new red blood cells (see chapter 18, fig. 18.23). The production of red blood cells and synthesis of hemoglobin depends on the supply of iron, along with that of vitamin B_{12} and folic acid.

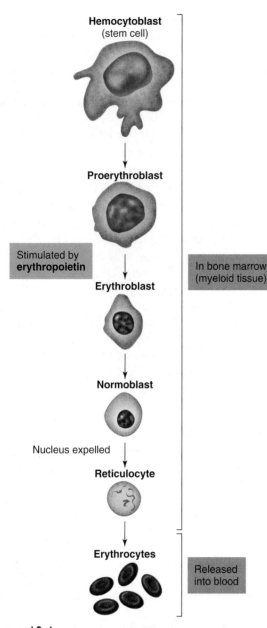

Hemocytoblast
(stem cell)

Proerythroblast

Stimulated by
erythropoietin

In bone marrow
(myeloid tissue)

Erythroblast

Normoblast

Nucleus expelled

Reticulocyte

Erythrocytes

Released
into blood

■ **Figure 13.4** **The stages of erythropoiesis.** The proliferation and differentiation of cells that will become mature erythrocytes (red blood cells) occurs in the bone marrow and is stimulated by the hormone erythropoietin, secreted by the kidneys.

Because of the recycling of iron, dietary requirements for iron are usually quite small. Males (and women after menopause) have a dietary iron requirement of only about 10 mg/day. Women with average menstrual blood loss need 15 mg/day, and pregnant women need 30 mg/day. Because of these relatively small dietary requirements, iron-deficiency anemia in adults is usually not due to a dietary deficiency but rather to blood loss, which reduces the amount of iron that can be recycled.

Red Blood Cell Antigens and Blood Typing

There are certain molecules on the surfaces of all cells in the body that can be recognized as foreign by the immune system of another individual. These molecules are known as *antigens.* As part of the immune response, particular lymphocytes secrete a class of proteins called *antibodies* that bond in a specific fashion with antigens. The specificity of antibodies for antigens is analogous to the specificity of enzymes for their substrates, and of receptor proteins for neurotransmitters and hormones. A complete description of antibodies and antigens is provided in chapter 15.

ABO System

The distinguishing antigens on other cells are far more varied than the antigens on red blood cells. Red blood cell antigens, however, are of extreme clinical importance because their types must be matched between donors and recipients for blood transfusions. There are several groups of red blood cell antigens, but the major group is known as the **ABO system.** In terms of the antigens present on the red blood cell surface, a person may be *type A* (with only A antigens), *type B* (with only B antigens), *type AB* (with both A and B antigens), or *type O* (with neither A nor B antigens). Each person's blood type—A, B, or O—denotes the antigens present on the red blood cell surface, which are the products of the genes (located on chromosome number 9) that code for these antigens.

Each person inherits two genes (one from each parent) that control the production of the ABO antigens. The genes for A or B antigens are dominant to the gene for O, since O simply means the absence of A or B. The genes for A and B are often shown as I^A and I^B, and the recessive gene for O is shown as the lowercase i. A person who is type A, therefore, may have inherited the A gene from each parent (may have the genotype $I^A I^A$), or the A gene from one parent and the O gene from the other parent (and thus have the genotype $I^A i$). Likewise, a person who is type B may have the genotype $I^B I^B$ or $I^B i$. It follows that a type O person inherited the O gene from each parent (has the genotype ii), whereas a type AB person inherited the A gene from one parent and the B gene from the other (there is no dominant-recessive relationship between A and B).

The immune system exhibits tolerance to its own red blood cell antigens. People who are type A, for example, do not produce anti-A antibodies. Surprisingly, however, they do make antibodies against the B antigen and, conversely, people with blood type B make antibodies against the A antigen (fig. 13.5). This is believed to result from the fact that antibodies made in response to some common bacteria cross-react with the A or B antigens. People who are type A, therefore, acquire antibodies that can react with B antigens by exposure to these bacteria, but they do not develop antibodies that can react with A antigens because tolerance mechanisms prevent this.

People who are type AB develop tolerance to both of these antigens, and thus do not produce either anti-A or anti-B antibodies. Those who are type O, by contrast, do not develop tolerance

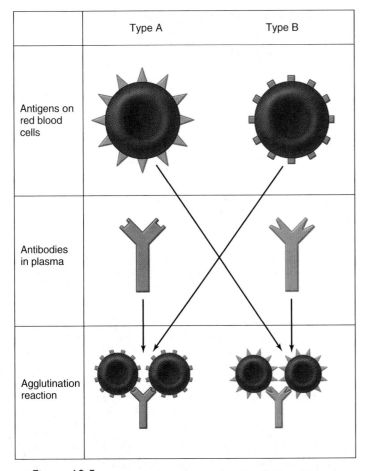

	Type A	Type B
Antigens on red blood cells		
Antibodies in plasma		
Agglutination reaction		

Figure 13.5 *Agglutination reaction.* People with type A blood have type A antigens on their red blood cells and antibodies in their plasma against the type B antigen. People with type B blood have type B antigens on their red blood cells and antibodies in their plasma against the type A antigen. Therefore, if red blood cells from one blood type are mixed with antibodies from the plasma of the other blood type, an agglutination reaction occurs. In this reaction, red blood cells stick together because of antigen-antibody binding.

Table 13.3 The ABO System of Red Blood Cell Antigens

Genotype	Antigen on RBCs	Antibody in Plasma
$I^A I^A$; $I^A i$	A	Anti-B
$I^B I^B$; $I^B i$	B	Anti-A
ii	O	Anti-A and anti-B
$I^A I^B$	AB	Neither anti-A nor anti-B

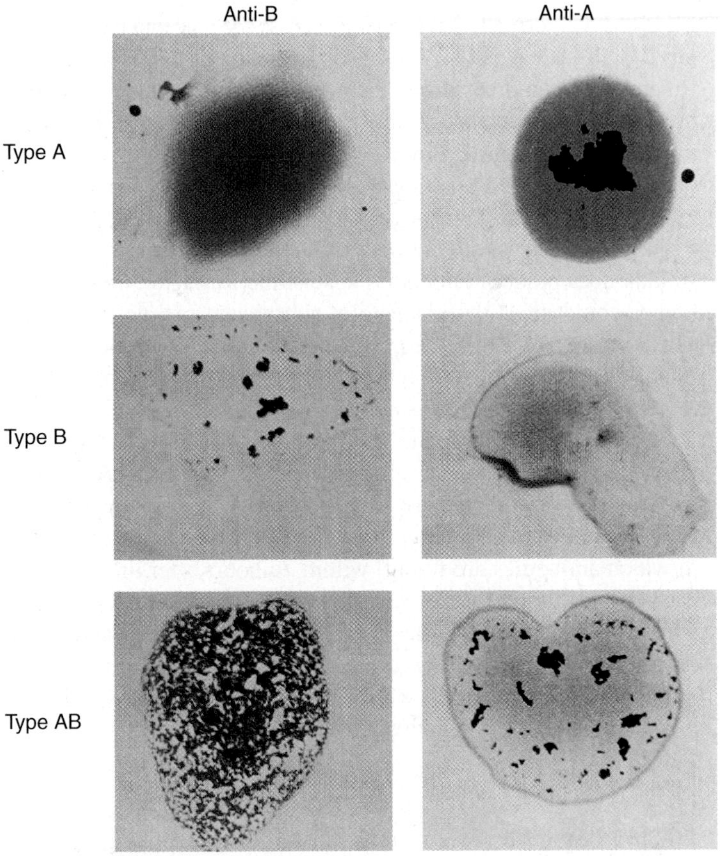

Figure 13.6 *Blood typing.* Agglutination (clumping) of red blood cells occurs when cells with A-type antigens are mixed with anti-A antibodies and when cells with B-type antigens are mixed with anti-B antibodies. No agglutination would occur with type O blood (not shown).

to either antigen; therefore, they have both anti-A and anti-B antibodies in their plasma (table 13.3).

Transfusion Reactions

Before transfusions are performed, a *major crossmatch* is made by mixing serum from the recipient with blood cells from the donor. If the types do not match—if the donor is type A, for example, and the recipient is type B—the recipient's antibodies attach to the donor's red blood cells and form bridges that cause the cells to clump together, or **agglutinate** (figs. 13.5 and 13.6). Because of this agglutination reaction, the A and B antigens are sometimes called *agglutinogens,* and the antibodies against them are called *agglutinins.* Transfusion errors that result in such agglutination can lead to blockage of small blood vessels and cause hemolysis (rupture of red blood cells), which may damage the kidneys and other organs.

In emergencies, type O blood has been given to people who are type A, B, AB, or O. Since type O red blood cells lack

A and B antigens, the recipient's antibodies cannot cause agglutination of the donor red blood cells. Type O is, therefore, a *universal donor*—but only as long as the volume of plasma donated is small, since plasma from a type O person would agglutinate type A, type B, and type AB red blood cells. Likewise, type AB people are *universal recipients* because they lack anti-A and anti-B antibodies, and thus cannot agglutinate donor red blood cells. (Donor plasma could agglutinate recipient red blood cells if the transfusion volume were too large.) Because of the dangers involved, use of the universal donor and recipient concept is strongly discouraged in practice.

Rh Factor

Another group of antigens found on the red blood cells of most people is the **Rh factor** (named for the rhesus monkey, in which these antigens were first discovered). There are a number of different antigens in this group, but one stands out because of its medical significance. This Rh antigen is termed D, and is often indicated as Rho(D). If this Rh antigen is present on a person's red blood cells, the person is **Rh positive;** if it is absent, the person is **Rh negative.** The Rh-positive condition is by far the more common (with a frequency of 85% in the Caucasian population, for example).

The Rh factor is of particular significance when Rh-negative mothers give birth to Rh-positive babies. Since the fetal and maternal blood are normally kept separate across the placenta (see chapter 20), the Rh-negative mother is not usually exposed to the Rh antigen of the fetus during the pregnancy. At the time of birth, however, a variable degree of exposure may occur, and the mother's immune system may become sensitized and produce antibodies against the Rh antigen. This does not always occur, however, because the exposure may be minimal and because Rh-negative women vary in their sensitivity to the Rh factor. If the woman does produce antibodies against the Rh factor, these antibodies could cross the placenta in subsequent pregnancies and cause hemolysis of the Rh-positive red blood cells of the fetus. Therefore, the baby could be born anemic, with a condition called *erythroblastosis fetalis,* or *hemolytic disease of the newborn.*

Erythroblastosis fetalis can be prevented by injecting the Rh-negative mother with an antibody preparation against the Rh factor (a trade name for this preparation is RhoGAM—the GAM is short for gamma globulin, the class of plasma proteins in which antibodies are found) within 72 hours after the birth of each Rh-positive baby. This is a type of passive immunization in which the injected antibodies inactivate the Rh antigens and thus prevent the mother from becoming actively immunized to them. Some physicians now give RhoGAM throughout the Rh-positive pregnancy of any Rh-negative woman.

Blood Clotting

When a blood vessel is injured, a number of physiological mechanisms are activated that promote **hemostasis,** or the cessation of bleeding (*hemo* = blood; *stasis* = standing). Breakage of the endothelial lining of a vessel exposes collagen proteins from the subendothelial connective tissue to the blood. This initiates three separate, but overlapping, hemostatic mechanisms: (1) vasoconstriction, (2) the formation of a platelet plug, and (3) the production of a web of fibrin proteins that penetrates and surrounds the platelet plug.

Functions of Platelets

In the absence of vessel damage, platelets are repelled from each other and from the endothelial lining of vessels. The repulsion of platelets from an intact endothelium is believed to be due to *prostacyclin,* a type of prostaglandin (see chapter 11, fig. 11.34), produced within the endothelium. Mechanisms that prevent platelets from sticking to the blood vessels and to each other are obviously needed to prevent inappropriate blood clotting.

Damage to the endothelium of vessels exposes subendothelial tissue to the blood. Platelets are able to stick to exposed collagen proteins that have become coated with a protein (*von Willebrand factor*) secreted by endothelial cells. Platelets contain secretory granules; when platelets stick to collagen, they *degranulate* as the secretory granules release their products. These products include *adenosine diphosphate (ADP), serotonin,* and a prostaglandin called *thromboxane A_2.* This event is known as the **platelet release reaction.**

Serotonin and thromboxane A_2 stimulate vasoconstriction, which helps to decrease blood flow to the injured vessel. Phospholipids that are exposed on the platelet membrane participate in the activation of clotting factors.

The release of ADP and thromboxane A_2 from platelets that are stuck to exposed collagen makes other platelets in the vicinity "sticky," so that they adhere to those stuck to the collagen. The second layer of platelets, in turn, undergoes a platelet release reaction, and the ADP and thromboxane A_2 that are secreted cause additional platelets to aggregate at the site of injury. This produces a **platelet plug** in the damaged vessel, which is strengthened by the activation of plasma clotting factors.

In order to undergo a release reaction, the platelets must produce prostaglandins. **Aspirin** inhibits the cyclooxygenase enzyme that catalyzes the conversion of arachidonic acid (a cyclic fatty acid) into prostaglandins (chapter 11; see fig. 11.34), and thereby inhibits the release reaction and consequent formation of a platelet plug. Since platelets lack nuclei and are not complete cells, they cannot regenerate new enzymes. Therefore, the enzymes remain inhibited for the life of the platelets. The ingestion of excessive amounts of aspirin can thus significantly prolong bleeding time for several days, which is why blood donors and women in the last trimester of pregnancy are advised to avoid aspirin. Slight inhibition of platelet aggregation by low doses of aspirin, however, can reduce the risk of atherosclerotic heart disease, and such a regimen is often recommended for patients diagnosed with this condition.

Clotting Factors: Formation of Fibrin

The platelet plug is strengthened by a meshwork of insoluble protein fibers known as **fibrin** (fig. 13.7). Blood clots therefore contain platelets and fibrin, and they usually contain trapped red blood cells that give the clot a red color (clots formed in arteries, where the blood flow is more rapid, generally lack red blood cells and thus appear gray). Finally, contraction of the platelet mass in the process of *clot retraction* forms a more compact and effective plug. Fluid squeezed from the clot as it retracts is called *serum,* which is plasma without fibrinogen, the soluble precursor of fibrin. (Serum is obtained in laboratories by allowing blood to clot in a test tube and then centrifuging the tube so that the clot and blood cells become packed at the bottom of the tube.)

The conversion of fibrinogen into fibrin may occur via either of two pathways. Blood left in a test tube will clot without the addition of any external chemicals; the pathway that produces this clot is thus called the **intrinsic pathway.** The intrinsic pathway also produces clots in damaged blood vessels when collagen is exposed to plasma. Damaged tissues, however, release a chemical that initiates a "shortcut" to the formation of fibrin. Since this chemical is not part of blood, the shorter pathway is called the **extrinsic pathway.**

The intrinsic pathway is initiated by the exposure of plasma to a negatively charged surface, such as that provided by collagen at the site of a wound or by the glass of a test tube. This activates a plasma protein called factor XII (table 13.4), which is a protein-digesting enzyme (a protease). Active factor XII in turn activates another clotting factor, which activates yet another. The plasma clotting factors are numbered in order of their discovery, which does not reflect the actual sequence of reactions.

The next steps in the sequence require the presence of Ca^{2+} and phospholipids, the latter provided by platelets. These steps result in the conversion of an inactive glycoprotein, called **prothrombin,** into the active enzyme **thrombin.** Thrombin converts the soluble protein **fibrinogen** into **fibrin** monomers. These

monomers are joined together to produce the insoluble fibrin polymers that form a meshwork supporting the platelet plug. The intrinsic clotting sequence is shown on the right side of figure 13.8.

A number of hereditary diseases involve the clotting system. Examples of hereditary clotting disorders include two different genetic defects in factor VIII. A defect in one subunit of factor VIII prevents this factor from participating in the intrinsic clotting pathway. This genetic disease, called **hemophilia A,** is an X-linked recessive trait that is prevalent in the royal families of Europe. A defect in another subunit of factor VIII results in **von Willebrand's disease.** In this disease, rapidly circulating platelets are unable to stick to collagen, and a platelet plug cannot be formed. Some acquired and inherited defects in the clotting system are summarized in table 13.5.

The formation of fibrin can occur more rapidly as a result of the release of **tissue thromboplastin** from damaged tissue cells. This extrinsic clotting pathway is shown on the left side of figure 13.8. Notice that the intrinsic and extrinsic clotting pathways eventually merge to form a final common pathway that results in the formation of insoluble fibrin polymers.

Dissolution of Clots

As the damaged blood vessel wall is repaired, activated factor XII promotes the conversion of an inactive molecule in plasma into the active form called *kallikrein.* Kallikrein, in turn, catalyzes the conversion of inactive *plasminogen* into the active molecule **plasmin.** Plasmin is an enzyme that digests fibrin into "split products," thus promoting dissolution of the clot.

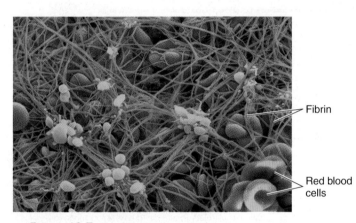

Figure 13.7 Colorized scanning electron micrograph of a blood clot. The threads of fibrin have been colored green, the erythrocytes are shown red, and the platelets have been colored purple.

Table 13.4 The Plasma Clotting Factors

Factor	Name	Function	Pathway
I	Fibrinogen	Converted to fibrin	Common
II	Prothrombin	Converted to thrombin (enzyme)	Common
III	Tissue thromboplastin	Cofactor	Extrinsic
IV	Calcium ions (Ca^{2+})	Cofactor	Intrinsic, extrinsic, and common
V	Proaccelerin	Cofactor	Common
VII*	Proconvertin	Enzyme	Extrinsic
VIII	Antihemophilic factor	Cofactor	Intrinsic
IX	Plasma thromboplastin component; Christmas factor	Enzyme	Intrinsic
X	Stuart-Prower factor	Enzyme	Common
XI	Plasma thromboplastin antecedent	Enzyme	Intrinsic
XII	Hageman factor	Enzyme	Intrinsic
XIII	Fibrin stabilizing factor	Enzyme	Common

*Factor VI is no longer referenced; it is now believed to be the same substance as activated factor V.

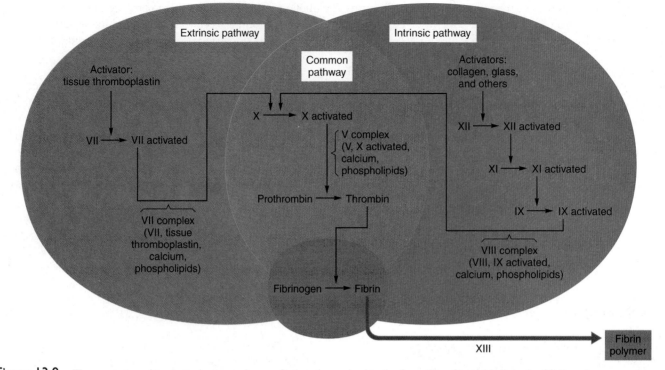

Figure 13.8 **The extrinsic and intrinsic clotting pathways.** Both pathways lead to the formation of insoluble threads of fibrin polymers.

Table 13.5 Some Acquired and Inherited Clotting Disorders and a Listing of Anticoagulant Drugs

Category	Cause of Disorder	Comments
Acquired clotting disorders	Vitamin K deficiency	Inadequate formation of prothrombin and other clotting factors in the liver
Inherited clotting disorders	Hemophilia A (defective factor VIII$_{AHF}$)	Recessive trait carried on X chromosome; results in delayed formation of fibrin
	von Willebrand's disease (defective factor VIII$_{VWF}$)	Dominant trait carried on autosomal chromosome; impaired ability of platelets to adhere to collagen in subendothelial connective tissue
	Hemophilia B (defective factor IX); also called Christmas disease	Recessive trait carried on X chromosome; results in delayed formation of fibrin

Anticoagulants	
Aspirin	Inhibits prostaglandin production, resulting in a defective platelet release reaction
Coumarin	Inhibits activation of vitamin K
Heparin	Inhibits activity of thrombin
Citrate	Combines with Ca^{2+}, and thus inhibits the activity of many clotting factors

In addition to kallikrein, a number of other plasminogen activators are used clinically to promote dissolution of clots. An exciting development in genetic engineering technology is the commercial availability of an endogenous compound, **tissue plasminogen activator (TPA),** which is the product of human genes introduced into bacteria. **Streptokinase,** a natural bacterial product, is a potent and more widely used activator of plasminogen. Streptokinase and TPA may be injected into the general circulation or injected specifically into a coronary vessel that has become occluded by a thrombus (blood clot).

Anticoagulants

Clotting of blood in test tubes can be prevented by the addition of *sodium citrate* or *ethylenediaminetetraacetic acid (EDTA),* both of which chelate (bind to) calcium. By this means, Ca^{2+} levels in the blood that can participate in the clotting sequence are lowered, and clotting is inhibited. A mucoprotein called *heparin* can also be added to the tube to prevent clotting. Heparin activates *antithrombin III,* a plasma protein that combines with and inactivates thrombin. Heparin is also given intravenously during certain medical procedures to prevent clotting. The *coumarin* drugs, whose mechanism of action is different from that of heparin, are also used as anticoagulants. These drugs (dicumarol and warfarin) prevent blood

Table 13.6 Terms Used to Describe Acid-Base Balance

Term	Definition
Acidosis, respiratory	Increased CO_2 retention (due to hypoventilation), which can result in the accumulation of carbonic acid and thus a fall in blood pH to below normal
Acidosis, metabolic	Increased production of "nonvolatile" acids, such as lactic acid, fatty acids, and ketone bodies, or loss of blood bicarbonate (such as by diarrhea), resulting in a fall in blood pH to below normal
Alkalosis, respiratory	A rise in blood pH due to loss of CO_2 and carbonic acid (through hyperventilation)
Alkalosis, metabolic	A rise in blood pH produced by loss of nonvolatile acids (such as excessive vomiting) or by excessive accumulation of bicarbonate base
Compensated acidosis or alkalosis	Metabolic acidosis or alkalosis are partially compensated for by opposite changes in blood carbonic acid levels (through changes in ventilation). Respiratory acidosis or alkalosis are partially compensated for by increased retention or excretion of bicarbonate in the urine.

clotting by inhibiting the cellular activation of vitamin K, thereby causing a vitamin K deficiency at the cellular level.

Vitamin K is needed for the conversion of glutamate, an amino acid found in many of the clotting factor proteins, into a derivative called *gamma-carboxyglutamate.* This derivative is more effective than glutamate at bonding to Ca^{2+} and such bonding is needed for proper function of clotting factors II, VII, IX, and X. Because of the indirect action of vitamin K on blood clotting, coumarin must be given to a patient for several days before it becomes effective as an anticoagulant.

Test Yourself Before You Continue

1. Distinguish between the different types of formed elements of the blood in terms of their origin, appearance, and function.

2. Describe how the rate of erythropoiesis is regulated.

3. Explain what is meant by "type A positive" and describe what can happen in a blood transfusion if donor and recipient are not properly matched.

4. Explain the meaning "intrinsic" and "extrinsic" as applied to the clotting pathways. How do the two pathways differ from each other? Which steps are common to both?

Acid-Base Balance of the Blood

The pH of blood plasma is maintained within a narrow range of values through the functions of the lungs and kidneys. The lungs regulate the carbon dioxide concentration of the blood, and the kidneys regulate the bicarbonate concentration.

The blood plasma within arteries normally has a pH between 7.35 and 7.45, with an average of 7.40. Using the definition of pH described in chapter 2, this means that arterial blood has a H^+ concentration of about $10^{-7.4}$ molar. Some of these hydrogen ions are derived from carbonic acid, which is formed in the blood plasma from carbon dioxide and which can ionize, as indicated in these equations:

$$CO_2 + H_2O \rightleftarrows H_2CO_3$$

$$H_2CO_3 \rightleftarrows H^+ + HCO_3^-$$

Carbon dioxide is produced by tissue cells through aerobic cell respiration and is transported by the blood to the lungs, where it can be exhaled. As will be described in more detail in chapter 16, carbonic acid can be reconverted to carbon dioxide, which is a gas. Because it can be converted to a gas, carbonic acid is referred to as a *volatile acid,* and its concentration in the blood is controlled by the lungs through proper ventilation (breathing). All other acids in the blood—including lactic acid, fatty acids, ketone bodies, and so on—are *nonvolatile acids.*

Under normal conditions, the H^+ released by nonvolatile acids does not affect the blood pH because these hydrogen ions are bound to molecules that function as *buffers.* The major buffer in the plasma is the *bicarbonate* (HCO_3^-) ion, and it buffers H^+ as described in this equation:

$$HCO_3^- + H^+ \rightarrow H_2CO_3$$

This buffering reaction could not go on forever because the free HCO_3^- would eventually disappear. If this were to occur, the H^+ concentration would increase and the pH of the blood would decrease. Under normal conditions, however, excessive H^+ is eliminated in the urine by the kidneys. Through this action, and through their ability to produce bicarbonate, the kidneys are responsible for maintaining a normal concentration of free bicarbonate in the plasma. The role of the kidneys in acid-base balance is described in chapter 17.

A fall in blood pH below 7.35 is called **acidosis** because the pH is to the acid side of normal. Acidosis does not mean acidic (pH less than 7); a blood pH of 7.2, for example, represents serious acidosis. Similarly, a rise in blood pH above 7.45 is called **alkalosis.** Both of these conditions are categorized into respiratory and metabolic components of acid-base balance (table 13.6).

Respiratory acidosis is caused by inadequate ventilation (hypoventilation), which results in a rise in the plasma concentration of carbon dioxide, and thus carbonic acid. **Respiratory alkalosis,** by contrast, is caused by excessive ventilation (hyperventilation). **Metabolic acidosis** can result from excessive production of nonvolatile acids; for example, it can result from excessive production of ketone bodies in uncontrolled diabetes mellitus (see chapter 19). It can also result from the loss of bicarbonate, in which case there would not be sufficient free bicarbonate to buffer the nonvolatile acids. (This occurs in diarrhea because of the loss of bicarbonate derived from pancreatic juice—see chapter 18.) **Metabolic alkalosis,** by contrast, can be caused by either too much bicarbonate (perhaps from an

Table 13.7   Classification of Metabolic and Respiratory Components of Acidosis and Alkalosis

Plasma CO₂	Plasma HCO₃⁻	Condition	Causes
Normal	Low	Metabolic acidosis	Increased production of "nonvolatile" acids (lactic acids, ketone bodies, and others), or loss of HCO_3^- in diarrhea)
Normal	High	Metabolic alkalosis	Vomiting of gastric acid; hypokalemia; excessive steroid administration
Low	Low	Respiratory alkalosis	Hyperventilation
High	High	Respiratory acidosis	Hypoventilation

intravenous infusion) or inadequate nonvolatile acids (perhaps as a result of excessive vomiting). Excessive vomiting may cause metabolic alkalosis through loss of the acid in gastric juice, which is normally absorbed from the intestine into the blood.

Since the *respiratory component* of acid-base balance is represented by the plasma carbon dioxide concentration and the *metabolic component* is represented by the free bicarbonate concentration, the study of acid-base balance can be simplified. A normal arterial blood pH is obtained when there is a proper ratio of bicarbonate to carbon dioxide. Indeed, the pH can be calculated given these values, and a normal pH is obtained when the ratio of these concentrations is 20 to 1. This is given by the **Henderson–Hasselbalch equation:**

$$pH = 6.1 + \log \frac{[HCO_3^-]}{0.03 P_{CO_2}}$$

where P_{CO_2} = partial pressure of CO_2, which is proportional to its concentration.

Respiratory acidosis or alkalosis occurs when the carbon dioxide concentrations are abnormal. Metabolic acidosis and alkalosis occur when the bicarbonate concentrations are abnormal (table 13.7). Often, however, a primary disturbance in one area (for example, metabolic acidosis) will be accompanied by secondary changes in another area (for example, respiratory alkalosis). It is important for hospital personnel to identify and treat the area of primary disturbance, but such analysis lies outside the scope of this discussion.

A more complete description of the respiratory and metabolic components of acid-base balance requires the study of pulmonary and renal function, and so will be presented with these topics in chapters 16 and 17.

Test Yourself Before You Continue

1. State the normal pH range of arterial blood plasma and explain how it is affected by the concentration of carbon dioxide in the blood. Explain how the plasma carbon dioxide concentration is regulated.

2. Explain how bicarbonate helps to maintain acid-base balance and describe the conditions that may result in metabolic acidosis or alkalosis.

Structure of the Heart

The heart contains four chambers: two atria, which receive venous blood, and two ventricles, which eject blood into arteries. The right ventricle pumps blood to the lungs, where the blood becomes oxygenated; the left ventricle pumps oxygenated blood to the entire body. The proper flow of blood within the heart is aided by two pairs of one-way valves.

About the size of a fist, the hollow, cone-shaped **heart** is divided into four chambers. The right and left **atria** (singular, *atrium*) receive blood from the venous system; the right and left **ventricles** pump blood into the arterial system. The right atrium and ventricle (sometimes called the *right pump*) are separated from the left atrium and ventricle (the *left pump*) by a muscular wall, or *septum*. This septum normally prevents mixture of the blood from the two sides of the heart.

Clinical Investigation Clue

Remember that Jason has a ventricular septal defect (a hole in the septum that separates the ventricles).

Given that the blood is under higher pressure in the left ventricle than the right during ventricular contraction, what effect could Jason's septal defect have on the blood in his heart?

Between the atria and ventricles, there is a layer of dense connective tissue known as the **fibrous skeleton** of the heart. Bundles of myocardial cells (described in chapter 12) in the atria attach to the upper margin of this fibrous skeleton and form a single functioning unit, or *myocardium*. The myocardial cell bundles of the ventricles attach to the lower margin and form a different myocardium. As a result, the myocardia of the atria and ventricles are structurally and functionally separated from each other, and special conducting tissue is needed to carry action potentials from the atria to the ventricles. The connective tissue of the fibrous skeleton also forms rings, called *annuli fibrosi,* around the four heart valves, providing a foundation for the support of the valve flaps.

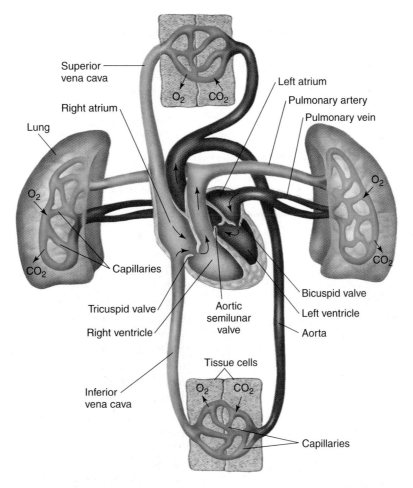

■ **Figure 13.9** **A diagram of the circulatory system.** The systemic circulation includes the aorta and venae cavae; the pulmonary circulation includes the pulmonary arteries and pulmonary veins.

Pulmonary and Systemic Circulations

Blood whose oxygen content has become partially depleted and whose carbon dioxide content has increased as a result of tissue metabolism returns to the right atrium. This blood then enters the right ventricle, which pumps it into the *pulmonary trunk* and *pulmonary arteries.* The pulmonary arteries branch to transport blood to the lungs, where gas exchange occurs between the lung capillaries and the air sacs (alveoli) of the lungs. Oxygen diffuses from the air to the capillary blood, while carbon dioxide diffuses in the opposite direction.

The blood that returns to the left atrium by way of the *pulmonary veins* is therefore enriched in oxygen and partially depleted of carbon dioxide. The path of blood from the heart (right ventricle), through the lungs, and back to the heart (left atrium) completes one circuit: the **pulmonary circulation.**

Oxygen-rich blood in the left atrium enters the left ventricle and is pumped into a very large, elastic artery—the *aorta.* The aorta ascends for a short distance, makes a U-turn, and then descends through the thoracic (chest) and abdominal cavities. Arterial branches from the aorta supply oxygen-rich blood to all of the organ systems and are thus part of the **systemic circulation.**

As a result of cellular respiration, the oxygen concentration is lower and the carbon dioxide concentration is higher in the tissues than in the capillary blood. Blood that drains into the systemic veins is thus partially depleted of oxygen and increased in carbon dioxide content. These veins ultimately empty into two large veins—the *superior* and *inferior venae cavae*—that return the oxygen-poor blood to the right atrium. This completes the systemic circulation: from the heart (left ventricle), through the organ systems, and back to the heart (right atrium). The systemic and pulmonary circulations are illustrated in figure 13.9, and their characteristics are summarized in table 13.8.

The numerous small muscular arteries and arterioles of the systemic circulation present greater resistance to blood flow than that in the pulmonary circulation. Despite the differences in resistance, the rate of blood flow through the systemic circulation must be matched to the flow rate of the pulmonary circulation. Since the amount of work performed by the left ventricle is greater (by a factor of 5 to 7) than that performed by the right ventricle, it is not surprising that the muscular wall of the left ventricle is thicker (8 to 10 mm) than that of the right ventricle (2 to 3 mm).

Table 13.8 Summary of the Pulmonary and Systemic Circulations

	Source	Arteries	O₂ Content of Arteries	Veins	O₂ Content of Veins	Termination
Pulmonary Circulation	Right ventricle	Pulmonary arteries	Low	Pulmonary veins	High	Left atrium
Systemic Circulation	Left ventricle	Aorta and its branches	High	Superior and inferior venae cavae and their branches*	Low	Right atrium

*Blood from the coronary circulation does not enter the venae cavae, but instead returns directly to the right atrium via the coronary sinus.

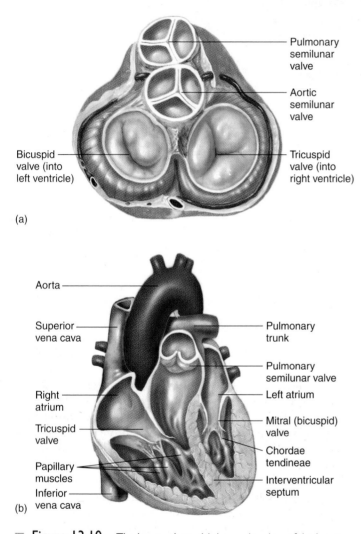

(a)

(b)

Figure 13.10 **The heart valves.** (*a*) A superior view of the heart valves. (*b*) A sagittal section through the heart, showing both AV valves and the pulmonary semilunar valve (the aortic semilunar valve is not visible in this view).

Atrioventricular and Semilunar Valves

Although adjacent myocardial cells are joined together mechanically and electrically by intercalated discs (chapter 12; see figs. 12.31 and 12.32), the atria and ventricles are separated into two functional units by a sheet of connective tissue—the fibrous skeleton previously mentioned. Embedded within this sheet of tissue are one-way **atrioventricular (AV) valves.** The AV valve located between the right atrium and right ventricle

has three flaps, and is therefore called the *tricuspid valve.* The AV valve between the left atrium and left ventricle has two flaps and is thus called the *bicuspid valve,* or, alternatively, the *mitral valve* (fig. 13.10).

The AV valves allow blood to flow from the atria to the ventricles, but they normally prevent the backflow of blood into the atria. Opening and closing of these valves occur as a result of pressure differences between the atria and ventricles. When the ventricles are relaxed, the venous return of blood to the atria causes the pressure in the atria to exceed that in the ventricles. The AV valves therefore open, allowing blood to enter the ventricles. As the ventricles contract, the intraventricular pressure rises above the pressure in the atria and pushes the AV valves closed.

There is a danger, however, that the high pressure produced by contraction of the ventricles could push the valve flaps too much and evert them. This is normally prevented by contraction of the *papillary muscles* within the ventricles, which are connected to the AV valve flaps by strong tendinous cords called the *chordae tendineae* (fig. 13.10). Contraction of the papillary muscles occurs at the same time as contraction of the muscular walls of the ventricles and serves to keep the valve flaps tightly closed.

One-way **semilunar valves** (fig. 13.11) are located at the origin of the pulmonary artery and aorta. These valves open during ventricular contraction, allowing blood to enter the pulmonary and systemic circulations. During ventricular relaxation, when the pressure in the arteries is greater than the pressure in the ventricles, the semilunar valves snap shut, thus preventing the backflow of blood into the ventricles.

Test Yourself Before You Continue

1. Using a flow diagram (arrows), describe the pathway of the pulmonary circulation. Indicate the relative amounts of oxygen and carbon dioxide in the vessels involved.

2. Use a flow diagram to describe the systemic circulation and indicate the relative amounts of oxygen and carbon dioxide in the blood vessels.

3. List the AV valves and the valves of the pulmonary artery and aorta. How do these valves ensure a one-way flow of blood?

4. Describe the structure of the fibrous skeleton of the heart. What is the significance of this structure?

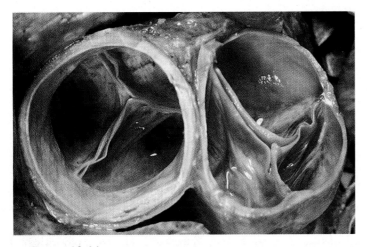

■ Figure 13.11 A photograph of the aortic and pulmonary semilunar valves. The valve flaps are shown in their closed position.

Cardiac Cycle and Heart Sounds

The two atria fill with blood and then contract simultaneously. This is followed by simultaneous contraction of both ventricles, which sends blood through the pulmonary and systemic circulations. Contraction of the ventricles closes the AV valves and opens the semilunar valves; relaxation of the ventricles causes the semilunar valves to close. The closing of first the AV valves and then the semilunar valves produces the "lub-dub" sounds heard with a stethoscope.

The **cardiac cycle** refers to the repeating pattern of contraction and relaxation of the heart. The phase of contraction is called **systole,** and the phase of relaxation is called **diastole.** When these terms are used without reference to specific chambers, they refer to contraction and relaxation of the ventricles. It should be noted, however, that the atria also contract and relax. There is an atrial systole and diastole. Atrial contraction occurs toward the end of diastole, when the ventricles are relaxed; when the ventricles contract during systole, the atria are relaxed (fig. 13.12).

The heart thus has a two-step pumping action. The right and left atria contract almost simultaneously, followed by contraction of the right and left ventricles 0.1 to 0.2 second later. During the time when both the atria and ventricles are relaxed, the venous return of blood fills the atria. The buildup of pressure that results causes the AV valves to open and blood to flow from atria to ventricles. It has been estimated that the ventricles are about 80% filled with blood even before the atria contract. Contraction of the atria adds the final 20% to the *end-diastolic volume,* which is the total volume of blood in the ventricles at the end of diastole.

Contraction of the ventricles in systole ejects about two-thirds of the blood they contain—an amount called the *stroke volume*—leaving one-third of the initial amount left in the ven-

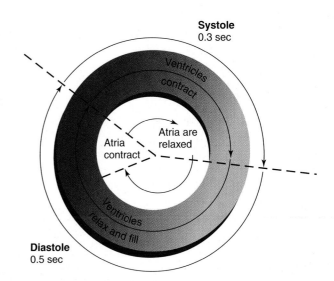

■ Figure 13.12 The cardiac cycle of ventricular systole and diastole. Contraction of the atria occurs in the last 0.1 second of ventricular diastole. Relaxation of the atria occurs during ventricular systole. The durations given for systole and diastole relate to a cardiac rate of 75 beats per minute.

tricles as the *end-systolic volume.* The ventricles then fill with blood during the next cycle. At an average *cardiac rate* of 75 beats per minute, each cycle lasts 0.8 second; 0.5 second is spent in diastole, and systole takes 0.3 second (fig. 13.12).

Interestingly, the blood contributed by contraction of the atria does not appear to be essential for life. Elderly people who have **atrial fibrillation** (a condition in which the atria fail to contract) do not appear to have a higher mortality than those who have normally functioning atria. People with atrial fibrillation, however, become fatigued more easily during exercise because the reduced filling of the ventricles compromises the ability of the heart to sufficiently increase its output during exercise. (Cardiac output and blood flow during rest and exercise are discussed in chapter 14.)

Pressure Changes During the Cardiac Cycle

When the heart is in diastole, pressure in the systemic arteries averages about 80 mmHg (millimeters of mercury). These events in the cardiac cycle then occur:

1. As the ventricles begin their contraction, the intraventricular pressure rises, causing the AV valves to snap shut. At this time, the ventricles are neither being filled with blood (because the AV valves are closed) nor ejecting blood (because the intraventricular pressure has not risen sufficiently to open the semilunar valves). This is the phase of *isovolumetric contraction.*

2. When the pressure in the left ventricle becomes greater than the pressure in the aorta, the phase of *ejection* begins

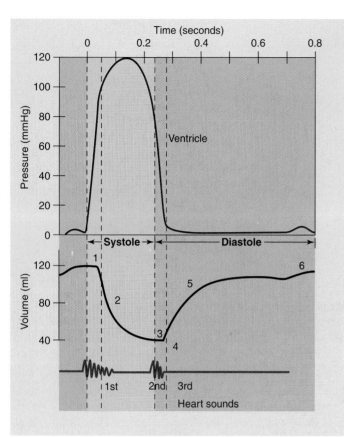

■ **Figure 13.13** The relationship between heart sounds and the intraventricular pressure and volume. The numbers refer to the events described in the text.

as the semilunar valves open. The pressure in the left ventricle and aorta rises to about 120 mmHg (fig. 13.13) when ejection begins and the ventricular volume decreases.

3. As the pressure in the left ventricle falls below the pressure in the aorta, the back pressure causes the semilunar valves to snap shut. The pressure in the aorta falls to 80 mmHg, while pressure in the left ventricle falls to 0 mmHg.

4. During *isovolumetric relaxation,* the AV and semilunar valves are closed. This phase lasts until the pressure in the ventricles falls below the pressure in the atria.

5. When the pressure in the ventricles falls below the pressure in the atria, the AV valves open and a phase of *rapid filling* of the ventricles occurs.

6. *Atrial contraction (atrial systole)* empties the final amount of blood into the ventricles immediately prior to the next phase of isovolumetric contraction of the ventricles.

Similar events occur in the right ventricle and pulmonary circulation, but the pressures are lower. The maximum pressure produced at systole in the right ventricle is 25 mmHg, which falls to a low of 8 mmHg at diastole.

Heart Sounds

Closing of the AV and semilunar valves produces sounds that can be heard by listening through a stethoscope placed on the chest. These sounds are often verbalized as "lub-dub." The

"lub," or **first sound,** is produced by closing of the AV valves during isovolumetric contraction of the ventricles. The "dub," or **second sound,** is produced by closing of the semilunar valves when the pressure in the ventricles falls below the pressure in the arteries. The first sound is thus heard when the ventricles contract at systole, and the second sound is heard when the ventricles relax at the beginning of diastole.

The first sound may be heard to split into tricuspid and mitral components, particularly during inhalation. Closing of the tricuspid is best heard at the fifth intercostal space (between the ribs), just to the right of the sternum; closing of the mitral valve is best heard at the fifth left intercostal space at the apex of the heart (fig. 13.14). The second sound also may be split under certain conditions. Closing of the pulmonary and aortic semilunar valves is best heard at the second left and right intercostal spaces, respectively.

Heart Murmurs

Murmurs are abnormal heart sounds produced by abnormal patterns of blood flow in the heart. Many murmurs are caused by defective heart valves. Defective heart valves may be congenital, or they may occur as a result of *rheumatic endocarditis,* associated with rheumatic fever. In this disease, the valves become damaged by antibodies made in response to an infection caused by streptococcus bacteria (the same bacteria that produce strep throat). Many people have small defects that produce detectable murmurs but do not seriously compromise the pumping ability of the heart. Larger defects, however, may have dangerous consequences and thus may require surgical correction.

In *mitral stenosis,* for example, the mitral valve becomes thickened and calcified. This can impair the blood flow from the left atrium to the left ventricle. An accumulation of blood in the left atrium may cause a rise in left atrial and pulmonary vein pressure, resulting in pulmonary hypertension. To compensate for the increased pulmonary pressure, the right ventricle grows thicker and stronger.

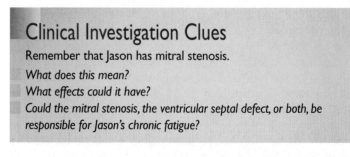

Clinical Investigation Clues

Remember that Jason has mitral stenosis.
What does this mean?
What effects could it have?
Could the mitral stenosis, the ventricular septal defect, or both, be responsible for Jason's chronic fatigue?

Valves are said to be *incompetent* when they do not close properly, and murmurs may be produced as blood regurgitates through the valve flaps. One important cause of incompetent AV valves is damage to the papillary muscles (see fig. 13.10). When this occurs, the tension in the chordae tendineae may not be sufficient to prevent the valve from everting as pressure in the ventricle rises during systole.

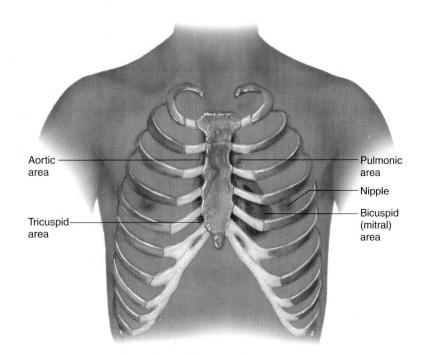

Aortic area

Pulmonic area

Nipple

Bicuspid (mitral) area

Tricuspid area

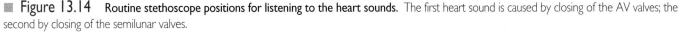 **Figure 13.14** Routine stethoscope positions for listening to the heart sounds. The first heart sound is caused by closing of the AV valves; the second by closing of the semilunar valves.

Murmurs also can be produced by the flow of blood through *septal defects*—holes in the septum between the right and left sides of the heart. These are usually congenital and may occur either in the interatrial or interventricular septum (fig. 13.15). When a septal defect is not accompanied by other abnormalities, blood will usually pass through the defect from the left to the right side, due to the higher pressure on the left side. The buildup of blood and pressure on the right side of the heart that results may lead to pulmonary hypertension and edema (fluid in the lungs).

 The lungs of a fetus are collapsed, and blood is routed away from the pulmonary circulation by an opening in the interatrial septum called the **foramen ovale** (fig. 13.15) and by a connection between the pulmonary trunk and aorta called the **ductus arteriosus** (fig. 13.16). These shunts normally close after birth, but when they remain open (are *patent*), murmurs can result. Since blood usually goes from left to right through these shunts, the left ventricle still pumps blood that is high in oxygen. When other defects are present that increase the pressure in the right pump (as in the *tetralogy of Fallot*), however, a significant amount of oxygen-depleted blood from the right side of the heart may enter the left side. The mixture of oxygen-poor blood from the right side with oxygen-rich blood in the left side of the heart lowers the oxygen concentration of the blood ejected into the systemic circulation. Since blood low in oxygen imparts a bluish tinge to the skin, the baby may be born *cyanotic* (blue).

Test Yourself Before You Continue

1. Using a drawing or flow chart, describe the sequence of events that occurs during the cardiac cycle. Indicate when atrial and ventricular filling occur and when atrial and ventricular contraction occur.

2. Describe how the pressure in the left ventricle and in the systemic arteries varies during the cardiac cycle.

3. Draw a figure to illustrate the pressure variations described in question no. 2, and indicate in your figure when the AV and semilunar valves close. Discuss the origin of the heart sounds.

4. Explain why blood usually flows from left to right through a septal defect. Under what conditions would a septal defect produce cyanosis?

Electrical Activity of the Heart and the Electrocardiogram

The pacemaker region of the heart (SA node) exhibits a spontaneous depolarization that causes action potentials, resulting in the automatic beating of the heart. Electrical impulses are conducted by myocardial cells in the atria and are transmitted to the ventricles by specialized conducting tissue. Electrocardiogram waves correspond to the electrical events in the heart as follows: P wave (depolarization of the atria); QRS wave (depolarization of the ventricles); and T wave (repolarization of the ventricles).

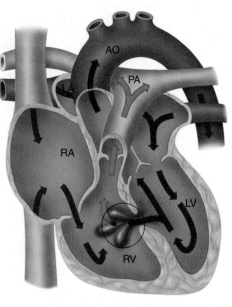

Septal defect
in atria

Septal defect
in ventricles

■ **Figure 13.15** **Abnormal patterns of blood flow due to septal defects.** Left-to-right shunting of blood is shown (*circled areas*) because the left pump is at a higher pressure than the right pump. Under certain conditions, however, the pressure in the right atrium may exceed that of the left, causing right-to-left shunting of blood through a septal defect in the atria (patent foramen ovale). (RA = right atrium; RV = right ventricle; LV = left ventricle; AO = aorta; PA = pulmonary arteries.)

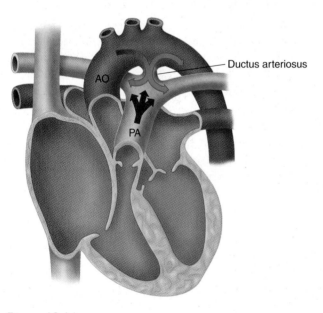

Ductus arteriosus

■ **Figure 13.16** **The flow of blood through a patent (open) ductus arteriosus.** The ductus is normally open in a fetus but closes after birth, eventually becoming the ligamentum arteriosum. (AO = aorta; PA = pulmonary arteries.)

As described in chapter 12, myocardial cells are short, branched, and interconnected by gap junctions. Gap junctions function as electrical synapses, and have been described in chapter 7 (see fig. 7.19) and chapter 12 (see fig. 12.31). The entire mass of cells interconnected by gap junctions is known as a *myocardium*. A myocardium is a single functioning unit, or

functional syncitium, since action potentials that originate in any cell in the mass can be transmitted to all the other cells. The myocardia of the atria and ventricles are separated from each other by the fibrous skeleton of the heart, as previously described. Since impulses normally originate in the atria, the atrial myocardium is excited before that of the ventricles.

Electrical Activity of the Heart

If the heart of a frog is removed from the body and all neural innervations are severed, it will still continue to beat as long as the myocardial cells remain alive. The automatic nature of the heartbeat is referred to as *automaticity*. As a result of experiments with isolated myocardial cells and clinical experience with patients who have specific heart disorders, many regions within the heart have been shown to be capable of originating action potentials and functioning as pacemakers. In a normal heart, however, only one region demonstrates spontaneous electrical activity and by this means functions as a pacemaker. This pacemaker region is called the **sinoatrial node,** or **SA node.** The SA node is located in the right atrium, near the opening of the superior vena cava.

Pacemaker Potential

The cells of the SA node do not maintain a resting membrane potential in the manner of resting neurons or skeletal muscle cells. Instead, during the period of diastole, the SA node exhibits a slow spontaneous depolarization called the **pacemaker potential.** The membrane potential begins at about –60 mV and gradually depolarizes to –40 mV, which is the threshold for producing an action potential in these cells. This spontaneous depolarization is produced by the diffusion of Ca^{2+} through openings in the membrane called *slow calcium channels*. At the threshold level of depolarization, other channels, called *fast calcium channels,* open, and Ca^{2+} rapidly diffuses into the cells. The opening of voltage-regulated Na^+ gates, and the inward diffusion of Na^+ that results, may also contribute to the upshoot phase of the action potential in pacemaker cells (fig. 13.17). Repolarization is produced by the opening of K^+ gates and outward diffusion of K^+, as in the other excitable tissues previously discussed. Once repolarization to –60 mV has been achieved, a new pacemaker potential begins, again culminating with a new action potential at the end of diastole.

Some other regions of the heart, including the area around the SA node and the atrioventricular bundle, can potentially produce pacemaker potentials. The rate of spontaneous depolarization of these cells, however, is slower than that of the SA node. Thus, the potential pacemaker cells are stimulated by action potentials from the SA node before they can stimulate themselves through their own pacemaker potentials. If action potentials from the SA node are prevented from reaching these areas (through blockage of conduction), they will generate pacemaker potentials at their own rate and serve as sites for the origin of action potentials; they will function as pacemakers. A pacemaker other than the SA node is called an *ectopic pacemaker,* or alternatively, an *ectopic focus.* From this discussion, it is clear that the rhythm set by such an ectopic pacemaker is usually slower than that normally set by the SA node.

Myocardial Action Potential

Once another myocardial cell has been stimulated by action potentials originating in the SA node, it produces its own action potentials. The majority of myocardial cells have resting membrane potentials of about –90 mV. When stimulated by action potentials from a pacemaker region, these cells become depolarized to threshold, at which point their voltage-regulated Na^+ gates open. The upshoot phase of the action potential of nonpacemaker cells is due to the inward diffusion of Na^+. Following the rapid reversal of the membrane polarity, the membrane potential quickly declines to about –15 mV. Unlike the action potential of other cells, however, this level of depolarization is maintained for 200 to 300 msec before repolarization (fig. 13.18). This *plateau phase* results from a slow inward diffusion of Ca^{2+}, which balances a slow outward diffusion of cations. Rapid repolarization at the end of the plateau phase is achieved, as in other cells, by the opening of K^+ channels and the rapid outward diffusion of K^+ that results.

Conducting Tissues of the Heart

Action potentials that originate in the SA node spread to adjacent myocardial cells of the right and left atria through the gap junctions between these cells. Since the myocardium of the atria is separated from the myocardium of the ventricles by the fibrous skeleton of the heart, however, the impulse cannot be conducted directly from the atria to the ventricles. Specialized conducting tissue, composed of modified myocardial cells, is thus required. These specialized myocardial cells form the *AV node, bundle of His,* and *Purkinje fibers.*

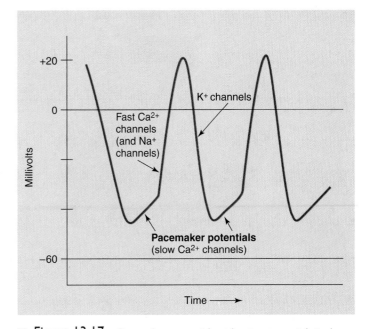

Figure 13.17 Pacemaker potentials and action potentials in the SA node. The pacemaker potentials are spontaneous depolarizations. When they reach threshold, they trigger action potentials.

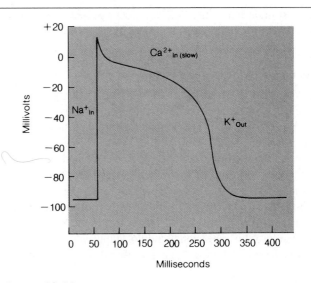

Figure 13.18 An action potential in a myocardial cell from the ventricles. The plateau phase of the action potential is maintained by a slow inward diffusion of Ca^{2+} The cardiac action potential, as a result, is about 100 times longer in duration than the "spike potential" of an axon.

Once the impulse has spread through the atria, it passes to the **atrioventricular node (AV node),** which is located on the inferior portion of the interatrial septum (fig. 13.19). From here, the impulse continues through the **atrioventricular bundle,** or **bundle of His** (pronounced "hiss"), beginning at the top of the interventricular septum. This conducting tissue pierces the fibrous skeleton of the heart and continues to descend along the interventricular septum. The atrioventricular bundle divides into right and left bundle branches, which are continuous with the **Purkinje fibers** within the ventricular

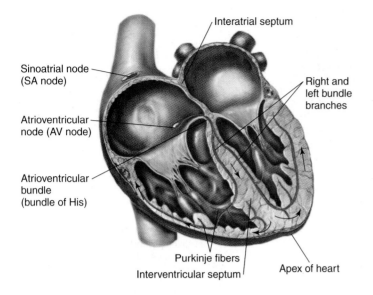

- Interatrial septum
- Sinoatrial node (SA node)
- Atrioventricular node (AV node)
- Atrioventricular bundle (bundle of His)
- Right and left bundle branches
- Purkinje fibers
- Interventricular septum
- Apex of heart

■ **Figure 13.19** **The conduction system of the heart.** The conduction system consists of specialized myocardial cells that rapidly conduct the impulses from the atria into the ventricles.

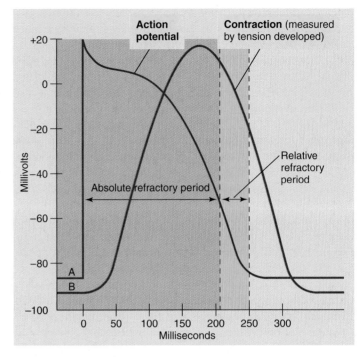

■ **Figure 13.20** Correlation of the myocardial action potential with myocardial contraction. The time course for the myocardial action potential (A) is compared with the duration of contraction (B). Notice that the long action potential results in a correspondingly long absolute refractory period and relative refractory period. These refractory periods last almost as long as the contraction, so that the myocardial cells cannot be stimulated a second time until they have completed their contraction from the first stimulus.

walls. Stimulation of the Purkinje fibers causes both ventricles to contract simultaneously and eject blood into the pulmonary and systemic circulation.

Conduction of the Impulse

Action potentials from the SA node spread very quickly—at a rate of 0.8 to 1.0 meter per second (m/sec)—across the myocardial cells of both atria. The conduction rate then slows considerably as the impulse passes into the AV node. Slow conduction of impulses (0.03 to 0.05 m/sec) through the AV node accounts for over half of the time delay between excitation of the atria and ventricles. After the impulses spread through the AV node, the conduction rate increases greatly in the atrioventricular bundle and reaches very high velocities (5 m/sec) in the Purkinje fibers. As a result of this rapid conduction of impulses, ventricular contraction begins 0.1 to 0.2 second after the contraction of the atria.

Excitation-Contraction Coupling in Heart Muscle

Depolarization of myocardial cells stimulates the opening of voltage-gated Ca^{2+} channels in the sarcolemma (plasma membrane of the myocardial cells). This allows Ca^{2+} to diffuse down its concentration gradient into the cell. The Ca^{2+} that enters the cytoplasm from the extracellular fluid serves as a stimulus for the opening of Ca^{2+} release channels in the sarcoplasmic reticulum, which stores Ca^{2+} (by active transport) during muscle relaxation. Since the Ca^{2+} release channels in the sarcoplasmic reticulum are opened by the increased Ca^{2+} concentration in the cytoplasm, this mechanism is called **calcium-stimulated-calcium-release.** The Ca^{2+} entering across the sarcolemma serves mainly as a stimulus, while the Ca^{2+} released from the sarcoplasmic reticulum contributes most to the rise in cytoplasmic Ca^{2+} concentration during depolarization of the myocardial cell.

Once Ca^{2+} is in the cytoplasm, it binds to troponin and stimulates contraction (described in chapter 12). As a result, myocardial cells contract when they are depolarized (fig. 13.20). During repolarization, the cytoplasmic concentration of Ca^{2+} is lowered by active transport of Ca^{2+} out of the cell across the sarcolemma (using a Na^+-Ca^{2+} exchanger), and by active transport of Ca^{2+} into the cisternae of the sarcoplasmic reticulum. This allows relaxation to occur during repolarization (fig. 13.20).

Unlike skeletal muscles, the heart cannot sustain a contraction. This is because the atria and ventricles behave as if each were composed of only one muscle cell; the entire myocardium of each is electrically stimulated as a single unit and contracts as a unit. This contraction, corresponding in time to the long action potential of myocardial cells and lasting almost 300 msec, is analogous to the twitch produced by a single skeletal muscle fiber (which lasts only 20 to 100 msec in comparison). The heart normally cannot be stimulated again until after it has relaxed from its previous contraction because myocardial cells have *long refractory periods* (fig. 13.20) that correspond to the long duration of their action potentials. Summation of contractions is thus prevented, and the myocardium must relax after each contraction. By this means, the rhythmic pumping action of the heart is ensured.

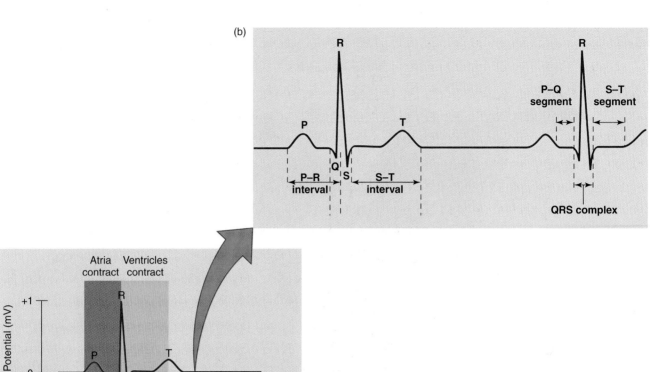

Figure 13.21   The electrocardiogram (ECG). The ECG indicates the conduction of electrical impulses through the heart (a) and measures and records both the intensity of this electrical activity (in millivolts) and the time intervals involved (b).

Abnormal patterns of electrical conduction in the heart can produce abnormalities of the cardiac cycle and seriously compromise the function of the heart. These **arrhythmias** may be treated with a variety of drugs that inhibit specific aspects of the cardiac action potentials and thereby inhibit the production or conduction of impulses along abnormal pathways. Drugs used to treat arrhythmias may (1) block the fast Na^{2+} channel (quinidine, procainamide, lidocaine); (2) block the slow Ca^{2+} channel (verapamil); or (3) block β-adrenergic receptors (propranolol, atenolol). By this means, the latter drugs block the ability of catecholamines to stimulate the heart.

The Electrocardiogram

The body is a good conductor of electricity because tissue fluids have a high concentration of ions that move (creating a current) in response to potential differences. Potential differences generated by the heart are thus conducted to the body surface, where they can be recorded by surface electrodes placed on the skin.

The recording thus obtained is called an **electrocardiogram** (**ECG** or **EKG**) (fig. 13.21); the recording device is called an *electrocardiograph*. Note that the ECG is not a recording of action potentials, but it does result from the production and conduction of action potentials in the heart. The correlation of an action potential produced in the ventricles to the waves of the ECG is shown in figure 13.21. This figure shows that the spread of depolarization through the ventricles corresponds to the plateau phase of the action potential, and thus to contraction of the ventricles.

A pair of surface electrodes placed directly on the heart will record a repeating pattern of potential changes. As action potentials spread from the atria to the ventricles, the voltage measured between these two electrodes will vary in a way that provides a "picture" of the electrical activity of the heart. By changing the position of the ECG recording electrodes on the body surface, a more complete picture of the electrical events can be obtained.

There are two types of ECG recording electrodes, or "leads." The *bipolar limb leads* record the voltage between electrodes placed on the wrists and legs. These bipolar leads include lead I

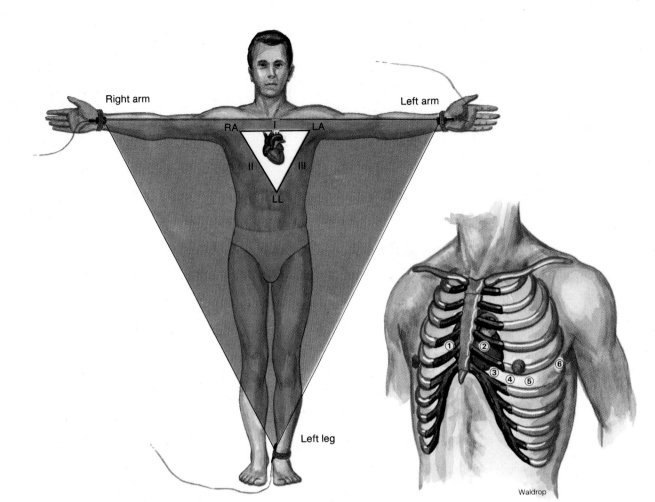

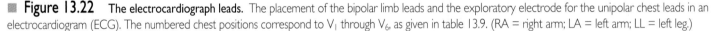

■ **Figure 13.22** **The electrocardiograph leads.** The placement of the bipolar limb leads and the exploratory electrode for the unipolar chest leads in an electrocardiogram (ECG). The numbered chest positions correspond to V₁ through V₆, as given in table 13.9. (RA = right arm; LA = left arm; LL = left leg.)

(right arm to left arm), lead II (right arm to left leg), and lead III (left arm to left leg). The right leg is used as a ground lead. In the *unipolar leads,* voltage is recorded between a single "exploratory electrode" placed on the body and an electrode that is built into the electrocardiograph and maintained at zero potential (ground).

The unipolar limb leads are placed on the right arm, left arm, and left leg, and are abbreviated AVR, AVL, and AVF, respectively. The unipolar chest leads are labeled 1 through 6, starting from the midline position (fig. 13.22). There are thus a total of twelve standard ECG leads that "view" the changing pattern of the heart's electrical activity from different perspectives (table 13.9). This is important because certain abnormalities are best seen with particular leads and may not be visible at all with other leads.

Each cardiac cycle produces three distinct ECG waves, designated P, QRS, and T. It should be noted that these waves represent changes in potential between two regions on the surface of the heart that are produced by the composite effects of action potentials in numerous myocardial cells. For example, the spread of depolarization through the atria causes a potential

Table 13.9 Electrocardiograph (ECG) Leads

Name of Lead	Placement of Electrodes
Bipolar limb leads	
I	Right arm and left arm
II	Right arm and left leg
III	Left arm and left leg
Unipolar limb leads	
AVR	Right arm
AVL	Left arm
AVF	Left leg
Unipolar chest leads	
V₁	4th intercostal space to the right of the sternum
V₂	4th intercostal space to the left of the sternum
V₃	5th intercostal space to the left of the sternum
V₄	5th intercostal space in line with the middle of the clavicle (collarbone)
V₅	5th intercostal space to the left of V₄
V₆	5th intercostal space in line with the middle of the axilla (underarm)

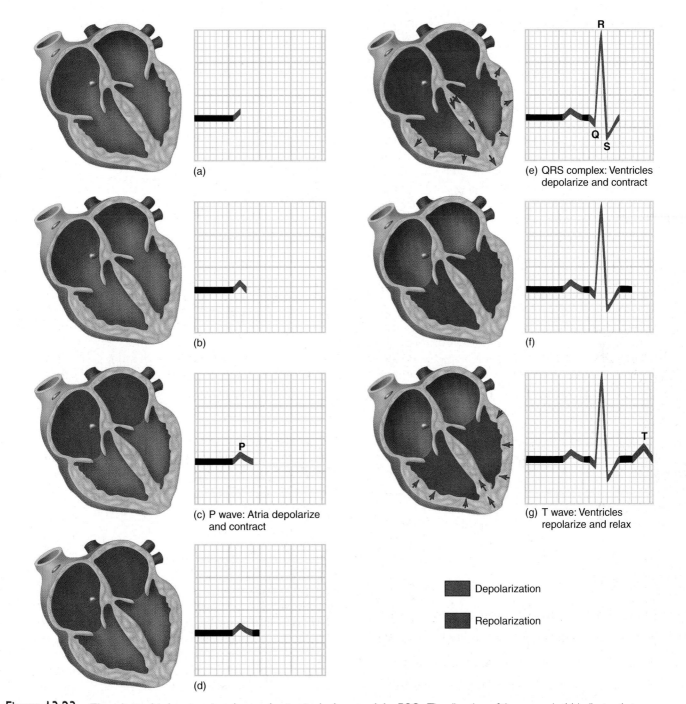

(a)

(b)

(c) P wave: Atria depolarize
and contract

(d)

(e) QRS complex: Ventricles
depolarize and contract

(f)

(g) T wave: Ventricles
repolarize and relax

▮ Depolarization

▮ Repolarization

■ **Figure 13.23** **The relationship between impulse conduction in the heart and the ECG.** The direction of the arrows in (e) indicates that depolarization of the ventricles occurs from the inside (endocardium) out (to the epicardium). The arrows in (g), by contrast, indicate that repolarization of the ventricles occurs in the opposite direction.

difference that is indicated by an upward deflection of the ECG line. When about half the mass of the atria is depolarized, this upward deflection reaches a maximum value because the potential difference between the depolarized and unstimulated portions of the atria is at a maximum. When the entire mass of the atria is depolarized, the ECG returns to baseline because all regions of the atria have the same polarity. The spread of atrial depolarization thus creates the **P wave.**

Conduction of the impulse into the ventricles similarly creates a potential difference that results in a sharp upward deflection of the ECG line, which then returns to the baseline as the entire mass of the ventricles becomes depolarized. The spread of the depolarization into the ventricles is thus represented by the **QRS wave.** The plateau phase of the cardiac action potential is related to the *S-T segment* of the ECG (see fig. 13.21). Finally, repolarization of the ventricles produces the **T wave** (fig. 13.23).

Correlation of the ECG with Heart Sounds

Depolarization of the ventricles, as indicated by the QRS wave, stimulates contraction by promoting the uptake of Ca^{2+} into the regions of the sarcomeres. The QRS wave is thus seen to occur at the beginning of systole. The rise in intraventricular pressure that results causes the AV valves to close, so that the first heart sound (S_1, or lub) is produced immediately after the QRS wave (fig. 13.24).

Repolarization of the ventricles, as indicated by the T wave, occurs at the same time that the ventricles relax at the beginning of diastole. The resulting fall in intraventricular pressure causes the aortic and pulmonary semilunar valves to close, so that the second heart sound (S_2, or dub) is produced shortly after the T wave begins in an electrocardiogram.

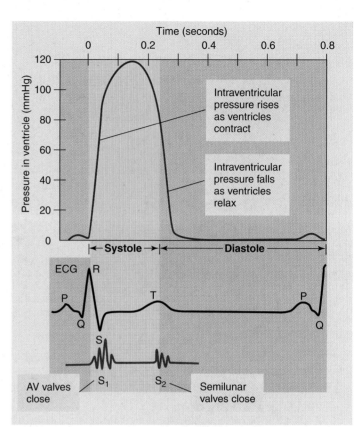

Figure 13.24 The relationship between changes in intraventricular pressure and the ECG. The QRS wave (representing depolarization of the ventricles) occurs at the beginning of systole, whereas the T wave (representing repolarization of the ventricles) occurs at the beginning of diastole.

Test Yourself Before You Continue

1. Describe the electrical activity of the cells of the SA node and explain how the SA node functions as the normal pacemaker.

2. Using a line diagram, illustrate a myocardial action potential and the time course for myocardial contraction. Explain how the relationship between these two events prevents the heart from sustaining a contraction and how it normally prevents abnormal rhythms of electrical activity.

3. Draw an ECG and label the waves. Indicate the electrical events in the heart that produce these waves.

4. Draw a figure that shows the relationship between ECG waves and the heart sounds. Explain this relationship.

5. Describe the pathway of electrical conduction of the heart, starting with the SA node. How does damage to the AV node affect this conduction pathway and the ECG?

Blood Vessels

The thick muscle layer of arteries allows them to transport blood ejected from the heart under high pressure, and the elastic recoil of the large arteries further contributes to blood flow. The thinner muscle layer of veins allows them to distend when an increased amount of blood enters them, and their one-way valves ensure that blood flows back to the heart. Capillaries are composed of only one layer of endothelium, which facilitates the rapid exchange of materials between the blood and tissue fluid.

Blood vessels form a tubular network throughout the body that permits blood to flow from the heart to all the living cells of the body and then back to the heart. Blood leaving the heart passes through vessels of progressively smaller diameters, referred to as *arteries, arterioles,* and *capillaries.* Capillaries are microscopic vessels that join the arterial flow to the venous flow. Blood returning to the heart from the capillaries passes through vessels of progressively larger diameters, called *venules* and *veins.*

The walls of arteries and veins are composed of three coats, or "tunics." The outermost layer is the **tunica externa,** the middle layer is the **tunica media,** and the inner layer is the **tunica interna.** The tunica externa is composed of connective tissue, whereas the tunica media is composed primarily of smooth muscle. The tunica interna consists of three parts: (1) an innermost simple squamous epithelium, the *endothelium,* which lines the lumina of all blood vessels; (2) the basement membrane (a layer of glycoproteins) overlying some connective tissue fibers; and (3) a layer of elastic fibers, or *elastin,* forming an *internal elastic lamina.*

Although arteries and veins have the same basic structure (fig. 13.25), there are some significant differences between them. Arteries have more muscle for their diameters than do comparably sized veins. As a result, arteries appear more rounded in cross section, whereas veins are usually partially collapsed. In addition, many veins have valves, which are absent in arteries.

Arteries

In the aorta and other large arteries, there are numerous layers of elastin fibers between the smooth muscle cells of the tunica

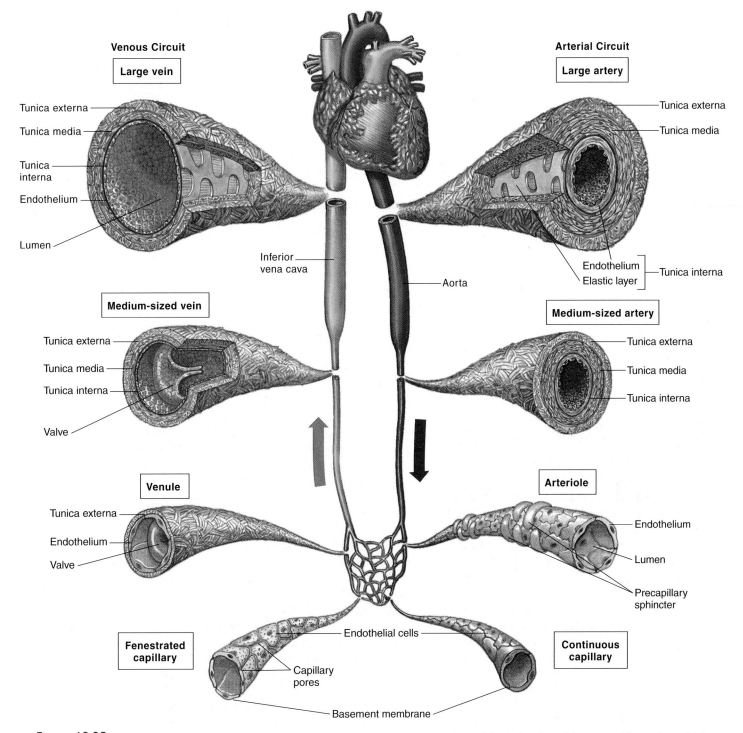

■ Figure 13.25 **Figure 13.25** **The structure of blood vessels.** Notice the relative thickness and composition of the tunicas (layers) in comparable arteries and veins.

media. These large **elastic arteries** expand when the pressure of the blood rises as a result of the heart's contraction; they recoil, like a stretched rubber band, when the blood pressure falls during relaxation of the ventricles. This elastic recoil drives the blood during the diastolic phase—the longest phase of the cardiac cycle—when the heart is resting and not providing a driving pressure.

The small arteries and arterioles are less elastic than the larger arteries and have a thicker layer of smooth muscle for their diameters. Unlike the larger elastic arteries, therefore, the diameter of the smaller **muscular arteries** changes only slightly as the pressure of the blood rises and falls during the heart's pumping activity. Since arterioles and small muscular arteries have narrow lumina, they provide the greatest resistance to blood flow through the arterial system.

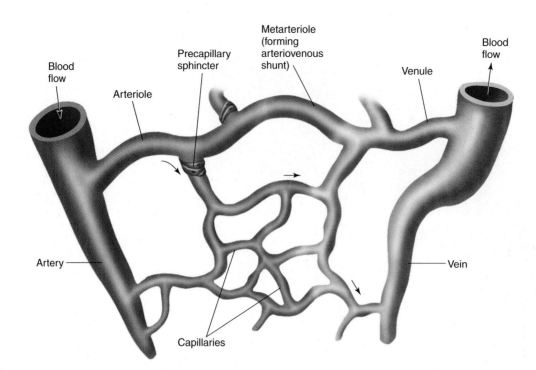

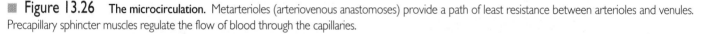

Figure 13.26 **The microcirculation.** Metarterioles (arteriovenous anastomoses) provide a path of least resistance between arterioles and venules. Precapillary sphincter muscles regulate the flow of blood through the capillaries.

Small muscular arteries that are 100 μm or less in diameter branch to form smaller **arterioles** (20 to 30 μm in diameter). In some tissues, blood from the arterioles can enter the venules directly through *arteriovenous anastomoses.* In most cases, however, blood from arterioles passes into capillaries (fig. 13.26). Capillaries are the narrowest of blood vessels (7 to 10 μm in diameter). They serve as the "business end" of the circulatory system, in which exchanges of gases and nutrients between the blood and the tissues occur.

Capillaries

The arterial system branches extensively (table 13.10) to deliver blood to over 40 billion capillaries in the body. As evidence of the extensiveness of these branchings, consider the fact that scarcely any cell in the body is more than 60 to 80 μm away from any capillary. The tiny capillaries provide a total surface area of 1,000 square miles for exchanges between blood and tissue fluid.

The amount of blood flowing through a particular capillary bed depends primarily on the resistance to blood flow in the small arteries and arterioles that supply blood to that capillary bed. Vasoconstriction in these vessels thus decreases blood flow to the capillary bed, whereas vasodilation increases blood flow.

The relatively high resistance in the small arteries and arterioles in resting skeletal muscles, for example, reduces capillary blood flow to only about 5% to 10% of its maximum capacity. In some organs (such as the intestine), blood flow may also be regulated by circular muscle bands called *precapillary sphincters* at the origin of the capillaries (fig. 13.26).

Unlike the vessels of the arterial and venous systems, the walls of capillaries are composed of just one cell layer—a simple squamous epithelium, or endothelium (fig. 13.27). The absence of smooth muscle and connective tissue layers permits a more rapid exchange of materials between the blood and the tissues.

Types of Capillaries

Different organs have different types of capillaries, distinguished by significant differences in structure. In terms of their endothelial lining, these capillary types include those that are *continuous,* those that are *fenestrated,* and those that are *discontinuous.*

Continuous capillaries are those in which adjacent endothelial cells are closely joined together. These are found in muscles, lungs, adipose tissue, and in the central nervous system. The lack of intercellular channels in continuous capillaries in the CNS contributes to the blood-brain barrier (chapter 7). Continuous capillaries in other organs have narrow intercellular channels (from 40 to 45 Å in width) that permit the passage of molecules other than protein between the capillary blood and tissue fluid (fig. 13.27).

Examination of endothelial cells with an electron microscope has revealed the presence of pinocytotic vesicles (fig. 13.27), which suggests that the intracellular transport of material may occur across the capillary walls. This type of transport appears to be the only mechanism of capillary exchange available within the central nervous system and may account, in part, for the selective nature of the blood-brain barrier.

Table 13.10 Characteristics of the Vascular Supply to the Mesenteries in a Dog

Kind of Vessels	Diameter (mm)	Number	Total Cross-Sectional Area (cm²)	Length (cm)	Total Volume (cm³)
Aorta	10	1	0.8	40	30
Large arteries	3	40	3.0	20	60
Main artery branches	1	600	5.0	10	50
Terminal branches	.06	1,800	5.0	1	25
Arterioles	0.02	40,000,000	125	0.2	25
Capillaries	0.008	1,200,000,000	600	0.1	60
Venules	0.03	80,000,000	570	0.2	110
Terminal veins	1.5	1,800	30	1	30
Main venous branches	2.4	600	27	10	270
Large veins	6.0	40	11	20	220
Vena cava	12.5	1	1.2	40	50
					930

Note: The pattern of vascular supply is similar in dogs and humans.

Source: Animal Physiology, 4th ed. by Gordon et al., © 1982. Adapted by permission of Prentice-Hall, Inc., Upper Saddle River, NJ.

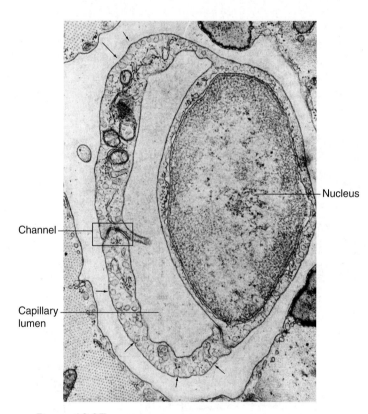

Figure 13.27 An electron micrograph of a capillary in the heart. Notice the thin intercellular channel (*middle left*) and the capillary wall, composed of only one cell layer. Arrows show some of the many pinocytotic vesicles.

(labels: Nucleus; Channel; Capillary lumen)

Fenestrated capillaries occur in the kidneys, endocrine glands, and intestines. These capillaries are characterized by wide intercellular pores (800 to 1,000 Å) that are covered by a layer of mucoprotein, which serves as a basement membrane over the capillary endothelium. This mucoprotein layer restricts the passage of certain molecules (particularly proteins) that might otherwise be able to pass through the large capillary pores. **Discontinuous cap-**

illaries are found in the bone marrow, liver, and spleen. The distance between endothelial cells is so great that these capillaries look like little cavities (*sinusoids*) in the organ.

Angiogenesis refers to the formation of new blood vessels from preexisting vessels, which are usually venules. Since all living cells must be within 100 µm of a capillary, angiogenesis is required during tissue growth. Angiogenesis is thus involved in the pathogenesis of *neoplasms* (tumors), and of the blindness caused by neovascularization of the retina in *diabetic retinopathy* and *age-related macular degeneration* (the most common cause of blindness). The treatment of these diseases may therefore be improved by inhibiting angiogenesis. Treatment for *ischemic heart disease,* on the other hand, may be improved by promoting angiogenesis in the coronary circulation. These therapies may manipulate paracrine regulators known to promote angiogenesis, including **vascular endothelial growth factor (VEGF)** and **fibroblast growth factor (FGF).**

Veins

Most of the total blood volume is contained in the venous system. Unlike arteries, which provide resistance to the flow of blood from the heart, veins are able to expand as they accumulate additional amounts of blood. The average pressure in the veins is only 2 mmHg, compared to a much higher average arterial pressure of about 100 mmHg. These values, expressed in millimeters of mercury, represent the hydrostatic pressure that the blood exerts on the walls of the vessels.

The low venous pressure is insufficient to return blood to the heart, particularly from the lower limbs. Veins, however, pass between skeletal muscle groups that provide a massaging action as they contract (fig. 13.28). As the veins are squeezed by contracting skeletal muscles, a one-way flow of blood to the heart is ensured by the presence of **venous valves.** The ability of these valves to prevent the flow of blood away from the heart was demonstrated in the seventeenth century by William Harvey (fig. 13.29). After applying a tourniquet to a subject's arm, Harvey found that he could push the blood in a bulging vein toward the heart, but not in the reverse direction.

The effect of the massaging action of skeletal muscles on venous blood flow is often described as the **skeletal muscle pump.** The rate of venous return to the heart is dependent, in large part, on the action of skeletal muscle pumps. When these pumps are less active, as when a person stands still or is bedridden, blood accumulates in the veins and causes them to bulge. When a person is more active, blood returns to the heart at a faster rate and less is left in the venous system.

Action of the skeletal muscle pumps aid the return of venous blood from the lower limbs to the large abdominal veins. Movement of venous blood from abdominal to thoracic veins, however, is aided by an additional mechanism—breathing. When a person inhales, the diaphragm—a muscular sheet separating the thoracic and abdominal cavities—contracts. Contraction of the dome-shaped diaphragm causes it to flatten and descend inferiorly into the abdomen. This has the dual effect of increasing the pressure in the abdomen, thus squeezing the abdominal veins, and decreasing the pressure in the thoracic cavity. The pressure difference in the veins created by this inspiratory movement of the diaphragm forces blood into the thoracic veins that return the venous blood to the heart.

The accumulation of blood in the veins of the legs over a long period of time, as may occur in people with occupations that require standing still all day, can cause the veins to stretch to the point where the venous valves are no longer efficient. This can also result from the compression of abdominal veins by a fetus during pregnancy. Venous congestion and stretching produced in this way can result in **varicose veins.** Venous congestion in the lower limbs is reduced during walking, when movements of the foot activate the soleus muscle pump. This effect can be produced in bedridden people by extending and flexing the ankle joints.

Test Yourself Before You Continue

1. Describe the basic structural pattern of arteries and veins. Explain how arteries and veins differ in structure and how these differences contribute to their differences in function.

2. Describe the functional significance of the skeletal muscle pump and illustrate the action of venous valves.

3. Explain the functions of capillaries and describe the structural differences between capillaries in different organs.

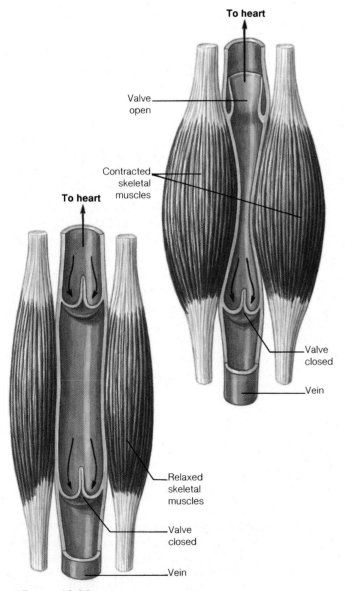

■ **Figure 13.28**   The action of the one-way venous valves. Contraction of skeletal muscles helps to pump blood toward the heart, but the flow of blood away from the heart is prevented by closure of the venous valves.

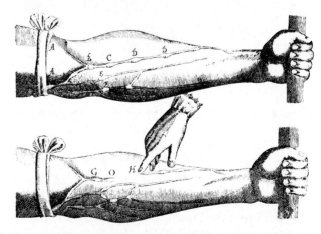

■ **Figure 13.29**   A demonstration of venous valves by William Harvey. By blocking venous drainage with a tourniquet, Harvey showed that the blood in the bulged vein was not permitted to move away from the heart, thereby demonstrating the action of venous valves.

After William Harvey, On the Motion of the Heart and Blood in Animals, 1628.

Atherosclerosis and Cardiac Arrhythmias

Atherosclerosis is a disease process that can lead to obstruction of coronary blood flow. As a result, the electrical properties of the heart and its ability to function as a pump may be seriously compromised. Abnormal cardiac rhythms, or arrhythmias, can be detected by the abnormal electrocardiogram patterns they produce.

Atherosclerosis

Atherosclerosis is the most common form of arteriosclerosis (hardening of the arteries) and, through its contribution to heart disease and stroke, is responsible for about 50% of the deaths in the United States, Europe, and Japan. In atherosclerosis, localized **plaques,** or *atheromas,* protrude into the lumen of the artery and thus reduce blood flow. The atheromas additionally serve as sites for *thrombus* (blood clot) formation, which can further occlude the blood supply to an organ (fig. 13.30).

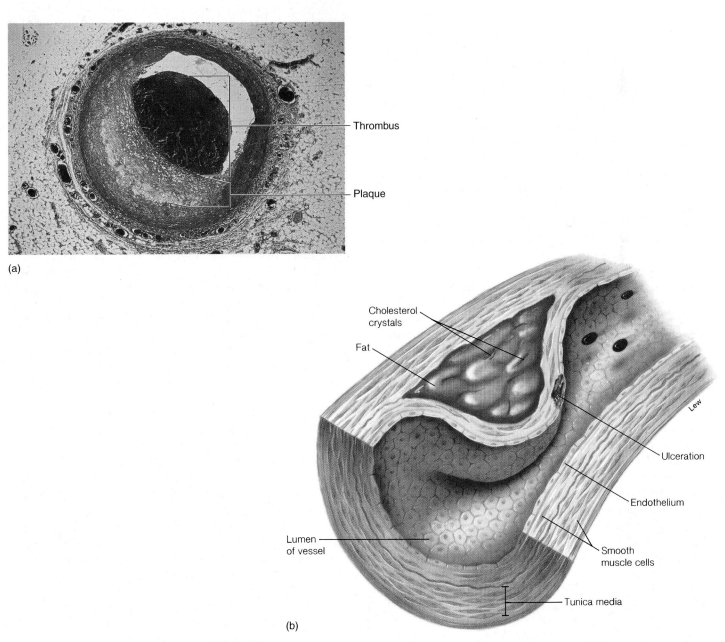

Figure 13.30 **Atherosclerosis.** *(a)* A photograph of the lumen (cavity) of a human coronary artery that is partially occluded by an atherosclerotic plaque and a thrombus. *(b)* A diagram of the structure of an atherosclerotic plaque.

It is currently believed that the process of atherosclerosis begins as a result of damage, or "insult," to the endothelium. Such insults are produced by smoking, hypertension (high blood pressure), high blood cholesterol, and diabetes. The first anatomically recognized change is the appearance of "fatty streaks," which are gray-white areas that protrude into the lumen of arteries, particularly at arterial branch points. These are aggregations of lipid-filled macrophages and lymphocytes within the tunica interna. They are present to a small degree in the aorta and coronary arteries of children aged 10 to 14, but progress to more advanced stages at different rates in different people. In the intermediate stage, the area contains layers of macrophages and smooth muscle cells. The more advanced lesions, called *fibrous plaques*, consist of a cap of connective tissue with smooth muscle cells over accumulated lipid and debris, macrophages that have been derived from monocytes (see chapter 15), and lymphocytes.

The disease process may be instigated by damage to the endothelium, but its progression appears to result from the action of a wide variety of cytokines and other paracrine regulators secreted by the endothelium and by the other participating cells, including platelets, macrophages, and lymphocytes. Some of these regulators attract monocytes and lymphocytes to the damaged endothelium and cause them to penetrate into the tunica interna. The monocytes then become macrophages, engulf lipids, and take on the appearance of "foamy cells." Smooth muscle cells change from a contractile state to a "synthetic" state, in which they produce and secrete connective tissue matrix proteins. (This is unique; in other tissues, connective tissue matrix is secreted by cells called fibroblasts.) The changed smooth muscle cells respond to chemical attractants and migrate from the tunica media to the tunica interna, where they can proliferate.

Endothelial cells normally prevent the progression just described by presenting a physical barrier to the penetration of monocytes and lymphocytes and by producing paracrine regulators such as nitric oxide. The vasodilator action of nitric oxide helps to counter the vasoconstrictor effects of another paracrine regulator, endothelin-1, which is increased in atherosclerosis. Hypertension, smoking, and high blood cholesterol, among other risk factors, interfere with this protective function.

Cholesterol and Plasma Lipoproteins

There is considerable evidence that high blood cholesterol is associated with an increased risk of atherosclerosis. This high blood cholesterol can be produced by a diet rich in cholesterol and saturated fat, or it may be the result of an inherited condition known as *familial hypercholesteremia.* This condition is inherited as a single dominant gene; individuals who inherit two of these genes have extremely high cholesterol concentrations (regardless of diet) and usually suffer heart attacks during childhood.

Lipids, including cholesterol, are carried in the blood attached to protein carriers (this topic is covered in detail in chapter 18). Cholesterol is carried to the arteries by plasma proteins called **low-density lipoproteins (LDLs).** LDLs, produced by the liver, are small protein-coated droplets of cholesterol, neutral fat, free fatty acids, and phospholipids. Cells in various organs

contain receptors for the proteins in LDLs; when LDL proteins attach to their receptors, the cell engulfs the LDL by receptor-mediated endocytosis (chapter 3; see fig. 3.4) and utilizes the cholesterol for different purposes. Most of the LDL particles in the blood are removed in this way by the liver.

People who eat a diet high in cholesterol and saturated fat, and people with familial hypercholesteremia, have a high blood LDL concentration because their livers have a low number of LDL receptors. With fewer LDL receptors, the liver is less able to remove the LDL from the blood, and thus more LDL is available to enter the endothelial cells of arteries.

Many people with dangerously high LDL-cholesterol concentrations take drugs known as **statins.** These drugs function as inhibitors of the enzyme *HMG-coenzyme A reductase,* which catalyzes the rate-limiting step in cholesterol synthesis. The statins therefore decrease the ability of the liver to produce its own cholesterol. The lowered intracellular cholesterol then stimulates the production of LDL receptors, allowing the liver cells to engulf more LDL-cholesterol. When a person takes a statin drug, therefore, the liver cells remove more LDL-cholesterol from the blood and thus decrease the amount of blood LDL-cholesterol that can enter the endothelial cells of arteries.

When endothelial cells engulf LDL, they oxidize it to a product called *oxidized LDL.* Recent evidence suggests that oxidized LDL contributes to endothelial cell injury, migration of monocytes and lymphocytes into the tunica interna, conversion of monocytes into macrophages, and other events that occur in the progression of atherosclerosis.

Since oxidized LDL seems to be so important in the progression of atherosclerosis, it would appear that antioxidant compounds could be used to treat this condition or help to prevent it. The antioxidant drug *probucol,* as well as *vitamin C, vitamin E,* and *beta-carotene,* which are antioxidants (see chapter 19), have been shown to be effective in this regard.

Excessive cholesterol may be released from cells and travel in the blood as **high-density lipoproteins (HDLs),** which are removed by the liver. The cholesterol in HDL is not taken into the artery wall because these cells lack the membrane receptor required for endocytosis of the HDL particles. For this reason, HDL-cholesterol does not contribute to atherosclerosis. Indeed, a high proportion of HDL-cholesterol as compared to LDL-cholesterol is beneficial, since it indicates that cholesterol may be traveling away from the blood vessels to the liver. The concentration of HDL-cholesterol appears to be higher and the risk of atherosclerosis lower in people who exercise regularly. The HDL-cholesterol concentration, for example, is higher in marathon runners than in joggers and is higher in joggers than in sedentary individuals. Women in general have higher HDL-cholesterol concentrations and a lower risk of atherosclerosis than men.

Clinical Investigation Clues

Remember that Jason had a high blood cholesterol and a high LDL/HDL ratio.

What dangers are indicated by the lab results?

What can Jason do about reducing these dangers?

Ischemic Heart Disease

A tissue is said to be **ischemic** when its oxygen supply is deficient because of inadequate blood flow. The most common cause of myocardial ischemia is atherosclerosis of the coronary arteries. The adequacy of blood flow is relative—it depends on the tissue's metabolic requirements for oxygen. An obstruction in a coronary artery, for example, may allow sufficient coronary blood flow at rest but not when the heart is stressed by exercise or emotional conditions. In these cases, the increased activity of the sympathoadrenal system causes the heart rate and blood pressure to rise, increasing the work of the heart and raising its oxygen requirements. Recent evidence also suggests that mental stress can cause constriction of atherosclerotic coronary arteries, leading to ischemia of the heart muscle. The vasoconstriction is believed to result from abnormal function of a damaged endothelium, which normally prevents constriction (through secretion of paracrine regulators) in response to mental stress. The control of vasoconstriction and vasodilation is discussed more fully in chapter 14.

Myocardial ischemia is associated with increased concentrations of blood lactic acid produced by anaerobic respiration of the ischemic tissue. This condition often causes substernal pain, which may also be referred to the left shoulder and arm, as well as to other areas. This referred pain is called **angina pectoris.** People with angina frequently take nitroglycerin or related drugs that help to relieve the ischemia and pain. These drugs are effective because they produce vasodilation, which improves circulation to the heart and decreases the work that the ventricles must perform to eject blood into the arteries.

Myocardial cells are adapted to respire aerobically and cannot respire anaerobically for more than a few minutes. If ischemia and anaerobic respiration are prolonged, *necrosis* (cellular death) may occur in the areas most deprived of oxygen. A

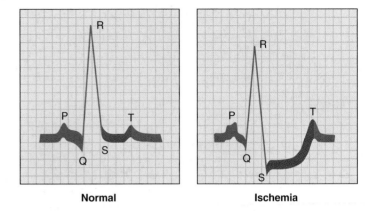

Figure 13.31 Depression of the S-T segment as a result of myocardial ischemia. This is but one of many ECG changes that alert trained personnel to the existence of heart problems.

sudden, irreversible injury of this kind is called a **myocardial infarction,** or **MI.** The lay term "heart attack," though imprecise, usually refers to a myocardial infarction.

Myocardial ischemia may be detected by changes in the S-T segment of the electrocardiogram (fig. 13.31). The diagnosis of myocardial infarction is aided by measurement of the blood levels of enzymes released by the infarcted tissue. Plasma concentrations of *creatine phosphokinase (CPK),* for example, increase within 3 to 6 hours after the onset of symptoms and return to normal after 3 days. Plasma levels of *lactate dehydrogenase (LDH)* reach a peak within 48 to 72 hours after the onset of symptoms and remain elevated for about 11 days. Plasma concentrations of *troponins T* and *I* (regulatory muscle proteins released by damaged myocardial cells; see p. 354) are also now important for the diagnosis of myocardial infarction.

Arrhythmias Detected by the Electrocardiograph

Arrhythmias, or abnormal heart rhythms, can be detected and described by the abnormal ECG tracings they produce. Although proper clinical interpretation of electrocardiograms requires information not covered in this chapter, some knowledge of abnormal rhythms is interesting in itself and is useful in gaining an understanding of normal physiology.

Since a heartbeat occurs whenever a normal QRS complex is seen, and since the ECG chart paper moves at a known speed so that its *x*-axis indicates time, the cardiac rate (beats per minute) can be easily obtained from an ECG recording. A cardiac rate slower than 60 beats per minute indicates **bradycardia;** a rate faster than 100 beats per minute is described as **tachycardia** (fig. 13.32).

Both bradycardia and tachycardia can occur normally. Endurance-trained athletes, for example, often have heart rates ranging from 40 to 60 beats per minute. This *athlete's bradycardia* occurs as a result of higher levels of parasympathetic inhibition of the SA node and is a beneficial adaptation. Activation of

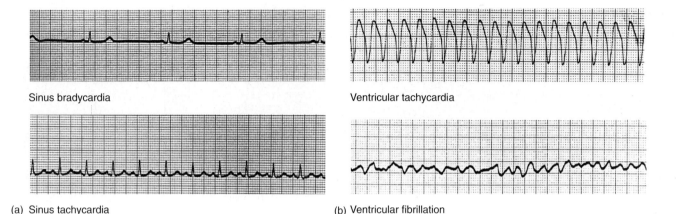

Sinus bradycardia

Ventricular tachycardia

(a) Sinus tachycardia

(b) Ventricular fibrillation

■ **Figure 13.32** **Some arrhythmias detected by the ECG.** In *(a)* the heartbeat is paced by the normal pacemaker—the SA node (hence the name *sinus rhythm*). This can be abnormally slow (bradycardia—42 beats per minute in this example) or fast (tachycardia—125 beats per minute in this example). Compare the pattern of tachycardia in *(a)* with the tachycardia in *(b)*. Ventricular tachycardia is produced by an ectopic pacemaker in the ventricles. This dangerous condition can quickly lead to ventricular fibrillation, also shown in *(b)*.

the sympathetic division of the ANS during exercise or emergencies ("fight or flight"), causes a normal tachycardia.

Abnormal tachycardia occurs if the heart rate increases when the person is at rest. This may be due to abnormally fast pacing by the atria (caused, for example, by drugs), or to the development of abnormally fast *ectopic pacemakers*—cells located outside the SA node that assume a pacemaker function. This abnormal atrial tachycardia thus differs from normal, or *sinus,* (SA node) *tachycardia. Ventricular tachycardia* results when abnormally fast ectopic pacemakers in the ventricles cause them to beat rapidly and independently of the atria. This is very dangerous because it can quickly degenerate into a lethal condition known as *ventricular fibrillation.*

Clinical Investigation Clues

Remember that Jason's ECG showed sinus tachycardia.

What does this mean?

What might produce this condition?

How would this ECG finding be related to the doctor's observations of Jason's radial pulse?

Flutter and Fibrillation

Extremely rapid rates of electrical excitation and contraction of either the atria or the ventricles may produce flutter or fibrillation. In **flutter,** the contractions are very rapid (200 to 300 per minute) but are coordinated. In **fibrillation,** contractions of different groups of myocardial fibers occur at different times, so that a coordinated pumping action of the chambers is impossible.

Atrial flutter usually degenerates quickly into *atrial fibrillation.* This causes the pumping action of the atria to stop. Since the ventricles fill to about 80% of their end-diastolic volume before atrial contraction normally occurs, however, the heart is still able to eject a sufficient quantity of blood into the circulation. People who have atrial fibrillation can thus

live for many years. People who have *ventricular fibrillation* (fig. 13.32), by contrast, can live for only a few minutes before the brain and heart—which are very dependent upon oxygen for their metabolism—cease to function.

Fibrillation is caused by a continuous recycling of electrical waves, known as **circus rhythms,** through the myocardium. This recycling is normally prevented by the fact that the entire myocardium enters a refractory period (due to the long duration of action potentials, as previously discussed). If some cells emerge from their refractory period before others, however, electrical waves can be continuously regenerated and conducted. Recycling of electrical waves along continuously changing pathways produces uncoordinated contraction and an impotent pumping action.

Circus rhythms are thus produced whenever impulses can be conducted without interruption by nonrefractory tissue. This may occur when the conduction pathway is longer than normal, as in a dilated heart. It can also be produced by an electric shock delivered at the middle of the T wave, when different myocardial cells are in different stages of recovery from their refractory period. Finally, circus rhythms and fibrillation may be produced by damage to the myocardium, which slows the normal rate of impulse conduction.

Fibrillation can sometimes be stopped by a strong electric shock delivered to the chest. This procedure is called **electrical defibrillation.** The electric shock depolarizes all of the myocardial cells at the same time, causing them all to enter a refractory state. Conduction of circus rhythms thus stops, and the SA node can begin to stimulate contraction in a normal fashion. This does not correct the initial problem that caused circus rhythms and fibrillation, but it does keep the person alive long enough to take other corrective measures.

AV Node Block

The time interval between the beginning of atrial depolarization—indicated by the P wave—and the beginning of ventricular depolarization (as shown by the Q part of the QRS complex) is called the *P-R interval* (see fig. 13.21). In the normal heart, this time

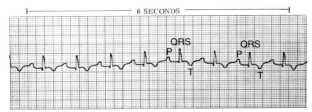

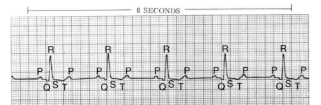

First-degree AV block

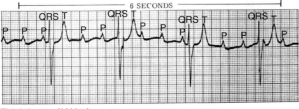

Second-degree AV block

Third-degree AV block

Figure 13.33 Atrioventricular (AV) node block. In first-degree block, the P-R interval is greater than 0.20 second (in the example here, the P-R interval is 0.26–0.28 second). In second-degree block, P waves are seen that are not accompanied by QRS waves. In this example, the atria are beating 90 times per minute (as represented by the P waves), while the ventricles are beating 50 times per minute (as represented by the QRS waves). In third-degree block, the ventricles are paced independently of the atria by an ectopic pacemaker. Ventricular depolarization (QRS) and repolarization (T) therefore have a variable position in the electrocardiogram relative to the P waves (atrial depolarization).

interval is 0.12 to 0.20 second in duration. Damage to the AV node causes slowing of impulse conduction and is reflected by changes in the P-R interval. This condition is known as *AV node block* (fig. 13.33).

First-degree AV node block occurs when the rate of impulse conduction through the AV node (as reflected by the P-R interval) exceeds 0.20 second. **Second-degree AV node block** occurs when the AV node is damaged so severely that only one out of every two, three, or four atrial electrical waves can pass through to the ventricles. This is indicated in an ECG by the presence of P waves without associated QRS waves.

In **third-degree,** or **complete, AV node block,** none of the atrial waves can pass through the AV node to the ventricles. The atria are paced by the SA node (follow a normal "sinus rhythm"), but the ventricles are paced by an ectopic pacemaker (usually located in the bundle of His or Purkinje fibers). Since the SA node is the normal pacemaker by virtue of the fact that it has the fastest cycle of electrical activity, the ectopic pacemaker in the ventricles causes them to beat at an abnormally slow rate. The bradycardia that results is usually corrected by insertion of an artificial pacemaker.

A number of abnormal conditions, including a blockage in conduction of the impulse along the bundle of His, require the insertion of an **artificial pacemaker.** This is a battery-powered device, about the size of a locket, which may be placed in permanent position under the skin. The electrodes from the pacemaker are guided through a vein to the right atrium, through the tricuspid valve, and into the right ventricle. The electrodes are fixed to the trabeculae carnae and are in contact with the wall of the ventricle. When these electrodes deliver shocks—either at a continuous pace or on demand (when the heart's own impulse doesn't arrive on time)—both ventricles are depolarized and contract, and then repolarize and relax, just as they do in response to endogenous stimulation.

Test Yourself Before You Continue

1. Explain how cholesterol is carried in the plasma and how the concentrations of cholesterol carriers are related to the risk for developing atherosclerosis.

2. Explain how angina pectoris is produced and discuss the significance of this symptom.

3. Define *bradycardia* and *tachycardia,* and give normal and pathological examples of each condition. Also, describe how flutter and fibrillation are produced.

4. Explain the effects of first-, second-, and third-degree AV node block on the electrocardiogram.

Lymphatic System

Lymphatic vessels absorb excess interstitial fluid and transport this fluid—now called lymph—to ducts that drain into veins. Lymph nodes, and lymphoid tissue in the thymus, spleen, and tonsils produce lymphocytes, which are white blood cells involved in immunity.

The **lymphatic system** has three basic functions: (1) it transports interstitial (tissue) fluid, initially formed as a blood filtrate, back to the blood; (2) it transports absorbed fat from the small intestine to the blood; and (3) its cells—called *lymphocytes*—help to provide immunological defenses against disease-causing agents.

The smallest vessels of the lymphatic system are the **lymphatic capillaries** (fig. 13.34). Lymphatic capillaries are microscopic closed-ended tubes that form vast networks in the intercellular spaces within most organs. Because the walls of lymphatic capillaries are composed of endothelial cells with porous junctions, interstitial fluid, proteins, microorganisms, and absorbed fat (in the intestine) can easily enter. Once fluid enters the lymphatic capillaries, it is referred to as **lymph.**

From merging lymphatic capillaries, the lymph is carried into larger lymphatic vessels called **lymph ducts.** The walls of

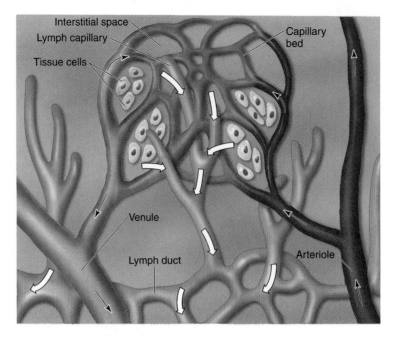

■ Figure 13.34 The relationship between blood capillaries and lymphatic capillaries. Notice that lymphatic capillaries are blind-ended. They are, however, highly permeable, so that excess fluid and protein within the interstitial space can drain into the lymphatic system.

lymph ducts are similar to those of veins. They have the same three layers and also contain valves to prevent backflow. Fluid movement within these vessels occurs as a result of peristaltic waves of contraction (chapter 12). The smooth muscle within the lymph ducts contains a pacemaker that initiates action potentials associated with the entry of Ca^{2+}, which stimulates contraction. The activity of the pacemaker, and hence the peristaltic waves of contraction, are increased in response to stretch of the vessel. The lymph ducts eventually empty into one of two principal vessels: the *thoracic duct* or the *right lymphatic duct*. These ducts drain the lymph into the left and right subclavian veins, respectively. Thus interstitial fluid, which is formed by filtration of plasma out of blood capillaries (a process described in chapter 14), is ultimately returned to the cardiovascular system (fig. 13.35).

Before the lymph is returned to the cardiovascular system, it is filtered through **lymph nodes** (fig. 13.36). Lymph nodes contain phagocytic cells, which help to remove pathogens, and *germinal centers*, which are sites of lymphocyte production. The tonsils, thymus, and spleen—together called **lymphoid organs**—likewise contain germinal centers and are sites of lymphocyte production. Lymphocytes are the cells of the immune system that respond in a specific fashion to antigens, and their functions are described as part of the immune system in chapter 15.

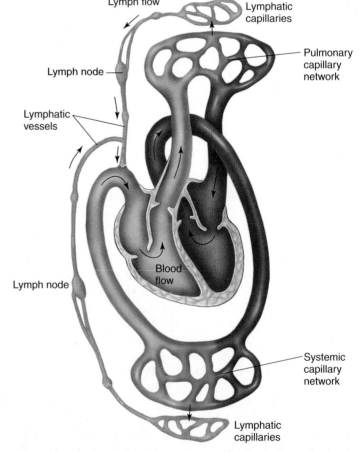

■ Figure 13.35 The relationship between the circulatory and lymphatic systems. This schematic illustrates that the lymphatic system transports fluid from the interstitial space back to the blood through a system of lymphatic vessels. Lymph is eventually returned to the vascular system at the subclavian veins.

Test Yourself Before You Continue

1. Compare the composition of lymph and blood and describe the relationship between blood capillaries and lymphatic capillaries.
2. Explain how the lymphatic system and the cardiovascular system are related. How do these systems differ?
3. Describe the functions of lymph nodes and lymphoid organs.

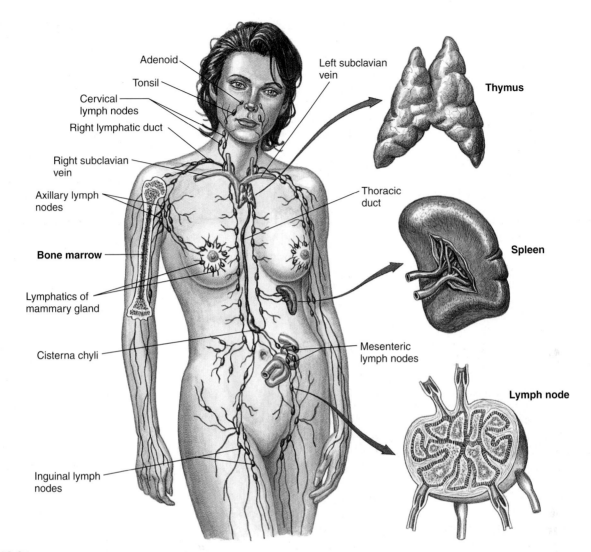

Adenoid

Tonsil

Cervical
lymph nodes

Right lymphatic duct

Right subclavian
vein

Axillary lymph
nodes

Bone marrow

Lymphatics of
mammary gland

Cisterna chyli

Inguinal lymph
nodes

Left subclavian
vein

Thymus

Thoracic
duct

Spleen

Mesenteric
lymph nodes

Lymph node

■ Figure 13.36   **The location of lymph nodes along the lymphatic pathways.** Lymph nodes are small bean-shaped bodies, enclosed within dense connective tissue capsules.

Summary

Functions and Components of the Circulatory System   366

I. The blood transports oxygen and nutrients to all the cells of the body and removes waste products from the tissues. It also serves a regulatory function through its transport of hormones.

   A. Oxygen is carried by red blood cells, or erythrocytes.

   B. White blood cells, or leukocytes, serve to protect the body from disease.

II. The circulatory system consists of the cardiovascular system (heart and blood vessels) and the lymphatic system.

Composition of the Blood   367

I. Plasma is the fluid part of the blood, containing dissolved ions and various organic molecules.

   A. Hormones are found in the plasma portion of the blood.

   B. Plasma proteins include albumins; globulins (alpha, beta, and gamma); and fibrinogen.

II. The formed elements of the blood include erythrocytes, leukocytes, and platelets.

   A. Erythrocytes, or red blood cells, contain hemoglobin and transport oxygen.

   B. Leukocytes may be granular (also called polymorphonuclear) or agranular. They function in immunity.

   C. Platelets, or thrombocytes, are required for blood clotting.

III. The production of red blood cells is stimulated by the hormone erythropoietin, and the development of different kinds of white blood cells is controlled by chemicals called lymphokines.

IV. The major blood typing groups are the ABO system and the Rh system.

    A. Blood type refers to the kind of antigens found on the surface of red blood cells.

    B. When different types of blood are mixed, antibodies against the red blood cell antigens cause the red blood cells to agglutinate.

V. When a blood vessel is damaged, platelets adhere to the exposed subendothelial collagen proteins.

    A. Platelets that stick to collagen undergo a release reaction in which they secrete ADP, serotonin, and thromboxane A_2.

    B. Serotonin and thromboxane A_2 cause vasoconstriction. ADP and thromboxane A_2 attract other platelets and cause them to stick to the growing mass of platelets that are stuck to the collagen in the broken vessel.

VI. In the formation of a blood clot, a soluble protein called fibrinogen is converted into insoluble threads of fibrin.

    A. This reaction is catalyzed by the enzyme thrombin.

    B. Thrombin is derived from prothrombin, its inactive precursor, by either an intrinsic or an extrinsic pathway.

        1. The intrinsic pathway, the longer of the two, requires the activation of more clotting factors.

        2. The shorter extrinsic pathway is initiated by the secretion of tissue thromboplastin.

    C. The clotting sequence requires Ca^{2+} as a cofactor and phospholipids present in the platelet cell membranes.

VII. Dissolution of the clot eventually occurs by the action of plasmin, which cleaves fibrin into split products.

Acid-Base Balance of the Blood   377

I. The normal pH of arterial blood is 7.40, with a range of 7.35 to 7.45.

    A. Carbonic acid is formed from carbon dioxide and contributes to the blood pH. It is referred to as a volatile acid because it can be eliminated in the exhaled breath.

    B. Nonvolatile acids, such as lactic acid and the ketone bodies, are buffered by bicarbonate.

II. The blood pH is maintained by a proper ratio of carbon dioxide to bicarbonate.

    A. The lungs maintain the correct carbon dioxide concentration. An increase in carbon dioxide, due to inadequate ventilation, produces respiratory acidosis.

    B. The kidneys maintain the free-bicarbonate concentration. An abnormally low plasma bicarbonate concentration produces metabolic acidosis.

Structure of the Heart   378

I. The right and left sides of the heart pump blood through the pulmonary and systemic circulations.

    A. The right ventricle pumps blood to the lungs. This blood then returns to the left atrium.

    B. The left ventricle pumps blood into the aorta and systemic arteries. This blood then returns to the right atrium.

II. The heart contains two pairs of one-way valves.

    A. The atrioventricular valves allow blood to flow from the atria to the ventricles, but not in the reverse direction.

    B. The semilunar valves allow blood to leave the ventricles and enter the pulmonary and systemic circulations, but they prevent blood from returning from the arteries to the ventricles.

III. The electrical impulse begins in the sinoatrial node and spreads through both atria by electrical conduction from one myocardial cell to another.

    A. The impulse then excites the atrioventricular node, from which it is conducted by the bundle of His into the ventricles.

    B. The Purkinje fibers transmit the impulse into the ventricular muscle and cause it to contract.

Cardiac Cycle and Heart Sounds   381

I. The heart is a two-step pump. The atria contract first, and then the ventricles.

    A. During diastole, first the atria and then the ventricles fill with blood.

    B. The ventricles are about 80% filled before the atria contract and add the final 20% to the end-diastolic volume.

    C. Contraction of the ventricles ejects about two-thirds of their blood, leaving about one-third as the end-systolic volume.

II. When the ventricles contract at systole, the pressure within them first rises sufficiently to close the AV valves and then rises sufficiently to open the semilunar valves.

    A. Blood is ejected from the ventricles until the pressure within them falls below the pressure in the arteries. At this point, the semilunar valves close and the ventricles begin relaxation.

    B. When the pressure in the ventricles falls below the pressure in the atria, a phase of rapid filling of the ventricles occurs, followed by the final filling caused by contraction of the atria.

III. Closing of the AV valves produces the first heart sound, or "lub," at systole. Closing of the semilunar valves produces the second heart sound, or "dub," at diastole. Abnormal valves can cause abnormal sounds called murmurs.

Electrical Activity of the Heart and the Electrocardiogram   383

I. In the normal heart, action potentials originate in the SA node as a result of spontaneous depolarization called the pacemaker potential.

    A. When this spontaneous depolarization reaches a threshold value, opening of the voltage-regulated Na^+ gates and fast Ca^{2+} channels produces action potential.

    B. Repolarization is produced by the outward diffusion of K^+, but a stable resting membrane potential is not attained because spontaneous depolarization once again occurs.

    C. Other myocardial cells are capable of spontaneous activity, but the SA node is the normal pacemaker because its rate of spontaneous depolarization is the fastest.

    D. When the action potential produced by the SA node reaches other myocardial cells, they produce action potentials with a long plateau phase because of the slow inward diffusion of Ca^{2+}.

E. The long action potential and long refractory period of myocardial cells allows the entire mass of cells to be in a refractory period while it contracts. This prevents the myocardium from being stimulated again until after it relaxes.

II. The regular pattern of conduction in the heart produces a changing pattern of potential differences between two points on the body surface.

A. The recording of this changing pattern caused by the heart's electrical activity is called an electrocardiogram (ECG).

B. The P wave is caused by depolarization of the atria; the QRS wave is caused by depolarization of the ventricles; and the T wave is produced by repolarization of the ventricles.

Blood Vessels   390

I. Arteries contain three layers, or tunics: the interna, media, and externa.

A. The tunica interna consists of a layer of endothelium, which is separated from the tunica media by a band of elastin fibers.

B. The tunica media consists of smooth muscle.

C. The tunica externa is the outermost layer.

D. Large arteries, containing many layers of elastin, can expand and recoil with rising and falling blood pressure. Medium and small arteries and arterioles are less distensible, and thus provide greater resistance to blood flow.

II. Capillaries are the narrowest but the most numerous of the blood vessels.

A. Capillary walls consist of just one layer of endothelial cells. They provide for the exchange of molecules between the blood and the surrounding tissues.

B. The flow of blood from arterioles to capillaries is regulated by precapillary sphincter muscles.

C. The capillary wall may be continuous, fenestrated, or discontinuous.

III. Veins have the same three tunics as arteries, but they generally have a thinner muscular layer than comparably sized arteries.

A. Veins are more distensible than arteries and can expand to hold a larger quantity of blood.

B. Many veins have venous valves that ensure a one-way flow of blood to the heart.

C. The flow of blood back to the heart is aided by contraction of the skeletal muscles that surround veins. The effect of this action is called the skeletal muscle pump.

Atherosclerosis and Cardiac Arrhythmias   395

I. Atherosclerosis of arteries can occlude blood flow to the heart and brain and is a causative factor in about 50% of all deaths in the United States, Europe, and Japan.

A. Atherosclerosis begins with injury to the endothelium, the movement of monocytes and lymphocytes into the tunica interna, and the conversion of monocytes into macrophages that engulf lipids. Smooth muscle cells then proliferate and secrete extracellular matrix.

B. Atherosclerosis is promoted by such risk factors as smoking, hypertension, and high plasma cholesterol concentration. Low-density lipoproteins (LDLs), which carry cholesterol into the artery wall, are oxidized by the endothelium and are a major contributor to atherosclerosis.

II. Occlusion of blood flow in the coronary arteries by atherosclerosis may produce ischemia of the heart muscle and angina pectoris, which may lead to myocardial infarction.

III. The ECG can be used to detect abnormal cardiac rates, abnormal conduction between the atria and ventricles, and other abnormal patterns of electrical conduction in the heart.

Lymphatic System   399

I. Lymphatic capillaries are blind-ended but highly permeable. They drain excess tissue fluid into lymph ducts.

II. Lymph passes through lymph nodes and is returned by way of the lymph ducts to the venous blood.

Review Activities

Test Your Knowledge of Terms and Facts

1. Which of these statements is *false?*
   a. Most of the total blood volume is contained in veins.
   b. Capillaries have a greater total surface area than any other type of vessel.
   c. Exchanges between blood and tissue fluid occur across the walls of venules.
   d. Small arteries and arterioles present great resistance to blood flow.

2. All arteries in the body contain oxygen-rich blood with the exception of
   a. the aorta.
   b. the pulmonary artery.
   c. the renal artery.
   d. the coronary arteries.

3. The "lub," or first heart sound, is produced by closing of
   a. the aortic semilunar valve.
   b. the pulmonary semilunar valve.
   c. the tricuspid valve.
   d. the bicuspid valve.
   e. both AV valves.

4. The first heart sound is produced at
   a. the beginning of systole.
   b. the end of systole.
   c. the beginning of diastole.
   d. the end of diastole.

5. Changes in the cardiac rate primarily reflect changes in the duration of
   a. systole.
   b. diastole.

6. The QRS wave of an ECG is produced by
   a. depolarization of the atria.
   b. repolarization of the atria.
   c. depolarization of the ventricles.
   d. repolarization of the ventricles.

7. The second heart sound immediately follows the occurrence of
   a. the P wave.
   b. the QRS wave.
   c. the T wave.

8. The cells that normally have the fastest rate of spontaneous diastolic depolarization are located in
   a. the SA node.
   b. the AV node.
   c. the bundle of His.
   d. the Purkinje fibers.

9. Which of these statements is *true?*
   a. The heart can produce a graded contraction.
   b. The heart can produce a sustained contraction.
   c. The action potentials produced at each cardiac cycle normally travel around the heart in circus rhythms.
   d. All of the myocardial cells in the ventricles are normally in a refractory period at the same time.

10. An ischemic injury to the heart that destroys myocardial cells is
    a. angina pectoris.

    b. a myocardial infarction.
    c. fibrillation.
    d. heart block.

11. The activation of factor X occurs in
    a. the intrinsic pathway only.
    b. the extrinsic pathway only.
    c. both the intrinsic and extrinsic pathways.
    d. neither the intrinsic nor extrinsic pathway.

12. Platelets
    a. form a plug by sticking to each other.
    b. release chemicals that stimulate vasoconstriction.
    c. provide phospholipids needed for the intrinsic pathway.
    d. serve all of these functions.

13. Antibodies against both type A and type B antigens are found in the plasma of a person who is
    a. type A.
    b. type B.
    c. type AB.
    d. type O.
    e. any of these types.

14. Production of which of the following blood cells is stimulated by a hormone secreted by the kidneys?
    a. lymphocytes
    b. monocytes
    c. erythrocytes

    d. neutrophils
    e. thrombocytes

15. Which of these statements about plasmin is *true?*
    a. It is involved in the intrinsic clotting system.
    b. It is involved in the extrinsic clotting system.
    c. It functions in fibrinolysis.
    d. It promotes the formation of emboli.

16. During the phase of isovolumetric relaxation of the ventricles, the pressure in the ventricles is
    a. rising.
    b. falling.
    c. first rising, then falling.
    d. constant.

17. Peristaltic waves of contraction move fluid within which of these vessels?
    a. arteries
    b. veins
    c. capillaries
    d. lymphatic vessels
    e. all of these

18. Excessive diarrhea may cause
    a. respiratory acidosis.
    b. respiratory alkalosis.
    c. metabolic acidosis.
    d. metabolic alkalosis.

Test Your Understanding of Concepts and Principles

1. Explain why the beat of the heart is automatic and why the SA node functions as the normal pacemaker.[1]

2. Compare the duration of the heart's contraction with the myocardial action potential and refractory period. Explain the significance of these relationships.

3. Describe the pressure changes that occur during the cardiac cycle and relate these changes to the occurrence of the heart sounds.

4. Can a defective valve be detected by an ECG? Can a partially damaged AV node be detected by auscultation (listening) with a stethoscope? Explain.

5. Describe the causes of the P, QRS, and T waves of an ECG and indicate at which point in the cardiac cycle each of these waves occurs. Explain why the first heart sound occurs immediately after the QRS wave and why the second sound occurs at the time of the T wave.

6. Explain how a cut in the skin initiates both the intrinsic and extrinsic clotting pathways. Which pathway is shorter? Why?

7. Distinguish between the respiratory and metabolic components of acid-base balance. What are some of the causes of acid-base disturbances?

8. Explain how aspirin, coumarin drugs, EDTA, and heparin function as anticoagulants. Which of these are effective when added to a test tube? Which are not? Why?

9. Explain how blood moves through arteries, capillaries, and veins. How does exercise affect this movement?

10. Explain the processes involved in the development of atherosclerosis. How might antioxidants help to retard the progression of this disease? How might exercise help? What other changes in lifestyle might help to prevent or reduce atherosclerotic plaques?

[1] *Note:* This question is answered in the Chapter 13 Study Guide found on the Online Learning Center at www.mhhe.com/fox8.

Test Your Ability to Analyze and Apply Your Knowledge

1. Hematopoietic stem cells account for less than 1% of the cells in the bone marrow. These cells can be separated from the others prior to bone marrow transplantation, but it is better to first inject the donor with recombinant cytokines. Identify the cytokines that might be used and describe their effects.

2. A patient has a low red blood cell count, and microscopic examination of his blood reveals an abnormally high proportion of circulating reticulocytes. Upon subsequent examination, the patient is diagnosed with a bleeding ulcer. This is surgically corrected, and in due course his blood measurements return to normal. What was the reason for the low red blood cell count and high proportion of reticulocytes?

3. A chemical called EDTA, like citrate, binds to (or "chelates") Ca^{2+}. Suppose a person had EDTA infused into the blood. What effect would this have on the intrinsic and extrinsic clotting pathways? How would these effects differ from the effects of aspirin on blood clotting?

4. During the course of a physiology laboratory, a student finds that her PR interval is 0.24 second. Concerned, she takes her own ECG again an hour later and sees an area of the ECG strip where the PR interval becomes longer and longer. Performing an ECG measurement on herself for a third time, she sees an area of the strip where a P wave is not followed by a QRS or T; further along in the strip, however, a normal pattern reappears. What do you think these recordings indicate?

5. A newborn baby with a patent foramen ovale or a ventricular septal defect might be cyanotic (blue). Will a two-year-old with these defects also be cyanotic? Explain your answer.

Related Websites

Check out the Links Library at www.mhhe.com/fox8 for links to sites containing resources related to the heart and circulation. These links are monitored to ensure current URLs.

14

Cardiac Output, Blood Flow, and Blood Pressure

Objectives

After studying this chapter, you should be able to . . .

1. define *cardiac output* and explain how it is calculated.

2. explain how autonomic nerves regulate cardiac rate and the strength of ventricular contraction.

3. explain the intrinsic regulation of stroke volume (the Frank–Starling Law of the Heart).

4. list the factors that affect the venous return of blood to the heart.

5. explain how interstitial fluid is formed and how it is returned to the capillary blood.

6. describe the conditions that may lead to edema.

7. explain how antidiuretic hormone helps to regulate blood volume, plasma osmolality, and blood pressure.

8. explain the role of aldosterone in the regulation of blood volume and blood pressure.

9. describe the renin-angiotensin-aldosterone system and discuss its significance in cardiovascular regulation.

10. use Poiseuille's law to explain how blood flow is regulated.

11. define *total peripheral resistance* and explain how vascular resistance is regulated by extrinsic control mechanisms.

12. describe the functions of nitric oxide and endothelin-1 in the paracrine regulation of blood flow.

13. describe the intrinsic mechanisms involved in the autoregulation of blood flow.

14. explain the mechanisms by which blood flow to the heart and skeletal muscles is regulated.

15. describe the changes that occur in the cardiac output and distribution of blood flow during exercise.

16. describe the cutaneous circulation and explain how circulation in the skin is regulated.

17. list the factors that regulate the arterial blood pressure.

18. describe the baroreceptor reflex and explain its significance in blood pressure regulation.

19. explain how the sounds of Korotkoff are produced and how these sounds are used to measure blood pressure.

20. explain how the pulse pressure and mean arterial pressure are calculated and discuss the significance of these measurements.

21. explain the mechanisms that contribute to and that help compensate for the conditions of hypertension, circulatory shock, and congestive heart failure.

Chapter at a Glance

Take Advantage of the Technology

*Visit the Online Learning Center for
these additional study resources.*

- Interactive quizzing
- Online study guide
- Current news feeds
- Crossword puzzles
- Vocabulary flashcards
- Labeling activities

www.mhhe.com/fox8

Clinical Investigation

Charlie is a college student on a biology field trip to study the ecology of the desert. Unfortunately, he gets so engrossed in his subject he wanders away from the group and gets lost. Thirty-six hours later, he is found crawling along a seldom-used one-lane road. He is very weak, and when he is brought to the hospital he has a weak, rapid pulse, low blood pressure, and cold skin. His urine output is low, and analysis reveals that it has a high osmolality (concentration) but a virtual absence of sodium. Charlie is given intravenous fluid containing albumin.

What caused Charlie's symptoms and laboratory findings? Why was he given intravenous fluid containing albumin?

Cardiac Output

The pumping ability of the heart is a function of the beats per minute (cardiac rate) and the volume of blood ejected per beat (stroke volume). The cardiac rate and stroke volume are regulated by autonomic nerves and by mechanisms intrinsic to the cardiovascular system.

The **cardiac output** is the volume of blood pumped per minute by each ventricle. The average resting **cardiac rate** in an adult is 70 beats per minute; the **average stroke volume** (volume of blood pumped per beat by each ventricle) is 70 to 80 ml per beat. The product of these two variables gives an average cardiac output of 5,500 ml (5.5 L) per minute:

$$\text{Cardiac output} = \text{stroke volume} \times \text{cardiac rate}$$
$$\text{(ml/min)} \qquad \text{(ml/beat)} \qquad \text{(beats/min)}$$

The **total blood volume** also averages about 5.5 L. This means that each ventricle pumps the equivalent of the total blood volume each minute under resting conditions. Put another way, it takes about a minute for a drop of blood to complete the systemic and pulmonary circuits. An increase in cardiac output, as occurs during exercise, must thus be accompanied by an increased rate of blood flow through the circulation. This is accomplished by factors that regulate the cardiac rate and stroke volume.

Regulation of Cardiac Rate

In the complete absence of neural influences, the heart will continue to beat according to the rhythm set by the SA node. This automatic rhythm is produced by the spontaneous depolarization of the resting membrane potential to a threshold level, at which point voltage-regulated membrane gates are opened and action potentials are produced. As described in chapter 13, Ca^{2+} enters the myocardial cytoplasm during the action potential, attaches to troponin, and causes contraction.

Normally, however, sympathetic and vagus (parasympathetic) nerve fibers to the heart are continuously active and modify the rate of spontaneous depolarization of the SA node. Norepinephrine, released primarily by sympathetic nerve end-

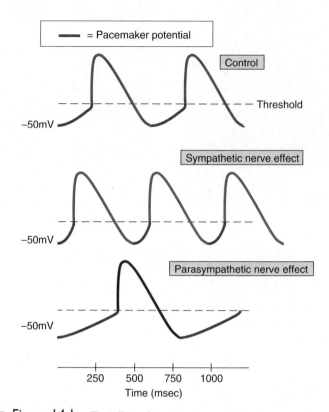

Figure 14.1 The effect of autonomic nerves on the pacemaker potentials in the SA node. The heart's rhythm is set by the rate of spontaneous depolarization in the SA node. This spontaneous depolarization is known as the pacemaker potential, and its rate is increased by sympathetic nerve stimulation and decreased by parasympathetic nerve inhibition.

ings, and epinephrine, secreted by the adrenal medulla, stimulate the opening of Na^+ and Ca^{2+} channels in the plasma membrane of pacemaker cells of the SA node. This increases the rate of the spontaneous depolarization (the *pacemaker potential*), and thereby stimulates an increased rate of firing of the SA node (fig. 14.1). Acetylcholine, released from parasympathetic endings, promotes the opening of K^+ channels in the pacemaker cells (see chapter 7, fig. 7.24). This hyperpolarizes the SA node and thus decreases the rate of its spontaneous firing (fig. 14.1). The actual pace set by the SA node at any time depends on the net effect of these antagonistic influences. Mechanisms that affect the cardiac rate are said to have a **chronotropic effect** (*chrono* = time). Those that increase cardiac rate have a positive chronotropic effect; those that decrease the rate have a negative chronotropic effect.

Autonomic innervation of the SA node represents the major means by which cardiac rate is regulated. However, other autonomic control mechanisms also affect cardiac rate to a lesser degree. Sympathetic endings in the musculature of the atria and ventricles increase the strength of contraction and cause a slight decrease in the time spent in systole when the cardiac rate is high (table 14.1).

During exercise, the cardiac rate increases as a result of decreased vagus nerve inhibition of the SA node. Further increases

Table 14.1 Effects of Autonomic Nerve Activity on the Heart

Region Affected	Sympathetic Nerve Effects	Parasympathetic Nerve Effects
SA node	Increased rate of diastolic depolarization; increased cardiac rate	Decreased rate of diastolic depolarization; decreased cardiac rate
AV node	Increased conduction rate	Decreased conduction rate
Atrial muscle	Increased strength of contraction	Decreased strength of contraction
Ventricular muscle	Increased strength of contraction	No significant effect

in cardiac rate are achieved by increased sympathetic nerve stimulation. The resting bradycardia (slow heart rate) of endurance-trained athletes is due largely to high vagus nerve activity.

The activity of the autonomic innervation of the heart is coordinated by the **cardiac control center** in the medulla oblongata of the brain stem. The question of whether there are separate cardioaccelerator and cardioinhibitory centers in the medulla is currently controversial. The cardiac control center, in turn, is affected by higher brain areas and by sensory feedback from pressure receptors, or *baroreceptors,* in the aorta and carotid arteries. In this way, a fall in blood pressure can produce a reflex increase in the heart rate. This baroreceptor reflex is discussed in more detail in relation to blood pressure regulation later in this chapter.

Clinical Investigation Clue

Remember that when they found Charlie crawling along the road, he had a fast, weak pulse.

What physiological mechanism was responsible for Charlie's rapid pulse?

Regulation of Stroke Volume

The stroke volume is regulated by three variables: (1) the **end-diastolic volume (EDV),** which is the volume of blood in the ventricles at the end of diastole; (2) the **total peripheral resistance,** which is the frictional resistance, or impedance to blood flow, in the arteries; and (3) the **contractility,** or strength, of ventricular contraction.

The end-diastolic volume is the amount of blood in the ventricles immediately before they begin to contract. This is a workload imposed on the ventricles prior to contraction, and thus is sometimes called a **preload.** The stroke volume is directly proportional to the preload; an increase in EDV results in an increase in stroke volume. (This relationship is known as the Frank–Starling Law of the Heart, discussed shortly). The stroke volume is also directly proportional to contractility; when the ventricles contract more forcefully, they pump more blood.

In order to eject blood, the pressure generated in a ventricle when it contracts must be greater than the pressure in the arteries (since blood flows only from higher pressure to lower pressure). The pressure in the arterial system before the ventricle contracts is, in turn, a function of the total peripheral resistance—the higher the peripheral resistance, the higher the pressure. As blood begins to be ejected from the ventricle, the added volume of blood in the arteries causes a rise in mean arterial pressure against the "bottleneck" presented by the peripheral resistance; ejection of blood stops shortly after the aortic pressure becomes equal to the intraventricular pressure. The total peripheral resistance thus presents an impedance to the ejection of blood from the ventricle, or an **afterload** imposed on the ventricle after contraction has begun.

In summary, the stroke volume is inversely proportional to the total peripheral resistance; the greater the peripheral resistance, the lower the stroke volume. It should be noted that this lowering of stroke volume in response to a raised peripheral resistance occurs for only a few beats. Thereafter, a healthy heart is able to compensate for the increased peripheral resistance by beating more strongly. This compensation occurs by means of a mechanism described in the next section (Frank–Starling Law of the Heart).

The proportion of the end-diastolic volume that is ejected against a given afterload depends on the strength of ventricular contraction. Normally, contraction strength is sufficient to eject 70 to 80 ml of blood out of a total end-diastolic volume of 110 to 130 ml. The **ejection fraction** is thus about 60%. More blood is pumped per beat as the EDV increases, and thus the ejection fraction remains relatively constant over a range of end-diastolic volumes. In order for this to be true, the strength of ventricular contraction must increase as the end-diastolic volume increases.

Frank–Starling Law of the Heart

Two physiologists, Otto Frank and Ernest Starling, demonstrated that the strength of ventricular contraction varies directly with the end-diastolic volume (fig. 14.2). Even in experiments where the heart is removed from the body (and is thus not subject to neural or hormonal regulation) and where the still-beating heart is filled with blood flowing from a reservoir, an increase in EDV within the physiological range results in increased contraction strength and, therefore, in increased stroke volume. This relationship between EDV, contraction strength, and stroke volume is thus a built-in, or *intrinsic,* property of heart muscle, and is known as the **Frank–Starling Law of the Heart.**

Intrinsic Control of Contraction Strength

The intrinsic control of contraction strength and stroke volume is due to variations in the degree to which the myocardium is stretched by the end-diastolic volume. As the EDV rises within the physiological range, the myocardium is increasingly stretched and, as a result, contracts more forcefully.

As discussed in chapter 12, stretch can also increase the contraction strength of skeletal muscles (see fig. 12.20). The resting length of skeletal muscles, however, is close to ideal, so that significant stretching decreases contraction strength. This is not true of the heart. Prior to filling with blood during diastole, the sarcomere lengths of myocardial cells are only about 1.5 μm. At this length, the actin filaments from each side overlap in the middle of the sarcomeres, and the cells can contract only weakly (fig. 14.3).

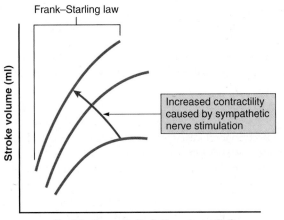

Figure 14.2   The Frank–Starling law and sympathetic nerve **effects.** The graphs demonstrate the Frank–Starling law: As the end-diastolic volume is increased, the stroke volume is increased. The graphs also demonstrate, by comparing the three curves, that the stroke volume is higher at any given end-diastolic volume when the ventricle is stimulated by sympathetic nerves. This is shown by the steeper curves to the left (see the red arrow).

As the ventricles fill with blood, the myocardium stretches so that the actin filaments overlap with myosin only at the edges of the A bands (fig. 14.3). This increases the number of interactions between actin and myosin, allowing more force to be developed during contraction. Since this more advantageous overlapping of actin and myosin is produced by stretching of the ventricles, and since the degree of stretching is controlled by the degree of filling (the end-diastolic volume), the strength of contraction is intrinsically adjusted by the end-diastolic volume.

As shown in figure 14.4, muscle length has a more pronounced effect on contraction strength in cardiac muscle than in skeletal muscle. That is, a particular increase in sarcomere length will stimulate contraction strength more in cardiac muscle than in skeletal muscle. This is believed to be due to an increased sensitivity of stretched cardiac muscle to the stimulatory effects of Ca^{2+}.

The Frank–Starling law explains how the heart can adjust to a rise in total peripheral resistance: (1) a rise in peripheral resistance causes a decrease in the stroke volume of the ventricle, so that (2) more blood remains in the ventricle and the end-diastolic volume is greater for the next cycle; as a result, (3) the ventricle is stretched to a greater degree in the next cycle and contracts more strongly to eject more blood. This allows a healthy ventricle to sustain a normal cardiac output.

A very important consequence of these events is that the cardiac output of the left ventricle, which pumps blood into the systemic circulation with its ever-changing resistances, can be adjusted to match the output of the right ventricle (which pumps blood into the pulmonary circulation). Clearly, the rate of blood

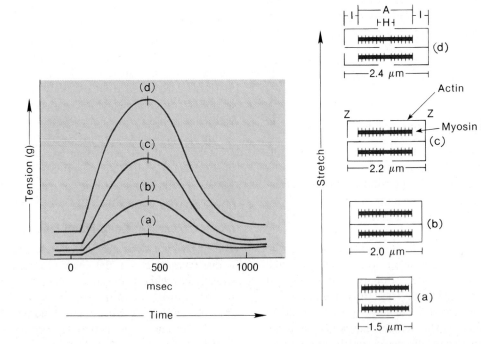

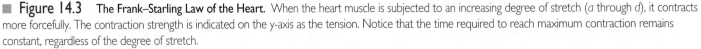

Figure 14.3   The Frank–Starling Law of the Heart.   When the heart muscle is subjected to an increasing degree of stretch (a through d), it contracts more forcefully. The contraction strength is indicated on the y-axis as the tension. Notice that the time required to reach maximum contraction remains constant, regardless of the degree of stretch.

flow through the pulmonary and systemic circulations must be equal in order to prevent fluid accumulation in the lungs and to deliver fully oxygenated blood to the body.

Extrinsic Control of Contractility

The *contractility* is the strength of contraction at any given fiber length. At any given degree of stretch, the strength of ventricular contraction depends on the activity of the sympathoadrenal system. Norepinephrine from sympathetic nerve endings and epinephrine from the adrenal medulla produce an increase in contraction strength (see figs. 14.2 and 14.4). This **positive inotropic effect** results from an increase in the amount of Ca^{2+} available to the sarcomeres.

The cardiac output is thus affected in two ways by the activity of the sympathoadrenal system: (1) through a positive inotropic effect on contractility and (2) through a positive chronotropic effect on cardiac rate (fig. 14.5). Stimulation through parasympathetic nerve endings to the SA node and conducting tissue has a negative chronotropic effect but does not directly affect the contraction strength of the ventricles. However, the increased EDV that results from a slower cardiac rate can increase contraction strength through the mechanism described by the Frank–Starling Law of the Heart.

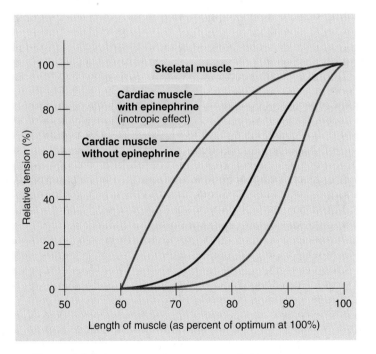

■ **Figure 14.4**   The effect of muscle length and epinephrine on contraction strength. In this schematic comparison, all three curves demonstrate that each muscle contracts with its maximum force (100% relative tension) at its own optimum length (100% optimum length). As the length is decreased from optimum, each curve demonstrates a decreased contraction strength. Notice that the decline is steeper for cardiac muscle than for skeletal muscle, demonstrating the importance of the Frank–Starling relationship in heart physiology. At any length, however, epinephrine increases the strength of myocardial contraction, demonstrating a positive inotropic effect.

Venous Return

The end-diastolic volume—and thus the stroke volume and cardiac output—is controlled by factors that affect the **venous return,** which is the return of blood to the heart via veins. The rate at which the atria and ventricles are filled with venous blood depends on the total blood volume and the venous pressure (pressure in the veins). It is the venous pressure that serves as the driving force for the return of blood to the heart.

Veins have thinner, less muscular walls than do arteries; thus, they have a higher **compliance.** This means that a given amount of pressure will cause more distension (expansion) in veins than in arteries, so that the veins can hold more blood. Approximately two-thirds of the total blood volume is located in the veins (fig. 14.6). Veins are therefore called *capacitance vessels,* after electronic devices called capacitors that store electrical charges. Muscular arteries and arterioles expand less under pressure (are less compliant), and thus are called *resistance vessels.*

Although veins contain almost 70% of the total blood volume, the mean venous pressure is only 2 mmHg, compared to a mean arterial pressure of 90 to 100 mmHg. The lower venous pressure is due in part to a pressure drop between arteries and capillaries and in part to the high venous compliance.

The venous pressure is highest in the venules (10 mmHg) and lowest at the junction of the venae cavae with the right atrium (0 mmHg). In addition to this pressure difference, the venous return to the heart is aided by (1) sympathetic nerve activity, which stimulates smooth muscle contraction in the venous walls and thus reduces compliance; (2) the skeletal muscle pump, which squeezes veins during muscle contraction; and (3) the pressure difference between the thoracic and

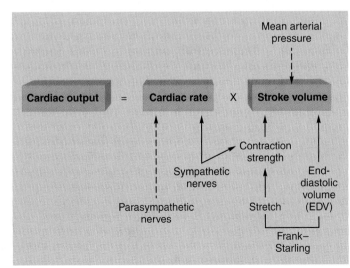

■ **Figure 14.5**   The regulation of cardiac output. Factors that stimulate cardiac output are shown as solid arrows; factors that inhibit cardiac output are shown as dashed arrows.

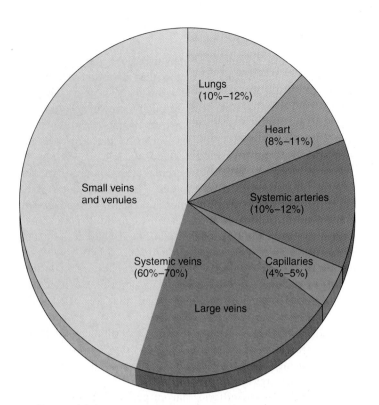

■ **Figure 14.6**   The distribution of blood within the circulatory
system at rest. Notice that the venous system contains most of the blood;
it functions as a reservoir from which more blood can be added to the
circulation under appropriate conditions (such as exercise).

abdominal cavities, which promotes the flow of venous blood
back to the heart.

  Contraction of the skeletal muscles functions as a "pump"
by virtue of its squeezing action on veins (described in chapter 13;
see fig. 13.28). Contraction of the diaphragm during inhalation
also improves venous return. The diaphragm lowers as it contracts,
thus increasing the thoracic volume and decreasing the abdominal
volume. This creates a partial vacuum in the thoracic cavity and a
higher pressure in the abdominal cavity. The pressure difference
thus produced favors blood flow from abdominal to thoracic veins
(fig. 14.7).

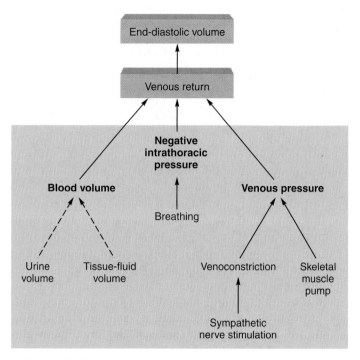

■ **Figure 14.7**   Variables that affect venous return and thus end-
diastolic volume. Direct relationships are indicated by solid arrows; inverse
relationships are shown with dashed arrows.

Blood Volume

Fluid in the extracellular environment of the body is distributed
between the blood and the interstitial fluid compartments by filtration
and osmotic forces acting across the walls of capillaries. The function
of the kidneys influences blood volume because urine is derived from
blood plasma. The hormones ADH and aldosterone act on the
kidneys to help regulate the blood volume.

  Blood volume represents one part, or compartment, of the
total body water. Approximately two-thirds of the total body
water is contained within cells—in the intracellular compartment.
The remaining one-third is in the **extracellular compartment.**
This extracellular fluid is normally distributed so that about 80%
is contained in the tissues—as *tissue,* or *interstitial, fluid*—with
the blood plasma accounting for the remaining 20% (fig. 14.8).

  The distribution of water between the tissue fluid and the
blood plasma is determined by a balance between opposing forces
acting at the capillaries. Blood pressure, for example, promotes the
formation of tissue fluid from plasma, whereas osmotic forces draw
water from the tissues into the vascular system. The total volume of
intracellular and extracellular fluid is normally maintained constant
by a balance between water loss and water gain. Mechanisms that
affect drinking, urine volume, and the distribution of water between
plasma and tissue fluid thus help to regulate blood volume and, by
this means, help to regulate cardiac output and blood flow.

Test Yourself Before You Continue

1. Describe how the stroke volume is intrinsically regulated by the
    end-diastolic volume. Why is this regulation significant?
2. Describe the effects of autonomic nerve stimulation on the
    cardiac rate and stroke volume.
3. Define the terms *preload* and *afterload* and explain how these
    factors affect the cardiac output.
4. List the factors that affect venous return. Using a flowchart, show
    how an increased venous return can result in an increased
    cardiac output.

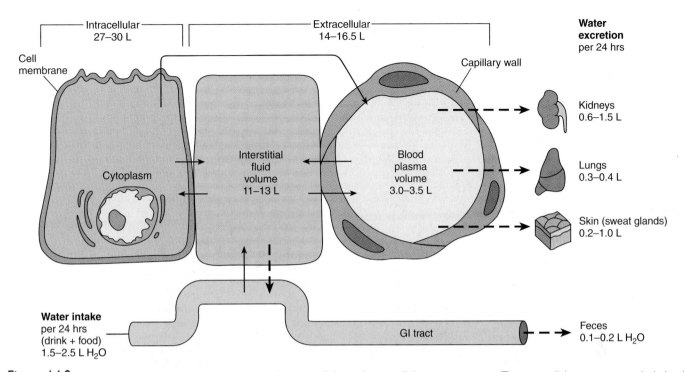

Figure 14.8 The distribution of body water between the intracellular and extracellular compartments. The extracellular compartment includes the blood plasma and the interstitial (tissue) fluid.

Exchange of Fluid Between Capillaries and Tissues

The distribution of extracellular fluid between the plasma and interstitial compartments is in a state of dynamic equilibrium. Tissue fluid is not normally a "stagnant pond"; rather, it is a continuously circulating medium, formed from and returning to the vascular system. In this way, the tissue cells receive a continuously fresh supply of glucose and other plasma solutes that are filtered through tiny endothelial channels in the capillary walls.

Filtration results from blood pressure within the capillaries. This hydrostatic pressure, which is exerted against the inner capillary wall, is equal to about 37 mmHg at the arteriolar end of systemic capillaries and drops to about 17 mmHg at the venular end of the capillaries. The **net filtration pressure** is equal to the hydrostatic pressure of the blood in the capillaries minus the hydrostatic pressure of tissue fluid outside the capillaries, which opposes filtration. If, as an extreme example, these two values were equal, there would be no filtration. The magnitude of the tissue hydrostatic pressure varies from organ to organ. With a hydrostatic pressure in the tissue fluid of 1 mmHg, as it is outside the capillaries of skeletal muscles, the net filtration pressure would be 37 − 1 = 36 mmHg at the arteriolar end of the capillary and 17 − 1 = 16 mmHg at the venular end.

Glucose, comparably sized organic molecules, inorganic salts, and ions are filtered along with water through the capillary channels. The concentrations of these substances in tissue fluid are thus the same as in plasma. The protein concentration of tissue fluid (2 g/100 ml), however, is less than the protein concentration of plasma (6 to 8 g/100 ml). This difference is due to the restricted filtration of proteins through the capillary pores. The osmotic pres-

sure exerted by plasma proteins—called the **colloid osmotic pressure** of the plasma (because proteins are present as a colloidal suspension)—is therefore much greater than the colloid osmotic pressure of tissue fluid. The difference between these two osmotic pressures is called the **oncotic pressure.** Since the colloid osmotic pressure of the tissue fluid is sufficiently low to be neglected, the oncotic pressure is essentially equal to the colloid osmotic pressure of the plasma. This value has been estimated to be 25 mmHg. Since water will move by osmosis from the solution of lower to the solution of higher osmotic pressure (chapter 6), this oncotic pressure favors the movement of water into the capillaries.

Whether fluid will move out of or into the capillary depends on the magnitude of the net filtration pressure, which varies from the arteriolar to the venular end of the capillary, and on the oncotic pressure. These opposing forces that affect the distribution of fluid across the capillary are known as **Starling forces,** and their effects can be calculated according to this equation:

Fluid movement is proportional to:

$$\underbrace{(P_\text{c} + \pi_\text{i})}_{} - \underbrace{(P_\text{i} + \pi_\text{p})}_{}$$

(Fluid out) − (Fluid in)

where

P_c = hydrostatic pressure in the capillary
π_i = colloid osmotic pressure of the interstitial (tissue) fluid
P_i = hydrostatic pressure of interstitial fluid
π_p = colloid osmotic pressure of the blood plasma

The expression to the left of the minus sign represents the sum of forces acting to move fluid out of the capillary. The expression to the right represents the sum of forces acting to move

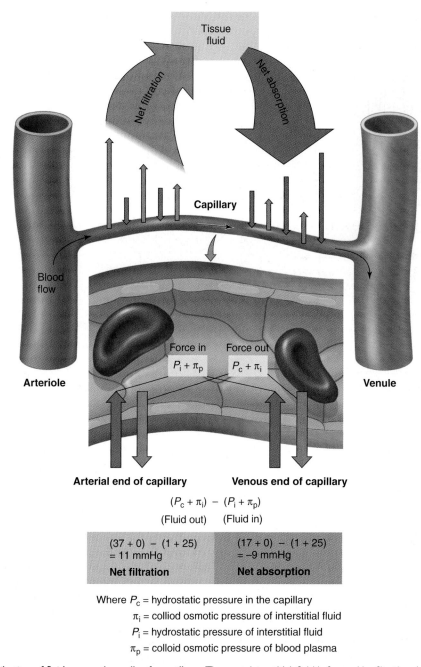

$$(P_c + \pi_i) - (P_i + \pi_p)$$

(Fluid out)   (Fluid in)

(37 + 0) − (1 + 25)	(17 + 0) − (1 + 25)
= 11 mmHg	= −9 mmHg
Net filtration	**Net absorption**

Where P_c = hydrostatic pressure in the capillary

π_i = colliod osmotic pressure of interstitial fluid

P_i = hydrostatic pressure of interstitial fluid

π_p = colloid osmotic pressure of blood plasma

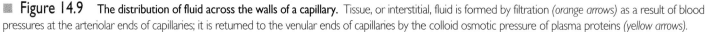 **Figure 14.9**   **The distribution of fluid across the walls of a capillary.** Tissue, or interstitial, fluid is formed by filtration *(orange arrows)* as a result of blood pressures at the arteriolar ends of capillaries; it is returned to the venular ends of capillaries by the colloid osmotic pressure of plasma proteins *(yellow arrows)*.

fluid into the capillary. Figure 14.9 provides typical values for blood capillaries in skeletal muscles. Notice that the sum of the forces acting on the capillary is a positive number at the arteriolar end and a negative number at the venular end of the capillary. The positive value at the arteriolar end indicates that the Starling forces that favor the extrusion of fluid from the capillary predominates. The negative value at the venular end indicates that the net Starling forces favor the return of fluid to the capillary. Fluid thus leaves the capillaries at the arteriolar end and returns to the capillaries at the venular end (fig. 14.9, *top*).

Clinical Investigation Clues

Remember that Charlie was given intravenous fluid containing albumin.

Why was Charlie given albumin?

What advantage does intravenous albumin have over intravenous saline or dextrose (glucose)?

This "classic" view of capillary dynamics has been modified in recent years by the realization that the balance of filtration and reabsorption varies in different tissues and under different conditions in a particular capillary. For example, a capillary may be open or closed off by precapillary muscles that function as sphincters. When the capillary is open, blood flow is high and the net filtration force exceeds the force for the osmotic return of water throughout the length of the capillary. The opposite is true if the precapillary sphincter closes and the blood flow through the capillary is reduced.

Through the action of the Starling forces, plasma and tissue fluid are continuously interchanged. The return of fluid to the vascular system at the venular ends of the capillaries, however, does not exactly equal the amount filtered at the arteriolar ends. According to some estimates, approximately 85% of the capillary filtrate is returned directly to the capillaries; the remaining 15% (amounting to at least 2 L per day) is returned to the vascular system by way of the lymphatic system. Lymphatic capillaries, it may be recalled from chapter 13 (see fig. 13.34), drain excess tissue fluid and proteins and, by way of lymphatic vessels, ultimately return this fluid to the venous system.

In the tropical disease *filariasis,* mosquitoes transmit a nematode worm parasite to humans. The larvae of these worms invade lymphatic vessels and block lymphatic drainage. The edema that results can be so severe that the tissues swell to produce an elephant-like appearance, with thickening and cracking of the skin. This condition is thus aptly named **elephantiasis** (fig. 14.10). The World Health Organization estimates that this disease currently affects at least 120 million people, primarily in India and Africa. A new drug regimen has been found to be 99% effective against the filariasis parasite, and a worldwide effort to treat this disease is now underway.

Causes of Edema

Excessive accumulation of tissue fluid is known as **edema.** This condition is normally prevented by a proper balance between capillary filtration and osmotic uptake of water and by proper lymphatic drainage. Edema may thus result from

1. *high arterial blood pressure,* which increases capillary pressure and causes excessive filtration;
2. *venous obstruction*—as in phlebitis (where a thrombus forms in a vein) or mechanical compression of veins (during pregnancy, for example)—which produces a congestive increase in capillary pressure;
3. *leakage of plasma proteins into interstitial fluid,* which causes reduced osmotic flow of water into the capillaries (this occurs during inflammation and allergic reactions as a result of increased capillary permeability);
4. *myxedema*—the excessive production of particular glycoproteins (mucin) in the extracellular matrix caused by hypothyroidism;

■ **Figure 14.10   The severe edema of elephantiasis.** Parasitic larvae that block lymphatic drainage produce tissue edema and the tremendous enlargement of the limbs and external genitalia in elephantiasis.

Table 14.2   Causes of Edema

Cause	Comments
Increased blood pressure or venous obstruction	Increases capillary filtration pressure so that more tissue fluid is formed at the arteriolar ends of capillaries.
Increased tissue protein concentration	Decreases osmosis of water into the venular ends of capillaries. Usually a localized tissue edema due to leakage of plasma proteins through capillaries during inflammation and allergic reactions. Myxedema due to hypothyroidism is also in this category.
Decreased plasma protein concentration	Decreases osmosis of water into the venular ends of capillaries. May be caused by liver disease (which can be associated with insufficient plasma protein production), kidney disease (due to leakage of plasma protein into urine), or protein malnutrition.
Obstruction of lymphatic vessels	Infections by filaria roundworms (nematodes) transmitted by a certain species of mosquito block lymphatic drainage, causing edema and tremendous swelling of the affected areas.

5. *decreased plasma protein concentration,* as a result of liver disease (the liver makes most of the plasma proteins) or kidney disease where plasma proteins are excreted in the urine;
6. *obstruction of the lymphatic drainage* (table 14.2).

Regulation of Blood Volume by the Kidneys

The formation of urine begins in the same manner as the formation of tissue fluid—by filtration of plasma through capillary pores. These capillaries are known as *glomeruli*, and the filtrate they produce enters a system of tubules that transports and modifies the filtrate (by mechanisms discussed in chapter 17). The kidneys produce about 180 L per day of blood filtrate, but since there is only 5.5 L of blood in the body, it is clear that most of this filtrate must be returned to the vascular system and recycled. Only about 1.5 L of urine is excreted daily; 98% to 99% of the amount filtered is **reabsorbed** back into the vascular system.

The volume of urine excreted can be varied by changes in the reabsorption of filtrate. If 99% of the filtrate is reabsorbed, for example, 1% must be excreted. Decreasing the reabsorption by only 1%—from 99% to 98%—would double the volume of urine excreted (an increase to 2% of the amount filtered). Carrying the logic further, a doubling of urine volume from, for example, 1 to 2 liters, would result in the loss of an additional liter of blood volume. The percentage of the glomerular filtrate reabsorbed—and thus the urine volume and blood volume—is adjusted according to the needs of the body by the action of specific hormones on the kidneys. Through their effects on the kidneys, and the resulting changes in blood volume, these hormones serve important functions in the regulation of the cardiovascular system.

Regulation by Antidiuretic Hormone (ADH)

One of the major hormones involved in the regulation of blood volume is **antidiuretic hormone (ADH),** also known as *vasopressin.* As described in chapter 11, this hormone is produced by neurons in the hypothalamus, transported by axons into the posterior pituitary, and released from this storage gland in response to hypothalamic stimulation. The release of ADH from the posterior pituitary occurs when neurons in the hypothalamus called **osmoreceptors** detect an increase in plasma osmolality (osmotic pressure).

An increase in plasma osmolality occurs when the plasma becomes more concentrated (chapter 6). This can be produced either by *dehydration* or by excessive *salt intake.* Stimulation of osmoreceptors produces sensations of thirst, leading to increased water intake and an increase in the amount of ADH released from the posterior pituitary. Through mechanisms that will be discussed in conjunction with kidney physiology in chapter 17, ADH stimulates water reabsorption from the filtrate. A smaller volume of urine is thus excreted as a result of the action of ADH (fig. 14.11).

A person who is dehydrated or who consumes excessive amounts of salt thus drinks more and urinates less. This raises the blood volume and, in the process, dilutes the plasma to lower its previously elevated osmolality. The rise in blood volume that results from these mechanisms is extremely important in stabilizing the condition of a dehydrated person with low blood volume and pressure.

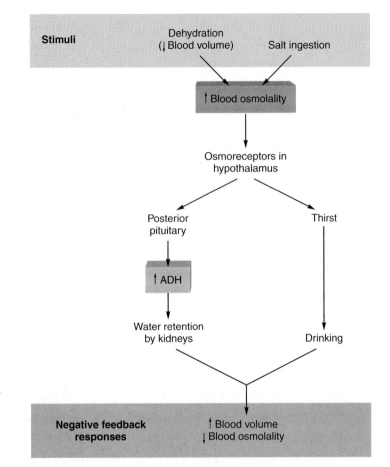

Figure 14.11 The negative feedback control of blood volume and blood osmolality. Thirst and ADH secretion are triggered by a rise in plasma osmolality. Homeostasis is maintained by countermeasures, including drinking and conservation of water by the kidneys.

Clinical Investigation Clues

Remember that Charlie had a low urine output and that his urine had a high osmolality (concentration).

What physiological mechanism could be responsible for this?
What benefit does Charlie derive from this mechanism?

Drinking excessive amounts of water without excessive amounts of salt does not result in a prolonged increase in blood volume and pressure. The water does enter the blood from the intestine and momentarily raises the blood volume; at the same time, however, it dilutes the blood. Dilution of the blood decreases the plasma osmolality and thus inhibits the release of ADH. With less ADH there is less reabsorption of filtrate in the kidneys—a larger volume of urine is excreted. Water is therefore a *diuretic*—a substance that promotes urine formation—because it inhibits the release of antidiuretic hormone.

An increase in blood volume can thus be compensated by a fall in ADH secretion. However, expanded blood volume also stimulates stretch receptors in the left atrium of the heart, causing the increased secretion of a different hormone. This is a polypeptide known as *atrial natriuretic peptide* (discussed in a separate section shortly). This hormone promotes the increased excretion of salt and water in the urine, thereby helping to lower the blood volume.

During prolonged exercise, particularly on a warm day, a substantial amount of water (up to 900 ml per hour) may be lost from the body through sweating. The lowering of blood volume that results decreases the ability of the body to dissipate heat, and the consequent over-heating of the body can cause ill effects and put an end to the exercise. The need for athletes to remain well hydrated is commonly recognized, but drinking pure water may not be the answer. This is because blood sodium is lost in sweat, so that a lesser amount of water is required to dilute the blood osmolality back to normal. When the blood osmolality is normal, the urge to drink is extinguished. For these reasons, athletes performing prolonged endurance exercise should drink solutions containing sodium (as well as carbohydrates for energy), and they should drink at a predetermined rate rather than at a rate determined only by thirst.

Regulation by Aldosterone

From the preceding discussion, it is clear that a certain amount of dietary salt is required to maintain blood volume and pressure. Since Na^+ and Cl^- are easily filtered in the kidneys, a mechanism must exist to promote the reabsorption and retention of salt when the dietary salt intake is too low. **Aldosterone,** a steroid hormone secreted by the adrenal cortex, stimulates the reabsorption of salt by the kidneys. Aldosterone is thus a "salt-retaining hormone." Retention of salt indirectly promotes retention of water (in part, by the action of ADH, as previously discussed). The action of aldosterone produces an increase in blood volume, but, unlike ADH, it does not produce a change in plasma osmolality. This is because aldosterone promotes the reabsorption of salt and water in proportionate amounts, whereas ADH promotes only the reabsorption of water. Thus, unlike ADH, aldosterone does not act to dilute the blood.

The secretion of aldosterone is stimulated during salt deprivation, when the blood volume and pressure are reduced. The adrenal cortex, however, is not directly stimulated to secrete aldosterone by these conditions. Instead, a decrease in blood volume and pressure activates an intermediate mechanism, described in the next section.

Throughout most of human history, salt was in short supply and was therefore highly valued. Moorish merchants in the sixth century traded an ounce of salt for an ounce of gold, and salt cakes were used as money in Abyssinia. Part of a Roman soldier's pay was given in salt—a practice from which the word *salary* (*sal* = salt) derives. Salt was also used to purchase slaves—hence the phrase "worth his salt."

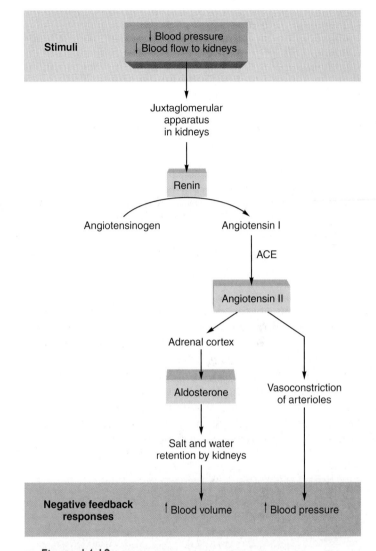

Figure 14.12 The renin-angiotensin-aldosterone system. This system helps to maintain homeostasis through the negative feedback control of blood volume and pressure. (ACE = angiotensin-converting enzyme.)

Renin-Angiotensin-Aldosterone System

When the blood flow and pressure are reduced in the renal artery (as they would be in the low-blood-volume state of salt deprivation), a group of cells in the kidneys called the *juxtaglomerular apparatus* secretes the enzyme **renin** into the blood. This enzyme cleaves a ten-amino-acid polypeptide called *angiotensin I* from a plasma protein called *angiotensinogen*. As angiotensin I passes through the capillaries of the lungs, an *angiotensin-converting enzyme (ACE)* removes two amino acids. This leaves an eight-amino-acid polypeptide called **angiotensin II** (fig. 14.12). Conditions of salt deprivation, low blood volume, and low blood pressure, in summary, cause increased production of angiotensin II in the blood.

Angiotensin II exerts numerous effects that produce a rise in blood pressure. This rise in pressure is partly due to vasoconstriction and partly to increases in blood volume. Vasoconstriction of arterioles and small muscular arteries is produced directly by the effects of angiotensin II on the smooth muscle layers of these vessels. The increased blood volume is an indirect effect of angiotensin II.

Angiotensin II promotes a rise in blood volume by means of two mechanisms: (1) thirst centers in the hypothalamus are stimulated by angiotensin II, and thus more water is ingested, and (2) secretion of aldosterone from the adrenal cortex is stimulated by angiotensin II, and higher aldosterone secretion causes more salt and water to be retained by the kidneys. The relationship between angiotensin II and aldosterone is sometimes described as the **renin-angiotensin-aldosterone system.**

Clinical Investigation Clues

Remember that Charlie's urine had virtually no Na^+ in it.
What physiological mechanism is responsible for this?
What benefits does Charlie derive from this mechanism?

The renin-angiotensin-aldosterone system can also work in the opposite direction: high salt intake, leading to high blood volume and pressure, normally inhibits renin secretion. With less angiotensin II formation and less aldosterone secretion, less salt is retained by the kidneys and more is excreted in the urine. Unfortunately, many people with chronically high blood pressure may have normal or even elevated levels of renin secretion. In these cases, the intake of salt must be lowered to match the impaired ability to excrete salt in the urine.

One of the newer classes of drugs that can be used to treat hypertension (high blood pressure) are the **angiotensin-converting enzyme, or ACE, inhibitors.** These drugs (such as captopril) block the formation of angiotensin II, thus reducing its vasoconstrictor effect. The ACE inhibitors also increase the activity of bradykinin, a polypeptide that promotes vasodilation. The reduced formation of angiotensin II and increased action of bradykinin result in vasodilation, which decreases the total peripheral resistance. Because this reduces the afterload of the heart, the ACE inhibitors are also used to treat left ventricular hypertrophy and congestive heart failure. Another new class of antihypertensive drugs allows angiotensin II to be formed but selectively blocks the angiotensin II receptors.

Atrial Natriuretic Peptide

As described in the previous section, a fall in blood volume is compensated for by renal retention of fluid through activation of the renin-angiotensin-aldosterone system. An increase in blood volume, conversely, is compensated for by renal excretion of a larger volume of urine. Experiments suggest that the increase in water excretion under conditions of high blood volume is at least partly due to an increase in the excretion of Na^+ in the urine, or *natriuresis* (*natrium* = sodium; *uresis* = making water).

Increased Na^+ excretion (natriuresis) may be produced by a decline in aldosterone secretion, but there is evidence that there is a separate hormone that stimulates natriuresis. This *natriuretic hormone* would thus be antagonistic to aldosterone and would promote Na^+ and water excretion in the urine in response to a rise in blood volume. A polypeptide hormone with these properties, identified as **atrial natriuretic peptide (ANP),** is produced by the atria of the heart. By promoting salt and water excretion in the urine, ANP can act to lower the blood volume and pressure. This is analogous to the action of diuretic drugs taken by people with hypertension, as described later in this chapter.

In addition to its stimulation of salt and water excretion by the kidneys. ANP also antagonizes various actions of angiotensin II. As a result of this action, ANP decreases the secretion of aldosterone and promotes vasodilation.

Test Yourself Before You Continue

1. Describe the composition of tissue (interstitial) fluid. Using a flow diagram, explain how tissue fluid is formed and how it is returned to the vascular system.
2. Define the term *edema* and describe four different mechanisms that can produce this condition.
3. Describe the effects of dehydration on blood and urine volumes. What cause-and-effect mechanism is involved?
4. Explain why salt deprivation causes increased salt and water retention by the kidneys.
5. Describe the actions of atrial natriuretic peptide and explain their significance.

Vascular Resistance to Blood Flow

The rate of blood flow to an organ is related to the resistance to flow in the small arteries and arterioles that serve the organ. Vasodilation decreases resistance and increases flow, whereas vasoconstriction increases resistance and decreases flow. Vasodilation and vasoconstriction occur in response to intrinsic and extrinsic regulatory mechanisms.

The amount of blood that the heart pumps per minute is equal to the rate of venous return, and thus is equal to the rate of blood flow through the entire circulation. The cardiac output of 5 to 6 L per minute is distributed unequally to the different organs. At rest, blood flow is about 2,500 ml/min through the liver, kidneys, and gastrointestinal tract; 1,200 ml/min through the skeletal muscles; 750 ml/min through the brain; and 250 ml/min through the coronary arteries of the heart. The balance of the cardiac output (500 to 1,100 ml/min) is distributed to the other organs (table 14.3).

Table 14.3 Estimated Distribution of the Cardiac Output at Rest

Organs	Blood Flow	
	Milliliters per Minute	Percent Total
Gastrointestinal tract and liver	1,400	24
Kidneys	1,100	19
Brain	750	13
Heart	250	4
Skeletal muscles	1,200	21
Skin	500	9
Other organs	600	10
Total organs	5,800	100

Source: From O.L. Wade and J.M. Bishop, *Cardiac Output and Regional Blood Flow.* Copyright © 1962 Blackwell Science, Ltd. Used with permission.

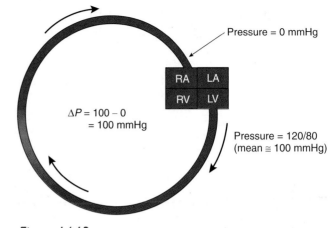

■ **Figure 14.13** Blood flow is produced by a pressure difference. The flow of blood in the systemic circulation is ultimately dependent on the pressure difference (ΔP) between the mean pressure of about 100 mmHg at the origin of flow in the aorta and the pressure at the end of the circuit—0 mmHg in the vena cava, where it joins the right atrium (RA). (LA = left atrium; RV = right ventricle; LV = left ventricle.)

Physical Laws Describing Blood Flow

The flow of blood through the vascular system, like the flow of any fluid through a tube, depends in part on the difference in pressure at the two ends of the tube. If the pressure at both ends of the tube is the same, there will be no flow. If the pressure at one end is greater than at the other, blood will flow from the region of higher to the region of lower pressure. The rate of blood flow is proportional to the pressure difference ($P_1 - P_2$) between the two ends of the tube. The term **pressure difference** is abbreviated ΔP, in which the Greek letter Δ *(delta)* means "change in."

If the systemic circulation is pictured as a single tube leading from and back to the heart (fig. 14.13), blood flow through this system would occur as a result of the pressure difference between the beginning of the tube (the aorta) and the end of the tube (the junction of the venae cavae with the right atrium). The average pressure, or **mean arterial pressure (MAP),** is about 100 mmHg; the pressure at the right atrium is 0 mmHg. The "pressure head," or driving force (ΔP), is therefore about $100 - 0 = 100$ mmHg.

Blood flow is directly proportional to the pressure difference between the two ends of the tube (ΔP) but is *inversely proportional* to the frictional resistance to blood flow through the vessels. Inverse proportionality is expressed by showing one of the factors in the denominator of a fraction, since a fraction decreases when the denominator increases:

$$\text{Blood flow} \propto \frac{\Delta P}{\text{resistance}}$$

The **resistance** to blood flow through a vessel is directly proportional to the length of the vessel and to the viscosity of the blood (the "thickness," or ability of molecules to "slip over" each other; for example, honey is quite viscous). Of particular physiological importance, the vascular resistance is inversely proportional to the fourth power of the radius of the vessel:

$$\text{Resistance} \propto \frac{L\eta}{r^4}$$

where

L = length of vessel
η = viscosity of blood
r = radius of vessel

For example, if one vessel has half the radius of another and if all other factors are the same, the smaller vessel will have sixteen times (2^4) the resistance of the larger vessel. Blood flow through the larger vessel, as a result, will be sixteen times greater than in the smaller vessel (fig. 14.14).

When physical constants are added to this relationship, the rate of blood flow can be calculated according to **Poiseuille's** *(pwă-zuh'yez)* **law:**

$$\text{Blood flow} = \frac{\Delta P r^4 (\pi)}{\eta L (8)}$$

Vessel length (L) and blood viscosity (the Greek letter *eta,* written η) do not vary significantly in normal physiology, although blood viscosity is increased in severe dehydration and in the polycythemia (high red blood cell count) that occurs as an adaptation to life at high altitudes. The major physiological regulators of blood flow through an organ are the mean arterial pressure (P, driving the flow) and the vascular resistance to flow. At a given mean arterial pressure, blood can be diverted from one organ to another by variations in the degree of vasoconstriction and vasodilation of small arteries and arterioles (that is, by variations in vessel radius, r). Vasoconstriction in one organ and vasodilation in another result in a diversion, or *shunting,* of blood to the second organ. Since arterioles are the smallest arteries and can become narrower by vasoconstriction, they provide the greatest resistance to blood flow (fig. 14.15). Blood flow to an organ is thus largely determined by the degree of vasoconstriction or vasodilation of its arterioles. The rate of blood flow to an organ can be increased by dilation of its arterioles and can be decreased by constriction of its arterioles.

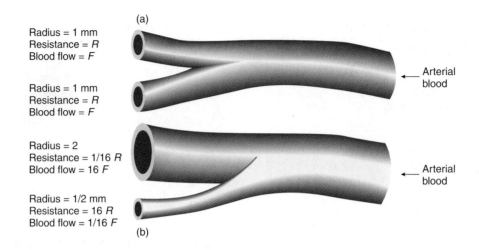

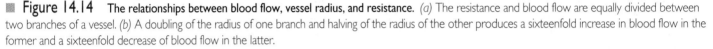

■ Figure 14.14 The relationships between blood flow, vessel radius, and resistance. (a) The resistance and blood flow are equally divided between two branches of a vessel. (b) A doubling of the radius of one branch and halving of the radius of the other produces a sixteenfold increase in blood flow in the former and a sixteenfold decrease of blood flow in the latter.

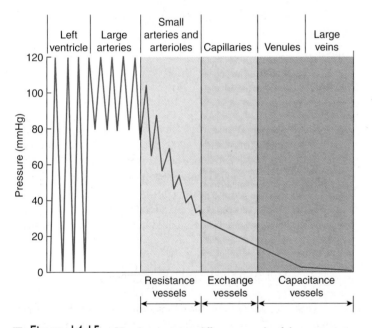

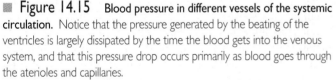

■ Figure 14.15 Blood pressure in different vessels of the systemic circulation. Notice that the pressure generated by the beating of the ventricles is largely dissipated by the time the blood gets into the venous system, and that this pressure drop occurs primarily as blood goes through the aterioles and capillaries.

Total Peripheral Resistance

The sum of all the vascular resistances within the systemic circulation is called the **total peripheral resistance.** The arteries that supply blood to the organs are generally in parallel rather than in series with each other. That is, arterial blood passes through only one set of resistance vessels (arterioles) before returning to the heart (fig. 14.16). Since one organ is not "downstream" from another in terms of its arterial supply, changes in resistance within one organ directly affect blood flow in that organ only.

Vasodilation in a large organ might, however, significantly decrease the total peripheral resistance and, by this means, might decrease the mean arterial pressure. In the absence of compensatory mechanisms, the driving force for blood flow through all organs might be reduced. This situation is normally prevented by an increase in the cardiac output and by vasoconstriction in other areas. During exercise of the large muscles, for example, the arterioles in the exercising muscles are dilated. This would cause a great fall in mean arterial pressure if there were no compensations. The blood pressure actually rises during exercise, however, because the cardiac output is increased and because there is constriction of arterioles in the viscera and skin.

Extrinsic Regulation of Blood Flow

The term *extrinsic regulation* refers to control by the autonomic nervous system and endocrine system. Angiotensin II, for example, directly stimulates vascular smooth muscle to produce generalized vasoconstriction. Antidiuretic hormone (ADH) also has a vasoconstrictor effect at high concentrations; this is why it is also called *vasopressin*. This vasopressor effect of ADH is not believed to be significant under physiological conditions in humans.

Regulation by Sympathetic Nerves

Stimulation of the sympathoadrenal system produces an increase in the cardiac output (as previously discussed) and an increase in total peripheral resistance. The latter effect is due to alpha-adrenergic stimulation (chapter 9; see fig. 9.10) of vascular smooth muscle by norepinephrine and, to a lesser degree, by epinephrine. This produces vasoconstriction of the arterioles in the viscera and skin.

Even when a person is calm, the sympathoadrenal system is active to a certain degree and helps set the "tone" of vascular smooth muscles. In this case, **adrenergic sympathetic fibers** (those that release norepinephrine) activate alpha-adrenergic receptors to cause a basal level of vasoconstriction throughout the

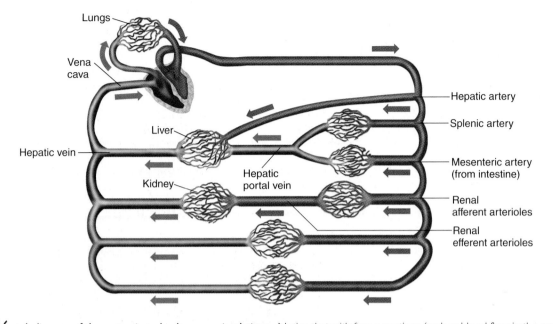

Figure 14.16 **A diagram of the systemic and pulmonary circulations.** Notice that with few exceptions (such as blood flow in the renal circulation) the flow of arterial blood is in parallel rather than in series (arterial blood does not usually flow from one organ to another).

body. During the fight-or-flight reaction, an increase in the activity of adrenergic fibers produces vasoconstriction in the digestive tract, kidneys, and skin.

Arterioles in skeletal muscles receive **cholinergic sympathetic fibers,** which release acetylcholine as a neurotransmitter. During the fight-or-flight reaction, the activity of these cholinergic fibers increases. This causes vasodilation. Vasodilation in skeletal muscles is also produced by epinephrine secreted by the adrenal medulla, which stimulates beta-adrenergic receptors. During the fight-or-flight reaction, therefore, blood flow is decreased to the viscera and skin because of the alpha-adrenergic effects of vasoconstriction in these organs, whereas blood flow to the skeletal muscles is increased. This diversion of blood flow to the skeletal muscles during emergency conditions may give these muscles an "extra edge" in responding to the emergency. Once exercise begins, however, the blood flow to skeletal muscles increases far more due to other mechanisms (described shortly under Intrinsic Regulation of Blood Flow).

Cocaine inhibits the reuptake of norepinephrine into the adrenergic axons, resulting in enhanced sympathetic-induced vasoconstriction. Chest pain, as a result of myocardial ischemia produced in this way, is a common cocaine-related problem. The nicotine from cigarette smoke acts synergistically with cocaine to induce vasoconstriction.

Parasympathetic Control of Blood Flow

Parasympathetic endings in arterioles are always cholinergic and always promote vasodilation. Parasympathetic innervation of blood vessels, however, is limited to the digestive tract, external genitalia, and salivary glands. Because of this limited distribu-

tion, the parasympathetic system is less important than the sympathetic system in the control of total peripheral resistance.

The extrinsic control of blood flow is summarized in table 14.4.

Paracrine Regulation of Blood Flow

Paracrine regulators, as described in chapter 11, are molecules produced by one tissue that help to regulate another tissue of the same organ. Blood vessels are particularly subject to paracrine regulation. Specifically, the endothelium of the tunica interna produces a number of paracrine regulators that cause the smooth muscle of the tunica media to either relax or contract.

The endothelium produces several molecules that promote smooth muscle relaxation, including **nitric oxide, bradykinin,** and **prostacyclin** (chapter 11). The endothelium-derived relaxation factor that earlier research had shown to be required for the vasodilation response to nerve stimulation appears to be nitric oxide.

The endothelium of arterioles contains an enzyme, *endothelial nitric oxide synthase (eNOS),* which produces nitric oxide (NO) from L-arginine. The NO diffuses into the smooth muscle cells of the tunica media of arterioles and activates the enzyme guanylate cyclase, which converts GTP into cyclic GMP (cGMP) and pyrophosphate (PP_i). The cGMP serves as a second messenger that, through a variety of mechanisms, acts to lower the cytoplasmic Ca^{2+} concentration. This leads to smooth muscle relaxation and thus vasodilation (see chapter 20, fig. 20.23). In many arterioles, a baseline level of NO production helps regulate the resting "tone" (degree of vasoconstriction/vasodilation) of the arterioles. The production of NO can be increased by ACh released from parasympathetic axons, which acts via the Ca^{2+}-calmodulin system (chapter 11) to stimulate nitric oxide synthase in the endothelial cells of blood vessels.

Table 14.4 Extrinsic Control of Vascular Resistance and Blood Flow

Extrinsic Agent	Effect	Comments
Sympathetic nerves		
Alpha-adrenergic	Vasoconstriction	Vasoconstriction is the dominant effect of sympathetic nerve stimulation on the vascular system, and it occurs throughout the body.
Beta-adrenergic	Vasodilation	There is some activity in arterioles in skeletal muscles and in coronary vessels, but effects are masked by dominant alpha-receptor-mediated constriction.
Cholinergic	Vasodilation	Effects are localized to arterioles in skeletal muscles and are produced only during defense (fight-or-flight) reactions.
Parasympathetic nerves	Vasodilation	Effects are restricted primarily to the gastrointestinal tract, external genitalia, and salivary glands and have little effect on total peripheral resistance.
Angiotensin II	Vasoconstriction	A powerful vasoconstrictor produced as a result of secretion of renin from the kidneys, it may function to help maintain adequate filtration pressure in the kidneys when systemic blood flow and pressure are reduced.
ADH (vasopressin)	Vasoconstriction	Although the effects of this hormone on vascular resistance and blood pressure in anesthetized animals are well documented, the importance of these effects in conscious humans is controversial.
Histamine	Vasodilation	Histamine promotes localized vasodilation during inflammation and allergic reactions.
Bradykinins	Vasodilation	Bradykinins are polypeptides secreted by sweat glands and by the endothelium of blood vessels; they promote local vasodilation.
Prostaglandins	Vasodilation or vasoconstriction	Prostaglandins are cyclic fatty acids that can be produced by most tissues, including blood vessel walls. Prostaglandin I_2 is a vasodilator, whereas thromboxane A_2 is a vasoconstrictor. The physiological significance of these effects is presently controversial.

In addition to serving as a paracrine regulator locally within a blood vessel, nitric oxide can also be carried by red blood cells because it binds to the sulfur atoms of the cysteines in hemoglobin. Red blood cells carry nitric oxide from areas where the oxygen concentration is high (nitric oxide is produced in the lungs) to areas where it is low. The lower the oxygen concentration, the more nitric oxide will be released from the red blood cells to cause vasodilation. The vasodilation then produces increased blood flow and delivery of oxygen to the tissue. This effect benefits ischemic tissues. Indeed, vasodilator drugs given to treat angina pectoris—such as *nitroglycerin*—promote vasodilation indirectly through their conversion to nitric oxide.

The endothelium also produces paracrine regulators that promote vasoconstriction. Notable among these is the polypeptide **endothelin-1.** This paracrine regulator stimulates vasoconstriction of arterioles, thus raising the total peripheral resistance. In normal physiology, this action may work together with those regulators that promote vasodilation to help regulate blood pressure.

Intrinsic Regulation of Blood Flow

Intrinsic, or "built-in," mechanisms within individual organs provide a localized regulation of vascular resistance and blood flow. Intrinsic mechanisms are classified as *myogenic* or *metabolic.* Some organs, the brain and kidneys in particular, utilize these intrinsic mechanisms to maintain relatively constant flow rates despite wide fluctuations in blood pressure. This ability is termed **autoregulation.**

Myogenic Control Mechanisms

If the arterial blood pressure and flow through an organ are inadequate—if the organ is inadequately *perfused* with blood—the metabolism of the organ cannot be maintained beyond a limited time period. Excessively high blood pressure can also be dangerous, par-

ticularly in the brain, because this may result in the rupture of fine blood vessels (causing cerebrovascular accident—CVA, or stroke).

Changes in systemic arterial pressure are compensated for in the brain and some other organs by the appropriate responses of vascular smooth muscle. A decrease in arterial pressure causes cerebral vessels to dilate, so that adequate rates of blood flow can be maintained despite the decreased pressure. High blood pressure, by contrast, causes cerebral vessels to constrict, so that finer vessels downstream are protected from the elevated pressure. These responses are myogenic; they are direct responses by the vascular smooth muscle to changes in pressure.

Metabolic Control Mechanisms

Local vasodilation within an organ can occur as a result of the chemical environment created by the organ's metabolism. The localized chemical conditions that promote vasodilation include (1) *decreased oxygen concentrations* that result from increased metabolic rate; (2) *increased carbon dioxide concentrations;* (3) *decreased tissue pH* (due to CO_2, lactic acid, and other metabolic products); and (4) the *release of adenosine or K^+* from the tissue cells. Through these chemical changes, the organ signals its blood vessels of its need for increased oxygen delivery.

The vasodilation that occurs in response to tissue metabolism can be demonstrated by constricting the blood supply to an area for a short time and then removing the constriction. The constriction allows metabolic products to accumulate by preventing venous drainage of the area. When the constriction is removed and blood flow resumes, the metabolic products that have accumulated cause vasodilation. The tissue thus appears red. This response is called **reactive hyperemia.** A similar increase in blood flow occurs in skeletal muscles and other organs as a result of increased metabolism. This is called **active hyperemia.** The increased blood flow can wash out the vasodilator metabolites, so that blood flow can fall to pre-exercise levels a few minutes after exercise ends.

Blood Flow to the Heart and Skeletal Muscles

Blood flow to the heart and skeletal muscles is regulated by both extrinsic and intrinsic mechanisms. These mechanisms provide increased blood flow when the metabolic requirements of these tissues are raised during exercise.

Survival requires that the heart and brain receive an adequate supply of blood at all times. The ability of skeletal muscles to respond quickly in emergencies and to maintain continued high levels of activity also may be critically important for survival. During such times, high rates of blood flow to the skeletal muscles must be maintained without compromising blood flow to the heart and brain. This is accomplished by mechanisms that increase the cardiac output and divert the blood away from the viscera and skin so that the heart, skeletal muscles, and brain receive a greater proportion of the total blood flow.

Clinical Investigation Clues

Remember that Charlie was very weak when he was found.
How did his dehydration affect his cardiac output?
How would this effect cause him to be weak?

Aerobic Requirements of the Heart

The coronary arteries supply an enormous number of capillaries, which are packed within the myocardium at a density ranging from 2,500 to 4,000 per cubic millimeter of tissue. Fast-twitch skeletal muscles, by contrast, have a capillary density of 300 to 400 per cubic millimeter of tissue. Each myocardial cell, as a consequence, is within 10 μm of a capillary (compared to an average distance in other organs of 70 μm). The exchange of gases by diffusion between myocardial cells and capillary blood thus occurs very quickly.

Contraction of the myocardium squeezes the coronary arteries. Unlike blood flow in all other organs, flow in the coronary vessels thus decreases in systole and increases during diastole. The myocardium, however, contains large amounts of *myoglobin,* a pigment related to hemoglobin (the molecules in red blood cells that carry oxygen). Myoglobin in the myocardium stores oxygen during diastole and releases its oxygen during systole. In this way, the myocardial cells can receive a continuous supply of oxygen even though coronary blood flow is temporarily reduced during systole.

In addition to containing large amounts of myoglobin, heart muscle contains numerous mitochondria and aerobic respiratory enzymes. This indicates that—even more than slow-twitch skeletal muscles—the heart is extremely specialized for aerobic respiration. The normal heart always respires aerobically, even during heavy exercise when the metabolic demand for oxygen can rise to five times resting levels. This increased oxygen requirement is met by a corresponding increase in coronary blood flow, from about 80 ml at rest to about 400 ml per minute per 100 g tissue during heavy exercise.

Regulation of Coronary Blood Flow

The coronary arterioles contain both alpha and beta adrenergic receptors, which promote vasoconstriction and vasodilation, respectively. Norepinephrine released by sympathetic nerve fibers stimulates alpha-adrenergic receptors to raise vascular resistance at rest. Epinephrine released by the adrenal medulla can stimulate the beta-adrenergic receptors to produce vasodilation when the sympathoadrenal system is activated during the fight-or-flight reaction.

Most of the vasodilation that occurs during exercise, however, is due to intrinsic metabolic control mechanisms. The intrinsic mechanisms occur as follows: (1) as the metabolism of the myocardium increases, there are local accumulations of carbon dioxide, K^+, and adenosine in the tissue, together with depletion of oxygen; (2) these localized changes act directly on the vascular smooth muscle to cause relaxation and vasodilation.

Under abnormal conditions blood flow to the myocardium may be inadequate, resulting in myocardial ischemia (chapter 13). The inadequate flow may be due to blockage by atheromas and/or blood clots or to muscular spasm of a coronary artery (fig. 14.17). Occlusion of a coronary artery can be visualized by inserting a catheter (plastic tube) into a brachial or femoral artery all the way to the opening of the coronary arteries in the aorta and then injecting a radiographic contrast material. The picture thus obtained is called an **angiogram.**

In a technique called **balloon angioplasty,** an inflatable balloon is used to open the coronary arteries. However, *restenosis* (recurrence of narrowing) often occurs. For this reason, a cylindrical support called a **stent** may be inserted to help keep the artery open. If the occlusion is sufficiently great, a **coronary bypass** may be performed. In this procedure, a length of blood vessel, usually taken from the saphenous vein in the leg, is sutured to the aorta and to the coronary artery at a location beyond the site of the occlusion (fig. 14.18).

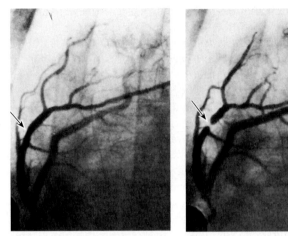

(a)   (b)

Figure 14.17   Angiograms of the left coronary artery of a heart patient. These angiograms were taken (*a*) when the patient's ECG was normal and (*b*) when the ECG showed evidence of myocardial ischemia. Notice that a coronary artery spasm (see arrow in [b]) appears to accompany the ischemia.

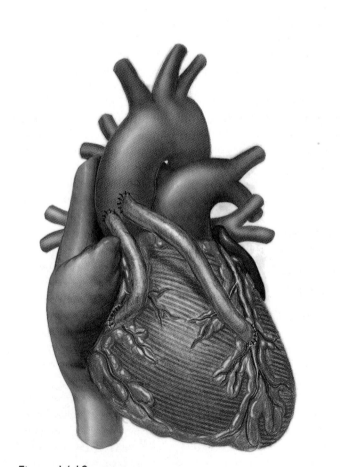

■ **Figure 14.18**   A diagram of coronary artery bypass surgery. Segments of the saphenous vein of the patient are commonly used as coronary bypass vessels.

Regulation of Blood Flow Through Skeletal Muscles

The arterioles in skeletal muscles, like those of the coronary circulation, have a high vascular resistance at rest as a result of alpha-adrenergic sympathetic stimulation. This produces a relatively low blood flow. Because muscles have such a large mass, however, they still receive from 20% to 25% of the total blood flow in the body at rest. Also, as in the heart, blood flow in a skeletal muscle decreases when the muscle contracts and squeezes its arterioles, and in fact blood flow stops entirely when the muscle contracts beyond about 70% of its maximum. Pain and fatigue thus occur much more quickly when an isometric contraction is sustained than when rhythmic isotonic contractions are performed.

In addition to adrenergic fibers, which promote vasoconstriction by stimulation of alpha-adrenergic receptors, there are also sympathetic cholinergic fibers in skeletal muscles. These cholinergic fibers, together with the stimulation of beta-adrenergic receptors by the hormone epinephrine, stimulate vasodilation as part of the fight-or-flight response to any stressful state, including that existing just prior to exercise (table 14.5). These extrinsic controls have been previously discussed and function to regulate blood flow through muscles at rest and upon anticipation of exercise.

As exercise progresses, the vasodilation and increased skeletal muscle blood flow that occur are almost entirely due to intrinsic metabolic control. The high metabolic rate of skeletal muscles during exercise causes local changes, such as increased carbon dioxide concentrations, decreased pH (due to carbonic acid and lactic acid), decreased oxygen, increased extracellular K^+, and the secretion of adenosine. As in the intrinsic control of the coronary circulation, these changes cause vasodilation of arterioles in skeletal muscles. This decreases the vascular resistance and increases the rate of blood flow. As a result of these changes, skeletal muscles can receive as much as 85% of the total blood flow in the body during maximal exercise.

Circulatory Changes During Exercise

While the vascular resistance in skeletal muscles decreases during exercise, the resistance to flow through the visceral organs and skin increases. This increased resistance occurs because of vasoconstriction stimulated by adrenergic sympathetic fibers, and it results in decreased rates of blood flow through these organs. During exercise, therefore, the blood flow to skeletal muscles increases because of three simultaneous changes: (1) increased total blood flow (cardiac output); (2) metabolic vasodilation in the exercising muscles; and (3) the diversion of blood away from the viscera and skin. Blood flow to the heart also increases during exercise, whereas blood flow to the brain does not appear to change significantly (fig. 14.19).

During exercise, the cardiac output can increase fivefold—from about 5 L per minute to about 25 L per minute. This is primarily due to an increase in cardiac rate. The cardiac rate,

Table 14.5 Changes in Skeletal Muscle Blood Flow Under Conditions of Rest and Exercise

Condition	Blood Flow (ml/min)	Mechanism
Rest	1,000	High adrenergic sympathetic stimulation of vascular alpha receptors, causing vasoconstriction
Beginning exercise	Increased	Dilation of arterioles in skeletal muscles due to cholinergic sympathetic nerve activity and stimulation of beta-adrenergic receptors by the hormone epinephrine
Heavy exercise	20,000	Fall in alpha-adrenergic activity Increased sympathetic cholinergic activity Increased metabolic rate of exercising muscles, producing intrinsic vasodilation

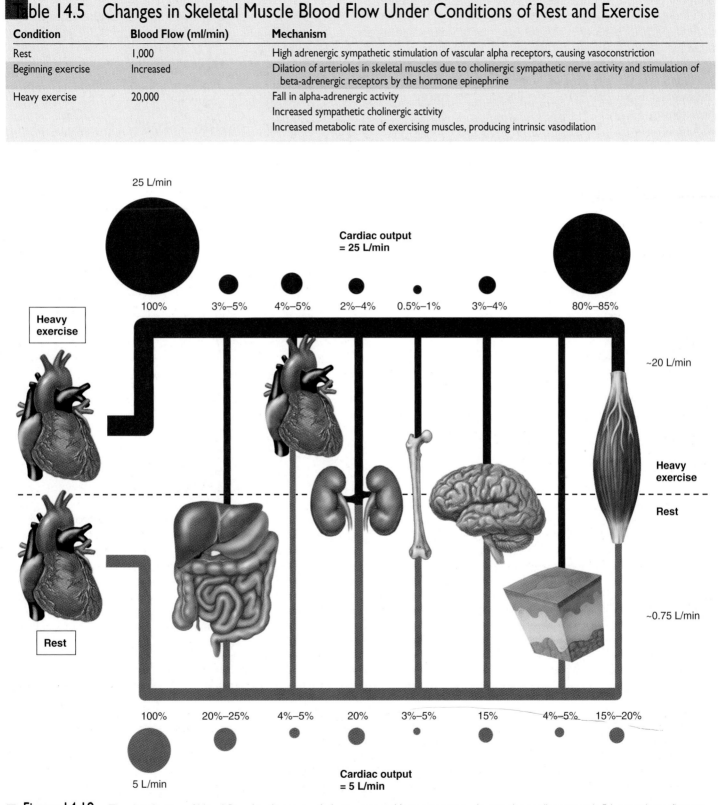

Figure 14.19 The distribution of blood flow (cardiac output) during rest and heavy exercise. At rest, the cardiac output is 5 L per minute *(bottom of figure)*; during heavy exercise the cardiac output increases to 25 L per minute *(top of figure)*. At rest, for example, the brain receives 15% of 5 L per minute (= 750 ml/min), whereas during exercise it receives 3% to 4% of 25 L per minute (0.03 25 = 750 ml/min). Flow to the skeletal muscles increases more than twentyfold because the total cardiac output increases (from 5 L/min to 25 L/min) and because the percentage of the total received by the muscles increases from 15% to 80%.

however, can increase only up to a maximum value (table 14.6), which is determined mainly by a person's age. In well-trained athletes, the stroke volume can also increase significantly, allowing these individuals to achieve cardiac outputs during strenuous exercise up to six or seven times greater than their resting values. This high cardiac output results in increased oxygen delivery to the exercising muscles; this is the major reason for the much higher than average maximal oxygen uptake ($\dot{V}_{O_2}$max) of elite athletes (chapter 12).

In most people, the increase in stroke volume that occurs during exercise will not exceed 35%. The fact that the stroke volume can increase at all during exercise may at first be surprising, given that the heart has less time to fill with blood between beats when it is pumping faster. Despite the faster beat, however, the end-diastolic volume during exercise is not decreased. This is because the venous return is aided by the improved action of the skeletal muscle pumps and by increased respiratory movements during exercise (fig. 14.20). Since the end-diastolic volume is not significantly changed during exercise, any increase in stroke volume that occurs must be due to an increase in the proportion of blood ejected per stroke.

The proportion of the end-diastolic volume ejected per stroke can increase from 60% at rest to as much as 90% dur-

ing heavy exercise. This increased *ejection fraction* is produced by the increased contractility that results from sympathoadrenal stimulation. There also may be a decrease in total peripheral resistance as a result of vasodilation in the exercising skeletal muscles, which decreases the afterload and thus further augments the increase in stroke volume. The cardiovascular changes that occur during exercise are summarized in table 14.7.

Endurance training often results in a lowering of the resting cardiac rate and an increase in the resting stroke volume. The lowering of the resting cardiac rate results from a greater degree of inhibition of the SA node by the vagus nerve. The increased resting stroke volume is believed to be due to an increase in blood volume; indeed, studies have shown that the blood volume can increase by about 500 ml after only 8 days

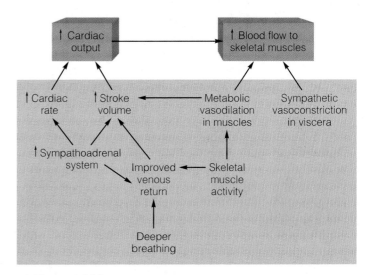

■ **Figure 14.20** **Cardiovascular adaptations to exercise.** These adaptations (1) increase the cardiac output, and thus the total blood flow; and (2) cause vasodilation in the exercising muscles, thereby diverting a higher proportion of the blood flow to those muscles.

Table 14.6 Relationship Between Age and Average Maximum Cardiac Rate*

Age	Maximum Cardiac Rate
20–29	190 beats/min
30–39	160 beats/min
40–49	150 beats/min
50–59	140 beats/min
60+	130 beats/min

*Maximum cardiac rate can be estimated by subtracting your age from 220.

Table 14.7 Cardiovascular Changes During Moderate Exercise

Variable	Change	Mechanisms
Cardiac output	Increased	Increased cardiac rate and stroke volume
Cardiac rate	Increased	Increased sympathetic nerve activity; decreased activity of the vagus nerve
Stroke volume	Increased	Increased myocardial contractility due to stimulation by sympathoadrenal system; decreased total peripheral resistance
Total peripheral resistance	Decreased	Vasodilation of arterioles in skeletal muscles (and in skin when thermoregulatory adjustments are needed)
Arterial blood pressure	Increased	Increased systolic and pulse pressure due primarily to increased cardiac output; diastolic pressure rises less due to decreased total peripheral resistance
End-diastolic volume	Unchanged	Decreased filling time at high cardiac rates is compensated for by increased venous pressure, increased activity of the skeletal muscle pump, and decreased intrathoracic pressure aiding the venous return
Blood flow to heart and muscles	Increased	Increased muscle metabolism produces intrinsic vasodilation; aided by increased cardiac output and increased vascular resistance in visceral organs
Blood flow to visceral organs	Decreased	Vasoconstriction in digestive tract, liver, and kidneys due to sympathetic nerve stimulation
Blood flow to skin	Increased	Metabolic heat produced by exercising muscles produces reflex (involving hypothalamus) that reduces sympathetic constriction of arteriovenous shunts and arterioles
Blood flow to brain	Unchanged	Autoregulation of cerebral vessels, which maintains constant cerebral blood flow despite increased arterial blood pressure

of training. These adaptations enable the trained athlete to produce a larger proportionate increase in cardiac output and achieve a higher absolute cardiac output during exercise. This large cardiac output is the major factor in the improved oxygen delivery to skeletal muscles that occurs as a result of endurance training.

Test Yourself Before You Continue

1. Describe blood flow and oxygen delivery to the myocardium during systole and diastole.
2. State how blood flow to the heart is affected by exercise. Explain how blood flow to the heart is regulated at rest and during exercise.
3. Describe the mechanisms that produce vasodilation of the arterioles in skeletal muscles during exercise. Give two other reasons for the increased blood flow to muscles during exercise.
4. Explain how the stroke volume can increase during exercise despite the fact that the filling times are reduced at high cardiac rates.

Blood Flow to the Brain and Skin

Intrinsic control mechanisms help to maintain a relatively constant blood flow to the brain. Blood flow to the skin, by contrast, can vary tremendously in response to regulation by sympathetic nerve stimulation.

The examination of cerebral and cutaneous blood flow is a study in contrasts. Cerebral blood flow is regulated primarily by intrinsic mechanisms; cutaneous blood flow is regulated by extrinsic mechanisms. Cerebral blood flow is relatively constant; cutaneous blood flow exhibits more variation than blood flow in any other organ. The brain is the organ that can least tolerate low rates of blood flow, whereas the skin can tolerate low rates of blood flow better than any other organ.

Cerebral Circulation

When the brain is deprived of oxygen for just a few seconds, a person loses consciousness; irreversible brain injury may occur after a few minutes. For these reasons, the cerebral blood flow is held remarkably constant at about 750 ml per minute. This amounts to about 15% of the total cardiac output at rest.

Unlike the coronary and skeletal muscle blood flow, cerebral blood flow is not normally influenced by sympathetic nerve activity. Only when the mean arterial pressure rises to about 200 mmHg do sympathetic nerves cause a significant degree of vasoconstriction in the cerebral circulation. This vasoconstriction helps to protect small, thin-walled arterioles from bursting under the pressure, and thus helps to prevent cerebrovascular accident (stroke).

In the normal range of arterial pressures, cerebral blood flow is regulated almost exclusively by local intrinsic mechanisms—a process called *autoregulation,* as previously mentioned. These mechanisms help to ensure a constant rate of blood flow despite changes in systemic arterial pressure. The autoregulation of cerebral blood flow is achieved by both myogenic and metabolic mechanisms.

Myogenic Regulation

Myogenic regulation occurs when there is variation in systemic arterial pressure. When the blood pressure falls, the cerebral arteries automatically dilate; when the pressure rises, they constrict. This helps to maintain a constant flow rate during the normal pressure variations that occur during rest, exercise, and emotional states.

The cerebral vessels are also sensitive to the carbon dioxide concentration of arterial blood. When the carbon dioxide concentration rises as a result of inadequate ventilation (hypoventilation), the cerebral arterioles dilate. This is believed to be due to decreases in the pH of cerebrospinal fluid rather than to a direct effect of CO_2 on the cerebral vessels. Conversely, when the arterial CO_2 falls below normal during hyperventilation, the cerebral vessels constrict. The resulting decrease in cerebral blood flow is responsible for the dizziness that occurs during hyperventilation.

Metabolic Regulation

The cerebral arterioles are exquisitely sensitive to local changes in metabolic activity, so that those brain regions with the highest metabolic activity receive the most blood. Indeed, areas of the brain that control specific processes have been mapped by the changing patterns of blood flow that result when these areas are activated. Visual and auditory stimuli, for example, increase blood flow to the appropriate sensory areas of the cerebral cortex, whereas motor activities, such as movements of the eyes, arms, and organs of speech, result in different patterns of blood flow (fig. 14.21).

The exact mechanisms by which increases in neural activity in a particular area of the brain elicit local vasodilation are not completely understood. There is evidence, however, that local cerebral vasodilation may be caused by K^+ which is released from active neurons during repolarization. It has been proposed that astrocytes may take up this extruded K^+ near the active neurons and then release the K^+ through their vascular processes (chapter 7; see fig. 7.10) that surround arterioles, thereby causing the arterioles to dilate.

Cutaneous Blood Flow

The skin is the outer covering of the body and as such serves as the first line of defense against invasion by disease-causing organisms. The skin, as the interface between the internal and external environments, also helps to maintain a constant deep-body temperature despite changes in the ambient (external) temperature—a process called *thermoregulation.* The thinness and large area of the skin (1.0 to 1.5 mm thick; 1.7 to

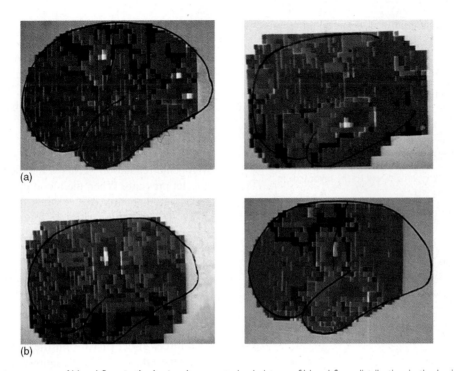

(a)

(b)

Figure 14.21 **Changing patterns of blood flow in the brain.** A computerized picture of blood-flow distribution in the brain after injecting the carotid artery with a radioactive isotope. In (a), on the left, the subject followed a moving object with his eyes. High activity is seen over the occipital lobe of the brain. In (a), on the right, the subject listened to spoken words. Notice that the high activity is seen over the temporal lobe (the auditory cortex). In (b), on the left, the subject moved his fingers on the side of the body opposite to the cerebral hemisphere being studied. In (b), on the right, the subject counted to 20. High activity is seen over the mouth area of the motor cortex, the supplementary motor area, and the auditory cortex.

1.8 square meters in surface area) make it an effective radiator of heat when the body temperature rises above the ambient temperature. The transfer of heat from the body to the external environment is aided by the flow of warm blood through capillary loops near the surface of the skin.

Blood flow through the skin is adjusted to maintain deep-body temperature at about 37° C (98.6° F). These adjustments are made by variations in the degree of constriction or dilation of ordinary arterioles and of unique **arteriovenous anastomoses** (fig. 14.22). These latter vessels, found predominantly in the fingertips, palms of the hands, toes, soles of the feet, ears, nose, and lips, shunt (divert) blood directly from arterioles to deep venules, thus bypassing superficial capillary loops. Both the ordinary arterioles and the arteriovenous anastomoses are innervated by sympathetic nerve fibers. When the ambient temperature is low, sympathetic nerves stimulate cutaneous vasoconstriction; cutaneous blood flow is thus decreased, so that less heat will be lost from the body. Since the arteriovenous anastomoses also constrict, the skin may appear rosy because the blood is diverted to the superficial capillary loops. In spite of this rosy appearance, however, the total cutaneous blood flow and rate of heat loss is lower than under usual conditions.

Skin can tolerate an extremely low blood flow in cold weather because its metabolic rate decreases when the ambient temperature decreases. In cold weather, therefore, the skin requires less blood. As a result of exposure to extreme cold, however, blood flow to the skin can be so severely restricted that

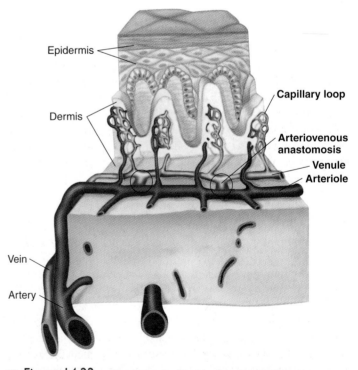

Figure 14.22 **Circulation in the skin showing arteriovenous anastomoses.** These vessels function as shunts, allowing blood to be diverted directly from the arteriole to the venule, and thus to bypass superficial capillary loops.

the tissue dies—a condition known as *frostbite*. Blood flow to the skin can vary from less than 20 ml per minute at maximal vasoconstriction to as much as 3 to 4 L per minute at maximal vasodilation.

As the temperature warms, cutaneous arterioles in the hands and feet dilate as a result of decreased sympathetic nerve activity. Continued warming causes dilation of arterioles in other areas of the skin. If the resulting increase in cutaneous blood flow is not sufficient to cool the body, sweat gland secretion may be stimulated. Perspiration helps to cool the body as it evaporates from the surface of the skin. The sweat gland also secrete **bradykinin,** a polypeptide that stimulates vasodilation.

Under the usual conditions of ambient temperature, the cutaneous vascular resistance is high and the blood flow is low when a person is not exercising. In the pre-exercise state of fight or flight, sympathetic nerve activity reduces cutaneous blood flow still further. During exercise, however, the need to maintain a deep-body temperature takes precedence over the need to maintain an adequate systemic blood pressure. As the body temperature rises during exercise, vasodilation in cutaneous vessels occurs together with vasodilation in the exercising muscles. This can produce an even greater lowering of total peripheral resistance. If exercise is performed in hot and humid weather, and if restrictive clothing increases skin temperature and cutaneous vasodilation, a dangerously low blood pressure may be produced after exercise has ceased and the cardiac output has declined. People have lost consciousness and have even died as a result.

Changes in cutaneous blood flow occur as a result of changes in sympathetic nerve activity. Since the activity of the sympathetic nervous system is controlled by the brain, emotional states, acting through control centers in the medulla oblongata, can affect sympathetic activity and cutaneous blood flow. During fear reactions, for example, vasoconstriction in the skin, along with activation of the sweat glands, can produce a pallor and a "cold sweat." Other emotions may cause vasodilation and blushing.

Blood Pressure

The pressure of the arterial blood is regulated by the blood volume, total peripheral resistance, and the cardiac rate. Regulatory mechanisms adjust these factors in a negative feedback manner to compensate for deviations. Arterial pressure rises and falls as the heart goes through systole and diastole.

Resistance to flow in the arterial system is greatest in the arterioles because these vessels have the smallest diameters. Although the total blood flow through a system of arterioles must be equal to the flow in the larger vessel that gave rise to those arterioles, the narrow diameter of each arteriole reduces the flow in each according to Poiseuille's law. Blood flow and pressure are thus reduced in the capillaries, which are located downstream of the high resistance imposed by the arterioles. (The slow velocity of blood flow through capillaries enhances diffusion across the capillary wall.) The blood pressure upstream of the arterioles—in the medium and large arteries—is correspondingly increased (fig. 14.23).

The blood pressure and flow within the capillaries are further reduced by the fact that their total cross-sectional area is much greater, due to their large number, than the cross-sectional areas of the arteries and arterioles (fig. 14.24). Thus, although each capillary is much narrower than each arteriole, the capillary beds served by arterioles do not provide as great a resistance to blood flow as do the arterioles.

Variations in the diameter of arterioles as a result of vasoconstriction and vasodilation thus affect blood flow through capillaries and, simultaneously, the *arterial blood pressure* "upstream" from the capillaries. In this way, an increase in total peripheral resistance due to vasoconstriction of arterioles can raise arterial blood pressure. Blood pressure can also be raised

Clinical Investigation Clues

Remember that Charlie's skin was cold to the touch.

What does this indicate about his cutaneous blood flow?

What produced this effect?

What benefit does Charlie derive from this mechanism?

Test Yourself before You Continue

1. Define the term *autoregulation* and describe how this process is accomplished in the cerebral circulation.

2. Explain how hyperventilation can cause dizziness.

3. Explain how cutaneous blood flow is adjusted to maintain a constant deep-body temperature.

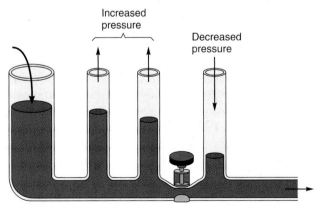

■ **Figure 14.23** The effect of vasoconstriction on blood pressure. A constriction increases blood pressure upstream (analogous to the arterial pressure) and decreases pressure downstream (analogous to capillary and venous pressure).

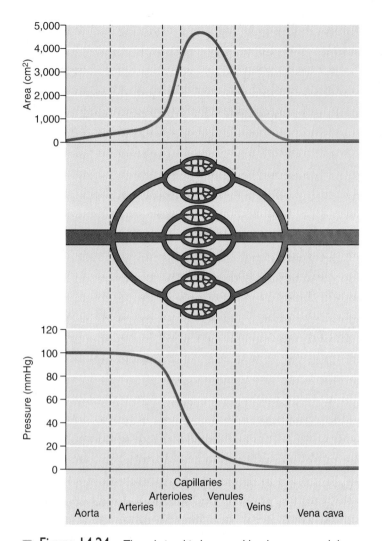

Figure 14.24 The relationship between blood pressure and the cross-sectional area of vessels. As blood passes from the aorta to the smaller arteries, arterioles, and capillaries, the cross-sectional area increases as the pressure decreases.

by an increase in the cardiac output. This may be due to elevations in cardiac rate or in stroke volume, which in turn are affected by other factors. The three most important variables affecting blood pressure are the **cardiac rate, stroke volume** (determined primarily by the **blood volume**), and total **peripheral resistance.** An increase in any of these, if not compensated for by a decrease in another variable, will result in an increased blood pressure.

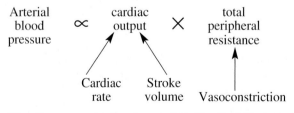

Blood pressure can thus be regulated by the kidneys, which control blood volume and thus stroke volume, and by the sympathoadrenal system. Increased activity of the sympathoadrenal

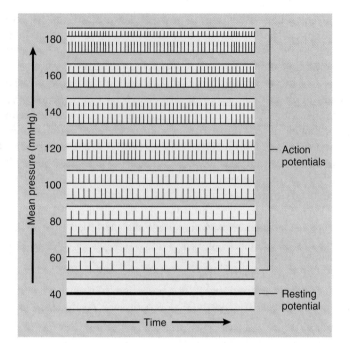

Figure 14.25 The effect of blood pressure on the baroreceptor response. This is a recording of the action potential frequency in sensory nerve fibers from baroreceptors in the carotid sinus and aortic arch. As the blood pressure increases, the baroreceptors become increasingly stretched. This results in a higher frequency of action potentials transmitted to the cardiac and vasomotor control centers in the medulla oblongata.

system can raise blood pressure by stimulating vasoconstriction of arterioles (thus raising total peripheral resistance) and by promoting an increased cardiac output. Sympathetic stimulation can also affect blood volume indirectly, by stimulating constriction of renal blood vessels and thus reducing urine output.

Blood pressure is measured in units of **millimeters of mercury (mm Hg).** In performing this measurement, the blood pushes on one surface of a U-shaped column of mercury while the atmosphere pushes on the other surface (see chapter 16, fig. 16.19). If the blood pressure were equal to the atmospheric pressure, the measurement would be zero mm Hg. For the same reason, a mean arterial pressure of 100 mm Hg indicates that the blood pressure is 100 mm Hg higher than the atmospheric pressure. Instruments used to measure blood pressure, called **sphygmomanometers,** thus contain mercury or are spring-loaded devices that are calibrated against mercurial instruments.

Baroreceptor Reflex

In order for blood pressure to be maintained within limits, specialized receptors for pressure are needed. These **baroreceptors** are stretch receptors located in the *aortic arch* and in the *carotid sinuses.* An increase in pressure causes the walls of these arterial regions to stretch, increasing the frequency of action potentials along sensory nerve fibers (fig. 14.25). A fall in pressure below the normal range, by contrast, causes a decrease in the frequency of action potentials produced by these sensory nerve fibers.

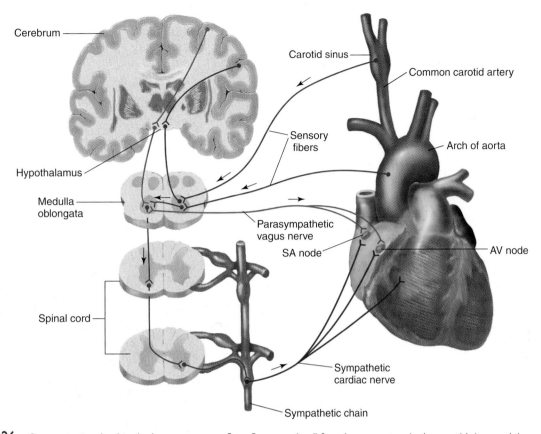

Figure 14.26 Structures involved in the baroreceptor reflex. Sensory stimuli from baroreceptors in the carotid sinus and the aortic arch, acting via control centers in the medulla oblongata, affect the activity of sympathetic and parasympathetic nerve fibers in the heart.

Sensory nerve activity from the baroreceptors ascends, via the vagus and glossopharyngeal nerves, to the medulla oblongata, which directs the autonomic system to respond appropriately. The **vasomotor control centers** in the medulla controls vasoconstriction/vasodilation, and hence helps to regulate total peripheral resistance. The **cardiac control centers** in the medulla regulates the cardiac rate (fig. 14.26). Acting through the activity of motor fibers within the vagus and sympathetic nerves controlled by these brain centers, the baroreceptors function to counteract blood pressure changes so that fluctuations in pressure are minimized.

The baroreceptor reflex is activated whenever blood pressure increases or decreases. The reflex is somewhat more sensitive to decreases in pressure than to increases, and is more sensitive to sudden changes in pressure than to more gradual changes. A good example of the importance of the baroreceptor reflex in normal physiology is its activation whenever a person goes from a lying to a standing position.

When a person goes from a lying to a standing position, there is a shift of 500 to 700 ml of blood from the veins of the thoracic cavity to veins in the lower extremities, which expand to contain the extra volume of blood. This pooling of blood in the lower extremities reduces the venous return and cardiac output, but the resulting fall in blood pressure is almost immediately compensated for by the baroreceptor reflex. A decrease in baroreceptor sensory information, traveling in the glossopharyngeal nerve (IX) and the vagus nerve (X) to the medulla oblongata, inhibits parasympa-

thetic activity and promotes sympathetic nerve activity. This produces an increase in cardiac rate and vasoconstriction, which help to maintain an adequate blood pressure upon standing (fig. 14.27).

Clinical Investigation Clue

Remember that Charlie had low blood pressure, a rapid pulse, and cold skin.

How can knowledge of the baroreceptor reflex be used to understand the relationships among these observations?

Input from baroreceptors can also mediate the opposite response. When the blood pressure rises above an individual's normal range, the baroreceptor reflex causes a slowing of the cardiac rate and vasodilation. Manual massage of the carotid sinus, a procedure sometimes employed by physicians to reduce tachycardia and lower blood pressure, also evokes this reflex. Such carotid massage should be used cautiously, however, because the intense vagus-nerve-induced slowing of the cardiac rate could cause loss of consciousness (as occurs in emotional fainting). Manual massage of both carotid sinuses simultaneously can even cause cardiac arrest in susceptible people.

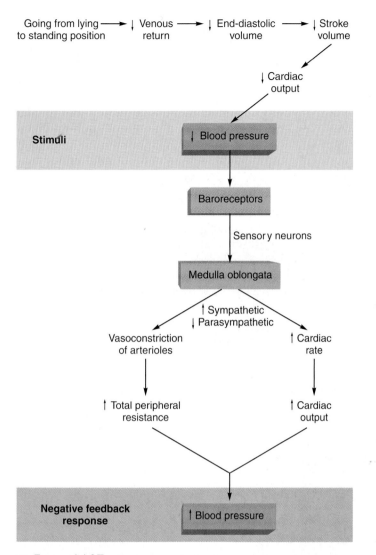

Going from lying → ↓ Venous → ↓ End-diastolic → ↓ Stroke
to standing position   return   volume   volume

↓ Cardiac output

Stimuli   ↓ Blood pressure

Baroreceptors

Sensory neurons

Medulla oblongata

↑ Sympathetic
↓ Parasympathetic

Vasoconstriction   ↑ Cardiac
of arterioles   rate

↑ Total peripheral   ↑ Cardiac
resistance   output

Negative feedback response   ↑ Blood pressure

■ **Figure 14.27**   The negative feedback control of blood pressure by the baroreceptor reflex. This reflex helps to maintain an adequate blood pressure upon standing.

 Since the baroreceptor reflex may require a few seconds before it is fully effective, many people feel dizzy and disoriented if they stand up too quickly. If the baroreceptor sensitivity is abnormally reduced, perhaps by atherosclerosis, an uncompensated fall in pressure may occur upon standing. This condition—called **postural, or orthostatic, hypotension** (*hypotension* = low blood pressure)—can make a person feel extremely dizzy or even faint because of inadequate perfusion of the brain.

Atrial Stretch Reflexes

In addition to the baroreceptor reflex, several other reflexes help to regulate blood pressure. The reflex control of ADH release by osmoreceptors in the hypothalamus and the control of angiotensin II production and aldosterone secretion by the juxta-

glomerular apparatus of the kidneys have been previously discussed. Antidiuretic hormone and aldosterone increase blood pressure by increasing blood volume, and angiotensin II stimulates vasoconstriction to cause an increase in blood pressure.

Other reflexes important to blood pressure regulation are initiated by **atrial stretch receptors** located in the atria of the heart. These receptors are activated by increased venous return to the heart and, in response (1) stimulate reflex tachycardia, as a result of increased sympathetic nerve activity; (2) inhibit ADH release, resulting in the excretion of larger volumes of urine and a lowering of blood volume; and (3) promote increased secretion of atrial natriuretic peptide (ANP). The ANP, as previously discussed, lowers blood volume by increasing urinary salt and water excretion and by antagonizing the actions of angiotensin II.

 Valsalva's maneuver is the term used to describe an expiratory effort against a closed glottis (which prevents the air from escaping—see chapter 16). This maneuver, commonly performed during forceful defecation or when lifting heavy weights, increases the intrathoracic pressure. Compression of the thoracic veins causes a fall in venous return and cardiac output, thus lowering arterial blood pressure. The lowering of arterial pressure then stimulates the baroreceptor reflex, resulting in tachycardia and increased total peripheral resistance. When the glottis is finally opened and the air is exhaled, the cardiac output returns to normal. The total peripheral resistance is still elevated, however, causing a rise in blood pressure. The blood pressure is then brought back to normal by the baroreceptor reflex, which causes a slowing of the heart rate. These fluctuations in cardiac output and blood pressure can be dangerous in people with cardiovascular disease. Even healthy people are advised to exhale normally when lifting weights.

Measurement of Blood Pressure

The first documented measurement of blood pressure was accomplished by Stephen Hales (1677–1761), an English clergyman and physiologist. Hales inserted a cannula into the artery of a horse and measured the heights to which blood would rise in the vertical tube. The height of this blood column bounced between the **systolic pressure** at its highest and the **diastolic pressure** at its lowest, as the heart went through its cycle of systole and diastole. Modern clinical blood pressure measurements, fortunately, are less direct. The indirect, or **auscultatory,** method is based on the correlation of blood pressure and arterial sounds.

In the auscultatory method, an inflatable rubber bladder within a cloth cuff is wrapped around the upper arm, and a stethoscope is applied over the brachial artery (fig. 14.28). The artery is normally silent before inflation of the cuff because blood normally travels in a smooth, *laminar flow* through the arteries. The term *laminar* means "layered"— blood in the central axial stream moves the fastest, and blood

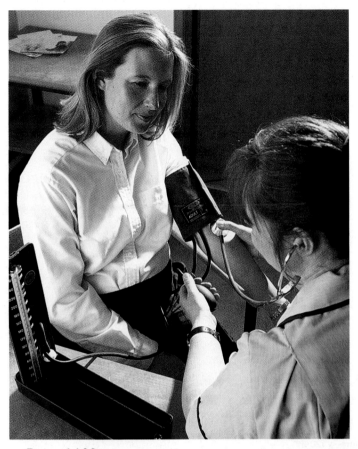

Figure 14.28 A pressure cuff and sphygmomanometer are used to measure blood pressure. The examiner is listening for the Korotkoff sounds.

flowing closer to the artery wall moves more slowly. There is little transverse movement between these layers that would produce mixing.

The laminar flow that normally occurs in arteries is smooth and silent. When the artery is pinched, however, blood flow through the constriction becomes turbulent. This causes the artery to produce sounds, much like the sounds produced by water flowing through a kink in a garden hose. The tendency of the cuff pressure to constrict the artery is opposed by the blood pressure. Thus, in order to constrict the artery, the cuff pressure must be greater than the diastolic blood pressure. If the cuff pressure is also greater than the systolic blood pressure, the artery will be pinched off and silent. *Turbulent flow* and sounds produced by the artery as a result of this flow, therefore, occur only when the cuff pressure is greater than the diastolic blood pressure and lower than the systolic pressure.

Let's say that a person has a systolic pressure of 120 mmHg and a diastolic pressure of 80 mmHg (the average normal values). When the cuff pressure is between 80 and 120 mmHg, the artery will be closed during diastole and open during systole. As the artery begins to open with every systole, turbulent flow of blood through the constriction will create vibrations that are known as the **sounds of Korotkoff,** as shown in figure 14.29. These are usually "tapping" sounds because the artery becomes constricted, blood flow stops, and silence is restored with every diastole. It

should be understood that the sounds of Korotkoff are *not* "lub-dub" sounds produced by closing of the heart valves (those sounds can be heard only on the chest, not on the brachial artery).

Initially, the cuff is usually inflated to produce a pressure greater than the systolic pressure, so that the artery is pinched off and silent. The pressure in the cuff is read from an attached meter called a *sphygmomanometer.* A valve is then turned to allow the release of air from the cuff, causing a gradual decrease in cuff pressure. When the cuff pressure is equal to the systolic pressure, the **first Korotkoff sound** is heard as blood passes in a turbulent flow through the constricted opening of the artery.

Korotkoff sounds will continue to be heard at every systole as long as the cuff pressure remains greater than the diastolic pressure. When the cuff pressure becomes equal to or less than the diastolic pressure, the sounds disappear because the artery remains open and laminar flow resumes. (fig. 14.30). The **last Korotkoff sound** thus occurs when the cuff pressure is equal to the diastolic pressure.

Different phases in the measurement of blood pressure are identified on the basis of the quality of the Korotkoff sounds (fig. 14.31). In some people, the Korotkoff sounds do not disappear even when the cuff pressure is reduced to zero (zero pressure means that it is equal to atmospheric pressure). In these cases—and often routinely—the onset of muffling of the sounds (phase 4 in fig. 14.31) is used as an indication of diastolic pressure rather than the onset of silence (phase 5).

The average arterial blood pressure in the systemic circulation is 120/80 mmHg, whereas the average pulmonary arterial blood pressure is only 22/8 mmHg. Because of the Frank–Starling relationship, the cardiac output from the right ventricle into the pulmonary circulation is matched to that of the left ventricle into the systemic circulation. Since the cardiac outputs are the same, the lower pulmonary blood pressure must be caused by a lower peripheral resistance in the pulmonary circulation. Because the right ventricle pumps blood against a lower resistance, it has a lighter workload and its walls are thinner than those of the left ventricle.

Pulse Pressure and Mean Arterial Pressure

When someone "takes a pulse," he or she palpates an artery (for example, the radial artery) and feels the expansion of the artery occur in response to the beating of the heart; the pulse rate is thus a measure of the cardiac rate. The expansion of the artery with each pulse occurs as a result of the rise in blood pressure within the artery as the artery receives the volume of blood ejected by a stroke of the left ventricle.

Since the pulse is produced by the rise in pressure from diastolic to systolic levels, the difference between these two pressures is known as the **pulse pressure.** A person with a blood pressure of 120/80 (systolic/diastolic) would therefore have a pulse pressure of 40 mmHg.

Pulse pressure = systolic pressure – diastolic pressure

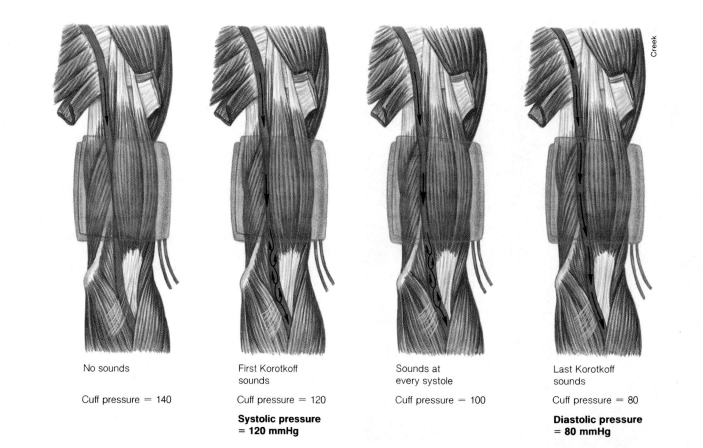

No sounds

Cuff pressure = 140

First Korotkoff
sounds

Cuff pressure = 120

**Systolic pressure
= 120 mmHg**

Sounds at
every systole

Cuff pressure = 100

Last Korotkoff
sounds

Cuff pressure = 80

**Diastolic pressure
= 80 mmHg**

Blood pressure = 120/80

■ **Figure 14.29** **The blood flow and Korotkoff sounds during a blood pressure measurement.** When the cuff pressure is above the systolic pressure, the artery is constricted. When the cuff pressure is below the diastolic pressure, the artery is open and flow is laminar. When the cuff pressure is between the diastolic and systolic pressure, blood flow is turbulent and the Korotkoff sounds are heard with each systole.

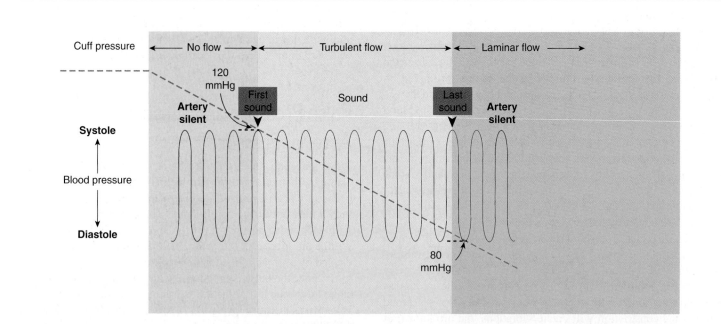

■ **Figure 14.30** **The indirect, or auscultatory, method of blood pressure measurement.** The first Korotkoff sound is heard when the cuff pressure is equal to the systolic blood pressure, and the last sound is heard when the cuff pressure is equal to the diastolic pressure. The dashed line indicates the cuff pressure.

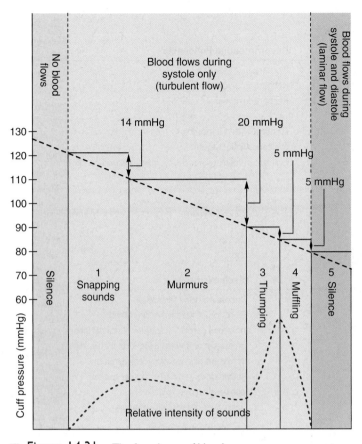

Figure 14.31 The five phases of blood pressure measurement. Not all phases are heard in all people. The cuff pressure is indicated by the falling dashed line.

riod of diastole is longer than the period of systole. Mean arterial pressure can be approximated by adding one-third of the pulse pressure to the diastolic pressure. For a person with a blood pressure of 120/80, for example, the mean arterial pressure would be approximately 80 + 1/3 (40) = 93 mmHg.

Mean arterial pressure = diastolic pressure + 1/3 pulse pressure

A rise in total peripheral resistance and cardiac rate increases the diastolic pressure more than it increases the systolic pressure. When the baroreceptor reflex is activated by going from a lying to a standing position, for example, the diastolic pressure usually increases by 5 to 10 mmHg, whereas the systolic pressure either remains unchanged or is slightly reduced (as a result of decreased venous return). People with hypertension (high blood pressure), who usually have elevated total peripheral resistance and cardiac rates, likewise have a greater increase in diastolic than in systolic pressure. Dehydration or blood loss results in decreased cardiac output, and thus also produces a decrease in pulse pressure.

An increase in cardiac output, by contrast, raises the systolic pressure more than it raises the diastolic pressure (although both pressures do rise). This occurs during exercise, for example, when the blood pressure may rise to values as high as 200/100 (yielding a pulse pressure of 100 mmHg).

At diastole in this example, the aortic pressure equals 80 mmHg. When the left ventricle contracts, the intraventricular pressure rises above 80 mmHg and ejection begins. As a result, the amount of blood in the aorta increases by the amount ejected from the left ventricle (the stroke volume). Due to the increase in volume, there is an increase in blood pressure. The pressure in the brachial artery, where blood pressure measurements are commonly taken, therefore increases to 120 mmHg in this example. The rise in pressure from diastolic to systolic levels (pulse pressure) is thus a reflection of the stroke volume.

Clinical Investigation Clue

Remember that Charlie's pulse was weak.

What is the sequence of effects that caused Charlie's pulse to be weak?

The **mean arterial pressure** represents the average arterial pressure during the cardiac cycle. This value is significant because it is the difference between this pressure and the venous pressure that drives blood through the capillary beds of organs. The mean arterial pressure is not a simple arithmetic average because the pe-

Test Yourself Before You Continue

1. Describe the relationship between blood pressure and the total cross-sectional area of arteries, arterioles, and capillaries. Describe how arterioles influence blood flow through capillaries and arterial blood pressure.

2. Explain how the baroreceptor reflex helps to compensate for a fall in blood pressure. Why will a person who is severely dehydrated have a rapid pulse?

3. Describe how the sounds of Korotkoff are produced and explain how these sounds are used to measure blood pressure.

4. Define *pulse pressure* and explain the physiological significance of this measurement.

Hypertension, Shock, and Congestive Heart Failure

An understanding of the normal physiology of the cardiovascular system is prerequisite to the study of its pathophysiology, or mechanisms of abnormal function. Since the mechanisms that regulate cardiac output, blood flow, and blood pressure are highlighted in particular disease states, a study of pathophysiology at this time can augment your understanding of the mechanisms involved in normal function.

Table 14.8 Blood Pressure Classification in Adults*

Category	Systolic mmHg		Diastolic mmHg	Recommended Follow-Up
Optimal	< 120	and	< 80	Recheck in 2 years
Normal	< 130	and	< 85	Recheck in 2 years
High normal	130–139	or	85–89	Recheck in 1 year
Hypertension:				
Stage 1—mild	140–159	or	90–99	Confirm within 2 months
Stage 2—moderate	160–179	or	100–109	Evaluate within 1 month
Stage 3—severe	≥ 180	or	≥ 110	Evaluate immediately or within 1 week based on clinical situation

Note: Diagnosis of high blood pressure is based on the average of two or more readings taken at each of two or more visits after initial screening. Unusually low readings should be evaluated for clinical significance. © 1997 NIH

*The National Institutes of Health (NIH), from the Sixth Report of the Joint Committee on Detection, Evaluation, and Treatment of High Blood Pressure.

Table 14.9 Possible Causes of Secondary Hypertension

System Involved	Examples	Mechanisms
Kidneys	Kidney disease	Decreased urine formation
	Renal artery disease	Secretion of vasoactive chemicals
Endocrine	Excess catecholamines (tumor of adrenal medulla)	Increased cardiac output and total peripheral resistance
	Excess aldosterone (Conn's syndrome)	Excess salt and water retention by the kidneys
Nervous	Increased intracranial pressure	Activation of sympathoadrenal system
	Damage to vasomotor center	Activation of sympathoadrenal system
Cardiovascular	Complete heart block; patent ductus arteriosus	Increased stroke volume
	Arteriosclerosis of aorta; coarctation of aorta	Decreased distensibility of aorta

Hypertension

Approximately 20% of all adults in the United States have *hypertension*—blood pressure in excess of the normal range for a person's age and sex. Hypertension that is a result of (secondary to) known disease processes is logically called **secondary hypertension.** Of the hypertensive population, secondary hypertension accounts for only about 5%. Hypertension that is the result of complex and poorly understood processes is not so logically called **primary,** or **essential, hypertension.** Hypertension in adults is defined by a systolic pressure greater than 140 mmHg and/or a diastolic pressure greater than 90 mmHg (table 14.8).

Diseases of the kidneys and arteriosclerosis of the renal arteries can cause secondary hypertension because of high blood volume. More commonly, the reduction of renal blood flow can raise blood pressure by stimulating the secretion of vasoactive chemicals from the kidneys. Experiments in which the renal artery is pinched, for example, produce hypertension that is associated (at least initially) with elevated renin secretion. These and other causes of secondary hypertension are summarized in table 14.9.

Essential Hypertension

The vast majority of people with hypertension have essential hypertension. An increased total peripheral resistance is a universal characteristic of this condition. Cardiac rate and the cardiac output are elevated in many, but not all, of these cases.

The secretion of renin, which is correlated with angiotensin II production and aldosterone secretion, is likewise variable. Although some people with essential hypertension have low renin secretion, most have either normal or elevated levels of renin secretion. Renin secretion in the normal range is inappropriate for people with hypertension, since high blood pressure should inhibit renin secretion and, through a lowering of aldosterone, result in greater excretion of salt and water. Inappropriately high levels of renin secretion could thus contribute to hypertension by promoting (via stimulation of aldosterone secretion) salt and water retention and high blood volume.

Sustained high stress (acting via the sympathetic nervous system) and high salt intake appear to act synergistically in the development of hypertension. There is some evidence that Na^+ enhances the vascular response to sympathetic stimulation. Further, sympathetic nerve stimulation can cause constriction of the renal blood vessels and thus decrease the excretion of salt and water.

As an adaptive response to prolonged high blood pressure, the arterial wall becomes thickened. This response can lead to arteriosclerosis and results in an even greater increase in total peripheral resistance, thus raising blood pressure still more in a positive feedback fashion.

The interactions between salt intake, sympathetic nerve activity, cardiovascular responses to sympathetic nerve activity, kidney function, and genetics make it difficult to sort out the cause-and-effect sequence that leads to essential hypertension. Current evidence suggests that the inability of the kidneys to properly eliminate salt and water is a shared characteristic in all cases of essential hypertension. Further, there is evidence that salt intake may be the single most important factor. Chimpanzees with their natural, low-salt diet, have low blood pressure. When given human levels of dietary salt, however, their blood pressure rises. "Pre-literate" people whose diet is natural and low in salt similarly exhibit low blood pressure that does not

Table 14.10 Mechanisms of Action of Selected Antihypertensive Drugs

Category of Drugs	Examples	Mechanisms
Diuretics	Thiazide; furosemide	Increase volume of urine excreted, thus lowering blood volume
Sympathoadrenal system inhibitors	Clonidine; alpha-methyldopa	Act to decrease sympathoadrenal stimulation by bonding to α_2-adrenergic receptors in the brain
	Guanethidine; reserpine	Deplete norepinephrine from sympathetic nerve endings
	Atenolol	Blocks beta-adrenergic receptors, decreasing cardiac output and/or renin secretion
	Phentolamine	Blocks alpha-adrenergic receptors, decreasing sympathetic vasoconstriction
Direct vasodilators	Hydralazine; minoxidil sodium nitroprusside	Cause vasodilation by acting directly on vascular smooth muscle
Calcium channel blockers	Verapamil; diltiazem	Inhibit diffusion of Ca^{2+} into vascular smooth muscle cells, causing vasodilation and reduced peripheral resistance
Angiotensin-converting enzyme (ACE) inhibitors	Captopril; enalapril	Inhibit the conversion of angiotensin I into angiotensin II
Angiotensin II–receptor antagonists	Losartan	Blocks the binding of angiotensin II to its receptor

rise with age. Even though some people may be more salt-sensitive than others, these findings suggest that everyone with hypertension should restrict their intake of dietary salt.

Dangers of Hypertension

If other factors remain constant, blood flow increases as arterial blood pressure increases. The organs of people with hypertension are thus adequately perfused with blood until the excessively high pressure causes vascular damage. Because most patients are asymptomatic (without symptoms) until substantial vascular damage has occurred, hypertension is often referred to as a silent killer.

Hypertension is dangerous for a number of reasons. First, high arterial pressure increases the afterload, making it more difficult for the ventricles to eject blood. The heart, then, must work harder, which can result in pathological changes in heart structure and function, leading to congestive heart failure. Additionally, high pressure may damage cerebral blood vessels, leading to cerebrovascular accident, or "stroke." (Stroke is the third-leading cause of death in the United States.) Finally, hypertension contributes to the development of atherosclerosis, which can itself lead to heart disease and stroke as previously described.

Preeclampsia is a toxemia of late pregnancy characterized by high blood pressure, proteinuria (the presence of proteins in the urine), and edema. For reasons discussed in chapter 17, only negligible amounts of proteins are normally found in urine, and the excretion of plasma proteins in the urine can cause edema. In preeclampsia, the sensitivity of blood vessels to pressor agents (which cause vasoconstriction) is increased, resulting in decreased organ perfusion and increased blood pressure. The danger of preeclampsia is that it can quickly degenerate into a state called *eclampsia,* in which seizures occur. This can be life-threatening, and so the woman with preeclampsia is immediately treated for her symptoms and the fetus is delivered as quickly as possible.

Treatment of Hypertension

The first form of treatment that is usually attempted is modification of lifestyle. This modification includes cessation of smoking, moderation of alcohol intake, and weight reduction, if applicable. It can also include programmed exercise and a reduction in sodium intake. People with essential hypertension may have a potassium deficiency, and there is evidence that eating food that is rich in potassium may help to lower blood pressure. There is also evidence that supplementing the diet with Ca^{2+} may be of benefit, but this is more controversial.

If lifestyle modifications alone are insufficient, various drugs may be prescribed. Most commonly, these are *diuretics* that increase urine volume, thus decreasing blood volume and pressure. Drugs that block β_1-adrenergic receptors (such as atenolol) lower blood pressure by decreasing the cardiac rate and are also frequently prescribed. ACE inhibitors, calcium antagonists, and various vasodilators (table 14.10) may also be used in particular situations. A new class of drugs, angiotensin II-receptor antagonists, is now also available.

Circulatory Shock

Circulatory shock occurs when there is inadequate blood flow and/or oxygen utilization by the tissues. Some of the signs of shock (table 14.11) are a result of inadequate tissue perfusion; other signs of shock are produced by cardiovascular responses that help to compensate for the poor tissue perfusion (table 14.12). When these compensations are effective, they (together with emergency medical care) are able to reestablish adequate tissue perfusion. In some cases, however, and for reasons that are not clearly understood, the shock may progress to an irreversible stage, and death may result.

Hypovolemic Shock

The term **hypovolemic shock** refers to circulatory shock that is due to low blood volume, as might be caused by hemorrhage (bleeding), dehydration, or burns. This is accompanied by

Table 14.11 Signs of Shock

	Early Sign	Late Sign
Blood pressure	Decreased pulse pressure Increased diastolic pressure	Decreased systolic pressure
Urine	Decreased Na$^+$ concentration Increased osmolality	Decreased volume
Blood pH	Increased pH (alkalosis) due to hyperventilation	Decreased pH (acidosis) due to "metabolic" acids
Effects of poor tissue perfusion	Slight restlessness; occasionally warm, dry skin	Cold, clammy skin; "cloudy" senses

Source: From *Principles and Techniques of Critical Care*, Vol. 1, edited by R. F. Wilson. Copyright © 1977 F. A. Davis Company, Philadelphia, PA. Used by permission.

Table 14.12 Cardiovascular Reflexes That Help to Compensate for Circulatory Shock

Organ(s)	Compensatory Mechanisms
Heart	Sympathoadrenal stimulation produces increased cardiac rate and increased stroke volume due to "positive inotropic effect" on myocardial contractility
Digestive tract and skin	Decreased blood flow due to vasoconstriction as a result of sympathetic nerve stimulation (alpha-adrenergic effect)
Kidneys	Decreased urine production as a result of sympathetic-nerve-induced constriction of renal arterioles; increased salt and water retention due to increased plasma levels of aldosterone and antidiuretic hormone (ADH)

decreased blood pressure and decreased cardiac output. In response to these changes, the sympathoadrenal system is activated by means of the baroreceptor reflex. As a result, tachycardia is produced and vasoconstriction occurs in the skin, digestive tract, kidneys, and muscles. Decreased blood flow through the kidneys stimulates renin secretion and activation of the renin-angiotensin-aldosterone system. A person in hypovolemic shock thus has low blood pressure; a rapid pulse; cold, clammy skin; and a reduced urine output.

Since the resistance in the coronary and cerebral circulations is not increased, blood is diverted to the heart and brain at the expense of other organs. Interestingly, a similar response occurs in diving mammals and, to a lesser degree, in Japanese pearl divers during prolonged submersion. These responses help to deliver blood to the two organs that have the highest requirements for aerobic metabolism.

Vasoconstriction in organs other than the brain and heart raises total peripheral resistance, which helps (along with the reflex increase in cardiac rate) to compensate for the drop in blood pressure due to low blood volume. Constriction of arterioles also decreases capillary blood flow and capillary filtration pressure. As a result, less filtrate is formed. At the same time, the osmotic return of fluid to the capillaries is either unchanged or increased (during dehydration). The blood volume is thus raised at the expense of tissue fluid volume. Blood volume is also conserved by decreased urine production, which occurs as a result of vasoconstriction in the kidneys and the water-conserving effects of ADH and aldosterone, which are secreted in increased amounts during shock.

Septic Shock

Septic shock refers to a dangerously low blood pressure (hypotension) that may result from sepsis, or infection. This can occur through the action of a bacterial lipopolysaccharide called *endotoxin*. The mortality associated with septic shock is presently very high, estimated at 50% to 70%. According to recent information, endotoxin activates the enzyme nitric oxide synthase within macrophages—cells that play an important role in the immune response (see chapter 15). As previously discussed, nitric oxide synthase produces nitric oxide, which promotes vasodilation and, as a result, a fall in blood pressure. Septic shock has recently been treated effectively with drugs that inhibit the production of nitric oxide.

Other Causes of Circulatory Shock

A rapid fall in blood pressure occurs in **anaphylactic shock** as a result of a severe allergic reaction (usually to bee stings or penicillin). This results from the widespread release of histamine, which causes vasodilation and thus decreases total peripheral resistance. A rapid fall in blood pressure also occurs in **neurogenic shock,** in which sympathetic tone is decreased, usually because of upper spinal cord damage or spinal anesthesia. **Cardiogenic shock** results from cardiac failure, as defined by a cardiac output inadequate to maintain tissue perfusion. This commonly results from infarction that causes the loss of a significant proportion of the myocardium.

Congestive Heart Failure

Cardiac failure occurs when the cardiac output is insufficient to maintain the blood flow required by the body. This may be due to heart disease—resulting from myocardial infarction or congenital defects—or to hypertension, which increases the afterload of the heart. The most common causes of left ventricular heart failure are myocardial infarction, aortic valve stenosis, and incompetence of the aortic and bicuspid (mitral) valves. Failure of the right ventricle is usually caused by prior failure of the left ventricle.

Heart failure can also result from disturbance in the electrolyte concentrations of the blood. Excessive plasma K^+ concentration decreases the resting membrane potential of myocardial cells, and low blood Ca^{2+} reduces excitation-contraction coupling. High blood K^+ and low blood Ca^{2+} can thus cause the heart to stop in diastole. Conversely, low blood K^+ and high blood Ca^{2+} can arrest the heart in systole.

The term *congestive* is often used in describing heart failure because of the increased venous volume and pressure that results. Failure of the left ventricle, for example, raises the left atrial pressure and produces pulmonary congestion and edema. This causes shortness of breath and fatigue; if severe, pulmonary edema can be fatal. Failure of the right ventricle results in increased right atrial pressure, which produces congestion and edema in the systemic circulation.

The compensatory responses that occur during congestive heart failure are similar to those that occur during hypovolemic shock. Activation of the sympathoadrenal system stimulates cardiac rate, contractility of the ventricles, and constriction of arterioles. As in hypovolemic shock, renin secretion is increased and urine output is reduced. The increased secretion of renin and consequent activation of the renin-angiotensin-aldosterone system causes salt and water retention. This occurs despite an increased secretion of atrial natriuretic peptide (which would have the compensatory effect of promoting salt and water excretion).

As a result of these compensations, chronically low cardiac output is associated with elevated blood volume and dilation and hypertrophy of the ventricles. These changes can themselves be dangerous. Elevated blood volume places a work overload on the heart, and the enlarged ventricles have a higher metabolic requirement for oxygen. These problems are often treated with drugs that increase myocardial contractility (such as digitalis), drugs that are vasodilators (such as nitroglycerin), and diuretic drugs that lower blood volume by increasing the volume of urine excreted.

People with congestive heart failure are often treated with the drug **digitalis.** Digitalis appears to bind to and inhibit the action of Na^+/K^+ pumps in the cell membranes, causing a rise in the intracellular concentrations of Na^+. The increased availability of Na^+, in turn, stimulates the activity of another membrane transport carrier, which exchanges Na^+ for extracellular Ca^{2+}. As a result, the intracellular concentrations of Ca^{2+} are increased which strengthens the contractions of the heart.

Test Yourself Before You Continue

1. Explain how stress and a high-salt diet can contribute to hypertension. Also, explain how different drugs may act to lower blood pressure.

2. Using a flowchart to show cause and effect, explain why a person in hypovolemic shock may have a fast pulse and cold, clammy skin.

3. Describe the compensatory mechanisms that act to raise blood volume during cardiovascular shock.

4. Explain how septic shock may be produced.

5. Describe congestive heart failure and explain the compensatory responses that occur during this condition.

HPer Links of the Circulatory System with Other Body Systems

Integumentary System

- The skin helps to protect the body from pathogens ....(p. 446)
- The skin provides a site for thermo-regulation ....(p. 427)
- The circulatory system delivers blood for exchange of gases, nutrients, and wastes with all of the body organs, including the skin ....(p. 366)
- Blood clotting occurs if the skin is broken ....(p. 374)

Skeletal System

- Hematopoiesis occurs in the bone marrow ....(p. 370)
- The rib cage protects heart and thoracic vessels ....(p. 379)
- The blood delivers calcium and phosphate for deposition of bone and removes calcium and phosphate during bone resorption ....(p. 623)
- The blood delivers parathyroid hormone and other hormones that regulate bone growth and maintenance ....(p. 624)

Muscular System

- Cardiac muscle function is central to the activity of the heart ....(p. 378)
- Smooth muscle function in blood vessels regulates the blood flow and blood pressure ....(p. 391)
- Skeletal muscle contractions squeeze veins and thus aid venous blood flow ....(p. 394)
- The blood removes lactic acid and heat from active muscles ....(p. 394)

Nervous System

- Autonomic nerves help to regulate the cardiac output ....(p. 408)
- Autonomic nerves help to regulate the vascular resistance, blood flow, and blood pressure ....(p. 420)

- Cerebral capillaries participate in the blood-brain barrier ....(p. 159)

Endocrine System

- Epinephrine and norepinephrine from the adrenal medulla help to regulate cardiac function and vascular resistance ..(p. 230)
- Thyroxine and other hormones influence the blood pressure ....(p. 309)
- The blood transports hormones to their target organs ....(p. 144)

Immune System

- The immune system protects against infections ....(p. 446)
- Lymphatic vessels drain tissue fluid and return it to the venous system ...(p. 399)
- Lymphocytes from the bone marrow and lymphoid organs circulate in the blood ....(p. 369)
- Neutrophils leave the vascular system to participate in aspects of the immune response ....(p. 447)
- The circulation carries chemical regulators of the immune response ....(p. 315)

Respiratory System

- The lungs provide oxygen for transport by blood and provide for elimination of carbon dioxide ....(p. 480)
- Ventilation helps to regulate the pH of the blood ....(p. 511)
- The blood transports gases between the lungs and tissue cells ....(p. 494)

Urinary System

- The kidneys regulate the volume, pH, and electrolyte balance of blood ....(p. 524)
- The kidneys excrete waste products, derived from blood plasma, in the urine ....(p. 539)
- Blood pressure is required for kidney function ....(p. 529)

Digestive System

- Intestinal absorption of nutrients, including iron and particular B vitamins, is needed for red blood cell production ....(p. 370)
- The hepatic portal vein permits the enterohepatic circulation of some absorbed molecules ....(p. 576)
- The circulation transports nutrients from the GI tract to all the tissues in the body ....(p. 366)

Reproductive System

- Gonadal hormones, particularly testosterone, stimulate red blood cell production ....(p. 370)
- The placenta permits exchanges of gases, nutrients, and waste products between the maternal and fetal blood ....(p. 672)
- Erection of the penis and clitoris results from vasodilation of blood vessels (p. 643)

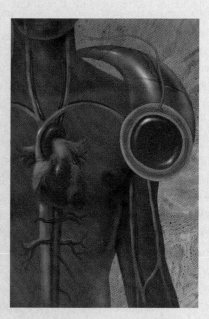

Summary

Cardiac Output 408

I. Cardiac rate is increased by sympathoadrenal stimulation and decreased by the effects of parasympathetic fibers that innervate the SA node.

II. Stroke volume is regulated both extrinsically and intrinsically.

    A. The Frank–Starling Law of the Heart describes the way the end-diastolic volume, through various degrees of myocardial stretching, influences the contraction strength of the myocardium and thus the stroke volume.

    B. The end-diastolic volume is called the preload. The total peripheral resistance, through its effect on arterial blood pressure, provides an afterload that acts to reduce the stroke volume.

    C. At a given end-diastolic volume, the amount of blood ejected depends on contractility. Strength of contraction is increased by sympathoadrenal stimulation.

III. The venous return of blood to the heart is dependent largely on the total blood volume and mechanisms that improve the flow of blood in veins.

    A. The total blood volume is regulated by the kidneys.

    B. The venous flow of blood to the heart is aided by the action of skeletal muscle pumps and the effects of breathing.

Blood Volume 412

I. Tissue fluid is formed from and returns to the blood.

    A. The hydrostatic pressure of the blood forces fluid from the arteriolar ends of capillaries into the interstitial spaces of the tissues.

    B. Since the colloid osmotic pressure of plasma is greater than that of tissue fluid, water returns by osmosis to the venular ends of capillaries.

    C. Excess tissue fluid is returned to the venous system by lymphatic vessels.

    D. Edema occurs when excess tissue fluid accumulates.

II. The kidneys control the blood volume by regulating the amount of filtered fluid that will be reabsorbed.

    A. Antidiuretic hormone stimulates reabsorption of water from the kidney filtrate, and thus acts to maintain the blood volume.

    B. A decrease in blood flow through the kidneys activates the renin-angiotensin-aldosterone system.

    C. Angiotensin II stimulates vasoconstriction and the secretion of aldosterone by the adrenal cortex.

    D. Aldosterone acts on the kidneys to promote the retention of salt and water.

Vascular Resistance to Blood Flow 418

I. According to Poiseuille's law, blood flow is directly related to the pressure difference between the two ends of a vessel and inversely related to the resistance to blood flow through the vessel.

II. Extrinsic regulation of vascular resistance is provided mainly by the sympathetic nervous system, which stimulates vasoconstriction of arterioles in the viscera and skin.

III. Intrinsic control of vascular resistance allows organs to autoregulate their blood flow rates.

    A. Myogenic regulation occurs when vessels constrict or dilate as a direct response to a rise or fall in blood pressure.

    B. Metabolic regulation occurs when vessels dilate in response to the local chemical environment within the organ.

Blood Flow to the Heart and Skeletal Muscles 423

I. The heart normally respires aerobically because of its extensive capillary supply and high myoglobin and enzyme content.

II. During exercise, when the heart's metabolism increases, intrinsic metabolic mechanisms stimulate vasodilation of the coronary vessels, and thus increase coronary blood flow.

III. Just prior to exercise and at the start of exercise, blood flow through skeletal muscles increases because of vasodilation caused by the activity of cholinergic sympathetic nerve fibers. During exercise, intrinsic metabolic vasodilation occurs.

IV. Since cardiac output can increase by a factor of five or more during exercise, the heart and skeletal muscles receive an increased proportion of a higher total blood flow.

    A. The cardiac rate increases because of lower activity of the vagus nerve and higher activity of sympathetic nerves.

    B. The venous return is greater because of higher activity of the skeletal muscle pumps and increased breathing.

    C. Increased contractility of the heart, combined with a decrease in total peripheral resistance, can result in a higher stroke volume.

Blood Flow to the Brain and Skin 427

I. Cerebral blood flow is regulated both myogenically and metabolically.

    A. Cerebral vessels automatically constrict if the systemic blood pressure rises too high.

    B. Metabolic products cause local vessels to dilate and supply more-active areas with more blood.

II. The skin contains unique arteriovenous anastomoses that can shunt the blood away from surface capillary loops.

    A. The activity of sympathetic nerve fibers causes constriction of cutaneous arterioles.

    B. As a thermoregulatory response, cutaneous blood flow and blood flow through surface capillary loops increase when the body temperature rises.

Blood Pressure 429

I. Baroreceptors in the aortic arch and carotid sinuses affect the cardiac rate and the total peripheral resistance via the sympathetic nervous system.

    A. The baroreceptor reflex causes pressure to be maintained when an upright posture is assumed. This reflex can cause a lowered pressure when the carotid sinuses are massaged.

B. Other mechanisms that affect blood volume help to regulate blood pressure.

II. Blood pressure is commonly measured indirectly by auscultation of the brachial artery when a pressure cuff is inflated and deflated.

   A. The first sound of Korotkoff, caused by turbulent flow of blood through a constriction in the artery, occurs when the cuff pressure equals the systolic pressure.

   B. The last sound of Korotkoff is heard when the cuff pressure equals the diastolic blood pressure.

III. The mean arterial pressure represents the driving force for blood flow through the arterial system.

Hypertension, Shock, and Congestive Heart Failure   435

I. Hypertension, or high blood pressure, is classified as either primary or secondary.

   A. Primary hypertension, also called essential hypertension, may result from the interaction of numerous mechanisms that raise the blood volume, cardiac output, and/or peripheral resistance.

   B. Secondary hypertension is the direct result of known specific diseases.

II. Circulatory shock occurs when delivery of oxygen to the organs of the body is inadequate.

   A. In hypovolemic shock, low blood volume causes low blood pressure that may progress to an irreversible state.

   B. The fall in blood volume and pressure stimulates various reflexes that produce a rise in cardiac rate, a shift of fluid from the tissues into the vascular system, a decrease in urine volume, and vasoconstriction.

III. Congestive heart failure occurs when the cardiac output is insufficient to supply the blood flow required by the body. The term *congestive* is used to describe the increased venous volume and pressure that result.

Review Activities

Test Your Knowledge of Terms and Facts

1. According to the Frank–Starling Law of the Heart, the strength of ventricular contraction is
   a. directly proportional to the end-diastolic volume.
   b. inversely proportional to the end-diastolic volume.
   c. independent of the end-diastolic volume.

2. In the absence of compensations, the stroke volume will decrease when
   a. blood volume increases.
   b. venous return increases.
   c. contractility increases.
   d. arterial blood pressure increases.

3. Which of these statements about tissue fluid is *false?*
   a. It contains the same glucose and salt concentration as plasma.
   b. It contains a lower protein concentration than plasma.
   c. Its colloid osmotic pressure is greater than that of plasma.
   d. Its hydrostatic pressure is lower than that of plasma.

4. Edema may be caused by
   a. high blood pressure.
   b. decreased plasma protein concentration.
   c. leakage of plasma protein into tissue fluid.
   d. blockage of lymphatic vessels.
   e. all of these.

5. Both ADH and aldosterone act to
   a. increase urine volume.
   b. increase blood volume.

   c. increase total peripheral resistance.
   d. produce all of these effects.

6. The greatest resistance to blood flow occurs in
   a. large arteries.
   b. medium-sized arteries.
   c. arterioles.
   d. capillaries.

7. If a vessel were to dilate to twice its previous radius, and if pressure remained constant, blood flow through this vessel would
   a. increase by a factor of 16.
   b. increase by a factor of 4.
   c. increase by a factor of 2.
   d. decrease by a factor of 2.

8. The sounds of Korotkoff are produced by
   a. closing of the semilunar valves.
   b. closing of the AV valves.
   c. the turbulent flow of blood through an artery.
   d. elastic recoil of the aorta.

9. Vasodilation in the heart and skeletal muscles during exercise is primarily due to the effects of
   a. alpha-adrenergic stimulation.
   b. beta-adrenergic stimulation.
   c. cholinergic stimulation.
   d. products released by the exercising muscle cells.

10. Blood flow in the coronary circulation
   a. increases during systole.
   b. increases during diastole.
   c. remains constant throughout the cardiac cycle.

11. Blood flow in the cerebral circulation
   a. varies with systemic arterial pressure.
   b. is regulated primarily by the sympathetic system.
   c. is maintained constant within physiological limits.
   d. increases during exercise.

12. Which of these organs is able to tolerate the greatest reduction in blood flow?
   a. brain
   b. heart
   c. skeletal muscles
   d. skin

13. Which of these statements about arteriovenous shunts in the skin is *true?*
   a. They divert blood to superficial capillary loops.
   b. They are closed when the ambient temperature is very low.
   c. They are closed when the deep-body temperature rises much above 37° C.
   d. All of these are true.

14. An increase in blood volume will cause
   a. a decrease in ADH secretion.
   b. an increase in Na^+ excretion in the urine.
   c. a decrease in renin secretion.
   d. all of these.

15. The volume of blood pumped per minute by the left ventricle is
   a. greater than the volume pumped by the right ventricle.
   b. less than the volume pumped by the right ventricle.

c. the same as the volume pumped by the right ventricle.

d. either less or greater than the volume pumped by the right ventricle, depending on the strength of contraction.

16. Blood pressure is lowest in

  a. arteries.
  b. arterioles.
  c. capillaries.
  d. venules.
  e. veins.

17. Stretch receptors in the aortic arch and carotid sinus

  a. stimulate secretion of atrial natriuretic peptide.

  b. serve as baroreceptors that affect activity of the vagus and sympathetic nerves.
  c. serve as osmoreceptors that stimulate the release of ADH.
  d. stimulate renin secretion, thus increasing angiotensin II formation.

18. Angiotensin II

  a. stimulates vasoconstriction.
  b. stimulates the adrenal cortex to secrete aldosterone.
  c. inhibits the action of bradykinin.
  d. does all of these.

19. Which of these is a paracrine regulator that stimulates vasoconstriction?

  a. nitric oxide
  b. prostacyclin
  c. bradykinin
  d. endothelin-1

20. The pulse pressure is a measure of

  a. the number of heartbeats per minute.
  b. the sum of the diastolic and systolic pressures.
  c. the difference between the systolic and diastolic pressures.
  d. the difference between the arterial and venous pressures.

Test Your Understanding of Concepts and Principles

1. Define the terms *contractility, preload,* and *afterload,* and explain how these factors affect the cardiac output.[1]

2. Using the Frank–Starling Law of the Heart, explain how the stroke volume is affected by (a) bradycardia and (b) a "missed beat."

3. Which part of the cardiovascular system contains the most blood? Which part provides the greatest resistance to blood flow? Which part provides the greatest cross-sectional area? Explain.

4. Explain how the kidneys regulate blood volume.

5. A person who is dehydrated drinks more and urinates less. Explain the mechanisms involved.

6. Using Poiseuille's law, explain how arterial blood flow can be diverted from one organ system to another.

7. Describe the mechanisms that increase the cardiac output during exercise and that increase the rate of blood flow to the heart and skeletal muscles.

8. Explain why an anxious person may have a cold, clammy skin and why the skin becomes hot and flushed on a hot, humid day.

9. Explain the different ways in which a drug that acts as an inhibitor of angiotensin-converting enzyme (ACE) can lower the blood pressure. Also, explain how diuretics and β_1-adrenergic-blocking drugs work to lower the blood pressure.

10. Explain how hypotension may be produced in (a) hypovolemic shock and (b) septic shock. Also, explain the mechanisms whereby people in shock have a rapid but weak pulse, cold and clammy skin, and low urine output.

Test Your Ability to Analyze and Apply Your Knowledge

1. One consequence of the Frank–Starling Law of the Heart is that the outputs of the right and left ventricles are matched. Explain why this is important and how this matching is accomplished.

2. An elderly man who is taking digoxin for a weak heart complains that his feet hurt. Upon examination, his feet are found to be swollen and discolored, with purple splotches and expanded veins. He is told to keep his feet raised and is given a prescription for Lasix, a

powerful diuretic. Discuss this man's condition and the rationale for his treatment.

3. You are bicycling in a 100-mile benefit race because you want to help the cause, but you didn't count on such a hot, humid day. You've gone through both water bottles and, in the last 10 miles, you are thirsty again. Should you accept the water that one bystander offers or the sports drink offered by another? Explain your choice.

4. As the leader of a revolution to take over a large country, you direct your followers to seize the salt mines. Why is this important? When the revolution succeeds and you become president, you ask your surgeon general to wage a health campaign urging citizens to reduce their salt intake. Why?

5. Which type of exercise, isotonic contractions or isometric contractions, puts more of a "strain" on the heart? Explain.

Related Websites

Check out the Links Library at www.mhhe.com/fox8 for links to sites related to cardiac output, blood flow, and blood pressure. These links are monitored to ensure current URLs.

---

[1]*Note:* This question is answered in the chapter 14 Study Guide found on the Online Learning Center at www.mhhe.com/fox8.

15 The Immune System

Objectives   *After studying this chapter, you should be able to . . .*

1. describe some of the mechanisms of nonspecific immunity and distinguish between nonspecific and specific immune defenses.

2. describe how B lymphocytes respond to antigens and define the terms *memory cell* and *plasma cell.*

3. describe the structure and classification of antibodies and discuss the nature of antigens.

4. describe the complement system and explain how antigen-antibody reactions indirectly lead to the destruction of an invading pathogen.

5. describe the events that occur during a local inflammation.

6. describe the process of active immunity and explain how the clonal selection theory may account for this process.

7. describe the mechanisms of passive immunity and give natural and clinical examples of this form of immunization.

8. explain how monoclonal antibodies are produced and describe some of their clinical uses.

9. explain how T lymphocytes are classified and describe the function of the thymus.

10. define the term *lymphokines* and list some of these molecules, together with their functions.

11. describe the histocompatibility antigens and explain their importance in the function of the T cell receptor proteins.

12. describe the interaction between macrophages and helper T lymphocytes and explain how the helper T cells affect immunological defense by killer T cells and B cells.

13. describe the possible role of suppressor T lymphocytes in the negative feedback control of the immune response.

14. describe differences in the immune response to viruses and bacteria.

15. describe the possible mechanisms responsible for tolerance of self-antigens.

16. describe some of the characteristics of cancer and explain how natural killer cells and killer T lymphocytes provide immunological surveillance against cancer.

17. define the term *autoimmune disease,* give examples of different kinds of autoimmune diseases, and explain some of the mechanisms by which these diseases are produced.

18. explain how immune complex diseases may be produced and give examples of these diseases.

19. distinguish between immediate hypersensitivity and delayed hypersensitivity and describe the mechanisms responsible for each form of allergy.

Take Advantage of the Technology

Visit the Online Learning Center for these additional study resources.

- Interactive quizzing
- Online study guide
- Current news feeds
- Crossword puzzles
- Vocabulary flashcards
- Labeling activities

www.mhhe.com/fox8

Clinical Investigation

Gary, an active 8-year-old, was playing by crawling through the underbrush in the surrounding hills while his parents were picnicking. When he returned to them he was covered in dirt and had a bloody thumb, which he said had been cut by an unseen bit of barbed wire that had been left on the ground. They bandaged the thumb, but, while he was eating a sandwich a wasp stung him on his injured hand! It was his first bee sting, and this event provided a tearful conclusion to the family's outing. The next day, the pain of the cut and bee sting subsided, but Gary developed a rash on his abdomen.

The family physician prescribed a cortisone cream for Gary's rash and said that antihistamines wouldn't help the rash. The doctor told Gary's mom not to worry about the cut, because the boy had fortunately been given a tetanus vaccine only 6 months earlier. However, Gary was stung by a bee 2 months later, and this time developed a severe swelling. The doctor prescribed antihistamine treatment for this, which did resolve the swelling.

How can you explain Gary's symptoms and medical treatment?

Defense Mechanisms

Nonspecific immune protection is provided by such mechanisms as phagocytosis, fever, and the release of interferons. Specific immunity, which involves the functions of lymphocytes, is directed at specific molecules, or parts of molecules, known as antigens.

The immune system includes all of the structures and processes that provide a defense against potential pathogens (disease-causing agents). These defenses can be grouped in two categories: **innate** (or **nonspecific**) **immunity** and **adaptive** (or **specific**) **immunity.** Although these two categories refer to different defense mechanisms, there are areas in which they overlap.

Innate, or nonspecific, defense mechanisms are inherited as part of the structure of each organism. Epithelial membranes that cover the body surfaces, for example, restrict infection by most pathogens. The strong acidity of gastric juice (pH 1–2) also helps to kill many microorganisms before they can invade the body. These external defenses are backed by internal defenses, such as phagocytosis, which function in both a specific and nonspecific manner (table 15.1).

Each individual can acquire the ability to defend against specific pathogens by prior exposure to those pathogens. This adaptive, or specific, immune response is a function of lymphocytes. Internal specific and nonspecific defense mechanisms function together to combat infection, with lymphocytes interacting in a coordinated effort with phagocytic cells.

The genes required for *innate immunity* are inherited. Since this limits the number of genes that can be devoted to this task, innate immune mechanisms combat whole categories of pathogens. A category of bacteria (called gram-negative), for example, can be recognized by the presence of particular molecules (called lipopolysaccharide) on their surfaces. In *adaptive immunity,* by contrast, specific features of pathogens are recognized. The enormous number of different genes required for this task is too large to be inherited. Instead, the variation is produced by genetic changes in lymphocytes during the life of each person after birth.

Innate (Nonspecific) Immunity

Innate immunity includes both external and internal defenses. These defenses are always present in the body and represent the first line of defense against invasion by potential pathogens.

Invading pathogens, such as bacteria, that have crossed epithelial barriers enter connective tissues. These invaders—or chemicals, called *toxins,* secreted from them—may enter blood or lymphatic capillaries and be carried to other areas of the body. Innate immunological defenses are the first employed to counter the invasion and spread of infection. If these defenses are not sufficient to destroy the pathogens, lymphocytes may be recruited and their specific actions used to reinforce the nonspecific immune defenses.

Phagocytosis

The innate defense mechanisms distinguish between the kinds of carbohydrates that are produced by mammalian cells and those produced by bacteria. The bacterial carbohydrates that "flag" the cell for phagocytic attack are part of the glycoproteins and lipopolysaccharides on the bacterial cell wall.

There are three major groups of phagocytic cells: (1) **neutrophils;** (2) the cells of the **mononuclear phagocyte system,** including *monocytes* in the blood and *macrophages* (derived from monocytes) in the connective tissues; and (3) **organ-specific phagocytes** in the liver, spleen, lymph nodes, lungs, and brain

Table 15.1 Structures and Defense Mechanisms of Nonspecific (Innate) Immunity

	Structure	Mechanisms
External	Skin	Physical barrier to penetration by pathogens; secretions contain lysozyme (enzyme that destroys bacteria)
	Digestive tract	High acidity of stomach; protection by normal bacterial population of colon
	Respiratory tract	Secretion of mucus; movement of mucus by cilia; alveolar macrophages
	Genitourinary tract	Acidity of urine; vaginal lactic acid
Internal	Phagocytic cells	Ingest and destroy bacteria, cellular debris, denatured proteins, and toxins
	Interferons	Inhibit replication of viruses
	Complement proteins	Promote destruction of bacteria; enhance inflammatory response
	Endogenous pyrogen	Secreted by leukocytes and other cells; produces fever

(table 15.2). Organ-specific phagocytes, such as the microglia of the brain, are embryologically and functionally related to macrophages and may be considered part of the mononuclear phagocyte system.

The *Kupffer cells* in the liver, as well as phagocytic cells in the spleen and lymph nodes, are **fixed phagocytes.** This term refers to the fact that these cells are immobile ("fixed") in the walls of the sinusoids (chapter 13) within these organs. As blood flows through these wide capillaries of the liver and spleen, foreign chemicals and debris are removed by phagocytosis and chemically inactivated within the phagocytic cells. Invading pathogens are very effectively removed in this manner, so that blood is usually sterile after a few passes through the liver and spleen. Fixed phagocytes in lymph nodes similarly help to remove foreign particles from the lymph.

Connective tissues have a resident population of all leukocyte types. Neutrophils and monocytes in particular can be highly mobile within connective tissues as they scavenge for invaders and cellular debris. These leukocytes are recruited to the site of an infection by a process called *chemotaxis*—movement toward chemical attractants. The chemical attractants are a subclass of cytokines (autocrine/paracrine regulators—see chapter 11) known as **chemokines.** Neutrophils are the first to arrive at the site of an infection; monocytes arrive later and can be transformed into macrophages as the battle progresses.

If the infection has spread, new phagocytic cells from the blood may join those already in the connective tissue. These new neutrophils and monocytes are able to squeeze through the tiny gaps between adjacent endothelial cells in the capillary wall and enter the connective tissues. This process, called **extravasation** (or **diapedesis**) is illustrated in figure 15.1.

Phagocytic cells engulf particles in a manner similar to the way an amoeba eats. The particle becomes surrounded by cytoplasmic extensions called pseudopods, which ultimately fuse. The particle thus becomes surrounded by a membrane derived from the plasma membrane (fig. 15.2) and contained within an organelle analogous to a food vacuole in an amoeba. This vacuole then fuses with lysosomes (organelles that contain digestive enzymes), so that the ingested particle and the digestive enzymes still are separated from the cytoplasm by a continuous membrane. Often, however, lysosomal enzymes are released before the food vacuole has completely formed. When this occurs, free lysosomal enzymes may be released into the infected area and contribute to inflammation.

Table 15.2 Phagocytic Cells and Their Locations

Phagocyte	Location
Neutrophils	Blood and all tissues
Monocytes	Blood
Tissue macrophages (histiocytes)	All tissues (including spleen, lymph nodes, bone marrow)
Kupffer cells	Liver
Alveolar macrophages	Lungs
Microglia	Central nervous system

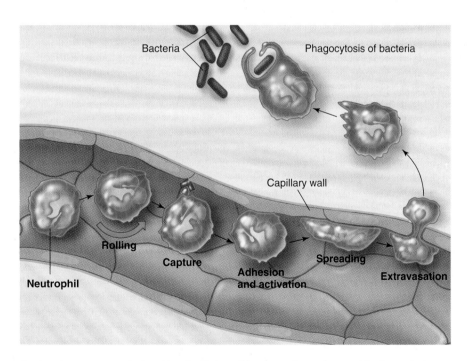

■ **Figure 15.1**   **Stages involved in the migration of white blood cells from blood vessels into tissues.** The figure depicts a neutrophil that goes through the stages of rolling, capture, adhesion and activation, and finally extravasation (diapedesis) through the blood vessel wall. This process is set in motion when the invading bacteria secrete certain chemicals, which attract and activate the white blood cells. The steps of extravasation require the binding of particular molecules on the white blood cell surface to receptor molecules on the surface of endothelial cells.

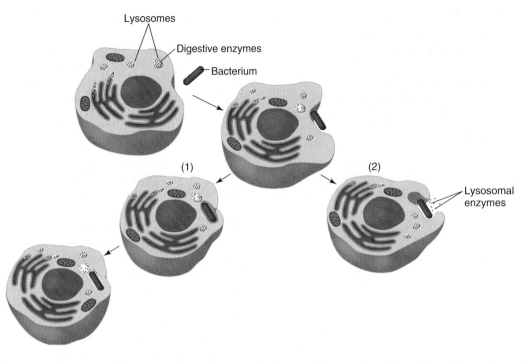

■ **Figure 15.2** **Phagocytosis by a neutrophil or macrophage.** A phagocytic cell extends its pseudopods around the object to be engulfed (such as a bacterium). (Blue dots represent lysosomal enzymes.) *(1)* If the pseudopods fuse to form a complete food vacuole, lysosomal enzymes are restricted to the organelle formed by the lysosome and food vacuole. *(2)* If the lysosome fuses with the vacuole before fusion of the pseudopods is complete, lysosomal enzymes are released into the infected area of tissue.

Fever

Fever may be a component of the nonspecific defense system. Body temperature is regulated by the hypothalamus, which contains a thermoregulatory control center (a "thermostat") that coordinates skeletal muscle shivering and the activity of the sympathoadrenal system to maintain body temperature at about 37° C. This thermostat is reset upward in response to a chemical called **endogenous pyrogen.** In at least some infections, the endogenous pyrogen has been identified as interleukin-1β, which is first produced as a cytokine by leukocytes and is then produced by the brain itself.

The cell wall of gram-negative bacteria contains **endotoxin,** a lipopolysaccharide that stimulates monocytes and macrophages to release various cytokines. These cytokines, including *interleukin-1, interleukin-6,* and *tumor necrosis factor,* act to produce fever, increased sleepiness, and a fall in the plasma iron concentration.

Although high fevers are definitely dangerous, a mild to moderate fever may be a beneficial response that aids recovery from bacterial infections. The fall in plasma iron concentrations that accompany a fever can inhibit bacterial activity and represents one possible benefit of a fever; others include increased activity of neutrophils and increased production of interferon.

Interferons

In 1957, researchers demonstrated that cells infected with a virus produced polypeptides that interfered with the ability of a second, unrelated strain of virus to infect other cells in the same culture. These **interferons,** as they were called, thus produced a nonspecific, short-acting resistance to viral infection. Although this discovery generated a great deal of excitement, further research was hindered by the fact that human interferons could be obtained only in very small quantities; moreover, animal interferons were shown to have little effect in humans. In 1980, however, a technique called *genetic recombination* (chapter 3) made it possible to introduce human interferon genes into bacteria, enabling the bacteria to act as interferon factories.

There are three major categories of interferons: *alpha, beta,* and *gamma interferons.* Almost all cells in the body make alpha and beta interferons. These polypeptides act as messengers that protect other cells in the vicinity from viral infection. The viruses are still able to penetrate these other cells, but the ability of the viruses to replicate and assemble new virus particles is inhibited. Viral infection, replication, and dispersal are illustrated in figure 15.3, using the virus that causes AIDS as an example. Gamma interferon is produced only by particular lymphocytes and a related type of cell called a natural killer cell. The secretion of gamma interferon by these cells is part of the immunological defense against infection and cancer. Some of the effects of interferons are summarized in table 15.3.

The Food and Drug Administration (FDA) has recently approved the use of interferons to treat a number of diseases. Alpha interferon, for example, is now being used to treat hepatitis C, hairy-cell leukemia, virally induced genital warts, and Kaposi's sarcoma. The FDA has also approved the use of beta interferon to treat relapsing-remitting multiple sclerosis and the

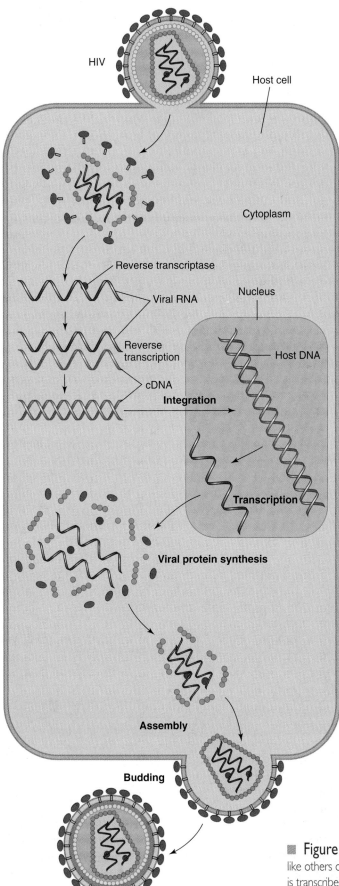

Table 15.3	Effects of Interferons	
Stimulation		**Inhibition**
Macrophage phagocytosis		Cell division
Activity of cytotoxic ("killer") T cells		Tumor growth
Activity of natural killer cells		Maturation of adipose cells
Production of antibodies		Maturation of erythrocytes

use of gamma interferon to treat chronic granulomatous disease. Treatment of numerous forms of cancer with interferon is currently in various stages of clinical trials.

Adaptive (Specific) Immunity

A German bacteriologist, Emil Adolf von Behring, demonstrated in 1890 that a guinea pig previously injected with a sublethal dose of diphtheria toxin could survive subsequent injections of otherwise lethal doses of that toxin. Further, von Behring showed that this immunity could be transferred to a second, nonexposed animal by injections of serum from the immunized guinea pig. He concluded that the immunized animal had chemicals in its serum—which he called **antibodies**—that were responsible for the immunity. He also showed that these antibodies conferred immunity only to diphtheria infections; the antibodies were *specific* in their actions. It was later learned that antibodies are proteins produced by a particular type of lymphocyte.

Antigens

Antigens are molecules that stimulate the production of specific antibodies and combine specifically with the antibodies produced. Most antigens are large molecules (such as proteins) with a molecular weight greater than about 10,000, although there are important exceptions. Also, most antigens are foreign to the blood and other body fluids. This is because the immune system can distinguish its own "self" molecules from those of any other organism ("nonself") and normally mounts an immune response only against nonself antigens. The ability of a molecule to function as an antigen depends not only on its size but also on the complexity of its structure. The plastics used in artificial implants are composed of large molecules, but they are not very antigenic because of their simple, repeating structures.

A large, complex molecule can have a number of different **antigenic determinant sites,** which are areas of the molecule that stimulate production of, and combine with, different antibodies. Most naturally occurring antigens have many antigenic determinant sites and stimulate the production of different antibodies with specificities for these sites.

■ **Figure 15.3** The life cycle of the human immunodeficiency virus (HIV). This virus, like others of its family, contains RNA instead of DNA. Once inside the host cell, the viral RNA is transcribed by reverse transcriptase into complementary DNA (cDNA). The genes in the cDNA then direct the synthesis of new virus particles.

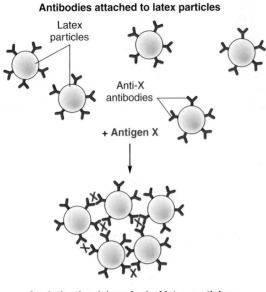

Antibodies attached to latex particles

Latex particles

Anti-X antibodies

+ Antigen X

Agglutination (clumping) of latex particles

■ **Figure 15.4** An immunoassay using the agglutination technique. Antibodies against a particular antigen are adsorbed to latex particles. When these are mixed with a solution that contains the appropriate antigen, the formation of the antigen-antibody complexes produces clumping (agglutination) that can be seen with the unaided eye.

Haptens

Many small organic molecules are not antigenic by themselves but can become antigens if they bind to proteins (and thus become antigenic determinant sites on the proteins). This discovery was made by Karl Landsteiner, who also discovered the ABO blood groups (chapter 13). By bonding these small molecules—which Landsteiner called **haptens**—to proteins in the laboratory, new antigens could be created for research or diagnostic purposes. The bonding of foreign haptens to a person's own proteins can also occur in the body. By this means, derivatives of penicillin, for example, that would otherwise be harmless can produce fatal allergic reactions in susceptible people.

Immunoassays

When the antigen or antibody is attached to the surface of a cell or to particles of latex rubber (in commercial diagnostic tests), the antigen-antibody reaction becomes visible because the particles *agglutinate* (clump) as a result of antigen-antibody bonding (fig. 15.4). These agglutinated particles can be used to assay a variety of antigens, and tests that utilize this procedure are called immunoassays. Blood typing and modern pregnancy tests are examples of such **immunoassays.** In order to increase their sensitivity, modern immunoassays generally use antibodies that exhibit specificity for just one antigenic determinant site. The technique for generating such uniformly specific antibodies is described in a later section on monoclonal antibodies.

Lymphocytes and Lymphoid Organs

Leukocytes, erythrocytes, and blood platelets are all ultimately derived from ("stem from") unspecialized cells in the bone marrow. These *stem cells* produce the specialized blood cells, and they replace themselves by cell division so that the stem cell population is not exhausted. Lymphocytes produced in this manner seed the thymus, spleen, and lymph nodes, producing self-replacing lymphocyte colonies in these organs.

The lymphocytes that seed the thymus become **T lymphocytes,** or **T cells** (the letter *T* stands for thymus-dependent). These cells have surface characteristics and an immunological function that differ from those of other lymphocytes. The thymus, in turn, seeds other organs; about 65% to 85% of the lymphocytes in blood and most of the lymphocytes in the germinal centers of the lymph nodes and spleen are T lymphocytes. T lymphocytes, therefore, either come from or had an ancestor that came from the thymus.

Most of the lymphocytes that are not T lymphocytes are called **B lymphocytes,** or **B cells.** The letter *B* derives from immunological research performed in chickens. Chickens have an organ called the *bursa of Fabricius* that processes B lymphocytes. Since mammals do not have a bursa, the *B* is often translated as the "bursa equivalent" for humans and other mammals. It is currently believed that the B lymphocytes in mammals are processed in the bone marrow, which conveniently also begins with the letter *B*. Since the bone marrow produces B lymphocytes and the thymus produces T lymphocytes, the bone marrow and thymus are considered to be **primary lymphoid organs.**

Both B and T lymphocytes function in specific immunity. The B lymphocytes combat bacterial infections, as well as some viral infections, by secreting antibodies into the blood and lymph. Because blood and lymph are body fluids (humors), the B lymphocytes are said to provide **humoral immunity,** although the term *antibody-mediated immunity* is also used. T lymphocytes attack host cells that have become infected with viruses or fungi, transplanted human cells, and cancerous cells. The T lymphocytes do not secrete antibodies; they must come in close proximity to the victim cell, or have actual physical contact with the cell, in order to destroy it. T lymphocytes are therefore said to provide **cell-mediated immunity** (table 15.4).

Thymus

The **thymus** extends from below the thyroid in the neck into the thoracic cavity. As mentioned in chapter 11, this organ grows during childhood but gradually regresses after puberty. Lymphocytes from the fetal liver and spleen, and from the bone marrow postnatally, seed the thymus and become transformed into T cells. These lymphocytes, in turn, enter the blood and seed lymph nodes and other organs, where they divide to produce new T cells when stimulated by antigens.

Small T lymphocytes that have not yet been stimulated by antigens have very long life spans—months or perhaps years. Still, new T cells must be continuously produced to provide efficient cell-mediated immunity. This is particularly important following cancer chemotherapy and during HIV infection (in AIDS),

Table 15.4 Comparison of B and T Lymphocytes

Characteristic	B Lymphocytes	T Lymphocytes
Site where processed	Bone marrow	Thymus
Type of immunity	Humoral (secretes antibodies)	Cell-mediated
Subpopulations	Memory cells and plasma cells	Cytotoxic (killer) T cells, helper cells, suppressor cells
Presence of surface antibodies	Yes—IgM or IgD	Not detectable
Receptors for antigens	Present—are surface antibodies	Present—are related to immunoglobulins
Life span	Short	Long
Tissue distribution	High in spleen, low in blood	High in blood and lymph
Percentage of blood lymphocytes	10%–15%	75%–80%
Transformed by antigens to	Plasma cells	Activated lymphocytes
Secretory product	Antibodies	Lymphokines
Immunity to viral infections	Enteroviruses, poliomyelitis	Most others
Immunity to bacterial infections	*Streptococcus, Staphylococcus,* many others	Tuberculosis, leprosy
Immunity to fungal infections	None known	Many
Immunity to parasitic infections	Trypanosomiasis, maybe to malaria	Most others

Table 15.5 Summary of Events in a Local Inflammation

Category	Events
Innate (Nonspecific) Immunity	Bacteria enter a break in the skin.
	Resident phagocytic cells—neutrophils and macrophages—engulf the bacteria.
	Nonspecific activation of complement proteins occurs.
Adaptive (Specific) Immunity	B cells are stimulated to produce specific antibodies.
	Phagocytosis is enhanced by antibodies attached to bacterial surface antigens (opsonization).
	Specific activation of complement proteins occurs, which stimulates phagocytosis, chemotaxis of new phagocytes to the infected area, and secretion of histamine from tissue mast cells.
	Extravasation (diapedesis) allows new phagocytic leukocytes (neutrophils and monocytes) to invade the infected area.
	Vasodilation and increased capillary permeability (as a result of histamine secretion) produce redness and edema.

when the population of T lymphocytes has been depleted. Under these conditions, the thymus can replenish the T lymphocyte population through late childhood. Repopulation of T lymphocytes occurs more slowly in adulthood, and appears to be accomplished mostly by production of T lymphocytes in the secondary lymphoid organs rather than in the thymus. This is because the thymus of adults becomes more of a fatty organ, although production of lymphocytes in the thymus has been demonstrated to occur to some extent even in people over the age of 70.

Secondary Lymphoid Organs

The **secondary lymphoid organs** include the lymph nodes, spleen, tonsils, and areas called Peyer's patches under the mucosa of the intestine (see chapter 13, fig. 13.36). These organs are strategically located across epithelial membranes in areas where antigens could gain entry to the blood or lymph. The spleen filters blood, whereas the other secondary lymphoid organs filter lymph received from lymphatic vessels.

Lymphocytes migrate from the primary lymphoid organs—the bone marrow and thymus—to the secondary lymphoid organs. Indeed, lymphocytes move constantly through the blood and lymph, going from lymphoid organ to lymphoid organ. This ceaseless travel increases the likelihood that a given lympho-

cyte, specific for a particular antigen, will be able to encounter that antigen. This process is aided, particularly in the case of T lymphocytes, by other cells that are known as *antigen-presenting cells* (see fig. 15.15). Secretion of chemokines (chemical attractants) by these cells increases the chances that the appropriate lymphocyte will encounter its specific antigen.

Local Inflammation

Aspects of the innate and adaptive immune responses and their interactions are well illustrated by the events that occur when bacteria enter a break in the skin and produce a **local inflammation** (table 15.5). The inflammatory reaction is initiated by the nonspecific mechanisms of phagocytosis and complement activation. (Complement proteins are activated during humoral immunity by B lymphocytes, as described in a later section.) Activated complement further increases this nonspecific response by attracting new phagocytes to the area and by stimulating their activity.

After some time, B lymphocytes are stimulated to produce antibodies against specific antigens that are part of the invading bacteria. Attachment of these antibodies to antigens in the bacteria greatly amplifies the previously nonspecific response. This occurs because of greater activation of complement, which directly

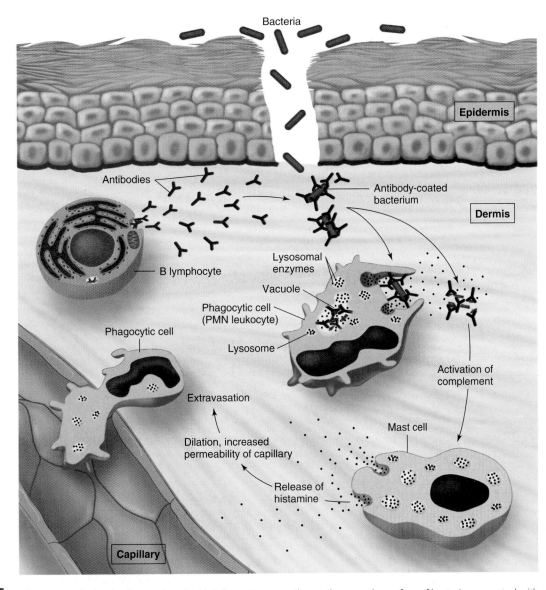

Figure 15.5 **The events of a local inflammation.** In this inflammatory reaction, antigens on the surface of bacteria are coated with antibodies and ingested by phagocytic cells. Symptoms of inflammation are produced by the release of lysosomal enzymes and by the secretion of histamine and other chemicals from tissue mast cells.

destroys the bacteria and which also—together with the antibodies themselves—promotes the phagocytic activity of neutrophils, macrophages, and monocytes (fig. 15.5). The ability of antibodies to promote phagocytosis is called *opsonization.*

Leukocytes within vessels in the inflamed area stick to the endothelial cells of the vessels through interactions between *adhesion molecules* on the two surfaces. The leukocytes can then roll along the wall of the vessel toward particular chemicals. As mentioned earlier, this movement, called *chemotaxis,* is produced by molecules called chemokines. Complement proteins and bacterial products may serve as chemokines, drawing the leukocytes toward the site of infection.

The leukocytes squeeze between adjacent endothelial cells (the process of extravasation, discussed earlier) and enter the subendothelial connective tissue. There, particular molecules on the leukocyte membrane interact with surrounding molecules that guide the leukocytes to the infection. The first to arrive are

the neutrophils, followed by monocytes (which can change into macrophages) and T lymphocytes (fig. 15.6). Most of the phagocytic leukocytes (neutrophils and monocytes) die in the course of the infection, but lymphocytes can travel through the lymphatic system and re-enter the circulation.

The adherence of monocytes to extracellular matrix proteins (chapter 6) promotes their conversion into macrophages. Macrophages ingest microorganisms and fragments of the extracellular matrix by phagocytosis. The macrophages also produce nitric oxide, which aids the destruction of bacteria. However, as inflammation progresses, the release of lysosomal enzymes from macrophages into the extracellular matrix causes the destruction of leukocytes and other tissue cells.

Mast cells are found in most tissues, but are especially concentrated in the skin, bronchioles (airways in the lungs), and intestinal mucosa. They are identified by their content of *heparin,* a molecule of clinical importance because of its

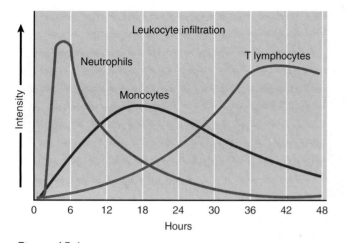

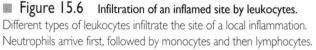

Figure 15.6 Infiltration of an inflamed site by leukocytes. Different types of leukocytes infiltrate the site of a local inflammation. Neutrophils arrive first, followed by monocytes and then lymphocytes.

anticoagulant ability (chapter 13). However, mast cells produce a variety of other molecules that play important roles in inflammation (and in allergy, discussed in a later section).

Mast cells release **histamine** which is stored in intracellular granules and secreted during inflammation and allergy. Histamine binds to its H_1 histamine receptors in the smooth muscle of bronchioles to stimulate bronchiolar constriction (as in asthma), but produces relaxation of the smooth muscles in blood vessels (causing vasodilation). Histamine also promotes increased capillary permeability, bringing more leukocytes to the infected area.

With a time delay, mast cells release inflammatory prostaglandins and leukotrienes (chapter 11; see fig. 11.34), as well as a variety of cytokines that promote inflammation. In addition, mast cells secrete *tumor necrosis factor$_\alpha$ (TNF$_a$),* which acts as a chemokine to recruit neutrophils to the infected site.

These effects produce the characteristic symptoms of a local inflammation: *redness* and *warmth* (due to histamine-stimulated vasodilation); *swelling* (edema) and *pus* (the accumulation of dead leukocytes); and *pain.* If the infection continues, the release of endogenous pyrogen from leukocytes and macrophages may also produce a fever, as previously discussed.

Test Yourself Before You Continue

1. List the phagocytic cells found in blood and lymph, and indicate which organs contain fixed phagocytes.
2. Describe the actions of interferons.
3. Distinguish between innate and adaptive immunity, and describe the properties of antigens.
4. Distinguish between B and T lymphocytes in terms of their origins and immune functions.
5. Identify the primary and secondary lymphoid organs and describe their functions.
6. Describe the events that occur during a local inflammation.

Functions of B Lymphocytes

B lymphocytes secrete antibodies that can bind to antigens in a specific fashion. This bonding stimulates a cascade of reactions whereby a system of plasma proteins called complement is activated. Some of the activated complement proteins kill the cells containing the antigen; others promote phagocytosis, resulting in a more effective defense against pathogens.

Exposure of a B lymphocyte to the appropriate antigen results in cell growth followed by many cell divisions. Some of the progeny become **memory cells;** these are visually indistinguishable from the original cell and are important in active immunity. Others are transformed into **plasma cells** (fig. 15.7). Plasma cells are protein factories that produce about 2,000 antibody proteins per second.

The antibodies that are produced by plasma cells when B lymphocytes are exposed to a particular antigen react specifically with that antigen. Such antigens may be isolated molecules, as illustrated in figure 15.7, or they may be molecules at the surface of an invading foreign cell (fig. 15.8). The specific bonding of antibodies to antigens serves to identify the enemy and to activate defense mechanisms that lead to the invader's destruction.

Antibodies

Antibody proteins are also known as **immunoglobulins.** They are found in the gamma globulin class of plasma proteins, as identified by a technique called *electrophoresis* in which different types of plasma proteins are separated by their movement in an electric field (fig. 15.9). The five distinct bands of proteins that appear are albumin, alpha-1 globulin, alpha-2 globulin, beta globulin, and gamma globulin.

The gamma globulin band is wide and diffuse because it represents a heterogeneous class of molecules. Since antibodies are specific in their actions, it follows that different types of antibodies should have different structures. An antibody against smallpox, for example, does not confer immunity to poliomyelitis and, therefore, must have a slightly different structure than an antibody against polio. Despite these differences, antibodies are structurally related and form only a few classes.

There are five immunoglobulin (abbreviated Ig) subclasses: *IgG, IgA, IgM, IgD,* and *IgE.* Most of the antibodies in serum are in the IgG subclass, whereas most of the antibodies in external secretions (saliva and milk) are IgA (table 15.6). Antibodies in the IgE subclass are involved in certain allergic reactions.

Antibody Structure

All antibody molecules consist of four interconnected polypeptide chains. Two long, heavy chains (the *H chains*) are joined to two shorter, lighter *L chains.* Research has shown that these four

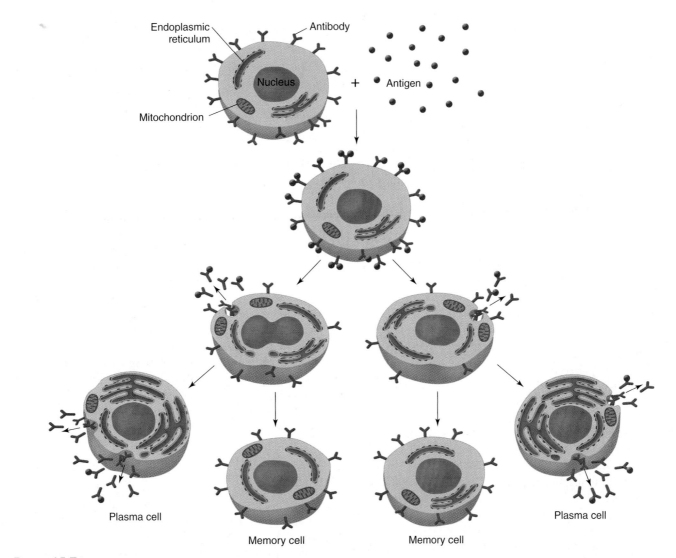

Figure 15.7 **B lymphocytes are stimulated to become plasma cells and memory cells.** B lymphocytes have antibodies on their surface that function as receptors for specific antigens. The interaction of antigens and antibodies on the surface stimulates cell division and the maturation of the B cell progeny into memory cells and plasma cells. Plasma cells produce and secrete large amounts of the antibody. (Note the extensive rough endoplasmic reticulum in these cells.)

Figure 15.8 **A pathogen, such as a bacterium, has many different antigens on its surface.** Each of these antigens interacts with a specific B-cell receptor protein, thereby activating those B cells that can produce antibodies against those specific antigens.

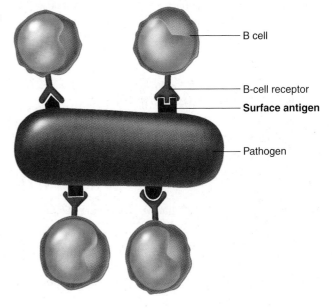

chains are arranged in the form of a Y. The stalk of the Y has been called the "crystallizable fragment" (abbreviated F_c), whereas the top of the Y is the "antigen-binding fragment" (F_{ab}). This structure is shown in figure 15.10.

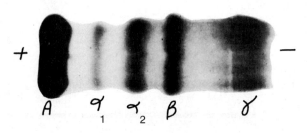

Figure 15.9 **The separation of serum protein by electrophoresis.** This technique separates different groups of proteins on the basis of their electric charges and sizes. (A = albumin; α_1 = alpha-1 globulin; α_2 = alpha-2 globulin; β = beta globulin; γ = gamma globulin.)

Table 15.6   The Immunoglobulins

Immunoglobulin	Functions
IgG	Main form of antibodies in circulation: production increased after immunization; secreted during secondary response
IgA	Main antibody type in external secretions, such as saliva and mother's milk
IgE	Responsible for allergic symptoms in immediate hypersensitivity reactions
IgM	Function as antigen receptors on lymphocyte surface prior to immunization; secreted during primary response
IgD	Function as antigen receptors on lymphocyte surface prior to immunization; other functions unknown

The amino acid sequences of some antibodies have been determined through the analysis of antibodies sampled from people with multiple myelomas. These lymphocyte tumors arise from the division of a single B lymphocyte, forming a population of genetically identical cells (a clone) that secretes identical antibodies. Clones and the antibodies they secrete are different, however, from one patient to another. Analyses of these antibodies have shown that the F_c regions of different antibodies are the same (are constant), whereas the F_{ab} regions are variable. Variability of the antigen-binding regions is required for the specificity of antibodies for antigens. Thus, it is the F_{ab} region of an antibody that provides a specific site for bonding with a particular antigen (fig. 15.11).

B lymphocytes have antibodies on their plasma membrane that serve as **receptors** for antigens. Combination of antigens with these antibody receptors stimulates the B cell to divide and produce more of these antibodies, which are secreted. Exposure to a given antigen thus results in increased amounts of the specific type of antibody that can attack that antigen. This provides active immunity, as described in the next major section.

Diversity of Antibodies

It is estimated that there are about 100 million trillion (10^{20}) antibody molecules in each individual, representing a few million different specificities for different antigens. Considering that antibodies that bind to particular antigens can cross-react with closely related antigens to some extent, this tremendous antibody diversity usually ensures that there will be some antibodies that can combine with almost any antigen a person might encounter. These observations evoke a question that has long fascinated scientists: How can a few million different antibodies be

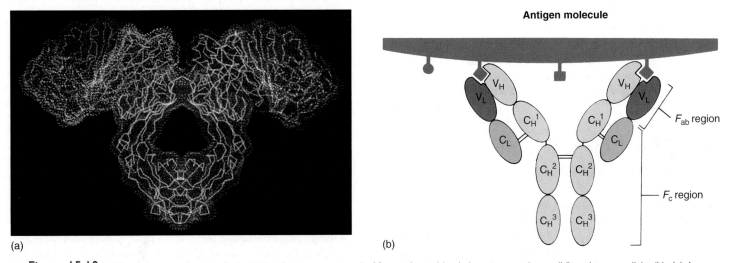

(a)   (b)

Figure 15.10 **The structure of antibodies.** Antibodies are composed of four polypeptide chains—two are heavy (H) and two are light (L). (a) A computer-generated model of antibody structure. (b) A simplified diagram showing the constant and variable regions. (The variable regions are abbreviated V, and the constant regions are abbreviated C.) Antigens combine with the variable regions. Each antibody molecule is divided into an F_{ab} (antigen-binding) fragment and an F_c (crystallizable) fragment.

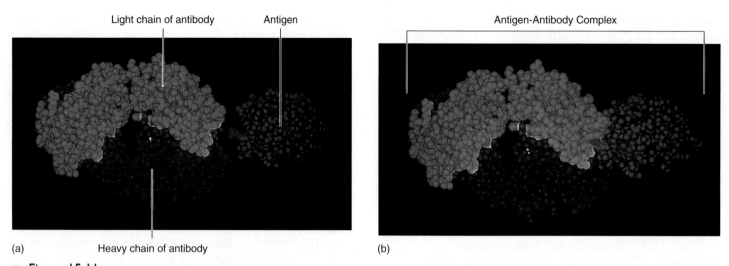

■ **Figure 15.11**   **The antigen-binding site of an antibody.** The structure of the F_{ab} portion of an antibody molecule and the antigen with which it combines as determined by X-ray diffraction. *(a)* The heavy and light chains of the antibody are shown in blue and yellow, respectively, and the antigen is shown in green. Note the complementary shape at the region where the two join together in *(b)*.

Reprinted with permission from A. G. Amit, "Three Dimensional Structure of an Antigen-Antibody Complex at 2.8 A Resolution" in *Science*, vol. 233, August 15, 1986. Copyright © 1986 American Association for the Advancement of Science.

produced? A person cannot possibly inherit a correspondingly large number of genes devoted to antibody production.

Two mechanisms have been proposed to explain antibody diversity. First, since different combinations of heavy and light chains can produce different antibody specificities, a person does not have to inherit a million different genes to code for a million different antibodies. If a few hundred genes code for different H chains and a few hundred code for different L chains, different combinations of these polypeptide chains could produce millions of different antibodies. The number of possible combinations is made even greater by the fact that different segments of DNA code for different segments of the heavy and light chains. Three segments in the antigen-combining region of a heavy chain and two in a light chain are coded by different segments of DNA and can be combined in different ways to make an antibody molecule.

Second, the diversity of antibodies could increase during development if, when some lymphocytes divided, the progeny received antibody genes that had been slightly altered by mutations. Such mutations are called *somatic mutations* because they occur in body cells rather than in sperm cells or ova. Antibody diversity would thus increase with age as the lymphocyte population increased.

The Complement System

The combination of antibodies with antigens does not itself cause destruction of the antigens or the pathogenic organisms that contain these antigens. Antibodies, rather, serve to identify the targets for immunological attack and to activate nonspecific immune processes that destroy the invader. Bacteria that are "buttered" with antibodies, for example, are better targets for phagocytosis by neutrophils and macrophages. The ability of antibodies to stimulate phagocytosis is termed **opsonization.** Immune destruction of bacteria is also promoted by antibody-induced activation of a system of serum proteins known as *complement.*

In the early part of the twentieth century, it was learned that rabbit antibodies that bind to the red blood cell antigens of sheep could not lyse (destroy) these cells unless certain protein components of serum were present. These proteins, called **complement,** constitute a nonspecific defense system that is activated by the bonding of antibodies to antigens, and by this means is directed against specific invaders that have been identified by antibodies.

The complement proteins are designated C1 through C9. These proteins are present in an inactive state within plasma and other body fluids and become activated by the attachment of antibodies to antigens. In terms of their functions, the complement proteins can be subdivided into three components: (1) recognition (C1); (2) activation (C4, C2, and C3, in that order); and (3) attack (C5 through C9). The attack phase consists of **complement fixation,** in which complement proteins attach to the cell membrane and destroy the victim cell.

There are two pathways of complement activation. The **classic pathway** is initiated by the binding of antibodies of the IgG and IgM subclasses to antigens on the invading cell's membrane. This is more rapid and efficient than the **alternative pathway,** which is initiated by the unique polysaccharides that coat bacterial cells.

In the classic pathway, IgG and IgM antibodies activate C1, which catalyzes the hydrolysis of C4 into two fragments, $C4_a$ and $C4_b$ (fig. 15.12). The $C4_b$ fragment binds to the cell membrane (is "fixed") and becomes an active enzyme. Then, through an intermediate step involving the splitting of C2, C3 is cleaved into $C3_a$ and $C3_b$. Acting through a different sequence of events, the alternative pathway of complement activation also results in the conversion of C3 into $C3_a$ and $C3_b$, so that the two pathways converge at this point.

The $C3_b$ converts C5 into $C5_a$ and $C5_b$. The $C3_a$ and $C5_a$ stimulate mast cells to release histamine; $C5_a$ additionally serves as a chemokine to attract neutrophils and monocytes to the site

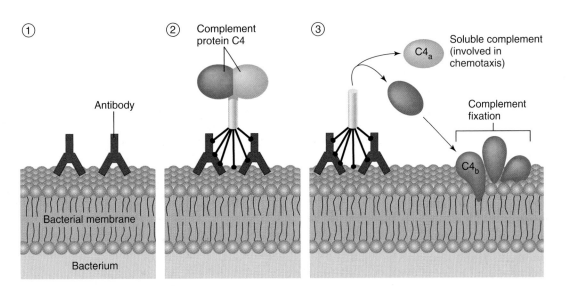

■ **Figure 15.12** **The fixation of complement proteins.** The formation of an antibody-antigen complex causes complement protein C4 to be split into two subunits—$C4_a$ and $C4_b$. The $C4_b$ subunit attaches (is fixed) to the membrane of the cell to be destroyed (such as a bacterium). This event triggers the activation of other complement proteins, some of which attach to the $C4_b$ on the membrane surface.

of infection. Meanwhile, C5 through C9 are inserted into the bacterial cell membrane to form a **membrane attack complex** (fig. 15.13). The attack complex is a large pore that can kill the bacterial cell through the osmotic influx of water. Note that the complement proteins, not the antibodies directly, kill the cell; antibodies serve only as activators of this process in the classic pathway.

Complement fragments that are liberated into the surrounding fluid rather than becoming fixed have a number of effects. These effects include (1) *chemotaxis*—the liberated complement fragments attract phagocytic cells to the site of complement activation; (2) *opsonization*—phagocytic cells have receptors for $C3_b$, so that this fragment may form bridges between the phagocyte and the victim cell, thus facilitating phagocytosis; and (3) *stimulation of the release of histamine* from mast cells and basophils by fragments $C3_a$ and $C5_a$. As a result of histamine release, blood flow to the infected area is increased because of vasodilation and increased capillary permeability. This helps to bring in more phagocytic cells to combat the infection, but the increased capillary permeability can also result in edema through leakage of plasma proteins into the surrounding tissue fluid.

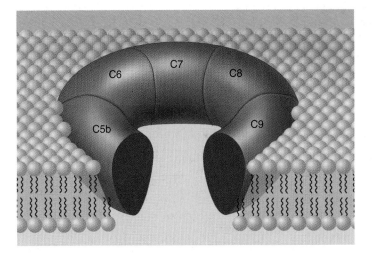

■ **Figure 15.13** **The membrane attack complex.** Fixed complement proteins C5 through C9 assemble in the plasma membrane of the victim cell as a membrane attack complex. This complex forms a large pore that punctures the membrane and thereby promotes the destruction of the cell.

Test Yourself Before You Continue

1. Illustrate the structure of an antibody molecule. Label the constant and variable regions, the F_c and F_{ab} parts, and the heavy and light chains.
2. Define *opsonization* and identify two types of molecules that promote this process.
3. Describe complement fixation and explain the roles of complement fragments that do not become fixed.

Functions of T Lymphocytes

Each subpopulation of T lymphocytes has specific immune functions. Killer T cells effect cell-mediated destruction of specific victim cells, and helper and suppressor T cells play supporting roles. T cells are activated only by antigens presented to them on the surface of particular antigen-presenting cells. Activated helper T cells produce lymphokines that stimulate other cells of the immune system.

The thymus processes lymphocytes in such a way that their functions become quite distinct from those of B cells. Lymphocytes residing in the thymus or originating from the thymus, or those derived from cells that came from the thymus, are all T lymphocytes. These cells can be distinguished from B cells by specialized techniques. Unlike B cells, the T lymphocytes provide specific immune protection without secreting antibodies. This is accomplished in different ways by the three subpopulations of T lymphocytes.

Killer, Helper, and Suppressor T Lymphocytes

The **killer,** or **cytotoxic, T lymphocytes** can be identified in the laboratory by a surface molecule called *CD8*. Their function is to destroy body cells that harbor foreign molecules. These are usually molecules from an invading microorganism, but they can also be molecules produced by the cell's genome because of a malignant transformation, or they may simply be body molecules that had never been presented before to the immune system.

In contrast to the action of B lymphocytes, which kill at a distance through humoral immunity (the secretion of antibodies), killer, or cytotoxic, T lymphocytes kill their victim cells by *cell-mediated destruction.* This means that they must be in actual physical contact with the victim cells. When this occurs, the killer cells secrete molecules called **perforins** and enzymes called **granzymes.** The perforins enter the plasma membrane of the victim cell and polymerize to form a very large pore. This is similar to the pore formed by the membrane attack complex of complement proteins, and results in the osmotic destruction of the victim cell. The granzymes enter the victim cell and, through the activation of caspases (enzymes involved in apoptosis—see chapter 3), cause the destruction of the victim cell's DNA.

The killer T lymphocytes defend against viral and fungal infections and are also responsible for transplant rejection reactions and for immunological surveillance against cancer. Although most bacterial infections are fought by B lymphocytes, some are the targets of cell-mediated attack by killer T lymphocytes. This is the case with the tubercle bacilli that cause tuberculosis. Injections of some of these bacteria under the skin produce inflammation after a latent period of 48 to 72 hours. This *delayed hypersensitivity reaction* is cell-mediated rather than humoral, as shown by the fact that it can be induced in an unexposed guinea pig by an infusion of lymphocytes, but not of serum, from an exposed animal.

The **helper T lymphocytes** (identified in the laboratory by the surface molecule *CD4*), and **suppressor T lymphocytes** indirectly participate in the specific immune response by regulating the responses of the B cells (fig. 15.14) and the killer T cells. The activity of B cells and killer T cells is increased by helper T lymphocytes and decreased by suppressor T lymphocytes. The amount of antibodies secreted in response to antigens is thus affected by the relative numbers of helper to suppressor T cells that develop in response to a given antigen.

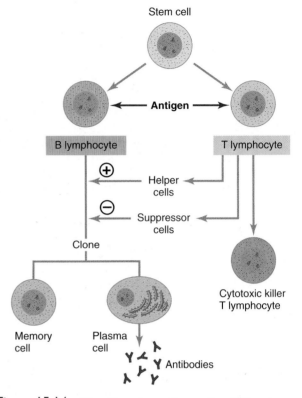

Figure 15.14   The effect of an antigen on B and T lymphocytes. A given antigen can stimulate the production of both B and T lymphocyte clones. The ability to produce B lymphocyte clones, however, is also influenced by the relative effects of helper and suppressor T lymphocytes.

Acquired immune deficiency syndrome (AIDS) has killed approximately 22 million people worldwide. Today, more Americans have died of AIDS than were killed in World Wars I and II combined. About 36 million people worldwide are currently infected, and since AIDS has been shown to have a latency period of approximately 8 years, most will display symptoms of the disease in the near future. AIDS is caused by the **human immunodeficiency virus (HIV)** (see fig. 15.3), which specifically destroys the helper T lymphocytes. This results in decreased immunological function and greater susceptibility to opportunistic infections, including *Pneumocystis carinii pneumonia.* Many people with AIDS also develop a previously rare form of cancer known as *Kaposi's sarcoma.*

Current treatment for AIDS includes the use of drugs that inhibit reverse transcriptase, the enzyme used by the virus to replicate its RNA (see fig. 15.3). Recently, two different reverse transcriptase inhibitors have been combined with a protease inhibitor (protease enzymes are needed to cut viral protein into segments for assembly of the viral coat) to produce a "cocktail" that has proved to be an effective treatment. New drugs are also under development that inhibit the fusion of HIV with its victim cells by targeting the part of the glycoprotein "spokes" that protrude from the HIV particles (see fig. 15.3), which serve to anchor the virus to the plasma membrane of its victim. Hopefully, these and other new drugs will provide better treatment until a safe and effective vaccine might be developed.

Lymphokines

The T lymphocytes, as well as some other cells such as macrophages, secrete a number of polypeptides that serve in an autocrine fashion (chapter 11) to regulate many aspects of the immune system. These products are generally called **cytokines;** the term **lymphokine** is often used to refer to the cytokines of lymphocytes. When a cytokine is first discovered, it is named according to its biological activity (e.g., *B cell-stimulating factor*). Since each cytokine has many different actions (table 15.7), however, such names can be misleading. Scientists have thus agreed to use the name *interleukin,* followed by a number, to indicate a cytokine once its amino acid sequence has been determined.

Interleukin-1, for example, is secreted by macrophages and other cells and can activate the T cell system. B cell-stimulating factor, now called *interleukin-4,* is secreted by T lymphocytes and is required for the proliferation and clone development of B cells. *Interleukin-2* is released by helper T lymphocytes and is required for activation of killer T lymphocytes, among other functions. It is now used in the treatment of certain cancers. *Granulocyte colony-stimulating factor (G-CSF)* and *granulocyte-monocyte colony-stimulating factor (GM-CSF)* are lymphokines that promote leukocyte development and are now available for use in medical treatments (discussed in chapter 13).

Current research has demonstrated that there are two subtypes of helper T lymphocytes, designated **T$_H$1** and **T$_H$2.** Helper T lymphocytes of the T$_H$1 subtype produce interleukin-2 and gamma interferon. Because they secrete these lymphokines, T$_H$1 cells activate killer T cells and promote cell-mediated immunity. The lymphokines secreted by the T$_H$1 lymphocytes also stimulate nitric oxide production in macrophages, increasing their activity. The T$_H$2 lymphocytes secrete interleukin-4, interleukin-5, interleukin-10, and other lymphokines that stimulate B lymphocytes to promote humoral immunity. The lymphokines secreted by T$_H$2 cells, particularly interleukin-4, can also activate mast cells and other agents that promote an allergic immune response.

Scientists have discovered that "uncommitted" helper T lymphocytes are changed into the T$_H$1 subtype in response to a cytokine called interleukin-12, which is secreted by macrophages and dendritic cells (discussed shortly) under appropriate conditions. This process could thus provide a switch for determining how much of the immune response to an antigen will be cell-mediated and how much will be humoral.

In response to **endotoxin,** a molecule released by bacteria, and to cytokines such as interleukin-1 and gamma interferon, production of the enzyme nitric oxide synthase is induced within macrophages. As discussed in chapter 14, this enzyme catalyzes the formation of nitric oxide, which in excessive amounts may produce the hypotension of septic shock. A normal amount of nitric oxide, however, is required for macrophages to destroy bacteria and tumor cells.

Table 15.7 Some Cytokines That Regulate the Immune System

Cytokine	Biological Functions
Interleukin-1 (IL-1)	Induces proliferation and activation of T lymphocytes
Interleukin-2 (IL-2)	Induces proliferation of activated T lymphocytes
Interleukin-3 (IL-3)	Stimulates proliferation of bone marrow stem cells and mast cells
Interleukin-4 (IL-4)	Stimulates proliferation of activated B cells; promotes production of IgE antibodies; increase activity of cytotoxic T cells
Interleukin-5 (IL-5)	Induces activation of cytotoxic T cells; promotes eosinophil differentiation and serves as chemokine for eosinophils
Interleukin-6 (IL-6)	Stimulates proliferation and activation of T and B lymphocytes
Granulocyte/monocyte-macrophage colony-stimulating factor (GM-CSF)	Stimulates proliferation and differentiation of neutrophils, eosinophils, monocytes, and macrophages

T Cell Receptor Proteins

The antigens recognized by B lymphocytes may be either proteins or carbohydrates, but only protein antigens are recognized by most T lymphocytes. Unlike B cells, T cells do not make antibodies and thus do not have antibodies on their surfaces to serve as receptors for these antigens. The T cells do, however, have a different type of antigen receptor on their membrane surfaces, and these T cell receptors have been identified as molecules closely related to the immunoglobulins. The T cell receptors differ from the antibody receptors on B cells in a very important respect: the T cell receptors *cannot bind to free antigens.* In order for T lymphocytes to respond to foreign antigens, the antigens must be presented to the T cells on the membrane of **antigen-presenting cells.**

The chief antigen-presenting cells are macrophages and stellate-shaped **dendritic cells** (fig. 15.15) Dendritic cells arise from stem cells in the bone marrow, but migrate through the blood and lymph to almost every tissue. They are especially concentrated at potential sites where antigen-bearing microorganisms might enter, such as the skin, intestinal mucosa, and lungs. For example, the basal layer of the epidermis contains **Langerhans cells,** which are immature dendritic cells. These cells engulf protein antigens by pinocytosis, partially digest these proteins into shorter polypeptides, and then move these polypeptides to the cell surface. At the cell surface, the foreign polypeptides are associated with molecules called *histocompatability antigens* (discussed in the next section). This allows the antigen-presenting cells to activate the T lymphocytes (fig. 15.15)

In order to interact with the correct T lymphocytes (those that have specificity for the antigen), however, the dendritic cells must migrate through lymphatic vessels to the secondary lymphoid organs, where they secrete chemokines to attract T lymphocytes. This migration affords the antigen-presenting cells the opportunity for a close encounter with the correct T lymphocytes.

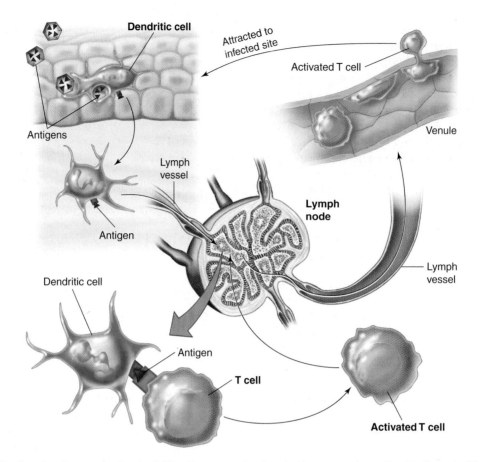

Attracted to infected site

Activated T cell

Antigens

Venule

Lymph vessel

Lymph node

Antigen

Lymph vessel

Dendritic cell

Antigen

T cell

Activated T cell

Figure 15.15 Migration of antigen-presenting dendritic cells to secondary lymphoid organs activates T cells. Once the T cells have been activated by antigens presented to them by the dendritic cells, the activated cells divide to produce a clone. Some of these cells then migrate from the lymphoid organ into the blood. Once in the blood, these activated T cells can home in on the site of the infection because of chemoattractant molecules produced during the inflammation.

Histocompatibility Antigens

Tissue that is transplanted from one person to another contains antigens that are foreign to the host. This is because all tissue cells, with the exception of mature red blood cells, are genetically marked with a characteristic combination of **histocompatibility antigens** on the membrane surface. The greater the variance in these antigens between the donor and the recipient in a transplant, the greater will be the chance of transplant rejection. Prior to organ transplantation, therefore, the "tissue type" of the recipient is matched to that of potential donors. Since the person's white blood cells are used for this purpose, histocompatibility antigens in humans are also called **human leukocyte antigens (HLAs).** They are also called *MHC molecules,* after the name of the genes that code for them.

The histocompatibility antigens are proteins that are coded by a group of genes called the **major histocompatibility complex (MHC),** located on chromosome number 6. These four genes are labeled A, B, C, and D. Each of them can code for only one protein in a given individual, but because each gene has multiple alleles (forms), this protein can be different in different people. Two people, for example, could both have antigen A3, but one might have antigen B17 and the other antigen B21.

The closer two people are related, the closer the match between their histocompatibility antigens.

Interactions Between Antigen-Presenting Cells and T Lymphocytes

The major histocompatibility complex of genes produces two classes of MHC molecules, designated *class 1* and *class 2,* that are found on the cell surface. The class-1 molecules are produced by all cells in the body except red blood cells. Class-2 MHC molecules are produced only by antigen-presenting cells—macrophages, dendritic cells, and B lymphocytes. These cells present their class-2 MHC molecules together with the foreign polypeptide antigen to helper T lymphocytes. This activates the helper T lymphocytes, so that they can promote the B-cell immune response.

The helper T lymphocytes can only be activated by antigens presented to them in association with class-2 MHC molecules. Killer (cytotoxic) T lymphocytes, by contrast, can be activated to destroy a victim cell only if the cell presents antigens to them in association with class-1 MHC molecules. The different require-

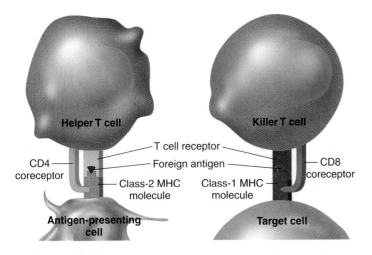

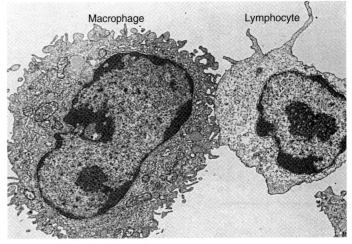

Figure 15.16 Coreceptors on helper and killer T cells. A foreign antigen is presented to T lymphocytes in association with MHC molecules. The CD4, on helper T cells, and CD8 coreceptors on killer T cells, permit each type of T cell to interact only with a specific class of MHC molecule.

(a)

ments for class-1 or class-2 MHC molecules result from the presence of *coreceptors,* which are proteins associated with the T cell receptors. The coreceptor known as *CD8* is associated with the killer T lymphocyte receptor and interacts only with the class-1 MHC molecules; the coreceptor known as *CD4* is associated with the helper T lymphocyte receptor and interacts only with the class-2 MHC molecules. These structures are illustrated in figure 15.16.

T Lymphocyte Response to a Virus

When a foreign particle, such as a virus, infects the body, it is taken up by macrophages (or dendritic cells) via phagocytosis and partially digested. Within the macrophage, the partially digested virus particles provide foreign antigens that are moved to the surface of the cell membrane. At the membrane, these foreign antigens form a complex with the class-2 MHC molecules. This combination of MHC molecules and foreign antigens is required for interaction with the receptors on the surface of helper T cells. The macrophages thus "present" the antigens to the helper T cells and, in this way, stimulate activation of the T cells (fig. 15.17). It should be remembered that T cells are "blind" to free antigens; they can respond only to antigens presented to them by dendritic cells and macrophages in combination with class-2 MHC molecules.

The first phase of macrophage-T cell interaction then occurs: the macrophage is stimulated to secrete the cytokine known as interleukin-1. As previously discussed, interleukin-1 stimulates cell division and proliferation of T lymphocytes. The activated helper T cells, in turn, secrete macrophage colony-stimulating factor and gamma interferon, which promote the activity of macrophages. In addition, interleukin-2 is secreted by the T lymphocytes and stimulates the macrophages to secrete *tumor necrosis factor,* which is particularly effective in killing cancer cells.

Killer T cells can destroy infected cells only if those cells display the foreign antigen together with their class-1 MHC molecules (fig. 15.18). Such interaction of killer T cells with the

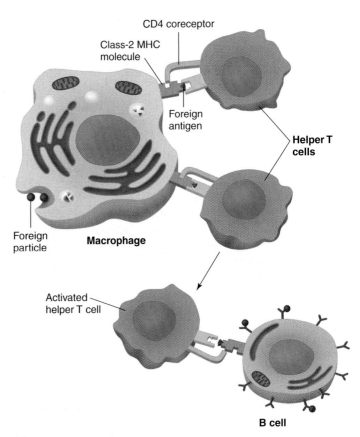

(b)

Figure 15.17 The interaction of antigen-presenting cells, T cells, and B cells. *(a)* An electron micrograph showing contact between a macrophage *(left)* and a lymphocyte *(right).* As illustrated in *(b),* such contact between a macrophage (or other antigen-presenting cell) and a T cell requires that the helper T cell interact with both the foreign antigen and the class-2 MHC molecule on the surface of the macrophage. In this figure, the helper T cell is now activated and able to interact with a B cell.

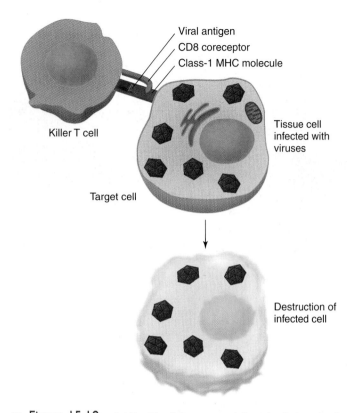

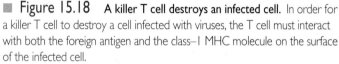

Viral antigen
CD8 coreceptor
Class-1 MHC molecule

Killer T cell

Tissue cell infected with viruses

Target cell

Destruction of infected cell

■ **Figure 15.18** A killer T cell destroys an infected cell. In order for a killer T cell to destroy a cell infected with viruses, the T cell must interact with both the foreign antigen and the class–1 MHC molecule on the surface of the infected cell.

foreign antigen-MHC class-1 complex also stimulates proliferation of those killer T cells. In addition, proliferation of the killer T lymphocytes is stimulated by interleukin-2 secreted by the helper T lymphocytes that were activated by macrophages, as previously described (fig. 15.19).

The network of interactions among the different cell types of the immune system now spreads outward. Helper T cells, activated to an antigen by macrophages or other antigen-presenting cells, can also promote the humoral immune response of B cells. In order to do this, the membrane receptor proteins on the surface of the helper T lymphocytes must interact with molecules on the surface of the B cells. This occurs when the foreign antigen attaches to the immunoglobulin receptors on the B cells, so that the B cells can present this antigen together with its class-2 MHC molecules to the receptors on the helper T cells (fig. 15.20). This interaction stimulates proliferation of the B cells, their conversion to plasma cells, and their secretion of antibodies against the foreign antigens.

Destruction of T Lymphocytes

The activated T lymphocytes must be destroyed after the infection has been cleared. This occurs because T cells produce a surface receptor called **FAS.** Production of FAS increases during the infection and, after a few days, the activated T lymphocytes begin to produce another surface molecule called **FAS ligand.** The binding of FAS to FAS ligand, on the same or on different cells, triggers the apoptosis (cell suicide) of the lymphocytes.

This mechanism also helps to maintain certain parts of the body—such as the inner region of the eye and the tubules of the testis—as *immunologically privileged sites.* These sites harbor molecules that the immune system would mistakenly treat as foreign antigens if the site were not somehow protected. The Sertoli cells of the testicular tubules (chapter 20; see fig. 20.17), for example, protect developing sperm from immune attack through two mechanisms. First, the tight junctions between adjacent Sertoli cells form a barrier that normally prevents exposure of the immune system to the developing sperm. Second, the Sertoli cells produce FAS ligand, which triggers apoptosis of any T lymphocytes that may enter the area.

Some tumor cells, unfortunately, have also been found to produce FAS ligand, which may defend the tumor from immune attack by triggering the apoptosis of lymphocytes. The role of the immune system in the defense against cancer is discussed in a later section.

CLINICAL Glucocorticoids (such as hydrocortisone) secreted by the adrenal cortex can act to inhibit the activity of the immune system and suppress inflammation. This is why **cortisone** and its analogues (such as **Prednisone**) are used clinically to treat inflammatory disorders and to inhibit the immune rejection of transplanted organs. The immunosuppressive effect of these hormones may be due to the fact that they inhibit secretion of the cytokines. It is interesting in this regard that interleukin-1 (IL-1), which can be produced by microglia in the brain, has been shown to stimulate the pituitary-adrenal axis by promoting CRH, ACTH, and glucocorticoid secretion (chapter 11). In a negative feedback fashion, the glucocorticoids then inhibit the immune system and suppress the production of the inflammatory cytokines, including IL-1, IL-2, and TNF$_\alpha$. These and related observations have opened up a new scientific field devoted to the study of interactions between the nervous, endocrine, and immune systems.

Clinical Investigation Clue

Remember that Gary was given cortisone for his rash.
Why would cortisone be useful for treating Gary's rash?

Test Yourself Before You Continue

1. Describe the role of the thymus in cell-mediated immunity.
2. Define the term *cytokines,* state the origin of these molecules, and describe their different functions.
3. Define the term *histocompatibility antigens* and explain the importance of class-1 and class-2 MHC molecules in the function of T cells.
4. Describe the requirements for activation of helper T cells by macrophages. Explain how helper T cells promote the immunological defenses provided by killer T cells and by B cells.

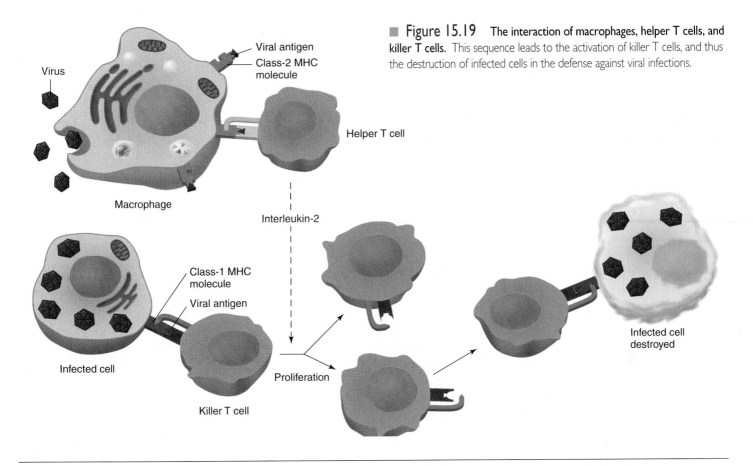

■ **Figure 15.19** The interaction of macrophages, helper T cells, and killer T cells. This sequence leads to the activation of killer T cells, and thus the destruction of infected cells in the defense against viral infections.

Virus

Viral antigen

Class-2 MHC molecule

Macrophage

Helper T cell

Interleukin-2

Class-1 MHC molecule

Viral antigen

Infected cell

Killer T cell

Proliferation

Infected cell destroyed

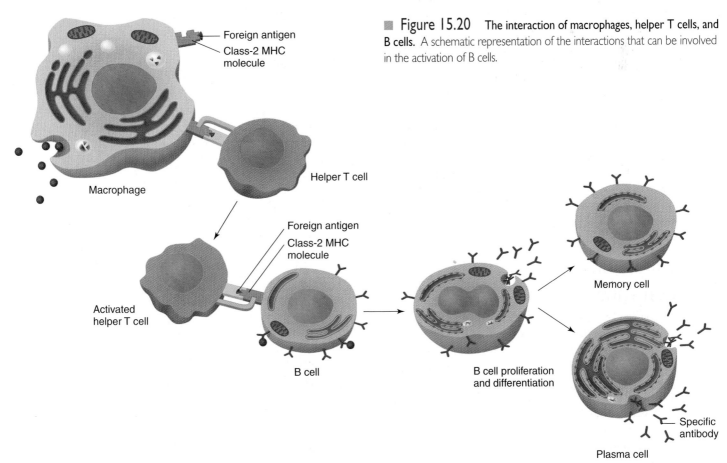

■ **Figure 15.20** The interaction of macrophages, helper T cells, and B cells. A schematic representation of the interactions that can be involved in the activation of B cells.

Foreign antigen

Class-2 MHC molecule

Macrophage

Helper T cell

Activated helper T cell

Foreign antigen

Class-2 MHC molecule

B cell

B cell proliferation and differentiation

Memory cell

Specific antibody

Plasma cell

Active and Passive Immunity

When a person is first exposed to a pathogen, the immune response may be insufficient to combat the disease. In the process, however, the lymphocytes that have specificity for that antigen are stimulated to divide many times and produce a clone. This is active immunity, and it can protect the person from getting the disease upon subsequent exposures.

It first became known in Western Europe in the mid-eighteenth century that the fatal effects of smallpox could be prevented by inducing mild cases of the disease. This was accomplished at that time by rubbing needles into the pustules of people who had mild forms of smallpox and injecting these needles into healthy people. Understandably, this method of immunization did not gain wide acceptance.

Acting on the observation that milkmaids who contracted cowpox—a disease similar to smallpox but less *virulent* (less pathogenic)—were immune to smallpox, an English physician named Edward Jenner inoculated a healthy boy with cowpox. When the boy recovered, Jenner inoculated him with what was considered a deadly amount of smallpox, to which the boy proved to be immune. (This was fortunate for both the boy—who was an orphan—and Jenner; Jenner's fame spread, and as the boy grew into manhood he proudly gave testimonials on Jenner's behalf.) This experiment, performed in 1796, began the first widespread immunization program.

A similar, but more sophisticated, demonstration of the effectiveness of immunizations was performed by Louis Pasteur almost a century later. Pasteur isolated the bacteria that cause anthrax and heated them until their *virulence* (ability to cause disease) was greatly reduced (or *attenuated*), although their *antigenicity* (the nature of their antigens) was not significantly changed (fig. 15.21). He then injected these attenuated bacteria into twenty-five cows, leaving twenty-five unimmunized. Several weeks later, before a gathering of scientists, he injected all fifty cows with the completely active anthrax bacteria. All twenty-five of the unimmunized cows died—all twenty-five of the immunized animals survived.

Active Immunity and the Clonal Selection Theory

When a person is exposed to a particular pathogen for the first time, there is a latent period of 5 to 10 days before measurable amounts of specific antibodies appear in the blood. This sluggish **primary response** may not be sufficient to protect the person against the disease caused by the pathogen. Antibody concentrations in the blood during this primary response reach a plateau in a few days and decline after a few weeks.

A subsequent exposure of that person to the same antigen results in a **secondary response** (fig. 15.22). Compared to the primary response, antibody production during the secondary response

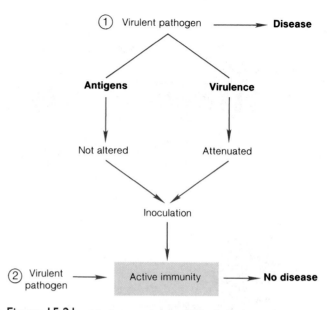

Figure 15.21   Virulence and antigenicity. Active immunity to a pathogen can be gained by exposure to the fully virulent form or by inoculation with a pathogen whose virulence (ability to cause disease) has been attenuated (reduced) without altering its antigenicity (nature of its antigens).

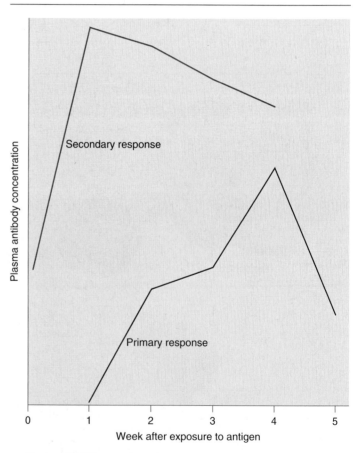

Figure 15.22   The primary and secondary immune responses. A comparison of antibody production in the primary response (upon first exposure to an antigen) to antibody production in the secondary response (upon subsequent exposure to the antigen). The more rapid production of antibodies in the secondary response is believed to be due to the development of lymphocyte clones produced during the primary response.

is much more rapid. Maximum antibody concentrations in the blood are reached in less than 2 hours and are maintained for a longer time than in the primary response. This rapid rise in antibody production is usually sufficient to prevent the disease.

Clonal Selection Theory

The immunization procedures of Jenner and Pasteur were effective because the people who were inoculated produced a secondary rather than a primary response when exposed to the virulent pathogens. The type of protection they were afforded does not depend on accumulations of antibodies in the blood, since secondary responses occur even after antibodies produced by the primary response have disappeared. Immunizations, therefore, seem to produce a type of "learning" in which the ability of the immune system to combat a particular pathogen is improved by prior exposure.

The mechanisms by which secondary responses are produced are not completely understood; the **clonal selection theory,** however, accounts for most of the evidence. According to this theory, B lymphocytes *inherit* the ability to produce particular antibodies (and T lymphocytes inherit the ability to respond to particular antigens). A given B lymphocyte can produce only one type of antibody, with specificity for one antigen. Since this ability is genetically inherited rather than acquired, some lymphocytes can respond to smallpox, for example, and produce antibodies against it even if the person has never been previously exposed to this disease.

The inherited specificity of each lymphocyte is reflected in the antigen receptor proteins on the surface of the lymphocyte's plasma membrane. Exposure to smallpox antigens thus stimulates these specific lymphocytes to divide many times until a large population of genetically identical cells—a *clone*—is produced. Some of these cells become plasma cells that secrete antibodies for the primary response; others become memory cells that can be stimulated to secrete antibodies during the secondary response (fig. 15.23).

Notice that, according to the clonal selection theory (table 15.8), antigens do not induce lymphocytes to make the appropriate antibodies. Rather, antigens select lymphocytes (through interaction with surface receptors) that are already able to make antibodies against that antigen. This is analogous to evolution by natural selection. An environmental agent (in this case, antigens) acts on the genetic diversity already present in a population of organisms (lymphocytes) to cause an increase in number of the individuals selected.

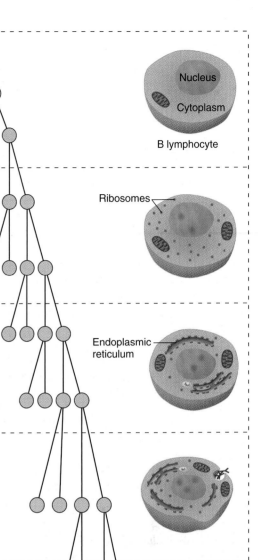

Figure 15.23 The clonal selection theory as applied to B lymphocytes. Most members of the B lymphocyte clone become memory cells, but some antibody-secreting plasma cells.

Table 15.8 Summary of the Clonal Selection Theory (As Applied to B Cells)

Process	Results
Lymphocytes inherit the ability to produce specific antibodies.	Prior to antigen exposure, lymphocytes that can make the appropriate antibodies are already present in the body.
Antigens interact with antibody receptors on the lymphocyte surface.	Antigen-antibody interaction stimulates cell division and the development of lymphocyte clones that contain memory cells and plasma cells that secrete antibodies.
Subsequent exposure to the specific antigens produces a more efficient response.	Exposure of lymphocyte clones to specific antigens results in greater and more rapid production of specific antibodies.

Active Immunity

The development of a secondary response provides **active immunity** against the specific pathogens. The development of active immunity requires prior exposure to the specific antigens, at which time the sluggishness of the primary response may cause the person to develop the disease. Some parents, for example, deliberately expose their children to others who have measles, chickenpox, or mumps so that their children will be immune to these diseases in later life, when the diseases are potentially more serious.

Clinical immunization programs induce primary responses by inoculating people with pathogens whose virulence has been attenuated or destroyed (such as Pasteur's heat-inactivated anthrax bacteria) or by using closely related strains of microorganisms that are antigenically similar but less pathogenic (such as Jenner's cowpox inoculations). The name for these procedures—**vaccinations** (after the Latin word *vacca,* meaning "cow")—reflects the history of this technique. All of these procedures cause the development of lymphocyte clones that can combat the virulent pathogens by producing secondary responses.

The first successful polio vaccine (the Salk vaccine) was composed of viruses that had been inactivated by treatment with formaldehyde. These "killed" viruses were injected into the body, in contrast to the currently used oral (Sabin) vaccine. The oral vaccine contains "living" viruses that have attenuated virulence. These viruses invade the epithelial lining of the intestine and multiply, but do not invade nerve tissue. The immune system can, therefore, become sensitized to polio antigens and produce a secondary response if polio viruses that attack the nervous system are later encountered.

Clinical Investigation Clue

Remember that Gary had recently been given a tetanus vaccine.
How did the vaccine prevent Gary from getting tetanus from his wound?

Immunological Tolerance

The ability to produce antibodies against **non-self (foreign) antigens,** while tolerating (not producing antibodies against) **self-antigens** occurs during the first month or so of postnatal life, when immunological competence is established. If a fetal mouse of one strain receives transplanted antigens from a different strain, therefore, it will not recognize tissue transplanted later in life from the other strain as foreign; consequently, it will not immunologically reject the transplant.

The ability of an individual's immune system to recognize and tolerate self-antigens requires continuous exposure of the immune system to those antigens. If this exposure begins when the immune system is weak—such as in fetal and early postnatal life—tolerance is more complete and long lasting than that produced by exposure beginning later in life. Some self-antigens, however, are normally hidden from the blood, such as thyroglobulin within the thyroid gland and lens protein in the eye. An exposure to these self-antigens results in antibody produc-

tion just as if these proteins were foreign. Antibodies made against self-antigens are called **autoantibodies.** Killer T cells that attack self-antigens are called **autoreactive T cells.**

Although the mechanisms are not well understood, two general theories have been proposed to account for immunological tolerance: **clonal deletion** and **clonal anergy.** According to the *clonal deletion theory,* tolerance to self-antigens is achieved by destruction of the lymphocytes that recognize self-antigens. This occurs primarily during fetal life, when those lymphocytes that have receptors on their surface for self-antigens are recognized and destroyed. There is much evidence for clonal deletion in the thymus, and this mechanism is believed to be largely responsible for T cell tolerance. *Anergy* (which means "without working") occurs when lymphocytes directed against self-antigens are present throughout life but, for complex and poorly understood reasons, do not attack those antigens. Clonal anergy is believed to be largely responsible for tolerance in B cells, and there is some evidence that it may also contribute to tolerance in T cells.

Passive Immunity

The term **passive immunity** refers to the immune protection that can be produced by the transfer of antibodies to a recipient from a human or animal donor. The donor has been actively immunized, as explained by the clonal selection theory. The person who receives these ready-made antibodies is thus passively immunized to the same antigens. Passive immunity also occurs naturally in the transfer of immunity from mother to fetus during pregnancy and from mother to baby during nursing.

The ability to mount a specific immune response—called **immunological competence**—does not develop until about a month after birth. The fetus, therefore, cannot immunologically reject its mother. The immune system of the mother is fully competent but does not usually respond to fetal antigens for reasons that are not completely understood. Some IgG antibodies from the mother do cross the placenta and enter the fetal circulation, however, and these serve to confer passive immunity to the fetus.

The fetus and the newborn baby are thus immune to the same antigens as the mother. However, since the baby did not itself produce the lymphocyte clones needed to form these antibodies, such passive immunity gradually disappears. If the baby is breast-fed it can receive additional antibodies of the IgA subclass in its mother's milk and *colostrum* (the secretion an infant feeds on for the first 2 or 3 days until the onset of true lactation). This provides additional passive immunity until the baby can produce its own antibodies through active immunity (see chapter 20, fig. 20.56).

Passive immunizations are used clinically to protect people who have been exposed to extremely virulent infections or toxins, such as tetanus, hepatitis, rabies, and snake venom. In these cases, the affected person is injected with *antiserum* (serum containing antibodies), also called *antitoxin,* from an animal that has been previously exposed to the pathogen. The animal develops the lymphocyte clones and active immunity, and thus has a high concentration of antibodies in its blood. Since the person who is injected with these antibodies does not develop active immunity, he or she must again be injected with antitoxin upon subsequent exposures.

Active and passive immunity are compared in table 15.9.

Table 15.9 Comparison of Active and Passive Immunity

Characteristic	Active Immunity	Passive Immunity
Injection of person with	Antigens	Antibodies
Source of antibodies	The person inoculated	Natural—the mother; artificial—injection with antibodies
Method	Injection with killed or attenuated pathogens or their toxins	Natural—transfer of antibodies across the placenta; artificial—injection with antibodies
Time to develop resistance	5 to 14 days	Immediately after injection
Duration of resistance	Long (perhaps years)	Short (days to weeks)
When used	Before exposure to pathogen	Before or after exposure to pathogen

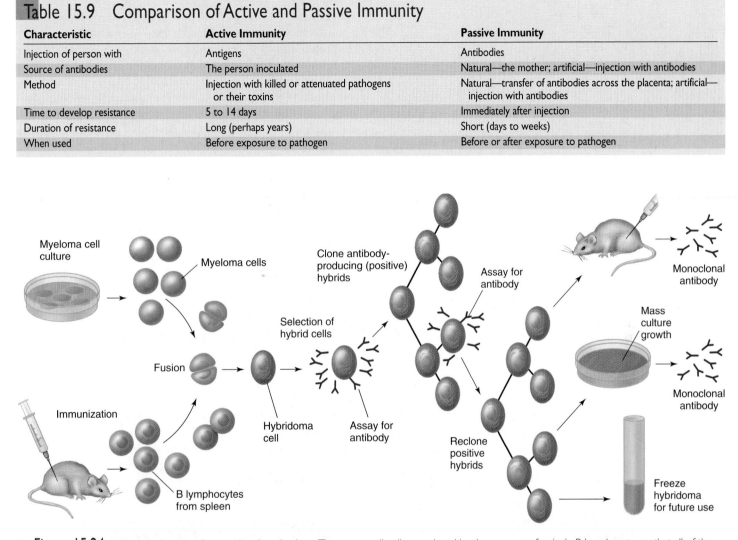

Figure 15.24 **The production of monoclonal antibodies.** These are antibodies produced by the progeny of a single B lymphocyte, so that all of the antibodies are directed against a specific antigen.

Monoclonal Antibodies

In addition to their use in passive immunity, antibodies are also commercially prepared for use in research and clinical laboratory tests. In the past, antibodies were obtained by chemically purifying a specific antigen and then injecting this antigen into animals. Since an antigen typically has many different antigenic determinant sites, however, the antibodies obtained by this method were polyclonal; they had different specificities. This decreased their sensitivity to a particular antigenic site and resulted in some degree of cross-reaction with closely related antigen molecules.

Monoclonal antibodies, by contrast, exhibit specificity for only one antigenic determinant site. In the preparation of monoclonal antibodies, an animal (frequently, a mouse) is injected with an antigen and subsequently killed. B lymphocytes are then obtained from the animal's spleen and placed in thousands of different *in vitro* incubation vessels. These cells soon die, however, unless they are hybridized with cancerous multi-

ple myeloma cells. The fusion of a B lymphocyte with a cancerous cell produces a potentially immortal hybrid that undergoes cell division and produces a clone, called a *hybridoma*. Each hybridoma secretes large amounts of identical monoclonal antibodies. From among the thousands of hybridomas produced in this way, the one that produces the desired antibody is cultured for large-scale production and the rest are discarded (fig. 15.24).

The availability of large quantities of pure monoclonal antibodies has resulted in the development of much more sensitive clinical laboratory tests (for pregnancy, for example). These pure antibodies have also been used to pick one molecule (the specific antigen interferon, for example) out of a solution of many molecules so as to isolate and concentrate it. In the future, monoclonal antibodies against specific tumor antigens may aid the diagnosis of cancer. Even more exciting, drugs that can kill normal as well as cancerous cells might be aimed directly at a tumor by combining these drugs with monoclonal antibodies against specific tumor antigens.

Tumor Immunology

Tumor cells can reveal antigens that stimulate the destruction of the tumor. When cancers develop, this immunological surveillance system—primarily the function of T cells and natural killer cells—has failed to prevent the growth and metastasis of the tumor.

Oncology (the study of tumors) has revealed that tumor biology is similar to and interrelated with the functions of the immune system. Most tumors appear to be clones of single cells that have become transformed in a process similar to the development of lymphocyte clones in response to specific antigens. Lymphocyte clones, however, are under complex inhibitory control systems—such as those exerted by suppressor T lymphocytes and negative feedback by antibodies. The division of tumor cells, by contrast, is not effectively controlled by normal inhibitory mechanisms. Tumor cells are also relatively unspecialized—they *dedifferentiate,* which means that they become similar to the less specialized cells of an embryo.

Tumors are described as *benign* when they are relatively slow growing and limited to a specific location (warts, for example). *Malignant* tumors grow more rapidly and undergo **metastasis,** a term that refers to the dispersion of tumor cells and the resultant seeding of new tumors in different locations. The term **cancer,** as it is generally applied, refers to malignant tumors.

As tumor cells dedifferentiate, they reveal surface antigens that can stimulate the immune destruction of the tumor. Consistent with the concept of dedifferentiation, some of these antigens are proteins produced in embryonic or fetal life and not normally produced postnatally. Since they are absent at the time immunological competence is established, they are treated as foreign and fit subjects for immunological attack when they are produced by cancerous cells. The release of two such antigens into the blood has provided the basis for laboratory diagnosis of some cancers. *Carcinoembryonic antigen tests* are useful in the diagnosis of colon cancer, for example, and tests for *alphafetoprotein* (normally produced only by the fetal liver) help in the diagnosis of liver cancer.

Tumor antigens activate the immune system, initiating an attack primarily by killer T lymphocytes (fig. 15.25) and natural killer cells (described in the next section). The concept of **immunological surveillance** against cancer was introduced in the early 1970s to describe the proposed role of the immune system in fighting cancer. According to this concept, tumor cells frequently appear in the body but are normally recognized and destroyed by the immune system

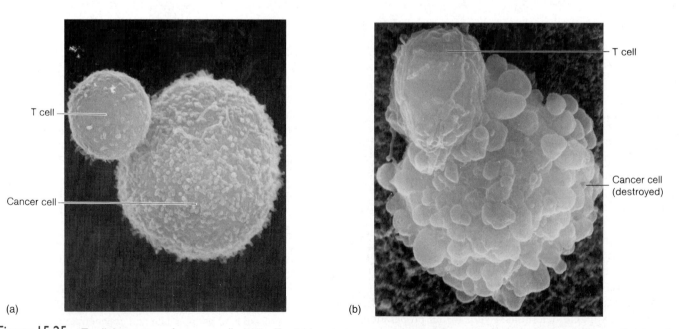

(a)

(b)

■ Figure 15.25 **T cell destruction of a cancer cell.** A killer T cell *(a)* contacts a cancer cell (the larger cell), in a manner that requires specific interaction with antigens on the cancer cell. The killer T cell releases lymphokines, including toxins that cause the death of the cancer cell, as shown in *(b)*.
Scanning electron micrographs © Andrejs Liepens.

before they can cause cancer. There is evidence that immunological surveillance does prevent some types of cancer; this explains why, for example, people with AIDS (who have a depressed immune system) have a high incidence of Kaposi's sarcoma. It is not clear, however, why all types of cancers do not appear with high frequency in AIDS patients and others whose immune systems are suppressed. For these reasons, the generality of the immunological surveillance system concept is currently controversial.

Natural Killer Cells

A particular strain of hairless mice genetically lack a thymus and T lymphocytes, yet these mice do not appear to have an especially high incidence of tumor production. This surprising observation led to the discovery of **natural killer (NK) cells,** which are lymphocytes that are related to, but distinct from, T lymphocytes. Unlike killer T cells, NK cells destroy tumors in a nonspecific fashion and do not require prior exposure for sensitization to the tumor antigens. The NK cells thus provide a first line of cell-mediated defense, which is subsequently backed up by a specific response mediated by killer T cells. These two cell types interact, however; the activity of NK cells is stimulated by interferon, released as one of the lymphokines from T lymphocytes.

Recent evidence suggests that NK cells particularly attack cells that lack class-1 MHC antigens. As previously mentioned, all of a person's normal tissue cells display this antigen. The method of attack is similar to that of the killer (cytotoxic) T lymphocytes: they release perforin proteins and the granzyme enzyme. Perforins insert into the victim plasma membrane, polymerize, and thereby form a large pore in the membrane. Granzyme is taken into the victim cell and indirectly leads to the destruction of its DNA.

Immunotherapy for Cancer

The production of human interferons by genetically engineered bacteria has made large amounts of these substances available for the experimental treatment of cancer. Thus far, interferons have proven to be a useful addition to the treatment of particular forms of cancer, including some types of lymphomas, renal carcinoma, melanoma, Kaposi's sarcoma, and breast cancer. They have not, however, proved to be the "magic bullet" against cancer (a term coined by Paul Ehrlich) as had previously been hoped.

A team of scientists led by Dr. S.A. Rosenberg at the National Cancer Institute has pioneered the use of another lymphokine that is now available through genetic engineering techniques. This is *interleukin-2 (IL-2),* which activates both killer T lymphocytes and B lymphocytes. The investigators removed some of the blood from cancer patients who could not be successfully treated by conventional means and isolated a population of their lymphocytes. They treated these lymphocytes with IL-2 to produce *lymphokine-activated killer (LAK) cells* and then reinfused these cells, together with IL-2 and interferons, into the patients. Depending on the combinations and dosages, they obtained remarkable success (but not a complete cure for all cancers) in many of these patients.

The research team next identified a subpopulation of lymphocytes that had invaded solid tumors in mice. These *tumor-infiltrating lymphocyte (TIL)* cells were allowed to replicate in tissue culture, whereupon they were reintroduced into the mice with excellent results. Recently, the same techniques were used to treat an experimental group of people with metastatic melanoma, a cancer that claims the lives of 6,000 Americans annually. The patients were first given conventional chemotherapy and radiation therapy. They were then treated with their own TIL cells and interleukin-2. Some of the preliminary results of this treatment seem promising, but, like gamma interferon, IL-2 is not a magic bullet against cancer.

Besides interleukin-2 and gamma interferon, other cytokines may be useful in the treatment of cancer and are currently undergoing experimental investigations. Interleukin-12, for example, seems promising because it is needed for the changing of uncommitted helper T lymphocytes into the T_H1 subtype that bolsters cell-mediated immunity. Scientists are also attempting to identify specific antigens that may be uniquely expressed in cancer cells in an effort to help the immune system target cancer cells for destruction.

Effects of Aging and Stress

Susceptibility to cancer varies greatly. The Epstein-Barr virus that causes Burkitt's lymphoma in some individuals (mainly in Africa), for example, can also be found in healthy people throughout the world. Most often the virus is harmless; in some cases, it causes mononucleosis (involving a limited proliferation of white blood cells). Only rarely does this virus cause the uncontrolled proliferation of leukocytes characteristic of Burkitt's lymphoma. The reasons for these differences in response to the Epstein-Barr virus, and indeed for the differing susceptibilities of people to other forms of cancer, are not well understood.

It is known that cancer risk increases with age. According to one theory, this is due to the fact that aging lymphocytes gradually accumulate genetic errors that decrease their effectiveness. The functions of the thymus also decline with age in parallel with a decrease in cell-mediated immune competence. Both of these changes, and perhaps others not yet discovered, could increase susceptibility to cancer.

Numerous experiments have demonstrated that tumors grow faster in experimental animals subjected to stress than they do in unstressed control animals. This is generally attributed to the fact that stressed animals, including humans, exhibit increased secretion of corticosteroid hormones that act to suppress the immune system (which is why cortisone is given to people who receive organ transplants and to people with chronic inflammatory diseases). Some recent experiments, however, suggest that the stress-induced suppression of the immune system may also be due to other factors that do not involve the adrenal cortex. Future advances in cancer therapy may incorporate methods of strengthening the immune system into protocols aimed at directly destroying tumors.

Diseases Caused by the Immune System

Immune mechanisms that normally protect the body are very complex and subject to errors that can result in diseases. Autoimmune diseases and allergies are two categories of disease that are not caused by an invading pathogen, but rather by a derangement in the normal functions of the immune system.

The ability of the normal immune system to tolerate self-antigens while it identifies and attacks foreign antigens provides a specific defense against invading pathogens. In every individual, however, this system of defense against invaders at times commits domestic offenses. This can result in diseases that range in severity from the sniffles to sudden death.

Diseases caused by the immune system can be grouped into three interrelated categories: (1) *autoimmune diseases,* (2) *immune complex diseases,* and (3) *allergy,* or *hypersensitivity.* It is important to remember that these diseases are not caused by foreign pathogens but by abnormal responses of the immune system.

Autoimmunity

Autoimmune diseases are those produced by failure of the immune system to recognize and tolerate self-antigens. This failure results in the activation of autoreactive T cells and the production of autoantibodies by B cells, causing inflammation and organ damage (table 15.10). There are over forty known or suspected autoimmune diseases that affect 5% to 7% of the population. Two-thirds of those affected are women.

There are at least five reasons why self-tolerance may fail:

1. **An antigen that does not normally circulate in the blood may become exposed to the immune system.** Thyroglobulin protein that is normally trapped within the thyroid follicles, for example, can stimulate the production of autoantibodies that cause the destruction of the thyroid; this occurs in *Hashimoto's thyroiditis.* Similarly, autoantibodies developed against lens protein in a damaged eye may cause the destruction of a healthy eye (in *sympathetic ophthalmia*).

2. **A self-antigen that is otherwise tolerated may be altered by combining with a foreign hapten.** The disease *thrombocytopenia* (low platelet count), for example, can be caused by the autoimmune destruction of thrombocytes (platelets). This occurs when drugs such as aspirin, sulfonamide, antihistamines, digoxin, and others combine with platelet proteins to produce new antigens. The symptoms of this disease usually disappear when the person stops taking these drugs.

3. **Antibodies may be produced that are directed against other antibodies.** Such interactions may be necessary for the prevention of autoimmunity, but imbalances may actually cause autoimmune diseases. *Rheumatoid arthritis,* for example, is an autoimmune disease associated with the abnormal production of one group of antibodies (of the IgM type) that attack other antibodies (of the IgG type). This contributes to an inflammation reaction of the joints characteristic of the disease.

Table 15.10   Some Examples of Autoimmune Diseases

Disease	Antigen
Postvaccinal and postinfectious encephalomyelitis	Myelin, cross-reactive
Aspermatogenesis	Sperm
Sympathetic ophthalmia	Uvea
Hashimoto's disease	Thyroglobulin
Graves' disease	Receptor proteins for TSH
Autoimmune hemolytic disease	I, Rh, and others on surface of RBCs
Thrombocytopenic purpura	Hapten-platelet or hapten-absorbed antigen complex
Myasthenia gravis	Acetylcholine receptors
Rheumatic fever	Streptococcal, cross-reactive with heart valves
Glomerulonephritis	Streptococcal, cross-reactive with kidney
Rheumatoid arthritis	IgG
Systemic lupus erythematosus	DNA, nucleoprotein, RNA, etc.
Diabetes mellitus (type I)	Beta cells in pancreatic islets
Multiple sclerosis	Components of myelin sheaths

Source: Modified from James T. Barrett, *Textbook of Immunology,* 5th ed. Copyright © 1988 Mosby-Yearbook. Reprinted by permission.

4. **Antibodies produced against foreign antigens may cross-react with self-antigens.** Autoimmune diseases of this sort can occur, for example, as a result of *Streptococcus* bacterial infections. Antibodies produced in response to antigens in this bacterium may cross-react with self-antigens in the heart and kidneys. The inflammation induced by such autoantibodies can produce heart damage (including the valve defects characteristic of *rheumatic fever*) and damage to the glomerular capillaries in the kidneys (*glomerulonephritis*).

5. **Self-antigens, such as receptor proteins, may be presented to the helper T lymphocytes together with class-2 MHC molecules.** Normally, only antigen-presenting cells (macrophages, dendritic cells, and antigen-activated B cells) produce class-2 MHC molecules, which are associated with foreign antigens and recognized by helper T cells. Perhaps as a result of viral infection, however, cells that do not normally produce class-2 MHC molecules may start to do so and, in this way, present a self-antigen to the helper T cells. In *Graves' disease*, for example, the thyroid cells produce class-2 MHC molecules, and the immune system produces autoantibodies against the TSH receptor proteins in the thyroid cells. These autoantibodies, called *TSAb's* for "thyroid-stimulating antibody," interact with the TSH receptors and overstimulate the thyroid gland. Similarly, in *type I diabetes mellitus*, the beta cells of the pancreatic islets abnormally produce class-2 MHC molecules, resulting in autoimmune destruction of the insulin-producing cells.

Immune Complex Diseases

The term *immune complexes* refers to antigen-antibody combinations that are free rather than attached to bacterial or other cells. The formation of such complexes activates complement proteins and promotes inflammation. This inflammation is normally self-limiting because the immune complexes are removed by phagocytic cells. When large numbers of immune complexes are continuously formed, however, the inflammation may be prolonged. Also, the dispersion of immune complexes to other sites can lead to widespread inflammation and organ damage. The damage produced by this inflammatory response is called immune complex disease.

Immune complex diseases can result from infections by bacteria, parasites, and viruses. In hepatitis B, for example, an immune complex that consists of viral antigens and antibodies can cause widespread inflammation of arteries (*periarteritis*). Arterial damage is not caused by the hepatitis virus itself but by the inflammatory process.

Immune complex disease can also result from the formation of complexes between self-antigens and autoantibodies. This produces systemic autoimmune disease, where the inflammation and tissue damage is not limited to a particular location. Two examples of this are the autoimmune diseases **rheumatoid arthritis** and **systemic lupus erythematosus (SLE).**

In rheumatoid arthritis, immune complexes in the synovial joints induce the activation of complement proteins and the se-

cretion of inflammatory cytokines. This leads to inflammation of the synovial joint and often leads to the progressive destruction of cartilage and bone. Such destruction is mediated by matrix metalloproteinase enzymes (chapter 6), released into the extracellular matrix in response to inflammatory cytokines secreted by helper T lymphocytes. Interestingly, antibodies of the IgM type are produced against the F_c portion of the person's own IgG antibodies! Such IgM autoantibodies are known as *rheumatoid factor,* and are diagnostic of rheumatoid arthritis.

People with SLE produce IgG autoantibodies against their own nuclear constituents, such as chromatin (DNA and protein), small nuclear ribonucleoprotein (snRNP)—described in chapter 3—and others. This can result in the formation of immune complexes throughout the body. In the glomerular capillaries (the filtering units of the kidneys, described in chapter 17), the inflammation provoked by the immune complexes can produce *glomerulonephritis.*

Allergy

The term *allergy,* often used interchangeably with *hypersensitivity,* refers to particular types of abnormal immune responses to antigens, which are called *allergens* in these cases. There are two major forms of allergy: (1) *immediate hypersensitivity,* which is due to an abnormal B lymphocyte response to an allergen that produces symptoms within seconds or minutes, and (2) *delayed hypersensitivity,* which is an abnormal T cell response that produces symptoms between 24 and 72 hours after exposure to an allergen. These two types of hypersensitivity are compared in table 15.11.

Immediate Hypersensitivity

Immediate hypersensitivity can produce allergic rhinitis (chronic runny or stuffy nose); conjunctivitis (red eyes); allergic asthma; atopic dermatitis (urticaria, or hives); and other symptoms. These symptoms result from the immune response to the allergen. In people who are not allergic, the allergen stimulates one type of helper T lymphocyte, the T_H1 cells, to secrete interferon-γ and interleukin-2. In people who are allergic, dendritic cells stimulate the other type of helper T lymphocytes, the T_H2 cells, to secrete other lymphokines, including interleukin-4 and interleukin-13. These, in

Table 15.11 Allergy: Comparison of Immediate and Delayed Hypersensitivity Reactions

Characteristic	Immediate Reaction	Delayed Reaction
Time for onset of symptoms	Within several minutes	Within 1 to 3 days
Lymphocytes involved	B cells	T cells
Immune effector	IgE antibodies	Cell-mediated immunity
Allergies most commonly produced	Hay fever, asthma, and most other allergic conditions	Contact dermatitis (such as to poison ivy and poison oak)
Therapy	Antihistamines and adrenergic drugs	Corticosteroids (such as cortisone)

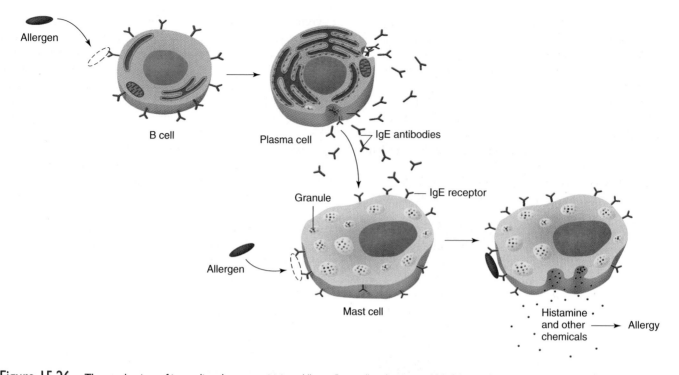

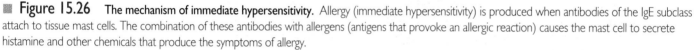

■ Figure 15.26 **The mechanism of immediate hypersensitivity.** Allergy (immediate hypersensitivity) is produced when antibodies of the IgE subclass attach to tissue mast cells. The combination of these antibodies with allergens (antigens that provoke an allergic reaction) causes the mast cell to secrete histamine and other chemicals that produce the symptoms of allergy.

turn, stimulate B lymphocytes and plasma cells to secrete antibodies of the IgE subclass instead of the normal IgG antibodies.

Unlike IgG antibodies, IgE antibodies do not circulate in the blood. Instead they attach to tissue mast cells and basophils, which have membrane receptors for these antibodies. When the person is again exposed to the same allergen, the allergen binds to the antibodies attached to the mast cells and basophils. This stimulates these cells to secrete various chemicals, including **histamine** (fig. 15.26). During this process, leukocytes may also secrete **prostaglandin D** and related molecules called **leukotrienes.** These chemicals produce the symptoms of the allergic reactions. It should be noted that histamine stimulates smooth muscle contraction in the respiratory tract but causes smooth muscle relaxation in the walls of blood vessels. The different effects are due to differences in the histamine receptors of these target tissues.

The symptoms of hay fever (itching, sneezing, tearing, runny nose) are produced largely by histamine and can be treated effectively by antihistamine drugs that block the H_1-histamine receptor. In asthma, the difficulty in breathing is caused by inflammation and smooth muscle constriction in the bronchioles as a result of chemicals released by mast cells and eosinophils. In particular, the bronchoconstriction in asthma is produced by leukotrienes, which are mainly secreted by activated eosinophils. Asthma is treated with epinephrine and more specific β-adrenergic stimulating drugs (chapter 9), which cause bronchodilation, and with corticosteroids, which inhibit inflammation and leukotriene synthesis. Asthma and its treatment are discussed more fully in chapter 16. Regarding food allergies (to milk, eggs, peanuts, soy, wheat, and others), no specific therapy is currently available. People with a food allergy must thus be very diligent about avoiding the particular food.

Clinical Investigation Clues

Remember that Gary's response to the bee sting was greater with the second bee sting than with the first, and was treated with antihistamines.

Why was his reaction greater to the second bee sting than to the first?

Why were antihistamines useful in treating the effects of the bee sting?

Immediate hypersensitivity to a particular antigen is commonly tested for by injecting various antigens under the skin (fig. 15.27). Within a short time a *flare-and-wheal reaction* is produced if the person is allergic to that antigen. This reaction is due to the release of histamine and other chemical mediators: the flare (spreading flush) is due to vasodilation, and the wheal (elevated area) results from local edema.

Allergens that provoke immediate hypersensitivity include various foods, bee stings, and pollen grains. The most common allergy of this type is seasonal hay fever, which may be provoked by ragweed (*Ambrosia*) pollen grains (fig. 15.28*a*). People who have chronic allergic rhinitis and asthma because of an allergy to dust or feathers are usually allergic to a tiny mite (fig. 15.28*b*) that lives in dust and eats the scales of skin that are constantly shed from the body. Actually, most of the antigens from the dust mite are not in its body but rather in its feces—tiny particles that can enter the nasal mucosa, much like pollen grains. There can be more than 100,000 mite feces per gram of house dust!

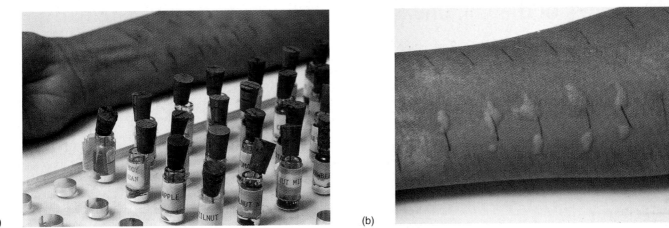

■ Figure 15.27 **A skin test for allergy.** If an allergen (a) is injected into the skin of a sensitive individual, a typical flare-and-wheal response (b) occurs within several minutes.

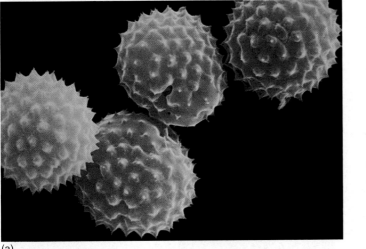

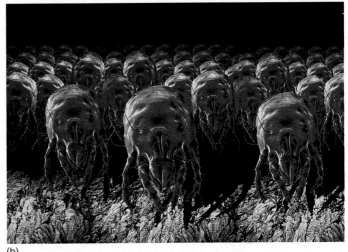

■ Figure 15.28 **Common allergens.** (a) A scanning electron micrograph of ragweed (Ambrosia), which is responsible for hay fever. (b) A scanning electron micrograph of house dust mites (Dermatophagoides farinae). Waste-product particles produced by the dust mite are often responsible for chronic allergic rhinitis and asthma.

Part (a): Reproduced by permission from R. G. Kessel and C. Y. Shih, Scanning Electron Microscopy, Springer-Verlag, 1976.

Delayed Hypersensitivity

In **delayed hypersensitivity,** as the name implies, symptoms take a longer time (hours to days) to develop than in immediate hypersensitivity. This may be because immediate hypersensitivity is mediated by antibodies, whereas delayed hypersensitivity is a cell-mediated T lymphocyte response. Since the symptoms are caused by the secretion of lymphokines rather than by the secretion of histamine, treatment with antihistamines provides little benefit. At present, corticosteroids are the only drugs that can effectively treat delayed hypersensitivity.

One of the best-known examples of delayed hypersensitivity is **contact dermatitis,** caused by poison ivy, poison oak, and poison sumac. The skin tests for tuberculosis—the tine test and the Mantoux test—also rely on delayed hypersensitivity reactions. If a person has been exposed to the tubercle bacillus and consequently has developed T cell clones, skin reactions appear within a few days after the tubercle antigens are rubbed into the skin with small needles (tine test) or are injected under the skin (Mantoux test).

Clinical Investigation Clues

Remember that Gary developed a rash on his abdomen the day following his time crawling through the underbrush in the hills.

What may have caused Gary's rash?

Why was it treated with cortisone rather than antihistamines?

Test Yourself Before You Continue

1. Explain the mechanisms that may be responsible for autoimmune diseases.

2. Distinguish between immediate and delayed hypersensitivity.

3. Describe the sequence of events by which allergens can produce symptoms of runny nose, skin rash, and asthma.

HPer Links of the Immune System with Other Body Systems

Integumentary System

Skeletal System

Muscular System

Nervous System

Endocrine System

Circulatory System

Respiratory System

Urinary System

Digestive System

Reproductive System

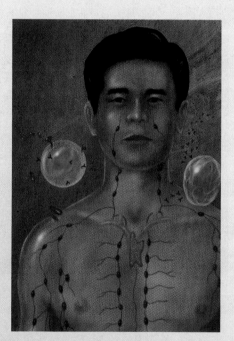

Summary

Defense Mechanisms 446

I. Nonspecific defense mechanisms include barriers to penetration of the body, as well as internal defenses.

    A. Phagocytic cells engulf invading pathogens.

    B. Interferons are polypeptides secreted by cells infected with viruses that help to protect other cells from viral infections.

II. Specific immune responses are directed against antigens.

    A. Antigens are molecules or parts of molecules that are usually large, complex, and foreign.

    B. A given molecule can have a number of antigenic determinant sites that stimulate the production of different antibodies.

III. Specific immunity is a function of lymphocytes.

    A. B lymphocytes secrete antibodies and provide humoral immunity.

    B. T lymphocytes provide cell-mediated immunity.

    C. The thymus and bone marrow are the primary lymphoid organs, which produce lymphocytes that seed the secondary lymphoid organs.

IV. Specific and nonspecific immune mechanisms cooperate in the development of local inflammation.

Functions of B Lymphocytes 453

I. There are five subclasses of antibodies, or immunoglobulins: IgG, IgA, IgM, IgD, and IgE.

    A. These subclasses differ with respect to the polypeptides in the constant region of the heavy chains.

    B. Each type of antibody has two variable regions that combine with specific antigens.

    C. The combination of antibodies with antigens promotes phagocytosis.

II. Antigen-antibody complexes activate a system of proteins called the complement system.

    A. This results in complement fixation, in which complement proteins attach to a cell membrane and promote the destruction of the cell.

    B. Free complement proteins promote opsonization and chemotaxis and stimulate the release of histamine from tissue mast cells.

Functions of T Lymphocytes 457

I. The thymus processes T lymphocytes and secretes hormones that are believed to be required for an effective immune response of T lymphocytes throughout the body.

II. There are three subcategories of T lymphocytes.

    A. Killer T lymphocytes kill victim cells by a mechanism that does not involve antibodies but that does require close contact between the killer T cell and the victim cell.

    B. Killer T lymphocytes are responsible for transplant rejection and for the immunological defense against fungal and viral infections, as well as for defense against some bacterial infections.

    C. Helper T lymphocytes stimulate, and suppressor T lymphocytes suppress, the function of B lymphocytes and killer T lymphocytes.

    D. The T lymphocytes secrete a family of compounds called lymphokines that promote the action of lymphocytes and macrophages.

    E. Receptor proteins on the cell membrane of T lymphocytes must bind to a foreign antigen in combination with a histocompatibility antigen in order for the T cell to become activated.

    F. Histocompatibility antigens, or MHC molecules, are a family of molecules on the membranes of cells that are present in different combinations in different individuals.

III. Antigen-presenting cells, such as macrophages and dendritic cells, partially digest a foreign protein, such as a virus, and present the antigens to the lymphocytes on the surface in combination with class-2 MHC antigens.

    A. Helper T lymphocytes require such interaction with antigen-presenting cells in order to be activated by a foreign antigen; when activated in this way, the helper T cells secrete interleukin-2.

    B. Interleukin-2 stimulates proliferation of killer T lymphocytes that are specific for the foreign antigen.

    C. In order for the killer T lymphocytes to attack a victim cell, the victim cell must present the foreign antigen in combination with a class-1 MHC molecule.

    D. Interleukin-2 also stimulates proliferation of B lymphocytes, and thus promotes the secretion of antibodies in response to the foreign antigen.

Active and Passive Immunity 464

I. A primary response is produced when a person is first exposed to a pathogen. A subsequent exposure results in a secondary response.

    A. During the secondary response, IgM antibodies are produced slowly and the person is likely to get sick.

    B. During the secondary response, IgG antibodies are produced quickly and the person is able to resist the pathogen.

    C. In active immunizations, the person is exposed to pathogens of attenuated virulence that have the same antigenecity as the virulent pathogen.

    D. The secondary response is believed to be due to the development of lymphocyte clones as a result of the antigen-stimulated proliferation of appropriate lymphocytes.

II. Tolerance to self-antigens occurs in prenatal development by destruction of T lymphocytes in the thymus that have specificity for the self-antigens.

    A. This is called clonal deletion.

    B. Clonal energy, or the suppression of lymphocytes, may also occur and may be responsible for tolerance to self-antigens by B lymphocytes.

C. When tolerance mechanisms are ineffective, the immune system may attack self-antigens to cause autoimmune diseases.

III. Passive immunity is provided by the transfer of antibodies from an immune to a nonimmune organism.

    A. Passive immunity occurs naturally in the transfer of antibodies from mother to fetus.

    B. Injections of antiserum provide passive immunity to some pathogenic organisms and toxins.

IV. Monoclonal antibodies are made by hybridomas, which are formed artificially by the fusion of B lymphocytes and multiple myeloma cells.

Tumor Immunology   468

I. Immunological surveillance against cancer is provided mainly by killer T lymphocytes and natural killer cells.

A. Cancerous cells dedifferentiate and may produce fetal antigens. These or other antigens may be presented to lymphocytes in association with abnormally produced class-2 MHC antigens.

B. Natural killer cells are nonspecific, whereas T lymphocytes are directed against specific antigens on the cancer cell surface.

C. Immunological surveillance against cancer is weakened by stress.

Diseases Caused by the Immune System   470

I. Autoimmune diseases may be caused by the production of autoantibodies against self-antigens, or they may result from the development of autoreactive T lymphocytes.

II. Immune complex diseases are those caused by the inflammation that results

when free antigens are bound to antibodies.

III. There are two types of allergic responses: immediate hypersensitivity and delayed hypersensitivity.

    A. Immediate hypersensitivity results when an allergen provokes the production of antibodies in the IgE class. These antibodies attach to tissue mast cells and stimulate the release of chemicals from the mast cells.

    B. Mast cells secrete histamine, leukotrienes, and prostaglandins, which are believed to produce the symptoms of allergy.

    C. Delayed hypersensitivity, as in contact dermatitis, is a cell-mediated response of T lymphocytes.

Review Activities

Test Your Knowledge of Terms and Facts

1. Which of these offers a nonspecific defense against viral infection?

    a. antibodies
    b. leukotrienes
    c. interferon
    d. histamine

Match the cell type with its secretion.

    2. killer T cells   **a.** antibodies
    3. mast cells   **b.** perforins
    4. plasma cells   **c.** lysosomal enzymes
    5. macrophages   **d.** histamine

6. Which of these statements about the F_{ab} portion of antibodies is *true?*

    a. It binds to antigens.
    b. Its amino acid sequences are variable.
    c. It consists of both H and L chains.
    d. All of these are true.

7. Which of these statements about complement proteins $C3_a$ and $C5_a$ is *false?*

    a. They are released during the complement fixation process.
    b. They stimulate chemotaxis of phagocytic cells.

    c. They promote the activity of phagocytic cells.
    d. They produce pores in the victim cell membrane.

8. Mast cell secretion during an immediate hypersensitivity reaction is stimulated when antigens combine with

    a. IgG antibodies.
    b. IgE antibodies.
    c. IgM antibodies.
    d. IgA antibodies.

9. During a secondary immune response,

    a. antibodies are made quickly and in great amounts.
    b. antibody production lasts longer than in a primary response.
    c. antibodies of the IgG class are produced.
    d. lymphocyte clones are believed to develop.
    e. all of these apply.

10. Which of these cell types aids the activation of T lymphocytes by antigens?

    a. macrophages
    b. neutrophils

    c. mast cells
    d. natural killer cells

11. Which of these statements about T lymphocytes is *false?*

    a. Some T cells promote the activity of B cells.
    b. Some T cells suppress the activity of B cells.
    c. Some T cells secrete interferon.
    d. Some T cells produce antibodies.

12. Delayed hypersensitivity is mediated by

    a. T cells.
    b. B cells.
    c. plasma cells.
    d. natural killer cells.

13. Active immunity may be produced by

    a. contracting a disease.
    b. receiving a vaccine.
    c. receiving gamma globulin injections.
    d. both *a* and *b*.
    e. both *b* and *c*.

14. Which of these statements about class-2 MHC molecules is *false?*

   a. They are found on the surface of B lymphocytes.

   b. They are found on the surface of macrophages.

   c. They are required for B cell activation by a foreign antigen.

   d. They are needed for interaction of helper and killer T cells.

   e. They are presented together with foreign antigens by macrophages.

Match the cytokine with its description.

15. interleukin-1

16. interleukin-2

17. interleukin-12

   a. stimulates formation of T_H1 helper T lymphocytes

   b. stimulates ACTH secretion

   c. stimulates proliferation of killer T lymphocytes

   d. stimulates proliferation of B lymphocytes

18. Which of these statements about gamma interferon is *false?*

   a. It is a polypeptide autocrine regulator.

   b. It can be produced in response to viral infections.

   c. It stimulates the immune system to attack infected cells and tumors.

   d. It is produced by almost all cells in the body.

Test Your Understanding of Concepts and Principles

1. Explain how antibodies help to destroy invading bacterial cells.[1]

2. Identify the different types of interferons and describe their origin and actions.

3. Distinguish between the class-1 and class-2 MHC molecules in terms of their locations and functions.

4. Describe the role of macrophages in activating the specific immune response to antigens.

5. Distinguish between the two subtypes of helper T lymphocytes and explain how they may be produced.

6. Describe how plasma cells attack antigens and how they can destroy an invading foreign cell. Compare this mechanism with that by which killer T lymphocytes destroy a target cell.

7. Explain how tolerance to self-antigens may be produced. Also, give two examples of autoimmune diseases and explain their possible causes.

8. Use the clonal selection theory to explain how active immunity is produced by vaccinations.

9. Describe the nature of passive immunity and explain how antitoxins are produced and used.

10. Distinguish between immediate and delayed hypersensitivity. What drugs are used to treat immediate hypersensitivity and how do these drugs work? Why don't these compounds work in treating delayed hypersensitivity?

Test Your Ability to Analyze and Apply Your Knowledge

1. The specific T lymphocyte immune response is usually directed against proteins, whereas nonspecific immune mechanisms are generally directed against foreign carbohydrates in the form of glycoproteins and lipopolysaccharides. How might these differences in target molecules be explained?

2. Lizards are cold-blooded; their body temperature is largely determined by the ambient temperature. Devise an experiment using lizards to test whether an elevated body temperature, as in a fever, can be beneficial to an organism with an infection.

3. Why are antibodies composed of different chains, and why are there several genes that encode the parts of a particular antibody molecule? What would happen if each antibody were coded for by only one gene?

4. As a scientist trying to cure allergy, you are elated to discover a drug that destroys all mast cells. How might this drug help to prevent allergy? What negative side effects might this drug have?

5. The part of the placenta that invades the mother's uterine lining (the endometrium) has recently been found to produce FAS ligand. What might this accomplish, and why might this action be necessary?

Related Websites

Check out the Links Library at www.mhhe.com/fox8 for links to sites containing resources related to the immune system. These links are monitored to ensure current URLs.

---

[1]*Note:* This question is answered in the chapter 15 Study Guide found on the Online Learning Center at www.mhhe.com/fox8.

16 Respiratory Physiology

Objectives

After studying this chapter, you should be able to . . .

1. describe the functions of the respiratory system, distinguish between the conducting and respiratory zone structures, and discuss the significance of the thoracic membranes.

2. explain how the intrapulmonary and intrapleural pressures vary during ventilation and relate these pressure changes to Boyle's law.

3. define the terms *compliance* and *elasticity,* and explain how these lung properties affect ventilation.

4. discuss the significance of surface tension in lung mechanics, explain how the law of Laplace applies to lung function, and describe the role of pulmonary surfactant.

5. explain how inspiration and expiration are accomplished in unforced breathing and describe the accessory respiratory muscles used in forced breathing.

6. define the various lung volumes and capacities that can be measured by spirometry and explain how obstructive diseases can be detected by the FEV_1 test.

7. describe the nature of asthma, bronchitis, emphysema, and pulmonary fibrosis.

8. explain Dalton's law and illustrate how the partial pressure of a gas in a mixture of gases is calculated.

9. explain Henry's law, describe how blood P_{O_2} and P_{CO_2} are measured, and discuss the clinical significance of these measurements.

10. describe the roles of the medulla oblongata, pons, and cerebral cortex in the regulation of breathing.

11. explain why changes in the P_{CO_2} and pH of blood, rather than in its oxygen content, serve as the primary stimuli in the control of breathing.

12. explain how chemoreceptors in the medulla oblongata and the peripheral chemoreceptors in the aortic and carotid bodies respond to changes in P_{CO_2}, pH, and P_{O_2}.

13. describe the Hering-Breuer reflex and explain its significance.

14. describe the different forms of hemoglobin and discuss the significance of these different forms.

15. describe the loading and unloading reactions and explain how the extent of these reactions is influenced by the P_{O_2} and affinity of hemoglobin for oxygen.

16. describe the oxyhemoglobin dissociation curve, discuss the significance of its shape, and demonstrate how the curve is used to derive the percent unloading of oxygen.

17. explain how oxygen transport is influenced by changes in blood pH and temperature, and explain the effect and physiological significance of 2,3-DPG on oxygen transport.

18. list the different forms in which carbon dioxide is carried by the blood, and explain the chloride shift in the tissues and the reverse chloride shift in the lungs.

19. explain how carbon dioxide affects blood pH, and how hypoventilation and hyperventilation affect acid-base balance.

20. describe the hyperpnea of exercise and explain how the anaerobic threshold is affected by endurance training.

21. explain the respiratory adjustments to life at a high altitude.

Refresh Your Memory

Before you begin this chapter, you may want to review these concepts from previous chapters:

Take Advantage of the Technology

Visit the Online Learning Center for these additional study resources.

■ Interactive quizzing
■ Online study guide
■ Current news feeds
■ Crossword puzzles
■ Vocabulary flashcards
■ Labeling activities

www.mhhe.com/fox8

Clinical Investigation

Harry, a taxi cab driver, is found parked in his taxi and in great pain. He has a puncture wound in his right thorax (where he had been stabbed by an assailant), and the paramedics correctly determine that his lung is collapsed. When he is taken to the hospital and blood gas measurements are taken, he is found to have a high arterial P_{CO_2} and a pH of 7.15. He is treated surgically, and during his recovery smokes numerous cigarettes despite his physician's warning. When blood gas measurements are again performed, his arterial P_{CO_2} and pH are found to be normal, but he shows a carboxyhemoglobin saturation of 18%. Pulmonary function tests are ordered, and they reveal that he has a vital capacity slightly below normal and an FEV_1 that is significantly lower than normal.

What condition did Harry have when he was rushed to the hospital? What was the cause of the initial blood gas results, and of his later carboxyhemoglobin levels? Also, what do the pulmonary function tests suggest about Harrry's health?

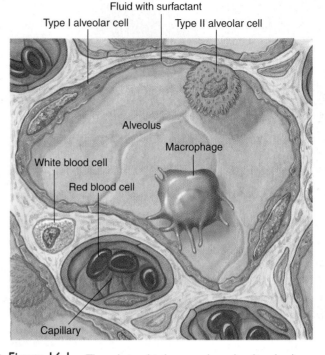

Figure 16.1 The relationship between lung alveoli and pulmonary capillaries. Notice that alveolar walls are quite narrow and lined with type I and type II alveolar cells. Pulmonary macrophages can phagocytose particles that enter the lungs.

The Respiratory System

The respiratory system is divided into a respiratory zone, which is the site of gas exchange between air and blood, and a conducting zone, which conducts the air to the respiratory zone. The exchange of gases between air and blood occurs across the walls of respiratory alveoli. These tiny air sacs, only a single cell layer thick, permit rapid rates of gas diffusion.

The term *respiration* includes three separate but related functions: (1) **ventilation** (breathing); (2) **gas exchange,** which occurs between the air and blood in the lungs and between the blood and other tissues of the body; and (3) **oxygen utilization** by the tissues in the energy-liberating reactions of cell respiration. Ventilation and the exchange of gases (oxygen and carbon dioxide) between the air and blood are collectively called *external respiration.* Gas exchange between the blood and other tissues and oxygen utilization by the tissues are collectively known as *internal respiration.*

Ventilation is the mechanical process that moves air into and out of the lungs. Since the oxygen concentration of air is higher in the lungs than in the blood, oxygen diffuses from air to blood. Carbon dioxide, conversely, moves from the blood to the air within the lungs by diffusing down its concentration gradient. As a result of this gas exchange, the inspired air contains more oxygen and less carbon dioxide than the expired air. More importantly, blood leaving the lungs (in the pulmonary veins) has a higher oxygen and a lower carbon dioxide concentration than the blood delivered to the lungs in the pulmonary arteries. This is because the lungs function to bring the blood into gaseous equilibrium with the air.

Gas exchange between the air and blood occurs entirely by diffusion through lung tissue. This diffusion occurs very rapidly because of the large surface area within the lungs and the very small diffusion distance between blood and air.

Structure of the Respiratory System

Gas exchange in the lungs occurs across an estimated 300 million tiny (0.25 to 0.50 mm in diameter) air sacs, known as **alveoli.** Their enormous number provides a large surface area (60 to 80 square meters, or about 760 square feet) for diffusion of gases. The diffusion rate is further increased by the fact that each alveolus is only one cell-layer thick, so that the total "air-blood barrier" is only two cells across (an alveolar cell and a capillary endothelial cell), or about 2 μm. This is an average distance because there are two types of cells in the alveolar wall, *type I* and *type II,* and the type II alveolar cells are thicker than the type I cells (fig. 16.1). Where the basement membranes of capillary endothelial cells fuse with those of type I alveolar cells, the diffusion distance can be as small as 0.3 μm (fig. 16.2), which is about 1/100th the width of a human hair.

Alveoli are polyhedral in shape and are usually clustered, like the units of a honeycomb. Air within one member of a cluster can enter other members through tiny pores (fig. 16.3). These clusters of alveoli usually occur at the ends of *respiratory bronchioles,* the very thin air tubes that end blindly in alveolar sacs. Individual alveoli also occur as separate outpouchings along the length of respiratory bronchioles. Although the distance between each respiratory bronchiole and its terminal alveoli is only about 0.5 mm, these units together constitute most of the mass of the lungs.

The air passages of the respiratory system are divided into two functional zones. The **respiratory zone** is the region where gas exchange occurs, and it therefore includes the respiratory bronchioles (because they contain separate outpouchings of alveoli)

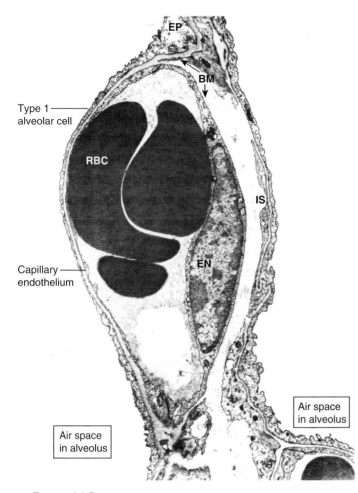

Figure 16.2 An electron micrograph of a capillary within the alveolar wall. Notice the short distance separating the alveolar space on one side (left, in this figure) from the capillary. (EP = epithelial cell of alveolus; RBC = red blood cell; BM = basement membrane; IS = interstitial connective tissue.)

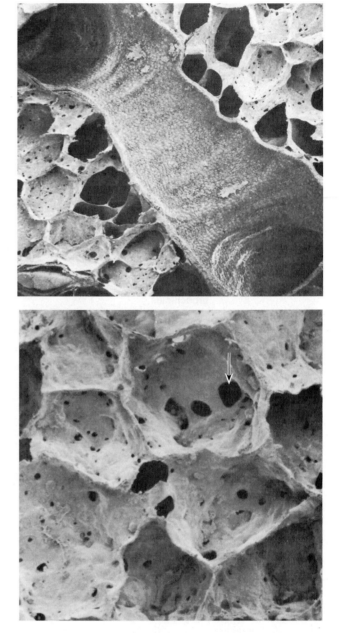

(a)

(b)

Figure 16.3 A scanning electron micrograph of lung tissue. (a) A small bronchiole passes between many alveoli. (b) The alveoli are seen under higher power, with an arrow indicating an alveolar pore through which air can pass from one alveolus to another.

and the terminal alveolar sacs. The **conducting zone** includes all of the anatomical structures through which air passes before reaching the respiratory zone (fig. 16.4; see also fig. 16.21).

Air enters the respiratory bronchioles from *terminal bronchioles,* which are narrow airways formed from many successive divisions of the *right* and *left primary bronchi.* These two large air passages, in turn, are continuous with the *trachea,* or windpipe, which is located in the neck in front of the esophagus (a muscular tube that carries food to the stomach). The trachea is a sturdy tube supported by rings of cartilage (fig. 16.5).

Air enters the trachea from the *pharynx,* which is the cavity behind the palate that receives the contents of both the oral and nasal passages. In order for air to enter or leave the trachea and lungs, however, it must pass through a valvelike opening called the *glottis* between the vocal folds. The ventricular and vocal folds are part of the *larynx,* or voice box, which guards the entrance to the trachea (fig. 16.6). The projection at the front of the throat, commonly called the "Adam's apple," is formed by the largest cartilage of the larynx.

If the trachea becomes occluded through inflammation, excessive secretion, trauma, or aspiration of a foreign object, it may be necessary to create an emergency opening into this tube so that ventilation can still occur. A **tracheotomy** is the procedure of surgically opening the trachea, and a **tracheostomy** involves the insertion of a tube into the trachea to permit breathing and to keep the passageway open. A tracheotomy should be performed only by a competent physician because of the great risk of cutting a recurrent laryngeal nerve or the common carotid artery.

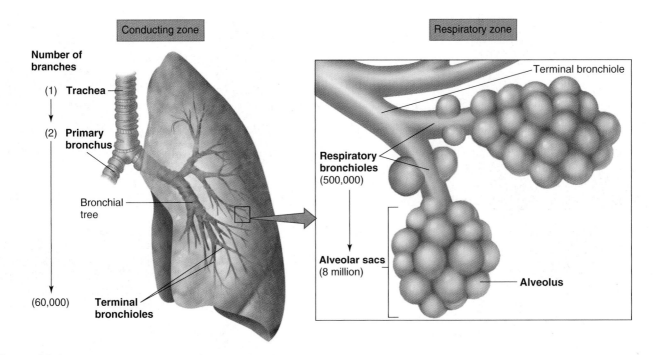

Figure 16.4 **The conducting and respiratory zones of the respiratory system.** The conducting zone consists of airways that conduct the air to the respiratory zone, which is the region where gas exchange occurs. The numbers of each member of the airways and the total number of alveolar sacs are shown in parentheses.

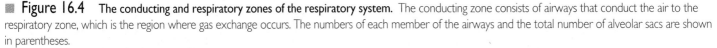

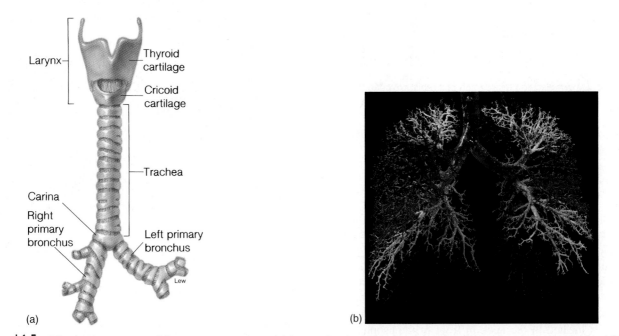

Figure 16.5 **The conducting zone of the respiratory system.** (a) An anterior view extending from the larynx to the terminal bronchi and (b) the airway from the trachea to the terminal bronchioles, as represented by a plastic cast.

The conducting zone of the respiratory system, in summary, consists of the mouth, nose, pharynx, larynx, trachea, primary bronchi, and all successive branchings of the bronchioles up to and including the terminal bronchioles. In addition to conducting air into the respiratory zone, these structures serve additional functions: *warming* and *humidification* of the inspired air and *filtration* and *cleaning*.

Regardless of the temperature and humidity of the ambient air, when the inspired air reaches the respiratory zone it is at a temperature of 37° C (body temperature), and it is saturated with water vapor as it flows over the warm, wet mucous membranes that line the respiratory airways. This ensures that a constant internal body temperature will be maintained and that delicate lung tissue will be protected from desiccation.

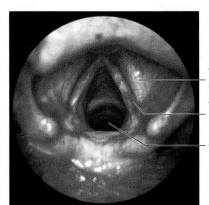

Ventricular fold
(false vocal cord)

Vocal fold
(true vocal cord)

Glottis

Figure 16.6 A photograph of the larynx showing the true and false vocal cords and the glottis. The vocal folds (true vocal cords) function in sound production, whereas the ventricular folds (false vocal cords) do not.

Mucus secreted by cells of the conducting zone structures serves to trap small particles in the inspired air and thereby performs a filtration function. This mucus is moved along at a rate of 1 to 2 centimeters per minute by cilia projecting from the tops of epithelial cells that line the conducting zone (fig. 16.7). There are about 300 cilia per cell that beat in a coordinated fashion to move mucus toward the pharynx, where it can either be swallowed or expectorated.

As a result of this filtration function, particles larger than about 6 μm do not normally enter the respiratory zone of the lungs. The importance of this function is evidenced by *black lung,* a disease that occurs in miners who inhale large amounts of carbon dust from coal, which causes them to develop pulmonary fibrosis. The alveoli themselves are normally kept clean by the action of resident macrophages (see fig. 16.1). The cleansing action of cilia and macrophages in the lungs is diminished by cigarette smoke.

Thoracic Cavity

The *diaphragm,* a dome-shaped sheet of striated muscle, divides the anterior body cavity into two parts. The area below the diaphragm, the *abdominopelvic cavity,* contains the liver, pancreas, gastrointestinal tract, spleen, genitourinary tract, and other organs. Above the diaphragm, the *thoracic cavity* contains the heart, large blood vessels, trachea, esophagus, and thymus in the central region, and is filled elsewhere by the right and left lungs.

The structures in the central region—or *mediastinum*—are enveloped by two layers of wet epithelial membrane collectively called the *pleural membranes.* The superficial layer, or *parietal pleura,* lines the inside of the thoracic wall. The deep layer, or *visceral pleura,* covers the surface of the lungs (fig. 16.8).

The lungs normally fill the thoracic cavity so that the visceral pleura covering each lung is pushed against the parietal pleura lining the thoracic wall. There is thus, under normal conditions, little or no air between the visceral and parietal pleura. There is, however, a "potential space"—called the *intrapleural space*—that can become a real space if the visceral and parietal pleurae separate when a lung collapses. The normal position of the lungs in the thoracic cavity is shown in the radiograph in figure 16.9.

Figure 16.7 A scanning electron micrograph of cilia in a bronchial wall. The cilia that project from the tops of the epithelial cells help to cleanse the lung by moving trapped particles.

Test Yourself Before You Continue

1. Describe the structures involved in gas exchange in the lungs and explain how gas exchange occurs.
2. Describe the structures and functions of the conducting zone of the respiratory system.
3. Describe how each lung is compartmentalized by the pleural membranes. What is the relationship between the visceral and parietal pleurae?

Physical Aspects of Ventilation

The movement of air into and out of the lungs occurs as a result of pressure differences induced by changes in lung volumes. Ventilation is thus influenced by the physical properties of the lungs, including their compliance, elasticity, and surface tension.

Movement of air, from higher to lower pressure, between the conducting zone and the terminal bronchioles occurs as a result of the pressure difference between the two ends of the airways. Air flow through bronchioles, like blood flow through blood vessels, is directly proportional to the pressure difference and inversely proportional to the frictional resistance to flow. The pressure differences in the pulmonary system are induced by changes in lung volumes. The compliance, elasticity, and surface tension of the lungs are physical properties that affect their functioning.

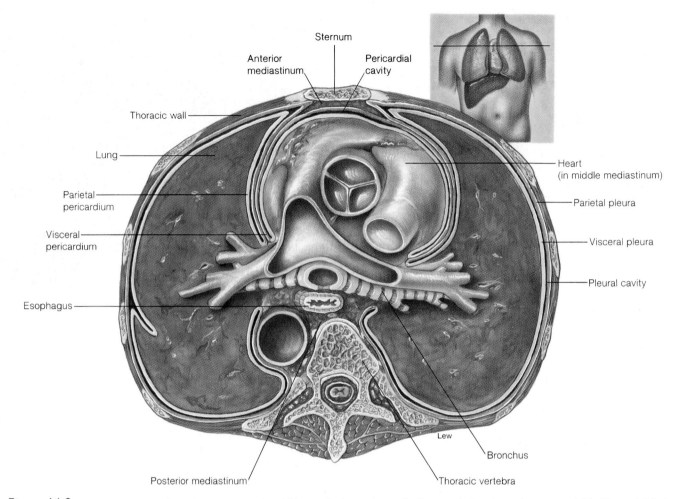

■ **Figure 16.8** **A cross section of the thoracic cavity.** In addition to the lungs, the mediastinum and pleural membranes are visible. The parietal pleura is shown in green, and the visceral pleura in blue.

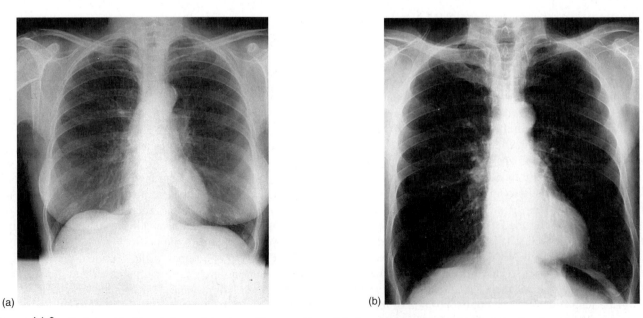

(a)   (b)

■ **Figure 16.9** **Radiographic (X-ray) views of the chest.** These are X rays (a) of a normal female and (b) of a normal male.

Intrapulmonary and Intrapleural Pressures

The visceral and parietal pleurae are normally flush against each other, so that the lungs are stuck to the chest wall in the same manner as two wet pieces of glass sticking to each other. The *intrapleural space* contains only a film of fluid secreted by the two membranes. The pleural cavity in a healthy person is thus potential rather than real; it can become real only in abnormal situations when air enters the intrapleural space. Since the lungs normally remain in contact with the chest wall, they expand and contract along with the thoracic cavity during respiratory movements.

Air enters the lungs during inspiration because the atmospheric pressure is greater than the **intrapulmonary, or intra-alveolar, pressure.** Since the atmospheric pressure does not usually change, the intrapulmonary pressure must fall below atmospheric pressure to cause inspiration. A pressure below that of the atmosphere is called a *subatmospheric pressure,* or *negative pressure.* During quiet inspiration, for example, the intrapulmonary pressure may decrease to 3 mmHg below the pressure of the atmosphere. This subatmospheric pressure is shown as –3 mmHg. Expiration, conversely, occurs when the intrapulmonary pressure is greater than the atmospheric pressure. During quiet expiration, for example, the intrapulmonary pressure may rise to at least +3 mmHg over the atmospheric pressure.

The lack of air in the intrapleural space produces a subatmospheric **intrapleural pressure** that is lower than the intrapulmonary pressure (table 16.1). There is thus a pressure difference across the wall of the lung—called the **transpulmonary** (or **transmural**) **pressure**—which is the difference between the intrapulmonary pressure and the intrapleural pressure. Since the pressure within the lungs (intrapulmonary pressure) is greater than that outside the lungs (intrapleural pressure), the difference in pressure (transpulmonary pressure) keeps the lungs against the chest wall. Thus, the changes in lung volume parallel changes in thoracic volume during inspiration and expiration.

Boyle's Law

Changes in intrapulmonary pressure occur as a result of changes in lung volume. This follows from **Boyle's law,** which states that the pressure of a given quantity of gas is inversely proportional to its volume. An increase in lung volume during inspiration decreases intrapulmonary pressure to subatmospheric levels; air therefore goes in. A decrease in lung volume, conversely, raises the intrapulmonary pressure above that of the atmosphere, expelling air from the lungs. These changes in lung volume occur as a consequence of changes in thoracic volume, as will be described in a later section on the mechanics of breathing.

Physical Properties of the Lungs

In order for inspiration to occur, the lungs must be able to expand when stretched; they must have high *compliance.* In order for expiration to occur, the lungs must get smaller when this tension is released: they must have *elasticity.* The tendency to get smaller is also aided by *surface tension* forces within the alveoli.

Table 16.1 Pressure Changes in Normal, Quiet Breathing

	Inspiration	Expiration
Intrapulmonary pressure (mmHg)	–3	+3
Intrapleural pressure (mmHg)	–6	–3
Transpulmonary pressure (mmHg)	+3	+6

Note: Pressures indicate mmHg below or above atmospheric pressure.

Compliance

The lungs are very distensible (stretchable)—they are, in fact, about a hundred times more distensible than a toy balloon. Another term for distensibility is *compliance,* which here refers to the ease with which the lungs can expand under pressure. **Lung compliance** can be defined as the change in lung volume per change in transpulmonary pressure, expressed symbolically as $\Delta V / \Delta P$. A given transpulmonary pressure, in other words, will cause greater or lesser expansion, depending on the compliance of the lungs.

The compliance of the lungs is reduced by factors that produce a resistance to distension. If the lungs were filled with concrete (as an extreme example), a given transpulmonary pressure would produce no increase in lung volume and no air would enter; the compliance would be zero. The infiltration of lung tissue with connective tissue proteins, a condition called *pulmonary fibrosis,* similarly decreases lung compliance.

Elasticity

The term **elasticity** refers to the tendency of a structure to return to its initial size after being distended. Because of their high content of elastin proteins, the lungs are very elastic and resist distension. Since the lungs are normally stuck to the chest wall, they are always in a state of elastic tension. This tension increases during inspiration when the lungs are stretched and is reduced by elastic recoil during expiration.

The elastic nature of lung tissue is revealed when air enters the intrapleural space (as a result of an open chest wound, for example). This condition, called a **pneumothorax,** is shown in figure 16.10. As air enters the intrapleural space, the intrapleural pressure rises until it is equal to the atmospheric pressure. When the intrapleural pressure is the same as the intrapulmonary pressure, the lung can no longer expand. Not only does the lung not expand during inspiration, it actually collapses away from the chest wall as a result of elastic recoil, a condition called *atelectasis.* Fortunately, a pneumothorax usually causes only one lung to collapse, since each lung is contained in a separate pleural compartment.

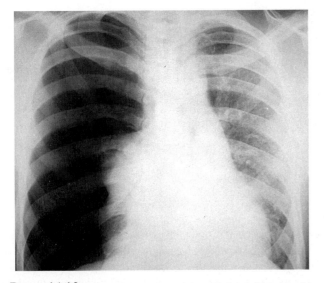

■ **Figure 16.10** **A pneumothorax of the right lung.** The right side of the thorax appears uniformly dark because it is filled with air. The spaces between the ribs are also greater on the right side due to release from the elastic tension of the lungs. The left lung appears denser (less dark) because of shunting of blood from the right to the left lung.

Clinical Investigation Clues

Remember that Harry's stab wound caused the collapse of the lung.

What condition does Harry have?

How can he still be alive when the paramedics found him?

What physical factors caused the lung to collapse?

Surface Tension

The forces that act to resist distension include elastic resistance and the **surface tension** that is exerted by fluid in the alveoli. The lungs both secrete and absorb fluid in two antagonistic processes that normally leave only a very thin film of fluid on the alveolar surface. Fluid absorption is driven (through osmosis) by the active transport of Na^+, while fluid secretion is driven by the active transport of Cl^- out of the alveolar epithelial cells. Research has demonstrated that people with cystic fibrosis have a genetic defect in one of the Cl^- carriers (called the *cystic fibrosis transmembrane regulator,* or *CFTR,* as described in chapter 6). This results in an imbalance of fluid absorption and secretion, so that the airway fluid becomes excessively viscous (with a lower water content) and difficult to clear.

The thin film of fluid normally present in the alveolus has a surface tension, which is due to the fact that water molecules at the surface are attracted more to other water molecules than to air. As a result, the surface water molecules are pulled tightly together by attractive forces from underneath. This surface tension produces a force that is directed inward, raising the pressure within the alveolus. As described by the **law of Laplace,** the pressure thus created is directly proportional to the surface tension and in-

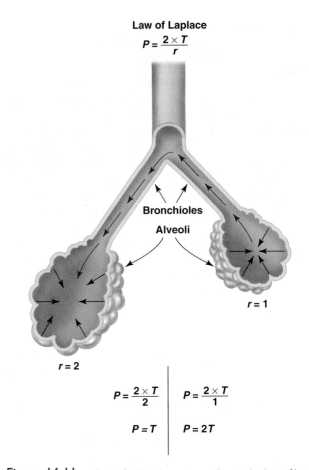

■ **Figure 16.11** **The law of Laplace.** According to the law of Laplace, the pressure created by surface tension should be greater in the smaller alveolus than in the larger alveolus. This implies that (without surfactant) smaller alveoli would collapse and empty their air into larger alveoli.

versely proportional to the radius of the alveolus (fig. 16.11). According to this law, the pressure in a smaller alveolus would be greater than in a larger alveolus if the surface tension were the same in both. The greater pressure of the smaller alveolus would then cause it to empty its air into the larger one. This does not normally occur because, as an alveolus decreases in size, its surface tension (the numerator in the equation) is decreased at the same time that its radius (the denominator) is reduced. The reason for the decreased surface tension, which prevents the alveoli from collapsing, is described in the next section.

Surfactant and the Respiratory Distress Syndrome

Alveolar fluid contains a phospholipid known as *dipalmitoyl lecithin* (probably attached to a protein) that functions to lower surface tension. This compound is called **surfactant**—a contraction of the term *surface active agent.* The surfactant molecules become interspersed between water molecules at the water-air interface of the alveoli, thereby reducing the attractive forces (hydrogen bonds, described in chapter 2) between water molecules that produce the surface tension. Thus, because of

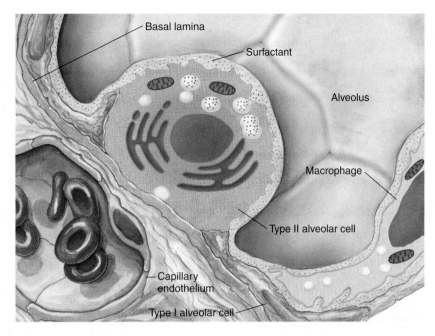

Basal lamina

Surfactant

Alveolus

Macrophage

Type II alveolar cell

Capillary endothelium

Type I alveolar cell

■ **Figure 16.12**   **The production of pulmonary surfactant.** Produced by type II alveolar cells, surfactant appears to be composed of a derivative of lecithin combined with protein.

surfactant, the surface tension in the alveoli is reduced. Further, the ability of surfactant to lower surface tension improves as the alveoli get smaller during expiration. This may be because the surfactant molecules become more concentrated as the alveoli get smaller. Surfactant thus prevents the alveoli from collapsing during expiration, as would be predicted from the law of Laplace. Even after a forceful expiration, the alveoli remain open and a *residual volume* of air remains in the lungs. Since the alveoli do not collapse, less surface tension has to be overcome to inflate them at the next inspiration.

Surfactant is produced by type II alveolar cells (fig. 16.12) in late fetal life. Premature babies are sometimes born with lungs that lack sufficient surfactant, and their alveoli are collapsed as a result. This condition is called **respiratory distress syndrome (RDS).** Considering that a full-term pregnancy lasts 37 to 42 weeks, RDS occurs in about 60% of babies born at less than 28 weeks, 30% of babies born at 28 to 34 weeks, and less than 5% of babies born after 34 weeks of gestation. The risk of RDS can be assessed by analysis of amniotic fluid (surrounding the fetus), and mothers can be given exogenous corticosteroids to accelerate the maturation of their fetus's lungs.

People with septic shock (a fall in blood pressure due to widespread vasodilation, which occurs as a result of a systemic infection) may develop a condition called **acute respiratory distress syndrome (ARDS).** In this condition, inflammation causes increased capillary and alveolar permeability that lead to the accumulation of a protein-rich fluid in the lungs. This decreases lung compliance and is accompanied by a reduced surfactant, which further lowers compliance. The blood leaving the lungs, as a result, has an abnormally low oxygen concentration (a condition called *hypoxemia*).

Even under normal conditions, the first breath of life is a difficult one because the newborn must overcome great surface tension forces in order to inflate its partially collapsed alveoli. The transpulmonary pressure required for the first breath is fifteen to twenty times that required for subsequent breaths, and an infant with respiratory distress syndrome must duplicate this effort with every breath. Fortunately, many babies with this condition can be saved by mechanical ventilators and by exogenous surfactant delivered to the baby's lungs by means of an endotracheal tube. The exogenous surfactant may be a synthetic mixture of phospholipids, or it may be surfactant obtained from bovine lungs. The mechanical ventilator and exogenous surfactant help to keep the baby alive long enough for its lungs to mature, so that it can manufacture sufficient surfactant on its own.

Test Yourself Before You Continue

1. Describe the changes in the intrapulmonary and intrapleural pressures that occur during inspiration and use Boyle's law to explain the reasons for these changes.

2. Explain how the compliance and elasticity of the lungs affect inspiration and expiration.

3. Describe pulmonary surfactant and discuss its significance.

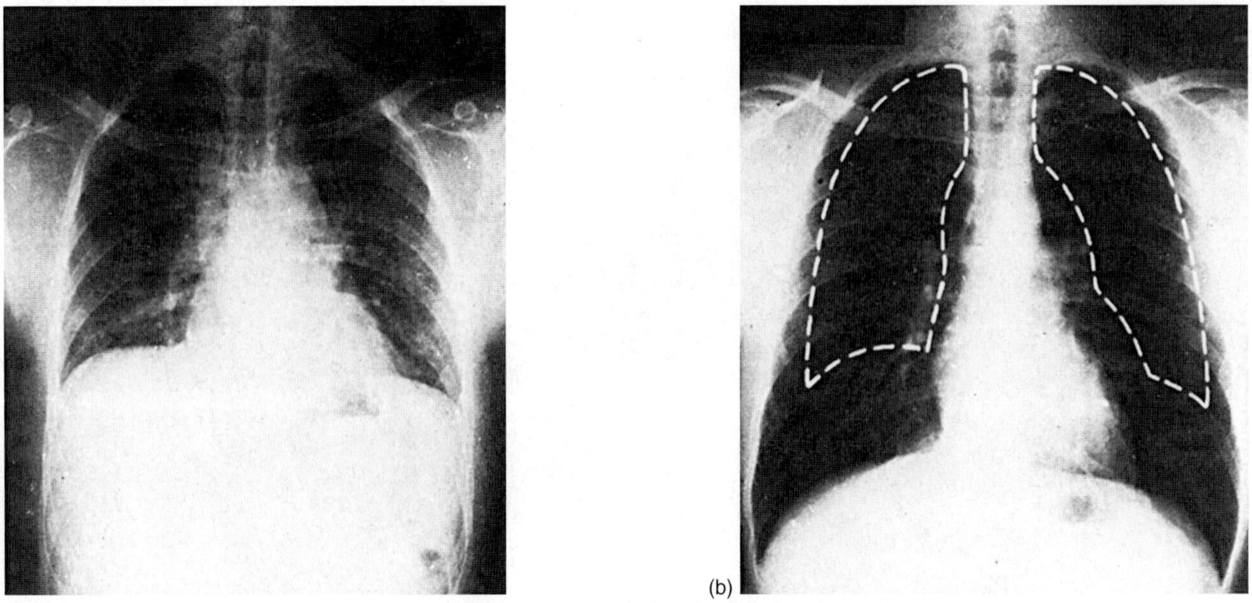

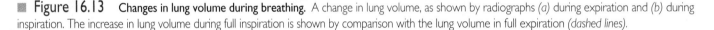

(a)   (b)

■ **Figure 16.13** **Changes in lung volume during breathing.** A change in lung volume, as shown by radiographs *(a)* during expiration and *(b)* during inspiration. The increase in lung volume during full inspiration is shown by comparison with the lung volume in full expiration *(dashed lines)*.

Mechanics of Breathing

Normal, quiet inspiration results from muscle contraction, and normal expiration from muscle relaxation and elastic recoil. These actions can be forced by contractions of the accessory respiratory muscles. The amount of air inspired and expired can be measured in a number of ways to test pulmonary function.

The thorax must be sufficiently rigid to protect vital organs and provide attachments for a number of short, powerful muscles. However, breathing, or **pulmonary ventilation,** also requires a flexible thorax that can function as a bellows during the ventilation cycle. The structure of the rib cage and associated cartilages provides continuous elastic tension, so that when stretched by muscle contraction during inspiration, the rib cage can return passively to its resting dimensions when the muscles relax. This elastic recoil is greatly aided by the elasticity of the lungs.

Pulmonary ventilation consists of two phases: *inspiration* and *expiration*. Inspiration (inhalation) and expiration (exhalation) are accomplished by alternately increasing and decreasing the volumes of the thorax and lungs (fig. 16.13).

Inspiration and Expiration

Between the bony portions of the rib cage are two layers of intercostal muscles: the **external intercostal muscles** and the **internal intercostal muscles** (fig. 16.14). Between the costal cartilages, however, there is only one muscle layer, and its fibers are oriented in a manner similar to those of the internal intercostals. These muscles are therefore called the *interchondral part* of the internal intercostals. Another name for them is the **parasternal intercostals.**

An unforced, or quiet, inspiration results primarily from contraction of the dome-shaped diaphragm, which lowers and flattens when it contracts. This increases thoracic volume in a vertical direction. Inspiration is aided by contraction of the parasternal and external intercostals, which raise the ribs when they contract and increase thoracic volume laterally. Other thoracic muscles become involved in forced (deep) inspiration. The most important of these are the *scalenes,* followed by the *pectoralis minor,* and in extreme cases the *sternocleidomastoid muscles.* Contraction of these muscles elevates the ribs in an anteroposterior direction; at the same time, the upper rib cage is stabilized so that the intercostals become more effective. The increase in thoracic volume produced by these muscle contractions decreases intrapulmonary (intra-alveolar) pressure, thereby causing air to flow into the lungs.

Quiet expiration is a passive process. After becoming stretched by contractions of the diaphragm and thoracic muscles, the thorax and lungs recoil as a result of their elastic tension when the respiratory muscles relax. The decrease in lung volume raises the pressure within the alveoli above the atmospheric pressure and pushes the air out. During forced expiration, the internal intercostal muscles (excluding the interchondral part) contract and depress the rib cage. The abdominal muscles also aid expiration because, when they contract, they force abdominal organs up against the diaphragm and further decrease the volume of the thorax. By this means, the intrapulmonary pressure can rise 20 or 30 mmHg above the atmospheric pressure. The events that occur during inspiration

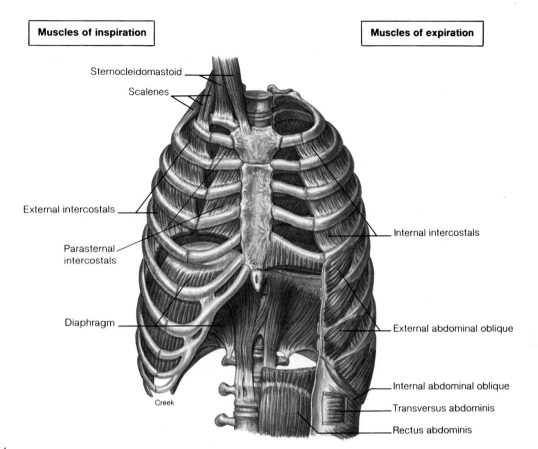

| Muscles of inspiration | | Muscles of expiration |

Figure 16.14 **The muscles involved in breathing.** The principal muscles of inspiration are shown on the left, and those of expiration are shown on the right.

Table 16.2 Mechanisms Involved in Normal, Quiet Ventilation and Forced Ventilation

	Inspiration	Expiration
Normal, quiet breathing	Contraction of the diaphragm and external intercostal muscles increases the thoracic and lung volume, decreasing intrapulmonary pressure to about –3 mmHg.	Relaxation of the diaphragm and external intercostals, plus elastic recoil of lungs, decreases lung volume and increases intrapulmonary pressure to about +3 mmHg.
Forced ventilation	Inspiration, aided by contraction of accessory muscles such as the scalenes and sternocleidomastoid, decreases intrapulmonary pressure to –20 mmHg or lower.	Expiration, aided by contraction of abdominal muscles and internal intercostal muscles, increases intrapulmonary pressure to +30 mmHg or higher.

and expiration are summarized in table 16.2 and shown in figure 16.15.

Pulmonary Function Tests

Pulmonary function may be assessed clinically by means of a technique known as *spirometry.* In this procedure, a subject breathes in a closed system in which air is trapped within a light plastic bell floating in water. The bell moves up when the subject exhales and down when the subject inhales. The movements of the bell cause corresponding movements of a pen, which traces a record of the breathing called a *spirogram* (fig. 16.16).

More sophisticated computerized devices are also commonly employed to assess lung function.

Lung Volumes and Capacities

An example of a spirogram is shown in figure 16.16, and the various lung volumes and capacities are defined in table 16.3. A lung capacity is equal to the sum of two or more lung volumes. During quiet breathing, for example, the amount of air expired in each breath is the **tidal volume.** The maximum amount of air that can be forcefully exhaled after a maximum inhalation is called the **vital capacity,** which is equal to the sum of the **inspiratory reserve volume, tidal volume,** and **expiratory reserve**

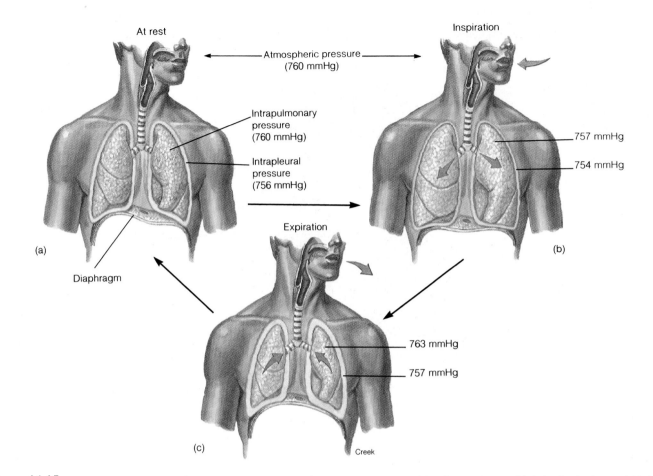

■ Figure 16.15 **The mechanics of pulmonary ventilation.** Pressures (at sea level) are shown (a) before inspiration, (b) during inspiration, and (c) during expiration. During inspiration, the intrapulmonary pressure is lower than the atmospheric pressure, and during expiration it is greater than the atmospheric pressure.

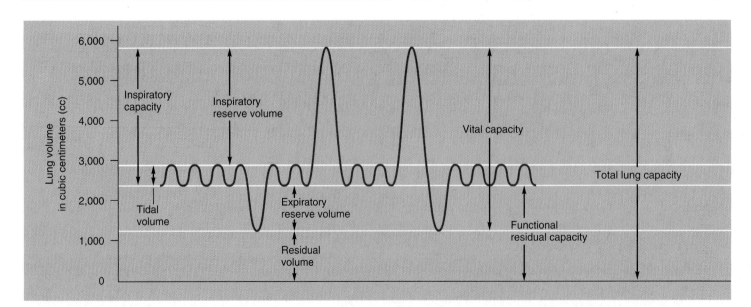

■ Figure 16.16 **A spirogram showing lung volumes and capacities.** A lung capacity is the sum of two or more lung volumes. The vital capacity, for example, is the sum of the tidal volume, the inspiratory reserve volume, and the expiratory reserve volume. Note that residual volume cannot be measured with a spirometer because it is air that cannot be exhaled. Therefore, the total lung capacity (the sum of the vital capacity and the residual volume) also cannot be measured with a spirometer.

Table 16.3 Terms Used to Describe Lung Volumes and Capacities

Term	Definition
Lung Volumes	The four nonoverlapping components of the total lung capacity
Tidal volume	The volume of gas inspired or expired in an unforced respiratory cycle
Inspiratory reserve volume	The maximum volume of gas that can be inspired during forced breathing in addition to tidal volume
Expiratory reserve volume	The maximum volume of gas that can be expired during forced breathing in addition to tidal volume
Residual volume	The volume of gas remaining in the lungs after a maximum expiration
Lung Capacities	Measurements that are the sum of two or more lung volumes
Total lung capacity	The total amount of gas in the lungs after a maximum inspiration
Vital capacity	The maximum amount of gas that can be expired after a maximum inspiration
Inspiratory capacity	The maximum amount of gas that can be inspired after a normal tidal expiration
Functional residual capacity	The amount of gas remaining in the lungs after a normal tidal expiration

Table 16.4 Ventilation Terminology

Term	Definition
Air spaces	Alveolar ducts, alveolar sacs, and alveoli
Airways	Structures that conduct air from the mouth and nose to the respiratory bronchioles
Alveolar ventilation	Removal and replacement of gas in alveoli; equal to the tidal volume minus the volume of dead space times the breathing rate
Anatomical dead space	Volume of the conducting airways to the zone where gas exchange occurs
Apnea	Cessation of breathing
Dyspnea	Unpleasant, subjective feeling of difficult or labored breathing
Eupnea	Normal, comfortable breathing at rest
Hyperventilation	Alveolar ventilation that is excessive in relation to metabolic rate; results in abnormally low alveolar CO_2
Hypoventilation	Alveolar ventilation that is low in relation to metabolic rate; results in abnormally high alveolar CO_2
Physiological dead space	Combination of anatomical dead space and underventilated or underperfused alveoli that do not contribute normally to blood gas exchange
Pneumothorax	Presence of gas in the intrapleural space (the space between the visceral and parietal pleurae) causing lung collapse
Torr	Unit of pressure nearly equal to the millimeter of mercury (760 mmHg = 760 torr)

volume (fig. 16.16). Multiplying the tidal volume at rest by the number of breaths per minute yields a **total minute volume** of about 6 L per minute. During exercise, the tidal volume and the number of breaths per minute increase to produce a total minute volume as high as 100 to 200 L per minute.

It should be noted that not all of the inspired volume reaches the alveoli with each breath. As fresh air is inhaled, it is mixed with air in the **anatomical dead space** (table 16.4). This dead space comprises the conducting zone of the respiratory system—nose, mouth, larynx, trachea, bronchi, and bronchioles—where no gas exchange occurs. Air within the anatomical dead space has a lower oxygen concentration and a higher carbon dioxide concentration than the external air. Since the air in the dead space enters the alveoli first, the amount of fresh air reaching the alveoli with each breath is less than the tidal volume. But, since the volume of air in the dead space is an anatomical constant, the percentage of fresh air entering the alveoli is increased with increasing tidal volumes. For example, if the anatomical dead space is 150 ml and the tidal volume is 500 ml, the percentage of fresh air reaching the alveoli is $350/500 \times 100\% = 70\%$. If the tidal volume is increased to 2,000 ml, and the anatomical dead space is still 150 ml, the percentage of fresh air reaching the

alveoli is increased to $1,850/2,000 \times 100\% = 93\%$. An increase in tidal volume can thus be a factor in the respiratory adaptations to exercise and high altitude.

Restrictive and Obstructive Disorders

Spirometry is useful in the diagnosis of lung diseases. On the basis of pulmonary function tests, lung disorders can be classified as *restrictive* or *obstructive*. In **restrictive disorders,** such as pulmonary fibrosis, the vital capacity is reduced to below normal. The rate at which the vital capacity can be forcibly exhaled, however, is normal. In disorders that are exclusively obstructive, by contrast, the vital capacity is normal because lung tissue is not damaged. In asthma for example, the vital capacity is normal, but expiration is more difficult and takes a longer time because bronchoconstriction increases the resistance to air flow. **Obstructive disorders** are therefore diagnosed by tests that measure the rate of expiration. One such test is the **forced expiratory volume (FEV),** in which the percentage of the vital capacity that can be exhaled in the first second (FEV_1) is measured (fig. 16.17). An FEV_1 that is significantly less than 80% suggests the presence of obstructive pulmonary disease.

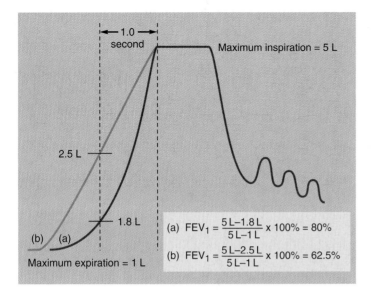

Figure 16.17 The one-second forced expiratory volume (FEV$_1$) **test.** The percentage in (a) is normal, whereas that in (b) may indicate an obstructive pulmonary disorder such as asthma or bronchitis.

Bronchoconstriction often occurs in response to inhalation of noxious agents present in smoke or smog. The FEV$_1$ has therefore been used by researchers to determine the effects of various components of smog and passive cigarette smoke on pulmonary function. These studies have shown that it is unhealthy to exercise on very smoggy days and that inhalation of smoke from other people's cigarettes in a closed environment can adversely affect pulmonary function.

There is normally a decline in the FEV$_1$ with age, but research suggests that this decline may be accelerated in cigarette smokers. Smokers under the age of 35 who quit have improved lung function; those who quit after the age of 35 slow their age-related decline in FEV$_1$ to normal rates.

Pulmonary Disorders

People with pulmonary disorders frequently complain of **dyspnea,** a subjective feeling of "shortness of breath." Dyspnea may occur even when ventilation is normal, however, and may not occur even when total minute volume is very high, as in exercise. Some of the terms associated with ventilation are defined in table 16.4.

Asthma

The dyspnea, wheezing, and other symptoms of **asthma** are produced by an obstruction of air flow through the bronchioles that occurs in episodes, or "attacks." This obstruction is caused by inflammation, mucous secretion, and bronchoconstriction. Inflammation of the airways is characteristic of asthma, and itself contributes to increased airway responsiveness to agents that promote bronchiolar constriction. Bronchoconstriction further increases airway resistance and makes breathing difficult. The increased airway resistance of asthma may be provoked by allergic

reactions in which immunoglobulin E (IgE) is produced (see chapter 15), by exercise (in exercise-induced bronchoconstriction), by breathing cold, dry air, or by aspirin (in a minority of asthmatics).

Asthma is often treated with glucocorticoid drugs, which inhibit inflammation. A new antileukotriene drug also is now available to suppress the inflammatory response. Epinephrine and related compounds stimulate beta-adrenergic receptors in the bronchioles, and by this means promote bronchodilation. Therefore, epinephrine was frequently used as an inhaled spray to relieve the symptoms of an asthma attack. It has since been learned that there are two subtypes of beta receptors for epinephrine, and that the subtype in the heart (called β_1) is different from the one in the bronchioles (β_2). Capitalizing on these differences, compounds such as *terbutaline* have been developed. These compounds can more selectively stimulate the β_2-adrenergic receptors and cause bronchodilation without affecting the heart to the extent that epinephrine does.

These conditions result in the secretion of several substances by tissue mast cells and eosinophils (chapter 15). Chief among these are histamine and the leukotrienes (derived from the same parent molecule, arachidonic acid, as are the prostaglandins—see chapter 11, fig. 11.34). Drugs that block leukotriene synthesis or action are among the newest compounds available for the treatment of asthma.

Emphysema

Alveolar tissue is destroyed in the chronic, progressive condition called **emphysema,** which results in fewer but larger alveoli (fig. 16.18). This reduces the surface area for gas exchange and decreases the ability of the bronchioles to remain open during expiration. Collapse of the bronchioles as a result of the compression of the lungs during expiration produces *air trapping,* which further decreases the efficiency of gas exchange in the alveoli.

Among the different types of emphysema, the most common occurs almost exclusively in people who have smoked cigarettes heavily over a period of years. A component of cigarette smoke apparently stimulates the macrophages and leukocytes to secrete proteolytic (protein-digesting) enzymes that destroy lung tissues. A less common type of emphysema results from the genetic inability to produce a plasma protein called α_1-antitrypsin. This protein normally inhibits proteolytic enzymes such as trypsin, and thus normally protects the lungs against the effects of enzymes that are released from alveolar macrophages.

Chronic bronchitis, asthma, and emphysema, the most common causes of respiratory failure, are together called **chronic obstructive pulmonary disease (COPD).** In addition to the more direct obstructive and restrictive aspects of these conditions, other pathological changes may occur. These include edema, inflammation, hyperplasia (an increase in the number of cells), zones of pulmonary fibrosis, pneumonia, pulmonary emboli (traveling blood clots), and heart failure. Patients with severe chronic bronchitis or

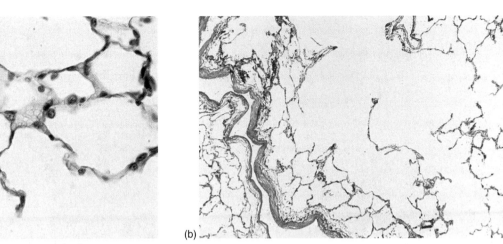

(a)

(b)

 Figure 16.18 **Emphysema destroys lung tissue.** These are photomicrographs of tissue (*a*) from a normal lung and (*b*) from the lung of a person with emphysema. The destruction of lung tissue in emphysema results in fewer and larger alveoli.

emphysema may develop *cor pulmonale*—pulmonary hypertension with hypertrophy and the eventual failure of the right ventricle. COPD is the fifth leading cause of death in the United States.

Clinical Investigation Clues

Remember that Harry has a slightly low vital capacity and a significantly reduced FEV_1.

What is the likely cause of the lowered FEV_1?
What does it signify?
What can he do to improve his FEV_1?
What might be responsible for his lowered vital capacity?

Pulmonary Fibrosis

Under certain conditions, for reasons that are poorly understood, lung damage leads to **pulmonary fibrosis** instead of emphysema. In this condition, the normal structure of the lungs is disrupted by the accumulation of fibrous connective tissue proteins. Fibrosis can result, for example, from the inhalation of particles less than 6 μm in size that can accumulate in the respiratory zone of the lungs. Included in this category is *anthracosis,* or black lung, which is produced by the inhalation of carbon particles from coal dust.

Test Yourself Before You Continue

1. Describe the actions of the diaphragm and external intercostal muscles during inspiration. How is quiet expiration produced?
2. Explain how forced inspiration and forced expiration are produced.
3. Define the terms *tidal volume* and *vital capacity*. Explain how the total minute volume is calculated and how this value is affected by exercise.
4. How are the vital capacity and the forced expiratory volume measurements affected by asthma and pulmonary fibrosis? Give the reasons for these effects.

Gas Exchange in the Lungs

Gas exchange between the alveolar air and the blood in pulmonary capillaries results in an increased oxygen concentration and a decreased carbon dioxide concentration in the blood leaving the lungs. This blood enters the systemic arteries, where blood gas measurements are taken to assess the effectiveness of lung function.

The atmosphere is an ocean of gas that exerts pressure on all objects within it. This pressure can be measured with a glass U-tube filled with fluid. One end of the U-tube is exposed to the atmosphere, while the other side is continuous with a sealed vacuum tube. Since the atmosphere presses on the open-ended side, but not on the side connected to the vacuum tube, atmospheric pressure pushes fluid in the U-tube up on the vacuum side to a height determined by the atmospheric pressure and the density of the fluid. Water, for example, will be pushed up to a height of 33.9 feet (10,332 mm) at sea level, whereas mercury (Hg)—which is more dense—will be raised to a height of 760 mm. As a matter of convenience, therefore, devices used to measure atmospheric pressure (barometers) use mercury rather than water. The atmospheric pressure at sea level is thus said to be equal to 760 mmHg (or 760 *torr*), which is also described as a pressure of *one atmosphere* (fig. 16.19).

According to **Dalton's law,** the total pressure of a gas mixture (such as air) is equal to the sum of the pressures that each gas in the mixture would exert independently. The pressure that a particular gas in a mixture exerts independently is the **partial pressure** of that gas, which is equal to the product of the total pressure and the fraction of that gas in the mixture.

The total pressure of the gas mixture is thus equal to the sum of the partial pressures of the constituent gases. Since oxygen constitutes about 21% of the atmosphere, for example, its partial pressure (abbreviated P_{O_2}) is 21% of 760, or about 159 mmHg. Since nitrogen constitutes about 78% of the atmosphere, its partial pressure is equal to $0.78 \times 760 = 593$ mmHg.

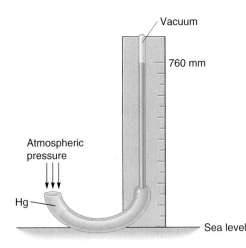

■ **Figure 16.19** The measurement of atmospheric pressure.
Atmospheric pressure at sea level can push a column of mercury to a height
of 760 millimeters. This is also described as 760 torr, or one atmospheric
pressure.

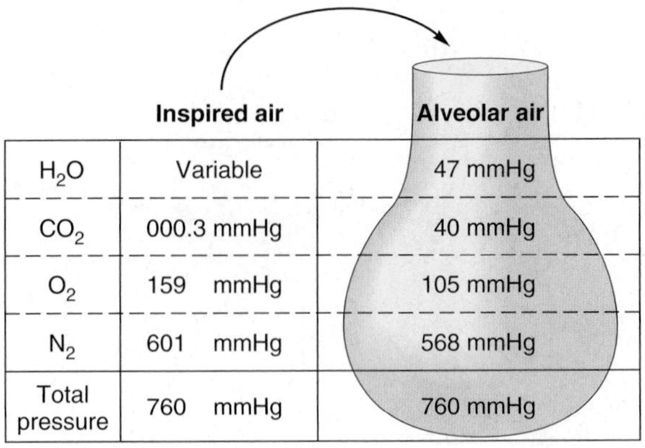

	Inspired air	Alveolar air
H_2O	Variable	47 mmHg
CO_2	000.3 mmHg	40 mmHg
O_2	159 mmHg	105 mmHg
N_2	601 mmHg	568 mmHg
Total pressure	760 mmHg	760 mmHg

■ **Figure 16.20** Partial pressures of gases in inspired air and
alveolar air at sea level. Notice that as air enters the alveoli its oxygen
content decreases and its carbon dioxide content increases. Also notice that
air in the alveoli is saturated with water vapor (giving it a partial pressure of
47 mmHg), which dilutes the contribution of other gases to the total
pressure.

These two gases thus contribute about 99% of the total pressure
of 760 mmHg:

$$P_{\text{dry atmosphere}} = P_{N_2} + P_{O_2} + P_{CO_2} = 760 \text{ mmHg}$$

Calculation of P_{O_2}

With increasing altitude, the total atmospheric pressure and the
partial pressure of the constituent gases decrease (table 16.5). At
Denver, for example (5,000 feet above sea level), the atmospheric pressure is decreased to 619 mmHg, and the P_{O_2} is
therefore reduced to $619 \times 0.21 = 130$ mmHg. At the peak of
Mount Everest (at 29,000 feet), the P_{O_2} is only 42 mmHg. As
one descends below sea level, as in ocean diving, the total pressure increases by one atmosphere for every 33 feet. At 33 feet
therefore, the pressure equals $2 \times 760 = 1,520$ mmHg. At
66 feet, the pressure equals three atmospheres.

Inspired air contains variable amounts of moisture. By the
time the air has passed into the respiratory zone of the lungs,
however, it is normally saturated with water vapor (has a relative
humidity of 100%). The capacity of air to contain water vapor
depends on its temperature; since the temperature of the respiratory zone is constant at 37° C, its water vapor pressure is also
constant (at 47 mmHg).

Water vapor, like the other constituent gases, contributes a
partial pressure to the total atmospheric pressure. Since the total
atmospheric pressure is constant (depending only on the height
of the air mass), the water vapor "dilutes" the contribution of
other gases to the total pressure:

$$P_{\text{wet atmosphere}} = P_{N_2} + P_{O_2} + P_{CO_2} + P_{H_2O}$$

When the effect of water vapor pressure is considered, the
partial pressure of oxygen in the inspired air is decreased at sea
level to

$$P_{O_2} \text{ (sea level)} = 0.21 (760 - 47) = 150 \text{ mmHg}$$

As a result of gas exchange in the alveoli, the P_{O_2} of alveolar air is further diminished to about 105 mmHg. The partial
pressures of the inspired air and the partial pressures of alveolar
air are compared in figure 16.20.

Partial Pressures of Gases in Blood

The enormous surface area of alveoli and the short diffusion distance between alveolar air and the capillary blood quickly help
to bring the blood into gaseous equilibrium with the alveolar air.
This function is further aided by the tremendous number of capillaries that surround each alveolus, forming an almost continuous sheet of blood around the alveoli (fig. 16.21).

When a liquid and a gas, such as blood and alveolar air,
are at equilibrium, the amount of gas dissolved in the fluid
reaches a maximum value. According to **Henry's law,** this value
depends on (1) the solubility of the gas in the fluid, which is a
physical constant; (2) the temperature of the fluid—more gas
can be dissolved in cold water than warm water; and (3) the partial pressure of the gas. Since the temperature of the blood does
not vary significantly, *the concentration of a gas dissolved in a
fluid (such as plasma) depends directly on its partial pressure in
the gas mixture.* When water—or plasma—is brought into equilibrium with air at a P_{O_2} of 100 mmHg, for example, the fluid
will contain 0.3 ml of O_2 per 100 ml fluid at 37° C. If the P_{O_2} of
the gas were reduced by half, the amount of dissolved oxygen
would also be reduced by half.

Blood Gas Measurements

Measurement of the oxygen content of blood (in ml O_2 per
100 ml blood) is a laborious procedure. Fortunately, an **oxygen
electrode** that produces an electric current in proportion to the
concentration of *dissolved oxygen* has been developed. If this
electrode is placed in a fluid while oxygen is artificially bubbled
into it, the current produced by the oxygen electrode will in-

Table 16.5 Effect of Altitude on Partial Oxygen Pressure (P_{O_2})

Altitude (Feet Above Sea Level)	Atmospheric Pressure (mmHg)	P_{O_2} in Air (mmHg)	P_{O_2} in Alveoli (mmHg)	P_{O_2} in Arterial Blood (mmHg)
0	760	159	105	100
2,000	707	148	97	92
4,000	656	137	90	85
6,000	609	127	84	79
8,000	564	118	79	74
10,000	523	109	74	69
20,000	349	73	40	35
30,000	226	47	21	19

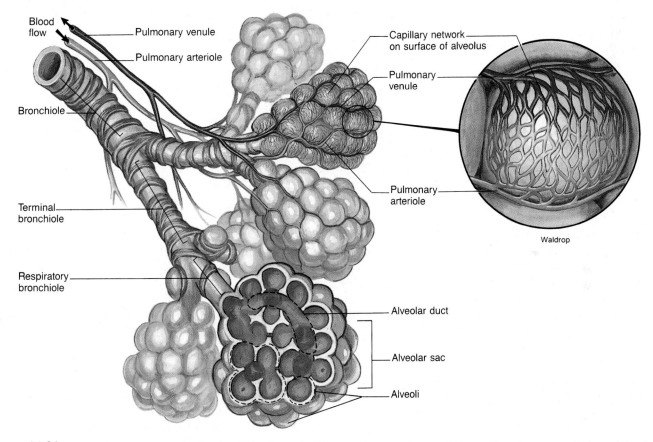

Waldrop

■ **Figure 16.21** **The relationship between alveoli and blood vessels.** The extensive area of contact between the pulmonary capillaries and the alveoli allows for rapid exchange of gases between the air and blood.

crease up to a maximum value. At this maximum value, the fluid is saturated with oxygen—that is, all of the oxygen that can be dissolved at that temperature and P_{O_2} is dissolved. At a constant temperature, the amount dissolved, and thus the electric current, depend only on the P_{O_2} of the gas.

As a matter of convenience, it can now be said that *the fluid has the same P_{O_2} as the gas.* If it is known that the gas has a P_{O_2} of 152 mmHg, for example, the deflection of a needle by the oxygen electrode can be calibrated on a scale at 152 mmHg (fig. 16.22). The actual amount of dissolved oxygen under these

circumstances is not particularly important (it can be looked up in solubility tables, if desired); it is simply a linear function of the P_{O_2}. A lower P_{O_2} indicates that less oxygen is dissolved; a higher P_{O_2} indicates that more oxygen is dissolved.

If the oxygen electrode is next inserted into an unknown sample of blood, the P_{O_2} of that sample can be read directly from the previously calibrated scale. Suppose, as illustrated in figure 16.22, the blood sample has a P_{O_2} of 100 mmHg. Since alveolar air has a P_{O_2} of about 105 mmHg, this reading indicates that the blood is almost in complete equilibrium with the alveolar air.

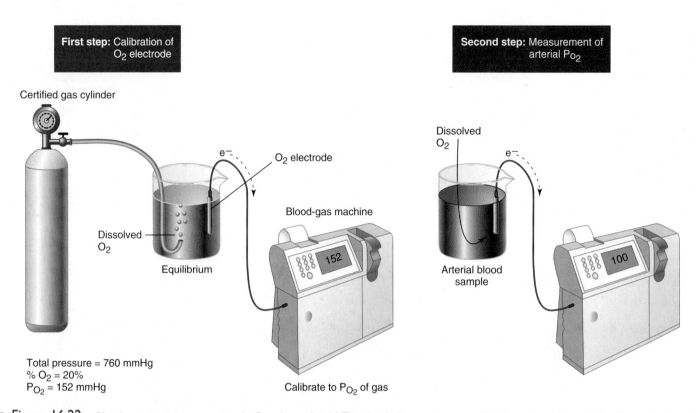

First step: Calibration of O₂ electrode

Certified gas cylinder

e⁻   O₂ electrode

Dissolved O₂

Equilibrium

Blood-gas machine

152

Total pressure = 760 mmHg
% O₂ = 20%
P_O₂ = 152 mmHg

Calibrate to P_O₂ of gas

Second step: Measurement of arterial P_O₂

Dissolved O₂

e⁻

Arterial blood sample

100

■ **Figure 16.22** **Blood gas measurements using the P_O₂ electrode.** (a) The electrical current generated by the oxygen electrode is calibrated so that the needle of the blood gas machine points to the P_O₂ of the gas with which the fluid is in equilibrium. (b) Once standardized in this way, the electrode can be inserted into a fluid such as blood, and the P_O₂ of this solution can be measured.

The oxygen electrode responds only to oxygen dissolved in water or plasma; it cannot respond to oxygen that is bound to hemoglobin in red blood cells. Most of the oxygen in blood, however, is located in the red blood cells attached to hemoglobin. The oxygen content of whole blood thus depends on both its P_{O_2} and its red blood cell and hemoglobin content. At a P_{O_2} of about 100 mmHg, whole blood normally contains almost 20 ml O_2 per 100 ml blood; of this amount, only 0.3 ml of O_2 is dissolved in the plasma and 19.7 ml of O_2 is found within the red blood cells (see fig. 16.32). Since only the 0.3 ml of O_2 affects the P_{O_2} measurement, this measurement would be unchanged if the red blood cells were removed from the sample.

Significance of Blood P_O₂ and P_CO₂ Measurements

Since blood P_{O_2} measurements are not directly affected by the oxygen in red blood cells, the P_{O_2} does not provide a measurement of the total oxygen content of whole blood. It does, however, provide a good index of *lung function*. If the inspired air had a normal P_{O_2} but the arterial P_{O_2} was below normal, for example, you could conclude that gas exchange in the lungs was impaired. Measurements of arterial P_{O_2} thus provide valuable information in treating people with pulmonary diseases, in performing surgery (when breathing may be depressed by anesthesia), and in caring for premature babies with respiratory distress syndrome.

When the lungs are functioning properly, the P_{O_2} of systemic arterial blood is only 5 mmHg less than the P_{O_2} of alveolar air. At a normal P_{O_2} of about 100 mmHg, hemoglobin is almost completely loaded with oxygen. Thus an increase in blood P_{O_2}—produced, for example, by breathing 100% oxygen from a gas tank—cannot significantly increase the amount of oxygen contained in the red blood cells. It can, however, significantly increase the amount of oxygen dissolved in the plasma (because the amount dissolved is directly determined by the P_{O_2}). If the P_{O_2} doubles, the amount of oxygen dissolved in the plasma also doubles, but the total oxygen content of whole blood increases only slightly. This is because the plasma contains relatively little oxygen compared to the red blood cells.

Since the oxygen carried by red blood cells must first dissolve in plasma before it can diffuse to the tissue cells, however, a doubling of the blood P_{O_2} means that the *rate of oxygen diffusion* to the tissues would double under these conditions. For this reason, breathing from a tank of 100% oxygen (with a P_{O_2} of 760 mmHg) would significantly increase oxygen delivery to the tissues, although it would have little effect on the total oxygen content of blood.

An electrode that produces a current in response to dissolved carbon dioxide is also used, so that the P_{CO_2} of blood can be measured together with its P_{O_2}. Blood in the systemic veins, which is delivered to the lungs by the pulmonary arteries, usually has a P_{O_2} of 40 mmHg and a P_{CO_2} of 46 mmHg. After gas exchange in the alveoli of the lungs, blood in the pulmonary veins and systemic arteries has a P_{O_2} of about 100 mmHg and a

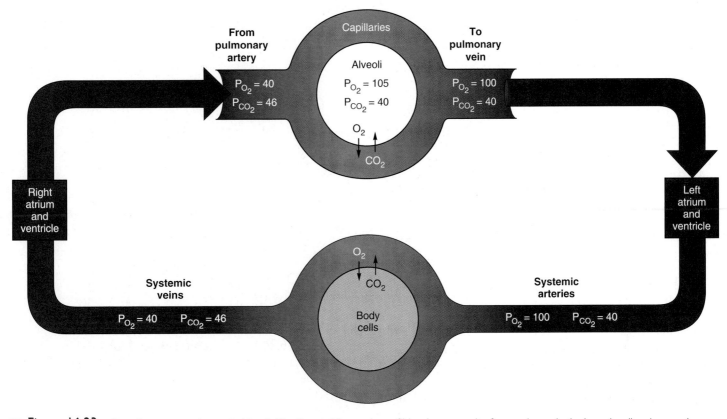

■ **Figure 16.23** **Partial pressures of gases in blood.** The P_{O_2} and P_{CO_2} values of blood are a result of gas exchange in the lung alveoli and gas exchange between systemic capillaries and body cells.

P_{CO_2} of 40 mmHg (fig. 16.23). The values in arterial blood are relatively constant and clinically significant because they reflect lung function. Blood gas measurements of venous blood are not as useful because these values are far more variable. Venous P_{O_2} is much lower and P_{CO_2} much higher after exercise, for example, than at rest, whereas arterial values are not significantly affected by moderate physical activity.

Clinical Investigation Clues

Remember that Harry had an abnormally high arterial P_{CO_2} when he was brought to the hospital. The later measurement, however, was normal.

What was the cause of his elevated arterial P_{CO_2}?
What was done to Harry that caused the later P_{CO_2} measurement to be normal?

Pulmonary Circulation and Ventilation/Perfusion Ratios

In a fetus, the pulmonary circulation has a high vascular resistance because the lungs are partially collapsed. This high vascular resistance helps to shunt blood from the right to the left atrium through the foramen ovale, and from the pulmonary ar-

tery to the aorta through the ductus arteriosus (described in chapter 13). After birth, the foramen ovale and ductus arteriosus close, and the vascular resistance of the pulmonary circulation falls sharply. This fall in vascular resistance at birth is due to (1) opening of the vessels as a result of the subatmospheric intrapulmonary pressure and physical stretching of the lungs during inspiration and (2) dilation of the pulmonary arterioles in response to increased alveolar P_{O_2}.

In the adult, the right ventricle (like the left) has a cardiac output of about 5.5 L per minute. The rate of blood flow through the pulmonary circulation is thus equal to the flow rate through the systemic circulation. Blood flow, as described in chapter 14, is directly proportional to the pressure difference between the two ends of a vessel and inversely proportional to the vascular resistance. In the systemic circulation, the mean arterial pressure is 90 to 100 mmHg and the pressure of the right atrium is 0 mmHg; therefore, the pressure difference is about 100 mmHg. The mean pressure of the pulmonary artery, by contrast, is only 15 mmHg and the pressure of the left atrium is 5 mmHg. The driving pressure in the pulmonary circulation is thus 15 – 5, or 10 mmHg.

Since the driving pressure in the pulmonary circulation is only one-tenth that of the systemic circulation, and since the flow rates are equal, it follows that the pulmonary vascular resistance must be one-tenth that of the systemic vascular resistance. The pulmonary circulation, in other words, is a low-resistance, low-pressure pathway. The low pulmonary blood pressure produces less filtration pressure (chapter 14; see fig. 14.9) than that produced in the systemic capillaries, and thus affords protection against *pulmonary edema.* This is a dangerous condition in which excessive fluid can enter the interstitial spaces of the lungs and then the alveoli, impeding ventilation and gas exchange. Pulmonary edema occurs when there is pulmonary hypertension, which may be produced by left ventricular heart failure.

Pulmonary arterioles constrict when the alveolar P_{O_2} is low and dilate as the alveolar P_{O_2} is raised. This response is opposite to that of systemic arterioles, which dilate in response to low tissue P_{O_2} (as described in chapter 14). Dilation of the systemic arterioles when the P_{O_2} is low helps to supply more blood and oxygen to the tissues; constriction of the pulmonary arterioles when the alveolar P_{O_2} is low helps to decrease blood flow to alveoli that are inadequately ventilated.

Constriction of the pulmonary arterioles where the alveolar P_{O_2} is low and their dilation where the alveolar P_{O_2} is high helps to *match ventilation to perfusion* (the term *perfusion* refers to blood flow). If this did not occur, blood from poorly ventilated alveoli would mix with blood from well-ventilated alveoli, and the blood leaving the lungs would have a lowered P_{O_2} as a result of this dilution effect.

Dilution of the P_{O_2} of pulmonary vein blood actually does occur to some degree despite these regulatory mechanisms. When a person stands upright, the force of gravity causes a greater blood flow to the base of the lungs than to the apex (top). Ventilation likewise increases from apex to base, but this increase is not proportionate to the increase in blood flow. The *ventilation/perfusion ratio* at the apex is thus high (0.24 L air divided by 0.07 L blood per minute gives a ratio of 3.4/1.0), while at the base of the lungs it is low (0.82 L air divided by 1.29 L blood per minute gives a ratio of 0.6/1.0). This is illustrated in figure 16.24.

Functionally, the alveoli at the apex of the lungs are thus overventilated (or underperfused) and are actually larger than alveoli at the base. This mismatch of ventilation/perfusion ratios is normal, but is largely responsible for the 5 mmHg difference in P_{O_2} between alveolar air and arterial blood. Abnormally large mismatches of ventilation/perfusion ratios can occur in cases of pneumonia, pulmonary emboli, edema, and other pulmonary disorders.

Disorders Caused by High Partial Pressures of Gases

The total atmospheric pressure increases by one atmosphere (760 mmHg) for every 10 m (33 ft) below sea level. If a diver descends 10 meters below sea level, therefore, the partial pressures and amounts of dissolved gases in the plasma will be twice those values at sea level. At 20 meters, they are three times, and at 30 meters they are four times the values at sea level. The increased amounts of nitrogen and oxygen dissolved in the blood plasma under these conditions can have serious effects on the body.

Oxygen Toxicity

Although breathing 100% oxygen at one or two atmospheres pressure can be safely tolerated for a few hours, higher partial

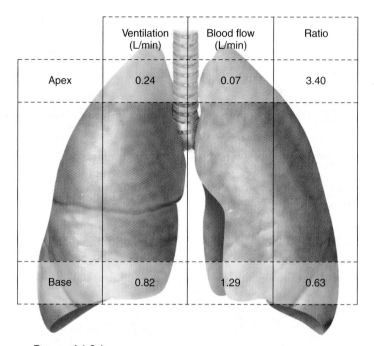

	Ventilation (L/min)	Blood flow (L/min)	Ratio
Apex	0.24	0.07	3.40
Base	0.82	1.29	0.63

■ **Figure 16.24**   Lung ventilation/perfusion ratios. The ventilation, blood flow, and ventilation/perfusion ratios are indicated for the apex and base of the lungs. The ratios indicate that the apex is relatively overventilated and the base underventilated in relation to their blood flows. As a result of such uneven matching of ventilation to perfusion, the blood leaving the lungs has a P_{O_2} that is slightly lower (by about 5 mmHg) than the P_{O_2} of alveolar air.

oxygen pressures can be very dangerous. Oxygen toxicity may develop rapidly when the P_{O_2} rises above about 2.5 atmospheres. This is apparently caused by the oxidation of enzymes and other destructive changes that can damage the nervous system and lead to coma and death. For these reasons, deep-sea divers commonly use gas mixtures in which oxygen is diluted with inert gases such as nitrogen (as in ordinary air) or helium.

Hyperbaric oxygen therapy, in which a patient is given 100% oxygen gas at 2 to 3 atmospheres pressure to breathe for varying lengths of time, is used to treat carbon monoxide poisoning, decompression sickness, severe traumatic injury (such as crush injury), infections that could lead to gas gangrene, and other conditions. While normal plasma oxygen concentration is 0.3 ml O_2/100 ml blood (as previously described), breathing 100% oxygen at a pressure of 3 atmospheres raises the plasma concentration to about 6 ml O_2/100 ml blood. This helps to kill anaerobic bacteria, such as those that cause gangrene; promote wound healing; reduce the size of gas bubbles (in the case of decompression sickness); and to quickly eliminate carbon monoxide from the body. Although hyperbaric oxygen was formerly used to treat premature infants for respiratory distress, the practice was discontinued because it caused a fibrotic deterioration of the retina that frequently resulted in blindness.

Nitrogen Narcosis

Although at sea level nitrogen is physiologically inert, larger amounts of dissolved nitrogen under hyperbaric (high-pressure) conditions have deleterious effects. Since it takes time for the nitrogen to dissolve, these effects usually do not appear until a diver has been submerged for more than an hour. **Nitrogen narcosis** resembles alcohol intoxication; depending on the depth of the dive, the diver may experience what Jacques Cousteau termed "rapture of the deep." Dizziness and extreme drowsiness are other narcotizing effects.

Decompression Sickness

The amount of nitrogen dissolved in the plasma as the diver ascends to sea level decreases, as a result of the progressive decrease in the P_{N_2}. If the diver surfaces slowly, a large amount of nitrogen can diffuse through the alveoli and be eliminated in the expired breath. If decompression occurs too rapidly, however, bubbles of nitrogen gas (N_2) can form in the tissue fluids and enter the blood. This process is analogous to the formation of carbon dioxide bubbles in a champagne bottle when the cork is removed. The bubbles of N_2 gas in the blood can block small blood channels, producing muscle and joint pain as well as more serious damage. These effects are known as **decompression sickness,** commonly called "the bends." The primary treatment for decompression sickness is hyperbaric oxygen treatment.

Airplanes that fly long distances at high altitudes (30,000 to 40,000 ft) have pressurized cabins so that the passengers and crew do not experience the very low atmospheric pressures of these altitudes. If a cabin were to become rapidly depressurized at high altitude, much less nitrogen could remain dissolved at the greatly lowered pressure. People in this situation, like the divers that ascend too rapidly, would thus experience decompression sickness.

Test Yourself Before You Continue

1. Explain how the P_{O_2} of air is calculated and how this value is affected by altitude, diving, and water vapor pressure.
2. Explain how blood P_{O_2} measurements are taken and discuss the physiological and clinical significance of these measurements.
3. Explain how the arterial P_{O_2} and the oxygen content of whole blood are affected by (a) hyperventilation, (b) breathing from a tank containing 100% oxygen, (c) anemia (low red blood cell count and hemoglobin concentration), and (d) high altitude.
4. Describe the ventilation/perfusion ratios of the lungs and explain why systemic arterial blood has a slightly lower P_{O_2} than alveolar air.
5. Explain how decompression sickness is produced in divers who ascend too rapidly.

Regulation of Breathing

The motor neurons that stimulate the respiratory muscles are controlled by two major descending pathways: one that controls voluntary breathing and another that controls involuntary breathing. The unconscious rhythmic control of breathing is influenced by sensory feedback from receptors sensitive to the P_{CO_2}, pH, and P_{O_2} of arterial blood.

Inspiration and expiration are produced by the contraction and relaxation of skeletal muscles in response to activity in somatic motor neurons in the spinal cord. The activity of these motor neurons is controlled, in turn, by descending tracts from neurons in the respiratory control centers in the medulla oblongata and from neurons in the cerebral cortex.

Brain Stem Respiratory Centers

Medulla Oblongata and Pons

A loose aggregation of neurons in the reticular formation of the *medulla oblongata* forms the **rhythmicity center** that controls automatic breathing. The rhythmicity center consists of interacting pools of neurons that fire either during inspiration (**inspiratory, or I, neurons**) or expiration (**expiratory, or E, neurons**). The I neurons project to and stimulate spinal motoneurons that innervate the respiratory muscles. Expiration is a passive process that occurs when the I neurons are inhibited, presumably by the activity of the E neurons.

The inspiratory neurons are located primarily in the *dorsal respiratory group,* and the expiratory neurons in the *ventral respiratory group.* These form two parallel columns within the medulla oblongata. The dorsal group of neurons regulates the activity of the phrenic nerves to the diaphragm, and the ventral group controls the motor neurons to the internal intercostal muscles.

The activity of the I and E neurons varies in a reciprocal way to produce a rhythmic pattern of breathing. There is evidence that the rhythmicity of I and E neurons may be driven by the cyclic activity of particular pacemaker neurons within the medulla. These pacemaker neurons display spontaneous, cyclic changes in the membrane potential, somewhat like the pacemaker cells of the heart (chapter 13).

The activity of the medullary rhythmicity center is influenced by centers in the *pons.* As a result of research in which the brain stem is destroyed at different levels, two respiratory control centers have been identified in the pons. One area—the **apneustic center**—appears to promote inspiration by stimulating the I neurons in the medulla. The other area—the **pneumotaxic center**—seems to antagonize the apneustic center and inhibit inspiration (fig. 16.25).

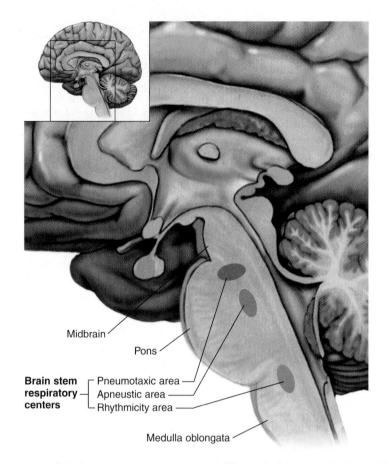

Midbrain

Pons

**Brain stem
respiratory
centers** ⎰ Pneumotaxic area
        ⎱ Apneustic area
        ⎱ Rhythmicity area

Medulla oblongata

 Figure 16.25 Approximate locations of the brain stem respiratory centers. The rhythmicity center in the medulla oblongata directly controls breathing, but it receives input from the control centers in the pons and from chemoreceptors.

Chemoreceptors

The automatic control of breathing is also influenced by input from receptors sensitive to the chemical composition of the blood. There are two groups of *chemoreceptors* that respond to changes in blood P_{CO_2}, pH, and P_{O_2}. These are the **central chemoreceptors** in the medulla oblongata and the **peripheral chemoreceptors.** The peripheral chemoreceptors are contained within small nodules associated with the aorta and the carotid arteries, and they receive blood from these critical arteries via small arterial branches. The peripheral chemoreceptors include the **aortic bodies,** located around the aortic arch, and the **carotid bodies,** located in each common carotid artery at the point where it branches into the internal and external carotid arteries (fig. 16.26). The aortic and carotid bodies should not be confused with the aortic and carotid sinuses (chapter 14) that are located within these arteries. The aortic and carotid sinuses contain receptors that monitor the blood pressure.

The peripheral chemoreceptors control breathing indirectly via sensory nerve fibers to the medulla. The aortic bodies send sensory information to the medulla in the vagus nerve (X); the carotid bodies stimulate sensory fibers in the glossopharyngeal nerve (IX). The neural and sensory control of ventilation is summarized in figure 16.27.

The automatic control of breathing is regulated by nerve fibers that descend in the lateral and ventral white matter of the spinal cord from the medulla oblongata. The voluntary control of breathing is a function of the cerebral cortex and involves nerve fibers that descend in the corticospinal tracts (chapter 8). The separation of the voluntary and involuntary pathways is dramatically illustrated in the condition called **Ondine's curse** (the term is taken from a German fairy tale). In this condition, neurological damage abolishes the automatic but not the voluntary control of breathing. People with Ondine's curse must remind themselves to breathe and they cannot go to sleep without the aid of a mechanical respirator.

Effects of Blood P_{CO_2} and pH on Ventilation

Chemoreceptor input to the brain stem modifies the rate and depth of breathing so that, under normal conditions, arterial P_{CO_2}, pH, and P_{O_2} remain relatively constant. If hypoventilation (inadequate ventilation) occurs, P_{CO_2} quickly rises and pH falls. The fall in pH is due to the fact that carbon dioxide can combine

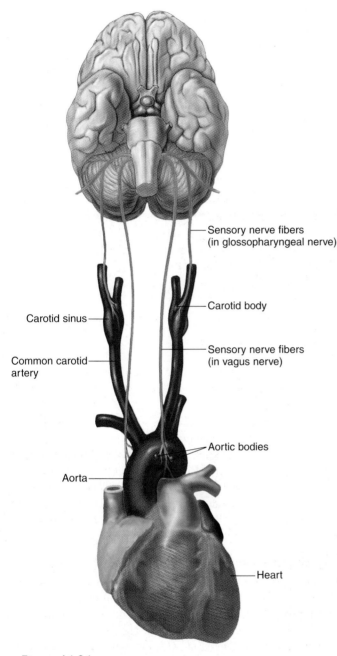

■ **Figure 16.26** Sensory input from the aortic and carotid bodies. The peripheral chemoreceptors (aortic and carotid bodies) regulate the brain stem respiratory centers by means of sensory nerve stimulation.

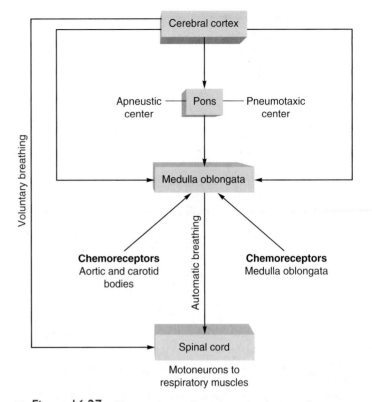

■ **Figure 16.27** The regulation of ventilation by the central nervous system. The feedback effects of pulmonary stretch receptors and "irritant" receptors on the control of breathing are not shown in this flowchart.

with water to form carbonic acid, which, as a weak acid, can release H$^+$ into the solution. This is shown in these equations:

$$CO_2 + H_2O \rightarrow H_2CO_3$$

$$H_2CO_3 \rightarrow H^+ + HCO_3^-$$

The oxygen content of the blood decreases much more slowly because of the large "reservoir" of oxygen attached to hemoglobin. During hyperventilation, conversely, blood P$_{CO_2}$ quickly falls and pH rises because of the excessive elimination of carbonic acid. The oxygen content of blood, on the other hand, is not significantly increased by hyperventilation (hemoglobin in arterial blood is 97% saturated with oxygen during normal ventilation).

For the reasons just stated, the blood P$_{CO_2}$ and pH are more immediately affected by changes in ventilation than is the oxygen content. Indeed, changes in P$_{CO_2}$ provide a sensitive index of ventilation, as shown in figure 16.28. In view of these facts, it is not surprising that changes in P$_{CO_2}$ provide the most potent stimulus for the reflex control of ventilation. Ventilation, in other words, is adjusted to maintain a constant P$_{CO_2}$; proper oxygenation of the blood occurs naturally as a side product of this reflex control.

The rate and depth of ventilation are normally adjusted to maintain an arterial P$_{CO_2}$ of 40 mmHg. Hypoventilation causes a rise in P$_{CO_2}$—a condition called *hypercapnia*. Hyperventilation, conversely, results in *hypocapnia*. Chemoreceptor regulation of breathing in response to changes in P$_{CO_2}$ is illustrated in figure 16.29.

Chemoreceptors in the Medulla

The chemoreceptors most sensitive to changes in the arterial P$_{CO_2}$ are located in the ventral area of the medulla oblongata, near the exit of the ninth and tenth cranial nerves. These chemoreceptor neurons are anatomically separate from, but synaptically communicate with, the neurons of the respiratory control center in the medulla.

An increase in arterial P$_{CO_2}$ causes a rise in the H$^+$ concentration of the blood as a result of increased carbonic acid concentrations. The H$^+$ in the blood, however, cannot cross the blood-brain barrier, and thus cannot influence the medullary chemoreceptors. Carbon dioxide in the arterial blood *can* cross the blood-brain barrier and, through the formation of carbonic acid, can lower the pH of cerebrospinal fluid (fig. 16.30). This

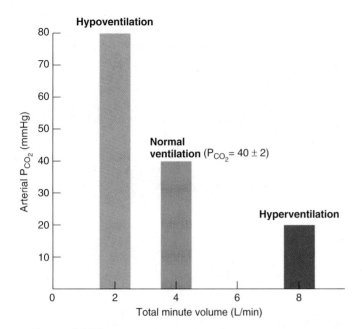

Figure 16.28 The relationship between total minute volume and arterial P$_{CO_2}$. Notice that these are inversely related: when the total minute volume increases by a factor of 2, the arterial P$_{CO_2}$ decreases by half. The total minute volume measures breathing, and is equal to the amount of air in each breath (the tidal volume) multiplied by the number of breaths per minute. The P$_{CO_2}$ measures the CO$_2$ concentration of arterial blood plasma.

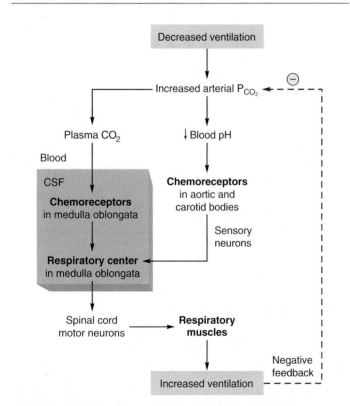

Figure 16.29 Chemoreceptor control of breathing. This figure depicts the negative feedback control of ventilation through changes in blood P$_{CO_2}$ and pH. The blood-brain barrier, represented by the orange box, allows CO$_2$ to pass into the cerebrospinal fluid but prevents the passage of H$^+$.

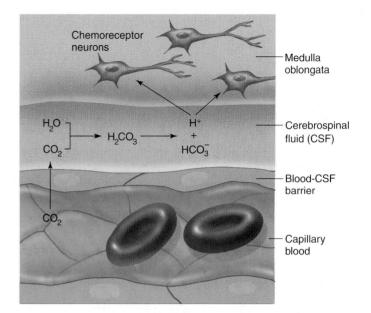

Figure 16.30 How blood CO$_2$ affects chemoreceptors in the medulla oblongata. An increase in blood CO$_2$ stimulates breathing indirectly by lowering the pH of blood and cerebrospinal fluid (CSF). This figure illustrates how a rise in blood CO$_2$ increases the H$^+$ concentration (lowers the pH) of CSF and thereby stimulates chemoreceptor neurons in the medulla oblongata.

fall in cerebrospinal fluid pH directly stimulates the chemoreceptors in the medulla when there is a rise in arterial P$_{CO_2}$.

The chemoreceptors in the medulla are ultimately responsible for 70% to 80% of the increased ventilation that occurs in response to a sustained rise in arterial P$_{CO_2}$. This response, however, takes several minutes. The immediate increase in ventilation that occurs when P$_{CO_2}$ rises is produced by stimulation of the peripheral chemoreceptors.

Peripheral Chemoreceptors

The aortic and carotid bodies are not stimulated directly by blood CO$_2$. Instead, they are stimulated by a rise in the H$^+$ concentration (fall in pH) of arterial blood, which occurs when the blood CO$_2$, and thus carbonic acid, is raised. The retention of CO$_2$ during hypoventilation thus stimulates the medullary chemoreceptors through a lowering of cerebrospinal fluid pH and stimulates peripheral chemoreceptors through a lowering of blood pH.

People who hyperventilate during psychological stress are sometimes told to breathe into a paper bag so that they rebreathe their expired air, enriched with CO$_2$. This procedure helps to raise their blood P$_{CO_2}$ back up to the normal range. This is needed because hypocapnia causes cerebral vasoconstriction, reducing brain perfusion and producing ischemia. The cerebral ischemia causes dizziness and can lead to an acidotic condition in the brain, which, through stimulation of the medullary chemoreceptors, causes further hyperventilation. Breathing into a paper bag can thus relieve the hypocapnia and stop the hyperventilation.

Effects of Blood P_{O_2} on Ventilation

Under normal conditions, blood P_{O_2} affects breathing only indirectly, by influencing the chemoreceptor sensitivity to changes in P_{CO_2}. Chemoreceptor sensitivity to P_{CO_2} is augmented by a low P_{O_2} (so ventilation is increased at a high altitude, for example) and is decreased by a high P_{O_2}. If the blood P_{O_2} is raised by breathing 100% oxygen, therefore, the breath can be held longer because the response to increased P_{CO_2} is blunted.

When the blood P_{CO_2} is held constant by experimental techniques, the P_{O_2} of arterial blood must fall from 100 mmHg to below 50 mmHg before ventilation is significantly stimulated (fig. 16.31). This stimulation is apparently due to a direct effect of P_{O_2} on the carotid bodies. Since this degree of *hypoxemia,* or low blood oxygen, does not normally occur at sea level, P_{O_2} does not normally exert this direct effect on breathing.

In emphysema, when there is a chronic retention of carbon dioxide, the chemoreceptor response to carbon dioxide becomes blunted. This is because the choroid plexus in the brain (chapter 8) secretes more bicarbonate into the cerebrospinal fluid, buffering the fall in cerebrospinal fluid pH. The abnormally high P_{CO_2}, however, enhances the sensitivity of the carotid bodies to a fall in P_{O_2}. For people with emphysema, therefore, breathing may be stimulated by a **hypoxic drive** rather than by increases in blood P_{CO_2}. Over a long period, however, the chronic hypoxia reduces the sensitivity of the carotid bodies in people with emphysema or other forms of chronic obstructive pulmonary disease, exacerbating their breathing problems.

The effects of changes in the blood P_{CO_2}, pH, and P_{O_2} on chemoreceptors and the regulation of ventilation are summarized in table 16.6.

A variety of disease processes can produce cessation of breathing during sleep, or *sleep apnea.* **Sudden infant death syndrome (SIDS)** is an especially tragic condition that claims about 1 in 1,000 babies under 12 months in the United States annually. Victims are apparently healthy 2-to-5-month-old babies who die in their sleep for no obvious reason—hence, the layperson's term "crib death." These deaths seem to be caused by failure of the respiratory control mechanisms in the brain stem and/or by failure of the carotid bodies to be stimulated by reduced arterial oxygen. Since 1992, when the American Academy of Pediatrics began a campaign recommending that parents put infants to sleep on their backs rather than on their stomachs, the number of infants dying from SIDS has dropped by 38%.

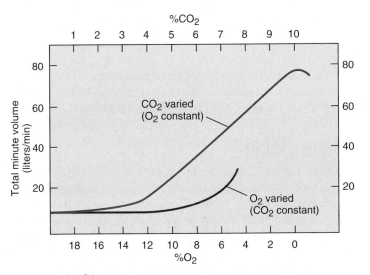

Figure 16.31 **Comparing the effects of blood CO_2 and O_2 on breathing.** The graph depicts the effects of increasing blood concentrations of CO_2 (see the scale at the top of the graph) on breathing, as measured by the total minute volume. The effects of decreasing concentrations of blood O_2 (see the scale at the bottom of the graph) on breathing are also shown for comparison. Notice that breathing increases linearly with increasing CO_2 concentration, whereas O_2 concentrations must decrease to half the normal value before breathing is stimulated.

Effects of Pulmonary Receptors on Ventilation

The lungs contain various types of receptors that influence the brain stem respiratory control centers via sensory fibers in the vagus nerves. **Unmyelinated C fibers** are sensory neurons in the lungs that can be stimulated by *capsaicin,* the chemical in hot peppers that creates the burning sensation. These receptors produce an initial apnea, followed by rapid, shallow breathing when a person eats these peppers. The unmyelinated C fibers are also stimulated by histamine and bradykinin, which are released in response to noxious agents. **Irritant receptors** in the wall of the larynx, and receptors in the lungs identified as **rapidly adapting receptors,** can cause a person to cough in response to

Table 16.6 Sensitivity of Chemoreceptors to Changes in Blood Gases and pH

Stimulus	Chemoreceptor	Comments
↑P_{CO_2}	Medullary chemoreceptors; aortic and carotid bodies	Medullary chemoreceptors are sensitive to the pH of cerebrospinal fluid (CSF). Diffusion of CO_2 from the blood into the CSF lowers the pH of CSF by forming carbonic acid. Similarly, the aortic and carotid bodies are stimulated by a fall in blood pH induced by increases in blood CO_2.
↓pH	Aortic and carotid bodies	Peripheral chemoreceptors are stimulated by decreased blood pH independent of the effect of blood CO_2. Chemoreceptors in the medulla are not affected by changes in blood pH because H$^+$ cannot cross the blood-brain barrier.
↓P_{O_2}	Carotid bodies	Low blood P_{O_2} (hypoxemia) augments the chemoreceptor response to increases in blood P_{CO_2} and can stimulate ventilation directly when the P_{O_2} falls below 50 mmHg.

components of smoke and smog, and to inhaled particulates. The rapidly adapting receptors in the lungs are stimulated most directly by an increase in the amount of fluid in the pulmonary interstitial tissue. Since the same chemicals that stimulate the unmyelinated C fibers can cause increased pulmonary interstitial fluid (due to extravasation from pulmonary capillaries—see chapter 14), a person may also cough after eating hot peppers.

The **Hering-Breuer reflex** is stimulated by **pulmonary stretch receptors.** The activation of these receptors during inspiration inhibits the respiratory control centers, making further inspiration increasingly difficult. This helps to prevent undue distension of the lungs and may contribute to the smoothness of the ventilation cycles. A similar inhibitory reflex may occur during expiration. The Hering-Breuer reflex appears to be important in maintaining normal ventilation in the newborn. Pulmonary stretch receptors in adults, however, are probably not active at normal resting tidal volumes (500 ml per breath) but may contribute to respiratory control at high tidal volumes, as during exercise.

Test Yourself Before You Continue

1. Describe the effects of voluntary hyperventilation and breath holding on arterial P_{CO_2}, pH, and oxygen content. Indicate the relative degree of changes in these values.

2. Using a flowchart to show a negative feedback loop, explain the relationship between ventilation and arterial P_{CO_2}.

3. Explain the effect of increased arterial P_{CO_2} on (a) chemoreceptors in the medulla oblongata and (b) chemoreceptors in the aortic and carotid bodies.

4. Explain the role of arterial P_{O_2} in the regulation of breathing. Why does ventilation increase when a person goes to a high altitude?

Hemoglobin and Oxygen Transport

Hemoglobin without oxygen, or deoxyhemoglobin, can bond with oxygen to form oxyhemoglobin. This "loading" reaction occurs in the capillaries of the lungs. The dissociation of oxyhemoglobin, or "unloading" reaction, occurs in the tissue capillaries. The bond strength between hemoglobin and oxygen, and thus the extent of the unloading reaction, is adjusted by various factors to ensure an adequate delivery of oxygen to the tissues.

If the lungs are functioning properly, blood leaving in the pulmonary veins and traveling in the systemic arteries has a P_{O_2} of about 100 mmHg, indicating a plasma oxygen concentration of about 0.3 ml O_2 per 100 ml blood. The total oxygen content of the blood, however, cannot be derived if only the P_{O_2} of plasma is known. The total oxygen content depends not only on

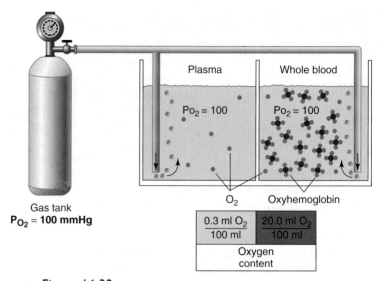

Figure 16.32   **The oxygen content of blood.** Plasma and whole blood that are brought into equilibrium with the same gas mixture have the same P_{O_2}, and thus the same number of dissolved oxygen molecules (shown as black dots). The oxygen content of whole blood, however, is much higher than that of plasma because of the binding of oxygen to hemoglobin.

the P_{O_2} but also on the hemoglobin concentration. If the P_{O_2} and hemoglobin concentration are normal, arterial blood contains about 20 ml of O_2 per 100 ml blood (fig. 16.32).

Hemoglobin

Most of the oxygen in the blood is contained within the red blood cells, where it is chemically bonded to **hemoglobin.** As described in chapter 13, each hemoglobin molecule consists of four polypeptide chains called *globins* and four nitrogen-containing, disc-shaped organic pigment molecules called *hemes* (fig. 16.33).

The protein part of hemoglobin is composed of two identical *alpha chains,* each 141 amino acids long, and two identical *beta chains,* each 146 amino acids long. Each of the four polypeptide chains is combined with one heme group. In the center of each heme group is one atom of iron, which can combine with one molecule of oxygen. One hemoglobin molecule can thus combine with four molecules of oxygen—and since there are about 280 million hemoglobin molecules per red blood cell, each red blood cell can carry over a billion molecules of oxygen.

Normal heme contains iron in the reduced form (Fe^{2+}, or ferrous iron). In this form, the iron can share electrons and bond with oxygen to form **oxyhemoglobin.** When oxyhemoglobin dissociates to release oxygen to the tissues, the heme iron is still in the reduced (Fe^{2+}) form and the hemoglobin is called **deoxyhemoglobin,** or **reduced hemoglobin.** The term *oxyhemoglobin* is thus not equivalent to *oxidized* hemoglobin; hemoglobin does not lose an electron (and become oxidized) when it combines with oxygen. Oxidized hemoglobin, or **methemoglobin,** has iron in the oxidized (Fe^{3+}, or ferric) state. Methemoglobin thus lacks the electron it needs to form a bond with oxygen and cannot

Hemoglobin

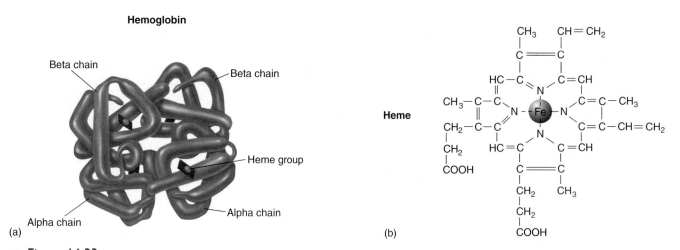

Figure 16.33 **The structure of hemoglobin.** *(a)* An illustration of the three-dimensional structure of hemoglobin in which the two alpha and two beta polypeptide chains are shown. The four heme groups are represented as flat structures with atoms of iron *(spheres)* in the centers. *(b)* The structural formula for heme.

participate in oxygen transport. Blood normally contains only a small amount of methemoglobin, but certain drugs can increase this amount.

In **carboxyhemoglobin,** another abnormal form of hemoglobin, the reduced heme is combined with *carbon monoxide* instead of oxygen. Since the bond with carbon monoxide is about 210 times stronger than the bond with oxygen, carbon monoxide tends to displace oxygen in hemoglobin and remains attached to hemoglobin as the blood passes through systemic capillaries. In carbon monoxide poisoning—which, in severe form results primarily from smoke inhalation and suicide attempts, and in milder forms from breathing smoggy air and smoking cigarettes—the transport of oxygen to the tissues is reduced.

According to federal standards, the percentage of carboxyhemoglobin in the blood of active nonsmokers should be no higher than 1.5%. However, concentrations of 3% in nonsmokers and 10% in smokers have been reported in some cities. Although these high levels may not cause immediate problems in healthy people, long-term adverse effects on health are possible. People with respiratory or cardiovascular diseases would be particularly vulnerable to the negative effects of carboxyhemoglobin on oxygen transport.

Clinical Investigation Clues

Remember that Harry smoked cigarettes, drove a taxi, and had a carboxyhemoglobin saturation of 18%

How are these observations related?

What danger does this pose to Harry (note—be sure to take his other problems into account)?

What can he do to lower his carboxyhemoglobin saturation?

Hemoglobin Concentration

The *oxygen-carrying capacity* of whole blood is determined by its concentration of hemoglobin. If the hemoglobin concentration is below normal—in a condition called **anemia**—the oxygen content of the blood will be abnormally low. Conversely, when the hemoglobin concentration rises above the normal range—as occurs in **polycythemia** (high red blood cell count)—the oxygen-carrying capacity of blood is increased accordingly. This can occur as an adaptation to life at a high altitude.

The production of hemoglobin and red blood cells in bone marrow is controlled by a hormone called **erythropoietin,** produced by the kidneys. The secretion of erythropoietin—and thus the production of red blood cells—is stimulated when the amount of oxygen delivered to the kidneys is lower than normal. Red blood cell production is also promoted by androgens, which explains why the hemoglobin concentration in men is from 1 to 2 g per 100 ml higher than in women.

The Loading and Unloading Reactions

Deoxyhemoglobin and oxygen combine to form oxyhemoglobin; this is called the **loading reaction.** Oxyhemoglobin, in turn, dissociates to yield deoxyhemoglobin and free oxygen molecules; this is the **unloading reaction.** The loading reaction occurs in the lungs and the unloading reaction occurs in the systemic capillaries.

Loading and unloading can thus be shown as a reversible reaction:

$$\text{Deoxyhemoglobin} + O_2 \underset{\text{(tissues)}}{\overset{\text{(lungs)}}{\rightleftharpoons}} \text{Oxyhemoglobin}$$

The extent to which the reaction will go in each direction depends on two factors: (1) the P_{O_2} of the environment and (2) the *affinity,* or bond strength, between hemoglobin and oxygen. High P_{O_2} drives the equation to the right (favors the loading reaction); at the high P_{O_2} of the pulmonary capillaries, almost all

Table 16.7   Relationship Between Percent Oxyhemoglobin Saturation and P_{O_2} (at pH of 7.40 and Temperature of 37° C)

P_{O_2} (mmHg)	100	80	61	45	40	36	30	26	23	21	19
Percent Oxyhemoglobin	97	95	90	80	75	70	60	50	40	35	30
	Arterial Blood				Venous Blood						

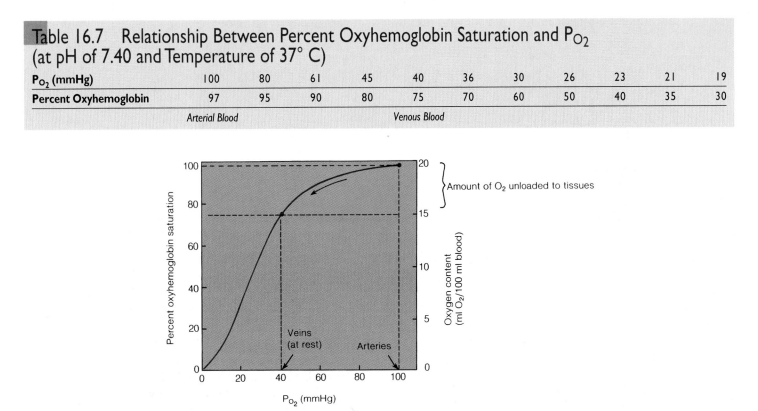

■ **Figure 16.34**   **The oxyhemoglobin dissociation curve.** The percentage of oxyhemoglobin saturation and the blood oxygen content are shown at different values of P_{O_2}. Notice that the percent oxyhemoglobin decreases by about 25% as the blood passes through the tissue from arteries to veins, resulting in the unloading of approximately 5 ml of O_2 per 100 ml of blood to the tissues.

the deoxyhemoglobin molecules combine with oxygen. Low P_{O_2} in the systemic capillaries drives the reaction in the opposite direction to promote unloading. The extent of this unloading depends on how low the P_{O_2} values are.

The affinity between hemoglobin and oxygen also influences the loading and unloading reactions. A very strong bond would favor loading but inhibit unloading; a weak bond would hinder loading but improve unloading. The bond strength between hemoglobin and oxygen is normally strong enough so that 97% of the hemoglobin leaving the lungs is in the form of oxyhemoglobin, yet the bond is sufficiently weak so that adequate amounts of oxygen are unloaded to sustain aerobic respiration in the tissues.

The Oxyhemoglobin Dissociation Curve

Blood in the systemic arteries, at a P_{O_2} of 100 mmHg, has a *percent oxyhemoglobin saturation* of 97% (which means that 97% of the hemoglobin is in the form of oxyhemoglobin). This blood is delivered to the systemic capillaries, where oxygen diffuses into the cells and is consumed in aerobic respiration. Blood leaving in the systemic veins is thus reduced in oxygen; it has a P_{O_2} of about 40 mmHg and a percent oxyhemoglobin saturation of about 75% when a person is at rest (table 16.7). Expressed another way, blood entering the tissues contains 20 ml O_2 per 100 ml blood, and blood leaving the tissues contains 15.5 ml O_2

per 100 ml blood (fig. 16.34). Thus, 22%, or 4.5 ml of O_2 out of the 20 ml of O_2 per 100 ml blood, is unloaded to the tissues.

A graphic illustration of the percent oxyhemoglobin saturation at different values of P_{O_2} is called an **oxyhemoglobin dissociation curve** (fig. 16.34). The values in this graph are obtained by subjecting samples of blood *in vitro* to different partial oxygen pressures. The percent oxyhemoglobin saturations obtained, however, can be used to predict what the unloading percentages would be *in vivo* with a given difference in arterial and venous P_{O_2} values.

Figure 16.34 shows the difference between the arterial and venous P_{O_2} and the percent oxyhemoglobin saturation at rest. The relatively large amount of oxyhemoglobin remaining in the venous blood at rest serves as an oxygen reserve. If a person stops breathing, a sufficient reserve of oxygen in the blood will keep the brain and heart alive for about 4 to 5 minutes without using cardiopulmonary resuscitation (CPR) techniques. This reserve supply of oxygen can also be tapped when a tissue's requirements for oxygen are raised.

The oxyhemoglobin dissociation curve is S-shaped, or *sigmoidal*. The fact that it is relatively flat at high P_{O_2} values indicates that changes in P_{O_2} within this range have little effect on the loading reaction. One would have to ascend as high as 10,000 feet, for example, before the oxyhemoglobin saturation of arterial blood would decrease from 97% to 93%. At more common elevations, the percent oxyhemoglobin saturation would not be significantly different from the 97% value at sea level.

Table 16.8 Effect of pH on Hemoglobin Affinity for Oxygen and Unloading of Oxygen to the Tissues

pH	Affinity	Arterial O$_2$ Content per 100 ml	Venous O$_2$ Content per 100 ml	O$_2$ Unloaded to Tissues per 100 ml
7.40	Normal	19.8 ml O$_2$	14.8 ml O$_2$	5.0 ml O$_2$
7.60	Increased	20.0 ml O$_2$	17.0 ml O$_2$	3.0 ml O$_2$
7.20	Decreased	19.2 ml O$_2$	12.6 ml O$_2$	6.6 ml O$_2$

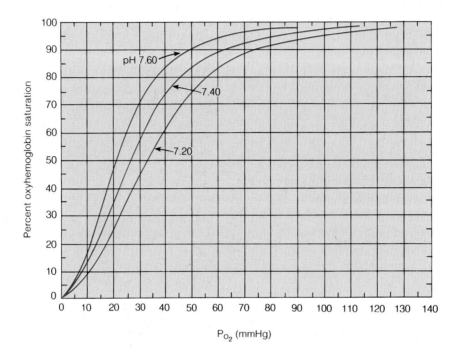

■ **Figure 16.35** **The effect of pH on the oxyhemoglobin dissociation curve.** A decrease in blood pH (an increase in H$^+$ concentration) decreases the affinity of hemoglobin for oxygen at each P$_{O_2}$ value, resulting in a "shift to the right" of the oxyhemoglobin dissociation curve. This is called the Bohr effect. A curve that is shifted to the right has a lower percent oxyhemoglobin saturation at each P$_{O_2}$.

At the steep part of the sigmoidal curve, however, small changes in P$_{O_2}$ values produce large differences in percent saturation. A decrease in *venous* P$_{O_2}$ from 40 mmHg to 30 mmHg, as might occur during mild exercise, corresponds to a change in percent saturation from 75% to 58%. Since the *arterial* percent saturation is usually still 97% during exercise, the lowered venous percent saturation indicates that more oxygen has been unloaded to the tissues. The difference between the arterial and venous percent saturations indicates the percent unloading. In the preceding example, 97% − 75% = 22% unloading at rest, and 97% − 58% = 39% unloading during mild exercise. During heavier exercise, the venous P$_{O_2}$ can drop to 20 mmHg or lower, indicating a percent unloading of about 80%.

Effect of pH and Temperature on Oxygen Transport

In addition to changes in P$_{O_2}$, the loading and unloading reactions are influenced by changes in the affinity of hemoglobin for oxygen. Such changes ensure that active skeletal muscles will receive more oxygen from the blood than they do at rest. This occurs as a result of the lowered pH and increased temperature in exercising muscles.

The affinity is decreased when the pH is lowered and increased when the pH is raised; this is called the **Bohr effect.** When the affinity of hemoglobin for oxygen is reduced, there is slightly less loading of the blood with oxygen in the lungs but greater unloading of oxygen in the tissues. The net effect is that the tissues receive more oxygen when the blood pH is lowered (table 16.8). Since the pH can be decreased by carbon dioxide (through the formation of carbonic acid), the Bohr effect helps to provide more oxygen to the tissues when their carbon dioxide output is increased by a faster metabolism.

When oxyhemoglobin dissociation curves are graphed at different pH values, the dissociation curve is seen to be shifted to the right by a lowering of pH and shifted to the left by a rise in pH (fig. 16.35). If the percent unloading is calculated (by subtracting the percent oxyhemoglobin saturation for arterial and venous blood), it will be seen that a *shift to the right* of the curve indicates a greater unloading of oxygen. A *shift to the left,* conversely, indicates less unloading but slightly more oxygen loading in the lungs.

When oxyhemoglobin dissociation curves are constructed at different temperatures, the curve moves rightward as the temperature increases. The rightward shift of the curve indicates that the affinity of hemoglobin for oxygen is decreased by a rise in temperature. An increase in temperature weakens the bond between hemoglobin and oxygen and thus has the same effect as a fall in pH. At higher temperatures, therefore, more oxygen is unloaded to the tissues than would be the case if the bond strength were constant. This effect can significantly enhance the delivery of oxygen to muscles that are warmed during exercise.

Effect of 2,3-DPG on Oxygen Transport

Mature red blood cells lack both nuclei and mitochondria. Without mitochondria they cannot respire aerobically; the very cells that carry oxygen are the only cells in the body that cannot use it! Red blood cells, therefore, must obtain energy through the anaerobic respiration of glucose. At a certain point in the glycolytic pathway, a "side reaction" occurs in the red blood cells that results in a unique product—**2,3-diphosphoglyceric acid (2,3-DPG).**

The enzyme that produces 2,3-DPG is inhibited by oxyhemoglobin. When the oxyhemoglobin concentration is decreased, therefore, the production of 2,3-DPG is increased. This increase in 2,3-DPG production can occur when the total hemoglobin concentration is low (in anemia) or when the P_{O_2} is low (at a high altitude, for example). The bonding of 2,3-DPG with deoxyhemoglobin makes the deoxyhemoglobin more stable. Therefore, a higher proportion of the oxyhemoglobin will be converted to deoxyhemoglobin by the unloading of its oxygen. An increased concentration of 2,3-DPG in red blood cells thus increases oxygen unloading (table 16.9) and shifts the oxyhemoglobin dissociation curve to the right.

The importance of 2,3-DPG in red blood cells is now recognized in blood banking. Red blood cells that have been stored for some time can lose their ability to produce 2,3-DPG as they lose their ability to metabolize glucose. Modern techniques for blood storage, therefore, include the addition of energy substrates for respiration and phosphate sources needed for the production of 2,3-DPG.

Anemia

When the total blood hemoglobin concentration falls below normal in anemia, each red blood cell produces increased amounts of 2,3-DPG. A normal hemoglobin concentration of 15 g per 100 ml unloads about 4.5 ml O_2 per 100 ml at rest, as previously described. If the hemoglobin concentration were reduced by half, you might expect that the tissues would receive only half the normal amount of oxygen (2.25 ml O_2 per 100 ml). It has been shown, however, that an amount as great as 3.3 ml O_2 per 100 ml is unloaded to the tissues under these conditions. This occurs as a result of a rise in 2,3-DPG production that causes a decrease in the affinity of hemoglobin for oxygen.

Fetal Hemoglobin

The effects of 2,3-DPG are also important in the transfer of oxygen from maternal to fetal blood. In an adult, hemoglobin molecules are composed of two alpha and two beta chains as previously described, whereas fetal hemoglobin has two *gamma* chains in place of the beta chains (gamma chains differ from beta chains in thirty-seven of their amino acids). Normal **adult hemoglobin** in the mother (**hemoglobin A**) is able to bind to 2,3-DPG. **Fetal hemoglobin,** or **hemoglobin F,** by contrast, cannot bind to 2,3-DPG, and thus has a higher affinity for oxygen than does hemoglobin A. Since hemoglobin F can have a higher percent oxyhemoglobin than hemoglobin A at a given P_{O_2}, oxygen is transferred from the maternal to the fetal blood as these two come into close proximity in the placenta.

Inherited Defects in Hemoglobin Structure and Function

A number of hemoglobin diseases are produced by inherited (congenital) defects in the protein part of hemoglobin. **Sickle-cell anemia**—a disease occurring almost exclusively in blacks and carried in a recessive state by 8% to 11% of the African-American population—for example, is caused by an abnormal form of hemoglobin called *hemoglobin S*. Hemoglobin S differs from normal hemoglobin A in only one amino acid: valine is substituted for glutamic acid in position 6 on the beta chains. This amino acid substitution is caused by a single base change in the region of DNA that codes for the beta chains.

Under conditions of low blood P_{O_2}, hemoglobin S comes out of solution and cross-links to form a "paracrystalline gel" within the red blood cells. This causes the characteristic sickle shape of

Table 16.9 Factors That Affect the Affinity of Hemoglobin for Oxygen and the Position of the Oxyhemoglobin Dissociation Curve

Factor	Affinity	Position of Curve	Comments
$\downarrow$pH	Decreased	Shift to the right	Called the Bohr effect; increases oxygen delivery during hypercapnia
$\uparrow$Temperature	Decreased	Shift to the right	Increases oxygen unloading during exercise and fever
$\uparrow$2,3-DPG	Decreased	Shift to the right	Increases oxygen unloading when there is a decrease in total hemoglobin or total oxygen content; an adaptation to anemia and high-altitude living

red blood cells (fig. 16.36) and makes them less flexible and more fragile. Since red blood cells must be able to bend in the middle to pass through many narrow capillaries, a decrease in their flexibility may cause them to block small blood channels and produce organ ischemia. The decreased solubility of hemoglobin S in solutions of low P_{O_2} is used in the diagnosis of sickle-cell anemia and sickle-cell trait (the carrier state, in which a person has the genes for both hemoglobin A and hemoglobin S). Sickle-cell anemia is treated with the drug *hydroxyurea,* which stimulates the production of gamma chains (characteristic of hemoglobin F) in place of the defective beta chains of hemoglobin S.

Thalassemia is any of a family of hemoglobin diseases found predominantly among people of Mediterranean ancestry. In *alpha thalassemia,* there is decreased synthesis of the alpha chains of hemoglobin, whereas in *beta thalassemia* the synthesis of the beta chains is impaired. One of the compensations for thalassemia is increased synthesis of gamma chains, resulting in the retention of large amounts of hemoglobin F (fetal hemoglobin) into adulthood.

Some types of abnormal hemoglobins have been shown to be advantageous in the environments in which they evolved. A person who is a carrier for sickle-cell anemia, for example (and who therefore has both hemoglobin A and hemoglobin S), has a high resistance to malaria. This is because the parasite that causes malaria cannot live in red blood cells that contain hemoglobin S.

Muscle Myoglobin

As described in chapter 12, **myoglobin** is a red pigment found exclusively in striated muscle cells. In particular, slow-twitch, aerobically respiring skeletal fibers and cardiac muscle cells are rich in myoglobin. Myoglobin is similar to hemoglobin, but it has one heme rather than four; therefore, it can combine with only one molecule of oxygen.

Myoglobin has a higher affinity for oxygen than does hemoglobin, and its dissociation curve is therefore to the left of the oxyhemoglobin dissociation curve (fig. 16.37). The shape of the myoglobin curve is also different from the oxyhemoglobin

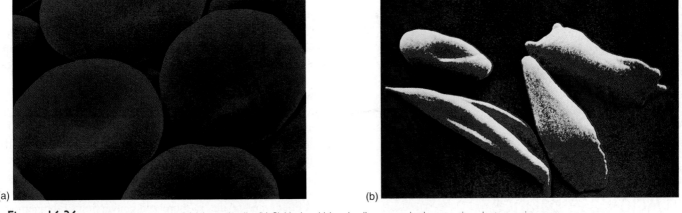

(a)   (b)

■ **Figure 16.36**   Sickle-cell anemia. *(a)* Normal cells. *(b)* Sickled red blood cells as seen in the scanning electron microscope.

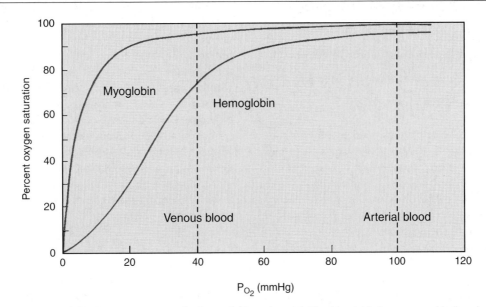

■ **Figure 16.37**   A comparison of the dissociation curves for hemoglobin and myoglobin. Myoglobin is an oxygen-binding pigment in skeletal muscles. At the P_{O_2} of venous blood, the myoglobin retains almost all of its oxygen, indicating that it has a higher affinity than hemoglobin for oxygen. The myoglobin, however, does release its oxygen at the very low P_{O_2} values found inside the mitochondria.

dissociation curve. The myoglobin curve is rectangular, indicating that oxygen will be released only when the P_{O_2} becomes very low.

Since the P_{O_2} in mitochondria is very low (because oxygen is incorporated into water here), myoglobin may act as a "go-between" in the transfer of oxygen from blood to the mitochondria within muscle cells. Myoglobin may also have an oxygen-storage function, which is of particular importance in the heart. During diastole, when the coronary blood flow is greatest, myoglobin can load up with oxygen. This stored oxygen can then be released during systole, when the coronary arteries are squeezed closed by the contracting myocardium.

Test Yourself Before You Continue

1. Use a graph to illustrate the effects of P_{O_2} on the loading and unloading reactions.
2. Draw an oxyhemoglobin dissociation curve and label the P_{O_2} values for arterial and venous blood under resting conditions. Use this graph to show the changes in unloading that occur during exercise.
3. Explain how changes in pH and temperature affect oxygen transport and state when such changes occur.
4. Explain how a person who is anemic or a person at high altitude could have an increase in the percent unloading of oxygen by hemoglobin.

Carbon Dioxide Transport and Acid-Base Balance

Carbon dioxide is transported in the blood primarily in the form of bicarbonate (HCO_3^-), which is released when carbonic acid dissociates. Bicarbonate can buffer H^+, and thus helps to maintain a normal arterial pH. Hypoventilation raises, and hyperventilation lowers, the carbonic acid concentration of the blood.

Carbon dioxide is carried by the blood in three forms: (1) as *dissolved* CO_2—carbon dioxide is about twenty-one times more soluble than oxygen in water, and about one-tenth of the total blood CO_2 is dissolved in plasma; (2) as *carbaminohemoglobin*—about one-fifth of the total blood CO_2 is carried attached to an amino acid in hemoglobin (carbaminohemoglobin should not be confused with carboxyhemoglobin, which is a combination of hemoglobin and carbon monoxide); and (3) as *bicarbonate ion,* which accounts for most of the CO_2 carried by the blood.

Carbon dioxide is able to combine with water to form carbonic acid. This reaction occurs spontaneously in the plasma at a slow rate, but it occurs much more rapidly within the red blood cells because of the catalytic action of the enzyme **carbonic anhydrase.** Since this enzyme is confined to the red blood cells,

most of the carbonic acid is produced there rather than in the plasma. The formation of carbonic acid from CO_2 and water is favored by the high P_{CO_2} found in tissue capillaries (this is an example of the *law of mass action,* described in chapter 4).

$$CO_2 + H_2O \xrightarrow[\text{high } P_{CO_2}]{\text{carbonic anhydrase}} H_2CO_3$$

The Chloride Shift

As a result of catalysis by carbonic anhydrase within the red blood cells, large amounts of carbonic acid are produced as blood passes through the systemic capillaries. The buildup of carbonic acid concentrations within the red blood cells favors the dissociation of these molecules into hydrogen ions (protons, which contribute to the acidity of a solution) and HCO_3^- (bicarbonate), as shown by this equation:

$$H_2CO_3 \rightarrow H^+ + HCO_3^-$$

The hydrogen ions (H^+) released by the dissociation of carbonic acid are largely buffered by their combination with deoxyhemoglobin within the red blood cells. Although the unbuffered hydrogen ions are free to diffuse out of the red blood cells, more bicarbonate diffuses outward into the plasma than does H^+. As a result of the "trapping" of hydrogen ions within the red blood cells by their attachment to hemoglobin and the outward diffusion of bicarbonate, the inside of the red blood cell gains a net positive charge. This attracts chloride ions (Cl^-), which move into the red blood cells as HCO_3^- moves out. This exchange of anions as blood travels through the tissue capillaries is called the **chloride shift** (fig. 16.38).

The unloading of oxygen is increased by the bonding of H^+ (released from carbonic acid) to oxyhemoglobin. This is the Bohr effect, and results in increased conversion of oxyhemoglobin to deoxyhemoglobin. Now, deoxyhemoglobin bonds H^+ more strongly than does oxyhemoglobin, so the act of unloading its oxygen improves the ability of hemoglobin to buffer the H^+ released by carbonic acid. Removal of H^+ from solution by combining with hemoglobin (through the law of mass action), in turn, favors the continued production of carbonic acid and thereby improves the ability of the blood to transport carbon dioxide. Thus, carbon dioxide increases oxygen unloading, and oxygen unloading increases carbon dioxide transport.

When blood reaches the pulmonary capillaries (fig. 16.39), deoxyhemoglobin is converted to oxyhemoglobin. Since oxyhemoglobin has a weaker affinity for H^+ than does deoxyhemoglobin, hydrogen ions are released within the red blood cells. This attracts HCO_3^- from the plasma, which combines with H^+ to form carbonic acid:

$$H^+ + HCO_3^- \rightarrow H_2CO_3$$

Under conditions of lower P_{CO_2}, as occurs in the pulmonary capillaries, carbonic anhydrase catalyzes the conversion of carbonic acid to carbon dioxide and water:

$$H_2CO_3 \xrightarrow[\text{low } P_{CO_2}]{\text{carbonic anhydrase}} CO_2 + H_2O$$

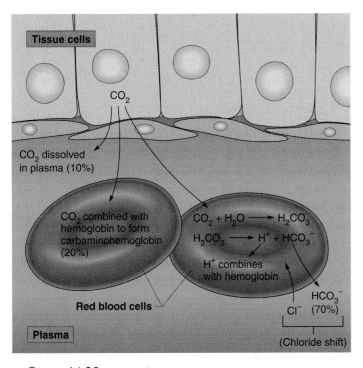

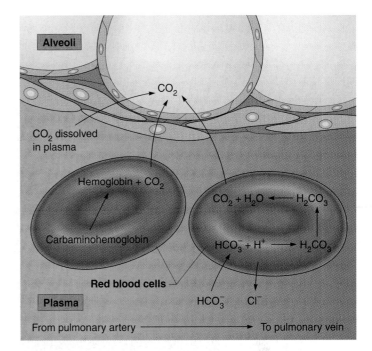

■ **Figure 16.38** **Carbon dioxide transport and the chloride shift.**
Carbon dioxide is transported in three forms: as dissolved CO_2 gas, attached to hemoglobin as carbaminohemoglobin, and as carbonic acid and bicarbonate. Percentages indicate the proportion of CO_2 in each of the forms. Notice that when bicarbonate (HCO_3^-) diffuses out of the red blood cells, Cl^- diffuses in to retain electrical neutrality. This exchange is the chloride shift.

■ **Figure 16.39** **The reverse chloride shift in the lungs.** Carbon dioxide is released from the blood as it travels through the pulmonary capillaries. A "reverse chloride shift" occurs during this time, and carbonic acid is transformed into CO_2 and H_2O. The CO_2 is eliminated in the exhaled air.

In summary, the carbon dioxide produced by the cells is converted within the systemic capillaries, mostly through the action of carbonic anhydrase in the red blood cells, to carbonic acid. With the buildup of carbonic acid concentrations in the red blood cells, the carbonic acid dissociates into bicarbonate and H^+, which results in the chloride shift. A *reverse chloride shift* operates in the pulmonary capillaries to convert carbonic acid to H_2O and CO_2 gas, which is eliminated in the expired breath (fig. 16.39). The P_{CO_2}, carbonic acid, H^+, and bicarbonate concentrations in the systemic arteries are thus maintained relatively constant by normal ventilation. This is required to maintain the acid-base balance of the blood (fig. 16.40), as discussed in chapter 13 and in the next section.

Ventilation and Acid-Base Balance

The basic concepts and terminology relating to the acid-base balance of the blood were introduced in chapter 13. In brief review, *acidosis* refers to an arterial pH below 7.35, and *alkalosis* refers to an arterial pH above 7.45. There are two components of each: respiratory and metabolic. The *respiratory component* refers to the carbon dioxide concentration of the blood, as measured by the P_{CO_2}. As implied by its name, the respiratory component is regulated by the respiratory system. The *metabolic component* is controlled by the kidneys, and is discussed in chapter 17.

Ventilation is normally adjusted to keep pace with the metabolic rate, so that the arterial P_{CO_2} remains in the normal range. In **hypoventilation,** the ventilation is insufficient to "blow off" carbon dioxide and maintain a normal P_{CO_2}. Indeed,

hypoventilation can be operationally defined as an abnormally high arterial P_{CO_2}. Under these conditions, carbonic acid production is excessively high and **respiratory acidosis** occurs.

In **hyperventilation,** conversely, the rate of ventilation is greater than the rate of CO_2 production. Arterial P_{CO_2} therefore decreases, so that less carbonic acid is formed than under normal conditions. The depletion of carbonic acid raises the pH, and **respiratory alkalosis** occurs.

A change in blood pH, produced by alterations in either the respiratory or metabolic component of acid-base balance, can be partially compensated for by a change in the other component. For example, a person with metabolic acidosis will hyperventilate. This is because the aortic and carotid bodies are stimulated by an increased blood H^+ concentration (fall in pH). As a result of the hyperventilation, a secondary respiratory alkalosis is produced. The person is still acidotic, but not as much so as would be the case without the compensation. People with partially compensated metabolic acidosis would thus have a low pH, which would be accompanied by a low blood P_{CO_2} as a result of the hyperventilation. Metabolic alkalosis, similarly, is partially compensated for by the retention of carbonic acid due to hypoventilation (table 16.10).

Clinical Investigation Clues

Remember that Harry had a high arterial P_{CO_2} and a pH of 7.15.

What does the pH of 7.15 represent, and what is the relationship between this pH and the arterial P_{CO_2}?

What caused these conditions to occur, and how were they corrected?

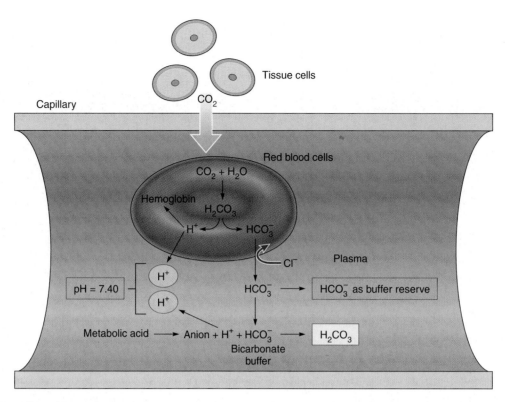

■ Figure 16.40 **The effect of bicarbonate on blood pH.** Bicarbonate released into the plasma from red blood cells buffers the H⁺ produced by the ionization of metabolic acids (lactic acid, fatty acids, ketone bodies, and others). Binding of H⁺ to hemoglobin also promotes the unloading of O_2.

Table 16.10   Effect of Lung Function on Blood Acid-Base Balance

Condition	pH	P_{CO_2}	Ventilation	Cause of Compensation
Normal	7.35–7.45	39–41 mmHg	Normal	Not applicable
Respiratory acidosis	Low	High	Hypoventilation	Cause of the acidosis
Respiratory alkalosis	High	Low	Hyperventilation	Cause of the alkalosis
Metabolic acidosis	Low	Low	Hyperventilation	Compensation for acidosis
Metabolic alkalosis	High	High	Hypoventilation	Compensation for alkalosis

Test Yourself Before You Continue

1. List the ways in which carbon dioxide is carried by the blood. Using equations, show how carbonic acid and bicarbonate are formed.

2. Describe the events that occur in the chloride shift in the systemic capillaries; also describe the reverse chloride shift that occurs in the pulmonary capillaries.

3. Describe the functions of bicarbonate and carbonic acid in blood.

4. Describe the effects of hyperventilation and hypoventilation on the blood pH and explain the mechanisms involved.

5. Explain why a person with ketoacidosis hyperventilates. What are the potential benefits of hyperventilation under these conditions?

Effect of Exercise and High Altitude on Respiratory Function

The arterial blood gases and pH do not significantly change during moderate exercise because ventilation increases during exercise to keep pace with the increased metabolism. Adjustments are also made at high altitude in both the control of ventilation and the oxygen transport ability of the blood to permit adequate delivery of oxygen to the tissues.

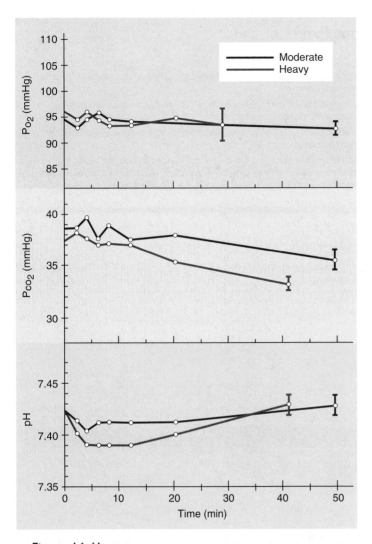

Figure 16.41 The effect of exercise on arterial blood gases and pH. Notice that there are no consistent or significant changes in these measurements during the first several minutes of moderate and heavy exercise, and that only the P_{CO_2} changes (actually decreases) during more prolonged exercise.

Changes in ventilation and oxygen delivery occur during exercise and during acclimatization to a high altitude. These changes help to compensate for the increased metabolic rate during exercise and for the decreased arterial P_{O_2} at high altitudes.

Ventilation During Exercise

As soon as a person begins to exercise, breathing becomes deeper and more rapid to produce a total minute volume that is many times the resting value. This increased ventilation, particularly in well-trained athletes, is exquisitely matched to the simultaneous increase in oxygen consumption and carbon dioxide production by the exercising muscles. The arterial blood P_{O_2}, P_{CO_2}, and pH thus remain surprisingly constant during exercise (fig. 16.41).

It is tempting to suppose that ventilation increases during exercise as a result of the increased CO_2 production by the exer-

cising muscles. Ventilation and CO_2 production increase simultaneously, however, so that blood measurements of P_{CO_2} during exercise are not significantly higher than at rest. The mechanisms responsible for the increased ventilation during exercise must therefore be more complex.

Two kinds of mechanisms—*neurogenic* and *humoral*—have been proposed to explain the increased ventilation that occurs during exercise. Possible neurogenic mechanisms include the following: (1) sensory nerve activity from the exercising limbs may stimulate the respiratory muscles, either through spinal reflexes or via the brain stem respiratory centers, and/or (2) input from the cerebral cortex may stimulate the brain stem centers to modify ventilation. These neurogenic theories help to explain the immediate increase in ventilation that occurs as exercise begins.

Rapid and deep ventilation continues after exercise has stopped, suggesting that humoral (chemical) factors in the blood may also stimulate ventilation during exercise. Since the P_{O_2}, P_{CO_2}, and pH of the blood samples from exercising subjects are within the resting range, these humoral theories propose that (1) the P_{CO_2} and pH in the region of the chemoreceptors may be different from these values "downstream," where blood samples are taken, and/or (2) cyclic variations in these values that cannot be detected by blood samples may stimulate the chemoreceptors. The evidence suggests that both neurogenic and humoral mechanisms are involved in the **hyperpnea,** or increased total minute volume, of exercise. (Note that hyperpnea differs from hyperventilation in that the blood P_{CO_2} is decreased in hyperventilation.)

Lactate Threshold and Endurance Training

The ability of the cardiopulmonary system to deliver adequate amounts of oxygen to the exercising muscles at the beginning of exercise may be insufficient because of the time lag required to make proper cardiovascular adjustments. During this time, therefore, the muscles respire anaerobically and a "stitch in the side"—possibly due to hypoxia of the diaphragm—may develop. After numerous cardiovascular and pulmonary adjustments have been made, a person may experience a "second wind" when the muscles are receiving sufficient oxygen for their needs.

Continued heavy exercise can cause a person to reach the **lactate threshold,** which is the maximum rate of oxygen consumption that can be attained before blood lactic acid levels rise as a result of anaerobic respiration. This occurs when 50% to 70% of the person's maximal oxygen uptake has been reached. The rise in lactic acid levels is due to the aerobic limitations of the muscles; it is not due to a malfunction of the cardiopulmonary system. Indeed, the arterial oxygen hemoglobin saturation remains at 97%, and venous blood draining the muscles contains unused oxygen.

The lactate threshold, however, is higher in endurance-trained athletes than it is in other people. These athletes, because of their higher cardiac output, have a higher rate of oxygen delivery to their muscles. As mentioned in chapter 12, endurance training also increases the skeletal muscle content of mitochondria and Krebs cycle enzymes, enabling the muscles to utilize

Table 16.11   Changes in Respiratory Function During Exercise

Variable	Change	Comments
Ventilation	Increased	In moderate exercise, ventilation is matched to increased metabolic rate. Mechanisms responsible for increased ventilation are not well understood.
Blood gases	No change	Blood gas measurements during light and moderate exercise show little change because ventilation is increased to match increased muscle oxygen consumption and carbon dioxide production.
Oxygen delivery to muscles	Increased	Although the total oxygen content and P_{O_2} do not increase during exercise, there is an increased rate of blood flow to the exercising muscles.
Oxygen extraction by muscles	Increased	Increased oxygen consumption lowers the tissue P_{O_2} and lowers the affinity of hemoglobin for oxygen (due to the effect of increased temperature). More oxygen, as a result, is unloaded so that venous blood contains a lower oxyhemoglobin saturation than at rest. This effect is enhanced by endurance training.

Table 16.12   Blood Gas Measurements at Different Altitudes

Altitude	Arterial P_{O_2} (mmHg)	Percent Oxyhemoglobin Saturation	Arterial P_{CO_2} (mmHg)
Sea level	90–95	96%	40
1,524 m (5,000 ft)	75–81	95%	32–33
2,286 m (7,500 ft)	69–74	92%–93%	31–33
4,572 m (15,000 ft)	48–53	86%	25
6,096 m (20,000 ft)	37–45	76%	20
7,620 m (25,000 ft)	32–39	68%	13
8,848 m (29,029 ft)	26–33	58%	9.5–13.8

Source: From P. H. Hackett et al., "High Altitude Medicine" in *Management of Wilderness and Environmental Emergencies,* 2d ed., edited by Paul S. Auerbach and Edward C. Geehr. Copyright © 1989 Mosby-Yearbook. Reprinted by permission.

more of the oxygen delivered to them by the arterial blood. The effects of exercise and endurance training on respiratory function are summarized in table 16.11.

Acclimatization to High Altitude

When a person from a region near sea level moves to a significantly higher elevation, several adjustments in respiratory function must be made to compensate for the decreased P_{O_2} at the higher altitude. These adjustments include changes in ventilation, in hemoglobin affinity for oxygen, and in total hemoglobin concentration.

Reference to table 16.12 indicates that at an altitude of 7,500 feet, for example, the P_{O_2} of arterial blood is 69 to 74 mmHg (compared to 90 to 95 mmHg at sea level). This table also indicates that the percent oxyhemoglobin saturation at this altitude is between 92% and 93%, compared to about 96% at sea level. The amount of oxygen attached to hemoglobin, and thus the total oxygen content of blood, is therefore decreased. In addition, the rate at which oxygen can be delivered to the cells (by the plasma-derived tissue fluid) after it dissociates from oxyhemoglobin is reduced at the higher altitude. This is because the maximum concentration of oxygen that can be dissolved in the plasma decreases in a linear fashion with the fall in P_{O_2}. People may thus experience rapid fatigue even at more moderate elevations (for example, 5,000 to 6,000 feet), at which the oxyhemoglobin saturation is only slightly decreased. Compensations made by the respiratory system gradually reduce the amount of fatigue caused by a given amount of exertion at high altitudes.

Changes in Ventilation

Starting at altitudes as low as 1,500 meters (5,000 feet), the decreased arterial P_{O_2} stimulates an increase in ventilation. This **hypoxic ventilatory response** produces hyperventilation, which lowers the arterial P_{CO_2} (table 16.12) and thus produces a respiratory alkalosis. The rise in arterial pH helps to blunt the hyperventilation, and within a few days the total minute volume becomes stabilized at a level 2.5 L/min higher than that at sea level.

Hyperventilation at high altitude increases tidal volume, thus reducing the proportionate contribution of air from the anatomical dead space and increasing the proportion of fresh air brought to the alveoli. This improves the oxygenation of the blood over what it would be in the absence of the hyperventilation. Hyperventilation, however, cannot increase blood P_{O_2} above that of the inspired air. The P_{O_2} of arterial blood decreases with increasing altitude, regardless of the ventilation. In the Peruvian Andes, for example, the normal arterial P_{O_2} is reduced from about 100 mmHg (at sea level) to 45 mmHg. The loading of hemoglobin with oxygen is therefore incomplete, producing an oxyhemoglobin saturation that is decreased from 97% (at sea level) to 81%.

Nitric oxide (NO) is produced in the lungs, and a recent study demonstrated increased NO concentration in the lungs of chronically hypoxic people who live at high altitude. Since NO is a vasodilator (chapter 14), this could increase pulmonary blood flow and perhaps thus improve the oxygenation of the blood in these people.

Furthermore, NO is loaded onto the protein portion of hemoglobin in the lungs and is unloaded in the tissue capillaries, where it can cause vasodilation. Therefore, the NO produced in the lungs

Table 16.13 Changes in Respiratory Function During Acclimatization to High Altitude

Variable	Change	Comments
Partial pressure of oxygen	Decreased	Due to decreased total atmospheric pressure
Partial pressure of carbon dioxide	Decreased	Due to hyperventilation in response to low arterial P_{O_2}
Percent oxyhemoglobin saturation	Decreased	Due to lower P_{O_2} in pulmonary capillaries
Ventilation	Increased	Due to lower P_{O_2}
Total hemoglobin	Increased	Due to stimulation by erythropoietin; raises oxygen capacity of blood to partially or completely compensate for the reduced partial pressure
Oxyhemoglobin affinity	Decreased	Due to increased DPG within the red blood cells; results in a higher percent unloading of oxygen to the tissues

and carried by hemoglobin to the systematic capillaries could promote increased blood flow and oxygen delivery to the tissues.

Finally, NO bound to sulfur atoms (and therefore abbreviated *SNOs*), in the cysteine groups of hemoglobin and other proteins, can be transferred from the blood to the respiratory control center, where it stimulates breathing. Thus, SNOs may contribute to the hypoxic ventilatory response (increased breathing when the arterial P_{O2} is low, as described earlier). All of these mechanisms of NO action would provide partial compensations for the chronic hypoxia of life at high altitude.

Acute mountain sickness (AMS) is common in people who arrive at altitudes in excess of 5,000 feet. Cardinal symptoms of AMS are headache, malaise, anorexia, nausea, and fragmented sleep. Headache, the most common symptom, may result from changes in blood flow to the brain. Low arterial P_{O_2} stimulates vasodilation of vessels in the pia mater, increasing blood flow and pressure within the skull. The hypocapnia produced by hyperventilation, however, causes cerebral vasoconstriction. Whether there is a net cerebral vasoconstriction or vasodilation depends on the balance between these two antagonistic effects. Pulmonary edema, common at altitudes above 9,000 feet, can produce shortness of breath, coughing, and a mild fever. Cerebral edema, which generally occurs above an altitude of 10,000 feet, can produce mental confusion and even hallucinations. Pulmonary and cerebral edema are potentially dangerous and should be alleviated by descending to a lower altitude.

The Affinity of Hemoglobin for Oxygen

Normal arterial blood at sea level unloads only about 22% of its oxygen to the tissues at rest; the percent saturation is reduced from 97% in arterial blood to 75% in venous blood. As a partial compensation for the decrease in oxygen content at high altitude, the affinity of hemoglobin for oxygen is reduced, so that a higher proportion of oxygen is unloaded. This occurs because the low oxyhemoglobin content of red blood cells stimulates the production of 2,3-DPG, which in turn decreases the affinity of hemoglobin for oxygen.

The action of 2,3-DPG to decrease the affinity of hemoglobin for oxygen thus predominates over the action of respiratory alkalosis (caused by the hyperventilation) to increase the affinity. At very high altitudes, however, the story becomes more complex. In one study, the very low arterial P_{O_2} (28 mmHg) of subjects at the summit of Mount Everest stimulated intense hyperventilation, so that the arterial P_{CO_2} was decreased to 7.5 mmHg. The resultant respiratory alkalosis (in this case, arterial pH greater than 7.7) caused the oxyhemoglobin dissociation curve to shift to the left (indicating greater affinity of hemoglobin for oxygen) despite the antagonistic effect of increased 2,3-DPG concentrations. It was suggested that the increased affinity of hemoglobin for oxygen caused by the respiratory alkalosis may have been beneficial at such a high altitude, since it increased the loading of hemoglobin with oxygen in the lungs.

Increased Hemoglobin and Red Blood Cell Production

In response to tissue hypoxia, the kidneys secrete the hormone erythropoietin (chapter 13). Erythropoietin stimulates the bone marrow to increase its production of hemoglobin and red blood cells. In the Peruvian Andes, for example, people have a total hemoglobin concentration that is increased from 15 g per 100 ml (at sea level) to 19.8 g per 100 ml. Although the percent oxyhemoglobin saturation is still lower than at sea level, the total oxygen content of the blood is actually greater—22.4 ml O_2 per 100 ml compared to a sea-level value of about 20 ml O_2 per 100 ml. These adjustments of the respiratory system to high altitude are summarized in table 16.13.

It should be noted that these changes are not unalloyed benefits. Polycythemia (high red blood cell count) increases the viscosity of blood; hematocrits of 55% to 60% have been measured in people who live in the Himalayas, and higher values are reached if dehydration accompanies the polycythemia. The increased blood viscosity contributes to pulmonary hypertension, which can cause accompanying edema and ventricular hypertrophy that can lead to heart failure.

Test Yourself Before You Continue

1. Describe the effect of exercise on the P_{O_2}, P_{CO_2}, and pH blood values and explain how ventilation might be increased during exercise.

2. Explain why endurance-trained athletes have a higher than average anaerobic threshold.

3. Describe the changes that occur in the respiratory system during acclimatization to life at a high altitude.

HPer Links of the Respiratory System with Other Body Systems

Integumentary System

- Nasal hairs and mucus prevent dust and other foreign material from damaging respiratory passageways . . . .**(p. 482)**

Skeletal System

- The lungs are protected by the rib cage, and bones of the rib cage serve as levers for the action of respiratory muscles . . . .**(p. 488)**
- Red blood cells, needed for oxygen transport, are produced in the bone marrow . . . .**(p. 371)**
- The respiratory system provides all organs, including the bones, with oxygen and eliminates carbon dioxide . . . .**(p. 480)**

Muscular System

- Contractions of skeletal muscles are needed for ventilation . . . .**(p. 488)**
- Muscles consume large amounts of oxygen and produce large amounts of carbon dioxide during exercise . . . .**(p. 342)**

Nervous System

- The nervous system regulates the rate and depth of breathing . . . .**(p. 499)**
- Autonomic nerves regulate blood flow, and hence the delivery of blood to tissues for gas exchange . . . .**(p. 420)**

Endocrine System

- Epinephrine dilates bronchioles, reducing airway resistance . . . .**(p. 231)**

- Thyroxine and epinephrine stimulate the rate of cell respiration . . . .**(p. 609)**

Circulatory System

- The heart and arterial system delivers oxygen from the lungs to the body tissues, and veins transport carbon dioxide from the body tissues to the lungs . . . . .**(p. 366)**
- Blood capillaries allow gas exchange for cell respiration in the tissues and lungs **(p. 392)**

Immune System

- The immune system protects against infections that could damage the respiratory system . . . .**(p. 446)**
- Alveolar macrophages and the action of cilia in the airways help to protect the lungs from infection . . . .**(p. 483)**

Urinary System

- The kidneys regulate the volume and electrolyte balance of the blood . .**(p. 524)**
- The kidneys participate with the lungs in the regulation of blood pH . . . . . .**(p. 548)**

Digestive System

- The GI tract provides nutrients to be used by cells of the lungs and other organs . . . .**(p. 561)**
- The respiratory system provides oxygen for cell respiration of glucose and other nutrients brought into the blood by the digestive system . . . .**(p. 480)**

Reproductive System

- The lungs provide oxygen for cell respiration of reproductive organs and eliminate carbon dioxide produced by these organs . . . .**(p. 480)**
- Changes in breathing and cell respiration occur during sexual arousal . . . . .**(p. 643)**

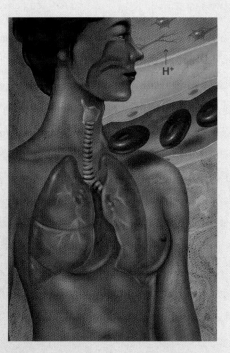

Summary

The Respiratory System 480

I. Alveoli are microscopic thin-walled air sacs that provide an enormous surface area for gas diffusion.
  A. The region of the lungs where gas exchange with the blood occurs is known as the respiratory zone.
  B. The trachea, bronchi, and bronchioles that deliver air to the respiratory zone constitute the conducting zone.

II. The thoracic cavity is delimited by the chest wall and diaphragm.
  A. The structures of the thoracic cavity are covered by thin, wet pleurae.
  B. The lungs are covered by a visceral pleura that is normally flush against the parietal pleura that lines the chest wall.
  C. The potential space between the visceral and parietal plurae is called the intrapleural space.

Physical Aspects of Ventilation 483

I. The intrapleural and intrapulmonary pressures vary during ventilation.
  A. The intrapleural pressure is always less than the intrapulmonary pressure.
  B. The intrapulmonary pressure is subatmospheric during inspiration and greater than the atmospheric pressure during expiration.
  C. Pressure changes in the lungs are produced by variations in lung volume in accordance with the inverse relationship between the volume and pressure of a gas described by Boyle's law.

II. The mechanics of ventilation are influenced by the physical properties of the lungs.
  A. The compliance of the lungs, or the ease with which they expand, refers specifically to the change in lung volume per change in transpulmonary pressure (the difference between intrapulmonary pressure and intrapleural pressure).
  B. The elasticity of the lungs refers to their tendency to recoil after distension.
  C. The surface tension of the fluid in the alveoli exerts a force directed inward, which acts to resist distension.

III. On first consideration, it would seem that the surface tension in the alveoli would create a pressure that would cause small alveoli to collapse and empty their air into larger alveoli.
  A. This would occur because the pressure caused by a given amount of surface tension would be greater in smaller alveoli than in larger alveoli, as described by the law of Laplace.
  B. Surface tension does not normally cause the collapse of alveoli, however, because pulmonary surfactant (a combination of phospholipid and protein) lowers the surface tension sufficiently.
  C. In hyaline membrane disease, the lungs of premature infants collapse because of a lack of surfactant.

Mechanics of Breathing 488

I. Inspiration and expiration are accomplished by contraction and relaxation of striated muscles.
  A. During quiet inspiration, the diaphragm and external intercostal muscles contract, and thus increase the volume of the thorax.
  B. During quiet expiration, these muscles relax, and the elastic recoil of the lungs and thorax causes a decrease in thoracic volume.
  C. Forced inspiration and expiration are aided by contraction of the accessory respiratory muscles.

II. Spirometry aids the diagnosis of a number of pulmonary disorders.
  A. In restrictive disease, such as pulmonary fibrosis, the vital capacity measurement is decreased to below normal.
  B. In obstructive disease, such as asthma and bronchitis, the forced expiratory volume is reduced to below normal because of increased airway resistance to air flow.

III. Asthma results from bronchoconstriction; emphysema, asthma, and chronic bronchitis are frequently referred to collectively as chronic obstructive pulmonary disease.

Gas Exchange in the Lungs 493

I. According to Dalton's law, the total pressure of a gas mixture is equal to the sum of the pressures that each gas in the mixture would exert independently.
  A. The partial pressure of a gas in a dry gas mixture is thus equal to the total pressure times the percent composition of that gas in the mixture.
  B. Since the total pressure of a gas mixture decreases with altitude above sea level, the partial pressures of the constituent gases likewise decrease with altitude.
  C. When the partial pressure of a gas in a wet gas mixture is calculated, the water vapor pressure must be taken into account.

II. According to Henry's law, the amount of gas that can be dissolved in a fluid is directly proportional to the partial pressure of that gas in contact with the fluid.
  A. The concentrations of oxygen and carbon dioxide that are dissolved in plasma are proportional to an electric current generated by special electrodes that react with these gases.
  B. Normal arterial blood has a P_{O_2} of 100 mmHg, indicating a concentration of dissolved oxygen of 0.3 ml per 100 ml of blood; the oxygen contained in red blood cells (about 19.7 ml per 100 ml of blood) does not affect the P_{O_2} measurement.

III. The P_{O_2} and P_{CO_2} measurements of arterial blood provide information about lung function.

IV. In addition to proper ventilation of the lungs, blood flow (perfusion) in the lungs must be adequate and matched to air flow (ventilation) in order for adequate gas exchange to occur.

V. Abnormally high partial pressures of gases in blood can cause a variety of disorders, including oxygen toxicity, nitrogen narcosis, and decompression sickness.

Regulation of Breathing   499

I. The rhythmicity center in the medulla oblongata directly controls the muscles of respiration.

   A. Activity of the inspiratory and expiratory neurons varies in a reciprocal way to produce an automatic breathing cycle.

   B. Activity in the medulla is influenced by the apneustic and pneumotaxic centers in the pons, as well as by sensory feedback information.

   C. Conscious breathing involves direct control by the cerebral cortex via corticospinal tracts.

II. Breathing is affected by chemoreceptors sensitive to the P_{CO_2}, pH, and P_{O_2} of the blood.

   A. The P_{CO_2} of the blood and consequent changes in pH are usually of greater importance than the blood P_{O_2} in the regulation of breathing.

   B. Central chemoreceptors in the medulla oblongata are sensitive to changes in blood P_{CO_2} because of the resultant changes in the pH of cerebrospinal fluid.

   C. The peripheral chemoreceptors in the aortic and carotid bodies are sensitive to changes in blood P_{CO_2} indirectly, because of consequent changes in blood pH.

III. Decreases in blood P_{O_2} directly stimulate breathing only when the blood P_{O_2} is lower than 50 mmHg. A drop in P_{O_2} also stimulates breathing indirectly, by making the chemoreceptors more sensitive to changes in P_{CO_2} and pH.

IV. At tidal volumes of 1 L or more, inspiration is inhibited by stretch receptors in the lungs (the Hering-Breuer reflex). A similar reflex may act to inhibit expiration.

Hemoglobin and Oxygen Transport   504

I. Hemoglobin is composed of two alpha and two beta polypeptide chains and four heme groups each containing a central atom of iron.

   A. When the iron is in the reduced form and not attached to oxygen, the hemoglobin is called deoxyhemoglobin, or reduced hemoglobin; when it is attached to oxygen, it is called oxyhemoglobin.

   B. If the iron is attached to carbon monoxide, the hemoglobin is called carboxyhemoglobin. When the iron is in an oxidized state and unable to transport any gas, the hemoglobin is called methemoglobin.

   C. Deoxyhemoglobin combines with oxygen in the lungs (the loading reaction) and breaks its bonds with oxygen in the tissue capillaries (the unloading reaction). The extent of each reaction is determined by the P_{O_2} and the affinity of hemoglobin for oxygen.

II. A graph of percent oxyhemoglobin saturation at different values of P_{O_2} is called an oxyhemoglobin dissociation curve.

   A. At rest, the difference between arterial and venous oxyhemoglobin saturations indicates that about 22% of the oxyhemoglobin unloads its oxygen to the tissues.

   B. During exercise, the venous P_{O_2} and percent oxyhemoglobin saturation are decreased, indicating that a higher percentage of the oxyhemoglobin has unloaded its oxygen to the tissues.

III. The pH and temperature of the blood influence the affinity of hemoglobin for oxygen, and thus the extent of loading and unloading.

   A. A fall in pH decreases the affinity of hemoglobin for oxygen, and a

rise in pH increases the affinity. This is called the Bohr effect.

   B. A rise in temperature decreases the affinity of hemoglobin for oxygen.

   C. When the affinity is decreased, the oxyhemoglobin dissociation curve is shifted to the right. This indicates a greater unloading percentage of oxygen to the tissues.

IV. The affinity of hemoglobin for oxygen is also decreased by an organic molecule in the red blood cells called 2,3-diphosphoglyceric acid (2,3-DPG).

   A. Since oxyhemoglobin inhibits 2,3-DPG production, 2,3-DPG concentrations will be higher when anemia or low P_{O_2} (as in high altitude) cause a decrease in oxyhemoglobin.

   B. If a person is anemic, the lowered hemoglobin concentration is partially compensated for because a higher percentage of the oxyhemoglobin will unload its oxygen as a result of the effect of 2,3-DPG.

   C. Fetal hemoglobin cannot bind to 2,3-DPG, and thus it has a higher affinity for oxygen than the mother's hemoglobin. This facilitates the transfer of oxygen to the fetus.

V. Inherited defects in the amino acid composition of hemoglobin are responsible for such diseases as sickle-cell anemia and thalassemia.

VI. Striated muscles contain myoglobin, a pigment related to hemoglobin that can combine with oxygen and deliver it to the muscle cell mitochondria at low P_{O_2} values.

Carbon Dioxide Transport and Acid-Base Balance   510

I. Red blood cells contain an enzyme called carbonic anhydrase that catalyzes the reversible reaction whereby carbon dioxide and water are used to form carbonic acid.

   A. This reaction is favored by the high P_{CO_2} in the tissue capillaries, and as a result, carbon dioxide produced by the tissues is converted into carbonic acid in the red blood cells.

B. Carbonic acid then ionizes to form H^+ and HCO_3^- (bicarbonate).

C. Since much of the H^+ is buffered by hemoglobin, but more bicarbonate is free to diffuse outward, an electrical gradient is established that draws Cl^- into the red blood cells. This is called the chloride shift.

D. A reverse chloride shift occurs in the lungs. In this process, the low P_{CO_2} favors the conversion of carbonic acid to carbon dioxide, which can be exhaled.

II. By adjusting the blood concentration of carbon dioxide, and thus of carbonic acid, the process of ventilation helps to maintain proper acid-base balance of the blood.

A. Normal arterial blood pH is 7.40. A pH below 7.35 is termed

acidosis; a pH above 7.45 is termed alkalosis.

B. Hyperventilation causes respiratory alkalosis, and hypoventilation causes respiratory acidosis.

C. Metabolic acidosis stimulates hyperventilation, which can cause a respiratory alkalosis as a partial compensation.

Effect of Exercise and High Altitude on Respiratory Function 512

I. During exercise there is increased ventilation, or hyperpnea, which is matched to the increased metabolic rate so that the arterial blood P_{CO_2} remains normal.

A. This hyperpnea may be caused by proprioceptor information, cerebral input, and/or changes in arterial P_{CO_2} and pH.

B. During heavy exercise, the anaerobic threshold may be reached at 50% to 70% of the

maximal oxygen uptake. At this point, lactic acid is released into the blood by the muscles.

C. Endurance training enables the muscles to utilize oxygen more effectively, so that greater levels of exercise can be performed before the anaerobic threshold is reached.

II. Acclimatization to a high altitude involves changes that help to deliver oxygen more effectively to the tissues, despite reduced arterial P_{O_2}.

A. Hyperventilation occurs in response to the low P_{O_2}.

B. The red blood cells produce more 2,3-DPG, which lowers the affinity of hemoglobin for oxygen and improves the unloading reaction.

C. The kidneys produce the hormone erythropoietin, which stimulates the bone marrow to increase its production of red blood cells, so that more oxygen can be carried by the blood at given values of P_{O_2}.

Review Activities

Test Your Knowledge of Terms and Facts

1. Which of these statements about intrapulmonary pressure and intrapleural pressure is *true?*

  a. The intrapulmonary pressure is always subatmospheric.

  b. The intrapleural pressure is always greater than the intrapulmonary pressure.

  c. The intrapulmonary pressure is greater than the intrapleural pressure.

  d. The intrapleural pressure equals the atmospheric pressure.

2. If the transpulmonary pressure equals zero,

  a. a pneumothorax has probably occurred.

  b. the lungs cannot inflate.

  c. elastic recoil causes the lungs to collapse.

  d. all of these apply.

3. The maximum amount of air that can be expired after a maximum inspiration is

  a. the tidal volume.

  b. the forced expiratory volume.

  c. the vital capacity.

  d. the maximum expiratory flow rate.

4. If the blood lacked red blood cells but the lungs were functioning normally,

  a. the arterial P_{O_2} would be normal.

  b. the oxygen content of arterial blood would be normal.

  c. both *a* and *b* would apply.

  d. neither *a* nor *b* would apply.

5. If a person were to dive with scuba equipment to a depth of 66 feet, which of these statements would be *false?*

  a. The arterial P_{O_2} would be three times normal.

  b. The oxygen content of plasma would be three times normal.

  c. The oxygen content of whole blood would be three times normal.

6. Which of these would be most affected by a decrease in the affinity of hemoglobin for oxygen?

  a. arterial P_{O_2}

  b. arterial percent oxyhemoglobin saturation

  c. venous oxyhemoglobin saturation

  d. arterial P_{CO_2}

7. If a person with normal lung function were to hyperventilate for several seconds, there would be a significant
   a. increase in the arterial P_{O_2}.
   b. decrease in the arterial P_{CO_2}.
   c. increase in the arterial percent oxyhemoglobin saturation.
   d. decrease in the arterial pH.

8. Erythropoietin is produced by
   a. the kidneys.
   b. the liver.
   c. the lungs.
   d. the bone marrow.

9. The affinity of hemoglobin for oxygen is decreased under conditions of
   a. acidosis.
   b. fever.
   c. anemia.
   d. acclimatization to a high altitude.
   e. all of these.

10. Most of the carbon dioxide in the blood is carried in the form of
    a. dissolved CO_2.
    b. carbaminohemoglobin.
    c. bicarbonate.
    d. carboxyhemoglobin.

11. The bicarbonate concentration of the blood would be decreased during
    a. metabolic acidosis.
    b. respiratory acidosis.
    c. metabolic alkalosis.
    d. respiratory alkalosis.

12. The chemoreceptors in the medulla are directly stimulated by
    a. CO_2 from the blood.
    b. H^+ from the blood.
    c. H^+ in cerebrospinal fluid that is derived from blood CO_2.
    d. decreased arterial P_{O_2}.

13. The rhythmic control of breathing is produced by the activity of inspiratory and expiratory neurons in
    a. the medulla oblongata.
    b. the apneustic center of the pons.
    c. the pneumotaxic center of the pons.
    d. the cerebral cortex.

14. Which of these occur(s) during hypoxemia?
    a. increased ventilation
    b. increased production of 2,3-DPG
    c. increased production of erythropoietin
    d. all of these

15. During exercise, which of these statements is *true?*
    a. The arterial percent oxyhemoglobin saturation is decreased.
    b. The venous percent oxyhemoglobin saturation is decreased.
    c. The arterial P_{CO_2} is measurably increased.
    d. The arterial pH is measurably decreased.

16. All of these can bond with hemoglobin *except*
    a. HCO_3^-.
    b. O_2.
    c. H^+.
    d. CO_2.
    e. NO

17. Which of these statements about the partial pressure of carbon dioxide is true?
    a. It is higher in the alveoli than in the pulmonary arteries.
    b. It is higher in the systemic arteries than in the tissues.
    c. It is higher in the systemic veins than in the systemic arteries.
    d. It is higher in the pulmonary veins than in the pulmonary arteries.

Test Your Understanding of Concepts and Principles

1. Using a flow diagram to show cause and effect, explain how contraction of the diaphragm produces inspiration.[1]

2. Radiographic (X-ray) pictures show that the rib cage of a person with a pneumothorax is expanded and the ribs are farther apart. Explain why this should be so.

3. Explain, using a flowchart, how a rise in blood P_{CO_2} stimulates breathing. Include both the central and peripheral chemoreceptors in your answer.

4. Explain why a person with ketoacidosis may hyperventilate. What benefit might it provide? Also explain why this hyperventilation can be stopped by an intravenous fluid containing bicarbonate.

5. What blood measurements can be performed to detect (a) anemia, (b) carbon monoxide poisoning, and (c) poor lung function?

6. Explain how measurements of blood P_{CO_2}, bicarbonate, and pH are affected by hypoventilation and hyperventilation.

7. Describe the changes in ventilation that occur during exercise. How are these changes produced and how do they affect arterial blood gases and pH?

8. How would an increase in the red blood cell content of 2,3-DPG affect the P_{O_2} of venous blood? Explain your answer.

9. Explain the mechanisms that produce changes in ventilation at a high altitude.

Why are these changes beneficial? Under what conditions could they be detrimental. What other factors operate at a high altitude to improve oxygen delivery to the tissues?

10. Compare asthma and emphysema in terms of their characteristics and the effects they have on pulmonary function tests.

11. Explain the mechanisms involved in quiet inspiration and in forced inspiration, and in quiet expiration and forced expiration. What muscles are involved in each case?

12. Describe the formation, composition, and function of pulmonary surfactant. What happens when surfactant is absent? How is this condition treated?

---

[1]*Note:* This question is answered in the chapter 16 Study Guide found on the Online Learning Center at www.mhhe.com/fox8.

Test Your Ability to Analyze and Apply Your Knowledge

1. The nature of the sounds produced by percussion (tapping) a patient's chest can tell a physician a great deal about the condition of the organs within the thoracic cavity. Healthy, air-filled lungs resonate, or sound hollow. How do you think the lungs of a person with emphysema would sound in comparison to healthy lungs? What kind of sounds would be produced by a collapsed lung, or one that was partially filled with fluid?

2. Explain why the first breath of a healthy neonate is more difficult than subsequent breaths and why premature infants often require respiratory assistance (a mechanical ventilator) to keep their lungs inflated. How else is this condition treated?

3. Nicotine from cigarette smoke causes the buildup of mucus and paralyzes the cilia that line the respiratory tract. How might these conditions affect pulmonary function tests? If smoking has led to emphysema, how would the pulmonary function tests change?

4. Carbon monoxide poisoning from smoke inhalation and suicide attempts is the most common cause of death from poisoning in the United States. How would carbon monoxide poisoning affect a person's coloring, particularly of the mucous membranes? How would it affect the hemoglobin concentration, hematocrit, and percent oxyhemoglobin saturation? How would chronic carbon monoxide poisoning affect the person's red blood cell content of 2,3-DPG?

5. After driving from sea level to a trail head in the high Sierras, you get out of your car and feel dizzy. What do you suppose is causing your dizziness? How is this beneficial and how is it detrimental? What may eventually happen to help to reduce the cause of the dizziness?

Related Websites

Check out the Links Library at www.mhhe.com/fox8 for links to sites containing resources related to respiratory physiology. These links are monitored to ensure current URLs.

17

Physiology of the Kidneys

Objectives

After studying this chapter, you should be able to . . .

1. describe the different regions of the nephron tubules and the location of the tubules in the kidney.

2. describe the structural and functional relationships between the nephron tubules and their associated blood vessels.

3. describe the composition of glomerular ultrafiltrate and explain how it is produced.

4. explain how the proximal convoluted tubule reabsorbs salt and water.

5. describe active transport and osmosis in the loop of Henle and explain how these processes produce a countercurrent multiplier system.

6. explain how the vasa recta function in countercurrent exchange.

7. describe the role of antidiuretic hormone (ADH) in regulating the final urine volume.

8. describe the mechanisms of glucose reabsorption and define the terms *transport maximum* and *renal plasma threshold.*

9. define the term *renal plasma clearance* and explain why the clearance of inulin is equal to the glomerular filtration rate.

10. explain how the clearance of different molecules is determined and how the processes of reabsorption and secretion affect the clearance measurement.

11. describe the mechanism of Na^+ reabsorption in the distal tubule and explain why this reabsorption occurs together with the secretion of K^+.

12. describe the effects of aldosterone on the cortical portion of the collecting duct and explain how aldosterone secretion is regulated.

13. explain how activation of the renin-angiotensin-aldosterone system results in the stimulation of aldosterone secretion.

14. explain how the interaction between plasma K^+ and H^+ concentrations affects the tubular secretion of these ions.

15. describe the role of the kidneys in the regulation of acid-base balance.

16. describe the different mechanisms by which substances can act as diuretics and explain why some diuretics cause excessive loss of K^+.

Chapter at a Glance

Clinical Investigation

Emily, a high school senior, visits her family physician complaining of pain in her lower back, somewhere between the twelfth rib and the lumbar vertebrae. Her urine is noticeably discolored, and urinalysis reveals that she has hematuria (blood in the urine). However, she does not report pain during urination. The physician is relieved to see that her urine has only trace amounts of protein. Further examination and tests show that Emily has mild oliguria (reduced urine production), some edema, and an elevated plasma creatinine concentration.

Emily tells the doctor that she is still competing on her cross-country racing team, even though her throat has been sore for almost a month. A throat culture reveals that she has a streptococcus infection. Emily is given antibiotics and hydrochlorothiazide and, within a few weeks, her symptoms are gone.

What was responsible for Emily's symptoms, and why did they disappear with this treatment?

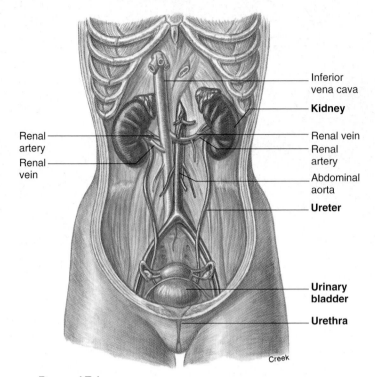

Figure 17.1 **The organs of the urinary system.** The urinary system of a female is shown; that of a male is the same, except that the urethra runs through the penis.

Structure and Function of the Kidneys

Each kidney contains many tiny tubules that empty into a cavity drained by the ureter. Each of the tubules receives a blood filtrate from a capillary bed called the glomerulus. The filtrate is similar to interstitial fluid, but it is modified as it passes through different regions of the tubule and is thereby changed into urine. The tubules and associated blood vessels thus form the functioning units of the kidneys, known as nephrons.

The primary function of the kidneys is regulation of the extracellular fluid (plasma and interstitial fluid) environment in the body. This is accomplished through the formation of urine, which is a modified filtrate of plasma. In the process of urine formation, the kidneys regulate:

1. the volume of blood plasma (and thus contribute significantly to the regulation of blood pressure);
2. the concentration of waste products in the blood;
3. the concentration of electrolytes (Na^+, K^+, HCO_3^- and other ions) in the plasma; and
4. the pH of plasma.

In order to understand how the kidneys perform these functions, a knowledge of kidney structure is required.

Gross Structure of the Urinary System

The paired **kidneys** lie on either side of the vertebral column below the diaphragm and liver. Each adult kidney weighs about 160 g and is about 11 cm (4 in.) long and 5 to 7 cm (2 to 3 in.) wide—about the size of a fist. Urine produced in the kidneys is drained into a cavity known as the *renal pelvis* (= basin), and then it is channeled from each kidney via long ducts—the **ureters**—to the **urinary bladder** (fig. 17.1).

A coronal section of the kidney shows two distinct regions (fig. 17.2). The outer *cortex* is reddish brown and granular in appearance because of its many capillaries. The deeper region, or *medulla,* is striped in appearance due to the presence of microscopic tubules and blood vessels. The medulla is composed of eight to fifteen conical *renal pyramids* separated by *renal columns.*

Kidney stones (*renal calculi*) are composed of crystals and proteins that grow until they break loose and pass into the urine collection system. Small stones that are anchored in place are not usually noticed, but large stones in the calyces or pelvis may obstruct the flow of urine. When a stone breaks loose and passes into a ureter, it produces a steadily increasing sensation of pain. The pain often becomes so intense that the patient requires narcotic drugs. Most kidney stones contain crystals of calcium oxalate, but stones may also be composed of crystals of calcium phosphate, uric acid, or cystine. These substances are normally present in urine in a supersaturated state, from which they can crystalize for a variety of reasons. The stones may be removed surgically or broken up by a noninvasive procedure called *shock-wave lithotripsy.*

Clinical Investigation Clues

Remember that Emily had pain in her lower back, between the twelfth rib and the lumbar vertebrae.

From which organ might the pain originate?
Is it likely that Emily has a kidney stone?

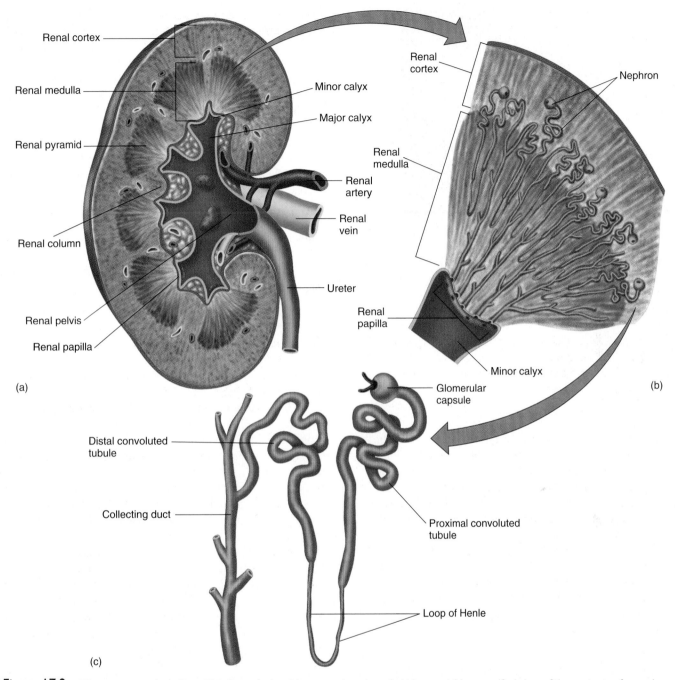

■ **Figure 17.2** **The structure of a kidney.** The figure depicts (a) a coronal section of a kidney and (b) a magnified view of the contents of a renal pyramid. (c) A single nephron tubule, microscopic in actual size, is shown isolated.

The cavity of the kidney is divided into several portions. Each pyramid projects into a small depression called a *minor calyx* (the plural form is *calyces*). Several minor calyces unite to form a *major calyx.* The major calyces then join to form the funnel-shaped *renal pelvis.* The renal pelvis collects urine from the calyces and transports it to the ureters and urinary bladder (fig. 17.3).

The **urinary bladder** is a storage sac for urine, and its shape is determined by the amount of urine it contains. An empty urinary bladder is pyramidal; as it fills, it becomes ovoid and bulges upward into the abdominal cavity. The urinary bladder is drained inferiorly by the tubular **urethra.** In females, the urethra is 4 cm (1.5 in.) long

and opens into the space between the labia minora (see chapter 20). In males, the urethra is about 20 cm (8 in.) long and opens at the tip of the penis, from which it can discharge either urine or semen.

Micturition Reflex

Two muscular sphincters surround the urethra. The upper sphincter, composed of smooth muscle, is called the *internal urethral sphincter;* the lower sphincter, composed of voluntary skeletal muscle, is called the *external urethral sphincter.* The actions of these sphincters are regulated in the process of urination, which is also known as **micturition.**

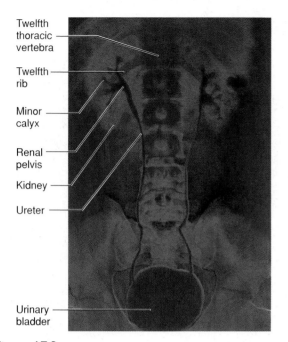

Twelfth
thoracic
vertebra

Twelfth
rib

Minor
calyx

Renal
pelvis

Kidney

Ureter

Urinary
bladder

■ **Figure 17.3** A pseudocolor radiograph of the urinary system. In this photograph, shades of gray are assigned colors. The calyces of the kidneys, the renal pelvises, the ureters, and the urinary bladder are visible.

Micturition is controlled by a reflex center located in the second, third, and fourth sacral levels of the spinal cord. Filling of the urinary bladder activates stretch receptors that send impulses to this micturition center. As a result, parasympathetic neurons are activated, causing rhythmic contractions of the detrusor muscle of the urinary bladder and relaxation of the internal urethral sphincter. At this point, a sense of urgency is perceived by the brain, but there is still voluntary control over the external urethral sphincter. When urination is consciously allowed to occur, descending motor tracts to the micturition center inhibit somatic motor fibers to the external urethral sphincter. This muscle then relaxes, and urine is expelled. The ability to voluntarily inhibit micturition generally develops between the ages of 2 and 3.

Microscopic Structure of the Kidney

The **nephron** is the functional unit of the kidney responsible for the formation of urine. Each kidney contains more than a million nephrons. A nephron consists of small tubes, or **tubules,** and associated small blood vessels. Fluid formed by capillary filtration enters the tubules and is subsequently modified by transport processes; the resulting fluid that leaves the tubules is urine.

Renal Blood Vessels

Arterial blood enters the kidney through the *renal artery,* which divides into *interlobar arteries* (fig. 17.4) that pass between the pyramids through the renal columns. *Arcuate arteries* branch from the interlobar arteries at the boundary of the cortex and medulla. A number of *interlobular arteries* radiate

from the arcuate arteries into the cortex and subdivide into numerous **afferent arterioles** (fig. 17.5), which are microscopic. The afferent arterioles deliver blood into **glomeruli**—capillary networks that produce a blood filtrate that enters the urinary tubules. The blood remaining in a glomerulus leaves through an **efferent arteriole,** which delivers the blood into another capillary network—the **peritubular capillaries** surrounding the renal tubules.

This arrangement of blood vessels is unique. It is the only one in the body in which a capillary bed (the glomerulus) is drained by an arteriole rather than by a venule and delivered to a second capillary bed located downstream (the peritubular capillaries). Blood from the peritubular capillaries is drained into veins that parallel the course of the arteries in the kidney. These veins are called the *interlobular veins, arcuate veins,* and *interlobar veins.* The interlobar veins descend between the pyramids, converge, and leave the kidney as a single *renal vein,* which empties into the inferior vena cava.

Nephron Tubules

The tubular portion of a nephron consists of a *glomerular capsule, a proximal convoluted tubule, a descending limb of the loop of Henle, an ascending limb of the loop of Henle,* and a *distal convoluted tubule* (fig. 17.5).

The **glomerular (Bowman's) capsule** surrounds the glomerulus. The glomerular capsule and its associated glomerulus are located in the cortex of the kidney and together constitute the *renal corpuscle.* The glomerular capsule contains an inner visceral layer of epithelium around the glomerular capillaries and an outer parietal layer. The space between these two layers is continuous with the lumen of the tubule and receives the glomerular filtrate, as will be described in the next section.

Filtrate that enters the glomerular capsule passes into the lumen of the **proximal convoluted tubule.** The wall of the proximal convoluted tubule consists of a single layer of cuboidal cells containing millions of microvilli; these microvilli increase the surface area for reabsorption. In the process of reabsorption, salt, water, and other molecules needed by the body are transported from the lumen, through the tubular cells and into the surrounding peritubular capillaries.

The glomerulus, glomerular capsule, and convoluted tubule are located in the renal cortex. Fluid passes from the proximal convoluted tubule to the **loop of Henle.** This fluid is carried into the medulla in the **descending limb** of the loop and returns to the cortex in the **ascending limb** of the loop. Back in the cortex, the tubule again becomes coiled and is called the **distal convoluted tubule.** The distal convoluted tubule is shorter than the proximal tubule and has relatively few microvilli. The distal convoluted tubule terminates as it empties into a collecting duct.

The two principal types of nephrons are classified according to their position in the kidney and the lengths of their loops of Henle. Nephrons that originate in the inner one-third of the cortex—called *juxtamedullary nephrons* because they are

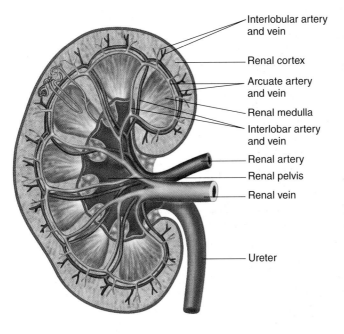

Interlobular artery
and vein

Renal cortex

Arcuate artery
and vein

Renal medulla

Interlobar artery
and vein

Renal artery

Renal pelvis

Renal vein

Ureter

(a)

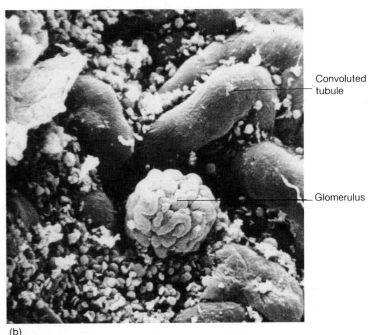

Convoluted
tubule

Glomerulus

(b)

■ **Figure 17.4**   **The vascular structure of the kidneys.** *(a)* An illustration of the major arterial supply and *(b)* a scanning electron micrograph of a glomerulus (300×).

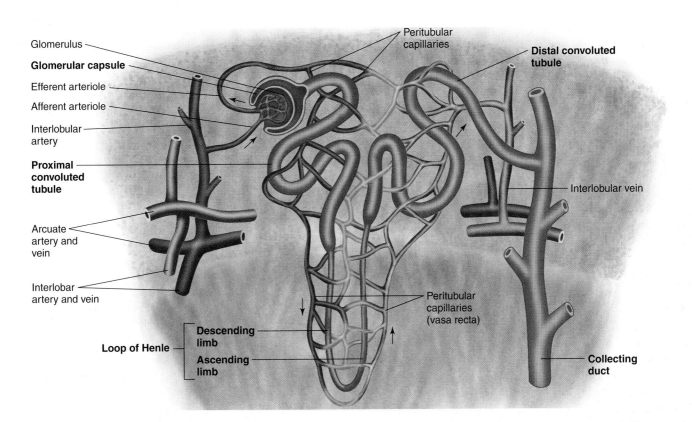

Glomerulus

Glomerular capsule

Efferent arteriole

Afferent arteriole

Interlobular
artery

**Proximal
convoluted
tubule**

Arcuate
artery and
vein

Interlobar
artery and vein

Peritubular
capillaries

**Distal convoluted
tubule**

Interlobular vein

Peritubular
capillaries
(vasa recta)

**Descending
limb**

Loop of Henle

**Ascending
limb**

**Collecting
duct**

■ **Figure 17.5**   **The nephron tubules and associated blood vessels.** In this simplified illustration, the blood flow from a glomerulus to an efferent arteriole, to the peritubular capillaries, and to the venous drainage of the kidneys is indicated with arrows. The names for the different regions of the nephron tubules are indicated with boldface type.

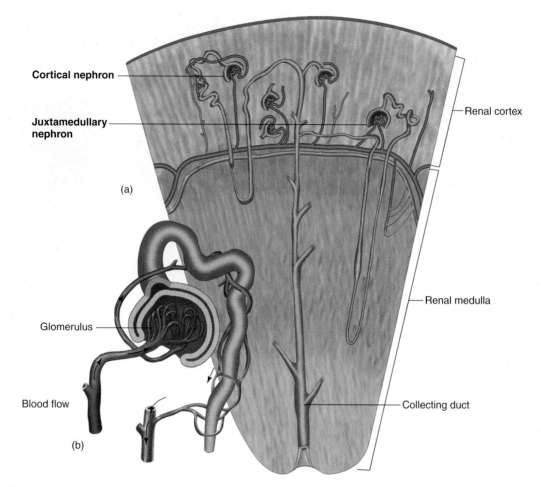

Cortical nephron

Juxtamedullary nephron

Renal cortex

(a)

Glomerulus

Renal medulla

Blood flow

Collecting duct

(b)

■ **Figure 17.6**   **The contents of a renal pyramid.** *(a)* The position of cortical and juxtamedullary nephrons is shown within the renal pyramid of the kidney. *(b)* The direction of blood flow in the vessels of the nephron is indicated with arrows.

next to the medulla—have longer loops of Henle than the more numerous *cortical nephrons,* which originate in the outer two-thirds of the cortex (fig. 17.6). The juxtamedullary nephrons play an important role in the ability of the kidney to produce a concentrated urine.

A **collecting duct** receives fluid from the distal convoluted tubules of several nephrons. Fluid is then drained by the collecting duct from the cortex to the medulla as the collecting duct passes through a renal pyramid. This fluid, now called urine, passes into a minor calyx. Urine is then funneled through the renal pelvis and out of the kidney in the ureter.

Polycystic kidney disease is a condition inherited as an autosomal dominant trait (see chapter 20) that affects 1 in every 600 to 1,000 people. This disease is thus more common than sickle-cell anemia, cystic fibrosis, or muscular dystrophy, which are also genetic diseases. In 50% of the people who inherit the defective gene (located on the short arm of chromosome 16), progressive renal failure develops during middle age to the point that dialysis or kidney transplants are required. The cysts that develop are expanded portions of the renal tubule. Cysts that originate in the proximal tubule contain fluid that resembles glomerular filtrate and plasma. Cysts that originate in the distal tubule contain fluid with a lower NaCl concentration and a higher potassium and urea concentration than plasma as a result of the transport processes that occur during the passage of fluid through the tubules.

Test Yourself Before You Continue

1. Describe the "theme" of kidney function in a single sentence. List the components of this functional theme.
2. Trace the course of blood flow through the kidney from the renal artery to the renal vein.
3. Trace the course of tubular fluid from the glomerular capsules to the ureter.
4. Draw a diagram of the tubular component of a nephron. Label the segments and indicate which parts are in the cortex and which are in the medulla.

Glomerular Filtration

The glomerular capillaries have large pores in their walls, and the layer of Bowman's capsule in contact with the glomerulus has filtration slits. Water, together with dissolved solutes (but not proteins), can thus pass from the blood plasma to the inside of the capsule and the nephron tubules. The volume of this filtrate produced by both kidneys per minute is called the glomerular filtration rate (GFR).

Endothelial cells of the glomerular capillaries have large pores (200 to 500 Å in diameter) called fenestrae; thus, the glomerular endothelium is said to be *fenestrated*. As a result of these large pores, glomerular capillaries are 100 to 400 times more permeable to plasma water and dissolved solutes than are the capillaries of skeletal muscles. Although the pores of glomerular capillaries are large, they are still small enough to prevent the passage of red blood cells, white blood cells, and platelets into the filtrate.

Before the filtrate can enter the interior of the glomerular capsule, it must pass through the capillary pores, the basement membrane (a thin layer of glycoproteins lying immediately outside the endothelial cells), and the inner (visceral) layer of the glomerular capsule. The inner layer of the glomerular capsule is composed of unique cells called *podocytes.* Each podocyte is shaped somewhat like an octopus, with a bulbous cell body and several thick arms. Each arm has thousands of cytoplasmic extensions known as *pedicels,* or *foot processes* (fig. 17.7). These pedicels interdigitate, like the fingers of clasped hands, as they wrap around the glomerular capillaries. The narrow slits between adjacent pedicels provide the passageways through which filtered molecules must pass to enter the interior of the glomerular capsule (fig. 17.8).

Although the glomerular capillary pores are apparently large enough to permit the passage of proteins, the fluid that enters the capsular space contains only a small amount of plasma proteins. This relative exclusion of plasma proteins from the filtrate is partially a result of their negative charges, which hinder their passage through the negatively charged glycoproteins in the basement membrane of the capillaries (fig. 17.9). The large size and negative charges of plasma proteins may also restrict their movement through the filtration slits between pedicels.

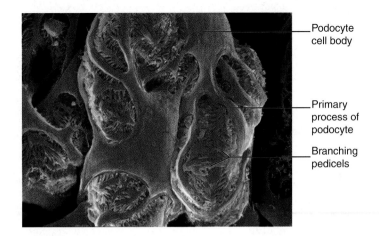

■ **Figure 17.7** **A scanning electron micrograph of the glomerular capillaries and capsule.** The inner (visceral) layer of the glomerular (Bowman's) capsule is composed of podocytes, as shown in this scanning electron micrograph. Very fine extensions of these podocytes form foot processes, or pedicels, that interdigitate around the glomerular capillaries. Spaces between adjacent pedicels form the "filtration slits" (see also fig. 17.8).

Glomerular Ultrafiltrate

The fluid that enters the glomerular capsule is called **ultrafiltrate** (fig. 17.10) because it is formed under pressure—the hydrostatic pressure of the blood. This process is similar to the formation of tissue fluid by other capillary beds in the body in response to Starling forces (chapter 14; see fig. 14.9). The force favoring filtration is opposed by a counterforce developed by the hydrostatic pressure of fluid in the glomerular capsule. Also, since the protein concentration of the tubular fluid is low (less than 2 to 5 mg per 100 ml) compared to that of plasma (6 to 8 g per 100 ml), the greater colloid osmotic pressure of plasma promotes the osmotic return of filtered water. When these opposing forces are subtracted from the hydrostatic pressure of the glomerular capillaries, a *net filtration pressure* of only about 10 mmHg is obtained.

Because glomerular capillaries are extremely permeable and have an extensive surface area, this modest net filtration pressure produces an extraordinarily large volume of filtrate. The **glomerular filtration rate (GFR)** is the volume of filtrate produced by both kidneys per minute. The GFR averages 115 ml per minute in women and 125 ml per minute in men. This is equivalent to 7.5 L per hour or 180 L per day (about 45 gallons)! Since the total blood volume averages about 5.5 L, this means that the total blood volume is filtered into the urinary tubules every 40 minutes. Most of the filtered water must obviously be returned immediately to the vascular system, or a person would literally urinate to death within minutes.

Clinical Investigation Clues

Remember that the physician was relieved that Emily had only trace amounts of protein in her urine.

Why didn't her urine contain a higher amount of protein?

If it did, what might that condition (called proteinuria) indicate?

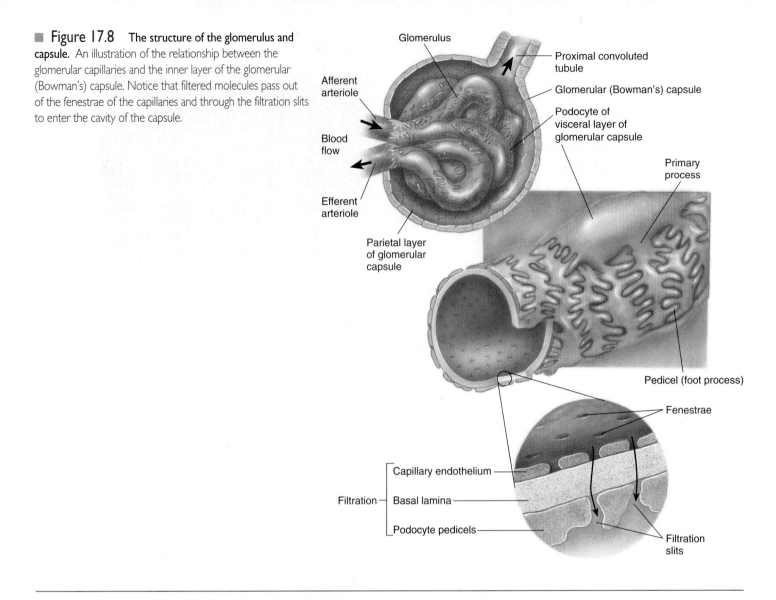

■ **Figure 17.8** **The structure of the glomerulus and capsule.** An illustration of the relationship between the glomerular capillaries and the inner layer of the glomerular (Bowman's) capsule. Notice that filtered molecules pass out of the fenestrae of the capillaries and through the filtration slits to enter the cavity of the capsule.

Glomerulus

Afferent arteriole

Blood flow

Efferent arteriole

Parietal layer of glomerular capsule

Proximal convoluted tubule

Glomerular (Bowman's) capsule

Podocyte of visceral layer of glomerular capsule

Primary process

Pedicel (foot process)

Fenestrae

Capillary endothelium

Filtration — Basal lamina

Podocyte pedicels

Filtration slits

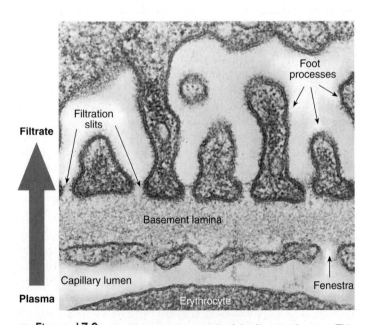

Filtrate

Plasma

Foot processes

Filtration slits

Basement lamina

Capillary lumen

Fenestra

Erythrocyte

■ **Figure 17.9** **An electron micrograph of the filtration barrier.** This electron micrograph shows the barrier separating the capillary lumen from the cavity of the glomerular (Bowman's) capsule.

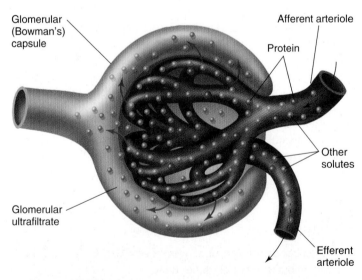

Glomerular (Bowman's) capsule

Afferent arteriole

Protein

Other solutes

Glomerular ultrafiltrate

Efferent arteriole

■ **Figure 17.10** **The formation of glomerular ultrafiltrate.** Only a very small proportion of plasma proteins (*green spheres*) are filtered, but smaller plasma solutes (*purple spheres*) easily enter the glomerular ultrafiltrate. Arrows indicate the direction of filtration.

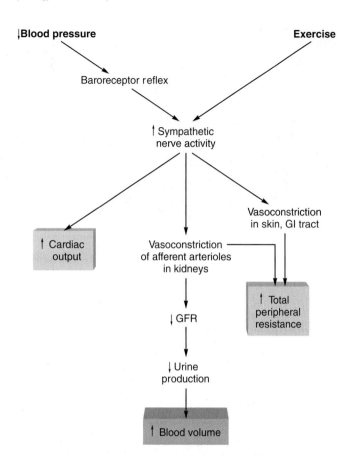

Figure 17.11   **Sympathetic nerve effects.** The effect of increased sympathetic nerve activity on kidney function and other physiological processes is illustrated.

Regulation of Glomerular Filtration Rate

Vasoconstriction or dilation of afferent arterioles affects the rate of blood flow to the glomerulus, and thus affects the glomerular filtration rate. Changes in the diameter of the afferent arterioles result from both extrinsic regulatory mechanisms (produced by sympathetic nerve innervation), and intrinsic regulatory mechanisms (those within the kidneys, also termed *renal autoregulation*). These mechanisms are needed to ensure that the GFR will be high enough to allow the kidneys to eliminate wastes and regulate blood pressure, but not so high as to cause excessive water loss.

Sympathetic Nerve Effects

An increase in sympathetic nerve activity, as occurs during the fight-or-flight reaction and exercise, stimulates constriction of afferent arterioles. This helps to preserve blood volume and to divert blood to the muscles and heart. A similar effect occurs during cardiovascular shock, when sympathetic nerve activity stimulates vasoconstriction. The decreased GFR and the resulting decreased rate of urine formation help to compensate for the rapid drop of blood pressure under these circumstances (fig. 17.11).

Table 17.1   Regulation of the Glomerular Filtration Rate (GFR)

Regulation	Stimulus	Afferent Arteriole	GFR
Sympathetic nerves	Activation by baroreceptor reflex or by higher brain centers	Constricts	Decreases
Autoregulation	Decreased blood pressure	Dilates	No change
Autoregulation	Increased blood pressure	Constricts	No change

Renal Autoregulation

When the direct effect of sympathetic stimulation is experimentally removed, the effect of systemic blood pressure on GFR can be observed. Under these conditions, surprisingly, the GFR remains relatively constant despite changes in mean arterial pressure within a range of 70 to 180 mmHg (normal mean arterial pressure is 100 mmHg). The ability of the kidneys to maintain a relatively constant GFR in the face of fluctuating blood pressures is called **renal autoregulation.**

Renal autoregulation is achieved through the effects of locally produced chemicals on the afferent arterioles (effects on the efferent arterioles are believed to be of secondary importance). When systemic arterial pressure falls toward a mean of 70 mmHg, the afferent arterioles dilate, and when the pressure rises, the afferent arterioles constrict. Blood flow to the glomeruli and GFR can thus remain relatively constant within the autoregulatory range of blood pressure values. The effects of different regulatory mechanisms on the GFR are summarized in table 17.1.

Autoregulation is also achieved through a negative feedback relationship between the afferent arterioles and the volume of fluid in the filtrate. An increased flow of filtrate is sensed by a special group of cells called the *macula densa* in the thick portion of the ascending limb (see fig. 17.25). When the macula densa senses an increased flow of filtrate, it signals the afferent arterioles to constrict. This lowers the GFR, thereby decreasing the formation of filtrate in a process called **tubuloglomerular feedback.**

Test Yourself Before You Continue

1. Describe the structures that plasma fluid must pass through before entering the glomerular capsule. Explain how proteins are excluded from the filtrate.

2. Describe the forces that affect the formation of glomerular ultrafiltrate.

3. Describe the effect of sympathetic innervation on the glomerular filtration rate and explain what is meant by renal autoregulation.

Reabsorption of Salt and Water

Most of the salt and water filtered from the blood is returned to the blood through the wall of the proximal tubule. The reabsorption of water occurs by osmosis, in which water follows the transport of NaCl from the tubule into the surrounding capillaries. Most of the water remaining in the filtrate is reabsorbed across the wall of the collecting duct in the renal medulla. This occurs as a result of the high osmotic pressure of the surrounding tissue fluid, which is produced by transport processes in the loop of Henle.

Although about 180 L of glomerular ultrafiltrate are produced each day, the kidneys normally excrete only 1 to 2 L of urine in a 24-hour period. Approximately 99% of the filtrate must thus be returned to the vascular system, while 1% is excreted in the urine. The urine volume, however, varies according to the needs of the body. When a well-hydrated person drinks a liter or more of water, urine production increases to 16 ml per minute (the equivalent of 23 L per day if this were to continue for 24 hours). In severe dehydration, when the body needs to conserve water, only 0.3 ml of urine per minute, or 400 ml per day, are produced. A volume of 400 ml of urine per day is the minimum needed to excrete the metabolic wastes produced by the body; this is called the **obligatory water loss.** When water in excess of this amount is excreted, the urine becomes increasingly diluted as its volume is increased.

Regardless of the body's state of hydration, it is clear that most of the filtered water must be returned to the vascular system to maintain blood volume and pressure. The return of filtered molecules from the tubules to the blood is called **reabsorption** (fig. 17.12). It is important to realize that the transport of water always occurs passively by osmosis; there is no such thing as active transport of water. A concentration gradient must thus be created between tubular fluid and blood that favors the osmotic return of water to the vascular system.

Reabsorption in the Proximal Tubule

Since all plasma solutes, with the exception of proteins, are able to enter the glomerular ultrafiltrate freely, the total solute concentration (osmolality) of the filtrate is essentially the same as that of plasma. This total solute concentration is equal to 300 milliosmoles per liter (300 mOsm). The filtrate is thus said to be *isosmotic* with the plasma (chapter 6). Reabsorption by osmosis cannot occur unless the solute concentrations of plasma in the peritubular capillaries and the filtrate are altered by active transport processes. This is achieved by the active transport of Na^+ from the filtrate to the peritubular blood.

Active and Passive Transport

The epithelial cells that compose the wall of the proximal tubule are joined together by tight junctions only on their apical sides—

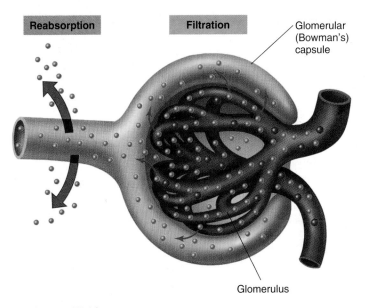

Reabsorption   **Filtration**   Glomerular (Bowman's) capsule

Glomerulus

Figure 17.12   Filtration and reabsorption. Plasma water and its dissolved solutes (except proteins) enter the glomerular ultrafiltrate by filtration, but most of these filtered molecules are reabsorbed. The term *reabsorption* refers to the transport of molecules out of the tubular filtrate back into the blood.

that is, the sides of each cell that are closest to the lumen of the tubule (fig. 17.13). Each cell therefore has four exposed surfaces: the apical side facing the lumen, which contains microvilli; the basal side facing the peritubular capillaries; and the lateral sides facing the narrow clefts between adjacent epithelial cells.

The concentration of Na^+ in the glomerular ultrafiltrate—and thus in the fluid entering the proximal tubule—is the same as that in plasma. The epithelial cells of the tubule, however, have a much lower Na^+ concentration. This lower Na^+ concentration is partially due to the low permeability of the plasma membrane to Na^+ and partially due to the active transport of Na^+ out of the cell by Na^+/K^+ pumps, as described in chapter 6. In the cells of the proximal tubule, the Na^+/K^+ pumps are located in the basal and lateral sides of the plasma membrane but not in the apical membrane. As a result of the action of these active transport pumps, a concentration gradient is created that favors the diffusion of Na^+ from the tubular fluid across the apical plasma membranes and into the epithelial cells of the proximal tubule. The Na^+ is then extruded into the surrounding tissue fluid by the Na^+/K^+ pumps.

The transport of Na^+ from the tubular fluid to the interstitial (tissue) fluid surrounding the proximal tubule creates a potential difference across the wall of the tubule, with the lumen as the negative pole. This electrical gradient favors the passive transport of Cl^- toward the higher Na^+ concentration in the interstitial fluid. Chloride ions, therefore, passively follow sodium ions out of the filtrate into the interstitial fluid. As a result of the accumulation of NaCl, the osmolality and osmotic pressure of the interstitial fluid surrounding the epithelial cells are increased above those of the tubular fluid. This is particularly true of the interstitial fluid between the lateral membranes of adjacent epithelial cells, where

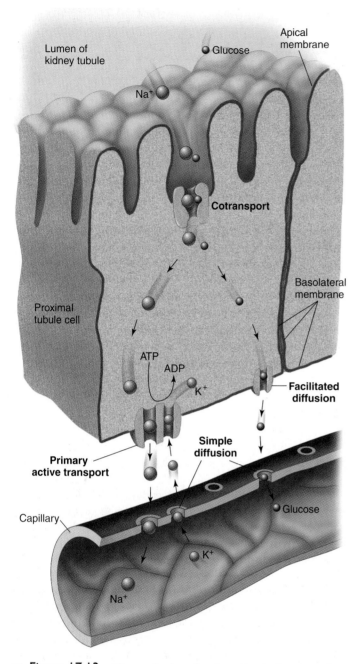

Figure 17.13 The mechanism of reabsorption in the proximal tubule. The appearance of proximal tubule cells in the electron microscope is illustrated. Molecules that are reabsorbed pass through the tubule cells from the apical membrane (facing the filtrate) to the basolateral membrane (facing the blood). There is coupled transport (a type of active transport) of glucose and Na$^+$ into the cytoplasm, and primary active transport of Na$^+$ across the basolateral membrane by the Na$^+$/K$^+$ pump.

the narrow spaces permit the accumulated NaCl to achieve a higher concentration.

An osmotic gradient is thus created between the tubular fluid and the interstitial fluid surrounding the proximal tubule. Since the cells of the proximal tubule are permeable to water, water moves by osmosis from the tubular fluid into the epithelial cells and then across the basal and lateral sides of the epithelial

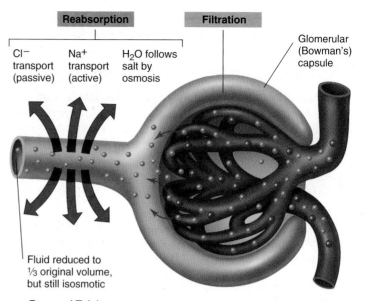

Figure 17.14 Salt and water reabsorption in the proximal tubule. Sodium is actively transported out of the filtrate (see fig. 17.13) and chloride follows passively by electrical attraction. Water follows the salt out of the tubular filtrate into the peritubular capillaries by osmosis.

cells into the interstitial fluid. The salt and water that were reabsorbed from the tubular fluid can then move passively into the surrounding peritubular capillaries, and in this way be returned to the blood (fig. 17.14).

Significance of Proximal Tubule Reabsorption

Approximately 65% of the salt and water in the original glomerular ultrafiltrate is reabsorbed across the proximal tubule and returned to the vascular system. The volume of tubular fluid remaining is reduced accordingly, but this fluid is still isosmotic with the blood, which has a concentration of 300 mOsm. This is because the plasma membranes in the proximal tubule are freely permeable to water, so that water and salt are removed in proportionate amounts.

An additional smaller amount of salt and water (about 20%) is returned to the vascular system by reabsorption through the descending limb of the loop of Henle. This reabsorption, like that in the proximal tubule, occurs constantly, regardless of the person's state of hydration. Unlike reabsorption in later regions of the nephron (distal tubule and collecting duct), it is not subject to hormonal regulation. Therefore, approximately 85% of the filtered salt and water is reabsorbed in a constant fashion in the early regions of the nephron (proximal tubule and loop of Henle). This reabsorption is very costly in terms of energy expenditures, accounting for as much as 6% of the calories consumed by the body at rest.

Since 85% of the original glomerular ultrafiltrate is reabsorbed in the early regions of the nephron, only 15% of the initial filtrate remains to enter the distal convoluted tubule and collecting duct. This is still a large volume of fluid—15% × GFR (180 L per day) = 27 L per day—that must be reabsorbed to varying degrees in accordance with the body's state of hydration. This "fine tuning" of

the percentage of reabsorption and urine volume is accomplished by the action of hormones on the later regions of the nephron.

The Countercurrent Multiplier System

Water cannot be actively transported across the tubule wall, and osmosis of water cannot occur if the tubular fluid and surrounding interstitial fluid are isotonic to each other. In order for water to be reabsorbed by osmosis, the surrounding interstitial fluid must be hypertonic. The osmotic pressure of the interstitial fluid in the renal medulla is, in fact, raised to over four times that of plasma by the juxtamedullary nephrons. This results partly from the fact that the tubule bends; the geometry of the loop of Henle permits interaction between the descending and ascending limbs. Since the ascending limb is the active partner in this interaction, its properties will be described before those of the descending limb.

Ascending Limb of the Loop of Henle

Salt (NaCl) is actively extruded from the ascending limb into the surrounding interstitial fluid. This is not accomplished, however, by the same process that occurs in the proximal tubule. Instead, Na^+ diffuses from the filtrate into the cells of the thick portion of the ascending limb, accompanied by the secondary active transport of K^+ and Cl^-. This occurs in a ratio of 1 Na^+ to 1 K^+ to 2 Cl^-. The Na^+ is then actively transported across the basolateral membrane to the interstitial fluid by the Na^+/K^+ pumps. Cl^- follows the Na^+ passively because of electrical attraction, and K^+ passively diffuses back into the filtrate (fig. 17.15).

The ascending limb is structurally divisible into two regions: a *thin segment,* nearest the tip of the loop, and a *thick segment* of varying lengths, which carries the filtrate outward into the cortex and into the distal convoluted tubule. It is currently believed that only the cells of the thick segments of the ascending limb are capable of actively transporting NaCl from the filtrate into the surrounding interstitial fluid.

Although the mechanism of NaCl transport is different in the ascending limb than in the proximal tubule, the net effect is the same: salt (NaCl) is extruded into the surrounding fluid. Unlike the epithelial walls of the proximal tubule, however, the walls of the ascending limb of the loop of Henle are *not permeable to water.* The tubular fluid thus becomes increasingly dilute as it ascends toward the cortex, whereas the interstitial fluid around the loops of Henle in the medulla becomes increasingly more concentrated. By means of these processes, the tubular fluid that enters the distal tubule in the cortex is made hypotonic (with a concentration of about 100 mOsm), whereas the interstitial fluid in the medulla is made hypertonic.

Descending Limb of the Loop of Henle

The deeper regions of the medulla, around the tips of the loops of juxtamedullary nephrons, reach a concentration of 1,200 to 1,400 mOsm. In order to reach this high a concentration, the salt pumped out of the ascending limb must accumulate in the interstitial fluid of the medulla. This occurs because of the properties of the descending limb, to be discussed next, and because blood vessels around the loop do not carry back all of the extruded salt to the general circulation. The capillaries in the medulla are

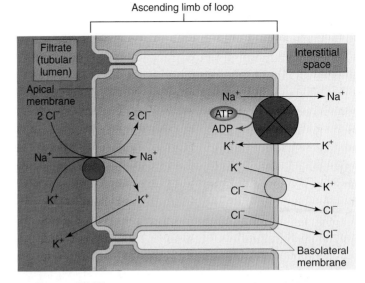

Figure 17.15 **The transport of ions in the ascending limb.** In the thick segment of the ascending limb of the loop, Na^+ and K^+ together with two Cl^- enter the tubule cells. Na^+ is then actively transported out into the interstitial space and Cl^- follows passively. The K^+ diffuses back into the filtrate, and some also enters the interstitial space.

uniquely arranged to trap NaCl in the interstitial fluid, as will be discussed shortly.

The descending limb does not actively transport salt, and indeed is believed to be impermeable to the passive diffusion of salt. It is, however, permeable to water. Since the surrounding interstitial fluid is hypertonic to the filtrate in the descending limb, water is drawn out of the descending limb by osmosis and enters blood capillaries. The concentration of tubular fluid is thus increased, and its volume is decreased, as it descends toward the tips of the loops.

As a result of these passive transport processes in the descending limb, the fluid that "rounds the bend" at the tip of the loop has the same osmolality as that of the surrounding interstitial fluid (1,200 to 1,400 mOsm). There is, therefore, a higher salt concentration arriving in the ascending limb than there would be if the descending limb simply delivered isotonic fluid. Salt transport by the ascending limb is increased accordingly, so that the "saltiness" (NaCl concentration) of the interstitial fluid is multiplied (fig. 17.16).

Countercurrent Multiplication

Countercurrent flow (flow in opposite directions) in the ascending and descending limbs and the close proximity of the two limbs allow for interaction between them. Since the concentration of the tubular fluid in the descending limb reflects the concentration of surrounding interstitial fluid, and since the concentration of this fluid is raised by the active extrusion of salt from the ascending limb, a *positive feedback mechanism* is created. The more salt the ascending limb extrudes, the more concentrated will be the fluid that is delivered to it from the descending limb. This positive feedback mechanism multiplies the concentration of interstitial fluid and descending limb fluid, and is thus called the **countercurrent multiplier system.**

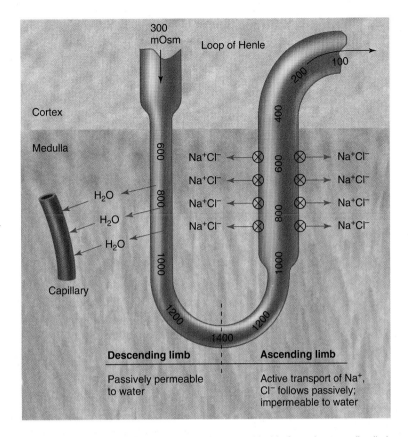

Figure 17.16 **The countercurrent multiplier system.** The extrusion of sodium chloride from the ascending limb makes the surrounding interstitial fluid more concentrated. Multiplication of this concentration is due to the fact that the descending limb is passively permeable to water, which causes its fluid to increase in concentration as the surrounding interstitial fluid becomes more concentrated. The values of these changes in osmolality, together with the effect on surrounding interstitial fluid concentration, are shown in milliosmolal units.

The countercurrent multiplier system recirculates salt and thus traps some of the salt that enters the loop of Henle in the interstitial fluid of the renal medulla. This system results in a gradually increasing concentration of renal interstitial fluid from the cortex to the inner medulla; the osmolality of interstitial fluid increases from 300 mOsm (isotonic) in the cortex to between 1,200 and 1,400 mOsm in the deepest part of the medulla. This hypertonicity is required for water reabsorption, as will be explained shortly.

Vasa Recta

In order for the countercurrent multiplier system to be effective, most of the salt that is extruded from the ascending limbs must remain in the interstitial fluid of the medulla, while most of the water that leaves the descending limbs must be removed by the blood. This is accomplished by the *vasa recta*—long, thin-walled vessels that parallel the loops of Henle of the juxtamedullary nephrons (see fig. 17.19). The descending vasa recta have characteristics of both capillaries and arterioles because their continuous endothelium is surrounded by smooth muscle remnants. These vessels have *urea transporters* (for facilitative diffusion) and *aquaporin proteins,* which function as water channels through the membrane (chapter 6). The ascending vasa recta are capillaries with a fenestrated endothelium. As

described in chapter 13, the wide gaps between endothelial cells in such capillaries permit rapid rates of diffusion.

The vasa recta maintain the hypertonicity of the renal medulla by means of a mechanism known as **countercurrent exchange.** Salt and other dissolved solutes (primarily urea, described in the next section) that are present at high concentrations in the interstitial fluid diffuse into the descending vasa recta. However, these same solutes then passively diffuse out of the ascending vasa recta and back into the interstitial fluid to complete the countercurrent exchange. They do this because, at each level of the medulla, the concentration of solutes is higher in the ascending vessels than in the interstitial fluid, and higher in the interstitial fluid than in the descending vessels. Solutes are thus recirculated and trapped within the medulla.

The walls of the vasa recta are freely permeable to water and to dissolved NaCl and urea. Plasma proteins, however, do not easily pass through the capillary walls of the vasa recta. The colloid osmotic pressure (oncotic pressure) within the vasa recta, therefore, is higher than in the surrounding interstitial fluid. This is similar to the situation in other capillary beds (chapter 14) and results in the osmotic movement of water into both the descending and ascending limbs of the vasa recta. The vasa recta thus trap salt and urea within the interstitial fluid but transport water out of the renal medulla (fig. 17.17).

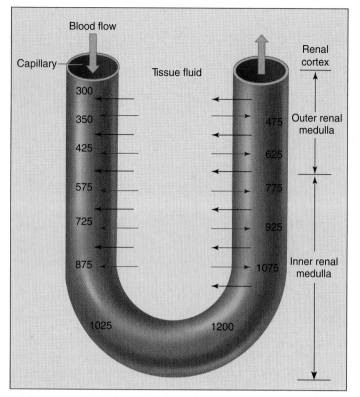

Black arrows = diffusion of NaCl and urea
Blue arrows = movement of water by osmosis

■ **Figure 17.17**   **Countercurrent exchange in the vasa recta.** The diffusion of salt and water first into and then out of these blood vessels helps to maintain the "saltiness" (hypertonicity) of the interstitial fluid in the renal medulla. (Numbers indicate osmolality.)

Effects of Urea

Countercurrent multiplication of the NaCl concentration is the mechanism that contributes most to the hypertonicity of the interstitial fluid in the medulla. However, **urea,** a waste product of amino acid metabolism (chapter 5; see fig. 5.15), also contributes significantly to the total osmolality of the interstitial fluid.

The role of urea was inferred from experimental evidence showing that active transport of Na$^+$ occurs only in the thick segments of the ascending limbs. The thin segments of the ascending limbs, which are located in the deeper regions of the medulla, are not able to extrude salt actively. But since salt does indeed leave the thin segments, a diffusion gradient for salt must exist, despite the fact that the surrounding interstitial fluid has the same osmolality as the tubular fluid. Investigators therefore concluded that molecules other than salt—specifically urea—contribute to the hypertonicity of the interstitial fluid.

It was later shown that the ascending limb of the loop of Henle and the terminal portion of the collecting duct in the inner medulla are permeable to urea. Indeed, the region of the collecting duct in the inner medulla has specific urea transporters that permit a very high rate of diffusion into the surrounding interstitial fluid. Urea can thus diffuse out of this portion of the collect-

ing duct and into the ascending limb (fig. 17.18). In this way, a certain amount of urea is recycled through these two segments of the nephron. The urea is thereby trapped in the interstitial fluid where it can contribute significantly to the high osmolality of the medulla. This relates to the ability to produce a concentrated urine, as will be described in the next section.

The transport properties of different tubule segments are summarized in table 17.2.

Collecting Duct: Effect of Antidiuretic Hormone (ADH)

As a result of the recycling of salt between the ascending and descending limbs and the recycling of urea between the collecting duct and the loop of Henle, the interstitial fluid is made very hypertonic. The collecting ducts must channel their fluid through this hypertonic environment in order to empty their contents of urine into the calyces. Whereas the fluid surrounding the collecting ducts in the medulla is hypertonic, the fluid that passes into the collecting ducts in the cortex is hypotonic as a result of the active extrusion of salt by the ascending limbs of the loops.

The medullary region of the collecting duct is impermeable to the high concentration of NaCl that surrounds it. The wall of the collecting duct, however, is permeable to water. Since the surrounding interstitial fluid in the renal medulla is very hypertonic because of the countercurrent multiplier system, water is drawn out of the collecting ducts by osmosis. This water does not dilute the surrounding interstitial fluid because it is transported by capillaries to the general circulation. In this way, most of the water remaining in the filtrate is returned to the vascular system (fig. 17.19).

Note that it is the osmotic gradient created by the countercurrent multiplier system that provides the force for water reabsorption through the collecting ducts. The rate at which this osmotic movement occurs, however, is determined by the permeability of the collecting duct to water. This depends on the number of **aquaporins** (water channels) in the plasma membranes of the collecting duct epithelial cells.

Aquaporins are produced as proteins within the membranes of vesicles that bud from the Golgi apparatus (chapter 3). In the absence of stimulation, these vesicles are present in the cytoplasm of the collecting duct cells. When **antidiuretic hormone (ADH)** binds to its membrane receptors on the collecting duct, it acts (via cAMP as a second messenger) to stimulate the fusion of these vesicles with the cell membrane (chapter 6; see fig. 6.15). This is identical to exocytosis, except that here there is no secretion of product. The importance of this process in the collecting duct is that the water channels are incorporated into the plasma membrane when the vesicles and membrane fuse. In response to ADH, therefore, the collecting duct becomes more permeable to water. When ADH is no longer available to bind to its membrane receptors, the water channels are removed from the plasma membrane by a process of endocytosis. Endocytosis is the opposite of exocytosis;

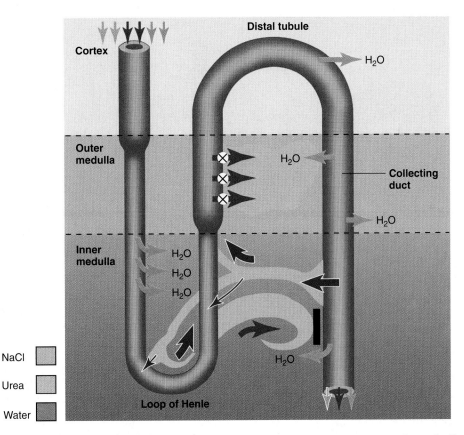

Figure 17.18 **The role of urea in urine concentration.** Urea diffuses out of the inner collecting duct and contributes significantly to the concentration of the interstitial fluid in the renal medulla. The active transport of Na^+ out of the thick segments of the ascending limbs also contributes to the hypertonicity of the medulla, so that water is reabsorbed by osmosis from the collecting ducts.

Table 17.2 **Transport Properties of Different Segments of the Renal Tubules and the Collecting Ducts**

Nephron Segment	Active Transport	Passive Transport		
		Salt	Water	Urea
Proximal tubule	Na^+	Cl^-	Yes	Yes
Descending limb of Henle's loop	None	Maybe	Yes	No
Thin segment of ascending limb	None	NaCl	No	Yes
Thick segment of ascending limb	Na^+	Cl^-	No	No
Distal tubule	Na^+	Cl^-	No**	No
Collecting duct*	Slight Na^+	No	Yes (ADH) or slight (no ADH)	Yes

*The permeability of the collecting duct to water depends on the presence of ADH.

**The last part of the distal tubule, however, is permeable to water.

the plasma membrane invaginates to reform vesicles that again contain the water channels. Alternating exocytosis and endocytosis in response to the presence and absence of ADH, respectively, is believed to result in the recycling of water channels within the cell.

When the concentration of ADH is increased, the collecting ducts become more permeable to water and more water is reabsorbed. A decrease in ADH, conversely, results in less reabsorption of water and thus in the excretion of a larger volume of more dilute urine (fig. 17.20). ADH is produced by neurons in the hypothalamus and is released from the posterior pituitary (chapter 11; see fig. 11.13). The secretion of ADH is stimulated when osmoreceptors in the hypothalamus respond to an increase in blood osmolality. During dehydration, therefore, when the plasma becomes more concentrated, increased secretion of ADH promotes increased permeability of the collecting ducts to water. In severe dehydration only the minimal amount of water needed to eliminate the body's wastes is excreted. This minimum, an *obligatory water loss* of about 400 ml per day, is limited by the

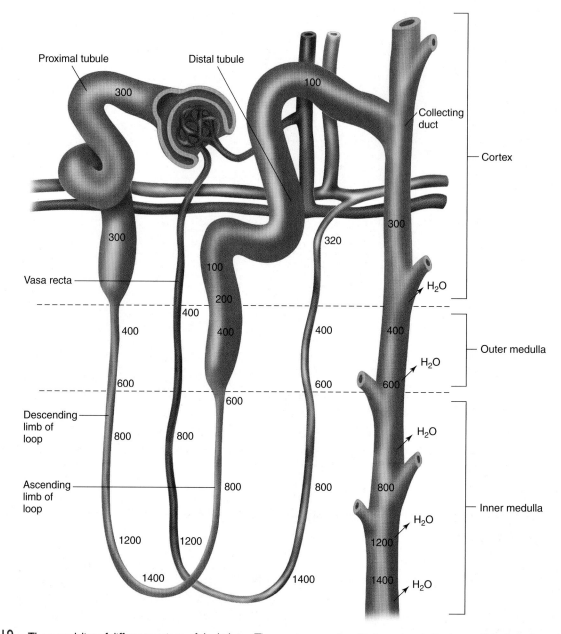

Figure 17.19 **The osmolality of different regions of the kidney.** The countercurrent multiplier system in the loop of Henle and countercurrent exchange in the vasa recta help to create a hypertonic renal medulla. Under the influence of antidiuretic hormone (ADH), the collecting duct becomes more permeable to water, and thus more water is drawn out by osmosis into the hypertonic renal medulla and peritubular capillaries.

fact that urine cannot become more concentrated than the medullary tissue fluid surrounding the collecting ducts. Under these conditions about 99.8% of the initial glomerular ultrafiltrate is reabsorbed.

A person in a state of normal hydration excretes about 1.5 L of urine per day, indicating that 99.2% of the glomerular ultrafiltrate volume is reabsorbed. Notice that small changes in percent reabsorption translate into large changes in urine volume. Drinking more water—and thus decreasing ADH secretion (fig. 17.20 and table 17.3)—results in correspondingly larger volumes of urine excretion. It should be noted, however, that even in the complete absence of ADH some water is still reabsorbed through the collecting ducts.

Diabetes insipidus is a disease associated with the inadequate secretion or action of ADH. When the secretion of ADH is adequate, but a genetic defect in the ADH receptors or the aquaporin channels renders the kidneys incapable of responding to ADH, the condition is called *nephrogenetic diabetes insipidus.* Without proper ADH secretion or action, the collecting ducts are not very permeable to water, and so a large volume (5 to 10 L per day) of dilute urine is produced. The dehydration that results causes intense thirst, but a person with this condition has difficulty drinking enough to compensate for the large volumes of water lost in the urine.

Table 17.3 Antidiuretic Hormone Secretion and Action

Stimulus	Receptors	Secretion of ADH	Effects on Urine Volume	Effects on Blood
↑Osmolality (dehydration)	Osmoreceptors in hypothalamus	Increased	Decreased	Increased water retention; decreased blood osmolality
↓Osmolality	Osmoreceptors in hypothalamus	Decreased	Increased	Water loss increases blood osmolality
↑Blood volume	Stretch receptors in left atrium	Decreased	Increased	Decreased blood volume
↓Blood volume	Stretch receptors in left atrium	Increased	Decreased	Increased blood volume

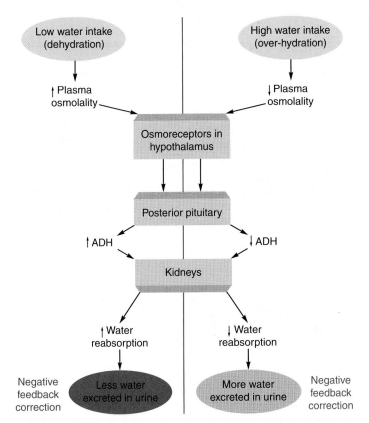

Figure 17.20 Homeostasis of plasma concentration is maintained by ADH. In dehydration (left side of figure), a rise in ADH secretion results in a reduction in the excretion of water in the urine. In overhydration (right side of figure), the excess water is eliminated through a decrease in ADH secretion. These changes provide negative feedback correction, maintaining homeostasis of plasma osmolality and, indirectly, blood volume.

Test Yourself Before You Continue

1. Describe the mechanisms for salt and water reabsorption in the proximal tubule.
2. Compare the transport of Na⁺, Cl⁻, and water across the walls of the proximal tubule, ascending and descending limbs of the loop of Henle, and collecting duct.
3. Describe the interaction between the ascending and descending limbs of the loop and explain how this interaction results in a hypertonic renal medulla.
4. Explain how ADH helps the body to conserve water. How do variations in ADH secretion affect the volume and concentration of urine?

Renal Plasma Clearance

As blood passes through the kidneys, some of the constituents of the plasma are removed and excreted in the urine. The blood is thus "cleared," to some extent, of particular solutes in the process of urine formation. These solutes may be removed from the blood by filtration through the glomerular capillaries or by secretion by the tubular cells into the filtrate. At the same time, certain molecules in the tubular fluid can be reabsorbed back into the blood.

One of the major functions of the kidneys is to eliminate excess ions and waste products from the blood. *Clearing* the blood of these substances is accomplished through their excretion in the urine. Because of renal clearance, the concentrations of these substances in the blood leaving the kidneys (in the renal vein) is lower than their concentrations in the blood entering the kidneys (in the renal artery).

Transport Process Affecting Renal Clearance

Renal clearance refers to the ability of the kidneys to remove molecules from the blood plasma by excreting them in the urine. Molecules and ions dissolved in the plasma can be filtered through the glomerular capillaries and enter the glomerular capsules. Then, those that are not reabsorbed will be eliminated in the urine; they will be "cleared" from the blood.

The process of filtration, a type of bulk transport through capillaries, promotes renal clearance. The process of reabsorption—involving membrane transport by means of carrier proteins—moves particular molecules and ions from the filtrate into the blood, and thus reduces the renal clearance of these molecules from the blood.

There is another process that affects renal clearance, a membrane transport process called **secretion** (fig. 17.21). In terms of its direction of transport, secretion is the opposite of reabsorption—secreted molecules and ions move out of the peritubular capillaries into the interstitial fluid, and then are transported across the basolateral membrane of the tubular epithelial

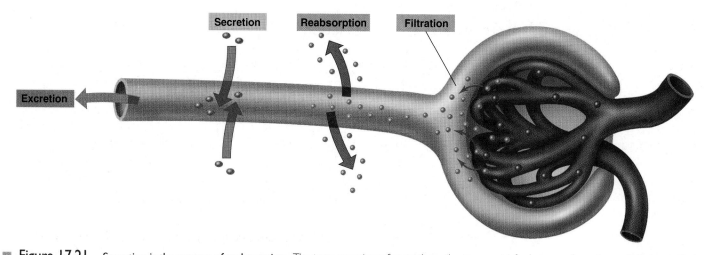

Figure 17.21 **Secretion is the reverse of reabsorption.** The term *secretion* refers to the active transport of substances from the peritubular capillaries into the tubular fluid. This transport is opposite in direction to that which occurs in reabsorption.

cells and into the lumen of the nephron tubule. Molecules that are both filtered and secreted are thus eliminated in the urine more rapidly (are cleared from the blood more rapidly) than molecules that are not secreted. In summary, the process of reabsorption decreases renal clearance, while the process of secretion increases renal clearance.

Tubular Secretion of Drugs

Many molecules foreign to the body—known generally as *xenobiotics* and including toxins and drugs—are eliminated in the urine more rapidly than would be possible by just glomerular filtration. This implies that they are secreted by membrane carriers that somehow recognize them as foreign to the body. Considering that membrane carriers are specific and that there are so many possible xenobiotic molecules, how is this accomplished?

Scientists have discovered that there are a large number of different carriers, usually in the basolateral membrane of the proximal tubule, that are **organic anion transporters.** These carriers are each specific for a broad range of molecules; they are described as being *polyspecific*. The specificity of one type

of carrier overlaps with the specificity of other carriers, so that they can transport a wide variety of exogenous ("originating outside") and endogenous ("originating inside") molecules across the nephron tubules. This allows the kidneys to rapidly eliminate potentially toxic molecules from the blood. However, tubular secretion of therapeutic drugs can interfere with the ability of those drugs to work.

Renal Clearance of Inulin: Measurement of GFR

If a substance is neither reabsorbed nor secreted by the tubules, the amount excreted in the urine per minute will be equal to the amount that is filtered out of the glomeruli per minute. There does not seem to be a single substance produced by the body, however, that is not reabsorbed or secreted to some degree. Plants such as artichokes, dahlias, onions, and garlic, fortunately, do produce such a compound. This compound, a polymer of the monosaccharide fructose, is **inulin.** Once injected into the blood, inulin is filtered by the glomeruli, and the amount of inulin excreted per minute is exactly equal to the amount that was filtered per minute (fig. 17.22).

If the concentration of inulin in urine is measured and the rate of urine formation is determined, the rate of inulin excretion can easily be calculated:

$$\text{Quantity excreted per minute} = \underset{\left(\frac{ml}{min}\right)}{V} \times \underset{\left(\frac{mg}{ml}\right)}{U}$$
$$(mg/min)$$

where

V = rate of urine formation
U = inulin concentration in urine

The rate at which a substance is filtered by the glomeruli (in milligrams per minute) can be calculated by multiplying the milliliters of plasma filtered per minute (the **glomerular**

Many antibiotics are secreted by the renal tubules and thus are rapidly cleared from the body. **Penicillin,** for example, is secreted into the tubular filtrate and, because of this, large amounts of the drug have to be administered. When penicillin was first used during World War II, however, it was in short supply. Scientists then discovered that a different organic anion (benzoic acid) would compete with penicillin for the carrier proteins and thus prevent penicillin from being too rapidly eliminated from the body. A newer competitor for these carriers, *probenecid,* is sometimes used to inhibit the tubular secretion of certain antibiotics. This improves the effectiveness of the antibiotic and reduces its potential toxicity to the kidneys (nephrotoxicity).

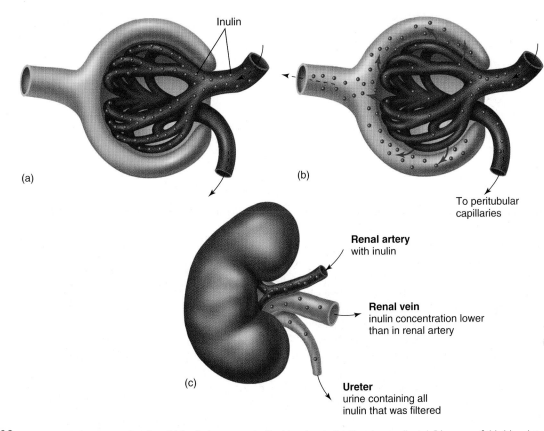

Inulin

(a)

(b)

To peritubular
capillaries

Renal artery
with inulin

Renal vein
inulin concentration lower
than in renal artery

(c)

Ureter
urine containing all
inulin that was filtered

■ **Figure 17.22** **The renal clearance of inulin.** *(a)* Inulin is present in the blood entering the glomeruli, and *(b)* some of this blood, together with its dissolved inulin, is filtered. All of this filtered inulin enters the urine, whereas most of the filtered water is returned to the vascular system (is reabsorbed). *(c)* The blood leaving the kidneys in the renal vein, therefore, contains less inulin than the blood that entered the kidneys in the renal artery. Since inulin is filtered but neither reabsorbed nor secreted, the inulin clearance rate equals the glomerular filtration rate (GFR).

filtration rate, or **GFR**) by the concentration of that substance in the plasma, as shown in this equation:

$$\text{Quantity filtered per minute} = \underset{\left(\dfrac{ml}{min}\right)}{GFR} \times \underset{\left(\dfrac{mg}{ml}\right)}{P}$$
$$(mg/min)$$

where

P = inulin concentration in plasma

Since inulin is neither reabsorbed nor secreted, the amount filtered equals the amount excreted:

$$\underset{\text{(amount filtered)}}{GFR \times P} = \underset{\text{(amount excreted)}}{V \times U}$$

If the preceding equation is now solved for the glomerular filtration rate,

$$GFR_{(ml/min)} = \frac{V_{(ml/min)} \times U_{(mg/ml)}}{P_{(mg/ml)}}$$

Suppose, for example, that inulin is infused into a vein and its concentrations in the urine and plasma are found to be 30 mg per ml and 0.5 mg per ml, respectively. If the rate of urine formation is 2 ml per minute, the GFR can be calculated as:

$$GFR = \frac{2 \text{ ml/min} \times 30 \text{ mg/ml}}{0.5 \text{ mg/ml}} = 120 \text{ ml/min}$$

This equation states that 120 ml of plasma must have been filtered each minute in order to excrete the measured amount of inulin that appeared in the urine. The glomerular filtration rate is thus 120 ml per minute in this example.

Measurements of the plasma concentration of **creatinine** are often used clinically as an index of kidney function. Creatinine, produced as a waste product of muscle creatine, is secreted to a slight degree by the renal tubules so that its excretion rate is a little above that of inulin. Since it is released into the blood at a constant rate, and since its excretion is closely matched to the GFR, an abnormal decrease in GFR causes the plasma creatinine concentration to rise. Thus, a simple measurement of blood creatinine concentration can indicate whether the GFR is normal and provide information about the health of the kidneys.

Clinical Investigation Clues

Remember that Emily had mild oliguria, edema, and an elevated plasma creatinine concentration.

What does an elevated plasma creatinine concentration suggest?

How might this be related to Emily's oliguria and edema?

Clearance Calculations

The **renal plasma clearance** is the volume of plasma from which a substance is completely removed in one minute by excretion in the urine. Notice that the units for renal plasma clearance are ml/min. In the case of inulin, which is filtered but neither reabsorbed nor secreted, the amount of inulin that enters the urine is that which is contained in the volume of plasma filtered. The clearance of inulin is thus equal to the GFR (120 ml/min in the previous example). This volume of filtered plasma, however, also contains other solutes that may be reabsorbed to varying degrees. If a portion of a filtered solute is reabsorbed, the amount excreted in the urine is less than that which was contained in the 120 ml of plasma filtered. Thus, *the renal plasma clearance of a substance that is reabsorbed must be less than the GFR* (table 17.4).

If a substance is not reabsorbed, all of the filtered amount will be cleared. If this substance is, in addition, secreted by active transport into the renal tubules from the peritubular blood, an additional amount of plasma can be cleared of that substance. Therefore, *the renal plasma clearance of a substance that is filtered and secreted is greater than the GFR* (table 17.5). In order to compare the renal "handling" of various substances in terms of their reabsorption or secretion, the renal plasma

clearance is calculated using the same formula used for determining the GFR:

$$\text{Renal plasma clearance} = \frac{V \times U}{P}$$

where

V = urine volume per minute
U = concentration of substance in urine
P = concentration of substance in plasma

Clearance of Urea

Urea may be used as an example of how the clearance calculations can reveal the way the kidneys handle a molecule. Urea is a waste product of amino acid metabolism that is secreted by the liver into the blood and filtered into the glomerular capsules. Using the formula for renal clearance previously described and these sample values, the urea clearance can be obtained:

$$V = 2 \text{ ml/min}$$
$$U = 7.5 \text{ mg/ml of urea}$$
$$P = 0.2 \text{ mg/ml of urea}$$

$$\text{Urea clearance} = \frac{(2 \text{ ml/min})(7.5 \text{ mg/ml})}{0.2 \text{ mg/ml}} = 75 \text{ ml/min}$$

The clearance of urea in this example (75 ml/min) is less than the clearance of inulin (120 ml/min). Thus, even though 120 ml of plasma filtrate entered the nephrons per minute, only the amount of urea contained in 75 ml of filtrate is excreted. The kidneys therefore must reabsorb some of the urea that is filtered. Despite the fact that it is a waste product, a significant portion of the filtered urea (ranging from 40% to 60%) is always reabsorbed. This is a passive transport process that occurs because of the presence in the tubules of carriers for the facilitative diffusion of urea.

Table 17.4 Effects of Filtration, Reabsorption, and Secretion on Renal Plasma Clearance

Term	Definition	Effect on Renal Clearance
Filtration	A substance enters the glomerular ultrafiltrate.	Some or all of a filtered substance may enter the urine and be "cleared" from the blood.
Reabsorption	A substance is transported from the filtrate, through tubular cells, and into the blood.	Reabsorption decreases the rate at which a substance is cleared; clearance rate is less than the glomerular filtration rate (GFR).
Secretion	A substance is transported from peritubular blood, through tubular cells, and into the filtrate.	When a substance is secreted by the nephrons, its renal plasma clearance is greater than the GFR.

Table 17.5 Renal "Handling" of Different Plasma Molecules

If Substance Is:	Example	Concentration in Renal Vein	Renal Clearance Rate
Not filtered	Proteins	Same as in renal artery	Zero
Filtered, not reabsorbed or secreted	Inulin	Less than in renal artery	Equal to GFR (115–125 ml/min)
Filtered, partially reabsorbed	Urea	Less than in renal artery	Less than GFR
Filtered, completely reabsorbed	Glucose	Same as in renal artery	Zero
Filtered and secreted	PAH	Less than in renal artery; approaches zero	Greater than GFR; up to total plasma flow rate (~625 ml/min)
Filtered, reabsorbed, and secreted	K+	Variable	Variable

Clearance of PAH: Measurement of Renal Blood Flow

Not all of the blood delivered to the glomeruli is filtered into the glomerular capsules; most of the glomerular blood passes through to the efferent arterioles and peritubular capillaries. The inulin and urea in this unfiltered blood are not excreted but instead return to the general circulation. Blood must therefore make many passes through the kidneys before it can be completely cleared of a given amount of inulin or urea.

For compounds in the unfiltered renal blood to be cleared, they must be secreted into the tubules by active transport from the peritubular capillaries. In this way, all of the blood going to the kidneys can potentially be cleared of a secreted compound in a single pass. This is the case for an exogenous molecule called **para-aminohippuric acid,** or **PAH** (fig. 17.23), that can be infused into the blood. The clearance (in ml/min) of PAH can be used to measure the **total renal blood flow.** The normal PAH clearance has been found to average 625 ml/min. Since the glomerular filtration rate averages about 120 ml/min, this indicates that only about 120/625, or roughly 20%, of the renal plasma flow is filtered. The remaining 80% passes on to the efferent arterioles.

Since filtration and secretion clear only the molecules dissolved in plasma, the PAH clearance actually measures the renal plasma flow. In order to convert this to the total renal blood flow, the volume of blood occupied by erythrocytes must be taken into account. If the hematocrit (chapter 13) is 45, for example, erythrocytes occupy 45% of the blood volume and plasma accounts for the remaining 55%. The total renal blood flow is calculated by dividing the PAH clearance by the fractional blood volume occupied by plasma (0.55, in this example). The total renal blood flow in this example is thus 625 ml/min divided by 0.55, or 1.1 L/min.

Reabsorption of Glucose

Glucose and amino acids in the blood are easily filtered by the glomeruli into the renal tubules. These molecules, however, are usually not present in the urine. It can therefore be concluded that filtered glucose and amino acids are normally completely reabsorbed by the nephrons. This occurs in the proximal tubule by secondary active transport, which is mediated by membrane carriers that cotransport glucose and Na^+ (see fig. 17.13), or amino acids and Na^+.

Carrier-mediated transport displays the property of *saturation.* This means that when the transported molecule (such as glucose) is present in sufficiently high concentrations, all of the carriers become occupied and the transport rate reaches a maximal value. The concentration of transported molecules needed to just saturate the carriers and achieve the maximal transport rate is called the **transport maximum** (abbreviated T_m).

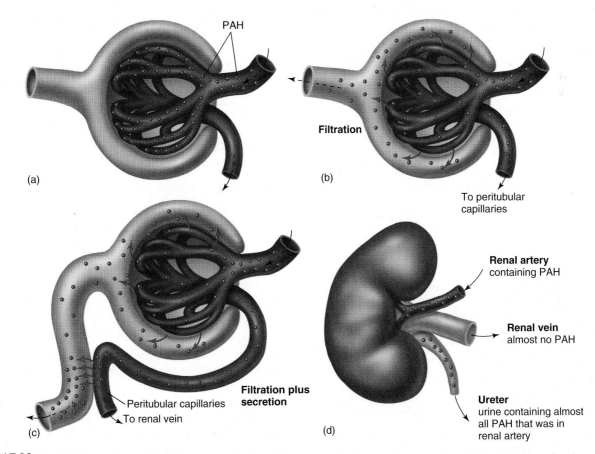

■ **Figure 17.23** **The renal clearance of PAH.** Some of the para-aminohippuric acid (PAH) in glomerular blood *(a)* is filtered into the glomerular (Bowman's) capsules *(b)*. The PAH present in the unfiltered blood is secreted from the peritubular capillaries into the nephron *(c)*, so that all of the blood leaving the kidneys is free of PAH *(d)*. The clearance of PAH therefore equals the total renal blood flow.

The carriers for glucose and amino acids in the renal tubules are not normally saturated and so are able to remove the filtered molecules completely. The T_m for glucose, for example, averages 375 mg per minute, which is well above the normal rate at which glucose is delivered to the tubules. The rate of glucose delivery can be calculated by multiplying the plasma glucose concentration (about 1 mg per ml in the fasting state) by the GFR (about 125 ml per minute). Approximately 125 mg per minute are thus delivered to the tubules, whereas a rate of 375 mg per minute is required to reach saturation.

Glycosuria

Glucose appears in the urine—a condition called **glycosuria**—when more glucose passes through the tubules than can be reabsorbed. This occurs when the plasma glucose concentration reaches 180 to 200 mg per 100 ml. Since the rate of glucose delivery under these conditions is still below the average T_m for glucose, we must conclude that some nephrons have considerably lower T_m values than the average.

The **renal plasma threshold** is the minimum plasma concentration of a substance that results in the excretion of that substance in the urine. The renal plasma threshold for glucose, for example, is 180 to 200 mg per 100 ml. Glucose is normally absent from urine because plasma glucose concentrations normally remain below this threshold value. Fasting plasma glucose is about 100 mg per 100 ml, for example, and the plasma glucose concentration following meals does not usually exceed 150 mg per 100 ml. The appearance of glucose in the urine (glycosuria) occurs only when the plasma glucose concentration is abnormally high (*hyperglycemia*) and exceeds the renal plasma threshold.

Fasting hyperglycemia is caused by the inadequate secretion or action of insulin. When this hyperglycemia results in glycosuria, the disease is called **diabetes mellitus.** A person with uncontrolled diabetes mellitus also excretes a large volume of urine because the excreted glucose carries water with it as a result of the osmotic pressure it generates in the tubules. This condition should not be confused with diabetes insipidus (discussed previously), in which a large volume of dilute urine is excreted as a result of inadequate ADH secretion.

Test Yourself Before You Continue

1. Define *renal plasma clearance* and describe how this volume is measured. Explain why the glomerular filtration rate is equal to the clearance rate of inulin.
2. Define the terms *reabsorption* and *secretion.* Using examples, describe how the renal plasma clearance is affected by the processes of reabsorption and secretion.
3. Explain why the total renal blood flow can be measured by the renal plasma clearance of PAH.
4. Define *transport maximum* and *renal plasma threshold.* Explain why people with diabetes mellitus have glycosuria.

Renal Control of Electrolyte and Acid-Base Balance

The kidneys regulate the blood concentration of Na^+, K^+, HCO_3^-, and H^+. Aldosterone stimulates the reabsorption of Na^+ in exchange for K^+ in the tubule. Aldosterone thus promotes the renal retention of Na^+ and the excretion of K^+. Secretion of aldosterone from the adrenal cortex is stimulated directly by a high blood K^+ concentration and indirectly by a low Na^+ concentration via the renin-angiotensin system.

The kidneys help to regulate the concentrations of plasma electrolytes—sodium, potassium, chloride, bicarbonate, and phosphate—by matching the urinary excretion of these compounds to the amounts ingested. The control of plasma Na^+ is important in the regulation of blood volume and pressure; the control of plasma K^+ is required to maintain proper function of cardiac and skeletal muscles.

Role of Aldosterone in Na^+/K^+ Balance

Approximately 90% of the filtered Na^+ and K^+ is reabsorbed in the early part of the nephron before the filtrate reaches the distal tubule. This reabsorption occurs at a constant rate and is not subject to hormonal regulation. The final concentration of Na^+ and K^+ in the urine is varied according to the needs of the body by processes that occur in the late distal tubule and in the cortical region of the collecting duct (the portion of the collecting duct within the medulla does not participate in this regulation). Renal reabsorption of Na^+ and secretion of K^+ are regulated by **aldosterone,** the principal mineralocorticoid secreted by the adrenal cortex (chapter 11).

Sodium Reabsorption

Although 90% of the filtered sodium is reabsorbed in the early region of the nephron, the amount left in the filtrate delivered to the distal convoluted tubule is still quite large. In the absence of aldosterone, 80% of this remaining amount is reabsorbed through the wall of the tubule into the peritubular blood; this represents 8% of the amount filtered. The amount of sodium excreted without aldosterone is thus 2% of the amount filtered. Although this percentage seems small, the actual amount it represents is an impressive 30 g of sodium excreted in the urine each day. When aldosterone is secreted in maximal amounts, by contrast, all of the sodium delivered to the distal tubule is reabsorbed. In this case urine contains no Na^+ at all.

Aldosterone stimulates Na^+ reabsorption to some degree in the late distal convoluted tubule, but the primary site of aldosterone action is in the **cortical collecting duct.** This is the initial portion of the collecting duct, located in the renal cortex, which has different permeability properties than the terminal portion of the collecting duct, located in the renal medulla.

Potassium Secretion

About 90% of the filtered potassium is reabsorbed in the early regions of the nephron (mainly from the proximal tubule). In order for potassium to appear in the urine, it must be secreted into later regions of the nephron tubule. Secretion of potassium occurs in the parts of the nephron that are sensitive to aldosterone—that is, in the late distal tubule and cortical collecting duct (fig. 17.24).

As Na^+ is reabsorbed in these regions of the nephron, the lumen of the tubule becomes more negatively charged (–50 mV) compared to the basolateral side. This potential difference then drives the secretion of K^+ into the tubule. The transport carrier for Na^+ is separate from the transport carrier for K^+, so, although Na^+ is reabsorbed in exchange for the secretion of K^+, there is not a 1:1 exchange of these ions.

The amount of K^+ secretion into the late distal tubule and cortical collecting duct depends on: (1) the amount of Na^+ delivered to these regions of the nephron; and (2) the amount of aldosterone secreted. If the blood concentration of K^+ rises, this will stimulate increased aldosterone secretion from the adrenal cortex. The aldosterone then stimulates increased reabsorption of Na^+ and, as a result, increased secretion of K^+.

Some diuretic drugs inhibit Na^+ reabsorption in the loop of Henle and therefore increase the delivery of Na^+ to the distal tubule. This results in an increased reabsorption of Na^+ and secretion of K^+ in the cortical collecting duct. People who take these diuretics, therefore, tend to have excessive K^+ loss in the urine. The actions of different types of diuretics are discussed in the section "Clinical Applications" at the end of this chapter.

The body cannot get rid of excess K^+ in the absence of aldosterone-stimulated secretion of K^+ into the cortical collecting ducts. Indeed, when both adrenal glands are removed from an experimental animal, the **hyperkalemia** (high blood K^+) that results can produce fatal cardiac arrhythmias. Abnormally low plasma K^+ concentrations **(hypokalemia)**, as might result from excessive aldosterone secretion or from diuretic drugs, can produce arrhythmias as well as muscle weakness.

Control of Aldosterone Secretion

Since aldosterone promotes Na^+ retention and K^+ loss, one might predict (on the basis of negative feedback) that aldosterone secretion would be increased when there was a low Na^+ or a high K^+ concentration in the blood. This indeed is the case. A rise in blood K^+ *directly* stimulates the secretion of aldosterone from the adrenal cortex. A decrease in plasma Na^+ concentration, if it causes a fall in blood volume, also promotes aldosterone secretion. However, the stimulatory effect of a fall in blood volume on aldosterone secretion is indirect, as described in the next section.

Juxtaglomerular Apparatus

The **juxtaglomerular apparatus** is the region in each nephron where the afferent arteriole comes into contact with the last portion of the thick ascending limb of the loop (fig. 17.25). Under the microscope, the afferent arteriole and tubule in this small region have a different appearance than in other regions. *Granular cells* within the afferent arteriole secrete the enzyme **renin** into the blood; this enzyme catalyzes the conversion of *angiotensinogen* (a protein) into *angiotensin I* (a ten-amino-acid polypeptide).

Secretion of renin into the blood thus results in the formation of angiotensin I, which is then converted to **angiotensin II** (an eight-amino-acid polypeptide) by *angiotensin-converting enzyme (ACE)*. This conversion occurs primarily as blood passes through the capillaries of the lungs, where most of the converting enzyme is present. Angiotensin II, in addition to its other effects (described in chapter 14), stimulates the adrenal cortex to secrete aldosterone. Thus, secretion of renin from the granular cells of the juxtaglomerular apparatus initiates the **renin-angiotensin-aldosterone system.** Conditions that result in increased renin secretion cause increased aldosterone secretion and, by this means, promote the reabsorption of Na^+ from cortical collecting duct into the blood.

Regulation of Renin Secretion

An inadequate dietary intake of salt (NaCl) is always accompanied by a fall in blood volume. This is because the decreased plasma concentration (osmolality) inhibits ADH secretion. With

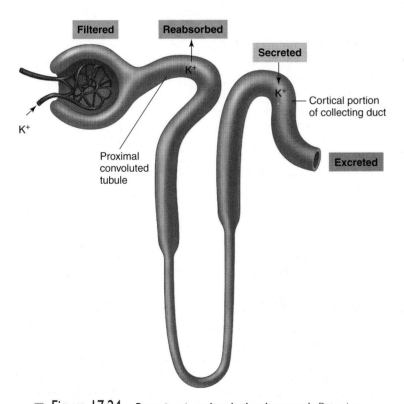

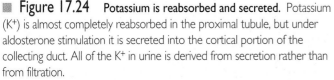

■ **Figure 17.24** Potassium is reabsorbed and secreted. Potassium (K^+) is almost completely reabsorbed in the proximal tubule, but under aldosterone stimulation it is secreted into the cortical portion of the collecting duct. All of the K^+ in urine is derived from secretion rather than from filtration.

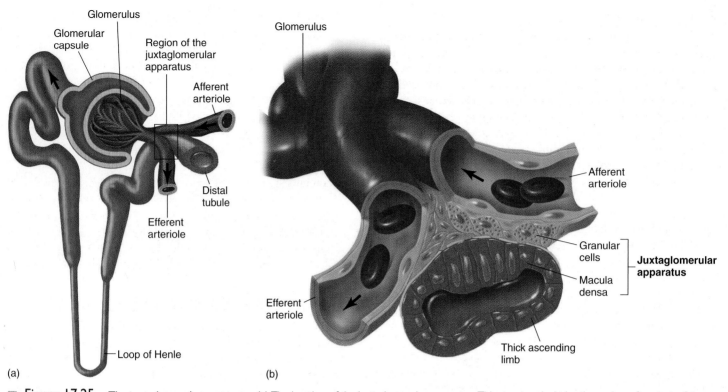

Figure 17.25   **The juxtaglomerular apparatus.** *(a)* The location of the juxtaglomerular apparatus. This structure includes the region of contact of the afferent arteriole with the last portion of the thick ascending limb of the loop. The afferent arterioles in this region contain granular cells that secrete renin, and the tubule cells in contact with the granular cells form an area called the macula densa, seen in *(b)*.

less ADH, less water is reabsorbed through the collecting ducts and more is excreted in the urine. The fall in blood volume and the fall in renal blood flow that result cause increased renin secretion. Increased renin secretion is believed to be due in part to the direct effect of blood pressure on the granular cells, which may function as baroreceptors in the afferent arterioles. Renin secretion is also stimulated by sympathetic nerve activity, which is increased by the baroreceptor reflex (chapter 14) when the blood volume and pressure fall.

An increased secretion of renin acts, via the increased production of angiotensin II, to stimulate aldosterone secretion. Consequently, less sodium is excreted in the urine and more is retained in the blood. This negative feedback system is illustrated in figure 17.26.

Role of the Macula Densa

The region of the ascending limb in contact with the granular cells of the afferent arteriole is called the **macula densa** (fig. 17.25). There is evidence that this region helps to inhibit renin secretion when the blood Na⁺ concentration is raised.

The cells of the macula densa respond to the Na^+ in the filtrate delivered to the distal tubule. When the plasma Na^+ concentration is raised, or when the GFR is increased, the rate of Na^+ delivered to the distal tubule is also increased. Through an effect on the macula densa, this increase in filtered Na^+ inhibits the granular cells from secreting renin. Aldosterone secretion thus decreases, and since less Na^+ is reabsorbed in the cortical collecting duct, more Na^+ is excreted in the urine. The regulation of renin and aldosterone secretion is summarized in table 17.6.

Atrial Natriuretic Peptide

Expansion of the blood volume causes increased salt and water excretion in the urine. This is partly due to an inhibition of aldosterone secretion, as previously described. However, it is also caused by increased secretion of a *natriuretic hormone,* a hormone that stimulates salt excretion (*natrium* is Latin for sodium)—an action opposite to that of aldosterone. The natriuretic hormone has been identified as a 28-amino-acid polypeptide called **atrial natriuretic peptide (ANP),** also called *atrial natriuretic factor.* Atrial natriuretic peptide is produced by the atria of the heart and secreted in response to the stretching of the atrial walls by increased blood volume. In response to ANP action, the kidneys lower the blood volume by excreting more of the salt and water filtered out of the blood by the glomeruli. Atrial natriuretic peptide thus functions as an endogenous diuretic.

Relationship Between Na⁺, K⁺, and H⁺

The plasma K^+ concentration indirectly affects the plasma H^+ concentration (pH). Changes in plasma pH likewise affect the K^+ concentration of the blood. When the extracellular H^+ concentration increases, for example, some of the H^+ moves into the cells and causes cellular K^+ to diffuse outward into the extracellular fluid. The plasma concentration of H^+ is thus decreased while the K^+ increases, helping to reestablish the proper ratio of these ions in the extracellular fluid. A similar effect occurs in the cells of the distal region of the nephron.

In the cells of the late distal tubule and cortical collecting duct, positively charged ions (K^+ and H^+) are secreted in

Table 17.6 Regulation of Renin and Aldosterone Secretion

Stimulus	Effect on Renin Secretion	Angiotensin II Production	Aldosterone Secretion	Mechanisms
↓Blood volume	Increased	Increased	Increased	Low blood volume stimulates renal baroreceptors; granular cells release renin.
↑Blood volume	Decreased	Decreased	Decreased	Increased blood volume inhibits baroreceptors; increased Na⁺ in distal tubule acts via macula densa to inhibit release of renin from granular cells.
↑K⁺	None	Not changed	Increased	Direct stimulation of adrenal cortex
↑Sympathetic nerve activity	Increased	Increased	Increased	α-adrenergic effect stimulates constriction of afferent arterioles; β-adrenergic effect stimulates renin secretion directly.

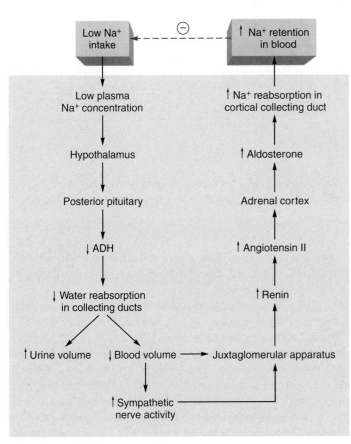

Figure 17.26 Homeostasis of plasma Na⁺. This is the sequence of events by which a low sodium (salt) intake leads to increased sodium reabsorption by the kidneys. The dashed arrow and negative sign indicate the completion of the negative feedback loop.

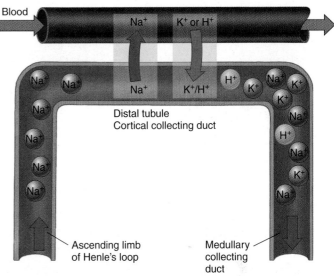

Figure 17.27 The reabsorption of Na⁺ and secretion of K⁺. In the distal tubule, K⁺ and H⁺ are secreted in response to the potential difference produced by the reabsorption of Na⁺. High concentrations of H⁺ may therefore decrease K⁺ secretion, and vice versa.

response to the negative polarity produced by reabsorption of Na⁺ (fig. 17.27). When a person has severe acidosis, there is an increased amount of H⁺ secretion at the expense of a decrease in the amount of K⁺ secreted. Acidosis may thus be accompanied by a rise in blood K⁺. If, on the other hand, hyperkalemia is the primary problem, there is an increased secretion of K⁺ and thus a decreased secretion of H⁺. Hyperkalemia can thus cause an increase in the blood concentration of H⁺ and acidosis.

Aldosterone indirectly stimulates the secretion of H⁺, as well as K⁺, into the cortical collecting ducts. Therefore, abnormally high aldosterone secretion, as occurs in **primary aldosteronism,** or **Conn's syndrome,** results in both hypokalemia and metabolic alkalosis. Conversely, abnormally low aldosterone secretion, as occurs in **Addison's disease,** can produce hyperkalemia accompanied by metabolic acidosis.

If a person is suffering from potassium deprivation, according to recent evidence, the collecting duct may be able to partially compensate by reabsorbing some K⁺. This occurs in the outer medulla, and results in the reabsorption of some of the K⁺ that was secreted into the cortical collecting duct.

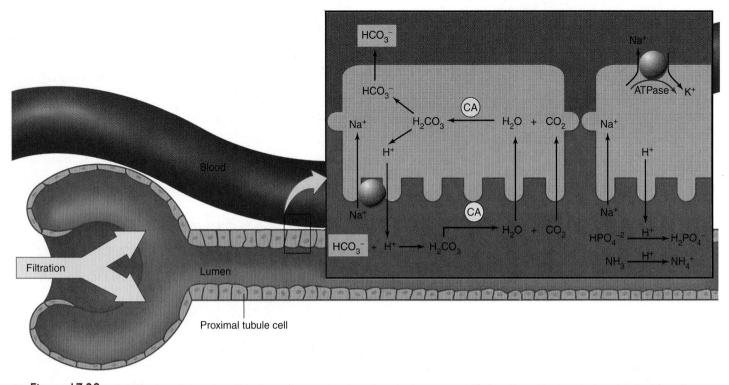

Figure 17.28 **Acidification of the urine.** This diagram summarizes how the urine becomes acidified and how bicarbonate is reabsorbed from the filtrate. It also depicts the buffering of the urine by phosphate and ammonium buffers. (CA = Carbonic anhydrase.) The inset depicts an expanded view of proximal tubule cells.

Renal Acid-Base Regulation

The kidneys help to regulate the blood pH by excreting H^+ in the urine and by reabsorbing bicarbonate. The H^+ enters the filtrate in two ways: by filtration through the glomeruli and by secretion into the tubules. Most of the H^+ secretion occurs across the wall of the proximal tubule in exchange for the reabsorption of Na^+. This exchange is performed by a transport carrier described as "antiport," because it moves the Na^+ and H^+ in opposite directions (see chapter 6).

Since the kidneys normally reabsorb almost all of the filtered bicarbonate and excrete H^+, normal urine contains little bicarbonate and is slightly acidic (with a pH range between 5 and 7). The mechanisms involved in acidification of the urine and reabsorption of bicarbonate are summarized in figure 17.28.

Reabsorption of Bicarbonate in the Proximal Tubule

The apical membranes of the tubule cells (facing the lumen) are impermeable to bicarbonate. The reabsorption of bicarbonate must therefore occur indirectly. When the urine is acidic, HCO_3^- combines with H^+ to form carbonic acid. Carbonic acid in the filtrate is then converted to CO_2 and H_2O in a reaction catalyzed by **carbonic anhydrase.** This enzyme is located in the apical cell membrane of the proximal tubule in contact with the filtrate.

Notice that the reaction that occurs in the filtrate is the same one that occurs within red blood cells in pulmonary capillaries (as discussed in chapter 16).

The tubule cell cytoplasm also contains carbonic anhydrase. As CO_2 concentrations increase in the filtrate, the CO_2 diffuses into the tubule cells. Within the tubule cell cytoplasm, carbonic anhydrase catalyzes the reaction in which CO_2 and H_2O form carbonic acid. The carbonic acid then dissociates to HCO_3^- and H^+ within the tubule cells. (These are the same events that occur in the red blood cells of tissue capillaries.) The bicarbonate within the tubule cell can then diffuse through the basolateral membrane and enter the blood (fig. 17.28). When conditions are normal, the same amount of HCO_3^- passes into the blood as was removed from the filtrate. The H^+, which was produced at the same time as HCO_3^- in the cytoplasm of the tubule cell, can either pass back into the filtrate or pass into the blood. Under acidotic conditions, almost all of the H^+ goes back into the filtrate and is used to help reabsorb all of the filtered bicarbonate.

During alkalosis, less H^+ is secreted into the filtrate. Since the reabsorption of filtered bicarbonate requires that HCO_3^- combine with H^+ to form carbonic acid, less bicarbonate is reabsorbed. This results in urinary excretion of bicarbonate, which helps to partially compensate for the alkalosis.

By these mechanisms, disturbances in acid-base balance caused by respiratory problems can be partially compensated for

Table 17.7 Categories of Disturbances in Acid-Base Balance

| P_{CO_2} (mmHg) | Bicarbonate (mEq/L)* | | |
	Less than 21	21–26	More than 26
More than 45	Combined metabolic and respiratory acidosis	Respiratory acidosis	Metabolic alkalosis and respiratory acidosis
35–45	Metabolic acidosis	Normal	Metabolic alkalosis
Less than 35	Metabolic acidosis and respiratory alkalosis	Respiratory alkalosis	Combined metabolic and respiratory alkalosis

* mEq/L = milliequivalents per liter. This is the millimolar concentration of HCO_3^- multiplied by its valence (×1).

by changes in plasma bicarbonate concentrations. Metabolic acidosis or alkalosis—in which changes in bicarbonate concentrations occur as the primary disturbance—similarly can be partially compensated for by changes in ventilation. These interactions of the respiratory and metabolic components of acid-base balance are summarized in table 17.7.

When people go to the high elevations of the mountains, they hyperventilate (as discussed in chapter 16). This lowers the arterial P_{CO_2} and produces a respiratory alkalosis. The kidneys participate in this acclimatization by excreting a larger amount of bicarbonate. This helps to compensate for the alkalosis and bring the blood pH back down toward normal. It is interesting in this regard that the drug *acetazolamide,* which inhibits renal carbonic anhydrase, is often used to treat **acute mountain sickness** (see page 515). The inhibition of renal carbonic anhydrase causes the loss of bicarbonate and water in the urine, producing a metabolic acidosis and diuresis that help to alleviate the symptoms.

Urinary Buffers

When a person has a blood pH of less than 7.35 (acidosis), the urine pH almost always falls below 5.5. The nephron, however, cannot produce a urine pH that is significantly less than 4.5. In order for more H^+ to be excreted, the acid must be buffered. (Actually, even in normal urine, most of the H^+ excreted is in a buffered form.) Bicarbonate cannot serve this buffering function because it is normally completely reabsorbed. Instead, the buffering action of phosphates (mainly HPO_4^{2-}) and ammonia (NH_3) provide the means for excreting most of the H^+ in the urine. Phosphate enters the urine by filtration. Ammonia (whose presence is strongly evident in a diaper pail or kitty litter box) is produced in the tubule cells by deamination of amino acids. These molecules buffer H^+ as described in these equations:

$$NH_3 + H^+ \rightarrow NH_4^+ \text{ (ammonium ion)}$$

$$HPO_4^{2-} + H^+ \rightarrow H_2PO_4^-$$

Test Yourself Before You Continue

1. Describe the effects of aldosterone on the renal nephrons and explain how aldosterone secretion is regulated.
2. Explain how changes in blood volume regulate renin secretion and how the secretion of renin helps to regulate the blood volume.
3. Explain the mechanisms by which the cortical collecting duct secretes K^+ and H^+. How might hyperkalemia affect the blood pH?
4. Explain how the kidneys reabsorb filtered bicarbonate and how this process is affected by acidosis and alkalosis.
5. Suppose a person with diabetes mellitus had an arterial pH of 7.30, an abnormally low arterial P_{CO_2}, and an abnormally low bicarbonate concentration. What type of acid-base disturbance would this be? What might have caused the imbalances?

Clinical Applications

Different types of diuretic drugs act on specific segments of the nephron tubule to indirectly inhibit the reabsorption of water and thus promote the lowering of blood volume. A knowledge of how diuretics exert their effects thus enhances understanding of the physiology of the nephron. Clinical analysis of the urine, similarly, is meaningful only when the mechanisms that produce normal urine composition are understood.

The importance of renal function in maintaining homeostasis and the ease with which urine can be collected and used as a mirror of the plasma's chemical composition make the clinical study of renal function and urine composition particularly useful. Further, the ability of the kidneys to regulate blood volume is exploited clinically in the management of high blood pressure.

Use of Diuretics

People who need to lower their blood volume because of hypertension, congestive heart failure, or edema take medications called **diuretics** that increase the volume of urine excreted. Diuretics directly lower blood volume (and hence, blood pressure) by increasing the proportion of the glomerular filtrate that is excreted as urine. These drugs also decrease the interstitial fluid volume (and hence, relieve edema) by a more indirect route. By lowering plasma volume, diuretic drugs increase the concentration, and thus the oncotic pressure, of the plasma within blood capillaries (chapter 14; see fig. 14.9). This promotes the osmosis of tissue fluid into the capillary blood, helping to reduce the edema.

The various diuretic drugs act on the renal nephron in different ways (table 17.8; fig. 17.29). On the basis of their chemical structure or aspects of their actions, commonly used diuretics are categorized as *loop diuretics, thiazides, carbonic anhydrase inhibitors, osmotic diuretics,* or *potassium-sparing diuretics.*

The most powerful diuretics, which inhibit salt and water reabsorption by as much as 25%, are the drugs that act to inhibit

Table 17.8 Actions of Different Classes of Diuretics

Category of Diuretic	Example	Mechanism of Action	Major Site of Action
Loop diuretics	Furosemide	Inhibits sodium transport	Thick segments of ascending limbs
Thiazides	Hydrochlorothiazide	Inhibits sodium transport	Last part of ascending limb and first part of distal tubule
Carbonic anhydrase inhibitors	Acetazolamide	Inhibits reabsorption of bicarbonate	Proximal tubule
Osmotic diuretics	Mannitol	Reduces osmotic reabsorption of water by reducing osmotic gradient	Last part of distal tubule and cortical collecting duct
Postassium-sparing diuretics	Spironolactone	Inhibits action of aldosterone	Last part of distal tubule and cortical collecting duct
	Triamterene	Inhibits Na⁺ reabsorption and K⁺ secretion	Last part of distal tubule and cortical collecting duct

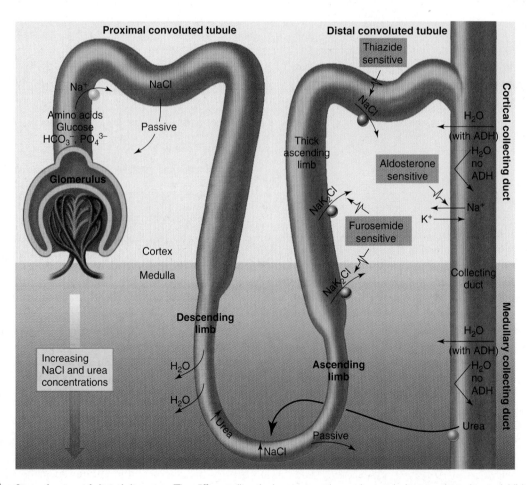

■ **Figure 17.29**   **Sites of action of clinical diuretics.** The different diuretic drugs act on the nephron tubules at various sites to inhibit the reabsorption of water. As a result of these actions, less water is reabsorbed into the blood and more is excreted in the urine. This lowers the blood volume and pressure.

active salt transport out of the ascending limb of the loop of Henle. Examples of these **loop diuretics** are *furosemide (Lasix)* and *ethacrynic acid*. The **thiazide diuretics** (e.g., *hydrochlorothiazide*) inhibit salt and water reabsorption by as much as 8% through inhibition of salt transport by the first segment of the distal convoluted tubule. The **carbonic anhydrase inhibitors** (e.g., *acetazolamide*) are much weaker diuretics; they act primarily in the proximal tubule to prevent the water reabsorption that occurs when bicarbonate is reabsorbed.

When extra solutes are present in the filtrate, they increase the osmotic pressure of the filtrate and in this way decrease the reabsorption of water by osmosis. The extra solutes thus act as **osmotic diuretics.** *Mannitol* is sometimes used clinically for this purpose. Osmotic diuresis can occur in diabetes mellitus because glucose is present in the filtrate and urine; this extra solute causes the excretion of excessive amounts of water in the urine and can result in severe dehydration of a person with uncontrolled diabetes.

Complications may arise from the use of diuretics that cause an excessive loss of K$^+$ in the urine. If K$^+$ secretion into the tubules is significantly increased, **hypokalemia** (abnormally low blood K$^+$ levels) may result. This condition can lead to neuromuscular disorders and to electrocardiographic abnormalities. People who take diuretics for the treatment of high blood pressure are usually on a low-sodium diet, and they often must supplement their meals with potassium chloride (KCl) to offset the loss of K$^+$.

The previously mentioned diuretics can, as a result of increased Na$^+$ delivery to the cortical collecting duct, result in the excessive secretion of K$^+$ into the filtrate and its excessive elimination in the urine. For this reason, **potassium-sparing diuretics** are sometimes used. *Spironolactones (Aldactone)* are aldosterone antagonists that compete with aldosterone for cytoplasmic receptor proteins in the cells of the cortical collecting duct. These drugs thus block the aldosterone stimulation of Na$^+$ reabsorption and K$^+$ secretion. *Triamterene (Dyrenium)* is a different type of potassium-sparing diuretic that appears to act on the tubule more directly to block Na$^+$ reabsorption and K$^+$ secretion. Combinations of spironolactone or triamterene together with hydrochlorothiazide (*Aldactazide* and *Dyazide,* respectively) are sometimes also prescribed for the diuretic treatment of hypertension.

Clinical Investigation Clue

Remember that Emily had edema and was given hydrochlorothiazide.

What is the action of hydrochlorothiazide, and how would this drug benefit Emily?

Renal Function Tests and Kidney Disease

Renal function can be tested by techniques that include the renal plasma clearance of PAH, which measures total blood flow to the kidneys, and the measurement of the GFR by the inulin clearance. The plasma creatinine concentration, as previously described, also provides an index of renal function. These tests aid the diagnosis of kidney diseases such as glomerulonephritis and renal insufficiency. The *urinary albumin excretion rate* is a commonly performed test that can detect an excretion rate of blood albumin that is slightly above normal. This condition, called **microalbuminuria,** is often the first manifestation of renal damage caused by diabetes or hypertension.

Acute Renal Failure

In **acute renal failure,** the ability of the kidneys to excrete wastes and regulate the homeostasis of blood volume, pH, and electrolytes deteriorates over a relatively short period of time (hours to days). There is a rise in blood creatinine concentration and a decrease in the renal plasma clearance of creatinine. This may be due to a reduced blood flow through the kidneys, perhaps as a result of atherosclerosis or inflammation of the renal tubules. The compromised kidney function may be the result of ischemia caused by the reduced blood flow, but it may also result from excessive use of certain drugs, including nonsteroidal anti-inflammatory drugs (NSAIDs) such as phenacetin.

Glomerulonephritis

Inflammation of the glomeruli, or **glomerulonephritis,** is believed to be an *autoimmune disease*—a disease that involves the person's own antibodies (as described in chapter 15). These antibodies may have been raised against the basement membrane of the glomerular capillaries. More commonly, however, they appear to have been produced in response to streptococcus infections (such as strep throat). A variable number of glomeruli are destroyed in this condition, and the remaining glomeruli become more permeable to plasma proteins. Leakage of proteins into the urine results in decreased plasma colloid osmotic pressure and can therefore lead to edema.

Clinical Investigation Clue

Remember that Emily had a month-long streptococcus infection and that her symptoms disappeared after taking an antibiotic (plus hydrochlorothiazide).

How might the strep infection be related to kidney function and to Emily's symptoms?

Renal Insufficiency

When nephrons are destroyed—as in chronic glomerulonephritis, infection of the renal pelvis and nephrons (*pyelonephritis*), or loss of a kidney—or when kidney function is reduced by damage caused by diabetes mellitus, arteriosclerosis, or blockage by kidney stones, a condition of **renal insufficiency** may develop. This can cause hypertension, which is due primarily to the retention of salt and water, and **uremia** (high plasma urea concentrations). The inability to excrete urea is accompanied by an elevated plasma H^+ concentration (acidosis) and an elevated K^+ concentration, which are more immediately dangerous than the high levels of urea. Uremic coma appears to result from these associated changes.

Patients with uremia or the potential for developing uremia are often placed on a *dialysis* machine. The term *dialysis* refers to the separation of molecules on the basis of their ability to diffuse through an artificial selectively permeable membrane. This principle is used in the "artificial kidney machine" for **hemodialysis.** Urea and other wastes in the patient's blood can easily pass through the membrane pores, whereas plasma proteins are left behind (just as occurs across glomerular capillaries). The plasma is thus cleansed of these wastes as they pass from the blood into the dialysis fluid. Unlike the tubules, however, the dialysis membrane cannot reabsorb Na^+, K^+, glucose, and other needed molecules. These substances are kept in the blood by including them in the dialysis fluid so that there is no concentration gradient that would favor their diffusion through the membrane. Hemodialysis is commonly performed three times a week for several hours each session.

More recent techniques include the use of the patient's own peritoneal membranes (which line the abdominal cavity) for dialysis. Dialysis fluid is introduced into the peritoneal cavity, and then, after a period of time, discarded after wastes have accumulated. This procedure, called **continuous ambulatory peritoneal dialysis (CAPD),** can be performed several times a day by the patients themselves on an outpatient basis. Although CAPD is more convenient and less expensive for patients than hemodialysis, it is less efficient in removing wastes and it is more often complicated by infection.

The many dangers presented by renal insufficiency and the difficulties encountered in attempting to compensate for this condition are stark reminders of the importance of renal function in maintaining homeostasis. The ability of the kidneys to regulate blood volume and chemical composition in accordance with the body's changing needs requires great complexity of function. Homeostasis is maintained in large part by coordination of renal functions with those of the cardiovascular and pulmonary systems, as described in the preceding chapters.

Test Yourself Before You Continue

1. List the different categories of clinical diuretics and explain how each exerts its diuretic effect.

2. Explain why most diuretics can cause excessive loss of K^+. How is this prevented by the potassium-sparing diuretics?

3. Define *uremia* and discuss the dangers associated with this condition. Explain how uremia can be corrected through the use of renal dialysis.

HPer Links of the Urinary System with Other Body Systems

Integumentary System

- Evaporative water loss from the skin helps to control body temperature, but effects on blood volume must be compensated for by the kidneys . . . .(p. 427)
- The skin produces vitamin D_3, which is activated in the kidneys . . . .(p. 625)
- The kidneys maintain homeostasis of blood volume, pressure, and composition, which is needed for the health of the integumentary and other systems . . . .(p. 524)

Skeletal System

- The pelvic girdle supports and protects some organs of the urinary system . . . . . .(p. 524)
- Bones store calcium and phosphate, and thus cooperate with the kidneys to regulate the blood levels of these ions . . . . .(p. 623)

Muscular System

- Muscles in the urinary tract assist the storage and voiding of urine . . . . .(p. 525)
- Smooth muscles in the renal blood vessels regulate renal blood flow, and thus the glomerular filtration rate . . . .(p. 531)

Nervous System

- Autonomic nerves help to regulate renal blood flow, and hence glomerular filtration . . . .(p. 531)
- The nervous system provides motor control of micturition . . . .(p. 525)

Endocrine System

- Antidiuretic hormone stimulates reabsorption of water from the renal tubules . . . .(p. 536)
- Aldosterone stimulates sodium reabsorption and potassium secretion by the kidneys . . . .(p. 544)

- Natriuretic hormones stimulate sodium excretion by the kidneys . . . .(p. 418)
- The kidneys produce the hormone erythropoietin . . . .(p. 371)
- The kidneys secrete renin, which activates the renin-angiotensin-aldosterone system . . . .(p. 417)

Circulatory System

- The blood transports oxygen and nutrients to all systems, including the urinary system, and removes wastes . . . .(p. 366)
- The heart secretes atrial natriuretic peptide, which helps to regulate the kidneys . . . .(p. 418)
- Erythropoietin from the kidneys stimulates red blood cell production . . . .(p. 371)
- The kidneys filter the blood to produce urine, while regulating blood volume, composition, and pressure . . . . . .(p. 524)

Immune System

- The immune system protects all systems, including the urinary system, against infections . . . .(p. 446)
- Lymphatic vessels help to maintain a balance between blood and interstitial fluid . . . .(p. 399)
- The acidity of urine provides a nonspecific defense against urinary tract infection . . . .(p. 548)

Respiratory System

- The lungs provide oxygen and eliminate carbon dioxide for all systems, including the urinary system . . . .(p. 480)
- The lungs and kidneys cooperate in the regulation of blood pH . . . .(p. 377)

Digestive System

- The GI tract provides nutrients for all tissues, including those of the urinary system . . . .(p. 561)
- The GI tract, like the urinary system, helps to eliminate waste products . . . . .(p. 579)

Reproductive System

- The urethra of a male passes through the penis and can eject either urine or semen . . . .(p. 651)
- The kidneys participate in the regulation of blood volume and pressure, which is required for functioning of the reproductive system . . . .(p. 416)
- The mother's urinary system eliminates metabolic wastes from the fetus . .(p. 672)

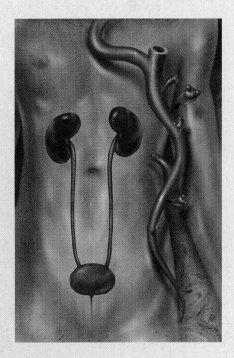

Summary

Structure and Function of the Kidneys 524

I. The kidney is divided into an outer cortex and inner medulla.

  A. The medulla is composed of renal pyramids, separated by renal columns.

  B. The renal pyramids empty urine into the calyces that drain into the renal pelvis. From there, urine flows into the ureter and is transported to the bladder to be stored.

II. Each kidney contains more than a million microscopic functional units called nephrons. Nephrons consist of vascular and tubular components.

  A. Filtration occurs in the glomerulus, which receives blood from an afferent arteriole.

  B. Glomerular blood is drained by an efferent arteriole, which delivers blood to peritubular capillaries that surround the nephron tubules.

  C. The glomerular (Bowman's) capsule and the proximal and distal convoluted tubules are located in the cortex.

  D. The loop of Henle is located in the medulla.

  E. Filtrate from the distal convoluted tubule is drained into collecting ducts, which plunge through the medulla to empty urine into the calyces.

Glomerular Filtration 529

I. A filtrate derived from plasma in the glomerulus must pass through a basement membrane of the glomerular capillaries and through slits in the processes of the podocytes—the cells that compose the inner layer of the glomerular (Bowman's) capsule.

  A. The glomerular ultrafiltrate, formed under the force of blood pressure, has a low protein concentration.

  B. The glomerular filtration rate (GFR) is the volume of filtrate produced by both kidneys each minute. It ranges from 115 to 125 ml/min.

II. The GFR can be regulated by constriction or dilation of the afferent arterioles.

  A. Sympathetic innervation causes constriction of the afferent arterioles.

  B. Intrinsic mechanisms help to autoregulate the rate of renal blood flow and the GFR.

Reabsorption of Salt and Water 532

I. Approximately 65% of the filtered salt and water is reabsorbed across the proximal convoluted tubules.

  A. Sodium is actively transported, chloride follows passively by electrical attraction, and water follows the salt out of the proximal tubule.

  B. Salt transport in the proximal tubules is not under hormonal regulation.

II. The reabsorption of most of the remaining water occurs as a result of the action of the countercurrent multiplier system.

  A. Sodium is actively extruded from the ascending limb, followed passively by chloride.

  B. Since the ascending limb is impermeable to water, the remaining filtrate becomes hypotonic.

  C. Because of this salt transport and because of countercurrent exchange in the vasa recta, the interstitial fluid of the medulla becomes hypertonic.

  D. The hypertonicity of the medulla is multiplied by a positive feedback mechanism involving the descending limb, which is passively permeable to water and perhaps to salt.

III. The collecting duct is permeable to water but not to salt.

  A. As the collecting ducts pass through the hypertonic renal medulla, water leaves by osmosis and is carried away in surrounding capillaries.

  B. The permeability of the collecting ducts to water is stimulated by antidiuretic hormone (ADH).

Renal Plasma Clearance 539

I. Inulin is filtered but neither reabsorbed nor secreted. Its clearance is thus equal to the glomerular filtration rate.

II. Some of the filtered urea is reabsorbed. Its clearance is therefore less than the glomerular filtration rate.

III. Since almost all the PAH in blood going through the kidneys is cleared by filtration and secretion, the PAH clearance is a measure of the total renal blood flow.

IV. Normally all of the filtered glucose is reabsorbed. Glycosuria occurs when the transport carriers for glucose become saturated as a result of hyperglycemia.

Renal Control of Electrolyte and Acid-Base Balance 544

I. Aldosterone stimulates sodium reabsorption and potassium secretion in the distal convoluted tubule.

II. Aldosterone secretion is stimulated directly by a rise in blood potassium and indirectly by a fall in blood volume.

  A. Decreased blood flow and pressure through the kidneys stimulates the secretion of the enzyme renin from the juxtaglomerular apparatus.

  B. Renin catalyzes the formation of angiotensin I, which is then converted to angiotensin II.

  C. Angiotensin II stimulates the adrenal cortex to secrete aldosterone.

III. Aldosterone stimulates the secretion of H^+, as well as potassium, into the filtrate in exchange for sodium.

IV. The nephrons filter bicarbonate and reabsorb the amount required to maintain acid-base balance. Reabsorption of bicarbonate, however, is indirect.

    A. Filtered bicarbonate combines with H^+ to form carbonic acid in the filtrate.

    B. Carbonic anhydrase in the membranes of microvilli in the tubules catalyzes the conversion of carbonic acid to carbon dioxide and water.

    C. Carbon dioxide is reabsorbed and converted in either the tubule cells or the red blood cells to carbonic acid, which dissociates to bicarbonate and H^+.

    D. In addition to reabsorbing bicarbonate, the nephrons filter and secrete H^+, which is excreted in the urine buffered by ammonium and phosphate buffers.

Clinical Applications 549

I. Diuretic drugs are used clinically to increase the urine volume and thus to lower the blood volume and pressure.

    A. Loop diuretics and the thiazides inhibit active Na^+ transport in the ascending limb and early portion of the distal tubule, respectively.

    B. Osmotic diuretics are extra solutes in the filtrate that increase the osmotic pressure of the filtrate and inhibit the osmotic reabsorption of water.

    C. The potassium-sparing diuretics act on the distal tubule to inhibit the reabsorption of Na^+ and secretion of K^+.

II. In glomerulonephritis, the glomeruli can permit the leakage of plasma proteins into the urine.

III. The technique of renal dialysis is used to treat people with renal insufficiency.

Review Activities

Test Your Knowledge of Terms and Facts

1. Which of these statements about the renal pyramids is *false?*
   a. They are located in the medulla.
   b. They contain glomeruli.
   c. They contain collecting ducts.
   d. They empty urine into the calyces.

Match these:
2. Active transport of sodium; water follows passively
3. Active transport sodium; water impermeable to water
4. Passively permeable to water only
5. Passively permeable to water and urea

   a. proximal tubule
   b. descending limb of loop
   c. ascending limb of loop
   d. distal tubule
   e. medullary collecting duct

6. Antidiuretic hormone promotes the retention of water by stimulating
   a. the active transport of water.
   b. the active transport of chloride.
   c. the active transport of sodium.
   d. the permeability of the collecting duct to water.

7. Aldosterone stimulates sodium reabsorption and potassium secretion in
   a. the proximal convoluted tubule.
   b. the descending limb of the loop.
   c. the ascending limb of the loop.
   d. the cortical collecting duct.

8. Substance X has a clearance greater than zero but less than that of inulin. What can you conclude about substance X?
   a. It is not filtered.
   b. It is filtered, but neither reabsorbed nor secreted.
   c. It is filtered and partially reabsorbed.
   d. It is filtered and secreted.

9. Substance Y has a clearance greater than that of inulin. What can you conclude about substance Y?
   a. It is not filtered.
   b. It is filtered, but neither reabsorbed nor secreted.
   c. It is filtered and partially reabsorbed.
   d. It is filtered and secreted.

10. About 65% of the glomerular ultrafiltrate is reabsorbed in
   a. the proximal tubule.
   b. the distal tubule.
   c. the loop of Henle.
   d. the collecting duct.

11. Diuretic drugs that act in the loop of Henle
   a. inhibit active sodium transport.
   b. cause an increased flow of filtrate to the distal convoluted tubule.
   c. cause an increased secretion of potassium into the tubule.
   d. promote the excretion of salt and water.
   e. do all of these.

12. The appearance of glucose in the urine
   a. occurs normally.
   b. indicates the presence of kidney disease.
   c. occurs only when the transport carriers for glucose become saturated.
   d. is a result of hypoglycemia.

13. Reabsorption of water through the tubules occurs by
    a. osmosis.
    b. active transport.
    c. facilitated diffusion.
    d. all of these.

14. Which of these factors oppose(s) filtration from the glomerulus?
    a. plasma oncotic pressure
    b. hydrostatic pressure in glomerular (Bowman's) capsule
    c. plasma hydrostatic pressure
    d. both *a* and *b*
    e. both *b* and *c*

15. The countercurrent exchange in the vasa recta
    a. removes Na^+ from the extracellular fluid.
    b. maintains high concentrations of NaCl in the extracellular fluid.
    c. raises the concentration of Na^+ in the blood leaving the kidneys.
    d. causes large quantities of Na^+ to enter the filtrate.
    e. does all of these.

16. The kidneys help to maintain acid-base balance by
    a. the secretion of H^+ in the distal regions of the nephron.
    b. the action of carbonic anhydrase within the apical cell membranes.
    c. the action of carbonic anhydrase within the cytoplasm of the tubule cells.
    d. the buffering action of phosphates and ammonia in the urine.
    e. all of these means.

Test Your Understanding of Concepts and Principles

1. Explain how glomerular ultrafiltrate is produced and why it has a low protein concentration.[1]

2. Explain how the countercurrent multiplier system works and discuss its functional significance.

3. Explain how countercurrent exchange occurs in the vasa recta and discuss the functional significance of this mechanism.

4. Explain how an increase in ADH secretion promotes increased water reabsorption and how water reabsorption decreases when ADH secretion is decreased.

5. Explain how the structure of the epithelial wall of the proximal tubule and the distribution of Na^+/K^+ pumps in the epithelial cell membranes contribute to the ability of the proximal tubule to reabsorb salt and water.

6. Describe how the thiazide diuretics, loop diuretics, and osmotic diuretics work. How do these substances cause hypokalemia?

7. Which diuretic drugs do not produce hypokalemia? How do these drugs work?

8. What happens to urinary bicarbonate excretion when a person hyperventilates? How might this response be helpful?

9. Describe the location of the macula densa and explain its role in the regulation of renin secretion and in tubuloglomerular feedback.

10. Describe how the nephron handles K^+, how the urinary excretion of K^+ changes under different conditions, and how this process is regulated by aldosterone.

---

[1] *Note:* This question is answered in the chapter 17 Study Guide found on the Online Learning Center at www.mhhe.com/fox8.

Test Your Ability to Analyze and Apply Your Knowledge

1. The very high rates of urea transport in the region of the collecting duct in the inner medulla are due to the presence of specific urea transporters that are stimulated by ADH. Suppose you collect urine from two patients who have been deprived of water overnight. One has normally functioning kidneys, and the other has a genetic defect in the urea transporters. How would the two urine samples differ? Explain.

2. Two men are diagnosed with diabetes insipidus. One didn't have the disorder until he suffered a stroke. The other had withstood the condition all his life, and it had never responded to exogenous ADH despite the presence of normal ADH receptors. What might be the cause of the diabetes insipidus in the two men?

3. Suppose a woman with a family history of polycystic kidney disease develops proteinuria. She has elevated blood creatinine levels and a reduced inulin clearance. What might these lab results indicate? Explain.

4. You love to spend hours fishing in a float tube in a lake, where the lower half of your body is submerged and the upper half is supported by an inner tube. However, you always have to leave the lake sooner than you'd like because you produce urine at a faster than usual rate. Using your knowledge about the regulation of urine volume, propose an explanation as to why a person might produce more urine under these conditions.

5. You have an infection, and you see that the physician is about to inject you with millions of units of penicillin. What do you think will happen to your urine production as a result? Explain. In the hope of speeding your recovery, you gobble extra amounts of vitamin C. How will this affect your urine output?

Related Websites

Check out the Links Library at www.mhhe.com/fox8 for links to sites containing resources related to the physiology of the kidneys. These links are monitored to ensure current URLs.

18 The Digestive System

Objectives

After studying this chapter, you should be able to . . .

1. describe the functions of the digestive system and list its structures and regions.

2. explain how one-way transport is accomplished in the digestive tract.

3. describe the layers of the gastrointestinal tract and state the function(s) of each.

4. describe the structure of the gastric mucosa, list the secretions of the mucosa and their functions, and identify the cells that produce each of these secretions.

5. describe the roles of HCl and pepsin in digestion and explain why the stomach does not normally digest itself.

6. describe the structure and function of the villi, microvilli, and crypts in the small intestine.

7. describe the location and functions of the brush border enzymes of the intestine.

8. explain the electrical activity that occurs in the intestine and describe the nature of peristalsis and segmentation.

9. explain how the large intestine absorbs fluid and electrolytes.

10. describe the flow of blood in the liver and explain how the liver modifies the chemical composition of the blood.

11. describe the composition and functions of bile and explain how bile is kept separate from blood in the liver.

12. trace the pathway of the formation, conjugation, and excretion of bilirubin and explain how jaundice may be produced.

13. explain the significance of the enterohepatic circulation of various compounds and describe the enterohepatic circulation of bile pigment.

14. identify the endocrine and exocrine structures of the pancreas and describe the composition and functions of pancreatic juice.

15. explain how gastric secretion is regulated during the cephalic, gastric, and intestinal phases.

16. describe the structure and function of the enteric nervous system.

17. explain how pancreatic juice and bile secretion is regulated by nerves and hormones.

18. discuss the nature and actions of the different gastrointestinal hormones.

19. describe the enzymes involved in the digestion of carbohydrates, lipids, and proteins, and explain how monosaccharides and amino acids are absorbed.

20. describe the roles of bile and pancreatic lipase in fat digestion and trace the pathways and structures involved in the absorption of lipids.

Refresh Your Memory

Before you begin this chapter, you may want to review these concepts from previous chapters:

■ Divisions of the Autonomic Nervous System 222

■ Functions of the Autonomic Nervous System 227

■ Endocrine Glands and Hormones 286

■ Autocrine and Paracrine Regulation 316

Take Advantage of the Technology

Visit the Online Learning Center for these additional study resources.

■ Interactive quizzing
■ Online study guide
■ Current news feeds
■ Crossword puzzles
■ Vocabulary flashcards
■ Labeling activities

www.mhhe.com/fox8

Clinical Investigation

Alan is a 23-year-old who appears at the college health center complaining of severe but transient pains that are provoked at particular times. He says that he gets a sharp pain in his stomach when he drinks wine, which he admits to doing on occasion. However, he also says that he gets a pain below his right scapula whenever he eats particular foods, such as peanut butter and bacon. This pain is not provoked when he eats fish or skinned chicken, or when he drinks alcohol. The doctor notes that the sclera of Alan's eyes are markedly yellow.

Laboratory tests reveal that Alan has fatty stools (though no blood in the stools), a prolonged clotting time, and elevated levels of conjugated bilirubin in the blood. Other blood tests, however, are normal, including those for ammonia, urea, free bilirubin, and pancreatic amylase in the plasma. His red and white blood cell counts are normal and he does not have a fever.

What do these observations and laboratory results reveal about the possible causes of Alan's symptoms?

Introduction to the Digestive System

Within the lumen of the gastrointestinal tract, large food molecules are hydrolyzed into their monomers (subunits). These monomers pass through the inner layer, or mucosa, of the small intestine to enter the blood or lymph in a process called absorption. Digestion and absorption are aided by specializations of the mucosa and by characteristic movements caused by contractions of the muscle layers of the gastrointestinal tract.

Unlike plants, which can form organic molecules using inorganic compounds such as carbon dioxide, water, and ammonia, humans and other animals must obtain their basic organic

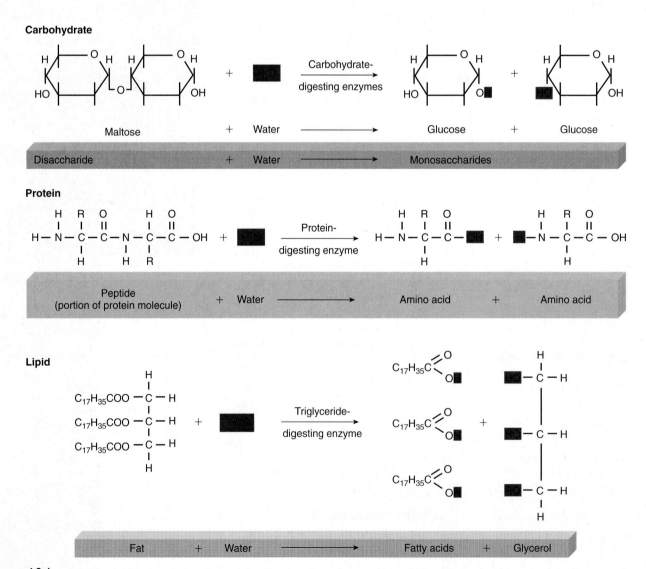

Carbohydrate

Maltose + Water → Glucose + Glucose

Disaccharide + Water → Monosaccharides

Protein

Peptide (portion of protein molecule) + Water → Amino acid + Amino acid

Lipid

Fat + Water → Fatty acids + Glycerol

■ Figure 18.1 **The digestion of food molecules through hydrolysis reactions.** These reactions ultimately release the subunit molecules of each food category.

molecules from food. Some of the ingested food molecules are needed for their energy (caloric) value—obtained by the reactions of cell respiration and used in the production of ATP—and the balance is used to make additional tissue.

Most of the organic molecules that are ingested are similar to the molecules that form human tissues. These are generally large molecules (*polymers*), which are composed of subunits (*monomers*). Within the gastrointestinal tract, the **digestion** of these large molecules into their monomers occurs by means of *hydrolysis reactions* (reviewed in fig. 18.1). The monomers thus formed are transported across the wall of the small intestine into the blood and lymph in the process of **absorption.** Digestion and absorption are the primary functions of the digestive system.

Since the composition of food is similar to the composition of body tissues, enzymes that digest food are also capable of digesting a person's own tissues. This does not normally occur, however, because a variety of protective devices inactivate digestive enzymes in the body and keep them away from the cytoplasm of the cells. The fully active digestive enzymes are normally confined to the lumen (cavity) of the gastrointestinal tract.

The lumen of the gastrointestinal tract is open at both ends (mouth and anus), and is thus continuous with the environment. In this sense, the harsh conditions required for digestion occur *outside* the body. Indigestible materials, such as cellulose from plant walls, pass from one end to the other without crossing the epithelial lining of the digestive tract; since they are not absorbed, they do not enter the body.

In *Planaria* (a type of flatworm), the gastrointestinal tract has only one opening—the mouth is also the anus. Each cell that lines the gastrointestinal tract is thus exposed to food, absorbable digestion products, and waste products. The two open ends of the digestive tract of higher organisms, by contrast, permit one-way transport, which is ensured by wavelike muscle contractions and by the action of sphincter muscles. This one-way transport allows different regions of the gastrointestinal tract to be specialized for different functions, as a "dis-assembly line." These functions of the digestive system include:

1. **Motility.** This refers to the movement of food through the digestive tract through the processes of
   a. *Ingestion:* Taking food into the mouth.
   b. *Mastication:* Chewing the food and mixing it with saliva.
   c. *Deglutition:* Swallowing food.
   d. *Peristalsis:* Rhythmic, wavelike contractions that move food through the gastrointestinal tract.
2. **Secretion.** This includes both exocrine and endocrine secretions.
   a. *Exocrine secretions:* Water, hydrochloric acid, bicarbonate, and many digestive enzymes are secreted into the lumen of the gastrointestinal tract. The stomach alone, for example, secretes 2 to 3 liters of gastric juice a day.
   b. *Endocrine secretions:* The stomach and small intestine secrete a number of hormones that help to regulate the digestive system.
3. **Digestion.** This refers to the breakdown of food molecules into their smaller subunits, which can be absorbed.
4. **Absorption.** This refers to the passage of digested end products into the blood or lymph.
5. **Storage and elimination.** This refers to the temporary storage and subsequent elimination of indigestible food molecules.

Anatomically and functionally, the digestive system can be divided into the tubular **gastrointestinal (GI) tract,** or *alimentary canal,* and **accessory digestive organs.** The GI tract is approximately 9 m (30 ft) long and extends from the mouth to the anus. It traverses the thoracic cavity and enters the abdominal cavity at the level of the diaphragm. The anus is located at the inferior portion of the pelvic cavity. The organs of the GI tract include the *oral cavity, pharynx, esophagus, stomach, small intestine,* and *large intestine* (fig. 18.2). The accessory digestive organs include the *teeth, tongue, salivary glands, liver,*

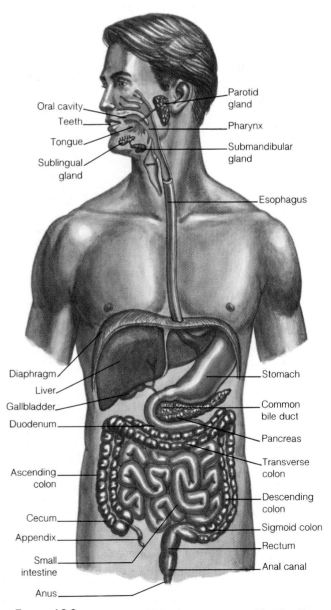

Figure 18.2 **The organs of the digestive system.** The digestive system includes the gastrointestinal tract and the accessory digestive organs.

gallbladder, and *pancreas.* The term *viscera* is frequently used to refer to the abdominal organs of digestion, but it also can be used in reference to any of the organs in the thoracic and abdominal cavities.

Layers of the Gastrointestinal Tract

The GI tract from the esophagus to the anal canal is composed of four layers, or *tunics.* Each tunic contains a dominant tissue type that performs specific functions in the digestive process. The four tunics of the GI tract, from the inside out, are the *mucosa, submucosa, muscularis,* and *serosa* (fig. 18.3a).

Mucosa

The **mucosa,** which lines the lumen of the GI tract, is the absorptive and major secretory layer. It consists of a simple columnar epithelium supported by the *lamina propria,* a thin layer of areolar connective tissue containing numerous lymph nodules, which are important in protecting against disease (fig. 18.3b). External to the lamina propria is a thin layer of smooth muscle called the *muscularis mucosae.* This is the muscle layer responsible for the numerous small folds in certain portions of the GI tract. These folds greatly increase the absorptive surface area. Specialized goblet cells in the mucosa secrete mucus throughout most of the GI tract.

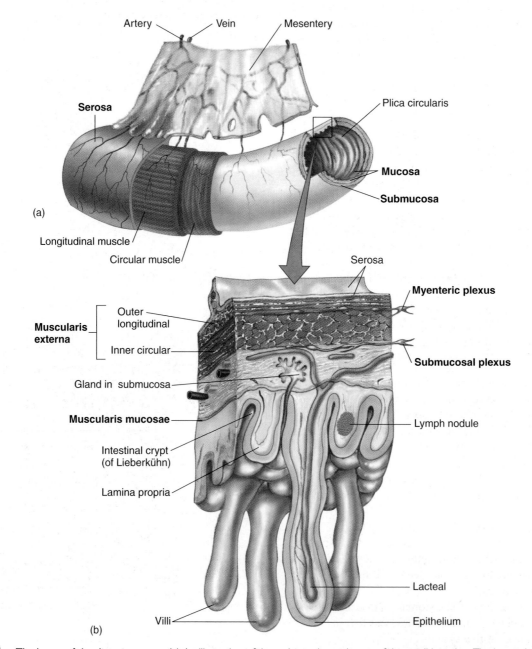

Artery   Vein   Mesentery

Serosa

Plica circularis

Mucosa

Submucosa

(a)

Longitudinal muscle

Circular muscle

Serosa

Myenteric plexus

Muscularis externa

Outer longitudinal

Inner circular

Submucosal plexus

Gland in submucosa

Muscularis mucosae

Lymph nodule

Intestinal crypt (of Lieberkühn)

Lamina propria

Lacteal

Villi

Epithelium

(b)

Figure 18.3   The layers of the digestive tract. *(a)* An illustration of the major tunics, or layers, of the small intestine. The insert shows how folds of mucosa form projections called villi in the small intestine. *(b)* An illustration of a cross section of the small intestine showing layers and glands.

Submucosa

The relatively thick **submucosa** is a highly vascular layer of connective tissue that serves the mucosa. Absorbed molecules that pass through the columnar epithelial cells of the mucosa enter into blood and lymphatic vessels of the submucosa. In addition to blood vessels, the submucosa contains glands and nerve plexuses. The **submucosal plexus** (*Meissner's plexus*) (fig. 18.3*b*) provides an autonomic nerve supply to the muscularis mucosae.

Muscularis

The **muscularis** (also called the *muscularis externa*) is responsible for segmental contractions and peristaltic movement through the GI tract. The muscularis has an inner circular and an outer longitudinal layer of smooth muscle. Contractions of these layers move the food through the tract and physically pulverize and mix the food with digestive enzymes. The **myenteric plexus** (*Auerbach's plexus*), located between the two muscle layers, provides the major nerve supply to the GI tract. It includes fibers and ganglia from both the sympathetic and parasympathetic divisions of the autonomic nervous system.

Serosa

The outer **serosa** completes the wall of the GI tract. It is a binding and protective layer consisting of areolar connective tissue covered with a layer of simple squamous epithelium.

Regulation of the Gastrointestinal Tract

The GI tract is innervated by the sympathetic and parasympathetic divisions of the **autonomic nervous system.** As discussed in chapter 9, **parasympathetic nerves** in general stimulate motility and secretions of the gastrointestinal tract. The **vagus nerve** is the source of parasympathetic activity in the esophagus, stomach, pancreas, gallbladder, small intestine, and upper portion of the large intestine. The lower portion of the large intestine receives parasympathetic innervation from spinal nerves in the sacral region. The submucosal plexus and myenteric plexus are the sites where parasympathetic preganglionic fibers synapse with postganglionic neurons that innervate the smooth muscle of the GI tract.

Postganglionic sympathetic fibers pass through the submucosal and myenteric plexuses and innervate the GI tract. The effects of the **sympathetic nerves** reduce peristalsis and secretory activity and stimulate the contraction of sphincter muscles along the GI tract; therefore, they are antagonistic to the effects of parasympathetic nerve stimulation.

Autonomic regulation, which is "extrinsic" to the gastrointestinal tract, is superimposed on "intrinsic" modes of regulation. The gastrointestinal tract contains **intrinsic sensory neurons** that have their cell bodies within the gut wall and are not part of the autonomic system. These help in the local regulation of the digestive tract by a complex neural network within the wall of the gut called the **enteric nervous system,** or **enteric brain** (discussed later in this chapter). Regulation by the enteric nervous system complements **paracrine regulation** by molecules acting locally within the tissues of the GI tract, as well as **hormonal regulation** by hormones secreted by the mucosa.

In summary, the digestive system is regulated extrinsically by the autonomic nervous system and endocrine system, and intrinsically by the enteric nervous system and various paracrine regulators. The details of this regulation will be described in subsequent sections.

Test Yourself Before You Continue

1. Define the terms *digestion* and *absorption*, describe how molecules are digested, and indicate which molecules are absorbed.
2. Describe the structure and function of the mucosa, submucosa, and muscularis.
3. Describe the location and composition of the submucosal and myenteric plexuses and explain the actions of autonomic nerves on the gastrointestinal tract.

From Mouth to Stomach

Swallowed food is passed through the esophagus to the stomach by wavelike contractions known as peristalsis. The mucosa of the stomach secretes hydrochloric acid and pepsinogen. Upon entering the lumen of the stomach, pepsinogen is converted into the active protein-digesting enzyme known as pepsin. The stomach partially digests proteins and functions to store its contents, called chyme, for later processing by the small intestine.

Mastication (chewing) of food mixes it with saliva, secreted by the salivary glands. In addition to mucus and various antimicrobial agents, saliva contains *salivary amylase,* an enzyme that can catalyze the partial digestion of starch. **Deglutition** (swallowing) begins as a voluntary activity in which the larynx is raised so that the epiglottis covers the entrance to the respiratory tract (chapter 16), preventing ingested material from entering. Swallowing involves three phases—*oral, pharyngeal,* and *esophageal*—that require the coordinated contraction of 25 pairs of muscles in the mouth, pharynx, larynx, and esophagus. The formation of a *bolus* (a mass to be swallowed) of food in the mouth is under voluntary control, while the pharyngeal and esophageal phases are involuntary and cannot be stopped once they have begun. The pharyngeal phase involves striated muscles of the larynx, pharynx, and mouth (the tongue and suprahyoid muscles), which are innervated by somatic motor neurons. The lower esophagus contains smooth muscles, innervated by autonomic neurons. The pattern of contractions required for swallowing is coordinated by interacting neurons in the medulla oblongata, which function as a **swallowing center.**

Once in the stomach, the ingested material is churned and mixed with hydrochloric acid and the protein-digesting enzyme pepsin. The mixture thus produced is pushed by muscular contractions of the stomach past the pyloric sphincter (*pylorus* = gatekeeper), which guards the junction of the stomach and the duodenum of the small intestine.

Esophagus

The **esophagus** is that portion of the GI tract which connects the pharynx to the stomach. It is a muscular tube approximately 25 cm (10 in.) long, located posterior to the trachea within the mediastinum of the thorax. Before terminating in the stomach, the esophagus passes through the diaphragm by means of an opening called the *esophageal hiatus.* The esophagus is lined with a nonkeratinized stratified squamous epithelium; its walls contain either skeletal or smooth muscle, depending on the location. The upper third of the esophagus contains skeletal muscle, the middle third contains a mixture of skeletal and smooth muscle, and the terminal portion contains only smooth muscle.

Swallowed food is pushed from the oral to the anal end of the esophagus (and, afterward, of the intestine) by a wavelike muscular contraction called **peristalsis** (fig. 18.4). Movement of the bolus along the digestive tract occurs because the circular smooth muscle contracts behind, and relaxes in front of, the bolus. This is followed by shortening of the tube by longitudinal muscle contraction. These contractions progress from the superior end of the esophagus to the *gastroesophageal junction* at a rate of 2 to 4 cm per second as they empty the contents of the esophagus into the cardiac region of the stomach.

The lumen of the terminal portion of the esophagus is slightly narrowed because of a thickening of the circular muscle fibers in its wall. This portion is referred to as the **lower esophageal (gastroesophageal) sphincter.** After food passes into the stomach, constriction of the muscle fibers of this region help to prevent the stomach contents from regurgitating into the esophagus. Regurgitation would occur because the pressure in the abdominal cavity is greater than the pressure in the thoracic cavity as a result of respiratory movements. The lower esophageal sphincter must therefore remain closed until food is pushed through it by peristalsis into the stomach.

The lower esophageal sphincter is not a true sphincter muscle that can be identified histologically, and it does at times permit the acidic contents of the stomach to enter the esophagus. This can create a burning sensation commonly called **heartburn,** although the heart is not involved. In infants under a year of age, the lower esophageal sphincter may function erratically, causing them to "spit up" following meals. Certain mammals, such as rodents, have a true gastroesophageal sphincter and thus cannot regurgitate. This is why poison grains are effective in killing mice and rats.

Stomach

The J-shaped **stomach** is the most distensible part of the GI tract. It is continuous with the esophagus superiorly and empties into the duodenum of the small intestine inferiorly. The functions of the stomach are to store food, to initiate the digestion of proteins, to kill bacteria with the strong acidity of gastric juice, and to move the food into the small intestine as a pasty material called **chyme.**

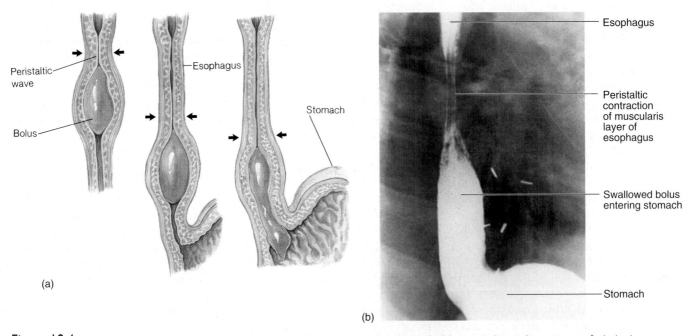

■ **Figure 18.4**   **Peristalsis in the esophagus.** *(a)* A diagram and *(b)* a radiograph showing peristaltic contraction and movement of a bolus into the stomach.

Swallowed food is delivered from the esophagus to the *cardiac region* of the stomach (figs. 18.5 and 18.6). An imaginary horizontal line drawn through the cardiac region divides the stomach into an upper *fundus* and a lower *body,* which together compose about two-thirds of the stomach. The distal portion of the stomach is called the *pyloric region.* The pyloric region begins in a somewhat widened area, the *antrum,* and ends at the *pyloric sphincter.* Contractions of the stomach churn the chyme, mixing it more thoroughly with the gastric secretions. These contractions also push partially digested food from the antrum through the pyloric sphincter and into the first part of the small intestine.

The inner surface of the stomach is thrown into long folds called *rugae,* which can be seen with the unaided eye. Microscopic examination of the gastric mucosa shows that it is likewise folded. The openings of these folds into the stomach lumen are called **gastric pits.** The cells that line the folds deeper in the

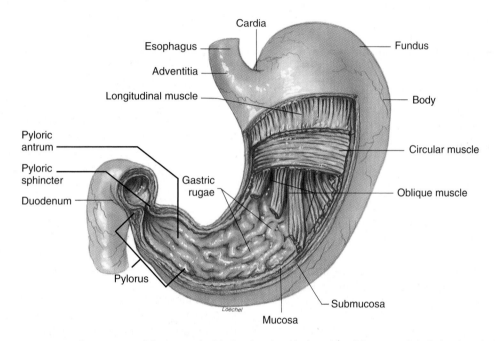

■ **Figure 18.5** **Primary regions and structures of the stomach.** Notice that the pyloric region of the stomach includes the pyloric antrum (the wider portion of the pylorus) as well as the pyloric sphincter.

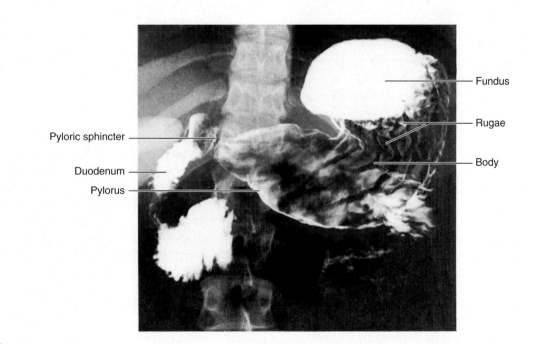

■ **Figure 18.6** **A radiograph of the stomach.** Note the rugae, which are foldings of the inner wall of the stomach (including the submucosa and mucosa).

mucosa secrete various products into the stomach; these cells form the exocrine **gastric glands** (fig. 18.7).

Gastric glands contain several types of cells that secrete different products:

1. **goblet cells,** which secrete *mucus;*
2. **parietal cells,** which secrete *hydrochloric acid (HCl);*
3. **chief** (or **zymogenic**) **cells,** which secrete *pepsinogen,* an inactive form of the protein-digesting enzyme *pepsin;*
4. **enterochromaffin-like (ECL) cells,** found in the stomach and intestine, which secrete *histamine* and *5-hydroxytryptamine* (also called *serotonin*) as paracrine regulators of the GI tract;
5. **G cells,** which secrete the hormone *gastrin* into the blood; and
6. **D cells,** which secrete the hormone *somatostatin.*

In addition to these products, the gastric mucosa (probably the parietal cells) secretes a polypeptide called **intrinsic factor,** which is required for the intestinal absorption of vitamin B_{12}. Vitamin B_{12} is necessary for the production of red blood cells in the bone marrow (see the next boxed clinical application). Also, the stomach has recently been shown to secrete a hormone named **ghrelin.** Secretion of this newly discovered hormone rises before meals and falls after meals. This may serve as a signal from the stomach to the brain that helps regulate hunger, as described in chapter 19.

The exocrine secretions of the gastric cells, together with a large amount of water (2 to 4 L/day), form a highly acidic solution known as **gastric juice.**

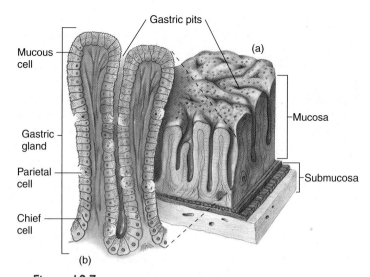

Figure 18.7 Gastric pits and gastric glands of the mucosa. *(a)* Gastric pits are the openings of the gastric glands. *(b)* Gastric glands consist of several types of cells (including mucous cells, chief cells, and parietal cells), each of which produces a specific secretion.

The only stomach function that appears to be essential for life is the secretion of *intrinsic factor.* This polypeptide is needed for the absorption of vitamin B_{12} in the terminal portion of the ileum in the small intestine, and vitamin B_{12} is required for maturation of red blood cells in the bone marrow. Following surgical removal of the stomach (gastrectomy), a patient has to receive B_{12} injections or take B_{12} orally, together with intrinsic factor. Without vitamin B_{12}, **pernicious anemia** will develop.

Pepsin and Hydrochloric Acid Secretion

The parietal cells secrete H^+, at a pH as low as 0.8, into the gastric lumen by primary active transport (involving carriers that function as an ATPase). These carriers, known as **H^+/K^+ ATPase pumps,** transport H^+ uphill against a million-to-one concentration gradient into the lumen of the stomach while they transport K^+ in the opposite direction (fig. 18.8).

At the same time, the parietal cell's basolateral membrane (facing the blood in capillaries of the lamina propria) take in Cl^- against its electrochemical gradient by coupling its transport to the downhill movement of bicarbonate (HCO_3^-). The bicarbonate ion is produced within the parietal cell by the dissociation of carbonic acid, formed from CO_2 and H_2O by the enzyme carbonic anhydrase. Therefore, the parietal cell can secrete Cl^- (by facilitative diffusion) as well as H^+ into the gastric juice while it secretes bicarbonate into the blood (fig. 18.8).

The secretion of HCl by the parietal cells is stimulated by a variety of factors, including the hormone gastrin, secreted by the G cells, and acetylcholine (ACh), released by axons of the vagus nerve. Most of the effects of gastrin and ACh on acid secretion, however, are currently believed to be indirect. Gastrin and ACh from vagal axons stimulate the release of histamine from the ECL cells of the gastric mucosa, and histamine, in turn, acts as a paracrine regulator to stimulate the parietal cells to secrete HCl (see fig. 18.30). The endocrine regulation of the digestive system is discussed in detail later in this chapter.

People with **gastroesophageal reflux disease,** a common disorder involving the reflux of acidic gastric juice into the esophagus, are often treated with specific drugs (including *Prilosec* and *Prevacid*) that inhibit the K^+/H^+ pumps in the gastric mucosa. Since gastric acid secretion is stimulated by histamine released from the ECL cells, people with **peptic ulcers** can be treated with drugs that block histamine action. Drugs in this category, such as *Tagamet* and *Zantac,* specifically block the H_2 histamine receptors in the gastric mucosa. This is a different receptor subtype than that blocked by antihistamines commonly used to treat cold and allergy symptoms.

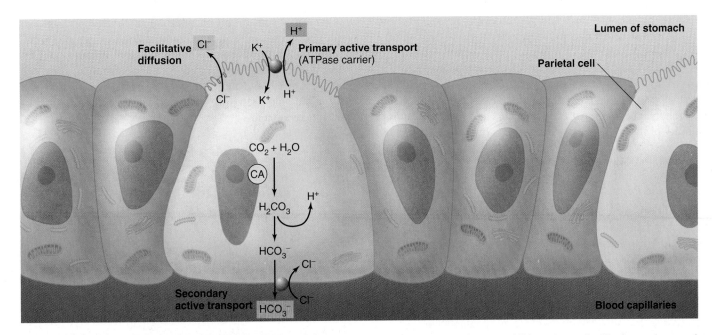

■ **Figure 18.8** **Secretion of gastric acid by parietal cells.** The apical membrane (facing the lumen) secretes H^+ in exchange for K^+ using a primary active transport carrier that is powered by the hydrolysis of ATP. The basolateral membrane (facing the blood) secretes bicarbonate (HCO_3^-) in exchange for Cl^-. The Cl^- moves into the cell against its electrochemical gradient, powered by the downhill movement of HCO_3^- out of the cell. This HCO_3^- is produced by the dissociation of carbonic acid (H_2CO_3), which is formed from CO_2 and H_2O by the action of the enzyme carbonic anhydrase (abbreviated CA). The Cl^- then leaves the apical portion of the membrane by diffusion through a membrane channel. The parietal cells thus secrete HCl into the stomach lumen as they secrete HCO_3^- into the blood.

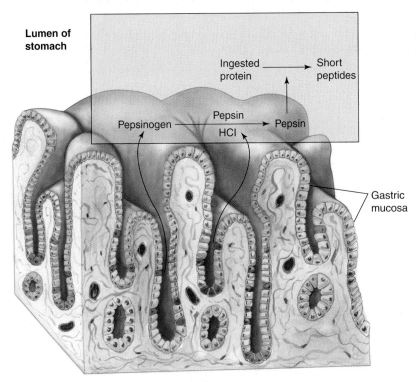

■ **Figure 18.9** **The activation of pepsin.** The gastric mucosa secretes the inactive enzyme pepsinogen and hydrochloric acid (HCl). In the presence of HCl, the active enzyme pepsin is produced. Pepsin digests proteins into shorter polypeptides.

The high concentration of HCl from the parietal cells makes gastric juice very acidic, with a pH of less than 2. This strong acidity serves three functions:

1. Ingested proteins are denatured at low pH—that is, their tertiary structure (chapter 2) is altered so that they become more digestible.
2. Under acidic conditions, weak pepsinogen enzymes partially digest each other—this frees the fully active pepsin enzyme as small inhibitory fragments are removed (fig. 18.9).
3. Pepsin is more active under acidic conditions—it has a pH optimum (chapter 4) of about 2.0.

As a result of the activation of pepsin under acidic conditions, the fully active pepsin is able to catalyze the hydrolysis of peptide bonds in the ingested protein. Thus, the cooperative activities of pepsin and HCl permit the partial digestion of food protein in the stomach.

Digestion and Absorption in the Stomach

Proteins are only partially digested in the stomach by the action of pepsin, while carbohydrates and fats are not digested at all by pepsin. (Digestion of starch begins in the mouth with the action of salivary amylase and continues for a time when the food enters the stomach, but amylase soon becomes inactivated by the strong acidity of gastric juice.) The complete digestion of food molecules occurs later, when chyme enters the small intestine.

Therefore, people who have had partial gastric resections—and even those who have had complete gastrectomies—can still adequately digest and absorb their food.

Almost all of the products of digestion are absorbed through the wall of the small intestine; the only commonly ingested substances that can be absorbed across the stomach wall are alcohol and aspirin. Absorption occurs as a result of the lipid solubility of these molecules. The passage of aspirin through the gastric mucosa has been shown to cause bleeding, which may be significant if aspirin is taken in large doses.

Gastritis and Peptic Ulcers

Peptic ulcers are erosions of the mucous membranes of the stomach or duodenum produced by the action of HCl. In *Zollinger-Ellison syndrome,* ulcers of the duodenum are produced by excessive gastric acid secretion in response to very high levels of the hormone gastrin. Gastrin is normally produced by the stomach but, in this case, it may be secreted by a pancreatic tumor. This is a rare condition, but it does demonstrate that excessive gastric acid can cause ulcers of the duodenum. Ulcers of the stomach, however, are not believed to be due to excessive acid secretion, but rather to mechanisms that reduce the barriers of the gastric mucosa to self-digestion.

It has been known for some time that most people who have peptic ulcers are infected with a bacterium known as *Helicobacter pylori,* which resides in the gastrointestinal tract of almost half the adult population worldwide. Also, clinical trials have demonstrated that antibiotics that eliminate this infection help in the treatment of the peptic ulcers. Thus, many people with ulcers can be effectively treated with a combination of antibiotics and K^+/H^+ pump inhibitors (such as *Prilosec*).

Experiments demonstrate that the plasma membranes of the parietal and chief cells of the gastric mucosa are highly impermeable to the acid in the gastric lumen. Other protective mechanisms include a layer of alkaline mucus, containing bicarbonate, covering the gastric mucosa; tight junctions between adjacent epithelial cells, preventing acid from leaking into the submucosa; a rapid rate of cell division, allowing damaged cells to be replaced (the entire epithelium is replaced every 3 days); and several protective effects provided by prostaglandins produced by the gastric mucosa. Indeed, a common cause of gastric ulcers is believed to be the use of nonsteroidal anti-inflammatory drugs (NSAIDs). This class of drugs, including aspirin and ibuprofen, acts to inhibit the production of prostaglandins (as discussed in chapter 11).

When the gastric barriers to self-digestion are broken down, acid can leak through the mucosa to the submucosa, causing direct damage and stimulating inflammation. The histamine released from mast cells during inflammation may stimulate further acid secretion (see fig. 18.30) and result in further damage to the mucosa. The inflammation that occurs during these events is called **acute gastritis.** This is why drugs that block the H_2 histamine receptors (such as *Tagamet* and *Zantac*) may be used to treat the gastritis.

The duodenum is normally protected from gastric acid by the buffering action of bicarbonate in alkaline pancreatic juice, as well as by secretion of bicarbonate by Brunner's glands in the submucosa of the duodenum. However, people who develop duodenal ulcers produce excessive amounts of gastric acid that are not neutralized by the bicarbonate ion. People with gastritis and peptic ulcers must avoid substances that stimulate acid secretion, including coffee and wine, and often must take antacids.

Clinical Investigation Clues

Remember that Alan got a sharp pain in his stomach whenever he drank wine.

What may have caused Alan's pain?

What medicine might help reduce this pain?

Test Yourself Before You Continue

1. Describe the structure and function of the lower esophageal sphincter.
2. List the secretory cells of the gastric mucosa and the products they secrete.
3. Describe the functions of hydrochloric acid in the stomach.
4. Explain how peptic ulcers are produced and why they are more likely to occur in the duodenum than in the stomach.
5. Explain how gastrin and vagus nerve stimulation cause the parietal cells to secrete HCl.

Small Intestine

The mucosa of the small intestine is folded into villi that project into the lumen. In addition, the cells that line these villi have foldings of their plasma membrane called microvilli. This arrangement greatly increases the surface area for absorption. It also improves digestion, since the digestive enzymes of the small intestine are embedded within the plasma membrane of the microvilli.

The **small intestine** (fig. 18.10) is that portion of the GI tract between the pyloric sphincter of the stomach and the ileocecal valve opening into the large intestine. It is called "small" because of its relatively small diameter compared to that of the large intestine. The small intestine is the longest part of the GI tract, however. It is approximately 3 m (12 ft) long in a living person, but it will measure nearly twice this length in a cadaver when the muscle wall is relaxed. The first 20 to 30 cm (10 in.)

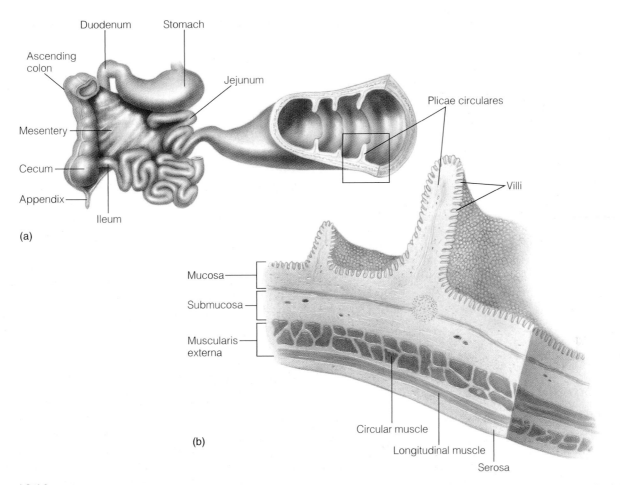

(a)

(b)

■ **Figure 18.10**   **The small intestine.** *(a)* The regions of the small intestine. *(b)* A section of the intestinal wall showing the tissue layers, plicae circulates, and villi.

extending from the pyloric sphincter is the **duodenum.** The next two-fifths of the small intestine is the **jejunum,** and the last three-fifths is the **ileum.** The ileum empties into the large intestine through the ileocecal valve.

The products of digestion are absorbed across the epithelial lining of intestinal mucosa. Absorption of carbohydrates, lipids, amino acids, calcium, and iron occurs primarily in the duodenum and jejunum. Bile salts, vitamin B_{12}, water, and electrolytes are absorbed primarily in the ileum. Absorption occurs at a rapid rate as a result of extensive foldings of the intestinal mucosa, which greatly increase its absorptive surface area. The mucosa and submucosa form large folds, called *plicae circulares,* which can be observed with the unaided eye. The surface area is further increased by microscopic folds of mucosa, called *villi,* and by foldings of the apical plasma membrane of epithelial cells (which can be seen only with an electron microscope), called *microvilli.*

Villi and Microvilli

Each **villus** is a fingerlike fold of mucosa that projects into the intestinal lumen (fig. 18.11). The villi are covered with columnar epithelial cells, among which are interspersed mucus-secreting

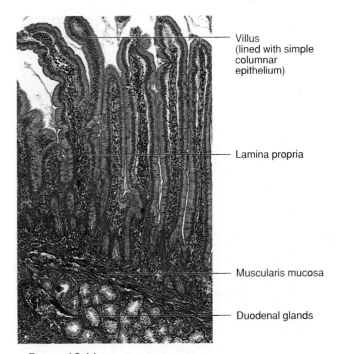

■ **Figure 18.11**   **The histology of the duodenum.** Note the duodenal (Brunner's) glands. These exocrine glands, unique to the duodenum, extend into the submucosa and produce a bicarbonate-rich, alkaline secretion.

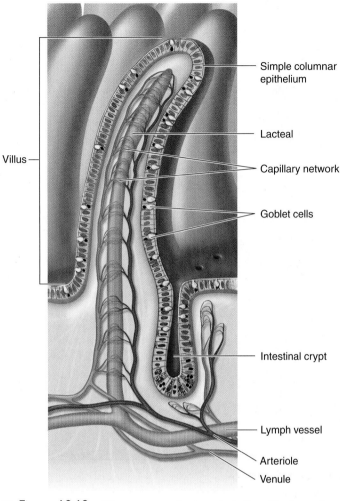

Villus

Simple columnar epithelium

Lacteal

Capillary network

Goblet cells

Intestinal crypt

Lymph vessel

Arteriole

Venule

Figure 18.12   **The structure of an intestinal villus.** The figure also depicts an intestinal crypt (crypt of Lieberkühn), in which new epithelial cells are produced by mitosis.

goblet cells. The lamina propria, which forms the connective tissue core of each villus, contains numerous lymphocytes, blood capillaries, and a lymphatic vessel called the *central lacteal* (fig. 18.12). Absorbed monosaccharides and amino acids enter the blood capillaries; absorbed fat enters the central lacteals.

Epithelial cells at the tips of the villi are continuously exfoliated (shed) and are replaced by cells that are pushed up from the bases of the villi. The epithelium at the base of the villi invaginates downward at various points to form narrow pouches that open through pores to the intestinal lumen. These structures are called **intestinal crypts,** or *crypts of Lieberkühn* (fig. 18.12).

Microvilli are formed by foldings at the apical surface of each epithelial cell membrane. These minute projections can be seen clearly only in an electron microscope. In a light microscope, the microvilli produce a somewhat vague **brush border** on the edges of the columnar epithelial cells. The terms *brush border* and *microvilli* are thus often used interchangeably in describing the small intestine (fig. 18.13).

Intestinal Enzymes

In addition to providing a large surface area for absorption, the plasma membranes of the microvilli contain digestive enzymes that hydrolyze disaccharides, polypeptides, and other substrates (table 18.1). These **brush border enzymes** are not secreted into the lumen, but instead remain attached to the cell membrane with their active sites exposed to the chyme. One brush border enzyme, **enterokinase,** is required for activation of the protein-digesting enzyme *trypsin,* which enters the small intestine in pancreatic juice.

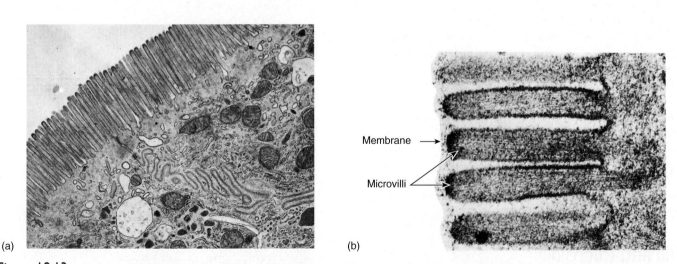

Membrane

Microvilli

(a)   (b)

Figure 18.13   **Electron micrographs of microvilli.** Microvilli are evident at the apical surface of the columnar epithelial cells in the small intestine. These are seen here *(a)* at lower magnification and *(b)* at higher magnification. Microvilli increase the surface area for absorption and also have the brush border digestive enzymes embedded in their plasma membranes.

Table 18.1 Brush Border Enzymes Attached to the Cell Membrane of Microvilli in the Small Intestine

Category	Enzyme	Comments
Disaccharidase	Sucrase	Digests sucrose to glucose and fructose; deficiency produces gastrointestinal disturbances
	Maltase	Digests maltose to glucose
	Lactase	Digests lactose to glucose and galactose; deficiency produces gastrointestinal disturbances (lactose intolerance)
Peptidase	Aminopeptidase	Produces free amino acids, dipeptides, and tripeptides
	Enterokinase	Activates trypsin (and indirectly other pancreatic juice enzymes); deficiency results in protein malnutrition
Phosphatase	Ca^{2+}, Mg^{2+}-ATPase	Needed for absorption of dietary calcium; enzyme activity regulated by vitamin D
	Alkaline phosphatase	Removes phosphate groups from organic molecules; enzyme activity may be regulated by vitamin D

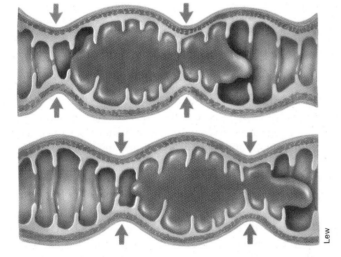

■ **Figure 18.14** **Segmentation of the small intestine.** Simultaneous contractions of numerous segments of the intestine help to mix the chyme with digestive enzymes and mucus.

The ability to digest milk sugar, or lactose, depends on the presence of a brush border enzyme called *lactase.* This enzyme is present in all children under the age of 4 but becomes inactive to some degree in most adults (people of Asian or African heritage are more often lactase deficient than Caucasians). A deficiency of lactase can result in **lactose intolerance,** a condition in which too much undigested lactose in the intestine causes diarrhea, gas, cramps, and other unpleasant symptoms. Yogurt is better tolerated than milk because it contains lactase (produced by the yogurt bacteria), which becomes activated in the duodenum and digests lactose.

Intestinal Contractions and Motility

Two major types of contractions occur in the small intestine: *peristalsis* and *segmentation.* Peristalsis is much weaker in the small intestine than in the esophagus and stomach. Intestinal motility—the movement of chyme through the intestine—is relatively slow and is due primarily to the fact that the pressure at the pyloric end of the small intestine is greater than at the distal end.

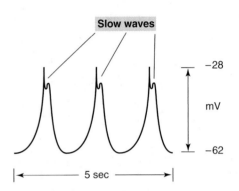

■ **Figure 18.15** **Slow waves in the intestine.** The slow waves are produced by the interstitial cells of Cajal (ICC), not by smooth muscle cells, and are apparently conducted by networks of ICC that are electrically joined together within the intestinal wall. Smooth muscle cells respond to this depolarization by producing action potentials and contracting. Note that the slow waves occur much slower (with a rate measured in seconds) than do the pacemaker potentials in the heart.

The major contractile activity of the small intestine is **segmentation.** This term refers to muscular constrictions of the lumen, which occur simultaneously at different intestinal segments (fig. 18.14). This action serves to mix the chyme more thoroughly.

Contractions of intestinal smooth muscles occur automatically in response to endogenous pacemaker activity, somewhat analogous to the automatic beating of the heart. In intestinal smooth muscle, however, the rhythm of contractions is paced by graded depolarizations called **slow waves** (fig. 18.15). Current evidence suggests that the slow waves are produced by unique cells, often associated with autonomic nerve endings. However, these pacemaker cells are neither neurons nor smooth muscle cells; they are the cells identified histologically as the **interstitial cells of Cajal.** These cells have long processes joined to each other and to smooth muscle cells by gap junctions, which permit the spread of depolarization from one cell to the next (fig. 18.16).

The slow waves are conducted between interconnected interstitial cells of Cajal through electrical synapses between these cells. Current evidence suggests that only the interstitial cells of Cajal can produce and conduct the slow waves. Smooth muscle cells, though electrically joined to each other (through electrical synapses termed *nexuses*) and to the interstitial cells of Cajal, cannot do this.

The slow waves produced and conducted by the interstitial cells of Cajal serve to depolarize the adjacent smooth muscle cells.

Structures	Functions

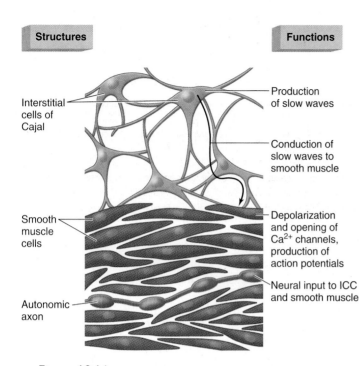

Interstitial cells of Cajal

Smooth muscle cells

Autonomic axon

Production of slow waves

Conduction of slow waves to smooth muscle

Depolarization and opening of Ca^{2+} channels, production of action potentials

Neural input to ICC and smooth muscle

Figure 18.16 Cells responsible for the electrical events within the muscularis. The interstitial cells of Cajal (ICC) generate the slow waves, which pace the contractions of the intestine. Slow waves are conducted into the smooth muscle cells, where they can stimulate opening of Ca^{2+} channels. This produces action potentials and stimulates contraction. Autonomic axons have varicosities that release neurotransmitters, which modify the inherent electrical activity of the interstitial cells of Cajal and smooth muscle cells.

When the slow-wave depolarization exceeds a threshold value, it triggers action potentials in the smooth muscle cells by opening voltage-gated Ca^{2+} channels. The inward flow of Ca^{2+} has two effects: (1) it produces the upward depolarization phase of the action potential (repolarization is produced by outward flow of K^+); and (2) it stimulates contraction (as described in chapter 12). Contraction may then be aided by additional calcium released from the sarcoplasmic reticulum through calcium-induced calcium release.

Autonomic nerves modify these automatic contractions of the intestine. When acetylcholine (from parasympathetic axons) stimulates its muscarinic ACh receptors in the smooth muscle cells, it increases the amplitude and duration of the slow waves. Thus, it increases the production of action potentials and promotes contractions and motility of the intestine. Inhibitory neurotransmitters, by contrast, hyperpolarize the smooth muscle membrane and thereby decrease the activity of the intestine.

Test Yourself Before You Continue

1. Describe the structures that increase the surface area of the small intestine and explain the function of the intestinal crypts.

2. Explain what is meant by the term *brush border* and give some examples of brush border enzymes. Why is it that many adults cannot tolerate milk?

3. Explain how smooth muscle contraction in the small intestine is regulated. What is the function of segmentation?

Large Intestine

The large intestine absorbs water, electrolytes, and certain vitamins from the chyme it receives from the small intestine. In a process regulated by the action of sphincter muscles, the large intestine then passes waste products out of the body through the rectum and anal canal.

The **large intestine,** or **colon,** extends from the ileocecal valve to the anus, framing the small intestine on three sides. Chyme from the ileum passes into the **cecum,** which is a blind pouch (open only at one end) at the beginning of the large intestine. Waste material then passes in sequence through the **ascending colon, transverse colon, descending colon, sigmoid colon, rectum,** and **anal canal** (fig. 18.17). Waste material (feces) is excreted through the *anus,* the external opening of the anal canal.

The mucosa of the large intestine, like that of the small intestine, contains many scattered lymphocytes and lymphatic nodules and is covered by columnar epithelial cells and mucus-secreting goblet cells. Although this epithelium does form crypts (fig. 18.18), there are no villi in the large intestine—the intestinal mucosa therefore appears flat. The outer surface of the colon bulges outward to form pouches, or **haustra** (fig. 18.17 and 18.19). Occasionally, the muscularis externa of the haustra may become so weakened that the wall forms a more elongated outpouching, or diverticulum (*divert* = turned aside). Inflammation of one or more of these structures is called *diverticulitis.*

The *appendix* is a short, thin outpouching of the cecum. It does not function in digestion, but like the tonsils, it contains numerous lymphatic nodules (fig. 18.18) and is subject to inflammation—a condition called **appendicitis.** This is commonly detected in its later stages by pain in the lower right quadrant of the abdomen. If the appendix ruptures, infectious material can spread throughout the surrounding body cavity, causing inflammation of the peritoneum—*peritonitis.* This dangerous event may be prevented by surgical removal of the inflamed appendix (*appendectomy*).

Clinical Investigation Clues

Remember that Alan had pains in particular locations only when provoked by certain food and drinks; also, he did not have an elevated white blood cell count or a fever.

Is it likely that Alan has appendicitis?
Why or why not?

The large intestine has little or no digestive function, but it does absorb water and electrolytes from the remaining chyme, as well as several B complex vitamins and vitamin K. Bacteria residing in the intestine, primarily the colon—collectively referred

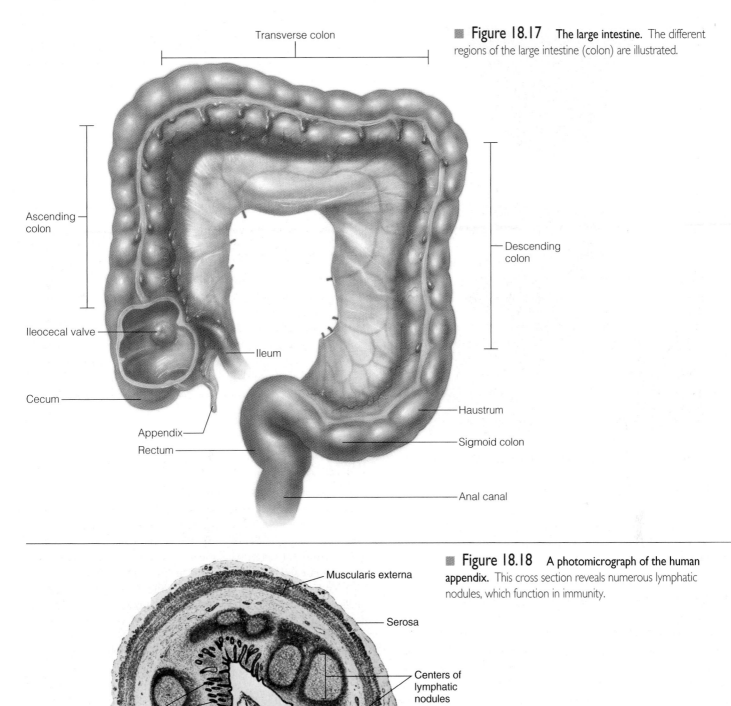

Transverse colon

Ascending colon

Ileocecal valve

Ileum

Cecum

Appendix

Rectum

Descending colon

Haustrum

Sigmoid colon

Anal canal

■ Figure 18.17   The large intestine. The different regions of the large intestine (colon) are illustrated.

Muscularis externa

Serosa

Centers of lymphatic nodules

Crypts of Lieberkühn

Lumen with feces

Submucosa

■ Figure 18.18   A photomicrograph of the human appendix. This cross section reveals numerous lymphatic nodules, which function in immunity.

to as the **intestinal microbiota** or **microflora**—produce significant amounts of vitamin K and folic acid (see chapter 19), which are absorbed in the large intestine.

The number of bacterial cells in the human colon is said to exceed the total number of cells in the human body! This intes-

tinal microbiota originates at birth and performs a number of physiologically important functions. In addition to the production of B vitamins and vitamin K, bacteria in the colon ferment (through anaerobic respiration) some indigestible molecules in the chyme and secreted mucus. They produce *short-chain fatty acids*

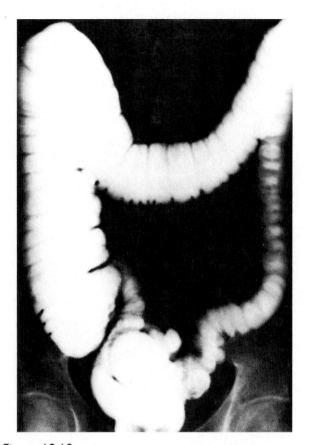

■ **Figure 18.19   A radiograph of the large intestine.** The large intestine is seen after a barium enema has been administered; the haustra are clearly visible.

(less than five carbons long), which are used for energy by the epithelial cells of the colon, and which aid the absorption of sodium, bicarbonate, calcium, magnesium, and iron in the large intestine.

Fluid and Electrolyte Absorption in the Intestine

Most of the fluid and electrolytes in the lumen of the GI tract are absorbed by the small intestine. Although a person may drink only about 1.5 L of water per day, the small intestine receives 7 to 9 L per day as a result of the fluid secreted into the GI tract by the salivary glands, stomach, pancreas, liver, and gallbladder. The small intestine absorbs most of this fluid and passes 1.5 to 2.0 L of fluid per day to the large intestine. The large intestine absorbs about 90% of this remaining volume, leaving less than 200 ml of fluid to be excreted in the feces.

Absorption of water in the intestine occurs passively as a result of the osmotic gradient created by the active transport of ions. The epithelial cells of the intestinal mucosa are joined together much like those of the kidney tubules and, like the kidney tubules, contain Na^+/K^+ pumps in the basolateral membrane. The analogy with kidney tubules is emphasized by the observation that aldosterone, which stimulates salt and water reabsorption in the renal tubules, also appears to stimulate salt and water absorption in the ileum.

The handling of salt and water transport in the large intestine is made more complex by the fact that the large intestine can secrete, as well as absorb, water. The secretion of water by the mucosa of the large intestine occurs by osmosis as a result of the active transport of Na^+ or Cl^- out of the epithelial cells into the intestinal lumen. Secretion in this way is normally minor compared to the far greater amount of salt and water absorption, but this balance may be altered in some disease states.

Diarrhea is characterized by excessive fluid excretion in the feces. Three different mechanisms, illustrated by three different diseases, can cause diarrhea. In *cholera*, severe diarrhea and dehydration result from *enterotoxin*, a chemical produced by the infecting bacteria. Release of enterotoxin stimulates active NaCl transport into the lumen of the intestine, followed by the osmotic movement of water. In *celiac sprue*, diarrhea is caused by damage to the intestinal mucosa produced in susceptible people by eating foods that contain gluten (proteins from grains such as wheat). In *lactose intolerance*, diarrhea is produced by the increased osmolarity of the contents of the intestinal lumen as a result of the presence of undigested lactose.

Defecation

After electrolytes and water have been absorbed, the waste material that is left passes to the rectum, leading to an increase in rectal pressure, relaxation of the internal anal sphincter, and the urge to defecate. If the urge to defecate is denied, feces are prevented from entering the anal canal by the external anal sphincter. In this case the feces remain in the rectum, and may even back up into the sigmoid colon. The **defecation reflex** normally occurs when the rectal pressure rises to a particular level that is determined, to a large degree, by habit. At this point, the external anal sphincter relaxes to admit feces into the anal canal.

During the act of defecation, the longitudinal rectal muscles contract to increase rectal pressure, and the internal and external anal sphincter muscles relax. Excretion is aided by contractions of abdominal and pelvic skeletal muscles, which raise the intra-abdominal pressure (this is part of Valsalva's maneuver, described in chapter 14). The raised pressure helps to push the feces from the rectum, through the anal canal, and out of the anus.

Test Yourself Before You Continue

1. Describe how electrolytes and water are absorbed in the large intestine and explain how diarrhea may be produced.
2. Identify the nature and significance of the intestinal microflora.
3. Describe the structures and mechanisms involved in defecation.

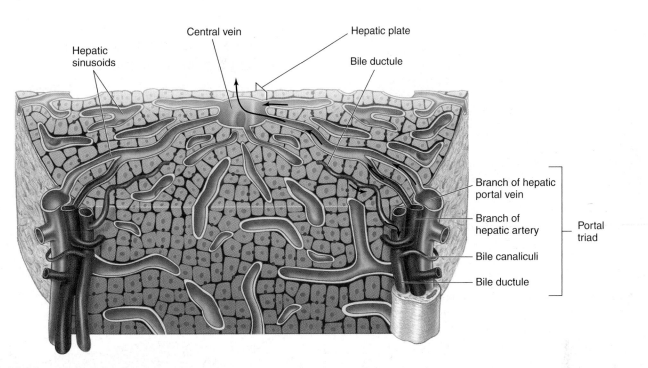

Central vein

Hepatic plate

Hepatic sinusoids

Bile ductule

Branch of hepatic portal vein

Branch of hepatic artery

Portal triad

Bile canaliculi

Bile ductule

■ **Figure 18.20**   *Microscopic structure of the liver.* Blood enters a liver lobule through the vessels in a portal triad, passes through hepatic sinusoids, and leaves the lobule through a central vein. The central veins converge to form hepatic veins that transport venous blood from the liver.

Liver, Gallbladder, and Pancreas

The liver regulates the chemical composition of the blood in numerous ways. In addition, the liver produces and secretes bile, which is stored and concentrated in the gallbladder prior to its discharge into the duodenum. The pancreas produces pancreatic juice, an exocrine secretion containing bicarbonate and important digestive enzymes, which is passed into the duodenum via the pancreatic duct.

The *liver* is positioned immediately beneath the diaphragm in the abdominal cavity. It is the largest internal organ, weighing about 1.3 kg (3.5 to 4.0 lb) in an adult. Attached to the inferior surface of the liver, between its right and quadrate lobes, is the pear-shaped *gallbladder.* This organ is approximately 7 to 10 cm (3 to 4 in.) long. The *pancreas,* which is about 12 to 15 cm (5 to 6 in.) long, is located behind the stomach along the posterior abdominal wall.

Structure of the Liver

Although the liver is the largest internal organ, it is, in a sense, only one to two cells thick. This is because the liver cells, or **hepatocytes,** form **hepatic plates** that are one to two cells thick. The plates are separated from each other by large capillary spaces called **sinusoids** (fig. 18.20).

The sinusoids have extremely large pores (called *fenestrae*) and, unlike other capillaries, lack a basement membrane. This makes the hepatic sinusoids much more permeable than other capillaries, even permitting the passage of plasma proteins with

protein-bound nonpolar molecules, such as fat and cholesterol. The sinusoids also contain phagocytic *Kupffer cells,* which are part of the reticuloendothelial system (chapter 15). The fenestrae, lack of a basement membrane, and plate structure of the liver provide intimate contact between the hepatocytes and the content of the blood.

Hepatic Portal System

The products of digestion that are absorbed into blood capillaries in the intestine do not directly enter the general circulation. Instead, this blood is delivered first to the liver. Capillaries in the digestive tract drain into the *hepatic portal vein,* which carries this blood to capillaries in the liver. It is not until the blood has passed through this second capillary bed that it enters the general circulation through the *hepatic vein* that drains the liver. The term **portal system** is used to describe this unique pattern of circulation: capillaries ⇒ vein ⇒ capillaries ⇒ vein. In addition to receiving venous blood from the intestine, the liver also receives arterial blood via the *hepatic artery.*

Liver Lobules

The hepatic plates are arranged into functional units called **liver lobules** (figs. 18.20 and 18.21). In the middle of each lobule is a *central vein,* and at the periphery of each lobule are branches of the hepatic portal vein and of the hepatic artery, which open into the sinusoids *between* hepatic plates. Arterial blood and portal venous blood, containing molecules absorbed in the GI tract, thus mix as the blood flows within the sinusoids from the periphery of the lobule to the central vein. The central veins of different liver lobules converge to form the hepatic vein, which carries blood from the liver to the inferior vena cava.

Bile is produced by the hepatocytes and secreted into thin channels called **bile canaliculi,** located *within* each hepatic plate

Branch of portal vein

Bile ductule   Bile canaliculus   Sinusoids

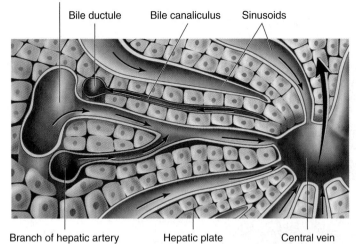

Branch of hepatic artery   Hepatic plate   Central vein

■ **Figure 18.21** **The flow of blood and bile in a liver lobule.** Blood flows within sinusoids from a portal vein to the central vein (from the periphery to the center of a lobule). Bile flows within hepatic plates from the center to bile ductules at the periphery of a lobule.

(fig. 18.21). These bile canaliculi are drained at the periphery of each lobule by *bile ducts,* which in turn drain into *hepatic ducts* that carry bile away from the liver. Since blood travels in the sinusoids and bile travels in the opposite direction within the hepatic plates, blood and bile do not mix in the liver lobules.

In **cirrhosis,** large numbers of liver lobules are destroyed and replaced with permanent connective tissue and "regenerative nodules" of hepatocytes. These regenerative nodules do not have the platelike structure of normal liver tissue, and are therefore less functional. One indication of this decreased function is the entry of ammonia (produced by intestinal bacteria) from the hepatic portal blood into the general circulation. Cirrhosis may be caused by chronic alcohol abuse, biliary obstruction, viral hepatitis, or by various chemicals that attack liver cells.

Enterohepatic Circulation

In addition to the normal constituents of bile, a wide variety of exogenous compounds (drugs) are secreted by the liver into the bile ducts (table 18.2). The liver can thus "clear" the blood of particular compounds by removing them from the blood and excreting them into the intestine with the bile. Molecules that are cleared from the blood by secretion into the bile are eliminated in the feces; this is analogous to renal clearance of blood through excretion in the urine (chapter 17).

Many compounds that are released with the bile into the intestine are not eliminated with the feces, however. Some of these can be absorbed through the small intestine and enter the hepatic portal blood. These molecules are thus carried back to the liver, where they can be again secreted by hepatocytes into the bile ducts. Compounds that recirculate between the liver and intes-

Table 18.2 Compounds Excreted by the Liver into the Bile Ducts

Category	Compound	Comments
Endogenous (Naturally Occurring)	Bile salts, urobilinogen, cholesterol	High percentage reabsorbed and has an enterohepatic circulation*
	Lecithin	Small percentage reabsorbed and has an enterohepatic circulation
	Bilirubin	No enterohepatic circulation
Exogenous (Drugs)	Ampicillin, streptomycin, tetracyline	High percentage reabsorbed and has an enterohepatic circulation
	Sulfonamides, penicillin	Small percentage reabsorbed and has an enterohepatic circulation

*Compounds with an enterohepatic circulation are absorbed to some degree by the intestine and are returned to the liver in the hepatic portal vein.

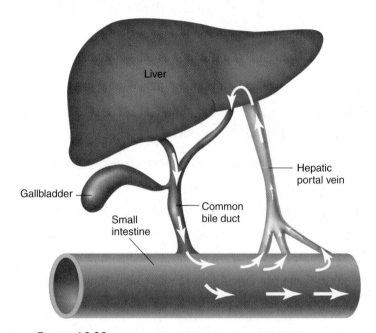

Liver

Hepatic portal vein

Gallbladder

Common bile duct

Small intestine

■ **Figure 18.22** **The enterohepatic circulation.** Substances secreted in the bile may be absorbed by the intestinal epithelium and recycled to the liver via the hepatic portal vein.

tine in this way are said to have an **enterohepatic circulation** (fig. 18.22). For example, a few grams of bile salts (discussed shortly) released into the intestine recirculate six to ten times a day, with only about 0.5 g of bile salts per day excreted in the feces.

Clinical Investigation Clue

Remember that Alan had normal blood levels of free bilirubin, ammonia, and urea.

What do these measurements suggest about the health of Alan's liver?

Table 18.3 Major Categories of Liver Function

Functional Category	Actions
Detoxication of Blood	Phagocytosis by Kupffer cells
	Chemical alteration of biologically active molecules (hormones and drugs)
	Production of urea, uric acid, and other molecules that are less toxic than parent compounds
	Excretion of molecules in bile
Carbohydrate Metabolism	Conversion of blood glucose to glycogen and fat
	Production of glucose from liver glycogen and from other molecules (amino acids, lactic acid) by gluconeogenesis
	Secretion of glucose into the blood
Lipid Metabolism	Synthesis of triglycerides and cholesterol
	Excretion of cholesterol in bile
	Production of ketone bodies from fatty acids
Protein Synthesis	Production of albumin
	Production of plasma transport proteins
	Production of clotting factors (fibrinogen, prothrombin, and others)
Secretion of Bile	Synthesis of bile salts
	Conjugation and excretion of bile pigment (bilirubin)

Functions of the Liver

As a result of its large and diverse enzymatic content and its unique structure, and because it receives venous blood from the intestine, the liver has a wider variety of functions than any other organ in the body. The major categories of liver function are summarized in table 18.3.

Bile Production and Secretion

The liver produces and secretes 250 to 1,500 ml of bile per day. The major constituents of bile are *bile pigment (bilirubin), bile salts, phospholipids* (mainly lecithin), *cholesterol,* and *inorganic ions.*

  Bile pigment, or **bilirubin,** is produced in the spleen, liver, and bone marrow as a derivative of the heme groups (minus the iron) from hemoglobin (fig. 18.23). The **free bilirubin** is not very water-soluble, and thus most is carried in the blood attached to albumin proteins. This protein-bound bilirubin can neither be filtered by the kidneys into the urine nor directly excreted by the liver into the bile.

  The liver can take some of the free bilirubin out of the blood and conjugate (combine) it with glucuronic acid. This **conjugated bilirubin** is water-soluble and can be secreted into the bile. Once in the bile, the conjugated bilirubin can enter the intestine where it is converted by bacteria into another pigment—**urobilinogen.** Derivatives of urobilinogen impart a brown color to the feces. About 30% to 50% of the urobilinogen, however, is absorbed by the intestine and enters the hepatic portal vein. Of the urobilinogen that enters the liver sinusoids, some is secreted into the bile and is thus returned to the intestine in an enterohepatic circulation; the rest enters the general circulation (fig. 18.24). The urobilinogen in plasma, unlike free bilirubin, is not attached to albumin. Urobilinogen is therefore easily filtered by the kidneys into the urine, where its derivatives produce an amber color.

Clinical Investigation Clues

Remember that Alan had high blood levels of conjugated bilirubin and had yellowing of his sclera.

What does the yellowing of the sclera indicate, and what is its cause?
What could cause his elevated blood levels of conjugated bilirubin?

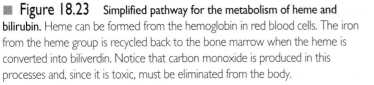

Figure 18.23 Simplified pathway for the metabolism of heme and bilirubin. Heme can be formed from the hemoglobin in red blood cells. The iron from the heme group is recycled back to the bone marrow when the heme is converted into biliverdin. Notice that carbon monoxide is produced in this processes and, since it is toxic, must be eliminated from the body.

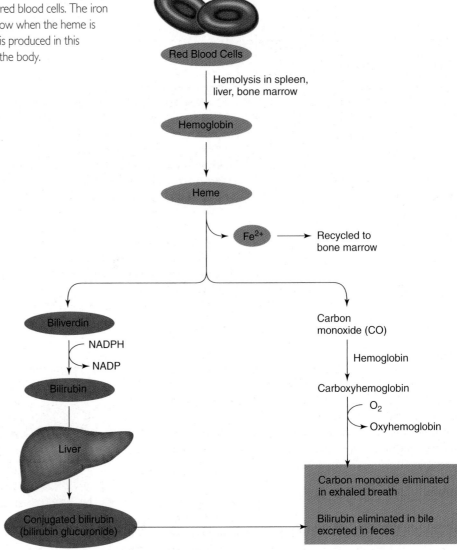

Figure 18.24 The enterohepatic circulation of urobilinogen. Bacteria in the intestine convert bilirubin (bile pigment) into urobilinogen. Some of this pigment leaves the body in the feces; some is absorbed by the intestine and is recycled through the liver. A portion of the urobilinogen that is absorbed enters the general circulation and is filtered by the kidneys into the urine.

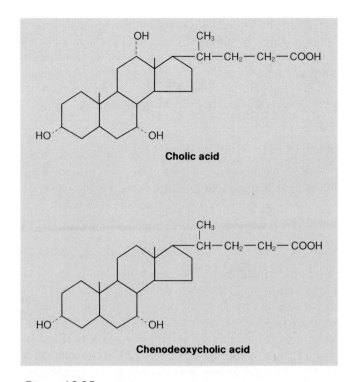

Cholic acid

Chenodeoxycholic acid

Figure 18.25 The two major bile acids in humans. These more polar derivatives of cholesterol form the bile salts.

Bile acids are derivatives of cholesterol that have two to four polar groups on each molecule. The principal bile acids in humans are *cholic acid* and *chenodeoxycholic acid* (fig. 18.25), conjugated with glycine or taurine to form the **bile salts.** In aqueous solutions these molecules "huddle" together to form aggregates known as **micelles.** As described in chapter 2, the nonpolar parts are located in the central region of the micelle (away from water), whereas the polar groups face water around the periphery of the micelle (see fig. 2.21). Lecithin, cholesterol, and other lipids in the small intestine enter these micelles, and the dual nature of the bile salts (part polar, part nonpolar) allows them to emulsify fat in the chyme.

The liver's production of bile acids from cholesterol is the major pathway of cholesterol breakdown in the body. This amounts to about half a gram of cholesterol converted into bile acids per day. No more than this is required, because approximately 95% of the bile acids released into the duodenum are absorbed in the ileum by means of specific carriers, and so have an enterohepatic circulation.

Detoxication of the Blood

The liver can remove hormones, drugs, and other biologically active molecules from the blood by (1) excretion of these compounds in the bile as previously described; (2) phagocytosis by the Kupffer cells that line the sinusoids; and (3) chemical alteration of these molecules within the hepatocytes.

Ammonia, for example, is a very toxic molecule produced by deamination of amino acids in the liver and by the action of bacteria in the intestine. Since the ammonia concentration of portal vein blood is four to fifty times greater than that of blood in the hepatic vein, it is clear that the ammonia is removed by the liver. The liver has the enzymes needed to convert ammonia into less toxic **urea** molecules, which are secreted by the liver into the blood and

excreted by the kidneys in the urine. Similarly, the liver converts toxic porphyrins into **bilirubin** and toxic purines into **uric acid.**

Steroid hormones and many drugs are inactivated in their passage through the liver by modifications of their chemical structures. The liver has enzymes that convert these nonpolar molecules into more polar (more water-soluble) forms by *hydroxylation* (the addition of OH^- groups) and by *conjugation* with highly polar groups such as sulfate and glucuronic acid. Polar derivatives of steroid hormones and drugs are less biologically active and, because of their increased water solubility, are more easily excreted by the kidneys into the urine.

Conjugation of steroid hormones and *xenobiotics* (foreign chemicals that are biologically active) makes them anionic (negatively charged) and hydrophilic (water-soluble). Thus changed, these compounds can be transported by liver cells into the bile canaliculi by **multispecific organic anion transport** carriers. These carriers have been cloned and identified as the same type that transports similar molecules into the nephron tubules. Through renal secretion (chapter 17) and secretion into the bile, therefore, these transport carriers help the body to eliminate potentially toxic molecules.

The liver cells contain enzymes for the metabolism of steroid hormones and other endogenous molecules, as well as for the detoxication of such exogenous toxic compounds as benzopyrene (a carcinogen from tobacco smoke and charbroiled meat), polychlorinated biphenyls (PCBs), and dioxin. The enzymes are members of a class called the **cytochrome P450 enzymes** (not related to the cytochromes of cell respiration) that comprises a few dozen enzymes with varying specificities. Together, these enzymes can metabolize thousands of toxic compounds. Since people vary in their hepatic content of the different cytochrome P450 enzymes, one person's sensitivity to a drug may be greater than another's because of a relative deficiency in the appropriate cytochrome P450 enzyme needed to metabolize that drug.

Production of the cytochrome P450 enzymes, needed for the hepatic metabolism of lipophilic compounds such as steroid hormones and drugs, is stimulated by the activation of a nuclear receptor. Nuclear receptors bind to particular molecular ligands and then activate specific genes (chapter 11; see fig. 11.7). The particular nuclear receptor that stimulates the production of cytochrome P450 enzymes is known as *SXR*—for *steroid and xenobiotic receptor.* A drug that activates SXR, and thereby induces the production of cytochrome P450 enzymes, would thus be expected to increase the hepatic metabolism of many other drugs. This is the mechanism responsible for many drug-drug interactions.

Secretion of Glucose, Triglycerides, and Ketone Bodies

As you may recall from chapter 5, the liver helps to regulate the blood glucose concentration by either removing glucose from the blood or adding glucose to it, according to the needs of the body. After a carbohydrate-rich meal, the liver can remove some glucose from the hepatic portal blood and convert it into glycogen and triglycerides through the processes of **glycogenesis** and

lipogenesis, respectively. During fasting, the liver secretes glucose into the blood. This glucose can be derived from the breakdown of stored glycogen in a process called **glycogenolysis,** or it can be produced by the conversion of noncarbohydrate molecules (such as amino acids) into glucose in a process called **gluconeogenesis.** The liver also contains the enzymes required to convert free fatty acids into ketone bodies **(ketogenesis),** which are secreted into the blood in large amounts during fasting. These processes are controlled by hormones and are explained further in chapter 19.

Production of Plasma Proteins

Plasma albumin and most of the plasma globulins (with the exception of immunoglobulins, or antibodies) are produced by the liver. Albumin constitutes about 70% of the total plasma protein and contributes most to the colloid osmotic pressure of the blood (chapter 14). The globulins produced by the liver have a wide variety of functions, including transport of cholesterol and triglycerides, transport of steroid and thyroid hormones, inhibition of trypsin activity, and blood clotting. Clotting factors I (fibrinogen), II (prothrombin), III, V, VII, IX, and XI, as well as angiotensinogen, are all produced by the liver.

Gallbladder

The **gallbladder** is a saclike organ attached to the inferior surface of the liver. This organ stores and concentrates bile, which drains to it from the liver by way of the bile ducts, hepatic ducts, and *cystic duct,* respectively. A sphincter valve at the neck of the gallbladder allows a 35- to 100-ml storage capacity. When the gallbladder fills with bile, it expands to the size and shape of a small pear. Bile is a yellowish green fluid containing bile salts, bilirubin, cholesterol, and other compounds, as previously discussed. Contraction of the muscularis layer of the gallbladder ejects bile through the cystic duct into the *common bile duct,* which conveys bile into the duodenum (fig. 18.26).

Approximately 20 million Americans have **gallstones**—small, hard mineral deposits (calculi) that can produce painful symptoms by obstructing the cystic or common bile ducts. Gallstones commonly contain cholesterol as their major component. Cholesterol normally has an extremely low water solubility (20 μg/L), but it can be present in bile at 2 million times its water solubility (40 g/L) because cholesterol molecules cluster together with bile salts and lecithin in the hydrophobic centers of micelles. In order for gallstones to be produced, the liver must secrete enough cholesterol to create a supersaturated solution, and some substance within the gallbladder must serve as a nucleus for the formation of cholesterol crystals. The gallstone is formed from cholesterol crystals that become hardened by the precipitation of inorganic salts (fig. 18.27). Gallstones may be removed surgically; cholesterol gallstones, however, may be dissolved by oral ingestion of bile acids. This may be combined with a newer treatment that involves fragmentation of the gallstones by high-energy shock waves delivered to a patient immersed in a water bath. This procedure is called *extracorporeal shock-wave lithotripsy.*

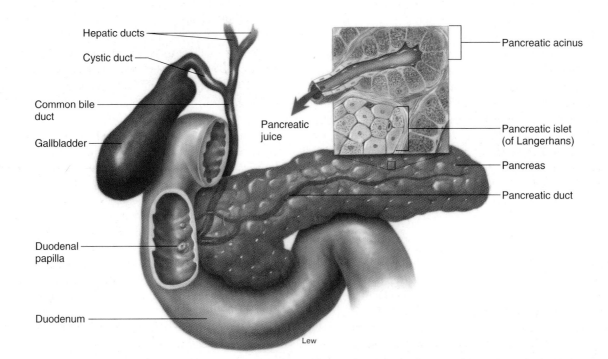

Figure 18.26 Pancreatic juice and bile are secreted into the duodenum. The pancreatic duct joins the common bile duct to empty its secretions through the duodenal papilla into the duodenum. The release of bile and pancreatic juice into the duodenum is controlled by the sphincter of ampulla (sphincter of Oddi).

Bile is continuously produced by the liver and drains through the hepatic and common bile ducts to the duodenum. When the small intestine is empty of food, the *sphincter of ampulla (sphincter of Oddi)* at the end of the common bile duct closes, and bile is forced up to the cystic duct and then to the gallbladder for storage.

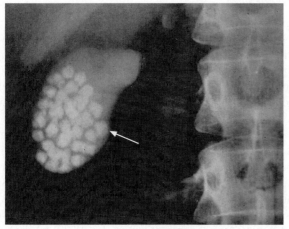

(a)

(b)

■ **Figure 18.27** Gallstones. (*a*) A radiograph of a gallbladder that contains gallstones (biliary calculi). (*b*) A posterior view of a gallbladder that has been surgically removed (cholecystectomy) and cut open to reveal its gallstones. (Note their size relative to that of a dime.)

Pancreas

The **pancreas** is a soft, glandular organ that has both exocrine and endocrine functions (fig. 18.28). The endocrine function is performed by clusters of cells called the **pancreatic islets,** or **islets of Langerhans** (fig. 18.28*a*), that secrete the hormones insulin and glucagon into the blood (see chapter 19). As an exocrine gland, the pancreas secretes pancreatic juice through the pancreatic duct into the duodenum. Within the lobules of the pancreas are the exocrine secretory units, called **acini** (fig. 18.28*b*). Each acinus consists of a single layer of epithelial cells surrounding a lumen, into which the constituents of pancreatic juice are secreted.

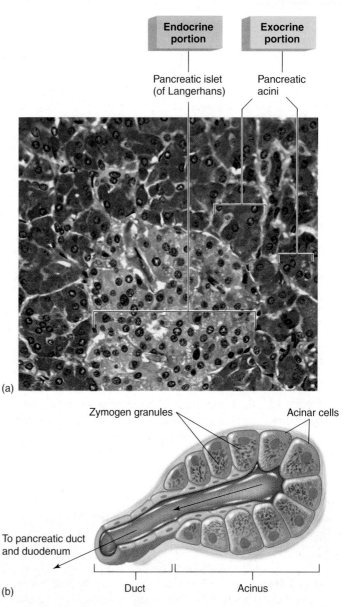

■ **Figure 18.28** The pancreas is both an exocrine and an endocrine gland. (*a*) A photomicrograph of the endocrine and exocrine portions of the pancreas. (*b*) An illustration depicting the exocrine pancreatic acini, where the acinar cells produce inactive enzymes stored in zymogen granules. The inactive enzymes are secreted by way of a duct system into the duodenum.

Table 18.4 Enzymes Contained in Pancreatic Juice

Enzyme	Zymogen	Activator	Action
Trypsin	Trypsinogen	Enterokinase	Cleaves internal peptide bonds
Chymotrypsin	Chymotrypsinogen	Trypsin	Cleaves internal peptide bonds
Elastase	Proelastase	Trypsin	Cleaves internal peptide bonds
Carboxypeptidase	Procarboxypeptidase	Trypsin	Cleaves last amino acid from carboxyl-terminal end of polypeptide
Phospholipase	Prophospholipase	Trypsin	Cleaves fatty acids from phospholipids such as lecithin
Lipase	None	None	Cleaves fatty acids from glycerol
Amylase	None	None	Digests starch to maltose and short chains of glucose molecules
Cholesterolesterase	None	None	Releases cholesterol from its bonds with other molecules
Ribonuclease	None	None	Cleaves RNA to form short chains
Deoxyribonuclease	None	None	Cleaves DNA to form short chains

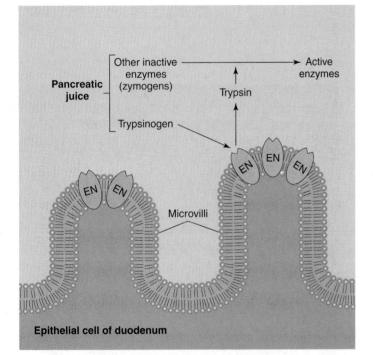

Figure 18.29 The activation of pancreatic juice enzymes. The pancreatic protein-digesting enzyme trypsin is secreted in an inactive form known as trypsinogen. This inactive enzyme (zymogen) is activated by a brush border enzyme, enterokinase (EN), located in the cell membrane of microvilli. Active trypsin in turn activates other zymogens in pancreatic juice.

Pancreatitis (inflammation of the pancreas) may result when conditions such as alcoholism, gallstones, traumatic injury, infections, or toxicosis from various drugs provoke activation of digestive enzymes within the pancreas. Leakage of trypsin into the blood also occurs, but trypsin is inactive in the blood because of the inhibitory action of two plasma proteins, α_1-antitrypsin and α_2-macroglobulin. Pancreatic amylase may also leak into the blood, but it is not active because its substrate (starch) is not present in blood. Pancreatic amylase activity can be measured *in vitro*, however, and these measurements are commonly performed to assess the health of the pancreas.

Pancreatic Juice

Pancreatic juice contains water, bicarbonate, and a wide variety of digestive enzymes. These enzymes include (1) **amylase,** which digests starch; (2) **trypsin,** which digests protein; and (3) **lipase,** which digests triglycerides. Other pancreatic enzymes are listed in table 18.4. It should be noted that the complete digestion of food molecules in the small intestine requires the action of both pancreatic enzymes and brush border enzymes.

Most pancreatic enzymes are produced as inactive molecules, or *zymogens,* so that the risk of self-digestion within the pancreas is minimized. The inactive form of trypsin, called trypsinogen, is activated within the small intestine by the catalytic action of the brush border enzyme *enterokinase.* Enterokinase converts trypsinogen to active trypsin. Trypsin, in turn, activates the other zymogens of pancreatic juice (fig. 18.29) by cleaving off polypeptide sequences that inhibit the activity of these enzymes.

The activation of trypsin, therefore, is the triggering event for the activation of other pancreatic enzymes. Actually, the pancreas does produce small amounts of active trypsin, but the other enzymes are not activated until pancreatic juice has entered the duodenum. This is because pancreatic juice also contains a small protein called *pancreatic trypsin inhibitor* that attaches to trypsin and inhibits its activity in the pancreas.

Test Yourself Before You Continue

1. Describe the structure of liver lobules and trace the pathways for the flow of blood and bile in the lobules.

2. Describe the composition and function of bile and trace the flow of bile from the liver and gallbladder to the duodenum.

3. Explain how the liver inactivates and excretes compounds such as hormones and drugs.

4. Describe the enterohepatic circulation of bilirubin and urobilinogen.

5. Explain how the liver helps to maintain a constant blood glucose concentration and how the pattern of venous blood flow permits this function.

6. Describe the endocrine and exocrine structures and functions of the pancreas. How is the pancreas protected against self-digestion?

Table 18.5 Effects of Gastrointestinal Hormones

Secreted by	Hormone	Effects
Stomach	Gastrin	Stimulates parietal cells to secrete HCl
		Stimulates chief cells to secrete pepsinogen
		Maintains structure of gastric mucosa
Small intestine	Secretin	Stimulates water and bicarbonate secretion in pancreatic juice
		Potentiates actions of cholecystokinin on pancreas
Small intestine	Cholecystokinin (CCK)	Stimulates contraction of gallbladder
		Stimulates secretion of pancreatic juice enzymes
		Inhibits gastric motility and secretion
		Maintains structure of exocrine pancreas (acini)
Small intestine	Gastric inhibitory peptide (GIP)	Inhibits gastric motility and secretion
		Stimulates secretion of insulin from pancreatic islets
Ileum and colon	Glucagon-like peptide-1 (GLP-1)	Inhibits gastric motility and secretion
		Stimulates secretion of insulin from pancreatic islets
	Guanylin	Stimulates intestinal secretion of Cl^-, causing elimination of NaCl and water in the feces

Neural and Endocrine Regulation of the Digestive System

The activities of different regions of the GI tract are coordinated by the actions of the vagus nerve and by various hormones. The stomach begins to increase its secretion in anticipation of a meal, and further increases its activities in response to the arrival of food. The entry of chyme into the duodenum stimulates the secretion of hormones that promote contractions of the gallbladder, the secretion of pancreatic juice, and the inhibition of gastric activity.

Neural and endocrine control mechanisms modify the activity of the digestive system. The sight, smell, or taste of food, for example, can stimulate salivary and gastric secretions via activation of the vagus nerve, which helps to "prime" the digestive system in preparation for a meal. Stimulation of the vagus, in this case, originates in the brain and is a conditioned reflex (as Pavlov demonstrated by training dogs to salivate in response to a bell). The vagus nerve is also involved in the reflex control of one part of the digestive system by another—these are "short reflexes," which do not involve the brain.

The GI tract is both an endocrine gland and a target for the action of various hormones. Indeed, the first hormones to be discovered were gastrointestinal hormones. In 1902 two English physiologists, Sir William Bayliss and Ernest Starling, discovered that the duodenum produced a chemical regulator. They named this substance **secretin** and proposed, in 1905, that it was but one of many yet undiscovered chemical regulators produced by the body. Bayliss and Starling coined the term *hormones* for this new class of regulators. In that same year, other investigators discovered that an extract from the stomach antrum stimulated gastric secretion. The hormone **gastrin** was thus the second hormone to be discovered.

The chemical structures of gastrin, secretin, and the duodenal hormone **cholecystokinin (CCK)** were determined in the 1960s. More recently, a fourth hormone produced by the small intestine, **gastric inhibitory peptide (GIP),** has been added to the list of proven GI tract hormones. The effects of these and other gastrointestinal hormones are summarized in table 18.5.

Regulation of Gastric Function

Gastric motility and secretion are, to some extent, automatic. Waves of contraction that serve to push chyme through the pyloric sphincter, for example, are initiated spontaneously by pacesetter cells in the greater curvature of the stomach. Likewise, the secretion of HCl from parietal cells and pepsinogen from chief cells can be stimulated in the absence of neural and hormonal influences by the presence of cooked or partially digested protein in the stomach. This action involves other cells in the gastric mucosa, including the G cells (that secrete the hormone gastrin); the enterochromaffin-like (ECL) cells, which secrete histamine; and the D cells, which secrete somatostatin.

The effects of autonomic nerves and hormones are superimposed on this automatic activity. This extrinsic control of gastric function is conveniently divided into three phases: (1) the *cephalic phase;* (2) the *gastric phase;* and (3) the *intestinal phase.* These are summarized in table 18.6.

Cephalic Phase

The **cephalic phase** of gastric regulation refers to control by the brain via the vagus nerve. As previously discussed, various conditioned stimuli can evoke gastric secretion. This conditioning in humans is, of course, more subtle than that exhibited by Pavlov's dogs in response to a bell. In fact, just talking about appetizing food is sometimes a more potent stimulus for gastric acid secretion than the actual sight and smell of food!

Activation of the vagus nerve (1) stimulates the chief cells to secrete pepsinogen; and (2) indirectly stimulates the parietal cells to secrete HCl. The vagus nerve endings directly stimulate G cells to secrete gastrin and the ECL cells to secrete histamine.

Table 18.6   The Three Phases of Gastric Secretion

Phase of Regulation	Description

Cephalic Phase	1. Sight, smell, and taste of food cause stimulation of vagus nuclei in brain
	2. Vagus stimulates acid secretion
	   a. Indirect stimulation of parietal cells (major effect)
	   b. Stimulation of gastrin secretion (lesser effect)

Gastric Phase	1. Distension of stomach stimulates vagus nerve; vagus stimulates acid secretion
	2. Amino acids and peptides in stomach lumen stimulate acid secretion
	   a. Direct stimulation of parietal cells (lesser effect)
	   b. Stimulation of gastrin secretion; gastrin stimulates acid secretion (major effect)
	3. Gastrin secretion inhibited when pH of gastric juice falls below 2.5

Intestinal Phase	1. Neural inhibition of gastric emptying and acid secretion
	   a. Arrival of chyme in duodenum causes distension, increase in osmotic pressure
	   b. These stimuli activate a neural reflex that inhibits gastric activity
	2. In response to fat in chyme, duodenum secretes a hormone that inhibits gastric acid secretion

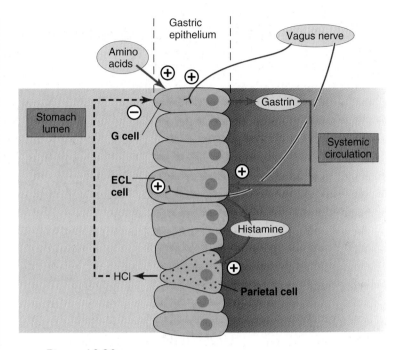

■ **Figure 18.30**   The regulation of gastric acid secretion. The presence of amino acids in the stomach lumen from partially digested proteins stimulates gastrin secretion. Gastrin secretion from G cells is also stimulated by vagus nerve activity. The secreted gastrin then acts as a hormone to stimulate histamine release from the ECL cells. The histamine, in turn, acts as a paracrine regulator to stimulate the parietal cells to secrete HCl. (⊕ = stimulation; ⊖ = inhibition.)

The gastrin secreted by G cells enters the systemic circulation and is carried back to the stomach, where it also stimulates the ECL cells to release histamine. Histamine, in turn, activates H_2 histamine receptors on the parietal cells to stimulate acid secretion (fig. 18.30).

This cephalic phase continues into the first 30 minutes of a meal, but then gradually declines in importance as the next phase becomes predominant.

Gastric Phase

The arrival of food into the stomach stimulates the **gastric phase** of regulation. Gastric secretion is stimulated in response to two factors: (1) distension of the stomach, which is determined by the amount of chyme, and (2) the chemical nature of the chyme.

While intact proteins in the chyme have little stimulatory effect, the partial digestion of proteins into shorter polypeptides and amino acids, particularly phenylalanine and tryptophan, stimulates the chief cells to secrete pepsinogen and the G cells to secrete gastrin. Gastrin, in turn, stimulates the secretion of pepsinogen from chief cells, but its effect on the parietal cells is primarily indirect. Gastrin stimulates the secretion of histamine from ECL cells, and the histamine then stimulates secretion of HCl from parietal cells, as previously described (fig. 18.30). A *positive feedback mechanism* thus develops. As more HCl and pepsinogen are secreted, more short polypeptides and amino acids are released from the ingested protein. This stimulates additional secretion of gastrin and, therefore, additional secretion of HCl and pepsinogen. It should be noted that glucose in the chyme has no effect on gastric secretion, and the presence of fat actually inhibits acid secretion.

Secretion of HCl during the gastric phase is also regulated by a *negative feedback mechanism*. As the pH of gastric juice drops, so does the secretion of gastrin—at a pH of 2.5, gastrin secretion is reduced, and at a pH of 1.0 gastrin secretion ceases. The secretion of HCl thus declines accordingly. This effect may be mediated by the hormone somatostatin, secreted by the D cells of the gastric mucosa. As the pH of gastric juice falls, the D cells are stimulated to secrete somatostatin, which then acts as a paracrine regulator to inhibit the secretion of gastrin from the G cells.

The presence of proteins and polypeptides in the stomach helps to buffer the acid and thus to prevent a rapid fall in gastric pH. More acid can thus be secreted when proteins are present than when they are absent. The arrival of protein into the stomach thus stimulates acid secretion in two ways—by the positive feedback mechanism previously discussed and by inhibition of the negative feedback control of acid secretion. Through these mechanisms, the amount of acid secreted is closely matched to the amount of protein ingested. As the stomach is emptied the protein buffers exit, the pH thus falls, and the secretion of gastrin and HCl is accordingly inhibited.

Intestinal Phase

The **intestinal phase** of gastric regulation refers to the inhibition of gastric activity when chyme enters the small intestine. Investigators in 1886 demonstrated that the addition of olive oil to a meal inhibits gastric emptying, and in 1929 it was shown that the presence of fat inhibits gastric juice secretion. This inhibitory intestinal phase of gastric regulation is due to both a neural reflex originating from the duodenum and to a chemical hormone secreted by the duodenum.

The arrival of chyme into the duodenum increases its osmolality. This stimulus, together with stretch of the duodenum and possibly other stimuli, activates sensory neurons of the vagus nerve and produces a neural reflex that results in the inhibition of gastric motility and secretion. The presence of fat in the chyme also stimulates the duodenum to secrete a hormone that inhibits gastric function. The general term for such an inhibitory hormone is an **enterogastrone.**

In the past, **gastric inhibitory peptide (GIP)** was thought to function as an enterogastrone—hence the name for this hormone. Many researchers, however, now believe that other intestinal hormones may serve this function. Other polypeptide hormones secreted by the small intestine that can inhibit gastric activity include **somatostatin** (produced by the intestine, as well as by the brain and stomach); **cholecystokinin (CCK),** secreted by the duodenum in response to the presence of chyme; and **glucagon-like peptide-1 (GLP-1),** secreted by the ileum and colon. GLP-1 is one of a family of peptides produced by the intestine that structurally resemble the hormone glucagon (secreted by the alpha cells of the pancreatic islets).

It could be that the only physiological role of GIP is stimulation of insulin secretion from the islets of Langerhans in response to the presence of glucose in the small intestine. Some scientists therefore propose that the name GIP be retained, but that it serve as an acronym for *glucose-dependent insulinotropic peptide.* It should be noted, however, that GLP-1 is also a very potent stimulator of insulin secretion. These intestinal hormones therefore stimulate the pancreas to "anticipate" a rise in blood glucose by secreting insulin (which acts to lower the blood glucose concentration) even before the glucose has been absorbed into the blood.

The inhibitory neural and endocrine mechanisms during the intestinal phase prevent the further passage of chyme from the stomach to the duodenum. This gives the duodenum time to process the load of chyme received previously. Since secretion of the enterogastrone is stimulated by fat in the chyme, a breakfast of bacon and eggs takes longer to pass through the stomach—and makes one feel "fuller" for a longer time—than does a breakfast of pancakes and syrup.

Regulation of Intestinal Function

Enteric Nervous System

The submucosal (Meissner's) and myenteric (Auerbach's) plexuses within the wall of the intestine contain 100 million neurons, about as many as are in the spinal cord! These include preganglionic parasympathetic axons, the ganglion cell bodies of postganglionic parasympathetic neurons, postganglionic sympathetic axons, and afferent (sensory) neurons. These plexuses also contain interneurons, as does the CNS. Also like the CNS, the **enteric nervous system** (or *enteric brain*) contains more glial cells than neurons, and these glial cells resemble the astrocytes of the brain.

Some sensory (afferent) neurons within the intestinal plexuses travel in the vagus nerve to deliver sensory information to the CNS. These are called *extrinsic afferents,* and they are involved in regulation by the autonomic nervous system. Other sensory neurons—called *intrinsic afferents*—have their cell bodies in the myenteric or submucosal plexuses and synapse with the interneurons of the enteric nervous system.

Peristalsis, for example, is regulated by the enteric nervous system. A bolus of chyme stimulates intrinsic afferents (with cell bodies in the myenteric plexus) that activate enteric interneurons, which in turn stimulate motor neurons. These motor neurons innervate both smooth muscle cells and interstitial cells of Cajal, where they release excitatory and inhibitory neurotransmitters. Smooth muscle contraction is stimulated by the neurotransmitters ACh and substance P above the bolus, and smooth muscle relaxation is stimulated by nitric oxide, vasoactive intestinal peptide (VIP), and ATP below the bolus (fig. 18.31).

Paracrine Regulators of the Intestine

There is evidence that the enterochromaffin-like cells (ECL cells) of the intestinal mucosa secrete **serotonin,** or **5-hydroxytryptamine,** in response to the stimuli of pressure and various chemicals. Serotonin then stimulates intrinsic afferents, which conduct impulses into the submucosal and myenteric plexuses and there activate motor neurons. Motor neurons that terminate in the muscularis can stimulate contractions; those that terminate in the intestinal crypts can stimulate the secretion of salt and water into the lumen. The ECL cells have also been shown to produce another paracrine regulator, termed **motilin,** which stimulates contraction in the duodenum and stomach antrum.

Guanylin is a recently discovered paracrine regulator produced by the ileum and colon. It derives its name from its ability to activate the enzyme guanylate cyclase, and thus to cause the production of cyclic GMP (cGMP) within the cytoplasm of intestinal epithelial cells. Acting through cGMP as a second messenger, guanylin stimulates the intestinal epithelial cells to secrete Cl^- and water and inhibits their absorption of Na^+. These actions increase the amount of salt and water lost from the body in the feces. A related polypeptide, called **uroguanylin,** has been found in the urine. This polypeptide appears to be produced by the intestine, and may therefore function as a hormone that stimulates the kidneys to excrete salt in the urine.

Certain *Escherichia coli* bacteria produce *heat-stable enterotoxins* that are responsible for **traveler's diarrhea.** The enterotoxins act by stimulating the same receptors on the apical membranes of the intestinal epithelial cells that are activated by guanylin. By mimicking the actions of guanylin, the enterotoxins stimulate intestinal Cl^- and water secretion to produce diarrhea.

Intestinal Reflexes

There are several intestinal reflexes that are controlled locally, by means of the enteric nervous system and paracrine regulators,

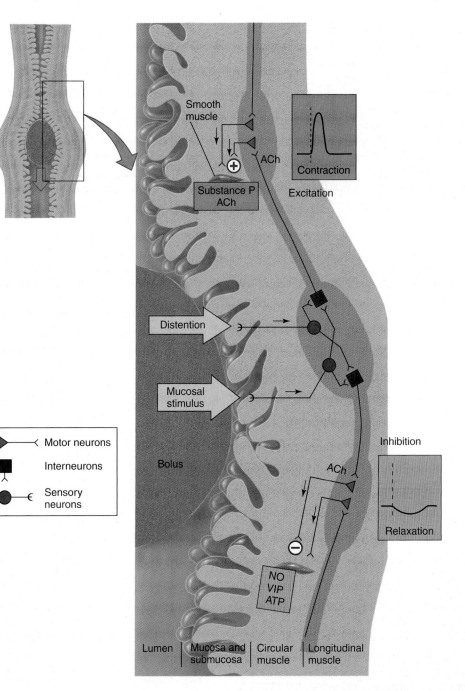

■ **Figure 18.31**   **The enteric nervous system.** Peristalsis is produced by local reflexes involving the enteric nervous system. Notice that the enteric nervous system consists of motor neurons, interneurons, and sensory neurons. The neurotransmitters that stimulate smooth muscle contraction are indicated with a ⊕, while those that produce smooth muscle relaxation are indicated with a ⊖. (NO = nitric oxide; VIP = vasoactive intestinal peptide.)

and extrinsically through the actions of the nerves and hormones previously discussed. These reflexes include:

1. the **gastroileal reflex,** in which increased gastric activity causes increased motility of the ileum and increased movements of chyme through the ileocecal sphincter;
2. the **ileogastric reflex,** in which distension of the ileum causes a decrease in gastric motility;
3. the **intestino-intestinal reflexes,** in which overdistension of one intestinal segment causes relaxation throughout the rest of the intestine.

Regulation of Pancreatic Juice and Bile Secretion

The arrival of chyme into the duodenum stimulates the intestinal phase of gastric regulation and, at the same time, stimulates reflex secretion of pancreatic juice and bile. The entry of new chyme is thus retarded as the previous load is digested. The secretion of pancreatic juice and bile is stimulated both by neural reflexes initiated in the duodenum and by secretion of the duodenal hormones cholecystokinin and secretin.

Secretion of Pancreatic Juice

The secretion of pancreatic juice is stimulated by both secretin and CCK. These two hormones, however, are secreted in response to different stimuli, and they have different effects on the composition of pancreatic juice. The release of secretin occurs in response to a fall in duodenal pH to below 4.5; this pH fall occurs for only a short time, however, because the acidic chyme is rapidly neutralized by alkaline pancreatic juice. The secretion of CCK, by contrast, occurs in response to the protein and fat content of chyme in the duodenum.

Secretin stimulates the production of bicarbonate by the pancreas. Since bicarbonate neutralizes the acidic chyme and since secretin is released in response to the low pH of chyme, a negative feedback loop is completed. CCK, by contrast, stimulates the production of pancreatic enzymes such as trypsin, lipase, and amylase. Partially digested protein and fat are the most potent stimulators of CCK secretion, and CCK secretion continues until the chyme has passed through the duodenum and early region of the jejunum.

Secretin and CCK can have different effects on the same cells (the pancreatic acinar cells) because their actions are mediated by different second messengers. The second messenger of secretin action is cyclic AMP, whereas the second messenger for CCK is Ca^{2+}.

Secretion of Bile

The liver secretes bile continuously, but this secretion is greatly augmented following a meal. The increased secretion is due to the release of secretin and CCK from the duodenum. Secretin stimulates the liver to secrete bicarbonate into the bile, and CCK enhances this effect. The arrival of chyme in the duodenum also causes the gallbladder to contract and eject bile. Contraction of the gallbladder occurs in response to neural reflexes from the duodenum and to hormonal stimulation by CCK.

Clinical Investigation Clues

Remember that Alan's pain below his right scapula was triggered by eating peanut butter and bacon, but not by eating fish or skinned chicken.

What component of the food triggered Alan's pain?

What physiological mechanism is involved in this process?

Trophic Effects of Gastrointestinal Hormones

Patients with tumors of the stomach pylorus exhibit excessive acid secretion and hyperplasia (growth) of the gastric mucosa. Surgical removal of the pylorus reduces gastric secretion and prevents growth of the gastric mucosa. Patients with peptic ulcers are sometimes treated by vagotomy—cutting of the portion of the vagus nerve that innervates the stomach. Vagotomy also reduces acid secretion but has no effect on the gastric mucosa.

These observations suggest that the hormone gastrin, secreted by the pyloric mucosa, may exert stimulatory, or *trophic*, effects on the gastric mucosa. The structure of the gastric mucosa, in other words, is dependent upon the effects of gastrin.

In the same way, the structure of the acinar (exocrine) cells of the pancreas is dependent upon the trophic effects of CCK. Perhaps this explains why the pancreas, as well as the GI tract, atrophies during starvation. Since neural reflexes appear to be capable of regulating digestion, perhaps the primary function of the GI hormones is trophic—that is, maintenance of the structure of their target organs.

Test Yourself Before You Continue

1. Describe the positive and negative feedback mechanisms that operate during the gastric phase of HCl and pepsinogen secretion.
2. Describe the mechanisms involved in the intestinal phase of gastric regulation and explain why a fatty meal takes longer to leave the stomach than a meal low in fat.
3. Explain the hormonal mechanisms involved in the production and release of pancreatic juice and bile.
4. Describe the enteric nervous system and identify some of the short reflexes that regulate intestinal function.

Digestion and Absorption of Carbohydrates, Lipids, and Proteins

Polysaccharides and polypeptides are hydrolyzed into their subunits. These subunits enter the epithelial cells of the intestinal villi and are secreted into blood capillaries. Fat is emulsified by the action of bile salts, hydrolyzed into fatty acids and monoglycerides, and absorbed into the intestinal epithelial cells. Once inside the cells, triglycerides are resynthesized and combined with proteins to form particles that are secreted into the lymphatic fluid.

The caloric (energy) value of food is derived mainly from its content of carbohydrates, lipids, and proteins. In the average American diet, carbohydrates account for approximately 50% of the total calories, protein accounts for 11% to 14%, and lipids make up the balance. These food molecules consist primarily of long combinations of subunits (monomers) that must be digested by hydrolysis reactions into free monomers before absorption can occur. The characteristics of the major digestive enzymes are summarized in table 18.7.

Table 18.7   Characteristics of the Major Digestive Enzymes

Enzyme	Site of Action	Source	Substrate	Optimum pH	Product(s)
Salivary amylase	Mouth	Saliva	Starch	6.7	Maltose
Pepsin	Stomach	Gastric glands	Protein	1.6–2.4	Shorter polypeptides
Pancreatic amylase	Duodenum	Pancreatic juice	Starch	6.7–7.0	Maltose, maltriose, and oligosaccharides
Trypsin, chymotrypsin, carboxypeptidase	Small intestine	Pancreatic juice	Polypeptides	8.0	Amino acids, dipeptides, and tripeptides
Pancreatic lipase	Small intestine	Pancreatic juice	Triglycerides	8.0	Fatty acids and monoglycerides
Maltase	Small intestine	Brush border of epithelial cells	Maltose	5.0–7.0	Glucose
Sucrase	Small intestine	Brush border of epithelial cells	Sucrose	5.0–7.0	Glucose + fructose
Lactase	Small intestine	Brush border of epithelial cells	Lactose	5.8–6.2	Glucose + galactose
Aminopeptidase	Small intestine	Brush border of epithelial cells	Polypeptides	8.0	Amino acids, dipeptides, tripeptides

Digestion and Absorption of Carbohydrates

Most carbohydrates are ingested as starch, which is a long polysaccharide of glucose in the form of straight chains with occasional branchings. The most commonly ingested sugars are sucrose (table sugar, a disaccharide consisting of glucose and fructose) and lactose (milk sugar, a disaccharide consisting of glucose and galactose). The digestion of starch begins in the mouth with the action of **salivary amylase.** This enzyme cleaves some of the bonds between adjacent glucose molecules, but most people don't chew their food long enough for sufficient digestion to occur in the mouth. The digestive action of salivary amylase stops some time after the swallowed bolus enters the stomach because this enzyme is inactivated at the low pH of gastric juice.

The digestion of starch, therefore, occurs mainly in the duodenum as a result of the action of **pancreatic amylase.** This enzyme cleaves the straight chains of starch to produce the disaccharide *maltose* and the trisaccharide *maltriose*. Pancreatic amylase, however, cannot hydrolyze the bond between glucose molecules at the branch points in the starch. As a result, short, branched chains of glucose molecules called *oligosaccharides* are released together with maltose and maltriose by the activity of this enzyme (fig. 18.32).

Maltose, maltriose, and oligosaccharides are hydrolyzed to their monosaccharides by brush border enzymes located on the microvilli of the epithelial cells in the small intestine. The brush border enzymes also hydrolyze the disaccharides sucrose and lactose into their component monosaccharides. These monosaccharides are then moved across the epithelial cell membrane by secondary active transport, in which the glucose shares a common membrane carrier with Na^+ (chapter 6; see fig. 6.18). Finally, glucose is secreted from the epithelial cells into blood capillaries within the intestinal villi.

Digestion and Absorption of Proteins

Protein digestion begins in the stomach with the action of pepsin. Some amino acids are liberated in the stomach, but the major products of pepsin digestion are short-chain polypeptides.

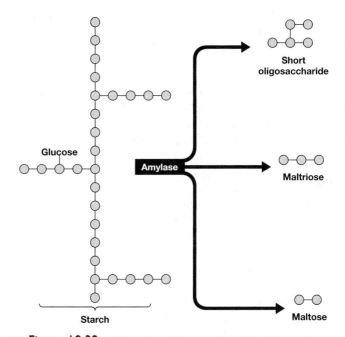

■ **Figure 18.32**   The action of pancreatic amylase. Pancreatic amylase digests starch into maltose, maltriose, and short oligosaccharides containing branch points in the chain of glucose molecules.

Pepsin digestion helps to produce a more homogenous chyme, but it is not essential for the complete digestion of protein that occurs—even in people with total gastrectomies—in the small intestine.

Most protein digestion occurs in the duodenum and jejunum. The pancreatic juice enzymes **trypsin, chymotrypsin, and elastase** cleave peptide bonds in the interior of the polypeptide chains. These enzymes are thus grouped together as *endopeptidases*. Enzymes that remove amino acids from the ends of polypeptide chains, by contrast, are *exopeptidases*. These include the pancreatic juice enzyme **carboxypeptidase,** which removes amino acids from the carboxyl-terminal end of polypeptide chains, and the brush border enzyme **aminopeptidase.** Aminopeptidase cleaves amino acids from the amino-terminal end of polypeptide chains.

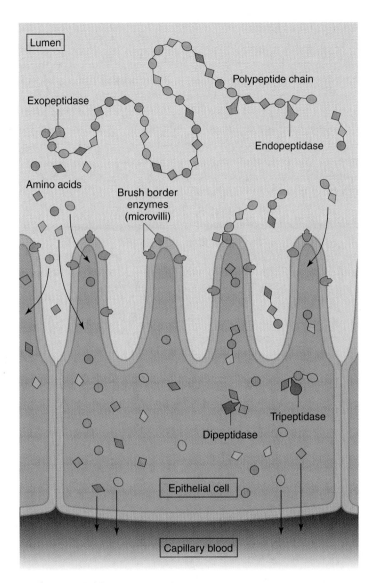

Figure 18.33 The digestion and absorption of proteins. Polypeptide chains of proteins are digested into free amino acids, peptides, and tripeptides by the action of pancreatic juice enzymes and brush border enzymes. The amino acids, dipeptides, and tripeptides enter duodenal epithelial cells. Dipeptides and tripeptides are hydrolyzed into free amino acids within the epithelial cells, and these products are secreted into capillaries that carry them to the hepatic portal vein.

As a result of the action of these enzymes, polypeptide chains are digested into free amino acids, dipeptides, and tripeptides. The free amino acids are absorbed by cotransport with Na^+ into the epithelial cells and secreted into blood capillaries. The dipeptides and tripeptides enter epithelial cells by the action of a single membrane carrier that has recently been characterized. This carrier functions in secondary active transport using a H^+ gradient to transport dipeptides and tripeptides into the cell cytoplasm. Within the cytoplasm, the dipeptides and tripeptides are hydrolyzed into free amino acids, which are then secreted into the blood (fig. 18.33).

Newborn babies appear to be capable of absorbing a substantial amount of undigested proteins (hence they can absorb some antibodies from their mother's first milk); in adults, however, only the free amino acids enter the portal vein. Foreign food protein, which would be very antigenic, does not normally enter the blood. An interesting exception is the protein toxin that causes botulism, produced by the bacterium *Clostridium botulinum*. This protein is resistant to digestion and is thus intact when it is absorbed into the blood.

Digestion and Absorption of Lipids

The salivary glands and stomach of neonates (newborns) produce lipases. In adults, however, very little lipid digestion occurs until the lipid globules in chyme arrive in the duodenum. Through mechanisms described in the next section, the arrival of lipids (primarily triglyceride, or fat) in the duodenum serves as a stimulus for the secretion of bile. In a process called **emulsification,** bile salt micelles are secreted into the duodenum and act to break up the fat droplets into tiny *emulsification droplets* of triglycerides. Note that emulsification is not chemical digestion—the bonds joining glycerol and fatty acids are not hydrolyzed by this process.

Clinical Investigation Clues

Remember that Alan had fatty stools and a prolonged clotting time.

■ Given what you have previously determined about Alan's condition, what could have caused his fatty stools?

■ How is this related to his prolonged clotting time? (Hint—vitamin K, needed for the formation of some clotting factors, is a fat-soluble vitamin.)

Digestion of Lipids

The emulsification of fat aids digestion because the smaller and more numerous emulsification droplets present a greater surface area than the unemulsified fat droplets that originally entered the duodenum. Fat digestion occurs at the surface of the droplets through the enzymatic action of **pancreatic lipase,** which is aided in its action by a protein called *colipase* (also secreted by the pancreas) that coats the emulsification droplets and "anchors" the lipase enzyme to them. Through hydrolysis, lipase removes two of the three fatty acids from each triglyceride molecule and thus liberates *free fatty acids* and *monoglycerides* (fig. 18.34). **Phospholipase A** likewise digests phospholipids such as lecithin into fatty acids and *lysolecithin* (the remainder of the lecithin molecule after the removal of two fatty acids).

Free fatty acids, monoglycerides, and lysolecithin, which are more polar than the undigested lipids, quickly become associated with micelles of bile salts, lecithin, and cholesterol to form "mixed micelles" (fig. 18.35). These micelles then move to the brush border of the intestinal epithelium where absorption occurs.

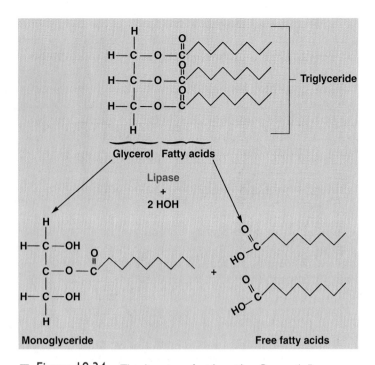

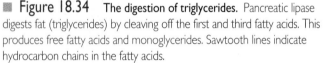

■ **Figure 18.34** **The digestion of triglycerides.** Pancreatic lipase digests fat (triglycerides) by cleaving off the first and third fatty acids. This produces free fatty acids and monoglycerides. Sawtooth lines indicate hydrocarbon chains in the fatty acids.

Absorption of Lipids

Free fatty acids, monoglycerides, and lysolecithin can leave the micelles and pass through the membrane of the microvilli to enter the intestinal epithelial cells. There is also some evidence that the micelles may be transported intact into the epithelial cells and that the lipid digestion products may be removed intracellularly from the micelles. In either event, these products are used to *resynthesize* triglycerides and phospholipids within the epithelial cells. This process is different from the absorption of amino acids and monosaccharides, which pass through the epithelial cells without being altered.

Triglycerides, phospholipids, and cholesterol are then combined with protein inside the epithelial cells to form small particles called **chylomicrons.** These tiny lipid and protein combinations are secreted into the central lacteals (lymphatic capillaries) of the intestinal villi (fig. 18.36). Absorbed lipids thus pass through the lymphatic system, eventually entering the venous blood by way of the thoracic duct (chapter 13). By contrast, amino acids and monosaccharides enter the hepatic portal vein.

Transport of Lipids in the Blood

Once the chylomicrons are in the blood, their triglyceride content is removed by the enzyme *lipoprotein lipase,* which is attached to the endothelium of blood vessels. This enzyme hydrolyzes triglycerides and thus provides free fatty acids and

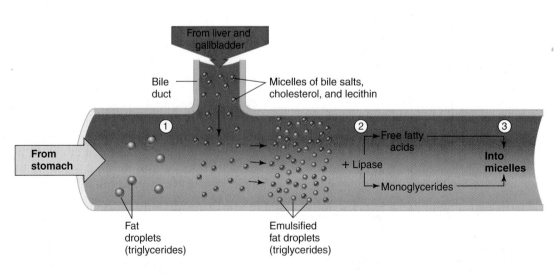

Step 1: Emulsification of fat droplets by bile salts

Step 2: Hydrolysis of triglycerides in emulsified fat droplets into fatty acid and monoglycerides

Step 3: Dissolving of fatty acids and monoglycerides into micelles to produce "mixed micelles"

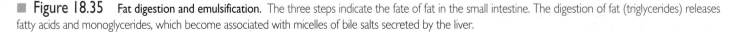

■ **Figure 18.35** **Fat digestion and emulsification.** The three steps indicate the fate of fat in the small intestine. The digestion of fat (triglycerides) releases fatty acids and monoglycerides, which become associated with micelles of bile salts secreted by the liver.

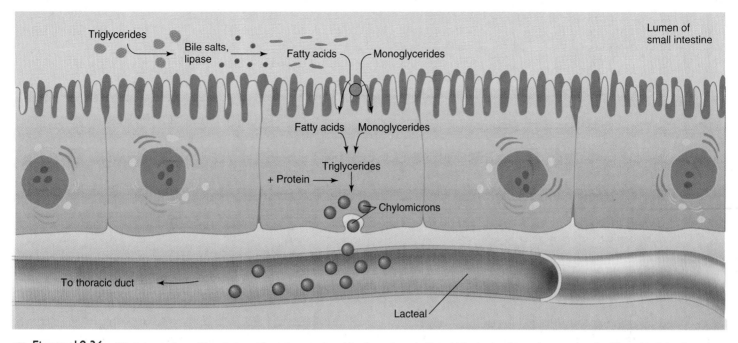

■ **Figure 18.36** **The absorption of fat.** Fatty acids and monoglycerides from the micelles within the small intestine are absorbed by epithelial cells and converted intracellularly into triglycerides. These are then combined with protein to form chylomicrons, which enter the lymphatic vessels (lacteals) of the villi. These lymphatic vessels transport the chylomicrons to the thoracic duct, which empties them into the venous blood (of the left subclavian vein).

Table 18.8 Characteristics of the Lipid Carrier Proteins (Lipoproteins) Found in Plasma

Lipoprotein Class	Origin	Destination	Major Lipids	Functions
Chylomicrons	Intestine	Many organs	Triglycerides, other lipids	Deliver lipids of dietary origin to body cells
Very-low-density lipoproteins (VLDLs)	Liver	Many organs	Triglycerides, cholesterol	Deliver endogenously produced triglycerides to body cells
Low-density lipoproteins (LDLs)	Intravascular removal of triglycerides from VLDLs	Blood vessels, liver	Cholesterol	Deliver endogenously produced cholesterol to various organs
High-density lipoproteins (HDLs)	Liver and intestine	Liver and steroid-hormone-producing glands	Cholesterol	Remove and degrade cholesterol

glycerol for use by the cells. The remaining *remnant particles,* containing cholesterol, are taken up by the liver. This is a process of endocytosis (chapter 3; see fig. 3.4) that requires membrane receptors for the protein part (or *apoprotein*) of the remnant particle.

Cholesterol and triglycerides produced by the liver are combined with other apoproteins and secreted into the blood as **very-low-density lipoproteins (VLDLs),** which deliver triglycerides to different organs. Once the triglycerides are removed, the VLDL particles are converted to **low-density lipoproteins (LDLs),** which transport cholesterol to various organs, including blood vessels. This can contribute to the development of atherosclerosis (chapter 13). Excess cholesterol is returned from these organs to the liver attached to **high-density lipoproteins (HDLs).** A high ratio of HDL-cholesterol to total cholesterol is believed to afford protection against atherosclerosis. The characteristics of these lipoproteins are summarized in table 18.8.

Test Yourself Before You Continue

1. List the enzymes involved in carbohydrate digestion, indicating their origins, sites of action, substrates, and products.

2. List each enzyme involved in protein digestion, indicating its origin and site of action. Also, indicate whether the enzyme is an endopeptidase or exopeptidase.

3. Describe how bile aids both the digestion and absorption of fats. Explain how the absorption of fat differs from the absorption of amino acids and monosaccharides.

4. Trace the pathway and fate of a molecule of triglyceride and a molecule of cholesterol in a chylomicron within an intestinal epithelial cell.

5. Cholesterol in the blood may be attached to any of four possible lipoproteins. Distinguish among these proteins in terms of the origin and destination of the cholesterol they carry.

INTERACTIONS

HPer Links of the Digestive System with Other Body Systems

Integumentary System

- The skin produces vitamin D, which indirectly helps to regulate the intestinal absorption of Ca^{2+} ....(p. 625)
- Adipose tissue in the hypodermis of the skin stores triglycerides ....(p. 605)
- The digestive system provides nutrients for all systems, including the integumentary system ....(p. 561)

Skeletal System

- The extracellular matrix of bones stores calcium phosphate ....(p. 623)
- The small intestine absorbs Ca^{2+} and PO_4^{3-}, which are needed for deposition of bone ....(p. 625)

Muscular System

- Muscle contractions are needed for chewing, swallowing, peristalsis, and segmentation ....(p. 561)
- Sphincter muscles help to regulate the passage of material along the GI tract ....(p. 565)
- The liver removes lactic acid produced by exercising skeletal muscles ....(p. 108)

Nervous System

- Autonomic nerves help to regulate the digestive system ....(p. 563)
- The enteric nervous system functions like the CNS to regulate the intestine .(p. 585)

Endocrine System

- Gastrin, produced by the stomach, helps to regulate the secretion of gastric juice ....(p. 584)
- Several hormones secreted by the small intestine regulate different aspects of the digestive system ....(p. 583)
- A hormone produced by the intestine stimulates the pancreatic islets to secrete insulin ....(p. 585)

- Adipose tissue secretes leptin, which helps to regulate hunger ....(p. 606)
- The liver removes some hormones from the blood, changes them chemically, and excretes them in the bile ....(p. 579)

Immune System

- The immune system protects all organs against infections, including those of the digestive system ....(p. 446)
- Lymphatic vessels carry absorbed fat from the small intestine to the venous system ....(p. 590)
- The liver aids the immune system by metabolizing certain toxins and excreting them in the bile ....(p. 579)
- The mucosa of the GI tract contains lymph nodules that protect against disease ....(p. 451)
- Acids and enzymes secreted by the GI tract provide nonspecific defense against microbes ....(p. 446)

Circulatory System

- The blood transports absorbed amino acids, monosaccharides, and other molecules from the intestine to liver, and then to other organs ....(p. 576)
- The hepatic portal vein allows some absorbed molecules to have an enterohepatic circulation ....(p. 576)
- The intestinal absorption of vitamin B_{12} (needed for red blood cell production) requires intrinsic factor, secreted by the stomach ....(p. 566)
- Iron must be absorbed through the intestine to allow a normal rate of hemoglobin production ....(p. 371)
- The liver synthesizes clotting proteins, plasma albumin, and all other plasma proteins except antibodies ......(p. 580)

Respiratory System

- The lungs provide oxygen for the metabolism of all organs, including those of the digestive system ....(p. 480)
- The oxygen provided by the respiratory system is used to metabolize food molecules brought into the body by the digestive system ....(p. 108)

Urinary System

- The kidneys eliminate metabolic wastes from all organs, including those of the digestive system ....(p. 524)
- The kidneys help to convert vitamin D into the active form required for calcium absorption in the intestine ......(p. 625)

Reproductive System

- Sex steroids, particularly androgens, stimulate the rate of fuel consumption by the body ....(p. 609)
- During pregnancy, the GI tract of the mother helps to provide nutrients that pass through the placenta to the embryo and fetus ....(p. 672)

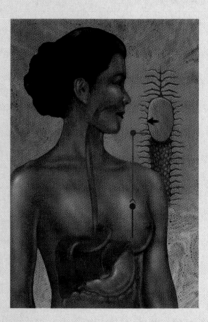

Summary

Introduction to the Digestive System 560

I. The digestion of food molecules involves the hydrolysis of these molecules into their subunits.

   A. The digestion of food occurs in the lumen of the GI tract and is catalyzed by specific enzymes.

   B. The digestion products are absorbed through the intestinal mucosa and enter the blood or lymph.

II. The layers (tunics) of the GI tract are, from the inside outward, the mucosa, submucosa, muscularis, and serosa.

   A. The mucosa consists of a simple columnar epithelium, a layer of connective tissue called the lamina propria, and a thin layer of smooth muscle called the muscularis mucosae.

   B. The submucosa is composed of connective tissue; the muscularis consists of layers of smooth muscles; the serosa is connective tissue covered by the visceral peritoneum.

   C. The submucosa contains the submucosal plexus, and the muscularis contains the myenteric plexus of autonomic nerves.

From Mouth to Stomach 563

I. Peristaltic waves of contraction push food through the lower esophageal sphincter into the stomach.

II. The stomach consists of a cardia, fundus, body, and pylorus (antrum). The pylorus terminates with the pyloric sphincter.

   A. The lining of the stomach is thrown into folds, or rugae, and the mucosal surface forms gastric pits that lead into gastric glands.

   B. The parietal cells of the gastric glands secrete HCl; the chief cells secrete pepsinogen.

   C. In the acidic environment of gastric juice, pepsinogen is converted into the active protein-digesting enzyme called pepsin.

   D. Some digestion of protein occurs in the stomach, but the most important function of the stomach is the secretion of intrinsic factor, which is needed for the absorption of vitamin B_{12} in the intestine.

Small Intestine 568

I. The small intestine is divided into the duodenum, jejunum, and ileum. The common bile duct and pancreatic duct empty into the duodenum.

II. Fingerlike extensions of mucosa called villi project into the lumen, and at the bases of the villi the mucosa forms narrow pouches called the crypts of Lieberkühn.

   A. New epithelial cells are formed in the crypts.

   B. The membrane of intestinal epithelial cells is folded to form microvilli. This brush border of the mucosa increases the surface area.

III. Digestive enzymes, called brush border enzymes, are located in the membranes of the microvilli.

IV. The small intestine exhibits two major types of movements—peristalsis and segmentation.

Large Intestine 572

I. The large intestine is divided into the cecum, colon, rectum, and anal canal.

   A. The appendix is attached to the inferior medial margin of the cecum.

   B. The colon consists of ascending, transverse, descending, and sigmoid portions.

   C. Bulges in the walls of the large intestine are called haustra.

II. The large intestine absorbs water and electrolytes.

   A. Although most of the water that enters the GI tract is absorbed in the small intestine, 1.5 to 2.0 L pass to the large intestine each day. The large intestine absorbs about 90% of this amount.

   B. Na^+ is actively absorbed and water follows passively, in a manner analogous to the reabsorption of NaCl and water in the renal tubules.

III. Defecation occurs when the anal sphincters relax and contraction of other muscles raises the rectal pressure.

Liver, Gallbladder, and Pancreas 575

I. The liver, the largest internal organ, is composed of functional units called lobules.

   A. Liver lobules consist of plates of hepatic cells separated by capillary sinusoids.

   B. Blood flows from the periphery of each lobule, where the hepatic artery and portal vein empty, through the sinusoids and out the central vein.

   C. Bile flows within the hepatocyte plates, in canaliculi, to the bile ducts.

   D. Substances excreted in the bile can be returned to the liver in the hepatic portal blood. This is called an enterohepatic circulation.

   E. Bile consists of a pigment called bilirubin, bile salts, cholesterol, and other molecules.

   F. The liver detoxifies the blood by excreting substances in the bile, by phagocytosis, and by chemical inactivation.

   G. The liver modifies the plasma concentrations of proteins, glucose, triglycerides, and ketone bodies.

II. The gallbladder stores and concentrates the bile. It releases bile through the cystic duct and common bile duct to the duodenum.

III. The pancreas is both an exocrine and an endocrine gland.

   A. The endocrine portion, known as the islets of Langerhans, secretes the hormones insulin and glucagon.

   B. The exocrine acini of the pancreas produce pancreatic juice, which contains various digestive enzymes and bicarbonate.

Neural and Endocrine Regulation of the Digestive System 583

I. The regulation of gastric function occurs in three phases.

   A. In the cephalic phase, the activity of higher brain centers, acting via the vagus nerve, stimulates gastric juice secretion.

   B. In the gastric phase, the secretion of HCl and pepsin is controlled by the gastric contents and by the hormone gastrin, secreted by the gastric mucosa.

  C. In the intestinal phase, the activity of the stomach is inhibited by neural reflexes and hormonal secretion from the duodenum.

II. Intestinal function is regulated, at least in part, by short, local reflexes coordinated by the enteric nervous system.

  A. The enteric nervous system contains interneurons, intrinsic sensory neurons, and autonomic motor neurons.

  B. Peristalis is coordinated by the enteric nervous system, which produces smooth muscle contraction above the bolus and relaxation below the bolus of chyme.

  C. Short reflexes include the gastroileal reflex, ileogastric reflex, and intestino-intestinal reflexes.

III. The secretion of the hormones secretin and cholecystokinin (CCK) regulates pancreatic juice and bile secretion.

  A. Secretin secretion is stimulated by the arrival of acidic chyme into the duodenum.

  B. CCK secretion is stimulated by the presence of fat in the chyme arriving in the duodenum.

  C. Contraction of the gallbladder occurs in response to a neural reflex and to the secretion of CCK by the duodenum.

IV. Gastrointestinal hormones may be needed for the maintenance of the GI tract and accessory digestive organs.

Digestion and Absorption of Carbohydrates, Lipids, and Proteins   587

I. The digestion of starch begins in the mouth through the action of salivary amylase.

  A. Pancreatic amylase digests starch into disaccharides and short-chain oligosaccharides.

  B. Complete digestion into monosaccharides is accomplished by brush border enzymes.

II. Protein digestion begins in the stomach through the action of pepsin.

  A. Pancreatic juice contains the protein-digesting enzymes trypsin and chymotrypsin, among others.

  B. The brush border contains digestive enzymes that help to complete the digestion of proteins into amino acids.

  C. Amino acids, like monosaccharides, are absorbed and secreted into capillary blood entering the portal vein.

III. Lipids are digested in the small intestine after being emulsified by bile salts.

  A. Free fatty acids and monoglycerides enter particles called micelles, formed in large part by bile salts, and they are absorbed in this form or as free molecules.

  B. Once inside the mucosal epithelial cells, these subunits are used to resynthesize triglycerides.

  C. Triglycerides in the epithelial cells, together with proteins, form chylomicrons, which are secreted into the central lacteals of the villi.

  D. Chylomicrons are transported by lymph to the thoracic duct and from there enter the blood.

Review Activities

Test Your Knowledge of Terms and Facts

1. Which of these statements about intrinsic factor is *true?*
  a. It is secreted by the stomach.
  b. It is a polypeptide.
  c. It promotes absorption of vitamin B$_{12}$ in the intestine.
  d. It helps prevent pernicious anemia.
  e. All of these are true.

2. Intestinal enzymes such as lactase are
  a. secreted by the intestine into the chyme.
  b. produced by the intestinal crypts (of Lieberkühn.)
  c. produced by the pancreas.
  d. attached to the cell membrane of microvilli in the epithelial cells of the mucosa.

3. Which of these statements about gastric secretion of HCl is *false?*
  a. HCl is secreted by parietal cells.
  b. HCl hydrolyzes peptide bonds.
  c. HCl is needed for the conversion of pepsinogen to pepsin.
  d. HCl is needed for maximum activity of pepsin.

4. Most digestion occurs in
  a. the mouth.
  b. the stomach.
  c. the small intestine.
  d. the large intestine.

5. Which of these statements about trypsin is *true?*
  a. Trypsin is derived from trypsinogen by the digestive action of pepsin.
  b. Active trypsin is secreted into the pancreatic acini.
  c. Trypsin is produced in the crypts (of Lieberkühn).
  d. Trypsinogen is converted to trypsin by the brush border enzyme enterokinase.

6. During the gastric phase, the secretion of HCl and pepsinogen is stimulated by
  a. vagus nerve stimulation that originates in the brain.
  b. polypeptides in the gastric lumen and by gastrin secretion.
  c. secretin and cholecystokinin from the duodenum.
  d. all of these.

7. The secretion of HCl by the stomach mucosa is inhibited by
  a. neural reflexes from the duodenum.
  b. the secretion of an enterogastrone from the duodenum.
  c. the lowering of gastric pH.
  d. all of these.

8. The first organ to receive the blood-borne products of digestion is
  a. the liver.
  b. the pancreas.
  c. the heart.
  d. the brain.

9. Which of these statements about hepatic portal blood is *true?*
  a. It contains absorbed fat.
  b. It contains ingested proteins.
  c. It is mixed with bile in the liver.
  d. It is mixed with blood from the hepatic artery in the liver.

10. Absorption of salt and water is the principal function of which region of the GI tract?
  a. esophagus
  b. stomach

c. duodenum
d. jejunum
e. large intestine

11. Cholecystokinin (CCK) is a hormone that stimulates
   a. bile production.
   b. release of pancreatic enzymes.
   c. contraction of the gallbladder.
   d. both *a* and *b*.
   e. both *b* and *c*.

12. Which of these statements about vitamin B_{12} is *false?*
   a. Lack of this vitamin can produce pernicious anemia.
   b. Intrinsic factor is needed for absorption of vitamin B_{12}.

c. Damage to the gastric mucosa may lead to a deficiency in vitamin B_{12}.
d. Vitamin B_{12} is absorbed primarily in the jejunum.

13. Which of these statements about starch digestion is *false?*
   a. It begins in the mouth.
   b. It occurs in the stomach.
   c. It requires the action of pancreatic amylase.
   d. It requires brush border enzymes for completion.

14. Which of these statements about fat digestion and absorption is *false?*
   a. Emulsification by bile salts increases the rate of fat digestion.

b. Triglycerides are hydrolyzed by the action of pancreatic lipase.
c. Triglycerides are resynthesized from monoglycerides and fatty acids in the intestinal epithelial cells.
d. Triglycerides, as particles called chylomicrons, are absorbed into blood capillaries within the villi.

15. Which of these statements about contraction of intestinal smooth muscle is *true?*
   a. It occurs automatically.
   b. It is increased by parasympathetic nerve stimulation.
   c. It produces segmentation.
   d. All of these are true.

Test Your Understanding of Concepts and Principles

1. Explain how the gastric secretion of HCl and pepsin is regulated during the cephalic, gastric, and intestinal phases.[1]

2. Describe how pancreatic enzymes become activated in the lumen of the intestine. Why are these mechanisms needed?

3. Explain the function of bicarbonate in pancreatic juice. How may peptic ulcers in the duodenum be produced?

4. Describe the mechanisms that are believed to protect the gastric mucosa from self-digestion. What factors might be responsible for the development of a peptic ulcer in the stomach?

5. Explain why the pancreas is considered both an exocrine and an endocrine gland. Given this information, predict what effects tying of the pancreatic duct would have on pancreatic structure and function.

6. Explain how jaundice is produced when (a) the person has gallstones, (b) the person has a high rate of red blood cell destruction, and (c) the person has liver disease. In which case(s) would phototherapy for the jaundice be effective? Explain.

7. Describe the steps involved in the digestion and absorption of fat.

8. Distinguish between chylomicrons, very-low-density lipoproteins, low-density lipoproteins, and high-density lipoproteins.

9. Identify the different neurons present in the wall of the intestine and explain how these neurons are involved in "short reflexes." Why is the enteric nervous system sometimes described as an "enteric brain"?

10. Trace the course of blood flow through the liver and discuss the significance of this pattern in terms of the detoxication of the blood. Describe the enzymes and the reactions involved in this detoxication.

Test Your Ability to Analyze and Apply Your Knowledge

1. Which surgery do you think would have the most profound effect on digestion: (a) removal of the stomach (gastrectomy), (b) removal of the pancreas (pancreatectomy), or (c) removal of the gallbladder (cholecystectomy)? Explain your reasoning.

2. Describe the adaptations of the GI tract that make it more efficient by either increasing the surface area for absorption or increasing the contact

between food particles and digestive enzymes.

3. Discuss how the ECL cells of the gastric mucosa function as a final common pathway for the neural, endocrine, and paracrine regulation of gastric acid secretion. What does this imply about the effectiveness of drug intervention to block excessive acid secretion?

4. Bacterial heat-stable enterotoxins can cause a type of diarrhea by stimulating

the enzyme guanylate cyclase, which raises cyclic GMP levels within intestinal cells. Why might this be considered an example of mimicry? How does it cause diarrhea?

5. The hormone insulin is secreted by the pancreatic islets in response to a rise in blood glucose concentration. Surprisingly, however, the insulin secretion is greater in response to oral glucose than to intravenous glucose. Explain why this is so.

Related Websites

Check out the Links Library at www.mhhe.com/fox8 for links to sites containing resources related to the digestive system. These links are monitored to ensure current URLs.

---

[1]*Note:* This question is answered in the chapter 18 Study Guide found on the Online Learning Center at www.mhhe.com/fox8.

19 Regulation of Metabolism

Objectives

After studying this chapter, you should be able to:

1. identify factors that influence the metabolic rate and explain the significance of the basal metabolic rate.

2. distinguish between the caloric and anabolic requirements for food and define the terms *essential amino acids* and *essential fatty acids*.

3. distinguish between fat-soluble and water-soluble vitamins and describe some of the functions of different vitamins.

4. define the terms *energy reserves* and *circulating energy substrates* and explain how these sources of energy interact during anabolism and catabolism.

5. describe the regulation of eating and discuss the endocrine control of metabolism in general terms.

6. describe the regulation of adipocyte development and the roles of adipocytes in the regulation of hunger and tissue responsiveness to insulin.

7. describe the actions of insulin and glucagon, and explain how the secretion of these hormones is regulated.

8. explain how insulin and glucagon regulate metabolism during feeding and fasting.

9. describe the symptoms of type 1 and type 2 diabetes mellitus and explain how these conditions are produced.

10. describe the metabolic effects of epinephrine and the glucocorticoids.

11. describe the effects of thyroxine on cell respiration and explain the relationship between thyroxine levels and the basal metabolic rate.

12. describe the symptoms of hypothyroidism and hyperthyroidism, and explain how these conditions are produced.

13. describe the metabolic effects of growth hormone and explain why growth hormone and thyroxine are needed for proper body growth.

14. describe the actions of parathyroid hormone, 1,25-dihydroxyvitamin D_3, and calcitonin, and explain how the secretion of these hormones is regulated.

15. describe how 1,25-dihydroxyvitamin D_3 is produced and explain why this compound is needed to prevent osteomalacia and rickets.

Chapter at a Glance

Take Advantage of the Technology

Visit the Online Learning Center for
these additional study resources.

- Interactive quizzing
- Online study guide
- Current news feeds
- Crossword puzzles
- Vocabulary flashcards
- Labeling activities

    www.mhhe.com/fox8

Clinical Investigation

Phyllis is a 44-year-old legal secretary who is moderately overweight. After weeks of complaining to her friends and family of her nausea, headaches, frequent urination, and continuous thirst, she decides to go to a physician. In the course of reporting her medical history, she mentions that both her mother and uncle are diabetics. She provides a sample of urine, which does not give evidence of glycosuria. She is told to return the next day to provide a fasting blood sample. When this sample is analyzed, her fasting blood glucose concentration is measured at 150 mg/dl. An oral glucose tolerance test is subsequently performed, and a blood glucose concentration of 220 mg/dl is measured 2 hours following the ingestion of the glucose solution.

The physician places her on a weight-reduction program and advises her to begin a mild but regular exercise regimen. He tells her that he will prescribe pills for her if the diet and exercise are not effective in relieving her symptoms. What diagnosis did the physician make? Why did he make this diagnosis and subsequent recommendations? What pills might he prescribe?

Nutritional Requirements

The body's energy requirements must be met by the caloric value of food to prevent catabolism of the body's own fat, carbohydrates, and protein. Additionally, food molecules—particularly the essential amino acids and fatty acids—are needed for replacement of molecules in the body that are continuously degraded. Vitamins and minerals do not directly provide energy but instead are required for diverse enzymatic reactions.

Living tissue is maintained by the constant expenditure of energy. This energy is obtained directly from ATP and indirectly from the cellular respiration of glucose, fatty acids, ketone bodies, amino acids, and other organic molecules. These molecules are ultimately obtained from food, but they can also be obtained from the glycogen, fat, and protein stored in the body.

The energy value of food is commonly measured in **kilocalories,** which are also called "big calories" and spelled with a capital letter (Calories). One kilocalorie (kcal) is equal to 1,000 calories; one calorie is defined as the amount of heat required to raise the temperature of one cubic centimeter of water from 14.5° to 15.5° C. As described in chapter 5, the amount of energy released as heat when a quantity of food is combusted *in vitro* is equal to the amount of energy released within cells through the process of aerobic respiration. This is 4 kilocalories per gram for carbohydrates or proteins and 9 kilocalories per gram for fat. When this energy is released by cell respiration, some is transferred to the high-energy bonds of ATP and some is lost as heat.

Metabolic Rate and Caloric Requirements

The total rate of body metabolism, or the **metabolic rate,** can be measured by either the amount of heat generated by the body or by the amount of oxygen consumed by the body per minute. This rate is influenced by a variety of factors. For example, the metabolic rate is increased by physical activity and by eating. The increased rate of metabolism that accompanies the assimilation of food can last more than 6 hours after a meal.

Body temperature is also an important factor in determining metabolic rate. The reasons for this are twofold: (1) temperature itself influences the rate of chemical reactions and (2) the hypothalamus contains *temperature control centers*, as well as temperature-sensitive cells that act as sensors for changes in body temperature. In response to deviations from a "set point" for body temperature (chapter 1), the control areas of the hypothalamus can direct physiological responses that help to correct the deviations and maintain a constant body temperature. Changes in body temperature are thus accompanied by physiological responses that influence the total metabolic rate.

Hypothermia (low body temperature)—where the core body temperature is lowered to between 26° and 32.5° C (78° and 90° F)—is often induced during open heart or brain surgery. Compensatory responses to the lowered temperature are dampened by the general anesthetic, and the lower body temperature drastically reduces the needs of the tissues for oxygen. Under these conditions, the heart can be stopped and bleeding is significantly reduced.

The metabolic rate (measured by the rate of oxygen consumption) of an awake, relaxed person 12 to 14 hours after eating and at a comfortable temperature is known as the **basal metabolic rate (BMR).** The BMR is determined primarily by a person's age, sex, and body surface area, but it is also strongly influenced by the level of thyroid secretion. A person with hyperthyroidism has an abnormally high BMR, and a person with hypothyroidism has a low BMR. An interesting recent finding is that the BMR may be influenced by genetic inheritance; it appears that at least some families that are prone to obesity may have a genetically determined low BMR.

In general, however, individual differences in energy requirements are due primarily to differences in physical activity. Daily energy expenditures may range from 1,300 to 5,000 kilocalories per day. The average values for people not engaged in heavy manual labor but who are active during their leisure time are about 2,900 kilocalories per day for men and 2,100 kilocalories per day for women. People engaged in office work, the professions, sales, and comparable occupations consume up to 5 kilocalories per minute during work. More physically demanding occupations may require energy expenditures of 7.5 to 10 kilocalories per minute.

Table 19.1   Energy Consumed (in Kilocalories per Minute) in Different Types of Activities

Activity	Weight in Pounds			
	105–115	127–137	160–170	182–192
Bicycling				
10 mph	5.41	6.16	7.33	7.91
Stationary, 10 mph	5.50	6.25	7.41	8.16
Calisthenics	3.91	4.50	7.33	7.91
Dancing				
Aerobic	5.83	6.58	7.83	8.58
Square	5.50	6.25	7.41	8.00
Gardening, Weeding, and Digging	5.08	5.75	6.83	7.50
Jogging				
5.5 mph	8.58	9.75	11.50	12.66
6.5 mph	8.90	10.20	12.00	13.20
8.0 mph	10.40	11.90	14.10	15.50
9.0 mph	12.00	13.80	16.20	17.80
Rowing, Machine				
Easily	3.91	4.50	5.25	5.83
Vigorously	8.58	9.75	11.50	12.66
Skiing				
Downhill	7.75	8.83	10.41	11.50
Cross-country, 5 mph	9.16	10.41	12.25	13.33
Cross-country, 9 mph	13.08	14.83	17.58	19.33
Swimming, Crawl				
20 yards per minute	3.91	4.50	5.25	5.83
40 yards per minute	7.83	8.91	10.50	11.58
55 yards per minute	11.00	12.50	14.75	16.25
Walking				
2 mph	2.40	2.80	3.30	3.60
3 mph	3.90	4.50	6.30	6.80
4 mph	4.50	5.20	6.10	6.80

When the caloric intake is greater than the energy expenditures, excess calories are stored primarily as fat. This is true regardless of the source of the calories—carbohydrates, protein, or fat—because these molecules can be converted to fat by the metabolic pathways described in chapter 5 (see fig. 5.17).

Weight is lost when the caloric value of the food ingested is less than the amount required in cell respiration over a period of time. Weight loss, therefore, can be achieved by dieting alone or in combination with an exercise program to raise the metabolic rate. A summary of the caloric expenditure associated with different forms of exercise is provided in table 19.1. Recent experiments, however, demonstrate why it is often so difficult to lose (or gain) weight. When subjects were maintained at 10% less than their usual weight, their metabolic rate decreased, and when they were maintained at 10% greater than their usual body weight, their metabolic rate increased. The body, it seems, tends to defend its usual weight by altering the energy expenditure as well as by regulating the food intake.

Anabolic Requirements

In addition to providing the body with energy, food also supplies the raw materials for synthesis reactions—collectively termed **anabolism**—that occur constantly within the cells of the body.

Anabolic reactions include those that synthesize DNA and RNA, protein, glycogen, triglycerides, and other polymers. These anabolic reactions must occur constantly to replace those molecules that are hydrolyzed into their component monomers. These hydrolysis reactions, together with the reactions of cell respiration that ultimately break the monomers down to carbon dioxide and water, are collectively termed **catabolism.**

Acting through changes in hormonal secretion, exercise and fasting increase the catabolism of stored glycogen, fat, and body protein. These molecules are also broken down at a certain rate in a person who is neither exercising nor fasting. Some of the monomers thus formed (amino acids, glucose, and fatty acids) are used immediately to resynthesize body protein, glycogen, and fat. However, some of the glucose derived from stored glycogen, for example, or fatty acids derived from stored triglycerides, are used to provide energy in the process of cell respiration. For this reason, new monomers must be obtained from food to prevent a continual decline in the amount of protein, glycogen, and fat in the body.

The *turnover rate* of a particular molecule is the rate at which it is broken down and resynthesized. For example, the average daily turnover for carbohydrates is 250 g/day. Since some of the glucose in the body is reused to form glycogen, the average daily dietary requirement for carbohydrate is somewhat less than this amount—about 150 g/day. The average daily turnover for protein is 150 g/day, but since many of the amino acids derived from the catabolism of body proteins can be reused in protein synthesis, a person needs only about 35 g/day of protein in the diet. It should be noted that these are average figures and will vary in accordance with individual differences in size, sex, age, genetics, and physical activity. The average daily turnover for fat is about 100 g/day, but very little is required in the diet (other than that which supplies fat-soluble vitamins and essential fatty acids), since fat can be produced from excess carbohydrates.

The minimal amounts of dietary protein and fat required to meet the turnover rate are adequate only if they supply sufficient amounts of the essential amino acids and fatty acids. These molecules are termed *essential* because they cannot be synthesized by the body and must be obtained in the diet. The nine **essential amino acids** are lysine, tryptophan, phenylalanine, threonine, valine, methionine, leucine, isoleucine, and histidine. The **essential fatty acids** are linoleic acid and linolenic acid.

Unsaturated fatty acids—those with double bonds between the carbons—are characterized by the location of the first double bond. Linoleic acid, found in corn oil, contains eighteen carbons and two double bonds. It has its first double bond on the sixth carbon from the methyl (CH_3) end, and is therefore designated as an n-6 (or omega-6) fatty acid. Linolenic acid, found in canola oil, also has eighteen carbons, but it has three double bonds. More significantly for health, its first double bond is on the third carbon from the methyl end; linolenic acid is an n-3 (also called omega-3) fatty acid. Several studies suggest that n-3 fatty acids may offer protection from cardiovascular disease.

Eskimos who eat a traditional diet of meat and fish have a surprisingly low blood concentration of triglycerides and cholesterol, and a low incidence of ischemic heart disease, despite the high fat and cholesterol content of their food. Several studies suggest that the n-3 fatty acids of the cold-water fish are the source of the apparent protective effect. The n-3 fatty acids of fish include *eicosapentaenoic acid*, or *EPA* (with twenty carbons), and *docosahexaenoic acid*, or *DHA* (with twenty-two carbons). The n-3 fatty acids may help to inhibit platelet function in thrombus formation, the progression of atherosclerosis, and/or ventricular arrhythmias. Several studies have confirmed the protective effect of fish and fish oil in the diet, and on the basis of this evidence it seems prudent to eat fish at least once or twice a week on a continuing basis.

Table 19.2   Recommended Dietary Allowances for Vitamins and Minerals[1]

Category	Age (Years) or Condition	Weight[2] (kg)	Weight[2] (lb)	Height[2] (cm)	Height[2] (in)	Protein (g)	Fat-Soluble Vitamins Vitamin A (μg RE)[3]	Vitamin D (μg)[4]	Vitamin E (mg α-TE)[5]	Vitamin K (μg)
Infants	0.0–0.05	6	13	60	24	13	375	7.5	3	5
	0.5–1	9	20	71	28	14	375	10	4	10
Children	1–3	13	29	90	35	16	400	10	6	15
	4–6	20	44	112	44	24	500	10	7	20
	7–10	28	62	132	52	28	700	10	7	30
Males	11–14	45	99	157	62	45	1,000	10	10	45
	15–18	66	145	176	69	59	1,000	10	10	65
	19–24	72	160	177	70	58	1,000	10	10	70
	25–50	79	174	176	70	63	1,000	5	10	80
	51+	77	170	173	68	63	1,000	5	10	80
Females	11–14	46	101	157	62	45	800	10	8	45
	15–18	55	120	163	64	44	800	10	8	55
	19–24	58	128	164	65	46	800	10	8	60
	25–50	63	138	163	64	50	800	5	8	65
	51+	65	143	160	63	50	800	5	8	65
Pregnant						60	800	10	10	65
Lactating	1st 6 months					65	1,300	10	12	65
	2nd 6 months					62	1,200	10	11	65

Source: Reprinted with permission from *Recommended Dietary Allowances*, 10th Edition. Copyright 1989 by the National Academy of Sciences. Courtesy of the National Academy Press, Washington, D.C.

[1] The allowances, expressed as average daily intakes over time, are intended to provide for individual variations among most normal persons as they live in the United States under usual environmental stresses. Diets should be based on a variety of common foods in order to provide other nutrients for which human requirements have been less well defined.

[2] Weights and heights of Reference Adults are actual medians for the U.S. population of the designated age, as reported by NHANES II. The use of these figures does not imply that the height-to-weight ratios are ideal.

[3] Retinol equivalents. 1 RE = 1 μg retinol or 6 μg β-carotene.

[4] As cholecalciferol. 10 μg cholecalciferol = 400 IU of vitamin D.

[5] α-tocopherol equivalents. 1 mg d-α-tocopherol = 1 α-TE.

Vitamins and Minerals

Vitamins are small organic molecules that serve as coenzymes in metabolic reactions or that have other highly specific functions. They must be obtained in the diet because the body either doesn't produce them, or it produces them in insufficient quantities. (Vitamin D is produced in limited quantities by the skin, and the B vitamins and vitamin K are produced by intestinal bacteria.) There are two classes of vitamins: fat-soluble and water-soluble. The **fat-soluble vitamins** include vitamins A, D, E, and K. The **water-soluble vitamins** include thiamine (B_1), riboflavin (B_2), niacin (B_3), pyridoxine (B_6), pantothenic acid, biotin, folic acid, vitamin B_{12}, and vitamin C (ascorbic acid). Recommended dietary allowances for these vitamins are listed in table 19.2.

Water-Soluble Vitamins

Derivatives of water-soluble vitamins serve as coenzymes in the metabolism of carbohydrates, lipids, and proteins. **Thiamine,** for example, is needed for the activity of the enzyme that converts pyruvic acid to acetyl coenzyme A. **Riboflavin** and **niacin** are needed for the production of FAD and NAD, respectively. FAD and NAD serve as coenzymes that transfer hydrogens during cell respiration (chapter 4; see fig. 4.17). **Pyridoxine** is a cofactor for the enzymes involved in amino acid metabolism. Deficiencies of the water-soluble vitamins can thus have widespread effects in the body (table 19.3).

Free radicals are highly reactive molecules that carry an unpaired electron. Such free radicals can damage tissues by removing an electron from, and thus oxidizing, other molecules. **Vitamin C** (a water-soluble vitamin) and vitamin E (a fat-soluble vitamin) function as *antioxidants* through their ability to inactivate free radicals. These vitamins may afford protection against some of the diseases that may be caused by free radicals.

Fat-Soluble Vitamins

Vitamin E has important antioxidant functions, as will be described shortly. Some fat-soluble vitamins have highly specialized functions. **Vitamin K,** for example, is required for the production of prothrombin and for clotting factors VII, IX, and X. Vitamins A and D also have functions unique to each, but these two vitamins overlap in their mechanisms of action.

Vitamin A is a collective term for a number of molecules that include *retinol* (the transport form of vitamin A), *retinal* (also known as retinaldehyde, used as the photopigment in the retina), and *retinoic acid.* Most of these molecules are ultimately derived

Water-Soluble Vitamins							Minerals						
Vitamin C (mg)	Thiamine (mg)	Riboflavin (mg)	Niacin (mg NE)[6]	Vitamin B_6 (mg)	Folate (µg)	Vitamin B_{12} (µg)	Calcium (mg)	Phosphorus (mg)	Magnesium (mg)	Iron (mg)	Zinc (mg)	Iodine (µg)	Selenium (µg)
30	0.3	0.4	5	0.3	25	0.3	400	300	40	6	5	40	10
35	0.4	0.5	6	0.6	35	0.5	600	500	60	10	5	50	15
40	0.7	0.8	9	1.0	50	0.7	800	800	80	10	10	70	20
45	0.9	1.1	12	1.1	75	1.0	800	800	120	10	10	90	20
45	1.0	1.2	13	1.4	100	1.4	800	800	170	10	10	120	30
50	1.3	1.5	17	1.7	150	2.0	1,200	1,200	270	12	15	150	40
60	1.5	1.8	20	2.0	200	2.0	1,200	1,200	400	12	15	150	50
60	1.5	1.7	19	2.0	200	2.0	1,200	1,200	350	10	15	150	70
60	1.5	1.7	19	2.0	200	2.0	800	800	350	10	15	150	70
60	1.2	1.4	15	2.0	200	2.0	800	800	350	10	15	150	70
50	1.1	1.3	15	1.4	150	2.0	1,200	1,200	280	15	12	150	45
60	1.1	1.3	15	1.5	180	2.0	1,200	1,200	300	15	12	150	50
60	1.1	1.3	15	1.6	180	2.0	1,200	1,200	280	15	12	150	55
60	1.1	1.3	15	1.6	180	2.0	800	800	280	15	12	150	55
60	1.0	1.2	13	1.6	180	2.0	800	800	280	10	12	150	55
70	1.5	1.6	17	2.2	400	2.2	1,200	1,200	300	30	15	175	65
95	1.6	1.8	20	2.1	280	2.6	1,200	1,200	355	15	19	200	75
90	1.6	1.7	20	2.1	260	2.6	1,200	1,200	340	15	16	200	75

[6] Niacin equivalents. 1 NE = 1 mg of niacin or 60 mg of dietary tryptophan.

Table 19.3 The Major Vitamins

Vitamin	Sources	Function	Deficiency Symptom(s)
A	Yellow vegetables and fruit	Constituent of visual pigment; strengthens epithelial membranes	Night blindness; dry skin
B_1 (Thiamine)	Liver, unrefined cereal grains	Cofactor for enzymes that catalyze decarboxylation	Beriberi; neuritis
B_2 (Riboflavin)	Liver, milk	Part of flavoproteins (such as FAD)	Glossitis; cheilosis
B_6 (Pyridoxine)	Liver, corn, wheat, and yeast	Coenzyme for decarboxylase and transaminase enzymes	Convulsions
B_{12} (Cyanocobalamin)	Liver, meat, eggs, milk	Coenzyme for amino acid metabolism; needed for erythropoiesis	Pernicious anemia
Biotin	Egg yolk, liver, tomatoes	Needed for fatty acid synthesis	Dermatitis; enteritis
C	Citrus fruits, green leafy vegetables	Needed for collagen synthesis in connective tissues	Scurvy
D	Fish liver	Needed for intestinal absorption of calcium and phosphate	Rickets; osteomalacia
E	Milk, eggs, meat, leafy vegetables	Antioxidant	Muscular dystrophy (nonhereditary)
Folate	Green leafy vegetables	Needed for reactions that transfer one carbon	Sprue; anemia
K	Green leafy vegetables	Promotes reactions needed for function of clotting factors	Hemorrhage; inability to form clot
Niacin	Liver, meat, yeast	Part of NAD and NADP	Pellagra
Pantothenic acid	Liver, eggs, yeast	Part of coenzyme A	Dermatitis; enteritis; adrenal insufficiency

from dietary β-carotene, present in such foods as carrots, leafy vegetables, and egg yolk. The β-carotene is converted by an enzyme in the intestine into two molecules of retinal. Most of the retinal is reduced to retinol, while some is oxidized to retinoic acid. It is the retinoic acid that binds to nuclear receptor proteins (chapter 11; see fig. 11.7) and directly produces the effects of vitamin A. Retinoic acid is involved, for example, in regulating embryonic development; vitamin A deficiency interferes with embryonic development, while excessive vitamin A during pregnancy can cause birth defects. Retinoic acid is also needed for the maintenance of epithelial membrane structure and function. Indeed, retinoids are now widely used to treat acne and other skin conditions.

Vitamin D is produced by the skin under the influence of ultraviolet light, but usually it is not produced in sufficient amounts for all of the body's needs. That is why we must eat food containing additional amounts of vitamin D, and why it is classified as a vitamin even though it can be produced by the body. The vitamin D secreted by the skin or consumed in the diet is inactive in its original form; it must first be converted into a derivative by enzymes in the liver and kidneys before it can be active in the body. Once the active derivative is produced, vitamin D helps to regulate calcium balance.

As may be recalled from chapter 11, a nuclear receptor for the active form of thyroid hormone or of vitamin D consists of two different polypeptides. One polypeptide binds to thyroid hormone (thyroid receptor, or *TR*) or vitamin D (vitamin D receptor, or *DR*), and one polypeptide binds to a form of retinoic acid (retinoic acid X receptor, or *RXR;* see fig. 11.7). This overlapping of receptors may permit "cross-talk" between the actions of thyroid hormone, vitamin D, and vitamin A. In view of this, it is not surprising that thyroxine, vitamin A, and vitamin D have overlapping functions—all three are involved in regulating gene expression and promoting differentiation (specialization) of tissues.

The best known function of vitamin D is regulation of calcium balance. Because it promotes the intestinal absorption of Ca^{2+} and PO_4^{3-}, vitamin D is needed for proper calcification of the bones. However, a vitamin D derivative called *calcipotriene* is now widely used for the treatment of **psoriasis,** a skin condition characterized by inflammation and excessive proliferation of keratinocytes (the cells of the epidermis that produce keratin). In this case, the vitamin D analogue inhibits proliferation and promotes differentiation of the keratinocytes. It has been suggested that vitamin D produced in the skin may function as an autocrine regulator of the epidermis.

Minerals (Elements)

Minerals (elements) are needed as cofactors for specific enzymes and for a wide variety of other critical functions. Those that are required daily in relatively large amounts include sodium, potassium, magnesium, calcium, phosphorus, and chlorine (see table 19.2). In addition, the following **trace elements** are recognized as essential: iron, zinc, manganese, fluorine, copper, molybdenum, chromium, and selenium. These must be ingested in microgram amounts up to 30 mg per day, for pregnant woman taking iron (tables 19.2 and 19.4).

Free Radicals and Antioxidants

The electrons in an atom are located in *orbitals,* with each orbital containing a maximum of two electrons. When an orbital has an unpaired electron, the molecule containing the unpaired electron is called a **free radical.** Free radicals are highly reactive in the body, oxidizing (removing an electron from) other

Table 19.4 Estimated Safe and Adequate Daily Dietary Intakes of Selected Vitamins and Minerals[1]

| Category | Age (Years) | Vitamins | | Trace Elements[2] | | | | |
		Biotin (µg)	Pantothenic Acid (mg)	Copper (mg)	Manganese (mg)	Fluoride (mg)	Chromium (µg)	Molybdenum (µg)
Infants	0–0.5	10	2	0.4–0.6	0.3–0.6	0.1–0.5	10–40	15–??
	0.5–1	15	3	0.6–0.7	0.6–1.0	0.2–1.0	20–60	20–40
Children and adolescents	1–3	20	3	0.7–1.0	1.0–1.5	0.5–1.5	20–80	25–50
	4–6	25	3–4	1.0–1.5	1.5–2.0	1.0–2.5	30–120	30–75
	7–10	30	4–5	1.0–2.0	2.0–3.0	1.5–2.5	50–200	50–150
	11+	30–100	4–7	1.5–2.5	2.0–5.0	1.5–2.5	50–200	75–250
Adults		30–100	4–7	1.5–3.0	2.0–5.0	1.5–4.0	50–200	75–250

Source: Reprinted with permission from *Recommended Dietary Allowances*, 10th Edition. Copyright 1989 by the National Academy of Sciences. Courtesy of the National Academy Press, Washington, D.C.

[1] Because there is less information on which to base allowances, these figures are not given in the main table of RDA and are provided here in the form of ranges of recommended intakes.

[2] Since the toxic levels for many trace elements may be only several times usual intakes, the upper levels for the trace elements given in this table should not be habitually exceeded.

atoms, or sometimes reducing (donating their electron to) other atoms. The major free radicals are referred to as **reactive oxygen species,** if they contain oxygen with an unpaired electron, or **reactive nitrogen species,** if they contain nitrogen with an unpaired electron.

The unpaired electron is symbolized with a dot superscript. Thus, reactive oxygen species include the *superoxide radical* (O_2^-), the *hydroxyl radical* ($HO^\bullet$), and others. Reactive nitrogen species include the *nitric oxide radical* ($NO^\bullet$) and others. These free radicals are produced by many cells in the body and serve some important physiological functions. The superoxide radical and nitric oxide radical produced in phagocytic cells such as neutrophils and macrophages, for example, help these cells to destroy bacteria. The superoxide radicals in phagocytic cells can be thought of as nonselective antibiotics, killing any infecting bacteria (as well as the neutrophils) and perhaps also injuring surrounding tissue cells, as these radicals contribute to the inflammation reaction. In addition, the superoxide radicals promote cellular proliferation (mitotic division) of fibroblasts, so that scar tissue can form. The superoxide radicals have similarly been shown to stimulate proliferation of lymphocytes in the process of clone production (chapter 15). The nitric oxide radical also has physiological actions, promoting relaxation of vascular smooth muscle and thus vasodilation (chapter 14), so that more blood can flow to the site of the inflammation. Thus, free radicals do serve useful physiological roles in the body. (fig. 19.1)

Excessive production of free radicals, however, can damage lipids, proteins, and DNA, and by this means exert an **oxidative stress** on the body. Oxidative stress has wide-ranging ill effects (fig. 19.1). It promotes cell death (apoptosis), contributes to aging and degenerative diseases associated with aging, promotes the malignant growth of cancers, and contributes to all inflammatory diseases (such as glomerulonephritis, rheumatoid arthritis, and lupus erythematosus). It has been implicated in ischemic heart disease, stroke, hyper-

tension, and a variety of neurological diseases, including multiple sclerosis, Alzheimer's disease, Parkinson's disease, and others. The wide range of diseases associated with oxidative stress stems from the widespread production of superoxide radicals in the mitochondria of all cells that undergo aerobic respiration.

The body protects itself against oxidative stress through various means, both enzymatic and nonenzymatic. The enzymes that help to prevent an excessive buildup of oxidants include *superoxide dismutase (SOD), catalase,* and *glutathione peroxidase.* The SOD enzyme catalyzes a reaction where two superoxide radicals form *hydrogen peroxide* (H_2O_2) and O_2. Hydrogen peroxide is not a free radical, but it is a potentially toxic oxidant (it can accept two more electrons), and so it must be eliminated. This is accomplished by catalase, in a reaction where two hydrogen peroxide molecules react to form H_2O and O_2; and by glutathione peroxidase, in a reaction where H_2O_2 reacts with NADPH + H$^+$ to form NADP and H_2O.

The body also protects itself from oxidative stress through nonenzymatic means (fig. 19.1). One of the most important protective mechanisms is the action of a tripeptide called **glutathione.** When it is in its reduced state, glutathione can react with certain free radicals and render them harmless. Thus, glutathione is said to be the major cellular **antioxidant.** *Ascorbic acid* (vitamin C), in the aqueous phase of cells, and α-*tocopherol* (the major form of vitamin E), in the lipid phase, help in this antioxidant function by picking up unpaired electrons from free radicals. This is said to "quench" the free radicals, although, in the reaction, vitamins C and E themselves gain an unpaired electron and thus become free radicals. Because of their chemical structures, however, they are weaker free radicals than those they quench. Many other molecules present in foods (primarily fruits and vegetables) have been shown to also have antioxidant properties, and research on the actions and potential health benefits of antioxidants is ongoing.

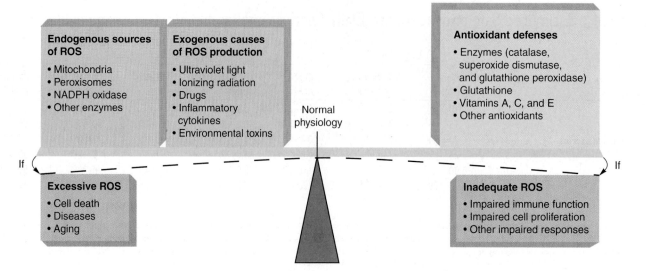

Figure 19.1 Reactive oxygen species (ROS) production and defense. Normal physiology requires that the reactive oxygen species (those that contain oxygen with an unpaired electron) be kept in balance.

Test Yourself Before You Continue

1. Explain how the metabolic rate is influenced by exercise, ambient temperature, and the assimilation of food.
2. Distinguish between the caloric and anabolic requirements of the diet.
3. List the water-soluble and fat-soluble vitamins and describe some of their functions.
4. Explain how vitamin D functions as a vitamin, hormone, and autocrine regulator.

Regulation of Energy Metabolism

The blood plasma contains circulating glucose, fatty acids, amino acids, and other molecules that can be used by the body tissues for cell respiration. These circulating molecules may be derived from food or from the breakdown of the body's own glycogen, fat, and protein. The building of the body's energy reserves following a meal and the utilization of these reserves between meals are regulated by the action of a number of hormones that act to promote either anabolism or catabolism.

The molecules that can be oxidized for energy by the processes of cell respiration may be derived from the **energy reserves** of glycogen, fat, or protein. Glycogen and fat function primarily as energy reserves; for proteins, by contrast, this represents a secondary, emergency function. Although body protein can provide amino acids for energy, it can do so only through the breakdown of proteins needed for muscle contraction, struc-

tural strength, enzymatic activity, and other functions. Alternatively, the molecules used for cell respiration can be derived from the products of digestion that are absorbed through the small intestine. Since these molecules—glucose, fatty acids, amino acids, and others—are carried by the blood to the cells for use in cell respiration, they can be called **circulating energy substrates** (fig. 19.2).

Because of differences in cellular enzyme content, different organs have different *preferred energy sources*. This concept was introduced in chapter 5. The brain has an almost absolute requirement for blood glucose as its energy source, for example. A fall in the plasma concentration of glucose to below about 50 mg per 100 ml can thus "starve" the brain and have disastrous consequences. Resting skeletal muscles, by contrast, use fatty acids as their preferred energy source. Similarly, ketone bodies (derived from fatty acids), lactic acid, and amino acids can be used to different degrees as energy sources by various organs. The plasma normally contains adequate concentrations of all of these circulating energy substrates to meet the energy needs of the body.

Eating

Ideally, one should eat the kinds and amounts of foods that provide adequate vitamins, minerals, essential amino acids and fatty acids, and calories. Proper caloric intake maintains energy reserves (primarily fat and glycogen) and results in a body weight within an optimum range for health.

Eating behavior appears to be at least partially controlled by areas of the hypothalamus. Lesions (destruction) in the ventromedial area of the hypothalamus produce *hyperphagia,* or overeating, and obesity in experimental animals. Lesions of the lateral hypothalamus, by contrast, produce *hypophagia* and weight loss. More recent experiments demonstrate that other brain regions are also involved in the control of eating behavior.

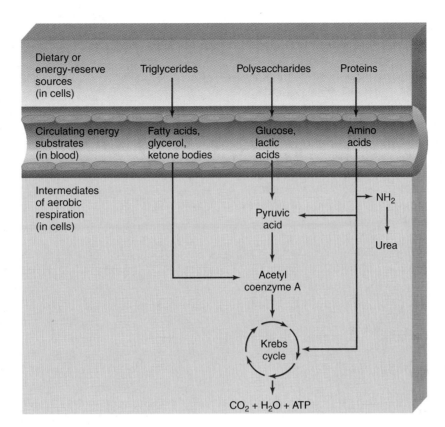

Figure 19.2 **A flowchart of energy pathways in the body.** The molecules indicated in the bottom rectangles are those found within cells, while the molecules in the middle rectangle are those that circulate in the blood.

The neurotransmitters that may be involved in neural pathways for eating behavior are being investigated. There is evidence, for example, that endorphins may be involved because injections of naloxone (a morphine-blocking drug) suppress overeating in rats. There is also evidence that the neurotransmitters norepinephrine and serotonin may be involved; injections of norepinephrine into the brain cause overeating in rats, whereas injections of serotonin have the opposite effect. Indeed, the diet pills *Redux* (D-fenfluramine) and *fen-phen* (L-fenfluramine and phentermine) work to reduce hunger by elevating brain levels of serotonin. (Both drugs have been taken off the market because of their association with heart valve problems.) The regulation of hunger is discussed shortly in a separate section.

Regulatory Functions of Adipose Tissue

It is difficult for a person to lose (or gain) weight, many scientists believe, because the body has negative feedback loops that act to "defend" a particular body weight, or more accurately, the amount of adipose tissue. This regulatory system has been called an *adipostat*. When a person eats more than is needed to maintain the set point of adipose tissue, the person's metabolic rate increases and hunger decreases, as previously described. Homeostasis of body weight implies negative feedback loops. Hunger and metabolism (acting through food and hormones) affect adi-

pose cells, so in terms of negative feedback, it seems logical that adipose cells should influence hunger and metabolism.

Adipose cells, or **adipocytes,** store fat within large vacuoles during times of plenty and serve as sites for the release of circulating energy substrates, primarily free fatty acids, during times of fasting. Since the synthesis and breakdown of fat is controlled by hormones that act on the adipocytes, the adipocytes traditionally have been viewed simply as passive storage depots of fat. Recent evidence suggests quite the opposite, however; adipocytes may themselves secrete hormones that play a pivotal role in the regulation of metabolism.

Development of Adipose Tissue

Some adipocytes appear during embryonic development, but their numbers increase greatly following birth. This increased number is due to both mitotic division of the adipocytes and the conversion of preadipocytes (derived from fibroblasts) into new adipocytes. This differentiation (specialization) is promoted by a high circulating level of fatty acids, particularly of saturated fatty acids. This represents a nice example of a negative feedback loop, where a rise in circulating fatty acids promotes processes that ultimately help to convert the fatty acids into stored fat.

The differentiation of adipocytes requires the action of a nuclear receptor protein—in the same family as the receptors for thyroid hormone, vitamin A, and vitamin D—known as **PPARγ.** (PPAR is an acronym for *peroxisome proliferator activated* receptor, and the γ is the Greek letter gamma, indicating the subtype of PPAR.) Just as the thyroid receptor is activated when it is bound to its ligand, PPARγ is activated when it is bound to its own specific ligand, a type of prostaglandin. When this ligand binds to the PPARγ receptor, it stimulates adipogenesis by promoting the development of preadipocytes into mature adipocytes. This occurs primarily in children, since the development of new adipocytes is more limited in adults.

Regulatory Functions of Adipocytes

In addition to storing fat (triglyceride, or triacylglycerol), adipocytes produce and secrete regulatory molecules. One of the most important of these is **leptin** (Greek *leptos* = thin), a hormone that signals the hypothalamus to indicate the level of fat storage. This hormone is involved in long-term regulation of eating and metabolism, as described in the next section. Another regulatory molecule produced by adipocytes is **tumor necrosis factor-alpha** (**TNFα**). TNFα is a cytokine that is also produced by macrophages and other cells of the immune system. When it is produced by adipocytes, the TNFα may act to reduce the sensitivity of cells (primarily skeletal muscle) to insulin. This may contribute to the *insulin resistance* that is observed in obese people.

The term "insulin resistance" refers to the fact that more insulin is required to maintain normal blood glucose concentrations. People with excessive amounts of large adipocytes (concentrated in the greater omentum) require more insulin to maintain normal blood glucose levels than do thinner people, who have smaller adipocytes.

Clinical Investigation Clue

Remember that Phyllis is moderately overweight and that the physician recommended a weight-reduction program to help alleviate her symptoms.

How does this relate to her sensitivity to insulin?

The importance of adipose tissue in the regulation of insulin sensitivity is shown by the action of a new class of drugs, the *thiazolidinediones*, for the treatment of **type 2 diabetes** (discussed later). These drugs currently include rosiglitazone (*Avandia*) and pioglitazone (*Actos*). They have one selective action: they bind to and activate the PPARγ nuclear receptors. This results in apoptosis of large adipocytes and an increase in the number of small adipocytes, among other effects. Accompanying these changes, there is decreased secretion of TNFα, leptin, and a newly discovered molecule named *resistin*, from the adipose tissue. Decreased adipose secretion of these products, perhaps together with other effects, may be responsible for the ability of the thiazolidinedione drugs to decrease insulin resistance in people with type 2 diabetes.

Regulation of Hunger

The possibility that adipose tissue secretes a hormonal *satiety factor* (a circulating chemical that decreases appetite) has been suspected for years on the basis of physiological evidence. According to this view, secretion of the satiety factor would increase following meals and decrease during fasting. Such a satiety factor could act through its regulation of the hunger centers in the hypothalamus.

The satiety factor secreted by adipose tissue has recently been identified. It is the product of a gene first observed in a strain of mice known as *ob/ob* (*ob* designates "obese"; the double symbol indicates that the mice are homozygous for this gene—they inherit it from both parents). Mice of this strain display hyperphagia (they eat too much) and decreased energy consumption. The *ob* gene has been cloned in mice and humans, and has been found to be expressed (produce mRNA) only in adipocytes. As expected, the expression of this gene is decreased during fasting and increased after feeding. The protein product of this gene, the presumed satiety factor, is a 167-amino-acid polypeptide now called *leptin*. The *ob* mice produce a mutated and ineffective form of leptin, and it is this defect that causes their obesity. When they are injected with normal leptin they stop eating and lose weight.

Scientists have also identified a few obese people with defective leptin genes. However, studies in humans show that the activity of the *ob* gene and the blood concentrations of leptin are raised in most obese people, and that weight loss results in a lowering of plasma leptin concentrations. Thus, unlike the case of the *ob/ob* mice, most cases of obesity in humans may be caused by a reduced sensitivity of the brain to the actions of leptin.

In the *ob/ob* mice, it was observed that injections of leptin caused a decreased amount of neuropeptide Y in the hypothalamus. This observation provides a clue about how leptin might work. As discussed in chapter 7, neuropeptide Y is a potent stimulator of appetite. It functions as a neurotransmitter of axons that extend within the hypothalamus from the arcuate nucleus to the paraventricular nucleus, two regions implicated in the control of eating behavior. When weight is lost, a reduced secretion of leptin from the adipocytes may result in increased production of neuropeptide Y, which then stimulates increased hunger and food intake and decreased expenditure of energy.

When weight is gained, conversely, an increased secretion of leptin may reduce hunger by inhibiting neuropeptide Y release in the hypothalamus. The control of hunger, however, appears to be more complex than this. Scientists have discovered that appetite can be suppressed by melanocyte-stimulating hormone (MSH) or by a related neuropeptide of the *melanocortin* family that binds to a specific melanocortin receptor in the hypothalamus. It has thus been proposed that when weight is gained, the rising levels of leptin may increase the activity of these melanocortin pathways, suppressing appetite and increasing energy expenditure.

In summary (fig 19.3), leptin is believed to target the arcuate nucleus of the hypothalamus, where it affects two populations of neurons. One population produces neuropeptide Y; these neurons are inhibited by leptin. The other population produces MSH and is stimulated by leptin. As a result, high leptin levels should suppress appetite, while lowered leptin levels should promote appetite. These effects are believed to help the body maintain its usual level of *adiposity* (fat storage).

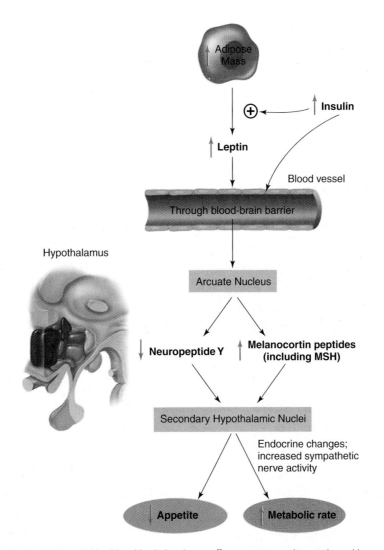

■ Figure 19.3 **The action of leptin.** Leptin crosses the blood-brain barrier to affect neurotransmitters released by neurons in the arcuate nucleus of the hypothalamus. This influences other hypothalamic nuclei, which in turn reduce appetite and increase metabolic rate. The figure also shows that insulin stimulates adipose cells to secrete leptin and is able to cross the blood-brain barrier and to act in a manner similar to leptin.

The actions of leptin and insulin (fig. 19.3) are important in the long-term regulation of eating and, by this means, in maintaining homeostasis of body weight. Other hormones are believed to be more involved in meal-to-meal feelings of hunger and satiety (a feeling of "fullness," and thus reduction of appetite). One of these is a recently discovered hormone called **ghrelin,** secreted by the stomach. Ghrelin secretion rises between meals, when the stomach is empty, and stimulates hunger. As the stomach fills during a meal, the secretion of ghrelin rapidly falls. Another hormone that regulates eating is the intestinal hormone **cholecystokinin (CCK).** Secretion of CCK rises during and immediately after a meal, and has been found to promote satiety. These two hormones thus act antagonistically on the arcuate nucleus of the hypothalamus (fig. 19.3) to promote feelings of hunger and satiety before and after a meal.

Intermediate between the long-term regulation of eating by leptin and insulin, and the short-term regulation by ghrelin and CCK, is a newly discovered hormone named PYY_{3-36}. This hormone is secreted by the small intestine in proportion to the calorie content of a meal. In a recent study, injections of this

hormone into humans was found to suppress appetite for up to 12 hours after a meal. Like leptin, PYY_{3-36} acts on the arcuate nucleus to decrease neuropeptide Y and increase melanocortin peptides (fig. 19.3). Through this action, PYY_{3-36} may regulate appetite to help determine the spacings between meals.

Low Adiposity: Starvation

Starvation and malnutrition are the leading causes of diminished immune capacity worldwide. People suffering from these conditions are thus more susceptible to infections. It is interesting in this regard that leptin receptors have been identified on the surface of helper T lymphocytes, which aid both humoral and cell-mediated immune responses (chapter 15).

People suffering from starvation have reduced adipose tissue and hence decreased leptin secretion. This can contribute to a decline in the ability of helper T lymphocytes to promote the immune response, and thus can—at least in part—account for the decline in immune competence in people who are starving.

The hypothalamus—a target of leptin action regulating appetite—is also involved in regulation of the reproductive system (chapter 20). There is evidence that leptin may be involved in regulating the onset of puberty and the menstrual cycle *(menarche)*. Adolescent girls who are excessively thin enter puberty later than the average age, and very thin women can experience amenorrhea (cessation of menstrual cycles). Adequate amounts of adipose tissue are thus required for proper functioning of the immune and reproductive systems.

 Anorexia nervosa and **bulimia nervosa** are eating disorders that affect mostly young women who are obsessively concerned about their weight and body shape. Anorexia is a potentially fatal condition caused by a compulsive pursuit of excessive thinness. There can be seriously low heart rate and blood pressure, decreased estrogen secretion and amenorrhea, and depression. In bulimia, the person engages in large, uncontrolled eating binges followed by methods to prevent weight gain, such as vomiting. Anorexia and bulimia are most common in societies where thinness is exalted but food is plentiful.

Obesity

Obesity is a risk factor for cardiovascular diseases, diabetes mellitus, gallbladder disease, and some malignancies (particularly endometrial and breast cancer). The distribution of fat in the body is also important; there is a greater risk of cardiovascular disease when the distribution of fat produces a high waist-to-hip ratio, or an "apple shape," as compared to a "pear shape." This is because the amount of intra-abdominal fat in the mesenteries and greater omentum is a better predictor of a health hazard than is the amount of subcutaneous fat. In terms of the risk of diabetes mellitus, the larger adipocytes of the "apple shape" are less sensitive to insulin than the smaller adipocytes of the "pear shape."

Clinical Investigation Clue

Remember that Phyllis is moderately overweight.

For which diseases does this place Phyllis in a higher than average risk category?

Obesity in childhood is due to an increase in both the size and number of adipocytes. Weight gain in adulthood is due mainly to an increase in adipocyte size in intra-abdominal fat, although the number of adipocytes elsewhere in the body can increase through activation of the PPARγ receptor (previously discussed). When weight is lost the adipocytes get smaller, but their number remains constant. It is thus important to prevent further increases in weight in all overweight people, but particularly so in children. This can best be achieved by a carefully chosen diet, low in saturated fat

(because of the effect of fatty acids on adipocyte growth and differentiation, as previously described), and exercise. Prolonged exercise of low to moderate intensity promotes weight loss because, under these conditions, skeletal muscles use fatty acids as their primary source of energy (chapter 12; see fig. 12.21).

Obesity is often diagnosed using a measurement called the **body mass index (BMI).** This measurement is calculated using the following formula:

$$\text{BMI} = \frac{w}{h^2}$$

where

w = weight in kilograms (pounds divided by 2.2)
h = height in meters (inches divided by 39.4)

Obesity has been defined by health agencies in different ways. The World Health Organization classifies people with a BMI of 30 or over as being at high risk for the diseases of obesity. According to the standards set by the National Institutes of Health, a healthy weight is indicated by a BMI between 19 and 25 (a BMI in the range of 25.0 to 29.9 is described as "overweight"; a BMI of over 30 is "obese"). According to a recent study, however, the lowest death rates from all causes occurred in men with a BMI in the range of 22.5 to 24.9, and in women with a BMI in the range of 22.0 to 23.4. Surveys suggest that most Americans have a BMI above these levels.

Caloric Expenditures

The caloric energy expenditure of the body has three components:

1. **Basal metabolic rate (BMR)** is the energy expenditure of a relaxed, resting person who is at a neutral ambient temperature (about 28° C) and who has not eaten in 8 to 12 hours. This comprises the majority (about 60%) of the total calorie expenditure in an average adult.
2. **Adaptive thermogenesis** is the heat energy expended in response to (a) changes in ambient temperature and (b) the digestion and absorption of food. This comprises about 10% of the total calorie expenditure, although this contribution can change in response to cold and diet.
3. **Physical activity** raises the metabolic rate and energy expenditure of skeletal muscles. This contribution to the total calorie expenditure is highly variable, depending on the type and intensity of the physical activity (see table 19.1).

In adaptive thermogenesis, a cold environment evokes shivering, which increases the metabolic rate and heat production of skeletal muscles. Since the skeletal muscles comprise about 40% of the total body weight, their metabolism has a profound effect on body temperature. Additionally, skeletal muscles may also contribute to **nonshivering thermogenesis** (heat production). There is evidence that skeletal muscles can produce an *uncoupling protein* that "uncouples" oxidative phosphorylation in mitochondria by causing the leakage of protons (H^+) from the intermembrane space (chapter 5; see fig. 5.10). This decreases the production of ATP, which in turn stimulates increased cell respiration and consequent heat production. Increased metabolism by brown fat (discussed in chapter 5) may also contribute to heat production in humans as a result of the production of uncoupling proteins in the mitochondria of brown adipocytes.

Diet is also an important regulator of adaptive thermogenesis, producing what is called the **thermic effect of food.** Interestingly, starvation decreases the metabolic rate by as much as 40%, and feeding increases the metabolic rate by about 25% to 40% in average adults, with corresponding increases in heat production. This diet-induced thermogenesis is most evident when diets are low in proteins.

The brain regulates adaptive thermogenesis. This is accomplished largely by activation of the sympathoadrenal system, where sympathetic innervation of skeletal muscles and brown fat, together with the effects of circulating epinephrine, cause increased metabolism in these tissues. Thyroid hormone secretion, controlled by the brain via TRH (thyrotropin-releasing hormone, which stimulates TSH secretion from the anterior pituitary—see chapter 11, fig. 11.16), is also needed for adaptive thermogenesis. Although thyroxine secretion is required for adaptive thermogenesis, the levels of thyroxine do not rise in response to cold or food, suggesting that the role of thyroid hormones is mainly a permissive one. In starvation, however, thyroxine levels do fall, suggesting that this decline may contribute to the slowdown in the metabolic rate during starvation.

During starvation, adipose tissue decreases its secretion of leptin, and this fall is needed for the fall in TRH secretion by the brain that occurs in starvation. Thus, the decline in leptin may be responsible for the decline in thryroxine secretion. There is also evidence that decreasing leptin during starvation may cause a decline in the sympathetic nerve stimulation of brown fat. Through both mechanisms, the decreased leptin levels that occur during starvation could cause a slowdown in the metabolic rate. This effect would help to conserve energy during starvation. Opposite responses when leptin levels are high, conversely, would help to raise the metabolic rate and put a brake on the growth of adipose tissue (fig. 19.3).

Hormonal Regulation of Metabolism

The absorption of energy carriers from the intestine is not continuous; it rises to high levels over a 4-hour period following each meal (the **absorptive state**) and tapers toward zero between meals, after each absorptive state has ended (the **postabsorptive,** or **fasting, state**). Despite this fluctuation, the plasma concentration of glucose and other energy substrates does not remain high during periods of absorption, nor it does normally fall below a certain level during periods of fasting. During the absorption of digestion products from the intestine, energy substrates are removed from the blood and deposited as energy reserves from which withdrawals can be made during times of fasting (fig. 19.4). This ensures an adequate plasma

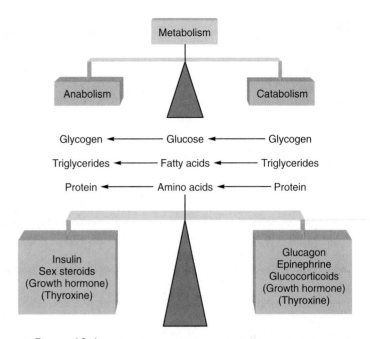

Figure 19.4 **The regulation of metabolic balance.** The balance of metabolism can be tilted toward anabolism (synthesis of energy reserves) or catabolism (utilization of energy reserves) by the combined actions of various hormones. Growth hormone and thyroxine have both anabolic and catabolic effects.

concentration of energy substrates to sustain tissue metabolism at all times.

The rate of deposit and withdrawal of energy substrates into and from the energy reserves and the conversion of one type of energy substrate into another are regulated by hormones. The balance between anabolism and catabolism is determined by the antagonistic effects of insulin, glucagon, growth hormone, thyroxine, and other hormones (fig. 19.4). The specific metabolic effects of these hormones are summarized in table 19.5, and some of their actions are illustrated in figure 19.5.

Table 19.5 Endocrine Regulation of Metabolism

Hormone	Blood Glucose	Carbohydrate Metabolism	Protein Metabolism	Lipid Metabolism
Insulin	Decreased	↑ Glycogen formation ↓ Glycogenolysis ↓ Gluconeogenesis	↑ Protein synthesis	↑ Lipogenesis ↓ Lipolysis ↓ Ketogenesis
Glucagon	Increased	↓ Glycogen formation ↑ Glycogenolysis ↑ Gluconeogenesis	No direct effect	↑ Lipolysis ↑ Ketogenesis
Growth hormone	Increased	↑ Glycogenolysis ↑ Gluconeogenesis ↓ Glucose utilization	↑ Protein synthesis	↓ Lipogenesis ↑ Lipolysis ↑ Ketogenesis
Glucocorticoids (hydrocortisone)	Increased	↑ Glycogen formation ↑ Gluconeogenesis	↓ Protein synthesis	↓ Lipogenesis ↑ Lipolysis ↑ Ketogenesis
Epinephrine	Increased	↓ Glycogen formation ↑ Glycogenolysis ↑ Gluconeogenesis	No direct effect	↑ Lipolysis ↑ Ketogenesis
Thyroid hormones	No effect	↑ Glucose utilization	↑ Protein synthesis	No direct effect

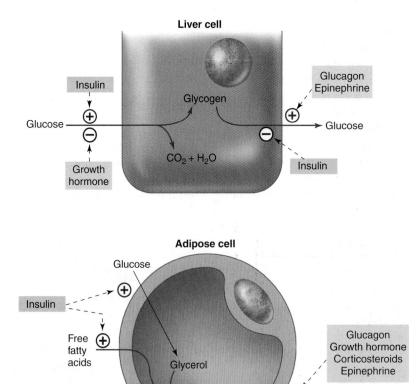

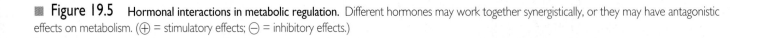

■ **Figure 19.5** **Hormonal interactions in metabolic regulation.** Different hormones may work together synergistically, or they may have antagonistic effects on metabolism. (⊕ = stimulatory effects; ⊖ = inhibitory effects.)

Energy Regulation by the Islets of Langerhans

Insulin secretion is stimulated by a rise in the blood glucose concentration, and insulin promotes the entry of blood glucose into tissue cells. Insulin thus increases the storage of glycogen and fat while causing the blood glucose concentration to fall. Glucagon secretion is stimulated by a fall in blood glucose, and glucagon acts to raise the blood glucose concentration by promoting glycogenolysis in the liver.

Scattered within a "sea" of pancreatic exocrine tissue (the acini) are islands of hormone-secreting cells (chapter 18; see fig. 18.28). These islets of Langerhans contain three distinct cell types that secrete different hormones. The most numerous are the *beta cells,* which secrete the hormone **insulin.** About 60% of each islet consists of beta cells. The *alpha cells* form about 25% of each islet and secrete the hormone **glucagon.** The least numerous cell type, the *delta cells,* produce **somatostatin,** the composition of which is identical to the somatostatin produced by the hypothalamus and the intestine.

All three pancreatic hormones are polypeptides. Insulin consists of two polypeptide chains—one that is twenty-one amino acids long and another that is thirty amino acids long—joined together by disulfide bonds. Glucagon contains twenty-one amino acids, and somatostatin contains fourteen. Insulin was the first of these hormones to be discovered (in 1921). The importance of insulin in diabetes mellitus was immediately recognized, and clinical use of insulin in the treatment of this disease began almost immediately after its discovery. The physiological role of glucagon was discovered later, but the physiological significance of islet-secreted somatostatin is still not well understood.

Regulation of Insulin and Glucagon Secretion

Insulin and glucagon secretion is largely regulated by the plasma concentrations of glucose and, to a lesser degree, of amino acids. The alpha and beta cells, therefore, act as both the sensors and effectors in this control system. Since the plasma concentration of glucose and amino acids rises during the absorption of a meal and falls during fasting, the secretion of insulin and glucagon likewise fluctuates between the absorptive and postabsorptive states. These changes in insulin and glucagon secretion, in turn, cause changes in plasma glucose and amino acid concentrations and thus help to maintain homeostasis via negative feedback loops (fig. 19.6).

As described in chapter 6, insulin stimulates the insertion of GLUT4 channels into the plasma membrane (due to the fusion of intracellular vesicles with the plasma membrane—see fig. 6.15) of its target cells, primarily in the skeletal and cardiac muscles, adipose tissue, and liver. This permits the entry of glucose into its target cells by facilitated diffusion. As a result, in-

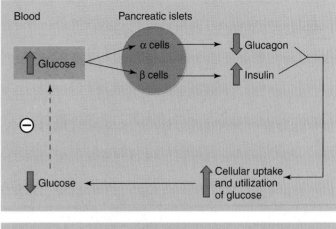

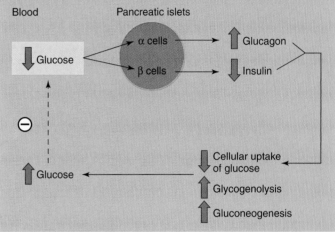

Figure 19.6 **The regulation of insulin and glucagon secretion.** The secretion from the β (beta) cells and α (alpha) cells of the pancreatic islets is regulated largely by the blood glucose concentration. (*a*) A high blood glucose concentration stimulates insulin and inhibits glucagon secretion. (*b*) A low blood glucose concentration stimulates glucagon and inhibits insulin secretion.

sulin promotes the production of the energy-storage molecules of glycogen and fat. Both actions decrease the plasma glucose concentration. Insulin also inhibits the breakdown of fat, induces the production of fat-forming enzymes, and inhibits the breakdown of muscle proteins. Thus, insulin promotes anabolism as it regulates the blood glucose concentration.

The mechanisms that regulate insulin and glucagon secretion and the actions of these hormones normally prevent the plasma glucose concentration from rising above 170 mg per 100 ml after a meal or from falling below about 50 mg per 100 ml between meals. This regulation is important because abnormally high blood glucose can damage certain tissues (as may occur in diabetes mellitus), and abnormally low blood glucose can damage the brain. The latter effect results from the fact that glucose enters the brain by facilitated diffusion; when the rate of this diffusion is too low, as a result of low plasma glucose concentrations, the supply of metabolic energy for the brain may become inadequate. This can result in weakness, dizziness, personality changes, and ultimately in coma and death.

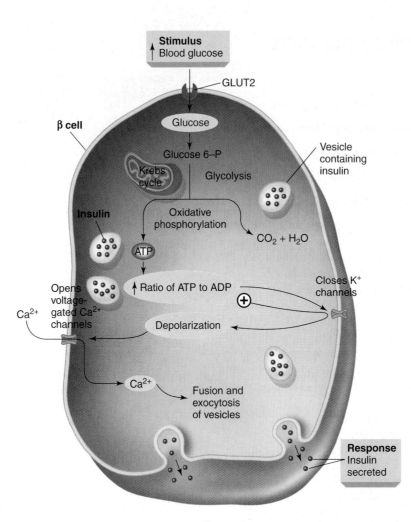

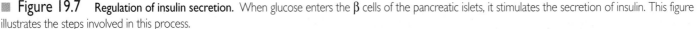

Figure 19.7 **Regulation of insulin secretion.** When glucose enters the β cells of the pancreatic islets, it stimulates the secretion of insulin. This figure illustrates the steps involved in this process.

Effects of Glucose and Amino Acids

The fasting plasma glucose concentration is in the range of 65 to 105 mg/dl. During the absorption of a meal, the plasma glucose concentration usually rises to a level between 140 and 150 mg/dl. This rise in plasma glucose (1) stimulates the beta cells to secrete insulin (fig 19.7), and (2) inhibits the secretion of glucagon from the alpha cells. Insulin then acts to stimulate the cellular uptake of plasma glucose. A rise in insulin secretion therefore lowers the plasma glucose concentration. Since glucagon has the antagonistic effect of raising the plasma glucose concentration by stimulating glycogenolysis in the liver, the inhibition of glucagon secretion complements the effect of increased insulin during the absorption of a carbohydrate meal. A rise in insulin and a fall in glucagon secretion thus help to lower the high plasma glucose concentration that occurs during periods of absorption.

During fasting, the plasma glucose concentration falls. At this time, therefore, (1) insulin secretion decreases and (2) glucagon secretion increases. These changes in hormone secretion prevent the cellular uptake of blood glucose into organs such as the muscles, liver, and adipose tissue and promote the release of glucose from the liver (through the stimulation of glycogen breakdown by glucagon). A negative feedback loop is therefore completed (fig. 19.6), helping to retard the fall in plasma glucose concentration that occurs during fasting.

The **oral glucose tolerance test** (fig. 19.8) is a measure of the ability of the beta cells to secrete insulin and of the ability of insulin to lower blood glucose. In this procedure, a person drinks a glucose solution and blood samples are taken periodically for plasma glucose measurements. In a normal person, the rise in blood glucose produced by drinking this solution is reversed to normal levels within 2 hours following glucose ingestion. In contrast, the plasma glucose concentration remains at 200 mg/dl or higher 2 hours after the oral glucose challenge in a person with diabetes mellitus.

Clinical Investigation Clues

Remember that Phyllis had a fasting blood glucose concentration of 150 mg/dl and a 2-hour measurement of 220 mg/dl in the oral glucose tolerance test.

What does her fasting blood glucose concentration indicate?

What additional information does her oral glucose tolerance test provide?

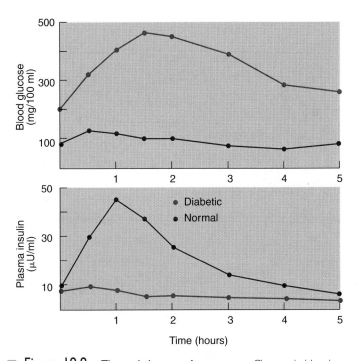

Figure 19.8 **The oral glucose tolerance test.** Changes in blood glucose and plasma insulin concentrations after the ingestion of 100 grams of glucose in an oral glucose tolerance test. The insulin is measured in activity units (U).

Insulin secretion is also stimulated by particular amino acids derived from dietary proteins. Meals that are high in protein, therefore, stimulate the secretion of insulin; if the meal is high in protein and low in carbohydrates, glucagon secretion will be stimulated as well. The increased glucagon secretion acts to raise the blood glucose, while the increased insulin promotes the entry of amino acids into tissue cells.

Effects of Autonomic Nerves

The islets of Langerhans receive both parasympathetic and sympathetic innervation. The activation of the parasympathetic system during meals stimulates insulin secretion at the same time that gastrointestinal function is stimulated. The activation of the sympathetic system, by contrast, stimulates glucagon secretion and inhibits insulin secretion. The effects of glucagon, together with those of epinephrine, produce a "stress hyperglycemia" when the sympathoadrenal system is activated.

Effects of Intestinal Hormones

Surprisingly, insulin secretion increases more rapidly following glucose ingestion than it does following an intravenous injection of glucose. This is due to the fact that the intestine, in response to glucose ingestion, secretes hormones that stimulate insulin secretion before the glucose has been absorbed. Insulin secretion thus begins to rise "in anticipation" of a rise in blood glucose. One of the intestinal hormones that mediates this effect is GIP—gastric inhibitory peptide, or, more appropriately in this context, *glucose-dependent insulinotropic peptide* (chapter 18). Other polypeptide hormones secreted by the intestine that have

similar effects are cholecystokinin (CCK) and glucagon-like peptide-1 (GLP-1), as described in chapter 18.

Insulin and Glucagon: Absorptive State

The lowering of plasma glucose by insulin is, in a sense, a side effect of the primary action of this hormone. Insulin is the major hormone that promotes anabolism in the body. During absorption of the products of digestion and the subsequent rise in the plasma concentrations of circulating energy substrates, insulin promotes the cellular uptake of plasma glucose and its incorporation into energy-reserve molecules of glycogen in the liver and muscles, and of triglycerides in adipose cells (chapter 11; see fig. 11.31). Quantitatively, skeletal muscles are responsible for most of the insulin-stimulated glucose uptake. Insulin also promotes the cellular uptake of amino acids and their incorporation into proteins. The stores of large energy-reserve molecules are thus increased while the plasma concentrations of glucose and amino acids are decreased.

A nonobese 70-kg (155-lb) man has approximately 10 kg (about 82,500 kcal) of stored fat. Since 250 g of fat can supply the energy requirements for 1 day, this reserve fuel is sufficient for about 40 days. Glycogen is less efficient as an energy reserve, .and less is stored in the body; there are about 100 g (400 kcal) of glycogen stored in the liver and 375 to 400 g (1,500 kcal) in skeletal muscles. Insulin promotes the cellular uptake of glucose into the liver and muscles and the conversion of glucose into glucose 6-phosphate. In the liver and muscles, this can be changed into glucose 1-phosphate, which is used as the precursor of glycogen. Once the stores of glycogen have been filled, the continued ingestion of excess calories results in the production of fat rather than of glycogen.

Insulin and Glucagon: Postabsorptive State

The plasma glucose concentration is maintained surprisingly constant during the fasting, or postabsorptive, state because of the secretion of glucose from the liver. This glucose is derived from the processes of glycogenolysis and gluconeogenesis, which are promoted by a high secretion of glucagon coupled with a low secretion of insulin.

Glucagon stimulates and insulin suppresses the hydrolysis of liver glycogen, or **glycogenolysis.** Thus during times of fasting, when glucagon secretion is high and insulin secretion is low, liver glycogen is used as a source of additional blood glucose. This results in the liberation of free glucose from glucose 6-phosphate by the action of an enzyme called *glucose 6-phosphatase* (chapter 5; see fig. 5.4). Only the liver has this enzyme, and therefore only the liver can use its stored glycogen as a source of additional blood glucose. Since muscles lack glucose 6-phosphatase, the glucose 6-phosphate produced from muscle glycogen can be used for glycolysis only by the muscle cells themselves.

Since there are only about 100 grams of stored glycogen in the liver, adequate blood glucose levels could not be maintained

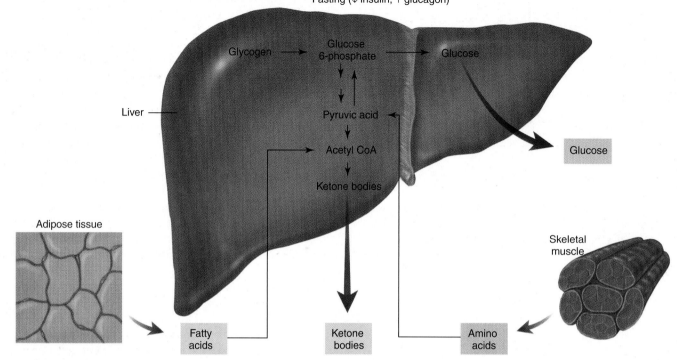

Fasting (↓ insulin, ↑ glucagon)

Figure 19.9 **Catabolism during fasting.** Increased glucagon secretion and decreased insulin secretion during fasting favors catabolism. These hormonal changes promote the release of glucose, fatty acids, ketone bodies, and amino acids into the blood. Notice that the liver secretes glucose that is derived both from the breakdown of liver glycogen and from the conversion of amino acids in gluconeogenesis.

for very long during fasting using this source alone. The low levels of insulin secretion during fasting, together with elevated glucagon secretion, however, promote **gluconeogenesis,** the formation of glucose from noncarbohydrate molecules. Low insulin allows the release of amino acids from skeletal muscles, while glucagon and cortisol (an adrenal hormone) stimulate the production of enzymes in the liver that convert amino acids to pyruvic acid and pyruvic acid into glucose. During prolonged fasting and exercise, gluconeogenesis in the liver using amino acids from muscles may be the only source of blood glucose.

The secretion of glucose from the liver during fasting compensates for the low blood glucose concentrations and helps to provide the brain with the glucose that it needs. But because insulin secretion is low during fasting, skeletal muscles cannot utilize blood glucose as an energy source. Instead, skeletal muscles—as well as the heart, liver, and kidneys—use free fatty acids as their major source of fuel. This helps to "spare" glucose for the brain.

The free fatty acids are made available by the action of glucagon. In the presence of low insulin levels, glucagon activates an enzyme in adipose cells called *hormone-sensitive lipase.* This enzyme catalyzes the hydrolysis of stored triglycerides and the release of free fatty acids and glycerol into the blood. Glucagon also activates enzymes in the liver that convert some of these fatty acids into ketone bodies, which are secreted into the blood (fig. 19.9). Several organs in the body can use ketone bodies, as well as fatty acids, as a source of acetyl CoA in aerobic respiration.

Through the stimulation of **lipolysis** (the breakdown of fat) and **ketogenesis** (the formation of ketone bodies), the high glucagon and low insulin levels that occur during fasting provide circulating energy substrates for use by the muscles, liver, and other organs. Through liver glycogenolysis and gluconeogenesis, these hormonal changes help to provide adequate levels of blood glucose to sustain the metabolism of the brain. The antagonistic action of insulin and glucagon (fig. 19.10) thus promotes appropriate metabolic responses during periods of fasting and periods of absorption.

Test Yourself Before You Continue

1. Describe how the secretions of insulin and glucagon change during periods of absorption and periods of fasting. How are these changes in hormone secretion produced?

2. Explain how the synthesis of fat in adipose cells is regulated by insulin. Also, explain how fat metabolism is regulated by insulin and glucagon during periods of absorption and fasting.

3. Define the following terms: *glycogenolysis, gluconeogenesis,* and *ketogenesis.* How do insulin and glucagon affect each of these processes during periods of absorption and fasting?

4. Describe two pathways used by the liver to produce glucose for secretion into the blood. Why can't skeletal muscles secrete glucose into the blood?

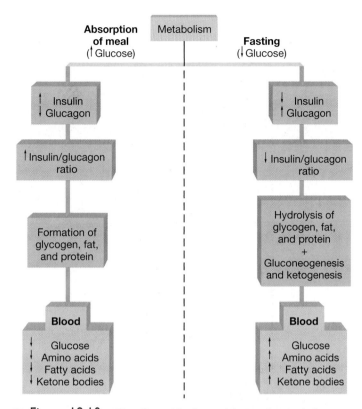

Figure 19.10 The effect of feeding and fasting on metabolism. Metabolic balance is tilted toward anabolism by feeding (absorption of a meal) and toward catabolism by fasting. This occurs because of an inverse relationship between insulin and glucagon secretion. Insulin secretion rises and glucagon secretion falls during food absorption, whereas the opposite occurs during fasting.

Table 19.6 Comparison of Type 1 and Type 2 Diabetes Mellitus

Feature	Type 1	Type 2
Usual age at onset	Under 20 years	Over 40 years
Development of symptoms	Rapid	Slow
Percentage of diabetic population	About 10%	About 90%
Development of ketoacidosis	Common	Rare
Association with obesity	Rare	Common
Beta cells of islets (at onset of disease)	Destroyed	Not destroyed
Insulin secretion	Decreased	Normal or increased
Autoantibodies to islet cells	Present	Absent
Associated with particular MHC antigens*	Yes	Unclear
Treatment	Insulin injections	Diet and exercise; oral stimulators of insulin sensitivity

*Discussed in chapter 15.

Diabetes Mellitus and Hypoglycemia

Inadequate secretion of insulin, or defects in the action of insulin, produce metabolic disturbances that are characteristic of diabetes mellitus. A person with type 1 diabetes requires injections of insulin; a person with type 2 diabetes can control this condition by other methods. In both types, hyperglycemia and glycosuria result from a deficiency and/or inadequate action of insulin. A person with reactive hypoglycemia, by contrast, secretes excessive amounts of insulin and thus experiences hypoglycemia in response to the stimulus of a carbohydrate meal.

Chronic high blood glucose, or hyperglycemia, is the hallmark of **diabetes mellitus.** The name of this disease is derived from the fact that glucose "spills over" into the urine when the blood glucose concentration is too high (*mellitus* is derived from a Latin word meaning "honeyed" or "sweet"). The general term *diabetes* comes from a Greek word meaning "siphon"; it refers to the frequent urination associated with this condition. The hyperglycemia of diabetes mellitus results from either the insufficient secretion of insulin by the beta cells of the islets of Langerhans or the inability of secreted insulin to stimulate the cellular uptake of glucose from the blood. Diabetes mellitus, in short, results from the inadequate secretion or action of insulin.

There are two major forms of diabetes mellitus. In **type 1** (*or insulin-dependent*) **diabetes,** the beta cells are progressively destroyed and secrete little or no insulin; injections of exogenous insulin are thus required to sustain the person's life. This form of the disease accounts for only about 10% of the known cases of diabetes. About 90% of the people who have diabetes have **type 2** (*non-insulin-dependent*) **diabetes.** Type 1 diabetes was once known as *juvenile-onset diabetes* because this condition is usually diagnosed in people under the age of 20. Type 2 diabetes has also been called *maturity-onset diabetes,* since it is usually diagnosed in people over the age of 40. Since the incidence of type 2 diabetes in children is rising (due to an increase in the frequency of obesity), however, these terms are no longer preferred. The two forms of diabetes mellitus are compared in table 19.6. (It should be noted that only the early stages of type 1 and type 2 diabetes mellitus are compared; some people with severe type 2 diabetes may also require insulin injections to control the hyperglycemia.)

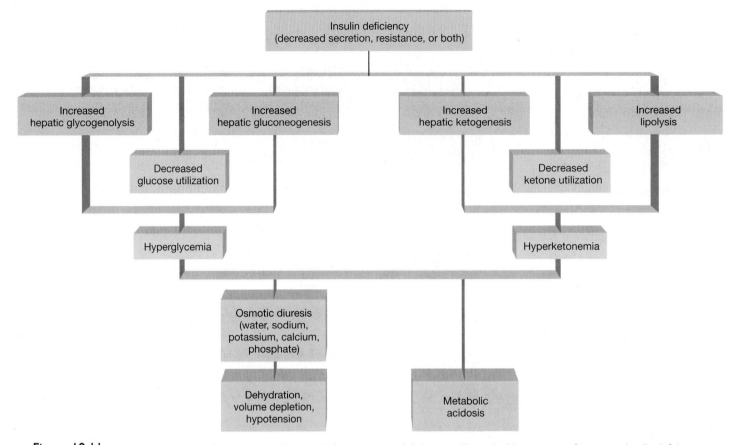

■ **Figure 19.11**   The consequences of an uncorrected insulin deficiency in type I diabetes mellitus. In this sequence of events, an insulin deficiency may lead to coma and death.

Type 1 Diabetes Mellitus

Type 1 diabetes mellitus results when the beta cells of the islets of Langerhans are progressively destroyed by autoimmune attack. Recent evidence in mice suggests that killer T lymphocytes (chapter 15) may target an enzyme known as glutamate decarboxylase in the beta cells. This autoimmune destruction of the beta cells may be provoked by an environmental agent, such as infection by viruses. In other cases, however, the cause is currently unknown. Removal of the insulin-secreting beta cells causes hyperglycemia and the appearance of glucose in the urine. Without insulin, glucose cannot enter the adipose cells; the rate of fat synthesis thus lags behind the rate of fat breakdown and large amounts of free fatty acids are released from the adipose cells.

In a person with uncontrolled type 1 diabetes, many of the fatty acids released from adipose cells are converted into ketone bodies in the liver. This may result in an elevated ketone body concentration in the blood (ketosis), and if the buffer reserve of bicarbonate is neutralized, it may also result in *ketoacidosis*. During this time, the glucose and excess ketone bodies that are excreted in the urine act as osmotic diuretics (chapter 17) and cause the excessive excretion of water in the urine. This can produce severe dehydration, which, together with ketoacidosis and

associated disturbances in electrolyte balance, may lead to coma and death (fig. 19.11).

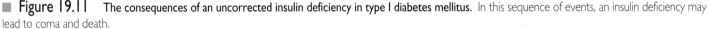

Clinical Investigation Clues

Remember that Phyllis did not have glycosuria when she gave a sample of urine to the physician. Yet she also complained of frequent urination and continuous thirst.

■ *What might cause Phyllis to have glycosuria at other times, which didn't show up in the urine sample?*

■ *Could Phyllis have ketonuria?*

■ *What might be the cause of her frequent urination and continuous thirst?*

In addition to the lack of insulin, people with type 1 diabetes have an abnormally high secretion of glucagon from the alpha cells of the islets. Glucagon stimulates glycogenolysis in the liver and thus helps to raise the blood glucose concentration. Glucagon also stimulates the production of enzymes in the liver that convert fatty acids into ketone bodies. The full range of symptoms of diabetes may result from high glucagon secretion as well as from the absence of insulin. The lack of insulin may be largely responsible for hyperglycemia and for the release of large amounts of fatty acids into the blood. The high glucagon secretion may contribute to the hyperglycemia and in large part causes the development of ketoacidosis.

Type 2 Diabetes Mellitus

The effects produced by insulin, or any hormone, depend on the concentration of that hormone in the blood and on the sensitivity of the target tissue to given amounts of the hormone. Tissue responsiveness to insulin, for example, varies under normal conditions. Exercise increases insulin sensitivity and obesity decreases insulin sensitivity of the target tissues. The islets of a nondiabetic obese person must therefore secrete high amounts of insulin to maintain the blood glucose concentration in the normal range. Conversely, nondiabetic people who are thin and who exercise regularly require lower amounts of insulin to maintain the proper blood glucose concentration.

Type 2 diabetes is usually slow to develop, is hereditary, and occurs most often in people who are overweight. Genetic factors are very significant; people at highest risk are those who have both parents with type 2 diabetes and those who are members of certain ethnic groups, particularly Native Americans of the Southwestern United States and Mexican-Americans. Unlike people with type 1 diabetes, those who have type 2 diabetes can have normal or even elevated levels of insulin in their blood. Despite this, people with type 2 diabetes have hyperglycemia if untreated. This must mean that, even though the insulin levels may be in the normal range, the amount of insulin secreted is insufficient to control blood glucose levels.

Much evidence has been obtained to show that people with type 2 diabetes have an abnormally low tissue sensitivity to insulin, or an **insulin resistance.** This is true even if the person is not obese, but the problem is compounded by the decreased tissue sensitivity that accompanies obesity, particularly of the "apple-shape" variety in which the adipose cells are enlarged. There is also evidence that the beta cells are not functioning correctly: whatever amount of insulin they secrete is inadequate to the task. People who are prediabetic may have *impaired glucose tolerance,* which is defined (in an oral glucose tolerance test) as a plasma glucose level of 140 to 200 mg/dl at 2 hours following the glucose ingestion. As previously mentioned, a value here of over 200 mg/dl indicates diabetes. Since impaired glucose tolerance is accompanied by higher levels of insulin (fig. 19.12), a state of insulin resistance is suggested. People with chronic type 2 diabetes, in contrast, can have both insulin resistance and a deficient secretion of insulin.

Since obesity decreases insulin sensitivity, people who are genetically predisposed to insulin resistance may develop symptoms of diabetes when they gain weight. Conversely, this type of diabetes mellitus can usually be controlled by increasing tissue sensitivity to insulin through diet and exercise. This is beneficial

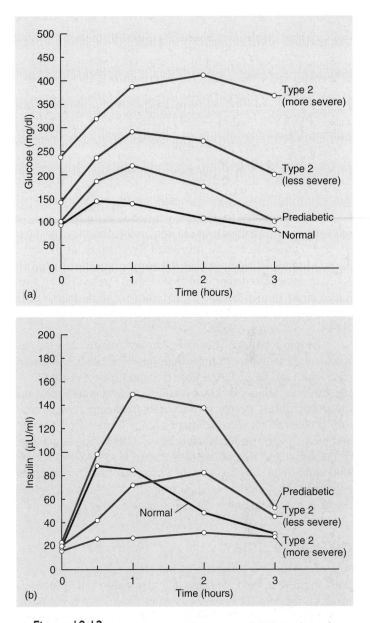

Figure 19.12 Oral glucose tolerance in prediabetes and type 2 diabetes. The oral glucose tolerance test showing (a) blood glucose concentrations and (b) insulin values following the ingestion of a glucose solution. Values are shown for people who are normal, prediabetic, and type 2 (non-insulin-dependent) diabetic. Prediabetics (those who demonstrate "insulin resistance") often show impaired glucose tolerance without fasting hyperglycemia.

Data from Simeon I. Taylor et al., "Insulin Resistance of Insulin Deficiency: Which is the Primary Cause of NIDDM?" in *Diabetes,* vol. 43, June 1994, p. 735.

because exercise, like insulin, increases the amount of membrane GLUT4 carriers (for the facilitative diffusion of glucose) in skeletal muscle cells. If diet and exercise are insufficient, oral drugs are available that increase insulin secretion from the beta cells (e.g., *sulfonylureas*) and that decrease the insulin resistance of the target tissues. These are the *thiazolidinediones* discussed earlier (see page 606).

Recent studies have demonstrated that, in most people with impaired glucose tolerance (who are prediabetic), the onset of type 2 diabetes can be prevented by changes in lifestyle. These changes include exercise and weight reduction, as previously mentioned, together with an increased intake of fiber, a reduced intake of total fat, and a reduced intake of saturated fat. In one recent study, these lifestyle changes decreased the risk of diabetes by 58% after 4 years.

People with type 2 diabetes do not usually develop ketoacidosis. The hyperglycemia itself, however, can be dangerous on a long-term basis. In the United States, diabetes is the leading cause of blindness, kidney failure, and amputation of the lower extremities. People with diabetes frequently have circulatory problems that increase the tendency to develop gangrene and increase the risk for atherosclerosis. The causes of damage to the retina and lens of the eyes and to blood vessels are not well understood. It is believed, however, that these problems result from a long-term exposure to high blood glucose, which damages tissues through a variety of mechanisms.

Hypoglycemia

A person with type 1 diabetes mellitus depends on insulin injections to prevent hyperglycemia and ketoacidosis. If inadequate insulin is injected, the person may enter a coma as a result of the ketoacidosis, electrolyte imbalance, and dehydration that develop. An overdose of insulin, however, can also produce a coma as a result of the hypoglycemia (abnormally low blood glucose levels) produced. The physical signs and symptoms of diabetic and hypoglycemic coma are sufficiently different to allow hospital personnel to distinguish between these two types.

Less severe symptoms of hypoglycemia are usually produced by an oversecretion of insulin from the islets of Langerhans after a carbohydrate meal. This **reactive hypoglycemia,** caused by an exaggerated response of the beta cells to a rise in blood glucose, is most commonly seen in adults who are genetically predisposed to type 2 diabetes. For this reason, people with reactive hypoglycemia must limit their intake of carbohydrates and eat small meals at frequent intervals, rather than two or three meals per day.

The symptoms of reactive hypoglycemia include tremor, hunger, weakness, blurred vision, and mental confusion. The appearance of some of these symptoms, however, does not necessarily indicate reactive hypoglycemia, and a given level of blood glucose does not always produce these symptoms. To confirm a diagnosis of reactive hypoglycemia, a number of tests must be

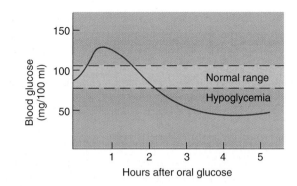

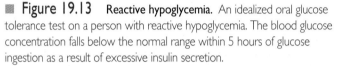

■ **Figure 19.13** **Reactive hypoglycemia.** An idealized oral glucose tolerance test on a person with reactive hypoglycemia. The blood glucose concentration falls below the normal range within 5 hours of glucose ingestion as a result of excessive insulin secretion.

performed. In the oral glucose tolerance test, for example, reactive hypoglycemia is shown when the initial rise in blood glucose produced by the ingestion of a glucose solution triggers excessive insulin secretion, so that the blood glucose levels fall below normal within 5 hours (fig. 19.13).

Metabolic Regulation by Adrenal Hormones, Thyroxine, and Growth Hormone

Epinephrine, the glucocorticoids, thyroxine, and growth hormone stimulate the catabolism of carbohydrates and lipids. These hormones are thus antagonistic to insulin in their regulation of carbohydrate and lipid metabolism. Thyroxine and growth hormone promote protein synthesis, however, and are needed for body growth and proper development of the central nervous system. The stimulatory effect of these hormones on protein synthesis is complementary to that of insulin.

The anabolic effects of insulin are antagonized by glucagon, as previously described, and by the actions of a variety of other hormones. The hormones of the adrenals, thyroid, and anterior pituitary (specifically growth hormone) antagonize the action of

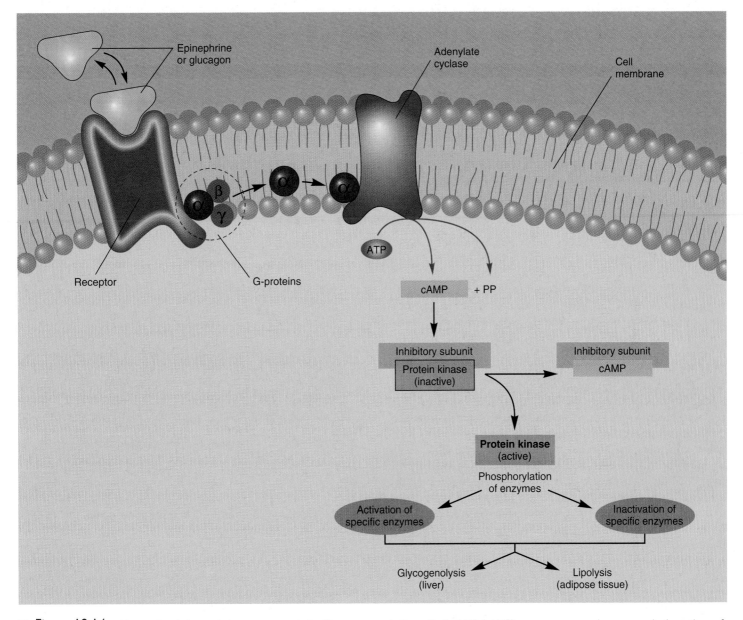

■ Figure 19.14   How epinephrine and glucagon exert their effects on metabolism.   Cyclic AMP (cAMP) serves as a second messenger in the actions of epinephrine and glucagon on liver and adipose tissue metabolism. (The mechanisms of hormone action are discussed in detail in chapter 11.)

insulin on carbohydrate and lipid metabolism. The actions of insulin, thyroxine, and growth hormone, however, can act synergistically in the stimulation of protein synthesis.

Adrenal Hormones

As described in chapter 11, the adrenal gland consists of two parts that function as separate glands. The two parts secrete different hormones and are regulated by different control systems. The **adrenal medulla** secretes catecholamine hormones—epinephrine and lesser amounts of norepinephrine—in response to sympathetic nerve stimulation. The **adrenal cortex** secretes corticosteroid hormones. These are grouped into two functional categories: **mineralocorticoids,** such as aldosterone, which act on the kidneys to regulate Na^+ and K^+ balance (chapter 17), and **glucocorticoids,** such as hydrocortisone (cortisol), which participate in metabolic regulation.

Metabolic Effects of Catecholamines

The metabolic effects of catecholamines (epinephrine and norepinephrine) are similar to those of glucagon. They stimulate glycogenolysis and the release of glucose from the liver, as well as lipolysis and the release of fatty acids from adipose tissue. These actions occur in response to glucagon during fasting, when low blood glucose stimulates glucagon secretion, and in response to catecholamines during the fight-or-flight reaction to stress. The latter effect provides circulating energy substrates in anticipation of the need for intense physical activity. Glucagon and epinephrine have similar mechanisms of action; the actions of both are mediated by cyclic AMP (fig. 19.14).

Sympathetic nerves, acting through the release of norepinephrine, can stimulate **β_3-adrenergic receptors** in brown adipose tissue (there appears to be few β_3 receptors in the ordinary, white fat of humans, and none in other tissues). As may be recalled from chapter 5, brown fat is a specialized tissue that contains an *uncoupling protein* that dissociates electron transport from the production of ATP. As a result, brown fat can have a very high rate of energy expenditure (unchecked by negative feedback from ATP) that is stimulated by catecholamines and thyroid hormones. Since human adults have less brown fat than is present in neonates, however, the significance of this effect is not fully understood. Even so, it has been shown that genetic defects in the β_3-adrenergic receptors of some people are associated with obesity and type 2 diabetes mellitus.

Metabolic Effects of Glucocorticoids

Hydrocortisone (cortisol) and other glucocorticoids are secreted by the adrenal cortex in response to ACTH stimulation. The secretion of ACTH from the anterior pituitary occurs as part of the general adaptation syndrome in response to stress (chapter 11). Since prolonged fasting or prolonged exercise certainly qualify as stressors, ACTH—and thus glucocorticoid secretion—is stimulated under these conditions. The increased secretion of glucocorticoids during prolonged fasting or exercise supports the effects of increased glucagon and decreased insulin secretion from the pancreatic islets.

Like glucagon, hydrocortisone promotes lipolysis and ketogenesis; it also stimulates the synthesis of hepatic enzymes that promote gluconeogenesis. Although hydrocortisone stimulates enzyme (protein) synthesis in the liver, it promotes protein breakdown in the muscles. This latter effect increases the blood levels of amino acids, and thus provides the substrates needed by the liver for gluconeogenesis. The release of circulating energy substrates—amino acids, glucose, fatty acids, and ketone bodies—into the blood in response to hydrocortisone (fig. 19.15) helps to compensate for a state of prolonged fasting or exercise. Whether these metabolic responses are beneficial in other stressful states is open to question.

Thyroxine

The thyroid follicles secrete **thyroxine,** also called **tetraiodothyronine (T_4),** in response to stimulation by thyroid-stimulating hormone (TSH) from the anterior pituitary (chapter 11; see fig. 11.23). The thyroid also secretes smaller amounts of **triiodothyronine (T_3)** in response to stimulation by TSH. Almost all organs in the body are targets of thyroxine action. Thyroxine itself, however, is not the active form of the hormone within the target cells; thyroxine is a prehormone that must first be converted to triiodothyronine (T_3) within the target cells to be active (chapter 11). Acting via its conversion to T_3, thyroxine (1) regulates the rate of cell respiration and (2) contributes to proper growth and development, particularly during early childhood.

Thyroxine (via its conversion to T_3) stimulates the rate of cell respiration in almost all cells in the body—an effect believed to be due to a lowering of cellular ATP concentrations. This effect is produced by (1) the production of uncoupling proteins (as in brown fat, discussed previously); and (2) stimulation of active transport Na^+/K^+ pumps, which serve as an energy sink to further lower ATP concentrations. As discussed in chapter 4, ATP exerts

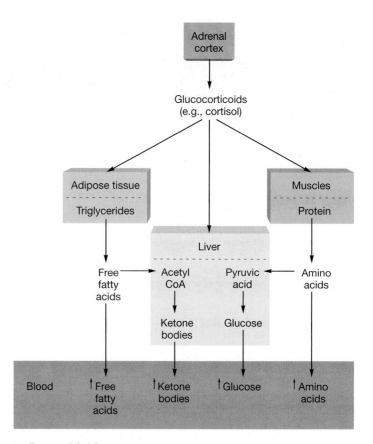

Figure 19.15 **The metabolic effects of glucocorticoids.** The catabolic actions of glucocorticoids help to raise the blood concentration of glucose and other energy-carrier molecules.

an end-product inhibition of cell respiration, so that when ATP concentrations decrease, the rate of cell respiration increases.

Much of the energy liberated during cell respiration escapes as heat, and uncoupling proteins increase the proportion of food energy that escapes as heat. Since thyroxine stimulates the production of uncoupling proteins and the rate of cell respiration, the actions of thyroxine increase the production of metabolic heat. The heat-producing, or *calorigenic* (*calor* = heat) *effects* of thyroxine are required for cold adaptation.

The metabolic rate under carefully defined resting conditions is known as the **basal metabolic rate (BMR).** This rate of basal metabolism has two components—one that is independent of thyroxine action and one that is regulated by thyroxine. In this way, thyroxine acts to "set" the BMR. The BMR can thus be used as an index of thyroid function. Indeed, such measurements were used clinically to evaluate the condition of the thyroid prior to the development of direct chemical determinations of T_4 and T_3 in the blood. A person who is hypothyroid may have a basal O_2 consumption about 30% lower than normal, while a person who is hyperthyroid may have a basal O_2 consumption up to 50% higher than normal.

A normal level of thyroxine secretion is required for growth and proper development of the central nervous system in children. This is why hypothyroidism in children can cause cretinism (see fig. 11.27). The symptoms of hypothyroidism and hyperthyroidism in adults are compared in table 11.8, p. 311.

A normal level of thyroxine secretion is required in order to maintain a balance of anabolism and catabolism. For reasons that are incompletely understood, both hypothyroidism and hyperthyroidism cause protein breakdown and muscle wasting.

Growth Hormone

The anterior pituitary secretes **growth hormone,** also called **somatotropin,** in larger amounts than any other of its hormones. As its name implies, growth hormone stimulates growth in children and adolescents. The continued high secretion of growth hormone in adults, particularly under the conditions of fasting and other forms of stress, implies that this hormone can have important metabolic effects even after the growing years have ended.

Regulation of Growth Hormone Secretion

The secretion of growth hormone is inhibited by somatostatin, which is produced by the hypothalamus and secreted into the hypothalamo-hypophyseal portal system (chapter 11). In addition, there is also a hypothalamic releasing hormone, GHRH, which stimulates growth hormone secretion. Growth hormone thus appears to be unique among the anterior pituitary hormones in that its secretion is controlled by both a releasing and an inhibiting hormone from the hypothalamus. The secretion of growth hormone follows a circadian ("about a day") pattern, increasing during sleep and decreasing during periods of wakefulness.

Growth hormone secretion is stimulated by an increase in the plasma concentration of amino acids and by a decrease in the plasma glucose concentration. These events occur during absorption of a high-protein meal, when amino acids are absorbed. The secretion of growth hormone is also increased during prolonged fasting, when plasma glucose is low and plasma amino acid concentration is raised by the breakdown of muscle protein.

Insulin-like Growth Factors

Insulin-like growth factors (IGFs), produced by many tissues, are polypeptides that are similar in structure to proinsulin (chapter 3; see fig. 3.25). They have insulin-like effects and serve as mediators for some of growth hormone's actions. The term **somatomedins** is often used to refer to two of these factors, designated *IGF-1* and *IGF-2,* because they mediate the actions of somatotropin (growth hormone). The liver produces and secretes IGF-1 in response to growth hormone stimulation, and this secreted IGF-1 then functions as a hormone in its own right, traveling in the blood to the target tissue. A major target is cartilage, where IGF-1 stimulates cell division and growth. IGF-1 also functions as an autocrine regulator (chapter 11), since the chondrocytes (cartilage cells) themselves produce IGF-1 in response to growth hormone stimulation. The growth-promoting actions of IGF-1, acting as both a hormone and an autocrine regulator, thus directly mediate the effects of growth hormone on cartilage. These actions are supported by IGF-2,

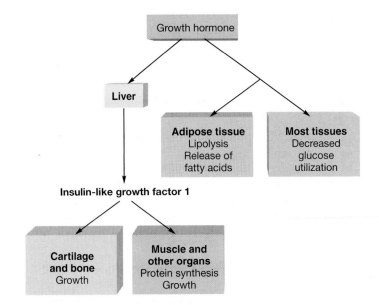

Figure 19.16 The metabolic effects of growth hormone. The growth-promoting, or anabolic, effects of growth hormone are mediated indirectly via stimulation of insulin-like growth factor 1 (also called somatomedin C) production by the liver.

which has more insulin-like actions. The action of growth hormone in stimulating lipolysis in adipose tissue and in decreasing glucose utilization is apparently not mediated by the somatomedins (fig. 19.16).

Effects of Growth Hormone on Metabolism

The fact that growth hormone secretion is increased during fasting and also during absorption of a protein meal reflects the complex nature of this hormone's action. Growth hormone has both anabolic and catabolic effects; it promotes protein synthesis (anabolism), and in this respect is similar to insulin. It also stimulates the catabolism of fat and the release of fatty acids from adipose tissue during periods of fasting (the postabsorptive state), as growth hormone secretion is increased at night. A rise in the plasma fatty acid concentration induced by growth hormone results in decreased rates of glycolysis in many organs. This inhibition of glycolysis by fatty acids, perhaps together with a more direct action of growth hormone, results in decreased glucose utilization by the tissues. Growth hormone thus acts to raise the blood glucose concentration, and for that reason is said to have a "diabetogenic" effect.

Growth hormone stimulates the cellular uptake of amino acids and protein synthesis in many organs of the body. These actions are useful when eating a protein-rich meal; amino acids are removed from the blood and used to form proteins, and the plasma concentration of glucose and fatty acids is increased to provide alternate energy sources (fig. 19.16). The anabolic effect of growth hormone is particularly important during the growing years, when it contributes to increases in bone length and in the mass of many soft tissues.

Effects of Growth Hormone on Body Growth

The stimulatory effects of growth hormone on skeletal growth results from stimulation of mitosis in the epiphyseal discs of cartilage present in the long bones of growing children and adolescents. This action is mediated by the somatomedins, IGF-1 and IGF-2, which stimulate the chondrocytes to divide and secrete more cartilage matrix. Part of this growing cartilage is converted to bone, enabling the bone to grow in length. This skeletal growth stops when the epiphyseal discs are converted to bone after the growth spurt during puberty, despite the fact that growth hormone secretion continues throughout adulthood.

An excessive secretion of growth hormone in children can produce **gigantism.** These children may grow up to 8 feet tall, at the same time maintaining normal body proportions. An excessive growth hormone secretion that occurs after the epiphyseal discs have sealed, however, cannot produce increases in height. In adults, the oversecretion of growth hormone results in an elongation of the jaw and deformities in the bones of the face, hands, and feet. This condition, called **acromegaly,** is accompanied by the growth of soft tissues and coarsening of the skin (fig. 19.17). It is interesting that athletes who take growth hormone supplements to increase their muscle mass may also experience body changes that resemble those of acromegaly.

An inadequate secretion of growth hormone during the growing years results in **dwarfism.** An interesting variant of this condition is *Laron dwarfism,* in which there is a genetic insensitivity to the effects of growth hormone. This insensitivity is associated with a reduction in the number of growth hormone receptors in the target cells. Genetic engineering has made available recombinant IGF-1, which has recently been approved by the FDA for the medical treatment of Laron dwarfism.

Age 9　　Age 16

Age 33　　Age 52

Figure 19.17 The progression of acromegaly in one individual. The coarsening of features and disfigurement are evident by age 33 and severe at age 52.

Test Yourself Before You Continue

1. Describe the effects of epinephrine and the glucocorticoids on the metabolism of carbohydrates and lipids. What is the significance of these effects as a response to stress?
2. Explain the actions of thyroxine on the basal metabolic rate. Why do people who are hypothyroid have a tendency to gain weight and why are they less resistant to cold stress?
3. Describe the effects of growth hormone on the metabolism of lipids, glucose, and amino acids.
4. Explain how growth hormone stimulates skeletal growth.

Table 19.7 Endocrine Regulation of Calcium and Phosphate Balance

Hormone	Effect on Intestine	Effect on Kidneys	Effect on Bone	Associated Diseases
Parathyroid hormone (PTH)	No direct effect	Stimulates Ca^{2+} reabsorption; inhibits PO_4^{3-} reabsorption	Stimulates resorption	Osteitis fibrosa cystica with hypercalcemia due to excess PTH
1,25-dihydroxyvitamin D_3	Stimulates absorption of Ca^{2+} and PO_4^{3-}	Stimulates reabsorption of Ca^{2+} and PO_4^{3-}	Stimulates resorption	Osteomalacia (adults) and rickets (children) due to deficiency of 1,25-dihydroxyvitamin D_3
Calcitonin	None	Inhibits resorption of Ca^{2+} and PO_4^{3-}	Stimulates deposition	None

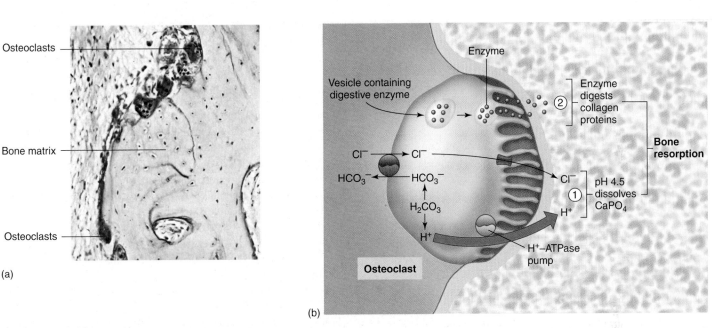

(a)

(b)

■ **Figure 19.18** **The resorption of bone by osteoclasts.** *(a)* A photomicrographs showing osteoclasts and bone matrix. *(b)* Figure depicting the mechanism of bone resorption. Notice that the bone is first demineralized by the dissolution of $CaPO_4$ from the matrix due to acid secretion by the osteoclast. After that, the organic component of the matrix (mainly collagen) is digested by the secretion of enzyme molecules (an enzyme called cathepsin K) from the osteoclast.

Regulation of Calcium and Phosphate Balance

A normal blood Ca^{2+} concentration is critically important for contraction of muscles and maintenance of proper membrane permeability. Parathyroid hormone promotes an elevation in blood Ca^{2+} by stimulating resorption of the calcium phosphate crystals from bone and renal excretion of phosphate. A derivative of vitamin D produced in the body, 1,25-dihydroxyvitamin D_3, promotes the intestinal absorption of calcium and phosphate.

The calcium and phosphate concentrations of plasma are affected by bone formation and resorption, intestinal absorption of Ca^{2+} and PO_4^{3-}, and urinary excretion of these ions. These processes are regulated by parathyroid hormone, 1,25-dihydroxy vitamin D_3, and calcitonin, as summarized in table 19.7.

Bone Deposition and Resorption

The skeleton, in addition to providing support for the body, serves as a large store of calcium and phosphate in the form of crystals called *hydroxyapatite,* which has the formula $Ca_{10}(PO_4)_6(OH)_2$. The calcium phosphate in these hydroxyapatite crystals is derived from the blood by the action of bone-forming cells, or **osteoblasts.** The osteoblasts secrete an organic matrix composed largely of collagen protein, which becomes hardened by deposits of hydroxyapatite. This process is called **bone deposition. Bone resorption** (dissolution of hydroxyapatite), produced by the action of **osteoclasts** (fig. 19.18*a*), results in the return of bone calcium and phosphate to the blood.

Bone resorption begins when the osteoclast attaches to the bone matrix and forms a "ruffled membrane" (see fig. 19.18*b*).

Since the bone matrix contains both an inorganic component (the calcium phospate crystals) and an organic component (collagen and other proteins), the osteoclast must secrete products that both dissolve calcium phosphate and digest the proteins of the bone matrix. The dissolution of calcium phosphate is accomplished by transport of H^+ by a H^+-ATPase pump in the ruffled membrane, thereby acidifying the bone matrix (to a pH of about 4.5) immediately adjacent to the osteoclast. A channel for Cl^- allows Cl^- to follow the H^+, preserving electrical neutrality. Finally, despite the extrusion of H^+ from the osteoclast, the cytoplasm is prevented from becoming too basic by the action of an active transport Cl^-/HCO_3^--pump on the opposite surface of the osteoclast (fig. 19.18b).

The protein component of the bone matrix is digested by enzymes, primarily one called cathepsin K, released by the osteoclasts. The osteoclast can then move to another site and begin the resorption process again, or be eliminated. Interestingly, there is evidence that estrogen, often given to treat osteoporosis in postmenopausal women, works in part by stimulating the apoptosis (cell suicide) of osteoclasts.

The formation and resorption of bone occur constantly at rates determined by the relative activity of osteoblasts and osteoclasts. Body growth during the first two decades of life occurs because bone formation proceeds at a faster rate than bone resorption. By age 50 or 60, the rate of bone resorption often exceeds the rate of bone deposition. The constant activity of osteoblasts and osteoclasts allows bone to be remodeled throughout life. The position of the teeth, for example, can be changed by orthodontic appliances (braces), which cause bone resorption on the pressure-bearing side and bone formation on the opposite side of the alveolar sockets.

Despite the changing rates of bone formation and resorption, the plasma concentrations of calcium and phosphate are maintained by hormonal control of the intestinal absorption and urinary excretion of these ions. These hormonal control mechanisms are very effective in maintaining the plasma calcium and phosphate concentrations within narrow limits. Plasma calcium, for example, is normally maintained at about 2.5 millimolar, or 5 milliequivalents per liter (a milliequivalent equals a millimole multiplied by the valence of the ion; in this case, ×2).

The maintenance of normal plasma calcium concentrations is important because of the wide variety of effects that calcium has in the body. Calcium is needed for blood clotting, for example, and for a variety of cell signaling functions. These include the role of calcium as a second messenger of hormone action (chapter 11), as a signal for neurotransmitter release from axons in response to action potentials (chapter 7), and as the stimulus for muscle contraction in response to electrical excitation (chapter 12).

Calcium is also needed to maintain proper membrane permeability. An abnormally low plasma calcium concentration increases the permeability of the cell membranes to Na^+ and other ions. Hypocalcemia, therefore, enhances the excitability of nerves and muscles and can result in muscle spasm (tetany).

 The rate of bone deposition equals the rate of bone resorption in healthy people on earth. In the **microgravity** (essentially, weightlessness) of space, however, astronauts have suffered from a slow, progressive loss of calcium from the weight-bearing bones of the legs and spine. For reasons that are not presently understood, about 100 mg of calcium are lost per day, which has reduced bone mineral density up to 20% in some astronauts who have been in space for several months. This loss cannot be countered simply by giving astronauts calcium, since hypercalcemia may cause kidney stones and other problems. The exercise machines that have been used in space have helped to prevent loss of muscle mass in astronauts, but they have not been effective in countering the problem of bone resorption.

Parathyroid Hormone and Calcitonin

Whenever the plasma concentration of Ca^{2+} begins to fall, the parathyroid glands are stimulated to secrete increased amounts of **parathyroid hormone (PTH),** which acts to raise the blood Ca^{2+} back to normal levels. As might be predicted from this action of PTH, people who have their parathyroid glands removed (as may occur accidentally during surgical removal of the thyroid) will experience hypocalcemia. This can cause severe muscle tetany, for reasons previously discussed, and serves as a dramatic reminder of the importance of PTH.

Parathyroid hormone helps to raise the blood Ca^{2+} concentration primarily by stimulating the activity of osteoclasts to resorb bone. In addition, PTH stimulates the kidneys to reabsorb Ca^{2+} from the glomerular filtrate while inhibiting the reabsorption of PO_4^{3-}. This raises blood Ca^{2+} levels without promoting the deposition of calcium phosphate crystals in bone. Finally, PTH promotes the formation of 1,25-dihydroxyvitamin D_3 (as described in the next section), and so it also helps to raise the blood calcium levels indirectly through the effects of this other hormone.

Many cancers secrete a hormone known as *parathyroid hormone-related protein.* This molecule causes hypercalcemia by interacting with the PTH receptors, and thus increasing bone resorption, stimulating the renal reabsorption of Ca^{2+}, and promoting the renal excretion of PO_4^{3-}.

As mentioned in chapter 11, the thyroid gland secretes a hormone called **calcitonin.** Though its physiological significance in humans is questionable, its pharmacological action (as a drug) can be useful—it inhibits the resorption of bone. People with stress fractures of vertebrae due to osteoporosis (discussed in the next Clinical box), may be helped by injections or nasal sprays of calcitonin.

Another hormone needed for regulation of the skeletal system is **estrogen.** As may be recalled from chapter 11, estrogen is derived from androgen. For both men and women, estrogen produced within the epiphyseal discs (the cartilage "growth plates" in growing children) is needed for the discs to "seal" (become bone); this stops growth. Also, proper bone mineralization, and the prevention of osteoporosis, requires the production of estrogen within the bone. Men are less prone to osteoporosis than are postmenopausal women, because men have higher blood levels of testosterone that can be converted by the bone into estrogen. Postmenopausal women, by contrast, have low blood concentrations of androgens (secreted from the adrenal cortex), and so cannot produce as much estrogen locally within the bones.

The most common bone disorder in elderly people is **osteoporosis.** Osteoporosis is characterized by parallel losses of mineral and organic matrix from bone, reducing bone mass and density (fig. 19.19) and increasing the risk of fractures. Although the causes of osteoporosis are not well understood, age-related bone loss occurs more rapidly in women than men (osteoporosis is almost ten times more common in women after menopause than in men at comparable ages), suggesting that the fall in estrogen secretion at menopause contributes to this condition. The withdrawal of sex steroids causes increased formation of osteoclasts, producing an imbalance between bone resorption and bone deposition. Premenopausal women who have a very low percentage of body fat and amenorrhea can also have osteoporosis.

Physicians advise teenage girls, who are attaining their maximum bone mass, to eat such calcium-rich foods as milk and other dairy products. This may reduce the progression of osteoporosis when they get older. Additionally, calcium supplementation and other dietary changes are recommended for women prior to menopause. Estrogen replacement therapy for postmenopausal women is common because it helps to prevent bone loss and reduce other symptoms of menopause. Osteoporosis may also be treated with drugs that inhibit bone resorption, including calcitonin (generally derived from salmon) administered by injection or nasal spray.

1,25-Dihydroxyvitamin D₃

The production of **1,25-dihydroxyvitamin D₃** begins in the skin, where vitamin D₃ is produced from its precursor molecule (7-dehydrocholesterol) under the influence of sunlight. In equatorial regions of the globe, exposure to sunlight can allow sufficient cutaneous production of vitamin D₃. In more northerly or southerly latitudes, however, exposure to the winter sun may not allow sufficient production of vitamin D₃. When the skin does not make sufficient amounts of vitamin D₃, this compound must be ingested in the diet—that is why it is called a vitamin. Whether this compound is secreted into the blood from the skin or enters the blood after being absorbed from the intestine, vitamin D₃ functions as a *prehormone;* in order to be biologically active, it must be chemically changed (chapter 11).

An enzyme in the liver adds a hydroxyl group (OH) to carbon number 25, which converts vitamin D₃ into 25-hydroxy vitamin D₃. In order to be active, however, another hydroxyl group must be added to carbon number 1. Hydroxylation of the first carbon is accomplished by an enzyme in the kidneys, which converts the molecule to 1,25-dihydroxyvitamin D₃ (fig. 19.20).

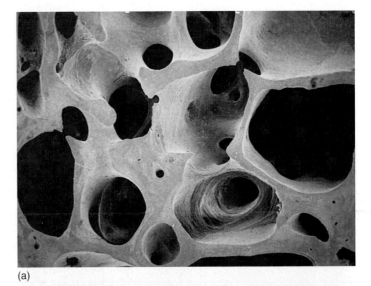

(a)

(b)

 Figure 19.19 Scanning electron micrographs of bone. These biopsy specimens were taken from the iliac crest. Compare bone thickness in (a) a normal specimen and (b) a specimen from a person with osteoporosis.

From L. G. Raisz, S. W. Dempster, et al., "Mechanisms of Disease" in *New England Journal of Medicine,* Vol. 218 (13):818. Copyright © 1988 Massachusetts Medical Society. All rights reserved.

The activity of this enzyme in the kidneys is stimulated by parathyroid hormone (fig. 19.21). Increased secretion of PTH, stimulated by low blood Ca^{2+}, is thus accompanied by the increased production of 1,25-dihydroxyvitamin D₃.

The hormone 1,25-dihydroxyvitamin D₃ helps to raise the plasma concentrations of calcium and phosphate by stimulating (1) the intestinal absorption of calcium and phosphate, (2) the resorption of bones, and (3) the renal reabsorption of calcium and phosphate so that less is excreted in the urine. Notice that 1,25-dihydroxyvitamin D₃, but not parathyroid hormone, directly stimulates intestinal absorption of calcium and phosphate and promotes the reabsorption of phosphate in the kidneys. The

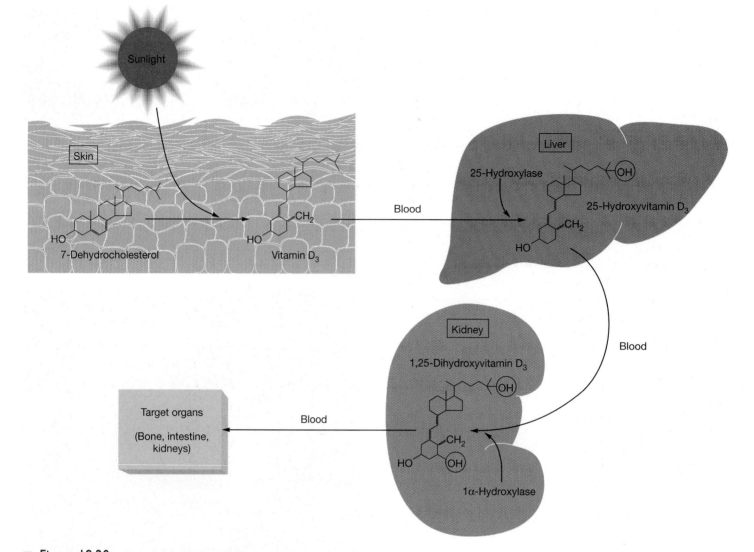

■ **Figure 19.20**   **The production of 1,25-dihydroxyvitamin D₃**. This hormone is produced in the kidneys from the inactive precursor 25-hydroxyvitamin D₃ (formed in the liver). The latter molecule is produced from vitamin D₃ secreted by the skin.

■ **Figure 19.21**   **The negative feedback control of parathyroid hormone secretion.** A decrease in plasma Ca²⁺ directly stimulates the secretion of parathyroid hormone (PTH). The production of 1,25-dihydroxyvitamin D₃ also rises when Ca²⁺ is low because PTH stimulates the final hydroxylation step in the formation of this compound in the kidneys.

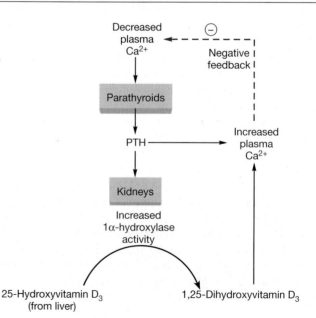

effect of simultaneously raising the blood concentrations of Ca^{2+} and PO_4^{3-} results in the increased tendency of these two ions to precipitate as hydroxyapatite crystals in bone.

Since 1,25-dihydroxyvitamin D_3 directly stimulates bone resorption, it seems paradoxical that this hormone is needed for proper bone deposition and, in fact, that inadequate amounts of 1,25-dihydroxyvitamin D_3 result in the bone demineralization of osteomalacia and rickets. This apparent paradox may be explained by the fact that the primary function of 1,25-dihydroxyvitamin D_3 is stimulation of intestinal Ca^{2+} and PO_4^{3-} absorption. When calcium intake is adequate, the major result of 1,25-dihydroxyvitamin D_3 action is the availability of Ca^{2+} and PO_4^{3-} in sufficient amounts to promote bone deposition. Only when calcium intake is inadequate does the direct effect of 1,25-dihydroxyvitamin D_3 on bone resorption become significant, acting to ensure proper blood Ca^{2+} levels.

In addition to osteoporosis, a number of other bone disorders are associated with abnormal calcium and phosphate balance. In **osteomalacia** (in adults) and **rickets** (in children), inadequate intake of vitamin D results in inadequate calcification of the organic matrix in bones. Excessive secretion of parathyroid hormone results in **osteitis fibrosa cystica,** in which excessive osteoclast activity causes resorption of both the mineral and organic components of bone, which are then replaced by fibrous tissue.

Negative Feedback Control of Calcium and Phosphate Balance

The secretion of parathyroid hormone is controlled by the plasma calcium concentrations. Its secretion is stimulated by low calcium concentrations and inhibited by high calcium concentrations. Since parathyroid hormone stimulates the final hydroxylation step in the formation of 1,25-dihydroxyvitamin D_3, a rise in parathyroid hormone results in an increase in production of 1,25-dihydroxyvitamin D_3. Low blood calcium can thus be corrected by the effects of increased parathyroid hormone and 1,25-dihydroxyvitamin D_3 (fig. 19.22).

It is possible for plasma calcium levels to fall while phosphate levels remain normal. In this case, the increased secretion of parathyroid hormone and the production of 1,25-dihydroxyvitamin D_3 that result could abnormally raise phosphate levels while acting to restore normal calcium levels. This is prevented by the inhibition of phosphate reabsorption in the kidneys by parathyroid hormone, so that more phosphate is excreted in the urine (fig. 19.20). In this way, blood calcium can be raised to normal levels without excessively raising blood phosphate concentrations.

Experiments in the 1960s revealed that high blood calcium in dogs may be lowered by a hormone secreted from the thyroid gland. This hormone thus has an effect opposite to that of

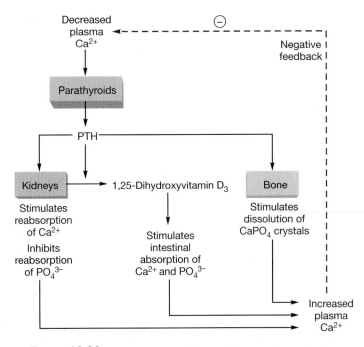

Figure 19.22 Homeostasis of plasma Ca^{2+} concentrations. A negative feedback loop returns low blood Ca^{2+} concentrations to normal without simultaneously raising blood phosphate levels above normal.

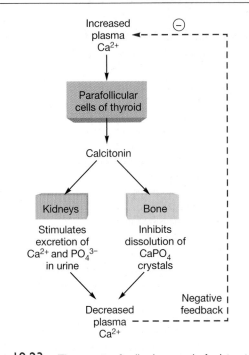

Figure 19.23 The negative feedback control of calcitonin secretion. The action of calcitonin is antagonistic to that of parathyroid hormone.

parathyroid hormone and 1,25-dihydroxyvitamin D_3. The calcium-lowering hormone, called *calcitonin,* was found to be secreted by the *parafollicular cells,* or *C cells,* of the thyroid. These cells are scattered among the follicular cells that secrete thyroxine.

The secretion of calcitonin is stimulated by high plasma calcium levels and acts to lower calcium levels by (1) inhibiting the activity of osteoclasts, thus reducing bone resorption, and (2) stimulating the urinary excretion of calcium and phosphate by inhibiting their reabsorption in the kidneys (fig. 19.23).

Although it is attractive to think that calcium balance is regulated by the effects of antagonistic hormones, the significance of calcitonin in human physiology remains unclear. Patients who have had their thyroid gland surgically removed (as for thyroid cancer) are *not* hypercalcemic, as one might expect them to be if calcitonin were needed to lower blood calcium levels. The ability of very large pharmacological doses of calcitonin to inhibit osteoclast activity and bone resorption, however, is clinically useful in the treatment of *Paget's disease,* in which osteoclast activity causes softening of bone. It is sometimes also used to treat osteoporosis, as previously described.

Test Yourself Before You Continue

1. Describe the mechanisms by which the secretion of parathyroid hormone and of calcitonin is regulated.
2. List the steps involved in the formation of 1,25-dihydroxyvitamin D_3 and state how this formation is influenced by parathyroid hormone.
3. Describe the actions of parathyroid hormone, 1,25-dihydroxyvitamin D_3, and calcitonin on the intestine, skeletal system, and kidneys, and explain how these actions affect the blood levels of calcium.
4. Account for the different effects of 1,25-dihydroxyvitamin D_3 on bones according to whether calcium intake is adequate or inadequate.

Summary

Nutritional Requirements   598

I. Food provides molecules used in cell respiration for energy.
  A. The metabolic rate is influenced by physical activity, temperature and eating. The basal metabolic rate is measured as the rate of oxygen consumption when such influences are standardized and minimal.
  B. The energy provided in food and the energy consumed by the body are measured in units of kilocalories.
  C. When the caloric intake is greater than the energy expenditure over a period of time, the excess calories are stored primarily as fat.
II. Vitamins and elements serve primarily as cofactors and coenzymes.
  A. Vitamins are divided into those that are fat-soluble (A, D, E, and K) and those that are water-soluble.
  B. Many water-soluble vitamins are needed for the activity of the enzymes involved in cell respiration.
  C. The fat-soluble vitamins A and D have specific functions but share similar mechanisms of action, activating nuclear receptors and regulating genetic expression.

Regulation of Energy Metabolism   604

I. The body tissues can use circulating energy substrates, including glucose, fatty acids, ketone bodies, lactic acid, amino acid, and others, for cell respiration.
  A. Different organs have different preferred energy sources.
  B. Circulating energy substrates can be obtained from food or from the energy reserves of glycogen, fat, and protein in the body.
II. Eating behavior is regulated, at least in part, by the hypothalamus.
  A. Lesions of the ventromedial area of the hypothalamus produce hyperphagia, whereas lesions of the lateral hypothalamus produce hypophagia.
  B. A variety of neurotransmitters have been implicated in the control of eating behavior. These include the endorphins, norepinephrine, serotonin, cholecystokinin, and neuropeptide Y.
III. Adipose cells, or adipocytes, are both the targets of hormonal regulation and themselves endocrine in nature.
  A. In children, circulating saturated fatty acids promote cell division and differentiation of new adipocytes. This activity involves the bonding of a prostaglandin with a nuclear receptor known as PPARγ.
  B. Adipocytes secrete leptin, which regulates food intake and metabolism, and TNFα, which may help to regulate the sensitivity of skeletal muscles to insulin.
IV. The control of energy balance in the body is regulated by the anabolic and catabolic effects of a variety of hormones.

Energy Regulation by the Islets of Langerhans   611

I. A rise in plasma glucose concentration stimulates insulin and inhibits glucagon secretion.

   A. Amino acids stimulate the secretion of both insulin and glucagon.

   B. Insulin secretion is also stimulated by parasympathetic innervation of the islets and by the action of intestinal hormones such as gastric inhibitory peptide (GIP).

II. During the intestinal absorption of a meal, insulin promotes the uptake of blood glucose into skeletal muscle and other tissues.

   A. This lowers the blood glucose concentration and increases the energy reserves of glycogen, fat, and protein.

   B. Skeletal muscles are the major organs that remove blood glucose in response to insulin stimulation.

III. During periods of fasting, insulin secretion decreases and glucagon secretion increases.

   A. Glucagon stimulates glycogenolysis in the liver, gluconeogenesis, lipolysis, and ketogenesis.

   B. These effects help to maintain adequate levels of blood glucose for the brain and provide alternate energy sources for other organs.

Diabetes Mellitus and Hypoglycemia   615

I. Diabetes mellitus and reactive hypoglycemia represent disorders of the islets of Langerhans.

   A. Type 1 diabetes mellitus occurs when the beta cells are destroyed; the resulting lack of insulin and excessive glucagon secretion produce the symptoms of this disease.

   B. Type 2 diabetes mellitus occurs as a result of a relative tissue insensitivity to insulin and inadequate insulin secretion; this condition is aggravated by obesity and improved by exercise.

   C. Reactive hypoglycemia occurs when the islets secrete excessive amounts of insulin in response to a rise in blood glucose concentration.

Metabolic Regulation by Adrenal Hormones, Thyroxine, and Growth Hormone   618

I. The adrenal hormones involved in energy regulation include epinephrine from the adrenal medulla and glucocorticoids (mainly hydrocortisone) from the adrenal cortex.

   A. The effects of epinephrine are similar to those of glucagon. Epinephrine stimulates glycogenolysis and lipolysis, and activates increased metabolism of brown fat.

   B. Glucocorticoids promote the breakdown of muscle protein and the conversion of amino acids to glucose in the liver.

II. Thyroxine stimulates the rate of cell respiration in almost all cells in the body.

   A. Thyroxine sets the basal metabolic rate (BMR), which is the rate at which energy (and oxygen) is consumed by the body under resting conditions.

   B. Thyroxine also promotes protein synthesis and is needed for proper body growth and development, particularly of the central nervous system.

III. The secretion of growth hormone is regulated by releasing and inhibiting hormones from the hypothalamus.

   A. The secretion of growth hormone is stimulated by a protein meal and by a fall in glucose, as occurs during fasting.

   B. Growth hormone stimulates catabolism of lipids and inhibits glucose utilization.

   C. Growth hormone also stimulates protein synthesis, and thus promotes body growth.

   D. The anabolic effects of growth hormone, including the stimulation of bone growth in childhood, are produced indirectly via polypeptides called insulin-like growth factors, or somatomedins.

Regulation of Calcium and Phosphate Balance   623

I. Bone contains calcium and phosphate in the form of hydroxyapatite crystals. This serves as a reserve supply of calcium and phosphate for the blood.

   A. The formation and resorption of bone are produced by the action of osteoblasts and osteoclasts, respectively.

   B. The plasma concentrations of calcium and phosphate are also affected by absorption from the intestine and by the urinary excretion of these ions.

II. Parathyroid hormone stimulates bone resorption and calcium reabsorption in the kidneys. This hormone thus acts to raise the blood calcium concentration.

   A. The secretion of parathyroid hormone is stimulated by a fall in blood calcium levels.

   B. Parathyroid hormone also inhibits reabsorption of phosphate in the kidneys, so that more phosphate is excreted in the urine.

III. 1,25-dihydroxyvitamin D_3 is derived from vitamin D by hydroxylation reactions in the liver and kidneys.

   A. The last hydroxylation step is stimulated by parathyroid hormone.

   B. 1,25-dihydroxyvitamin D_3 stimulates the intestinal absorption of calcium and phosphate, resorption of bone, and renal reabsorption of phosphate.

IV. A rise in parathyroid hormone, accompanied by the increased production of 1,25-dihydroxyvitamin D_3, helps to maintain proper blood levels of calcium and phosphate in response to a fall in calcium levels.

V. Calcitonin is secreted by the parafollicular cells of the thyroid gland.

   A. Calcitonin secretion is stimulated by a rise in blood calcium levels.

   B. Calcitonin, at least at pharmacological levels, acts to lower blood calcium by inhibiting bone resorption and stimulating the urinary excretion of calcium and phosphate.

Review Activities

Test Your Knowledge of Terms and Facts

Match these:

1. absorption of a carbohydrate meal
2. fasting

   a. rise in insulin; rise in glucagon
   b. fall in insulin; rise in glucagon
   c. rise in insulin; fall in glucagon
   d. fall in insulin; fall in glucagon

Match these:

3. growth hormone
4. thyroxine
5. hydrocortisone

   a. increased protein synthesis; increased cell respiration
   b. protein catabolism in muscles; gluconeogenesis in liver
   c. protein synthesis in muscles; decreased glucose utilization
   d. fall in blood glucose; increased fat synthesis

6. A lowering of blood glucose concentration promotes
   a. decreased lipogenesis.
   b. increased lipolysis.
   c. increased glycogenolysis.
   d. all of these.
7. Glucose can be secreted into the blood by
   a. the liver.
   b. the muscles.
   c. the liver and muscles.
   d. the liver, muscles, and brain.

8. The basal metabolic rate is determined primarily by
   a. hydrocortisone.
   b. insulin.
   c. growth hormone.
   d. thyroxine.
9. Somatomedins are required for the anabolic effects of
   a. hydrocortisone.
   b. insulin.
   c. growth hormone.
   d. thyroxine.
10. The increased intestinal absorption of calcium is stimulated directly by
    a. parathyroid hormone.
    b. 1,25-dihydroxyvitamin D_3.
    c. calcitonin.
    d. all of these.
11. A rise in blood calcium levels directly stimulates
    a. parathyroid hormone secretion.
    b. calcitonin secretion.
    c. 1,25-dihydroxyvitamin D_3 formation.
    d. all of these.
12. At rest, about 12% of the total calories consumed are used for
    a. protein synthesis.
    b. cell transport.
    c. the Na^+/K^+ pumps.
    d. DNA replication.
13. Which of these hormones stimulates anabolism of proteins and catabolism of fat?
    a. growth hormone
    b. thyroxine
    c. insulin

    d. glucagon
    e. epinephrine
14. If a person eats 600 kilocalories of protein in a meal, which of these statements will be *false?*
    a. Insulin secretion will be increased.
    b. The metabolic rate will be increased over basal conditions.
    c. The tissue cells will use some of the amino acids for resynthesis of body proteins.
    d. The tissue cells will obtain 600 kilocalories worth of energy.
    e. Body-heat production and oxygen consumption will be increased over basal conditions.
15. Ketoacidosis in untreated diabetes mellitus is due to
    a. excessive fluid loss.
    b. hypoventilation.
    c. excessive eating and obesity.
    d. excessive fat catabolism.
16. Which of these statements about leptin is *false?*
    a. It is secreted by adipocytes.
    b. It increases the energy expenditure of the body.
    c. It stimulates the release of neuropeptide Y in the hypothalamus.
    d. It promotes feelings of satiety, decreasing food intake.

Test Your Understanding of Concepts and Principles

1. Compare the metabolic effects of fasting to the state of uncontrolled type 1 diabetes mellitus. Explain the hormonal similarities of these conditions.[1]
2. Glucocorticoids stimulate the breakdown of protein in muscles but the synthesis of protein in the liver. Explain the significance of these different effects.

3. Describe how thyroxine affects cell respiration. Why does a person who is hypothyroid have a tendency to gain weight and less tolerance for cold?
4. Compare and contrast the metabolic effects of thyroxine and growth hormone.
5. Why is vitamin D considered both a vitamin and a prehormone? Explain why people with osteoporosis might be

helped by taking controlled amounts of vitamin D.
6. Define the term *insulin resistance.* Explain the relationship between insulin resistance, obesity, exercise, and non-insulin-dependent diabetes mellitus.
7. Describe the chemical nature and origin of the somatomedins and explain the physiological significance of these growth factors.

---

[1]*Note:* This question is answered in the chapter 19 Study Guide found on the Online Learning Center at www.mhhe.com/fox8.

8. Explain how insulin secretion and glucagon secretion are influenced by (a) fasting, (b) a meal that is high in carbohydrate and low in protein, and (c) a meal that is high in protein and high in carbohydrate. Also, explain how the changes in insulin and glucagon secretion under these conditions function to maintain homeostasis.

9. Using a cause-and-effect sequence, explain how an inadequate intake of dietary calcium or vitamin D can cause bone resorption. Also, describe the cause-and-effect sequence whereby an adequate intake of calcium and vitamin D may promote bone deposition.

10. Describe the conditions of gigantism, acromegaly, Laron dwarfism, and kwashiorkor, and explain how these conditions relate to blood levels of growth hormone and IGF-1.

Test Your Ability to Analyze and Apply Your Knowledge

1. Your friend is trying to lose weight and at first is very successful. After a time, however, she complains that it seems to take more exercise and a far more stringent diet to lose even one more pound. What might explain her difficulties?

2. How can a high-fat diet in childhood lead to increased numbers of adipocytes? Explain how this process may be related to the ability of adipocytes to regulate the insulin sensitivity of skeletal muscles in adults.

3. Discuss the role of GLUT4 in glucose metabolism and use this concept to explain why exercise helps to control type 2 diabetes mellitus.

4. You are running in a 10-K race and, to keep your mind occupied, you try to remember which physiological processes regulate blood glucose levels during exercise. Step by step, what are these processes?

5. Discuss the location and physiological significance of the β_3 adrenergic receptors and explain how a hypothetical β_3 adrenergic agonist drug might help in the treatment of obesity.

6. A person with type 1 diabetes mellitus accidentally overdoses on insulin. What symptoms might she experience, and why? If she remains conscious, what treatment might be offered to adjust her blood glucose level?

Related Websites

Check out the Links Library at www.mhhe.com/fox8 for links to sites containing resources related to the regulation of metabolism. These links are monitored to ensure current URLs.

20 Reproduction

Objectives

After studying this chapter, you should be able to . . .

1. describe how the chromosomal content determines the sex of an embryo and how this relates to the development of testes or ovaries.

2. explain how the development of accessory sex organs and external genitalia is affected by the presence or absence of testes in the embryo.

3. describe the hormonal changes that occur during puberty, the mechanisms that may control the onset of puberty, and the secondary sex characteristics that develop during puberty.

4. explain how the secretions of FSH and LH are regulated in the male and describe the actions of these hormones on the testis.

5. describe the structure of the testis and the interaction between the interstitial Leydig cells and seminiferous tubules.

6. describe the stages of spermatogenesis and the functions of Sertoli cells in this process.

7. explain the hormonal control of spermatogenesis and describe the effects of androgens on the male accessory sex organs.

8. describe the composition of semen, explain the physiology of erection and ejaculation, and discuss the various factors that affect male fertility.

9. describe the four phases of the human sexual response.

10. describe oogenesis and the stages of follicle development through ovulation and the formation of a corpus luteum.

11. explain the hormonal interactions involved in the control of ovulation.

12. describe the changes in the secretion of ovarian sex steroids during a nonfertile cycle and explain the function and fate of the corpus luteum.

13. explain how the secretion of FSH and LH is controlled through negative and positive feedback mechanisms during a menstrual cycle.

14. explain how contraceptive pills prevent ovulation.

15. describe the cyclic changes that occur in the endometrium and the hormonal mechanisms that cause these changes.

16. describe the acrosomal reaction and the events that occur at fertilization, blastocyst formation, and implantation.

17. explain how menstruation and further ovulation are normally prevented during pregnancy.

18. describe the structure and functions of the placenta.

19. list the hormones secreted by the placenta and describe their actions.

20. discuss the factors that stimulate uterine contractions during labor and parturition, and explain how the onset of labor may be regulated.

21. describe the hormonal requirements for development of the mammary glands during pregnancy and explain how lactation is prevented during pregnancy.

22. describe the milk-ejection reflex.

Refresh Your Memory

Before you begin this chapter, you may want to review these concepts from previous chapters:

■ DNA Synthesis and Cell Division   69

■ Meiosis   75

■ Mechanism of Steroid Hormone Action   293

■ Pituitary Gland   299

■ Autocrine and Paracrine Regulation   316

Chapter at a Glance

Take Advantage of the Technology

Visit the Online Learning Center for these additional study resources.

■ Interactive quizzing

■ Online study guide

■ Current news feeds

■ Crossword puzzles

■ Vocabulary flashcards

■ Labeling activities

www.mhhe.com/fox8

Clinical Investigation

Gloria, a second-year college student, goes to the student health center complaining of amenorrhea of several months' duration. Before that, her periods had always been normal. Gloria's body weight is below average for her height, but she exhibits normal development of secondary sexual characteristics. She had been diagnosed a few years earlier as hypothyroid, and thyroxine pills were prescribed for this condition. Gloria teaches an aerobics class at a local gym, and she exercises at least an hour a day. She states that she is not taking birth control pills.

A pregnancy test is performed and is negative. Blood is drawn, and the subsequent laboratory report indicates normal thyroxine levels; in fact, all measurements are within the normal range (her blood cholesterol is even on the low side). What is the most likely cause of Gloria's condition? What should she do about it?

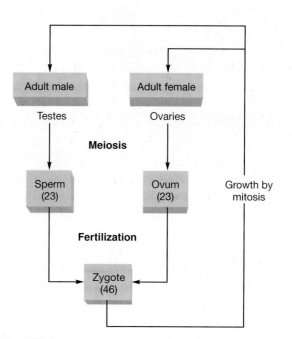

■ **Figure 20.1** **The human life cycle.** Numbers in parentheses indicate the haploid state (twenty-three chromosomes) or diploid state (forty-six chromosomes).

Sexual Reproduction

Early embryonic gonads can become either testes or ovaries. A particular gene on the Y chromosome induces the embryonic gonads to become testes. Females lack a Y chromosome, and the absence of this gene causes the development of ovaries. The embryonic testes secrete testosterone, which induces the development of male accessory sex organs and external genitalia. The absence of testes (rather than the presence of ovaries) in a female embryo causes the development of the female accessory sex organs.

"A chicken is an egg's way of making another egg." Phrased in more modern terms, genes are "selfish." Genes, according to this view, do not exist in order to make a well-functioning chicken (or other organism). The organism, rather, exists and functions so that the genes can survive beyond the mortal life of individual members of a species. Whether or not one accepts this rather cynical view, it is clear that reproduction is one of life's essential functions. The incredible complexity of structure and function in living organisms could not be produced in successive generations by chance; mechanisms must exist to transmit the blueprint (genetic code) from one generation to the next. Sexual reproduction, in which genes from two individuals are combined in random and novel ways with each new generation, offers the further advantage of introducing great variability into a population. This diversity of genetic constitution helps to ensure that some members of a population will survive changes in the environment over evolutionary time.

In sexual reproduction, **germ cells,** or **gametes** (sperm and ova), are formed within the *gonads* (testes and ovaries) by a process of reduction division, or *meiosis* (chapter 3). During this type of cell division, the normal number of chromosomes in human cells—forty-six—is halved, so that each gamete receives twenty-three chromosomes. Fusion of a sperm cell and ovum

(egg cell) in the act of **fertilization** results in restoration of the original chromosome number of forty-six in the **zygote,** or fertilized egg. Growth of the zygote into an adult member of the next generation occurs by means of mitotic cell divisions, as described in chapter 3. When this individual reaches puberty, mature sperm or ova will be formed by meiosis within the gonads so that the life cycle can be continued (fig. 20.1).

Sex Determination

Each zygote inherits twenty-three chromosomes from its mother and twenty-three chromosomes from its father. This does not produce forty-six different chromosomes, but rather twenty-three pairs of *homologous chromosomes*. The members of a homologous pair, with the important exception of the sex chromosomes, look like each other and contain similar genes (such as those coding for eye color, height, and so on). These homologous pairs of chromosomes can be photographed and numbered (as shown in fig. 20.2). Each cell that contains forty-six chromosomes (that is *diploid*) has two number 1 chromosomes, two number 2 chromosomes, and so on through pair number 22. The first twenty-two pairs of chromosomes are called **autosomal chromosomes.**

The twenty-third pair of chromosomes are the **sex chromosomes.** In a female, these consist of two X chromosomes, whereas in a male there is one X chromosome and one Y chromosome. The X and Y chromosomes look different and contain different genes. This is the exceptional pair of homologous chromosomes mentioned earlier.

When a diploid cell (with forty-six chromosomes) undergoes meiotic division, its daughter cells receive only one chromosome from each homologous pair of chromosomes. The

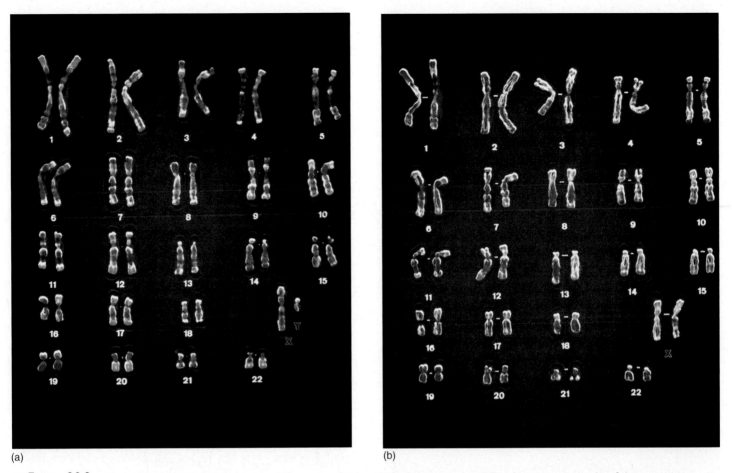

(a)
(b)

Figure 20.2   Homologous pairs of chromosomes.   These were obtained from diploid human cells. The first twenty-two pairs of chromosomes are called the autosomal chromosomes. The sex chromosomes are (a) XY for a male and (b) XX for a female.

gametes are therefore said to be *haploid* (they contain only half the number of chromosomes in the diploid parent cell). Each sperm cell, for example, will receive only one chromosome of homologous pair number 5—either the one originally contributed by the mother, or the one originally contributed by the father (modified by the effects of crossing-over; see chapter 3, fig. 3.34). Which of the two chromosomes—maternal or paternal—ends up in a given sperm cell is completely random. This is also true for the sex chromosomes, so that approximately half of the sperm produced will contain an X and approximately half will contain a Y chromosome.

The egg cells (ova) in a woman's ovary will receive a similar random assortment of maternal and paternal chromosomes. Since the body cells of females have two X chromosomes, however, all of the ova will normally contain one X chromosome. Because all ova contain one X chromosome, whereas some sperm are X-bearing and others are Y-bearing, *the chromosomal sex of the zygote is determined by the fertilizing sperm cell.* If a Y-bearing sperm cell fertilizes the ovum, the zygote will be XY and male; if an X-bearing sperm cell fertilizes the ovum, the zygote will be XX and female.

Although each diploid cell in a woman's body inherits two X chromosomes, it appears that only one of each pair of X chromosomes remains active. The other X chromosome forms a clump of inactive heterochromatin, which can often be seen as a dark spot, called a *Barr body,* in the nucleus of cheek cells (fig. 20.3). This provides a convenient test for chromosomal sex in cases where it is suspected that the chromosomal sex may differ from the apparent ("phenotypic") sex of the individual. Also, some of the nuclei in the neutrophils of females have a "drumstick" appendage not seen in neutrophils from males.

Formation of Testes and Ovaries

Following conception, the gonads of males and females are similar in appearance for the first forty or so days of development. During this time, cells that will give rise to sperm (called *spermatogonia*) and cells that will give rise to ova (called *oogonia*) migrate from the yolk sac to the developing embryonic gonads. At this stage, the embryonic structures have the potential to become either **testes** or **ovaries.** The hypothetical substance that promotes their conversion to testes (fig. 20.4) has been called the **testis-determining factor (TDF).**

Although it has long been recognized that male sex is determined by the presence of a Y chromosome and female sex by the absence of the Y chromosome, the genes involved have only recently been localized. In rare male babies with XX genotypes, scientists have discovered that one of the X chromosomes contains a segment of the Y chromosome—the result of an error

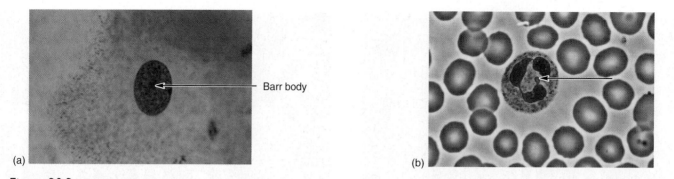

Figure 20.3 **Barr bodies.** The nuclei of cheek cells from females *(a)* have Barr bodies *(arrow)*. These are formed from one of the X chromosomes, which is inactive. No Barr body is present in the cell obtained from a male because males have only one X chromosome, which remains active. Some neutrophils obtained from females *(b)* have a "drumstick-like" appendage *(arrow)* that is not found in the neutrophils of males.

that occurred during the meiotic cell division that formed the sperm cell. Similarly, rare female babies with XY genotypes were found to be missing the same portion of the Y chromosome erroneously inserted into the X chromosome of XX males.

Through these and other observations, it has been shown that the gene for the testis-determining factor is located on the short arm of the Y chromosome. Evidence suggests that it may be a particular gene known as **SRY** (for *s*ex-determining *r*egion of the *Y*). This gene is found in the Y chromosome of all mammals and is highly conserved, meaning that it shows little variation in structure over evolutionary time.

 Notice that it is normally the presence or absence of the Y chromosome that determines whether the embryo will have testes or ovaries. This point is well illustrated by two genetic abnormalities. In **Klinefelter's syndrome,** the affected person has forty-seven instead of forty-six chromosomes because of the presence of an extra X chromosome. This person, with an XXY genotype, will develop testes and have a male phenotype despite the presence of two X chromosomes. Patients with **Turner's syndrome,** who have the genotype XO (and therefore have only forty-five chromosomes), have poorly developed ("streak") gonads and are phenotypically female.

The structures that will eventually produce sperm within the testes, the **seminiferous tubules,** appear very early in embryonic development—between 43 and 50 days following conception. The tubules contain two major cell types: germinal and nongerminal. The **germinal cells** are those that will eventually become sperm through meiosis and subsequent specialization. The nongerminal cells are called **Sertoli** (or **sustentacular**) **cells.** The Sertoli cells appear at about day 42. At about day 65, the **Leydig** (or **interstitial**) **cells** appear in the embryonic testes. The Leydig cells are clustered in the *interstitial tissue* that surrounds the seminiferous tubules. The interstitial Leydig cells constitute the endocrine tissue of the testes. In contrast to the rapid development of the testes, the functional units of the ovaries—called the **ovarian follicles**—do not appear until the second trimester of pregnancy (at about day 105).

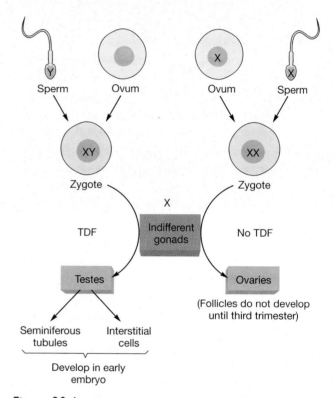

Figure 20.4 **The chromosomal sex and the development of embryonic gonads.** The very early embryo has "indifferent gonads" that can develop into either testes or ovaries. The testis-determining factor (TDF) is a gene located on the Y chromosome. In the absence of TDF, ovaries will develop.

The early-appearing Leydig cells in the embryonic testes secrete large amounts of male sex hormones, or *androgens* (*andro* = man; *gen* = forming). The major androgen secreted by these cells is **testosterone.** Testosterone secretion begins as early as 8 weeks after conception, reaches a peak at 12 to 14 weeks, and then declines to very low levels by the end of the second trimester (at about 21 weeks). Testosterone secretion during embryonic development in the male serves the very important function of masculinizing the embryonic structures; similarly high levels of testosterone will not appear again in the life of the individual until the time of puberty.

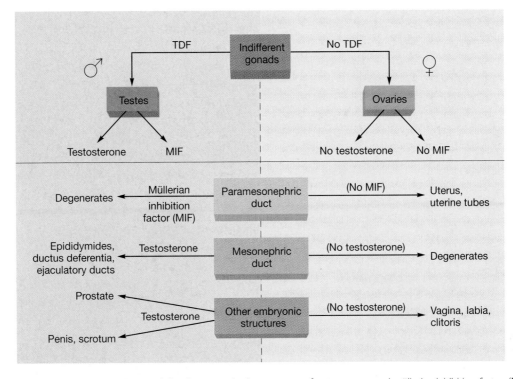

■ **Figure 20.5** The regulation of embryonic sexual development. In the presence of testosterone and müllerian inhibition factor (MIF) secreted by the testes, male external genitalia and accessory sex organs develop. In the absence of these secretions, female structures develop.

As the testes develop, they move within the abdominal cavity and gradually descend into the *scrotum*. Descent of the testes is sometimes not complete until shortly after birth. The temperature of the scrotum is maintained at about 35° C—about 3° C below normal body temperature. This cooler temperature is needed for spermatogenesis. The fact that spermatogenesis does not occur in males with undescended testes—a condition called *cryptorchidism* (*crypt* = hidden; *orchid* = testes)—demonstrates this requirement.

Associated with each spermatic cord is a strand of skeletal muscle called the cremaster muscle. In cold weather, the cremaster muscles contract and elevate the testes, bringing them closer to the warmth of the trunk. The **cremasteric reflex** produces the same effect when the inside of a man's thigh is stroked. In a baby, however, this stimulation can cause the testes to be drawn up through the inguinal canal into the body cavity. The testes can also be drawn up into the body cavity voluntarily by trained Sumo wrestlers.

Development of Accessory Sex Organs and External Genitalia

In addition to testes and ovaries, various internal **accessory sex organs** are needed for reproductive function. Most of these are derived from two systems of embryonic ducts. Male accessory organs are derived from the **wolffian (mesonephric) ducts,** and female accessory organs are derived from the **müllerian (paramesonephric) ducts** (fig. 20.5). Interestingly, the two duct systems are present in both male and female embryos between day 25 and day 50, and so embryos of both sexes have the potential to form the accessory organs characteristic of either sex.

Experimental removal of the testes (castration) from male embryonic animals results in regression of the wolffian ducts and development of the müllerian ducts into *female accessory organs:* the **uterus** and **uterine (fallopian) tubes.** Female accessory sex organs, therefore, develop as a result of the absence of testes rather than as a result of the presence of ovaries.

In a male, the Sertoli cells of the seminiferous tubules secrete *müllerian inhibition factor (MIF),* a polypeptide that causes regression of the müllerian ducts beginning at about day 60. The secretion of testosterone by the Leydig cells of the testes subsequently causes growth and development of the wolffian ducts into *male accessory sex organs:* the **epididymis, ductus (vas) deferens, seminal vesicles,** and **ejaculatory duct.**

The **external genitalia** of males and females are essentially identical during the first 6 weeks of development, sharing in common a *urogenital sinus, genital tubercle, urethral folds,* and a pair of *labioscrotal swellings.* The secretions of the testes masculinize these structures to form the **penis** and spongy (penile) urethra, **prostate,** and **scrotum.** In the absence of secreted testosterone, the genital tubercle that forms the penis in a male will become the **clitoris** in a female. The penis and clitoris are thus said to be *homologous structures.* Similarly, the labioscrotal swellings form the scrotum in a male or the **labia majora** in a female; these structures are therefore homologous also (fig. 20.6).

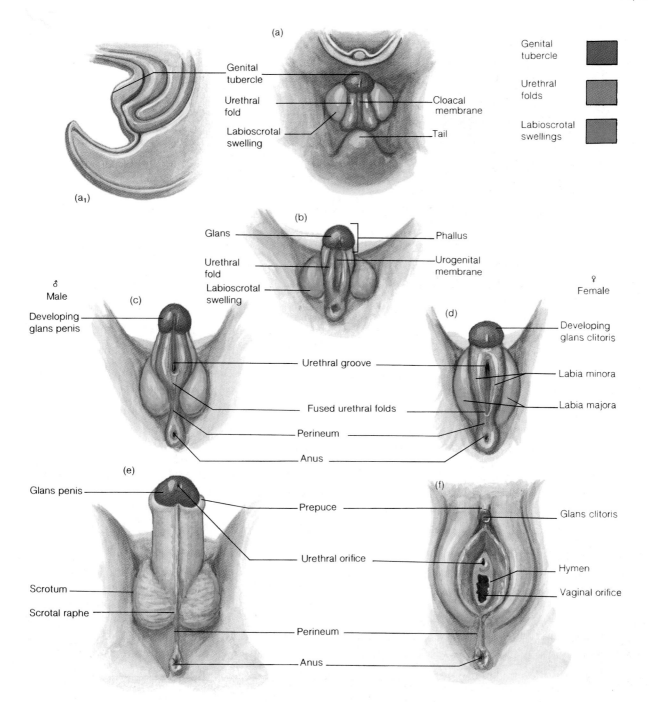

■ Figure 20.6   The development of the external genitalia in the male and female. (*a* [*a₁*, sagittal view]) At 6 weeks, the urethral fold and labioscrotal swelling have differentiated from the genital tubercle. *(b)* At 8 weeks, a distinct phallus is present during the indifferent stage. By week 12, the genitalia have become distinctly male *(c)* or female *(d),* being derived from homologous structures. *(e, f)* At 16 weeks, the genitalia are formed.

Masculinization of the embryonic structures described occurs as a result of testosterone secreted by the embryonic testes. Testosterone itself, however, is not the active agent within all of the target organs. Once inside particular target cells, testosterone is converted by the enzyme 5α-*reductase* into the active hormone known as **dihydrotestosterone (DHT)** (fig. 20.7). DHT is needed for the development and maintenance of the penis, spongy urethra, scrotum, and prostate. Evidence suggests that testosterone itself directly stimulates the wolffian duct derivatives—epididymis, ductus deferens, ejaculatory duct, and seminal vesicles.

In summary, the genetic sex is determined by whether a Y-bearing or an X-bearing sperm cell fertilizes the ovum; the presence or absence of a Y chromosome in turn determines whether the gonads of the embryo will be testes or ovaries; the presence or absence of testes, finally, determines whether the accessory sex organs and external genitalia will be male or female (table 20.1). This regulatory pattern of sex determina-tion makes sense in light of the fact that both male and female embryos develop within an environment high in estrogen, which is secreted by the mother's ovaries and the placenta. If the secretions of the ovaries determined the sex, all embryos would be female.

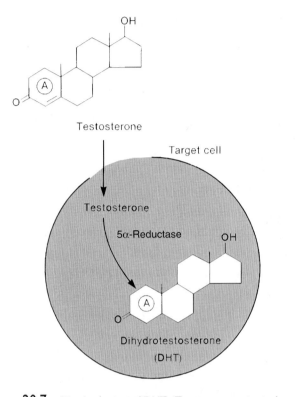

Figure 20.7 The formation of DHT. Testosterone, secreted by the interstitial (Leydig) cells of the testes, is converted into dihydrotestosterone (DHT) within the target cells. This reaction involves the addition of a hydrogen (and the removal of the double carbon bond) in the first (A) ring of the steroid.

Disorders of Embryonic Sexual Development

Hermaphroditism is a condition in which both ovarian and testicular tissue is present in the body. About 34% of her-maphrodites have an ovary on one side and a testis on the other. About 20% have ovotestes—part testis and part ovary—on both sides. The remaining 46% have an ovotestis on one side and an ovary or testis on the other. This condition is ex-tremely rare and appears to be caused by the fact that some embryonic cells receive the short arm of the Y chromosome, with its SRY gene, whereas others do not. More common (though still rare) disorders of sex determination involve indi-viduals with either testes or ovaries, but not both, who have accessory sex organs and external genitalia that are incom-pletely developed or that are inappropriate for their chromoso-mal sex. These individuals are called *pseudohermaphrodites* (*pseudo* = false).

The most common cause of female pseudohermaphro-ditism is *congenital adrenal hyperplasia*. This condition, which is inherited as a recessive trait, is caused by the exces-sive secretion of androgens from the adrenal cortex. Because the cortex does not secrete müllerian inhibition factor, a fe-male with this condition would have müllerian duct deriva-tives (uterus and fallopian tubes), but she would also have wolffian duct derivatives and partially masculinized external genitalia.

Table 20.1 A Developmental Timetable for the Reproductive System

Approximate Time After Fertilization			Developmental Changes	
Days	**Trimester**	**Indifferent**	**Male**	**Female**
19	First	Germ cells migrate from yolk sac.		
25–30		Wolffian ducts begin development.		
44–48		Müllerian ducts begin development.		
50–52		Urogenital sinus and tubercle develop.		
53–60			Tubules and Sertoli cells appear. Müllerian ducts begin to regress.	
60–75			Leydig cells appear and begin testosterone production.	Formation of vagina begins.
			Wolffian ducts grow.	Regression of wolffian ducts begins.
105	Second			Development of ovarian follicles begins.
120				Uterus is formed.
160–260	Third		Testes descend into scrotum. Growth of external genitalia occurs.	Formation of vagina is complete.

Source: Reproduced, with permission, from the *Annual Review of Physiology*, Volume 40, p. 279. Copyright © 1978 by Annual Reviews, Inc.

An interesting cause of male pseudohermaphroditism is *testicular feminization syndrome.* Individuals with this condition have normally functioning testes but lack receptors for testosterone. Thus, although large amounts of testosterone are secreted, the embryonic tissues cannot respond to this hormone. Female genitalia therefore develop, but the vagina ends blindly (a uterus and fallopian tubes do not develop because of the secretion of müllerian inhibition factor). Male accessory sex organs likewise cannot develop because the wolffian ducts lack testosterone receptors. A child with this condition appears externally to be a normal prepubertal girl, but she has testes in her body cavity and no accessory sex organs. These testes secrete an exceedingly large amount of testosterone at puberty because of the absence of negative feedback inhibition. This abnormally large amount of testosterone is converted by the liver and adipose tissue into estrogens. As a result, the person with testicular feminization syndrome develops into a female with well-developed breasts who never menstruates (and who, of course, can never become pregnant).

Clinical Investigation Clue

Remember that Gloria has normal secondary sexual characteristics and used to have regular periods.

Do you think it likely that Gloria may suffer from any of the problems in sexual development discussed in this section?

Some male pseudohermaphrodites have normally functioning testes and normal testosterone receptors, but they genetically lack the ability to produce the enzyme 5α-reductase. Individuals with *5α-reductase deficiency* have normal epididymides, ductus (vasa) deferentia, seminal vesicles, and ejaculatory ducts because the development of these structures is stimulated directly by testosterone. The external genitalia are poorly developed and more female in appearance, however, because DHT, which cannot be produced from testosterone in the absence of 5α-reductase, is required for the development of male external genitalia.

Test Yourself Before You Continue

1. Define the terms *diploid* and *haploid,* and explain how the chromosomal sex of an individual is determined.
2. Explain how the chromosomal sex determines whether testes or ovaries will be formed.
3. List the male and female accessory sex organs and explain how the development of one or the other set of organs is determined.
4. Describe the abnormalities characteristic of testicular feminization syndrome and of 5α-reductase deficiency and explain how these abnormalities are produced.

Endocrine Regulation of Reproduction

The functions of the testes and ovaries are regulated by gonadotropic hormones secreted by the anterior pituitary. The gonadotropic hormones stimulate the gonads to secrete their sex steroid hormones, and these steroid hormones, in turn, have an inhibitory effect on the secretion of the gonadotropic hormones. This interaction between the anterior pituitary and the gonads forms a negative feedback loop.

The embryonic testes during the first trimester of pregnancy are active endocrine glands, secreting the high amounts of testosterone needed to masculinize the male embryo's external genitalia and accessory sex organs. Ovaries, by contrast, do not mature until the third trimester of pregnancy. Testosterone secretion in the male fetus declines during the second trimester of pregnancy, however, so that the gonads of both sexes are relatively inactive at the time of birth.

Before puberty, there are equally low blood concentrations of *sex steroids*—androgens and estrogens—in both males and females. Apparently, this is not due to deficiencies in the ability of the gonads to produce these hormones, but rather to lack of sufficient stimulation. During *puberty,* the gonads secrete increased amounts of sex steroid hormones as a result of increased stimulation by **gonadotropic hormones** from the anterior pituitary.

Interactions Between the Hypothalamus, Pituitary Gland, and Gonads

The anterior pituitary produces and secretes two gonadotropic hormones—**FSH (follicle-stimulating hormone)** and **LH (luteinizing hormone).** Although these two hormones are named according to their actions in the female, the same hormones are secreted by the male's pituitary gland. The gonadotropic hormones of both sexes have three primary effects on the gonads: (1) stimulation of *spermatogenesis* or *oogenesis* (formation of sperm or ova); (2) stimulation of gonadal hormone secretion; and (3) maintenance of the structure of the gonads (the gonads atrophy if the pituitary gland is removed).

The secretion of both LH and FSH from the anterior pituitary is stimulated by a hormone produced by the hypothalamus and secreted into the hypothalamo-hypophyseal portal vessels (chapter 11). This releasing hormone is sometimes called *LHRH (luteinizing hormone-releasing hormone).* Since attempts to find a separate FSH-releasing hormone have thus far failed, and since LHRH stimulates FSH as well as LH secretion, LHRH is often referred to as **gonadotropin-releasing hormone (GnRH).**

If a male or female animal is castrated (has its gonads surgically removed), the secretion of FSH and LH increases to much higher levels than in the intact animal. This demonstrates

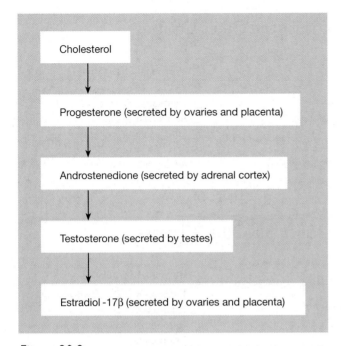

Figure 20.8 A simplified biosynthetic pathway for the steroid hormones. The sources of the sex hormones secreted in the blood are also indicated.

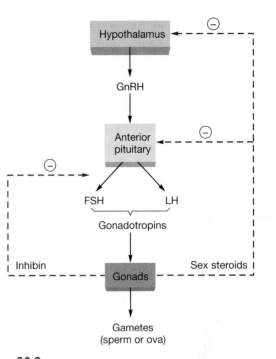

Figure 20.9 Interactions between the hypothalamus, anterior pituitary, and gonads. Sex steroids secreted by the gonads have a negative feedback effect on the secretion of GnRH (gonadotropin-releasing hormone) and on the secretion of gonadotropins. The gonads may also secrete a polypeptide hormone called inhibin that functions in the negative feedback control of FSH secretion.

that the gonads secrete products that exert a negative feedback effect on gonadotropin secretion. This negative feedback is exerted in large part by sex steroids: estrogen and progesterone in the female, and testosterone in the male. A biosynthetic pathway for these steroids is shown in figure 20.8.

The negative feedback effects of steroid hormones occurs by means of two mechanisms: (1) inhibition of GnRH secretion from the hypothalamus and (2) inhibition of the pituitary's response to a given amount of GnRH. In addition to steroid hormones, the testes and ovaries secrete a polypeptide hormone called **inhibin.** Inhibin is secreted by the Sertoli cells of the seminiferous tubules in males and by the granulosa cells of the ovarian follicles in females. This hormone specifically inhibits the anterior pituitary's secretion of FSH without affecting the secretion of LH.

Figure 20.9 illustrates the process of gonadal regulation. Although hypothalamus-pituitary-gonad interactions are similar in males and females, there are important differences. Secretion of gonadotropins and sex steroids is more or less constant in adult males. Secretion of gonadotropins and sex steroids in adult females, by contrast, shows cyclic variations (during the menstrual cycle). Also, during one phase of the female cycle—shortly before ovulation—estrogen exerts a positive feedback effect on LH secretion.

Studies have shown that secretion of GnRH from the hypothalamus is *pulsatile* rather than continuous, and thus the secretion of FSH and LH follows this pulsatile pattern. This **pulsatile secretion** is needed to prevent desensitization and downregulation of the target glands (discussed in chapter 11). It appears that the frequency of the pulses of secretion, as well as their amplitude (how much hormone is secreted per pulse), affects the target gland's response to the hormone. For example, it has been proposed that a slow frequency of GnRH pulses in women preferentially stimulates FSH secretion, while faster pulses of GnRH favor LH secretion.

If a powerful synthetic analogue of GnRH (such as *nafarelin*) is administered, the anterior pituitary first increases and then decreases its secretion of FSH and LH. This decrease, which is contrary to the normal stimulatory action of GnRH, is due to a desensitization of the anterior pituitary evoked by continuous exposure to GnRH. The decrease in LH causes a fall in testosterone secretion from the testes, or of estradiol secretion from the ovaries. The decreased testosterone secretion is useful in the treatment of men with **benign prostatic hyperplasia.** In this condition, common in men over the age of 60, testosterone supports abnormal growth of the prostate. The fall in estradiol secretion in women given synthetic GnRH analogues can be useful in the treatment of **endometriosis.** In this condition, ectopic endometrial tissue from the uterus (dependent on estradiol for growth) is found growing outside the uterus—for example, on the ovaries or on the peritoneum. These treatments illustrate the reasons why GnRH and the gonadotropins are normally secreted in a pulsatile fashion, and are particularly beneficial clinically because they are reversible.

The Onset of Puberty

Secretion of FSH and LH is high in the newborn, but falls to very low levels a few weeks after birth. Gonadotropin secretion remains low until the beginning of puberty, which is marked by rising levels of FSH followed by LH secretion. Experimental

Table 20.2 Development of Secondary Sex Characteristics and Other Changes That Occur During Puberty in Girls

Characteristic	Age of First Appearance	Hormonal Stimulation
Appearance of breast buds	8–13	Estrogen, progesterone, growth hormone, thyroxine, insulin, cortisol
Pubic hair	8–14	Adrenal androgens
Menarche (first menstrual flow)	10–16	Estrogen and progesterone
Axillary (underarm) hair	About 2 years after the appearance of pubic hair	Adrenal androgens
Eccrine sweat glands and sebaceous glands; acne (from blocked sebaceous glands)	About the same time as axillary hair growth	Adrenal androgens

Table 20.3 Development of Secondary Sex Characteristics and Other Changes That Occur During Puberty in Boys

Characteristic	Age of First Appearance	Hormonal Stimulation
Growth of testes	10–14	Testosterone, FSH, growth hormone
Pubic hair	10–15	Testosterone
Body growth	11–16	Testosterone, growth hormone
Growth of penis	11–15	Testosterone
Growth of larynx (voice lowers)	Same time as growth of penis	Testosterone
Facial and axillary (underarm) hair	About 2 years after the appearance of pubic hair	Testosterone
Eccrine sweat glands and sebaceous glands; acne (from blocked sebaceous glands)	About the same time as facial and axillary hair growth	Testosterone

evidence suggests that this rise in gonadotropin secretion is a result of two processes: (1) maturational changes in the brain that result in increased GnRH secretion by the hypothalamus and (2) decreased sensitivity of gonadotropin secretion to the negative feedback effects of sex steroid hormones.

The maturation of the brain that leads to increased GnRH secretion at the time of puberty appears to be programmed—children without gonads show increased FSH secretion at the normal time. Also during this period of time, a given amount of sex steroids has less of a suppressive effect on gonadotropin secretion than the same dose would have if administered prior to puberty. This suggests that the sensitivity of the hypothalamus and the pituitary to negative feedback effects decreases at puberty, which would also help to account for rising gonadotropin secretion at this time.

During late puberty there is a pulsatile secretion of gonadotropins—FSH and LH secretion increase during periods of sleep and decrease during periods of wakefulness. These pulses of increased gonadotropin secretion during puberty stimulate a rise in sex steroid secretion from the gonads. Increased secretion of testosterone from the testes and of **estradiol-17β** (estradiol is the major *estrogen,* or female sex steroid) from the ovaries during puberty, in turn, produces changes in body characteristic of the two sexes. Such **secondary sex characteristics** (tables 20.2 and 20.3) are the physical manifestations of the hormonal changes occurring during puberty. These changes are accompanied by a growth spurt, which begins at an earlier age in girls than in boys (fig. 20.10).

The age at which puberty begins is related to the amount of body fat and level of physical activity of the child. The aver-

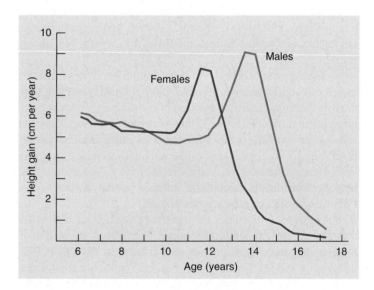

■ **Figure 20.10** **Growth as a function of sex and age.** Notice that the growth spurt during puberty occurs at an earlier age in females than in males.

age age of *menarche*—the first menstrual flow—is later (age 15) in girls who are very active physically than in the general population (age 12.6). This appears to be due to a requirement for a minimum percentage of body fat for menstruation to begin; this may represent a mechanism favored by natural selection to ensure that a woman can successfully complete a pregnancy and nurse the baby. Recent evidence suggests that the secretion of leptin from adipocytes (chapter 19) is required for puberty. Later in life, women who are very lean and physically active may have irregular cycles and *amenorrhea* (cessation of menstruation). This may also be related to the percentage of body fat. However, there is also evidence that physical exercise may act to inhibit GnRH and gonadotropin secretion.

Pineal Gland

The role of the **pineal gland** in human physiology is poorly understood. It is known that the pineal, a gland located deep within the brain, secretes the hormone **melatonin** as a derivative of the amino acid tryptophan (fig. 20.11) and that production of this hormone is influenced by light-dark cycles.

The pineal glands of some vertebrates have photoreceptors that are directly sensitive to environmental light. Although no such photoreceptors are present in the pineal glands of mammals, the secretion of melatonin has been shown to increase at night and decrease during daylight. The inhibitory effect of light on melatonin secretion in mammals is indirect. Pineal secretion is stimulated by postganglionic sympathetic neurons that originate in the superior cervical ganglion; activity of these neurons, in turn, is inhibited by nerve tracts that are activated by light striking the retina. The physiology of the pineal gland was discussed in chapter 11 (see fig. 11.32).

There is abundant experimental evidence that melatonin can inhibit gonadotropin secretion and thus have an "antigonad" effect in many vertebrates. However, the role of melatonin in the regulation of human reproduction has not yet been clearly established.

The Human Sexual Response

The sexual response, similar in both sexes, is often divided into four phases: excitation, plateau, orgasm, and resolution. The **excitation phase,** also known as **arousal,** is characterized by myotonia (increased muscle tone) and vasocongestion (the engorgement of a sexual organ with blood). This results in erection of the nipples in both sexes, although the effect is more intense and evident in females than in males. The clitoris swells (analogous to erection of the penis), and the labia minora swell to more than twice their previous size. Vasocongestion of the vagina leads to secretion of fluid, producing vaginal lubrication. Vasocongestion also causes considerable enlargement of the uterus, and in women who have not breast-fed a baby the breasts may enlarge as well.

During the **plateau phase,** the clitoris becomes partially hidden behind the labia minora because of the continued engorgement of the labia with blood. Similarly, the erected nipples become partially hidden by continued swelling of the *areolae*

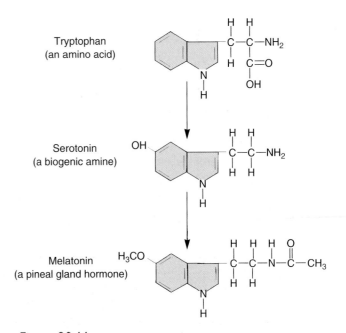

Figure 20.11 A simplified biosynthetic pathway for melatonin. Secretion of melatonin by the pineal gland follows a circadian (daily) rhythm tied to daily and seasonal changes in light.

(pigmented areas surrounding the nipples). Pronounced engorgement of the outer third of the vagina produces what Masters and Johnson, two scientists who performed pioneering studies of the human sexual response, called the "orgasmic platform."

In **orgasm,** which lasts only a few seconds, the uterus and orgasmic platform of the vagina contract several times. This is analogous to the contractions that accompany ejaculation in a male. Orgasm is followed by the **resolution phase,** in which the body returns to preexcitation conditions. Men, but not women, immediately enter a **refractory period** following orgasm, during which time they may produce an erection but are not able to ejaculate. Women, by contrast, lack a refractory period and are thus capable of multiple orgasms.

Male Reproductive System

The Leydig cells in the interstitial tissue of the testes are stimulated by LH to secrete testosterone, a potent androgen that acts to maintain the structure and function of the male accessory sex organs and to promote the development of male secondary sex characteristics. The Sertoli cells in the seminiferous tubules of the testes are stimulated by FSH. The cooperative actions of FSH and testosterone are required to initiate spermatogenesis.

The testes consist of two parts, or "compartments"—the seminiferous tubules, where spermatogenesis occurs, and the interstitial tissue, which contains the testosterone-secreting *Leydig cells* (fig. 20.12). The seminiferous tubules account for about 90% of the weight of an adult testis. The interstitial tissue is a thin web of connective tissue (containing Leydig cells) that fills the spaces between the tubules.

With regard to gonadotropin action, the testes are strictly compartmentalized. Cellular receptor proteins for FSH are located exclusively in the seminiferous tubules, where they are confined to the *Sertoli cells.* LH receptor proteins are located exclusively in the interstitial Leydig cells. Secretion of testosterone by the Leydig cells is stimulated by LH but not by FSH. Spermatogenesis in the tubules is stimulated by FSH. The apparent simplicity of this compartmentation is an illusion, however, because the two compartments can interact with each other in complex ways.

Control of Gonadotropin Secretion

Castration of a male animal results in an immediate rise in FSH and LH secretion. This demonstrates that hormones secreted by the testes exert negative feedback control of gonadotropin secretion. If testosterone is injected into the castrated animal, the secretion of LH can be returned to the previous (precastration) levels. This provides a classical example of negative feedback—LH stimulates testosterone secretion by the Leydig cells, and testosterone inhibits pituitary secretion of LH (fig. 20.13).

The amount of testosterone that is sufficient to suppress LH, however, is not sufficient to suppress the postcastration rise in FSH secretion in most experimental animals. In rams and bulls, a water-soluble (and, therefore, peptide rather than steroid) product of the seminiferous tubules specifically suppresses FSH secretion. This hormone, produced by the Sertoli cells, is called *inhibin.* The seminiferous tubules of the human testes also have been shown to produce inhibin, which inhibits FSH secretion in men. (There is also evidence that inhibin is produced by the ovaries, where it may function as a hormone and as a paracrine regulator of the ovaries.)

Testosterone Derivatives in the Brain

The brain contains testosterone receptors and is a target organ for this hormone. The effects of testosterone on the brain, such as the suppression of LH secretion, are not mediated directly by testosterone, however, but rather by its derivatives that are produced within the brain cells. Testosterone may be converted by the enzyme 5α-reductase to dihydrotestosterone (DHT), as previously described. The DHT, in turn, can be changed by other enzymes into other 5α-reduced androgens—abbreviated 3α-diol and 3β-diol (fig. 20.14). Alternatively,

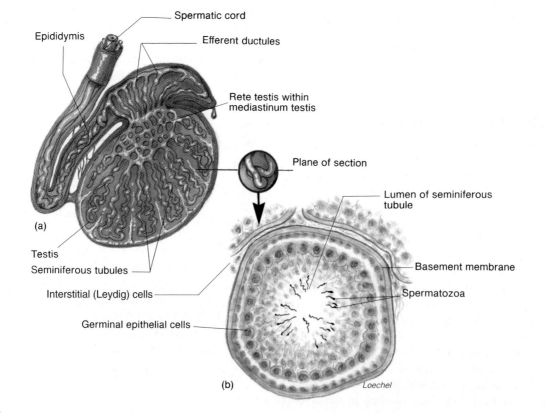

Spermatic cord
Epididymis
Efferent ductules
Rete testis within mediastinum testis
Plane of section
Lumen of seminiferous tubule
(a)
Testis
Seminiferous tubules
Interstitial (Leydig) cells
Germinal epithelial cells
Basement membrane
Spermatozoa
(b)
Loechel

■ Figure 20.12   The seminiferous tubules. *(a)* A sagittal section of a testis and *(b)* a transverse section of a seminiferous tubule.

testosterone may be converted within the brain to estradiol-17β. Although usually regarded as a female sex steroid, estradiol is therefore an active compound in normal male physiology! Estradiol is formed from testosterone by the action of an enzyme called *aromatase*. This reaction is known as *aromatization,* a term that refers to the presence of an aromatic carbon ring (chapter 2). The estradiol formed from

testosterone in the brain is required for the negative feedback effects of testosterone on LH secretion.

Testosterone Secretion and Age

The negative feedback effects of testosterone and inhibin help to maintain a relatively constant (that is, noncyclic) secretion of gonadotropins in males, resulting in relatively constant levels of androgen secretion from the testes. This contrasts with the cyclic secretion of gonadotropins and ovarian steroids in females. Women experience an abrupt cessation in sex steroid secretion during menopause. By contrast, the secretion of androgens declines only gradually and to varying degrees in men over 50 years of age. The causes of this age-related change in testicular function are not currently known. The decline in testosterone secretion cannot be due to decreasing gonadotropin secretion, since gonadotropin levels in the blood are, in fact, elevated (because of less negative feedback) at the time that testosterone levels are declining.

Endocrine Functions of the Testes

Testosterone is by far the major androgen secreted by the adult testis. This hormone and its derivatives (the 5α-reduced androgens) are responsible for initiation and maintenance of the body changes associated with puberty in males. Androgens are sometimes called *anabolic steroids* because they stimulate the growth of muscles and other structures (table 20.4). Increased testosterone secretion during puberty is also required for growth of the accessory sex organs—primarily the seminal vesicles and prostate. Removal of androgens by castration results in atrophy of these organs.

Androgens stimulate growth of the larynx (causing a lowering of the voice) and promote hemoglobin synthesis (males have higher hemoglobin levels than females) and bone growth. The effect of androgens on bone growth is self-limiting, however,

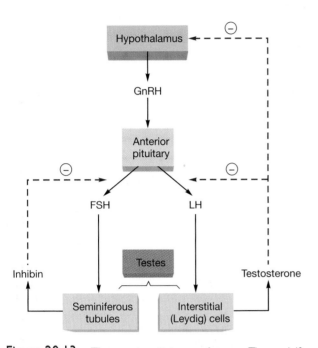

Figure 20.13 **The anterior pituitary and testes.** The seminiferous tubules are the targets of FSH action; the interstitial (Leydig) cells are targets of LH action. Testosterone secreted by the Leydig cells inhibits LH secretion; inhibin secreted by the tubules may inhibit FSH secretion.

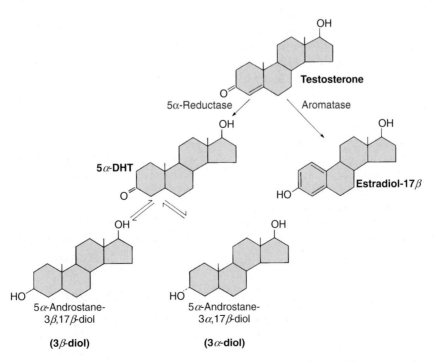

Figure 20.14 **Derivatives of testosterone.** Testosterone secreted by the interstitial (Leydig) cells of the testes can be converted into active metabolites in the brain and other target organs. These active metabolites include DHT and other 5α-reduced androgens and estradiol.

Table 20.4   Actions of Androgens in the Male

Category	Action
Sex Determination	Growth and development of wolffian ducts into epididymis, ductus deferens, seminal vesicles, and ejaculatory ducts
	Development of urogenital sinus into prostate
	Development of male external genitalia (penis and scrotum)
Spermatogenesis	At puberty: Completion of meiotic division and early maturation of spermatids
	After puberty: Maintenance of spermatogenesis
Secondary Sex Characteristics	Growth and maintenance of accessory sex organs
	Growth of penis
	Growth of facial and axillary hair
	Body growth
Anabolic Effects	Protein synthesis and muscle growth
	Growth of bones
	Growth of other organs (including larynx)
	Erythropoiesis (red blood cell formation)

because they ultimately cause replacement of cartilage by bone in the epiphyseal discs, thus "sealing" the discs and preventing further lengthening of the bones (as described in chapter 19).

Although androgens are by far the major endocrine products of the testes, there is evidence that both Sertoli cells and Leydig cells secrete small amounts of estradiol. Further, receptors for estradiol are found in Sertoli and Leydig cells, as well as in the cells lining the male reproductive tract (efferent ductules and epididymis) and accessory sex organs (prostate and seminal vesicles). Estrogen receptors have also been located in the developing sperm cells (spermatocytes and spermatids, described in the next section) of many species, including humans. This suggests a role for estrogens in spermatogenesis, and indeed knockout mice (chapter 3) missing an estrogen receptor gene are infertile. Further, men with a congenital deficiency in aromatase—the enzyme that converts androgens to estrogens (fig. 20.14)—are also infertile.

Estradiol, either secreted by the testes or produced locally as a paracrine regulator, may be responsible for a number of effects in men that have previously been attributed to androgens. For example, the importance of the conversion of testosterone into estradiol in the brain for negative feedback control was described earlier. Estrogen also may be responsible for sealing of the epiphyseal plates of cartilage; this is suggested by observations that men who lack the ability to produce estrogen or who lack estrogen receptors (due to rare genetic defects only recently discovered) maintain their epiphyseal plates and continue to grow.

The two compartments of the testes interact with each other in paracrine fashion (fig. 20.15). Paracrine regulation, as described in chapter 11, refers to chemical regulation that occurs among tissues within an organ. Testosterone from the Leydig cells is metabolized by the tubules into other active androgens and is required for spermatogenesis, for example. The tubules also secrete products that might influence Leydig cell function. Such interactions are suggested by evidence that exposure of pubertal male rats to FSH augments the responsiveness of the Leydig cells to LH. Since FSH can directly stimulate only the Sertoli cells of the tubules, the FSH-induced enhancement of LH responsiveness must be mediated by products secreted from the Sertoli cells.

Inhibin secreted by the Sertoli cells in response to FSH can facilitate the Leydig cells' response to LH, as measured by the

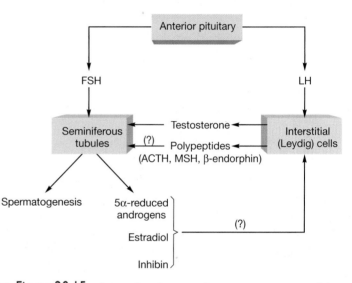

■ **Figure 20.15**   Interactions between the two compartments of the testes. Testosterone secreted by the interstitial (Leydig) cells stimulates spermatogenesis in the tubules. Leydig cells may also secrete ACTH, MSH, and β-endorphin. Secretion of inhibin by the tubules may affect the sensitivity of the Leydig cells to LH stimulation.

amount of testosterone secreted. Further, it has been shown that the Leydig cells are capable of producing a family of polypeptides previously associated only with the pituitary gland—ACTH, MSH, and β-endorphin. Experiments suggest that ACTH and MSH can stimulate Sertoli cell function, whereas β-endorphin can inhibit Sertoli function. The physiological significance of these fascinating paracrine interactions between the two compartments of the testes remains to be demonstrated.

Spermatogenesis

The germ cells that migrate from the yolk sac to the testes during early embryonic development become spermatogenic stem cells, called **spermatogonia,** within the outer region of the seminiferous tubules. Spermatogonia are diploid cells (with forty-six chromosomes) that ultimately give rise to mature

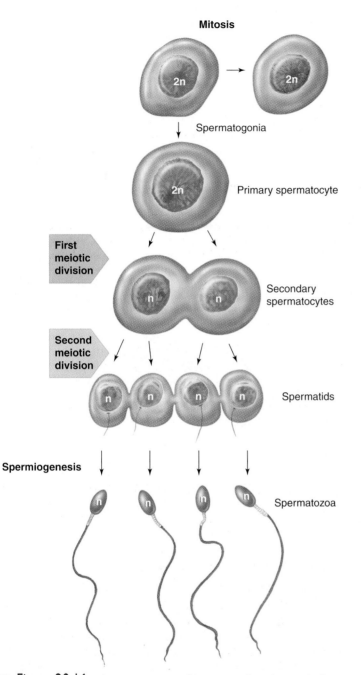

Mitosis

Spermatogonia

Primary spermatocyte

First meiotic division

Secondary spermatocytes

Second meiotic division

Spermatids

Spermiogenesis

Spermatozoa

■ **Figure 20.16** **Spermatogenesis.** Spermatogonia undergo mitotic division in which they replace themselves and produce a daughter cell that will undergo meiotic division. This cell is called a primary spermatocyte. Upon completion of the first meiotic division, the daughter cells are called secondary spermatocytes. Each of these completes a second meiotic division to form spermatids. Notice that the four spermatids produced by the meiosis of a primary spermatocyte are interconnected. Each spermatid forms a mature spermatozoon.

haploid gametes by a process of reductive cell division called *meiosis*. The steps of meiosis are summarized in chapter 3, figure 3.33.

Meiosis involves two nuclear divisions (see fig. 3.33). In the first part of this process, the DNA duplicates and homologous chromosomes are separated into two daughter cells.

Since each daughter cell contains only one of each homologous pair of chromosomes, the cells formed at the end of this first meiotic division contain twenty-three chromosomes each and are haploid. Each of the twenty-three chromosomes at this stage, however, consists of two strands (called *chromatids*) of identical DNA. During the second meiotic division, these duplicate chromatids are separated into daughter cells. Meiosis of one diploid spermatogonium cell therefore produces four haploid cells.

Actually, only about 1,000 to 2,000 stem cells migrate from the yolk sac into the embryonic testes. In order to produce many millions of sperm throughout adult life, these spermatogonia duplicate themselves by mitotic division and only one of the two cells—now called a **primary spermatocyte**—undergoes meiotic division (fig. 20.16). In this way, spermatogenesis can occur continuously without exhausting the number of spermatogonia.

When a diploid primary spermatocyte completes the first meiotic division (at telophase I), the two haploid cells thus produced are called **secondary spermatocytes.** At the end of the second meiotic division, each of the two secondary spermatocytes produces two haploid **spermatids.** One primary spermatocyte therefore produces four spermatids.

The sequence of events in spermatogenesis is reflected in the cellular arrangement of the wall of the seminiferous tubule. The spermatogonia and primary spermatocytes are located toward the outer side of the tubule, whereas spermatids and mature spermatozoa are located on the side of the tubule facing the lumen.

At the end of the second meiotic division, the four spermatids produced by meiosis of one primary spermatocyte are interconnected—their cytoplasm does not completely pinch off at the end of each division. Development of these interconnected spermatids into separate mature **spermatozoa** (singular, *spermatozoon*)—a process called *spermiogenesis*—requires the participation of the Sertoli cells (fig. 20.17).

Sertoli Cells

The nongerminal Sertoli cells form a continuous layer connected by tight junctions around the circumference of each tubule. In this way, they constitute a **blood-testis barrier;** molecules from the blood must pass through the cytoplasm of the Sertoli cells before entering germinal cells. Similarly, this barrier normally prevents the immune system from becoming sensitized to antigens in the developing sperm, and thus prevents autoimmune destruction of the sperm. The cytoplasm of the Sertoli cells extends from the periphery to the lumen of the tubule and envelops the developing germ cells, so that it is often difficult to tell where the cytoplasm of the Sertoli cells ends and that of germ cells begins.

The Sertoli cells help to make the seminiferous tubules an *immunologically privileged site* (protected from immune attack) through another mechanism as well. As described in chapter 15, the Sertoli cells produce **FAS ligand,** which binds to the *FAS* receptor on the surface of T lymphocytes. This triggers apoptosis (cell suicide) of the T lymphocytes and thus helps to prevent immune attack of the developing sperm.

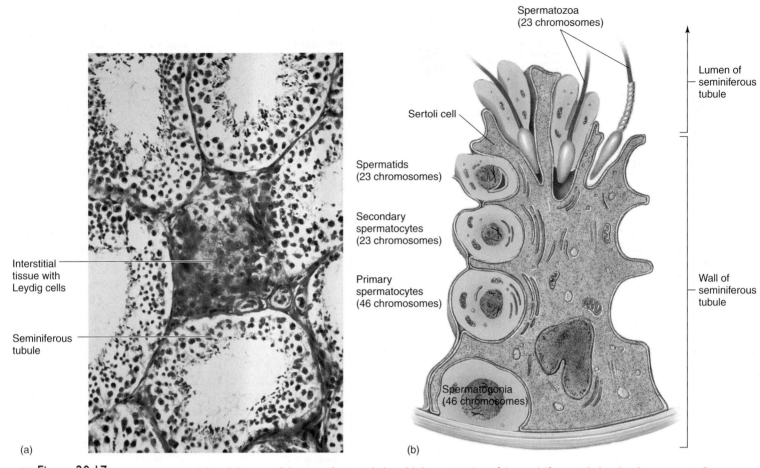

■ **Figure 20.17** A photomicrograph and diagram of the seminiferous tubules. *(a)* A cross section of the seminiferous tubules also shows surrounding interstitial tissue. *(b)* The stages of spermatogenesis are indicated within the germinal epithelium of a seminiferous tubule. The relationship between Sertoli cells and developing spermatozoa can also be seen.

In the process of spermiogenesis (conversion of spermatids to spermatozoa), most of the spermatid cytoplasm is eliminated. This occurs through phagocytosis by Sertoli cells of the "residual bodies" of cytoplasm from the spermatids (fig. 20.18). Phagocytosis of residual bodies may transmit informational molecules from germ cells to Sertoli cells. The Sertoli cells, in turn, may provide molecules needed by the germ cells. It is known, for example, that the X chromosome of germ cells is inactive during meiosis. Since this chromosome contains genes needed to produce many essential molecules, it is believed that these molecules are provided by the Sertoli cells during this time.

Sertoli cells secrete a protein called **androgen-binding protein (ABP)** into the lumen of the seminiferous tubules. This protein, as its name implies, binds to testosterone and thereby concentrates it within the tubules. The importance of Sertoli cells in tubular function is further evidenced by the fact that FSH receptors are confined to the Sertoli cells. Any effect of FSH on the tubules, therefore, must be mediated through the action of Sertoli cells. These include the FSH-induced stimulation of spermiogenesis and the autocrine interactions between Sertoli cells and Leydig cells previously described.

Hormonal Control of Spermatogenesis

The formation of primary spermatocytes and entry into early prophase I begins during embryonic development, but spermatogenesis is arrested at this point until puberty, when testosterone secretion rises. Testosterone is required for completion of meiotic division and for the early stages of spermatid maturation. This effect is probably not produced by testosterone directly, but rather by some of the molecules derived from testosterone (the 5α-reduced androgens and estrogens, described earlier) in the tubules. The testes also produce a wide variety of paracrine regulators—transforming growth factor, insulin-like growth factor-1, inhibin, and others—that may help to regulate spermatogenesis.

The later stages of spermatid maturation during puberty appear to require stimulation by FSH (fig. 20.19). This FSH effect is mediated by the Sertoli cells, as previously described. During puberty, therefore, both FSH and androgens are needed for the initiation of spermatogenesis.

FSH and testosterone (or its derivatives) stimulate sperm development indirectly, by acting on the Sertoli cells. It is currently believed that these hormones stimulate the Sertoli cells to secrete polypeptides, which in turn act as paracrine regulators to stimulate spermatogenesis.

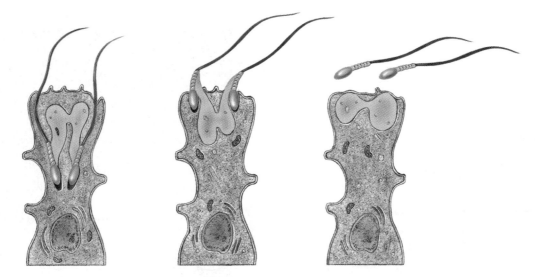

Figure 20.18   The processing of spermatids into spermatozoa (spermiogenesis). As the spermatids develop into spermatozoa, most of their cytoplasm is pinched off as residual bodies and ingested by the surrounding Sertoli cell cytoplasm.

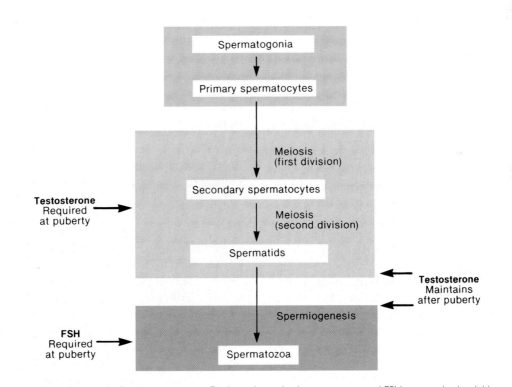

Figure 20.19   The endocrine control of spermatogenesis. During puberty, both testosterone and FSH are required to initiate spermatogenesis. In the adult, however, testosterone alone can maintain spermatogenesis.

At the conclusion of spermiogenesis, spermatozoa are released into the lumen of the seminiferous tubules. The spermatozoa consist of an oval-shaped *head* (which contains the DNA), *midpiece,* and a *tail* (fig. 20.20). Although the tail will ultimately be capable of flagellar movement, the sperm at this stage are nonmotile. They become motile and undergo other maturational changes outside the testis in the epididymis.

Male Accessory Sex Organs

The seminiferous tubules are connected at both ends to a tubular network called the *rete testis* (see fig. 20.12). Spermatozoa and tubular secretions are moved to this area of the testis and are drained via the *efferent ductules* into the **epididymis** (the plural is *epididymides*). The epididymis is a tightly coiled structure,

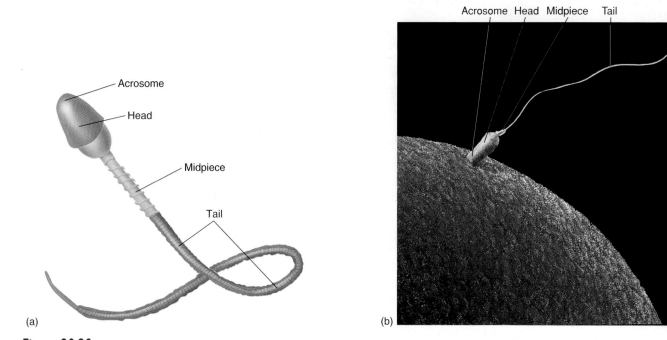

■ **Figure 20.20**   **A human spermatozoon.** *(a)* A diagrammatic representation and *(b)* a scanning electron micrograph in which a spermatozoon is seen in contact with an ovum.

about 5 meters (16 feet) long if stretched out, that receives the tubular products. Spermatozoa enter at the "head" of the epididymis and are drained from its "tail" by a single tube, the **ductus,** or **vas, deferens.**

Spermatozoa that enter the head of the epididymis are non-motile. This is partially due to the low pH of the fluid in the epididymis and ductus deferens, produced by the secretion of H^+ by active transport ATPase pumps. During their passage through the epididymis, the sperm undergo maturational changes that make them more resistant to changes in pH and temperature. The pH is neutralized by the alkaline prostatic fluid during ejaculation, so that the sperm are fully motile and become capable of fertilizing an ovum once they spend some time in the female reproductive tract. Sperm obtained from the seminiferous tubules, by contrast, cannot fertilize an ovum. The epididymis serves as a site for sperm maturation and for the storage of sperm between ejaculations.

The ductus deferens carries sperm from the epididymis out of the scrotum into the pelvic cavity. The **seminal vesicles** then add secretions that pass through their ducts; at this point, the ductus deferens becomes an **ejaculatory duct.** The ejaculatory duct is short (about 2 cm), however, because it enters the **prostate** and soon merges with the prostatic **urethra.** The prostate adds its secretions through numerous pores in the walls of the prostatic urethra, forming a fluid known as *semen* (fig. 20.21).

The seminal vesicles and prostate are androgen-dependent accessory sex organs—they will atrophy if androgen is withdrawn by castration. The seminal vesicles secrete fluid containing fructose, which serves as an energy source for the spermatozoa. This fluid secretion accounts for about 60% of the volume of the semen. The fluid contributed by the prostate contains citric acid, calcium, and coagulation proteins. Clotting proteins cause the

semen to coagulate after ejaculation, but the hydrolytic action of fibrinolysin later causes the coagulated semen to again assume a more liquid form, thereby freeing the sperm.

An immunoassay for *prostate-specific antigen (PSA)* is a common laboratory test for prostate disorders, including **prostate cancer.** A more common disorder, affecting most men over 60 to a greater or lesser degree, is **benign prostatic hyperplasia (BPH).** This disorder is responsible for most symptoms of bladder outlet obstruction, where there is difficulty in urination. BPH treatment may involve a surgical procedure called *transurethral resection (TUR),* or the use of drugs. Drugs used to treat BPH include α_1-*adrenergic receptor blockers,* which decrease the muscle tone of the prostate and bladder neck (making urination easier), and *5α-reductase inhibitors,* which inhibit the enzyme needed to convert testosterone into dihydrotestosterone (DHT). As previously described, DHT is required to maintain the structure of the prostate; therefore, a reduction in DHT may help to reduce the size of the prostate.

Erection, Emission, and Ejaculation

Erection, accompanied by increases in the length and width of the penis, is achieved as a result of blood flow into the "erectile tissues" of the penis. These erectile tissues include two paired structures—the *corpora cavernosa*—located on the dorsal side of the penis, and one unpaired *corpus spongiosum* on the ventral

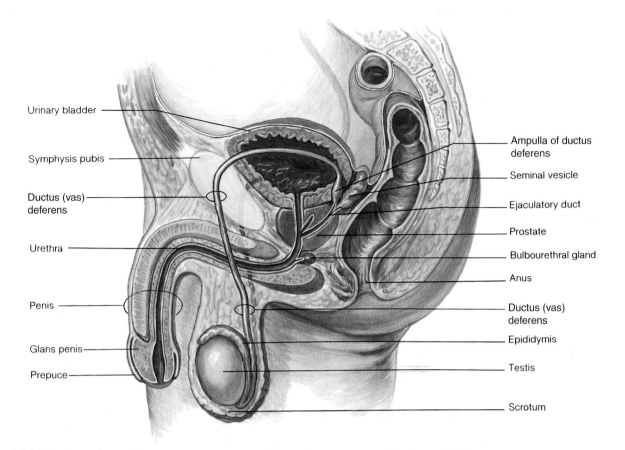

Urinary bladder

Symphysis pubis

Ductus (vas) deferens

Urethra

Penis

Glans penis

Prepuce

Ampulla of ductus deferens

Seminal vesicle

Ejaculatory duct

Prostate

Bulbourethral gland

Anus

Ductus (vas) deferens

Epididymis

Testis

Scrotum

■ **Figure 20.21** The organs of the male reproductive system. The male organs are seen here in a sagittal view.

side (fig. 20.22). The urethra runs through the center of the corpus spongiosum. The erectile tissue forms columns that extend the length of the penis, although the corpora cavernosa do not extend all the way to the tip.

Erection is achieved by parasympathetic nerve-induced vasodilation of arterioles that allows blood to flow into the corpora cavernosa of the penis. The neurotransmitter that mediates this increased blood flow is now believed to be nitric oxide (fig. 20.23). Nitric oxide released by parasympathetic axons and produced by the endothelial cells of penile blood vessels activates guanylate cyclase in the vascular smooth muscle cells. Guanylate cyclase catalyzes the production of cyclic GMP (cGMP), which closes Ca^{2+} channels in the plasma membrane (fig. 20.23). This decreases the cytoplasmic Ca^{2+} concentration, causing smooth muscle relaxation (chapter 12). The penile blood vessels thereby dilate to increase the blood flow into the erectile tissue, producing an erection.

As the erectile tissues become engorged with blood and the penis becomes turgid, venous outflow of blood is partially occluded, thus aiding erection. The term **emission** refers to the movement of semen into the urethra, and **ejaculation** refers to the forcible expulsion of semen from the urethra out of the penis. Emission and ejaculation are stimulated by sympathetic nerves, which cause peristaltic contractions of the tubular system, contractions of the seminal vesicles and prostate, and contractions of muscles at the base of the penis. Sexual function in the male thus requires the synergistic action (rather

than antagonistic action) of the parasympathetic and sympathetic systems.

Erection is controlled by two portions of the central nervous system—the hypothalamus in the brain and the sacral portion of the spinal cord. Conscious sexual thoughts originating in the cerebral cortex act via the hypothalamus to control the sacral region, which in turn increases parasympathetic nerve activity to promote vasodilation and erection of the penis. Conscious thought is not required for erection, however, because sensory stimulation of the penis can more directly activate the sacral region of the spinal cord and cause an erection.

 Nitric oxide, released in the penis in response to parasympathetic nerve stimulation, diffuses into the smooth muscle cells of blood vessels and stimulates the production of cyclic guanosine monophosphate (cGMP). The cGMP, in turn, causes the vascular smooth muscle to relax, so that blood can flow into the corpora cavernosa (fig. 20.23). This physiology is exploited by *sildenafil* (trade named *Viagra*), which can be taken as a pill to treat **erectile dysfunction**. Sildenafil blocks cGMP phosphodiesterase, an enzyme that functions to break down cGMP. This increases the concentration of cGMP and thus promotes vasodilation, leading to increased engorgement of the erectile spongy tissue with blood and consequently promoting erection (fig. 20.23).

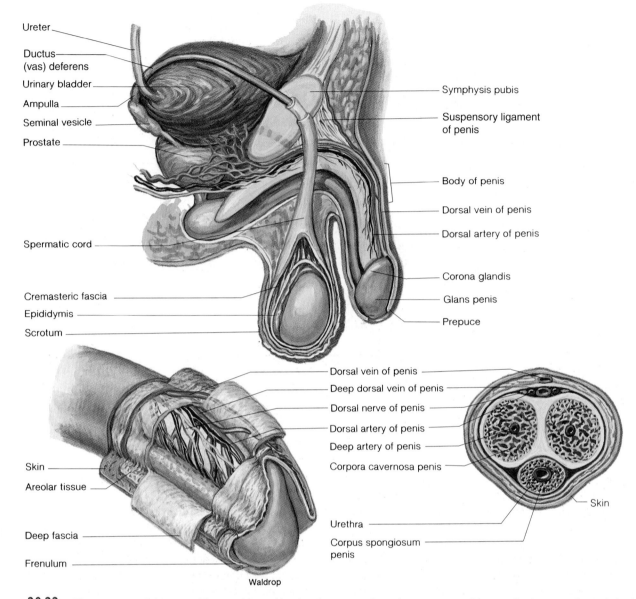

Ureter
Ductus (vas) deferens
Urinary bladder
Ampulla
Seminal vesicle
Prostate

Symphysis pubis
Suspensory ligament of penis

Body of penis
Dorsal vein of penis
Dorsal artery of penis

Spermatic cord

Corona glandis

Cremasteric fascia
Epididymis
Scrotum

Glans penis
Prepuce

Dorsal vein of penis
Deep dorsal vein of penis
Dorsal nerve of penis
Dorsal artery of penis
Deep artery of penis
Corpora cavernosa penis

Skin
Areolar tissue

Skin

Deep fascia
Frenulum

Urethra
Corpus spongiosum penis

Waldrop

■ **Figure 20.22** **The structure of the penis.** The attachment, blood and nerve supply, and arrangement of the erectile tissue are shown in both longitudinal and cross section.

Male Fertility

The approximate volume of semen for each ejaculation is 1.5 to 5.0 milliliters. The bulk of this fluid (45% to 80%) is produced by the seminal vesicles, and 15% to 30% is contributed by the prostate. There are usually between 60 and 150 million sperm per milliliter of ejaculate. Normal human semen values are summarized in table 20.5.

A sperm concentration below about 20 million per milliliter is termed *oligospermia* (*oligo* = few) and is associated with decreased fertility. A total sperm count below about 50 million per ejaculation is clinically significant in male infertility. Oligospermia may be caused by a variety of factors, including heat from a sauna or hot tub, various pharmaceutical drugs, lead and arsenic poisoning, and such illicit drugs as marijuana, cocaine, and anabolic steroids. It may be temporary or permanent. In addition to low sperm counts as a cause of in-

fertility, some men and women have antibodies against sperm antigens (this is very common in men with vasectomies). While such antibodies do not appear to affect health, they do reduce fertility.

Attempts have been made to develop new methods of male contraception. These have generally involved compounds that suppress gonadotropin secretion, such as testosterone or a combination of progesterone and a GnRH antagonist. Another compound, *gossypol*, which interferes with sperm development, has also been tried. These drugs can be effective but have unacceptable side effects. One of the most widely used methods of male contraception is a surgical procedure called a **vasectomy** (fig. 20.24). In this procedure, each ductus (vas) deferens is cut and tied or, in some cases, a valve or similar device is inserted. A vasectomy interferes with sperm transport but does not directly affect the secretion of androgens from Leydig cells in the interstitial tissue. Since spermatogenesis continues, the sperm produced cannot be drained from the

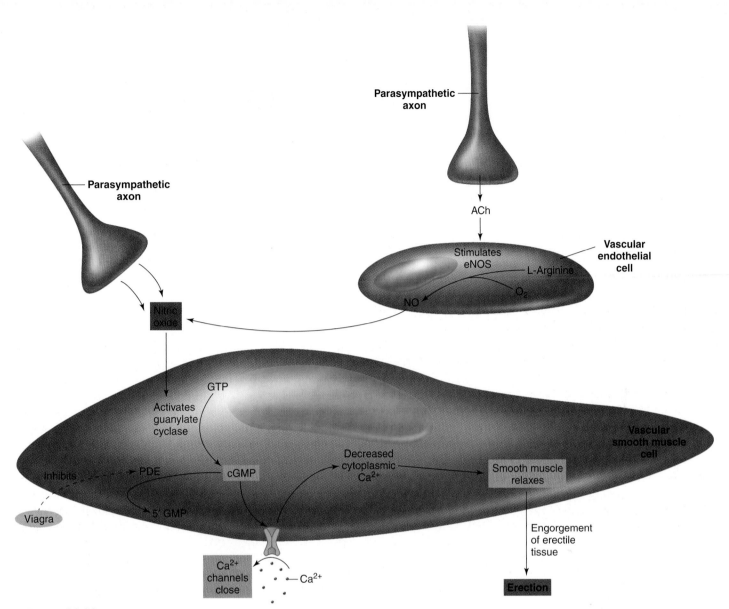

Figure 20.23 **Role of nitric oxide in penile erection and the action of Viagra.** Nitric oxide is released as a neurotransmitter by particular parasympathetic axons, and is secreted as a paracrine regulator by endothelial cells of vessels in the penis (eNOS = endothelial nitric oxide synthase). Nitric oxide stimulates the production of cyclic GMP (cGMP), which serves as the intracellular second messenger in the mechanism of smooth muscle relaxation that leads to erection. The enzyme phosphodiesterase (PDE) breaks down cGMP, and it is this enzyme that is inhibited by the drug sildenafil (Viagra).

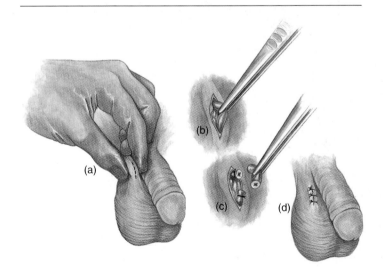

Figure 20.24 **A vasectomy.** In this surgical procedure a segment of the ductus (vas) deferens is removed through an incision in the scrotum.

Table 20.5   Semen Analysis

Characteristic	Reference Value
Volume of ejaculate	1.5–5.0 ml
Sperm count	40–250 million/ml
Sperm motility	
Percentage of motile forms:	
1 hour after ejaculation	70% or more
3 hours after ejaculation	60% or more
Leukocyte count	0–2,000/ml
pH	7.2–7.8
Fructose concentration	150–600 mg/100 ml

Source: Modified from L. Glasser, "Seminal Fluid and Subfertility," *Diagnostic Medicine,* July/August 1981, p. 28. Used by permission.

testes and instead accumulate in "crypts" that form in the seminiferous tubules, epididymis, and ductus deferens. These crypts present sites for inflammatory reactions in which spermatozoa are phagocytosed and destroyed by the immune system. It is thus not surprising that approximately 70% of men with vasectomies develop antisperm antibodies. These antibodies do not appear to cause autoimmune damage to the testes, but they do significantly diminish the possibility of reversing a vasectomy and restoring fertility.

Test Yourself Before You Continue

1. Describe the effects of castration on FSH and LH secretion in the male. Explain the experimental evidence suggesting that the testes produce a polypeptide that specifically inhibits FSH secretion.

2. Describe the two compartments of the testes with respect to (a) structure, (b) function, and (c) response to gonadotropin stimulation. Describe two ways in which these compartments interact.

3. Using a diagram, describe the stages of spermatogenesis. Why can spermatogenesis continue throughout life without using up all of the spermatogonia?

4. Describe the structure and proposed functions of the Sertoli cells in the seminiferous tubules.

5. Explain how FSH and androgens synergize to stimulate sperm production at puberty. Describe the hormonal requirements for spermatogenesis after puberty.

Female Reproductive System

The ovaries contain a large number of follicles, each of which encloses an ovum. Some of these follicles mature during the ovarian cycle, and the ova they contain progress to the secondary oocyte stage of meiosis. At ovulation, the largest follicle breaks open to extrude a secondary oocyte from the ovary. The empty follicle then becomes a corpus luteum, which ultimately degenerates at the end of a nonfertile cycle.

The two ovaries (fig. 20.25), about the size and shape of large almonds, are suspended by means of ligaments from the pelvic girdle. Extensions called *fimbriae* of the **uterine (fallopian) tubes** partially cover each ovary. Ova that are released from the ovary—in a process called *ovulation*—are normally drawn into the uterine tubes by the action of the ciliated epithelial lining of the tubes. The lumen of each uterine tube is continuous with the **uterus** (or womb), a pear-shaped muscular organ held in place within the pelvic cavity by ligaments.

The uterus consists of three layers. The outer layer of connective tissue is the **perimetrium,** the middle layer of smooth muscle is the **myometrium,** and the inner epithelial layer is the **endometrium.** The endometrium is a stratified, squamous,

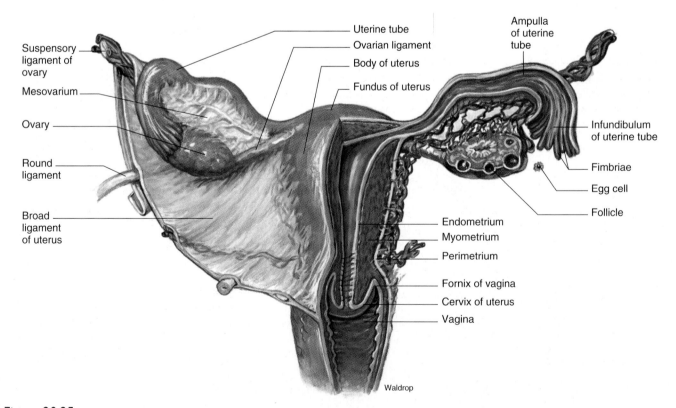

Figure 20.25   **The uterus, uterine tubes, and ovaries.** The supporting ligaments can also be seen in this posterior view.

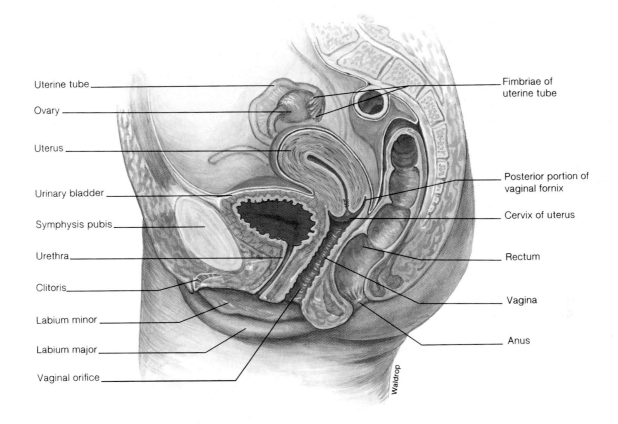

■ **Figure 20.26** **The organs of the female reproductive system.** These are shown in sagittal section.

nonkeratinized epithelium that consists of a *stratum basale* and a more superficial *stratum functionale*. The stratum functionale, which cyclically grows thicker as a result of estrogen and progesterone stimulation, is shed at menstruation.

The uterus narrows to form the *cervix* (= neck), which opens to the tubular **vagina.** The only physical barrier between the vagina and uterus is a plug of *cervical mucus.* These structures—the vagina, uterus, and fallopian tubes—constitute the accessory sex organs of the female (fig. 20.26). Like the accessory sex organs of the male, the female reproductive tract is affected by gonadal steroid hormones. Cyclic changes in ovarian secretion, as will be described in the next section, cause cyclic changes in the epithelial lining of the tract.

The vaginal opening is located immediately posterior to the opening of the urethra. Both openings are covered by longitudinal folds—the inner **labia minora** and outer **labia majora** (fig. 20.27). The **clitoris,** a small structure composed largely of erectile tissue, is located at the anterior margin of the labia minora.

Ovarian Cycle

The germ cells that migrate into the ovaries during early embryonic development multiply, so that by about 5 months of *gestation* (prenatal life) the ovaries contain approximately 6 million

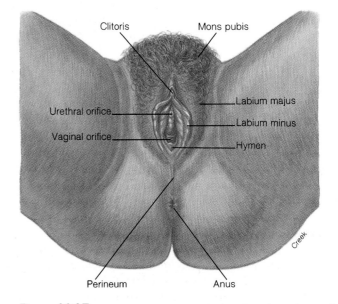

■ **Figure 20.27** **The external female genitalia.** The labia majora and clitoris in a female are homologous to the scrotum and penis, respectively, in a male.

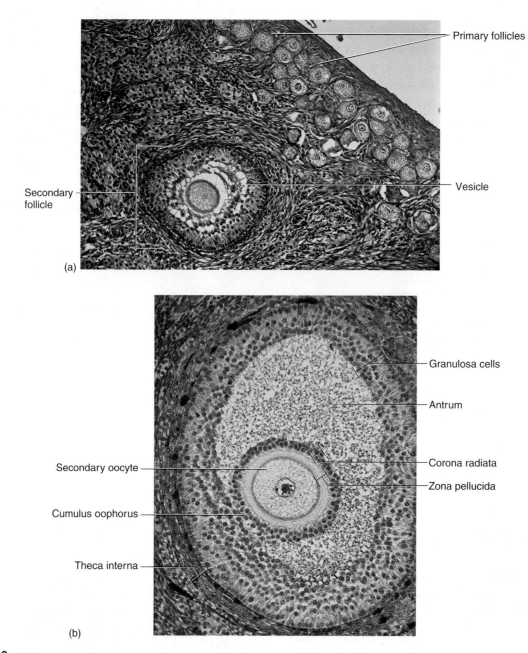

Primary follicles

Secondary follicle

Vesicle

(a)

Granulosa cells

Antrum

Secondary oocyte

Corona radiata

Zona pellucida

Cumulus oophorus

Theca interna

(b)

■ **Figure 20.28**   **Photomicrographs of the ovary.** (a) Primary follicles and one secondary follicle and (b) a graafian follicle are visible in these sections.

to 7 million oogonia. Most of these oogonia die prenatally through a process of apoptosis (chapter 3). The production of new oogonia stops at this point and never resumes again. The oogonia begin meiosis toward the end of gestation, at which time they are called **primary oocytes.** Like spermatogenesis in the prenatal male, oogenesis is arrested at prophase I of the first meiotic division. The primary oocytes are thus still diploid.

Primary oocytes decrease in number throughout a woman's life. The ovaries of a newborn girl contain about 2 million oocytes—all she will ever have. Each oocyte is contained within its own hollow ball of cells, the *ovarian follicle.* By the time a girl reaches puberty, the number of oocytes and follicles

has been reduced to 400,000. Only about 400 of these oocytes will ovulate during the woman's reproductive years, and the rest will die by apoptosis. Oogenesis ceases entirely at menopause (the time menstruation stops).

Primary oocytes that are not stimulated to complete the first meiotic division are contained within tiny **primary follicles** (fig. 20.28a). Immature primary follicles consist of only a single layer of follicle cells. In response to FSH stimulation, some of these oocytes and follicles get larger, and the follicular cells divide to produce numerous layers of **granulosa cells** that surround the oocyte and fill the follicle. Some primary follicles will be stimulated to grow still more, and they will develop a number of

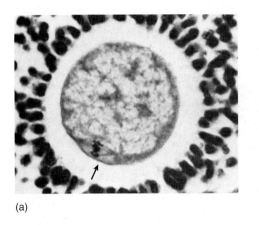

(a)

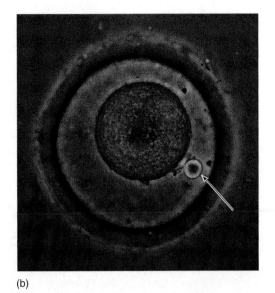

(b)

Figure 20.29 Photomicrographs of oocytes. *(a)* A primary oocyte at a metaphase I of meiosis. Notice the alignment of chromosomes *(arrow)*. *(b)* A human secondary oocyte formed at the end of the first meiotic division. Also shown is the first polar body *(arrow)*.

fluid-filled cavities called *vesicles;* at this point, they are called **secondary follicles** (fig. 20.28*a*). Continued growth of one of these follicles will be accompanied by the fusion of its vesicles to form a single fluid-filled cavity called an *antrum.* At this stage, the follicle is known as a **mature, or graafian, follicle** (fig. 20.28*b*).

As the follicle develops, the primary oocyte completes its first meiotic division. This does not form two complete cells, however, because only one cell—the **secondary oocyte**—gets all the cytoplasm. The other cell formed at this time becomes a small *polar body* (fig. 20.29), which eventually fragments and disappears. This unequal division of cytoplasm ensures that the ovum will be large enough to become a viable embryo should fertilization occur. The secondary oocyte then begins the second meiotic division, but meiosis is arrested at metaphase II. The second meiotic division is completed only by an oocyte that has been fertilized.

The secondary oocyte, arrested at metaphase II, is contained within a graafian follicle. The granulosa cells of this follicle form a ring around the oocyte and form a mound that supports the oocyte. This mound is called the *cumulus oophorus.* The ring of granulosa cells surrounding the oocyte is the *corona radiata.* Between the oocyte and the corona radiata is a thin gel-like layer of proteins and polysaccharides called the **zona pellucida** (see fig. 20.28*b*). The zona pellucida is significant because it presents a barrier to the ability of a sperm to fertilize an ovulated oocyte.

Under the stimulation of FSH from the anterior pituitary, the granulosa cells of the ovarian follicles secrete increasing amounts of estrogen as the follicles grow. Interestingly, the granulosa cells produce estrogen from its precursor testosterone, which is supplied by cells of the *theca interna,* the layer immediately outside the follicle (see fig. 20.28*b*).

Ovulation

Usually by the tenth to fourteenth day after the first day of menstruation only one follicle has continued its growth to become a fully mature graafian follicle (fig. 20.30). Other secondary follicles during that cycle regress and become *atretic*—a term that means "without an opening" in reference to their failure to rupture. Follicle atresia, or degeneration, is a type of apoptosis that results from a complex interplay of hormones and paracrine regulators. The gonadotropins (FSH and LH), as well as various paracrine regulators and estrogen act to protect follicles from atresia. By contrast, paracrine regulators that include androgens and FAS ligand (chapter 15) promote atresia of the follicles.

The follicle that is protected from atresia and that develops into a graafian follicle becomes so large that it forms a bulge on the surface of the ovary. Under proper hormonal stimulation, this follicle will rupture—much like the popping of a blister—and extrude its oocyte into the uterine tube in the process of **ovulation** (fig. 20.31).

The released cell is a secondary oocyte, surrounded by the zona pellucida and corona radiata. If it is not fertilized, it will degenerate in a couple of days. If a sperm passes through the corona radiata and zona pellucida and enters the cytoplasm of the secondary oocyte, the oocyte will then complete the second meiotic division. In this process, the cytoplasm is again not divided equally; most remains in the zygote (fertilized egg), leaving another polar body which, like the first, degenerates (fig. 20.32).

Changes continue in the ovary following ovulation. The empty follicle, under the influence of luteinizing hormone from the anterior pituitary, undergoes structural and biochemical

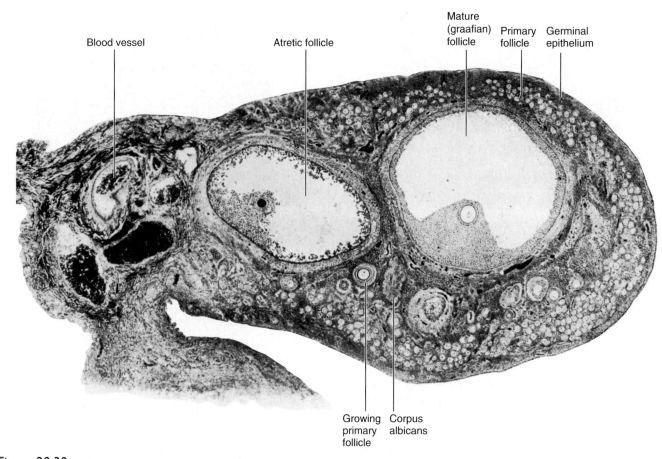

Blood vessel   Atretic follicle   Mature (graafian) follicle   Primary follicle   Germinal epithelium

Growing primary follicle   Corpus albicans

■ **Figure 20.30**   **An ovary containing follicles at different stages of development.** An atretic follicle is one that is dying by apoptosis. It will eventually become a corpus albicans.

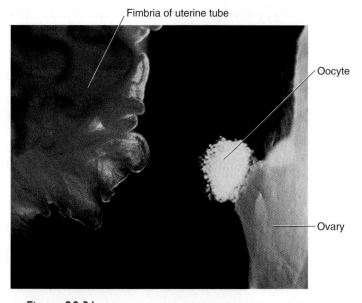

Fimbria of uterine tube

Oocyte

Ovary

■ **Figure 20.31**   **Ovulation from a human ovary.** Notice the cloud of fluid and granulosa cells surrounding the ovulated oocyte.

changes to become a **corpus luteum** (= yellow body). Unlike the ovarian follicles, which secrete only estrogen, the corpus luteum secretes two sex steroid hormones: estrogen and progesterone. Toward the end of a nonfertile cycle, the corpus luteum regresses to become a nonfunctional *corpus albicans*. These cyclic changes in the ovary are summarized in figure 20.33.

Pituitary-Ovarian Axis

The term **pituitary-ovarian axis** refers to the hormonal interactions between the anterior pituitary and the ovaries. The anterior pituitary secretes two gonadotropic hormones—follicle-stimulating hormone (FSH) and luteinizing hormone (LH)—both of which promote cyclic changes in the structure and function of the ovaries. The secretion of both gonadotropic hormones, as previously discussed, is controlled by a single releasing hormone from the hypothalamus—gonadotropin-releasing hormone (GnRH)—and by feedback effects from hormones secreted by the ovaries. The nature of these interactions will be described in detail in the next section.

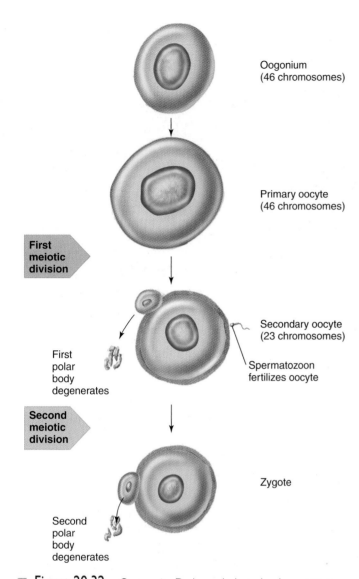

Oogonium
(46 chromosomes)

Primary oocyte
(46 chromosomes)

First meiotic division

First polar body degenerates

Secondary oocyte
(23 chromosomes)

Spermatozoon fertilizes oocyte

Second meiotic division

Zygote

Second polar body degenerates

■ **Figure 20.32** **Oogenesis.** During meiosis, each primary oocyte produces a single haploid gamete. If the secondary oocyte is fertilized, it forms a second polar body and its nucleus fuses with that of the sperm cell to become a zygote.

Test Yourself Before You Continue

1. Compare the structure and contents of a primary follicle, secondary follicle, and graafian follicle.
2. Define *ovulation* and describe the changes that occur in the ovary following ovulation in a nonfertile cycle.
3. Describe oogenesis and explain why only one mature ovum is produced by this process.
4. Compare the hormonal secretions of the ovarian follicles with those of a corpus luteum.

Menstrual Cycle

Cyclic changes in the secretion of gonadotropic hormones from the anterior pituitary cause the ovarian changes during a monthly cycle. The ovarian cycle is accompanied by cyclic changes in the secretion of estradiol and progesterone, which interact with the hypothalamus and pituitary to regulate gonadotropin secretion. The cyclic changes in ovarian hormone secretion also cause changes in the endometrium of the uterus during a menstrual cycle.

Humans, apes, and Old-World monkeys have cycles of ovarian activity that repeat at approximately one-month intervals; hence the name **menstrual cycle** (*menstru* = monthly). The term *menstruation* is used to indicate the periodic shedding of the stratum functionale of the endometrium, which becomes thickened prior to menstruation under the stimulation of ovarian steroid hormones. In primates (other than New-World monkeys) this shedding of the endometrium is accompanied by bleeding. There is no bleeding when most other mammals shed their endometrium, and therefore their cycles are not called menstrual cycles.

In human females and other primates that have menstrual cycles, coitus (sexual intercourse) may be permitted at any time of the cycle. Nonprimate female mammals, by contrast, are sexually receptive (in "heat" or "estrus") only at a particular time in their cycles, shortly before or after ovulation. These animals are therefore said to have *estrous cycles*. Bleeding occurs in some animals (such as dogs and cats) that have estrous cycles shortly before they permit coitus. This bleeding is a result of high estrogen secretion and is not associated with shedding of the endometrium. The bleeding that accompanies menstruation, by contrast, is caused by a fall in estrogen and progesterone secretion.

Since one releasing hormone can stimulate the secretion of both FSH and LH, one might expect always to see parallel changes in the secretion of these gonadotropins. This, however, is not the case. FSH secretion is slightly greater than LH secretion during an early phase of the menstrual cycle, whereas LH secretion greatly exceeds FSH secretion just prior to ovulation. These differences are believed to result from the feedback effects of ovarian sex steroids, which can change the amount of GnRH secreted, the pulse frequency of GnRH secretion, and the ability of the anterior pituitary cells to secrete FSH and LH. These complex interactions result in a pattern of hormone secretion that regulates the phases of the menstrual cycle.

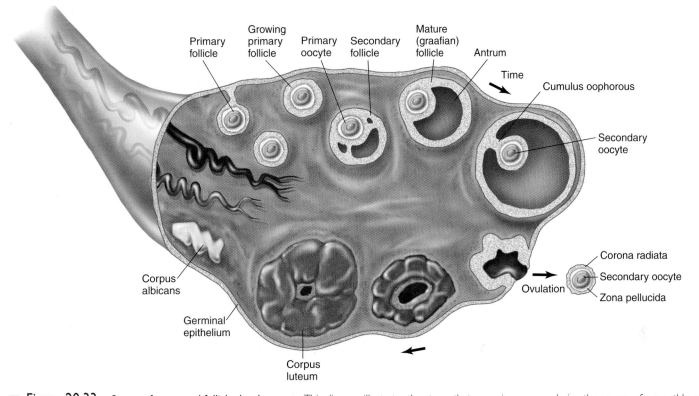

■ **Figure 20.33** **Stages of ovum and follicle development.** This diagram illustrates the stages that occur in an ovary during the course of a monthly cycle. The arrows indicate changes with time.

Phases of the Menstrual Cycle: Cyclic Changes in the Ovaries

The duration of the menstrual cycle is typically about 28 days. Since it is a cycle, there is no beginning or end, and the changes that occur are generally gradual. It is convenient, however, to call the first day of menstruation "day one" of the cycle, because the flow of menstrual blood is the most apparent of the changes that occur. It is also convenient to divide the cycle into phases based on changes that occur in the ovary and in the endometrium. The ovaries are in the *follicular phase* from the first day of menstruation until the day of ovulation. After ovulation, the ovaries are in the *luteal phase* until the first day of menstruation. The cyclic changes that occur in the endometrium are called the menstrual, proliferative, and secretory phases. These will be discussed separately. It should be noted that the time frames used for the following discussion are only averages. Individual cycles may exhibit considerable variation.

Follicular Phase

Menstruation lasts from day 1 to day 4 or 5 of the average cycle. During this time the secretions of ovarian steroid hormones are at their lowest, and the ovaries contain only primary follicles. During the **follicular phase** of the ovaries, which lasts from day 1 to about day 13 of the cycle (this duration is highly variable), some of the primary follicles grow, develop vesicles, and become secondary follicles. Toward the end of the follicular

phase, one follicle in one ovary reaches maturity and becomes a graafian follicle. As follicles grow, the granulosa cells secrete an increasing amount of **estradiol** (the principal estrogen), which reaches its highest concentration in the blood at about day 12 of the cycle, 2 days before ovulation.

The growth of the follicles and the secretion of estradiol are stimulated by, and dependent upon, FSH secreted from the anterior pituitary. The amount of FSH secreted during the early follicular phase is believed to be slightly greater than the amount secreted in the late follicular phase (fig. 20.34), although this can vary from cycle to cycle. FSH stimulates the production of FSH receptors in the granulosa cells, so that the follicles become increasingly sensitive to a given amount of FSH. This increased sensitivity is augmented by estradiol, which also stimulates the production of new FSH receptors in the follicles. As a result, the stimulatory effect of FSH on the follicles increases despite the fact that FSH levels in the blood do not increase throughout the follicular phase. Toward the end of the follicular phase, FSH and estradiol also stimulate the production of LH receptors in the graafian follicle. This prepares the graafian follicle for the next major event in the cycle.

The rapid rise in estradiol secretion from the granulosa cells during the follicular phase acts on the hypothalamus to increase the frequency of GnRH pulses. In addition, estradiol augments the ability of the pituitary to respond to GnRH with an increase in LH secretion. As a result of this stimulatory, or **positive feedback,** effect of estradiol on the pituitary, there is an increase in LH secretion in the late follicular phase that culminates in an **LH surge** (fig. 20.34).

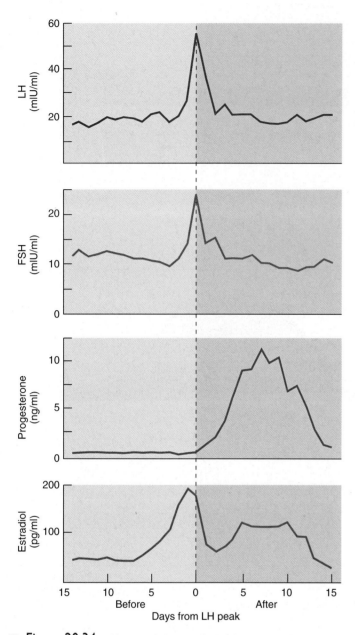

Figure 20.34 Hormonal changes during the menstrual cycle. Sample values are indicated for LH, FSH, progesterone, and estradiol during the menstrual cycle. The midcycle peak of LH is used as a reference day. (IU = international unit.)

The LH surge begins about 24 hours before ovulation and reaches its peak about 16 hours before ovulation. It is this surge that acts to trigger ovulation. Since GnRH stimulates the anterior pituitary to secrete both FSH and LH, there is a simultaneous, smaller surge in FSH secretion. Some investigators believe that this midcycle peak in FSH acts as a stimulus for the development of new follicles for the next month's cycle.

Ovulation

Under the influence of FSH stimulation, the graafian follicle grows so large that it becomes a thin-walled "blister" on the surface of the ovary. The growth of the follicle is accompanied by a rapid increase in the rate of estradiol secretion. This rapid increase in estradiol, in turn, triggers the LH surge at about day 13. Finally, the surge in LH secretion causes the wall of the graafian follicle to rupture at about day 14 (fig. 20.35, *top*). In the course of ovulation, a secondary oocyte, arrested at metaphase II of meiosis, is released from the ovary and swept by cilia into a uterine tube. The ovulated oocyte is still surrounded by a zona pellucida and corona radiata as it begins its journey to the uterus.

Ovulation occurs, therefore, as a result of the sequential effects of FSH and LH on the ovarian follicles. By means of the positive feedback effect of estradiol on LH secretion, the follicle in a sense sets the time for its own ovulation. This is because ovulation is triggered by an LH surge, and the LH surge is triggered by increased estradiol secretion that occurs while the follicle grows. In this way, the graafian follicle is not normally ovulated until it has reached the proper size and degree of maturation.

Luteal Phase

After ovulation, the empty follicle is stimulated by LH to become a new structure—the corpus luteum (fig. 20.36). This change in structure is accompanied by a change in function. Whereas the developing follicles secrete only estradiol, the corpus luteum secretes both estradiol and **progesterone.** Progesterone levels in the blood are negligible before ovulation but rise rapidly to a peak level during the **luteal phase,** approximately one week after ovulation (see figs. 20.34 and 20.35).

The high levels of progesterone combined with estradiol during the luteal phase exert an inhibitory, or **negative feedback,** effect on FSH and LH secretion. There is also evidence that the corpus luteum produces inhibin during the luteal phase, which may help to suppress FSH secretion or action. This serves to retard development of new follicles, so that further ovulation does not normally occur during that cycle. In this way, multiple ovulations (and possible pregnancies) on succeeding days of the cycle are prevented.

However, new follicles start to develop toward the end of one cycle in preparation for the next. This may be due to a decreased production of inhibin toward the end of the luteal phase. Estrogen and progesterone levels also fall during the late luteal phase (starting about day 22) because the corpus luteum regresses and stops functioning. In lower mammals, the decline in corpus luteum function is caused by a hormone called *luteolysin,* secreted by the uterus. There is evidence that the luteolysin in humans may be prostaglandin $F_{2\alpha}$ (see figs. 2.23 and 11.34), but the mechanisms of corpus luteum regression in humans is still incompletely understood. Luteolysis (breakdown of the corpus luteum) can be prevented by high levels of LH, but LH levels remain low during the luteal phase as a result of negative feedback exerted by ovarian steroids. In a sense, therefore, the corpus causes its own demise.

With the declining function of the corpus luteum, estrogen and progesterone fall to very low levels by day 28 of the cycle. The withdrawal of ovarian steroids causes menstruation and permits a new cycle of follicle development to progress.

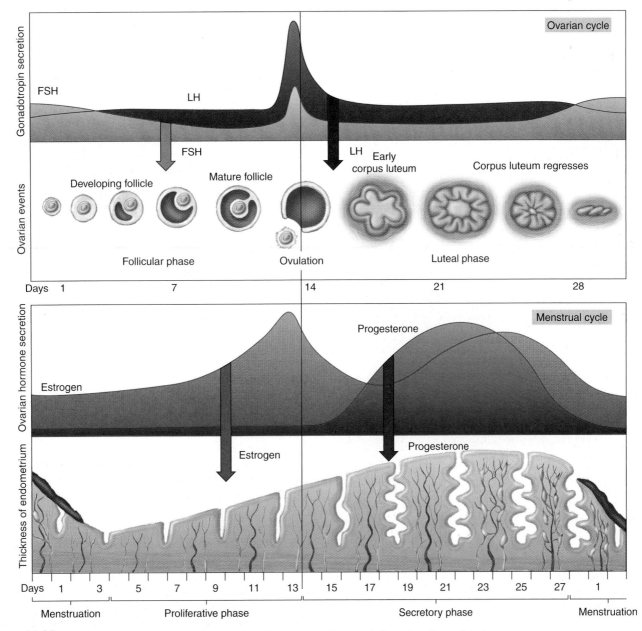

Figure 20.35 **The cycle of ovulation and menstruation.** The downward arrows indicate the effects of the hormones.

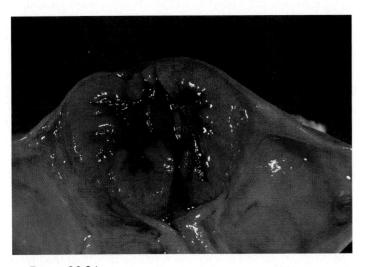

Figure 20.36 **A corpus luteum in a human ovary.** This structure is formed from the empty graafian follicle following ovulation.

Cyclic Changes in the Endometrium

In addition to a description of the female cycle in terms of ovarian function, the cycle can also be described in terms of the changes in the endometrium of the uterus. These changes occur because the development of the endometrium is timed by the cyclic changes in the secretion of estradiol and progesterone from the ovarian follicles. Three phases can be identified on the basis of changes in the endometrium: (1) the *proliferative phase;* (2) the *secretory phase;* and (3) the *menstrual phase* (fig. 20.35, *bottom*).

The **proliferative phase** of the endometrium occurs while the ovary is in its follicular phase. The increasing amounts of estradiol secreted by the developing follicles stimulate growth (proliferation) of the stratum functionale of the endometrium. In humans and other primates, coiled blood vessels called *spiral arteries* develop in the endometrium during this phase. Estradiol may also stimulate the production of receptor proteins for progesterone at this time, in preparation for the next phase of the cycle.

The **secretory phase** of the endometrium occurs when the ovary is in its luteal phase. In this phase, increased progesterone secretion stimulates the development of uterine glands. As a result of the combined actions of estradiol and progesterone, the endometrium becomes thick, vascular, and "spongy" in appearance, and the uterine glands become engorged with glycogen during the phase following ovulation. The endometrium is therefore well prepared to accept and nourish an embryo should fertilization occur.

The **menstrual phase** occurs as a result of the fall in ovarian hormone secretion during the late luteal phase. Necrosis (cellular death) and sloughing of the stratum functionale of the endometrium may be produced by constriction of the spiral arteries. It would seem that the spiral arteries are responsible for menstrual bleeding, since lower animals that lack these arteries do not bleed when they shed their endometrium. The phases of the menstrual cycle are summarized in figure 20.37 and in table 20.6.

The cyclic changes in ovarian secretion cause other cyclic changes in the female reproductive tract. High levels of estradiol secretion, for example, cause cornification of the vaginal epithelium (the upper cells die and become filled with keratin). High levels of estradiol also cause the production of a thin, watery cervical mucus that can easily be penetrated by spermatozoa. During the luteal phase of the cycle, the high levels of progesterone cause the cervical mucus to thicken and become sticky after ovulation has occurred.

 Abnormal menstruations are among the most common disorders of the female reproductive system. The term *amenorrhea* refers to the absence of menstruation. *Dysmenorrhea* refers to painful menstruation, which may be marked by severe cramping. In *menorrhagia*, menstrual flow is excessively profuse or prolonged, and in *metrorrhagia* uterine bleeding that is not associated with menstruation occurs at irregular intervals.

Clinical Investigation Clue

Remember that Gloria is experiencing amenorrhea.

Considering the previous clues, which is most likely to be responsible for her amenorrhea—decreased functioning of her endometrium, ovaries, anterior pituitary, or hypothalamus?

Contraceptive Methods

Contraceptive Pill

About 10 million women in the United States and 60 million women worldwide are currently using **oral contraceptives.** These contraceptives usually consist of a synthetic estrogen combined with a synthetic progesterone in the form of pills that

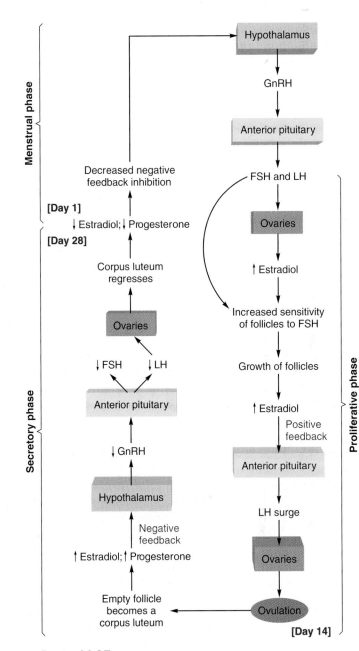

■ **Figure 20.37** Endocrine control of the ovarian cycle. This sequence of events is shown together with the associated phases of the endometrium during the menstrual cycle.

are taken once each day for 3 weeks after the last day of a menstrual period. This procedure causes an immediate increase in blood levels of ovarian steroids (from the pill), which is maintained for the normal duration of a monthly cycle. As a result of *negative feedback inhibition* of gonadotropin secretion, *ovulation never occurs.* The entire cycle is like a false luteal phase, with high levels of progesterone and estrogen and low levels of gonadotropins.

Since the contraceptive pills contain ovarian steroid hormones, the endometrium proliferates and becomes secretory just as it does during a normal cycle. In order to prevent an abnormal growth of the endometrium, women stop taking the steroid pills after 3 weeks (placebo pills are taken during the fourth week).

Table 20.6 Phases of the Menstrual Cycle

Phase of Cycle		Hormonal Changes			Tissue Changes	
Ovarian	Endometrial	Pituitary	Ovary	Ovarian		Endometrial
Follicular (days 1–4)	Menstrual	FSH and LH secretion low	Estradiol and progesterone remain low	Primary follicles grow		Outer two-thirds of endometrium is shed with accompanying bleeding
Follicular (days 5–13)	Proliferative	FSH slightly higher than LH secretion in early follicular phase	Estradiol secretion rises (due to FSH stimulation of follicles)	Follicles grow; graafian follicle develops (due to FSH stimulation)		Mitotic division increases thickness of endometrium; spiral arteries develop (due to estradiol stimulation)
Ovulatory (day 14)	Proliferative	LH surge (and increased FSH) stimulated by positive feedback from estradiol	Estradiol secretion falls	Graafian follicle ruptures and secondary oocyte is extruded into uterine tube		No change
Luteal (days 15–28)	Secretory	LH and FSH decrease (due to negative feedback from steroids)	Progesterone and estrogen secretion increase, then fall	Development of corpus luteum (due to LH stimulation); regression of corpus luteum		Glandular development in in endometrium (due to progesterone stimulation)

This causes estrogen and progesterone levels to fall, permitting menstruation to occur.

The side effects of earlier versions of the birth control pill have been reduced through a decrease in the content of estrogen and through the use of newer generations of progestogens (analogues of progesterone). The newer contraceptive pills are very effective and have a number of beneficial side effects, including a reduced risk for endometrial and ovarian cancer, and a reduction in osteoporosis. However, there may be an increased risk for breast cancer, and possibly cervical cancer, with oral contraceptives.

Newer systems for delivery of contraceptive steroids are designed so that the steroids are not taken orally, and as a result do not have to pass through the liver before entering the general circulation. (All drugs taken orally pass from the hepatic portal vein to the liver before they are delivered to any other organ, as described in chapter 18.) This permits lower doses of hormones to be effective. Such systems include a subcutaneous implant (Norplant), which need only be replaced after 5 years, and vaginal rings, which can be worn for three weeks. The long-term safety of these newer methods has not yet been established.

Clinical Investigation Clues

Remember that Gloria stated that she was not taking birth control pills.

If she were taking birth control pills, how could they be used to prevent periods?

Rhythm Method

Studies have demonstrated that the likelihood of a pregnancy is close to zero if coitus occurs more than 6 days prior to ovulation, and that the likelihood is very low if coitus occurs more than a day following ovulation. Conception is most likely to result when intercourse takes place 1 to 2 days prior to ovulation.

There is no evidence for differences in the sex ratio of babies conceived at these different times.

Cyclic changes in ovarian hormone secretion also cause cyclic changes in basal body temperature. In the **rhythm method** of birth control, a woman measures her oral basal body temperature upon waking to determine when ovulation has occurred. On the day of the LH peak, when estradiol secretion begins to decline, there is a slight drop in basal body temperature. Starting about 1 day after the LH peak, the basal body temperature sharply rises as a result of progesterone secretion, and it remains elevated throughout the luteal phase of the cycle (fig. 20.38). The day of ovulation for that month's cycle can be accurately determined by this method, making the method useful if conception is desired. Since the day of the cycle on which ovulation occurs is quite variable in many women, however, the rhythm method is not very reliable for contraception by predicting when the next ovulation will occur. The contraceptive pill is a statistically more effective means of birth control.

Menopause

The term **menopause** means literally "pause in the menses" and refers to the cessation of ovarian activity and menstruation that occurs at about the age of 50. During the postmenopausal years, which account for about a third of a woman's life span, the ovaries are depleted of follicles and stop secreting estradiol and inhibin. The fall in estradiol is due to changes in the ovaries, not in the pituitary; indeed, FSH and LH secretion by the pituitary is elevated because of a lack of negative feedback from estradiol and inhibin.

The only estrogen found in the blood of postmenopausal women is the weak estrogen *estrone,* formed by the mesenchymal cells in adipose tissue. Estrone is formed from weak androgens, such as *androstenedione* and *dehydroepiandrosterone (DHEA),* secreted from the adrenal cortex. Since adipose tissue is the only source of estrogen, postmenopausal women who have more adipose tissue have higher levels of estrogen and less propensity toward osteoporosis.

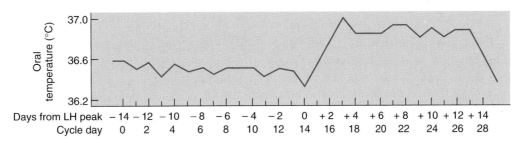

■ **Figure 20.38** Changes in basal body temperature during the menstrual cycle. Such changes can be used in the rhythm method of birth control.

It is the withdrawal of estradiol secretion from the ovaries that is most responsible for the many symptoms of menopause. These include vasomotor disturbances and urogenital atrophy. Vasomotor disturbances produce the "hot flashes" of menopause, where a fall in core body temperature is followed by feelings of heat and profuse perspiration. Atrophy of the urethra, vaginal wall, and vaginal glands occurs, with loss of lubrication. There is also increased risk of atherosclerotic cardiovascular disease and increased progression of osteoporosis (chapter 19).

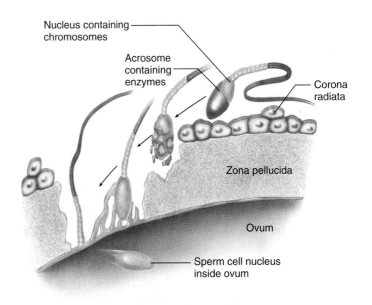

■ **Figure 20.39** The process of fertilization. As the head of the sperm cell encounters the gelatinous corona radiata of the secondary oocyte, the acrosomal vesicle ruptures and the sperm cell digests a path for itself by the action of the enzymes released from the acrosome. When the plasma membrane of the sperm cell contacts the plasma membrane of the ovum, they become continuous, and the nucleus of the sperm cell moves into the cytoplasm of the ovum.

Test Yourself Before You Continue

1. Describe the changes that occur in the ovary and endometrium during the follicular phase and explain how these changes are regulated by hormones.
2. Describe the hormonal regulation of ovulation.
3. Describe the formation, function, and fate of the corpus luteum. Also, describe the changes that occur in the endometrium during the luteal phase.
4. Explain the significance of negative feedback control during the luteal phase and describe the hormonal control of menstruation.

Fertilization, Pregnancy, and Parturition

Once fertilization has occurred, the secondary oocyte completes meiotic division. It then undergoes mitosis, first forming a ball of cells and then an early embryonic structure called a blastocyst. Cells of the blastocyst secrete human chorionic gonadotropin, a hormone that maintains the mother's corpus luteum and its production of estradiol and progesterone. This prevents menstruation so that the embryo can implant into the endometrium, develop, and form a placenta. Birth is dependent upon strong contractions of the uterus, which are stimulated by oxytocin from the posterior pituitary.

During the act of sexual intercourse, the male ejaculates an average of 300 million sperm into the vagina of the female. This tremendous number is needed because of the high sperm fatality—only about 100 survive to enter each fallopian tube. During their passage through the female reproductive tract, the sperm gain the ability to fertilize an ovum. This process is called **capacitation.** Although the changes that occur in capacitation are incompletely understood, experiments have shown that freshly ejaculated sperm are infertile; they must be present in the female tract for at least 7 hours before they can fertilize an ovum.

A woman usually ovulates only one ovum a month, for a total of less than 450 ova during her reproductive years. Each ovulation releases a secondary oocyte arrested at metaphase of the second meiotic division. The secondary oocyte, as previously described, enters the uterine tube surrounded by its zona pellucida (a thin transparent layer of protein and polysaccharides) and corona radiata of granulosa cells (fig. 20.39).

Fertilization

Fertilization normally occurs in the uterine tubes. Each sperm contains a large, enzyme-filled vesicle above its nucleus, known as an **acrosome,** that is central to this task (fig. 20.40). The interaction of sperm with particular molecules in the zona pellucida triggers an **acrosomal reaction.** This involves the progressive fusion of the acrosomal membrane with the plasma membrane of the sperm, creating pores through which the acrosomal enzymes can be released by exocytosis. These enzymes, including a protein-digesting enzyme and hyaluronidase (which digests hyaluronic acid, a constituent of the extracellular matrix), allow the sperm to digest a path through the zona pellucida to the oocyte.

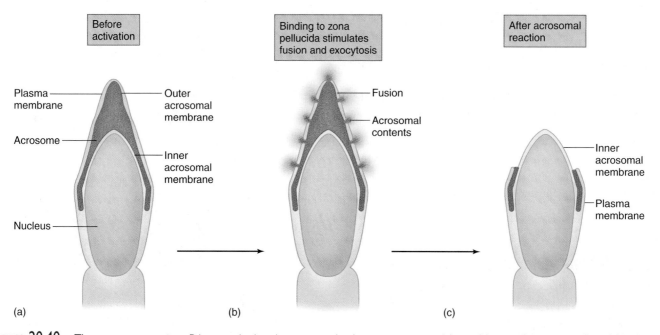

Figure 20.40 **The acrosome reaction.** Prior to activation, the acrosome is a large, enzyme-containing vesicle over the sperm nucleus. After the sperm binds to particular proteins in the zona pellucida surrounding the egg, the acrosomal membrane fuses with the plasma membrane in many locations, creating openings through which the acrosomal contents can be released by exocytosis. When the process is complete, the inner acrosomal membrane has become continuous with the plasma membrane.

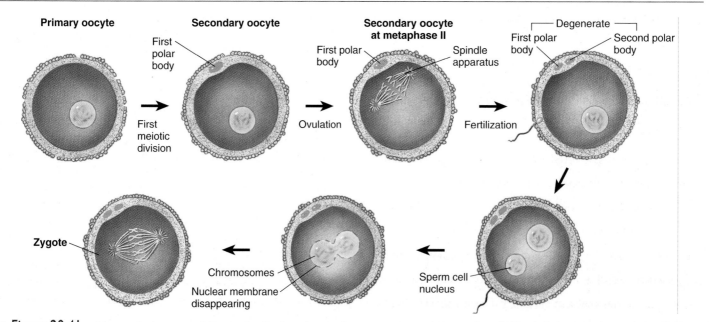

Figure 20.41 **Changes in the oocyte following fertilization.** A secondary oocyte, arrested at metaphase II of meiosis, is released at ovulation. If this cell is fertilized, it will complete its second meiotic division and produce a second polar body. The chromosomes of the two gametes are joined in the zygote.

As the first sperm tunnels its way through the zona pellucida and fuses with the plasma membrane of the oocyte, a number of changes occur that prevent other sperm from fertilizing the same oocyte. *Polyspermy* (the fertilization of an oocyte by many sperm) is thereby prevented; only one sperm can fertilize an oocyte. As fertilization occurs, the secondary oocyte is stimulated to complete its second meiotic division (fig. 20.41). Like the first meiotic division, the second produces one cell that contains all of the cytoplasm—the mature ovum—and one polar body. The second polar body, like the first, ultimately fragments and disintegrates.

At fertilization, the sperm cell enters the cytoplasm of the much larger egg cell. Within 12 hours, the nuclear membrane in the ovum disappears, and the haploid number of chromosomes (twenty-three) in the ovum is joined by the haploid number of chromosomes from the sperm cell. A fertilized egg, or **zygote,** containing the diploid number of chromosomes (forty-six) is thus formed (fig. 20.41).

It should be noted that the sperm cell contributes more than the paternal set of chromosomes to the zygote. Recent evidence demonstrates that the centrosome of the human zygote is derived from the sperm cell and not from the oocyte. As described in chapter 3, the centrosome is needed for the organization of microtubules into a spindle apparatus, so that duplicated chromosomes can be separated during mitosis. Without a centrosome to form the spindle apparatus, cell division (and hence embryonic development) cannot proceed.

A secondary oocyte that has been ovulated but not fertilized does not complete its second meiotic division, but instead disintegrates 12 to 24 hours after ovulation. Fertilization therefore cannot occur if intercourse takes place later than 1 day following ovulation. Sperm, by contrast, can survive up to 3 days in the female reproductive tract. Fertilization therefore can occur if intercourse takes place within a 3-day period prior to the day of ovulation.

The process of *in vitro* **fertilization** is sometimes used to produce pregnancies in women with absent or damaged uterine tubes or in women who are infertile for a variety of other reasons. A secondary oocyte may be collected by aspiration following ovulation (as estimated by waiting 36 to 38 hours after the LH surge). Alternatively, a woman may be treated with powerful FSH-like hormones that cause the development of multiple follicles, and preovulatory oocytes may be collected by aspiration guided by ultrasound and laparoscopy. The donor's sperm are treated so as to duplicate normal capacitation. The oocytes may be placed in a petri dish for 2 to 3 days, along with sperm collected from the donor, or newer techniques may be used to promote fertilization. These newer techniques include **ICSI** (for intracytoplasmic sperm injection), which involves the microinjection of sperm through the zona pellucida directly into the ovum (fig. 20.42). A number of embryos may be produced at the same time, and the surplus frozen in liquid nitrogen for later use. The embryos are usually transferred, three or more at a time, to the woman's uterus at their four-cell stage, 48 to 72 hours after fertilization. In some cases, the embryos may be transferred to the end of the uterine tube. The likelihood of a successful implantation is low (around 35%), and the procedure is expensive. The long-term safety of fertility drugs has also been questioned.

Cleavage and Blastocyst Formation

At about 30 to 36 hours after fertilization, the zygote divides by mitosis—a process called **cleavage**—into two smaller cells. The rate of cleavage is thereafter accelerated. A second cleavage, which occurs about 40 hours after fertilization, produces four cells. A third cleavage about 50 to 60 hours after fertilization produces a ball of eight cells called a **morula** (= mulberry). This very early embryo enters the uterus 3 days after ovulation has occurred (fig. 20.43).

Continued cleavage produces a morula consisting of thirty-two to sixty-four cells by the fourth day following fertilization. The embryo remains unattached to the uterine wall for the next 2 days, during which time it undergoes changes that convert it into a hollow structure called a **blastocyst** (fig. 20.44). The blastocyst consists of two parts: (1) an *inner cell mass,* which will become the fetus, and (2) a surrounding *chorion,* which will become part of the placenta. The cells that form the chorion are called *trophoblast cells.*

On the sixth day following fertilization, the blastocyst attaches to the uterine wall, with the side containing the inner cell mass positioned against the endometrium. The trophoblast cells produce enzymes that allow the blastocyst to "eat its way" into the thick endometrium. This begins the process of **implantation,** or **nidation,** and by the seventh to tenth day the blastocyst is completely buried in the endometrium (fig. 20.45). Approximately 75% of all lost pregnancies are due to a failure of implantation, and consequently are not recognized as pregnancies.

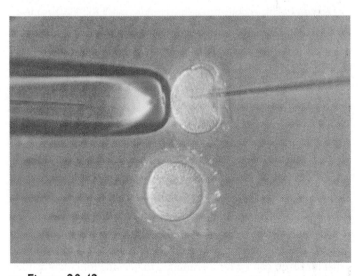

Figure 20.42 *In vitro* fertilization. A needle (the shadow on the right) is used to inject a single spermatozoon into a human oocyte *in vitro.*

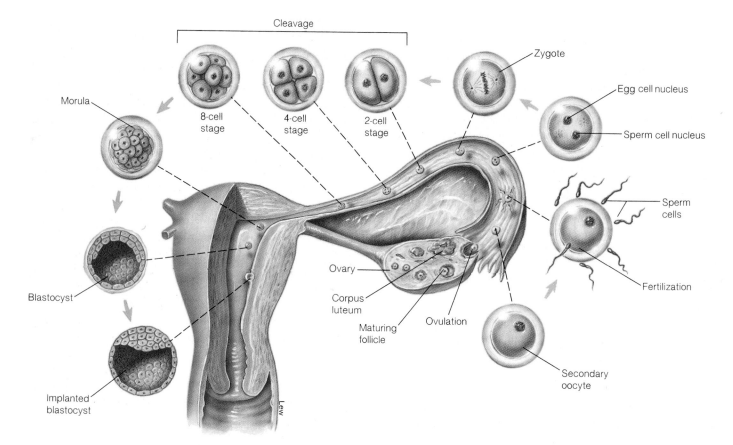

■ Figure 20.43 Fertilization, cleavage, and the formation of a blastocyst. A diagram showing the ovarian cycle, fertilization, and the events of the first week following fertilization. Implantation of the blastocyst begins between the fifth and seventh day and is generally complete by the tenth day.

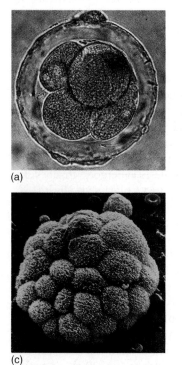

(a)

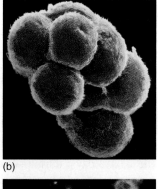

(b)

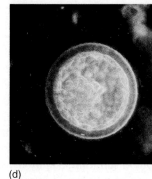

(c)   (d)

■ Figure 20.44 Scanning electron micrographs of preembryonic human development. A human ovum fertilized in a laboratory (*in vitro*) is seen at (*a*) the 4-cell stage. This is followed by (*b*) cleavage at the 16-cell stage and the formation of (*c*) a morula and (*d*) a blastocyst.

CLINICAL

Progesterone, secreted from the woman's corpus luteum, is required for the endometrium to support the implanted embryo and maintain the pregnancy. A drug developed in France, and recently approved for use in the United States, promotes abortion by blocking the progesterone receptors of the endometrial cells. This drug, called **RU486,** has the generic name **mifepristone.** When combined with a small amount of a prostaglandin, which stimulates contractions of the myometrium, RU486 can cause the endometrium to slough off, carrying the embryo with it. Sometimes called the "abortion pill," RU486 has generated bitter controversy in the United States. A recent study found mifepristone followed by prostaglandin treatment to be 96% to 99% effective at terminating pregnancies of 49 days or less.

Embryonic Stem Cells and Cloning

Only the fertilized egg cell and each of the early cleavage cells are **totipotent,** a term that refers to their ability to create the entire organism if implanted into a uterus. The nuclei of adult somatic cells, however, can be reprogrammed to become totipotent if they are transplanted into egg cell cytoplasm. Through such *somatic cell nuclear transfer,* the cloning of an entire adult organism (often called *reproductive cloning*) is possible, and indeed has

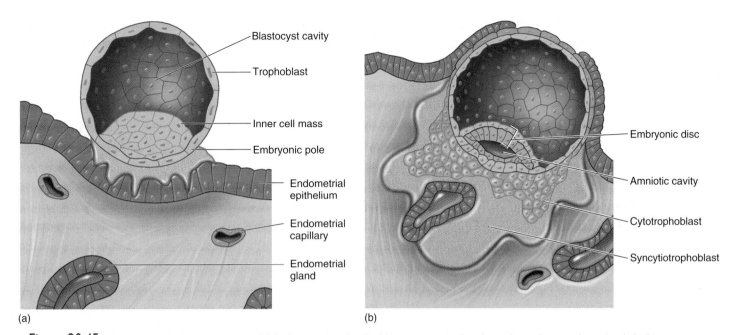

Figure 20.45 Implantation of the blastocyst. *(a)* A diagram showing the blastocyst attached to the endometrium on about the sixth day. *(b)* Implantation of the blastocyst at the ninth or tenth day.

been accomplished in sheep, cattle, cats, and other animals. The possible use of this technique to clone humans has been widely condemned by scientists and others for many reasons, including the low probability of producing healthy children.

This differs from the possibility of nuclear transplantation to produce stem cells, sometimes referred to as *therapeutic cloning,* for the purpose of growing specific tissues for the treatment of diseases. For example, nerve tissue produced by therapeutic cloning holds promise for the treatment of Parkinson's disease, multiple sclerosis, stroke, and spinal cord injury; cloning of islet of Langerhans beta cells may help treat diabetes mellitus; and other cloned tissues might offer new treatments for many other maladies. When nuclear transplantation is performed for the purpose of developing stem cells (therapeutic cloning), rather than for the purpose of reproductive cloning, the totipotent cell is not implanted into a uterus but is rather allowed to develop *in vitro* to the blastocyst stage.

Stem cell research uses cells that can be described as pluripotent or multipotent. Cells obtained from the inner cell mass of a blastocyst—termed **embryonic stem (ES) cells**—are **pluripotent.** Pluripotency refers to the ability to give rise to all tissues except the trophoblast cells of the placenta. This contrasts with adult stem cells, which have been described as **multipotent** because they can give rise to a number of differentiated cells. For example, neural stem cells (chapter 7) give rise to neurons and different types of glial cells, and hematopoietic stem cells (chapter 13) give rise to the different types of blood cells. There is also research suggesting that neural stem cells might be able to form blood and muscle cells, and that stem cells from the skin can be induced to develop into neurons, glial cells, smooth muscle cells, and adipocytes. The ability of adult stem cells to differentiate into such different tissue types, however, is incompletely understood and currently controversial.

In a recent report, scientists obtained neurons from cultured mouse ES cells and used these to reverse symptoms of Parkinson's disease in rats. However, the use of ES cells in this way does present some potential problems: the transplanted neurons derived from ES cells will likely be immunologically rejected by the host, and ES cells that are transplanted develop benign tumors containing different types of cells.

In another exciting recent report, scientists isolated what may be pluripotent stem cells from bone marrow cultures taken from adult humans, mice, and rats. When they injected the cells into mouse embryos, the descendants of those cells developed into almost every tissue type. It is not currently known if these cells normally exist in the bone marrow, or were created in the process of tissue culture. The scientists, hesitant at present to call these cells pluripotent, have named them "multipotent adult progenitor cells (MAPCs)." The potential health benefits of therapeutic cloning using ES cells, adult stem cells, and MAPCs have engendered excitement, hope, and ethical controversy.

Implantation of the Blastocyst and Formation of the Placenta

If fertilization does not take place, the corpus luteum begins to decrease its secretion of steroids about 10 days after ovulation. This withdrawal of steroids, as previously described, causes necrosis and sloughing of the endometrium following day 28 of the cycle. If fertilization and implantation have occurred, however, these events must obviously be prevented to maintain the pregnancy.

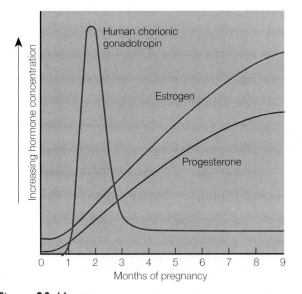

Figure 20.46 The secretion of human chorionic gonadotropin (hCG). This hormone is secreted by trophoblast cells during the first trimester of pregnancy, and it maintains the mother's corpus luteum for the first 5½ weeks. After that time, the placenta becomes the major sex-hormone-producing gland, secreting increasing amounts of estrogen and progesterone throughout pregnancy.

Chorionic Gonadotropin

The blastocyst saves itself from being eliminated with the endometrium by secreting a hormone that indirectly prevents menstruation. Even before the sixth day when implantation occurs, the trophoblast cells of the chorion secrete **chorionic gonadotropin,** or **hCG** (the *h* stands for "human"). This hormone is identical to LH in its effects and therefore is able to maintain the corpus luteum past the time when it would otherwise regress. The secretion of estradiol and progesterone is thus maintained and menstruation is normally prevented.

All **pregnancy tests** assay for the presence of hCG in blood or urine because this hormone is secreted by the blastocyst but not by the mother's endocrine glands. Modern pregnancy tests detect the beta subunit of hCG, which is unique to hCG and provides the least amount of cross-reaction with other hormones. Accurate and sensitive immunoassays for hCG in pregnancy tests employ antibodies that are produced by a clone of lymphocytes—termed *monoclonal antibodies* (chapter 15)—against the specific beta subunit of hCG. Home pregnancy kits that use these antibodies are generally accurate in the week following the first missed menstrual period.

The secretion of hCG declines by the tenth week of pregnancy (fig. 20.46). Actually, this hormone is required for only the first 5 to 6 weeks of pregnancy because the placenta itself becomes an active steroid hormone-secreting gland. By the fifth

to sixth week, the mother's corpus luteum begins to regress (even in the presence of hCG), but by this time the placenta is secreting more than sufficient amounts of steroids to maintain the endometrium and prevent menstruation.

Clinical Investigation Clues

Remember that Gloria's pregnancy test was negative.
What did they specifically test for?
If the test came out positive, what physiological mechanism would account for her amenorrhea?

Chorionic Membranes

Between days 7 and 12, as the blastocyst becomes completely embedded in the endometrium, the chorion becomes a two-cell-thick structure that consists of an inner *cytotrophoblast* layer and an outer *syncytiotrophoblast* layer (see fig. 20.45*b*). Meanwhile, the inner cell mass (which will become the fetus) also develops two cell layers. These are the *ectoderm* (which will form the nervous system and skin) and the *endoderm* (which will eventually form the gut and its derivatives). A third, middle embryonic layer—the *mesoderm*—is not yet seen at this stage. The embryo at this stage is a two-layer-thick disc separated from the cytotrophoblast of the chorion by an *amniotic cavity.*

As the syncytiotrophoblast invades the endometrium, it secretes protein-digesting enzymes that create numerous blood-filled cavities in the maternal tissue. The cytotrophoblast then forms projections, or *villi* (fig. 20.47), that grow into these pools of venous blood, producing a leafy-appearing structure called the *chorion frondosum* (*frond* = leaf). This occurs only on the side of the chorion that faces the uterine wall. As the embryonic structures grow, the other side of the chorion bulges into the cavity of the uterus, loses its villi, and takes on a smooth appearance.

Since the chorionic membrane is derived from the zygote, and since the zygote inherits paternal genes that produce proteins foreign to the mother, scientists have long wondered why the mother's immune system doesn't attack the embryonic tissues. The placenta, it seems, is an "immunologically privileged site." Recent studies suggest that this immune protection may be due to FAS ligand, which is produced by the cytotrophoblast. As you may recall from chapter 15, T lymphocytes produce a surface receptor called FAS. The binding of FAS to FAS ligand triggers the apoptosis (cell suicide) of those lymphocytes, thereby preventing them from attacking the placenta.

Formation of the Placenta and Amniotic Sac

As the blastocyst implants in the endometrium and the chorion develops, the cells of the endometrium also undergo changes. These changes, including cellular growth and the accumulation of glycogen, are collectively called the **decidual reaction.** The maternal tissue in contact with the chorion frondosum is called the *decidua basalis.* These two structures—chorion frondosum (fetal tissue) and decidua basalis (maternal tissue)—together form the functional unit known as the **placenta.**

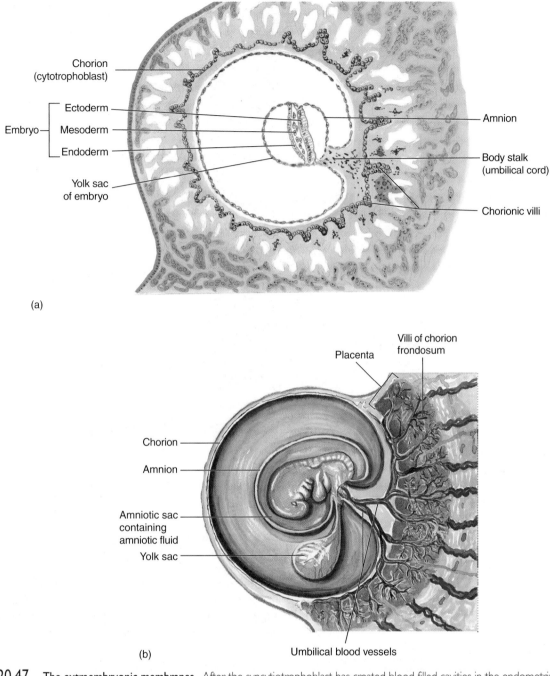

(a)

(b)

Figure 20.47 **The extraembryonic membranes.** After the syncytiotrophoblast has created blood-filled cavities in the endometrium, these cavities are invaded by extensions of the cytotrophoblast (a). These extensions, or villi, branch extensively to produce the chorion frondosum (b). The developing embryo is surrounded by a membrane called the amnion.

The disc-shaped human placenta is continuous at its outer surface with the smooth part of the chorion, which bulges into the uterine cavity. Immediately beneath the chorionic membrane is the amnion, which has grown to envelop the entire embryo (fig. 20.48). The embryo, together with its umbilical cord, is therefore located within the fluid-filled **amniotic sac.**

Amniotic fluid is formed initially as an isotonic secretion. Later, the volume is increased and the concentration changed by urine from the fetus. Amniotic fluid also contains cells that are sloughed off from the fetus, placenta, and amniotic sac. Since all of these cells are derived from the same fertilized ovum, all have the same genetic composition. Many genetic abnormalities can be detected by aspiration of this fluid and examination of the cells thus obtained. This procedure is called **amniocentesis** (fig. 20.49).

Amniocentesis is usually performed at about the sixteenth week of pregnancy. By this time the amniotic sac contains between 175 to 225 ml of fluid. Genetic diseases such as Down

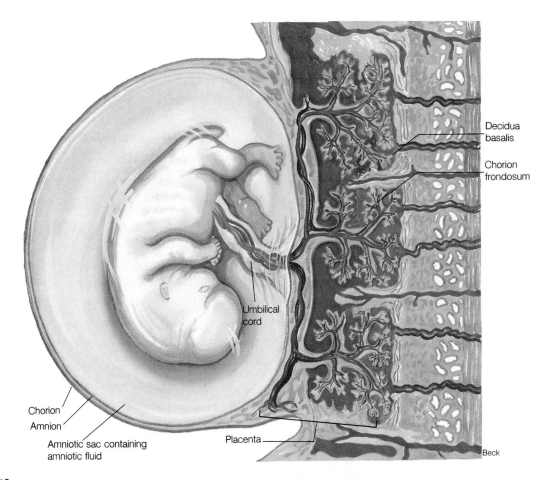

Decidua
basalis

Chorion
frondosum

Umbilical
cord

Chorion

Amnion

Amniotic sac containing
amniotic fluid

Placenta

Beck

■ **Figure 20.48** **The amniotic sac and placenta.** Blood from the embryo is carried to and from the chorion frondosum by umbilical arteries and veins. The maternal tissue between the chorionic villi is known as the decidua basalis; this tissue, together with the chorionic villi, forms the functioning placenta. The space between chorion and amnion is obliterated, and the fetus lies within the fluid-filled amniotic sac.

syndrome (characterized by three instead of two chromosomes number 21) can be detected by examining chromosomes; diseases such as Tay-Sachs disease, in which degeneration of myelin sheaths results from a defective enzyme, can be detected by biochemical techniques.

 The amniotic fluid that is withdrawn contains fetal cells at a concentration too low to permit direct determination of genetic or chromosomal disorders. These cells must therefore be cultured *in vitro* for 10 to 14 days before they are present in sufficient numbers for the laboratory tests required. A newer method, called **chorionic villus biopsy,** is now available to detect genetic disorders earlier than permitted by amniocentesis. In chorionic villus biopsy, a catheter is inserted through the cervix to the chorion and a sample of a chorionic villus is obtained by suction or cutting. Genetic tests can be performed directly on the villus sample because it contains much larger numbers of fetal cells than does a sample of amniotic fluid. Chorionic villus biopsy can provide genetic information at 12 weeks' gestation. Amniocentesis, by contrast, cannot provide such information before about 20 weeks.

Major structural abnormalities that may not be predictable from genetic analysis can often be detected by *ultrasound.* Sound-wave vibrations are reflected from the interface of tissues with different densities—such as the interface between the fetus and amniotic fluid—and used to produce an image. This technique is so sensitive that it can be used to detect a fetal heartbeat several weeks before it can be heard using a stethoscope.

Exchange of Molecules Across the Placenta

The *umbilical arteries* deliver fetal blood to vessels within the villi of the chorion frondosum of the placenta. This blood circulates within the villi and returns to the fetus via the *umbilical vein.* Maternal blood is delivered to and drained from the cavities within the decidua basalis that are located between the chorionic villi (fig. 20.50). In this way, maternal and fetal blood are brought close together but never mix within the placenta.

The placenta serves as a site for the exchange of gases and other molecules between the maternal and fetal blood. Oxygen diffuses from mother to fetus, and carbon dioxide diffuses in the opposite direction. Nutrient molecules and waste products likewise pass between maternal and fetal blood; the placenta is, after all, the only link between the fetus and the outside world.

The placenta is not merely a passive conduit for exchange between maternal and fetal blood, however. It has a very high metabolic rate, utilizing about a third of all the oxygen and glucose supplied by the maternal blood. The rate of protein synthesis is, in fact, higher in the placenta than in the liver. Like the liver, the placenta produces a great variety of enzymes capable of converting hormones and exogenous drugs into less active molecules. In this way potentially dangerous molecules in the maternal blood are often prevented from harming the fetus.

Endocrine Functions of the Placenta

The placenta secretes both steroid hormones and protein hormones. The protein hormones include **chorionic gonadotropin (hCG)** and **chorionic somatomammotropin (hCS),** both of which have actions similar to those of some anterior pituitary hormones (table 20.7). Chorionic gonadotropin has LH-like effects, as previously described; it also has thyroid-stimulating ability, like pituitary TSH. Chorionic somatomammotropin likewise has actions that are similar to two pituitary hormones: growth hormone and prolactin. The placental hormones hCG and hCS thus duplicate the actions of four anterior pituitary hormones.

Pituitary-like Hormones from the Placenta

The importance of chorionic gonadotropin in maintaining the mother's corpus luteum for the first 5½ weeks of pregnancy has been previously discussed. There is also some evidence that hCG may in some way help to prevent immunological rejection of the implanting embryo. Chorionic somatomammotropin acts together with growth hormone from the mother's pituitary to produce a diabetic-like effect in the pregnant woman. The effects of these two hormones promote (1) lipolysis and increased plasma fatty acid concentration; (2) glucose-sparing by maternal tissues and, therefore, increased blood glucose concentrations;

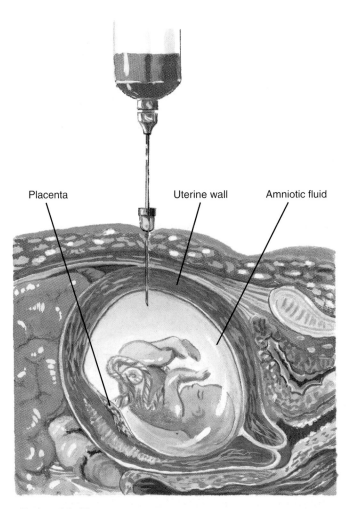

■ Figure 20.49 **Amniocentesis.** In this procedure, amniotic fluid containing suspended cells is withdrawn for examination. Various genetic diseases can be detected prenatally by this means.

Placenta   Uterine wall   Amniotic fluid

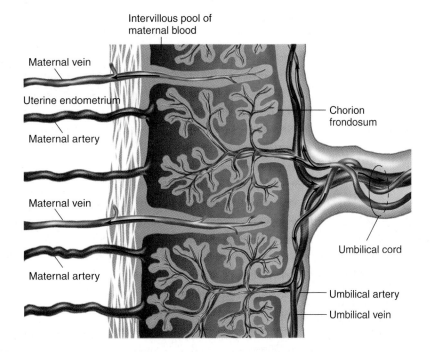

Intervillous pool of maternal blood

Maternal vein

Uterine endometrium

Maternal artery

Maternal vein

Maternal artery

Chorion frondosum

Umbilical cord

Umbilical artery

Umbilical vein

■ Figure 20.50 **The circulation of blood within the placenta.** Maternal blood is delivered to and drained from the spaces between the chorionic villi. Fetal blood is brought to blood vessels within the villi by branches of the umbilical artery and is drained by branches of the umbilical vein.

Table 20.7 Hormones Secreted by the Placenta

Hormones	Effects
Pituitary-like Hormones	
Chorionic gonadotropin (hCG)	Similar to LH; maintains mother's corpus luteum for first 5½ weeks of pregnancy; may be involved in suppressing immunological rejection of embryo; also exhibits TSH-like activity
Chorionic somatomammotropin (hCS)	Similar to prolactin and growth hormone; in the mother, hCS acts to promote increased fat breakdown and fatty acid release from adipose tissue and to promote the sparing of glucose for use by the fetus ("diabetic-like" effects)
Sex Steroids	
Progesterone	Helps maintain endometrium during pregnancy; helps suppress gonadotropin secretion; stimulates development of alveolar tissue in mammary glands
Estrogens	Help maintain endometrium during pregnancy; help suppress gonadotropin secretion; help stimulate mammary gland development; inhibit prolactin secretion; promote uterine sensitivity to oxytocin; stimulate duct development in mammary glands

and (3) polyuria (excretion of large volumes of urine), thereby producing a degree of dehydration and thirst. This diabetic-like effect in the mother helps to ensure a sufficient supply of glucose for the placenta and fetus, which (like the brain) use glucose as their primary energy source.

Steroid Hormones from the Placenta

After the first 5½ weeks of pregnancy, when the corpus luteum regresses, the placenta becomes the major sex-steroid-producing gland. The blood concentration of estrogens, as a result of placental secretion, rises to levels more than 100 times greater than those existing at the beginning of pregnancy. The placenta also secretes large amounts of progesterone, changing the estrogen/progesterone ratio in the blood from 100:1 at the beginning of pregnancy to close to 1:1 toward full-term.

The placenta, however, is an "incomplete endocrine gland" because it cannot produce estrogen and progesterone without the aid of precursors supplied to it by both the mother and the fetus. The placenta, for example, cannot produce cholesterol from acetate, and so it must be supplied with cholesterol from the mother's circulation. Cholesterol, which is a steroid containing twenty-seven carbons, can then be converted by enzymes in the placenta into steroids that contain twenty-one carbons—such as progesterone. The placenta, however, lacks the enzymes needed to convert progesterone into androgens (which have nineteen carbons). Therefore, androgens produced by the fetus are needed as substrates for the placenta to convert into estrogens (fig. 20.51), which have eighteen carbons.

In order for the placenta to produce estrogens, it needs to cooperate with the steroid-producing tissues (principally the adrenal cortex) in the fetus. Fetus and placenta thus form a single functioning system in terms of steroid hormone production. This system has been called the **fetal-placental unit** (fig. 20.51).

The ability of the placenta to convert androgens into estrogen helps to protect the female embryo from becoming masculinized by the androgens secreted from the mother's adrenal glands. In addition to producing estradiol, the placenta secretes large amounts of a weak estrogen called **estriol.** The production of estriol increases tenfold during pregnancy, so that by the third trimester estriol accounts for about 90% of the estrogens excreted in the mother's urine. Since almost all of this estriol comes from the placenta (rather than from maternal tissues), measurements of urinary estriol can be used clinically to assess the health of the placenta.

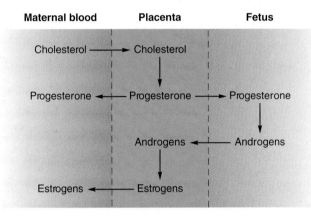

Figure 20.51 Interactions between the embryo and placenta produce the steroid hormones. The secretion of progesterone and estrogen from the placenta requires a supply of cholesterol from the mother's blood and the cooperation of fetal enzymes that convert progesterone to androgens.

Labor and Parturition

Powerful contractions of the uterus are needed to expel the fetus in the sequence of events called **labor.** These uterine contractions are known to be stimulated by two agents: (1) **oxytocin,** a polypeptide hormone produced in the hypothalamus and released by the posterior pituitary (and also produced by the uterus itself), and (2) **prostaglandins,** a class of cyclic fatty acids with paracrine functions produced within the uterus. The particular prostaglandins (PGs) involved are $PGF_{2\alpha}$ and PGE_2. Labor can indeed be induced artificially by injections of oxytocin or by insertion of prostaglandins into the vagina as a suppository.

Although labor is known to be stimulated by oxytocin and prostaglandins, the factors responsible for the initiation of labor are incompletely understood. In all mammals, labor is initiated by activation of the fetal adrenal cortex. In mammals other than primates, the fetal hypothalamus–anterior pituitary–adrenal cortex axis sets the time of labor. Corticosteroids secreted by the fetal adrenal cortex then stimulate the placenta to convert progesterone into estrogens. This is significant because progesterone inhibits activity of the myometrium, while estrogens stimulate the ability of the myometrium to contract. However, the initiation of labor in humans and other primates is more complex. Progesterone levels do not fall because the human placenta cannot

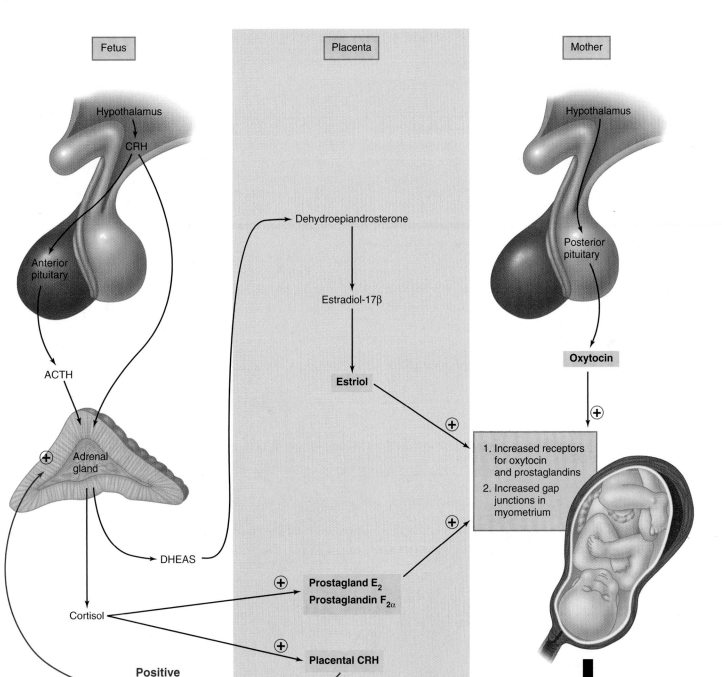

Figure 20.52 **Labor in humans.** The fetal adrenal gland secretes dehydroepiandrosterone sulfate (DHEAS) and cortisol upon stimulation by CRH (corticotropin releasing hormone) and ACTH (adrenocorticotropic hormone). In turn, cortisol stimulates the placenta to secrete CRH, producing a positive feedback loop. The DHEAS is converted by the placenta into estriol, which is needed, together with prostaglandins and oxytocin, to stimulate the myometrium of the mother's uterus to undergo changes leading to labor. The plus signs emphasize activation steps critical to this process.

convert progesterone into estrogens; it can only make estrogen when it is supplied with androgens from the fetus (fig. 20.51).

The fetal adrenal lacks a medulla, but the cortex itself is composed of two parts. The outer part secretes cortisol, as does the adult adrenal cortex. The inner part, called the *fetal adrenal zone,* secretes the androgen **dehydroepiandrosterone sulfate (DHEAS).** Once the DHEAS from the fetus travels to the pla-

centa, it is converted into estrogens. The rising secretion of estrogens (primarily estriol), in turn, stimulates the uterus to (1) produce receptors for oxytocin; (2) produce receptors for prostaglandins; and (3) produce gap junctions between myometrial cells in the uterus (fig. 20.52). The increase in oxytocin and prostaglandin receptors makes the myometrium more sensitive to these agents. The gap junctions (which function as electrical synapses—see chapter 7) help to synchronize and coordinate the contractions of the uterus.

This chain of events may be set in motion by the placenta, through its secretion of **corticotropin-releasing hormone (CRH).** The CRH produced by the placenta, like the CRH produced by the hypothalamus (chapter 11), stimulates the anterior pituitary to secrete ACTH (adrenocorticotropic hormone). There is also evidence for CRH receptors in the fetal adrenal gland, suggesting that the CRH produced by the placenta can itself stimulate adrenal secretion. Thus, CRH from the placenta directly and indirectly (via stimulation of ACTH secretion) stimulates the fetal adrenal cortex to secrete cortisol and DHEAS.

The secretion of cortisol from the fetal adrenal cortex helps to promote maturation of the fetus's lungs; it also stimulates the placenta to secrete CRH, resulting in a positive feedback loop that also increases secretion of DHEAS (fig. 20.52). The placenta can then convert the increased amounts of DHEAS into increased amounts of estriol. The estriol, in turn, activates the myometrium to become more sensitive to oxytocin and prostaglandins, as previously described. Thus, the chain of events culminates in parturition may be set in motion by the placenta's secretion of CRH. How this "placental clock" is timed, however, is not currently understood.

Studies in rhesus monkeys demonstrate that there is a rise in the oxytocin concentration of the mother's plasma during the night, but not during the day. The uterus also produces oxytocin, which may act as a paracrine regulator along with prostaglandins to stimulate contractions and supplement the actions of the oxytocin released by the posterior pituitary. The concentration of oxytocin receptors in the myometrium increases dramatically as a result of estrogen stimulation, as previously described, making the uterus more sensitive to oxytocin. These effects culminate in **parturition,** or childbirth.

Following delivery of the baby, oxytocin is needed to maintain the muscle tone of the myometrium and to reduce hemorrhaging from uterine arteries. Oxytocin may also play a role in promoting the involution (reduction in size) of the uterus following delivery; the uterus weighs about 1 kg (2.2 lb) at term but only about 60 g (2 oz) by the sixth week following delivery.

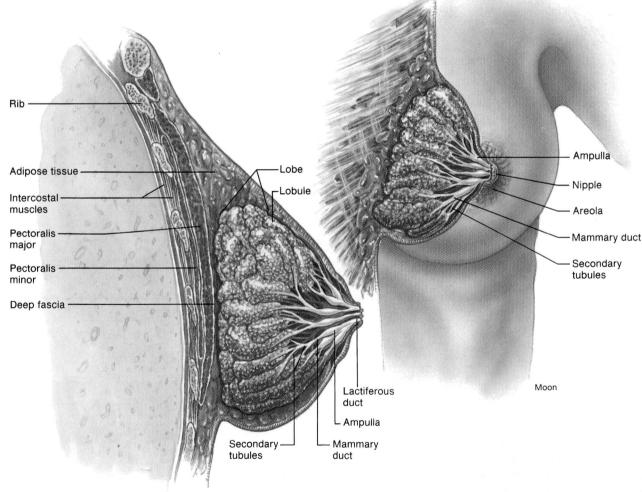

Rib

Adipose tissue

Intercostal muscles

Pectoralis major

Pectoralis minor

Deep fascia

Lobe

Lobule

Secondary tubules

Mammary duct

Ampulla

Lactiferous duct

Ampulla

Nipple

Areola

Mammary duct

Secondary tubules

Moon

(a)

(b)

Figure 20.53 The structure of the breast and mammary glands. (a) A sagittal section and (b) an anterior view partially sectioned.

Lactation

Each mammary gland is composed of fifteen to twenty *lobes,* divided by adipose tissue. The amount of adipose tissue determines the size and shape of the breast but has nothing to do with the ability of a woman to nurse. Each lobe is subdivided into *lobules,* which contain the glandular *alveoli* (fig. 20.53) that secrete the milk of a lactating female. The clustered alveoli secrete milk into a series of *secondary tubules.* These tubules converge to form a series of *mammary ducts,* which in turn converge to form a *lactiferous duct* that drains at the tip of the nipple. The lumen of each lactiferous duct expands just beneath the surface of the nipple to form an *ampulla,* where milk accumulates during nursing.

The changes that occur in the mammary glands during pregnancy and the regulation of lactation provide excellent examples of hormonal interactions and neuroendocrine regulation. Growth and development of the mammary glands during pregnancy requires the permissive actions of insulin, cortisol, and thyroid hormones. In the presence of adequate amounts of these hormones, high levels of progesterone stimulate the development of the mammary alveoli and estrogen stimulates proliferation of the tubules and ducts (fig. 20.54).

The production of milk proteins, including casein and lactalbumin, is stimulated after parturition by **prolactin,** a hormone secreted by the anterior pituitary. The secretion of prolactin is controlled primarily by *prolactin-inhibiting hormone (PIH),* which is believed to be dopamine produced by the hypothalamus and secreted into the portal blood vessels. The secretion of PIH is stimulated by high levels of estrogen. In addition, high levels of estrogen act directly on the mammary glands to block their stimulation by prolactin. During pregnancy, consequently, the high levels of estrogen prepare the breasts for lactation but prevent prolactin secretion and action.

After parturition, when the placenta is expelled as the *afterbirth,* declining levels of estrogen are accompanied by an increase in the secretion of prolactin. Milk production is therefore stimulated. If a woman does not wish to breast-feed her baby she may take oral estrogens to inhibit prolactin secretion. A different drug commonly given in these circumstances, and in other conditions in which it is desirable to inhibit prolactin secretion, is *bromocriptine.* This drug binds to dopamine receptors, and thus promotes the action of dopamine. The fact that this action inhibits prolactin secretion offers additional evidence that dopamine functions as the prolactin-inhibiting hormone (PIH).

The act of nursing helps to maintain high levels of prolactin secretion via a *neuroendocrine reflex* (fig. 20.55). Sensory endings in the breast, activated by the stimulus of suckling, relay impulses to the hypothalamus and inhibit the secretion of PIH. There is also indirect evidence that the stimulus of suckling may cause the secretion of a *prolactin-releasing hormone,* but this is controversial. Suckling thus results in the reflex secretion of high levels of prolactin that promotes the secretion of milk from the alveoli into the ducts. In order for the baby to get the milk, however, the action of another hormone is needed.

The stimulus of suckling also results in the reflex secretion of oxytocin from the posterior pituitary. This hormone is produced in the hypothalamus and stored in the posterior pituitary; its release results in the **milk-ejection reflex,** or **milk letdown.** This is because oxytocin stimulates contraction of the lactiferous ducts as well as of the uterus.

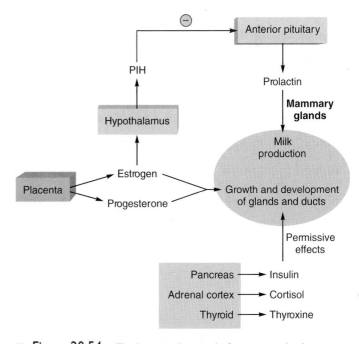

Figure 20.54 The hormonal control of mammary gland development and lactation. Notice that milk production is prevented during pregnancy by estrogen inhibition of prolactin secretion. This inhibition is accomplished by the stimulation of PIH (prolactin-inhibiting hormone) secretion from the hypothalamus.

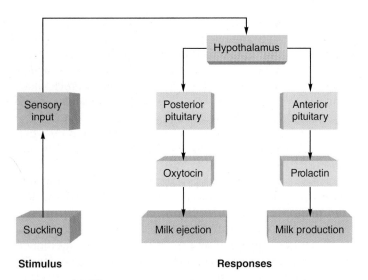

■ **Figure 20.55** Milk production and the milk-ejection reflex.
Lactation occurs in two stages: milk production (stimulated by prolactin) and milk ejection (stimulated by oxytocin). The stimulus of sucking triggers a neuroendocrine reflex that results in increased secretion of oxytocin and prolactin.

Beast-feeding supplements the immune protection given to the infant by its mother. While the fetus is *in utero,* immunoglobin G (IgG—chapter 15) antibodies cross the placenta from the maternal to the fetal blood. These antibodies provide passive immune protection to the baby for the first three to twelve months after birth (fig. 20.56). Infants that are breast-fed also receive IgA antibodies from the mother's milk, which provides additional passive immune protection within the baby's intestine. In addition, the mother's milk contains cytokines, lymphocytes, and antibodies that may promote the development of the baby's system of active immunity. Since the ability of the baby to produce its own antibodies is not well developed for several months after birth (fig. 20.56), the passive immunity provided by maternal antibodies in breast milk may be significant in protecting the baby from a variety of infections.

Breast-feeding, acting through reflex inhibition of GnRH secretion, can also inhibit the secretion of gonadotropins from the mother's anterior pituitary and thus inhibit ovulation. Breast-feeding is thus a natural contraceptive mechanism that helps to space births. This mechanism appears to be most effective in women with limited caloric intake and in those who breast-feed their babies at frequent intervals throughout the day and night. In the traditional societies of the less industrialized nations, therefore, breast-feeding is an effective contraceptive. Breast-feeding has much less of a contraceptive effect in women who are well nourished and who breast-feed their babies at more widely spaced intervals.

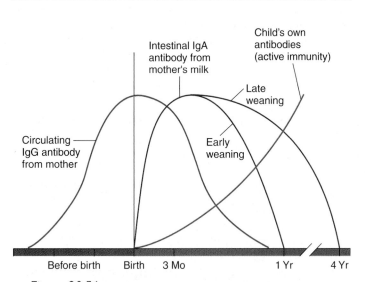

■ **Figure 20.56** Maternal antibodies that protect the baby.
Circulating IgG antibodies cross the placenta and protect the baby for 3 months to 1 year after birth. This passive immunity is supplemented by IgA antibodies in the baby's intestine obtained from the mother's milk. This protection lasts longer for babies weaned at a later age. Notice the inability of the baby to produce a large amount of its own antibodies until it is several months of age.

Adapted from R.M. Zinkernagel, "Advances in immunology: Maternal antibodies, childhood infections, and autoimmune diseases." *New England Journal of Medicine, 345:18,* pp. 1331–1335. Copyright © 2001 Massachusetts Medical Society. All rights reserved.

Milk letdown can become a conditioned reflex made in response to visual or auditory cues; the crying of a baby can elicit oxytocin secretion and the milk-ejection reflex. On the other hand, this reflex can be suppressed by the adrenergic effects produced in the fight-or-flight reaction. Thus, if a woman becomes nervous and anxious while breast-feeding, she will produce milk but it will not flow (there will be no milk letdown). This can cause increased pressure, intensifying her anxiety and frustration and further inhibiting the milk-ejection reflex. It is therefore important for mothers to nurse their babies in a quiet and calm environment. If needed, synthetic oxytocin can be given as a nasal spray to promote milk letdown.

Test Yourself Before You Continue

1. Describe the changes that occur in the sperm cell and ovum during fertilization.
2. Identify the source of hCG and explain why this hormone is needed to maintain pregnancy for the first 10 weeks.
3. List the fetal and maternal components of the placenta and describe the circulation in these two components. Explain how fetal and maternal gas exchange occurs.
4. List the protein hormones and sex steroids secreted by the placenta and describe their functions.
5. Identify the two agents that stimulate uterine contraction during labor and describe the proposed mechanisms that may initiate labor in humans.
6. Describe the hormonal interactions required for breast development during pregnancy and for lactation after delivery.

Concluding Remarks

It may seem strange to end a textbook on physiology with the topics of pregnancy and parturition. This is done in part for practical reasons; these topics are complex, and to understand them requires a grounding in subjects covered earlier. Also, it seems appropriate to end at the beginning, at the start of a new life. Although generations of researchers have accumulated an impressive body of knowledge, the study of physiology is still young and rapidly growing. I hope that this introductory textbook will serve students' immediate practical needs as a resource for understanding current applications, and that it will provide a good foundation for a lifetime of further study.

HPer Links of the Reproductive System with Other Body Systems

Integumentary System

- The skin serves as a sexual stimulant and helps to protect the body from pathogens ....(p. 446)
- Sex hormones affect the distribution of body hair, deposition of subcutaneous fat, and other secondary sexual characteristics ....(p. 642)

Skeletal System

- The pelvic girdle supports and protects some reproductive organs ....(p. 654)
- Sex hormones stimulate bone growth and maintenance ....(p. 625)

Muscular System

- Contractions of smooth muscles aid the movement of gametes ....(p. 651)
- Contractions of the myometrium aid labor and delivery ....(p. 674)
- Cremaster muscles help to maintain proper temperature of the testes ....(p. 637)
- Testosterone promotes an increase in muscle mass ....(p. 291)

Nervous System

- Autonomic nerves innervate the organs of male reproduction to stimulate erection and ejaculation ....(p. 651)
- Autonomic nerves promote aspects of the human sexual response ....(p. 643)
- The CNS, acting through the pituitary, coordinates different aspects of reproduction ....(p. 640)
- The limbic system of the brain is involved in sexual drive ....(p. 201)
- Gonadal sex hormones influence brain activity ....(p. 644)

Endocrine System

- The anterior pituitary controls the activity of the gonads ....(p. 640)
- Testosterone secreted by the testes maintains the structure and function of the male reproductive system ....(p. 637)
- Estradiol and progesterone secreted by the ovaries regulates the female accessory sex organs, including the endometrium of the uterus ....(p. 662)
- Hormones secreted by the placenta are needed to maintain a pregnancy ..(p. 670)
- Prolactin and oxytocin are required for production of breast milk and the milk-ejection reflex ....(p. 677)

Circulatory System

- The circulatory system transports oxygen and nutrients to the reproductive organs ....(p. 366)
- The fetal circulation permits the fetus to obtain oxygen and nutrients from the placenta ....(p. 672)
- Estrogen secreted by the ovaries helps to raise the level of HDL-cholesterol carriers in the blood, lowering the risk of atherosclerosis ....(p. 396)

Immune System

- The immune system protects the body, including the reproductive system, against infections ....(p. 446)
- The blood-testis barrier prevents the immune system from attacking sperm in the testes ....(p. 647)
- The placenta is an immunologically privileged site; it is protected against rejection by the mother's immune system ....(p. 670)

Respiratory System

- The lungs provide oxygen for all body systems, including the reproductive system, and provide for the elimination of carbon dioxide ....(p. 480)
- The red blood cells of a fetus contain hemoglobin F, which has a high affinity for oxygen ....(p. 508)

Urinary System

- The kidneys regulate the volume, pH, and electrolyte balance of the blood and eliminate wastes ....(p. 524)
- The male urethra transports urine as well as semen ....(p. 650)

Digestive System

- The GI tract provides nutrients for all of the organs of the body, including those of the reproductive system ....(p. 561)
- Nutrients obtained from the GI tract of the mother can cross the placenta to the embryo and fetus ....(p. 672)

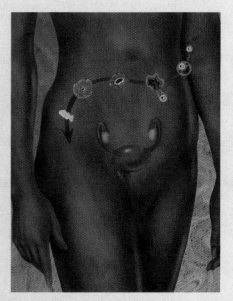

Summary

Sexual Reproduction 634

I. Sperm that bear X chromosomes produce XX zygotes when they fertilize an ovum; sperm that bear Y chromosomes produce XY zygotes.

  A. Embryos that have the XY genotype develop testes; those without a Y chromosome produce ovaries.

  B. The testes of a male embryo secrete testosterone and müllerian inhibition factor. MIF causes degeneration of female accessory sex organs, and testosterone promotes the formation of male accessory sex organs.

II. The male accessory sex organs are the epididymis, ductus (vas) deferens, seminal vesicles, prostate, and ejaculatory duct.

  A. The female accessory sex organs are the uterus and uterine (fallopian) tubes. They develop when testosterone and müllerian inhibition factor are absent.

  B. Testosterone indirectly (acting via conversion to dihydrotestosterone) promotes the formation of male external genitalia; female genitalia are formed when testosterone is absent.

III. Numerous disorders of embryonic sexual development can be understood in terms of the normal physiology of the developmental processes.

Endocrine Regulation of Reproduction 640

I. The gonads are stimulated by two anterior pituitary hormones: FSH (follicle-stimulating hormone) and LH (luteinizing hormone).

  A. The secretion of FSH and LH is stimulated by gonadotropin-releasing hormone (GnRH), which is secreted by the hypothalamus.

  B. The secretion of FSH and LH is also under the control of the gonads by means of negative feedback exerted by gonadal steroid hormones and by a peptide called inhibin.

II. The rise in FSH and LH secretion that occurs at puberty may be due to maturational changes in the brain and to decreased sensitivity of the hypothalamus and pituitary gland to the negative feedback effects of sex steroid hormones.

III. The pineal gland secretes melatonin. This hormone has an inhibitory effect on gonadal function in some species of mammals, but its role in human physiology is presently controversial.

IV. The human sexual response is divided into four phases: excitation, orgasm, plateau, and resolution. Both sexes follow a similar pattern.

Male Reproductive System 644

I. In the male, the pituitary secretion of LH is controlled by negative feedback from testosterone, whereas the secretion of FSH is controlled by the secretion of inhibin from the testes.

  A. The negative feedback effect of testosterone is actually produced by the conversion of testosterone to 5α-reduced androgens and to estradiol.

  B. The secretion of testosterone is relatively constant rather than cyclic, and it does not decline sharply at a particular age.

II. Testosterone promotes the growth of soft tissue and bones before the epiphyseal discs have sealed; thus, testosterone and related androgens are anabolic steroids.

  A. Testosterone is secreted by the interstitial Leydig cells under stimulation by LH.

  B. LH receptor proteins are located in the interstitial tissue. FSH receptors are located in the Sertoli cells within the seminiferous tubules.

  C. The Leydig cells of the interstitial compartment and the Sertoli cells of the tubular compartment of the testes secrete autocrine regulatory molecules that allow the two compartments to interact.

III. Diploid spermatogonia in the seminiferous tubules undergo meiotic cell division to produce haploid sperm.

  A. At the end of meiosis, four spermatids are formed. They develop into spermatozoa by a maturational process called spermiogenesis.

  B. Sertoli cells in the seminiferous tubules are required for spermatogenesis.

  C. At puberty, testosterone is required for the completion of meiosis, and FSH is required for spermiogenesis.

IV. Spermatozoa in the seminiferous tubules are conducted to the epididymis and drained from the epididymis into the ductus deferens. The prostate and seminal vesicles add fluid to the semen.

V. Penile erection is produced by parasympathetic-induced vasodilation. Ejaculation is produced by sympathetic nerve stimulation of peristaltic contraction of the male accessory sex organs.

Female Reproductive System 654

I. Primordial follicles in the ovary contain primary oocytes that have become arrested at prophase of the first meiotic division. Their number is maximal at birth and declines thereafter.

  A. A small number of oocytes in each cycle are stimulated to complete their first meiotic division and become secondary oocytes.

  B. At the completion of the first meiotic division, the secondary oocyte is the only complete cell formed. The other product of this division is a tiny polar body, which disintegrates.

II. One of the secondary follicles grows very large, becomes a graafian follicle, and is ovulated.

  A. Upon ovulation, the secondary oocyte is extruded from the ovary. It does not complete the second meiotic division unless it becomes fertilized.

  B. After ovulation, the empty follicle becomes a new endocrine gland called a corpus luteum.

  C. The ovarian follicles secrete only estradiol, whereas the corpus luteum secretes both estradiol and progesterone.

III. The hypothalamus secretes GnRH in a pulsatile fashion, causing pulsatile secretion of gonadotropins. This is needed to prevent desensitization and downregulation of the target glands.

Menstrual Cycle 659

I. During the follicular phase of the cycle, the ovarian follicles are stimulated by FSH from the anterior pituitary.

   A. Under FSH stimulation, the follicles grow, mature, and secrete increasing amounts of estradiol.

   B. At about day 13, the rapid rise in estradiol secretion stimulates a surge of LH from the anterior pituitary. This represents positive feedback.

   C. The LH surge stimulates ovulation at about day 14.

   D. After ovulation, the empty follicle is stimulated by LH to become a corpus luteum, at which point the ovary is in a luteal phase.

   E. The secretion of progesterone and estradiol rises during the first part of the luteal phase and exerts negative feedback on FSH and LH secretion.

   F. Without continued stimulation by LH, the corpus luteum regresses at the end of the luteal phase, and the secretion of estradiol and progesterone declines. This decline results in menstruation and the beginning of a new cycle.

II. The rising estradiol concentration during the follicular phase produces the proliferative phase of the endometrium. The secretion of progesterone during the luteal phase produces the secretory phase of the endometrium.

III. Oral contraceptive pills usually contain combinations of estrogen and progesterone that exert negative feedback control of FSH and LH secretion.

Fertilization, Pregnancy, and Parturition   665

I.   The sperm undergoes an acrosomal reaction, which allows it to penetrate the corona radiata and zona pellucida.

   A. Upon fertilization, the secondary oocyte completes meiotic division and produces a second polar body, which degenerates.

   B. The diploid zygote undergoes cleavage to form a morula and then a blastocyst. Implantation of the blastocyst in the endometrium begins between the fifth and seventh day.

II. The trophoblast cells of the blastocyst secrete human chorionic gonadotropin (hCG), which functions in the manner of LH and maintains the mother's corpus luteum for the first 10 weeks of pregnancy.

   A. The trophoblast cells provide the fetal contribution to the placenta. The placenta is also formed from adjacent maternal tissue in the endometrium.

   B. Oxygen, nutrients, and wastes are exchanged by diffusion between the fetal and maternal blood.

III. The placenta secretes chorionic somatomammotropin (hCS), chorionic gonadotropin (hCG), and steroid hormones.

   A. The action of hCS is similar to that of prolactin and growth hormone. The action of hCG is similar to that of LH and TSH.

   B. The major steroid hormone secreted by the placenta is estriol. The placenta and fetal glands

cooperate in the production of steroid hormones.

IV. Contraction of the uterus in labor is stimulated by oxytocin from the posterior pituitary and by prostaglandins, produced within the uterus.

   A. Androgens, primarily DHEAS, secreted by the fetal adrenal cortex are converted into estrogen by the placenta.

   B. Estrogen secreted by the placenta induces oxytocin synthesis, enhances uterine sensitivity to oxytocin, and promotes prostaglandin synthesis in the uterus. These events culminate in labor and delivery.

V.   The high levels of estrogen during pregnancy, acting synergistically with other hormones, stimulate growth and development of the mammary glands.

   A. Prolactin (and the prolactin-like effects of hCS) can stimulate the production of milk proteins. Prolactin secretion and action, however, are blocked during pregnancy by the high levels of estrogen secreted by the placenta.

   B. After delivery, when estrogen levels fall, prolactin stimulates milk production.

   C. The milk-ejection reflex is a neuroendocrine reflex. The stimulus of suckling causes reflex secretion of oxytocin. This stimulates contractions of the lactiferous ducts and the ejection of milk from the nipple.

Review Activities

Test Your Knowledge of Terms and Facts

Match these:

1. menstrual phase
2. follicular phase
3. luteal phase
4. ovulation

   a. high estrogen and progesterone; low FSH and LH
   b. low estrogen and progesterone
   c. LH surge
   d. increasing estrogen; low LH and low progesterone

5. A person with the genotype XO has
   a. ovaries.
   b. testes.
   c. both ovaries and testes.
   d. neither ovaries nor testes.

6. An embryo with the genotype XX develops female accessory sex organs because of
   a. androgens.
   b. estrogens.
   c. lack of androgens.
   d. lack of estrogens.

7. In the male,
   a. FSH is not secreted by the pituitary.
   b. FSH receptors are located in the Leydig cells.
   c. FSH receptors are located in the spermatogonia.
   d. FSH receptors are located in the Sertoli cells.

8. The secretion of FSH in a male is inhibited by negative feedback effects of
   a. inhibin secreted from the tubules.
   b. inhibin secreted from the Leydig cells.

c. testosterone secreted from the tubules.

d. testosterone secreted from the Leydig cells.

9. Which of these statements is *true?*

   a. Sperm are not motile until they pass through the epididymis.

   b. Sperm require capacitation in the female reproductive tract before they can fertilize an ovum.

   c. A secondary oocyte does not complete meiotic division until it has been fertilized.

   d. All of these are true.

10. The corpus luteum is maintained for the first 10 weeks of pregnancy by

    a. hCG.

    b. LH.

c. estrogen.

d. progesterone.

11. Fertilization normally occurs in

    a. the ovaries.

    b. the uterine tubes.

    c. the uterus.

    d. the vagina.

12. The placenta is formed from

    a. the fetal chorion frondosum.

    b. the maternal decidua basalis.

    c. both *a* and *b.*

    d. neither *a* nor *b.*

13. Uterine contractions are stimulated by

    a. oxytocin.

    b. prostaglandins.

    c. prolactin.

    d. both *a* and *b.*

    e. both *b* and *c.*

14. Contraction of the mammary glands and ducts during the milk-ejection reflex is stimulated by

    a. prolactin.

    b. oxytocin.

    c. estrogen.

    d. progesterone.

15. If GnRH were secreted in large amounts and at a constant rate rather than in a pulsatile fashion, which of these statements would be *true?*

    a. LH secretion will increase at first and then decrease.

    b. LH secretion will increase indefinitely.

    c. Testosterone secretion in a male will be continuously high.

    d. Estradiol secretion in a woman will be continuously high.

Test Your Understanding of Concepts and Principles

1. Identify the conversion products of testosterone and describe their functions in the brain, prostate, and seminiferous tubules.[1]

2. Explain why a testis is said to be composed of two separate compartments. Describe the interactions that may occur between these compartments.

3. Describe the roles of the Sertoli cells in the testes.

4. Describe the steps of spermatogenesis and explain its hormonal control.

5. Explain the hormonal interactions that control ovulation and cause it to occur at the proper time.

6. Compare menstrual bleeding and bleeding that occurs during the estrous

cycle of a dog in terms of hormonal control mechanisms and the ovarian cycle.

7. "The [contraceptive] pill tricks the brain into thinking you're pregnant." Interpret this popularized explanation in terms of physiological mechanisms.

8. Why does menstruation normally occur? Under what conditions does menstruation not occur? Explain.

9. Explain the proposed mechanisms whereby the act of a mother nursing her baby results in lactation. By what mechanisms might the sound of a baby crying elicit the milk-ejection reflex?

10. Describe the steps of oogenesis when fertilization occurs and when it does not occur. Why are polar bodies produced?

11. Identify the hormones secreted by the placenta. Why is the placenta considered an incomplete endocrine gland?

12. Describe the endocrine changes that occur at menopause and discuss the consequences of these changes. What are the benefits and risks associated with hormone replacement therapy?

13. Explain the sequence of events by which the male accessory sex organs and external genitalia are produced. What occurs when a male embryo lacks receptor proteins for testosterone? What occurs when a male embryo lacks the enzyme 5α-reductase?

14. Describe the mechanisms that have been proposed to time the onset of parturition in sheep and humans.

Test Your Ability to Analyze and Apply Your Knowledge

1. According to your friend, there is a female birth control pill and not a male birth control pill only because the medical establishment is run by men. Do you agree with her conspiracy theory? Provide physiological support for your answer.

2. Elderly men with benign prostatic hyperplasia are sometimes given estrogen treatments. How would this

help the condition? What other types of drugs may be given, and what would you predict their possible side effects to be?

3. Discuss the role of apoptosis and follicle atresia in ovarian physiology. How might this process be regulated?

4. Is it true that estrogen is an exclusively female hormone and that testosterone

is an exclusively male hormone? Explain your answer.

5. Surgical removal of a woman's ovaries (ovariectomy) can precipitate menstruation. Ovariectomy in a dog or cat, however, does not cause the discharge of uterine blood. How can you explain these different responses?

Related Websites

Check out the Links Library at www.mhhe.com/fox8 for links to sites containing resources related to reproduction. These links are monitored to ensure current URLs.

---

[1]*Note:* This question is answered in the chapter 20 Study Guide found on the Online Learning Center at www.mhhe.com/fox8.

Appendix A

Solutions to Clinical Investigations

Chapter 2

Since our enzymes can recognize only L-amino acids and D-sugars, the opposite stereoisomers that George was eating could not be used by his body. He was weak because he was literally starving. The ketonuria also may have contributed to his malaise. Since he was starving, his stored fat was being rapidly hydrolyzed into glycerol and fatty acids for use as energy sources. The excessive release of fatty acids from his adipose tissue resulted in the excessive production of ketone bodies by his liver; hence, his ketonuria.

Chapter 3

Timothy's past drug abuse could have resulted in the development of an extensive smooth endoplasmic reticulum, which contains many of the enzymes required to metabolize drugs. Liver disease could have been caused by the drug abuse, but there is an alternative explanation. The low amount of the enzyme that breaks down glycogen signals the presence of glycogen storage disease, a genetic condition in which a key lysosomal enzyme is lacking. This enzymatic evidence is supported by the observations of large amounts of glycogen granules and the lack of partially digested glycogen granules within secondary lysosomes. (In reality, such a genetic condition would more likely be diagnosed in early childhood.)

Chapter 4

The high blood concentrations of the MB isoenzyme form of creatine phosphokinase (CPK) following severe chest pains suggest that Tom experienced a myocardial infarction ("heart attack"—see chapter 13). His difficulty in urination, together with his high blood levels of acid phosphatase, suggest prostate disease. (The relationship between the prostate gland and the urinary system is described in chapter 20.) Further tests—including one for *prostate-specific antigen (PSA)*—can be performed to confirm this diagnosis. Tom "got the runs" when he ate ice cream probably because he has lactose intolerance—the lack of sufficient lactase to digest milk sugar (lactose).

Chapter 5

Brenda's great fatigue following workouts is partially related to the depletion of her glycogen reserves and extensive utilization of anaerobic respiration (with consequent production of lactic acid) for energy. Production of large amounts of lactic acid during exercise causes her need for extra oxygen to metabolize the lactic acid following exercise (the oxygen debt)—hence, her gasping and panting. Eating more carbohydrates would help Brenda to maintain the glycogen stores in her liver and muscles, and training more gradually could increase the ability of her muscles to obtain more of their energy through aerobic respiration, so that she would experience less pain and fatigue.

The pain in her arms and shoulders is probably the result of lactic acid production by the exercised skeletal muscles. However, the intense pain in her left pectoral region could be angina pectoris, caused by anaerobic respiration of the heart. If this is the case, it would indicate that the heart became ischemic because blood flow was inadequate for the demands placed upon it. Blood tests for particular enzymes released by damaged heart muscle (chapter 4) and an electrocardiogram (ECG) should be performed.

Chapter 6

Jessica's hyperglycemia caused her renal carrier proteins to become saturated, resulting in glycosuria (glucose in the urine). The elimination of glucose in the urine and its consequent osmotic effects caused the urinary excretion of an excessive amount of water, resulting in dehydration. This raised the plasma osmolality, stimulating the thirst center in the hypothalamus. (Hyperglycemia and excessive thirst and urination are cardinal signs of diabetes mellitus.) Further, the loss of plasma water (increased plasma osmolality) caused an increase in the concentration of plasma solutes, including K^+. The resulting hyperkalemia affected the membrane potential of myocardial cells of the heart, producing electrical abnormalities that were revealed in Jessica's electrocardiogram.

Chapter 7

The muscular paralysis and difficulty in breathing (due to paralysis of the diaphragm) could have been caused by saxitoxin poisoning from the shellfish, if they had been gathered during a red tide. A positive chemical analysis of Sandra's blood and of the shellfish for saxitoxin would confirm this diagnosis. The monoamine oxidase (MAO) inhibitor was probably prescribed to treat her depression. It turns out, however, that there are significant drug-food interactions with MAO inhibitors—in fact, shellfish is specifically contraindicated! Other drugs are now available to treat depression that have fewer side effects.

Chapter 8

Frank evidently suffered a cerebrovascular accident (CVA), otherwise known as a "stroke." The obstruction of blood flow in a cerebral artery damaged part of the precentral gyrus (motor cortex) in the left hemisphere. Since most corticospinal tracts decussate in the pyramids, this caused paralysis on the right side of his body. His spinal nerves were undamaged, so his knee-jerk reflex was intact. The damage to the left cerebral hemisphere apparently included damage to Broca's area, producing a characteristic aphasia that accompanied the paralysis of the right side of his body.

Chapter 9

Because of her final examinations, Cathy had been under prolonged stress, which overstimulated her sympathoadrenal system.

The increased sympathoadrenal activity could account for her rapid pulse (due to increased heart rate) and her hypertension (due to increased heart rate and vasoconstriction). Her headache was probably due to the fact that her pupils were dilated, thus admitting excessive amounts of light. Since she had been preparing drugs for a laboratory exercise on autonomic control, she may have been exposed to atropine, which would have caused dilation of her pupils. This possibility is strengthened by the fact that she felt her mouth to be excessively dry.

Chapter 10

Ed was on an international flight, so he was exposed to a long flight at high altitude (even though the airplane cabin is pressurized, it still is at a lower than sea-level pressure). Considering his head cold, his eustachean tube may not have been able to equalize pressure on both sides of the tympanic membrane, leading to pain and reduced hearing. If this is the explanation, the symptoms should resolve with time and the aid of a decongestant. Ed's visual problem suggests that he is experiencing presbyopia, which normally begins at about Ed's age.

Chapter 11

Rosemary's hyperglycemia cannot be attributed to diabetes mellitus because insulin activity is normal, as indicated by the glucose tolerance test. The symptoms might be due to hyperthyroidism, but this possibility is ruled out by the blood tests. The high blood levels of corticosteroids are not the result of ingestion of these compounds as drugs. However, the patient might have Cushing's syndrome, in which case an adrenal tumor could be responsible for the hypersecretion of corticosteroids and, as a result of negative feedback inhibition, a decrease in blood ACTH levels. This possibility is supported by Rosemary's low ACTH levels. Excessive corticosteroid levels cause the mobilization of glucose from the liver, thus increasing the blood glucose to hyperglycemic levels.

Chapter 12

Since Maria has a high maximal oxygen uptake, she should have good endurance with little fatigue and pain during exercise. The fact that her muscles are not large but have good tone supports her statement that she frequently engages in endurance-type exercise. The normal concentration of creatine phosphokinase suggests that her skeletal muscles and heart may not be damaged, but further tests should be done to confirm this, particularly since she has a history of hypertension. The fatigue and muscle pain might simply be due to excessive workouts, but the high blood Ca^{2+} concentration suggests another possibility. The high blood Ca^{2+} could be responsible for her excessively high muscle tone; this inability of her muscles to relax might, in fact, be responsible for the pain and fatigue. Maria should therefore undergo an endocrinological workup (for parathyroid hormone, for example) to determine the cause of her high blood Ca^{2+} levels.

Chapter 13

Jason has a heart murmur due to the ventricular septal defect and mitral stenosis, which were probably congenital. These conditions could reduce the amount of blood pumped by the left ventricle through the systemic arteries, and thus weaken his pulse. The reduced blood flow and consequent reduced oxygen delivery to the tissues could be the cause of his chronic fatigue. The lowered volume of blood pumped by the left ventricle could cause a reflex increase in the heart rate, as detected by his rapid pulse and the ECG tracing showing sinus tachycardia. Jason's high blood cholesterol and LDL/HDL ratio is probably unrelated to his symptoms. This condition could be dangerous, however, as it increases his risk for atherosclerosis. Jason should therefore be placed on a special diet, and perhaps medication, to lower his blood cholesterol.

Chapter 14

Charlie was suffering from dehydration, which lowered his blood volume and thus lowered his blood pressure. This stimulated the baroreceptor reflex, resulting in intense activation of sympathetic nerves. Sympathetic nerve activation caused vasoconstriction in cutaneous vessels—hence the cold skin—and an increase in cardiac rate (hence the high pulse rate). The intravenous albumin solution was given in the hospital to increase his blood volume and pressure. His urine output was low as a result of (1) sympathetic nerve-induced vasoconstriction of arterioles in the kidneys, which decreased blood flow to the kidneys; (2) water reabsorption in response to high ADH secretion, which resulted from stimulation of osmoreceptors in the hypothalamus; and (3) water and salt retention in response to aldosterone secretion, which was stimulated by activation of the renin-angiotensin system. The absence of sodium in his urine resulted from the high aldosterone secretion.

Chapter 15

While crawling through the underbrush, Gary may have been exposed to poison oak, causing a contact dermatitis. Since this is a delayed hypersensitivity response mediated by T cells, antihistamines would not have alleviated the symptoms. Cortisone helped, however, due to its immunosuppressive effect. The first bee sting did not have much of an effect, but it served to sensitize Gary (through the development of B cell clones) to the second bee sting. The second sting resulted in an immediate hypersensitivity response (mediated by IgE), which caused the release of histamine. This allergic reaction could thus be treated effectively with antihistamines. The previous tetanus vaccine that Gary had received had provided him with active immunity against tetanus.

Chapter 16

The puncture wound must have admitted air into the pleural cavity (pneumothorax), raising the intrapleural pressure and causing collapse of the right lung. Since the left lung is located in a separate pleural compartment, it was unaffected by the wound. As a result of the collapse of his right lung, Harry was hypoventilating. This caused retention of CO_2, thus raising his arterial P_{CO_2} and resulting in respiratory acidosis (as indicated by an arterial pH lower than 7.35). Upon recovery, analysis of his arterial blood revealed that he was breathing adequately but that he had a carboxyhemoglobin saturation of 18%. This very high level is probably due to a combination of smoking and driving in heavily congested areas, with much automobile exhaust. The high carboxyhemoglobin would reduce oxygen transport, thus aggravating any problems he might have with his cardiovascular or pulmonary system.

The significantly low FEV_1 indicates that Harry has an obstructive pulmonary problem, possibly caused by smoking and the inhalation of polluted air. A low FEV_1 could simply indicate bronchoconstriction, but the fact that Harry's vital capacity was a little low suggests that he may have early-stage lung damage, possibly emphysema. He should be strongly advised to quit smoking, and further pulmonary tests should be administered at regular intervals.

Chapter 17

The location of Emily's pain and the discoloration of her urine are indicative of a renal disorder. The elevated blood creatinine concentration could indicate a reduction in the glomerular filtration rate (GFR) as a result of the glomerulonephritis, and this reduced GFR could have been responsible for the fluid retention and observed edema. The presence of only trace amounts of protein in the urine, however, was encouraging, and could be explained by her running activity (proteinuria in this case would have been an ominous sign). The streptococcus infection, acting via an autoimmune reaction, was probably responsible for the glomerulonephritis. This was confirmed by the fact that the symptoms of glomerulonephritis disappeared after treatment with an antibiotic. Hydrochlorothiazide is a diuretic that helped to alleviate the edema by (1) promoting the excretion of larger amounts of urine and (2) shifting of edematous fluid from the interstitial to the vascular compartment.

Chapter 18

Alan may have gastritis or a peptic ulcer, as suggested by the sharp pain in his stomach when he drinks wine (a stimulator of gastric acid secretion). The lack of fever and the normal white blood cell count suggest that the inflammation associated with appendicitis is absent. The yellowing of the sclera indicates jaundice, and this symptom—together with the prolonged clotting time—could be caused by liver disease. However, liver disease would elevate the blood levels of free bilirubin, which were found to be normal. The normal levels of urea and ammonia in the blood likewise suggest normal liver function. Similarly, the normal pancreatic amylase levels suggest that the pancreas is not affected.

Alan's symptoms are most likely due to the presence of gallstones. Gallstones could obstruct the normal flow of bile, and thus prevent normal fat digestion. This would explain the fatty stools. The resulting loss of dietary fat could cause a deficiency in vitamin K, which is a fat-soluble vitamin required for the production of a number of clotting factors (chapter 13)—hence, the prolonged clotting time. The pain would be provoked by oily or fatty foods (peanut butter and bacon), which trigger a reflex contraction of the gallbladder once the fat arrives in the duodenum. Contraction of the gallbladder against an obstructed cystic duct or common bile duct often produces a severe referred pain below the right scapula.

Chapter 19

Phyllis' frequent urinations (polyuria) probably are causing her thirst and other symptoms. These symptoms and the fact that her mother and uncle were diabetics suggested that this woman might have diabetes mellitus. Indeed, polyuria, polyphagia (frequent eating), and polydipsia (frequent drinking)—the "three P's"—are cardinal symptoms of diabetes mellitus. The fasting hyperglycemia (blood glucose concentration of 150 mg/dl) confirmed the diagnosis of diabetes mellitus. This abnormally high fasting blood glucose is too low to result in glycosuria. She could have glycosuria after meals, however, which would be responsible for her polyuria. The oral glucose tolerance test further confirmed the diagnosis of diabetes mellitus, and the observations that this condition appeared to have begun in middle age and that it was not accompanied by ketosis and ketonuria suggested that it was type 2 diabetes mellitus. This being the case, she could increase her tissue sensitivity to insulin by diet and exercise. If this failed, she could probably control her symptoms with drugs that increase the tissue sensitivity to the effects of insulin.

Chapter 20

Gloria's lack of menstruation was not accompanied by pain, and she did not have a history of spotting or excessive menstrual bleeding. The fact that she had menstruated prior to her amenorrhea ruled out the possibility of primary amenorrhea. Her secondary amenorrhea could have been the result of pregnancy, but this was ruled out by the negative pregnancy test. The amenorrhea could have been caused by her hypothyroidism, but she stated that she took her thyroid pills regularly, and her blood test demonstrated normal thyroxine levels.

Gloria most likely has a secondary amenorrhea that is due to emotional stress, low body weight, and/or her strenuous exercise program. She should take steps to alleviate these conditions if she wants to resume her normal menstrual periods. If she refuses to gain weight and reduce her level of physical activity, her physician might recommend the use of oral contraceptives to help regulate her cycles.

Appendix B

Answers to Test Your Knowledge of Terms and Facts Questions

Chapter 1
1. d
2. d
3. b
4. b
5. d
6. c
7. b
8. b
9. a
10. c
11. c

Chapter 2
1. c
2. b
3. a
4. d
5. c
6. b
7. c
8. d
9. d
10. b
11. d
12. b
13. b
14. d

Chapter 3
1. d
2. b
3. a
4. c
5. d
6. b
7. a
8. c
9. a
10. b
11. d
12. e
13. b
14. a

Chapter 4
1. b
2. d
3. d
4. a
5. d
6. e
7. e
8. d
9. d
10. d

Chapter 5
1. b
2. a
3. c
4. e
5. d
6. c
7. a
8. c
9. a
10. d
11. b
12. d
13. b

Chapter 6
1. c
2. b
3. a
4. c
5. b
6. d
7. a
8. a
9. b
10. d
11. b
12. b

Chapter 7
1. c
2. d
3. a
4. a
5. c
6. d
7. d
8. a
9. c
10. c
11. b
12. d
13. d
14. b
15. a
16. c
17. a
18. e

Chapter 8
1. d
2. b
3. e
4. a
5. b
6. e
7. c
8. d
9. b
10. c
11. a
12. b
13. d
14. a

Chapter 9
1. c
2. c
3. c
4. a
5. c
6. b
7. b
8. e
9. c
10. c
11. b
12. c

Chapter 10
1. d
2. a
3. c
4. d
5. c
6. a
7. c
8. c
9. d
10. b
11. d
12. b
13. b
14. c
15. b
16. c
17. c
18. b

Chapter 11
1. d
2. d
3. e
4. e
5. d
6. a
7. b
8. e
9. d
10. a
11. d
12. c
13. b
14. d
15. c
16. b

Chapter 12
1. b
2. d
3. c
4. b
5. e
6. b
7. a
8. c
9. b
10. b
11. d
12. a
13. e
14. c
15. b

Chapter 13
1. c
2. b
3. e
4. a
5. b
6. c
7. c
8. a
9. d
10. b
11. c
12. d
13. d
14. c
15. c
16. b
17. d
18. c

Chapter 14
1. a
2. d
3. c
4. e
5. b
6. c
7. a
8. c
9. d
10. b
11. c
12. d
13. b
14. d
15. c
16. e
17. b
18. d
19. d
20. c

Chapter 15
1. c
2. b
3. d
4. a
5. c
6. d
7. d
8. b
9. e
10. a
11. d
12. a
13. d
14. c
15. b
16. c
17. a
18. d

Chapter 16
1. c
2. d
3. c
4. a
5. c
6. c
7. b
8. a
9. e
10. c
11. a
12. c
13. a
14. d
15. b
16. b
17. c

Chapter 17
1. b
2. a
3. c
4. b
5. e
6. d
7. d
8. c
9. d
10. a
11. e
12. c
13. a
14. d
15. b
16. e

Chapter 18
1. e
2. d
3. b
4. c
5. d
6. b
7. d
8. a
9. d
10. e
11. e
12. d
13. b
14. d
15. d

Chapter 19
1. c
2. b
3. c
4. a
5. b
6. d
7. a
8. d
9. c
10. b
11. b
12. c
13. a
14. d
15. d
16. c

Chapter 20
1. b
2. d
3. a
4. c
5. a
6. c
7. d
8. a
9. d
10. a
11. b
12. c
13. d
14. b
15. a

Glossary

A

a-, an- (Gk.) Not, without, lacking.

ab- (L.) Off, away from.

abdomen (*ab′ dŏ-men, ab-do men*) The region of the trunk between the diaphragm and pelvis.

abductor (*ab-duk′tor*) A muscle that moves the skeleton away from the midline of the body or away from the axial line of a limb.

ABO system The most common system of classification for red blood cell antigens. On the basis of antigens on the red blood cell surface, individuals can be type A, type B, type AB, or type O.

absorption (*ab-sorp′shun*) The transport of molecules across epithelial membranes into the body fluids.

accommodation (*ă-kom″ŏ-da′shun*) Adjustment; specifically, the process whereby the focal length of the eye is changed by automatic adjustment of the curvature of the lens to bring images of objects from various distances into focus on the retina.

acetyl (*as′ĭ-tl, ă-sĕt′l*) **CoA** Acetyl coenzyme A. An intermediate molecule in aerobic cell respiration that, together with oxaloacetic acid, begins the Krebs cycle. Acetyl CoA is also an intermediate in the synthesis of fatty acids.

acetylcholine (*ă-sĕt″l-ko′lēn*) **(ACh)** An acetic acid ester of choline—a substance that functions as a neurotransmitter chemical in somatic motor nerve and parasympathetic nerve fibers.

acetylcholinesterase (*ă-sĕt″l-ko″lĭ-nes′ tĕ-rās*) An enzyme in the membrane of postsynaptic cells that catalyzes the conversion of ACh into choline and acetic acid. This enzymatic reaction inactivates the neurotransmitter.

acidosis (*as″ĭ-do′sis*) An abnormal increase in the H⁺ concentration of the blood that lowers arterial pH below 7.35.

acromegaly (*ak″ro-meg′ă-le*) A condition caused by hypersecretion of growth hormone from the pituitary after maturity and characterized by enlargement of the extremities, such as the nose, jaws, fingers, and toes.

ACTH Adrenocorticotropic (*ă-dre″no-kor″ă-ko-trop′ik*) hormone A hormone secreted by the anterior pituitary that stimulates the adrenal cortex.

actin (*ak′tin*) A structural protein of muscle that, along with myosin, is responsible for muscle contraction.

action potential An all-or-none electrical event in an axon or muscle fiber in which the polarity of the membrane potential is rapidly reversed and reestablished.

active immunity Immunity involving sensitization, in which antibody production is stimulated by prior exposure to an antigen.

active transport The movement of molecules or ions across the cell membranes of epithelial cells by membrane carriers. An expenditure of cellular energy (ATP) is required.

ad- (L.) Toward, next to.

adductor (*ă-duk′tor*) A muscle that moves the skeleton toward the midline of the body or toward the axial plane of a limb.

adenohypophysis (*ad″n-o-hi-pof′ĭ-sis*) The anterior, glandular lobe of the pituitary gland that secretes FSH (follicle-stimulating hormone), LH (luteinizing hormone), ACTH (adrenocorticotropic hormone), TSH (thyroid-stimulating hormone), GH (growth hormone), and prolactin. Secretions of the anterior pituitary are controlled by hormones secreted by the hypothalamus.

adenylate cyclase (*ă-den′l-it si′klāse*) An enzyme found in cell membranes that catalyzes the conversion of ATP to cyclic AMP and pyrophosphate (PP₁). This enzyme is activated by an interaction between a specific hormone and its membrane receptor protein.

ADH Antidiuretic (*an″te-di″yŭ-ret′ik*) hormone, also known as *vasopressin.* A hormone produced by the hypothalamus and released from the posterior pituitary. It acts on the kidneys to promote water reabsorption, thus decreasing the urine volume.

adipose (*ad′ĭ-pōs*) **tissue** Fatty tissue. A type of connective tissue consisting of fat cells in a loose connective tissue matrix.

ADP Adenosine diphosphate (*ă-den′ŏ-sēn di-fos′făt*). A molecule that, together with inorganic phosphate, is used to make ATP (adenosine triphosphate).

adrenal cortex (*ă-dre′nal kor′teks*) The outer part of the adrenal gland. Derived from embryonic mesoderm, the adrenal cortex secretes corticosteroid hormones, including aldosterone and hydrocortisone.

adrenal medulla (*mĕdul′ă*) The inner part of the adrenal gland. Derived from embryonic postganglionic sympathetic neurons, the adrenal medulla secretes catecholamine hormones—epinephrine and (to a lesser degree) norepinephrine.

adrenergic (*ad″ rĕ-ner′jik*) Denoting the actions of epinephrine, norepinephrine, or other molecules with similar activity (as in *adrenergic receptor* and *adrenergic stimulation*).

aerobic (*ă-ro′bik*) **capacity** The ability of an organ to utilize oxygen and respire aerobically to meet its energy needs.

afferent (*af′er-ent*) Conveying or transmitting inward, toward a center. Afferent neurons, for example, conduct impulses toward the central nervous system; afferent arterioles carry blood toward the glomerulus.

agglutinate (*ă-gloot′n-āt*) A clumping of cells (usually erythrocytes) as a result of specific chemical interaction between surface antigens and antibodies.

agranular leukocytes (*a-gran′ yŭ-lar loo′kŏ-sīts*) While blood cells (leukocytes) with cytoplasmic granules that are too small to be clearly visible; specifically, lymphocytes and monocytes.

albumin (*al-byoo′min*) A water-soluble protein produced in the liver; the major component of the plasma proteins.

aldosterone (*al-dos′ter-ŏn*) The principal corticosteroid hormone involved in the regulation of electrolyte balance (mineralocorticoid).

alkalosis (*al″kă-lo′sis*) An abnormally high alkalinity of the blood and body fluids (blood pH > 7.45).

allergen (*al′-er-jen*) An antigen that evokes an allergic response rather than a normal immune response.

allergy (*al'er-je*) A state of hypersensitivity caused by exposure to allergens. It results in the liberation of histamine and other molecules with histamine-like effects.

all-or-none law The statement that a given response will be produced to its maximum extent in response to any stimulus equal to or greater than a threshold value. Action potentials obey an all-or-none law.

allosteric (*al"o-ster'ik*) Denoting the alteration of an enzyme's activity by its combination with a regulator molecule. Allosteric inhibition by an end product represents negative feedback control of an enzyme's activity.

alpha motoneuron (*al'fă mo"tŏ-noor'on*) The type of somatic motor neuron that stimulates extrafusal skeletal muscle fibers.

alveoli (*al-ve'ŏ-li*); sing., *alveolus*. Small, saclike dilations (as in *lung alveoli*).

amniocentesis (*am"ne-o-sen-te'sis*) A procedure for obtaining amniotic fluid and fetal cells in this fluid through transabdominal perforation of the uterus.

amnion (*am'ne-on*) A developmental membrane surrounding the fetus that contains amniotic fluid; commonly called the "bag of waters."

amphoteric (*am-fo-ter'ik*) Having both acidic and basic characteristics; used to denote a molecule that can be positively or negatively charged, depending on the pH of its environment.

amylase (*am'il-ās*) A digestive enzyme that hydrolyzes the bonds between glucose subunits in starch and glycogen. Salivary amylase is found in saliva and pancreatic amylase is found in pancreatic juice.

an- (Gk.) Without, not.

anabolic steroids (*an"ă-bol'ik ster'oidz*) Steroids with androgen-like stimulatory effects on protein synthesis.

anabolism (*ă-nab'ŏ-liz"em*) Chemical reactions within cells that result in the production of larger molecules from smaller ones; specifically, the synthesis of protein, glycogen, and fat.

anaerobic respiration (*an-ă-ro'bik res"pĭ-ra'shun*) A form of cell respiration involving the conversion of glucose to lactic acid in which energy is obtained without the use of molecular oxygen.

anaerobic threshold The maximum rate of oxygen consumption that can be attained before a significant amount of lactic acid is produced by the exercising skeletal muscles through anaerobic respiration. This generally occurs when about 60% of the person's total maximal oxygen uptake has been reached.

anaphylaxis (*an"ă-fĭ-lak'sis*) An unusually severe allergic reaction that can result in cardiovascular shock and death.

androgen (*an'drŏ-jen*) A steroid hormone that controls the development and maintenance of masculine characteristics; primarily testosterone secreted by the testes, although weaker androgens are secreted by the adrenal cortex.

anemia (*ă-ne'me-ă*) An abnormal reduction in the red blood cell count, hemoglobin concentration, or hematocrit, or any combination of these measurements. This condition is associated with a decreased ability of the blood to carry oxygen.

angina pectoris (*an-ji'nă pek'tŏ-ris*) A thoracic pain, often referred to the left pectoral and arm area, caused by myocardial ischemia.

angiogenesis (*an"je-o-jen'e-sis*) The growth of new blood vessels.

angiotensin II (*an"je-o-ten'sin*) An eight-amino-acid polypeptide formed from angiotensin I (a ten-amino-acid precursor), which in turn is formed from the cleavage of a protein (angiotensinogen) by the action of renin, an enzyme secreted by the kidneys. Angiotensin II is a powerful vasoconstrictor and a stimulator of aldosterone secretion from the adrenal cortex.

anion (*an'i-on*) An ion that is negatively charged, such as chloride, bicarbonate, or phosphate.

antagonistic effects Actions of regulators such as hormones or nerves that counteract the effects of other regulators. The actions of sympathetic and parasympathetic neurons on the heart, for example, are antagonistic.

anterior (*an-tēr'e-or*) At or toward the front of an organism, organ, or part; the ventral surface.

anterior pituitary (*pĭ-too'ĭ-ter-e*) See adenohypophysis.

antibodies (*an'tĭ-bod"ēz*) Immunoglobulin proteins secreted by B lymphocytes that have been transformed into plasma cells. Antibodies are responsible for humoral immunity. Their synthesis is induced by specific antigens, and they combine with these specific antigens but not with unrelated antigens.

anticoagulant (*an"te-ko-ag'yŭ-lant*) A substance that inhibits blood clotting.

anticodon (*an"te-ko'don*) A base triplet provided by three nucleotides within a loop of transfer RNA that is complementary in its base-pairing properties to a triplet (the codon in mRNA). The matching of codon to anticodon provides the mechanism for translation of the genetic code into a specific sequence of amino acids.

antigen (*an'tĭ-jen*) A molecule able to induce the production of antibodies and to react in a specific manner with antibodies.

antigenic (*an-tĭ-jen'ik*) **determinant site** The region of an antigen molecule that specifically reacts with particular antibodies. A large antigen molecule may have a number of such sites.

antioxidants Molecules that scavenge free radicals, thereby relieving the oxidative stress on the body.

antiport (*an'-tĭ-port*) A form of secondary active transport (coupled transport) in which a molecule or ion is moved together with, but in the opposite direction to, Na$^+$ ions; that is, out of the cell; also called *countertransport*.

antiserum (*an'tĭ-se"rum*) A serum containing antibodies that are specific for one or more antigens.

aphasia (*ă-fa'ze-ă*) Absent or defective speech, writing, or comprehension of written or spoken language caused by brain damage or disease. Broca's area, Wernicke's area, the arcuate fasciculus, or the angular gyrus may be involved.

apnea (*ap'ne-ă*) The temporary cessation of breathing.

apneustic (*ap-noo'stik*) **center** A collection of neurons in the brain stem that participates in the rhythmic control of breathing.

apoptosis (*ap"ŏ-to'sis*) Cellular death in which the cells show characteristic histological changes. It occurs as part of programmed cell death and other events in which cell death is a physiological response.

aquaporins (*ă-kwă-por'inz*) The protein channels in a cell (plasma) membrane that permit osmosis to occur across the membrane. In certain tissues, particularly the collecting ducts of the kidney, aquaporins are inserted into the cell membrane in response to stimulation by antidiuretic hormone.

aqueous humor (*a'kwe-us*) A fluid produced by the ciliary body that fills the anterior and posterior chambers of the eye.

arteriosclerosis (*ar-tir"e-o-sklĕ-ro'sis*) Any of a group of diseases characterized by thickening and hardening of the artery wall and narrowing of its lumen.

arteriovenous anastomosis (*ar-tir"e-o-ve'nus ă-nas"tŏ-mo'sis*) A direct connection between an artery and a vein that bypasses the capillary bed.

artery (*ar'tĕ-re*) A vessel that carries blood away from the heart.

astigmatism (*ă-stig'mă-tiz"em*) Unequal curvature of the refractive surfaces of the eye (cornea and/or lens), so that light that enters the eye along certain meridians does not focus on the retina.

atherosclerosis (*ath"ĕ-ro-sklĕ-ro'sis*) A common type of arteriosclerosis in which raised areas, or plaques, within the tunica interna of medium and large arteries are formed from smooth muscle cells, cholesterol, and other lipids. These plaques occlude the arteries and serve as sites for the formation of thrombi.

atomic number A whole number representing the number of positively charged protons in the nucleus of an atom.

atopic dermatitis (*ă-top'ik der"mă-ti'tis*) An allergic skin reaction to agents such as poison ivy and poison oak; a type of delayed hypersensitivity.

ATP Adenosine triphosphate (*ăden'ŏ-sēn tri-fos'făt*). The universal energy carrier of the cell.

atretic (*ă-tret'ik*) Without an opening. Atretic ovarian follicles are those that fail to rupture and release an oocyte.

atrial natriuretic (*a'tre-al na"trĭ-yoo-ret'ik*) **factor** A chemical secreted by the atria that acts as a natriuretic hormone (a hormone that promotes the urinary excretion of sodium).

atrioventricular node (*a"tre-o-ven-trik'yŭ-lar nōd*) A specialized mass of conducting tissue located in the right atrium near the junction of the interventricular septum. It transmits the impulse into the bundle of His; also called the *AV node*.

atrioventricular valves One-way valves located between the atria and ventricles. The AV valve on the right side of the heart is the tricuspid, and the AV valve on the left side is the bicuspid, or mitral, valve.

atrophy (*at'rŏ fe*) A gradual wasting away, or decrease in mass and size of an organ; the opposite of hypertrophy.

atropine (*at'rŏ-pēn*) An alkaloid drug, obtained from a plant of the species *Belladonna*, that acts as an anticholinergic agent. It is used medically to inhibit parasympathetic nerve effects, dilate the pupil of the eye, increase the heart rate, and inhibit intestinal movements.

auto- (Gk.) Self, same.

autoantibody (*aw"to-an'tĭ-bod"e*) An antibody that is formed in response to, and that reacts with, molecules that are part of one's own body.

autocrine (*aw'tŏ-krin*) **regulation** A type of regulation in which one part of an organ releases chemicals that help to regulate another part of the same organ. Prostaglandins, for example, are autocrine regulators.

autonomic (*aw"tŏ-nom'ik*) **nervous system** The part of the nervous system that involves control of smooth muscle, cardiac muscle, and glands. The autonomic nervous system is subdivided into the sympathetic and parasympathetic divisions.

autoregulation (*aw″to-reg′yŭ la′shun*) The ability of an organ to intrinsically modify the degree of constriction or dilation of its small arteries and arterioles, and thus to regulate the rate of its own blood flow. Autoregulation may occur through myogenic or metabolic mechanisms.

autosomal chromosomes (*aw″to-so′mal kro′mŏ-sōmz*) The paired chromosomes; those other than the sex chromosomes.

axon (*ak′son*) The process of a nerve cell that conducts impulses away from the cell body.

axonal (*ak′sŏ-nal, ak-son′al*) **transport** The transport of materials through the axon of a neuron. This usually occurs from the cell body to the end of the axon, but retrograde (backward) transport can also occur.

B

baroreceptors (*bar″o-re-sep′torz*) Receptors for arterial blood pressure located in the aortic arch and the carotid sinuses.

Barr body A microscopic structure in the cell nucleus produced from an inactive X chromosome in females.

basal ganglia (*ba′sal gang′gle-ă*) Gray matter, or nuclei, within the cerebral hemispheres, forming the corpus striatum, amygdaloid nucleus, and claustrum.

basal metabolic (*ba′sal met″ă-bol′ik*) **rate (BMR)** The rate of metabolism (expressed as oxygen consumption or heat production) under resting or basal conditions 8 to 12 hours after eating.

basophil (*ba′sŏ-fil*) The rarest type of leukocyte; a granular leukocyte with an affinity for blue stain in the standard staining procedure.

B cell lymphocytes (*lim′fŏ-sīts*) Lymphocytes that can be transformed by antigens into plasma cells that secrete antibodies (and are thus responsible for humoral immunity). The *B* stands for *bursa equivalent,* which is believed to be the bone marrow.

benign (*bĭ-nīn′*) Not malignant or life threatening.

bi- (L.) Two, twice.

bile (*bīl*) Fluid produced by the liver and stored in the gallbladder that contains bile salts, bile pigments, cholesterol, and other molecules. The bile is secreted into the small intestine.

bile salts Salts of derivatives of cholesterol in bile that are polar on one end and nonpolar on the other end of the molecule. Bile salts have detergent or surfactant effects and act to emulsify fat in the lumen of the small intestine.

bilirubin (*bil″ĭ-roo′bin*) Bile pigment derived from the breakdown of the heme portion of hemoglobin.

blastocyst (*blas′tŏ-sist*) The stage of early embryonic development that consists of an inner cell mass, which will become the embryo, and surrounding trophoblast cells, which will form part of the placenta. This is the form of the embryo that implants in the endometrium of the uterus beginning at about the fifth day following fertilization.

blood-brain barrier The structures and cells that selectively prevent particular molecules in the plasma from entering the central nervous system.

Bohr effect The effect of blood pH on the dissociation of oxyhemoglobin. Dissociation is promoted by a decrease in the pH.

Boyle's law The statement that the pressure of a given quantity of a gas is inversely proportional to its volume.

bradycardia (*brad″ĭ-kar′de-ă*) A slow cardiac rate; less than sixty beats per minute.

bradykinin (*brad″ĭ-ki′nin*) A short polypeptide that stimulates vasodilation and other cardiovascular changes.

bronchiole (*brong′ke-ōl*) the smallest of the air passages in the lungs, containing smooth muscle and cuboidal epithelial cells.

brown fat A type of fat most abundant at birth that provides a unique source of heat energy for infants, protecting them against hypothermia.

brush border enzymes Digestive enzymes that are located in the cell membrane of the microvilli of intestinal epithelial cells.

buffer A molecule that serves to prevent large changes in pH by either combining with H^+ or by releasing H^+ into solution.

bulk transport Transport of materials into a cell by endocytosis or phagocytosis, and out of a cell by exocytosis.

bundle of His (*hiss*) A band of rapidly conducting cardiac fibers originating in the AV node and extending down the atrioventricular septum to the apex of the heart. This tissue conducts action potentials from the atria into the ventricles.

C

cable properties A term that refers to the ability of neurons to conduct an electrical current. This occurs, for example, between nodes of Ranvier, where action potentials are produced in a myelinated fiber.

calcitonin (*kal″sĭ-to′nin*) Also called *thyrocalcitonin.* A polypeptide hormone produced by the parafollicular cells of the thyroid and secreted in response to hypercalcemia. It acts to lower blood calcium and phosphate concentrations and may serve as an antagonist of parathyroid hormone.

calmodulin (*kal″mod′yŭ-lin*) A receptor protein for Ca^{2+} located within the cytoplasm of target cells. It appears to mediate the effects of this ion on cellular activities.

calorie (*kal′ŏ-re*) A unit of heat equal to the amount of heat needed to raise the temperature of 1 gram of water by 1° C.

cAMP Cyclic adenosine monophosphate (*ă-den′ŏ-sēn mon″o-fos′făt*) A second messenger in the action of many hormones, including catecholamine, polypeptide, and glycoprotein hormones. It serves to mediate the effects of these hormones on their target cells.

cancer A tumor characterized by abnormally rapid cell division and the loss of specialized tissue characteristics. This term usually refers to malignant tumors.

capacitation (*kă-pas″ĭ-ta′shun*) Changes that occur within spermatozoa in the female reproductive tract that enable them to fertilize ova. Sperm that have not been capacitated in the female tract cannot fertilize ova.

capillary (*kap′ĭlar″e*) The smallest vessel in the vascular system. Capillary walls are only one cell thick, and all exchanges of molecules between the blood and tissue fluid occur across the capillary wall.

capsaicin (*kap-sa′ĭ-sin*) **receptor** Both an ion channel in cutaneous sensory dendrites and a receptor

for capsaicin—the molecule in chili peppers that causes sensations of heat and pain. In response to a noxiously high temperature, or to capsaicin in chili peppers, this ion channel opens, resulting in the perception of heat and pain.

carbohydrate (*kar″bo-hi′drāt*) An organic molecule containing carbon, hydrogen, and oxygen in a ratio of 1:2:1. The carbohydrate class of molecules is subdivided into monosaccharides, disaccharides, and polysaccharides.

carbonic anhydrase (*kar-bon′ik an-hi′drās*) An enzyme that catalyzes the formation or breakdown of carbonic acid. When carbon dioxide concentrations are relatively high, this enzyme catalyzes the formation of carbonic acid from CO_2 and H_2O. When carbon dioxide concentrations are low, the breakdown of carbonic acid to CO_2 and H_2O is catalyzed. These reactions aid the transport of carbon dioxide from tissues to alveolar air.

carboxyhemoglobin (*kar-bok″se-he″mŏglo′bin*) An abnormal form of hemoglobin in which the heme is bound to carbon monoxide.

cardiac (*kar′de-ak*) **muscle** Muscle of the heart, consisting of striated muscle cells. These cells are interconnected, forming a mass called the myocardium.

cardiac output The volume of blood pumped by either the right or the left ventricle each minute.

cardiogenic (*kar″de-o-jen′ik*) **shock** Shock that results from low cardiac output in heart disease.

carrier-mediated transport The transport of molecules or ions across a cell membrane by means of specific protein carriers. It includes both facilitated diffusion and active transport.

cast An accumulation of proteins molded from the kidney tubules that appear in urine sediment.

catabolism (*kă-tab′ŏ-liz-em*) Chemical reactions in a cell whereby larger, more complex molecules are converted into smaller molecules.

catalyst (*kata′-ă-list*) A substance that increases the rate of a chemical reaction without changing the nature of the reaction or being changed by the reaction.

catecholamines (*kat″ĕ-kol′ă-mēnz*) A group of molecules that includes epinephrine, norepinephrine, L-dopa, and related molecules. The effects of catecholamines are similar to those produced by activation of the sympathetic nervous system.

cations (*kat′i-ions*) Positively charged ions, such as sodium, potassium, calcium, and magnesium.

cell-mediated immunity Immunological defense provided by T cell lymphocytes that come into close proximity with their victim cells (as opposed to humoral immunity provided by the secretion of antibodies by plasma cells).

cellular respiration (*sel′yŭ-lar res″pĭ-ra′shun*) The energy-releasing metabolic pathways in a cell that oxidize organic molecules such as glucose and fatty acids.

centri- (L.) Center.

centriole (*sen′tre-ōl*) The cell organelle that forms the spindle apparatus during cell division.

centromere (*sen′trŏ-mēr*) The central region of a chromosome to which the chromosomal arms are attached.

cerebellum (*ser″ĕ-bel′um*) A part of the metencephalon of the brain that serves as a major center of control in the extrapyramidal motor system.

cerebral lateralization (ser'ĕ-bral lat"er-al-ĭ-za'shun) The specialization of function of each cerebral hemisphere. Language ability, for example, is lateralized to the left hemisphere in most people.

chemiosmotic (kem-e"os-mot'ik) **theory** The theory that oxidative phosphorylation within mitochondria is driven by the development of a H⁺ gradient across the inner mitochondrial membrane.

chemoreceptor (ke"mo-re-sep'tor) A neural receptor that is sensitive to chemical changes in blood and other body fluids.

chemotaxis (ke"mo-tak'sis) The movement of an organism or a cell, such as a leukocyte, toward a chemical stimulus.

Cheyne—Stokes (chān' stōks) **respiration** Breathing characterized by rhythmic waxing and waning of the depth of respiration, with regularly occurring periods of apnea (failure to breathe).

chloride (klor'īd) **shift** The diffusion of Cl⁻ into red blood cells as HCO₃⁻ diffuses out of the cells. This occurs in tissue capillaries as a result of the production of carbonic acid from carbon dioxide.

cholecystokinin (ko"lĭ-sis"to-ki'nin) **(CCK)** A hormone secreted by the duodenum that acts to stimulate contraction of the gallbladder and to promote the secretion of pancreatic juice.

cholesterol (kŏ-les'ter-ol) A twenty-seven-carbon steroid that serves as the precursor for steroid hormones.

cholinergic (ko"lĭ-ner'jik) Denoting nerve endings that, when stimulated, release acetylcholine as a neurotransmitter, such as those of the parasympathetic system.

chondrocyte (kon'dro-sīt) A cartilage-forming cell.

chorea (kŏ-re'ă) The occurrence of a wide variety of rapid, complex, jerky movements that appear to be well coordinated but that are performed involuntarily.

chromatids (kro'mă tidz) Duplicated chromosomes, joined together at the centromere, that separate during cell division.

chromatin (kro'mă-tin) Threadlike structures in the cell nucleus consisting primarily of DNA and protein. They represent the extended form of chromosomes during interphase.

chromosome (kro'mŏ-sōm) A structure in the cell nucleus, containing DNA and associated proteins, as well as RNA, that is made according to the genetic instructions in the DNA. Chromosomes are in a compact form during cell division; hence, they become visible as discrete structures in the light microscope at this time.

chylomicron (ki"lo-mi'kron) A particle of lipids and protein secreted by the intestinal epithelial cells into the lymph and transported by the lymphatic system to the blood.

chyme (kīm) A mixture of partially digested food and digestive juices that passes from the pylorus of the stomach into the duodenum.

cilia (sil'e-ă) sing., *cilium.* Tiny hairlike processes extending from the cell surface that beat in a coordinated fashion.

circadian (ser"kă-de'an) **rhythms** Physiological changes that repeat at approximately 24-hour periods. They are often synchronized to changes in the external environment, such as the day-night cycles.

cirrhosis (sĭ-ro'sis) Liver disease characterized by the loss of normal microscopic structure, which is replaced by fibrosis and nodular regeneration.

clonal (klōn'al) **selection theory** The theory in immunology that active immunity is produced by the development of clones of lymphocytes able to respond to a particular antigen.

clone (klōn) **1.** A group of cells derived from a single parent cell by mitotic cell division; since reproduction is asexual, the descendants of the parent cell are genetically identical. **2.** A term used to refer to cells as separate individuals (as in white blood cells) rather than as part of a growing organ.

CNS Central nervous system. That part of the nervous system consisting of the brain and spinal cord.

cochlea (kok'le-ă) The organ of hearing in the inner ear where nerve impulses are generated in response to sound waves.

codon (ko'don) The sequence of three nucleotide bases in mRNA that specifies a given amino acid and determines the position of that amino acid in a polypeptide chain through complementary base pairing with an anticodon in transfer RNA.

coenzyme (ko-en'zīm) An organic molecule, usually derived from a water-soluble vitamin, that combines with and activates specific enzyme proteins.

cofactor (ko'fac-tor) A substance needed for the catalytic action of an enzyme; generally used in reference to inorganic ions such as Ca²⁺ and Mg²⁺.

colloid osmotic (kol'oid oz-mot'ik) **pressure** Osmotic pressure exerted by plasma proteins that are present as a colloidal suspension; also called *oncotic pressure.*

com-, con- (L.) With, together.

compliance (kom-pli'ans) **1.** A measure of the ease with which a structure such as the lung expands under pressure. **2.** A measure of the change in volume as a function of pressure changes.

conducting zone The structures and airways that transmit inspired air into the respiratory zone of the lungs, where gas exchange occurs. The conducting zone includes such structures as the trachea, bronchi, and larger bronchioles.

cone Photoreceptor in the retina of the eye that provides color vision and high visual acuity.

congestive (kon-jes'tiv) **heart failure** The inability of the heart to deliver an adequate blood flow because of heart disease or hypertension. It is associated with breathlessness, salt and water retention, and edema.

conjunctivitis (kon-jungk"tĭ-vi'tis) Inflammation of the conjunctiva of the eye; sometimes called "pink eye."

connective tissue One of the four primary tissues, characterized by an abundance of extracellular material.

Conn's syndrome Primary hyperaldosteronism, in which excessive secretion of aldosterone produces electrolyte imbalances.

contralateral (kon"tră-lat'er-al) Taking place or originating in a corresponding part on the opposite side of the body.

cornea (kor'ne-ă) The transparent structure forming the anterior part of the connective tissue covering of the eye.

corpora quadrigemina (kor'por-ă kwad"rĭ-jem'ĭ-na) A region of the mesencephalon consisting of the superior and inferior colliculi. The superior colliculi are centers for the control of visual reflexes; the inferior colliculi are centers for the control of auditory reflexes.

corpus callosum (kor'pus kă-lo'sum) A large transverse tract of nerve fibers connecting the cerebral hemispheres.

cortex (kor'teks) **1.** The outer layer of an internal organ or body structure, as of the kidney or adrenal gland. **2.** The convoluted layer of gray matter that covers the surface of the cerebral hemispheres.

corticosteroid (kor"tĭ-ko-ster'oid) Any of a class of steroid hormones of the adrenal cortex, consisting of glucocorticoids (such as hydrocortisone) and mineralocorticoids (such as aldosterone).

cotransport Also called *coupled transport* or *secondary active transport.* Carrier-mediated transport in which a single carrier transports an ion (e.g., Na⁺) down its concentration gradient while transporting a specific molecule (e.g., glucose) against its concentration gradient. The hydrolysis of ATP is indirectly required for cotransport because it is needed to maintain the steep concentration gradient of the ion.

countercurrent exchange The process that occurs in the vasa recta of the renal medulla in which blood flows in U-shaped loops. This allows sodium chloride to be trapped in the interstitial fluid while water is carried away from the kidneys.

countercurrent multiplier system The interaction that occurs between the descending limb and the ascending limb of the loop of Henle in the kidney. This interaction results in the multiplication of the solute concentration in the interstitial fluid of the renal medulla.

creatine phosphate (kre'ă-tin fos'făt) An organic phosphate molecule in muscle cells that serves as a source of high-energy phosphate for the synthesis of ATP; also called *phosphocreatine.*

crenation (krĭ-na'shun) A notched or scalloped appearance of the red blood cell membrane caused by the osmotic loss of water from the cells.

cretinism (krēt'n-iz"em) A condition caused by insufficient thyroid secretion during prenatal development or the years of early childhood. It results in stunted growth and inadequate mental development.

crypt- (Gk.) Hidden, concealed.

cryptorchidism (krip-tor'kĭ-diz"em) A developmental defect in which the testes fail to descend into the scrotum, and instead remain in the body cavity.

curare (koo-ră re) A chemical derived from plant sources that causes flaccid paralysis by blocking ACh receptor proteins in muscle cell membranes.

Cushing's syndrome Symptoms caused by hypersecretion of adrenal steroid hormones as a result of tumors of the adrenal cortex or ACTH-secreting tumors of the anterior pituitary.

cyanosis (si'ă-no"sis) A bluish discoloration of the skin or mucous membranes due to excessive concentration of deoxyhemoglobin; indicative of inadequate oxygen concentration in the blood.

cyclins (si'klinz) A group of proteins that promote different phases of the cell cycle by activating enzymes called cyclin-dependant kinases.

cyto- (Gk.) Cell.

cytochrome (si'tŏ-krōm) A pigment in mitochondria that transports electrons in the process of aerobic respiration.

cytochrome P450 enzymes Enzymes of a particular kind, not related to the mitochondrial cytochromes, that metabolize a broad spectrum of biological molecules, including steroid hormones and toxic drugs. They are prominent in the liver, where they help in detoxication of the blood.

cytokine (*si'to-kīn*) An autocrine or paracrine regulator secreted by various tissues.

cytokinesis (*si"to-kĭ-ne'sis*) The division of the cytoplasm that occurs in mitosis and meiosis when a parent cell divides to produce two daughter cells.

cytoplasm (*si'tŏ-plaz"em*) The semifluid part of the cell between the cell membrane and the nucleus, exclusive of membrane-bound organelles. It contains many enzymes and structural proteins.

cytoskeleton (*si'to-skeI'ĕ-ton*) A latticework of structural proteins in the cytoplasm arranged in the form of microfilaments and microtubules.

D

Dalton's law The statement that the total pressure of a gas mixture is equal to the sum that each individual gas in the mixture would exert independently. The part contributed by each gas is known as the partial pressure of the gas.

dark adaptation The ability of the eyes to increase their sensitivity to low light levels over a period of time. Part of this adaptation involves increased amounts of visual pigment in the photoreceptors.

dark current The steady inward diffusion of Na^+ into the rods and cones when the photoreceptors are in the dark. Stimulation by light causes this dark current to be blocked, and thus hyperpolarizes the photoreceptors.

delayed hypersensitivity An allergic response in which the onset of symptoms may not occur until 2 or 3 days after exposure to an antigen. Produced by T cells, it is a type of cell-mediated immunity.

dendrite (*den'drīt*) A relatively short, highly branched neural process that carries electrical activity to the cell body.

dendritic (*den-drit'ik*) **cells** The most potent antigen-presenting cells for the activation of helper T lymphocytes. The dendritic cells originate in the bone marrow and migrate through the blood and lymph to lymphoid organs and to nonlymphoid organs such as the lungs and skin.

denervation (*de"ner-va'shun*) **hypersensitivity** The increased sensitivity of smooth muscles to neural stimulation after their innervation has been blocked or removed for a period of time.

dentin (*den'tin*) One of the hard tissues of the teeth. It covers the pulp cavity and is itself covered on its exposed surface by enamel and on its root surface by cementum.

deoxyhemoglobin (*de-ok"se-he"mō-glo'bin*) The form of hemoglobin in which the heme groups are in the normal reduced form but are not bound to a gas. Deoxyhemoglobin is produced when oxyhemoglobin releases oxygen.

depolarization (*de-po"lar-ĭ-za'shun*) The loss of membrane polarity in which the inside of the cell membrane becomes less negative in comparison to the outside of the membrane. The term is also used to indicate the reversal

of membrane polarity that occurs during the production of action potentials in nerve and muscle cells.

deposition (*dep-ŏ-zish'on*), **bone** The formation of the extracellular matrix of bone by osteoblasts. This process includes secretion of collagen and precipitation of calcium phosphate in the form of hydroxyapatite crystals.

detoxication (*de-tok"sĭ-ka'shun*) The reduction of the toxic properties of molecules. This occurs through chemical transformation of the molecules and takes place, to a large degree, in the liver.

diabetes insipidus (*di"ă-be'tēz in-sip'ĭ-dus*) A condition in which inadequate amounts of anti-diuretic hormone (ADH) are secreted by the posterior pituitary. It results in inadequate reabsorption of water by the kidney tubules, and thus in the excretion of a large volume of dilute urine.

diabetes mellitus (*mĕ-li'tus*) The appearance of glucose in the urine due to the presence of high plasma glucose concentrations, even in the fasting state. This disease is caused by either a lack of sufficient insulin secretion or by inadequate responsiveness of the target tissues to the effects of insulin.

dialysis (*di-al'ĭ-sis*) A method of removing unwanted elements from the blood by selective diffusion through a porous membrane.

diapedesis (*di"ă-pē-de'sis*) The migration of white blood cells through the endothelial walls of blood capillaries into the surrounding connective tissues.

diarrhea (*di"ă-re'ă*) Abnormal frequency of defecation accompanied by abnormal liquidity of the feces.

diastole (*di-as'tŏ-le*) The phase of relaxation in which the heart fills with blood. Unless accompanied by the modifier *atrial*, diastole refers to the resting phase of the ventricles.

diastolic (*di"ă-stol'ik*) **blood pressure** The minimum pressure in the arteries that is produced during the phase of diastole of the heart. It is indicated by the last sound of Korotkoff when taking a blood pressure measurement.

diffusion (*dĭ-fyoo'zhun*) The net movement of molecules or ions from regions of higher to regions of lower concentration.

digestion The process of converting food into molecules that can be absorbed through the intestine into the blood.

1,25-dihydroxyvitamin (*di"hi-drok"se-vi'tă-min*) D_3 The active form of vitamin D produced within the body by hydroxylation reactions in the liver and kidneys of vitamin D formed by the skin. This is a hormone that promotes the intestinal absorption of Ca^{2+}.

diploid (*dip'loid*) Denoting cells having two of each chromosome, or twice the number of chromosomes that are present in sperm or ova.

disaccharide (*di-sak'ă-rīd*) Any of a class of double sugars; carbohydrates that yield two simple sugars, or monosaccharides, upon hydrolysis.

diuretic (*di"yŭ-ret'ik*) A substance that increases the rate of urine production, thereby lowering the blood volume.

DNA Deoxyribonucleic (*de-ok"se-ri"bo-noo-kle'ik*) acid. A nucleic acid composed of nucleotide bases and deoxyribose sugar that contains the genetic code.

dopa (*do" pă*) Dihydroxyphenylalanine (*di"hi-drok"se-fen"al-ă-lă-nīn*). An amino acid formed in the liver from tyrosine and converted to dopamine in the brain. L-dopa is

used in the treatment of Parkinson's disease to stimulate dopamine production.

dopamine (*do' pă-mēn*) A type of neurotransmitter in the central nervous system; it is also the precursor of norepinephrine, another neurotransmitter molecule.

dopaminergic (*do"pă-mēn-er'jik*) **pathways** Neural pathways in the brain that release dopamine. The nigrostriatal pathway is involved in motor control, whereas the mesolimbic dopamine pathway is involved in mood and emotion.

2,3-DPG 2,3-diphosphoglyceric (*di-fos'fo-glis-er"ik*) acid. A product of red blood cells, 2,3-DPG bonds with the protein component of hemoglobin and increases the ability of oxyhemoglobin to dissociate and release its oxygen.

ductus arteriosus (*duk'tus ar-tir"e-o'sus*) A fetal blood vessel connecting the pulmonary artery directly to the aorta.

dwarfism A condition in which a person is undersized because of inadequate secretion of growth hormone.

dyspnea (*disp-ne'ă*) Subjective difficulty in breathing.

dystrophin (*dis-trof'in*) A protein associated with the sarcolemma of skeletal muscle cells that is produced by the defective gene of people with Duchenne's muscular dystrophy.

E

eccentric (*ek-sen'trik*) **contraction** A muscle contraction in which the muscle lengthens despite its contraction, due to a greater external stretching force applied to it. The contraction in this case can serve a shock absorbing function, for example, when the quadriceps muscles of the leg contract eccentrically upon landing when a person jumps from a heighth.

ECG Electrocardiogram (*ĕ-lek"tro-kar'de-ŏ-gram*) (also abbreviated EKG).A recording of electrical currents produced by the heart.

E. coli (*e ko'li*) A species of bacteria normally found in the human intestine; full name is *Escherichia* (*esh"ĭ-rik'e-ă*) *coli*.

ecto- (Gk.) Outside, outer.

-ectomy (Gk.) Surgical removal of a structure.

ectopic (*ek-top'ik*) Foreign, out of place.

ectopic focus An area of the heart other than the SA node that assumes pacemaker activity.

ectopic pregnancy Embryonic development that occurs anywhere other than in the uterus (as in the uterine tubes or body cavity).

edema (*ĕ-de'mă*) Swelling resulting from an increase in tissue fluid.

EEG Electroencephalogram (*ĕ-lek"tro-en-sef' ă-lŏ-gram*) A recording of the electrical activity of the brain from electrodes placed on the scalp.

effector (*ĕ-fek'tor*) **organs** A collective term for muscles and glands that are activated by motor neurons.

efferent (*ef'er-ent*) Conveying or transporting something away from a central location. Efferent nerve fibers conduct impulses away from the central nervous system, for example, and efferent arterioles transport blood away from the glomerulus.

eicosanoids (*i-ko'să-noidz*) The biologically active derivatives of arachidonic acid, a fatty acid found in cell membranes. The eicosanoids include prostaglandins and leukotrienes.

elasticity (*ĕ″las-tis′ĭ-te*) The tendency of a structure to recoil to its initial dimensions after being distended (stretched).

electrolyte (*ĕ-lek′tro-līt*) An ion or molecule that is able to ionize and thus carry an electric current. The most common electrolytes in the plasma are Na$^+$, HCO$_3^-$, and K$^+$.

electrophoresis (*ĕ-lek″tro-fŏ-re′sis*) A biochemical technique in which different molecules can be separated and identified by their rate of movement in an electric field.

element, chemical A substance that cannot be broken down by chemical means into simpler substances. An element is composed of atoms that all have the same atomic number. An element can, however, include different forms of a given atom (isotopes) that have different numbers of neutrons, and thus different atomic weights.

elephantiasis (*el″ĕ-fan-ti′ă-sis*) A disease in which the larvae of a nematode worm block lymphatic drainage and produce edema. The lower areas of the body can become enormously swollen as a result.

EMG Electromyogram (*ĕ-lek″tro-mi′ŏ-gram*) An electrical recording of the activity of skeletal muscles through the use of surface electrodes.

embryonic stem cells Also called ES cells, these are the cells of the inner cell mass of a blastocyst. Embryonic stem cells are pluripotent, and so are potentially capable of differentiating into all tissue types except the trophoblast cells of a placenta.

emmetropia (*em″ĭ-tro′pe-ă*) A condition of normal vision in which the image of objects is focused on the retina, as opposed to nearsightedness (myopia) or farsightedness (hypermetropia).

emphysema (*em″fĭ-se′mă, em″fĭ-ze′mă*) A lung disease in which alveoli are destroyed and the remaining alveoli become larger. It results in decreased vital capacity and increased airway resistance.

emulsification (*ĕ-mul″sĭ-fĭ-ka′shun*) The process of producing an emulsion or fine suspension. In the small intestine, fat globules are emulsified by the detergent action of bile.

end-diastolic (*di″ă-stol′ik*) **volume** The volume of blood in each ventricle at the end of diastole, immediately before the ventricles contract at systole.

endergonic (*en″der-gon′ik*) Denoting a chemical reaction that requires the input of energy from an external source in order to proceed.

endo- (Gk.) Within, inner.

endocrine (*en′dŏ-krin*) **glands** Glands that secrete hormones into the circulation rather than into a duct; also called *ductless glands.*

endocytosis (*en″do-si-to′sis*) The cellular uptake of particles that are too large to cross the cell membrane. This occurs by invagination of the cell membrane until a membrane-enclosed vesicle is pinched off within the cytoplasm.

endoderm (*en′dŏ-derm*) The innermost of the three primary germ layers of an embryo. It gives rise to the digestive tract and associated structures, the respiratory tract, the bladder, and the urethra.

endogenous (*en-doj′ĕ-nus*) Denoting a product or process arising from within the body (as opposed to exogenous products or influences, which arise from external sources).

endolymph (*en′dŏ-limf*) The fluid contained within the membranous labyrinth of the inner ear.

endometrium (*en″do-me′tre-um*) The mucous membrane of the uterus, the thickness and structure of which vary with the phases of the menstrual cycle.

endoplasmic reticulum (*en-do-plaz′mik rĕ-tik′yŭ-lum*) An extensive system of membrane-enclosed cavities within the cytoplasm of the cell. Those with ribosomes on their surface are called rough endoplasmic reticulum and participate in protein synthesis.

endorphin (*en-dor′fin*) Any of a group of endogenous opioid molecules that may act as a natural analgesic.

endothelin (*en″do-the′lin*) A polypeptide secreted by the endothelium of a blood vessel that serves as a paracrine regulator, promoting contraction of the smooth muscle and constriction of the vessel.

endothelium (*en″do-the′le-um*) The simple squamous epithelium that lines blood vessels and the heart.

endotoxin (*en″do-tok′sin*) A toxin found within certain types of bacteria that is able to stimulate the release of endogenous pyrogen and produce a fever.

end-plate potential The graded depolarization produced by ACh at the neuromuscular junction. This is equivalent to the excitatory postsynaptic potential produced at neuron-neuron synapses.

end-product inhibition The inhibition of enzymatic steps of a metabolic pathway by products formed at the end of that pathway.

enkephalin (*en-kef′ă-lin*) Either of two short polypeptides, containing five amino acids, that have analgesic effects. The two known enkephalins (which differ in only one amino acid) are endorphins, and may function as neurotransmitters in the brain.

enteric (*en-ter′ik*) A term referring to the intestine.

enterochromaffin (*en″ter-o-kro″maf′in*)-**like (ECL) cells** Cells of the gastric epithelium that secrete histamine. The ECL cells are stimulated by the hormone gastrin and by the vagus nerve; the histamine from ECL cells, in turn, stimulates gastric acid secretion from the parietal cells.

enteroglucagon (*en″ter-o-gloo′kă-gon*) One of a family of polypeptides secreted by the ileum and colon that structurally resemble the hormone glucagon. Enteroglucagons raise the blood glucose concentration, stimulate insulin secretion, and inhibit GI motility. An example is glucagon-like peptide-1.

enterohepatic (*en″ter-o-hĕ-pat′ik*) **circulation** The recirculation of a compound between the liver and small intestine. The compound is present in the bile secreted by the liver into the small intestine. It is then reabsorbed and returned to the liver via the hepatic portal vein.

entropy (*en′trŏ-pe*) The energy of a system that is not available to perform work. A measure of the degree of disorder in a system, entropy increases whenever energy is transformed.

enzyme (*en′zim*) A protein catalyst that increases the rate of specific chemical reactions.

epi- (Gk.) Upon, over, outer.

epidermis (*ep″ĭ-der′mis*) The stratified squamous epithelium of the skin, the outer layer of which is dead and filled with keratin.

epididymis (*ep″ĭ-did′ĭ-mis*); pl., *epididymides.* A tubelike structure outside the testes. Sperm pass from the seminiferous tubules into the head of the epididymis and then pass from the tail of the epididymis to the ductus (vas) deferens. The sperm mature, becoming motile, as they pass through the epididymis.

epinephrine (*ep″ĭ-nef′rin*) A catecholamine hormone secreted by the adrenal medulla in response to sympathetic nerve stimulation. It acts together with norepinephrine released from sympathetic nerve endings to prepare the organism for "fight or flight"; also known as *adrenaline.*

epithelium (*ep″ĭ-the′le-um*) One of the four primary tissue types; the type of tissue that covers and lines body surfaces and forms exocrine and endocrine glands.

EPSP Excitatory postsynaptic (*pōst″sĭ-nap′tik*) potential. A graded depolarization of a postsynaptic membrane in response to stimulation by a neurotransmitter chemical. EPSPs can be summated, but they can be transmitted only over short distances; they can stimulate the production of action potentials when a threshold level of depolarization is attained.

equilibrium (*e″kwĭ-lib′re-um*) **potential** The hypothetical membrane potential that would be created if only one ion were able to diffuse across a membrane and reach a stable, or equilibrium, state. In this stable state, the concentrations of the ion would remain constant inside and outside the membrane, and the membrane potential would be equal to a particular value.

erythroblastosis fetalis (*ĕ-rith″ro-blas-to′sis fe-tal′is*) Hemolytic anemia in an Rh-positive newborn caused by maternal antibodies against the Rh factor that have crossed the placenta.

erythrocyte (*ĕ-rith′rŏ-sīt*) A red blood cell. Erythrocytes are the formed elements of blood that contain hemoglobin and transport oxygen.

erythropoietin (*ĕ-rith″ro-poi′ĕ-tin*) A hormone secreted by the kidneys that stimulates the bone marrow to produce red blood cells.

essential amino acids The eight amino acids in adults or nine amino acids in children that cannot be made by the human body; therefore, they must be obtained in the diet.

estradiol (*es″tră-di′ol*) The major estrogen (female sex steroid hormone) secreted by the ovaries.

estrus (*es′trus*) **cycle** Cyclic changes in the structure and function of the ovaries and female reproductive tract, accompanied by periods of "heat" (estrus), or sexual receptivity; the lower mammalian equivalent of the menstrual cycle, but differing from the menstrual cycle in that the endometrium is not shed with accompanying bleeding.

ex- (L.) Out, off, from.

excitation-contraction coupling The means by which electrical excitation of a muscle results in muscle contraction. This coupling is achieved by Ca^{2+}, which enters the muscle cell cytoplasm in response to electrical excitation and which stimulates the events culminating in contraction.

exergonic (*ek″ser-gon′ik*) Denoting chemical reactions that liberate energy.

exo- (Gk.) Outside or outward.

exocrine (*ek′sŏ-krin*) **gland** A gland that discharges its secretion through a duct to the outside of an epithelial membrane.

exocytosis (*ek″so-si-to′sis*) The process of cellular secretion in which the secretory products are contained within a membrane-enclosed vesicle. The vesicle fuses with the cell membrane so that the lumen of the vesicle is open to the extracellular environment.

exon (*ek'son*) A nucleotide sequence in DNA that codes for the production of messenger RNA.

extensor (*ek-sten'sor*) A muscle that, upon contraction, increases the angle of a joint.

exteroceptor (*ek"ster-o-cep'tor*) A sensory receptor that is sensitive to changes in the external environment (as opposed to an interoceptor).

extra- (L.) Outside, beyond.

extrafusal (*ek"stră-fyooz'al*) **fibers** The ordinary muscle fibers within a skeletal muscle; not found within the muscle spindles.

extraocular (*ek"stră-ok'yŭ-lar*) **muscles** The muscles that insert into the sclera of the eye. They act to change the position of the eye in its orbit (as opposed to the intraocular muscles, such as those of the iris and ciliary body within the eye).

extrapyramidal (*ek"stră-pĭ-ram'ĭ-dl*) **tracts** Neural pathways that are situated outside of, or that are "independent of," pyramidal tracts. The major extrapyramidal tract is the reticulospinal tract, which originates in the reticular formation of the brain stem and receives excitatory and inhibitory input from both the cerebrum and the cerebellum. The extrapyramidal tracts are thus influenced by activity in the brain involving many synapses, and they appear to be required for fine control of voluntary movements.

facilitated (*fă-sil'ĭ-ta"tid*) **diffusion** The carrier-mediated transport of molecules through the cell membrane along the direction of their concentration gradients. It does not require the expenditure of metabolic energy.

FAD Flavin adenine dinucleotide (*fla'vin ad'n-ēn di-noo'kle-ō-tīd*). A coenzyme derived from riboflavin that participates in electron transport within the mitochondria.

FAS A surface receptor produced by T lymphocytes during an infection. After a few days, the activated T lymphocytes begin to produce another surface molecule, FAS ligand. The bonding of FAS with FAS ligand, on the same or on different cells, triggers apoptosis of the lymphocytes.

feces (*fe'sēz*) The excrement discharged from the large intestine.

fertilization (*fer'tĭ-lĭ-za"shun*) The fusion of an ovum and spermatozoon.

fiber, muscle A skeletal muscle cell.

fiber, nerve An axon of a motor neuron or the dendrite of a pseudounipolar sensory neuron in the PNS.

fibrillation (*fib"rĭ-la'shun*) A condition of cardiac muscle characterized electrically by random and continuously changing patterns of electrical activity and resulting in the inability of the myocardium to contract as a unit and pump blood. It can be fatal if it occurs in the ventricles.

fibrin (*fi'brin*) The insoluble protein formed from fibrinogen by the enzymatic action of thrombin during the process of blood clot formation.

fibrinogen (*fi-brin'ŏ-jen*) A soluble plasma protein that serves as the precursor of fibrin; also called *factor I.*

flaccid paralysis (*flak'sid pă-ral'ĭ-sis*) The inability to contract muscles, resulting in a loss of muscle tone. This may be due to damage to

lower motor neurons or to factors that block neuromuscular transmission.

flagellum (*flă-jel'um*) A whiplike structure that provides motility for sperm.

flare-and-wheal reaction A cutaneous reaction to skin injury or to the administration of antigens produced by release of histamine and related molecules; characterized by local edema and a red flare.

flavoprotein (*fla"vo-pro'te-in*) A conjugated protein containing a flavin pigment that is involved in electron transport within the mitochondria.

flexor (*flek'sor*) A muscle that decreases the angle of a joint when it contracts.

follicle (*fol'ĭ-k'l*) A microscopic hollow structure within an organ. Follicles are the functional units of the thyroid gland and of the ovary.

foramen ovale (*fŏ-ra'men o-val'e*) An opening normally present in the atrial septum of the fetal heart that allows direct communication between the right and left atria.

fovea centralis (*fo've-ă sen-tra'lis*) A tiny pit in the macula lutea of the retina that contains slim, elongated cones. It provides the highest visual acuity (clearest vision).

Frank–Starling Law of the Heart The statement describing the relationship between end-diastolic volume and stroke volume of the heart. A greater amount of blood in a ventricle prior to contraction results in greater stretch of the myocardium, and by this means produces a contraction of greater strength.

FSH Follicle-stimulating hormone. One of the two gonadotropic hormones secreted by the anterior pituitary. In females, FSH stimulates the development of the ovarian follicles; in males, it stimulates the production of sperm in the seminiferous tubules.

GABA Gamma-aminobutyric (*gam"ă-ă-me"no-byoo-tir'ik*) acid. An amino acid believed to function as an inhibitory neurotransmitter in the central nervous system.

gamete (*gam'ēt*)- collective term for haploid germ cells: sperm and ova.

gamma motoneuron (*gam' ă mo"tŏ-noor'on*) The type of somatic motor neuron that stimulates intrafusal fibers within the muscle spindles.

ganglion (*gang'gle-on*) A grouping of nerve cell bodies located outside the brain and spinal cord.

gap junctions Specialized regions of fusion between the cell membranes of two adjacent cells that permit the diffusion of ions and small molecules from one cell to the next. These regions serve as electrical synapses in certain areas, such as in cardiac muscle.

gas exchange The diffusion of oxygen and carbon dioxide down their concentration gradients that occurs between pulmonary capillaries and alveoli, and between systemic capillaries and the surrounding tissue cells.

gastric (*gas'trik*) **intrinsic factor** A glycoprotein secreted by the stomach that is needed for the absorption of vitamin B_{12}.

gastric juice The secretions of the gastric mucosa. Gastric juice contains water, hydrochloric acid, and pepsinogen as major components.

gastrin (*gas'trin*) A hormone secreted by the stomach that stimulates the gastric secretion of hydrochloric acid and pepsin.

gastroileal (*gas'tro-il"e-al*) **reflex** The reflex in which increased gastric activity causes increased motility of the ileum and increased movement of chyme through the ileocecal sphincter.

gates A term used to describe structures within the cell membrane that regulate the passage of ions through membrane channels. Gates may be chemically regulated (by neurotransmitters) or voltage regulated (in which case they open in response to a threshold level of depolarization).

gen- (Gk.) Producing.

generator (*jen'ĕ-ra"tor*) **potential** The graded depolarization produced by stimulation of a sensory receptor that results in the production of action potentials by a sensory neuron; also called the *receptor potential.*

genetic (*jĕ-net'ik*) **recombination** The formation of new combinations of genes, as by crossing-over between homologous chromosomes.

genetic transcription The process by which RNA is produced with a sequence of nucleotide bases that is complementary to a region of DNA.

genetic translation The process by which proteins are produced with amino acid sequences specified by the sequence of codons in messenger RNA.

gigantism (*ji-gan'tiz"em*) Abnormal body growth due to the excessive secretion of growth hormone.

glomerular (*glo-mer'yŭ-lar*) **filtration rate (GFR)** The volume of blood plasma filtered out of the glomeruli of both kidneys each minute. The GFR is measured by the renal plasma clearance of inulin.

glomerular ultrafiltrate Fluid filtered through the glomerular capillaries into the glomerular (Bowman's) capsule of the kidney tubules.

glomeruli (*glo-mer'yŭ-li*) The tufts of capillaries in the kidneys that filter fluid into the kidney tubules.

glomerulonephritis (*glo-mer"yŭ-lo-nĕ-fri'tis*) Inflammation of the renal glomeruli; associated with fluid retention, edema, hypertension, and the appearance of protein in the urine.

glucagon (*gloo'că-gon*) A polypeptide hormone secreted by the alpha cells of the islets of Langerhans in the pancreas that acts to promote glycogenolysis and raise the blood glucose levels.

glucocorticoid (*gloo"ko-kor'tĭ-koid*) Any of a class of steroid hormones secreted by the adrenal cortex (corticosteroids) that affects the metabolism of glucose, protein, and fat. These hormones also have anti-inflammatory and immunosuppressive effects. The major glucocorticoid in humans is hydrocortisone (cortisol).

gluconeogenesis (*gloo"ko-ne"ŏ-jen'ĭ-sis*) The formation of glucose from noncarbohydrate molecules, such as amino acids and lactic acid.

GLUT An acronym for *glucose transporters.* GLUT proteins promote the facilitated diffusion of glucose into cells. One isoform of GLUT, designated GLUT4, is inserted into the cell membranes of muscle and adipose cells in response to insulin stimulation and exercise.

glutamate (*gloo'tă-māt*) The ionized form of glutamic acid, an amino acid that serves as the major excitatory neurotransmitter of the CNS. *Glutamate* and *glutamic acid* are terms that can be used interchangeably.

glutathione (*gloo"tah-thi'on*) A tripeptide molecule that functions as the major cellular antioxidant.

glycogen (*gli"kŏ-jen*) A polysaccharide of glucose—also called *animal starch*—produced primarily in the liver and skeletal muscles. Similar to plant starch in composition, glycogen contains more highly branched chains of glucose subunits than does plant starch.

glycogenesis (*gli"kŏ-jen'ĭ-sis*) The formation of glycogen from glucose.

glycogenolysis (*gli"ko-jĕ-nol'ĭ-sis*) The hydrolysis of glycogen to glucose-1-phosphate, which can be converted to glucose-6-phosphate. The glucose-6-phosphate then may be oxidized via glycolysis or (in the liver) converted to free glucose.

glycolysis (*gli"kol'ĭ-sis*) The metabolic pathway that converts glucose to pyruvic acid. The final products are two molecules of pyruvic acid and two molecules of reduced NAD, with a net gain of two ATP molecules. In anaerobic respiration, the reduced NAD is oxidized by the conversion of pyruvic acid to lactic acid. In aerobic respiration, pyruvic acid enters the Krebs cycle in mitochondria, and reduced NAD is ultimately oxidized by oxygen to yield water.

glycosuria (*gli"kŏ-soor'e-ă*) The excretion of an abnormal amount of glucose in the urine (urine normally contains only trace amounts of glucose).

Golgi (*gol'je*) **apparatus** A network of stacked, flattened membranous sacs within the cytoplasm of cells. Its major function is to concentrate and package proteins within vesicles that bud off from it.

Golgi tendon organ A tension receptor in the tendons of muscles that becomes activated by the pull exerted by a muscle on its tendons; also called a *neurotendinous receptor*.

gonad (*go'nad*) A collective term for testes and ovaries.

gonadotropic (*go"nad-ŏ-trop'ik*) **hormones** Hormones of the anterior pituitary that stimulate gonadal function—the formation of gametes and secretion of sex steroids. The two gonadotropins are FSH (follicle-stimulating hormone) and LH (luteinizing hormone), which are essentially the same in males and females.

G-protein An association of three membrane-associated protein subunits, designated alpha, beta, and gamma, that is regulated by guanosine nucleotides (GDP and GTP). The G-protein subunits dissociate in response to a membrane signal and, in turn, activate other proteins in the cell.

graafian (*graf'e-an*) **follicle** A mature ovarian follicle, containing a single fluid-filled cavity, with the ovum located toward one side of the follicle and perched on top of a hill of granulosa cells.

granular leukocytes (*loo'kŏ-sūts*) Leukocytes with granules in the cytoplasm. On the basis of the staining properties of the granules, these cells are of three types: neutrophils, eosinophils, and basophils.

Graves' disease A hyperthyroid condition believed to be caused by excessive stimulation of the thyroid gland by autoantibodies. It is associated with exophthalmos (bulging eyes), high pulse rate, high metabolic rate, and other symptoms of hyperthyroidism.

gray matter The part of the central nervous system that contains neuron cell bodies and dendrites but few myelinated axons. It forms the cortex of the cerebrum, cerebral nuclei, and central region of the spinal cord.

growth hormone A hormone secreted by the anterior pituitary that stimulates growth of the skeleton and soft tissues during the growing years and that influences the metabolism of protein, carbohydrate, and fat throughout life.

gustducins (*gus-doo'sinz*) The G-proteins involved in the sense of taste, particularly of sweet and bitter tastes.

gyrus (*ji'rus*) A fold or convolution in the cerebrum.

H

haploid (*hap'loid*) Denoting cells that have one of each chromosome type and therefore half the number of chromosomes present in most other body cells. Only the gametes (sperm and ova) are haploid.

hapten (*hap'ten*) A small molecule that is not antigenic by itself, but which—when combined with proteins—becomes antigenic and thus capable of stimulating the production of specific antibodies.

haversian (*hă-ver'shan*) **system** A haversian canal and its concentrically arranged layers, or lamellae, of bone. It constitutes the basic structural unit of compact bone.

hay fever A seasonal type of allergic rhinitis caused by pollen. It is characterized by itching and tearing of the eyes, swelling of the nasal mucosa, attacks of sneezing, and often by asthma.

hCG Human chorionic gonadotropin (*kor'e-on-ik go-nad"ŏ-tro'pin*). A hormone secreted by the embryo that has LH-like actions and that is required for maintenance of the mother's corpus luteum for the first 10 weeks of pregnancy.

heart murmur An abnormal heart sound caused by an abnormal flow of blood in the heart. Murmurs are due to structural defects, usually of the valves or septum.

heart sounds The sounds produced by closing of the AV valves of the heart during systole (the first sound) and by closing of the semilunar valves of the aorta and pulmonary trunk during diastole (the second sound).

helper T cells A subpopulation of T cells (lymphocytes) that help to stimulate antibody production of B lymphocytes by antigens.

hematocrit (*he-mat'ŏ-krit*) The ratio of packed red blood cells to total blood volume in a centrifuged sample of blood, expressed as a percentage.

heme (*hēm*) The iron-containing red pigment that, together with the protein globin, forms hemoglobin.

hemoglobin (*he'mŏ-glo"bin*) The combination of heme pigment and protein within red blood cells that acts to transport oxygen and (to a lesser degree) carbon dioxide. Hemoglobin also serves as a weak buffer within red blood cells.

Henderson-Hasselbalch (*hen'der-son-has'el-balk*) **equation** A formula used to determine the blood pH produced by a given ratio of bicarbonate to carbon dioxide concentrations.

Henry's law The statement that the concentration of gas dissolved in a fluid is directly proportional to the partial pressure of that gas.

heparin (*hep'ar-in*) A mucopolysaccharide found in many tissues, but in greatest abundance in the lungs and liver. It is used medically as an anticoagulant.

hepatic (*hĕ-pat'ik*) Pertaining to the liver.

hepatitis (*hep"ă-it'tis*) Inflammation of the liver.

Hering-Breuer reflex A reflex in which distension of the lungs stimulates stretch receptors, which in turn act to inhibit further distension of the lungs.

hermaphrodite (*her-maf'rŏ-dīt*) An organism with both testicular and ovarian tissue.

hetero- (Gk.) Different, other.

heterochromatin (*het"ĕ-ro-kro'mă-tin*) A condensed, inactive form of chromatin.

hiatal hernia (*hi-a'tal her' ne-ă*) A protrusion of an abdominal structure through the esophageal hiatus of the diaphragm into the thoracic cavity.

high-density lipoproteins (*lip"o-pro'te-inz*) **(HDLs)** Combinations of lipids and proteins that migrate rapidly to the bottom of a test tube during centrifugation. HDLs are carrier proteins that are believed to transport cholesterol away from blood vessels to the liver, and thus to offer some protection from atherosclerosis.

higher motor neurons Neurons in the brain that, as part of the pyramidal or extrapyramidal system, influence the activity of the lower motor neurons in the spinal cord.

histamine (*his'tă-mēn*) A compound secreted by tissue mast cells and other connective tissue cells that stimulates vasodilation and increases capillary permeability. It is responsible for many of the symptoms of inflammation and allergy.

histocompatibility (*his"to-kom-pat"ĭ-bil'ĭ-te*) **antigens** A group of cell-surface antigens found on all cells of the body except mature red blood cells. They are important for the function of T lymphocytes, and the greater their variance, the greater will be the likelihood of transplant rejection.

histone (*his'toōn*) A basic protein associated with DNA that is believed to repress genetic expression.

homeo (Gk.) Same.

homeostasis (*ho"me-o-sta'sis*) The dynamic constancy of the internal environment, the maintenance of which is the principal function of physiological regulatory mechanisms. The concept of homeostasis provides a framework for understanding most physiological processes.

homologous (*hŏ-mol'-ŏ-gus*) **chromosomes** The matching pairs of chromosomes in a diploid cell.

hormone (*hor'mōn*) A regulatory chemical produced in an endocrine gland that is secreted into the blood and carried to target cells that respond to the hormone by an alteration in their metabolism.

hormone-response element A specific region of DNA that binds to a particular nuclear hormone receptor when that receptor is activated by bonding with its hormone. This stimulates genetic transcription (RNA synthesis).

humoral immunity (*hyoo'-mor-al ĭ-myoo'nĭ-te*) The form of acquired immunity in which antibody molecules are secreted in response to antigenic stimulation (as opposed to cell-mediated immunity).

hyaline (*hi'ă-līn*) **membrane disease** A disease affecting premature infants who lack pulmonary surfactant. It is characterized by collapse of the alveoli (atelectasis) and pulmonary edema; also called *respiratory distress syndrome*.

hydrocortisone (*hi″drŏ-kor′tĭ-sōn*) The principal corticosteroid hormone secreted by the adrenal cortex, with glucocorticoid action; also called *cortisol.*

hydrophilic (*hi″drŏ-fil′ik*) Denoting a substance that readily absorbs water; literally, "water loving."

hydrophobic (*hi″drŏ-fo′bik*) Denoting a substance that repels, and that is repelled by, water; literally, "water fearing."

hyper- (Gk.) Over, above, excessive.

hyperbaric (*hi″per-bar′ik*) **oxygen** Oxygen gas present at greater than atmospheric pressure.

hypercapnia (*hi″per-kap′ne-ă*) Excessive concentration of carbon dioxide in the blood.

hyperemia (*hi″per-e′-me-ă*) Excessive blood flow to a part of the body.

hyperglycemia (*hi″per-gli-se′me-ă*) An abnormally increased concentration of glucose in the blood.

hyperkalemia (*hi″per-kă-le′me-ă*) An abnormally high concentration of potassium in the blood.

hyperopia (*hi″per-o′pe-ă*) A refractive disorder in which rays of light are brought to a focus behind the retina as a result of the eyeball being too short; also called *farsightedness.*

hyperplasia (*hi″per-pla′ze-ă*) An increase in organ size because of an increase in the number of cells as a result of mitotic cell division.

hyperpnea (*hi″perp′ne-ă*) Increased total minute volume during exercise. Unlike hyperventilation, the arterial blood carbon dioxide values are not changed during hyperpnea because the increased ventilation is matched to an increased metabolic rate.

hyperpolarization (*hi″per-po″lar-ĭ-za′shun*) An increase in the negativity of the inside of a cell membrane with respect to the resting membrane potential.

hypersensitivity (*hi″per-sen″sĭ-tiv′ĭ-te*) Another name for *allergy;* an abnormal immune response that may be immediate (due to antibodies of the IgE class) or delayed (due to cell-mediated immunity).

hypertension (*hi″per-ten′shun*) High blood pressure. Classified as either primary, or essential, hypertension of unknown cause or secondary hypertension that develops as a result of other, known disease processes.

hypertonic (*hi″per-ton′ik*) Denoting a solution with a greater solute concentration, and thus a greater osmotic pressure, than plasma.

hypertrophy (*hi-per′trŏ-fe*) Growth of an organ because of an increase in the size of its cells.

hyperventilation (*hi-per-ven″tĭ-la′shun*) A high rate and depth of breathing that results in a decrease in the blood carbon dioxide concentration to below normal.

hypo- (Gk.) Under, below, less.

hypodermis (*hi″pŏ-der′mis*) A layer of fat beneath the dermis of the skin.

hypotension (*hi″po-ten′shun*) Abnormally low blood pressure.

hypothalamic (*hi″po-thă-lam′ik*) **hormones** Hormones produced by the hypothalamus. These include antidiuretic hormone and oxytocin, which are released from the posterior pituitary, and both releasing and inhibiting hormones that regulate the secretions of the anterior pituitary.

hypothalamo-hypophyseal (*hi″po-thă-lam′o-hi″po-fĭ-se′al*) **portal system** A vascular system that transports releasing and inhibiting hormones from the hypothalamus to the anterior pituitary.

hypothalamo-hypophyseal tract The tract of nerve fibers (axons) that transports antidiuretic hormone and oxytocin from the hypothalamus to the posterior pituitary.

hypothalamus (*hi″po-thal′ă-mus*) An area of the brain lying below the thalamus and above the pituitary gland. The hypothalamus regulates the pituitary gland and contributes to the regulation of the autonomic nervous system, among its many functions.

hypothermia (*hi″pŏ-ther′me-ă*) A low body temperature. This is a dangerous condition that is defended against by shivering and other physiological mechanisms that generate body heat.

hypovolemic (*hi″po-vo-le′mik*) **shock** A rapid fall in blood pressure as a result of diminished blood volume.

hypoxemia (*hi″pok-se′me-ă*) A low oxygen concentration of the arterial blood.

I

ileogastric (*il″e-o-gas′trik*) **reflex** The reflex in which distension of the ileum causes decreased gastric motility.

immediate hypersensitivity Hypersensitivity (allergy) that is mediated by antibodies of the IgE class and that results in the release of histamine and related compounds from tissue cells.

immunization (*im″yŭ-nĭ-za′shun*) The process of increasing one's resistance to pathogens. In active immunity, a person is injected with antigens that stimulate the development of clones of specific B or T lymphocytes; in passive immunity, a person is injected with antibodies made by another organism.

immunoassay (*im″yŭ-no-as′a*) Any of a number of laboratory or clinical techniques that employ specific bonding between an antigen and its homologous antibody in order to identify and quantify a substance in a sample.

immunoglobulins (*im″yŭ-no-glob′yŭ-linz*) Subclasses of the gamma globulin fraction of plasma proteins that have antibody functions, providing humoral immunity.

immunosurveillance (*im″yŭ-no-ser-va′lens*) The function of the immune system to recognize and attack malignant cells that produce antigens not recognized as "self." This function is believed to be cell mediated rather than humoral.

implantation (*im″plan-ta′shun*) The process by which a blastocyst attaches itself to and penetrates the endometrium of the uterus.

infarct (*in′farkt*) An area of necrotic (dead) tissue produced by inadequate blood flow (ischemia).

inhibin (*in-hib′in*) Believed to be a water-soluble hormone secreted by the seminiferous tubules of the testes that specifically exerts negative feedback control of FSH secretion from the anterior pituitary.

inositol triphosphate (*ĭ-no′sĭ-tol tri-fos′făt*) A second messenger in hormone action that is produced by the cell membrane of a target cell in response to the action of a hormone. This compound is believed to stimulate the release of Ca^{2+} from the endoplasmic reticulum of the cell.

insulin (*in′sŭ-lin*) A polypeptide hormone secreted by the beta cells of the islets of Langerhans in the pancreas that promotes the anabolism of carbohydrates, fat, and protein. Insulin acts to promote the cellular uptake of blood glucose and, therefore, to lower the blood glucose concentration; insulin deficiency produces hyperglycemia and diabetes mellitus.

integrins (*in-te′grinz*) A family of glycoproteins that extend from the cytoskeleton, through the plasma membrane of cells, and into the extracellular matrix. They serve to integrate different cells of a tissue and the extracellular matrix, and to bind cells to other cells, such as neutrophils to the endothelial cells of capillaries for extravasation.

inter- (L.) Between, among.

interferons (*in″ter-fer′unz*) Small proteins that inhibit the multiplication of viruses inside host cells and that also have antitumor properties.

interleukin-2 (*in″ter-loo′kin-2*) A lymphokine secreted by T lymphocytes that stimulates the proliferation of both B and T lymphocytes.

interneurons (*in″ter-noor′onz*) Those neurons within the central nervous system that do not extend into the peripheral nervous system. They are interposed between sensory (afferent) and motor (efferent) neurons; also called *association neurons.*

interoceptors (*in″ter-o-sep′torz*) Sensory receptors that respond to changes in the internal environment (as opposed to exteroceptors).

interphase The interval between successive cell divisions, during which time the chromosomes are in an extended state and are active in directing RNA synthesis.

intestino-intestinal (*in″tes′tĭ-no-in-tes′tĭ-nal*) **reflex** The reflex in which overdistension to one region of the intestine causes relaxation throughout the rest of the intestine.

intra- (L.) Within, inside.

intrafusal (*in″tră-fyoo′sal*) **fibers** Modified muscle fibers that are encapsulated to form muscle spindle organs, which are muscle stretch receptors.

intrapleural (*in″tră-ploor′al*) **space** An actual or potential space between the visceral pleura covering the lungs and the parietal pleura lining the thoracic wall. Normally, this is a potential space; it can become real only in abnormal situations.

intrapulmonary (*in″tră-pul′mŏ-nar″e*) **space** The space within the air sacs and airways of the lungs.

intron (*in′tron*) A noncoding nucleotide sequence in DNA that interrupts the coding regions (exons) for mRNA.

inulin (*in′yŭ-lin*) A polysaccharide of fructose, produced by certain plants, that is filtered by the human kidneys but neither reabsorbed nor secreted. The clearance rate of injected inulin is thus used to measure the glomerular filtration rate.

in vitro (*in ve′tro*) Occuring outside the body, in a test tube or other artificial environment.

in vivo (*in ve′vo*) Occuring within the body.

ion (*i′on*) An atom or a group of atoms that has a net positive or a net negative charge because of a loss or gain of electrons.

ionization (*i″on-ĭ-za′shun*) The dissociation of a solute to form ions.

ipsilateral (*ip″sĭ-lat′er-al*) On the same side (as opposed to contralateral).

IPSP Inhibitory postsynaptic potential. A hyperpolarization of the postsynaptic membrane in response to a particular neurotransmitter chemical, which makes it more difficult for the postsynaptic cell to attain the threshold level of depolarization required to produce action potentials. IPSPs are responsible for postsynaptic inhibition.

ischemia (*ĭ-ske′me-ă*) A rate of blood flow to an organ that is inadequate to supply sufficient oxygen and maintain aerobic respiration in that organ.

islets of Langerhans (*ī′letz of lang′er-hanz*) Encapsulated groupings of endocrine cells within the exocrine tissue of the pancreas, including alpha cells that secrete glucagon and beta cells that secrete insulin; also called *pancreatic islets.*

iso- (Gk.) Equal, same.

isoenzymes (*i″so-en′zimz*) Enzymes, usually produced by different organs, that catalyze the same reaction but that differ from each other in amino acid composition.

isometric (*i″sŏ-met′rik*) **contraction** Muscle contraction in which there is no appreciable shortening of the muscle.

isotonic (*i″sŏ-ton′ik*) **contraction** Muscle contraction in which the muscle shortens in length and maintains approximately the same amount of tension throughout the shortening process.

isotonic solution A solution having the same total solute concentration, osmolality, and osmotic pressure as the solution with which it is compared; a solution with the same solute concentration and osmotic pressure as plasma.

J

jaundice (*jawn′dis*) A condition characterized by high blood bilirubin levels and staining of the tissues with bilirubin, which imparts a yellow color to the skin and mucous membranes.

junctional (*jungk′shun-al*) **complexes** Structures that join adjacent epithelial cells together, including the zonula occludens, zonula adherens, and macula adherens (desmosome).

juxta- (L.) Near to, next to.

juxtaglomerular (*juk″stă-glo-mer′yŭ-lar* **apparatus** A renal structure in which regions of the nephron tubule and afferent arteriole are in contact with each other. Cells in the afferent arteriole of the juxtaglomerular apparatus secrete the enzyme renin into the blood, which activates the renin-angiotensin system.

K

keratin (*ker′ă-tin*) A protein that forms the principal component of the outer layer of the epidermis and of hair and nails.

ketoacidosis (*ke″to-ă-sĭ-do′sis*) A type of metabolic acidosis resulting from the excessive production of ketone bodies, as in diabetes mellitus.

ketogenesis (*ke″to-jen′ĭ-sis*) The production of ketone bodies.

ketone (*ke′tŏn*) **bodies** The substances derived from fatty acids via acetyl coenzyme A in the liver; namely, acetone, acetoacetic acid, and β-hydroxybutyric acid. Ketone bodies are oxidized by skeletal muscles for energy.

ketosis (*ke-to′sis*) An abnormal elevation in the blood concentration of ketone bodies. This condition does not necessarily produce acidosis.

kilocalorie (*kil′ŏ-kal″ŏ-re*) A unit of measurement equal to 1,000 calories, which are units of heat. (A kilocalorie is the amount of heat required to raise the temperature of 1 kilogram of water 1° C.) In nutrition, the kilocalorie is called a big calorie (Calorie).

kinase (*ki′nās*) Any of a class of enzymes that transfer phosphate groups to organic molecules. The activity of particular protein kinases may be promoted by hormones and other regulatory molecules. These enzymes can, in turn, phosphorylate other enzymes and thereby regulate their activities.

Klinefelter's (*klīn′fel-terz*) **syndrome** The syndrome produced in a male by the presence of an extra X chromosome (genotype XXY).

knockout mice Strains of mice in which a specific targeted gene has been inactivated by developing the mice from embryos injected with specifically mutated cells.

Krebs (*krebz*) **cycle** A cyclic metabolic pathway in the matrix of mitochondria by which the acetic acid part of acetyl CoA is oxidized and substrates provided for reactions that are coupled to the formation of ATP.

Kupffer (*koop′fer*) **cells** Phagocytic cells lining the sinusoids of the liver that are part of the reticuloendothelial system.

L

lactate threshold A measurement of the intensity of exercise. It is the percentage of a person's maximal oxygen uptake at which a rise in blood lactate levels occurs. The average lactate threshold occurs when exercise is performed at 50% to 70% of the maximal oxygen uptake (aerobic capacity).

lactose (*lak′tōs*) Milk sugar; a disaccharide of glucose and galactose.

lactose intolerance The inability of many adults to digest lactose because of a deficiency of the enzyme lactase.

Laplace, law of The statement that the pressure within an alveolus is directly proportional to its surface tension and inversely proportional to its radius.

larynx (*lar′ingks*) A structure consisting of epithelial tissue, muscle, and cartilage that serves as a sphincter guarding the entrance of the trachea. It is the organ responsible for voice production.

lateral inhibition The sharpening of perception that occurs in the neural processing of sensory input. Input from those receptors that are most greatly stimulated is enhanced, while input from other receptors is reduced. This results, for example, in improved pitch discrimination in hearing.

leptin (*lep′tin*) A hormone secreted by adipose tissue that acts as a satiety factor to reduce appetite. It also increases the body's caloric expenditure.

lesion (*le′zhun*) 1. A wounded or damaged area of tissue. 2. An injury or wound. 3. A single infected patch in a skin disease.

leukocyte (*loo′kŏ-sīt*) A white blood cell.

Leydig (*li′dig*) **cells** The interstitial cells of the testes that serve an endocrine function by secreting testosterone and other androgenic hormones.

ligament (*lig′ă-ment*) A tough cord or fibrous band of dense regular connective tissue that contains numerous parallel arrangements of collagen fibers. It connects bones or cartilages and serves to strengthen joints.

ligand (*li′gand, lig′and*) A smaller molecule that chemically binds to a larger molecule, which is usually a protein. Oxygen, for example, is the ligand for the heme in hemoglobin, and hormones or neurotransmitters can be the ligands for specific membrane proteins.

limbic (*lim′bik*) **system** A group of brain structures, including the hippocampus, cingulate gyrus, dentate gyrus, and amygdala. The limbic system appears to be important in memory, the control of autonomic function, and some aspects of emotion and behavior.

lipid (*lip′id*) An organic molecule that is nonpolar, and thus insoluble in water. Lipids include triglycerides, steroids, and phospholipids.

lipogenesis (*lip″ŏ-jen′ĕ-sis*) The formation of fat or triglycerides.

lipolysis (*li-pol′ĭ-sis*) The hydrolysis of triglycerides into free fatty acids and glycerol.

lipophilic (*lip″ŏ-fil′ik*) Pertaining to molecules that are nonpolar and thus soluble in lipids. The steroid hormones, thyroxine, and the lipid-soluble vitamins are examples of lipophilic molecules.

long-term potentiation (*pŏ-ten″she-a′shun*) The improved ability of a presynaptic neuron that has been stimulated at high frequency to subsequently stimulate a postsynaptic neuron over a period of weeks or even months. This may represent a mechanism of neural learning.

low-density lipoproteins (*lip″o-pro′te-inz*) **(LDLs)** Plasma proteins that transport triglycerides and cholesterol to the arteries. LDLs are believed to contribute to arteriosclerosis.

lower motor neuron The motor neuron that has its cell body in the gray matter of the spinal cord and that contributes axons to peripheral nerves. This neuron innervates muscles and glands.

lumen (*loo′men*) The cavity of a tube or hollow organ.

lung surfactant (*sur-fak′tant*) A mixture of lipoproteins (containing phospholipids) secreted by type II alveolar cells into the alveoli of the lungs. It lowers surface tension and prevents collapse of the lungs, as occurs in hyaline membrane disease when surfactant is absent.

luteinizing (*loo′te-ĭ-ni″zing*) **hormone (LH)** A gonadotropic hormone secreted by the anterior pituitary. In a female, LH stimulates ovulation and the development of a corpus luteum; in a male, it stimulates the Leydig cells to secrete androgens.

lymph (*limf*) A fluid derived from tissue fluid that flows through lymphatic vessels, returning to the venous bloodstream.

lymphatic (*lim-fat′ik*) **system** The lymphatic vessels and lymph nodes.

lymphocyte (*lim′fŏ-sīt*) A type of mononuclear leukocyte; the cell responsible for humoral and cell-mediated immunity.

lymphokine (*lim′fŏ-kīn*) Any of a group of chemicals released from T cells that contribute to cell-mediated immunity.

-lysis (Gk.) Breakage, disintegration.

lysosome (*li′sŏ-sŏm*) An organelle containing digestive enzymes that is responsible for intracellular digestion.

M

macro- *(Gk.)* Large.

macromolecule *(mak″rŏ-mol′ĭ-kyool)* A large molecule; a term commonly used to refer to protein, RNA, and DNA.

macrophage *(mak′rŏ-fāj)* A large phagocytic cell in connective tissue that contributes to both specific and nonspecific immunity.

macula densa *(mak′yŭ-lă den′să)* The region of the distal tubule of the renal nephron in contact with the afferent arteriole. This region functions as a sensory receptor for the amount of sodium excreted in the urine and acts to inhibit the secretion of renin from the juxtaglomerular apparatus.

macula lutea *(loo′te-ă)* A yellowish depression in the retina of the eye that contains the fovea centralis, the area of keenest vision.

malignant Denoting a structure or process that is life threatening. Of a tumor, tending to metastasize.

mast cell A type of connective tissue cell that produces and secretes histamine and heparin.

maximal oxygen uptake The maximum rate of oxygen consumption by the body per unit time during heavy exercise. Also called the *aerobic capacity,* the maximal oxygen uptake is commonly indicated with the symbol V_O2 max.

mean arterial pressure An adjusted average of the systolic and diastolic blood pressures. It averages about 100 mmHg in the systemic circulation and 10 mmHg in the pulmonary circulation.

mechanoreceptor *(mek″ă-no-re-sep′tor)* A sensory receptor that is stimulated by mechanical means. Mechanoreceptors include stretch receptors, hair cells in the inner ear, and pressure receptors.

medulla oblongata *(mĕ-dul′ă ob″long-gătă)* A part of the brain stem that contains neural centers for the control of breathing and for regulation of the cardiovascular system via autonomic nerves.

mega- *(Gk.)* Large, great.

megakaryocyte *(meg″ă-kar′e-o-sīt)* A bone marrow cell that gives rise to blood platelets.

meiosis *(mi-o′sis)* A type of cell division in which a diploid parent cell gives rise to haploid daughter cells. It occurs in the process of gamete production in the gonads.

melanin *(mel′ă-nin)* A dark pigment found in the skin, hair, choroid layer of the eye, and substantia nigra of the brain. It may also be present in certain tumors (melanomas).

melatonin *(mel′ă-to′nin)* A hormone secreted by the pineal gland that produces darkening of the skin in lower animals and that may contribute to the regulation of gonadal function in mammals. Secretion follows a circadian rhythm and peaks at night.

membrane potential The potential difference or voltage that exists between the two sides of a cell membrane. It exists in all cells but is capable of being changed by excitable cells (neurons and muscle cells).

membranous labyrinth *(mem′bră-nus lab′ĭ-rinth)* A system of communicating sacs and ducts within the bony labyrinth of the inner ear.

menarche *(mĕ-nar′ke)* The first menstrual discharge, normally occurring during puberty.

Ménière's *(mān-yarz′)* **disease** Deafness, tinnitus, and vertigo resulting from a disease of the labyrinth.

menopause *(men′ŏ-pawz)* The cessation of menstruation, usually occurring at about age 50.

menstrual *(men′stroo-al)* **cycle** The cyclic changes in the ovaries and endometrium of the uterus that lasts about a month. It is accompanied by shedding of the endometrium, with bleeding, and occurs only in humans and the higher primates.

menstruation *(men″stroo-a′shun)* Shedding of the outer two-thirds of the endometrium with accompanying bleeding as a result of a lowering of estrogen secretion by the ovaries at the end of the monthly cycle. The first day of menstruation is taken as day 1 of the menstrual cycle.

meso- *(Gk.)* Middle.

mesoderm *(mes′ŏ-derm)* The middle embryonic tissue layer that gives rise to connective tissue (including blood, bone, and cartilage); blood vessels; muscles; the adrenal cortex; and other organs.

messenger RNA (mRNA) A type of RNA that contains a base sequence complementary to a part of the DNA that specifies the synthesis of a particular protein.

meta- *(Gk.)* Change.

metabolic acidosis *(as″ĭ-do′sis)* **and alkalosis** *(al″kă-lo′sis)* Abnormal changes in arterial blood pH due to changes in nonvolatile acid concentration (for example, changes in lactic acid or ketone body concentrations) or to changes in blood bicarbonate concentration.

metabolism *(mĕ-tab′ŏ-liz-em)* All of the chemical reactions in the body. It includes those that result in energy storage (anabolism) and those that result in the liberation of energy (catabolism).

metastasis *(mĕ-tas′tă-sis)* A process whereby cells of a malignant tumor separate from the tumor, travel to a different site, and divide to produce a new tumor.

methemoglobin *(met-he′mŏ-glo″bin)* The abnormal form of hemoglobin in which the iron atoms in heme are oxidized to the ferrous form. Methemoglobin is incapable of bonding with oxygen.

micelle *(mi-sel′)* A colloidal particle formed by the aggregation of numerous molecules.

micro- *(L.)* Small; also, one-millionth.

microvilli *(mi″kro-vil′i)* Tiny fingerlike projections of a cell membrane. They occur on the apical (lumenal) surface of the cells of the small intestine and in the renal tubules.

micturition *(mik″tŭ-rish′un)* Urination.

milliequivalent *(mil″ĭ-e-kwiv′ă-lent)* The millimolar concentration of an ion multiplied by its number of charges.

mineralocorticoid *(min″er-al-o-kor′tĭ-koid)* Any of a class of steroid hormones of the adrenal cortex (corticosteroids) that regulate electrolyte balance.

mitosis *(mi-to′sis)* Cell division in which the two daughter cells receive the same number of chromosomes as the parent cell (both daughters and parent are diploid).

molal *(mo′lal)* Pertaining to the number of moles of solute per kilogram of solvent.

molar *(mo′lar)* Pertaining to the number of moles of solute per liter of solution.

mole *(mōl)* The number of grams of a chemical that is equal to its formula weight (atomic weight for an element or molecular weight for a compound).

mono- *(Gk.)* One, single.

monoamine *(mon″o-am′ēn)* Any of a class of neurotransmitter molecules containing one amino group. Examples are serotonin, dopamine, and norepinephrine.

monoamine oxidase *(mon′o-am′ēn ok′sĭ-dās)* **(MAO)** An enzyme that degrades monoamine neurotransmitters within presynaptic axon endings. Drugs that inhibit the action of this enzyme thus potentiate the pathways that use monoamines as neurotransmitters.

monoclonal *(mon″ŏ-klōn′al)* **antibodies** Identical antibodies derived from a clone of genetically identical plasma cells.

monocyte *(mon′o-sīt)* A mononuclear, nongranular, phagocytic leukocyte that can be transformed into a macrophage.

monomer *(mon′ŏ-mer)* A single molecular unit of a longer, more complex molecule. Monomers are joined together to form dimers, trimers, and polymers. The hydrolysis of polymers eventually yields separate monomers.

mononuclear leukocyte *(mon″o-noo′kle-ar loo′kŏ-sīt)* Any of a category of white blood cells that includes the lymphocytes and monocytes.

mononuclear phagocyte *(fag′ŏ-sīt)* **system** A term used to describe monocytes and tissue macrophages.

monosaccharide *(mon″ŏ-sak′ă-rīd)* The monomer of the more complex carbohydrates. Examples of monomers are glucose, fructose, and galactose. Also called a *simple sugar.*

-morph, morpho- *(Gk.)* Form, shape.

motile *(mo′til)* Capable of self-propelled movement.

motor cortex *(kor′teks)* The precentral gyrus of the frontal lobe of the cerebrum. Axons from this area form the descending pyramidal motor tracts.

motor neuron *(noor′on)* An efferent neuron that conducts action potentials away from the central nervous system to effector organs (muscles and glands). It forms the ventral roots of spinal nerves.

motor unit A lower motor neuron and all of the skeletal muscle fibers stimulated by branches of its axon. Larger motor units (more muscle fibers per neuron) produce more force when the unit is activated, but smaller motor units afford a finer degree of neural control over muscle contraction.

mucous *(myoo′kus)* **membrane** The layers of visceral organs that include the lining epithelium, submucosal connective tissue, and, (in some cases) a thin layer of smooth muscle (the muscularis mucosa).

muscarinic receptors *(mus″kă-rin′ik re-sep′torz)* Receptors for acetylcholine that are stimulated by postganglionic parasympathetic neurons. Their name is derived from the fact that they are also stimulated by the chemical muscarine, derived from a mushroom.

muscle spindle *(mus′el spin′d′l)* A sensory organ within skeletal muscle that is composed of intrafusal fibers. It is sensitive to muscle stretch and provides a length detector within muscles.

myelin *(mi′ĕ-lin)* **sheath** A sheath surrounding axons formed from the cell membrane of Schwann cells in the peripheral nervous system and from oligodendrocytes in the central nervous system.

myocardial infarction (*mi″ŏ-kar′de-al in-fark′shun*) An area of necrotic tissue in the myocardium that is filled in by scar (connective) tissue.

myofibril (*mi″ŏ-fi′bril*) A subunit of striated muscle fiber that consists of successive sarcomeres. Myofibrils run parallel to the long axis of the muscle fiber, and the pattern of their filaments provides the striations characteristic of striated muscle cells.

myogenic (*mi″ŏ-jen′ik*) Originating within muscle cells; this term is used to describe self-excitation by cardiac and smooth muscle cells.

myoglobin (*mi″ŏ-glo′bin*) A molecule composed of globin protein and heme pigment. It is related to hemoglobin but contains only one subunit (instead of the four in hemoglobin). Myoglobin is found in striated muscles, wherein it serves to store oxygen.

myoneural (*mi″ŏ-noor′al*) **junction** A synapse between a motor neuron and the muscle cell that it innervates; also called the *neuromuscular junction.*

myopia (*mi-o′pe-ă*) A condition of the eyes in which light is brought to a focus in front of the retina because the eye is too long; also called *nearsightedness.*

myosin (*mi′ŏ-sin*) The protein that forms the A bands of striated muscle cells. Together with the protein actin, myosin provides the basis for muscle contraction.

myxedema (*mik″sĭ-de′mă*) A type of edema associated with hypothyroidism. It is characterized by accumulation of mucoproteins in tissue fluid.

N

NAD Nicotinamide adenine dinucleotide (*nik″ŏ-tină-mĭd ad′nēn di-noo′kle-ŏ-tīd*) A coenzyme derived from niacin that functions to transport electrons in oxidation-reduction reactions. It helps to transport electrons to the electron-transport chain within mitochondria.

naloxone (*na″l ok-sōn, nă-lok′-sōn*) A drug that antagonizes the effects of morphine and endorphins.

natriuretic (*na″trĭ-yoo-ret′ik*) **hormone** A hormone that increases the urinary excretion of sodium. This hormone has been identified as atrial natriuretic peptide (ANP), produced by the atria of the heart.

necrosis (*nĕ-kro′sis*) Cellular death within tissues and organs as a result of pathological conditions. Necrosis differs histologically from the physiological cell death of apoptosis.

negative feedback A response mechanism that serves to maintain a state of internal constancy, or homeostasis. Effectors are activated by changes in the internal environment, and the inhibitory actions of the effectors serve to counteract these changes and maintain a state of balance.

neoplasm (*ne′ŏ-plazm*) A new, abnormal growth of tissue, as in a tumor.

nephron (*nef′ron*) The functional unit of the kidneys, consisting of a system of renal tubules and a vascular component that includes capillaries of the glomerulus and the peritubular capillaries.

Nernst equation The equation used to calculate the equilibrium membrane potential for given ions when the concentrations of those ions on each side of the membrane are known.

nerve A collection of motor axons and sensory dendrites in the peripheral nervous system.

neurilemma (*noor″ĭ-lem′ă*) The sheath of Schwann and its surrounding basement membrane that encircles nerve fibers in the peripheral nervous system.

neuroglia (*noo-rog′le-ă*) The supporting cells of the central nervous system that aid the functions of neurons. In addition to providing support, they participate in the metabolic and bioelectrical processes of the nervous system, also called *glial cells.*

neurohypophysis (*noor″o-hi-pof′ĭ-sis*) The posterior part of the pituitary gland that is derived from the brain. It releases vasopression (ADH) and oxytocin, both of which are produced in the hypothalamus.

neuron (*noor′on*) A nerve cell, consisting of a cell body that contains the nucleus; short branching processes called dendrites that carry electrical charges to the cell body; and a single fiber, or axon, that conducts nerve impulses away from the cell body.

neuropeptide (*noor″o-pep′tīd*) Any of various polypeptides found in neural tissue that are believed to function as neurotransmitters and neuromodulators. Neuropeptide Y, for example, is the most abundant polypeptide in the brain and has been implicated in a variety of processes, including the stimulation of appetite.

neurotransmitter (*noor″o-trans′mit-er*) A chemical contained in synaptic vesicles in nerve endings that is released into the synaptic cleft, where it causes the production of either excitatory or inhibitory postsynaptic potentials.

neurotrophin (*noor″ŏ-trof′in*) Any of a family of autocrine regulators secreted by neurons and neuroglial cells that promote axon growth and other effects. Nerve growth factor is an example.

neutron (*noo′tron*) An electrically neutral particle that exists together with positively charged protons in the nucleus of atoms.

nexus (*nek′sus*) A bond between members of a group; the type of intercellular connection found in single-unit smooth muscles.

niacin (*ni′-ă-sin*) A water-soluble B vitamin needed for the formation of NAD, which is a coenzyme that participates in the transfer of hydrogen atoms in many of the reactions of cell respiration.

nicotinic receptors (*nik″ŏ-tin′ik re-sep′torz*) Receptors for acetylcholine located in the autonomic ganglia and in neuromuscular junctions. Their name is derived from the fact that they can also be stimulated by nicotine, derived from the tobacco plant.

nidation (*ni-da′shun*) Implantation of the blastocyst in the endometrium of the uterus.

Nissl (*nis′l*) **bodies** Granular-appearing structures in the cell bodies of neurons that have an affinity for basic stain; they correspond to ribonucleoprotein; also called *chromatophilic substances.*

nitric oxide A gas that functions as a neurotransmitter in both the central nervous system and in peripheral autonomic neurons, and as an autocrine and paracrine regulator in many organs. It promotes vasodilation, intestinal relaxation, penile erection, and aids long-term potentiation in the brain.

nociceptor (*no″sĭ-sep′tor*) A receptor for pain that is stimulated by tissue damage.

nodes of Ranvier (*ran′ve-a*) Gaps in the myelin sheath of myelinated axons, located approximately 1 mm apart. Action potentials are produced only at the nodes of Ranvier in myelinated axons.

norepinephrine (*nor″ep-ĭ-nef′rin*) A catecholamine released as a neurotransmitter from postganglionic sympathetic nerve endings and as a hormone (together with epinephrine) by the adrenal medulla; also called *noradrenaline.*

nuclear receptors Receptors that bind to both a regulatory ligand (such as a hormone) and to DNA. The nuclear receptors, when activated by their ligands, regulate genetic expression (RNA synthesis).

nucleolus (*noo-kle′ŏ-lus*) A dark-staining area within a cell nucleus; the site of production of ribosomal RNA.

nucleoplasm (*noo′kle-ŏ-plaz″em*) The protoplasm of a nucleus.

nucleosome (*noo′kle-ŏ-sōm*) A complex of DNA and histone proteins that is believed to constitute an inactive form of DNA. In the electron microscope, the histones look like beads threaded on a string of chromatin.

nucleotide (*noo′kle-ŏ-tīd*) The subunit of DNA and RNA macromolecules. Each nucleotide is composed of a nitrogenous base (adenine, guanine, cytosine, and thymine or uracil); a sugar (deoxyribose or ribose); and a phosphate group.

nucleus (*noo′klē-us*) **brain** An aggregation of neuron cell bodies within the brain. Nuclei within the brain are surrounded by white matter and are located deep to the cerebral cortex.

nucleus, cell The organelle, surrounded by a double saclike membrane called the nuclear envelope (nuclear membrane), that contains the DNA and genetic information of the cell.

nystagmus (*nī-stag′mus*) Involuntary oscillatory movements of the eye.

O

obese (*o-bēs*) Excessively fat.

oligo- (Gk.) Few, small.

oligodendrocyte (*ol″ĭ-go-den′drŏ-sīt*) A type of glial cell that forms myelin sheaths around axons in the central nervous system.

oncogene (*on′kŏ-jēn*) A gene that contributes to cancer. Oncogenes are believed to be abnormal forms of genes that participate in normal cellular regulation.

oncology (*on-kol′ŏ-je*) The study of tumors.

oncotic (*on-kot′ik*) **pressure** The colloid osmotic pressure of solutions produced by proteins. In plasma, it serves to counterbalance the outward filtration of fluid from capillaries caused by hydrostatic pressure.

oo- (Gk.) Pertaining to an egg.

oocyte (*o′ŏ-sīt*) An immature egg cell (ovum). A primary oocyte has not yet completed the first meiotic division; a secondary oocyte has begun the second meiotic division. A secondary oocyte, arrested at metaphase II, is ovulated.

oogenesis (*o″ŏ-jen′ĕ-sis*) The formation of ova in the ovaries.

opsonization (*op″sŏ-nĭ-za′shun*) The process by which antibodies enhance the ability of phagocytic cells to attack bacteria.

optic (*op′tik*) **disc** The area of the retina where axons from ganglion cells gather to form the optic nerve and where blood vessels enter and leave the eye. It corresponds to the blind spot in the visual field caused by the absence of photoreceptors.

organ A structure in the body composed of two or more primary tissues that performs a specific function.

organelle (*or″gă-nel′*) A structure within cells that performs specialized tasks. Organelles include mitochondria, the Golgi apparatus, endoplasmic reticulum, nuclei, and lysosomes. The term is also used for some structures not enclosed by a membrane, such as ribosomes and centrioles.

organ of Corti (*kor′te*) The structure within the cochlea that constitutes the functional unit of hearing. It consists of hair cells and supporting cells on the basilar membrane that help to transduce sound waves into nerve impulses; also called the *spiral organ.*

organic anion transporters A family of transport proteins with a broad range of specificities (described as multispecific or polyspecific), able to transport many drugs and endogenous molecules across the cell membranes of the renal tubules and bile ductules. As a result of this action, the organic anion transporters are responsible for the renal secretion of many compounds (such as antibiotics) into the renal tubules, and for the secretion of many compounds into the bile, for elimination in the urine and feces, respectively.

osmolality (*oz″mŏ-lal′ĭ-te*) A measure of the total concentration of a solution; the number of moles of solute per kilogram of solvent.

osmoreceptor (*oz″mŏ-re-cep′tor*) A sensory neuron that responds to changes in the osmotic pressure of the surrounding fluid.

osmosis (*oz-mo′sis*) The passage of solvent (water) from a more dilute to a more concentrated solution through a membrane that is more permeable to water than to the solute.

osmotic (*oz-mot′ik*) **pressure** A measure of the tendency for a solution to gain water by osmosis when separated by a membrane from pure water. Directly related to the osmolality of the solution, it is the pressure required to just prevent osmosis.

osteo- (Gk.) Pertaining to bone.

osteoblast (*os′te-ŏ-blast*) A bone-forming cell.

osteoclast (*os′te-ŏ-klast*) A cell that resorbs bone by promoting the dissolution of calcium phosphate crystals.

osteocyte (*os′te-ŏ-sīt*) A mature bone cell that has become entrapped within a matrix of bone. This cell remains alive because it is nourished by means of canaliculi within the extracellular material of bone.

osteomalacia (*os″te-o-mă-la′shă*) Softening of bones due to a deficiency of vitamin D and calcium.

osteoporosis (*os″te-o-pŏ-ro′sis*) Demineralization of bone, seen most commonly in postmenopausal women and patients who are inactive or paralyzed. It may be accompanied by pain, loss of stature, and other deformities and fractures.

ovary (*o′vă-re*) The gonad of a female that produces ova and secretes female sex steroids.

ovi- (L.) Pertaining to an egg.

oviduct (*o′vĭ-dukt*) The part of the female reproductive tract that transports ova from the ovaries to the uterus. Also called the *uterine,* or *fallopian, tube.*

ovulation (*ov-yŭ-la′shun*) The extrusion of a secondary oocyte from the ovary.

oxidation-reduction (*ok″sĭ-da′shun-re-duk′shun*) The transfer of electrons or hydrogen atoms from one atom or molecule to another. The atom or molecule that loses the electrons or hydrogens is oxidized; the atom or molecule that gains the electrons or hydrogens is reduced.

oxidative phosphorylation (*ok″sĭ-da′tiv fos″for-ĭ-la′shun*) The formation of ATP by using energy derived from electron transport to oxygen. It occurs in the mitochondria.

oxidizing (*ok′sĭ-dizing*) **agent** An atom that accepts electrons in an oxidation-reduction reaction.

oxygen (*ok′sĭ-jen*) **debt** The extra amount of oxygen required by the body after exercise to metabolize lactic acid and to supply the higher metabolic rate of muscles warmed during exercise.

oxyhemoglobin (*ok″se-he″mŏ-glo′bin*) A compound formed by the bonding of molecular oxygen with hemoglobin.

oxyhemoglobin saturation The ratio, expressed as a percentage, of the amount of oxyhemoglobin compared to the total amount of hemoglobin in blood.

oxidative stress The damage to lipids, proteins, and DNA in the body produced by the excessive production of free radicals. Oxidative stress is believed to contribute to aging and a variety of diseases.

oxytocin (*ok″sĭ-to′sin*) One of the two hormones produced in the hypothalamus and released from the posterior pituitary (the other being vasopressin). Oxytocin stimulates the contraction of uterine smooth muscles and promotes milk letdown in females.

P

pacemaker The group of cells that has the fastest spontaneous rate of depolarization and contraction in a mass of electrically coupled cells; in the heart, this is the sinoatrial, or SA, node.

pacesetter potentials Changes in membrane potential produced spontaneously by pacemaker cells of single-unit smooth muscles.

pacinian corpuscle (*pă-sin′e-an kor′pus′l*) A cutaneous sensory receptor sensitive to pressure. It is characterized by an onionlike layering of cells around a central sensory dendrite.

PAH Para-aminohippuric (*par′ă-ă-me′no-hi-pyoor″ik*) acid A substance used to measure total renal plasma flow because its clearance rate is equal to the total rate of plasma flow to the kidneys. PAH is filtered and secreted by the renal nephrons but not reabsorbed.

pancreatic (*pan″kre-at′ik*) **islets** See islets of Langerhans.

pancreatic juice The secretions of the pancreas that are transported by the pancreatic duct to the duodenum. Pancreatic juice contains bicarbonate and the digestive enzymes trypsin, lipase, and amylase.

paracrine (*par′ă-krin*) **regulator** A regulatory molecule produced within one tissue that acts on a different tissue of the same organ. For

example, the endothelium of blood vessels secretes a number of paracrine regulators that act on the smooth muscle layer of the vessels to cause vasoconstriction or vasodilation.

parasympathetic (*par″ă-sim″pă-thet′ik*) Pertaining to the craniosacral division of the autonomic nervous system.

parathyroid (*par″ă-thi′roid*) **hormone (PTH)** A polypeptide hormone secreted by the parathyroid glands. PTH acts to raise the blood Ca^{2+} levels primarily by stimulating resorption of bone.

Parkinson's disease A tremor of the resting muscles and other symptoms caused by inadequate dopamine-producing neurons in the basal nuclei of the cerebrum. Also called *paralysis agitans.*

parturition (*par″tyoo-rish′un*) The process of giving birth; childbirth.

passive immunity Specific immunity granted by the administration of antibodies made by another organism.

Pasteur effect A decrease in the rate of glucose utilization and lactic acid production in tissues or organisms by their exposure to oxygen.

pathogen (*path′ŏ-jen*) Any disease-producing microorganism or substance.

pepsin (*pep′sin*) The protein-digesting enzyme secreted in gastric juice.

peptic ulcer (*pep′tik ul′ser*) An injury to the mucosa of the esophagus, stomach, or small intestine caused by the breakdown of gastric barriers to self-digestion or by excessive amounts of gastric acid.

perfusion (*per′fyoo″zhun*) The flow of blood through an organ.

peri- (Gk.) Around, surrounding.

perilymph (*per′ĭ-limf*) The fluid that fills the space between the membranous and bony labyrinths of the inner ear.

perimysium (*per″ĭ-mis′e-um*) The connective tissue surrounding a fascicle of skeletal muscle fibers.

periosteum (*per″e-os′te-um*) Connective tissue covering bones. It contains osteoblasts, and is therefore capable of forming new bone.

peripheral resistance The resistance to blood flow through the arterial system. Peripheral resistance is largely a function of the radius of small arteries and arterioles. The resistance to blood flow is proportional to the fourth power of the radius of the vessel.

peristalsis (*per″ĭ-stal′sis*) Waves of smooth muscle contraction in smooth muscles of the tubular digestive tract. It involves circular and longitudinal muscle fibers at successive locations along the tract and serves to propel the contents of the tract in one direction.

permissive effect The phenomenon in which the presence of one hormone "permits" the full exertion of the effects of another hormone. This may be due to promotion of the synthesis of the active form of the second hormone, or it may be due to an increase in the sensitivity of the target tissue to the effects of the second hormone.

pH The symbol (short for potential of hydrogen) used to describe the hydrogen ion (H^+) concentration of a solution. The pH scale in common use ranges from 0 to 14. Solutions with a pH of 7 are neutral; those with a pH lower than 7 are acidic; and those with a higher pH are basic.

phagocytosis (*fag″ŏ-si-to′sis*) Cellular eating; the ability of some cells (such as white blood

**cells) to engulf large particles (such as bacteria) and digest these particles by merging the food vacuole in which they are contained with a lysosome containing digestive enzymes.

phenylalanine (*fen″il-al′ă-nēn*) An amino acid that also serves as the precursor for L-dopa, dopamine, norepinephrine, and epinephrine.

phenylketonuria (*fen″il-kēt″n-oor′e-ă*) (**PKU**) An inborn error of metabolism that results in the inability to convert the amino acid phenylalanine into tyrosine. This defect can cause central nervous system damage if the child is not placed on a diet low in phenylalanine.

phonocardiogram (*fo″nŏ-kar′de-ŏ-gram*) A visual display of the heart sounds.

phosphatidylcholine (*fos-fat″i-dil-ko′len*) The chemical name for the molecule also called lecithin.

phosphodiesterase (*fos″fo-di-es′ter-ăs*) An enzyme that cleaves cyclic AMP into inactive products, thus inhibiting the action of cyclic AMP as a second messenger.

phospholipid (*fos″fo-lip′id*) A lipid containing a phosphate group. Phospholipid molecules (such as lecithin) are polar on one end and nonpolar on the other end. They make up a large part of the cell membrane and function in the lung alveoli as surfactants.

phosphorylation (*fos″for-ĭ-la′shun*) The addition of an inorganic phosphate group to an organic molecule; for example, the addition of a phosphate group to ADP to make ATP or the addition of a phosphate group to specific proteins as a result of the action of protein kinase enzymes.

photoreceptors (*fo″to-re-sep′torz*) Sensory cells (rods and cones) that respond electrically to light. They are located in the retina of the eyes.

pia mater (*pi′ă ma′ter*) The innermost of the connective tissue meninges that envelops the brain and spinal cord.

pineal (*pin′ e-al*) **gland** A gland within the brain that secretes the hormone melatonin. It is affected by sensory input from the photoreceptors of the eyes.

pinocytosis (*pin″ŏ-si-to′sis*) Cell drinking; invagination of the cell membrane to form narrow channels that pinch off into vacuoles. This permits cellular intake of extracellular fluid and dissolved molecules.

pituitary (*pĭ-too′ĭ-ter-e*) **gland** Also called the *hypophysis*. A small endocrine gland joined to the hypothalamus at the base of the brain. The pituitary gland is functionally divided into anterior and posterior portions. The anterior pituitary secretes ACTH, TSH, FSH, LH, growth hormone, and prolactin. The posterior pituitary releases oxytocin and antidiuretic hormone (ADH), which are produced by the hypothalamus.

plasma (*plaz′mă*) The fluid portion of the blood. Unlike serum (which lacks fibrinogen), plasma is capable of forming insoluble fibrin threads when in contact with test tubes.

plasma cells Cells derived from B lymphocytes that produce and secrete large amounts of antibodies. They are responsible for humoral immunity.

plasmalemma (*plaz″mă-lem′ă*) The cell membrane; an alternate term for the selectively permeable membrane that encloses the cytoplasm of a cell.

platelet (*plăt′let*) A disc-shaped structure, 2 to 4 micrometers in diameter, derived from bone marrow cells called megakaryocytes. Platelets circulate in the blood and participate (together with fibrin) in forming blood clots.

pluripotent (*ploo-rip′ŏ-tent*) A term used to describe the ability of early embryonic cells to specialize to produce all tissues except the trophoblast cells of the placenta.

pneumotaxic (*noo″mŏ-tak′sik*) **center** A neural center in the pons that rhythmically inhibits inspiration in a manner independent of sensory input.

pneumothorax (*noo″mo-thor′aks*) An abnormal condition in which air enters the intrapleural space, either through an open chest wound or from a tear in the lungs. This can lead to the collapse of a lung (atelectasis).

PNS The peripheral nervous system, including nerves and ganglia.

-pod, -podium (Gk.) Foot, leg, extension.

Poiseuille's (*pwă-zū′yez*) **law** The statement that the rate of blood flow through a vessel is directly proportional to the pressure difference between the two ends of the vessel and inversely proportional to the length of the vessel, the viscosity of the blood, and the fourth power of the radius of the vessel.

polar body A small daughter cell formed by meiosis that degenerates in the process of oocyte production.

polar molecule A molecule in which the shared electrons are not evenly distributed, so that one side of the molecule is negatively (or positively) charged in comparison with the other side. Polar molecules are soluble in polar solvents such as water.

poly- (Gk.) Many.

polycythemia (*pol″e-si-the′-me-ă*) An abnormally high red blood cell count.

polydipsia (*pol″e-dip′se-ă*) Excessive thirst.

polymer (*pol′ĭ-mer*) A large molecule formed by the combination of smaller subunits, or monomers.

polymorphonuclear (*pol″e-mor″fŏ-noo′kle-ar*) **leukocyte** A granular leukocyte containing a nucleus with a number of lobes connected by thin cytoplasmic strands. This term includes neutrophils, eosinophils, and basophils.

polypeptide (*pol″e-pep′tīd*) A chain of amino acids connected by covalent bonds called peptide bonds. A very large polypeptide is called a protein.

polyphagia (*pol′e-fa″je-ă*) Excessive eating.

polysaccharide (*pol″e-sak′ă-rīd*) A carbohydrate formed by covalent bonding of numerous monosaccharides. Examples are glycogen and starch.

polyuria (*pol″e-yoor′e-ă*) Excretion of an excessively large volume of urine in a given period.

portal (*por′tal*) **system** A system of vessels consisting of two capillary beds in series, where blood from the first is drained by veins into a second capillary bed, which in turn is drained by veins that return blood to the heart. The two major portal systems in the body are the hepatic portal system and the hypothalamo-hypophyseal portal system.

positive feedback A response mechanism that results in the amplification of an initial change. Positive feedback results in avalanche-like effects, as occur in the formation of a blood clot or in the production of the LH surge by the stimulatory effect of estrogen.

posterior (*pos-tēr′e-or*) At or toward the back of an organism, organ, or part; the dorsal surface.

posterior pituitary *See* neurohypophysis.

postsynaptic (*pŏst″sĭ-nap′tik*) **inhibition** The inhibition of a postsynaptic neuron by axon endings that release a neurotransmitter that induces hyperpolarization (inhibitory postsynaptic potentials).

potential (*pŏ-ten′shal*) **difference** In biology, the difference in charge between two solutions separated by a membrane. The potential difference is measured in voltage.

prehormone (*pre-hor′mōn*) An inactive form of a hormone secreted by an endocrine gland. The prehormone is converted within its target cells to the active form of the hormone.

presynaptic (*pre″sĭ-nap′tik*) **inhibition** Neural inhibition in which axoaxonic synapses inhibit the release of neurotransmitter chemicals from the presynaptic axon.

pro- (Gk.) Before, in front of, forward.

process (*pros′es, pro′ses*) **cell** Any thin cytoplasmic extension of a cell, such as the dendrites and axon of a neuron.

progesterone (*pro-jes′tĕ-rōn*) A steroid hormone secreted by the corpus luteum of the ovaries and by the placenta. Secretion of progesterone during the luteal phase of the menstrual cycle promotes the final maturation of the endometrium.

prohormone (*pro-hor′mōn*) The precursor of a polypeptide hormone that is larger and less active than the hormone. The prohormone is produced within the cells of an endocrine gland and is normally converted into the shorter, active hormone prior to secretion.

prolactin (*pro-lak′tin*) A hormone secreted by the anterior pituitary that stimulates lactation (acting together with other hormones) in the postpartum female. It may also participate (along with the gonadotropins) in regulating gonadal function in some mammals.

prophylaxis (*pro″fĭ-lak′sis*) Prevention or protection.

proprioceptor (*pro″pre-o-sep″tor*) A sensory receptor that provides information about body position and movement. Examples are receptors in muscles, tendons, and joints and in the semicircular canals of the inner ear.

prostaglandin (*pros″tă-glan′din*) Any of a family of fatty acids that serve numerous autocrine regulatory functions, including the stimulation of uterine contractions and of gastric acid secretion and the promotion of inflammation.

protein (*pro′te-in*) The class of organic molecules composed of large polypeptides, in which over a hundred amino acids are bonded together by peptide bonds.

protein kinase (*ki′nās*) The enzyme activated by cyclic AMP that catalyzes the phosphorylation of specific proteins (enzymes). Such phosphorylation may activate or inactivate enzymes.

proto- (Gk.) First, original.

proton (*pro′ton*) A unit of positive charge in the nucleus of atoms.

protoplasm (*pro′tŏ-plaz″em*)) A general term for the colloidal complex of protein that includes cytoplasm and nucleoplasm.

pseudo- (Gk.) False.

pseudohermaphrodite (*soo″dŏ-her-maf′rŏ-dīt*) An individual who has the gonads of one sex only, but some of the body features of the opposite sex. (A true hermaphrodite has both ovarian and testicular tissue.)

pseudopod (*soo″dŏ-pod*) A footlike extension of the cytoplasm that enables some cells (with amoeboid motion) to move across a substrate. Pseudopods also are used to surround food particles in the process of phagocytosis.

puberty (*pyoo'ber-te*) The period in an individual's life span when secondary sexual characteristics and fertility develop.

pulmonary (*pul'mŏ-ner"e*) **circulation** The part of the vascular system that includes the pulmonary arteries and pulmonary veins. It transports blood from the right ventricle of the heart through the lungs, and then back to the left atrium of the heart.

pupil The opening at the center of the iris of the eye.

Purkinje (*pur-kin'je*) **fibers** Specialized conducting tissue in the ventricles of the heart that carry impulses from the bundle of His to the myocardium of the ventricles.

pyramidal (*pĭ-ram'mŏ-dal*) **tracts** Motor tracts that descend without synaptic interruption from the cerebrum to the spinal cord, where they synapse either directly or indirectly (via spinal interneurons) with the lower motor neurons of the spinal cord; also called *corticospinal tracts*.

pyrogen (*pi'rŏ-jen*) A fever-producing substance.

Q

QRS complex The principal deflection of an electrocardiogram, produced by depolarization of the ventricles.

R

reabsorption (*re"ab-sorp'shun*) The transport of a substance from the lumen of the renal nephron into the peritubular capillaries.

receptive field An area of the body that, when stimulated by a sensory stimulus, activates a particular sensory receptor.

reciprocal innervation (*rĭ-sip'rŏ-kal in"er-va'shun*) The process whereby the motor neurons to an antagonistic muscle are inhibited when the motor neurons to an agonist muscle are stimulated. In this way, for example, the extensor muscle of the elbow joint is inhibited when the flexor muscles of this joint are stimulated to contract.

recruitment (*rĭ-kroot'ment*) In terms of muscle contraction, the successive stimulation of more and larger motor units in order to produce increasing strengths of muscle contraction.

reduced hemoglobin Hemoglobin with iron in the reduced ferrous state. It is able to bond with oxygen but is not combined with oxygen. Also called *deoxyhemoglobin*.

reducing agent An electron donor in a coupled oxidation-reduction reaction.

refraction (*re-frak'shun*) The bending of light rays when light passes from a medium of one density to a medium of another density. Refraction of light by the cornea and lens acts to focus the image on the retina of the eye.

refractory (*re-frak'tŏ-re*) **period** The period of time during which a region of axon or muscle cell membrane cannot be stimulated to produce an action potential (absolute refractory period), or when it can be stimulated only by a very strong stimulus (relative refractory period).

releasing hormones Polypeptide hormones secreted by neurons in the hypothalamus that travel in the hypothalamo-hypophyseal portal system to the anterior pituitary and stimulate the anterior pituitary to secrete specific hormones.

REM sleep The stage of sleep in which dreaming occurs. It is associated with rapid eye movements (REMs). REM sleep occurs three to four times each night and lasts from a few minutes to over an hour.

renal (*re'nal*) Pertaining to the kidneys.

renal plasma clearance The volume of plasma from which a particular solute is cleared each minute by the excretion of that solute in the urine. If there is no reabsorption or secretion of that solute by the nephron tubules, the renal plasma clearance is equal to the glomerular filtration rate.

renal pyramid (*pĭ'ră-mid*) One of a number of cone-shaped tissue masses that compose the renal medulla.

renin (*re'nin*) An enzyme secreted into the blood by the juxtaglomerular apparatus of the kidneys. Renin catalyzes the conversion of angiotensinogen into angiotensin II.

rennin (*ren'in*) A digestive enzyme secreted in the gastric juice of infants that catalyzes the digestion of the milk protein casein.

repolarization (*re-po"lar-ĭ-za'shun*) The reestablishment of the resting membrane potential after depolarization has occurred.

resorption (*re-sorp'shun*) **bone** The dissolution of the calcium phosphate crystals of bone by the action of osteoclasts.

respiratory acidosis (*rĭ-spīr'ă-tor-e as"ĭ-do'sis*) A lowering of the blood pH to below 7.35 as a result of the accumulation of CO_2 caused by hypoventilation.

respiratory alkalosis (*al"kă-lo'sis*) A rise in blood pH to above 7.45 as a result of the excessive elimination of blood CO_2 caused by hyperventilation.

respiratory distress syndrome A lung disease of the newborn, most frequently occurring in premature infants, that is caused by abnormally high alveolar surface tension as a result of a deficiency in lung surfactant; also called *hyaline membrane disease*.

respiratory zone The region of the lungs in which gas exchange between the inspired air and pulmonary blood occurs. It includes the respiratory bronchioles, in which individual alveoli are found, and the terminal alveoli.

resting potential The potential difference across a cell membrane when the cell is in an unstimulated state. The resting potential is always negatively charged on the inside of the membrane compared to the outside.

reticular (*rĕ-tik'yŭ-lar*) **activating system (RAS)** A complex network of nuclei and fiber tracts within the brain stem that produces nonspecific arousal of the cerebrum to incoming sensory information. The RAS thus maintains a state of alert consciousness and must be depressed during sleep.

retina (*ret'ĭ-nă*) The layer of the eye that contains neurons and photoreceptors (rods and cones).

retinoic (*ret"ĭ-no'ik*) **acid** The active form of vitamin A that binds to nuclear receptor proteins and directly produces the effects of vitamin A.

rhodopsin (*ro-dop'sin*) Visual purple. A pigment in rod cells that undergoes a photochemical dissociation in response to light and, in so doing, stimulates electrical activity in the photoreceptors.

riboflavin (*ri'bo-fla"vin*) Vitamin B_2. Riboflavin is a water-soluble vitamin that is used to form the coenzyme FAD, which participates in the transfer of hydrogen atoms.

ribosome (*ri'bo-sōm*) A cytoplasmic organelle composed of protein and ribosomal RNA that is responsible for the translation of messenger RNA and protein synthesis.

ribozymes (*ri'bo-zims*) RNA molecules that have catalytic ability.

rickets (*rik'ets*) A condition caused by a deficiency of vitamin D and associated with interference of the normal ossification of bone.

rigor mortis (*rig'or mor'tis*) The stiffening of a dead body due to the depletion of ATP and the production of rigor complexes between actin and myosin in muscles.

RNA Ribonucleic (*ri"bo-noo-kle'ik*) acid. A nucleic acid consisting of the nitrogenous bases adenine, guanine, cytosine, and uracil; the sugar ribose; and phosphate groups. There are three types of RNA found in cytoplasm: messenger RNA (mRNA), transfer RNA (tRNA), and ribosomal RNA (rRNA).

rods One of the two categories of photoreceptors (the other being cones) in the retina of the eye. Rods are responsible for black-and-white vision under low illumination.

S

saccadic (*să-kad'ik*) **eye movements** Very rapid eye movements that occur constantly and that change the focus on the retina from one point to another.

saltatory (*sal'tă-tor-e*) **conduction** The rapid passage of action potentials from one node of Ranvier to another in myelinated axons.

sarcolemma (*sar"cŏ-lem'ă*) The cell membrane of striated muscle cells.

sarcomere (*sar'kŏ-mēr*) The structural subunit of a myofibril in a striated muscle; equal to the distance between two successive Z lines.

sarcoplasm (*sar'kŏ-plaz"em*) The cytoplasm of striated muscle cells.

sarcoplasmic reticulum (*sar"kŏ-plaz'mik rĕ-tik'yŭ-lum*) The smooth or agranular endoplasmic reticulum of striated muscle cells. It surrounds each myofibril and serves to store Ca^{2+} when the muscle is at rest.

Schwann (*shvan*) **cell** A supporting cell of the peripheral nervous system that forms sheaths around peripheral nerve fibers. Schwann cells also direct regeneration of peripheral nerve fibers to their target cells.

sclera (*skler'ă*) The tough white outer coat of the eyeball, continuous anteriorly with the clear cornea.

second messenger A molecule or ion whose concentration within a target cell is increased by the action of a regulator molecule (e.g., a hormone or neurotransmitter) so as to stimulate the metabolism of that target cell in a way characteristic of the actions of the regulator molecule—that is, in a way that mediates the intracellular effects of the regulator molecule.

secretin (*sĕ-kre'tin*) A polypeptide hormone secreted by the small intestine in response to acidity of the intestinal lumen. Along with cholecystokinin, secretin stimulates the secretion of pancreatic juice into the small intestine.

secretion (*sĕ-kre'shun*) **renal** The transport of a substance from the blood through the wall of the nephron tubule into the urine.

semen (*se'men*) The fluid ejaculated by a male, containing sperm and additives from the prostate and seminal vesicles.

semicircular canals Three canals of the bony labyrinth that contain endolymph, which is continuous with the endolymph of the membranous labyrinth of the cochlea. The semicircular canals provide a sense of equilibrium.

semilunar (*sem″e-loo'nar*) **valves** The valve flaps of the aorta and pulmonary artery at their juncture with the ventricles.

seminal vesicles (*sem'ĭ-nal ves'ĭ-k'lz*) The paired organs located on the posterior border of the urinary bladder that empty their contents into the ejaculatory duct and thus contribute to the semen.

seminiferous tubules (*sem″ĭ-nif'er-us too'byoolz*) The tubules within the testes that produce spermatozoa by meiotic division of their germinal epithelium.

selectively permeable membrane A membrane with pores of a size that permit the passage of solvent and some solute molecules, while restricting the passage of other solute molecules.

sensory neuron (*noor'on*) An afferent neuron that conducts impulses from peripheral sensory organs into the central nervous system.

serosa (*sĭ-ro'să*) An outer epithelial membrane that covers the surface of a visceral organ.

serotonin (*ser″ŏ-to'nin*) monoamine neurotransmitter, chemically known as 5-hydroxytryptamine, derived from the amino acid L-tryptophan. Serotonin released at synapses in the brain have been associated with the regulation of mood and behavior, appetite, and cerebral circulation.

Sertoli (*ser-to'le*) **cells** Nongerminal supporting cells in the seminiferous tubules. Sertoli cells envelop spermatids and appear to participate in the transformation of spermatids into spermatozoa; also called *sustentacular cells.*

serum (*ser'um*) The fluid squeezed out of a clot as it retracts; supernatant when a sample of blood clots in a test tube and is centrifuged. Serum is plasma from which fibrinogen and other clotting proteins have been removed as a result of clotting.

sex chromosomes The X and Y chromosomes. These are the unequal pairs of chromosomes involved in sex determination (which depends on the presence or absence of a Y chromosome). Females lack a Y chromosome and normally have the genotype XX; males have a Y chromosome and normally have the genotype XY.

shock As it relates to the cardiovascular system, a rapid, uncontrolled fall in blood pressure, which in some cases becomes irreversible and leads to death.

sickle-cell anemia A hereditary autosomal recessive trait that occurs primarily in people of African ancestry, in whom it evolved apparently as a protection (in the carrier state) against malaria. In the homozygous state, hemoglobin S is made instead of hemoglobin A, which leads to the characteristic sickling of red blood cells, hemolytic anemia, and organ damage.

sinoatrial (*si″no-a'tre-al*) **node** A mass of specialized cardiac tissue in the wall of the right atrium that initiates the cardiac cycle; the SA node, also called the *pacemaker.*

sinus (*si'nus*) A cavity.

sinusoid (*si'nŭ-soid*) A modified capillary with a relatively large diameter that connects the arterioles and venules in the liver, bone marrow, lymphoid tissues, and some endocrine organs. In the liver, sinusoids are partially lined by phagocytic cells of the reticuloendothelial system.

skeletal muscle pump A term used with reference to the effect of skeletal muscle contraction on the flow of blood in veins. As the muscles contract, they squeeze the veins, and in this way help move the blood toward the heart.

sleep apnea A temporary cessation of breathing during sleep, usually lasting for several seconds.

sliding filament theory The theory that the thick and thin filaments of a myofibril slide past each other during muscle contraction, decreasing the length of the sarcomeres but maintaining their own initial length.

slow waves Pacemaker depolarizations in the intestine produced by pacemaker cells, the interstitial cells of Cajal; these produce action potentials and resulting smooth muscle contractions.

smooth muscle A specialized type of nonstriated muscle tissue composed of fusiform single-nucleated fibers. It contracts in an involuntary, rhythmic fashion in the walls of visceral organs.

sodium/potassium (*so'de-um/pŏ-tas'e-um*) **pump** An active transport carrier with ATPase enzymatic activity that acts to accumulate K⁺ within cells and extrude Na⁺ from cells, thus maintaining gradients for these ions across the cell membrane.

soma-, somato-, -some (Gk.) Body, unit.

somatesthetic (*so″mat-es-thet'ek*) **sensations** Sensations arising from cutaneous, muscle, tendon, and joint receptors. These sensations project to the postcentral gyrus of the cerebral cortex.

somatic (*so-mat'ik*) **motor neuron** A motor neuron in the spinal cord that innervates skeletal muscles. Somatic motor neurons are categorized as alpha and gamma motoneurons.

somatomammotropic (*so″mă-tŏ-mam″ŏ-trop'ik*) **hormone** A hormone secreted by the placenta that has actions similar to the pituitary hormones growth hormone and prolactin; also called *chorionic somatomammotropin (hCS).*

somatomedin (*so″mă-tŏ-med'n*) Any of a group of small polypeptides that are believed to be produced in the liver in response to growth hormone stimulation and to mediate the actions of growth hormone on the skeleton and other tissues.

somatostatin (*so″mă-tŏ-stat'n*) A polypeptide produced in the hypothalamus that acts to inhibit the secretion of growth hormone from the anterior pituitary. Somatostatin is also produced in the islets of Langerhans of the pancreas, but its function there has not been established.

somatotropic (*so″mă-tŏ-trop'ik*) **hormone** Growth hormone. An anabolic hormone secreted by the anterior pituitary that stimulates skeletal growth and protein synthesis in many organs.

sounds of Korotkoff (*kŏ-rot'kof*) The sounds heard when blood pressure measurements are taken. These sounds are produced by the turbulent flow of blood through an artery that has been partially constricted by a pressure cuff.

spastic paralysis (*spas'tik pă-ral'ĭ-sis*) Paralysis in which the muscles have such a high tone that they remain in a state of contracture. This may be caused by inability to degrade ACh released at the neuromuscular junction (as caused by certain drugs) or by damage to the spinal cord.

spermatid (*sper'mă-tid*) Any of the four haploid cells formed by meiosis in the seminiferous tubules that mature to become spermatozoa without further division.

spermatocyte (*sper-mat'ŏ-sīt*) A diploid cell of the seminiferous tubules in the testes that divides by meiosis to produce spermatids.

spermatogenesis (*sper″mă-to-jen'ĭ-sis*) The formation of spermatozoa, including meiosis and maturational processes in the seminiferous tubules.

spermatozoon (*sper'mă-to-zo'on*) pl., *spermatozoa* or, loosely, *sperm.* A mature sperm cell, formed from a spermatid.

spermiogenesis (*sper″me-ŏ-jen'ĭ-sis*) The maturational changes that transform spermatids into spermatozoa.

sphygmo- (Gk.) The pulse.

sphygmomanometer (*sfig″mo-mă-nom'ĭ-ter*) A manometer (pressure transducer) used to measure the blood pressure.

spindle fibers Filaments that extend from the poles of a cell to its equator and attach to chromosomes during the metaphase stage of cell division. Contraction of the spindle fibers pulls the chromosomes to opposite poles of the cell.

spironolactone (*spi-rŏ'no-lak'tōn*) A diuretic drug that acts as an aldosterone antagonist.

Starling forces The hydrostatic pressures and the colloid osmotic pressures of the blood and tissue fluid. The balance of these pressures determines the net movement of fluid out of or into blood capillaries.

stem cells Cells that are relatively undifferentiated (unspecialized) and able to divide and produce different specialized cells.

steroid (*ster'oid*) A lipid derived from cholesterol that has three six-sided carbon rings and one five-sided carbon ring. These form the steroid hormones of the adrenal cortex and gonads.

stretch reflex The monosynaptic reflex whereby stretching a muscle results in a reflex contraction. The knee-jerk reflex is an example of a stretch reflex.

striated (*stri'ăt-ed*) **muscle** Skeletal and cardiac muscle, the cells of which exhibit cross banding, or striations, because of the arrangement of thin and thick filaments.

stroke volume The amount of blood ejected from each ventricle at each heartbeat.

sub- (L.) Under, below.

substrate (*sub'strāt*) In enzymatic reactions, the molecules that combine with the active sites of an enzyme and that are converted to products by catalysis of the enzyme.

sulcus (*sul'kus*) A groove or furrow; a depression in the cerebrum that separates the folds, or gyri, of the cerebral cortex.

summation (*sŭ-ma'shun*) In neural physiology, the additive effects of graded synaptic potentials. In muscle physiology, the additive effects of contractions of different muscle fibers.

super-, supra- (L.) Above, over.

suppressor T cells A subpopulation of T lymphocytes that acts to inhibit the production of antibodies against specific antigens by B lymphocytes.

suprachiasmatic (*soo″pră-ki″az-mot′ik*) **nucleus(SCN)** The primary center for the regulation of circadian rhythms. Located in the hypothalamus, the SCN is believed to regulate circadian rhythms by means of its stimulation of melatonin secretion from the pineal gland.

surfactant (*sur-fak′tant*) In the lungs, a mixture of phospholipids and proteins produced by alveolar cells that reduces the surface tension of the alveoli and contributes to the elastic properties of the lungs.

sym-, syn- (Gk.) With, together.

symport (*sim′port*) A form of secondary active transport (coupled transport) in which a molecule or ion is moved together with, and in the same direction as, Na^+ ions; that is, into the cell.

synapse (*sin′aps*) The junction across which a nerve impulse is transmitted from an axon terminal to a neuron, a muscle cell, or a gland cell either directly or indirectly (via the release of chemical neurotransmitters).

synapses en passant The type of synapses formed by autonomic neurons with their target cells. Neurotransmitters are released into the extracellular fluid from a number of regions of the autonomic axons as they pass through the target tissue.

synapsin (*sĭ-nap′sin*) A protein within the membrane of the synaptic vesicles of axons. When activated by the arrival of action potentials, synapsins aid the fusion of the synaptic vesicles with the cell membrane so that the vesicles may undergo exocytosis and release their content of neurotransmitters.

synaptic plasticity (*sĭ-nap′tik plas-tis′ĭ-te*) The ability of synapses to change at a cellular or molecular level. At a cellular level, plasticity refers to the ability to form new synaptic associations. At a molecular level, plasticity refers to the ability of a presynaptic axon to release more than one type of neurotransmitter.

syncytium (*sin-sish′e-um*) The merging of cells in a tissue into a single functional unit. Because the atria and ventricles of the heart have gap junctions between their cells, these myocardia behave as syncytia.

synergistic (*sin″er-jis′tik*) Pertaining to regulatory processes or molecules (such as hormones) that have complementary or additive effects.

systemic (*sis-tem′ik*) **circulation** The circulation that carries oxygenated blood from the left ventricle via arteries to the tissue cells and that carries blood depleted of oxygen via veins to the right atrium; the general circulation, as compared to the pulmonary circulation.

systole (*sis′-tŏ-le*) The phase of contraction in the cardiac cycle. Used alone, this term refers to contraction of the ventricles; the term *atrial systole* refers to contraction of the atria.

T

tachycardia (*tak″ĭ-kar′de-ă*) An excessively rapid heart rate, usually applied to rates in excess of 100 beats per minute (in contrast to bradycardia, in which the heart rate is very slow—below 60 beats per minute).

target organ The organ that is specifically affected by the action of a hormone or other regulatory process.

T cell A type of lymphocyte that provides cell-mediated immunity (in contrast to B lymphocytes, which provide humoral immunity through the secretion of antibodies). There are three subpopulations of T cells: cytotoxic (killer), helper, and suppressor.

telo- (Gk.) An end, complete, final.

telomere (*tel′ŏ-mēr*) A DNA sequence at the end of a chromosome that is not copied by DNA polymerase during DNA replication. This inability to copy telomeres may contribute to cell aging and death. Germinal cells (that produce gametes) and cancer cells have an additional enzyme, telomerase, which copies the telomeres.

telophase (*tel′ŏ-fāz*) The last step of mitosis and the last step of the second division of meiosis.

tendon (*ten′dun*) The dense regular connective tissue that attaches a muscle to the bones of its origin and insertion.

testes (*tes′tēz*) sing; *testis*. Male gonads. Testes are also known as *testicles*.

testis-determining factor The product of a gene located on the short arm of the Y chromosome that causes the indeterminate embryonic gonads to develop into testes.

testosterone (*tes-tos′tĕ-rōn*) The major androgenic steroid secreted by the Leydig cells of the testes after puberty.

tetanus (*tet′n-us*) In physiology, a term used to denote a smooth, sustained contraction of a muscle, as opposed to muscle twitching.

tetraiodothyronine (*tet″rā-i″ŏ-dŏ-thi′ro-nēn*) (T_4) A hormone containing four iodine atoms; also known as *thyroxine*.

thalassemia (*thal″ă-se′me-ă*) Any of a group of hemolytic anemias caused by the hereditary inability to produce either the alpha or beta chain of hemoglobin. It is found primarily among Mediterranean people.

theophylline (*the-of′ĭ-lin*) A drug found in certain tea leaves that promotes dilation of the bronchioles by increasing the intracellular concentration of cyclic AMP (cAMP) in the smooth muscle cells. This effect is due to inhibition of the enzyme phosphodiesterase, which breaks down cAMP.

thermiogenesis (*ther-me-o-jen′ĭ-sis*) The production of heat by the body through mechanisms such as increased metabolic rate.

thorax (*thor′aks*) The part of the body cavity above the diaphragm; the chest.

threshold The minimum stimulus that just produces a response.

thrombin (*throm′bin*) A protein formed in blood plasma during clotting that enzymatically converts the soluble protein fibrinogen into insoluble fibrin.

thrombocyte (*throm′bŏ-sīt*) A blood platelet; a disc-shaped structure in blood that participates in clot formation.

thrombopoietin (*throm″bo-poi-e′tin*) A cytokine that stimulates the production of thrombocytes (blood platelets) from megakaryocytes in the bone marrow.

thrombosis (*throm-bo′sis*) The development or presence of a thrombus.

thrombus (*throm′bus*) A blood clot produced by the formation of fibrin threads around a platelet plug.

thymus (*thi′mus*) A lymphoid organ located in the superior portion of the anterior mediastinum. It processes T lymphocytes and secretes hormones that regulate the immune system.

thyroglobulin (*thi-ro-glob′yŭ-lin*) An iodine-containing protein in the colloid of the thyroid follicles that serves as a precursor for the thyroid hormones.

thyroxine (*thi-rok′sin*) Also called *tetraiodothyronine*, or T_4. The major hormone secreted by the thyroid gland. It regulates the basal metabolic rate and stimulates protein synthesis in many organs. A deficiency of this hormone in early childhood produces cretinism.

tinnitus (*tĭ-ni′tus*) The spontaneous sensation of a ringing sound or other noise without sound stimuli.

tolerance, immunological The ability of the immune system to distinguish self from nonself; thus, the immune system does not normally attack those antigens that are part of one's own tissues.

total minute volume The product of tidal volume (ml per breath) and ventilation rate (breaths per minute).

totipotent (*tōtip′ŏ-tent*) The ability of a cell to differentiate into all tissue types, and thus to form a new organism when appropriately stimulated and placed in the correct environment (a uterus).

toxin (*tok′sin*) A poison.

toxoid (*tok′soid*) A modified bacterial endotoxin that has lost toxicity but that still has the ability to act as an antigen and stimulate antibody production.

tracts A collection of axons within the central nervous system that forms the white matter of the CNS.

trans- (L.) Across, through.

transamination (*trans″am-ĭ-na′shun*) The transfer of an amino group from an amino acid to an alpha-keto acid, forming a new keto acid and a new amino acid without the appearance of free ammonia.

transcription (*tran-skrip′shun*). **genetic** The process by which messenger RNA is synthesized from a DNA template resulting in the transfer of genetic information from the DNA molecule to the mRNA.

transducins (*trans-doo′sinz*) The G-proteins involved in vision. When light causes the photodissociation of rhodopsin, the G-protein alpha subunit dissociates from the opsin and indirectly cause a reduction in the dark current of the photoreceptors.

translation (*trans-la′shun*), **genetic** The process by which messenger RNA directs the amino acid sequence of a growing polypeptide during protein synthesis.

transplantation (*trans″plan-ta′shun*) The grafting of tissue from one part of the body to another part, or from a donor to a recipient.

transpulmonary (*trans″pul′mŏ-ner″-e*) **pressure** The pressure difference across the wall of the lung; equal to the difference between intrapulmonary pressure and intrapleural pressure.

triiodothyronine (*tri″i-ŏ-thi′ro-nēn*) (T_3) A hormone secreted in small amounts by the thyroid; the active hormone in target cells formed from thyroxine.

tropomyosin (*tro′pŏ-mi′ŏ-sin*) A filamentous protein that attaches to actin in the thin filaments. Together with another protein called troponin, it acts to inhibit and regulate the attachment of myosin cross bridges to actin.

troponin (*tro′pŏ-nin*) A protein found in the thin filaments of the sarcomeres of skeletal muscle. A subunit of troponin binds to Ca^{2+}, and as a result causes tropomyosin to change position in the thin filament.

trypsin (*trip'sin*) A protein-digesting enzyme in pancreatic juice that is released into the small intestine.

tryptophan (*trip'tŏ-fān*) An amino acid that also serves as the precursor for the neurotransmitter molecule serotonin.

TSH Thyroid-stimulating hormone, also called *thyrotropin* (*thi"rŏ-tro'pin*) A hormone secreted by the anterior pituitary that stimulates the thyroid gland.

tubuloglomerular (*too"byŭ-lo-glo-mer'yŭ-lar*) **feedback** A control mechanism whereby an increased flow of fluid through the nephron tubules causes a reflex reduction in the glomerular filtration rate.

tumor necrosis (*nĕ-kro'sis*) **factor (TNF)** A cytokine released by immune cells and mast cells that causes destruction of tumors and migration of neutrophils toward the site of a bacterial infection. TNF is also secreted by adipose cells and may be a paracrine regulator of insulin sensitivity.

turgid (*tur'jid*) Swollen and congested.

twitch A rapid contraction and relaxation of a muscle fiber or a group of muscle fibers.

tympanic (*tim-pan'ik*) **membrane** The eardrum; a membrane separating the external from the middle ear that transduces sound waves into movements of the middle-ear ossicles.

tyrosine kinase (*ti'rŏ-sen ki'nās*) An enzyme that adds phosphate groups to tyrosine, an amino acid present in most proteins. The membrane receptor for insulin, for example, is a tyrosine kinase. When bound to insulin, the tyrosine kinase is activated, which leads to a cascade of effects that mediate insulin's action.

U

universal donor A person with blood type O, who is able to donate blood to people with other blood types in emergency blood transfusions.

universal recipient A person with blood type AB, who can receive blood of any type in emergency transfusions.

urea (*yoo-re'ă*) The chief nitrogenous waste product of protein catabolism in the urine, formed in the liver from amino acids.

uremia (*yoo-re'me-ă*) The retention of urea and other products of protein catabolism as a result of inadequate kidney function.

urobilinogen (*yoo"rŏ-bi-lin'ŏ-jen*) A compound formed from bilirubin in the intestine. Some is excreted in the feces and some is absorbed and enters the enterohepatic circulation, where it may be excreted either in the bile or in the urine.

V

vaccination (*vaks"sĭ-na'shun*) The clinical induction of active immunity by introducing antigens into the body so that the immune system becomes sensitized to them. The immune system will mount a secondary response to those antigens upon subsequent exposures.

vagina (*vă-ji'ă*) The tubular organ in the female leading from the external opening of the vulva to the cervix of the uterus.

vagus (*va'gus*) **nerve** The tenth cranial nerve, composed of sensory dendrites from visceral organs and preganglionic parasympathetic nerve fibers. The vagus is the major parasympathetic nerve in the body.

Valsalva's (*val-sal'vaz*) **maneuver** Exhalation against a closed glottis, so that intrathoracic pressure rises to the point that the veins returning blood to the heart are partially constricted. This produces circulatory and blood pressure changes that could be dangerous.

vasa-, vaso- (L.) Pertaining to blood vessels.

vasa vasora (*va'să va-sor'ă*) Blood vessels that supply blood to the walls of large blood vessels.

vasectomy (*vă-sek'tŏ-me, va-zek'tŏ-me*) Surgical removal of a portion of the ductus (vas) deferens to induce infertility.

vasoconstriction (*va"zo-kon-strik'shun*) A narrowing of the lumen of blood vessels as a result of contraction of the smooth muscles in their walls.

vasodilation (*va"zo-di-la'shun*) A widening of the lumen of blood vessels as a result of relaxation of the smooth muscles in their walls.

vasopressin (*va"zo-pres'in*) Another name for antidiuretic hormone (ADH), released from the posterior pituitary. The name *vasopressin* is derived from the fact that this hormone can stimulate constriction of blood vessels.

vein A blood vessel that returns blood to the heart.

ventilation (*ven"tĭ-la'shun*) Breathing; the process of moving air into and out of the lungs.

vertigo (*ver'tĭ-go*) A sensation of whirling motion, either of oneself or of external objects; dizziness, or loss of equilibrium.

vestibular (*vĕ-stib'yŭ-lar*) **apparatus** The parts of the inner ear, including the semicircular canals, otricle, and saccule, that function to provide a sense of equilibrium.

villi (*vil'i*) Fingerlike folds of the mucosa of the small intestine.

virulent (*vir'yŭ-lent*) Pathogenic, or able to cause disease.

vital capacity The maximum amount of air that can be forcibly expired after a maximal inspiration.

vitamin (*vi'tă-min*) Any of various unrelated organic molecules present in foods that are required in small amounts for normal metabolic function of the body. Vitamins are classified as water-soluble or fat-soluble.

W

white matter The portion of the central nervous system composed primarily of myelinated fiber tracts. This forms the region deep to the cerebral cortex in the brain and the outer portion of the spinal cord.

Z

zygote (*zi'gōt*) A fertilized ovum.

zymogen (*zi'mŏ-jen*) An inactive enzyme that becomes active when part of its structure is removed by another enzyme or by some other means.

Credits

Chapter 1
1.7, 1.8, 1.9, 1.10: © Ed Reschke; **1.11a:** © Ray Simons/ Photo Researchers; **1.11b:** © Photo Researchers; **1.11c, 1.12a, 1.16:** © Ed Reschke; **1.17b:** © Biology Media/Photo Researchers; **1.18a, 1.19b:** © Ed Reschke; **1.20:** Courtesy of Southern Illinois University, School of Medicine

Chapter 2
2.28: © Ed Reschke

Chapter 3
3.3ab: © Jeon-Visuals Unlimited; **3.4 (1–4):** From M.M. Perry and A.B. Gilbert. Journal of Cell Science 39:257-272, 1979. Reprinted by permission Company of Biologists, LTD.; **3.5a:** © Science Photo Library/Photo Researchers; **3.5b:** Courtesy Richard Chao; **3.6:** © Dr. Dennis Kunkel/PhotoTake; **3.7:** © K. G. Murtin/Visuals Unlimited; **3.9:** Courtesy Richard Chao; **3.10a:** Courtesy Keith R. Porter; **3.11:** Courtesy of Joacim Frank; **3.12a:** Courtesy of Keith Porter; **3.13a:** Courtesy of Mark S. Ladinsky and Kathryn D. Howell, University of Colorado **3.14:** Courtesy of Richard Chao; **3.15a:** Courtesy of Ed Pollock; **3.20:** © E. Kiselva-D. Fawcett/Visuals Unlimited; **3.29a–e:** © Ed Reschke; **3.30a:** © D.M. Phillips/ Visuals Unlimited; **3.31ab:** Images by Dr. Conly L. Rieder, Division of Molecular Medicine, Wadsworth Center, Albany, New York 12201-0509; **3.32:** © SPL/Photo Researchers

Chapter 7
7.6b: Constance Holden, "Versatile Cells Against Intractable Diseases." Science 297 (26 July 2002) p. 501. © AAAS; **7.7:** From H. Webster, The Vertebrate Peripheral Nervous System, John Hubbard (ed.), © 1974 Plenum Publishing Corporation; **7.20:** © John Heuser, Washington University School of Medicine, St. Louis, MO

Chapter 8
8.8: Courtesy of Dr. Llinas, New York University Medical Center; **8.9:** From W.T. Carpenter and R.W. Buchanan, "Medical Progress: Schizophrenia" in New England Journal of Medicine, 330:685, 1994, fig 1A. © 1994 Massachusetts Medical Society. All rights reserved; **8.16b:** © The McGraw-Hill Company, Inc./Karl Rubin, photographer

Chapter 10
10.10: © Discover Magazine. Reprinted with permission; **10.13a:** Courtesy of Dean E. Hillman; **10.21:** © Prof. P. Motta/Dept. of Anatomy/University "La Sapieza," Rome/Science Photo Library/Photo Researchers, Inc.; **10.29a:** © Thomas Simms; **10.32b:** Courtesy of P.N. Farnsworth, University of Medicine and Dentistry; **10.36b:** © Ominkron/Photo Researchers

Chapter 11
11.1b: © Ed Reschke; **11.21b:** © Nawrocki Stock Photo, Inc.; **11.22:** © Martin M. Rotker; **11.24, 11.26, 11.27:** © L. V. Bergman/The Bergman Collection

Chapter 12
12.2: © Ed Reschke; **12.3b:** © Victor B. Eichler; **12.6a, 12.7a:** © Dr. H. E. Huxley; **12.7b:** From R.G. Kessel and R.H. Kardon, Tissues and Organs: A Text-Atlas of Scanning Electron Microscopy, 1979, W. H. Freeman & Company; **12.9a:** © Dr. H. E. Huxley; **12.17:** © Stuart Fox; **12.23:** © Hans Hoppler, Respiratory Physiology, 44:94 (1981); **12.32,12.33a:** © Ed Reschke

Chapter 13
13.2: Reginald J. Poole/Polaroid International Photomicrograph Competition; **13.6 (6 images):** © Stuart Fox; **13.7:** From Nature, Volume 413, Issue 6855, 4 October 2001, cover. Photo by Dr. John W. Weisel, University of Pennsylvania; **13.8:** Adapted from A. Marchand, "Case of the Month, Circulating Anticoagulants: Chasing the Diagnosis" in Diagnostic Medicine, June 1983, p. 14. Copyright © 1984. Reprinted with permission of Medical Economics Company, Inc., Oradell, NJ; **13.11:** © Igaku-Shoin, Ltd., Tokyo, Japan; **13.27:** © Don W. Fawcett; **13.30a:** © Biophoto Associates/Photo Researchers, Inc.

Chapter 14
14.4: Source: F. Fuchs and S.H. Smith, "Calcium, Cross-Bridges, and the Frank–Starling Relationship." News in Physiological Sciences, 16 (Feb. 2001): 6; **14.10:** From E.K. Markell and M. Vogue, Medical Parasitology, 6th edition, 1986, W.B. Saunders Company; **14.17ab:** Courtesy of Donald S. Bain; **14.19:** Adapted from P. Astrand and K. Rodahl, Textbook of Work Physiology, 3rd ed. Copyright © 1986 McGraw-Hill, Inc., New York, NY. Used by permission of the authors; **14.21ab:** © Niels A. Lassen, Coppenhagen, Denmark; **14.28:** © Ruth Jenkinson/MIDIRS/Science Photo/Photo Researchers, Inc.

Chapter 15
15.9: © McGraw-Hill Company, Inc.; **15.10a:** Courtesy of Arthur J. Olson, Ph.D., The Scripps Research Institute, Molecular Graphics Laboratory; **15.11ab:** Reprinted with permission from A.G. Amit, "Three Dimensional Structure of an Antigen-Antibody Complex at 2.8 A Resolution" in Science, vol. 233, August 15, 1986. © American Association for the Advancement of Science; **15.17a:** From Alan S. Rosenthal, "Current Concepts: Regulation of the Immune Response—Role of the Macrophage" in New England Journal of Medicine, vol. 303:1153, Nov. 13, 1980, fig 2, © Massachusetts Medical Society. All rights reserved; **15.25ab:** © Dr. Andrejs Liepins; **15.27ab:** © SIU Biomed/ Custom Medical Stock; **15.28a:** © Stan Flegler/Visuals Unlimited; **15.28b:** © Dennis Kunkel/Phototake

Chapter 16
16.2: From John J. Murray, The Normal Lung, 2nd edition, 1986, © W.B. Saunders; **16.3ab:** Courtesy American Lung Association; **16.5b:** © John Watney Photo Library; **16.6:** © Phototake; **16.7:** © Manfred Kage/Peter Arnold; **16.9ab, 16.10:** Courtesy of Edward C. Vasquez, R.T.C.R.T., Dept. of Radiologic Technology, Los Angeles City College; **16.13ab:** From J.H. Comore, Jr., Physiology of Respiration: An Introductory Text, 2nd edition, 1974 © Yearbook Medical Publishers, Inc. Chicago; **16.18a:** © R. Calentine/Visuals Unlimited; **16.18b:** © M. Moore/Visuals Unlimited; **16.36a:** © Bill Longcore/Science Source/Photo Researchers; **16.36b:** © David Phillips/Photo Researchers

Chapter 17
17.3: © SPL/Photo Researchers; **17.4b:** © Biophoto Associates/Photo Researchers; **17.7:** © Professor P.M. Motta and M. Sastellucci/SPL/Photo Researchers; **17.9:** From Bloom and D.W. Fawcett, A Textbook of Histology, 11th edition. ©1986 W. B Saunders

Chapter 18
18.4b, 18.6: © Kent M. Van De Graaff; **18.13ab:** Two photos: From Keith R. Porter, D.H. Aplers and D. Seetharan, "Pathophysiology of Diseases Involving Intestinal Brush-Border Proteins" in New England Journal of Medicine, Vol. 296, 1997, p.1047, fig. 1. © 1977 Massachusetts Medical Society. All rights reserved; **18.18:** From Bloom and D.W. Fawcett, A Textbook of Histology, 11th edition. ©1986 W. B. Saunders; **18.19:** From W.A. Sodeman and T.M. Watson, Pathologic Physiology, 6th edition, 1969, © W.B. Saunders, Inc.; **18.27a:** © Carroll Weiss/Camera M.D. Studios; **18.27b:** © Dr. Sheril Burton; **18.28a:** © Ed Reschke

Chapter 19
19.17: Reprinted from "Acromegaly Diabetes, Hypermetabolism, Proteinuria and Heart Failure, Clinico-pathologic Conference" in American Journal of Medicine, 20, (1956) © 1956 with permission from Excerpta Media, Inc.; **19.18a:** From S.N. Bhasker (ed.), Orban's Oral Histology and Embryology, 8th edition. © 1976 C.V. Mosby & Co.; **19.19ab:** Photo from L.G. Raisz, S.W. Dempster, et al., New England Journal of Medicine, Vol. 218 (13):818. © 1988 Massachusetts Medical Society. All rights reserved.

Chapter 20
20.2a: © CNRI/Science Photo Library/Photo Researchers; **20.2b:** © Dept. of Clinical Cytogenetics, Addenbrooks Hospital, Cambridge/SPL/Photo Researchers; **20.3a:** © Visuals Unlimited; **20.3b:** © M. Abbey/Photo Researchers; **20.10:** From F. K. Shuttleworth, Monograph of the Society for Research in Child Development, Inc., Chicago, IL. Copyright © 1939 The Society for Research in Child Development, Inc. Reprinted by permission; **20.17a:** © Biophoto Associates/ Photo Researchers; **20.20b:** © F. Leroy/SPL/Photo Researchers, Inc; **20.28ab:** © Ed Reschke; **20.29a:** From Richard J. Blandau, A Textbook of Histology, 10th edition. © W.B. Saunders; **20.30:** From Bloom and D.W. Fawcett, A Textbook of Histology,10th edition. ©1975 W. B Saunders; **20.31:** © Dr. Landrum Shettles; **20.36:** © Martin M. Rotker; **20.42:** Courtesy of David L. Hill; **20.44a:** © Petit Format/Photo Researchers; **20.44b:** © SPL/Photo Researchers; **20.44c:** Dr. Landrum Shettles; **20.44d:** © Biophoto Associates/Photo Researchers.

Index

Note: Page references followed by the letters *f* and *t* indicate figures and tables, respectively.

D

D cells, 566, 583, 584
DAG. *See* Diacylglycerol
Dale, Henry, 7*t*
Dalton's law, 493
Dark adaptation, 269–270
Dark current, 271, 271*f*
Data, reproducible, 4
Deacetylation, 62, 63*f*
Dead space, anatomical, 491
Deafness
    conduction, 257, 260–261
    sensorineural, 260
Deamination, oxidative, 117–118, 118*f*
Decidua basalis, 670
Decidual reaction, 670
Decompression sickness, 499
Decussation, 197
Dedifferentiation, 468
Defecation, 574
Defecation reflex, 574
Defense mechanisms, 446–453. *See also*
    Immunity
Defibrillation, electrical, 398
Deglutition, 561, 563
Dehydration
    oral rehydration therapy for, 138
    physiological response to, 134, 134*f*, 416
Dehydration synthesis
    of amino acids, 38, 39*f*
    of carbohydrates, 32, 33*f*
Dehydroepiandrosterone sulfate
    (DHEAS), 675
Dehydrogenase(s)
    diseases associated with abnormal
        concentrations of, 88*t*
    function of, 88, 88*t*
Deletion, clonal, 466
Delta cells, 611
Delta waves, 196–197, 196*f*
Denaturation, of proteins, 39–40
Dendrites, 11, 11*f*
    function of, 153
    structure of, 152*f*, 153, 153*f*
Dendritic cells, 459, 460*f*
Dendritic spines, long-term potentiation
    and, 202
Denervation hypersensitivity, 221
Dense bodies, 355, 355*f*
Dense fibrous connective tissue, 14
Dense irregular connective tissue, 14, 16*f*
Dentin, 16, 17*f*
Dentinal tubules, 16
Deoxyhemoglobin, 504
Deoxyribonuclease, in pancreatic juice, 582*t*
Deoxyribonucleic acid (DNA)
    base triplets of, 65–66, 66*t*
    in chromatin, 62, 63*f*
    of mitochondria, 57, 58
    replication of, 69–70, 70*f*
        senescence and, 74
    in RNA synthesis, 62, 64
    structure of, 42, 43*f*, 62
    synthesis of, 69–78
Deoxyribonucleic acid (DNA) helicases, 69
Deoxyribonucleic acid (DNA) polymerases,
    70, 74
Deoxyribose, 42, 44*f*
Dephosphorylation, 90
Depolarization
    in action potentials, 162–163
    excitatory postsynaptic potential, 170, 170*f*
    of resting membrane potential, in axons,
        160*f*, 161, 162
    stimulus intensity for, 164
Depression, treatment for, 176, 177–178
Depressor muscles, 326*t*
Dermatitis, contact, 473
Dermatophagoides farinae, 472, 473*f*
Dermis, 18–19, 18*f*
Desensitization, to hormones, 291
Desmosomes, 138
Deuteranopia, 272
Deuterium, 25
Dextrose solution, 133
DHA. *See* Docosahexaenoic acid
DHEAS. *See* Dehydroepiandrosterone sulfate
DHT. *See* Dihydrotestosterone
Diabetes insipidus, 538
    nephrogenic, 538
Diabetes mellitus, 313, 615–618
    body shape and, 608
    and glycosuria, 135, 544
    ketoacidosis in, 36

metabolism in, 116
type 1 (insulin-dependent), 616–617
    as autoimmune disease, 470*t*, 471
    etiology of, 313, 616
    symptoms of, 616–617
    vs. type 2, 615, 615*t*
    uncorrected insulin deficiency in,
        616, 616*f*
type 2 (non-insulin-dependent), 615, 617–618
    etiology of, 313, 617
    oral glucose tolerance in, 617, 617*f*, 618
    symptoms of, 617
    treatment of, 606
    vs. type 1, 615, 615*t*
Diabetic retinopathy, 393
Diacylglycerol (DAG), 296
Dialysis, 128, 552
    continuous ambulatory peritoneal, 552
Diapedesis (extravasation), 369, 447, 447*f*
Diaphragm, 483, 488
    and venous return, 412
Diarrhea, 574
    etiology of, 574
    traveler's, 585
    treatment for, 138
Diastole, 381–382, 381*f*
Diastolic pressure, 432, 433
Diencephalon, 204, 204*f*
    development of, 191
Dieting, metabolism during, 116
Differentiation, cellular, 64
Diffusion, 127–130
    concentration gradients in, 128, 128*f*
    definition of, 128
    facilitated, 127, 135–136, 135*f*, 136*f*, 139*f*
    ion channels in, 129, 129*f*
    net (osmosis), 128, 130–133
    rate of, 129
    simple, 127
Digestion, 587–591
    of carbohydrates, 588
    definition of, 561
    enzymes in, 588*t*
    fat and, 585
    hydrolysis reactions in, 560*f*, 561
    of lipids, 589, 590*f*
    of proteins, 584, 588–589, 589*f*
    in stomach, 567–568
    transportation after, 366
Digestive system, 558–591. *See also*
    Gastrointestinal tract
    accessory organs in, 561–562
    cells of, 79
        apoptosis in, 72
    circulatory system and, 366, 440, 592
    components of, 19*t*, 561–562, 561*f*
    endocrine regulation of, 583–587, 584*f*
    endocrine secretions of, 561
    endocrine system and, 319, 592
    exocrine secretions of, 561
    functions of, 19*t*, 561
    glands of, 14
    immune system and, 474, 592
    integumentary system and, 592
    introduction to, 560–563
    membrane transport in, 145
    metabolism in, 120
    muscular system and, 359, 592
    nervous system and, 237, 592
    neural regulation of, 563, 583–587, 584*f*
    prostaglandins in, 318
    reproductive system and, 592, 680
    respiratory system and, 516, 592
    sensory system and, 277
    skeletal system and, 592
    urinary system and, 553, 592
Digitalis, 439
Dihydrotestosterone (DHT)
    formation of, 638, 639*f*, 650
    function of, 638
Dihydroxyphenylalanine (DOPA), 92
    chemical structure of, 229*f*
1,25-Dihydroxyvitamin D₃
    in calcium and phosphate balance, 625–627
    production of, 625, 626*f*
Diiodotyrosine (DIT), 309
Dimerization, 293
Dioxin, 579
Dipalmitoyl lecithin, 486
Dipeptides, 38
2,3-Diphosphoglyceric acid (2,3-DPG), and
    oxygen transport, 508, 508*t*
Diploid cells, 634–635
Direct phosphorylation, 113
Disaccharides, 31–32
    dehydration synthesis of, 33*f*
Discontinuous capillaries, 393

Dismutase
    as antioxidant, 111
    genetic defects in, 348
D-isomers, 31
Disulfide bonds, in proteins, 40, 41*f*
Disynaptic reflex, 351
DIT. *See* Diiodotyrosine
Diuretics, 549–551
    action of, 549–551, 550*t*
        site of, 550–551, 550*f*
    complications related to, 551
    for hypertension, 437, 437*t*
    loop, 550*t*, 551
    osmotic, 550*t*, 551
    potassium-sparing, 550*t*, 551
    thiazide, 550*t*, 551
    water as, 416
Divergence, 222
Diverticulitis, 572
DNA. *See* Deoxyribonucleic acid
DNA-binding domain, of hormone
    receptors, 292
Docosahexaenoic acid (DHA), 600
Domains, of hormone receptors, 292
DOPA. *See* Dihydroxyphenylalanine
L-Dopa, blood-brain barrier and, 159
Dopamine
    blood-brain barrier and, 159
    chemical structure of, 229*f*
    as neurotransmitter, 178, 180*t*, 206
Dopaminergic neurons, 178, 206
Dormitory effect, in menstruation, 305
Double helix, of DNA, 42, 43*f*
Downregulation, 291
2,3-DPG. *See* 2,3-Diphosphoglyceric acid
Drug(s)
    abuse of
        dopaminergic neurons and, 206
        synaptic effects of, 207*t*
    pharmaceutical, development of, 5–6
    tubular secretion of, 540
Dual innervation, of organs, 234–235
Duchenne's muscular dystrophy, 326
Duct(s), 14. *See also specific ducts*
Ductus arteriosus, 318, 383, 384*f*
Ductus (vas) deferens, 637, 650, 651*f*, 652*f*
Duodenal (Brunner's) glands, 568, 569*f*
Duodenum, 569
    ulcers of, 568
Dust mites, 472, 473*f*
Dwarfism, 622
    Laron, 622
    pituitary, 300
Dyazide, 551
Dynamic constancy, homeostasis as, 8, 8*f*
Dynorphin, 180
Dyrenium. *See* Triamterene
Dysmenorrhea, 663
Dyspnea, 492
Dysreflexia, autonomic, 235
Dystrophin, 326

E

E neurons. *See* Expiratory neurons
Ear(s), 255–261, 256*f*. *See also* Hearing
    and equilibrium, 251–255
    inner, 251–255, 251*f*, 256*f*, 257–261
    middle, 255–256, 256*f*
    outer, 255, 256*f*
    spiral organ of, 258–261, 259*f*
Eardrum, 255, 257
Eating, regulation of, 604–605
Eccentric contraction, 341
Eccles, John, 7*t*
Eccrine sweat glands, 14
ECG. *See* Electrocardiogram
ECL cells. *See* Enterochromaffin-like cells
Eclampsia, 437
ECS. *See* Electroconvulsive shock
Ectoderm, 190, 670
Ectopic pacemaker (ectopic focus), 385, 398
Edema
    altitude and, 515
    cerebral, 515
    etiology of, 131, 415, 415*t*
    pulmonary, 498, 515
EDTA. *See* Ethylenediaminetetraacetic acid
EDV. *See* End-diastolic volume
EEG. *See* Electroencephalogram
Effector organs, visceral, 221
Effectors
    antagonistic, 8
    in negative feedback loop, 6–7, 7*f*
    in positive feedback loop, 9

Efferent arterioles, 526
Efferent ductules, 649
Efferent neurons. *See* Motor neurons
Ehrlich, Paul, 469
Eicosanoids, 317
Eicosapentaenoic acid (EPA), 600
Ejaculation, 651
Ejaculatory duct, 650, 651*f*
    development of, 637
Ejection, in cardiac cycle, 381–382, 409
Ejection fraction, 409, 426
EKG. *See* Electrocardiogram
Elastase, in pancreatic juice, 582*t*, 588
Elastic arteries, 391
Elasticity, of lungs, 485, 488
Elastin, in extracellular matrix, 127
Electrocardiogram (ECG or EKG),
    387–390, 387*f*
    arrhythmias detected by, 397–399, 398*f*
    heart sounds and, 390, 390*f*
    impulse conduction and, 388–389, 389*f*
    leads of
        bipolar limb, 387–388, 388*f*, 388*t*
        unipolar, 388, 388*f*, 388*t*
    myocardial ischemia and, 397, 397*f*
Electrocardiograph, 387
Electroconvulsive shock (ECS) therapy, and
    memory, 201
Electroencephalogram (EEG), 194–197, 196*t*
    types of patterns in, 194–197, 196*f*
Electrolyte(s). *See also specific types*
    absorption of, in large intestine, 574
    balance, renal control of, 544–547
Electromagnetic spectrum, 262, 262*f*
Electromechanical release mechanism, 337–338
Electron(s)
    definition of, 24
    orbital of, 24
    valence, 24, 25
Electron acceptor, final, 104
    oxygen as, 104, 113
Electronegative atoms, 27, 27*f*
Electron-transport system, 110–111,
    111*f*, 112*f*
    and ATP production, 112–113
    chemiosmotic theory of, 112, 112*f*
Electrophoresis, 453, 455*f*
Elements
    chemical, 25
    trace, 601*t*, 602, 603*t*
Elephantiasis, 415, 415*f*
Elimination, in digestive system, 561
Embryonic development, 190–191, 191*f*
Embryonic stem cells, 64, 668–669
Emission, 651
Emmetropia, 267*f*
Emotion
    cerebral area related to, 200–201, 205
    olfaction and, 249
Emphysema, 492–493, 493*f*, 503
Emulsification, 589, 590*f*
Emulsification droplets, 589
Enamel, 16, 17*f*
Encephalomyelitis, postvaccinal and
    postinfectious, 470*t*
End-diastolic volume (EDV), 381
    definition of, 409
    and stroke volume, 409, 410*f*
    venous return and, 411, 412*f*
Endemic goiter, 309–310, 310*f*
Endergonic reactions, 94
End-feet, 159, 159*f*
Endocannabinoids, as neurotransmitters,
    180*t*, 181
Endocarditis, rheumatic, 382
Endocrine glands, 12, 286–287, 287*t*. *See
    also* Hormone(s); *specific glands and
    hormones*
    vs. exocrine glands, 14
    formation of, 15*f*
    location of major, 286*f*
    regulation by, 9
        vs. neural regulation, 289–290
    secretions of, in digestive system, 561
    signaling in, 144, 144*f*
Endocrine system, 286–287
    cells of, 79
    circulatory system and, 319, 440
    components of, 19*t*
    digestive system and, 319, 592
    functions of, 19*t*
    immune system and, 319, 474
    integumentary system and, 319
    membrane transport in, 145
    metabolism in, 120
    muscular system and, 319, 359
    nervous system and, 237, 319

N

n-3 fatty acids, 600
NAD. *See* Nicotinamide adenine dinucleotide
Nafarelin, 641
Naloxone, 180
National Library of Medicine, 5
Natriuresis, 418
Natriuretic hormone, 418, 546
Natriuretic peptide (factor), atrial, 546
Natural killer (NK) cells, 448, 469
Nature (journal), 5
Nature Medicine (journal), 5
Nearsightedness. *See* Myopia
Necrosis, 72, 397
Negative feedback loops. *See* Feedback control, negative
Neonate(s)
  hemolytic disease of newborn, 374
  immunity in, 466
  jaundice in, 577
  sudden infant death syndrome and, 503
  vascular resistance in, 497
Neoplasms, angiogenesis and, 393
Neostigmine, and neural control of skeletal muscles, 172t
Nephron(s), 526
  cortical, 528, 528f
  juxtamedullary, 526–528, 528f
  tubules of, 526–528, 527f
    reabsorption in, 532–534, 533f
Nernst equation, 141–142
Nerve(s)
  classification of, 154
  definition of, 153t, 154
  secretion by, mechanism for, 54
Nerve cells. *See* Neuron(s)
Nerve energies, law of specific, 242–243
Nerve gas, action of, 172t, 174
Nerve growth factor (NGF), 158
Nerve impulses, 143. *See also* Action potentials
Nerve tissue, 11, 11f
Nervous system. *See also* Autonomic nervous system; Central nervous system; Neuron(s)
  cells of, 79
  circulatory system and, 237, 440
  components of, 19t, 153t
  digestive system and, 237, 592
  endocrine system and, 237, 319
  enteric, 563, 585, 586f
  functions of, 19t
  immune system and, 237, 474
  integumentary system and, 237
  membrane transport in, 145
  metabolism in, 120
  muscular system and, 237, 359
  regulation by, 9
  reproductive system and, 237, 680
  respiratory system and, 237, 516
  sensory system and, 277
  skeletal system and, 237
  urinary system and, 237, 553
Neural crest, 190, 191f
Neural stem cells, 64, 156, 193, 669
  and learning, 203
  and memory, 203
Neural tube, 190, 191f
Neurilemma, 156
Neuroendocrine reflexes, 301, 677
Neurofibrillar tangles, in Alzheimer's disease, 175, 194
Neurogenesis, 193, 203
Neurogenic mechanism, of breathing and exercise, 513
Neurogenic shock, 439
Neuroglia, 152
Neuroglial cells, 11, 152, 155f
Neurohormones, 286
Neurohypophysis. *See* Pituitary gland, posterior
Neuroleptics, 178
Neuromodulators, 180
Neuromuscular junction, motor end plates at, 328, 328f
Neuromuscular junctions, 167, 174
Neuron(s), 152–154. *See also specific types*
  cable properties of, 164–165
  classification of
    functional, 154
    structural, 154
  components of, 11
  definition of, 152

functions of, 152
  long-term potentiation in, 182
  membrane potential of, alteration of, 160–161
  production of, 193
  regeneration of, 157–158, 158f
  structure of, 152–153, 152f, 153f
  synapses in, 167–170
Neuronal cell death, 182
Neuropeptide(s), 179–180
Neuropeptide Y, 180–181, 606
Neurotransmitters, 144. *See also specific substances*
  in autonomic nervous system, 175, 227–234
  early research on, 167
  and eating, 605
  excitatory, 179
  vs. hormones, 289–290
  inhibitory, 179
  parasympathetic release of, 229f
  receptor proteins for, 169
  release of, 168–169, 169f
  sympathetic release of, 229f
  in synaptic inhibition, 182–183
Neurotrophin(s), 158, 316–317, 316t
Neurotrophin-3, 158
Neurotrophin-4/5, 158
Neutral fats, 35
Neutral pH, 28
Neutrons, 24
Neutrophil(s), 369, 370f, 370t, 446
  abundance of, 369
  apoptosis of, 72
  functions of, 447, 448f
  in local inflammation, 452, 453f
  structure of, 369
New England Journal of Medicine, 5
Newborn. *See* Neonate(s)
News in Physiological Sciences, 5
NGF. *See* Nerve growth factor
Niacin, 96, 601, 602t
  recommended dietary allowance for, 601t
Nicotinamide adenine dinucleotide (NAD)
  action of, 98f
  in aerobic respiration, 108–114, 109f
  in glycolytic-lactic acid pathway, 104–105
  as hydrogen carrier, 96, 97f
  structure of, 97f
Nicotine
  dopaminergic neurons and, 178, 206
  synaptic effects of, 207t
Nicotinic ACh receptors, 171, 171f, 206, 232, 233f, 233t
Nidation, 667
NIDDM (non-insulin-dependent diabetes mellitus). *See* Diabetes mellitus, type 2
Niedergerde, R., 7t
Nifedipine, 357
Nigrostriatal dopamine system, 178, 206
Nissl bodies, 152
Nitric oxide
  altitude and, 514–515
  as autocrine regulator, 316t
  and erection, 651, 653f
  free radicals of, diseases associated with, 111
  as neurotransmitter, 144, 180t, 181, 233–234
  as paracrine regulator, 316–317
    of blood flow, 421–422
  as retrograde messenger, 202
  and septic shock, 438
Nitric oxide radical, 603
Nitric oxide synthase, 438, 459
Nitrogen
  balance, 116
  in blood, disorders associated with, 499
  partial pressure of, 493
  reactive species of, 603
  structure of, 24t
Nitrogen narcosis, 499
Nitrogenous base, of nucleotides, 42
Nitroglycerin, 422
NK. *See* Natural killer
N-Methyl-D-aspartate (NMDA) receptors, 179, 202
Nociceptors, 242, 243t, 244
Nodes of Ranvier, 157
  action potentials at, 165f, 166
Non-insulin-dependent diabetes mellitus (NIDDM). *See* Diabetes mellitus, type 2
Nonkeratinized membrane, 12, 12t, 13f
Nonshivering thermogenesis, 608–609

Nonsteroid family, of hormone receptors, 292
Nonsteroidal anti-inflammatory drugs (NSAIDs), 318
  excessive use of, 551
  and gastric ulcers, 568
Nonvolatile acids, 377–378
Norepinephrine
  and cardiac contraction strength, 411
  and cardiac rate, 408
  chemical structure of, 229f
  and eating, 605
  as hormone, 176
  hypersecretion of, 307
  metabolic regulation by, 619
  as neurotransmitter, 176, 177f, 178, 180t
    in autonomic nervous system, 228
    responses to, 230–231
  secretion of, 223, 289, 307
Normoblasts, 371, 372f
Norplant, 664
NSAIDs. *See* Nonsteroidal anti-inflammatory drugs
Nuclear bag fibers, 348
Nuclear chain fibers, 348
Nuclear envelope, 50t, 51f, 61
Nuclear hormone receptors, 292–294
Nuclear pore(s), 61, 62f
Nuclear pore complexes, 61, 62f
Nuclear receptors, 579
Nucleic acids, 38, 42–44
Nucleolus, 51f
  function of, 50t, 51, 61
  structure of, 50t
Nucleosomes, 62
Nucleotides, 42, 42f
  in DNA, 42, 44f, 62
  in RNA, 44, 44f
Nucleus(i) (cell), 51f, 61–64
  function of, 51
  structure of, 24, 51, 61, 61f
  transplantation of, 668–669
Nucleus(i) (nervous system), 152–153, 153t
  amygdaloid, 200, 202, 249
  basal, 197, 197f, 211, 211f, 352
  caudate, 197, 352
  composition of, 192
  geniculate
    lateral, 204, 274, 276
    medial, 204
  intralaminar, 204
  lentiform, 197
  paraventricular, 205, 205f, 301
  red, 206, 211, 211f
  septal, 200
  suprachiasmatic, 314
  supraoptic, 205, 205f, 301
  vagus, 208
  vestibular, 211, 211f
Nucleus accumbens, dopaminergic pathways and, 178
Null hypothesis, 5
Nutritional requirements, 598–603
Nystagmus, vestibular, 254–255

O

ob gene, 606
Obesity, 608
ob/ob mice, 606
Occipital lobe, 193, 194f, 194t
Oculomotor nerve, 212t, 225t
  parasympathetic fibers in, 223
Oddi, sphincter of, 581
Off-center fields, 275, 276f
Olfaction, 249–250
  cerebral area related to, 200, 249
  emotion and, 249
  molecular basis of, 249–250
  neural pathway for, 249, 250f
  receptors for, 248, 249–250
Olfactory bulb, 249, 250f
Olfactory nerve, 212t, 250f
Oligodendrocytes, 157f
  function of, 155, 156f
  in myelin sheath, 156
Oligosaccharides, production of, 588, 588f
Oligospermia, 652
Olive oil
  and gastric function, 584
  saturated fat content of, 34
Omega-3 fatty acids, 600
On-center fields, 275, 276f

Oncogenes, 71
Oncology, 468
Oncotic pressure, 413
Ondine's curse, 500
Oocytes, 657f
  after fertilization, 666f, 667
  primary, 656
  secondary, 657, 657f
Oogenesis, 640, 659f
Oogonia, 635, 656
Ophthalmia, sympathetic, 470, 470t
Opiates, synaptic effects of, 207t
Opioid(s)
  endogenous, as neurotransmitters, 180, 180t
  receptor proteins for, 180
Opsin, 269, 270f
Opsonization, 452, 456, 457
Optic chiasma, 274
Optic disc, 263
Optic nerve, 212t
Optic neuropathy, Leber's hereditary, 58
Optic radiation, 276
Optic tectum. *See* Colliculus(i), superior
Oral cavity, 561, 561f
Oral contraceptives, 663–664
Oral rehydration therapy, 138
Orbitals, 24
Organ(s). *See also specific organs*
  definition of, 10, 17
  dual innervation of, 234–235, 234t
  target, 9
Organ of Corti, 258–261, 259f
Organelles, 55–59
  definition of, 51
  digestion of, 56, 57
Organic acids, 29, 31f
Organic anion transporters, 540
Organic molecules, 29–31
Orgasm, 643
Origin, of skeletal muscle, 326
Orphan receptors, 293
Orthostatic hypotension, 432
Oscilloscope, 160, 160f
Osmolality, 132, 132f
  of blood, regulation of, 133–134, 134f, 416, 416f, 417
  of kidneys, 538f
  measurement of, 132
Osmoreceptors, 134, 416
Osmosis, 130–133
  definition of, 127, 128
  effects of, 130, 131f
  model of, 130f
  requirements for, 130
Osmotic pressure, 130–131, 131f
  colloid, 413
  ionization and, 132, 133f
Ossicles, middle-ear, 255
  damage to, 257
Osteitis deformans, alkaline phosphatase concentrations and, 88t
Osteitis fibrosa cystica, 627
Osteoblasts, 16, 623
Osteoclasts, 623, 623f
Osteocytes, 16, 17f
Osteomalacia, 627
Osteoporosis, 624–625, 625f
Otic ganglion, 224–225
Otitis media, 257
Otolith organs, 251, 251f, 252f, 253, 253f
Otolithic membrane, 253
Otosclerosis, 257
Outer ear, 255, 256f
Ova. *See* Ovum
Oval window, 255
Ovarian cycle, 655–657, 662f
  endocrine control of, 663f
Ovarian follicles, 315
  development of, 636, 656, 657, 658f, 660f
  mature, 657
  primary, 656–657
  secondary, 657
Ovary(ies), 654f, 655f, 656f
  cyclic changes in, 660–661
  endocrine functions of, 14, 287t
  exocrine functions of, 14
  formation of, 635–637, 636f
  sex steroid secretion by, 37
Overeating, 604, 605
Ovotestes, 639
Ovulation, 654, 657–658, 658f, 660f, 661, 662f
Ovulatory phase, of menstrual cycle, 664t
Ovum, 315
  development of, 660f
  mitochondria in, 58